Food, Pharmaceutical & Bioengineering Division

AA003035

Presentations at the 2008 AIChE Annual Meeting

Philadelphia, Pennsylvania
16 - 21 November 2008

Volume 1 of 2

ISBN: 978-1-61567-228-8

Printed from e-media with permission by:

Curran Associates, Inc.
57 Morehouse Lane
Red Hook, NY 12571

Some format issues inherent in the e-media version may also appear in this print version.

Printed by Curran Associates, Inc. (2009)

For permission requests, please contact AIChE
at the address below.

AIChE
3 Park Avenue
New York, NY 10016-5991

Phone: (203) 702-7660
Fax: (203) 775-5177

www.aiche.org

Additional copies of this publication are available from:

Curran Associates, Inc.
57 Morehouse Lane
Red Hook, NY 12571 USA
Phone: 845-758-0400
Fax: 845-758-2634
Email: curran@proceedings.com
Web: www.proceedings.com

TABLE OF CONTENTS

VOLUME 1

VOLUME 2

9a Optimization of Immobilization of Lipases on Various Supports for Transesterification of Tributyrin Using Dimethyl Carbonate

Mohamed A. Seyam, Richard C. Garner, Archana H. Trivedi, and Steven W. Peretti. Chemical and Biomolecular Engineering, North Carolina State University, Raleigh, NC 27695

The current methods of producing fatty acid methyl esters involves reacting triglycerides (the substrate) with methanol (the acyl acceptor), catalyzed by a soluble alkaline catalyst, resulting in glycerol as a byproduct. Since the value of glycerol is decreasing, this byproduct is then reacted to form glycerol carbonate derivatives. In this research project, the use of immobilized lipases as catalysts for these types of reactions is being studied. Immobilization of the lipase will allow for easy separation so that the lipase can be reused for multiple reactions. The current focus of the project is to evaluate loading and activity of a *Candida rugosa* lipase by using glass beads (150-212 μm, 212-300μm, and 425-600μm in diameter), Amberlite® Weakly Basic Anion Exchanger IRA-96, and Dowex® MWA-1 as supports, all containing an amine functional group. Immobilization onto these supports is carried out by physical adsorption and covalent binding. For covalent binding, glutaraldehyde and cyanuric chloride are used as cross-linking agents and are compared for their affect on loading and activity of the lipase. When running reactions, the substrate used is tributyrin and instead of using an alcohol, dimethyl carbonate (DMC) serves as the acyl acceptor. The reasons for using DMC instead of methanol are twofold: 1) Methanol is toxic while DMC is environmentally friendly, and 2) Using DMC forms glycerol carbonate derivatives readily as the byproduct, thus eliminating one step from the current process.

9b Yeast Surface Display Mini-Cellulosme for Cellulose Hydrolysis

Shen-Long Tsai[1], Jeongseok Oh[1], Ashok Mulchandani[2], and Wilfred Chen[1]. (1) Chemical and Environmental Engineering, University of California at Riverside, 900 University Ave, Riverside, CA 92521, (2) Department of Chemical and Environmental Engineering, University of California, Riverside, CA 92521

Cellulosomes are cellulolytic complexes found in many anaerobic microorganisms and have been shown to degrade cellulose efficiently. The major component of these macromolecule complexes is a structural scaffoldin consisting of repeating cohesin domains, which are docked individually with a cellulase tagged dockerin domain. The specific cohesin-dockerin interaction provides the mechanism for position-specific self-assembly. To emulate the success of a natural mechanism for efficient cellulose hydrolysis, a mini-cellulosome is assembled onto the yeast cell surface, enabling the ethanol-producing strain to utilize cellulose and concomitantly ferment it to ethanol. More importantly, by organizing these cellulases in an ordered structure, the enhanced synergy will increase the efficiency in hydrolysis, and thereby enhance ethanol production.

9c Rational Improvement of Simvastatin Synthase Solubility In Escherichia Coli Leads to Higher Whole-Cell Biocatalytic Activity

Xinkai Xie[1], Inna Pashkov[2], Xue Gao[1], Todd Yeates[2], and Yi Tang[1]. (1) Chemical and Biomolecular Engineering, University of California at Los Angeles, 420 Westwood Plaza, 5667 Boelter Hall, Los Angeles, CA 90095, (2) Chemistry and biology, University of California at Los Angeles, 420 Westwood Plaza, Los Angeles, CA 90095

Simvastatin is the active pharmaceutical ingredient of the blockbuster cholesterol lowering drug Zocor. We have previously developed an Escherichia coli based whole-cell biocatalytic platform towards the synthesis of simvastatin acid starting from the precursor monacolin J sodium salt (MJSS). The centerpiece of the biocatalytic approach is the simvastatin synthase LovD, which is highly prone to misfolding and aggregation when overexpressed from E. coli. Increasing the solubility of LovD without decreasing its catalytic activity can therefore elevate the performance of the whole-cell biocatalyst. Using a combination of homology structural prediction and site-directed mutagenesis, we identified two cysteine residues in LovD that are responsible for nonspecific intermolecular crosslinking, which leads to oligomer formation and protein aggregation. Replacement of Cys40 and Cys60 with alanine residues resulted in marked gain in both protein solubility and whole-cell biocatalytic activities. Further mutagenesis of these two residues with natural amino acids with similar size and polarity to cysteines showed that C40A and C60N are the most beneficial mutations at each position, affording 27% and 26% increase in whole cell activities, respectively. The double mutant C40A/C60N combines the individual improvements and displayed ~50% increases in protein solubility and whole-cell activity. Optimized fed-batch high-cell-density fermentation using YT2/C40A/C60N quantitatively (>99%) converted 45 mM MJSS to SS within 18 hours, which represents a significant improvement over the performance of wild type LovD under identical conditions. The high efficiency of the improved whole-cell platform renders the biocatalytic synthesis of simvastatin an attractive substitute over the existing semisynthetic routes.

9d Improvements In the Yield and Productivity of Nadph-Dependent, Whole-Cell Transformation In *Escherichia Coli*

Jonathan W. Chin and Patrick C. Cirino. Chemical Engineering, Penn State University, 222 Fenske Lab, University Park, PA 16802

Escherichia coli strain PC09 (Δ*xylB*, *crp**) expressing an NADPH-dependent xylose reductase (CbXR) produces xylitol from xylose while growing on glucose. Resting cells give molar yields of ~3-4 xylitol produced per glucose consumed, depending on the conditions used. Studies with elevated expression levels of CbXR and xylose transporters suggest that enzyme activity and xylose transport are not limiting xylitol production in PC09. Previously, we examined the effects of deletions in key metabolic pathways (e.g. Embden-Meyerhof-Parnas and pentose phosphate) and reactions (e.g. transhydrogenase and NADH dehydrogenase) on resting cell xylitol yield. The results demonstrated the importance of direct NADPH supply by $NADP^+$-utilizing enzymes in central metabolism for driving heterologous NADPH-dependent reactions, and suggest that the pool of reduced cofactors available for biotransformation is not readily interchangeable via transhydrogenase.

We will present two fundamentally different approaches to improving the coupling between glucose oxidation and NADPH-dependent xylose reduction. We first focus on improving the xylitol yield, where the NADPH requirement directly competes with growth. A metabolic model suggests that the maximum theoretical xylitol yield is 9.6 mole/mole glucose for non-growing cells. The oxidative pentose phosphate pathway (Zwf + Gnd) is predicted to supply 73% of the NADPH necessary for xylitol production by cycling glucose metabolism through the reverse reaction catalyzed by Pgi. Eliminating flux through transhydrogenase lowers the theoretical yield to 8.8. In contrast, when flux through the reverse Pgi reaction is restricted to a minimum level required for maintenance, the theoretical maximum yields drop to 9.2 and 3.6 for the wild-type and *pntA* deletion strains respectively. Experimentally, our approach to increase xylitol yield involves deletions in the phosphofructokinase genes and increasing expression of Zwf. Results from a variety of genetic modification scenarios will be presented.

An alternate strategy is to couple cell growth to xylitol production. Here, we begin with a growth-inhibited, NADPH-overproducing strain having deletions in *pgi* and *sthA*. Expression of NADPH-dependent xylose reductase then serves as a source of $NADP^+$ regeneration, and xylitol production correlates with growth. This growth coupling was not achieved in the *crp** strain. Co-expression of xylose transport genes therefore improves growth and xylitol production. This growth-coupled system allows for the evolution of strains with improved NADPH-dependent reaction productivity, and results will be presented in that context.

9e Directed Evolution of a Phloroglucinol Producing Type III Polyketide Synthase

Sheryl B. Rubin-Pitel, Wenjuan Zha, and Huimin Zhao. Chemical and Biomolecular Engineering, University of Illinois at Urbana-Champaign, 600 S. Mathews Avenue, Box C-3, M/C 712, Urbana, IL 61801

Phloroglucinol (1,3,5-trihydroxybenzene) is a process chemical utilized across a broad spectrum of industries. It is currently synthesized by a chemical method which includes a highly explosive substrate, trinitrotoluene (TNT), strong acid, and toxic chromium salt. Consequently, there has been interest in developing new syntheses for phloroglucinol, either by biosynthesis or an alternative chemical route. Biosynthesis of phloroglucinol by *Escherichia coli* would use glucose, a renewable feedstock, as a substrate, proceed under mild temperature and pH conditions, and would not lead to environmentally hazardous effluent.

We recently discovered that PhlD, a type III polyketide synthase in the diacetylphloroglucinol (DAPG) biosynthetic pathway, is in fact a phloroglucinol synthase. This finding creates the foundation for commercial biosynthesis of phloroglucinol either *in vivo* from glucose via fermentation by *E. coli*, or *in vitro* by combining substrate with purified protein or cell lysate. Characterization of the biochemical properties of PhlD revealed that like many enzymes from secondary metabolism, it suffers from slow catalytic activity and poor stability. Therefore we are performing directed evolution to engineer PhlD to become a practical industrial biocatalyst. Toward this end, a colorimetric screening system to quantify phloroglucinol production was developed and used to evaluate a library of PhlD mutants. In a first round of screening, the PhlD mutant library was generated by synthetic shuffling of the diversity present in 50 natural PhlD homologs. After screening approximately 40,000 clones, two significantly improved PhlD variants were discovered which possessed increases in both catalytic efficiency and thermostability. Based on these promising results, a second round of directed evolution is underway. We will apply the THR selection method, recently developed by Delcourt and coworkers, to further increase the thermostability of PhlD in an activity-independent approach by fusing PhlD to a thermostable kanamycin resistance gene and expressing the fusion protein in *Thermus thermophilus*.

9f Protein Engineering of a Thermostable Alcohol Dehydrogenase to Alter Cofactor and Substrate Specificities

Elliot Campbell and Scott Banta. Chemical Engineering, Columbia University, 801 Mudd, 500 W 120th St, New York, NY 10027

Pyrococcus furiosus alcohol dehydrogenase D (AdhD) is a highly thermostable member of the well-characterized aldo-keto reductase (AKR) superfamily. Its preference for secondary alcohols and NAD+ combined with promiscuous activity towards simple sugars, primary alcohols, and aldehydes makes it an attractive protein engineering candidate for applications such as biofuel cells and the development of artificial metabolic pathways. We have examined the mode of cofactor binding in this enzyme and explored the nature of its preference for NAD(H) vs NADP(H). Site-directed mutagenesis of the cofactor binding pocket identified those residues important for determining cofactor specificity and identified mutants with improved activity. Changing a histidine residue involved in binding the adenine half of the cofactor back to the consensus arginine residue (in NADP(H) binding AKRs) resulted in an order of magnitude improvement in activity with NADP(H). Additionally, a lysine to glycine mutation in the cofactor binding pocket increased activity with both cofactors. Mutagenesis was also performed within the substrate binding pocket in an attempt to improve activity with sugars or primary alcohols. Substrate specificity in the AKR superfamily is tailored through modification of surface loops near the active site. Multiple sequence alignments and homology modeling indicate these loops are severely truncated in AdhD, and likely contribute to its thermostability and broad specificity. Typically these loops are necessary to differentiate between various steroids and other large substrates, and their reduced size should not significantly impact the affinity for smaller primary alcohols and monosaccharides. We report on the kinetic parameters for wild-type AdhD and mutants with improved activities with these substrates. The results of this work will be applied to the continued evolution of novel specificities for this enzyme.

9g Cloning and Characterization of An α-Amino Ester Hydrolase from *Xanthomonas Campestris Pv. Campestris*

Janna Blum, Javier Chaparro-Riggers, and Andreas Bommarius. School of Chemical and Biomolecular Engineering, Georgia Institute of Technology, 311 Ferst Drive, NW, Atlanta, GA 30332

α-Amino ester hydrolases (αAEHs) are capable of the synthesis and hydrolysis of α-amino β-lactam antibiotics such as ampicllin and have only recently been cloned and over-expressed using *E. coli* [Barends:2003, Polderman-Tijmes:2002]. The enzymatic synthesis of ampicillin using α-amino ester hydrolases (αAEH) is an attractive alternative to using penicillin G acylase (PGA) for enzymatic synthesis of beta-lactam antibiotics. A 637 amino acids (~72 kDa) putative glutaryl 7-ACA acylase (gaa gene) from *Xanthomonas campestris pv. campestris* strain ATCC 33913 was identified in a genome study[daSilva:2002]. A homology search indicates that this protein is 93% identical to the known αAEH from *Xanthomonas citri* strain IFO 3835 and 59% identical to the *Acetobacter turbidans* strain ATCC 9325. Additionally, the catalytic triad (S174, D338, H370) and carboxylate cluster (D208, E340, D341) residues in the gaa protein are conserved, suggesting the putative glutaryl 7-ACA acylase from *X. campestris pv. campestris* strain ATCC 33913 belongs to the αAEH subclass of enzymes.

The gaa gene was cloned and expressed in E. coli. The *X. campestris pv. campestris* αAEH was characterized for activity, temperature, pH, and substrate specificity. The resulting kinetic parameters (kcat, KM) of ampicillin hydrolysis of this protein are comparable to other αAEH characterized from *A. turbidans* and *X. citri* proteins. Results on expression, optimal processing parameters and efforts to alter the substrate specificity of the *X. campestris pv. campestris* αAEH toward larger β-lactam resistant semi-synthetic antibiotics using site-directed mutagenesis around the active site will be discussed.

12a Ensemble Modeling of Metabolic Networks and Identification of Enzyme Targets for Metabolic Engineering

Matthew L. Rizk, Linh M. Tran, and James C. Liao. Chemical and Biomolecular Engineering, University of California, Los Angeles, 5531 Boelter Hall, 420 Westwood Plaza, Los Angeles, CA 90095

Complete modeling of metabolic networks is desirable but difficult for the lack of kinetics. As a step towards this goal, we develop an approach to build an ensemble of dynamic models which reach the same steady state. The models in the ensemble are based on the same mechanistic framework at the elementary reaction level, including known regulations, and span the space of all kinetics allowable by thermodynamics. This ensemble allows for the examination of possible phenotypes of the network upon perturbations, such as changes in enzyme expression levels. The size of the ensemble is reduced by acquiring data for such perturbation phenotypes. If the mechanistic framework is approximately accurate, the ensemble converges to a smaller set of models and becomes more predictive. This approach bypasses the need for detailed characterization of kinetic parameters and arrives at a set of models that describes relevant phenotypes upon enzyme perturbations.

Further, we demonstrate that such an ensemble modeling approach can be used to identify enzyme targets for overexpression for metabolic engineering strain design. We demonstrate the utility of this approach on the pathway to aromatics biosynthesis, succinate production, and lysine biosynthesis, identifying the same enzyme targets to increase production that have been shown in the literature.

12b Genome Sequencing Reveals Novel Features of Metabolism In *Agrobacterium* Sp

Anne Ruffing[1], Marlene Castro[2], Wei-Shou Hu[2], and Rachel Ruizhen Chen[1]. (1) Chemical and Biomolecular Engineering, Georgia Institute of Technology, 311 Ferst Dr., Atlanta, GA 30332-100, (2) Chemical Engineering and Materials Science, University of Minnesota, 421 Washington Avenue SE, Minneapolis, MN 55455-0132

Agrobacterium sp. ATCC 31749 is an important industrial organism due to its production of curdlan, a β1,3-glucan polysaccharide. We have genetically engineered ATCC 31749 for the production of up to 20 mM of *N*-acetyllactosamine. Although this organism is an advantageous host for oligosaccharide synthesis, little genetic information is available. In order to identify potential targets for optimizing oligosaccharide production, *Agrobacterium* sp. ATCC 31749 was sequenced using the Genome Sequencer FLX System from 454 Life Sciences.

The assembly resulted in 95 contigs with a total length of 5.4 Mb. Mapping to available genomes from the *Rhizobiaceae* family revealed that *Agrobacterium tumefaciens* presents the highest similarity to ATCC 31749. However, not all contigs mapped to *A. tumefaciens*, revealing differences between the two organisms.

Genes previously reported to influence curdlan production were identified in the ATCC 31749 sequences. The curdlan synthesis operon *crdASC*, which has homologous sequences in the linear chromosome of *A. tumefaciens*, was identified, as well as phosphatidylserine synthase (*pssAG*), with a homologous sequence in the circular chromosome of *A. tumefaciens*. Other relevant metabolic genes were identified, including genes for sucrose hydrolysis, nitrogen regulation and fixation.

Despite the presence of *crdASC* and *pssAG* homologs, curdlan production has not been detected in *A. tumefaciens*. Potential reasons for this include unexpressed genes, or lack of other, yet unidentified genes involved in curdlan production. Genes related to sugar import and transport were identified in ATCC 31749 sequences that do not map to the *A. tumefaciens* genome. The available sequences offer an opportunity to identify genes relevant to curdlan production, improve our understanding of the metabolic regulation in ATCC 31749, and reveal targets for future metabolic engineering endeavors.

12c Microfluidic Droplets as Nanobioreactors for Screening Metabolic Engineering Libraries

Benjamin L. Wang[1], Hang Zhou[1], David A. Weitz[2], and Gregory N. Stephanopoulos[1]. (1) Department of Chemical Engineering, Massachusetts Institute of Technology, 77 Massachusetts Ave., 56-422, Cambridge, MA 02139, (2) Department of Physics and School of Engineering and Applied Science, Harvard University, Pierce Hall, 29 Oxford St., Cambridge, MA 02138

Metabolic engineering has contributed significantly to the rational improvement of strains for industrial applications. In addition to enzymatic steps closely associated with the product-forming pathway, other, so-called distal genes may also impact production in a profound way due to (often unknown) kinetic and regulatory effects. Such genes are identified by combinatorial methods whereby libraries of the host strain are constructed harboring random variants of the basic strain genes, as well as random combinations of gene knock outs and over-expressions. Cells with superior properties are selected from these libraries and the specific genetic alteration identified in a process known as Inverse Metabolic Engineering. However, these approaches necessitate the use of high throughput screening methods to select the most desirable clones from these libraries of engineered strains. For many metabolic engineering libraries, the selection criterion is the production of a secreted metabolite or the consumption of a medium component. Thus, a strategy for compartmentalizing clones is necessary so that each clone grows in a separate environment allowing for the measurement of clone-specific metabolite concentrations. Traditional methods such as microwell plates for culturing and assaying can be utilized but are not sufficiently high throughput. Here, we will present our development of a high throughput screening platform which utilizes microfluidics to encapsulate yeast cells in nanoliter aqueous droplets surrounded by an immiscible fluorinated oil phase. This system contains modules for cell culturing, measurement of the metabolite of interest with a fluorescent enzymatic assay, and sorting. To demonstrate the capabilities of this platform, we screened an incoming population which contained equal numbers of two *Saccharomyces cerevisiae* strains and enriched for the higher xylose consuming strain by over 20 times. Furthermore, we screened a *S. cerevisiae* xylose transporter library for high xylose consumption and selected a mutant which consumes 5% more xylose than the library as a whole.

12d Determining the Molecular Mechanisms of Adaptive Evolution In Yeast

Katy Kao, Stanford, 300 Pasteur Dr, Stanford, CA 94305 and Gavin Sherlock, Genetics, Stanford, 300 Pasteur Dr, Stanford, CA 94305.

We present here the most detailed molecular characterization of an evolutionary series to date. Whole genome evolution of microbial strains to select for traits of interest via adaptive evolution has the potential to be a useful tool for biotechnological purposes. However, the population structure during microbial adaptation to an environment has largely been uncharacterized. We developed a method to better understand the population structure of yeast evolving under a nutrient-limited condition using fluorescently marked strains that are otherwise isogenic. This method allowed us to unequivocally demonstrate for the first time that clonal intereference and multiple mutations affect the population structure in a eukaryote. Our method allowed us to isolate the subpopulations containing the beneficial mutants and identify the exact mutations in five of the beneficial mutants throughout the course of a 440 generation evolution. Transcriptome profiles of these beneficial mutants correlated with the mutations observed.

12e Exhaustive Identification of All Engineering Interventions Leading to Targeted Overproductions

Sridhar Ranganathan[1], **Patrick F. Suthers**[2], and Costas D. Maranas[2]. (1) Industrial Engineering and Operations Research, The Pennsylvania State University, University Park, PA 16802, (2) Department of Chemical Engineering, The Pennsylvania State University, University Park, PA 16802

Existing computational methods for strain optimization generate engineering strategies only *one at a time*, out of the myriads of possibilities, thus limiting the array of choices presented to the biotechnologist. In addition, metabolic flux data (e.g., though MFA) obtained for the wild-type strain are not directly integrated with the strain optimization process. To remedy these limitations, we present a new computational framework that overcomes these issues. The key concept here is that instead of looking for specific engineering strategies one at a time, we look to classify all fluxes in the metabolic model depending upon whether or not they must increase, decrease, or become equal to zero to meet a pre-specified overproduction target. This classification is not limited to individual reactions but is extended for pairs, triplets, etc. while making use of all flux data available for the strain before engineering. The final output of this analysis is a logic tree-diagram that spans all possible sets of engineering interventions capable of meeting a pre-specified overproduction target. The developed methodology is tested by exhaustively identifying all engineering interventions for succinate production in a glucose-limited medium. The method recapitulates known engineering targets but also reveals new non-intuitive ones that boost succinate production by performing coordinated changes on pathways distant from the last steps of succinate synthesis.

Title: Design of efficient organisms using a combination of inverse metabolic engineering and metabolic evolution

Authors: Pornkamol (Apple) Unrean, Cong Trinh and Friedrich Srienc

Department of Chemical Engineering and Materials Science/BioTechnology Institute, University of Minnesota 1479 Gortner Avenue, Saint Paul, MN 55108, USA

Abstract

In this study, we demonstrate the use of combined inverse metabolic engineering and metabolic evolution for rational strain development. We have applied these two approaches for designing the efficient *E. coli* strains for the production of primary metabolites such as ethanol as well as for the production of complex secondary metabolites like lycopene. Inverse metabolic engineering based on Elementary Mode Analysis was employed to predict a set of multiple-gene deletion required for the cell to function according to an efficient pathway. Metabolic evolution of growth-based selection was used for co-selection of an increased production of the product. We have recently constructed the design strains and characterized their performances. Experimental results have shown that the designed strains achieved a better performance than the wild type under identical conditions. The results, therefore, confirm the validity of these applied strategies for rational strain development.

Introduction

The creation of organisms that are capable of performing in the most efficient way for the production of desirable metabolites is an intriguing goal in metabolic engineering. Developing organisms to achieve a better product yield and productivity often requires the simultaneous manipulating of pathways and optimizing of regulatory effects. Previously, these optimizations were often accomplished by random mutagenesis and high-throughput screening which are relatively time- and cost-consuming process and may lead to only suboptimal phenotype. In this study, we demonstrate a rational, systematic way of strain development using a combination of metabolic engineering based on Elementary Mode Analysis and metabolic evolution. In designing *E. coli* mutants for the most efficient production of desired products, we apply Elementary Mode Analysis for systematic optimization of biosynthesis pathway of the products. In combination with Elementary Mode Analysis, we exploit evolutionary process to overcome metabolic regulation limiting the synthesis of the desired products.

A metabolic network can function according to many different pathway options. Elementary Mode Analysis is a flux balance-based stoichiometric model that has emerged as a powerful tool for rigorously dissecting a metabolic network into its basic building blocks of the metabolic structure. We apply Elementary Mode Analysis for analyzing the metabolic network of an *E. coli* strain. Evaluation of all feasible fluxes identified by Elementary Mode Analysis permits us to identify potential genes deletion required for re-routing the operating pathway of an *E. coli* cell into the efficient one for production of desired products. Later, metabolic evolution in a growth-based selection scheme was employed on the knockout strains that are designed by Elementary Mode

Analysis. The evolutionary process allows for a co-selection of a mutant *E. coli* with an improved cell growth and synthesis of the product. Here, we describe the capacity of two optimized *E. coli* strains derived from the combined approaches of metabolic engineering and metabolic evolution. One *E. coli* strain is optimized for efficient production of ethanol from glycerol and another is optimized for efficient production of lycopene from glucose. The experimental results are also verified with the prediction from Elementary Mode Analysis.

Result

Strain design using Elementary Mode Analysis

The metabolic models of ethanol producing *E. coli* and lycopene producing *E. coli* are constructed. Elementary Mode Analysis on the constructed models, then, identifies total of available Elementary Modes (or pathway options). For a rational design of *E. coli* mutants for the most efficient production of ethanol from glycerol and for the most efficient production of lycopene from glucose, we use Elementary Mode Analysis to examine the effect of gene deletion on the production of the products in the cell. Gene knockout is simulated by removing the enzymatic reaction corresponding to that gene from the metabolic network model. The phenotype of that specific knockout mutant is, then, represented by a combination of remaining Elementary Modes. We sequentially screen the simulation result and identify an optimal set of gene knockouts that could restrict the mutants to operate according to the efficient-product producing modes. The designed mutant with these multiple gene deletions for efficient production of ethanol from glycerol is designated as TCS099/pLOI297 and for efficient production of lycopene from glucose is designated as LYC018/pACEBI. All Elementary Modes remain in both strains are directed to efficient ethanol production and to efficient lycopene production respectively as well as to maintain cell growth.

Strain Construction

To validate the prediction of strain design by Elementary Mode Analysis, we have constructed the designed multiple-genes knockout mutants TCS099 and LYC018. TCS099 and LYC018 have all targeted knockout genes as identified by Elementary Mode Analysis for efficient synthesis of ethanol and for efficient synthesis of lycopene respectively removed from their chromosomes. Gene disruption in the mutants is verified by PCR amplification using primers specific to that deleted genes and gel electrophoresis. The designed mutant TCS099 is transformed with plasmid pLOI297 and then tested for its performance on production of ethanol. Likewise, for production of lycopene, the designed strain LYC018 is transformed with plasmid pACEBI containing lycopene synthesis genes and tested for its capacity of lycopene synthesis.

Metabolic evolution

According to the strain design, TCS099 and LYC018 always coproduce biomass and the products, ethanol and lycopene respectively, in a growth-associated manner. This obligated couple between cell growth and the products provide a basis for strain improvement through metabolic evolution. That is, by selecting for improved growth would also co-select for improved production of the products. TCS099 mutant designed for efficient production of ethanol and LYC018 mutant designed for efficient production of lycopene are, therefore, evolved in a serial cultivation under a growth-based selection scheme where cell growth rate of the mutants in each batch culture was monitored during the serial dilution. The results reveal the continuous increase in cell growth of both mutants over the course of evolution. After several generations, the evolved strain of TCS099 and the evolved strain of LYC018 are isolated and renamed TCS099Rd50 and LYC018E1 correspondingly.

Strain characterization

Based on Elementary Mode Analysis, TCS099/pLOI297 was rationally designed for efficient ethanol production from glycerol and LYC018/pLOI297 was rationally designed for efficient lycopene production from glucose. The performance of the constructed strains based on the design of Elementary Mode Analysis is tested experimentally in comparison with the wild-type strain as a control in aerobic batch fermentation. The mutant LYC018/pACEBI outperforms the wild type MG1655/pACEBI for the conversion of glucose into lycopene. Similarly, TCS099/pLOI297 also outperforms the wild-type MG1655pLOI297 for the production of ethanol from glycerol. The production of ethanol and the production of lycopene are further improved in the evolved strains derived from growth-based selection evolution, TCS099Rd50/pLOI297 and LYC018E1/pACEBI comparing to their parental strains. In vivo phenotypes of all strains are also compared with predicted phenotype by Elementary Modes. The observed phenotype of all strains is in agreement with the predicted phenotypic space of Elementary Mode Analysis. These consistencies, therefore, confirm the accuracy of the Elementary Mode Model.

Discussion

In this study, we have shown an alternative way for rationally engineering efficient *E. coli* strain using the combined approach of inverse metabolic engineering and metabolic evolution. Inverse metabolic engineering via Elementary Mode Analysis has been applied for optimizing two *E. coli* strains, one for production of ethanol from glycerol and another for production of lycopene from glucose. Elementary Mode Analysis can be used to dissect a metabolic network into unique, non-decomposable pathways. A set of these pathways, therefore, presents all possible physiological states of cells at steady-state conditions. Elementary Mode Analysis is utilized to computationally search for the combination of expressed gene products that supports only a minimum number of possible pathways. A set of target gene knockouts is extracted through the sequential search of Elementary Mode Analysis aiming at minimizing the cell's pathway options while retaining the most efficiency of the desired products. The derived minimal strains can, therefore, only function on the basis of the most efficient production of the desired products. Both mutants with multiple gene deletions predicted by the model, TCS099/pLOI297 and LYC018/pACEBI, experimentally showed the improved performance on the production of the desired products, ethanol and lycopene respectively. This also confirms the validity of the Elementary Mode Analysis as a powerful tool for a rational strain design

Besides pathway optimization guided by Elementary Mode Analysis, another feasible alternative called metabolic evolution can be undertook to yield promising targets for further enhancement of ethanol or lycopene synthesis. In particular, the tightly couple between the products synthesis and the cell growth allows for a selection of an improved ethanol-producing mutant or an improved lycopene-producing mutant through a growth-based selection scheme. The improvement of cell growth observed during a serial dilution evolution suggests an occurrence of a new strain from the parent as a result of spontaneous mutation or mutations. The evolved strains, both TCS099Rd50/pLOI297 and LYC018E1/pACEBI that are obtained from growth based selection evolution are found to co-select for increased production of ethanol product and lycopene product respectively. Both evolved strains also show distinct kinetic of product accumulation which suggests that the different modes of action and different combinatorial genes mutation between the evolved strains and their parents. In vivo phenotype of evolved strains in comparison with Elementary Mode modeling result also reveals the shift toward a predicted optimal metabolic state over the course of evolution.

In conclusion, this study underscores the application of inverse metabolic engineering based on Elementary Mode Analysis in combination with metabolic evolution for developing of two efficient *E. coli* strains, one for the production of ethanol and another for the production of lycopene. The improved performance observed in the evolved strains suggested that *E. coli* can be metabolically engineered and evolved to increase fluxes to the production of the products. These applied strategies allow us to capture apparent limitations in obtaining the most efficient performance of the designed strain caused by both metabolic pathways and regulatory mechanisms. These combined approaches could aid in efficient strain design for production of many other useful products as well.

12g Isotopically Nonstationary ^{13}C Flux Analysis of *Synechocystis* Photoautotrophic Metabolism

Jamey D. Young[1], Avantika A. Shastri[2], Gregory Stephanopoulos[1], and John A. Morgan[2]. (1) Department of Chemical Engineering, Massachusetts Institute of Technology, 77 Massachusetts Ave., Building 56-439, Cambridge, MA 02139, (2) School of Chemical Engineering, Purdue University, 480 Stadium Mall Dr., West Lafayette, IN 47907

Photoautotrophic metabolism is the process by which plants and other photosynthetic organisms use light energy to fix freely available CO_2 into complex organic molecules. This represents the primary source of all food on earth as well as raw materials for bio-based production of fuels and chemicals. The ability to perform quantitative studies using isotope tracers and metabolic flux analysis (MFA) is critical for detecting pathway bottlenecks and deciphering flux regulation in these systems. Although ^{13}C is the preferred isotope tracer for studying central carbon metabolism in heterotrophic systems, photoautotrophs assimilate carbon solely from CO_2 and therefore produce a uniform **steady-state** ^{13}C-labeling pattern that is insensitive to fluxes. However, **transient** measurements of isotope incorporation following a step change from unlabeled to labeled CO_2 can be used to estimate fluxes successfully with newly developed techniques of isotopically nonstationary MFA (INST-MFA). We have assembled a package of computational routines that achieves more than 5000-fold speedup relative to previous INST-MFA approaches. These tools now permit comprehensive flux analysis of photoautotrophic metabolism, complementing previous studies that have been limited to heterotrophic or mixotrophic conditions.

We have applied the INST-MFA approach to study the metabolism of *Synechocystis* sp. PCC 6803, a model photosynthetic organism, under autotrophic conditions using both GC/MS and LC/MS/MS to quantify the trajectories of metabolite labeling that result from introduction of ^{13}C-labeled bicarbonate. The INST-MFA flux map was compared to values predicted by a linear programming (LP) method that does not require experimental measurements but instead assumes that carbon and light utilization are regulated to provide optimal growth. Although the LP predicts that there should be no flux through the oxidative pentose phosphate pathway, the experimental results indicate that around 10% of the fixed carbon is lost via this pathway. Due in part to these losses, 142 ± 12 moles of CO_2 must be fixed to yield a net gain of 100 C-moles of biomass. This is significantly more than the 111 moles of CO_2 predicted by the LP model, indicating that growth is suboptimal with respect to carbon utilization. Another notable result of the flux analysis is that *Synechocystis* appears to use transaldolase rather than aldolase as its primary route to regenerate S7P in the Calvin cycle. This is an interesting result, since plants are known to use aldolase exclusively for this purpose. Lastly, the flux analysis confirms that the oxygenation side reaction of RuBisCO is insignificant, as expected due to the potent CO_2-concentrating effect of carboxysomes in this organism.

14a Biobutanol Production Via Fermentation of Hot-Water Wood Extract Sugars

Shijie Liu, Lawrence Fitzgerald, Gary Steeves, Adam Smith, Shiuli Mahmud, Alan Shupe, and Som Smukher. Paper Science and Engineering, SUNY College of Environmental Science and Forestry, 1 Forestry Drive, Syracuse, NY 13210

Hot-water extraction of woody biomass produces a wood extract that is rich in hemicelluloses or xylan and other carbohydrates and extractives. This extracts are excellent feedstock for the production of biobutanol. Clostridium berjinckii is used as the fermentor grown in an anaerobic environment. It is found that the present hot-water extract production technique is adequate for the production of butanol.

14b Engineering *Escherichia Coli* for Improved Ethanol and Butanol Tolerance for Biofuel Production

Adam Smith, Manas Medisetty, Wen-Hsuan Huang, Joseph Priola, and **Dacheng Ren**. Department of Biomedical and Chemical Engineering, Syracuse University, 121 Link Hall, Syracuse, NY 13244

Economical production of biofuels from renewable biomass is critically important for reducing our dependence on the fast diminishing fossil fuels. *Escherichia coli* can ferment both pentose and hexose sugar constituents and has been engineered to produce ethanol and butanol. However, *E. coli* is not natively tolerant to these organic solvents. In this study, we seek to develop *E. coli* strains with enhanced solvent tolerance and to understand the genetic basis of such tolerance. Strains with significantly enhanced tolerance were obtained by challenging with increasing concentrations of ethanol/butanol. In addition, *E. coli* strains overexpressing certain genes were studied to identify the genes involved in solvent tolerance. Overexpression of the gene *rob* was found to increase the tolerance to ethanol. Our genetic engineering approach yielded an improved *E coli* strain that demonstrated a butanol MIC of more than double that of the wild type. Additionally, the engineered strain was capable of growth in medium containing 2.1% butanol with a growth yield that was 60% of that of the wild type grown in butanol-free medium. By increasing the solvent tolerance of *E. coli*, it is possible to improve alcohol production and reduce purification costs leading to higher product yield and profits.

14c Integrating Systems and Synthetic Biology to Optimize Biofuel Pathways In Microorganisms

Brian F. Pfleger[1], Wes D. Marner II[2], Sydnor T. Withers[2], and Robert Landick[3]. (1) Chemical and Biological Engineering, University of Wisconsin Madison, 3629 Engineering Hall, 1415 Engineering Dr, Madison, WI 53706, (2) Great Lakes Bioenergy Research Center, University of Wisconsin Madison, 3529 Microbial Sciences, 1550 Linden Dr., Madison, WI 53706, (3) Dept. of Biochemistry, University of Wisconsin Madison, 5441 Microbial Sciences, 1550 Linden Dr., Madison, WI 53706

The number of small molecules that can be produced by engineered microbes has rapidly increased in recent years. This increase can be attributed to the development of recombinant DNA technology and advances in metabolic engineering and directed evolution. The current trend in engineering microbes utilizes an iterative approach where genetic perturbations are made, metabolite production is evaluated, and new designs are implemented. The next generation is going to take advantage of both systems and synthetic biology to generate increasingly complex perturbations and clearer pictures of cellular physiology.

Within the Great Lakes Bioenergy Research Center (GLBRC), we are utilizing this strategy to develop the next generation of biofuel producers. Our strategy is to develop first-pass panels of biofuel producing strains based on a minimal set of mutations chosen with the assistance of metabolic models. These panels are screened for biofuel production, and the best strains are selected for REDIME (REiterative DIrected Microbial Evolution). The REDIME process entails growing the best candidate strains at larger scales, sampling the cultures for multi-omic analysis, identifying limiting genetic elements, designing/evolving new strains that circumvent the limitations, and evaluating the improvements before repeating the cycle. In order to implement this strategy, the GLBRC has assembled two facilities, a high throughput microbial phenotype screening facility, and a bioenergy chemostat facility. These facilities have the capability to screen panels of strains and generate sufficient biomass for multi-omic analysis. In addition, GLBRC researchers have developed pipelines for sequencing, transcriptomic, proteomic, metabolomic, and other systems-level studies. The goal of these pipelines is to decrease the turnaround between fermentation and analysis, such that the REDIME iteration time is as short as possible.

This talk will use examples from preliminary and published experiments to describe how the elements of synthetic and systems biology can be integrated into the REDIME process for biofuel production. Recent work has used systems biology to evaluate multi-omic data and to identify sources of toxicity during the production of biofuels. Subsequently, synthetic biology techniques afforded the ability to overcome these toxic effects by regulating metabolic circuits involved in the biofuel pathways. The combinations of systems biology and synthetic biology permit increasingly sophisticated levels of control over cellular processes. These methods are more robust than traditional gene overexpression and or deletions. I will also briefly discuss the development of new synthetic biology tools and their use in biofuel development.

14d Dynamic Modeling of Transesterification In a Simulated Moving Bed Chromatographic Reactor for Biodiesel Production

Mian Xu, Chemical Engg & Analytical Science, The University of Manchester, Manchester, United Kingdom, Jhuma Sadhukhan, Centre for Process Integration, Chemical Engineering & Analytical Science, The University of Manchester, Manchester, United Kingdom, and Karen Wilson, Department of Chemistry, University of York, Heslington, York, United Kingdom.

Biodiesel production is a very promising subject due to the relevance that it is environmental-friendly and an alternative diesel fuel for fossil fuel. Nowadays, most industrial applications of biodiesel production are performed by the transesterification of renewable biological sources based on heterogeneous catalysts, which requires separate reaction and separation processes. And the conversion of the reaction is restricted due to its equilibrium limitation. Thus, in this work, a simulated moving bed chromatographic reactor (SMBR) has been applied for the first time into the transesterification of biodiesel production, which allows us to carry out a simultaneous reaction and separation process and drive the transesterification reaction beyond its equilibrium. A detailed dynamic modeling of reaction as well as continuous adsorptive separation based on the surface reacting kinetic phenomena integrating with particle-level transport phenomena has been established firstly for a SMBR process design and optimization. This research offers a heuristic insight into the improvement of industrial production of biodiesel. A case study of the transesterification of tributyrin with methanol over hydrotalcite catalysts has been demonstrated to establish the effectiveness of this methodology.

Keywords: biodiesel production, SMBR, chromatography, heterogeneous catalysts

14e High Throughput Approach for Identifying Pentose Fermenting Microbes In the Environment

Valerie Lieu[1], Sharon L. Doty[2], and Daniel T. Schwartz[1]. (1) Department of Chemical Engineering, University of Washington, Box 351750, Seattle, WA 98195-1750, (2) College of Forest Resources, University of Washington, Seattle, WA 98195

Microbial populations in the environment are remarkably diverse, but sometimes difficult and tedious to culture and screen. We have recently obtained preliminary results suggesting that many fungi strains found in the tissue, surrounding soil, and fruit of plants may possess useful bioenergy production properties, namely, an ability to ferment xylose, the dominant five-carbon sugar of hemicellulose, to alcohol and other valuable biochemicals. Moreover, many of these plant-associated microbes are resistant to phytochemical inhibitors produced by the host. Saccharomyces cerevisiae, a common yeast used in industry for the production of bioethanol, cannot readily metabolize xylose. To date, we have used traditional bulk microbiology tools to characterize these fungi; our promising preliminary results point to the value of developing easy-to-use tools that will allow rapid screening of naturally occurring microbial populations for their potential impact on bioenergy and biochemical production from biomass. To characterize microbial populations requires high throughput methods to separate, grow, and analyze diversity in the population. Microfluidic devices have the ability to perform many of the reaction and separation operations chemical engineers are familiar with at the macroscale, but they do so in nanoliter volumes and they can be arrayed in parallel configurations to accelerate cell sorting. In this work, we describe a class of micro-eddy based microfluidic devices that can trap suspension cells in microscale continuously stirred fermenters.[1]-[3] Simulation and experiments are used to understand flow and mass transfer, as well as cell trapping, in the devices. Since microeddy-based devices have well characterized mixing, the production or consumption of chemicals within the eddies can be rapidly and quantitatively measured using optical techniques. Raman spectroscopy, fluorescence microscopy, and optical microscopy are all demonstrated. We also show that the devices can be scaled into an arrayed configuration, allowing higher throughput screening. We show how the arrayed trapping and study of single-cells from a mixed culture eliminates many of the bulk microbiology steps needed to create isolates from complex environmental samples, accelerating the screening process.

[1] B.R. Lutz, J. Chen, and D.T. Schwartz, Microfluidics without Microfabrication, PNAS (USA) 100, 4395-4398 (2003).

[2] B.R. Lutz, J. Chen, and D.T. Schwartz, Characterizing Homogeneous Chemistry Using Well-Mixed Microeddies, Anal. Chem. 78 1606-12 (2006).

[3] B.R. Lutz, J. Chen, and D.T. Schwartz, Hydrodynamic Tweezers: 1. Non-contact cell trapping in a laminar oscillating flow, Anal. Chem. 78 5429-5435 (2006).

14f Engineering a Fungal L-Arabinose Pathway towards the Utilization of Pentose Sugars for Production of Xylitol and Ethanol

Ryan P. Sullivan, Nikhil U. Nair, and Huimin Zhao. Chemical and Biomolecular Engineering, University of Illinois at Urbana-Champaign, 600 S Mathews Ave., MC-712 Box C-3, Urbana, IL 61801

Improved utilization of hemicellulose sugars (specifically D-xylose and L-arabinose) for biosynthesis of value-added products is crucial for fermentative processes to become more economically feasible. In yeast and filamentous fungi, both pentose sugars enter the pentose phosphate pathway (PPP) through a common intermediate, D-xylulose 5-phosphate. D-xylose is converted to D-xylulose 5-phosphate through two redox reactions, while L-arabinose conversion requires four redox reactions prior to entering the PPP. These redox reactions require alternative forms of nicotinamide cofactor, as the reductases (xylose reductase and L-xylulose reductase) typically favor reduced NADPH and the dehydrogenases (L-arabinitol 4-dehydrogenase and xylitol dehydrogenase) favor oxidized NAD+. This results in a cofactor imbalance that potentially prohibits the efficient utilization of these hemicellulose sugars, resulting in the excretion of sugar alcohol by-products formed by the initial xylose reductase.

This work describes the cloning, characterization, and engineering of an L-arabinose pathway comprised mainly of enzymes from the filamentous fungi Neurospora crassa for co-utilization of D-xylose and L-arabinose for improved production of xylitol, a five carbon sugar alcohol with a growing market as a sweetener. In particular, the rational design and directed evolution of L-arabinitol 4-dehydrogenase towards the utilization of NADP+ as a cofactor was accomplished in an attempt to partially relieve the cofactor imbalance involved in conversion of L-arabinose to xylitol, and is currently being investigated for improvement in xylitol production from both pentose sugars in a model organism Escherichia coli. The engineered strain generated for this work includes deletions of the endogenous bacterial pentose sugar pathway genes (xylA and araBAD) as well as a mutant carbon catabolite regulator (crp*) to allow for utilization of pentose substrates with glucose as a co-substrate.

Extending this concept even further, engineering both dehydrogenases for NADP+ utilization could conceivably create closed redox loops between the reductases and the dehydrogenases, resulting in the utilization of a single cofactor pair, NADP+/NADPH. Through rational design, N. crassa xylitol dehydrogenase cofactor specificity was completely reversed to NADP+, and the entire engineered pathway leading from L-arabinose and D-xylose to D-xylulose 5-phosphate prior to introduction into the PPP is being tested for improved fermentative ethanol production from the pentose sugars.

18a Thin Film Oxygen Sensor for Monitoring Cellular Oxygen Consumption

Peter C. Thomas[1], Samuel P. Forry[2], Srinivasa R. Raghavan[1], and Anne L. Plant[2]. (1) University of Maryland, Department of Bioengineering, College Park, MD 20742, (2) NIST, 100 Bureau Dr., STOP 8313, Gaithersburg, MD 20899

Cell culture techniques have continued to progress in order to achieve a more biomimetic environment for cell-based studies. However, oxygen concentration during cell culture is still a variable that has been largely overlooked. Conventional culturing condition is generally performed at an oxygen concentration of 20%. In comparison, cells in vivo exist at a much more hypoxic condition of 3% oxygen concentration. As more studies continue to demonstrate that oxygen concentration can affect cellular activities, monitoring local oxygen concentration as well as oxygen consumption rates by the cells will be crucial in creating a more biomimetic environment.

In the current work, an oxygen sensor, based on the quenching of a platinum fluorophore, is investigated as a technique to monitor the local oxygen concentration. The Stern-Volmer equation is used to characterize the change of fluorescent intensity due to oxygen concentration. Platinum-dye incorporated PDMS is spin-coated onto glass cover slips to produce thin film oxygen sensor. To facilitate the attachment of cells, the surface of the oxygen sensor is coated with ECM proteins. Combining with time-lapsed phase contrast microscopy, the current system allows simultaneous monitoring of local oxygen concentration and the effects on cellular morphology. Cell viability assay and other fluorescent techniques can be utilized simultaneously to provide information regarding cell viability and oxygen concentration. During normal metabolism, cells consume oxygen which subsequently alters the local oxygen concentration. This study will examine the changes of oxygen concentration during cell culture. In addition, conclusions regarding cell density and oxygen consumption rate will be presented. Finally, due to the large dynamic range observed by the current oxygen sensor, the feasibility of monitoring single cell oxygen consumption will also be discussed.

18b Model Based Monitoring and Control of a Cstr-Type Anaerobic Digester

Georgios Savoglidis, **Gerasimos Lyberatos**, and Costas Kravaris. University of Patras, 1 Karatheodori St., Patras, 26500, Greece

The process of anaerobic digestion is commonly used for sludge digestion in sewage treatment plants, for high organic load industrial wastewater treatment and for biogas production out of energy crops. One of the major issues addressed is maximizing the biogas production rate for energy purposes. At maximum biogas production conditions the operation of a CSTR digester becomes very sensitive to disturbances (such as organic overload or underload, entry of an inhibitor etc), leading many times to washout of the sensitive methanogenic biomass.

A mesophilic (35oC) CSTR-type 3 liter digester was constructed and properly equipped with the necessary monitoring and control equipment. The reactor temperature was maintained constant through an external jacket. The digester was started up with an anaerobic mixed culture from a wastewater treatment plant. A synthetic glucose-based feed was pumped to the reactor through a computer-controlled peristaltic pump. For monitoring of the process, a LabView monitoring environment was developed. The biogas production rate was measured in discrete time intervals using a standard volume device which was recording the time needed for a given volume of biogas to be generated. During these intervals the proportion of the methane in the biogas was also measured using an infrared ion detector.

The reactor was operated in a sequence of retention times, and the dynamic responses were followed until steady-state was reached. A two-state model (biomass and the limiting substrate as state variables) was developed for monitoring and control purposes. The model parameters were estimated from the steady-state data.

A nonlinear observer was then developed based on the exact error linearization method for the calculation of the observer gains. The observer was used for estimating the unknown states and parameters that cannot be measured and detection of the potential presence and magnitude of a disturbance. These disturbances could either be a change in the organic feed concentration, a change in the feed flow rate or entry of an inhibitor, which would be responsible for a reduction in the maximum specific growth rate of the bacteria. The ability of the observer to properly identify the type and magnitude of the imposed disturbance was demonstrated experimentally.

In the sequel, a control policy near optimal steady state, was approached through proportional output feedback control with a linear control law previously described in [1], and further examined for discrete time systems in [2] and [3]. The proposed control law globally stabilizes the process around the optimal steady state without leading to washout of the biomass. The robustness of the control law was also studied. The control law was extremely robust to variations of certain parameters of the model used.

References

[1]L. Syrou, I. Karafyllis, K. Stamatelatou, G. Lyberatos and C. Kravaris, Robust Global stabilization of continuous bioreactors, Proceedings 7th International Symposium on Dynamics and Control of Process Systems (DYCOPS 7), Cambridge , MA, July 2004.

[2]I. Karafyllis, G. Savvoglidis, L. Syrou, K. Stamatelatou, C. Kravaris and G. Lyberatos, Global stabilization of continuous bioreactors, Paper 15C08, EA64484, 2006 AIChE Annual Meeting, San Francisco, CA.

[3]I. Karafyllis, C. Kravaris, L. Syrou and G. Lyberatos, A vector Lyapunov function characterization of Input-to-State stability with application to robust global stabilization of the chemostat, European Journal of Control, Vol. 14, 2008, p. 47-61.

18c Measurement of Toluene and Trichloroethylene Using Oxygenase-Based Fiber-Optic Biosensors

Zhong Zhong[1], Sean Pieper[2], Manasi Katragadda[2], Thomas K. Wood[3], Kevin L. Lear[2], David S. Dandy[1], and Kenneth F. Reardon[1]. (1) Department of Chemical and Biological Engineering, Colorado State University, 1370 Campus Delivery, Fort Collins, CO 80523, (2) Department of Electrical and Computer Engineering, Colorado State University, 1370 Campus Delivery, Fort Collins, CO 80523, (3) Artie McFerrin Department of Chemical Engineering, Texas A&M University, 200 Jack E. Brown Building, MS 3122, College Station, TX 77843-3122

Groundwater monitoring plays an important role in safeguarding our water supply, especially drinking water resources. Toluene and trichloroethylene (TCE) are two common groundwater contaminants, as well as critical contaminants in the National Primary Drinking Water Regulations. Traditional monitoring for groundwater contaminants usually involves field sampling, transport, and laboratory measurements (usually by GC with an FID or MS detector, or by HPLC). These measurements are expensive and slow and require removal of the sample from the site, which adds to the cost and alters the analyte concentration through volatile losses and nonlinear volume averaging during the sampling process. Sensors, and especially biosensors, could avoid these disadvantages by providing real-time in situ measurement with highly specific and sensitive detection. A biosensor usually consists of a biological component that detects the analyte through an enzymatic reaction or biological binding, a transducer that translates the biochemical signal into an optical or electrical signal, and a detector that recognizes the signal and allows it to be correlated to the analyte concentration.

In this project, oxygenase-based fiber-optic biosensors are developed to measure toluene and TCE respectively. These biosensors are made by using oxygenase-expressing bacteria as the biocomponent and oxygen optodes as the transducer. The optical signal change (OSC) has been found to correlate to the analyte concentration. Employing different oxygenases as the biocomponent yielded biosensors with different responses to toluene and TCE. For example, biosensors constructed with ttoluene ortho-monooxygenase (TOM) exhibit an $OSC/\Delta C_{toluene}$ = 148 counts•L/mg with limits of detection of 300 μ g/L, while biosensors made with TOM-Green (a selected mutation of TOM derived from directed evolution) were found to have an $OSC/\Delta C_{TCE}$ = 694 counts•L/μ g with limits of detection of 0.2 μ g/L. The calibrations of these biosensors were verified by GC/MS. The average response time for these biosensor is around 1 hour. The biosensors could retain more than 75% of their original activity for at least a week and could be regenerated by exposure to formate or glucose.

18d Ligand-Responsive Micrornas: Toward Probing and Programming Cellular States

Chase L. Beisel, Chemical Engineering, California Institute of Technology, Pasadena, CA 91125 and Christina Smolke, Chemistry and Chemical Engineering, California Institute of Technology, 236A Spalding, Mail Code 210-41, Pasadena, CA 91125.

RNA is a flexible substrate for the construction of information processing devices that evaluate the cellular environment and regulate gene expression accordingly. These devices are principally composed of an aptamer, nucleic acids that bind target molecules with high affinity and specificity, and a gene regulatory element such that regulatory activity is modulated by a binding event between the aptamer and its cognate ligand. The gene regulatory element can act through a myriad of mechanisms that have unique dynamics and regulatory strength, although not every mechanism is functional in a given organism. Therefore, utilizing many of these mechanisms will greatly expand the applicability of RNA-based information processing devices.

In this work, we selected microRNAs (miRNAs) as the gene regulatory element. MiRNAs are endogenous small RNAs that silence gene expression through the RNA interference (RNAi) pathway and are present in most eukaryotes, including fungi, plants, and animals. The regulatory capacity of miRNAs is astounding, as recent studies have suggested that miRNAs regulate over 10% of human genes, including genes implicated in carcinogenesis. Recent studies have also revealed the structural requirements for miRNA processing, providing valuable insights that contribute to the design of ligand-responsive miRNAs.

We engineered ligand-responsive miRNAs by incorporating aptamers into the miRNA basal segment. Initial miRNA processing and subsequent activation of RNAi is sensitive to the basal segment secondary structure such that ligand binding inhibits processing, resulting in upregulation of the target gene. Through the development of ligand-responsive miRNAs, we established a modular composition framework that promotes independent replacement of the aptamer and miRNA sequences to change either the recognized ligand or the target gene. This design was extended by inserting multiple ligand-responsive miRNAs into the same transcript for multi-input control of multiple genes with variable extents of gene silencing. We also demonstrated that the dynamic range can be increased by inserting the ligand-responsive miRNA into the target transcript such that miRNA processing and RNAi-mediated silencing are both inhibited upon ligand binding.

Our work demonstrates that miRNAs can be engineered to serve as information processing devices to dynamically monitor and control cellular processes. These devices can serve as in vivo biosensors that provide a dynamic read-out of the intracellular environment or as autonomous control devices that evaluate the environment and regulate gene expression accordingly.

18e Immobilization Strategies for Single-Chain Antibody Microarrays

Shannon L. Servoss[1], Cheryl Baird[2], Keith Miller[2], Noah Pefaur[2], Rachel Gonzalez[2], David Apiyo[2], Heather Engelmann[2], Sudhir Srivastava[3], Jacob Kagan[3], Karin Rodland[2], and Richard Zangar[2]. (1) Chemical Engineering, University of Arkansas, Fayetteville, AR 72701, (2) Pacific Northwest National Laboratory, Richland, WA 99352, (3) National Cancer Institute, Bethesda, MD

Sandwich enzyme-linked immunosorbent assay (ELISA) microarrays have great potential for validating disease biomarkers. Each ELISA relies on the availability of two robust affinity reagents that retain activity when immobilized on a solid surface or when labeled for detection. Current procedures for the development of traditional immunoglobulin G (IgG) are slow and expensive. One alternative to IgG are single-chain antibodies (scFv), which are affinity reagents that have greater potential for high-throughput production. Unfortunately, scFv are typically less active than IgG following immobilization on a solid surface and not always suitable for use in sandwich ELISAs. We have investigated different immobilization strategies and scFv constructs to determine a more robust strategy for using scFv as ELISA reagents. Two promising strategies emerged from these studies: (1) the precapture of epitope-tagged scFv using an anti-epitope antibody and (2) the direct printing of a thioredoxin/scFv fusion protein on glass slides. Both strategies improved the stability of immobilized scFv and increased the sensitivity of the scFv ELISA microarray assays, although the anti-epitope precapture method introduced a risk of reagent transfer. Using the direct printing method, we show that scFv against prostate specific antigen (PSA) are highly specific when tested against 21 different IgG-based assays. In addition, the scFv microarray PSA assay gave comparable quantitative results to a commercial 96-well ELISA when tested using human serum samples. In addition, we find that thioredoxin-scFv fusions against other biomarkers of interest also have good lower limits of detection. Overall, these results suggest that minor modifications of the scFv construct are sufficient to produce reagents that are suitable for use in multiplex assay systems.

18f Performance Limits and Optimization for T-Rflp Sensors

Jeffrey Kantor, Chemical and Biomolecular Engineering, University of Notre Dame, Notre Dame, IN 46556

Terminal restriction fragment polymorphism (t-rflp) of rDNA for the16S small ribosomal subunit has become a widely used method for analyzing microbial communities, with the potential for a new class of low-cost sensors of bacterial organisms for medical and environmental applications. Large databases of fragment lengths are now available from experimental data in addition to in-silico digestions of over 500,000 bacterial organisms and up 90 restriction enzymes. Bacterial identification, however, is limited by the number of distinct fragment lengths produced by digestion with typical restriction enzymes, and the ability to experimentally resolve fragment lengths.

In this paper we demonstrate limits to performance of sensors based on t-rflp. The analysis uses Shannon and Renyi information entropy as a priori and a posteriori measures of bacterial probability density function. The conditional 'information gain' is then a metric measuring the effectiveness of a sensor based on either single or multiple restriction enzymes. We demonstrate this principle by computing information gain and the number of distinguishable bacterial 'equivalence classes' of typical restriction enzymes used for the phylogenetic analysis of bacterial colonies in the human gut.

Typical t-rflp instrumental methods combine digests from multiple restriction enzymes. For this case, we formulate and solve for the maximum entropy aposteriori probability distribution for bacterial organisms. We further show how to optimize information gain for two characteristic cases. The first is optimal combination of multiple digests to form a single t-rflp chromatogram which leads to a large scale convex optimization problem. The second case of multiple enzymes and multiple length chromatograms leads to an intractably large optimization problem, but for which we can propose performance metrics.

These ideas are demonstrated using literature data for t-rflp in medical and ecological applications.

Web Page: www.nd.edu/~jeff/Publications/Presentations/T-RFLP_Detection.pdf

41a Propofol Induces a GABA-Independent Expression of C-Fos and Egr-1 In Neuronal Cells

Srivatsan Kidambi[1], Joel Yarmush[2], Wayne Fong[2], Sangeetha Kamath[2], Joseph SchianodiCola[2], and Yaakov Nahmias[3]. (1) Center for Engineering in Medicine, Harvard Medical School, Massachusetts General Hospital, Shriners Hospital, New York Methodist Hospital, 51 Blossom Street, Boston, MA 02114, (2) Department of Anesthesiology, New York Methodist Hospital, Brooklyn, NY 11215, (3) Center for Engineering in Medicine, Massachusetts General Hospital, Harvard Medical School, 51 Blossom Street, Boston, MA 02114

Propofol is a commonly used intravenous anesthetic agent, which produce rapid induction of and recovery from general anesthesia. Numerous clinical studies reported that propofol can potentially cause amnesia and memory loss in human subjects. The underlying mechanism for this memory loss is unclear but may potentially be related to the induction of memory-associated genes such as c-Fos and Egr-1 by propofol.

Mouse neuroblastoma (N2A) cells were exposed to varying concentrations of propofol at multiple time intervals. Cellular viability was quantified by fluorescence microscopy and the transcription of the immediate early genes, c-Fos and Egr-1, was quantified using quantitative reverse transcriptase polymerase chain reaction (qRT-PCR). GABA-A receptor antagonists and MAPK/ERK inhibitors were used to investigate the mechanism of action. Propofol induces time and dose dependent transcription of c-Fos and Egr-1. At physiologically relevant concentration (3μ g/ml) propofol caused a 6-fold and 2.5-fold induction in c-Fos and Egr-1 transcription, respectively. The propofol-induced c-Fos and Egr-1 expression was unaffected by bicuculline, GABA receptor antagonist but was abolished by PD98059, a MAPK/ERK inhibitor, suggesting the possible involvement of the MAPK/ERK pathway. Our study suggests that at physiological concentrations propofol can induce long-term changes of neuronal function by changing gene transcription. Furthermore, the induction of c-Fos and Egr-1 by propofol is mediated via a non GABAergic mechanism involving the MAPK/ERK pathway. Future work will examine the effects of propofol in primary neurons and elucidate the role of the GABA pathway in a more physiologic model of the central nervous system.

41b Influence of Micrornas In the De-Differentiation of Primary Rat Hepatocyte Cells

Juan C. Lagos and Charles M. Roth. Chemical and Biochemical Engineering, Rutgers University, 59 Jackson Avenue, Princeton, NJ 08540

The development of a hepatocyte culture system stably expressing a spectrum of differentiated functions is a highly desired objective for pivotal applications in the fields of toxicology and bioartificial liver systems. Various approaches have been explored to prevent or delay phenotypic alterations attributable to changes in liver specific gene expression during hepatocyte isolation and cell culture processes. While a number of advances have been made in engineering substrates and media, a system for hepatocyte cell culture with specific cell functions at levels comparable to those observed in vivo does not yet exist.

MicroRNAs (miRNAs) are endogenously produced 21 to 23 nucleotide non-coding-RNAs that regulate expression of target genes via partial sequence complementarity. Roles for several miRNAs' have been demonstrated that encompass development, apoptosis, growth, fat metabolism and differentiation in diverse organism [1]. For example, Esau, et. al have shown that miR-122 regulates lipid metabolism in the adult liver in vivo [2]. However, regulation of microRNAs as a driving force for loss of differentiated functions in cultured hepatocytes has yet to be investigated.

In order to test the hypothesis that alterations of miRNAs play a significant role in the de-differentiation of hepatocytes during the first days of culture, we extracted the miRNAs from freshly isolated hepatocytes versus those cultured on a single layer of collagen for 4 days and profiled them by microarray analysis. Among the 238 rat miRNA probes, several miRNAs were consistently altered in their expression, including miR-122, which was induced approximately 3.5-fold in cultured vs. freshly isolated hepatocytes. Important targets of miR-122 were predicted computationally, establishing a biochemical network for miRNA regulation of primary rat hepatocytes. These include several liver-specific transcription factor genes. The mRNA levels of miR-122 and several of its targets were monitored and validated using TaqMan RT-PCR. These results demonstrate that microRNAs such as miR-122 are altered in their expression during early hepatocyte culture and suggest that they may play a role in coordinating gene expression programs.

In order to explore a possible causative role for miR-122 on hepatocyte differentiation status, we are currently manipulating its expression using miR-122 mimics and antagomirs. Liver morphology, urea and albumin production are being used in concert with miR-122 target gene expression as markers for hepatocyte differentiation. In addition, we are investigating whether manipulations known to improve performance of hepatocyte cultures – such as cultivation in double layer collagen configuration [3] or the addition of retinoic acid [4] – acts via a miRNA mediated mechanism. These findings set the stage for manipulation of function in hepatocyte culture using microRNAs as master molecular switches of gene expression programs.

References:

[1] Bartel, David P., 2004. MicroRNAs: Genomics, Biogenesis, Mechanism, and Function. Cell 116, 281 - 297

[2] Esau, C., Davis, S., Murray, S, Yu, X., Pandey, S., Pear, M., Watts, L., Booten, S., Graham, M., McKay, R., Subramaniam, A., Propp, S., Lollo, B., Freier, S., Bennett, F., Bhanot, S. and Monia, B., 2006. miR-122 regulation of lipid metabolism revealed by in vivo antisense targeting. Cell Metabolism 3, 87-98.

[3] Dunn, J., Tompkins, R., and Yarmush, M. 1991. Long-Term in Vitro Function of Adult Hepatocytes in a Collagen Snadwich Configuration. Biotechnol. Prog. 7, 237-245.

[4] Burley, M., Guzikowski, S. and Roth, C., 2007. Retinoic Acid Induces Morphological and Functional Changes in Primary Rat Hepatocytes. Submitted for publication.

41c Reduction of Adipocyte Triglyceride Accumulation through Targeted Perturbation of Pyruvate Metabolism

Hai Shi[1], Yaguang Si[2], and Kyongbum Lee[1]. (1) Chemical and Biological Engineering, Tufts University, 4 Colby Street, Room 284, Medford, MA 02155, (2) Obesity Research Center, Boston University School of Medicine, 650 Albany Street, Room 840, Boston, MA 02118

We previously reported that a redistribution of pyruvate flux from lactate fermentation into the TCA cycle significantly correlated with triglyceride (TG) accumulation in 3T3-L1 adipocytes. The goal of this study was to test the hypothesis that TG accumulation could be altered by specifically perturbing pyruvate metabolism. To address this hypothesis, we treated cultured 3T3-L1 adipocytes with inhibitors of lactate dehydrogenase (LDH) and pyruvate carboxylase (PC), and characterized their global effects on intermediary metabolism using metabolic flux analysis (MFA). Long-term inhibition of LDH or PC (over several days) did not alter the adipocyte differentiation program as assessed by the expression levels of peroxisome proliferator-activated receptor-γ and glycerol-3-phosphate dehydrogenase. Inhibiting LDH or PC up-regulated lipolysis and decreased TG accumulation. Inhibiting PC also up-regulated glycolysis. Metabolic flux analysis indicated that the reduction in TG accumulation was largely explained by decreased de novo fatty acid synthesis. Importantly, the increase in glycolysis and reduction in TG was observed even when fatty acids were added to the culture medium, albeit to a lesser extent than when glucose was the only major carbon source. In the absence of the inhibitors, addition of long chain free fatty acids (FFAs) to the culture medium did not affect net TG accumulation, as the increased uptake of FFAs depressed de novo synthesis. In conclusion, the results of this study support our hypothesis regarding pyruvate as a controlling metabolite of TG synthesis. Prospectively, the approach presented here could help formulate a new strategy to develop therapeutic agents for limiting excessive fat accumulation at the cellular level.

41d A Computational Study of 'inside-out' Signal Dynamics and Integrin Activation during Leukocyte Trafficking

Michael Beste, Chemical and Biomolecular Engineering, University of Pennsylvania, Philadelphia, PA 19104, Elena Lomakina, Pharmacology & Physiology, University of Rochester, Goergen Hall, Rochester, NY 14627, Richard E. Waugh, Biomedical Engineering, University of Rochester, Goergen Hall, Rochester, NY 14627, and Daniel A. Hammer, Bioengineering, University of Pennsylvania, 240 Skirkanich Hall, 210 South 33rd Street, Philadelphia, PA 19104.

Leukocytes homing to peripheral sites of immune challenge rely on rapid activation of 'inside-out' signaling pathways to trigger localized cell arrest and emigration into surrounding tissues. This cascade is initiated by endothelial-displayed chemokines which bind G-protein coupled receptors at the cell interface and culminates in the activation of resting surface integrins. Using a computational approach to model biochemical transduction events, we address how the intracellular organization of such 'inside-out' signaling networks impacts the efficiency of cell arrest. In particular, we focus on the regulation of the β_2 integrin, LFA-1, whose affinity is modulated by the small GTPase Rap1. Our stochastic framework explicitly simulates the temporal and spatial evolution of known molecular species downstream of the chemokine receptor leading to Rap1 and subsequent LFA-1 activation. The dynamics of integrin activation are synchronized to Adhesive Dynamics simulations which model the mechanics of adhesion and predict the overall likelihood for cell arrest.

Using this integrated modeling strategy, we demonstrate how feedback and amplification motifs within the signaling pathway confer leukocytes with a tunable sensitivity to chemotactic stimuli. These predictions are consistent with experimental measurements recently obtained from combined signaling and adhesion assays on human neutrophils. We extrapolate our general findings to consider the specialized case of lymphocytes which activate integrins through multiple chemokine receptors. When downstream signals cross-regulate two or more receptors, the model predicts that the 'inside-out' network can elicit functional synergy in the adhesion response and trigger cell arrest at sub-optimal levels of chemokine exposure.

41e Altered EGFR Trafficking and Signaling In Lung Cancers with Kinase-Inhibitor-Sensitizing Mutations

Matthew J. Lazzara, Chemical and Biomolecular Engineering, University of Pennsylvania, 311A Towne Building, 220 South 33rd Street, Philadelphia, PA 19104-6393 and Douglas A. Lauffenburger, Biological Engineering, Massachusetts Institute of Technology, 77 Massachusetts Ave, 56-341, Cambridge, MA 02139.

The majority of non-small-cell lung cancers involve EGFR overexpression, but clinical response to EGFR kinase inhibitors is limited almost exclusively to cells bearing a set of recently identified EGFR mutants. These mutations render cells "oncogene-addicted" to EGFR signaling, such that they readily undergo apoptosis in response to low concentrations of EGFR kinase inhibitors. Compared to WT EGFR counterparts, lung cancer cells with mutant EGFR display increased basal activity of the receptor, prolonged ligand-mediated receptor activation, and altered downstream activation dynamics of several key signaling intermediates. We have previously shown that these alterations result in part from a defect in the rate of ligand-mediated endocytosis of EGFR mutants. Here, we extend that work and demonstrate how altered EGFR endocytosis impacts downstream signaling dynamics in several key pathways. Specifically, we show that while the activities of some pathways are enhanced as a result of mutation, the activities of others are impaired. By overexpressing the EGFR mutants in cell lines of interest and working with cell lines with an inducible switch for the expression of a key protein involved in EGFR endocytosis, we have demonstrated that the alterations we observed in human lung cancer cell lines can be recapitulated in predictable ways and that they are tightly tied to receptor trafficking. In addition, we have shown that RNAi-mediated down-regulation of signaling intermediates impaired downstream of EGFR mutation in cells with WT EGFR increases cellular sensitivity to kinase inhibitors. In total, our results demonstrate that the EGFR mutations under consideration create a quantitative bifurcation of downstream signaling pathways, and that it is the net effect of these bifurcated signals that ultimately determines cellular response to EGFR kinase inhibition. Characterization of this bifurcation portends enhanced ability to successfully intervene pharmacologically in lung cancers with and without EGFR mutation.

41f Nicotinamide Increases the Megakaryocytic Maturation of Human Hematopoietic Stem Cells Primarily Due to Sirtuin Inhibition

Swapna Panuganti[1], Eleftherios T. Papoutsakis[2], and William M. Miller[1]. (1) Chemical and Biological Engineering, Northwestern University, 2145 Sheridan Road, Evanston, IL 60208, (2) Dept. of Chemical Engineering, Delaware Biotechnology Institute, 15 Innovation Way, Newark, DE 19711

Megakaryocytic cells (Mks), the precursors to platelets, are derived from hematopoietic stem cells (HSCs) and are among the least well-understood blood cell types. During differentiation, Mks increase in cytoplasmic volume and surface area, develop multi-lobated nuclei, become polyploid, form cytoplasmic extensions called proplatelets from which platelets are believed to arise, and undergo a complex apoptotic program. The process of becoming polyploid involves a variation of the mitotic cell cycle whereby the cells duplicate their DNA content without undergoing cytokinesis thus forming cells with 4N, 8N, 16N, etc. nuclei. Higher Mk DNA has been shown to be strongly correlated with greater platelet production. Thrombocytopenia, a deficiency in blood platelet levels, is associated with a variety of hematological malignancies and can also arise as a side effect of high-dose chemotherapy in cancer patients. In many cases, thrombocytopenia is associated with defective Mk maturation in terms of achieving high-ploidy Mk production *in vivo*. Elucidation of the factors that regulate Mk polyploidization will aid in developing treatments for Mk-related disorders. We have previously reported that nicotinamide, soluble vitamin B_3, increases the percentage of high-ploidy Mks ca. 3-fold and enhances proplatelet formation versus primary Mk cultures supplemented only with thrombopoietin. Nicotinamide is a well-established and potent inhibitor of sirtuin family of histone/protein deacetylases (SIRT) and has been found to be well tolerated by human subjects at high doses. The SIRT1/2 inhibitor cambinol increased Mk ploidy to a similar extent as nicotinamide without affecting Mk commitment or apoptosis. EX-527 and AGK2, selective inhibitors of SIRT1 and SIRT2, respectively, were approximately 30% as effective as nicotinamide at increasing Mk ploidy. This suggests that nicotinamide increases Mk ploidy via synergistic inhibition of SIRT1 and SIRT2. Nicotinamide also increases intracellular levels of NAD(H) 5-fold, but the application of an NAD^+ de novo pathway precursor had minimal impact on ploidy, suggesting that the increase in intracellular NAD(H) is not the primary factor impacting Mk ploidy. We are currently investigating further downstream roles of the SIRT homologs in Mks such as acetylation of apoptosis related factors. Future plans include knockdown of SIRT1/2 in primary murine HSC cultures.

41g Elasticity of Nsp1 and Its Binding with Kap95 Probed by Single Molecule Force Spectroscopy

Amit K. Dutta[1], Jacklyn Novatt[2], Tijana Jovanovic[2], Anna Sophia McKenney[2], Michael Rout[2], Brian T. Chait[2], and Georges Belfort[1]. (1) Chemical and Biological Engineering, Rensselaer Polytechnic Institute, 110 8th st, Troy, NY 12180-3590, (2) The Rockefeller University, 1230 York Avenue, New York, NY 10021

Nuclear pore complexes (NPC) facilitate receptor-mediated translocation of proteins and ribonucleoprotein complexes between the nucleus and the cytoplasm in eukaryotic cells. Small molecules can move through the NPC via passive diffusion, but the translocation of cargo proteins larger than ~40 kD generally requires specific transport receptor molecules collectively termed karyopherins (Kaps). The translocation of Kap-cargo complexes occurs due to specific binding between Kaps and integral NPC proteins called nucleoporins that contain degenerate multiple repeats of "Phe-Gly" and are called FG-Nups. Nsp1 is an FG-Nup located at the entrance and exit of the NPC. This work focuses on the preparation and nano-mechanical characterization of surface tethered Nsp1 using Atomic Force Microscopy. Using single molecule force spectroscopy, we probe the stretching, flexibility and Kap-binding characteristics of Nsp1. Knowing the elastic properties of Nsp1 should provide insight into the mechanisms of binding with Kap95 complexes and the transport of large cargo molecules through the Nuclear Pore Complex.

44a Controlling Orientations of Immobilized Oligopeptides Using N-Terminal Cysteine Labels

Xinyan Bi and Kun-Lin Yang. Department of Chemical and Biomolecular Engineering, National University of Singapore, 4 Engineering Dr 4, Singapore, 117576, Singapore

We present a strategy of using N-terminal cysteine labels for controlling the immobilization of oligopeptides on aldehyde-terminated surfaces through the formation of stable thiazolidine rings in this presentation. We study the effect of cysteine position (either N-terminal or C-terminal) and lysine residue on the immobilization of oligopeptides. Based on our ellipsometry and quartz crystal microbalance (QCM) results, we conclude that the proposed immobilization strategy is highly site-specific. It only works when cysteine is in the N-terminal position, and the formation of thiazolidine is much faster than the formation of imines between lysine residues and aldehydes, even in the presence of a reducing agent such as NaBH3CN. By labeling an oligopeptide CSNKTRIDEANNKATKML with an N-terminal cysteine, we immobilize this oligopeptide on an aldehyde-terminated surface and investigate the enzymatic activity of trypsin acting on the oligopeptide. It is found that trypsin is able to cleave the immobilized oligopeptide having a single anchoring point at the N-terminal cysteine. No cleavage is observed when the oligopeptide is immobilized through multiple anchoring points at lysine residues.

44b Giant Spacer Entropic Effect In Binding to Surface Attached Probes: DNA Microarrays

Arnold Vainrub, College of Veterinary Medicine, Auburn University, 109 Greene Hall, Auburn, AL 36849-5518, Xiaolian Gao, Department of Biology and Biochemistry, University of Houston, Houston, TX 77204-5001, and B. Montgomery Pettitt, Department of Chemistry, University of Houston, Houston, TX 77204-5003.

In DNA microarray a free in solution assayed genomic DNA hybridizes with an oligonucleotide probe attached to the surface and forms a DNA duplex. Restriction of translational and rotational motion of the duplex because of the surface attachment gives rise to the entropy decrease. The rotational entropy penalty is shown to be substantial and account for a giant linker effect in the hybridization on the surface. The translational motion entropy effect is shown to vanish. A simple gas model is used to elucidate the physics and calculate the reduction of equilibrium binding constant for surface attached biomolecular probes. The rotational entropy barrier decreases when the probe is attached to the surface through a spacer molecule (oligomer) and disappears when the spacer allows free rotation of the duplex. Experiments were done for various lengths of (dT)n oligomer spacer and different surface densities of oligonucleotide probes. The theory is shown to be in complete accord with the experiments.

Web Page: www.vetmed.auburn.edu/index.pl/vainrub

44c Genetic Identification of Malaria with Dielectrophoresis on Nano-Gentic Beads

Zachary R. Gagnon[1], Satyajyoti Senapati[2], and H.-C. Chang[1]. (1) Chemical and Biomolecular Engineering, University of Notre Dame, 182 Fitzpatrick Hall, Notre Dame, IN 46556, (2) Chemical & Biomolecular Engineering, University of Notre Dame, 182 Fitzpatrick Hall, Notre Dame, IN 46556

It is well known that silica particles on the order of 100 nm have an electrical conductivity $< 10^{-7}$ S/m, much less than even that of deionized water, ~ 20 μS/m, and still exhibit positive dielectrophoresis (DEP) when suspended in such a low conductivity electrolyte. Such behavior is commonly attributed to Stern layer adsorption of surface charges. This increased surface charge density on the particle surface facilitates ion and electron transport around the particle and can thus give rise to an increase in the effective conductivity of the particle enabling low conductive particles to still exhibit positive DEP in a media of greater conductivity.

The frequency at which the induced particle dipole goes to zero, known as the crossover frequency, is highly dependent on the surface conductance of the particle. We have shown previously that DNA hybridization on the surface of a 100 nm functionalized silica particle leads to detectable surface conduction changes which make it possible to detect DNA hybridization reactions by simply measuring changes in particle suspension crossover frequency (cof).

In this work we utilize the above mentioned phenomena to detect the malaria species Plasmodium falciparum in an infected red blood cell suspension. Additionally, it is shown that DEP can not only be used as a genetic sensor, but can also be utilized to optimize the conditions for which particle surface hybridization takes place. By varying the particle surface concentration of oligonucleotide it is shown that a suspension cof can determine the optimal surface concentration for which hybridization can occur. Additionally, since the cof is sensitive to the length of the DNA strand, we also demonstrate the ability to identify different species of malaria based on differences in their PCR amplified DNA lengths.

Identifying the Removal Mechanism of Fouling Proteins In a Surface Acoustic Wave Biosensor: A Numerical Study and Comparison to Experiments

Subramanian K.R.S. Sankaranarayanan and Venkat R. Bhethanabotla*
Sensors Research Laboratory, Department of Chemical and Biomedical Engineering,
University of South Florida, Tampa, Florida, 33620, USA

Abstract — Biosensors typically operate in liquid media for measurement of biomarkers and suffer from fouling mechanisms such as nonspecific binding of protein molecules to the device surface. Using a novel numerical technique as well as experiments, we have identified that fluid motion induced by high intensity sound waves, such as those propagating in these sensors, can lead to the removal of the nonspecifically bound proteins, thereby eliminating sensor fouling. We present a computational and experimental study of the acoustic-streaming phenomenon induced biofouling elimination by surface acoustic waves (SAWs). The transient solutions generated from the developed coupled field fluid solid interaction (FSI) model were utilized to predict trends in acoustic-streaming velocity for various design parameters such as voltage intensity, device frequency, fluid viscosity and density. The model predictions were utilized to compute the various interaction forces involved and thereby identify the possible mechanisms for removal of nonspecifically-bound proteins. Our study indicates that the SAW body force overcomes the adhesive forces of the fouling proteins to the device surface and the fluid-induced drag and lift forces prevent its re-attachment. The streaming velocity fields computed using the finite-element models in conjunction with the proposed mechanism were used to identify the conditions leading to improved removal efficiency. Our research findings have significant implications in designing reusable and highly sensitive biosensors.

Keywords – Acoustic streaming; SAW; Biofouling; Finite element method

I. Introduction

SAW and other transducer devices used in biological-species detection suffer from fouling that result from binding of nonspecific protein molecules to the device surface. Nonspecific binding dramatically reduces the sensitivity and selectivity of biosensors [1]. Our experimental and numerical study indicates that the acoustic streaming phenomenon, *i.e.*, fluid motion induced from high intensity sound waves, can be used effectively to remove nonspecifically bound proteins to allow reuse of SAW biosensors (Fig. 1). The acoustic-streaming phenomenon can be used to remove these nonspecifically-bound proteins to allow more accurate determinations and reuse of these devices [2]. The generated sound fields cause tangential fluid motion along the inter-phase boundaries. These motions exert steady viscous stress on the boundary layer leading to liquid circulation near the boundaries. Although these stresses are not large, they are still significant enough to remove loosely-bound material on the surface of the device. Understanding the fluid dynamics in such a system is useful for efficient removal of nonspecifically-bound proteins. The application of this work broadly applies to all transducers used for biological species sensing that suffer from fouling and nonspecific binding of protein molecules to the device surface.

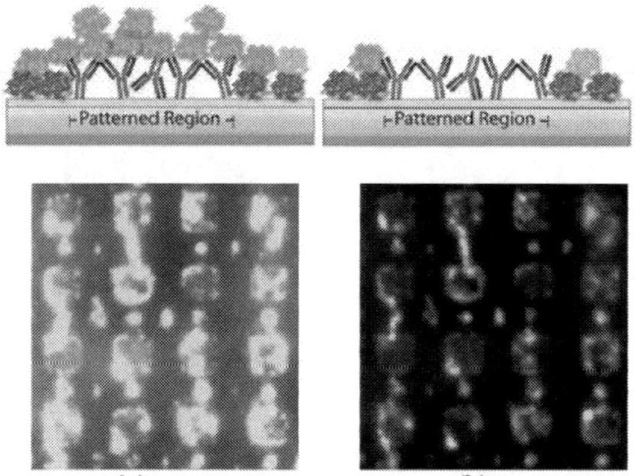

(a) (b)

Figure 1. Nonspecifically-bound bovine serum albumin (BSA) fluorescent image (pseudo-colored) with illustration above of SAW device surface at end of experiment (RF power dose = 12,500 mW sec). The illustrations in (a) and (b) depict the removal of nonspecifically-bound proteins accomplished by acoustic-streaming phenomenon induced by the Rayleigh wave interaction with the fluid medium.

In this work, a finite element fluid solid interaction (FE-FSI) model of acoustic streaming phenomenon resulting from the interaction of SAWs (Rayleigh mode) with liquid loading is developed (Fig. 2). Parameters studied using the FE-FSI model include voltage intensity, device frequency, fluid viscosity and density. The transient solutions generated from the model were used to predict trends in acoustic streaming velocity. The generated streaming velocity fields are then utilized to predict the various adhesive and removal forces experienced by the specific and nonspecifically bound protein molecules located at various regions along the SAW biosensor.

The predicted mechanism of removal of nonspecifically bound proteins is based on an order of magnitude comparison of the various forces involved. Typically, there are mainly three different mechanisms for particle removal, namely, sliding, rolling and lift-off. For the sake of efficient removal across the entire SAW delay path, it is necessary to identify the dominant mechanism involved. This can be established by comparing the relative magnitudes of the various removal forces. The forces responsible for removal of particles are mainly characterized as the direct SAW forces, the lift and the drag forces that result from the mean velocity field in the fluid, whereas, the dominant adhesive forces for particles immersed in a liquid are the van der Waals and double layer forces. Based on the calculated streaming velocity fields, the interaction forces are analyzed for varying input parameters to identify conditions that lead to improved removal efficiency.

II. THEORY

A coupled-field FSI model of a SAW device based on a micron-sized piezoelectric substrate ($YZ\text{-}LiNbO_3$) in contact with a liquid loading was developed to study surface-acoustic-wave interaction with fluid loading [3]. A system of four coupled wave equations for the electric potential and the three component of displacement in piezoelectric materials are solved for the piezoelectric substrate or the solid domain [4]:

$$-\rho\frac{\partial^2 u_i}{\partial t^2}+c^E_{ijkl}\frac{\partial^2 u_k}{\partial x_j \partial x_l}+e_{kij}\frac{\partial^2 \phi}{\partial x_k \partial x_j}=0 \qquad (3.1)$$

$$e_{ikl}\frac{\partial^2 u_k}{\partial x_i \partial x_l}-\varepsilon^s_{ik}\frac{\partial^2 \phi}{\partial x_i \partial x_k}=0 \qquad (3.2)$$

These coupled wave equations can be discretized and solved for generating displacement profiles and voltages at each element/node. The piezoelectric material displacements obtained from the above equations are applied to the fluid domain at each time step.

Fluid domain was modeled using the Navier-Stokes equation; the arbitrary-Lagrangian-Eulerian approach was employed to handle the mesh distortions arising from the motion of the solid substrate.

$$\rho\left(\frac{\partial v_f}{\partial t}\right)+v_f\cdot\nabla v_f+\nabla P-2\eta\nabla\cdot D=0 \qquad (3.3)$$

$$\nabla\cdot v_f=0 \qquad (3.4)$$

Here, v_f, P, ρ and η denote the fluid velocity, pressure, density, and viscosity, respectively. D is the rate of deformation tensor given by

$$D=\frac{1}{2}\left(\nabla v_f+(\nabla v_f)^t\right)$$

The fluid-solid coupling was established by maintaining stress and displacement continuity at the fluid-structure interface. The velocity v calculated from Eq. (3.3) and (3.4) contains harmonically varying terms and a "dc" term. The latter induces acoustic-streaming. When averaged over a relatively long time, the effect of the harmonically varying terms disappears and only the contributions from the dc part appear in the solution. The acoustic-streaming velocity ($\bar{v}_{a,i}$, i=x, y, and z) is therefore obtained by averaging v over a time period as follows:

$$\bar{v}_{a,i}=\frac{1}{T}\int_0^T v_i dt\text{, i=}x\text{, }y\text{ and }z \qquad (3.5)$$

---

* Author for correspondence at email: venkat@eng.usf.edu; This work was supported in part by NSF grants DGE-0221681, CHE-0722887 and ECCS-0801929, US Army Research Contract No. W81XWH-05-1-0585 and the USF Office of Research by an interdisciplinary research grant.

where T is the time period of the wave propagation.

III. COMPUTATIONAL DETAILS

Insights into the dominant mechanism responsible for efficient removal of nonspecifically bound proteins are obtained from the estimates of various forces predicted using perturbation methods [5] and trends in the streaming velocity fields. For simplicity, the protein molecules are modeled as spherical particles. The different SAW-induced forces, which include linear forces (added mass (F_{AM}), drag (F_D), lift (F_L) and Basset forces (F_B), and nonlinear ones due to radiation pressure and drag (F_{ST}), are computed. Additionally, the magnitudes of forces are evaluated for various particle sizes. The combined effect of the various interaction (adhesive and removal) forces is utilized to gain insights into the removal mechanisms of nonspecifically bound proteins. The principal adhesion forces (van der Waals (F_{vdW}) and electrical double layer (F_{el}) forces) are used to model the interaction of the specific and nonspecifically bound proteins to the biosensor surface. The forces responsible for removal of particles are mainly characterized as the acoustic radiation forces as well as the lift and drag forces that result from the mean velocity field in the fluid. The streaming velocity field (Fig. 3) computed using the FE-FSI model is utilized to estimate the fluid induced drag and lift forces for a range of operating conditions and fluid properties. Comparison of the various removal forces to the adhesive forces that bind the specific and nonspecific proteins is used to estimate the removal efficiency on the SAW biosensor and to optimize conditions.

IV. RESULTS AND DISCUSSION

A. Acoustic wave interaction with fluid medium

Simulation results predict strong coupling of ultrasonic surface waves on the piezoelectric substrate with the thin liquid layer causing wave mode conversion from Rayleigh to leaky SAW, which leads to acoustic-streaming as shown in Figs. 2 and 3. The leaky SAWs decay exponentially with distance from the source. Streaming velocity fields were studied for a range of input parameters such as input voltage, device frequency as well fluid properties such as density and viscosity. Our simulation results indicate that the induced streaming velocities typically vary from 1 µm/s to 1 cm/s with the exact values dictated by the device operating conditions as well as fluid properties.

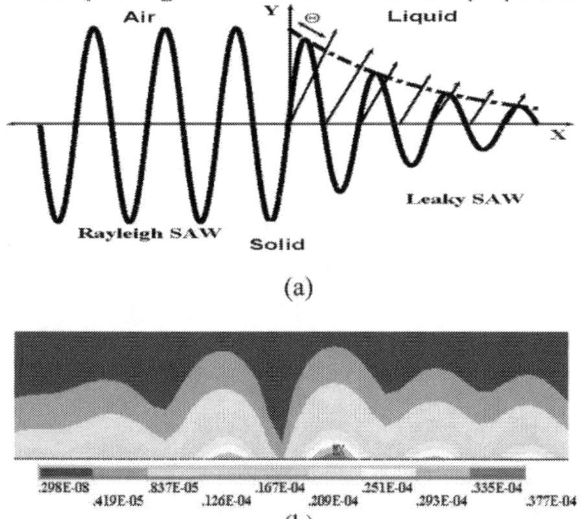

Figure 2. (a) Ultrasonic irradiation mechanism into the fluid medium (b) Simulated contours showing particle displacement profile (micrometers) in fluid region for an applied AC peak input voltage of 1 V in a SAW device operating at 100 MHz. The wave decays with distance from the interdigital transducer (IDT) fingers located at the center of the SAW device.

B. Acoustic streaming velocity

The SAW interaction with the fluid creates a net pressure gradient in the direction of sound propagation in the fluid which leads to an internal, acoustically induced streaming phenomenon [6-7]. The recirculation patterns resulting from the SAW motion gives rise to eddy formation as shown in Fig. 3. With increasing time, these eddies rise through the fluid and break into smaller ones, thereby dissipating their energy while new ones are created at the interface. Thus, the generated sound fields cause tangential motion along the inter-phase boundaries. The tangential motion results in fluid induced shear stresses, which if larger than the adhesive forces, can be used to remove the nonspecifically bound proteins. Apart from this force, the motion of the acoustic waves also induces a removal force on the fouling proteins. This force acts at a Rayleigh angle (θ) as shown in Fig. 2(a).

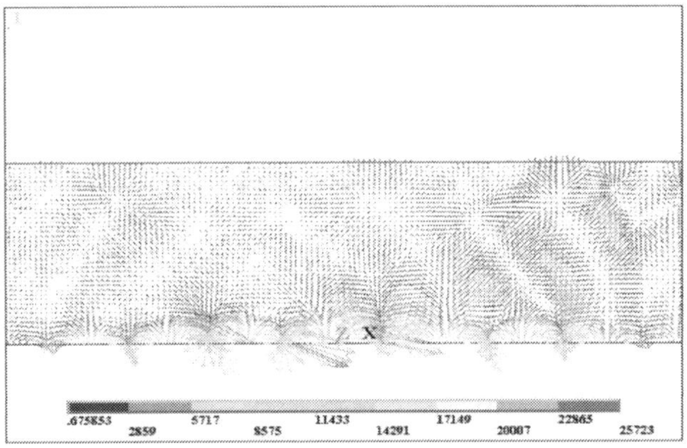

Figure 3. Induced streaming patterns in the fluid domain for a 100 MHz frequency SAW device. The calculated velocities are in μm/s. The displacements in the solid domain have been suppressed for clarity.

C. *Mechanism of nonspecifically bound protein removal*

An order of magnitude analysis of the various interaction forces was performed and the data for one such input condition are shown in Table I. Based on this analysis an approximate removal mechanism can be proposed. For micrometer- and sub-micrometer-sized protein agglomerates, the results in Table I illustrate that the fluid-induced removal forces are not significant enough to overcome the van der Waals forces. The SAW body force (F_{SAW}) is several orders of magnitude larger than the adhesion forces for the range of particle diameters considered in Table I.

TABLE I. FORCES (N) VERSUS PARTICLE RADIUS *R* (MICROMETERS) FOR A 100 MHZ SAW DEVICE AND AN APPLIED INPUT VOLTAGE OF 1 V. THE FLUID VISCOSITY IS 1 CP. THE CALCULATED FORCES ARE BASED ON THE STREAMING VELOCITY FIELD GENERATED AT THE IDT REGION.

R	0.1	1	10
F_{vdW}	-2×10^{-9}	-2×10^{-8}	-2×10^{-7}
F_{SAW}	4×10^{-6}	4×10^{-4}	4×10^{-2}
F_L	3×10^{-20}	3×10^{-18}	3×10^{-16}
F_{ST}	2×10^{-14}	2×10^{-13}	2×10^{-12}
F_{AM}	1.5×10^{-13}	1.5×10^{-10}	1.5×10^{-7}
F_B	6×10^{-14}	6×10^{-12}	6×10^{-10}

It therefore appears that the SAW direct force causes the removal or detachment of the nonspecific proteins from the SAW surface and moves the particle away from the region of influence of the adhesive forces. The van der Waals adhesion forces decrease rapidly with distance from the SAW surface. However, the SAW direct forces also decay rapidly with distance into the fluid. Hence, a steady force is required to cause the removal of the detached nonspecific proteins. The horizontal streaming-induced F_{ST} helps to push the nonspecific proteins away from the fouled area. The vertical streaming force or F_L helps prevent re-adhesion of the proteins to the surface. Thus, the fluid-induced drag and lift forces result in a net displacement of the detached proteins. The mechanism of removal is depicted in Fig. 4.

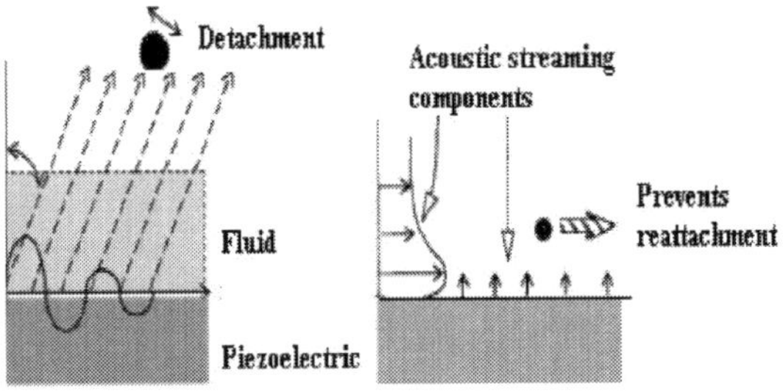

Figure 4. Predicted mechanism based on adhesive and removal forces calculated for known velocity fields. Body force arising from motion of SAW is responsible for detaching nonspecifically bound proteins. Streaming induced drag and lift forces prevent re-attachment of the detached nonspecific proteins to the SAW device surface.

D. *Efficiency of nonspecifically bound protein removal*

The main aim of this study is to achieve complete removal of nonspecifically bound proteins along the entire SAW delay path. Based on the removal mechanism predicted using our simulation model, the dominant force responsible for removal appears to be the SAW body force. The magnitude of this force decays rapidly as we move away from the IDT region, with the extent of decay depending on the applied voltages, device frequency and fluid properties such as density and viscosity. A decay length can be defined which represents the region beyond which removal forces are not strong enough to eliminate biofouling. Based on the forces calculated for various locations along a SAW delay path, it was found that the decay length for sub-micron sized particles is approximately 100 microns for the above device. Similar analysis was carried out for the SAW device operating at higher fluid viscosity of 10 cP. We find that at the MHz device frequencies simulated in the present work, the variation in streaming velocity for a fluid viscosity change from 1 to 10 cP is significant, especially at regions along the delay path located far away from the IDT region. We find that when the fluid viscosity is increased from 1 to 10 cP, the calculated decay length rapidly decreases from 100 microns to 60 microns. Therefore, our study indicates that for increasing fluid viscosity in the simulated range, the removal of nonspecifically-bound proteins that can be efficiently carried out by SAW-induced streaming phenomenon decreases rapidly with increasing distance along the SAW delay path.

V. CONCLUSIONS

Fluid-solid interaction finite element models were used to predict the mechanism of removal of nonspecifically bound proteins in SAW biosensing application. The generated streaming velocities were used to compute the various adhesive and removal forces involved. Our study indicates that the SAW body force overcomes the adhesive forces of the fouling proteins to the device surface and the fluid-induced drag and lift forces prevent its re-attachment. The streaming velocity fields computed using the finite-element models in conjunction with the proposed mechanism were used to identify the conditions leading to improved removal efficiency. In particular, the results of our numerical analysis show that higher amplitude and high frequency SAWs can clean biosensor surfaces most effectively in media with fluid properties similar to those of water, *i.e.*, lower viscosity and density, when (1) the SAW wavelength can be made comparable to the particle radius to promote effective acoustic-particle interaction; (2) the viscous boundary layer is thin and (3) the non-linear acoustic streaming forces exceed typical adhesion forces which is true for MHz frequencies. Based on the simulation analyses, possible mechanisms of nonspecifically bound protein removal are discussed and interpreted in terms of the experimental observations of SAW biosensor surface cleaning. Predictions of the model are in good agreement with those of simple analytical theories as well as the experimentally observed trends of nonspecifically bound protein removal in typical SAW biosensing operation.

ACKNOWLEDGMENT

The authors wish to thank the Academic Computing Center and the Engineering Computing Center at University of South Florida for providing the computational facilities. The authors thank Reetu Singh of USF for useful discussions.

REFERENCES

[1] S. Cular, D. W. Branch, V. R. Bhethanabotla, G. D. Meyer, and H. G. Craighead, "Removal of nonspecifically bound proteins on microarrays using surface acoustic waves." *IEEE Sensors Journal*, vol. 8(3), pp. 314-320, 2008.

[2] W. Nyborg, Acoustic Streaming. Academic press Inc, New York, NY, 1965.

[3] S. K. R. S.Sankaranarayanan, S. Cular, V. R. Bhethanabotla, and B. Joseph. "Flow induced by acoustic streaming on surface-acoustic-wave devices and its application in biofouling removal: A computational study and comparisons to experiment." *Physical Review E*, vol. 77, pp. 066308/1-066308/19, 2008.

[4] B. A. Auld, Acoustic Fields and Waves in Solids. John Wiley & Sons, NY, 1973.

[5] J. J. Campbell and W. R. Jones, "A method for estimating optimal crystal cuts and propagation directions for excitation of piezoelectric surface waves.," *IEEE Transactions on Sonics and Ultrasonics*, vol. 15, pp. 209-217, 1968.

[6] S. Shiokawa, Y. Matsui, and T. Ueda, *Proceedings of IEEE Ultrasonics Symposium*, Vol. 1, pp. 643-646, 1989.

[7] R. M. Moroney,(Univ. of California Berkeley,1995), Ph.D. Thesis.

44e Enzyme Modified Carbon Nanotube Biosensors for Pesticide Detection

Shankar Balasubramanian[1], Valber Pedrosa[1], Sheetal Paliwal[1], Virginia A. Davis[2], James Wild[3], and Aleksandr Simonian[1]. (1) Materials Research and Education Center, Auburn University, Wilmore, Auburn, AL 36849, (2) Department of Chemical Engineering, Auburn University, Auburn, AL 36849, (3) Biochemistry and Biophysics, Texas A & M University, 103 Biochemistry/Biophysics Building, Texas A & M University, 2128 TAMU, College Station, TX 77843

A simple approach is described for design and development of highly sensitive and stable electrochemical biosensor for direct detection of organophosphate pesticides. Organophosphorus hydrolase (OPH) functionalized single and multi-walled carbon nanotube (CNT) conjugates provided the platform for direct amperometric detection of Paraoxon, a model organophosphate. The catalytic hydrolysis of paraoxon produces equimoles of p-nitrophenol whose oxidation was monitored amperometrically in real time under flow-injection (FIA) mode. Buffer pH, flow rate and method of OPH immobilization were found to influence the sensitivity of paraoxon detection. The dynamic concentration range was 0.5 to 8.5 $\mu mol\ L^{-1}$ with a detection limit of 2.3 $\mu mol\ L^{-1}$ (S/N = 3). Our studies show that immobilized enzyme retains a significant degree of enzymatic activity, provided high sensitivity and showed remarkable stability with only 8% decrease in signal over 10-day period. These results suggest that covalent immobilization of OPH on CNT can be used for specific immobilization of OPH with advantages of low price, simplicity and biocompatibility.

44f Layer-by-Layer Assembly of Enzyme-Coated Single-Walled Carbon Nanotubes and Redox Polymers

Ta-Wei Tsai[1], Gustavo Heckert[1], and **David W. Schmidtke**[2]. (1) University of Oklahoma Bioengineering Center, University of Oklahoma, 100 East Boyd, Norman, OK 73019, (2) School of Chemical, Biological, and Materials Engineering, University of Oklahoma, 100 East Boyd, EC Rm. T-335, Norman, OK 73019

A sodium cholate suspension-dialysis method was used to adsorb two different redox enzymes, glucose oxidase (GOX) and horseradish peroxidase (HRP), onto single-walled carbon nanotubes (SWNT). By this method, solutions of highly dispersed and debundled SWNTs were prepared, that remained stable for 30 days and which retained 70% of the native enzymatic activity. We demonstrate incorporation of these enzyme-SWNT conjugates into amperometric biosensors through a layer-by-layer (LBL) self-assembly process. Incorporation of SWNT-enzyme conjugates resulted in a 5-fold increase in the sensitivity as compared to films without SWNTs. In the case of the GOX-based sensors, current densities as high as 400 &muA/cm2 could be reached, with the sensitivity of the sensors being controlled by varying the number of multilayers.

64a On the Role of Concentration Gradients In the Performance of Carrier-Bound Immobilized Biocatalysts: The Case of An Industrial Amidase

Caterina Boniello[1], Torsten Mayr[2], Ingo Klimant[2], Burghard König[3], Waander Riethorst[3], and Bernd Nidetzky[1]. (1) Institute of Biotechnology and Biochemical Engineering, Graz University of Technology, Research Centre Applied Biocatalysis, Petersgasse 12, Graz, Austria, (2) Institute of Analytical Chemistry and Radiochemistry, Graz University of Technology, Stremayrgasse 16, Graz, Austria, (3) Department of Biocatalysis - Development & Manufacturing Anti-Infectives, Sandoz GmbH, Biochemiestrasse 10, Kundl, Austria

Many industrial biocatalysts are used as immobilized enzyme preparations where the protein is tethered on the surface of an insoluble porous carrier. In this way the biocatalyst can be separated easily from the reaction solution and protein-free product is obtained. Recovery of the enzyme for multiple rounds of substrate conversion and improved operational stability of the biocatalyst are advantages often brought about by the immobilization, which may therefore contribute to a reduction of the process costs connected to the catalyst. There are however common disadvantages of rendering an enzyme insoluble through attachment on a carrier. A substantial decrease in specific enzymatic activity is often observed in the immobilized enzyme as compared to the soluble counterpart. A significant portion of the measured loss in catalytic effectiveness of the immobilized enzyme may be apparent and reflect the influence of mass transfer effects on the overall reaction rate. Hindered diffusion in the pores of the carrier particles may promote the formation of gradients in the concentrations of substrates and products along the radial dimension, which may strongly impact on the enzyme activity. Enzymatic reactions that lead to formation or uptake of protons may also produce substantial pH gradients along the characteristic length of the carrier pore, which in turn can alter the activity and stability of the immobilized enzyme. Although theory for the analysis of the performance of immobilized enzymes has been available for decades, there are only few examples published in the literature where actual measurements of intra-particle concentration gradients have been correlated with the kinetic properties of carrier-bound biocatalysts. The evidence gained from a detailed study is expected to have major implications on carrier selection and design of the immobilization process. Cost considerations dictate that the quest for the optimum carrier is a relevant issue of biocatalytic process development in the industry. We report here on the characterization of an industrially used amidase immobilized on epoxy-activated carriers (Sepabeads).

The relative rates of enzymatic reaction and mass transfer of substrate will be determined by the size of the (spherical) particles, the diameter of the pores, and the loading of enzyme activity per unit mass of carrier. Steady state kinetic analysis was used to evaluate the effect of variation in these three parameters on enzyme coefficients, the maximum initial rate (Vmax) and the Michaelis-Menten constant (Km). Using standard grade Sepabeads EC-EP which exhibit a pore diameter of between 30 and 40 nm we observed significant (up to 3-fold) increases in Km in response to an increase in enzyme loading and particle size. When using a carrier that shows a pore diameter about 6 times that of Sepabeads EC-EP, the Km of the immobilized amidase was independent of the enzyme loading in the range 7 - 70 mg protein / g dry carrier. The aggregate data analyzed by using the dimensionless Thiele modulus portray the diffusional effects on the enzymatic conversion rate in carriers featuring a relatively small and high pore diameter. The protein binding capacity increased with increasing specific surface area on the carrier, that is when the particle or pore diameter decreased. However, when using particles with relatively small pores, a substantial fraction of the available surface area appeared to be unavailable for binding of protein.

It was unexpected to observe a substantial effect of particle characteristics and enzyme loading on Vmax, which is recorded under conditions where the enzyme is saturated with substrate. Considering that the observable decrease in Vmax might be traceable to a variation in intra-particle pH resulting from the hydrolysis of an amide substrate and the hindered diffusion of the acidic product, we determined the difference between the pH of the bulk solution and the average pH in the carrier particle. Following impregnation of Sepabeads with a pH-responsive fluorescein label, fluorescence measurements were used to determine the pH in the particles in a time resolved manner while the enzymatic reaction took place. A significant difference (by about 1 pH unit) in the pH values measured inside and outside of the carrier indicates that product concentration gradients may have a large impact on the activity of the immobilized enzyme. The acidification of the carrier relative to the bulk increased with increasing enzyme loading and substrate concentration. Implications of these results for the design of a suitable carrier for amidase-catalyzed conversions are discussed.

64b Engineering Bacteria for the Production of Human-Like Metabolites of Diclofenac

Jamie E. Prior, Chemical Engineering, University of Colorado, 1111 Engineering Drive, Campus Box 424, Boulder, CO 80309, Touraj Shokati, Anesthesiology, University of Colorado, Campus Box B113, Denver, CO, Uwe Christians, Anethesiology, University of Colorado Health Sciences Center, Campus Box B113, Denver, CO, and Dr. Ryan T. Gill, University of Colorado, ECCH 111 Campus Box 424, Boulder, CO 80309.

Drug metabolism describes the process in which a drug is degraded or modified for elimination from the body. These degradation products, or metabolites, can be characterized by varying toxicity profiles compared to the original drug. As such, metabolites must also be studied in an effort to accurately depict the full effects of a specific drug. Microbial models for human drug metabolism have been an effective method to produce metabolites in quantities suitable for testing and are more attractive than other methods such as chemical synthesis. In both microbes and mammals, cytochrome P450 enzyme activity is responsible for the breakdown of drugs into metabolites. The bacterium, *Actinoplanes* sp. has been found to have similar drug metabolism profile for several drugs as humans. However, further engineering towards scaling up metabolite production has proven difficult. To address these concerns, we are working to identify the cytochrome P450s responsible for the biotransformation of these drugs for production of specific metabolites in order to optimize expression through metabolic engineering for increased production of drug metabolites.

To date we have identified at least 14 unique P450 genes in *Actinoplanes*, five of which we have the entire ORF. We have successfully cloned, expressed, and purified three of the P450s in the active form. In order to characterize the ability of each P450 to metabolize a range of drugs we are using a combination of heterologous expression as well as gene disruption. The two methods are useful when combined to work around the lack of native redox partners associated with the P450s in *E. coli* as well as to characterize genes not expressed well. Later optimization of redox partners will be possible once we have identified the specific P450s we are interested in working with to develop a system for *in vivo* drug metabolism in *E. coli*.

We have identified two of the P450s as possible targets for the observed drug metabolism in *Actinoplanes*. Substrate binding assays show one of the P450s binds diclofenac and lovastatin, while another P450 binds rapamycin. Studies are currently underway to identify the metabolite profile of each P450 using LC-MS. In summary, we have identified P450 genes in *Actinoplanes* that are responsible for the observed drug metabolism of diclofenac, and are working towards engineering a system in *E. coli* for the production of drug metabolites.

64c Deep Eutectic Solvents: Inexpensive, Biologically-Derived Media for Enzyme Catalysis

Johnathan T. Gorke[1], Friedrich Srienc[1], and Romas J. Kazlauskas[2]. (1) Chemical Engineering and Materials Science, University of Minnesota, 421 Washington Avenue SE, Minneapolis, MN 55455, (2) Biochemistry, Molecular Biology and Biophysics, University of Minnesota, 1479 Gortner Ave, Saint Paul, MN 55108

Room temperature ionic liquids (RTIL's) have the advantages of low vapor pressure and non-flammability, but high cost and potential toxicity limits their use. Deep eutectic solvents (DES's), eutectic mixtures of ammonium salts and hydrogen bond donors, such as 1:2 choline chloride/urea or choline chloride/glycerol are an alternative to conventional RTIL's[1]. In addition to having low vapor pressure and low flammability, DES's are composed of non-toxic and inexpensive materials including vitamins, amides, sugars, and alcohols. Many of these components are natural products or can be made from renewable materials, meaning that DES's are a sustainable alternative to common RTILs and to many petroleum-based organic solvents. Some DES's contain urea and/or high amounts of ammonium halide, so they do not appear to be likely solvents for biocatalysis. Surprisingly, we found that lipases are active in DES's for transesterification and aminolysis of ethyl valerate and butanol or butylamine[2]. The transesterification and aminolysis activities of immobilized *Candida antarctica* lipase B (CALB) in 1:2 mixtures of choline chloride/urea (ChCl:U) and 1:2 choline chloride/glycerol (ChCl:Gly) were comparable to those in toluene and as much as five-fold higher than the RTIL's 1-butyl-3-methylimidazolium tetrafluoroborate (BMIM[BF4]) and 1-butyl-methylimidazolum bis (trifluoromethane)sulfonamide (BMIM[Tf2N]). We found that the reactivity of the glycerol in ChCl:Gly was reduced by >200-fold compared to the transesterification of free glycerol by the lipase. Further, free CALB was at least 20 times more stable at 60 °C in a 1:2 mixture of choline chloride (5 M) and urea (10 M) than in aqueous 5 M choline chloride or 10 M urea. We hypothesize that the hydrogen-bonded structure of DES's prevents urea and halide from inactivating the enzyme and prevents the glycerol component of DES's from reacting. We also found that the addition of 10% to 25% ChCl:Gly to an aqueous medium could enhance esterase activity by as much as 3-fold for the pig liver esterase- and *Rhizopus oryzae* esterase-catalyzed hydrolysis of p-nitrophenyl acetate. We observed an even more dramatic 20-fold increase in conversion of styrene oxide to styrene glycol by an epoxide hydrolase from *Agrobacterium radiobacter* after addition of 25% ChCl:Gly to the reaction medium.

1. A. P. Abbott, G. Capper, D. L. Davies, R. K. Rasheed, and V. Tambyrajah, 2003, Chem. Commun., 70.

2. J. T. Gorke, F. Srienc, and R. J. Kazlauskas, 2008, Chem. Commun., 1235.

Web Page: www.rsc.org/Publishing/Journals/CC/article.asp?doi=b716317g

64d Matrix-Bound Multi-Enzyme System for Rapid and Effective Ethanol Metabolism

Matthew W. Eggert, Department of Chemical Engineering, Auburn University, 170 Wilmore, Auburn, AL 36849-5127, Mark E. Byrne, Biomimetic & Biohybrid Materials, Biomedical Devices, and Drug Delivery Laboratories, Department of Chemical Engineering, Auburn University, 324 Ross Hall, Auburn, AL 36849-5127, and Robert P. Chambers, Chemical Engineering, Auburn University, 240 Ross Hall, Auburn, AL 36849-5127.

For the purpose of improving the rate of alcohol metabolism within the human body, a multi-enzyme system containing alcohol (ADH), acetaldehyde (ALDH) and lactic (LDH) dehydrogenases, linked together by cofactor NAD, is currently being investigated. Three enzyme experiments show rapid metabolism of ethanol while maintaining potentially toxic acetaldehyde well below harmful concentrations. Efficient conversion of ethanol to non-toxic acetate, reduction of inhibition effects, and regeneration of cofactor NAD highlight the abilities of the multi-enzyme system. Using a multi-enzyme simulation model in conjunction with experimental data, effective relative amounts of each enzyme were determined for achieving a high overall rate of ethanol conversion.

Studies of covalently bound alcohol dehydrogenase onto a polymer matrix have been performed via UV free-radical polymerization following functionalization of the enzyme amino groups with acryloyl chloride. Enzyme activity, based on initial rate kinetics, beginning with the free enzyme through functionalization then polymerization is monitored. UV exposure is minimized to prevent denaturization of enzyme. Up to 40% ADH enzyme activity is retained in gel form. Experimental diffusion characteristics indicate sufficient transport for an effective matrix bound enzyme system.

64e Scaleable Production of Free Fatty Acids in *E. Coli*: Implications for Biodiesel Production

Harmit Vora[1], Xuefeng Lu[2], and Chaitan Khosla[1]. (1) Chemical Engineering, Stanford University, 381 North-South Mall, Stanford, CA 94305, (2) Metabolic Engineering, Qingdao Institute of Bioenergy and Bioprocess Technology, Chinese Academy of Sciences, No.100 Nanjing Road, Room 3204, Jinhua Apartment B, Qingdao, China

Whereas microbial fermentation processes for producing ethanol and related alcohol biofuels are well established, biodiesel (methyl esters of fatty acids) is exclusively derived from plant oils. Slow cycle times for engineering oilseed metabolism and the excessive accumulation of glycerol as a byproduct are two major drawbacks of deriving biodiesel from plants. Although most bacteria produce fatty acids as cell envelope precursors, the biosynthesis of fatty acids is tightly regulated at multiple levels. By introducing four distinct genetic changes into the *E. coli* genome, we have engineered this prototypical bacterium into an efficient producer of free fatty acids. The genetic alterations include: (a) knocking out the endogenous fadD gene, which encodes an acyl-CoA synthetase, to block fatty acid degradation; (b) heterologous expression of a plant thioesterase to increase the abundance of shorter chain fatty acids with an eye towards improving fuel quality; (c) increasing the supply of malonyl-CoA by over-expressing acetyl-CoA carboxylase; and (d) releasing feedback inhibition caused by long-chain fatty acyl-ACPs through over-expression of an endogenous thioesterase.

To evaluate the performance of the above biocatalyst in a process that mimics large-scale industrial fermentation processes, fed-batch fermentation runs were performed at a 6 L scale. Under defined media fermentation conditions, using glycerol as the primary carbon source, the metabolically engineered *E. coli* strain produced a final titer of about 2.5 g/L and a linear productivity of about 0.17 mg/ml/h sustained for 12 h post-induction. During this linear period, a carbon mass yield of about 4.8% was achieved. At least 50% of the fatty acids produced were present in the free acid form. The observation that *E. coli* can efficiently convert a cheap carbon source such as glycerol into free fatty acid without apparent toxicity opens the door for developing microbial biodiesel as an alternative to petroleum-based long-chain alkane fuels.

64f Enantioselective Bioreduction of Ketone with Cofactor Recycling by the Use of Coupled Permeabilized Microorganisms

Zhi Li, Department of Chemical and Biomolecular Engineering, National University of Singapore, 4 Engineering Drive 4, Singapore, 117576, Singapore, Wei Zhang, Singapore-MIT Alliance, National University of Singapore, 4 Engineering Drive 3, Singapore, 117576, Singapore, Kelvin O'Conner, Department of Industrial Microbiology, University College Dublin, Belfield, Dublin 4, Dublin, Ireland, and Daniel I. C. Wang, Department of Chemical Engineering, Massachusetts Institute of Technology, 77 Massachusetts Ave, Cambridge, MA 02139.

Biocatalytic oxidoreductions are key reactions for asymmetric syntheses of enantiopure compounds that are useful synthons and pharmaceutical intermediates. One of the challenges in the practical application of these reactions is the efficient recycling of the expensive nicotinamide cofactor NAD(P)H/NAD(P)+ which is often required in stoichiometric amount. We recently developed a novel cofactor recycling method for bioreduction based on the use of permeabilized microbial cells.[1-2] By using permeabilized cells of Bacillus pumilus Phe-C3 containing an enantioselective ketoreductase and a glucose-6-phosphate dehydrogenase as catalysts, bioreduction of ethyl 3-keto-4,4,4-trifluoro-butyrate in the presence of glucose-6-phosphate and trace amounts of NADP+ gave (R)-ethyl 3-hydroxy-4,4,4-trifluorobutyrate in 95% ee with NADPH recycling for 4200 times.[1] In a more general approach, permeabilized cells of B. pumilus Phe-C3 containing the ketoreductase were coupled with permeabilized cells of Bacillus subtilis BGSC 1A1 containing a glucose dehydrogenase (GDH), in the presence of glucose and trace amounts of NADP+, for the bioreduction of the 3-ketoester, giving rise to the recycling of NADPH for 1620 times.[2] To further improve the cofactor recycling in this general approach, we have been working on the development of highly active recombinant strain expressing GDH, the optimization of the permeabilization conditions for the cofactor-regenerating microorganisms, the optimal coupling of two permeabilized microorganisms for the bioreduction to achieve much higher turn over number of the cofactor recycling, and the long-term utilization of the permeabilized cells couple for biotransformation. Our recent progress will be presented.

REFERENCES:

[1]Zhang, J.; Witholt, B.; Li, Z. Adv. Syn. Catal. 2006, 348, 429-433.

[2]Zhang, J.; Witholt, B.; Li, Z. Chem. Commun. 2006, 398-400.

64g Changes of Cellulose Reactivity with Purified Enzyme Components during Hydrolysis

Jian Shi, College of Engineering-Center for Environmental Research and Technology, University of California, Riverside, 1084 Columbia Ave, Riverside, CA 92507, Bin Yang, University of California Riverside, 1084 Columbia Avenue, Riverside, CA 92507, Edmund Larenas, Genencor, A Danisco Divison, 925 Page Mill Rd.,, Palo Alto, CA 94304, Colin Mitchinson, Genencor International, 925 Page Mill Road, Palo Alto, CA 94304, and Charles E. Wyman, College of Engineering-Center for Environmental Research and Technology, University of California Riverside, 1084 Columbia Avenue, Riverside, CA 92507.

To better understand the mechanism of enzymatic hydrolysis of cellulose, we applied a restart protocol to explore how cellulose reactivity changed and the interaction of cellulose with major cellulase components over the course of enzymatic hydrolysis. This approach allowed us to accurately monitor the hydrolysis rate of partially converted cellulose over the time course of substrate digestion excluding the impact of cellulase changes. In this presentation, the effect of enzyme-substrate interactions on reaction rates was studied using purified key cellulase components (i.e., CBHI, EGI, and/or EGII) from wild type Trichoderma reesei, and their synergism was studied by comparing the interrupted enzymatic hydrolysis of pure cellulose as measured using the restart protocol with uninterrupted hydrolysis as well as with the hydrolysis behavior of individual key cellulase components. Important chemical and physical features, including DP, reducing ends, crystallinity index, accessible surface area, and cellulose reactivity with enzymes, were characterized over the time course of cellulose hydrolysis, and the effects of cellulose reactivity and cellulase processivity on enzymatic hydrolysis of cellulose will be discussed.

66a Biobutanol from Yeast. A Synergistic Genome and Protein Engineering Approach

Nikhil U. Nair[1], Zengyi Shao[1], Hua Zhao[1], Ryan P. Sullivan[1], Michael McLachlan[2], Tyler W. Johannes[1], and Huimin Zhao[1]. (1) Chemical & Biomolecular Engineering, University of Illinois at Urbana-Champaign, 221 RAL, Box C-3 RAL, MC-712, 600 S. Mathews Ave., Urbana, IL 61801, (2) Biophysics & Computation Biology, University of Illinois at Urbana-Champaign, 221 RAL, Box C-3 RAL, MC-712, 600 S. Mathews Ave., Urbana, IL 61801

Although bioethanol is currently the leading candidate to replace gasoline as transportation fuel, biobutanol has been recently gaining favor over ethanol. Having better properties as fuel such as higher energy content, higher octane number, and lower latent heat, it promises a better long-term solution to transportation fuel requirements. When other logistical concerns about long distance conveyance in pipelines are considered, butanol still comes out a winner over ethanol as it has lower solubility in water, is less corrosive, and has higher vapor pressure. Finally, unlike ethanol, butanol can be directly used in the current design of internal combustion engines.

In spite of its advantages, there are two major issues plaguing economical production of biobutanol. The first issue is that the natural producers of butanol are anaerobic gram-positive bacteria of the genus Clostridia, which are relatively difficult to culture. And while butanol fermentation using Clostridia has been studied extensively for decades, very few genetic tools are available to engineer them for improved titer and productivity. The second issue is that butanol is toxic to bacteria at concentrations over 20 g/L, far below its solubility in water (~70 g/L). This remains an unresolved issue with current Clostridial fermentations. Heterologous butanol production in more congenial organisms such as E. coli or S. cerevisiae (common yeast) could alleviate some problems associated with Clostridial fermentation. Given the choice between the two, S. cerevisiae might be the better option, although butanol production has already been demonstrated in E. coli. Having been evolved for high alcohol tolerance for hundreds of years, S. cerevisiae is a far more resilient organism than bacteria and therefore it may be possible to engineer it for higher tolerance to butanol than may be possible for any bacteria.

Expression of the entire six-step pathway from either Clostridia, or of homologs and orthologs of each enzyme from various organisms ranging from soil bacteria to protists, did not yield any butanol. Analysis revealed the inability to solubly express a few key proteins impeded butanol production. Directed evolution experiments were implemented to engineer soluble, functional mutants, which finally yielded a functional butanol biosynthetic pathway. In order to enhance productivity, a new method called MIRAGE (mutagenic inverted repeat assisted genome engineering) was developed to perform efficient, one-step genome modifications in yeast. Metabolic engineering using MIRAGE decreased byproduct formation and increased butanol productivity. Further engineering of enzymes by directed evolution and metabolic engineering by MIRAGE is currently underway to improve fermentative butanol yield.

66b Effects of Glucose and Insulin Levels on HepG2 Cell Metabolism

Vidya V. Iyer[1], Charles M. Roth[1], and M.G. Ierapetritou[2]. (1) Chemical and Biochemical Engineering, Rutgers University, 98 Brett Road, Piscataway, NJ 08854, (2) Chemical & Biochemical Engineering Department, Rutgers University, the State University of New Jersey, 98 Brett Road, Piscataway, NJ 08854

HepG2, hepatocellular carcinoma cells, are easily maintained and expanded in culture and have been shown to express a wide range of liver-specific functions. As such, HepG2 cells can be used as an efficient system for in vitro toxicology studies and possibly in bioartificial livers (Allen et al., 2001, Park & Lee, 2005). Although the molecular expression of HepG2 cells and biological phenotypes have been characterized extensively, relatively little data exists regarding quantitative metabolic information in response to simple medium perturbations.

In this study, HepG2 cells were cultured in vitro under varying levels of glucose and insulin for five days. Four different levels of glucose were used- Minimal essential media (MEM) 0.8 mg/ml glucose (Control), DMEM-5 mg/ml glucose (high glucose), DMEM-1 mg/ml glucose (low glucose) and DMEM-0 mg/ml glucose (no glucose). Each of the four media was studied at two levels of insulin- no insulin and low insulin (50 mU/ml, representing physiological levels). Cell growth was found to be highest for DMEM-1 and lowest for DMEM-0. We also observed that the urea production was highest on a per cell basis for DMEM-0, for both no and low insulin treatments. Further, it was seen that lactate acted as an alternative carbon source in the absence of glucose, whereas it acted as a sink for the high glucose media.

Other measurements that were performed included amino acid uptake, fatty acid production, and albumin production. The metabolic network consisted of reactions of central metabolism such as glycolysis, pentose phosphate pathway, lactate metabolism and TCA cycle, amino acid catabolism, urea cycle, fatty acid metabolism, lipid metabolism and albumin synthesis. Using the above mentioned network, extracellular measurements, and stoichiometric material balances on each metabolite in a flux balance analysis (FBA) framework, metabolic flux maps with quantitative values of the intracellular fluxes under varying levels of glucose and insulin were evaluated. Further, by comparing the intracellular metabolic fluxes of HepG2 cells among varying treatments, alterations in central metabolic pathways have been demonstrated.

We are currently using the HepG2 in vitro culture system to study drug and toxin metabolism. Previous studies with HepG2 have demonstrated the apoptotic effect of low concentrations of ethanol and corresponding identification of signal transduction pathways involved (Castaneda & Rosin-Steiner, 2006). Another study simultaneously exposed HepG2 cells to different concentrations of acetaminophen and N-acetylcysteine (Manov et al., 2004). They showed that while N-acetylcysteine prevented oxidative damage induced by acetaminophen, it could not prevent the induced apoptosis. Interestingly in vivo studies have also shown that N-acetylcysteine is protective only when administered early on after an acetaminophen overdose (Jones, 1998). Through combinatorial exposure of HepG2 cells to acetaminophen and either protective compounds (N-acetylcysteine) or a second xenobiotic (e.g., flutamide), we seek to identify metabolic alternatives that precede irreversible hepatic injury. To this end, we have an existing pathway model of hepatic metabolism with detoxification pathways. The interactions of the xenobiotic pathways with central metabolism are analyzed using FBA enabling the identification of corresponding regulatory pathways.

References

1 Allen J.W., Hassanein T. & Bhatia S.N. (2001) Advances in bioartificial liver devices. Hepatology, 34, 447-455.

2 Castaneda F. & Rosin-Steiner S. (2006) Low concentration of ethanol induce apoptosis in HepG2 cells: role of various signal transduction pathways. International Journal of Medical Sciences., 3, 160-167.

3 Jones A.L. (1998) Mechanism of action and value of N-acetylcysteine in the treatment of early and late acetaminophen poisoning: a critical review. Journal of Toxicology - Clinical Toxicology, 36, 277-285.

4 Manov I., Hirsh M. & Iancu T.C. (2004) N-Acetylcysteine does not Protect HepG2 Cells against Acetaminophen-Induced Apoptosis. Pharmacology and Toxicology, 94, 213-225.

5 Park J.K. & Lee D.H. (2005) Bioartificial liver systems: current status and future perspective. Journal of Bioscience and Bioengineering, 99, 311-319.

66c Distilling the Complexity of Metabolic Chemistry

Christopher Henry, Mathematics and Computer Science, Argonne National Laboratory, 9700 S. Cass Avenue, Argonne, IL 60439, Linda J. Broadbelt, Northwestern University, Department of Chemical Engineering, 2145 Sheridan Road E136, Evanston, IL 60208-3120, and Vassily Hatzimanikatis, Laboratory of Computational Systems Biotechnology, EPFL, CH-1015, Lausanne, Switzerland.

Recent advances in DNA sequencing technologies have reduced the time required to sequence an entire prokaryotic genome to a few days. Development of the Rapid Annotation using Subsystems Technology (RAST) server allows newly sequenced prokaryotic genomes to be annotated in less than 24 hours [1]. However, the conversion of newly annotated genomes into metabolic models that may be used to accurately predict cellular phenotypes, gene essentiality, and metabolic engineering strategies is still a process that can take months at the minimum [2]. While methods exist for automatically mapping biochemical reactions to the functional roles assigned to genes during the annotation process [3], these methods depend upon the EC classification system and databases of biochemical reactions such as the KEGG [4]. The metabolic models produced using these methods are plagued by errors arising from the misclassification of reactions within the EC system or by errors in the stoichiometry of reactions stored within the reaction databases. The manual effort needed to identify and correct such errors accounts for a significant portion of the time required to produce a functioning genome-scale metabolic model.

We have developed a new algorithm that utilizes the Biological Network Integrated Computational Explorer (BNICE) framework [5] to: (i) systematically assign EC classes to biochemical reactions, (ii) correct errors in reaction cofactor stoichiometry, and (iii) generate mappings from the reactant atoms to the product atoms of biochemical reactions. This algorithm utilizes a set of 86 reaction rules developed from a manual curation of the EC classification system. These reaction rules are applied iteratively to the substrates of each biochemical reaction in an attempt to reproduce the reaction. For each reaction that is reproduced using the reaction rules, errors in the reaction cofactor stoichiometry are identified and corrected, a mapping is generated between the atoms in the reactants and products of the reaction, and an EC number is assigned to the reaction based on the EC number associated with the reaction rules that reproduced the reaction. This methodology has been successfully applied to reproduce the majority of the eligible biochemical reactions contained in the KEGG database and in genome-scale metabolic models of *E. coli*, *S. cerevisiae*, and *B. subtilis*. The results of this work will help to improve the speed and efficiency with which new metabolic models may be assembled. The atom mappings generated by the method will be invaluable for interpreting the results of 13C tracer experiments being applied to measure the rates of intracellular reactions. Finally, this work indicates the areas of metabolism not covered by our current set of 86 reaction rules facilitating the continuing efforts to produce new reaction rules for the BNICE framework.

1. Aziz, R.K., et al., The RAST Server: rapid annotations using subsystems technology. BMC Genomics, 2008. 9: p. 75.

2. Price, N.D., J.L. Reed, and B.O. Palsson, Genome-scale models of microbial cells: evaluating the consequences of constraints. Nat Rev Microbiol, 2004. 2(11): p. 886-897.

3. DeJongh, M., et al., Toward the automated generation of genome-scale metabolic networks in the SEED. BMC Bioinformatics, 2007. 8: p. 139.

4. Kanehisa, M. and S. Goto, KEGG: Kyoto encyclopedia of genes and genomes. Nucleic Acids Research, 2000. 28(1): p. 27-30.

5. Hatzimanikatis, V., et al., Exploring the diversity of complex metabolic networks. Bioinformatics, 2004. 21(8): p. 1603-1609.

66d Coregulation Analysis of Highly Coexpressed and Non-Coexpressed Genes with a Novel Promoter Similarity Index

Nguyen Tung, BioMAPS Institute for Quantitative Biology, Rutgers University, Rutgers University, Piscataway, NJ 08540, Richard Nowakowski, Department of Neuroscience and Cell Biology, University of Medicine and Dentistry, Piscataway, NJ 08540, and Ioannis P. Androulakis, Biomedical and Chemical & Biochemical Engineering, Rutgers - The State University of New Jersey, 599 Taylor Road, Piscataway, NJ 08854.

Understanding gene regulation is a critical step towards deciphering the complexities of transcription [1, 2]. Although microarray technology can reveal patterns of gene expression [3, 4], the regulation mechanism driving gene expression is still unknown in most cases [5]. A reasonable hypothesis is that similar gene promoters may imply the possibility that similar transcription factors can bind and be potentially functional determinants of gene regulation [2, 6-8]. Thus, two critical questions emerge: (i) what constitutes a set of coexpressed versus non-coexpressed genes, and (ii) how do we handle the fact that a gene may have multiple promoters [7, 9] corresponding to different transcription start sites?

The first question is important since the silent underlying assumption of regulation analysis is that genes that have the same regulation mechanism ought to respond in a similar manner under similar stimuli. The second question is very critical as well. Since it is known that genes can be co-expressed under certain conditions and not under others it clearly implies that alternative regulation mechanisms must be available to each gene, otherwise the problem would be trivial. We therefore computationally examine possible combinations of alternative promoters with the hope of clarifying the mechanisms characterizing regulation.

In this presentation, we will first discuss a novel methodology based on a multi-clustering approach [10-13] which, instead of simply clustering the entire data set, attempts to identify within the data a subset of highly coexpressed and non-coexpressed genes i.e. with a confidence level δ identify those genes which are either co-expressed or non co-expressed. Subsequently, we consider whether some combinations from alternative promoter sequences for each gene [14] in this subset can disclose alternative regulation mechanisms. Under the assumption that the more similar the promoter sequences are, the more probable is that the corresponding genes are co-regulated [8, 15-17], we propose a novel index of promoter similarity that takes into account all possibilities of regulatory elements e.g. TFs, miRNAs [1, 18] to computationally represent the coregulation level between any two genes. Instead of scanning for common patterns e.g. conserved motifs [19, 20], TFBSs [21, 22] in promoter regions of coexpressed genes and then calculating the coregulation level between two genes [15-17], the measure here is estimated directly from their sequences [6, 8, 23, 24] regardless as to whether genes coexpressed or not. Using the proposed measure of promoter similarity, the pool of promoter sequences is hierarchically clustered [24, 25] with a similarity threshold calculated from the background set of sequences and mapped onto groups of coregulated genes. Clustering results are then rationalized in the context of expression under different experimental conditions to identify possible combinations of alternative promoters activated under a specific condition. Representative promoters in each cluster of coregulated genes now can be considered as the specific promoters for those genes involved in a specific regulation mechanism.

Consequently, our work aims at identifying the possibility that alternative promoters, thus regulation mechanisms, are activated under different conditions. Therefore, we wish to determine whether the combination of co-expression and co-regulation analyses can reveal condition-specific regulation mechanisms.

References

1. Chen K, Rajewsky N: The evolution of gene regulation by transcription factors and microRNAs. Nat Rev Genet 2007, 8(2):93-103.

2. Nguyen DH, D'Haeseleer P: Deciphering principles of transcription regulation in eukaryotic genomes. Mol Syst Biol 2006, 2:2006 0012.

3. Belacel N WQ, Cuperlovic-Culf M: Clustering methods for microarray gene expression data. OMICS 2006, 10(4):507-531.

4. Jiang D TC, Zhang A: Cluster Analysis for Gene Expression Data: A Survey. IEEE Trans on Knowledge and Data Eng 2004, 16(11):1370-1386.

5. Michalak P: Coexpression, coregulation, and cofunctionality of neighboring genes in eukaryotic genomes. Genomics 2008, 91(3):243-248.

6. Allocco DJ, Kohane IS, Butte AJ: Quantifying the relationship between co-expression, co-regulation and gene function. BMC Bioinformatics 2004, 5:18.

7. ENCODE Project Consortium: Identification and analysis of functional elements in 1% of the human genome by the ENCODE pilot project. Nature 2007, 447(7146):799-816.

8. Park PJ, Butte AJ, Kohane IS: Comparing expression profiles of genes with similar promoter regions. Bioinformatics 2002, 18(12):1576-1584.

9. Carninci P, Sandelin A, Lenhard B, Katayama S, Shimokawa K, Ponjavic J, Semple CA, Taylor MS, et al: Genome-wide analysis of mammalian promoter architecture and evolution. Nat Genet 2006, 38 (6):626-635.

10. Grotkjaer T, Winther O, Regenberg B, Nielsen J, Hansen LK: Robust multi-scale clustering of large DNA microarray datasets with the consensus algorithm. Bioinformatics 2006, 22(1):58-67.

11. Laderas T, McWeeney S: Consensus framework for exploring microarray data using multiple clustering methods. Omics 2007, 11(1):116-128.

12. Monti S TP, Mesirov J , Golub T: Consensus Clustering: A Resampling-Based Method for Class Discovery and Visualization of Gene Expression Microarray Data. Mach Learn 2003, 52:91-118.

13. Swift S, Tucker A, Vinciotti V, Martin N, Orengo C, Liu X, Kellam P: Consensus clustering and functional interpretation of gene-expression data. Genome Biol 2004, 5(11):R94.

14. Genomatix DB: http://www.genomatix.de/.

15. van Helden J: Metrics for comparing regulatory sequences on the basis of pattern counts. Bioinformatics 2004, 20(3):399-406.

16. van Helden J, Andre B, Collado-Vides J: Extracting regulatory sites from the upstream region of yeast genes by computational analysis of oligonucleotide frequencies. J Mol Biol 1998, 281(5):827-842.

17. van Helden J, Rios AF, Collado-Vides J: Discovering regulatory elements in non-coding sequences by analysis of spaced dyads. Nucleic Acids Res 2000, 28(8):1808-1818.

18. Hobert O: Gene regulation by transcription factors and microRNAs. Science 2008, 319(5871):1785-1786.

19. Chen X, Guo L, Fan Z, Jiang T: W-AlignACE: An improved Gibbs sampling algorithm based on more accurate position weight matrices learned from sequence and gene expression/ChIP-chip data. Bioinformatics 2008.

20. Elkon R, Linhart C, Sharan R, Shamir R, Shiloh Y: Genome-wide in silico identification of transcriptional regulators controlling the cell cycle in human cells. Genome Res 2003, 13(5):773-780.

21. Cartharius K, Frech K, Grote K, Klocke B, Haltmeier M, Klingenhoff A, Frisch M, Bayerlein M, Werner T: MatInspector and beyond: promoter analysis based on transcription factor binding sites. Bioinformatics 2005, 21(13):2933-2942.

22. Chekmenev DS, Haid C, Kel AE: P-Match: transcription factor binding site search by combining patterns and weight matrices. Nucleic Acids Res 2005, 33:W432-W437.

23. Blanco E, Messeguer X, Smith TF, Guigo R: Transcription factor map alignment of promoter regions. PLoS Comput Biol 2006, 2(5):e49.

24. Veerla S, Hoglund M: Analysis of promoter regions of co-expressed genes identified by microarray

analysis. BMC Bioinformatics 2006, 7:384.

25. Kankainen M, Loytynoja A: MATLIGN: a motif clustering, comparison and matching tool. BMC Bioinformatics 2007, 8:189.

66e Improving Cofactor Availability through Metabolic Engineering for Recombinant Phytochemicals Production

Joseph Chemler, Zachary Fowler, and Mattheos Koffas. Chemical and Biological Engineering, University at Buffalo, the State University of New York, 904 Furnas Hall, Buffalo, NY 14260

Many pharmaceuticals we use today originate from plant secondary metabolites such as Taxol and Artemisinin which are still extracted from natural sources. To solve the supply constraints encountered when using plant extracts, significant progress has been made in cloning the biosynthetic pathways into microorganisms to produce these compounds through fermentative processes. However, these approaches typically suffer from low yields. One issue that has received minimal attention is the availability of the NADPH, a necessary cofactor in many plant biosynthetic pathways, especially when cytochrome p450s are involved. Our group has been focusing on utilizing metabolic engineering to optimize NADPH availability. Specifically, our progress has covered the reconstruction of the NADPH-dependent (+)-catechin biosynthetic pathway within Escherichia coli. We have developed a constraint based model that uses an evolutionary algorithm to identify beneficial gene knockouts in order to improve NADPH yields. So far, we have increased (+)-catechin yields 10-fold using knockout strains and overexpression of NADPH-producing enzymes. In addition, we have applied our optimized strains to synthesis of novel flavonoid analogues using a mutational biosynthesis approach.

Web Page: www.eng.buffalo.edu/~mkoffas/

66f Identification of Optimal Measurement Sets for Isotopically Non-Stationary Mfa Experiments

YoungJung Chang[1], Patrick F. Suthers[2], and Costas D. Maranas[2]. (1) Chemical Engineering, The Pennsylvania State University, 147B Fenske Laboratory, The Pennsylvania State University, University Park, PA 16802, (2) Department of Chemical Engineering, The Pennsylvania State University, University Park, PA 16802

^{13}C-based stationary metabolic flux analysis (S-MFA) has been extensively used to estimate flux distributions from external flux and isotopic measurements. However, S-MFA is impractical for some cases because 1) isotopic steady state is very difficult to attain (e.g., miniaturized experiments), or 2) steady-state measurements are not enough to infer all intracellular fluxes. Isotopically nonstationary MFA (IN-MFA) can overcome some of these limitations by measuring isotopic distributions of internal metabolites at a number of different time points. In current IN-MFA methods, mass isotopomer distributions (MDVs) of internal metabolites are measured by GC-MS or LC-MS/MS at a set of fixed sampling times.

In this work, we extend our recent OptMeas procedure (Chang et al., 2008) developed for the identifiability analysis of S-MFA in order to account for the added concentration variables and differential isotopic balance equations. The extended OptMeas is used to identify the essential metabolites that need to be measured to ensure unique flux elucidation under isotopically nonstationary conditions. The identified essential measurements are then queried and refined to minimize the measurement cost while ensuring a unique flux distribution. It is important to note that this procedure calls for the repetitive solution of the inverse problem of IN-MFA, which is a dynamic optimization (DO) problem. The DO problem is converted to a nonlinear programming (NLP) problem by discretizing the time domain. In order to efficiently solve the resulting large-scale NLP problem, we combine a suitable network decomposition scheme with a Lagrangean decomposition based global optimization algorithm. The proposed approach was tested with a small network example involving eight metabolites and ten fluxes, and then applied to medium-scale demonstration examples including 1,3-propanediol producing *E. coli* strain. We found that the proposed approach is both scalable and reliable in predicting flux distribution and suggesting essential measurements.

Reference

(1) Chang, Y., Suthers, P. F., Maranas, C. D., Identification of optimal measurement sets for complete flux elucidation in MFA experiments, Biotech Bioeng, accepted, 2008.

66g Determining Metabolic Fluxes Using Experimental Measurements

Jennifer L. Reed, Chemical & Biological Engineering, UW Madison, 1415 Engineering Dr., 3639 Engineeering Hall, Madison, WI 53706, Qiang Hua, State Key Laboratory of Bioreactor Engineering, East China University of Science and Technology, Shanghai, China, and Bernhard O. Palsson, University of California, San Diego, 9500 Gilman Dr. Dept. 0412, La Jolla, CA 92093-0412.

Metabolic fluxes are useful for understanding cellular physiology and identifying ways to alter cell behavior. Metabolic fluxes are determined by incorporating experimental measurements into computational models and calculating the flux distributions which best match experimental observations. These experimental measurements include the (1) nutrient uptake and by-product secretion rates by cells, (2) biomass composition measurements, and (3) GC-MS or NMR measurements of ^{13}C labeled metabolites labeled arising from the processing of ^{13}C labeled substrates by the metabolic network. In order to quantify metabolic fluxes, experimental data needs to be analyzed by computational models. Metabolic flux analysis uses data on the relative abundance of ^{13}C labeled isotopomers (ie. metabolites which differ in the location of ^{13}C atoms) to calculate intracellular flux distributions (for review see [1]). Most models only include the central metabolic pathways for an organism, with some recent exceptions [2-4].

We investigated how sensitive flux distributions, calculated by isotopomer models, are to the inclusion of biosynthesis pathways as well as energy and redox balance constraints. We generated a biosynthetic isotopomer model for *Escherichia coli*, including central and biosynthetic metabolic pathways. Comparisons between model calculations show that inclusion of biosynthetic pathways leads to differences in global optimal solutions, as well as expanded confidence intervals compared to a central model.

We also investigated how balancing energy and redox carrier production and consumption rates, such as ATP, NAD(P) and NAD(P)H affects the calculated flux distributions. We found that incorporating these energy and redox balance constraints did not significantly affect how well the flux distributions fit the experimental data; however, these additional balances eliminate thermodynamically infeasible flux distributions and reduce confidence intervals for central metabolic fluxes. We also evaluated how well different types of data (metabolite uptake & secretion rates and GC-MS analysis of amino acids) eliminate flux distributions, and found that while often not measured in these types of experiments, oxygen uptake rates can significantly restrict flux values through the central metabolic pathways.

References

1. Wiechert W. 2001. ^{13}C Metabolic Flux Analysis. *Metabolic Engineering*. 3: 195-206.

2. Vo, TD; Lim, SK; Lee, WNP; Palsson, BO. 2006. Isotopomer analysis of cellular metabolism in

tissue culture: A comparative study between the pathway and network-based methods. *Metabolomics*. 2 (4): 243-256.

3. Vo, TD; Palsson, BO. 2006. Isotopomer analysis of myocardial substrate metabolism: A systems biology approach. *Biotechnology and Bioengineering*. 95(5): 972-983.

4. Suthers, PF; Burgard, AP; Dasika, MS; Nowroozi, F; Van Dien, S; Keasling, JD; Maranas, CD. 2007. Metabolic flux elucidation for large-scale models using C-13 labeled isotopes. *Metabolic Engineering*. 9 (5-6): 387-405.

68a Emulsion Function In a Simulated Gastric Environment. A Quantitative Study of Digestion Kinetics and Intestinal Permeability Enhancement

Fulden Buyukozturk[1], Hong Long[1], and **Rebecca L. Carrier**[2]. (1) Chemical Engineering, Northeastern University, 342 Snell Engineering Center, 360 Huntington Avenue, Boston, MA 02115, (2) Department of Chemical Engineering, Northeastern University, 342 Snell Engineering Center, 360 Huntington Avenue, Boston, MA 02115

The oral route is the most preferred method of administration of drugs. Unfortunately, this route is not possible for 50% of currently marketed drug compounds due to their low solubility in water and low oral bioavailability. Lipid-based drug delivery systems are commonly studied for the enhancement of oral bioavailability of hydrophobic drugs. However, these systems are not being commonly utilized commercially, partially due to the lack of guidance for formulation and lack of knowledge of how these systems function to enhance bioavailability [1, 2]. Self-emulsifying drug delivery systems (SEDDS) are oil-in-water emulsions, typically consisting of oil and one or more surfactants, that spontaneously emulsify under the gentle mixing conditions in the gastrointestinal environment. These systems, in particular, have shown great promise for enhancing oral bioavailability of low-solubility compounds as well as offering the advantages of minimal processing and inherent stability.

To gain a greater understanding of the dependence of emulsion function in the gastrointestinal (GI) tract on formulation design, a quantitative study of processes central to emulsion function, digestion and intestinal permeability enhancement, was undertaken. A broad range of formulations was studied: Oils from three different structural classes (long chain triglyceride (Soybean oil), medium chain trigylcerides (Neobee M5), and propylene glycol dicaprylate/dicaprate (Captex 200)) and surfactants varying in hydrophilic-lipophilic balance (HLB) value from 10-15 (Cremophor EL, Tween 80, a mixture of Capmul MCM and Labrasol) were combined at three different ratios. Naproxen was included to test the influence of drug on emulsion properties. The dependence of emulsion formation and emulsion characteristics on physicochemical emulsion formulation properties was investigated. An HLB value of 10 or higher was required for spontaneous emulsification; emulsions formed had particle sizes ranging from 60-500 nm. MTT assay results showed that only oil-surfactant mixtures that did not spontaneously form emulsions were toxic. Changes in transcellular permeability were assessed by measuring drug transport across epithelial layers, and epithelial tight junction integrity was assessed by measuring transepithelial electric resistance. Formulations composed of high levels of high HLB surfactant (Tween 80) significantly reduced epithelial tight junction integrity. Digestion kinetics were studied by monitoring the change in particle size and by quantitative analysis of digestion products by HPLC upon exposure of emulsions to digestive enzymes. Ultimately, the relation of emulsion formulation physical and chemical properties to relative kinetics of processes occurring in the GI tract (permeation enhancement, digestion, drug release) will be utilized to aid in rational design of emulsions and fulfillment of the tremendous potential of this technology for oral bioavailablility enhancement.

References

[1] Neslihan Gursoy, et al. Biomedicine Pharmacotherapy, 58; 173 (2004).

[2] Charman, William N., et al. Journal of Pharmaceutical Sciences,89; 967 (2000).

68b Manufacturing of Biodegradable Drug-Polymer Nanocomposites for Controlled Drug Release by Supercritical Fluid Extraction of Emulsions

Johannes Kluge[1], Francesco Fusaro[1], Nathalie Casas[1], Marco Mazzotti[1], **Ranjit Thakur**[2], and Gerhard Muhrer[2]. (1) Institute of Process Engineering, Department of Mechanical and Process Engineering, ETH Zurich, Sonneggstrasse 3, CH-8092 Zurich, Switzerland, (2) Novartis Pharma AG, CHBS, WSJ-145, Lichtstrasse 35, CH-4056 Basel, Switzerland

Biocompatible and biodegradable polymers such as poly-lactic-co-glycolic acid (PLGA) may be used as pharmaceutical excipients for pharmaceutical applications such as controlled drug release or targeted drug delivery. For such applications, the drug needs to be incorporated in the polymer in the form of composite micro- or nanoparticles for which narrow particle size distributions are desired. There is a need for the development of scalable processes that are compatible with the sensitive materials involved and with the high standards that are requested for product quality.

Supercritical fluids, mostly supercritical carbon dioxide (scCO2), have been applied in a number of different processes aiming at the production of pharmaceutical micro- or nanoparticles. Generally, such processes are characterized by a number of advantages that stem from the application of scCO2; these are typically mild operating conditions, a low consumption of organic solvents and high degree of product purity. Therefore it suggests itself to investigate whether and how such processes may also be applied for the production of drug-PLGA nanocomposites.

Supercritical Fluid Extraction of Emulsions (SFEE) is a novel technique that allows for the manufacturing of such particles. In this process, drug and polymer are mixed in an organic solution that is dispersed in an oil-in-water emulsion stabilized by a suitable surfactant. Then, the emulsion is mixed with scCO2 in order to remove the organic solvent from the emulsion droplets, leading to the co-precipitation of drug and PLGA. Thereby, stable and solvent-free suspensions in water are obtained, and the solidified polymer or composite particles may be recovered upon depressurization.

In the present study, we investigate different methods that may be applied for the incorporation of both lipophilic and hydrophilic drug substances. A further goal of the study is the characterization of particles with respect to the distribution of drug and polymer within the composite structure. Finally, the characterization of the kinetics of drug release from the co-formulation is carried out in vitro, thus assessing its potential performance in in-vivo applications.

68c Production and Stabilization of Organic Crystalline Sub-Micron and Nanoparticles Using High Pressure Homogenization for Drug Delivery

Frank Romanski, M. Silvina Tomassone, Eric Jayjock, and Fernando J. Muzzio. Chemical and Biochemical Engineering, Rutgers University, 98 Brett Rd., Piscataway, NJ 08854

Many newly discovered drugs are found to be poorly soluble in water. It has been known for over a decade that forming finer particles (micron and smaller) can improve the bio-availability and the dissolution rate of drugs due to their increased surface area. However, there is little knowledge in the area of short- and long-term stabilization of the particles following production. This issue is critical because nanoparticles in suspension agglomerate intensely, making them hard to process and in some cases defeating the purpose of using nanoparticles in the first place. Thus, the goal of this work is to obtain well stabilized and deagglomerated nanoparticle systems. Such stable nanosuspensions can the be used to incorporate nanoscopic drugs into many products including strip films, tablets, and multilayer coatings loaded with drug nanoparticles for controlled release performance.

At the present time, methods for preventing nanoparticle agglomeration are essentially empirical. In this work, we investigate the well known particle creation technique High Pressure Homogenization (HPH) as a "top down" approach to the creation of nanoparticles. The method has long been proven to create nanoparticles through the application of large amounts of shear and cavitation. We focus on the stabilization of the particles during creation rather than the creation technique itself. We have selected two hydrophobic drug materials as test cases; Fenofibrate (used to reduce the amount of cholesterol and triglycerides in blood) and Griseofulvin (antifungal type of antibiotic), both typically low dosage materials that are generally safe for laboratory use. Both drugs are Biopharmaceutics Classification System (BCS) Class II substances which are considered to have very poor water solubility, yet a high gastric permeability. The drugs are passed through a homogenizer to reduce the particle size by varying the pressure, solids loading, and concentration of surfactant. Several surfactants have been tested including Tween 80 and Poloxamer 407. However, better results were found using block co-polymers such as HPMC (hydroxypropyl-methylcellulose). Our experimental work suggests that using surfactants and polymers with a high affinity for the drug surface can effectively aid in stopping post production agglomeration of the particles both short-term and long-term. In addition to experimental results, several Molecular Dynamics (MD) studies have been conducted. The goal of these simulations is to test the molecular interactions between the surfactant and surface of the crystalline materials. The chemical functionality of a crystalline surface, the entropy of the lattice, and the energy of attachment derived from the MD simulations can indicate which stabilizers are optimal for the stabilization of the produced particles.

Surface Interactions and Microstructural Characterization of API Subjected to Cryogenic Milling

DIANA GUZMAN-VILLANUEVA[1], DEA HERRERA-RUIZ[2], M. TERESA CARVAJAL[1]

[1]Department of Industrial and Physical Pharmacy, School of Pharmacy, Purdue University, 575 Stadium Mall Drive, West Lafayette, IN 47907.

[2]Dea Herrera-Ruiz, Facultad de Farmacia, Universidad de Morelos, UAEM, Cuernavaca, Morelos, México.

INTRODUCTION

Use of crystalline active pharmaceutical ingredients (API) in solid oral dosage forms are a preferable form in the pharmaceutical industry because of good stability compared to amorphous. However, a large number of crystalline APIs has poor solubility being the rate limited for the bioavailability[1], this has become a challenge in pharmaceutical research to come out with creative solubilization strategies to enhance dissolution and hence bioavailability. One strategy is the particle size reduction to improve the dissolution rate by increasing the surface area of materials; therefore drugs with poor water solubility are often micronized. This approach offers advantages for both oral delivery and pulmonary drug administration.

There are different methods used to reduce the particle size the solids in pharmaceutical manufacturing, being the milling of larges particles a common way, but depending on the product to be milled and the required final particle size, the drug would be size reduced by pressure, friction or impact[2]. Although the milling is process frequently used in the course of drug formulation, one main concern is that the milled materials present characteristics such as poor control not only in the particle size, shape and electrostatic charge but on the surface interactions, microstructural and physicochemical properties. These in turn are responsible for the functionality and the stability upon storage of the materials.

During the course of the milling process, molecular compounds are subjected to a large variety of perturbations such as mechanical stress; the high energy input can cause a disruption of the crystal lattice such as formation of disordered amorphous solids or the creation of defect which alter the physicochemical properties of pharmaceutical solids with an increased chemical reactivity[3]. All these transformations have a direct influence on the stability and functionality of the compounds, due to thermodynamically activated solids can result in a greater susceptibility to spontaneous recrystallization during storage.For these reasons, there is need to study the microstructure from pharmaceutical solids when they are subjected to high energy milling processes.

METHODS

In this research, Nifedipine (NDP), Ketoconazole (KCZ), were chosen as a model drugs along with a common excipient: Lactose (LAC) widely used in the food and pharmacy industries.

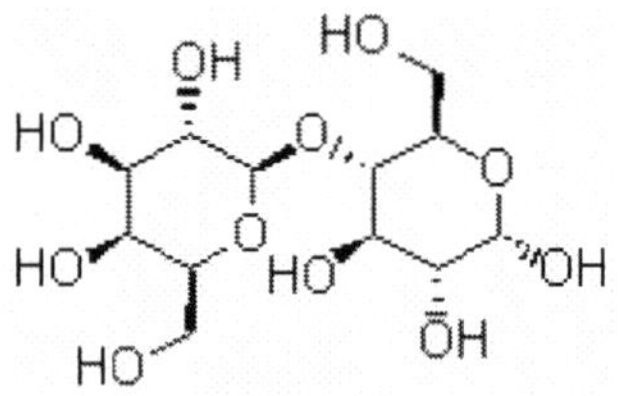

α-Lactose

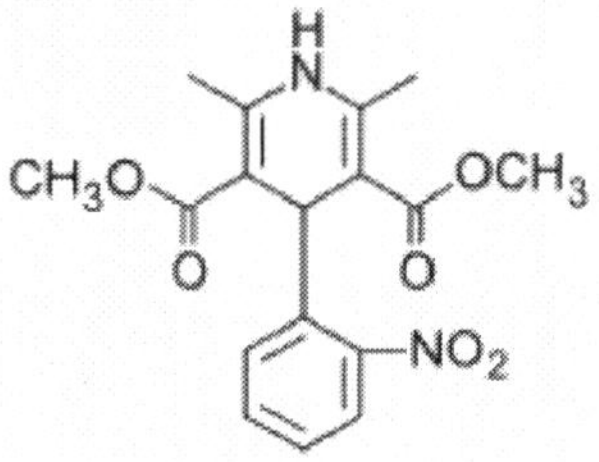

Nifedipine

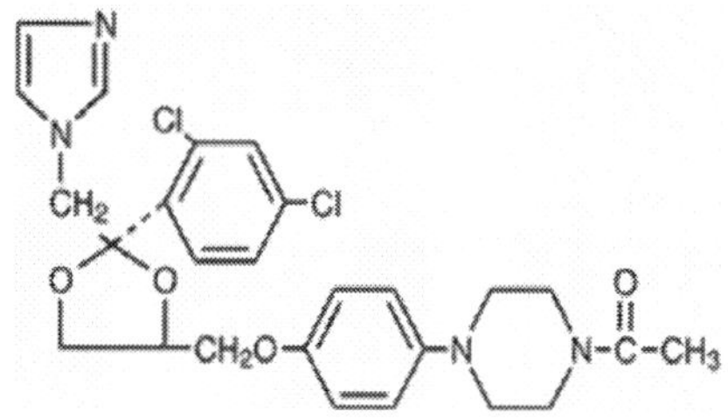

Ketoconazole

These compounds were subjected to cryogenic milling and stored at low relative humidity in desiccators until ready to be analyzed. Cryogenic milling was performed using a method employed by Feng et al[4]. Samples were milled for 10 y 30 min periods. Thermal and surface characterization was investigated. The structure of milled materials was compared with their respective counterparts of un-milled (crystalline) and amorphous (quench melt) using techniques such as Differential Scanning Calorimetry (DSC), Powder X-Ray Diffraction (PXRD), Optical Microscopy (OM) and Inverse Gas Chromatography (IGC).

Differential Scanning Calorimetry (DSC)

A Q10 DSC was used and the samples were analyzed in hermetic aluminum pans. To calibrate high purity indium was used. Nitrogen was used as purging gas. Approximately 3 mg of samples were weighed and heated at 10 °C/min until reach the melting point of the drugs, samples were carried out in triplicate. Different mixtures of crystalline, milled and amorphous samples were analyzed on the same DSC instrument under the same conditions and data were collected.

Powder X-Ray Diffraction (PXRD)

Samples were analyzed using a Siemens X-ray diffractrometer. All samples (un-milled, milled and quench melt) were placed in a circular silicon holder and then scanned from 4 to 40° 2θ with a scan rate of 4°/ min. The patterns obtained were compared to each other with their respective crystalline and quench melt form.

Optical Microscopy (OM)

A BH-2 Olympus Microscope was used to determine the presence of birefringence or double refraction as indicative of crystallinity in milled and un-milled samples. Approximately 2 mg of samples were placed at slide glasses and then were directly observed by optical microscope using a 10X-40X magnification in a black background.

RESULTS AND DISCUSSION

PXRD of un-milled samples were compared with the calculated patterns from the Cambridge Structural Database. These data indicate that un-milled samples correspond to the pure crystalline forms (Figure 1). The reduction in the intensity of crystallization peaks of NDP was confirmed by PXRD analysis. The PXRD of milled and quench melt samples also was

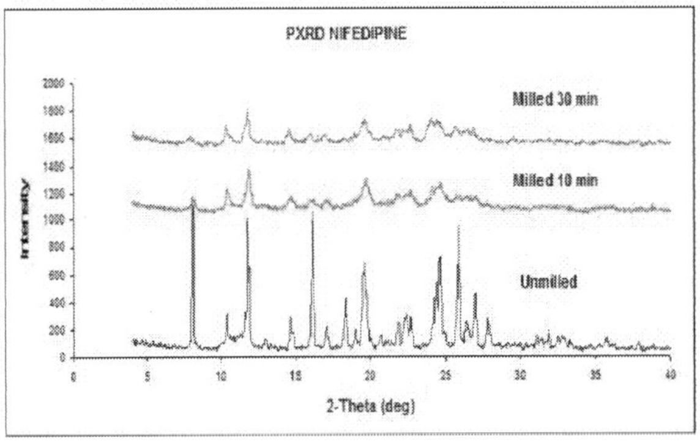

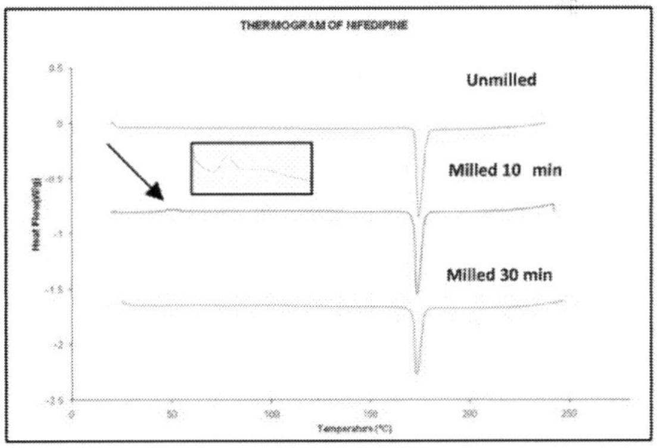

Figure 1. PXRD patters and DSC thermograms from Nifedipine

compared and they show a different behavior. Quench melt show the absence of crystallization peaks how was expected. DSC measurements were performed to study how the thermal properties are affected by cryogenic milling. Thermograms were obtained from un-milled crystalline samples, amorphous quench melt and samples subjected to different time periods. The thermograms show consistent melting points for crystalline materials and glass transition and crystallization events for amorphous materials. DSC measurements of un-milled, milled and quench melt NDP samples presented the consistent melting point peak. The melt quench sample also show the glass transition temperature ~38 °C as well as the crystallization event ~75°C respectively. However, the thermograms of milled samples 10 and 30 min show a new exothermic event approximately at 55°C overlapping the glass transition temperature. These thermograms are completely different to amorphous samples (Figure 1). The new exothermic event is only showed in milled samples. Both milling times present the sample this event being more prominent at 30 min. Therefore, the presence of exothermic peaks is a consequence of milling process. These exothermal peaks cannot be categorized as a crystallization peaks produced as a result of an amorphous state due to the absence of glass transition temperature but the creation of defects in the crystalline lattice which translates in the reduction of crystallinity. Furthermore, NDP milled show a decrease in the melting point as well as heat of fusion. These denote the presence of defect in the milled samples (Figure 1).

Under the microscope the un-milled and milled samples of NDP samples show a decrease of birefringence (Figure 2) and hence a decrease of crystallinity as shown by PXRD. However, milled samples still present broad regions with a high birefringence. That means that cryogenic milling reduces the crystallinity having the three-dimensional array of the crystalline lattice. In contrast, KCZ shows a different behavior. The thermograms of 10 and 30 min milling times show the characteristic events of an amorphous state. KCZ presents a glass transition temperature at ~ 36 °C. The crystallization peak was showed at ~ 50 °C for both times. The crystallization peak was increased as the milling time was longer. The analysis of PXRD of KCZ milled sample resulted in characteristic amorphous patterns without the presences of peaks (Figure not shown). These patterns denote a total loss of crystallinity compared with the un-milled materials. That shows that cryogenic milling is able to change the initial state of KTZ sample and produce the absence of long- range and three-dimensional molecular array characteristic of the crystalline state. Nevertheless, in the microscopy studies is observed a significant loss of birefringence at 10 min but there is still the presence of crystalline material. A total loss of birefringence at 30 min is noted (See figure 2). The results obtained show that the amorphous state take place due to the accumulation of perturbation resulting in a transformation from crystalline lattice to an amorphous state as is increasing the milling time. Therefore, the high mechanical energy input for the cryogenic milling not only generate the fragmentation of larger particles to reduce the particle size but it produce disruptions which alter the periodicity that characterizes the crystalline state, creating a disordered material.

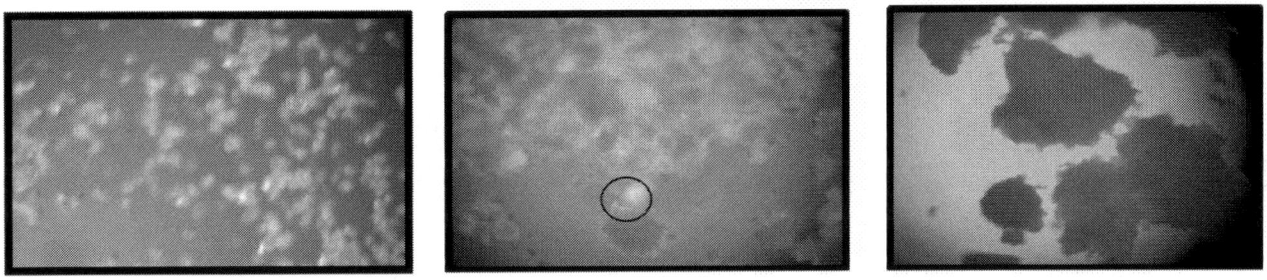

Ketoconazol Unmilled | Ketoconazol Milled 10 min | Ketoconazol Milled 30 min

Figure 2. Presence of crystalline material at after milling Ketoconazole showed by birefringence.

Similar behavior to that of the Ketaconazole is observed in LAC milled sample, a progressive deterioration of the crystalline lattice that ends with the accumulation of dislocations causing overall perturbations which contribute to the formation of amorphous regions as the milling time is increased. DSC measurements of un-milled LAC show the presence of melting point and dehydration events at ~219 ºC and ~ 150 ºC respectively. However, after 10 min milling LAC show the formation of two endothermic peaks approximately at dehydration and melting point events which correspond to the presence of a mixture of α and β-LAC. In contrast, the DSC thermogram of LAC milled 30 min show a thermal transition associated with the formation of an amorphous state represented for a sharp crystallization peak at ~50ºC (Figure 3).

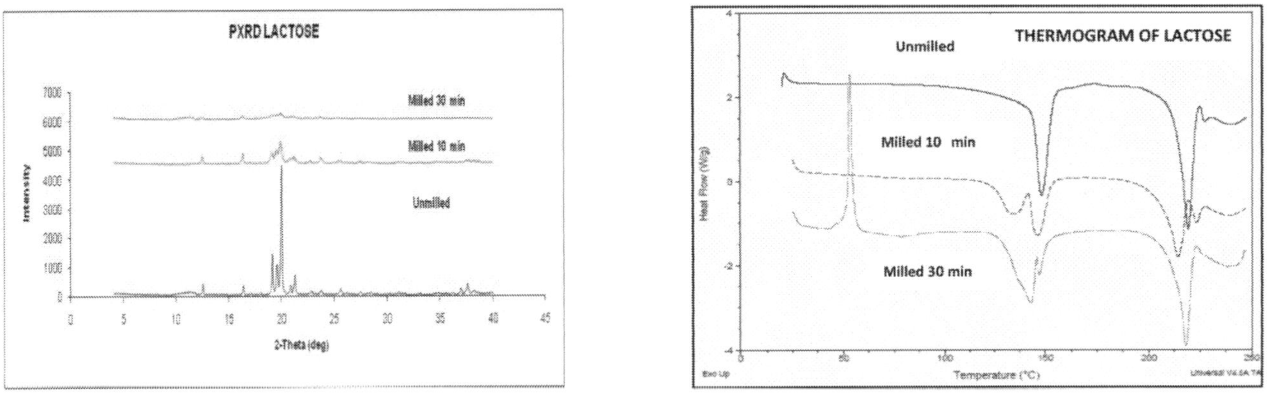

Figure 3. PXRD patters and DSC thermograms for Lactose.

The impact of the cryogenic milling in LAC samples is markedly observed by the reduction in the intensity of crystallization peaks in PXRD LAC patterns and by the loss of birefringence in microscopy studies.

CONCLUSIONS

The techniques used have demonstrated the physical transformation of the materials after being subjected to cryogenic milling. The presence of different events: *disorder*, showed in several orders of magnitude by LAC, the formation of *crystal defect* by NDP and the creation of *amorphous states* by KTZ. All of them manifest different behaviors according to their capacity of control the high mechanical energy input during milling process. Crystal defects were characterized by the presence of new exothermal peaks in milled samples which overlap the glass transition region present in the amorphous form by the DSC and as reduction of crystallinity in PXRD patterns. The amorphous form was represented by glass transition temperature and crystallizations peaks as typically reported in the literature. This work clearly demonstrated that cryogenic milling induced physical transformations in the crystalline lattice. The surface of the milled material is left with high surface free energy compared to that of the un-milled material. It is expected that the milled material will be prompt to interact more easily than the un-milled material. Work is still underway to understand the properties that govern these energetic states in order to take them into consideration when carrying out stability and performance studies.

REFERENCES

1. C. Bhugra, M. J. Pikal. Role of Thermodynamic, Molecular, and Kinetic Factors in Crystallization From the Amorphous State. *Journal of Pharmaceutical Sciencies* , Vol 97, NO. 4, 2008.
2. E. Shalaev, M. Shalaev, G Zografi. The Effect of Disorder on tehe Chemical Reactivity of on Organic Solid, Tetraglycine Methyl Ester: Change of the Reaction Mechanism. *Journal of Pharmaceutical Sciences* , Vol. 91, No. 2, 2002.
3. N. Resenack, B. W. Muller. Micron-Size Drug Particles: Common and Novel Micronization Techniques. *Pharmaceutical Development and Technology* , Vol 9, No. 1, pp 1-13, 2004.
4. T. Feng, R. Pinal, M. T. Carvajal. Process Induced Disorder in Crystalline Materials: Differentiating Defective Crystals form the Amorphous Form of Griseofulvin. *Journal of Pharmaceutical Sciences*, Vol. 97, No. 8, 2007.

68e Novel Methods of Granule Structuring and Release Control

Frantisek Stepanek, Department of Chemical Engineering, Institute of Chemical Technology, Prague, Technicka 5, Prague 6, 166 28, Czech Republic and Mansoor A. Ansari, Departiment of Chemical Engineering, Imperial College London, South Kensington Campus, London, SW7 2AZ, United Kingdom.

A granule is a composite mixture of one or more type of primary solid particles (API, excipient, etc.) held toghether by a binder, and represents a commonly used precursor or final dosage form for the delivery of solid active pharmaceutical ingredients. Granules are either transiotry, existing only from one process step to another (e.g., as a feed for tableting) or they are the final product on its own (e.g., when filled into capsules). In both considerations the spatial arrangement of formulation ingredients within the granule matrix is very important in regard to its mechanical properties (visco-elastic response during tabletting) and API release pattern during dissolution.

In our recent work [1, 2] we have established a novel methodology that can be used to control the spatial distribution of the API (or any other component of interest) within the granule structure by introducing the API as a solid dispersion with in-situ melt or aqueous polymeric binders. In the present work we examine the dissolution behaviour of tablets compressed from pre-designed granules of two types, i) granules with model API randomly distributed in the structure matrix and ii) granules with the model API encapsulated in its core.

We will also demonstrate the above mentioned approach to produce granules that contain compartmentalised reactive particles. These granules release an active ingredient produced in-situ as a result of the interaction between the reactive precursor particles contained within the granule. Since granule properties profoundly influence the tablet dissoultion charactersitics, the idea of comparmentalised granules would offer a range of applications in the area where in-situ API are essential formulation or physiological requirements.

References:
[1] Ansari M.A., and Stepanek F. 2006, "Design of granule Structure: computational methods and experimental realization" AIChE Journal , vol 52, no 11, pp: 3762-3774.
[2] Ansari, M.A., and Stepanek F. 2008, "The evolution of microstructure in threecomponent granulation and its effect on dissolution", Particulate Science and Technology , vol 26, no 1, pp: 55-66.

68f Practical Methods for Improving Flow Properties of Active Pharmaceutical Ingredients

Kalyana Pingali, Department of Chemical and Biochemical Engineering, Rutgers University, 98 Brett Road, Piscataway, NJ 08854, Kostas Saranteas, Chemical Process Research and Development, Sepracor Inc., 84 Waterford Drive, Marlborough, MA 01752, and Fernando J. Muzzio, Chemical and Biochemical Engineering, Rutgers University, 98 Brett Road, Piscataway, NJ 08854.

The essential aim of this paper is to develop effective methods for improving the flow properties of APIs without requiring particle size or shape modification. The 'formulation' approach used here focuses on enhancing flow properties of three chemically different drug powders (micronized acetaminophen, levalbuterol and didesmethylsibutramine) by using small amounts of lubricants, glidants, and other additives, both individually and in combination. Additives are intimately mixed using a laboratory-scale V-blender with an intensifier bar. Flow index, dilation and electrical impedance were measured for a total of 24 blends blend. The flow behavior of all three APIs improved with the addition of these additives. Relative effectiveness of different additive combinations displayed remarkable consistency for all three APIs. Simultaneous presence of $SiO2$, MgSt and talc led to substantial decreases in cohesiveness, causing major improvements in flowability of powder. All three properties showed a very tight correlation. Drug powders with improved flow were found to exhibit low dilation and low impedance values. A common linear correlation between flow index and impedance and also between dilation and impedance was observed for all three APIxs, indicating that electric properties play a substantial role in the cohesivity of all three APIs, and suggesting the presence of a common mechanism for the emergence (and mitigation) of cohesive phenomena.

68g Limiting Flow Rate Analysis of Direct Compression Grade Hypromellose and Its Relationship to Powder Flowability

Karl Jacob[1], Kacee Ender[2], Dave Wallick[2], and Harold Bernthal[2]. (1) The Dow Chemical Company, Dow Chemical Company, 1319 Bldg., Midland, MI 48667, (2) Water Soluble Polymers R&D, Dow Wulff Cellulosics, Larkin Lab, Midland, MI 48674

Direct compression (DC) is a tableting process in which a dry blend of ingredients is placed into a tablet hopper and compressed into tablets. Direct compression requires that the powder blend have high flowability. One issue with CR grade hypromellose is poor flowability and its tendency to arch in hoppers. A new grade of hypromellose with a larger mean particle size and improved flow has been developed for direct compression controlled-release applications. This is not the result of decreased particle cohesion (as measured by ring shear tests), but rather improved permeability of the material. This work introduces the concept of limiting flow rates and its importance as a key measure of flowability for fine materials. Limiting flow rate analysis was performed for both grades of hypromellose. Limiting flow rates are caused by the inability of the flowing solid to dilate in the conical section of the hopper as a result of low gas permeability of the bulk solid. The three key measurements used to determine the limiting flow rate are cohesive strength, permeability and compressibility. Cohesive strength is measured in a Schulze RST01.pc shear tester. Compressibility measures the bulk density as a function of load, while permeability measures the air permeation characteristics under similar loads. These data are then used in a program which solves a series of differential equations to determine the limiting flow rate. This paper will examine in detail the key relationships between the permeability, bulk density and the powder flow and how interstitial gas pressure in the conical region of the hopper plays a key role in determining the flow rate of DC mixtures.

As an example, a tablet hopper of specified size with a 6 inch diameter outlet that manufactures 500 mg tablets at a rate of 10,000 tablets per minute requires a flow rate of 660 lbs/hr of material out of the hopper. The limiting flow rate for the DC hypromellose was calculated to be 2400 lbs/hr for a 6 inch diameter outlet, while the flow rate for the CR grade hypromellose was calculated to be 100 lb/hr. The influence of low permeability will be shown to be dramatic with decreases in flow rates being several orders of magnitude, compared to the unimpeded flow rates of coarser materials. For the new DC hypromellose, the improvement in flow is approximately 25 times better than the conventional CR grade hypromellose.

69a Complete Microbiological Fuel Cell. A Photosynthetic Cathode Coupled to a Fermentative Anode

Erin E. Powell, John C. Bolster, Gordon A Hill, and Richard W Evitts. Chemical Engineering, University of Saskatchewan, 57 Campus Drive, Saskatoon, SK S7N 5A9, Canada

The culture of a photosynthetic microorganism using carbon dioxide as the growth substrate provides opportunities for bioremediation of an atmospheric pollutant and yield of valuable products. In addition, during the carbon dioxide capture and conversion process, it is possible to employ the photosynthetic culture in a microbial fuel cell (MFC) and generate electrical energy. In this research, the biokinetics of growth of a photosynthetic microorganism was studied and optimized. The growing cell culture was then combined with an ethanol fermentation culture, resulting in a complete microbial battery. The photosynthetic culture was employed as the electron acceptor in a cathodic half cell, while the fermentation culture operated as the anodic half cell, thus generating a potential difference across the cell. This new MFC is carbon dioxide neutral and produces ethanol in the anodic half cell chamber, and can reduce excess carbon at the cathode, in addition to generating electricity. The details of the complete MFC, along with power outputs will be presented. The effect of changing culture conditions on fuel cell operation will be reported.

69b Properties of Algal-Based Biodiesel

Nadia J. Abunasser[1], Mario Enrique Danton Garica Perez[2], Haiying Tang[3], Steven O. Salley[3], John Wilson[4], and K. Y. Simon Ng[3]. (1) Chemical Engineering and Materials Science, Wayne State University, 461 Burroughs, Detroit, MI 48202, (2) Alternative Energy Technology, Wayne State University, 461 Burroughs, Detroit, MI 48202, (3) Department of Chemical Engineering and Materials Science, Wayne State University, 5050 Anthony Wayne Drive, Detroit, MI 48202, (4) National Biofuel Energy Laboratory, NextEnergy, 461 Burroughs ST, Detroit, MI 48202

While interest in using algal oil as a source for biodiesel has increased in recent years, there are few comprehensive studies of the properties of the biofuels that would be produced from these oils. Dunaliella tertiolecta UTEX LB 999, grown in Erdschreiber's Medium and the Chlorella Minutissma UTEX 2219, grown in Modified 3N Bold Medium at room temperature with a 16-8 light/dark cycle, as well as Botryococcus braunii UTEX LB 572, grown in Waris+Soil Extract Medium at room temperature with a 12-12 light/dark cycle were evaluated. The algal oil was extracted using a chloroform/ methanol extraction, and the amount of oil collected from each strain was noted and compared to published values. The oil was then transesterfied to produce biodiesel. The properties such as the fatty acid methyl ester profiles, oxidative stability, viscosity, lubricity, and derived cetane number were measured and compared to those of biodiesel produced from traditional feedstock.

69c The Influence of Redox Potential and Sulfide Concentration in the Fermentation of Syngas to Ethanol

Peng Hu, Allyson Frankman, Dallin Glenn, and Randy S. Lewis. Department of Chemical Engineering, Brigham Young University, 329 CB, Provo, UT 84602

Abstract:

Many fermentation processes have been used to convert biomass to various fuels. Currently, the fermentation of syngas, which is a mixture of CO, H2 and N2, is being studied for the production of ethanol. In our studies, an anaerobic microbial catalyst Clostridium carboxidivorans was used and the key performance parameters, such as growth rate and ethanol/acetic acid production rates, were compared with regards to redox potential and sulfide concentration. Studies were carried out in both 100 ml bottles with daily replacement of headspace gas as well as in 1-liter bioreactors with continuous gas flow. During our studies, the reducing agent cysteine-sulfide was applied to maintain the anaerobic environment.

In all studies, microbial catalyst was observed to first produce acetic acid as a fermentative product while in the growth phase. As expected, a switch was observed from acetic acid to ethanol as the cell concentration reached steady state. Furthermore, it was observed that the acetic acid decreased, indicating that the cells could be converting the acid to ethanol. Interestingly, we observed that both the cell growth and the switch between acetic acid production and ethanol production corresponded with changes in the redox potential. Different redox potentials were observed at different growing or production rates. These observations indicated that redox potential may play important roles during the fermentation process. In order to improve cell growth and ethanol production, a model of redox control was built and experiments were conducted at different levels of redox potential.

In addition, we found sulfide was not only scavenging the residual oxygen, but also affecting the fermentative performance. Since there was an equilibrium of HS- and H2S between gas and liquid phases, the volatile H2S could leave the system with the continuous gas flow in reactor studies or with outlet gas during the headspace gas replacement in bottle studies. The sulfide loss with continuous gas flow in the reactors was much faster than that in the bottles. The different sulfide concentrations may cause different redox levels, and may affect the cell growth and ethanol production. Models of sulfide loss in both reactors and bottles agreed with experimental data. Based on the models, studies at different sulfide levels were conducted and the optimum sulfide concentrations for cells growth and ethanol production were found.

69d Bioenergetics of Succinic Acid Production by E.Coli

Amarjeet Singh, Chemical & Biological Engineering, University of Colorado at Boulder, 1111 Engineering Drive, Campus Box 424, Boulder, CO 80309 and Ryan T. Gill, Dept. of Chemical and Biological Engineering, University of Colorado, 1111 Engineering Drive, Campus Box 424, Boulder, CO 80309.

C4- dicarboxylic acids are potential building blocks for deriving commodity and specialty chemicals from biomass. E.coli strains engineered for succinic acid acids formation are, however, incapable of robust growth in minimal media. To elucidate the causal mechanisms for this observed growth defect, we employed a selection strategy to identify genes that improve growth of NZN111- a succinic acid production strain. Our studies suggest that altered internal redox balance is the primary cause for the growth inhibition of this strain. Using SCALEs, a genome wide approach, we identified a number of genetic changes that can balance the redox inside the cell and restore microaerobic growth of NZN111 in minimal media. Based on this genetic information, we employed genetic as well as process changes to achieve better anaerobic growth and subsequent higher succinic acid production.

69e Biological Production of Hydroxyacids from Renewable Sources

Collin Martin[1], Hsien-Chung Tseng[1], and **Kristala L. Jones Prather**[2]. (1) Department of Chemical Engineering, Massachusetts Institute of Technology, 77 Massachusetts Ave, Room 66-458, Cambridge, MA 02139, (2) Chemical Engineering, MIT, 77 Massachusetts Ave, Cambridge, MA 02139

Hydroxyacids represent an important class of compounds that see application in the production of polyesters, biodegradable plastics and antibiotics, and they serve as useful chiral synthetic building blocks for other fine chemicals and pharmaceuticals. We have developed processes for the microbial production of one such class of hydroxyacids, the hydroxyvalerates. 4-hydroxyvalerate (4HV) and 3-hydroxyvalerate (3HV) are produced from the inexpensive and renewable carbon source levulinic acid utilizing an engineered strain of *Pseudomonas putida*. We tested two strains of *P. putida*: a commercially available strain (KT2440) and the PHA synthase knockout strain GPp104. We also examined two enzyme systems for removing CoA acyl carriers from intracellular hydroxyacids: the *ptb/buk* system and *tesB*. Once a suitable strain and enzyme system was found, the process was modulated at the shake flask scale in minimal and rich media for the high-titer production of both 3HV and 4HV. Maximum observable titers were 14 g/L for 4HV produced in rich medium and 5.3 g/L for 3HV produced in minimal medium. To our knowledge, this work represents the first time that hydroxyvalerates have been produced from a feasible feedstock in shake flasks at such high concentrations. This microbial system, as well as recombinant *E. coli*, are currently being investigated for their potential to produce other, longer chain hydroxyacids at high titers.

69f Stimulation of Microbial Metabolism In Biofilms

Hung Duc Nguyen, School of Chemical Engineering and Bioengineering, Washington State University, PO Box 642710, Pullman, WA 99164-2710 and Haluk Beyenal, School of Chemical Engineering and Bioengineering, Washington State Univeristy, PO Box 642710, Pullman, WA 99164-2710.

Biofilms are used for many different bioprocesses. Although their use in biobased fuels or biochemical productions have been suggested, there have not been any detailed researches demonstrating how biofilms can be used. Our research group has extensive research experience on biofilm processes. Recently we developed a new biofilm reactor which can be used in biobased chemical or biofuel production. This presentation focuses on the design and use of biofilm reactors for biobased chemicals. Our experimental results demonstrated that biofilm processes can improve productivity of the selected process. We suggest biofilm technology as an effective alternative to classical suspended culture bioprocess for biofuel or biochemical production.

92a DNA Polymerization Dynamics of HIV-1 Reverse Transcriptase. A Single Molecule Study

Charles M. Schroeder[1], Sangjin Kim[2], and X. Sunney Xie[2]. (1) Chemical & Biomolecular Engineering, University of Illinois at Urbana-Champaign, Roger Adams Laboratory, 600 South Mathews Avenue, Urbana, IL 61801, (2) Chemistry and Chemical Biology, Harvard University, 12 Oxford Street, Cambridge, MA 02138

Human immunodeficiency virus-1 reverse transcriptase (HIV-1 RT) is a heterodimeric DNA polymerase responsible for synthesizing double-stranded proviral DNA from the single-stranded RNA viral genome. In this work, we study the DNA polymerization dynamics of HIV-1 RT at the single molecule level using a hydrodynamic flow-stretching assay. In this assay, bead-tethered DNA molecules are surface immobilized inside microfluidic flow cells, and DNA templates are stretched by hydrodynamic forces. In this manner, we directly observe polymerization dynamics as a function of template DNA sequence for single enzymes. We observe the primer extension activity of HIV-1 RT on single DNA templates with a force-dependent DNA synthesis rate. In the low force regime, the average synthesis rate is 22 nt/sec. Furthermore, we observe transient enzyme stalling at specific locations within the single-stranded DNA template, and we demonstrate that enzymatic pause sites are in high correlation with hairpins in the DNA template. Finally, we present the hydrodynamic flow-stretching assay as a useful single molecule tool to study both specific and non-specific nucleic acid-protein interactions in a multiplexed and high-throughput format.

92b Dicer-siRNA Binding: Effect on Silencing Efficacy of siRNAs

Hemant K. Kini and S. Patrick Walton. Chemical Engineering and Materials Science, Michigan State University, East Lansing, MI 48824

Dicer proteins are known to participate in RNA interference (RNAi) by cleaving double-stranded RNA (dsRNA) and pre-micro RNA (pre-miRNA) substrates into ~21-mer short interfering RNAs (siRNAs) and miRNAs, respectively. Studies have shown that Dicer alone forms stable complexes with siRNAs. It is also known that Dicer is one of the components of the active RNA-induced silencing complex (RISC). To enhance guide strand incorporation into RISC and reduce off-target silencing by the siRNA passenger strand, siRNAs are routinely synthesized with a mismatch at the guide strand 5' end. As we have shown that Dicer can interact directly with siRNAs, we examined the impact of terminal sequence and structure, including mismatches, on Dicer-siRNA interactions. We find that Dicer-siRNA binding is correlated with the silencing efficacy of the siRNAs. Besides supporting a role for Dicer in RISC formation and loading with the siRNA, these results also provide a basis for designing siRNAs with enhanced silencing.

92c Regulation of Virulence In Salmonella Typhimurium

Christopher V. Rao, Department of Chemical Engineering, University of Illinois, 600 South Matthews Ave., 211 Roger Adams Laboratory, Box C-3, Urbana, IL 61801 and Supreet Saini, Chemical and Biomolecular Engineering, University of Illinois, 600 South Matthews Ave., 211 Roger Adams Laboratory, Box C-3, Urbana, IL 61801.

Salmonella typhimurium causes a variety of diseases ranging from mild gastroenteritis to life threatening systemic infections. During the course of infection, Salmonella invades intestinal epithelial cells using a type III secretion system (TTSS) encoded on Salmonella pathogenicity island 1 (SPI1). The SPI1 TTSS forms a needle-like structure that injects proteins directly into the cytosol of the host cells. These injected proteins commandeer the host cell's intracellular machinery to promote bacterial internalization.

Through the combined use of experimental and mathematical analysis, we demonstrate that coupled negative and positive feedback loops are involved in the regulation of the SPI1 TTSS. These interlocking loops form a bistable switch with an environmentally-tuned activation threshold.

92d Identifying Genes Involved In Multidrug Resistance In Gram-Negative Bacteria

Miao Duo[1], Sheron Mark[2], Xi Chen[3], Shuyu Hou[2], and **Dacheng Ren**[2]. (1) Department of Civil and Environmental Engineering, Syracuse University, 121 Link Hall, Syracuse, NY 13244, (2) Department of Biomedical and Chemical Engineering, Syracuse University, 121 Link Hall, Syracuse, NY 13244, (3) Department of Biology, Syracuse University, 121 Link Hall, Syracuse, NY 13244

Bacterial multidrug resistance is a major cause of clinical failures in treating infections. In this study, *Escherichia coli* and *Pseudomonas aeruginosa* were used as model systems to explore the genetic basis of multidrug resistance. The random mutant libraries were constructed in *E. coli* EC100 and *P. aeruginosa* PAO1 using transposon mutagenesis. The libraries were screened to identify the mutants with enhance/reduced resistance to antibiotics. Out of the 4000 *E. coli* mutants screened up to date, 6 mutants were found to be more sensitive to chloramphenicol and 7 were more resistant compared to the wild-type EC100. Thirty-three drug sensitive mutants and 1 drug resistant mutant were identified based on 3000 *P. aeruginosa* mutants that have been screened. Thirty-four genes responsible for these phenotypic changes were identified by inverse PCR and DNA sequencing. The functions of these genes and their roles in biofilm multidrug resistance were explored. The gene rob of *E. coli* was found to be important for drug resistance in biofilms.

92e A "Cellular Oscilloscope" to Measure the Frequency Response of Gene Expression

Christopher J. Zopf and Narendra Maheshri. Chemical Engineering, MIT, 25 Ames St. Room 66-558, Cambridge, MA 02139

The kinetics of gene expression has been widely studied in response to step changes in signal. Frequency response studies represent a complementary approach to elucidating the kinetics of reaction systems involved in gene expression. We have developed a "cellular oscilloscope" in S. cerevisiae to measure the frequency response of gene expression "output" with respect to an oscillating "input" transcription factor concentration. This system relies on the fact that many eukaryotic transcription factors continuously shuttle between nucleus and cytoplasm and are regulated vis-à-vis their nuclear localization. We employ the phosphate responsive transcription factor Pho4p and show it can be forced in and out of the nucleus quickly (5-10 minutes) and reliably if one eliminates endogenous feedback control in the phosphate-sensing pathway. Microfluidic devices are employed to drive the Pho4p oscillations by varying the phosphate concentration. Using fluorescent proteins, we can measure both the oscillations in Pho4p localization and its effect on downstream gene expression in real-time. Using this system, we study how the promoter architecture of several Pho4p target promoters influences gene expression. Specifically, we ask whether a change in the nuclear concentration (thermodynamic affinity model) or in the nuclear residence time (kinetic proofreading model) of the transcription factor is critical for activation. Finally, we demonstrate the modularity of our system by isolating the minimal shuttling domain of Pho4p and fusing it to alternate DNA binding and activation domains. This will enable the study of promoters that are not part of the endogenous phosphate starvation pathway.

92f Evaluating Protein Interactions In the Endoplasmic Reticulum of *Saccharomyces cerevisiae*: Spatial Distribution of Molecular Chaperone and Co-Chaperones Evident at a Sub-Organelle Level

Carissa L. Young, David Raden, and Anne S. Robinson. Department of Chemical Engineering, University of Delaware, 150 Academy Street, Newark, DE 19716

The molecular chaperone BiP/Kar2 is a member of the Hsp70 family of chaperones that resides in the endoplasmic reticulum (ER) of *S. cerevisiae*. BiP is identified in multiple critical processes of the cell including ER translocation, protein folding/maturation, and ER-associated degradation (ERAD). Biochemical and genetic experiments have demonstrated BiP's association with selective co-chaperones for specific ER processes. We hypothesize that chaperone/co-chaperone interactions are a result of spatial inhomogeneity, regulated by co-chaperones, and this inhomogeneity serves as a means of dictating cellular functions. From a systems biology perspective, integration of experimental and computational design will allow us to elucidate the mechanisms associated with chaperone interactions in the ER.

Protein interactions of this system are highly dependent on the timescales of diffusion and kinetics. Diffusion calculations and estimates of species concentrations *in vivo* must incorporate the physiological conditions of the ER. The low resolution of traditional immunofluorescence techniques combined with BiP's relative high abundance has previously inhibited the determination of protein localization effects at a sub-organelle level as well as the transient behavior associated with co-chaperones. Advances in confocal light microscopy (CLM) combined with the use of green fluorescent protein (GFP) variants allow continuous monitoring of protein dynamics in living cells. Using CLM to visualize fluorescently tagged proteins *in vivo*, we are able to generate physiologically relevant data of protein dynamics in *S. cerevisiae*. We have created an *in vivo* system that investigates the spatial distribution, relative species concentration, and effective diffusion of molecular chaperone BiP and co-chaperones involved in multiple ER processes.

Fusion proteins of BiP and co-chaperone Sec63 have revealed that an inhomogeneous spatial distribution of proteins occurs at a sub-organelle level during translocation. Localization of BiP and Sec63 in the nuclear and peripheral ER has been reconstructed in three dimensions resulting in estimations of ER volume and species concentration. We have confirmed that the ER is continuous and structurally dynamic through Fluorescence Loss in Photobleaching (FLiP) experiments in conjunction with time series analysis of BiP, Sec63, and soluble GFP variants retained in the ER. The dynamics of freely soluble ER resident proteins have been captured at the millisecond timescale. Rapid diffusion of ER resident proteins in combination with an inhomogeneous spatial distribution of chaperone/co-chaperones indicates that we can potentially resolve transient interactions of ER proteins. Our experimental design has directed the development of mathematical models accurately reflecting spatial and temporal effects of chaperone/co-chaperone interactions[1,2]. The behavior of many systems, including multiple processes of the ER, could not be properly understood without this level of resolution[3].

[1] Marc Griesemer, Professor Linda Petzold, Department of Computer Science, University of California Santa Barbara

[2] Professor Francis Doyle III, Department of Chemical Engineering, University of California Santa Barbara

[3] Lemerle, C. *et al. Space as the final frontier in stochastic simulations of biological systems.* FEBS Letters. (2000) 579:1789-1794.

92g Quantitative Implications of Scaffolding In Signal Propagation

Anand R. Asthagiri and Stephen Chapman. Chemical Engineering, California Institute of Technology, 1200 E. California Blvd MC 210-41, Pasadena, CA 91125

Intracellular regulatory networks that control cell behaviors are remarkably intricate and involve numerous components. Amidst this apparent complexity is a key organizing feature: scaffolds. Protein scaffolds bind concomitantly to multiple components of a signaling pathway, thereby promoting signal transmission along a common physical "backbone". Natural systems broadly use scaffold proteins to propagate signals that regulate cell cycle, proliferation, differentiation and motility in species ranging from yeast to human. Furthermore, there is emerging interest in applying synthetic scaffolds to modulate regulatory and metabolic pathways in applications such as generating cellulosic biofuels. Despite its widespread occurrence in natural networks and its potential applications in synthetic circuits, how scaffolds quantitatively control signal transmission is unclear. In this talk, I will describe recent quantitative insights that we have gathered from studies of the Ste5 scaffold in yeast cells. Our results demonstrate that scaffolding allows tunability of signal throughput without disrupting the quality of signal transmission.

98a A Polymeric Electroporation Microneedle Array for Minimally Invasive and Highly Localized Electrogenetherapy and Electrochemotherapy

Yeu-Chun Kim[1], Seong-O Choi[1], Joshua Hutcheson[1], Mark Allen[2], and Mark R Prausnitz[1]. (1) School of Chemical and Biomolecular Engineering, Georgia Institute of Technology, 311 Ferst Drive, Atlanta, GA 30332-0100, (2) School of Electrical and Computer Engineering, Georgia Institute of Technology, 311 Ferst Drive, Atlanta, GA 30332-0100

1. Introduction

In recent years, biological and biomedical applications of microstructures have become one of the promising research areas of microelectromechanical systems (MEMS). These micro-scale devices for biological/biomedical applications are often referred to as "BioMEMS" [1]. Electroporation has been used for electrogenetherapy and electrochemotherapy [2]. For application of electroporation on skin, conventional macro-needle electrodes have been used [3], but they cause pain during insertion, can lead to tissue burning during electroporation, and require a high-voltage pulse generator.

The objective of this research is to develop a microdevice to deliver biomolecules such as DNA into skin by means of electroporation. This is particularly interesting for delivery of DNA vaccines into the skin's dendritic cells [4]. The basic requirements of the device are the need to penetrate across the *stratum corneum* layer and provide an electrical pathway to generate an electric field. In addition to that, the device should provide methodologies to reduce pain during injection and tissue burning, two issues that plague traditional therapies.

One promising approach to satisfy these requirements is to implement electrical functionality in a microneedle array (i.e., an array of electrically conductive needle s measuring hundreds of microns in length that are electrically isolated from each other to enable generation of strong field strengths across short electrode-electrode spacings). The microneedle array can provide minimally invasive, painless insertion into skin and can act as an electrode after proper metallization process. Due to the shallow insertion depth of the microneedles, pain during an application of an electric field by disturbing nerves can also be minimized. In addition, the voltage required for electroporation can be reduced due to the small size scale, avoiding the use of a high voltage source, which is not suitable for portable applications.

2. Experimental methods and results

The device contains a 10x10 array of microneedles, with adjacent microneedles electrically isolated, which gives a "checkerboard" pattern of electrical polarity (Figure 1). The height of the microneedles is 500μm, bottom is 200μm square, and spacing between microneedles is 200μm.

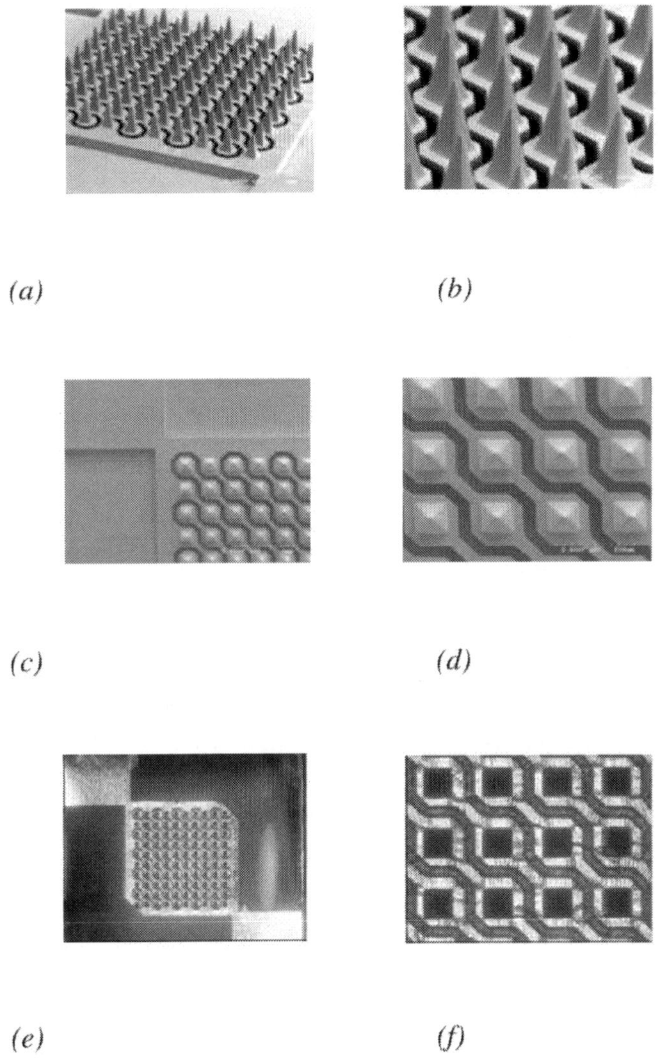

Figure 1. (a)-(d) SEM picture of the fabricated device; (e),(f) Photomicrograph of the device after metal transfer micromolding

To verify mechanical strength, an insertion test was performed with human cadaver skin. The fabricated microneedle array was inserted and removed from the skin, and the skin was dyed with trypan blue. Microscopic imaging showed blue stains in the microneedle array pattern, indicating that the microneedles pierced into the skin (Figure 2).

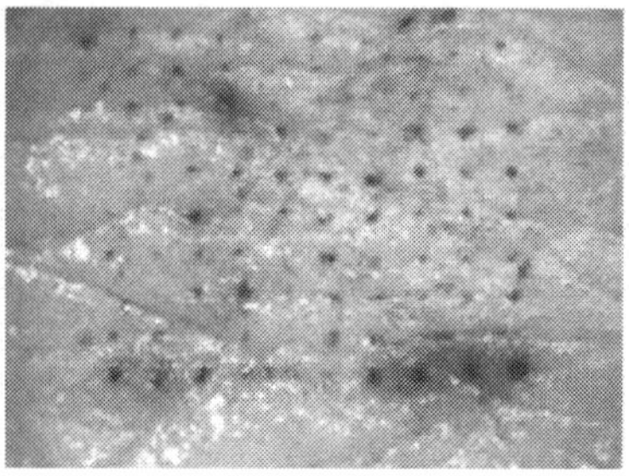

Figure 2. Photomicrograph of human cadaver skin after piercing with microneedle array and staining with blue dye

The capability of delivering molecules into cells using this electroporation microneedle array was examined *in vitro*. DU-145 prostate cancer cells were used for the electroporation experiment, and calcein was used as the model compound for delivery into the cell. A cell suspension was mixed with calcein and applied as a droplet over the microneedle array. After applying voltages of 12-91V for 2.5ms, the cells were collected, followed by adding propidium iodide (PI) to stain dead cells. When 12 V was applied for 2.5 ms, the average uptake of calcein was 7 % of the cells (Figure 3). The uptake of calcein increased until 69 V was applied and then decreased as the applied voltage increased. At 69 V, which corresponds to the nominal field strength (applied voltage divided by the distance between electrodes at the bottom of the device) of 3.45 kV/cm, a maximum uptake of 48 % of cells was achieved. However, cell viability suddenly dropped at that point, indicating that the field strength induced by the applied voltage started to cause cell death rather than uptake of calcein.

3. Conclusion

The ultimate goal of this research is to develop a micro-scale electroporation device that can deliver molecules into skin. We designed and fabricated electrically active and appropriately patterned microneedle arrays and showed that they were strong enough to insert into skin. We then observed that uptake of calcein occurred at conditions which correspond to macro-scale electroporation experiments, indicating that the fabricated device has the capability to deliver molecules into cells by electroporation. Overall, electroporation with a microneedle array can be a promising alternative method to deliver molecules into skin.

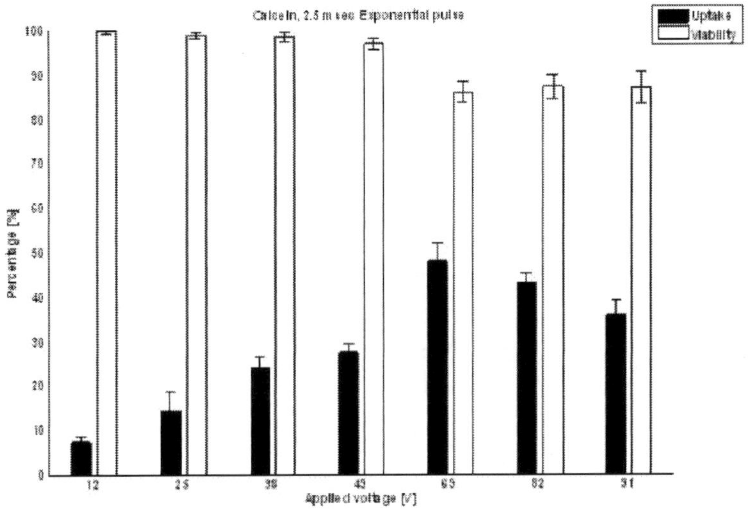

Figure 3. Uptake of calcein and cell viability at each applied voltage

References

1. R. S. Shawgo, A. C. R. Grayson, Y. W. Li, and M. J. Cima, *BioMEMS for drug delivery*, Current Opinion in Solid State & Materials Science (2002) 6, 329-334.

2. M. J. Jaroszeski et al., *Electrochemotherapy, Electrogenetherapy, and Transdermal Drug Delivery*, Humana Press, Inc., Totowa, 2000.

3. G.A. Hofmann et al., *Electroporation Therapy: A New Approach for the Treatment of Head and Neck Cancer*, IEEE Transactions on Biomedical Engineering (1999) 46, 752-759.

4. Roos AK, Moreno S, Leder C, Pavlenko M, King A, Pisa P, *Enhancement of cellular immune response to a prostate cancer DNA vaccine by intradermal electroporation,* Molecular Therapy (2006) 13, 320-327

98b Enhanced Microfluidic Electroporation for Oligonucleotides Delivery

Shengnian Wang, Institute for Micromanufacturing, Louisiana Tech University, 911 Hergot Avenue, Ruston, LA 71272, Xulang Zhang, NSF Center for Affordable Nanoengineering of Polymeric Biomedical Devices, The Ohio State University, 1381 Kinnear Road, Suite 100, Columbus, OH 43210, and Ly James Lee, Chemical and Biomolecular Engineering, The Ohio State University, Room 125A, Koffolt Labs., 140W. 19th Ave., Columbus, OH 43210.

Electroporation provides an easy way for intracellular delivery of biomolecules in most cell-related research. However, current electroporation instruments often require a tedious trial-and-error search process to identify a good protocol and many protocols recommended limited cell population (105–106). It would be very valuable if electroporation could be carried out in a continuous mode, under a mild, uniform electric field that is effective with every porated cell.

We have recently developed a Converging Flow Electroporation (CFE) device for in vitro biomolecules uptake. Cells mixed with biomolecules will continuously flow through a converging microchannel where a low AC/DC electric field is applied. When sequentially passing the throat region, each cell experiences an identical but highly focused electric pulse. In this way, it simplifies or even eliminates the optimization process for a new cell types and proper electroporation protocols can be quickly established. Using pWizGFP plasmid and K562 & mES cells as models, CFE showed significant improvement on transgene expression (~ 15%) compared to the Bio-Rad Gene Pulser XcellTM Electroporation System. Cell lysis during poration was mostly avoided with no observable cell debris and quantitative result by MTS assay confirmed our observation of cell conditions with the cell viability more than 50% or even higher. Similar results were also observed with pWizGFP transfection in (CCE).

With its continuous-mode operation, many DNA away from cells may be wasted, or have to be recycled afterwards. The relatively low concentration of DNA around cell also limits the further improvement of its working efficiency. To further enhance CFE efficiency and to reduce oligonucleotides consumptions, multi-functional nanoparticles which carry DNA and targeting molecules are used in further study. Through receptors on the cell surface, nanoparticles are conjugated onto cells and flow together through the electroporation zone. This largely overcomes barriers in the uptake of DNA when cell membranes are upset. We demonstrated this concept using transferring (TfR)-targeted lipoplex nanoparticle and found better transfection of eGFP plasmids and ODN in K562 cells. The cell proliferation assays also show better cell viability.

98c Microfluidic Integration of RNA Sample Preparation and Amplification for Global Gene Expression Profiling of Small Samples

Jason G. Kralj[1], Audrey Player[2], Ernest Kawasaki[2], Dan Edelman[2], and Laurie E. Locascio[1]. (1) Biochemical Sciences, National Institute of Standards and Technology, 100 Bureau Dr., MS 8313, Gaithersburg, MD 20899, (2) National Cancer Institute/National Institutes of Health, 8717 Grovemont Circle, Bethesda, MD 20877

We present the integration of the Eberwine process in a microfluidic system, incorporating small sample mRNA capture, reverse transcription, RNA digestion with second strand synthesis, and linear amplification using T7 RNA polymerase. Each step of the process has been shown to scale down to the single-cell level efficiently, and this is the first demonstration of all steps incorporated into a single device. Multilayer softlithography was used to fabricate 9-channel devices incorporating sieve valve structures, allowing functionalized microbeads to be used for immobilizing the nucleic acids. Here, Jurkat T-cell total RNA was used as a model sample with amounts from 100 ng down to approximately 100 pg. Amplified product from conventional benchtop and microfluidic methods was compared based on transcript length using electrophoresis, efficiency of transcription based on PCR of selected genes, and global gene expression profiles using microarrays. Early results indicate about 850 more genes were detected using the microfluidic process at the 1 ng level (100 cell equivalents), suggesting an order of magnitude improvement in yield of aRNA than conventional methods. Samples as small as 100 pg of total RNA (10 cell equivalents) have been successfully amplified. The main advantages of microfluidic processing include immobilization of the cDNA onto a stationary bead column, thereby eliminating sample transfer steps; RNAse contamination is minimized because the channel is sealed; and higher numbers of genes are detected on microarrays from samples 1 ng and smaller than amplification using benchtop methods.

98d A Micro Cell Culture Analog (μ CCA) with 3-D Hydrogel Culture to Assess Metabolism-Dependent Cytotoxicity of Anti-Cancer Drugs

Jong H. Sung and Michael Shuler. Chemical and Biomolecular Engineering, Cornell University, 120 Olin Hall, Ithaca, NY 14853

Development of a high-throughput in vitro system that can closely mimic human response to a drug is crucial for reducing the cost and the time of developing new drugs. Current cell-based assay system in a multi-well format is a static system which consists of a single cell type, and lack the important aspect of biotransformation and multi-organ interactions. In addition, 2-D monolayer cell culture does not provide a physiologically realistic environment and alter the behavior of cells, making the conventional cell-based assay systems inadequate for prediction of human response to a drug. A multi-chamber, microfluidic device utilizing 3-D constructs of cell-embedded hydrogel was developed and used to test the effect of anti-cancer drugs and their metabolites. A microscale cell culture analog (μ CCA) is a physical realization of a physiologically-based pharmacokinetic (PBPK) model, where chambers representing key organs are fabricated on a silicon chip and interconnected by microchannels for medium recirculation which emulates blood flow. We tested Tegafur, an oral prodrug of 5-FU which is a chemotherapeutic agent for colon cancer, using the μ CCA device with the HepG2/C3A cells and HCT-116 cells embedded in Matrigel constructs cultured in the liver and the tumor compartment, respectively. Tegafur is non-toxic to cells and is converted to 5-FU in the liver by cytochrome P450 enzymes, which then exerts a toxic effect on cells. The results show that the μ CCA device was able to capture the metabolism in the liver compartment and consequent damages to the tumor cells, which was not observed in a traditional, static 96-well plate assay. This 3-D μ CCA system was able to reproduce the action of a drug that could only be observed in animal or human model before. It can mimic the dynamic multiple organ interactions and enables the study of pharmacokinetic/pharmacodynamic properties of a drug in vitro in a more physiologically realistic microenvironment.

98e A Microfluidic Gradient Chamber for Spatially Defined Adipocyte Culture

Ning Lai and Kyongbum Lee. Chemical and Biological Engineering, Tufts University, 4 Colby Street, Science and Technology Center, Medford, MA 02155

Excessive expansion of adipose tissue and consequent obesity is a major risk factor for type 2 diabetes, which can further lead to stroke, hypertension and cardiovascular disease. In this light, there is a pressing need to identify factors and signaling events that regulate adipocyte differentiation and growth. Ideally, these factors and events are studied under controlled conditions in a precisely defined chemical environment. Here, we describe a microfluidic reactor-based adipose tissue model that mimics the contacting arrangement of adipocytes and capillaries. A key feature of the designed reactor is that it affords selective localization of micro-scale collagen gels, which serve as three-dimensional (3D) scaffolds for cell culture. The selective localization is accomplished by fabricating regions of high (convective) transport resistance as cell culture compartments. The culture compartment is lined by variably spaced, micro-fabricated posts. The posts define diffusion pores that connect the culture compartment to flanking medium channels. In this study, the culture compartments were seeded with 3T3-L1 preadipocytes, which have been shown to undergo consistent and uniform differentiation into adipocytes in conventional (static), Petri dish cultures. Reproducible positioning of cells in the culture compartment was obtained by first infusing a preadipocyte suspension in a collagen pre-polymer solution, and then initiating the cross-linking through a temperature and pH shift. Following localization and entrapment within the collagen gel matrix, the preadipocytes achieved their characteristic fibroblastic morphology, and proliferated at rates comparable to those observed in conventional Petri dish cultures. Long-term cell culture for up to two weeks was achieved without nutrient limitations and loss of viability by maintaining a continuous flow of medium. Morphological analysis and Oil Red O staining indicated that preadipocytes induced with a standard adipogenic cocktail underwent differentiation and formed microscopic intracellular lipid droplets. Leveraging the convection-free device design, stable diffusion gradients were generated across the culture compartment by introducing parallel media streams with different chemical compositions into each flow channel. A fluorescent dye (FITC) was used to verify and characterize the kinetics of gradient formation. The gradient feature was utilized to induce asymmetrical differentiation, where preadipocytes were exposed to the adipogenic cocktail from only one of the two flow channels. Lipid droplet formation was observed only for the preadipocytes on the induced side of the culture compartment, thus confirming spatially selective differentiation. Taken together, our results establish the proof-of-principle of a novel micro-fluidic adipose reactor that affords long-term and spatially defined co-culture of adipocytes and preadipocytes. Prospectively, the spatially defined co-culture could be used to conduct detailed studies on the recruitment of locally resident preadipocytes by mature and enlarged adipocytes.

98f Micropatterned Co-Culture Model of Bacteria and Epithelial Cells for Investigating Signal-Mediated Host-Pathogen Interactions

Jeongyun Kim, Fatih Senocak, Manjunath Hegde, and **Arul Jayaraman**. Artie McFerrin Department of Chemical Engineering, Texas A&M University, 3122 TAMU, College Station, TX 77843

The intestinal tract is colonized by hundreds of non-pathogenic bacterial species, including those belonging to the genus *Escherichia*, that are important in maintaining normal gastrointestinal (GI) tract function. Colonization of the GI tract by pathogenic bacteria such as *Escherichia coli* O157:H7 (EHEC) proceeds in three distinct steps: (i) migration of the pathogen to intestinal epithelial cell surface, (ii) colonization, and (iii) injection of virulence factors. While the driving forces underlying EHEC infections are not fully understood, it is becoming increasingly clear that soluble signal-mediated interactions are important in EHEC infections. Recent work from our lab has shown that phenotypes relevant in EHEC infections - chemotaxis, colonization, and attachment – are differentially regulated by eukaryotic hormones in the GI tract (e.g., norepinephrine) and bacterial quorum sensing signals (e.g., indole) [1, 2]. These results underscore the importance of GI tract signals in EHEC infections.

Since commensal bacteria residing in the GI tract are the source for most prokaryotic signals in the intestinal lumen, we hypothesized that the organization of commensal bacteria in the GI tract is a key determinant of pathogen colonization. In this work, we have adopted micropatterning techniques in order to manipulate and investigate the local microenvironment that EHEC encounters during colonization so that its role in EHEC infections can be determined. Conventional cell culture methods for studying EHEC attachment do not account for the effect of the GI tract microenvironment on EHEC colonization as it is difficult to localize bacteria in specific locations among eukaryotic cells. Using easy to operate pneumatic valves, we have fabricated several commensal bacterial "islands" ranging in diameter from 20 to 100 microns and surrounded them with eukaryotic cells so as to create an in-vivo like microenvironment. Microvalves were used to capture bacteria and localize them to specific regions where they colonize and form biofilms. The effect of commensal island size and bacterial attachment time was optimized as these two parameters are expected to directly influence the concentration of signaling molecules released into the microenvironment. The feasibility of localizing bacteria in specific regions was demonstrated using sequential introduction of dyes and bacteria expressing different fluorescent proteins. The ability to localize different bacterial species in different islands enables mimicking the in vivo GI tract heterogeneity. Following bacterial island formation, residual bacteria outside the island was lysed, followed by seeding of Hela cells to complete the co-culture system. Finally, EHEC was introduced into the micropatterned chamber and its colonization to the vicinity of specific bacterial islands studied. Our micropatterned co-culture of bacteria and epithelial cells is expected to enable investigation of signaling interactions between bacteria and host-cells that are crucial to infections.

REFERENCES

1. Bansal T, Englert DL, Lee J, Hegde M, Wood TK and Jayaraman A. Infection Immunity. 75: 4597 (2007).

2. Bansal T, Jesudhasan P, Pillai S, Wood TK, and Jayaraman A. Applied Microbiology & Biotechnology. 73: 4100 (2008).

98g Using Antibody-Modified Microelectrodes for Capture and Release of Cells

He Zhu, Jun Yan, and **Alexander Revzin**. Department of Biomedical Engineering, University of California at Davis, 451 East Health Sciences Dr., Davis, CA 95616

The present study sought to design a novel strategy for releasing antibody-bound cells through electrochemical disruption of the underlying antibody (Ab) layer. Poly (ethylene glycol) (PEG) hydrogel photolithography was employed to make the glass regions non-fouling, thus, ensuring selective localization of proteins and cells on the microelectrodes. The gold surfaces were decorated with anti-CD4 Ab molecules using standard alkanethiol self-assembly and carbodiimide coupling approaches. The Ab-functionalized electrodes selectively captured model T-lymphocytes (Molt-3 cells) expressing CD4 antigen while minimal cell adhesion was observed on PEG hydrogel-modified glass substrates. Importantly, application of a reductive potential (-1.2 V vs. Ag/AgCl reference electrode) resulted in release of surface-bound T-cells from the electrode surface. In the future, the cell sorting approach described here may be combined with microfluidic delivery to enable Ab-mediated capture of T-lymphocytes or other cell types followed by release of select cells for downstream gene expression studies or re-cultivation.

126a Early Steps In Aggregation of Amyloidogenic Peptide Gnnqqny

Allam S. Reddy[1], Manan Chopra[1], and Juan J. De Pablo[2]. (1) Department of Chemical and Biological Engineering, University of Wisconsin-Madison, 1415 Engineering Drive, Madison, WI 53706-1691, (2) Department of Chemical and Biological Engineering, University of Wisconsin, 1415 Engineering Drive, Madison, WI 53706

Protein aggregation has been implicated in the pathology of several neurodegenerative diseases. A comprehensive understanding of protein aggregation is vital to the development of therapeutic strategies for neurodegenerative diseases. Recently, amyloidogenic heptapeptide GNNQQNY has emerged as a molecule of choice for fundamental studies of protein aggregation. Despite its popularity of GNNQQNY in experiments, little theoretical work has been aimed at supporting and augmenting available experimental observations. In this work we present a rigorous theoretical analysis of the aggregation pathway of amyloidogenic heptapeptide GNNQQNY. The aggregation pathway of GNNQQNY involves a rare dimerization re-attraction, which is not easily studied in conventional simulations. We have used transition path sampling Monte Carlo simulations to study the pathways involved in that dimerization. Statistical analysis of the reaction pathways shows that the dimerization reaction proceeds via a zipping mechanism, initiated with the formation of contacts at the third residue (N). We also discuss the role of the water in the dimerization pathway. Our results can also explain the structural observations of GNNQQNY aggregates that have emerged from experiments.

126b New Twists to the Amyloid Folding Problem Revealed by Solid State NMR and Electron Microscopy of Alzheimer's Beta Amyloid Fibrils

Anant K. Paravastu[1], Isam Qahwash[2], Wa-Ming Yao[3], Richard D. Leapman[4], Steven C. Meredith[2], and Robert Tycko[3]. (1) Chemical and Biomedical Engineering, Florida A&M University and Florida State University, 2525 Pottsdamer St., Tallahassee, FL 32310, (2) Biochemistry and Molecular Biology, University of Chicago, 5841 S. Maryland Ave., Chicago, IL 60637, (3) Laboratory of Chemical Physics, National Institute of Diabetes & Digestive & Kidney Diseases, NIH, 9000 Rockville Pike, Building 5, Room 111, Bethesda, MD 20892, (4) National Institute of Biomedical Imaging and Bioengineering, NIH, 9000 Rockville Pike, Bethesda, MD 20892

Self-assembly of peptides into amyloid fibrils has received considerable attention because of the pathological conditions associated with amyloid deposition in vivo. The amyloid folding problem is especially interesting because of the variety of amino acid sequences that can self-assemble into the characteristic cross-beta structural motif. Recently, it was discovered that self-assembly of a single peptide could be driven towards different amyloid fibrils with distinct morphologies and underlying molecular structures (Petkova, A.T., Leapman, R.D., Tycko, R., et. al. Science 307, 262 (2005).). For the 40-residue Alzheimer's beta amyloid peptide (ABeta), we have characterized the range of possible fibril molecular structures by preparing fibrils using a variety of experimental protocols. Solid state NMR was used to assess secondary structures, structural order, and site-specific inter-atomic proximities. Electron microscopy was used to measure fibril dimensions, morphology, and mass. Together, these results are all consistent with folded beta strand molecular conformations within the cross-beta structural motif. Observed structural variations between different ABeta fibrils will be discussed in terms of interfaces between beta sheets within the fibril core, conformations of non-beta-strand regions, and two classes of overall fibril symmetry. This presentation will culminate in the description of a novel 3-fold symmetric ABeta fibril structure, and preliminary results on new ABeta fibrils seeded with human Alzheimer's brain material. The results indicate the importance of multiple distinct but energetically similar pathways for amyloid fibril formation.

126c Elucidating Therapeutic Monoclonal Antibody Aggregation

Veysel Kayser[1], Vladimir Voynov[1], Naresh Chennamsetty[1], Bernhard Helk[2], Kurt Forrer[2], and Bernhardt L. Trout[1]. (1) Chemical Engineering, Massachusetts Institute of Technology, 77 Massachusetts Ave, E19-531, Cambridge, MA 02139, (2) Novartis Pharma AG, Basel, Switzerland

A strong interest in protein aggregation has emerged in the pharmacology industry for therapeutic monoclonal antibodies (MAB). We report combined chromatographic and multi-spectroscopic studies of such MAB aggregation. Results from different spectroscopic methods agree well with each other, and they are used in a combinatorial way to get a complete picture of protein aggregation. SEC-HPLC was used to separate monomer from aggregates. Spectroscopic techniques consisting of intrinsic tryptophan fluorescence and extrinsic fluorophore binding were also applied, as well as static-light scattering. Both steady-state and time-resolved fluorescence spectroscopy, including fluorescence lifetime, anisotropy, quenching and FRET were used. Based on monomer peak loss in SEC-HPLC at elevated temperatures, aggregation mechanisms were elucidated, and a kinetic protein aggregation model, based on nucleation and growth polymerization, is proposed to predict in vitro therapeutic monoclonal antibody aggregation at low storage temperatures. According to our model, MAB first partially unfolds with at least one unfolding intermediate and forms an aggregation prone monomeric structure. Having said that, aggregation model is not limited to the formation of such structure, and in fact, aggregation could occur from any type of monomer. In addition, we found that aggregation of studied MABs show temperature dependent aggregation regimes. Based on accelerated studies, a kinetic model used for the prediction of MAB aggregation at low storage temperatures is proposed.

126d Air/water Interface Induced Folding and Self-Assembly of Amyloid-Beta Peptide Seed Fibrillogenesis

Eva Y. Chi[1], Amy Winans[1], Shelli L Frey[1], Jaroslaw Majewski[2], Kristian Kjaer[3], and Ka Yee C Lee[1]. (1) Department of Chemistry, The University of Chicago, 5735 S. Ellis Ave., Chicago, IL 60637, (2) Los Alamos Neutron Science Center, Los Alamos National Laboratory, Manuel Lujan Jr. Neutron Scattering Center, TA 53, Bld 622, MS H805, Los Alamos, NM 87545, (3) Niels Bohr Institute, University of Copenhagen, Copenhagen, Denmark

The amphipacity of the natively unstructured amyloid-β (Aβ40) peptide may play an important role in its aggregation into β-sheet rich fibrils that is associated with the pathogenesis of Alzheimer's disease. Using the air/water interface as an ideal hydrophobic interface, we characterized Aβ's surface activity and the structure, assembly, and morphology of Aβ adsorbed to the air/water interface. Aβ readily adsorbed to the air/water interface to form a contiguous film with a surface pressure of approximately 14 mN/m and showed an apparent critical micelle concentration of about 100 nM. Adsorbed Aβ was a single molecular layer extending ~ 20 Å into the aqueous subphase with in-plane ordering which gave rise to X-ray diffraction peaks. Analysis of the diffraction peaks showed that the air/water interface induced Aβ peptides to self-assemble into nano-size clusters with Aβ peptides folded in a β-sheet conformation. The presence of these clusters was further confirmed by imaging the morphology of the Aβ film using atomic force microscopy. The formation of these clusters were not affected by solution pH or ionic strength, or the presence of cosolutes sucrose and urea at concentrations known to stabilize and denature protein native structure in solution, suggesting that the hydrophobic interface-driven Aβ folding and assembly is robust and thereby strongly favorable. Furthermore, Aβ•adsorbed at the air/water interface can seed fibril growth in solution when re-introduced into the bulk. Our results implicate that that interface-induced Aβ folding and self-assembly may serve as a mechanism by which Aβ aggregation occurs in vivo.

126e Transthyretin and Beta-Amyloid. A Regulatory Network In Protein Misfolding Disease?

Regina M. Murphy and Jiali Du. Chemical and Biological Engineering, University of Wisconsin, 1415 Engineering Drive, Madison, WI 53706

Alzheimer's disease (AD) is the most common age-associated neurodegenerative disease, affecting approximately 5 million Americans. Most researchers believe that the deposition of beta-amyloid peptide (Aβ) as extracellular amyloid deposits in the brain, is causally linked to the neuronal dysfunction and death observed in the disease. Aβ is the proteolytic product of a precursor protein called APP. Transgenic mice, engineered to overexpress APP, develop a large number of Aβ deposits. Surprisingly, however, these mice suffer little neuronal death, thus calling into question the "amyloid cascade" hypothesis. Recently, it has been discovered that transgenic mice respond to the overexpression of APP by robustly upregulating synthesis of another protein, transthyretin (TTR). TTR serves as a transport protein in the blood and is the most common protein in cerebrospinal fluid. Remarkably, TTR itself can form amyloid deposits, and is responsible for causing a common age-related disease called senile systemic amyloidosis. We hypothesis that upregulation of TTR in transgenic mice serves a neuroprotective role, but the mechanism of this action is unknown. In this talk we present data showing that TTR interacts directly with Aβ, influencing both its aggregation kinetics and its toxicity. The nature of the interaction is complex, as TTR is a homotetrameric protein and Aβ is present as a mix of monomer, oligomer, and fibril. We produced and characterized recombinant wt TTR as well as mutants that differ in the stability of the tetramer. Our data show that TTR tetramer stability strongly influences the protein's regulation of Aβ aggregation kinetics. Interestingly, those TTR variants that accelerated Aβ aggregation were best at inhibiting Aβ toxicity in cell culture assays. The effect of TTR on Aβ aggregation has been characterized in detail using a variety of biophysical and biochemical tools. Our results delineate the importance of TTR and Aβ monomeric vs. oligomeric status in regulating their interaction. We present some strategies for adjusting TTR tetramer stability through the use of small-molecule drugs, as an indirect method for reducing Aβ toxicity by manipulating this protein-misfolding regulatory network.

126f Effect of Fermentation Conditions on Inclusion Body Quality in High Cell Density E. Coli Fermentations

Rishard H. Chen[1], Clarissa S. Lui, Matthew P. DeLisa[1], Gerd Ritter[2], Lloyd J. Old[2], Carl Batt, and Michael L. Shuler[3]. (1) Chemical Engineering, Cornell University, Olin Hall, Ithaca, NY 14853, (2) Ludwig Institute for Cancer Research, New York, NY 10021, (3) Chemical Engineering and Biomedical Engineering, Cornell University, 270 Olin Hall, Ithaca, NY 14853

Protein therapeutics produced in high cell density cultures of Escherichia coli using industrial-scale recombinant DNA processes commonly form insoluble aggregates. These inclusion bodies are advantageous because they are often very pure aggregates of the protein of interest, and their isolation is typically a first step in purification. This study shows that fermentation conditions can influence the size, shape, and density of inclusion bodies, which may facilitate the ease of their isolation and subsequent downstream purification.

Melan-A, a promising candidate for cancer immunotherapy, was used as a model protein therapeutic that forms inclusion bodies in E. coli. Two populations of inclusion bodies appeared depending upon fermentation temperature when their density was analyzed using cesium chloride gradient ultracentrifugation. Inclusion bodies produced at 30°C were mostly about 1.18 g/ml, while the density of inclusion bodies produced at 42°C were 1.27 g/ml. At 37°C approximately 80% of the inclusion bodies were in the lower density fractions. Using electron microscopy, it was observed that the physical structure of the inclusion bodies differed. Overall, inclusion bodies produced at 30°C and 37°C were 400 nm in diameter, but the 30°C inclusion bodies had a more amorphous structure with poorly defined boundaries between subunits. Inclusion bodies produced at 42°C were 300 nm and had a relatively smooth and spherical shape. Diffusion limitations of reactions with Bradford reagent, as are expected with inclusion bodies, were exacerbated with the 42°C inclusion bodies.

Currently under investigation are the effects of additional fermentation conditions on inclusion body quality, such as the adjustment of growth rates. In addition to electron microscopy, atomic force microscopy is also being attempted to determine the mechanical properties of these inclusion bodies. Furthermore, the ability to recover a His-tagged version of the protein on a nickel column is being studied. SSX-2, another cancer antigen, and GFP are also under similar investigations to show possible generalization of the observed phenomena.

126g Time Dependent Saxs Analysis of Insulin Oligomerization

Mirco Sorci, Rensselaer Polytechnic Institute, 110 8th st, Troy, NY 12180-3590, Joel E. Morgan, Biology, Rensselaer Polytechnic Institute, Troy, NY 12180, Kurt Erlacher, Bruker AXS Inc, Madison, WI, and **Georges Belfort**, Chemical and Biological Engineering, Rensselaer Polytechnic Institute (RPI), 110 8th St., Troy, NY 12180.

Small-angle X-ray scattering (SAXS) is a well known technique used for structural analysis of materials including macromolecules. In this work, we have used SAXS to measure the temporal formation of insulin oligomers, possible candidates as toxic agents in amyloid-related diseases. The X-rays form a scattering pattern which contains information on the size and shape of the samples (between 5 and 25 nm) and was first used here to characterize the native insulin system (predominantly dimeric at pH 1.6). The fits of the data using different methods (Guinier analysis and spherical model) were in good agreement and comparable with theoretical prediction. The technique was also used to deliver structural information of repeat distances and to follow the fibrillation process. Fibrillation comprises an initial lag phase, a subsequent growth step and saturation to a plateau. The results showed the formation of larger structures at the end of the oligomer process prior to fibrillation. The analysis showed good agreement with previous SANS results. In conclusion, SAXS is a useful tool for the investigation of oligomer assembly into fibrils. This process is critical to fully understand the molecular basis of amyloid-related diseases.

From raw cassava pulp to L-lactic acid via fermentation by *Rhizopus oryzae*

Nuttha Thongchul[1], Surapong Navankasattusas[1] and Shang-Tian Yang[2]*
[1]Institute of Biotechnology and Genetic Engineering, Chulalongkorn University, Bangkok, Thailand
[2]Department of Chemical and Biomolecular Engineering, The Ohio State University, OH, U.S.A.

Abstract

Currently the natural crude oil depletion has been brought into awareness. This affects many industries including petrochemical manufacturing. As a result, there are many attempts to develop novel processes to produce substitutes to such petrochemical products including plastics. Polylactic acid, made from polymerization of optically pure isomer of lactic acid, is one of such examples. In addition to the application in food and pharmaceutical industries, lactic acid can be used as a chemical building block for plastics and solvents. Nowadays lactic acid is produced via fermentation by lactic acid bacteria, which require complex media and produce a racemic mixture of lactic acid. In contrast, *Rhizopus oryzae* is capable of producing an optically pure L-lactic acid from simple and inexpensive media consisting of starchy materials and pentose sugars. Raw cassava pulp, containing approximately 50% starch (dry basis), is one of the starchy materials which are abundant in Thailand. Potentially, it can be used as a carbon substrate in L-lactic acid fermentation by *R. oryzae*. Therefore, in this study we investigated L-lactic acid fermentation from raw cassava pulp. Raw cassava pulp was treated by acid hydrolysis before fermentation. The effects of hydrolysis conditions, including acid type and concentration, and time, on starch recovery and glucose production were observed. Among the conditions studied, the maximum starch recovery (88%) and the highest glucose production (98 g/L) were obtained when hydrolyzing cassava pulp at 121°C for 15 min with 1.0 M HCl (1 g dry pulp to 9 mL HCl solution). The kinetics of L-lactic acid fermentation using acid treated cassava pulp hydrolysates (ACPH) from different conditions as the carbon source was also studied. It was found that free ions present in the hydrolysates affected L-lactic acid fermentation. *R. oryzae* was capable of growing in the fermentation medium containing HCl treated cassava pulp hydrolysate (HACPH). However, instead of L-lactic acid, ethanol was produced as a major product in the fermentation. While no growth and product formation were observed in the fermentation of H_2SO_4 treated cassava pulp hydrolysate (SACPH) and H_3PO_4 treated cassava pulp hydrolysate (PACPH).

Introduction

In addition to the application in food and pharmaceutical industries, lactic acid can be used as a chemical building block for plastics and solvents (Vink et al., 2004; Zhan et al., 2003). Nowadays lactic acid is produced via fermentation by lactic acid bacteria, which require complex media and produce a racemic mixture of lactic acid. This somewhat makes bacterial fermentation uneconomically attractive. Many works reported that *Rhizopus oryzae* was capable of utilizing various agricultural residues including solid agro-industrial wastes to produce a variety of products including L-lactic acid (Cata et al., 1999; Christen et al, 2000; Pandey et al., 2000; Yu and Hang, 1989). For examples, rice brans, wheat brans, sugarcane bagasse, apple pomaces, corncobs, sorghum, soyabean, potato wastes, paper pulp sulfite liquor, steam-exploded wood hydrolysate, and cassava pulp have been used as the carbon sources of *R. oryzae* in many studies (Aikat and Bhattacharyya, 2001; Huang et al., 2003; Ruengruglikit and Hang, 2003; Sun and Xu, 2008; Taherzadeh et al., 2003; Tani, et al., 1988; Woiciechowski et al., 1999). It is noted that the processing of agro-industrial raw material like cassava produces a large amount of wastes whose accumulation leads to the environmental pollution problem. At present, cassava pulp discharged from the cassava milling plant is used as an additive for making cassava pellets for animal feed although it can be used as a carbon substrate in bioprocessing (Nishise et al., 1988; Soccol and Vandenberghe, 2003). Cassava pulp is a starchy-rich lignocellulosic residue which contains approximately 50% starch (by weight, dry basis). Since it provides a rich organic nature and low ash content compared with other crop residues such as rice straw and wheat straw, it can be used as a substrate for bioconversion processes for the production of value-added products (Pandey et al., 2000; Sriroth et al., 2000). Therefore, application of cassava pulp in bioprocessing not only provides an alternated substrate, but also solves the pollution problem.

Thailand is one of the world's largest cassava exporters of cassava products. As a result, there has been a plenty of cassava pulp discharged daily from the cassava milling industry. Bioprocessing was used to convert starch remained in cassava pulp to value-added product, lactic acid. Due to the ability of *R. oryzae* to produce and secrete amylases, lactic acid fermentation of cassava pulp hydrolysates was investigated in this study. Before use in fermentation, raw cassava pulp was treated by acid. The effects of hydrolysis conditions, including acid type and concentration, and time, on starch recovery and glucose production were observed. In addition, the effect of the fermentation medium containing different acid treated cassava pulp hydrolysates (ACPH) on lactic acid fermentation by *R. oryzae* was investigated.

Materials and Methods

Starch recovery from raw cassava pulp

Oven-dried cassava pulp was hydrolyzed by 3 different mineral acids (HCl, H_2SO_4, and H_3PO_4) at 121°C and 15 psig. The effects of acid concentration, ratio of dry pulp to acid solution (g/mL), and hydrolysis time on reducing sugar concentration and yield were determined. After hydrolysis, the samples were centrifuged and the supernatants were collected for analysis. The reducing sugar present in the hydrolysate was analyzed by Dinitrosalicylic acid assay (Miller, 1959) and the yield of reducing sugar liberated from the pulp was determined.

Lactic acid fermentation from the hydrolysates of raw cassava pulp

Lactic acid fermentation from different ACPH by *R. oryzae* was carried out in free cell culture in the shaken flask culture at 30°C, 200 rpm. The fermentation kinetics as well as the cell growth and product formation were observed.

Microorganism

Rhizopus oryzae NRRL 395, a filamentous fungus producing L(+)-lactic acid obtained from the Northern Regional Research Center, Peroria, IL, was used in this study. The sporangiospores were collected from the 7-day culture on a potato dextrose agar (PDA) plate by shaving and extracting the spores with sterile water. The spore concentration was determined by spore counting using a haemacytometer. The spore suspension was then adjusted to desired concentration by dilution with sterile water.

Fermentation medium

Lactic acid production by *R. oryzae* consists of 2 phases, i.e. growth and production. During the growth phase, the medium consisted of 50 g/L carbon source and 5 g/L yeast extract was used for spore germination and initial cell growth where the control of pH was not necessary during this phase. After the growth phase, the growth medium was replaced by the production medium for enhancing lactic acid production. The production medium contained 70 g/L carbon source, 0.6 g/L KH_2PO_4, 0.25 g/L $MgSO_4 \cdot 7H_2O$, 0.088 g/L $ZnSO_4 \cdot 7H_2O$, and 0.3 g/L urea. 5 M NaOH was used for pH control (Thongchul and Yang, 2003). 2 groups of the carbon source were used in this study. Glucose was used as the typical carbon source in the fermentation. Another is the ACPH. The hydrolysates obtained from hydrolyzing the pulp with HCl, H_2SO_4, and H_3PO_4 were referred later as HACPH, SACPH, and PACPH, respectively.

Analytical methods

Moisture content and ash content of raw cassava pulp

The moisture content of raw cassava pulp was determined. Exact weight of raw pulp was recorded. The raw pulp was then dried at 105°C until constant weight. The moisture content (wet basis) was determined. Exact weight of moisture-free cassava pulp was recorded. Then the moisture-free pulp was put into the muffle at 600°C for 4h. After that the burnt pulp (ash) was weighed. The ash content (dry basis) was determined by dividing the weight of the ash by the exact weight of moisture-free cassava pulp.

Cell biomass

Cell biomass was harvested from the culture and washed with water to remove the residues. Washed biomass was dried at 105°C until constant weight was obtained.

Substrate and product concentrations

Glucose and L-lactic acid were analyzed by YSI 2700 glucose analyzer (Yellow Spring Instrument Co., Inc.). Before measurement, the sample was centrifuged and diluted with distilled water. High performance liquid chromatography (HPLC) was used to analyze the organic compounds (glucose, xylose, arabinose, lactic acid, fumaric acid, and ethanol) present in the samples. Samples were centrifuged, filtered through cellulose acetate membrane, and diluted with double distilled water. 20 μL diluted particle-free samples were injected into an organic acid analysis column (Biorad, Aminex HPX-87H ion exclusion organic acid column; 300mm×7.8mm) maintained at 45°C in a column oven. 0.005M H_2SO_4 was used as an eluant at 0.6 mL/min flow rate. A refractive index detector was used to detect the organic compounds. A standard containing 2 g/L of each component was injected as a reference to determine the sample concentration. The peak area was used for the comparison basis. It is noted that HPLC can detect both L(+) and (D)-lactic acids.

Dextrose equivalent (%DE)

Reducing sugar present in the hydrolysate was measured using dinitrosalicylic acid assay as described in Miller (1959). The reducing sugar was determined by comparing the absorbency at 540 nm of the reaction mixture to a standard curve of glucose solution. Dextrose equivalent was determined by the following eq.1

$$\%DE = \frac{[RS]}{[TS]} \times 100 \qquad \text{eq.1}$$

Where [*RS*] represents reducing sugar concentration liberated from raw cassava pulp (g/L) and [*TS*] is the total solid content in the hydrolysate (g/L). Total solid content was determined from drying hydrolysis solution at 105°C until constant weight. The weight amount of solid remained after drying is referred as total solid content.

Results and discussion

Raw cassava pulp

Cassava pulp obtained from cassava mill contains approximately 56.0% starch, 35.9% fiber, 5.3% protein, 2.7% ash, and 0.1% fat (dry basis) (Sriroth et al., 2000). The moisture content of raw pulp used in this study was 74.66% and the ash content was 1.21%.

Acid hydrolysis

Reducing sugar yield obtained from acid hydrolysis of oven-dried pulp at 121°C, 15 psig was observed. It was reported that steaming released uronic acids and acetyl groups in the form of acetic acid. Under these acid steaming conditions, the hemicelluloses in cassava pulp were hydrolyzed with concomitant removal of the hemicelluloses-lignin matrix (Agu et al., 1997). In addition, H^+ ion in acid molecule attacks not only starchy materials but also the lignocellulosic structure in the pulp, it should be noted that the reducing sugar as well as glucose liberated in the hydrolysate come from starch and lignocellulosic structure in cassava pulp (Urbaneja et al., 1996; Zhu et al., 2002).

Effects of reaction time and acid concentration

The reducing sugar in the hydrolysate obtained from acid hydrolysis of cassava pulp was determined. Cassava pulp was treated with 3 types of acid at different concentration under 121°C, 15 psig. Tables 1-3 show the effects of acid concentration and reaction time on reducing sugar yield. It was found that increasing acid concentration increased reducing sugar yield due to more H^+ ions dissociated; thus more chance of starch molecules in cassava pulp to be attacked (Chamy et al., 1994; Stryer, 1995). When excess H^+ ions were present at high acid concentration, reducing sugar concentration and yield were decreased indicating that reducing sugar was further hydrolyzed resulting in unwanted compounds which may inhibit microbial growth in fermentation (Malester et al., 1988; Woiciechowski et al., 1999). Potential fermentation inhibitor compounds include furan aldehydes formed by degradation of sugars, organic acid released from hemicellulose side-groups, and aldehydes and phenolics released from lignin (Nichols et al., 2008). The results also indicate that using less acid concentration required longer reaction time to reach high reducing sugar yield. It was found that the ratio of dry pulp to acid solution (g/mL) did not influence reducing sugar yield; however, the lower ratio gave lower reducing sugar concentration of which required concentration step before use in fermentation.

Type of acid and its dissociation constant

Among 3 acids, HCl provided the highest reducing sugar yield while H_3PO_4 gave the lowest yield (Tables 1-3). Comparing the dissociation constant (K_a) of these 3 acids, it is found that K_a of HCl is higher than H_2SO_4 and H_3PO_4, respectively. Therefore, more H^+ ions in HCl were readily dissociated leading to the higher chance of the starch molecules as well as the lignocellulosic material in cassava pulp to be attacked and broken down (Chamy et al., 1994; Urbaneja et al., 1996). This resulted in the higher reducing sugar concentration obtained and consequently the higher yield with less acid concentration used at the same reaction time.

Table 1 Reducing sugar present in the hydrolysate of oven-dried cassava pulp after hydrolysis by HCl under different conditions

Hydrolysis time (min)	HCl conc. (M)	Ratio of oven-dried cassava pulp to acid solution (g/mL)			
		1 g/9 mL		1 g/18 mL	
		Reducing sugar in hydrolysate (g/L)	Reducing sugar yield (% wt; g RS/g DP)	Reducing sugar in hydrolysate (g/L)	Reducing sugar yield (% wt; g RS/g DP)
15	0.25	86.28	77.66	46.35	83.44
	0.50	109.28	98.35	53.90	97.01
	0.75	108.32	97.49	**53.97**	97.15
	1.00	**111.50**	100.35	51.83	93.29
	1.25	101.66	91.50	50.87	91.56
	1.50	100.26	90.23	52.20	93.95
	1.75	103.59	93.23	42.81	77.05
	2.00	98.19	88.37	43.54	78.38
30	0.25	102.85	92.56	37.85	68.13
	0.50	**103.88**	93.49	47.09	84.77
	0.75	102.55	92.30	45.47	81.84
	1.00	98.93	89.04	**49.39**	88.90
	1.25	93.01	83.71	45.39	81.71
	1.50	90.50	81.45	44.00	79.18
	1.75	98.41	88.57	43.03	77.45
	2.00	90.05	81.05	39.11	70.39
45	0.25	83.40	75.06	42.07	75.72
	0.50	**102.55**	92.30	44.80	80.64
	0.75	91.09	81.98	45.76	82.37
	1.00	99.15	89.23	**47.69**	85.83
	1.25	88.21	79.39	46.43	83.57
	1.50	88.43	79.59	46.21	83.17
	1.75	91.83	82.65	38.00	68.40
	2.00	92.57	83.31	37.26	67.07
60	0.25	**110.46**	99.42	41.70	75.05
	0.50	108.91	98.02	41.62	74.92
	0.75	104.40	93.96	**47.09**	84.77
	1.00	98.85	88.97	46.58	83.84
	1.25	97.52	87.77	45.98	82.77
	1.50	93.68	84.31	44.51	80.11
	1.75	90.50	81.45	43.62	78.51
	2.00	94.05	84.64	37.33	67.20

Remark:

- 3 replicates of hydrolysis were conducted at each hydrolysis condition

Table 2 Reducing sugar present in the hydrolysate of oven-dried cassava pulp after hydrolysis by H_2SO_4 under different conditions

Hydrolysis time (min)	H_2SO_4 conc. (M)	Ratio of oven-dried cassava pulp to acid solution (g/mL)			
		1 g/9 mL		1 g/18 mL	
		Reducing sugar in hydrolysate (g/L)	Reducing sugar yield (% wt; g RS/g DP)	Reducing sugar in hydrolysate (g/L)	Reducing sugar yield (% wt; g RS/g DP)
15	0.25	83.10	74.79	33.85	60.93
	0.50	83.98	75.59	40.73	73.31
	0.75	83.98	75.59	41.84	75.31
	1.00	91.08	81.97	**43.61**	78.50
	1.25	90.64	81.57	39.17	70.51
	1.50	**93.08**	83.77	37.84	68.12
	1.75	87.09	78.38	37.84	68.12
	2.00	88.20	79.38	38.44	69.18
30	0.25	73.19	65.87	23.06	41.50
	0.50	78.22	70.39	35.63	64.13
	0.75	77.11	69.40	35.77	64.39
	1.00	74.08	66.67	38.07	68.52
	1.25	82.51	74.25	**47.01**	84.62
	1.50	**93.01**	83.70	45.16	81.30
	1.75	71.86	64.67	39.40	70.91
	2.00	70.38	63.34	38.44	69.18
45	0.25	70.67	63.61	26.31	47.36
	0.50	69.34	62.41	35.03	63.06
	0.75	75.11	67.60	36.96	66.52
	1.00	**75.78**	68.20	**39.91**	71.85
	1.25	73.56	66.20	20.39	36.71
	1.50	70.90	63.81	18.18	32.72
	1.75	69.64	62.68	35.33	63.59
	2.00	67.42	60.68	34.07	61.33
60	0.25	55.59	50.03	35.33	63.39
	0.50	66.46	59.81	37.54	67.59
	0.75	66.24	59.61	36.96	66.52
	1.00	66.83	60.15	36.66	65.99
	1.25	74.37	66.93	41.10	73.98
	1.50	76.44	68.80	33.70	60.67
	1.75	77.55	69.80	**41.39**	74.51
	2.00	**87.68**	78.91	34.89	62.80

Remark:

- 3 replicates of hydrolysis were conducted at each hydrolysis condition

Table 3 Reducing sugar present in the hydrolysate of oven-dried cassava pulp after hydrolysis by H_3PO_4 under different conditions

Hydrolysis time (min)	H_3PO_4 conc. (M)	Ratio of oven-dried cassava pulp to acid solution (g/mL)			
		1 g/9 mL		1 g/18 mL	
		Reducing sugar in hydrolysate (g/L)	Reducing sugar yield (% wt; g RS/g DP)	Reducing sugar in hydrolysate (g/L)	Reducing sugar yield (% wt; g RS/g DP)
15	0.25	46.58	41.92	26.39	47.50
	0.50	63.95	57.56	30.53	54.96
	0.75	86.65	77.99	38.74	69.73
	1.00	88.80	79.92	39.18	70.53
	1.25	85.17	76.66	42.07	75.72
	1.50	95.38	85.84	44.21	79.58
	1.75	**97.74**	87.97	**47.83**	86.10
	2.00	95.67	86.11	40.81	73.46
30	0.25	24.10	21.69	11.01	19.82
	0.50	37.26	33.53	17.89	32.20
	0.75	48.20	43.38	23.43	42.18
	1.00	54.71	49.24	27.13	48.83
	1.25	70.68	63.61	28.02	50.43
	1.50	68.46	61.62	**50.87**	91.56
	1.75	73.42	66.08	43.84	78.91
	2.00	**74.82**	67.34	49.98	89.96
45	0.25	15.89	14.30	20.03	36.06
	0.50	36.81	33.13	25.72	46.30
	0.75	46.28	41.65	29.94	53.89
	1.00	52.86	47.58	31.34	56.42
	1.25	58.48	52.63	33.49	60.28
	1.50	62.70	56.43	32.01	57.62
	1.75	**73.64**	66.28	33.34	60.01
	2.00	71.20	64.08	**36.89**	66.40
60	0.25	36.22	32.60	9.75	17.56
	0.50	37.26	33.53	26.98	48.57
	0.75	51.23	46.11	30.75	55.35
	1.00	59.22	53.30	33.12	59.61
	1.25	62.92	56.63	39.40	70.93
	1.50	**75.27**	**67.74**	**41.25**	74.25
	1.75	68.17	61.35	40.07	72.13
	2.00	73.05	65.74	36.22	65.20

Remark:

- 3 replicates of hydrolysis were conducted at each hydrolysis condition

Glucose production by acid hydrolysis

Reducing sugar yield obtained from acid and enzymatic hydrolysis of cassava pulp was compared. The highest reducing sugar concentration of 111.50 g/L with up to 100% reducing sugar yield was obtained when hydrolyzing cassava pulp at 121°C, 15 psig for 15 min using 1.0 M HCl solution at the ratio of 1 g dry pulp to 9 mL acid solution comparing with other conditions studied in acid hydrolysis (Tables 1-3). Nevertheless, these products from acid hydrolysis also contained plenty of acid salts which may harm the microorganisms and consequently inhibit desired product formation when directly be used as the carbon source in fermentation (Woiciechowski et al., 1999).

L-lactic acid fermentation

Different ACPH obtained from digesting the pulp with 1.0 M HCl, 1.5 M H_2SO_4, and 1.75 H_3PO_4 at 121°C, 15 psig, 15 min were used as the sole carbon source in the fermentation medium. The fermentation kinetics of *R. oryzae* cultivation in the media containing different ACPH was compared with the cultivation in the simple glucose medium.

L-lactic acid fermentation from glucose

Lactic acid fermentation consisted of 2 phases. In growth phase, yeast extract was used as a nitrogen source for promoting spore germination and initial cell growth. While during the production phase, urea was used as the nitrogen source with the help of $ZnSO_4$ for limiting growth and maintaining fungal activity to produce lactic acid. Fermentation kinetics with glucose as the carbon source is shown in Figure 1. It was found that ethanol was observed as a major product (0.2991 g ethanol/g glucose) in the growth phase instead of cell growth. While during the production phase, lactic acid was a major product (0.3097 g lactate/g glucose). Cell biomass and ethanol production was still found in the production phase resulting in less lactic acid yield and productivity than expected (Table 4).

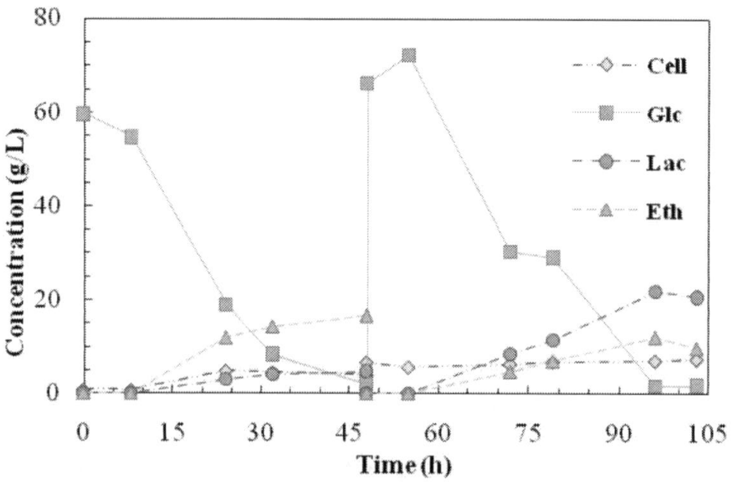

Figure 1 Fermentation kinetics of lactic acid production by *R. oryzae* at 30°C, 200 rpm using glucose as a carbon source

Table 4 Cell biomass and product formation in lactic acid fermentation using glucose as the carbon source

Phase	Max. lactate (g/L)	Yield (g/g)			Productivity (g/L·h)		
		Cell	Lactate	Ethanol	Cell	Lactate	Ethanol
Growth	4.77	0.1068	0.0835	0.2991	0.0873	0.1135	0.3998
Production	21.82	0.0139	0.3097	0.1606	0.0183	0.4322	0.2207

L-lactic acid fermentation from the acid hydrolysates

HACPH, SACPH, and PACPH were used in lactic acid fermentation. Glucose and reducing sugar concentrations in the hydrolysates were determined. The pH of the hydrolysates was approximately 0.3. Before fermentation, pH of the hydrolysate was adjusted to 6.0 (an optimal pH of *R. oryzae*) using concentrated NaOH solution. It was observed that the largest amount of NaOH was required in adjusting pH of PACPH whereas the smallest amount of NaOH was used in adjusting pH of HACPH. This was because the difference in HCl (1.0 M) and H_3PO_4 (1.75 M) concentrations were used in cassava pulp hydrolysis resulting in the difference in H^+ ions present in the hydrolysates. Therefore, the different amount of OH^- ions from NaOH solution was required in adjusting the pH of the hydrolysates. After adjusting the pH and diluting to the desired concentration, acid hydrolysates were used as the carbon source in the fermentation medium. It was found that growth and product formation were observed in fermentation of HACPH while no growth was found in fermentation of SACPH and PACPH. It is believed that a large amount of the remaining ions (Na^+, SO_4^{2-}, and PO_4^{3-}) in the hydrolysates caused growth inhibition resulting in no growth and product formation in the fermentation of SACPH and PACPH. By detoxification of these hydrolysates, it would help improve microbial growth and product formation in the fermentation process. The examples of the detoxification of the hydrolysates include adsorption on resins or activated charcoal, precipitation and filtration, and extraction with an organic solvent system such as Alamine 336 and Aliquat 336 dissolved in octanol (Gamez et al., 2006; Grzenia et al., 2008; Nichols et al., 2008; Qureshi et al., 2008).

Fermentation kinetics of *R. oryzae* grown in the medium containing HACPH is shown in Figure 2. Instead of lactic acid, ethanol was found as the major product in both growth and production phases. In this work, we were able to produce high cell density in the growth phase; however, in the production phase a large amount of cell biomass converted glucose to ethanol (Table 5). This was probably due to an operating condition in the shaken flask culture of *R. oryzae* grown in the HACPH medium favored ethanol production (Taherzadeh et al., 2003). In addition, oxygen limitation inside the mycelial clump in free cells caused metabolic shift to anaerobic route to yield ethanol as the major product. Previous study on lactic acid fermentation by pelletized *R. oryzae* indicated that lactic acid production was improved when culturing in a bioreactor due to better oxygen transfer and mixing compared to the culture in the shaken flask (Liu et al., 2006). Therefore, it is believed that with better oxygen transfer in the fermentation system, it is possible to achieve high cell production in the growth phase. Consequently with high cell biomass with good morphological control and sufficient oxygen supply, lactic acid production will be promoted in the production phase. Therefore, controlling oxygen supply is one of the major keys in lactic acid fermentation.

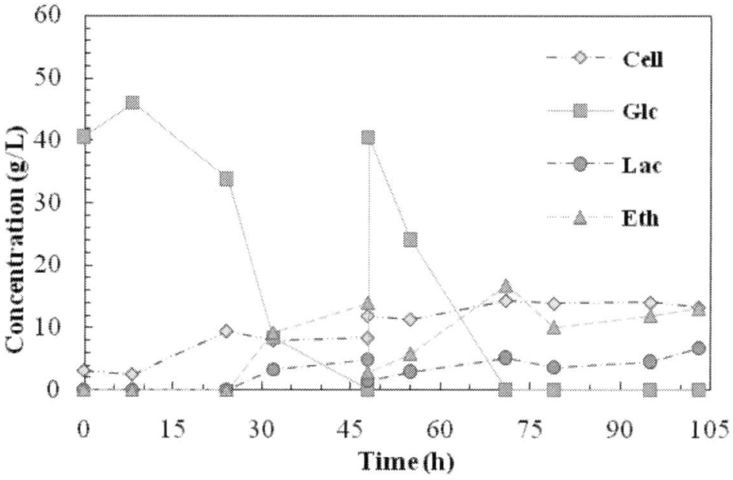

Figure 2 Fermentation kinetics of lactic acid production by *R. oryzae* at 30°C, 200 rpm using cassava pulp hydrolysate obtained from hydrolysis with 1.0 M HCl at 121°C, 15 psig, 15 min as a carbon source

Table 5 Cell biomass and product formation in lactic acid fermentation using HCl treated cassava pulp hydrolysate as the carbon source

Phase	Max. lactate (g/L)	Yield (g/g)			Productivity (g/L·h)		
		Cell	Lactate	Ethanol	Cell	Lactate	Ethanol
Growth	4.87	0.1090	0.1100	0.3150	0.1367	0.1068	0.3065
Production	6.67	0.0648	0.0926	0.3537	0.1193	0.1588	0.6201

Comparison of the sole carbon source used in lactic acid fermentation

It was observed that the lower cell biomass production was obtained in fermentation of glucose (the highest concentration of 7.38 g/L) as compared with the production from HACPH (14.23 g/L). High ethanol production was observed as a consequence of high cell biomass production in fermentation of HACPH (16.80 g/L during the growth phase). Taherzadeh et al. (2003) reported that poor medium compositions and cultivation conditions resulted in high lactic acid yield whereas ethanol and cell biomass yields were high in a rich medium. They found that high ethanol and cell biomass production was achieved when growing *R. oryzae* in paper pulp sulfite liquor containing 45% dry weight materials including 60% lignosulfonate, 12 g/L galactose, 21 g/L glucose, 75 g/L mannose, 25 g/L xylose, and 6 g/L acetic acid. They indicated that the rich medium resulted in faster sugar uptake and a higher ethanol yield whereas lactic acid yield was lower. In contrast, a poor medium (a synthetic medium containing glucose as the sole carbon source) resulted in slower sugar uptake, higher lactic acid yield, and lower ethanol yield. Thongchul (2005) reported that high cell biomass was produced with low lactic acid yield and productivity in the fermentation of corn fiber hydrolysate by *R. oryzae* NRRL395 in a Rotating Fibrous Bed bioreactor. Raw cassava pulp used in this study contained approximately 56% starch, 35.9% fiber, 5.3% protein, 2.7% ash, and 0.1% fat (dry basis) (Sriroth et al., 2000). Hydrolysis product of cassava pulp contained such materials in the soluble forms which probably promoted cell biomass and ethanol production; therefore, resulting in higher cell biomass and ethanol production but lower lactic acid production as compared with those obtained from culture in the glucose medium.

Conclusion

This study showed that the hydrolysis products of cassava pulp could be used as a sole carbon source by *R. oryzae* to produce cell biomass, lactic acid, and ethanol. Acid hydrolysis at high temperature and pressure yielded high concentration of reducing sugar liberated from cassava pulp within a short time. However, the presence of the remaining ions (Na^+, SO_4^{2-}, and PO_4^{3-}) in the hydrolysates caused growth inhibition resulting in no growth and product formation in the fermentation of SACPH and PACPH. By detoxification, it is believed that *R. oryzae* can grow in detoxified SACPH and PACPH. Instead of lactic acid, cell biomass and ethanol were the major products found in the fermentation of HACPH while *R. oryzae* produced lactic acid as the major product from the simple glucose medium. This was perhaps due to some proteins and fibers in the pulp remained soluble in the hydrolysates enriched the fermentation medium. Thus, the fermentation medium containing the hydrolysates favored cell biomass production and consequently led to ethanol formation due to excess growth and oxygen limitation. Oxygen limitation could be diminished by high agitation and aeration in the bioreactor and lactic acid production could be improved as compared to that in the shaken flask culture.

Acknowledgments

This work has been financially supported by Commission on Higher Education and Thailand Research Fund (Grant Number: MRG5080097). Raw cassava pulp was kindly provided by Chonjareon Cassava Mill, Thailand. The technical assistance in fermentation sample analysis provided by Ms.Sitanan Thitiprasert was appreciated.

References

1. Agu, R.C., Amadife, A.E., Ude, C.M., Onyia, A., Ogu, E.O., Okafor, M., 1997. Combined heat treatment and acid hydrolysis of cassava grate waste (CGW) biomass for ethanol production. Waste Management. 17, 91-96.
2. Aikat, K., Bhattacharyya, B.C., 2001. Protease production in solid state fermentation with liquid medium recycling in a stacked plate reactor and in a packed bed reactor by a local strain of *Rhizopus oryzae*. Process Biochemistry. 36, 1059-1068.
3. Cata, F.S., Soccol, C.R., Ramos, L.P., Fontana, J.D., 1999. Production of fumaric acid by fermentation of enzymatic hydrolysates derived from cassava bagasse. Bioresource Technology. 68, 23-28.
4. Chamy, R., Illanes, A., Aroca, G., Nunez, L., 1994. Acid hydrolysis of sugar beet pulp as pretreatment for fermentation. Bioresource Technology. 50, 149-152.

5. Christen, P., Bramorski, A., Revah, S., Soccol, C.R., 2000. Characterization of volatile compounds produced by *Rhizopus* strains grown on agro-industrial solid wastes. Bioresource Technology. 71, 211-215.
6. Gamez, S., Gonzalez-Cabriales, J.J., Ramirez, J.A., Garrote, G., Vazquez, M., 2006. Study of the hydrolysis of sugar cane bagasse using phosphoric acid. Journal of Food Engineering. 74,78-88.
7. Grzenia, D.L., Schell, D.J., Wickramasinghe, S.R., 2008. Membrane extraction for removal of acetic acid from biomass hydrolysates. Journal of Membrane Science. 322, 189-195.
8. Huang, L.P., Jin, B., Lant, P. and Zhou, J., 2003. Biotechnological production of lactic acid integrated with potato wastewater treatment by *Rhizopus arrhizus*. Journal of Chemical Technololgy and Biotechnology. 78, 899-906.
9. Liu, Y., Liao, W., Liu, C., Chen, S., 2006. Optimization of L-(+)-lactic acid production using pelletized filamentous *Rhizopus oryzae* NRRL395. Applied Biochemistry and Biotechnology. 129-132, 844-853.
10. Malester, A.I., Green, M., Kimchie, S., Shelef, G., 1988. The effect of the neutralizing capacity of cellulosic materials on the kinetics of cellulose dilute acid hydrolysis. Biological Waste. 26, 115-124.
11. Miller, G.L., 1959. Use of dinitrosalicylic acid reagent for determination of reducing sugar. Analytical Chemistry. 31, 426-428.
12. Nichols, N.N., Sharma, L.N., Mowery, R.A., Chambliss, C.K., 2008. Fungal metabolism of fermentation inhibitors present in corn stover dilute acid hydrolysate. Enzyme and Microbial Technology. 42, 624-630.
13. Nishise, H., Fuji, A., Ueno, M., Vongsuvanlert, V., Tani, Y., 1988. Production of raw cassava starch-digestive glucoamylase by *Rhizopus* sp. in liquid culture. Journal of Fermentation Technology. 66, 397-402.
14. Pandey, A., Soccol, C.R., Nigam, P., Soccol, V.T., Vandenberghe, L.P.S., Mohan, R., 2000. Biotechnological potential of agro-industrial residues. II: cassava bagasse. Bioresource Technology. 74, 81-87.
15. Qureshi, N., Ezeji, T.C., Ebener, J., Dien, B.S., Cotta, M.A., Blaschek, H.P., 2008. Butanol production by *Clostridium beijerinckii*. Part I: use of acid and enzyme hydrolyzed corn fiber. Bioresource Technology. 99, 5915-5922.
16. Ruengruglikit, C. and Hang, Y.D., 2003. L(+)-lactic acid production from corncobs by *Rhizopus oryzae* NRRL-395. LWT Food Science and Technology. 36, 573-575.
17. Soccol, C.R., Vandenberghe, L.P.S., 2003. Overview of applied solid-state fermentation in Brazil. Biochemical Engineering Journal. 13, 205-218.
18. Sriroth, K., Chollakup, R., Chotineeranat, S., Piyachomkwan, K. and Oates, C.G., 2000. Processing of cassava waste for improved biomass utilization. Bioresource Technology. 71, 63-69.
19. Stryer, B., 1995. Biochemistry, fourth ed. W.H. Freeman and Company, New York.
20. Sun, S.Y., Xu, Y., 2008. Solid-state fermentation for 'whole-cell synthetic lipase' production from *Rhizopus chinensis* and identification of the functional enzyme. Process Biochemistry. 43, 219-224.
21. Taherzadeh, M.J., Fox, M., Hjorth, H., Edebo, L., 2003. Production of mycelium biomass and ethanol from paper pulp sulfite liquor by *Rhizopus oryzae*. Bioresource Technology. 88, 167-177.
22. Tani, Y., Fuji, A., Nishise, H., 1988. Production of raw cassava starch-digestive glucoamylase by a 2-deoxyglucose-resistant mutant of *Rhizopus* sp. Journal of Fermentation Technology. 66, 545-551.
23. Thongchul, N. Lactic acid production by immobilized *Rhizopus oryzae* in a rotating fibrous bed bioreactor. PhD thesis. The Ohio State University. Columbus, Ohio. 2005.

24. Thongchul, N., Yang, S.-T., 2003. Controlling filamentous fungal morphology by immobilization on a rotating fibrous matrix to enhance oxygen transfer and L(+)-lactic acid production by *Rhizopus oryzae*, in: Saha, B.C. (Ed.), Fermentation Biotechnology. Oxford University Press, MA, 2003, pp. 36-51.
25. Urbaneja, G., Ferrer, J., Paez, G., Arenas, L., Colina, G., 1996. Acid hydrolysis and carbohydrates characterization of coffee pulp. Renewable Energy. 9, 1041-1044.
26. Vink, E.T.H., Rabago, K.R., Glassner, D.A., Springs, R., O'Corner, R.P. and Gruber, P.R., 2004. The sustainability of NatureWorks™ polylactide polymers and Ingeo™ polylactide fibers[a]: an update of the future. Macromolecular Bioscience. 4, 551-564.
27. Woiciechowski, A.L., Soccol, C.R., Ramos, L.P. and Pandey, A., 1999. Experimental design to enhance the production of L-(+)-lactic acid from steam-exploded wood hydrolysate using *Rhizopus oryzae* in a mixed-acid fermentation. Process Biochemistry. 34, 949-955.
28. Yu, R.-C., Hang, Y.D., 1989. Kinetics of direct fermentation of agricultural commodities to L(+)-lactic acid by *Rhizopus oryzae*. Biotechnology Letters. 11, 597-600.
29. Zhan, X., Wang, D., Tuinstra, M.R., Bean, S., Seib, P.A., Sun, X.S., 2003. Ethanol and lactic acid production as affected by sorghum genotype and location. Industrial Crop and Product. 18, 245-255.
30. Zhu, Y., Wu, Z. and Yang, S.-T., 2002. Butyric acid production from acid hydrolysate of corn fiber by *Clostridium tyrobutyricum* in a fibrous-bed bioreactor. Process Biochemistry. 38, 657-666.

146b Production of Platform Chemicals from Renewable Sugars: Low-Cost Succinic Acid Fermentation Using Engineered Escherichia Coli

Praveen V. Vadlani, Grain Science and Industry, Kansas State University, 1980 Kimball Avenue, 202 BIVAP Building, Manhattan, KS 66506, Ka Yiu San, Department of Bioengineering, Department of Chemical and Biomolecular Engineering, Rice University, Houston, TX 77005, and George N. Bennett, Biochemistry and Cell Biology, Rice University, 6100 Main Street MS#142, Houston, TX 77005.

Succinic acid is a platform chemical that can be used to make several industrial chemicals including non-toxic solvents, adhesives, paint-additives, and non-corrosive road deicers. A novel and cost-effective fermentation process is being developed to produce succinic acid using a metabolically engineered E. coli SBS550MG and renewable sugars derived from corn and grain sorghum. The constructed strain was characterized and evaluated with shake flask cultures. Fermentation experiments using 2% synthetic glucose were performed in a 2 L fermenter and a high succinic acid yield of 1.1 g/g glucose was obtained. Shake flask experiments to demonstrate the utilization of sorghum-glucose and corn steep liquor as fermentation media substitutes were also performed. A succinic acid yield of 1.1 g/g of glucose consumed was obtained, which was identical to the yield when synthetic glucose was used. Results from batch and fed-batch fermenter experiments using renewable sugars will be discussed in the presentation.

146c Fatty Acid and Lipid Accumulation In Municipal Wastewater Microorganisms Growing on a Nitrogen-Limited Synthethic Wastewater Medium

Andro Mondala, Rafael Hernandez, Todd French, Earl Alley, and Maria Paraschivescu. Chemical Engineering, Mississippi State University, Box 9595, Mississippi State, MS 39762

The potential for accumulating lipids and fatty acids in the indigenous microbial population in municipal wastewater was investigated. These lipids and fatty acids could then be used as feedstocks for the production of biofuels such as biodiesel and green diesel. A 5 % (v/v) inoculum of activated sludge from a wastewater treatment plant in Tuscaloosa, AL was grown under excess carbon and nitrogen-limited conditions in a batch culture using a synthetic wastewater medium. Glucose concentrations of 4, 8, 12, and 20 g/L in the medium were tested and samples were analyzed for cell mass (dry basis), glucose uptake, and ammonium-nitrogen concentration. The microbial cells were then harvested and the accumulated lipids and fatty acids in these cells were extracted using the method of Bligh and Dyer (1959). The extracts were then weighed to determine the gravimetric yield in terms of cell dry mass and then analyzed as their methyl esters by gas chromatography. The kinetics of oil accumulation and glucose uptake were determined and applied to design a wastewater treatment facility capable of generating oil for biofuels production.

146d Microbial Fermentation of Glycerol for the Production of Ethanol

Ashutosh Gupta, Chemical and Biomolecular Engineering, Rice University, MS-362, 6100 S Main St, Rice University, Houston, TX 77005 and Ramon Gonzalez, Departments of Chemical & Biomolecular Engineering and Bioengineering, Rice University, MS-362, P.O. Box 1892, Houston, TX 77251-1892.

Microbial fermentation of glycerol for the efficient production of ethanol

Unprecedented increase in crude oil prices in recent times has led to worldwide increase in the production of bio-based chemicals and fuels. Glycerol is an inevitable byproduct in the production of biodiesel. Its value for biodiesel manufactures has changed dramatically from being a valuable co-product to a waste product in last few years. This has created an urgent need to find new platforms for the production of value added products from glycerol.

We have recently identified several microorganisms, including E. coli and other enteric bacteria, which can efficiently ferment glycerol if cultivated under appropriate culture conditions. I all cases ethanol was found to be the main product of glycerol fermentation. These organisms have significant industrial advantages for glycerol utilization in terms of yield, productivity, low nutritional requirements, and ability to grow in presence/absence of oxygen.

In this work, we present the effect of pH on the fermentation of glycerol by these microorganisms in batch cultures. We show our result that changes in pH from acidic to basic medium, results in a transition from glycerol fermenting to glycerol non-fermenting behavior. We also report the effect of changes in media composition on glycerol utilization and growth. Effects of changing headspace environment from inert to carbon dioxide are also presented. Fermentation in these microorganisms is enabled by the production of 1,2 propanediol, which results in the consumption of excess reducing equivalents generated by the synthesis of cell biomass from glycerol.

146e Production of Butanol from Biodiesel Derived Crude Glycerol Using Pure Cultures of C. Pasteurianum (ATCC 6013)

Katherine A. Taconi[1], Keerthi Venkataramanan[1], and Duane Johnson[2]. (1) University of Alabama in Huntsville, 130 Engineering Building, Huntsville, AL 35899, (2) Alabama Biodiesel Corporation, Moundville, AL

The rapidly expanding market for biodiesel is dramatically altering the cost and availability of glycerol, as typical biodiesel production processes generate about 10 wt% glycerol. Biodiesel producers have little financial incentive to purify the crude glycerol, and since it contains methanol, crude glycerol is considered a hazardous waste. Therefore, it is critical that new and innovative processes for managing and utilizing the glycerol co-product be developed. In the longer term, as supplies continue to increase, glycerol will become a versatile building block chemical for the production of high value compounds within an integrated biorefinery.

The goal of this research is to investigate the value-added conversion of crude glycerol generated during biodiesel production using anaerobic fermentation. Pure cultures of Clostridium pasteurianum have previously been shown to produce significant amounts of butanol and 1,3-propanediol as well as some ethanol and lactic acid when utilizing purified glycerol as the sole substrate [1,2]. Butanol is of particular interest as a renewable biofuel, as it has a higher heating value, higher octane number, lower vapor pressure, and higher miscibility than ethanol, making butanol preferable to ethanol for blending with petroleum fuels.

This presentation will discuss the results of experiments evaluating growth and product formation of C. pasteurianum (ATCC®6013™) utilizing biodiesel-derived crude glycerol as the sole carbon source. Crude glycerol containing approximately 5 g/L methanol and various salts was obtained from Alabama Biodiesel Corporation (Moundville, AL). Batch cultures were grown on purified glycerol and crude glycerol for comparison. Cultures utilizing crude glycerol have demonstrated significant growth and production of butanol, ethanol, and 1,3-PDO as well as complete substrate utilization. To the best of the authors' knowledge, this is the first report of butanol production from crude glycerol using this organism.

To date, this work has produced butanol yields up to 32 wt%. In contrast to previous work, these experiments did not produce lactic acid and did accumulate butyric acid. This presentation will present additional results regarding the yields of butanol, ethanol, and 1,3-PDO as well as present preliminary data for the extraction of products from the fermentation broth to produce a biofuel blend.

[1] Biebl, H. "Fermentation of Glycerol by Clostridium pasteurianum – Batch and Continuous Culture Studies." Journal of Industrial Microbiology and Biotechnology. 2001. Vol. 27, 18-26.

[2] Dabrock, B.; Bahl, H.; Gottschalk, G. "Parameters Affecting Solvent Production by Clostridium pasteurianum." Applied and Environmental Microbiology. 1992. Vol. 58, No. 4, 1233-1239.

146f Sludges-to-Ethanol as a Supplementary Unit In Kraft Paper Mill

Li Kang[1], Wei Wang[2], and Y.Y. Lee[2]. (1) Chemical Engineering, Auburn University, 212 Ross Hall, Auburn University, Auburn, AL 36849, (2) Chemical Engineering Department, Auburn University, 212 Ross Hall, Auburn University, Auburn, AL 36849

A Kraft paper mill produces chemical pulp by treating wood chips with white liquor, a mixture of sodium hydroxide, sodium sulfide, and sodium carbonate at 130 - 180ºC. Some of the fine fibers are rejected in the paper making process and eventually become waste material, e.g., paper mill sludge. Its high carbohydrate and low lignin content makes the sludge an attractive feedstock for bioconversion into value-added products including ethanol. We have proven that Kraft paper mill sludge and recycle sludge are easily converted to ethanol by SSF or SSCF without any pretreatment. The sludges, however, are not appropriate for cellulase production because of high ash content. Kraft pulp itself is an ideal substrate for production of cellulase enzymes. Our laboratory data indicate that only about 20 kg of Kraft pulp is required to produce cellulase enzyme that can saccharify one dry-ton of the sludge. In this study, the Kraft pulp was used as the main carbon source for production of cellulase by Trichoderma Reesei, Rut C-30. Partially de-ashed Kraft paper mill sludges were tested as a substrate for production of cellulosic ethanol through fermentation using the cellulase procduced in-house and two different microorganisms - Recombinant E. coli KO-11 (SSCF) and Saccharomyces cerevisiae, NRE- D5A (SSF). The fermentation process proceeded without any major difficulty, and demonstrated ethanol yields comparable to those of pretreated corn stover. It is projected that 70 gallons of ethanol can be produced from these processes per ton of dry sludge. When the SSCF and SSF were operated in fed-batch mode, the ethanol concentration in the broth was increased to 48g/L and 62g/L, respectively.

CLOSED-LOOP CONTROL OF AN ARTIFICIAL PANCREATIC BETA CELL USING MULTI-PARAMETRIC MODEL PREDICTIVE CONTROL

Matthew W. Percival[†,‡], *Eyal Dassau*[†,‡], *Howard Zisser*[†,‡], *Lois Jovanovič*[†,‡], *and Francis J. Doyle III*[†,‡,*]
[†] *Department of Chemical Engineering, University of California, Santa Barbara*
[‡] *Sansum Diabetes Research Institute, Santa Barbara*
* *Corresponding author: frank.doyle@icb.ucsb.edu*

Abstract

A model predictive control algorithm was developed for regulating glycemia in subjects with type 1 diabetes mellitus using exogenous insulin infusion to reject disturbances from oral carbohydrate consumption. A clinically validated nonlinear physiological model of glucose metabolism was linearized and served as the basis for the controller. The control law was developed using multi-parametric programming techniques, so that the online optimization problem was reduced to the evaluation of an affine function. The controller was tuned for optimal response to mixed-meals of up to 40 g carbohydrate. The robustness of the controller was then evaluated for mismatch in the size (25%) and timing (15 minutes) of the meal. Due to the computational ease with which this control law was evaluated online, it would be suitable for implementation on a personal digital assistant with limited computational capacity.

Introduction

Diabetes mellitus is a disease characterized by insufficient endogenous insulin production, leading to poor regulation of plasma glucose concentration. The resulting chronic hyperglycemia can lead to macro- and micro-vascular complications, e.g., stroke, heart attacks, blindness, and kidney disease. An indication of the quality of glycemic regulation, and thus the likelihood of such complications is given by the quantity of glycosylated hemoglobin (A1C). Successful control consists of normalizing glycemia, and thus reducing A1C to below 6.0% [1].

Type 1 diabetes mellitus (T1DM) is an autoimmune disease that results in the destruction of the pancreatic beta cells found in the islets of Langerhans. These cells are responsible for the production of insulin, one of the hormones that regulate plasma glucose concentration. In order to regulate glycemia in T1DM, insulin must be delivered exogenously. Current therapeutics involve manually taking a capillary blood sample (a finger stick measurement) to determine the blood glucose concentration, then injecting an appropriate amount of insulin depending upon the subject and the situation, e.g., at meal time. For this method to be effective, this process must be repeated up to 12 times daily. Such a high level of discipline is burdensome, and thus patient adherence is typically lacking. As a result, glycemic control is inadequate and complications of T1DM occur. As an indication of the worldwide scale of the problem, there are 1.5 million people with T1DM in the United States alone [2].

Recent technological advances have brought continuous real-time subcutaneous glucose sensors and continuous insulin infusion pumps, which in combination with rapid-acting insulin analogs provide a framework for a closed-loop controller suitable for use in ambulatory conditions [3]. Real-time glucose sensors measure the glucose concentration in the subcutaneous tissue, and typically offer an updated glucose measurement every five minutes; this sample resolution is in line with that required for an algorithm to achieve efficient closed-loop control via subcutaneous insulin delivery [4].

The first algorithms for closed-loop control of glycemia were developed in the 1970s, most notably those used with the glucose controlled insulin infusion devise called the Biostator® [5]. These early algorithms relied on intravenous insulin and dextrose delivery, along with intravenous glucose measurements. Such invasive methods were therefore only suitable for use in an in-patient hospital setting. Currently, robust control algorithms for a closed-loop device suitable for use in ambulatory conditions are not available [6]. The main challenges involve the large lag time in insulin action of subcutaneously delivered insulin on the plasma glucose concentration, and sustained disturbances to glucose levels after the absorption of a meal. For these reasons, model predictive control (MPC) has been identified as a highly promising controller structure, and has been applied in a clinical environment [7]. The MPC framework also allows for the explicit inclusion of constraints [8], which are pertinent to the physical capabilities of the insulin pump.

In order to mitigate the problems involved with solving large optimization problems online, multi-parametric programming techniques can be applied which lead to the solution of a single optimization problem offline; the online problem thus reduces to reading a lookup table, and the evaluation of an affine function [9]. Such techniques are germane to the first generation of closed-loop devices for managing glycemia for two reasons:

(i) The first commercially available artificial beta cells will no doubt have limited computational capacity.

(ii) There should be an easy way to evaluate systematically controller outcomes under a range of expected conditions before application.

Such prior evaluation would be a requirement of the Food and Drug Administration (FDA) for a device to be approved for commercial use [10]. This algorithmic approach has previously been used to develop a multi-parametric MPC (mpMPC) algorithm in the context of diabetes, albeit for intravenous insulin delivery using a simplified model of insulin action [11].

In this paper, a simulation study was carried out to show the utility of mpMPC when applied to regulation of glycemia. An established physiological model of glucose metabolism [7, 12] characterizing orally ingested carbohydrate and subcutaneously injected insulin was used represent a person with diabetes. A control law was then developed using mpMPC. Simulations were carried out to determine optimal controller tuning in the case of meal (disturbance) rejection. The resulting controller was then tested for robustness by introducing elements of plant/model mismatch typical of a person with diabetes.

Experimental Methods

This section describes the design of the simulation study and the implementation of mpMPC. The simulation study was designed as an *in silico* closed-loop clinical trial. A physiological model represents the subject with T1DM, which is subject to meal ingestion and insulin delivery determined by the control algorithm.

Simulation model

The physiological models of glucose-insulin kinetics and subcutaneous insulin absorption presented by Hovorka et al. [7] and Wilinska et al. [12] formed the virtual T1DM subject. Orally ingested glucose is absorbed by the gut into plasma using a two-compartment fundamental model. Subcutaneously injected insulin is absorbed into the plasma from a fourth order model incorporating saturation effects. The glucose effects are described by fifth order model including three insulin sensitivity effects, endogenous glucose disposal and production, and insulin dependent and independent effects. The nonlinear effects of the model are assumed representative of those observed *in vivo*.

Controller development

The formulation of the control algorithm is a two part process:

i. Formulation of the objective function as a constrained optimization problem

ii. Reformulation of the objective function as a multi-parametric program and its solution

The objective function for the MPC problem includes quadratic penalties for the vector of predicted deviations from setpoint of glucose measurements (y) and the vector of future insulin delivery rates (u). The cost, J, is written as

$$\min_{u} J = y^T Q_y y + u^T R u. \tag{1}$$

The model predictions are given by a linearization of the plant around nominal operating conditions, which correspond to fasting and basal insulin delivery; the state-space representation of the model at time k is

$$\begin{aligned} x_{k+1} &= A x_k + B u_k \\ y_k &= C x_k \end{aligned}, \tag{2}$$

where x represents the state vector, u represents the inputs, and y the measurement; all are deviation variables. Physical restrictions on insulin delivery corresponding to the mechanical limitations of the insulin pump are included as constraints on the manipulated variable

$$u_{min} \leq u \leq u_{max}. \tag{3}$$

Constraints are also given on the region of state-space in which a solution is required:

$$x_{min} \leq x \leq x_{max}, \tag{4}$$

where x_{min} and x_{max} represent the minimum and maximum values permissible for each state in the state vector.

In order to apply multi-parametric programming techniques the objective function must be reformulated as a standard mpQP,

$$\begin{aligned} &\min_{z} \mu = \frac{1}{2} z^T H z \\ &s.t.\ Gz \leq W + GH^{-1}F^T x_0 \end{aligned} \tag{5}$$

where x_0 is the state vector at the current time, and z is a linear transformation of the manipulated variable, defined as

$$z = u + H^{-1}F^T x_0. \tag{6}$$

Matrices F, G, H and W are obtained from A, B, Q_y, and R. An explicit solution to this optimization problem exists [9]. The solution consists of multiple regions within state-space; within each region the optimal values of the manipulated variables, u, are an affine function of the current state vector x_0. The algorithm used to determine the regions in the solution and the corresponding control law is described by Pistikopoulos et al. [13]. Transformation of the parametric solution back into state-space gives an explicit control law for all possible initial conditions of the state vector. Full mathematical development of this method has been carried out several times in the literature [9, 11, 13].

Controller object

The solution to the multi-parametric programming problem was found using the Multi-Parametric Toolbox [14], and interfacing software, YALMIP [15]. The solution consists of a polytope defining the control law. The control law consists of regions defined in state-space, essentially a lookup table. Each region, or entry in the lookup table contains an affine function of the current state; the evaluation of this function gives the optimal control moves to be implemented.

In order to evaluate the control law, the state vector is evaluated using the latest measurements, thus determining the appropriate region, or lookup table entry. The affine function associated with that region is then evaluated with the current state vector, thus determining the control move. A graphical example of control law evaluation is given in Appendix A.

Simulated Results

Using the controller and simulated subject described in the previous section, controller tuning was performed and robustness due to expected levels of uncertainty in key variables were investigated. The goal was to investigate the rejection of a single meal and to ensure long-term controller stability over a period of 12 hours.

Simulations were performed with measured disturbances of 10-40 g CHO mixed-meals. For these announced disturbances, the prediction horizon and weighting matrices were tuned. In order to test for robustness, mismatches of 25% were introduced to the CHO measurement and of 15 minutes to the disturbance timing.

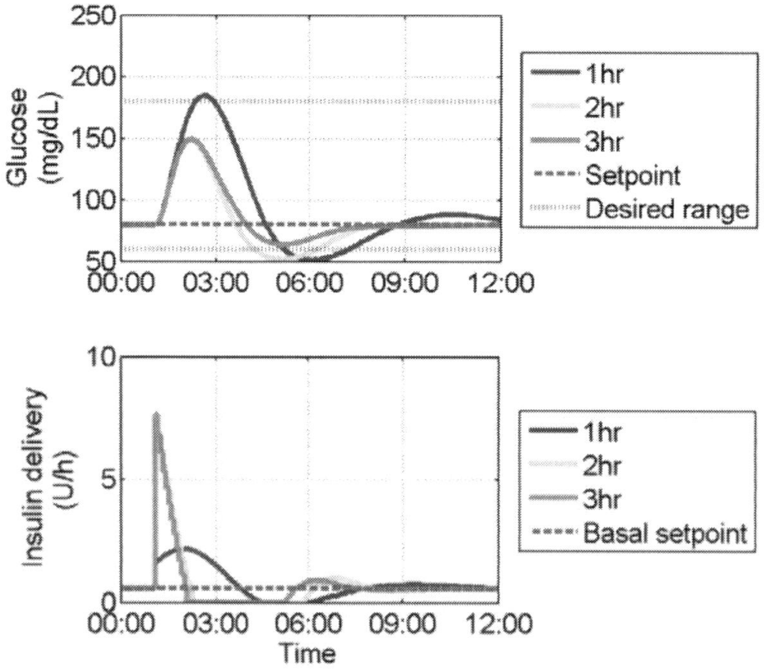

Figure 1: Closed-loop responses to an announced meal of 40 g CHO. TOP: glucose trajectory. Increasing the prediction reduces hyperglycemia and hypoglycemia. BOTTOM: insulin delivery. Increased prediction horizon allows more aggressive control action at meal time.

Prediction horizon

An appropriate choice of prediction horizon is one in which the full effects of current inputs are observed, i.e., the open-loop speed of response. Based on this principle, a prediction horizon of three hours is appropriate, as is corroborated by Figure 1.

Weighting matrices

The weighting matrices on glucose error, Q_y, and insulin delivery about basal, R, determine how aggressive the controller is; if large penalties are applied to glucose error, an aggressive controller will result; if large penalties are applied to insulin delivery about basal, a more conservative controller will result. In the case where Q_y and R are constant over time, the ratio Q_y/R is the single tuning parameter. As a general rule, a controller should be more conservative if there is known to be a significant plant/model mismatch, or if the consequences of large deviations from setpoint are severe.

In this nominal case with a model based on the Jacobian of the plant, the plant model mismatch should be small; hence, the deciding factor in tuning should be avoidance of hypoglycemia. Figure 2 shows the effects of various Q_y/R with a three hour prediction on controller performance. The best Q_y/R is equal to unity, since this setting avoids hypoglycemia and hyperglycemia.

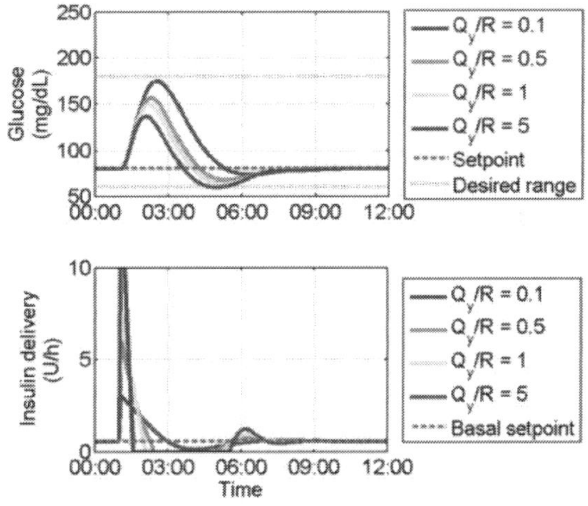

Figure 2: Closed-loop responses to an announced meal of 40 g CHO and the effect of varying the input/output weighting matrices. TOP: a less aggressive controller minimizes the approach to hypoglycemia while remaining below the hyperglycemic threshold. BOTTOM: the more aggressive controllers deliver insulin as a pseudo-bolus in order to minimize the glycemic excursion.

Robustness

Uncertainty is not only attributed to plant/model mismatch, but also to inputs. These input uncertainties arise whenever a meal is announced, specifically the quantity of carbohydrate in the meal and the exact time of the meal. It is not that meal size and time are impossible to determine exactly, but that the announcement itself is burdensome and a typical subject with T1DM does not record these figures precisely. Mismatches of 25% were introduced to the CHO measurement and of 15 minutes to the disturbance timing. Figures 3 and 4 show how time and size mismatches affect the controller tuned with $Q_y/R=1$. Combining these effects lead to hypoglycemic events with the $Q_y/R=1$, so the controller was further detuned. Figure 5 shows the combined effects of meal size and time mismatch with $Q_y/R=0.5$. Clearly detuning is necessary to avoid hypoglycemia in the case of uncertainty.

Based on the limits of uncertainty previously described, it is important to ascertain the greatest amount of CHO that can be consumed without inducing hypo- or hyperglycemia. Figure 6 shows the upper and lower bounds of the glycemic trajectory for meal sizes of 10, 25, and 40 g CHO. For this simulation model and these conditions of uncertainty, the maximum meal size is 35 g CHO. This meal size is very low in CHO, which serves to illustrate further the difficulties associated with rejection of meal-based disturbances.

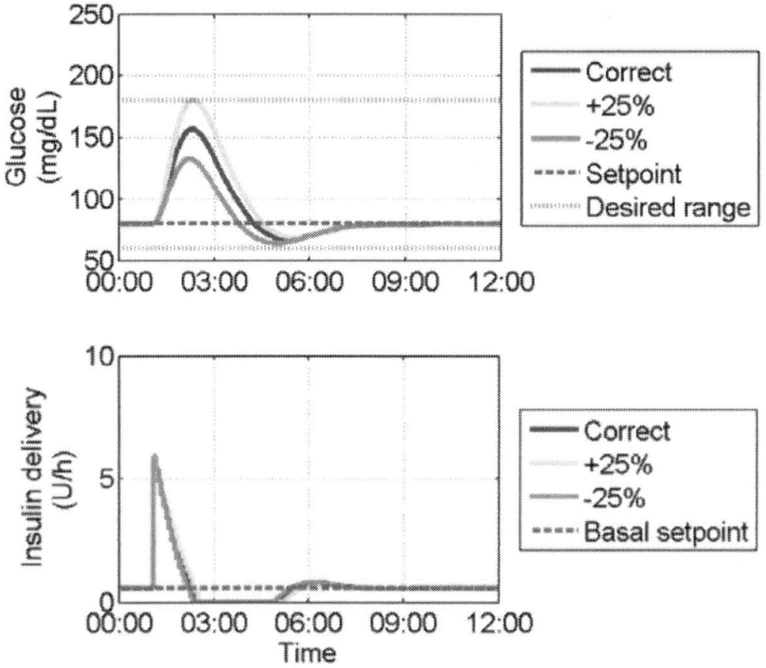

Figure 3: Closed-loop responses to an announced meal of 40 g CHO with the actual meal size of 40 g CHO ± 25%. TOP: glycemic excursions remain within desired range, thus avoiding hyperglycemia and hypoglycemia. BOTTOM: insulin delivery is initially the same in each case; feedback from glucose measurements adjusts the following delivery slightly to compensate for the differences in expected and actual glycemic excursions.

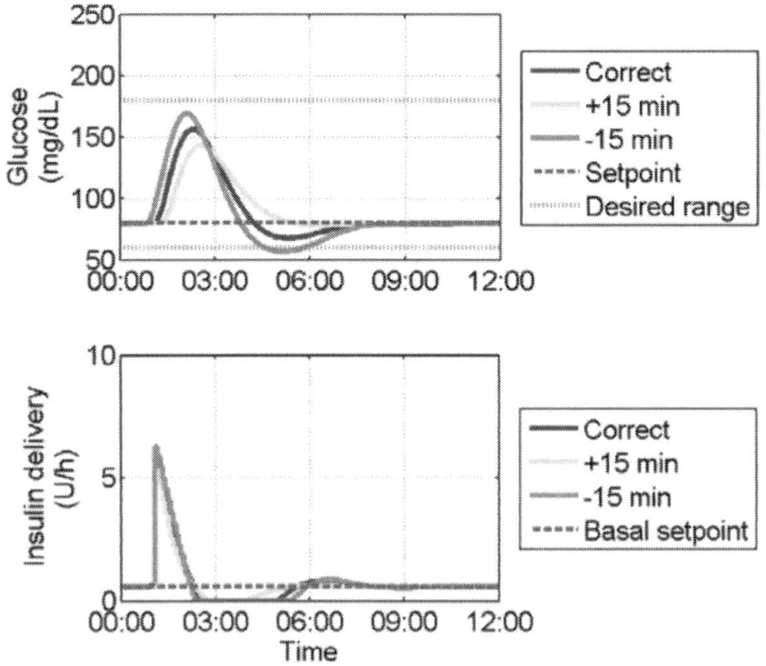

Figure 4: Closed-loop responses to an announced meal of 40 g CHO with the actual meal timing varied by ± 15 minutes. TOP: early consumption of the meal causes a minor hypoglycemic event four hours after meal time. BOTTOM: insulin delivery is initially the same in each case as a direct consequence of meal announcement; feedback from glucose measurements adjusts the following delivery slightly to compensate for the differences in expected and actual glycemic excursions.

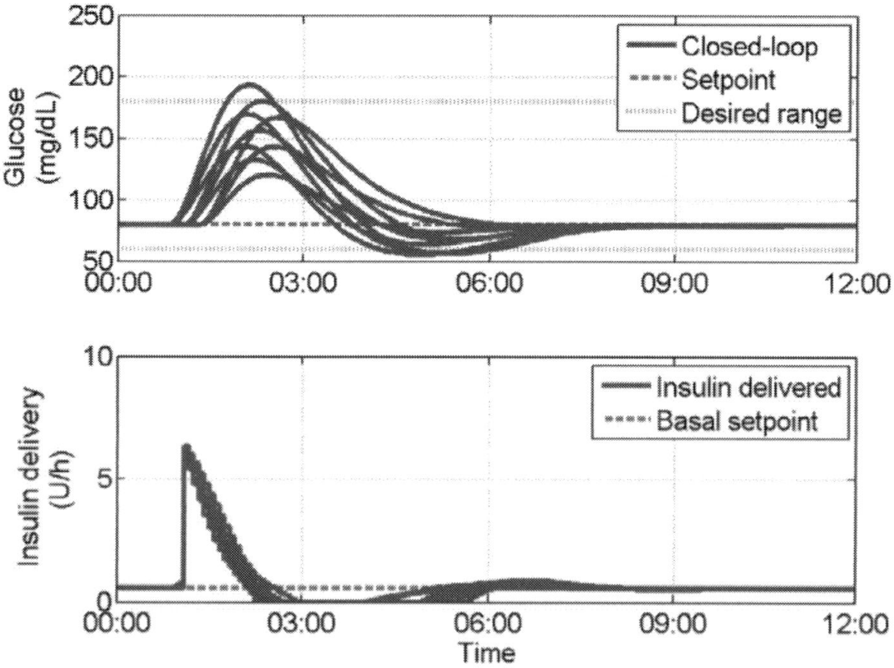

Figure 5: Closed-loop responses for when the mismatches of meal size and meal time are combined. TOP: in some cases minor hyperglycemic excursions and hypoglycemic events occur. BOTTOM: insulin delivery remains similar in all cases since the controller tuning is conservative, thus the major insulin delivery peak is due to the meal announcement.

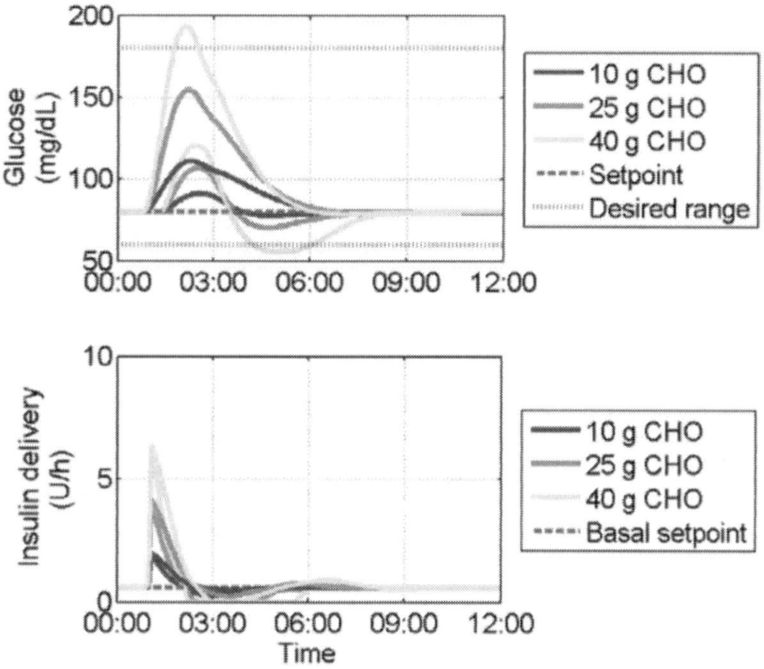

Figure 6: Maxima and minima of closed-loop responses under uncertainty for meals ranging from 10, 25, and 40 g CHO. TOP: the limits of the glucose trajectories show that the maximum meal size permissible in order to avoid hyperglycemia and hypoglycemia under these conditions of uncertainty is less than 40 g CHO.

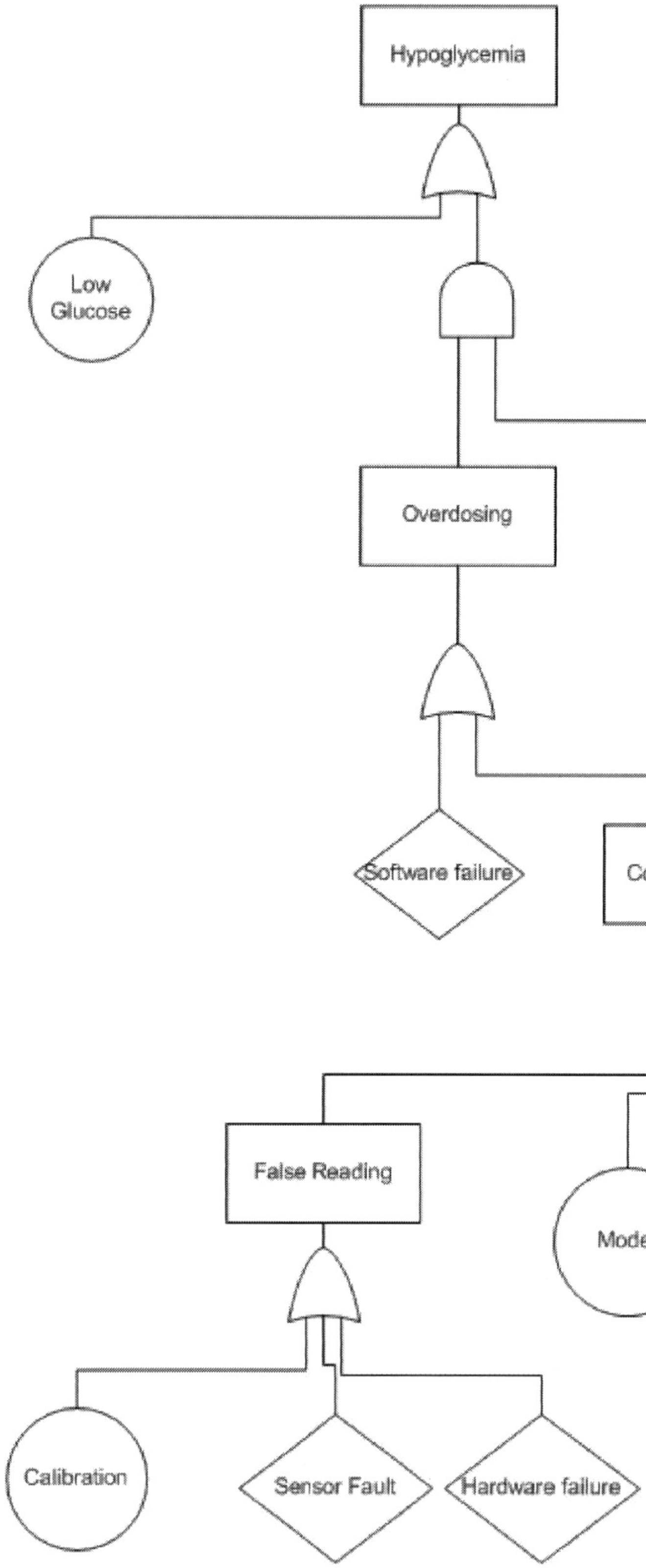

Figure 7: A simplified fault tree analysis for an artificial pancreatic beta cell with a top event defined as hypoglycemia. This analysis shows the causes of hypoglycemia by different components of the artificial pancreas.

Regulatory body approval of medical devices

An artificial pancreas (AP) for the regulation of glycemia is considered a high-risk device (class III) by regulatory bodies such as the FDA. As such, a thorough risk analysis and/or fault tree analysis (FTA) is required that demonstrates the safety of both the system and its subcomponents. In particular, the control algorithm and software that run the system need to demonstrate both safety and efficacy to be robust to both system disturbances other abnormalities [16, 17, 18].

The use of MPC as a controller of such a device opens the question of controller stability, for example, if constrained MPC is implemented or if an output constraint has been implemented there can be a situation of unsolvable optimization [19, 20]. As demonstrated on a simplified FTA (Figure 7), a top event defined as hypoglycemia can be developed due to overdosing of insulin that is the outcome of either software failure or a controller fault. MPC with real time optimization can therefore produce an error that can be a trigger for a catastrophic event.

The FDA has asked researchers conduct clinical trials to submit an Investigative Device Exemption (IDE) before clinical trials are carried out; this document should provide the necessary information for them to approve the device for clinical studies. As part of the requirements the control algorithm needs to be described in a way that will demonstrate its stability. This requirement is challenging if an optimization problem is solved in real time. One way to fulfill this requirement is to use unconstrained MPC or a PID algorithm, where controller output is then clipped in order to produce feasible instructions to the hardware; the control action is then no longer optimal. The use of mpMPC thus permits evaluation of controller stability and performance prior to any *in silico* evaluation of the algorithm based on an approve test simulator [21] and the IDE submission.

Conclusions

Development of a linear model from the Jacobian of the plant produced an efficacious control law. The use of a disturbance model is of paramount importance in this application, due to the large lag time for insulin action. Without such a model, controller performance is inadequate under conditions of model uncertainty similar to those likely to be experienced with intra-subject variation.

Evaluation of controller performance considering hypo- and hyperglycemia showed that these conditions could be avoided with meal sizes of up to 35 g CHO, even under the conditions of uncertainty expected. In the United States, a 70 kilogram man typically consumes over 70 grams of carbohydrate per meal plan. Thus this upper limit on meal size indicates that optimal therapy requires some form of patient adherence to the maximum amount of CHO consumed at one meal.

Offline optimization, although computationally expensive for high-order systems, has been shown to be a feasible technique for applying mpMPC to glycemic regulation. Indeed, there is an increase in the number of controller regions as the control law becomes more aggressive, and as more controller features, such as a reference trajectory, are included. For the controllers demonstrated here, the offline computation of the control law took a time of the order minutes.

Since the glucose-insulin kinetics in subjects with T1DM is known to be time-varying, regular updates of the offline optimization would be required to ensure a sufficiently accurate model. This would be feasible on a daily basis due to the time required for the offline optimization. Future work will include output constraints implemented through mixed integer programming, to satisfy hierarchical objectives on avoidance of hypoglycemic events and extended periods of hyperglycemia.

A distinctive feature of mpMPC is that it provides the necessary ability to ensure controller stability to different inputs as required by the FDA and demonstrated on several applications in [19] without the need to compromise the controller design to a less sophisticated one such as PID or unconstrained MPC. This implementation also provides means to analyze the algorithm before implementation, which is critical to satisfying FDA requirements concerning risk analysis.

Acknowledgements

This work was supported by the Juvenile Diabetes Research Foundation (JDRF) grant 22-2006-1115, the Institute for Collaborative Biotechnologies ICB, and the Otis Williams Fund at the Santa Barbara Foundation.

References

1. Hirsch IB, Brownlee M. Should minimal blood glucose variability become the gold standard of glycemic control? *J Diabetes Complications,* 2005; 19(3):178–181.
2. Eiselein L, Schwartz HJ, Rutledge JC. The challenge of type 1 diabetes mellitus. *ILAR J,* 2004; 45(3):231–236.
3. Bequette BW. A critical assessment of algorithms and challenges in the development of a closed–loop artificial pancreas. *Diabetes Technol Ther,* 2005; 7(1):28–47.
4. Hovorka, R. Continuous glucose monitoring and closed-loop systems. *Diabet Med,* 2005, *23,* 1-12
5. Clemens AH, Chang PH, Myers RW: The development of Biostator, a glucose controlled insulin infusion system (GCIIS). *Horm Metab Res*, 1977; S7:23–33.
6. Hovorka R, Wilinska ME, Chassin LJ, Dunger DB. Roadmap to the artificial pancreas. *Diabetes Res Clin Prac,* 2006, 74, S178-S182.
7. Hovorka R, Canonico V, Chassin LJ, Haueter U, Massi-Benedetti M, Federici MO, Pieber TR, Schaller HC, Schaupp L, Vering T, Wilinska ME. Nonlinear model predictive control of glucose concentration in subjects with type 1 diabetes. *Physiol Meas,* 2004; 25(4):905–20.
8. Garcia CE, Prett DM, Morari M. Model predictive control – theory and practice – a survey. *Automatica,* 1989; 25(3):335–348.
9. Bemporad A, Morari M, Dua V, Pistikopoulos EN. The explicit linear quadratic regulator for constrained systems. *Automatica,* 2002; 38(1):3–20.
10. General Principles of Software Validation; Final Guidance for Industry and FDA Staff. [Internet] Jan 11 2002; http://www.fda.gov/cdrh/comp/guidance/938.html.
11. Dua P, Doyle III FJ, Pistikopoulos EN. Model-based blood glucose control for type 1 diabetes via parametric programming. *IEEE Trans Biomed Eng*, 2006; 53(8):1478–1491.
12. Wilinska ME, Chassin LJ, Schaller HC, Schaupp L, Pieber TR, Hovorka R. Insulin kinetics in type 1 diabetes: continuous and bolus delivery of rapid acting insulin. *IEEE Trans Biomed Eng,* 2005; 52(1):3–12.
13. Pistikopoulos E, Georgiadis M, Dua V. Multi-parametric model-based control, *Wiley-VCH,* 2007.

14. Kvasnica M, Grieder P, Baotić M. Multi-Parametric Toolbox (MPT). 2004.
15. Löfberg J. YALMIP: A toolbox for modeling and optimization in MATLAB. *In proceedings of the CACSD Conference*. Taipei, Taiwan, 2004.
16. Fries RC, Reliable Design of Medical Devices: CRC Press, 2005.
17. Food and Drug Administration, Guide To Inspection Of Computerized Systems In Drug Processing, *Food and Drug Administration*, Washington, DC, 1983.
18. Food and Drug Administration, Guidance for the Content of Premarket Submissions for Software Contained in Medical Devices, *Food and Drug Administration*, Washington, DC, 2005.
19 Dua P, Kouramas K, Dua V, Pistikopoulos EN. MPC on a chip—Recent advances on the application of multi-parametric model-based control. *Computers & Chemical Engineering,* vol. 32, no. 4-5, pp. 754-765, April, 2008.
20. A. Bemporad, F. Borrelli, and M. Morari, On the optimal control law for linear discrete time hybrid systems, *Hybrid Systems: Computation and Control*, Lecture Notes in Computer Science, pp. 105-119, 2002.
21. Kovatchev BP, Breton MD, Dalla Man C, Cobelli C. In silico model and computer simulation environment approximating the human glucose/insulin utilization. *Food and Drug Administration Master File* MAF 1521, 2008.

Appendices

Appendix A: Graphical example of control law implementation

Consider an arbitrary second order linear state space model, with state vector given by

$$x = [x_1 \quad x_2]^T, \tag{A1}$$

where x_1 and x_2 represent the model states. The lookup table in this case is a two-dimensional polytope, with regions described graphically in Figure A1. The lookup table is searched until the region containing the state vector, x, is found. The next control move is calculated using the affine function associated with the current region using

$$I_D = Fx + g. \tag{A2}$$

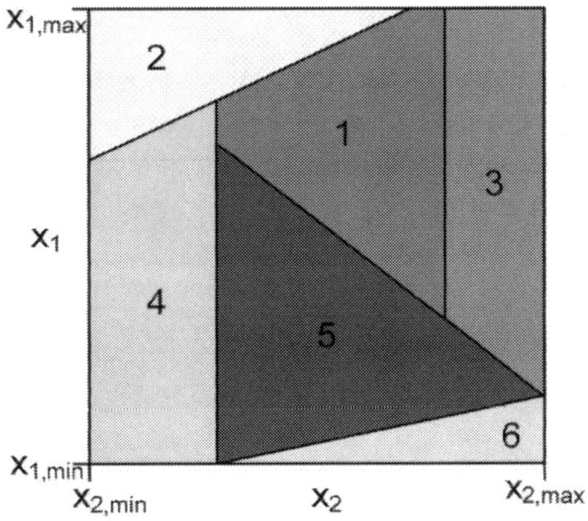

Figure A1: Graphical representation of the controller regions for a hypothetical controller across two states. Associated with each region is an affine function, representing the control law, which is evaluated at each time step using the state vector.

160b Developing Patient Specific Pharmacokinetic Models: Population Pharmacokinetic Analysis of Triazolam

Daniel Joseph Hines and Christos Georgakis. Chemical and Biological Engineering, Tufts University, 4 Colby Street, Medford, MA 02155

Prior population pharmacokinetic (PPK) modeling approaches have found limited success in developing dynamic models that accurately describe the variability that is observed in pharmacokinetic (PK) behavior among individual patients within a population. Many patient specific characteristics, for example, age, weight, or gender, are thought to be the significant factors that cause such variability between individuals. The overall goal of this research is to formulate a PK modeling methodology that will enable one to relate the variability of the model parameters to the appropriate patient specific characteristics. These models will benefit the field of pharmaceutical therapy in several ways: 1) the ability to tailor the administration of therapeutic drugs for the characteristics of a specific patient, and 2) gain insight that will advance the knowledge of the physiological processes that influence patient specific PK behavior.

In order to develop such a methodology the modeling of an example case has been undertaken. Drug plasma concentration vs. time data sets collected for 61 individuals who were orally administered a 0.5 mg dose of the common anesthetic, Triazolam, were provided by Dr. David Greenblatt of the Tufts School of Medicine. The individual factors: gender, age, weight, and height, were recorded for each of the patients to investigate the contribution of such factors to the observed inter-individual PK variability for this single occasion study.

Four different compartmental absorption models, quantifying the physiological processes relevant to drug adsorption, distribution, and excretion kinetics, were used to represent the PK data of each individual patient. The set of four models includes one-compartment and two-compartment models with and without a lag. Parameter estimation, using an ordinary or weighted least-squares approach along with constrained nonlinear optimization, resulted in satisfactory representation of the majority of the patient data. The models with a lag showed better representations of the data, and the one–compartment lag model was selected for further examination because of its smaller number of parameters. Partial Least Squares and Stepwise Multi-linear regression of linear and nonlinear terms is used to investigate the functional relationship between the model parameters and patient specific factors.

gmfELUCIDATING CHEMOTHERAPEUTIC DRUG TRANSPORT OF DIFFERENT AGENTS AND OF ANTIOANGIOGENIC THERAPY AGAINST A BRAIN TUMOR

Davis Yohanes Arifin, MEBCS, Singapore-MIT Alliance, 4 Engineering Drive 3, Singapore 117576
Kam Yiu Timothy Lee, Gleneagles Hospital Ltd, 6A Napier Road, Singapore 25850
Chi-Hwa Wang, Department of Chemical and Biomolecular Engineering, National University of Singapore, 4 Engineering Drive 4, Singapore, 117576
Kenneth A. Smith, Department of Chemical Engineering, Massachusetts Institute of Technology, Cambridge, MA 02139

Introduction

Treatment against brain tumors usually involves surgical removal, chemotherapy, and radiotherapy. For chemotherapy, the first FDA-approved treatment is the Gliadel® wafer (1). This wafer is able to provide sustained release of *carmustine* locally to the remnant tumor. However, results have been highly variable due to the significant loss of *carmustine* to blood capillaries, resulting in a limited (millimeter) penetration distance (2). This suggests that the cause of poor performance for this drug delivery device is related to the drug transport mechanism in the tumor tissue, not the local delivery concept.

The above example shows how drug transport properties are often overlooked. Though the drug is clinically effective against the tumor tissue, insufficient penetration will yield an ineffective therapy. Many new chemotherapeutic agents have been developed and might also be administered by local polymeric delivery. These drugs include *carmustine*, *paclitaxel*, *5-fluoroacil* (5-FU), and *methotrexate* (MTX). There is a need to analyze their transport mechanisms in brain tumor.

A related issue is the use of antiangiogenic agents to enhance the performance of chemotherapy. It has been hypothesized that this therapy could renormalize the tumor vasculature and, hence, improving the localization of the drug (3). This is achieved by increasing the drug convection within the tumor and decreasing the fluid convection out of the tumor margin (3). However, this concept needs to be further explored for the case of brain tumor, especially when surgery is performed.

To address these questions, a simulation analysis has been performed. This analysis has the advantages of being able to decouple and isolate different aspects of drug transport. Thus, we are able to elucidate (a) which chemotherapeutic agent is more favorable from a transport perspective, and (b) how the antiangiogenic agents could enhance the overall therapy. In addition, recent advances in patient-specific imaging techniques, e.g. magnetic resonance images (MRI) and computed tomography (CT), allow the extraction of brain tissue geometry as well as precise tumor size and location. Here, the simulation utilizes a three-dimensional geometry constructed from MRI of a brain tumor patient so that the possible effect of brain fluid flow can be examined.

Simulation and Mathematical Model

Patient-specific Model Geometry

A MRI of a brain tumor (courtesy of Brain and Spine Clinics, Gleneagles Hospital, Singapore) was reconstructed by Mimics 10 (Materialise Inc., 2006). This technique is able to reconstruct the brain geometry with a precision of +/- 3 mm (Fig. 1a). The tumor volume was 12.8 cc, giving an equivalent spherical mean radius (R) of 14.5 mm. The tumor was surgically removed, leaving a cavity in the core and a surrounding residual tumor with an average thickness (L) of 2.3 mm. Subsequently, eight wafers containing drug were inserted into the open cavity. Each is 14 mm in diameter and 1mm thick in thickness. The wafers are held in place by surgical cellulose (Surgicel). This is visualized in Fig. 1b.

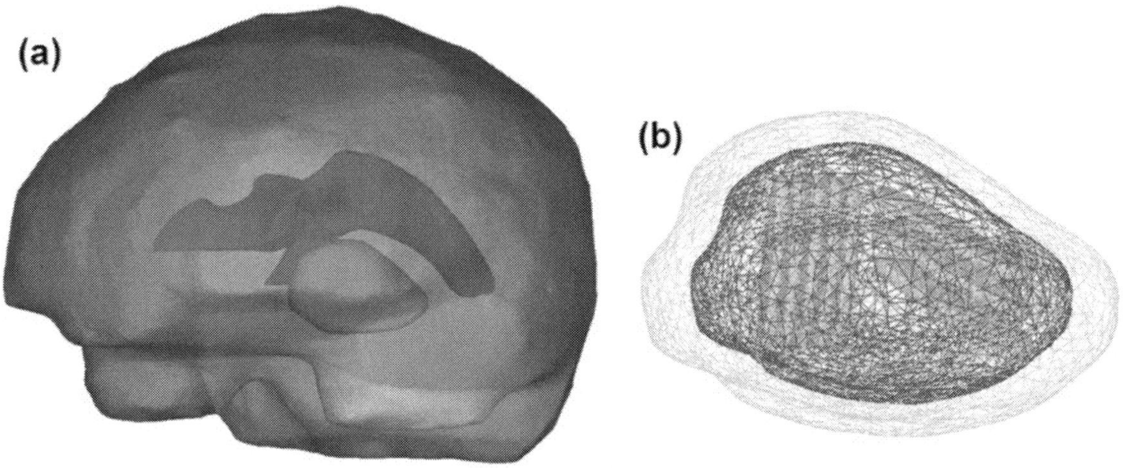

Fig. 1. (a) The model geometry used from a brain tumor patient, showing the presence of: ventricle (shown in blue), tumor (red) located in temporal lobe, and remaining normal brain tissue. (b) The 3D treatment domain with eight wafers containing drug (light green with blue line), cavity (dark red), and remnant tumor (gray).

Transport Model

The transport of interstitial fluid in the brain is described by coupling the modified continuity and momentum equations for fluid flow in a porous medium (4). The interstitial fluid is assumed to be water at 37 °C. Table I provides the baseline values of parameters related to the flow. The continuity equation for incompressible interstitial fluid in the brain tissue is:

$$\nabla \cdot \mathbf{v} = F_{\mathrm{V}} \tag{1}$$

where $\mathbf{v}$ is superficial fluid velocity vector and F_{V} is the rate of fluid gain from the capillary blood flow per unit volume of tissue. Fluid removal by the lymphatic system is not included because brain tissue lacks a well defined lymphatic system. The fluid gain is assumed to be a non-uniformly distributed source, depending on the pressure difference between blood vessels and interstitial fluid. The constitutive equation for F_{V} follows Starling's law (5):

$$F_{\mathrm{V}} = L_{\mathrm{p}}\left(\frac{S}{V}\right)\left[p_{\mathrm{V}} - p_{\mathrm{i}} - \sigma\left(\pi_{\mathrm{V}} - \pi_{\mathrm{i}}\right)\right] \tag{2}$$

where L_{p} is hydraulic conductivity of the microvascular wall, p_{V} is vascular pressure, p_{i} is interstitial fluid pressure, $\frac{S}{V}$ is available exchange area of the blood vessels per unit volume of tissue, σ is the osmotic reflection coefficient for plasma proteins, and π_{V} and π_{i} are osmotic pressures of blood plasma and interstitial fluid, respectively.

The brain tissue is assumed to be a rigid porous medium. The steady-state momentum equation for fluid flow through tissue is assumed to be:

$$-\nabla p_{\mathrm{i}} + \mu\nabla^2\mathbf{v} - \left(\frac{\mu}{K}\right)\mathbf{v} - \rho\mathbf{v}\cdot\nabla\mathbf{v} = 0 \qquad (4)$$

where ρ and μ are density and viscosity of the interstitial fluid and K is the Darcy permeability of the tissue.

The dimensionless forms of the equations of continuity and motion are then as follows:

$$\widetilde{\nabla}\cdot\tilde{\mathbf{v}} = \widetilde{F_{\mathrm{v}}} \qquad (5)$$

$$-\frac{\left(p_{\mathrm{ventricle}} - p_{\mathrm{outer}}\right)}{\rho v_{\mathrm{s}}^2}\widetilde{\nabla}\widetilde{p_{\mathrm{i}}} + \frac{1}{\mathrm{Re}}\widetilde{\nabla}^2\tilde{\mathbf{v}} - \frac{1}{\mathrm{Re}}\left(\frac{L_{\mathrm{brain}}^{\ 2}}{K}\right)\tilde{\mathbf{v}} - \tilde{\mathbf{v}}\cdot\widetilde{\nabla}\tilde{\mathbf{v}} = 0 \qquad (6)$$

where $\widetilde{\nabla} = L_{\mathrm{brain}}\nabla$ and L_{brain} is the mean radius of the brain, i.e. 70 mm, which has been chosen as the characteristic length scale), $\tilde{\mathbf{v}} = \mathbf{v}/v_{\mathrm{s}}$ ($v_{\mathrm{s}} = (K/\mu)\left(p_{\mathrm{ventricle}} - p_{\mathrm{outer}}\right)/L_{\mathrm{brain}} = 1\times10^{-7}$ m/s is the velocity scale), $\widetilde{p_{\mathrm{i}}} = \left(p_{\mathrm{i}} - p_{\mathrm{outer}}\right)/\left(p_{\mathrm{ventricle}} - p_{\mathrm{outer}}\right)$, and $\mathrm{Re} = \rho L_{\mathrm{brain}} v_{\mathrm{s}}/\mu$ is Reynolds number. Insertion of parameter values from Table I leads to the conclusion that Eq. 6 can be further simplified to:

$$\tilde{\mathbf{v}} = -\widetilde{\nabla}\widetilde{p_{\mathrm{i}}} \qquad (7)$$

Table I. Interstitial Fluid-related Parameter Values

Parameter		Cavity	Remnant Tumor	Normal Tissue
ρ	Density of the interstitial fluid (kg/m^3)		1,000 (7)	
μ	Viscosity of the interstitial fluid (Pa-s)		7×10^{-4} (7)	
p_{v}	Vascular pressure (Pa)	N/A	4,610 (6)	
S/V	Blood vessel exchange area (m^{-1})	N/A	20,000 (5)	7,000 (5)
π_{v}	Vascular osmotic pressure (Pa)	N/A	3,440 (12)	
π_{i}	Interstitial osmotic pressure (Pa)	N/A	1,110 (5)	740 (5)
σ	Osmotic reflection coefficient of plasma	N/A	0.82 (5)	0.91 (5)
L_{p}	Hydraulic conductivity (m/Pa/s)	N/A	1.1×10^{-12}	1.4×10^{-13}
K	Darcy's permeability (m^2)	1×10^{-11}	6.4×10^{-14}	6.4×10^{-15}

In the tissue, drug is always present in the two different forms: free and bound; the latter is usually not subject to metabolism (8). Furthermore, brain tissue consists of three phases: interstitial/extracellular space (ECS), intracellular space (ICS), and cell membrane (CM) (9-11). Both forms of drug are distributed between these three phases and can be expressed as the following:

$$C = \alpha C_{\mathrm{i}} + \beta C_{\mathrm{c}} + \left(1 - \alpha - \beta\right)C_{\mathrm{m}} \qquad (8)$$

$$B = \alpha B_{\mathrm{i}} + \beta B_{\mathrm{c}} + \left(1 - \alpha - \beta\right)B_{\mathrm{m}} \qquad (9)$$

where C is the free drug concentration per total brain volume; C_{i}, C_{c}, and C_{m} are the free drug concentrations per unit volume in the interstitial, intracellular, and membrane spaces; while α and β are the volume fractions of the interstitial and cellular spaces. The definition of bound drug is analogous to that for the free drug: B is the free drug concentration per total brain volume; B_{i}, B_{c}, and B_{m} are the bound drug concentrations per unit volume in the interstitial, intracellular, and membrane spaces. It is further assumed that the drug is neither eliminated nor bound in the membrane phase ($B_{\mathrm{m}} = 0$).

In tissue, drug availability is dependent on the rates of transport (diffusion and convection), elimination (via blood capillaries and metabolism/degradation), and local binding (9-11). In interstitial phase, the protein/fat-bound drug is assumed to be locally immobilized as it tends to immediately interact with cell/membrane surfaces due to its lipophilicity; therefore, only unbound drug is available for transport (8). Thus, the drug conservation equation, to be coupled to the flow equations, can be described as follows (9-11):

$$\frac{\partial C}{\partial t} = \alpha D_i \nabla^2 C_i - \nabla \cdot (\mathbf{v} C_i) - \alpha k_{bbb} C_i - \beta k_e C_c - \frac{\partial B}{\partial t} \quad (10)$$

where t is time, D_i is drug diffusivity in the interstitial space, k_{bbb} is the first-order elimination constant for drug transport through the BBB, and k_e is the enzymatic/non-enzymatic elimination constant due to metabolism/degradation. Table II depicts all parameter values related to the drug.

There are several assumptions made to simplify the drug transport equation as the following: (i) the free drug concentrations in interstitial (C_i) and intracellular phase (C_c) are directly proportional to those of bound drug (B_i and B_c), respectively, so that $K_i = B_i/C_i$ and $K_c = B_c/C_c$ where K_i and K_c are binding constant between bound and free drug in interstitial and intracellular space, respectively; and, (ii) while the membrane phase is inert, local equilibrium is present between interstitial, intracellular, and membrane phases so that $P_{c/i} = C_c/C_i$ and $P_{m/i} = C_m/C_i$, where $P_{c/i}$ is partition coefficient between intracellular and interstitial space, while $P_{m/i}$ is partition coefficient between cell membrane and interstitial phase. Note that, with these assumptions and transport data from Table II, the total concentration (C) becomes proportional to the concentration in the interstitial phase (C_i): $C = nC_i$, where n is a constant.

Table II. Drug-related parameter values

Parameter		BCNU	Taxol	5-FU	MTX
M_w	Molecular weight (kg/kmol)	214.06	853.9	130.08	454.44
$C_{i,eff}$	Effective concentration in interstitial phase (kg/m3)	1.3E-03(11)	7.6E-04(11)	2.6E-04(16)	2.7E-02(16)
$P_{c/i}$	Partition coefficient between cellular and interstitial phase	1(10)	1(10)	1(10)	1(10)
$P_{m/i}$	Partition coefficient between membrane and interstitial phase	10.3(10)	3162.3(11)	0.1(13)	0.01(13)
K_i, K_c	Binding constant between free and bound drugs in interstitial and cellular phase (assume K_i=K_c)	5(10)	5.1(12)	0.1(14)	0.7(15)
D_i	Diffusion coefficient in interstitial phase (m^2/s)	1.5E-09(10)	9.0E-10(11)	1.2E-09(13)	5.3E-10(13)
k_{bbb}	Drug elimination to blood capillaries (1/s)	1.4E-02(10)	1.4E-04(11)	1.8E-02(13)	2.8E-04(13)
k_e	Drug elimination due to enzymatic/non-enzymatic reactions (1/s)	1.1E-04(10)	6.8E-07(11)	5.6E-04(13)	1.5E-04(13)

Furthermore, the convection term, i.e. $\alpha \nabla \cdot (\mathbf{v} C_i)$, is comprised of two sub-terms: first containing the gradient of concentration is the usual convective term ($\alpha \mathbf{v} \cdot \nabla C_i$), while the second containing the gradient of fluid velocity is the dilution term ($\alpha C_i \nabla \cdot \mathbf{v}$). The latter can be simplified by Eq. 1 to become a sort of first-order elimination term, i.e. $\alpha F_v C_i$. Finally, the drug conservation equation becomes as follows:

$$\frac{\partial C_i}{\partial t}=\left(\frac{\alpha}{\alpha^*}\right)D_i\nabla^2C_i-\left(\frac{1}{\alpha^*}\right)\mathbf{v}\cdot\nabla C_i-\left(\frac{\alpha k_{bbb}+\beta k_e+\alpha F_v}{\alpha^*}\right)C_i \quad (11)$$

where $\alpha^*=\alpha(1+K_i)+\beta P_{c/i}(1+K_e)+(1-\alpha-\beta)P_{m/i}$ acts as a retardation constant for transport due to local binding and cell membrane partitioning. In the simplest form, this equation follows the simple diffusion/convection/reaction equation as follows:

$$\frac{\partial C_i}{\partial t}=D\nabla^2C_i-\mathbf{v}^*\cdot\nabla C_i-kC_i \quad (12)$$

where $D=(\alpha/\alpha^*)D_i$ is the apparent diffusion coefficient in the brain tissue, $\mathbf{v}^*=\mathbf{v}/\alpha^*$ is the apparent interstitial fluid velocity vector in the brain tissue, and $k=(\alpha k_{bbb}+\beta k_e+\alpha F_v)/\alpha^*$ is the apparent first-order elimination constant. In the dimensionless form, Eq. 12 can be rewritten as the following.

$$\frac{\partial \Gamma_i}{\partial \tau}=\tilde{\nabla}^2\Gamma_i-\text{Pe}\,\tilde{\mathbf{v}}\tilde{\nabla}\cdot\Gamma_i-\phi^2\Gamma_i \quad (13)$$

where $\Gamma_i=C_i/C_{i,eff}$ is dimensionless interstitial free drug concentration relative to the effective therapeutic concentration ($C_{i,eff}$), $\tilde{\mathbf{v}}=\mathbf{v}/v_s$ is the dimensionless interstitial fluid velocity, $\text{Pe}=v_sL_d/\alpha^*D$ is Peclet number that measures the relative importance of drug convection over diffusion in tissue, $\phi=L_d\sqrt{k/D}$ is Thiele modulus that is the ratio of drug elimination rate to diffusion transport, $L_d=\sqrt{D/k}$ is the diffusion/reaction length scale in the remnant tumor, and $\tau=t/k$ is the dimensionless time relative to the first-order elimination constant in the remnant tumor (k). The role of convection is determined by the magnitude of Pe. Note that $\phi=1$ as we standardize to the remnant tumor; in other domains, these dimensionless values will be different accordingly.

Drug release from the wafers

The drug release from the wafers is assumed to be an ideal constant (zero-order) rate, in which 61.6 mg of drug is released in a period of 30 days. Note that the different release mechanism of the drug is not of primary interest in this study.

With this constant release profile, the drug distribution will reach a quasi-steady state after a certain period of initial time. We are interested to study the penetration at this condition where the treatment prevails for most of the release period. Consequently, the quasi-steady form of Eq. 13 becomes:

$$0=\tilde{\nabla}^2\Gamma_i-\text{Pe}\,\tilde{\mathbf{v}}\tilde{\nabla}\cdot\Gamma_i-\phi^2\Gamma_i \quad (14)$$

Influence of the antiangiogenic therapy to interstitial transport

It is hypothesized that the influence of antiangiogenic therapy is to normalize the tumor vasculature (3). The normalized vasculature leads to a normalized fluid permeation of fluid from the vessels. This is mathematically modeled as the decrease of the hydraulic conductivity (L_p) and the blood vessel density (S/V) in the tumor; both parameters appear together as in the Starling's law, $L_p(S/V)$ (Eq. 2). In an ideal case, the tumor vasculature can be normalized perfectly (as in the normal tissue), the value will be about 21% (m) of its original value ($L_p(S/V)|_{normalized}=mL_p(S/V)|_{tumor}$). However, the normalization can rarely be ideal; therefore, this model accounts for the maximum benefit can be gained from the antiangiogenic therapy.

Results and Discussion

Minimal Interstitial Fluid Flow Changes by Antiangiogenic Treatment

Depending on the physiochemical properties of the drug, the role of convective flow to distribute the drug can be important. Therefore, it is important to understand the behavior of the interstitial fluid flow in the treatment domain, especially to account the influence of the antiangiogenic therapy. This is shown in Fig. 2. It basically shows that there is no significant difference in fluid flow under antiangiogenic therapy. In other words, the effect of the treatment to the interstitial fluid flow in the treatment domain is minimal. With the treatment, the mean interstitial pressure in the remnant tumor decreases minimally from 1,104 to 1,062 Pa while the mean fluid velocity remains at 3.18×10^{-7} m/s, confirming that the interstitial fluid flow is not affected by the treatment. Consequently, the drug distribution will no longer influenced by the treatment. This is because that upon the tumor removal surgery the volume portion of tumor is no longer significant; therefore, any changes related to the tumor permeation would be quite decent. At the same time, this also tells that the treatment might be best when big portion of tumor is still available.

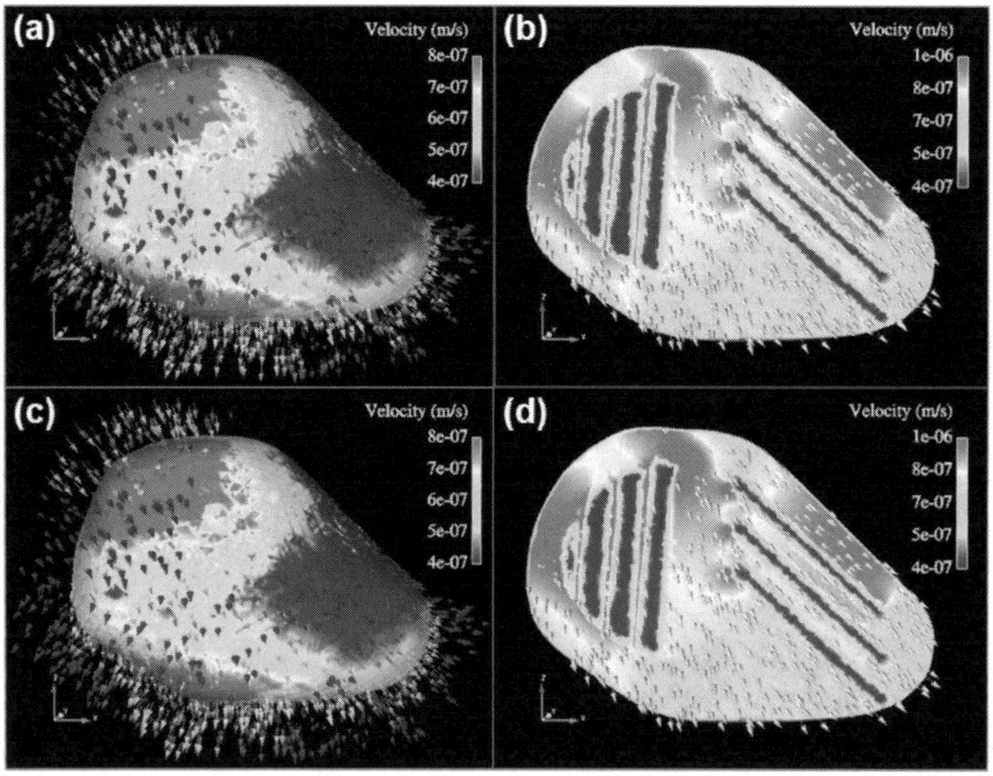

Fig. 2. The fluid flow around and inside the cavity domain for normal case (a and b, respectively) and for the case with antiangiogenic treatment (c and d, respectively). It shows that the antiangiogenic treatment provides minimal effect to the interstitial fluid in the treatment domain. As a result, the treatment will help very minimally the drug distribution to the remnant tumor.

Drug Distribution in the Cavity

While the antiangiogenic therapy might not help the drug distribution to the remnant tumor when surgery is performed, the way drug distributes still depends on its physiochemical properties. Fig. 3 shows the distribution of different chemotherapeutic agents on the cavity surface and its penetration to the remnant tumor (surrounding the cavity). In all cases, the drug concentration profile is along the pressure gradient between the ventricle and the brain periphery – the domain near brain periphery has a relatively higher drug concentration due to the direction of convective flow built by the pressure gradient between the ventricle and the brain periphery.

The physiochemical property of drug in the cavity leads to the availability of the drug on the cavity/remnant tumor interface. Higher drug concentration is observed for the case of *carmustine* and *paclitaxel* than those of 5-FU and MTX. This is because of low Thiele modulus ($\Phi = 0.08$ and 0.05) in the cavity for *carmustine* and *paclitaxel*, respectively. On the other hand, 5-FU and MTX has relatively higher Thiele modulus ($\Phi = 0.17$ and 0.49, respectively). In other words, for the former case (*carmustine* and *paclitaxel*), the elimination is minimal as compared to the drug transport rate by diffusion. In contrast, the elimination is more comparable to the diffusion as for 5-FU and MTX.

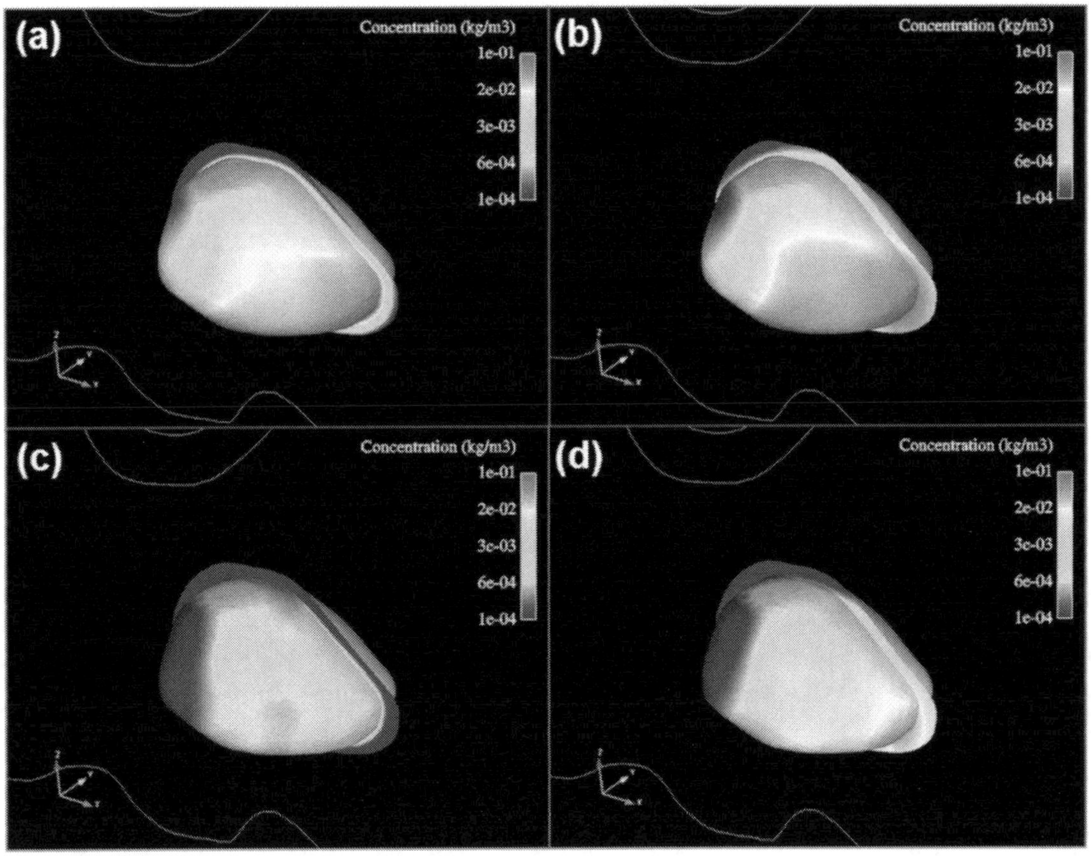

Fig. 3. Drug concentration contour on the surface of cavity and its surrounding remnant tumor (represented by a coronal plane in the middle section of the cavity), showed for four different chemotherapeutic agents: (a) *carmustine*, (b) *paclitaxel*, (c) 5-FU, and (d) MTX.

The convection appears to be more dominant for *paclitaxel* and MTX (Pe = 0.20 and 0.17, respectively) as compared to *carmustine* and 5-FU (Pe = 0.02 for both agents). This is the reason why MTX has higher interface concentration than 5-FU despite the former experiences higher Φ). This also explains the higher concentration on the interface for *paclitaxel* than *carmustine*. It is also important to note that the direction of convection based on the pressure gradient between the ventricle and brain periphery; thus, the drug will be more concentrated to the right bottom region when the convective transport is more dominant, as in the case of *paclitaxel* and MTX.

Drug penetration to the remnant tumor

Drug penetration to the remnant tumor is based on diffusion and convection. Among four, *paclitaxel* and MTX are the agents which are susceptible to the convective flow (Pe = 0.58 and 0.49, respectively). In contrast, the influence of convective flow can be considered minimal for *carmustine* and 5-FU (Pe = 0.06 for both agents). This suggests that both diffusion and convection are important for *paclitaxel* and MTX while only diffusion is important for *carmustine* and 5-FU. However, drug is also being eliminated when it penetrates the tumor. Hence, the drug penetration is again influenced by its diffusion/reaction length scale (L_d). Higher length scale indicates that drug can penetrate deeper to the tumor, and vice versa. As shown in Fig. 3, relative to the drug concentration on the cavity/remnant tumor interface, *paclitaxel* and MTX (L_d = 1.8 and 0.9 mm, respectively) penetrate farther than *carmustine* and 5-FU (L_d = 0.3 and 0.2 mm, respectively). The limited penetration of the latter two is primarily due to their rapid capillary elimination (high k_{bbb}).

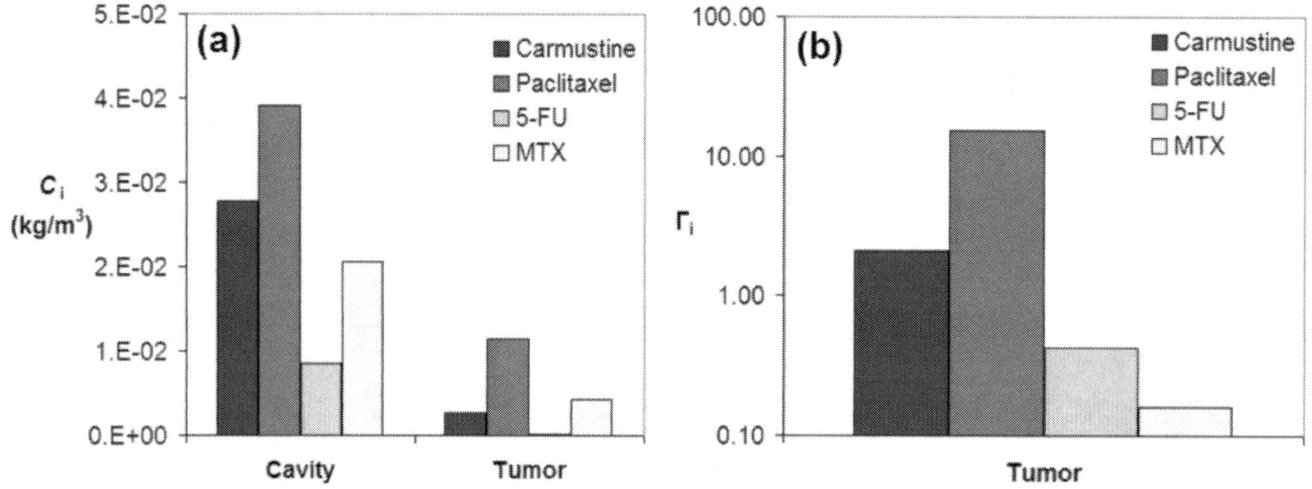

Fig. 4. The availability of each chemotherapeutic agent for a given dosage under constant release (61.6 mg continuously released in 30 days). (a) Shown is the mean concentration of the drug in the cavity and the remnant tumor. (b) Shown relative to the effective concentration in the tumor ($\Gamma_i = C_i / C_{i,eff}$).

The drug availability in the remnant tumor is quantitatively compared by its mean concentration. This is shown in Fig. 4a. In the cavity, the highest concentration is obtained by *paclitaxel*, followed by *carmustine*, MTX, and 5-FU. As discussed previously, this is because of the combination of diffusion and convection relative to elimination in the cavity. In the remnant tumor, *paclitaxel* is still available at the highest concentration. Interestingly, MTX surpasses *carmustine*. This is because, unlike MTX, *carmustine* is carried away significantly by capillary elimination. The availability of 5-FU also suffers due to the significant capillary elimination, and it is shown as the lowest available in the remnant tumor. However, the efficacy of each drug against brain tumor depends on its effective concentration ($C_{i,eff}$). Fig. 4b shows the normalized concentration relative to its effective concentration in the tumor ($\Gamma_i = C_i/C_{i,eff}$). For the same current dosage, *paclitaxel* again shows it superiority against other agents. The mean concentration in the tumor is about fifteen times of its effective concentration. This is followed by *carmustine*, 5-FU, and MTX, which provides about two, two-fifth, and one-fifth times of its individual effective concentration, respectively. Note that this prevails particularly for the dosage used in this study (61.6 mg of drug constantly released in 30 days). High Γ_i (>1) simply indicates that drug can penetrate well and be available above its effective concentration in the tumor. Consequently, lower dosage might be preferred to avoid any toxicity to the surrounding normal tissue. On the other hand, low Γ_i (<1) suggests the opposite: the drug penetrates poorly and is not available at sufficient concentration in the tumor. Consequently, higher dosage will be required to achieve a satisfactory treatment.

CONCLUSIONS

Clinical Implications

This study shows that the distribution of the drug in the brain tumor treatment relies on its physiochemical properties. Hence, while the therapeutic efficacy against the disease is essential, ones also need to carefully consider the amount of dosage and its release mechanism in order to achieve optimal drug delivery strategy. In this case, computer simulation has shown its advantages to predict the drug distribution for various scenarios. In addition, we predict that the effect of antiangiogenic therapy to enhance drug penetration in the case of brain tumor with resection will be quite minimal.

References

1. H. Brem and P. Gabikian (2001), Biodegradable polymer implants to treat brain tumors Journal of Controlled Release 74:63-67.
2. A.B. Fleming and W.M. Saltzman (2002), Pharmacokinetics of the carmustine implant. Clinical Pharmacokinetics 41:403-419.
3. R.K. Jain, R.T. Tong, L.L. Munn. Effect of vascular normalization by antiangiogenic therapy on interstitial hypertension, peritumor edema, and lymphatic metastasis: Insights from a mathematical model. Cancer Research 67:2729-2735.
4. J.F. Gross and A.S. Popel (1979), Mathematical models of transport phenomena in normal and neoplastic tissue In H.I. Peterson (ed.), Tumor Blood Circulation, CRC Press, Baca Raton, pp. 169-183.
5. L.T. Baxter and R.K. Jain (1989), Transport of fluid and macromolecules in tumors. I. Role of interstitial pressure and convection. Microvascular Research 37:77-104.
6. H.K. Kimelberg (2004), Water homeostasis in the brain: Basic concepts, Neuroscience 120:852-860.
7. R.H. Perry (1997), Perry's Chemical Engineers' Handbook (Ed.), McGraw-Hill, New York, pp. 59-71.
8. D.E. Golan, A.H. Tashjian, E. Armstrong, J.M. Galanter, A.W. Armstrong, R.A. Arnaout, H.S. Rose (2008), Principles of Pharmacology: The Pathophysiologic Basis of Drug Therapy, Lippincot Williams & Wilkins, Baltimore MD, pp. 58-59.
9. A.B. Fleming and W.M. Saltzman (2002), Pharmacokinetics of the *carmustine* implant. Clinical Pharmacokinetics 41:403-419.

10. L.K. Fung, M. Shin, B. Tyler, H. Brem, and W.M. Saltzman (1996), Chemotherapeutic drugs released from polymers: distribution of *1,3-bis(2-chloroethyl)-1-nitrosurea* in the rat brain. Pharmaceutical Research 13:671-682.
11. L.K. Fung, M.G. Ewend, A. Silis, E.P. Sipos, R. Thompson, M. Watts, O.M. Colvin, H. Brem, and W.M. Saltzman (1998), Pharmacokinetics of interstitial delivery of *carmustine*, *4-hydroperoxycyclophospamide*, and *paclitaxel* from a biodegradable polymer implant in the monkey brain. Cancer Research 58:672-684.
12. H.-J. Kuh, S.H. Jang, M.G. Wientjes, J.L.-S. Au (2000), Computational Model of Intracellular Pharmacokinetics of Paclitaxel. The Journal of Pharmacology and Experimental Therapeutics 293:761-770.
13. W.M. Saltzman and M.L. Radomsky (1991), Drugs released from polymers: diffusion and elimination in brain tissue. Chemical Engineering Science 46:2429-2444.
14. D.S. Wishart, C. Knox, A.C. Guo, S. Shrivastava, M. Hassanali, P. Stothard, Z. Chang, J. Woolsey, (2006) Drug Bank: A comprehensive resource for in silico drug discovery and exploration, Nucleic Acids Research 34:D668-672.
15. M.B. De Sousa Maia, S. Saivin, E. Chatelut, M.F. Malmary, G. Houin (1996), In vitro and in vivo protein binding of methotrexate assessed by microdialysis, International Journal of Clinical Pharamacology and Therapeutics 34:335:341.
16. C.M. Hand, J.R. Vender, P. Black (1988), Chemotherapy in experimental brain tumor, part 1: In vitro calorimetric MTT assay, Journal of Neuro-Oncologty 36:1-6.

160d Modeling the Effect of Chemotaxis In Glioblastoma Tumor Progression

Francisco G. Vital-Lopez[1], Antonios Armaou[1], Michelle Hutnik[2], and Costas D. Maranas[1]. (1) Department of Chemical Engineering, The Pennsylvania State University, University Park, PA 16802, (2) Medical Science Department, The Angiogenesis Foundation, 124 Mount Auburn Street, Suites 200N, Cambridge, MA 01238

The American Cancer Society estimates that 21,810 patients will be diagnosed with brain and other central nervous system cancers and 13,070 patients with this disease will die in the US during 2008. The most common and lethal form of this disease is glioblastoma multiforme (grade IV in the World Health Organization classification). Glioblastomas are intracranial neoplasms with uncontrolled proliferation, generally with a necrotic core, marked angiogenesis, bizarre infiltrating ability and high resistance to radio- and chemo-therapy. In developing therapeutics for this disease, it is critical to understand the intricate relationship between the many biological processes that take place during tumor progression.

We present a mathematical, agent-based, model that describes the progression of a brain tumor by capturing the interplay between processes occurring at the intracellular and tissue levels. Specifically, we focus on the temporal-spatial distribution of key biochemical cues (e.g., nutrient, growth factors) and the intracellular signaling pathways (e.g., PI3K-Akt and MAPK pathways) underlying the tumor growth, tumor cell migration and angiogenesis processes. We employ the model to explore the effect of possible migration mechanisms on the growth and invasion properties of the tumor. Simulation results show that the model is able to recapitulate in silico the morphology of brain tumors. We believe that the proposed modeling base (after further development and refinement) could ultimately help identify spatio-temporal dependent targets for tumor suppression that may otherwise remain concealed.

160e Investigation of Hemodynamics In Saccular Aneurysms

Marjan Rafat[1], Martin Heller[2], Howard A. Stone[1], and Debra T. Auguste[1]. (1) School of Engineering and Applied Sciences, Harvard University, 29 Oxford St., Cambridge, MA 02138, (2) Micro- and Nanotechnology, Technical University of Denmark, Building 345 East, DK-2800, Kongens Lyngby, Denmark

The mechanisms behind aneurysm growth and rupture are unknown. An increased understanding of hemodynamics, mechanical forces, and biological response within an aneurysm may enable aneurysm prediction, improve diagnosis, and allow development of new treatment strategies. We analyzed flow patterns in clinically relevant aneurysm geometries using the finite-element method to solve the incompressible Navier-Stokes equations numerically. Our analysis consists of a two-dimensional, steady-state flow over a range of Reynolds numbers ($0 < Re < 500$), which accounts for patient variation in blood pressure (velocity) and hematocrit (viscosity). The aneurysm geometry was varied systematically using height to width ratios. In addition, we studied aneurysms placed at the midpoint of a bifurcation, offset from the midpoint of a bifurcation, and on a lateral vessel. We observe the formation of symmetric and asymmetric flow profiles, with the appearance of closed and open eddies. The distinguishing features of the different flows are mapped as a function of Reynolds number and geometry. This work evaluates flow patterns for different aneurysm geometries, which may dictate effects caused by mass transfer and mechanical stress on the biological response. Small changes in the geometry or Re could result in distinct changes in the hemodynamic profile, which may (1) correlate with an increased incidence of rupture or (2) be used to manipulate the biological response to impede rupture or growth. Our nondimensional analysis may be broadly applied to a wide range of aneurysms and patient specific cases.

160f Effects of Bone Morphogenetic Protein Distribution and Shear Stress on 3-D Human Mesenchymal Stem Cell Construct Development In a 3-D Perfusion Bioreactor System

Teng Ma, Feng Zhao, Katelyn Sellgren, and Ravindran Chella. Chemical and Biomedical Engineering, Florida State University, 2525 Pottsdamer Street, Tallahassee, FL 32310

The process of creating functional 3-D tissues constructs in the bioreactor system using stem cells and scaffolds is influenced by the dynamic culture environment characterized by dynamic stresses and concentration gradients of essential nutrients and regulatory molecules. Among these microenvironmental parameters, media flow and molecule distributions are crucial regulators of 3-D hMSC tissue development. Our previous study has found that hMSCs actively respond to shear stresses at a range of 1×10^{-5} to 1×10^{-4} Pa, orders of magnitude lower than those observed on planar surfaces (Zhao et al., 2007). However, media flow also influences macromolecule transport and these two effects are strongly coupled in the 3-D constructs. In this study, we have combined the experimental and modeling approaches to delineate the effects of shear stress and distribution of bone morphogenetic proteins (BMP) by operating the modular perfusion bioreactor system developed in our lab (Zhao et al., 2005) in either transverse or parallel perfusion modes. BMPs are members of the TGF-β superfamily and are autologously secreted by hMSC, which have been demonstrated to be potent inducers of its osteogeneic differentiation. First, the effects of media flow on the developmental characteristics of hMSCs grown in 3-D poly (ethylene terephthalate) (PET) matrices were determined. The two perfusion modes have different hydrodynamic characteristics but identical volumetric flowrate of 0.2 mL/min, residence time, and initial cell seeding population. A 1.6 times higher proliferation rate, higher CFU-F formation, and stem cell gene expression at day 20 were observed in parallel perfusion in comparison with those under transverse perfusion. The transverse flow also upregulated the osteogenic differentiation potential at day 20 as measured by the osteonectin gene expression and calcium deposition in the matrices. In addition, ECM of hMSCs was patterned differently under the two perfusion modes, in which transverse perfusion directed a graded ECM distribution along the flow direction, as opposed to a denser and more uniform spatial patterning of ECM on both sides of 3-D constructs dictated by parallel perfusion flow mode. Hydrodynamic modeling results showed cells in transverse flow mode were exposed to media flow at ~ 3 μ m/s and shear stress in the range of 1×10^{-3} Pa. Concentration gradients of BMPs and the fraction of bound receptors in the constructs under various flow conditions are modeled to determine the spatial distribution of BMPs on hMSC proliferation and osteogenic differentiation. The spatial distribution of alkaline phosphatase in 3-D constructs will be determined to experimentally analyze the cellular response to long-term accumulation of BMP under various flow conditions. Combining modeling and experimental results will not only help to delineate the effects of the biomechanical microenvironment and macromolecule transport but also improve tissue engineering strategy using perfusion bioreactor systems.

Zhao F, Chella R and Ma T. Effects of shear stress on 3-D human mesenchymal stem cell construct development in a versatile perfusion bioreactor system: experiments and hydrodynamic modeling.

Biotechnology and Bioengineering, 2007;96:584-595.

Zhao F and Ma T. Perfusion bioreactor system for human mesenchymal stem cell tissue engineering: dynamic cell seeding and construct development. Biotechnology and Bioengineering, 2005;91:482-493.

160g Modeling Intracellular Dynamics of HIV Infection and Treatment

Samira Khalili and Antonios Armaou. Chemical Engineering, The Pennsylvania State University, University Park, PA 16802

Human Immunodeficiency Virus, the causative agent of acquired immunodeficiency syndrome (AIDS), has been classified in retrovirus family, and specifically in lentivirus group. Retroviruses are enveloped RNA viruses with specific characteristics and their replication cycle includes some uncommon features. The unusual morphological characteristics (cylindrical or cone-shape nuclei), several genes with regulatory function, and a biphasic course of gene expression distinguish lentiviruses form other retroviruses. HIV distinctively encodes six auxiliary genes (Tat, Rev, Nef, Vpr, Vif, and Vpu) with specific roles in infection cycle, leading to a complicated yet robust sequence of events that eventually leads to the creation of new virus particles and the lysis of the infected T-cell.

Like other biological systems, dynamics of HIV infection is very complicated. There are several mathematical models in the open literature that describe different aspects of the infection [1, 2, 3]; however most of them focus on the extracellular components of the infection. The interaction dynamics of intracellular components are extremely complicated and on the other hand, quantitative information from experimental studies does not include all parts of the cycle. These obstacles make developing intracellular mathematical models of HIV very difficult. In [4], a model for the entire infection cycle is proposed. However, some stages don't have the necessary level of detail. In other works [5, 6] precise models have been proposed, yet for a specific part of the cycle.

In this work we propose an intracellular model which covers the complete infection cycle yet is detailed enough that can provide a good insight into the HIV infection cycle. The current model includes feedback loops to capture the regulatory effects of important proteins such as Tat and Rev. Furthermore, this detailed model provides information about dynamics of viral components which can improve drug administration scheduling. For instance, Reverse Transcriptase Inhibitor (RTI) is most effective before the production of double stranded DNA from viral RNA. Decision about drug dosage and administration can be made with the knowledge of components trajectories over the time provided by this model. Ultimately, this model can be linked to a previously developed extracellular model [3] by coarsening the intracellular events and generate a multiscale model of HIV infection which considers dynamics of viral RNA and proteins production at the intracellular level, yet captures the disease progression at the animal level.

[1] A. S. Perelson and P. W. Nelson. Mathematical analysis of HIV-1 dynamics in vivo. SIAM review, 41(1): 3-44, 1999.

[2] J. M. Heffernan and L. M. Wahl. Natural variation in HIV-1 infection: Monte carlo estimates that include CD8 effector cells. J. Theor. Biol., 243: 191-204, 2006.

[3] S. Khalili and A. Armaou. Sensitivity analysis of HIV infection response to treatment via stochastic modeling. Chemical Engineering Science, 63(5):1330-1341, 2008.

[4] B. Reddy and J. Yin. Quantitative intracellular kinetics of HIV type 1. AIDS Res. Hum. Retroviruses. 15:273-283, 1999.

[5] K. Lim and J. Yin. Dynamic tradeoffs in the raft-mediated entry of Human Immunodeficiency Virus type 1 into cells. Biothech. & Bioeng. 93(2): 246-257, 2006.

[6] H. Kim and J. Yin. Robust growth of Human Immunodeficiency Virus Type. Biophysical Journal, 89: 2210-2221, 2005.

176a Establishing Multivariate Specifications for Incoming Materials Using Data from Multiple Scales

Salvador Garcia-Munoz, Process Modeling & Engineering Technology, Pfizer Global Research & Development, 445 Eastern Point Rd., Groton, CT 06340

In complex processes, the multiple properties that are measured from a raw material will have a combined impact onto the product's final quality attributes (FQA). The effects of the raw materials will be combined to those from the process and the scale of the operation. A new mathematical framework is developed to help understand and decouple the effects of raw materials, scale, and process conditions onto the FQA of a given product. Such framework allows the understanding of the process, the scale and the different control strategies available to operate the process since it decouples the effect of each of these onto the FQA. Ultimately it provides a mathematical framework to establish a multivariate specification purely driven by the FQA. The new technique is based on Latent variable models (LVM) coupled with optimization techniques. This proposed method can be used as the basis to quantitatively describe a design-space while taking into account the inherent variability in the FQA that is brought in by the process operation, the control system and the materials. The methodology is illustrated herein by applying it to establish a multivariate specification for the purchase of a polymeric excipient for the production of tablets.

176b The Concept of Qbd Applied to API Starting Materials

Timothy J. Watson, Chemical R&D, Pfizer, Inc, Groton, CT 06333

The concept and application of Quality by Design (QbD) principles has been, and will undoubtedly continue to be, an evolving topic in the pharmaceutical industry. In an effort to establish consensus and develop consistency, industry and regulators have frequently described QbD by dividing it into three fundamental, interrelated concepts; control strategy, design space, and criticality. This paper provides perspective of QbD applications for the assessment of criticality for starting materials (SM) used to manufacture API. Using a criticality decision tree, and examples from the torcetrapib manufacturing process, case studies for each SM's quality attributes (QA's) will be evaluated with proposed criticality assignments.

176c Quality by Design and Design Space Enabling Continuous Improvements for a Commercial Solid Oral Drug Product

Mary T. Am Ende[1], Georg Bernhard[2], Veronica Lubczyk[2], Ulrich Dressler[3], Thomas Katzschner[4], Andrew Prpich[1], and Holger Weyhers[5]. (1) Oral Products Center of Emphasis, Pfizer, Eastern Point Road, MS 8156-01, Groton, CT 06340, (2) RFT Program Office, Pfizer Global Manufacturing, Heinrich Mack Str 35, Illertissen, 89257, Germany, (3) DP Process Validation, Pfizer Global Manufacturing, Heinrich Mack Str 35, Illertissen, 89257, Germany, (4) Site Validation Committee, Pfizer Global Manufacturing, Heinrich Mack Str 35, Illertissen, 89257, Germany, (5) Site Leadership, Pfizer Global Manufacturing, Heinrich Mack Str 35, Illertissen, 89257, Germany

The purpose of this presentation is to demonstrate the risk management approach used in development of a commercial product filed using the Quality by Design and Design Space concepts during the FDA pilot program. The Critical Quality Attribute (CQA) identified through our Right First Time risk assessment work process clearly defined the product design space. With a successful launch campaign, the product demand exceeded the original forecasts. The manufacturing site was challenged to streamline processing to maximize production rates while maintaining quality.

In this presentation, we discuss two aspects for improvements in process efficiencies which demonstrate how commercial manufacturing translated the output of the risk management approach to significant cost savings and waste material reductions. First, a spike in the tablet potency at the end of the compression was revealed during technology transfer and determined to occur when 300 grams remained in the press (enough only to fill the die channels). The existing controls on the equipment allowed press shut-off to be automatically set at approximately 1.5 - 2 kg blend remaining in the press. Commercial manufacturing projects reduced the blend discharged quantity from ~1.5 - 2 kg to NLT 0.5 kg per batch, and resulted in significant cost savings in terms of improved efficiency, and reduced waste. The second efficiency improvement discussed here was aimed at eliminating a bottleneck in the film coating process by scaling up from 115 to 350kg. Two theoretical film coating models were employed to establish the process parameters that matched the smaller scale coating process using a design space approach, and thus eliminated the need for large scale trials. The final process produced quality product at the recommended operating ranges. These continuous improvement efforts resulted in a robust drug product process. These accomplishments are attributed to the Quality by Design and Continuous Improvement approaches utilized in the development and commercialization of this product.

176d Prospective Analysis of a Manufacturing Rig and Determination of Routine Operating Capability In Preparation for a New Product Launch Utilizing Quality by Design Principles

Claire Twomey, Technical Operations, Eli Lilly, Dunderrow, Co. Cork, Kinsale, Ireland, Helena Ryan, Science & Technology Team, Eli Lilly, Dunderrow, Co. Cork, Kinsale, Ireland, Niamh Warren, Process Engineering, Eli Lilly, Dunderrow, Co. Cork, Kinsale, Ireland, and Kevin D. Seibert, Cpr&d, Eli Lilly & Co., DC 4813, 1400 W. Raymond St., B110, Indianapolis, IN 46221.

Consistent with the science and risk based strategies associated with Quality by Design initiatives, routine process capability has been utilized as a means of establishing criticality of process parameters for a new product launch in a manufacturing facility. Prospective analysis of control systems was used to ensure assumptions made by development for routine process capability were consistent with actual system performance. Where necessary, control tuning was leveraged to reduce systematic variation and ensure equipment was adequately capable of the appropriate level of control required by the criticality assessment.

176e Process Modeling Based Approach towards Quality by Design for An API Synthetic Step

Shawn Brueggemeier, Emily Reiff, Olav Lyngberg, Lindsay Hobson, and Jose Tabora. Process Research and Development, Bristol-Myers Squibb, PO Box 191, New Brunswick, NJ 08903

Utilizing a model based approach, a multivariate design space was identified for the reaction and distillative crystallization unit operations of an API synthetic step. Initially, a kinetic model for the conversion of the final intermediate to API was developed from laboratory data as a function of concentration, temperature, and equivalents of reagent. In addition, a process model was also developed for the kinetics of API degradation in the subsequent distillative crystallization. The kinetic degradation model was coupled to a vapor-liquid-equilibrium model to evaluate process performance under different pressures and distillation times. The multidimensional process design space was then mapped by visualizing the results of the models as surface plots. When combined with quantitative API isolation information, these surface plots provided the foundation for a quality gate on the level of residual intermediate. This not only provided control of a critical quality attribute (CQA) but also compressed the reaction design space into a single output parameter. Model performance was verified in a late stage pilot plant campaign and the process models described here will be further utilized as a key portion of future Quality by Design efforts for this project.

176f Integrating Design for Six Sigma Tools and Qbd Concepts In Hot Melt Extrusion Process Development

Luke Schenck[1], Adam Chen[2], Michael Lowinger[3], Gregory M. Troup[4], Amanda Sinha[5], Jennifer Ho[2], Brit L. Rudeen[2], Stephanie Robison[2], and Celia Cruz[2]. (1) Chemical Process Development and Commercialization, Merck & Co, Inc., 126 E. Lincoln Highway, RY818-B324, Rahway, NJ 07065-0900, (2) Pharmaceutical Commercialization Technology, Merck & Co., Inc., WP78-110, P.O. Box 4, West Point, PA 19486, (3) Materials Characterization & Technology Assessment, Merck & Co. Inc., 770 Sumneytown Pike, West Point, PA 19486, (4) Process Analytics, Merck & Co. Inc., 770 Sumneytown Pike, West Point, PA 19486, (5) Analytical Development and Commercialization, Merck & Co, Inc., PO Box 4, WP78-210, West Point, PA 19486

This work summarizes how the Design for Six Sigma (DFSS) tool set was employed as a means to streamline delivery of Quality by Design (QbD) for a hot melt extrusion process. DFSS methodology was believed to provide an efficient means to capture the key components of both the systematic approach to development and risk management elements that are central to QbD. While various roadmaps exist for DFSS, the one that was utilized was CDOV or Concept-Design-Optimize-Verify.

DFSS Concept stage tools helped manage and translate existing knowledge - including experimental information, first principles understanding, models and best practices gleaned from peer reviewed literature - into a rationalized, systematic approach to development. This was achieved through process mapping activities combined with a quality function deployment grid that helped clarify and prioritize development targets. As such, a daunting, technically complex process became a more manageable set of potential critical process parameters to evaluate first. These parameters were thought to have the most impact in achieving the desired quality requirement of a molecularly dispersed, extruded product.

Successive DFSS Design stage tools utilized statistically designed experiments facilitated by first principles understanding and modeling. Initial screening experiments sought to help define the Critical Quality Attributes (CQA) and intentionally included perceived failure points. Even though these failure points were known to be sufficiently removed from the anticipated commercial operating space, the thought was a better understanding of failure would yield better definition of CQA's. The initial experiments successfully identified several key CQA's, and a single critical process parameter (CPP) having the most impact on product quality.

At this point, DFSS Optimization stage tools assisted in expanding process understanding and enhancing robustness, moving towards a defined design space. Response surface mapping experiments were completed to identify the curvature and interactions of the CPP in the vicinity of the anticipated commercial operating range. Throughout these experiments, PAT was heavily integrated. While a control strategy was not included in the scope of this work, the robustness testing that PAT facilitated provided a strong starting point for later formulating a control strategy.

Ultimately, the DFSS tools helped manage the technical complexity of developing this hot melt extrusion process and deliver on QbD. Decisions and assumptions throughout the development process were clearly spelled out, with CQA's and CPP's identified. The knowledge gained from this development exercise facilitated process optimization, and yielded a robust, well defined design space.

176g Design Space of Drug Product: Empirical Modelling Based on Multivariate Projection Methods

T. Kourti, GlaxoSmithKline & McMaster University, 1280 Main St. West., Hamilton, ON L8S 4L7, Canada

In this approach, the definition of design space is taken literally and Quality is expressed (in one model) as a function of the multidimensional combination of raw material (API and excipients), parameters of the process (Granulation, Drying, Blending, Compression, Coating) and IPC's.

A multivariate structure is also maintained for quality.

Separate models provide support for control strategy in a feed forward mode. Multivariate projection methods are used for modelling, while scientific and other knowledge is incorporated to create hybrid models.

The structure of the multi- factor DOE's required to derive such a model is discussed, as well as ways to develop control strategy based on empirical data.

178a The Importance of Using Limiting Flow Rates to Assess the Flowability of Pharmaceuticals, Excipients and Their Mixtures

Karl Jacob, The Dow Chemical Company, Dow Chemical Company, 1319 Bldg., Midland, MI 48667 and **James K. Prescott**, Jenike & Johanson, Inc., 400 Business Park Drive, Tyngsboro, MA 01879.

Reliable flow of pharmaceutical mixtures out of tablet press hoppers is a key to problem-free tablet press operation and consistent tablet properties. Failure to insure reliable flow can result in both considerable manual intervention for the tablet press hopper and poor tablet physical properties such as tablet weight, tablet hardness, etc. Numerous bench top tests are presently used to assess powder flowability such as rotating drums, powder flow out of funnels with variable outlet sizes, etc. In addition, laboratory-scale tablet presses are employed as a small scale model for full-scale operation. However, these tests fail to address the technical complexity, speed, and physics/mechanics of actual full-scale tablet press hopper operation. Many pharmaceutical systems with fine powders (for example, below 100 microns in particle size) exhibit significantly reduced flow rates on account of the two-phase flow which must occur in the conical section of a tablet press hopper. Solids must dilate or reduce in bulk density as they flow in the conical section of a tablet press hopper. If, in the case of fine materials, they are insufficiently permeable to gas flow, then this results in a limiting flow rate of solids. The reduction in flow can be significant and it is not unusual to observe decreases in flow of several orders of magnitude or complete stoppage of flow. This paper will develop the fundamental theory behind limiting flow rates, the key measurements required in order to calculate the limiting flow rate and present several case studies showing the importance of the limiting flow rate approach.

An investigation into the effects of binder viscosity, shear rate, mixing time, and primary particle size on the spreading of liquids in coating processes

Tyler Simmons and Richard Turton
Department of Chemical Engineering
West Virginia University, Morgantown 26506

Abstract

The purpose of the current study was to investigate the spreading of a liquid binder onto a particle bed. It was desired to characterize this liquid spreading in terms of common parameters including liquid viscosity, agitation intensity, mixing time, and primary particle size. A series of experiments was performed to determine the effects of these parameters on the uniformity of the coating on sugar cores (nonpareils). Silicone oils of viscosities ranging from 5 cSt to 10,000 cSt were used as the major component of the binder liquid. These silicone oils were blended with an isoparaffin to allow for the solubilizing of a blue dye in the silicone composition for visual marking purposes. Two shear rates were used for the runs. Mixing time was investigated with times ranging from 20 seconds to 200 seconds. Once the liquid spreading experiments were completed, the data were collected and analyzed. The coating uniformity was investigated by capturing images of blue-coated particles with a color camera and subsequently analyzing the color level intensity on each particle using Visilog® imaging software. The relative standard deviation (RSD) of the blue content on the particle samples was determined from the raw data, and this RSD was used as a measure of the liquid spreading onto the particle bed. A high value indicates poor mixing while a low value indicates good/uniform spreading. It was found that liquid viscosity, shear rate, mixing time, and primary particle size all have an effect on the spreading of the liquid in the bed; however, it was not possible to eliminate the occurrence of agglomeration in the system. This made it difficult to definitively characterize the spreading process alone. Nevertheless, after studying the physics of the system at a microscale level, promising results describing the agglomerate breakage step were obtained.

KEYWORDS: Binder, Coating, Agglomeration, Breakage, Uniformity.

I. INTRODUCTION

The distribution of the binder has an important effect on the granulation process and the distributed characteristics thereof. Indeed, the assumption of binder distribution and coverage is implicit to many agglomeration process models, for example, the microscale model of Ennis et al. [1]. If the wetting and spreading of the liquid is slow relative to the process time, its coverage will be poor and will lead to large variations in the inter-granule composition.

The uniformity of a batch of solids undergoing film coating using a solution of binder and solvent has been addressed previously for fluidized beds and perforated drum-coating devices by Cheng and Turton [2 and 3], Mann [4], Choi and Miessen [5], Sandadi et al. [6], and Sudsakorn and Turton [7]. The application of liquid binder solution simultaneously with a powder has also been addressed by Nakamura et al. [8]. However, the uniformity of coverage of core particles by a relatively viscous binder has received little attention in the literature. Important issues in this type of coating are the effects of binder properties (most importantly viscosity and surface tension), agitation intensity in the powder bed, mixing time, and primary particle size. In the current work, these parameters served as the focus of the investigation on the uniformity of the spreading of the binder on the surface of the core particles.

II. EXPERIMENTAL

The core particles used in this study were 18-20 mesh and 35-40 mesh nonpareils with average diameters of 920 microns and 460 microns, respectively. Silicone oils (polydimethylsiloxane) of viscosities ranging from 5 cSt to 10,000 cSt were used in this work. A blue dye tagging agent was combined with an isoparaffin and added to the silicone oils. The resulting series of binders had viscosities in the range 5 to 1700 cSt.

The experimental apparatus, shown in Figure 1, was comprised of a tilted aluminum disk 2 feet in diameter driven by a motor, the speed of which could be manipulated by the user from a control panel. A plexiglass™ enclosure was constructed and was held in place on top of the rotating disk by springs. This enclosure acted as the walls of the particle bed. The shear in the enclosure was achieved by the upward movement of particles near the surface of the disk due to the upward rotation of the disk and the cascading of particles down the bed once they reached the top wall.

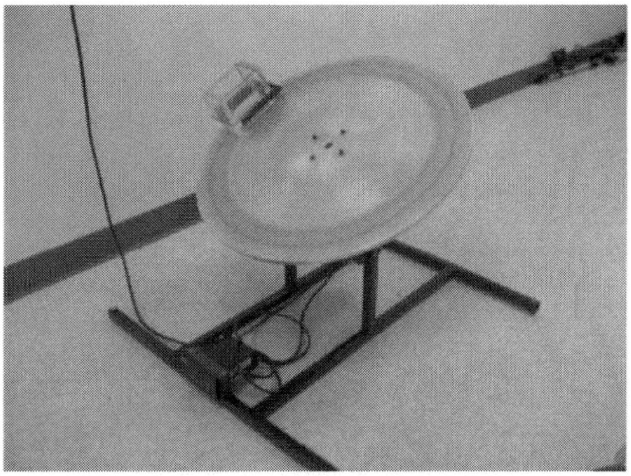

(a)

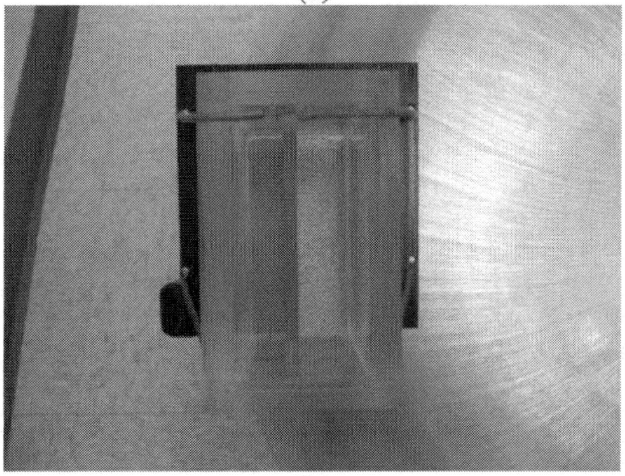

(b)

Figure 1: (a) Experimental apparatus (b) Top view of particle bed enclosure

A series of experiments was performed in which 20 drops of binder were added to equal-weight (10 g) batches of core particles over a time period of 20 seconds. Four different binder viscosities, two shear rates, and three mixing times were investigated for both the 18-20 mesh and 35-40 mesh particle sizes. The mixing time is reported as the time after all the liquid had been added to the particle bed, and all experimental runs were performed at ambient temperature. Table 1 shows the experimental matrix (for each of the primary particle sizes) used for this study.

Table 1: Experimental Matrix Used in the Current Work

Binder Viscosity (cSt)	Shear Rate	Mixing Time (s)
5	Disk Speed = 0.58 m/s Disk Speed = 5.00 m/s	20 60 200
24	Disk Speed = 0.58 m/s Disk Speed = 5.00 m/s	20 60 200
130	Disk Speed = 0.58 m/s Disk Speed = 5.00 m/s	20 60 200
1700	Disk Speed = 0.58 m/s Disk Speed = 5.00 m/s	20 60 200

In order to quantify the degree of spreading of the liquid binder on the cores, it was necessary to capture an image of the blue-coated particles. Multiple samples from each coating scenario were prepared, and a Pixelink firewire color CCD camera was used to capture the image of each sample. A typical image is shown as Figure 2. The imaging analysis algorithm was performed using Visilog™ software to determine the distribution of the binder solution on the particles. First the color image was split into its component colors (red, green, blue). Then, employing the red monochrome image, various edge detection tools were used to remove any "blobs" or "noise" from the image. A boundary was then set around each particle, and this perimeter enclosed the relevant area for the color level analysis. The software then analyzed the gray level, which was an indication of the blue intensity on the particle images, within each of these particle perimeters to produce results of the binder distribution on the population of the core particles.

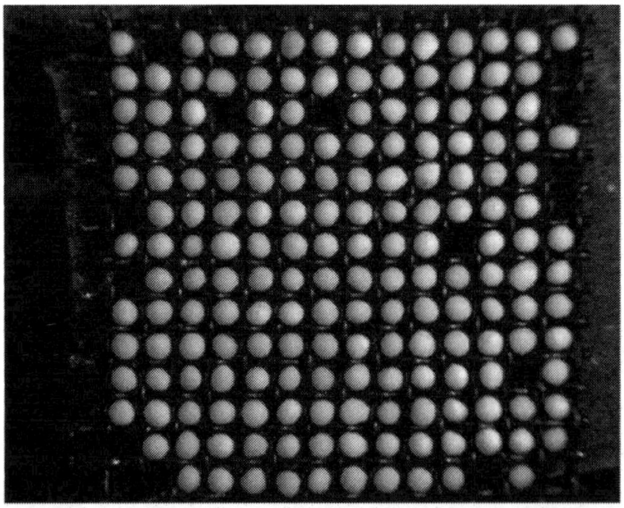

Figure 2: Color image of blue-coated particles, high shear setting and 60 s mixing time

III. RESULTS AND DISCUSSION

Coating Uniformity

Five samples (of approximately x particles each) were analyzed for the spread of binder solution for each experimental scenario, and the raw data were combined. This raw data comprised the relative color level, which was based on two standard values. One of the standard values represented a relative color value of 0 (no blue dye). This value was found by capturing images of uncoated white particles and determining the value of the whitest particle. The other represented a relative color value of 1 (large amount of blue dye), which was found in a similar fashion. Therefore, the relative blue level on each of the particles in the sample images was normalized. The mean ($\bar{x}$), standard deviation (σ_x), and relative standard deviation RSD ($\sigma_x/\bar{x}$) for these color level values were determined for all scenarios. Figure 3 illustrates the effects of binder viscosity, shear rate, and mixing time on RSD for the 18-20 mesh particle size. Low binder viscosities, high shear rates, and longer mixing times lead to a smaller RSD, and therefore, better spreading of the liquid binder.

Results were also obtained for the experiments using the 35-40 mesh primary particles; however, there appeared to be no significant trend of RSD with viscosity, shear rate, and mixing time. It is believed that because of the flow pattern observed in the powder bed for these experiments with the smaller particles, the data obtained is confounded. It was observed that, after the total amount of binder was added to these particles, the flow of the particles had a "slugging" motion. This "slugging" motion comprised of large groups of particles moving as a single entity down the length of the particle bed. In this situation, the flow field was quite different from that found with the larger particles. No longer did individual particles move separately in the bed but rather clumps of particles moved in a non-uniform motion. Not surprisingly, it was found that all RSD values for the smaller particle size were significantly larger than the corresponding values using the 18-20 mesh particles. This particle size effect may be overcome using higher shear rates in the apparatus, but such conditions could not be produced in the current equipment.

The mechanism by which the liquid binder spreads on the powder bed was also examined. Because the binder was added in a drop-wise fashion, agglomerates were initially formed in all experiments. Subsequently, due to the movement and impact of the dry bed particles, these agglomerates broke up and the liquid binder redistributed itself over the bed. The mechanism by which these initial agglomerates broke up was also investigated. The work of Simons and Fairbrother [9] and Willett and Seville [10] was used to develop a model to describe this behavior. The formulation of the model and results will be discussed in the presentation, if time allows.

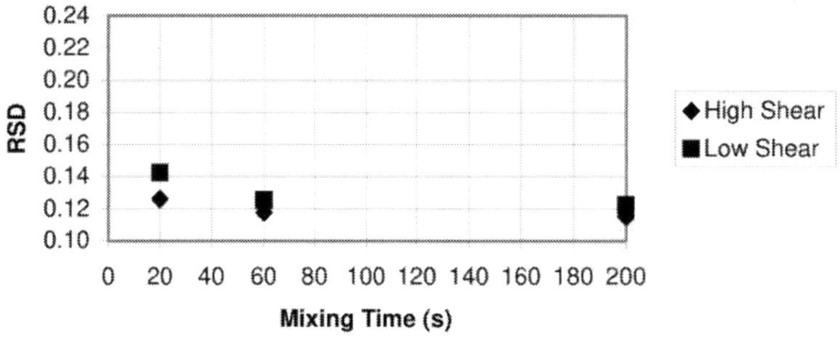

(a)

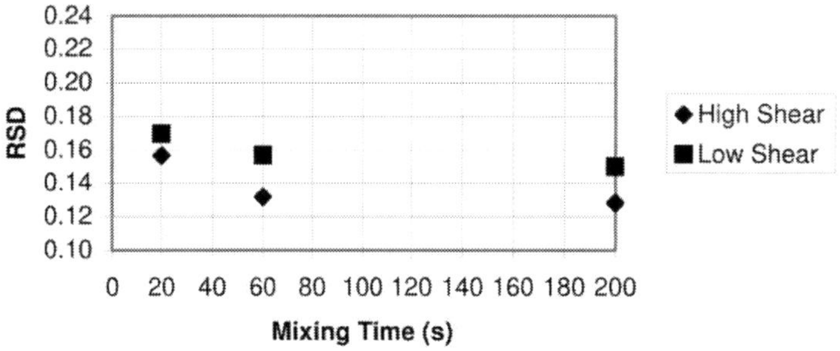

(b)

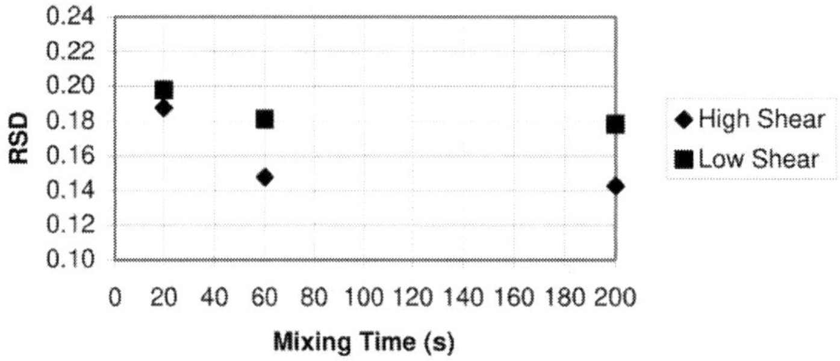

(c)

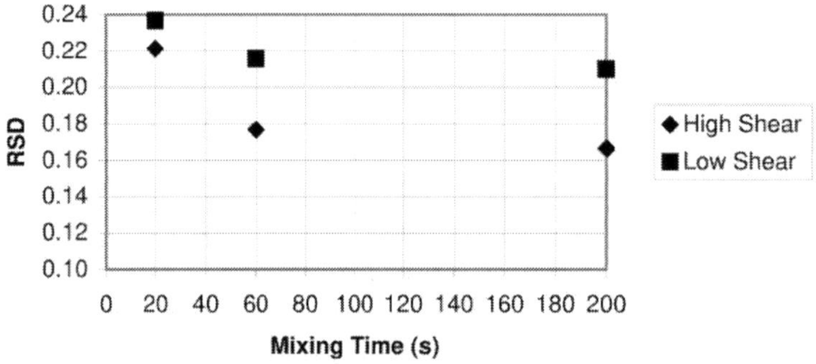

(d)

Figure 3: Effect of mixing time and shear rate for 18-20 mesh primary particle size at binder viscosity of (a) 5 cSt, (b) 24 cSt, (c) 130 cSt, and (d) 1700 cSt

IV. CONCLUSIONS

This work demonstrates a methodology to directly measure binder dispersion on particles in a mixer-granulator or coating process. From the coating uniformity results obtained from this work, it is hypothesized that parameters including binder viscosity, shear rate, mixing time, and primary particle size can be incorporated to model the spreading process. However, due to the presence of agglomeration in the system, it was not possible to definitively characterize the spreading mechanism alone. On the other hand, it is believed that there exists a strong coupling between binder dispersion and agglomeration, and the agglomeration and breakage mechanisms were investigated.

V. REFERENCES

1. Ennis, B.J., Tardos, G.I., and Pfeffer, R., A Microlevel-Based Characterization of Granulation Phenomena, Powder Technology, 65, 257-272 (1991)

2. Cheng, X.X., and R. Turton, The Prediction of Variability Occurring in Fluidized Bed Coating Equipment Part 1: The Measurement of Particle Circulation Rates in a Bottom Spray Fluidized Bed Coater, Pharmaceutical Development and Technology, 5(3), 311-322 (2000)

3. Cheng, X.X., and R. Turton, The Prediction of Variability Occurring in Fluidized Bed Coating Equipment Part 2: The Role of Non-Uniform Particle Coverage as Particles Pass Through the Spray Zone, Pharmaceutical Development and Technology, 5(3), 323-332 (2000)

4. U. Mann, E.J. Crosby, Can. J. Chem. Eng., 53, (1975) 579-581

5. M.S. Choi, A. Miessen, Chem. Eng. Sci., 52, (1997) 1073-1086

6. Sandadi, S., Pandey, P., and R. Turton, In-situ, Near Real-time Acquisition of Particle Motion in Rotating Pan Coating Equipment Using Imaging Techniques, Chem. Eng. Sci., 59, 5807-5817 (2004)

7. Sudsakorn, K., and R. Turton, Nonuniformity of Coating on a Size Distribution of Particles in a Fluidized Bed Coater, Powder Technology, 110, 37-43 (2000)

8. H. Nakamura, E. Abe, N. Yamada, Powder Technology, 99, 140-146 (1998)

9. S.J.R. Simons, R.J. Fairbrother, Direct Observations of Liquid Binder-Particle Interactions: The Role of Wetting Behavior in Agglomerate Growth, Powder Technology, 110, 44-58 (2000)

10. C.D. Willett, J.P.K. Seville, An Examination of the Effects of Contact Angle on the Properties of Exact and Toroidal Liquid Bridges, World Congress on Particle Technology 3, Paper 79 (1998) IChemE

Investigation on attrition strength of coated particles by response surface methodology

G. Perfetti[b,c], D.E.C. van der Meer[a], G.M.H. Meesters[b,c], W. J. Wildeboer[b]

[a]*Food Technology, School of Agriculture & Technology, INHOLLAND College, Kalfjeslaan 2, NL-2623 AA Delft, The Netherlands*
[b]*DSM Food Specialties, Alexander Fleminglaan 1, NL-2613 AX Delft, The Netherlands*
[c]*Particle Technology Group, Faculty of Applied Sciences, Delft University of Technology, Julianalaan 136, NL-2628 BL Delft, The Netherlands*

Abstract

The purpose of the present study is to obtain better understanding of the influence of different variables on the strength of coating, applied onto sodiumbenzoate particles. Through fluid bed coating of 1 mm sized particles, three experimental factors have been studied: the coating quantity related to the particle weight, the glass-transition temperature (T_g), and the fluid bed temperature (T_{FB}). Through repeated impact, the resistance to attrition of the coated particles was tested, which is a measure of the strength of the coating. The experiments have been designed according to the response surface methodology (RSM).
From statistical analysis, the coating quantity appears to have a significant effect on the resistance to attrition of the coated particle. This relationship is linear and positive, which means that an increasing quantity leads to more resistance to attrition. No direct effect on coating strength is detected of the fluid bed temperature.

1. Introduction: Reasons for coating

Particles can be coated for several reasons. In general, it is done to alter surface properties, or to add functionality for chemical or physical protection of sensitive ingredients. (Elversson et al., 2006)
It is possible to provide an active agent like aspirin with a protective coating, but usually tablets consist of several layers, shown in Fig. 1.

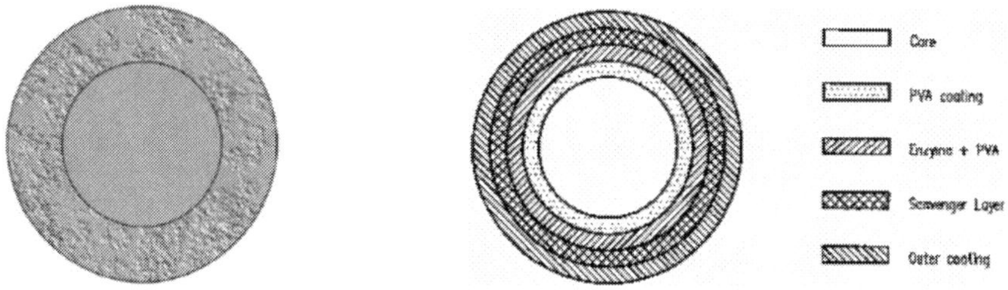

Figure 1. a) Cross-section of a thick-coated particle (Werner, et al., 2007[a]). b) Schematic diagram of an example of a multi-layered granule. (Arnold, et al., 1994)

1.1 Pharmaceutical and Food applications

In the pharmaceutical industry, the coating of particles with a fluid bed is very suitable for specialised dosage forms, like the dosage of a pharmaceutical protein. (Maa et al., 1996) The coating is required to provide desired release characteristics, depending on time or on environment of the tablet, such as the low pH in a stomach. The most common reasons for coating are (Jones, 1985, Guignon et al., 2002):

- taste masking;
- enteric drug release;
- isolation or barrier films;
- improved stability;
- facilitate aesthetics and recognisability: texture or colour;
- facilitate dosage and mixing: size, shape or surface texture.

The benefits of coating in the pharmaceutical industry rather correspond to the reasons for coating in the food-industry. The most common ones are (Dewettinck et al., 1998):

- increased shelf-life;
- taste masking;
- ease of handling;
- controlled release;
- improved aesthetics, taste and colour.

Improvement of food-quality and shelf-life can be obtained by opposing the diffusion of moisture, fat and aromas within the product or between the product and its reactive environment. Edible films and coatings are capable of preventing moisture migration, so they can improve shelf-life by preservation of flavour and freshness. (Dogan et al., 2007). Furthermore, coatings for food and enzyme granules need to prevent the manufacturers and end-users from exposure to these products. (Meesters, 2002).
Not only the coating process in general, but also the fluidised bed by itself knows several applications in food-processing. For example, it can be used for microencapsulation of flavours, micro-organisms and enzymes for the preparation of bread or cheese. Fluid beds could be put into service for puffing, freeze-drying, classification, blanching, spray-drying, granulation and agglomeration, for instance during the production process of milk powder or instant yeast. (Dewettinck et al., 1999[a,c]).

2. Experimental part

2.1 Materials

As coating agents aqueous-based solutions of the followings materials were used: two grades of Hydroxypropyl MethylCellulose HPMC (Pharmacoat® 603, 615, Syntapharm, Mülheim an der Ruhr, Germany) and one grade of PolyVinyl Alcohol PVA (Mowiol® 4-98, Sigma-Aldrich, UK). A list of studied materials with their characteristics and labelled properties is given in Tab. 1.

Table 1. Labelled Properties of the materials used in this work as delivered from the suppliers

Polymer	Brand Name	Grade	Degrees of substitution		Number Average Molecular Weight	Viscosity* [mPa·s]	Mol% Hydrolysis	Polymerization
			Methoxy	Hydroxypropoxyl				
Hydroxypropyl	Pharmacoat	603	28.0 - 30.0	7.0 - 12.0	13000	2.4 - 3.6		
Methylcellulose	Pharmacoat	615	28.0 - 30.0	7.0 - 12.0	65000	12.0 - 18.0		
PolyVinyl Alcohol	Mowiol	4-98			27000	4 - 5	98 -98.8	~ 600

* 4 % solution in water (20 °C)

As core material granules of sodiumbenzoate (Purox-S®, DSM, Geleen, The Netherlands). The particle size is in the range 850 - 1000 µm and they are rather spherical with several sharp edges.

2.2 Preparation of Solutions

Both HPMC and PVA aqueous solutions were prepared at 3% weight content. No additives compounds were used in this work.
The proper amount of HPMC is dispersed into 1/3 of the required amount of demi-water previously heated to more than 80 °C by magnetic heater, while strong stirring with a magnetic stirrer. The polymer gradually disperses to form uniform slurry. The stirring is continued until all particles are thoroughly wetted in the hot water. Later on the remaining water is added as ice water for rapid hydration and the resultant mixture is cooled down in an ice bath while vigorous stirring with a mechanical stirrer at 500 rpm for 30 min. The clear aqueous solution obtained were then allowed to hydrate overnight at 5 °C and allowed to de-foam and equilibrate to room temperature for several hours before use.

As a first step PVA is sprinkled into cold water during stirring and heated to 90 - 95 °C in a water b ath or by the use of live steam. The solution is heavily stirred during cooling in order to prevent skin formation. The speed of dissolution increases with increasing temperature and decreases with increasing molecule. The dissolving process is also made more difficult when there is a transition to higher concentrations. Foam production is minimised using suitable low-speed mechanical stirrer. The solutions are equilibrated to room temperature before use. As Polyvinyl alcohol solutions which have been stored for lengthy periods at high concentration and low temperatures may increase in viscosity they are produced and straightaway used for experiments.

2.3 Methods

2.3.1 Fluid Bed Coater

The equipment originally built for drying of yeast at lab-scale, has been modified by fastening a spray nozzle to the temperature sensor in central position in respect of fluid bed, connecting the nozzle to the central air-pressured system and adding pipes and peristaltic pump to supply the coating solution. Fig. 2 shows from left to right the empty glass tube, the running fluid bed coater, and a close-up of the spray nozzle and the fluid bed.

Figure 2. The fluid bed coating device.

The advantages of this particular coating device are the accurate temperature sensors and regulators, which are necessary to guarantee that the tunings correspond with the actual values. Unlike usual fluid bed coaters, this one has a rather small diameter of only 80 mm. The narrowness and the off-centre position of the spray nozzle promote wall-adhesion. When this occurs, the particles clinging to the wall must be removed quickly, or there will be particle loss and dissimilar distribution of the coating solution on the particle surface.
The fluid bed system is equipped with process software to regulate the settings. Such software enables us to control the fluid bed temperature, the inlet-air temperature or both of them in the same time. Air flow rate, temperature of the heating system, spray nozzle flow rate, coating solution's rate and relative humidity in the chamber are controlled and registered by the software.
Fluid bed coating processes have many different variables, involving the process, the product, or both. Each coating device has its own, particular values, which must be tuned specifically to create a stronger coating or better appearance of the coating, and to prevent unwanted side-effects like inter-particle agglomeration or wall adhesion.

The process parameters for fluid bed coating can be divided into three different process-steps: fluidisation, atomisation and drying, shown in Tab. 2. During fluid bed coating, the steps happen all at once and all three of them can affect film formation (Jones, 1988), but separating them is necessary to diminish the complexity of particle coating. (Werner et al., 2007[a])

Table 2. Process variables. (Jones, 1988, Maa et al., 1996, Werner et al., 2007[a])

Process-step	System design	Process parameters
Fluidisation	▪ batch size ▪ fluid bed system type ▪ equipment dimensions	▪ batch size ▪ drying air flow-rate ▪ inlet air temperature ▪ fluid bed temperature ▪ fluidisation air humidity
Atomisation	▪ spray mode (top, bottom) ▪ nozzle design (pneumatic...) ▪ number of spray nozzles,	▪ atomising air flow-rate ▪ droplet size ▪ nozzle distance from the bed ▪ coating solution rate
Drying	▪ batch-size ▪ equipment dimensions	▪ drying air flow-rate ▪ inlet air temperature ▪ fluid bed temperature ▪ relative humidity inside the fluid bed ▪ coating concentration in the solution

Tab. 3 outlines the product-related variables during fluidisation, atomisation and drying. These variables can be divided into physical core particle properties, and properties of the coating material.

Table 3. Product variables. (Jones, 1988, Maa et al., 1996, Werner et al., 2007[a])

Process-step	Particle properties	Coating material properties
Fluidisation	▪ particle size ▪ particle size distribution ▪ density of the particles ▪ substrate surface character	
Atomisation		▪ viscosity ▪ surface tension ▪ density ▪ coating concentration of solution ▪ solution temperature ▪ solution spray rate
Drying	▪ particle-size ▪ density of the particles	▪ glass-transition temperature

When creating a design of experiments, there are many possible choices for product parameters. According to Jones (1988), the surface area of the particles is a major factor in fluid bed coating. It is controlled by particle size, shape (sphericity), porosity, density, and friability. In practice, it means that the choice for a certain particle material has a large influence on the experiments.
Although there seems to be a connection between the glass-transition temperature T_g and the viscosity of the coating material, viscosity is not constant at T_g, because viscosity is strongly dependent on molecular weight, while T_g shows no molecular weight dependence in the regime of high molecular weight polymers like PVA or HPMC. Therefore, it is not the entire chain that is responsible for the glass-transition. (Inoue et al., 1995) T_g is an important factor during drying: combined with drying time and temperature it affects spreading and coalescence.
Tab. 4 shows all the process conditions and experimental values of all the performed spray-coating experiments.

Table 4. Used experimental design of coating experiments

Reactor	**Fluid bed Dryer + 2-way nozzle (spherical pattern)**
Nozzle	**Air-pressured (nozzle insert: 0.8 mm)[(1)]**
Nozzle's position	**Top-spray, ? 5 cm above the fluid bed (the spray cone did not undergo excessive contact with the reactor wall and temperature sensor): position can be adjusted**
Filter	**Standard**
Coating solution spray rate R_{sol}	**1.2 g/min** (3 rpm). Derived and calculated using a peristaltic pump[(2)] and a electronic balance[(3)]**
Mass of core material W_p	**150 g**
Inlet air Flow Rate $F_i^{(1)}$	**60 Kg/h constant**
Inlet Air Velocity	**2.7 m/s**
Fluid bed Temperature	**40* - 55* - 70 °C*, automatically kept constant by software**
Inlet air Temperature T_1	**Automatically determined by software**
Outlet air Temperature T_1	**Automatically recorded by Thermometer[(4)]**
Atomizing Air Pressure	**0.5 bar**
Heating System Temperature	**60 °C (F.B.T. = 40 °C), 95 °C (55 °C), 100 °C (70 °C)**
Fluid Bed Relative Humidity RH	**Recorded by means of the software**
Outlet Air Relative Humidity RH	**± 3%, recorded by means of the software**
Core particles	**Microcrystalline Cellulose, Cellets® 1000, 1.0 – 1.4 mm Sodium Benzoate, Purox® S 0.85 – 1.0 mm**
Coating Solution Dry Matter Content DM	**3% w/w**
Particles Coating Content w	**1 – 5 – 9 % w/w**
Mass of Coating Solution W_{cs}	**50.5 g (1%), 263.6 g (5%), 494.5 g (9%)**
Coating Solution Temperature T_s	**Room temperature**
Drying time (after spraying)	**10 min**
Drying Temperature (after spraying)	**40 °C (F.B.T. = 40 °C), 55 °C (55 °C), 70 °C (70 °C)**
Coating materials	**Pharmacoat HPMC 603 Pharmacoat HPMC 615 Poly Vinyl Alcohol PVA 4-98**
Coated Particles Storage Conditions	**Room Temperature (23 °C) Freezer (-18 °C)**
Ambient Air	
Relative Humidity	**Recorded by means of Thermometer[(4)]**
Temperature	**Recorded by means of Thermometer[(4)]**

* During the experiments Fluid bed temperature is kept constant by means of Control Software. To obtain it the software provide to set the corresponding values of Inlet Air Temperature and Heating System Temperature (keeping constant the coating solution flow rate)
** The stickiness of the coated particles must be taken into account: the limit of stickiness will represent the maximum amount of coating solution I can spray.
[(1)] Aeromatic Ltd. Bubendorf, Switzerland
[(2)] Watson Marlow 505U, VWR International, Amsterdam, The Netherlands
[(3)] Mettler Toledo PG8001-S, Mettler Toledo, Switzerland
[(4)] Hygro-Measuring System, Novasina ms1, Novasina AG, Lachen, Switzerland

2.3.2 Design of experiments

Coating strength is expected to depend on the state of the polymer during spraying, spreading and drying of the coating material. In this setting, the glass-transition temperature T_g of amorphous solids like polymers, is an interesting product characteristic constant. Below T_g the coating polymer has a glassy state: most bonds are intact, which makes a coating brittle, but strong. Because of the very low mobility of the macromolecules, tablets coated with this kind of polymers have extremely small diffusion rates (Siepmann et al., 2001). Above T_g the polymer has a rubbery state and is more elastic, due to weakening of secondary, non-covalent bonds (Picker, 2003).

When applying polymers on particles, the process temperature not only affects the state of the polymer, but is also responsible for drying velocity. If an aqueous coating solution is drying slowly, the molecular chains can arrange themselves neatly into each other, resulting in a firm network with little water left and therefore in a high T_g. Consequently a high process temperature results in fast drying where the chains are not allowed to arrange. Hence the water content between the chains is higher and T_g is lower. Tanaka et al. (2006) suggested water content in the air during spray drying might be responsible for lowered T_g, which is also related to relative humidity and process temperature. In fact when the temperature is too low, the drying of particles will be too slow, which might causes stickiness, agglomeration and wall adhesion. On the other hand, a process temperature that is too high might cause almost drying of the droplets before the impact with a particle. As a consequence, the droplets do not spread very well and this prevents the formation of a smooth coating layer.

Therefore it could be important not only to experiment with materials with different glass-transition temperatures, but also to combine it with varying fluid bed temperatures: above and beneath T_g of the coating material.

This is why the factors of the experiments contain low and high values. HPMC 615 has a naturally high T_g, around 150 ℃, while PVA has a lower T_g of circa 50 ℃ determined with differential scanni ng calorimetry (DSC). Fluid bed temperature is set on 40 ℃ and 70 ℃.

Because there is expected to be an optimum, a centrepoint is chosen between the upper values. The T_g of HPMC 603 is around 125 ℃, which is close enough to the ideal centerpoint of 100 ℃. The centrepoint of the fluid bed temperature is 55 ℃.

Van Laarhoven et al. (2008) already demonstrated a connection between a thicker coating layer and stronger resistance against wear. To compare strength due to the glassy or rubbery state of the coating material and strength due to coating thickness, these factors are put together in one design.

The quantity of applied coating equals a weight percentage of the coated particle that is 1 %, 5 % and 9 %, not taken into account possible enclosed water, losses of sprayed coating or particle losses during the coating process.

The central composite design

When all parameters can be kept constant during the different experiments it is possible to reduce the number of tests by using factorial design of experiments, a collective noun for a statistical method, that is able to analyse different factors and possible interaction between these factors. When there are many variables with low and high values, fractional factorial design can help to determine which variables have a major, minor or no influence on the outcome (Montgomery, 2005). In this case a central composite design (CCD) is a suitable option from the response surface methodology to reduce the number of experiments. In general there are three different types of CCD: circumscribed, inscribed (Fig. 3) and faced. All three of them contain eight corners, four axial points and one centre point (Myers, R.H. and D.C. Montgomery, 2002).

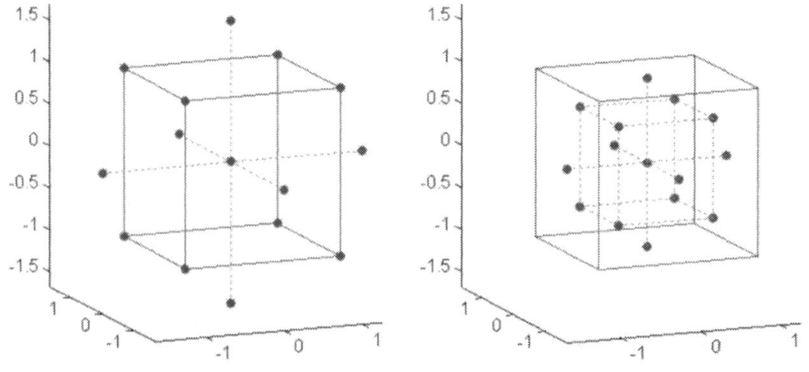

Figure 3. Response surface model: circumscribed and inscribed CCD. (MathWorks, 2008)

In order to diminish the different levels of each factor to three, a faced CCD is chosen, illustrated in Fig. 4. Unlike other CCD's there are no points inside or outside the walls of the cube in this design, except for the centrepoint that must be performed twice. Appendix IV shows the 16 necessary tests to determine a surface that could illustrate an optimum of values for the three parameters. Within the design there are three factors:

A: **the coating quantity;**

B: **the glass-transition temperature (T_g);**

C: **the fluid bed temperature (T_{FB}).**

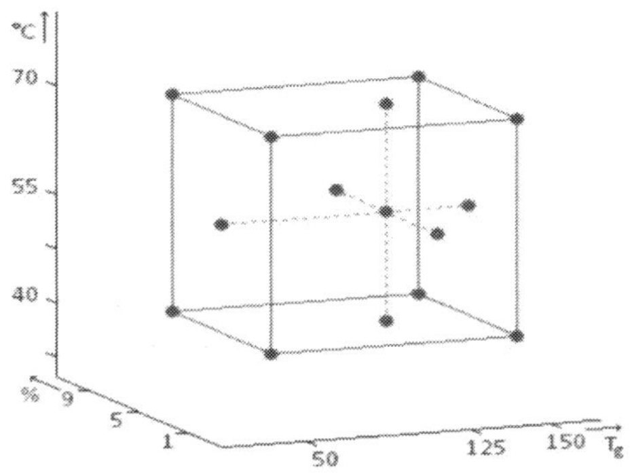

Figure 4. Projection of the chosen response surface model.

For the coating quantity, the input variable is the percentage of the total weight of the coated particle. Each factor has three levels: low, high, and a centre point. In this way, many combinations are possible, of which the 16 most important combinations are performed. The different coatings are tested on attrition with the repeated impact test, in order to assess coating quality.

2.3.3 Repeated Impact Tester RIT

During processing, handling and transportation, coated particles are liable to different forces, resulting in damage of the coating layer. This damage can be divided and isolated into different kinds:

- attrition by repeated impact;
- abrasion by tangential forces;
- breakage by compression.

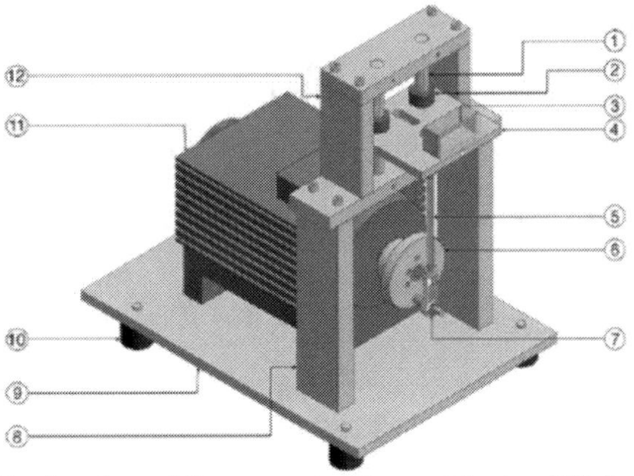

Figure 5. Schematic drawing of the repeated impact tester (Pitchumani et al., 2004)

Attrition by repeated impact can be tested by applying repetitive unidirectional forces on coated particles. For this purpose Pitchumani (2003) developed the repeated impact tester, illustrated in Fig. 5. The device causes an upward and downward movement to the particle chamber, providing collisions of the particles against the top and bottom walls of the chamber. The system is designed to have little or no tangential velocity component.

When the experiment is performed with a frequency of 40 Hertz, the particle chamber oscillates 40 times per second. Pitchumani et al. (2004) has proved that the particles collide twice in an oscillation with the top and bottom walls, which means there will be 51,200 collisions in 640 seconds. Coated particles have been undergoing 51,200; 102,400; 204,800; or 409,600 collisions. The criterion for dust formation: remaining mass after the

experiment (M_n), is calculated by division of the mass (M_i) after testing and sieving with a 450 µm sieve, by the initial particle mass (M_0).
The ratio of the remaining mass is the measure of the amount of dust formation, which equals the attrition of the coating. The less attrition takes place, the better the coating resists repeated impact. This is an indication for a stronger coating.

3 Results

3.1 Morphology of the coating

The different morphology of the coatings is clearly illustrated from scanning electron micrographs showed in Fig. 6, Fig. 7 and Fig. 8. Cross section of coated particles has been are shown only in the case of 9% coating thickness.

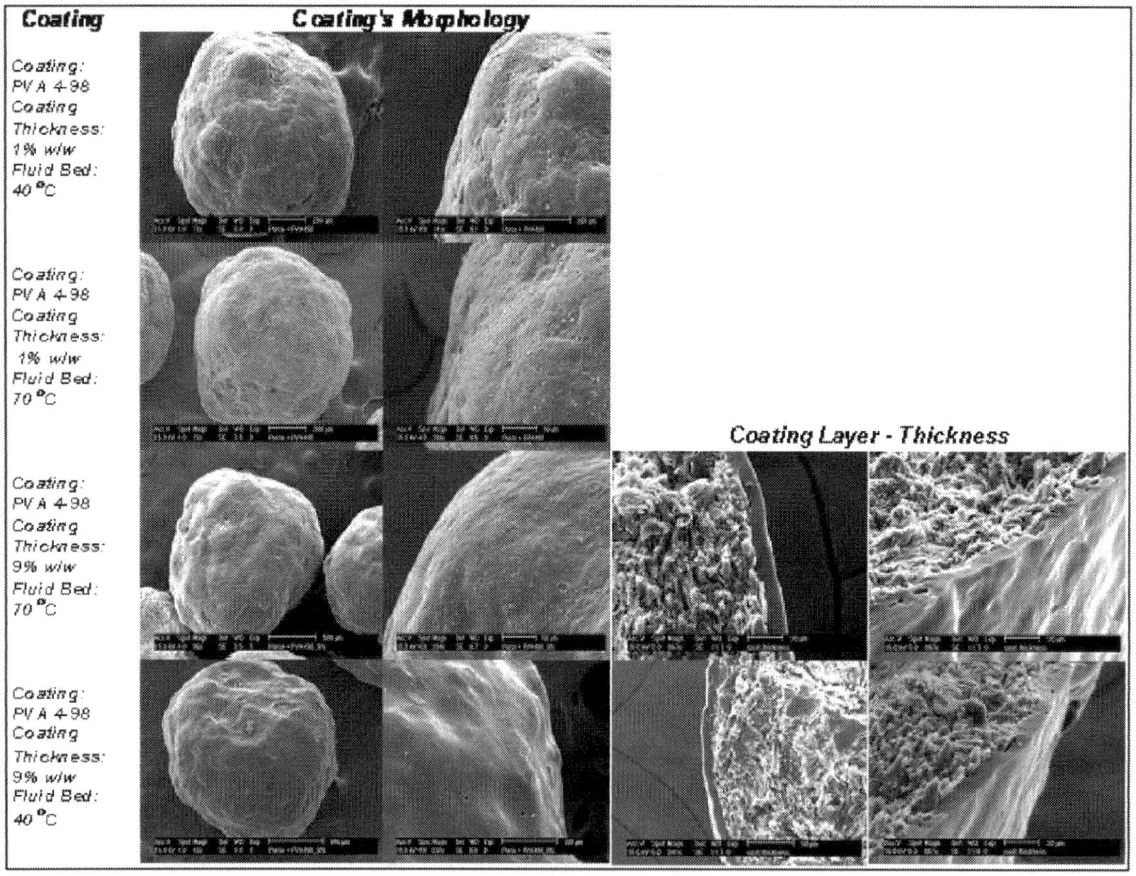

Figure 6: Scanning electron micrograph of Purox particles coated with PVA 4-98, 1% w/w (40 and 70 °C) and 9% w/w (40 and 70 °C) respectively

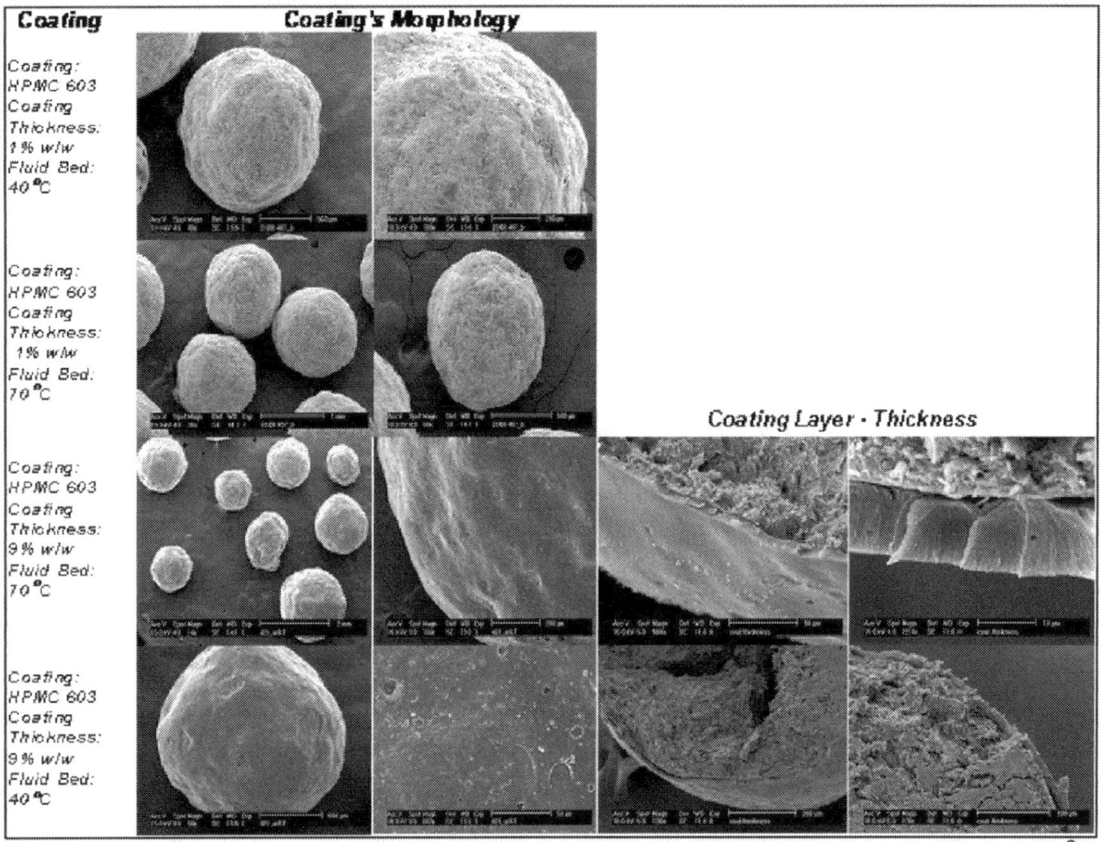

Fig. 7: Scanning electron micrograph of Purox particles coated with HPMC 603, 1% w/w (40 and 70 °C) and 9% w/w (40 and 70 °C) respectively

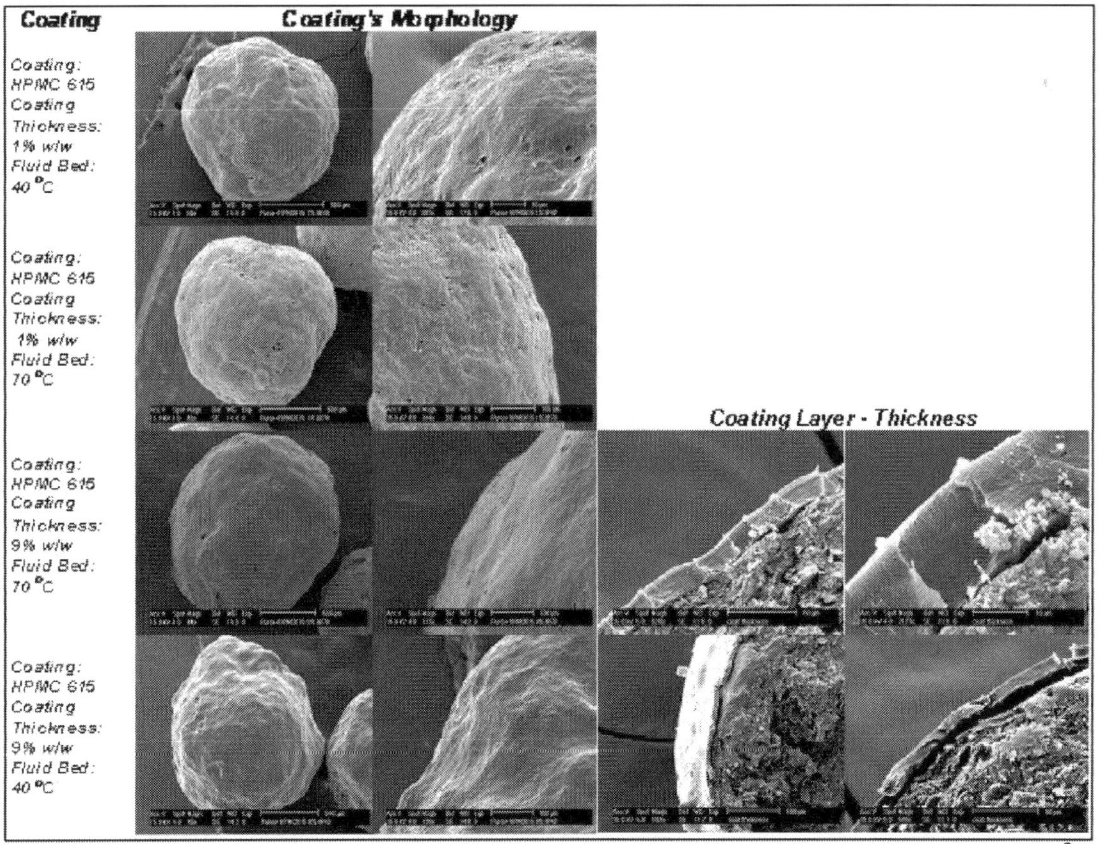

Fig. 8: Scanning electron micrograph of Purox particles coated with HPMC 615, 1% w/w (40 and 70 °C) and 9% w/w (40 and 70 °C) respectively

3.2 Repeated impact tests

The factors and the response variable resulting in repeated impact tests are listed in Tab. 5. Because the declination of the values is approximately linear, a trend line is drawn by calculating the shortest distance between the data points and the line, after which the line is extracted. At the domain between 0 and 50,000 collisions the lines are bent, because the tests always start with a remaining mass of 1.00. Remaining mass of 0.96, which corresponds to 4 % of mass loss, is chosen as limiting value: a minimum on which a coating must fulfil. The number of collisions at M_n = 0.96 per each batch is taken from repeated impact tests results and used as input for the statistic analysis: the response variable.

Table 5. Experimental factors A, B, and C, plus the response variable resistance against attrition.

No.	A: Coating quantity (%)	B: Coating material ~ T_g (℃)	C: Fluid bed temperature (℃)	Attrition ~ collisions at M_n = 0.96
14	9	150	40	2,130,000
16	9	150	70	560,000
12	5	150	55	508,000
6	9	50	70	413,000
7	9	50	40	342,000
2	1	50	70	339,000
13	9	125	55	338,000
1	1	50	40	243,000
9	5	125	40	232,000
8	5	125	70	204,000
11	5	125	55	183,000
3	5	50	55	178,000
4	5	125	55	125,000
15	1	150	40	91,000
10	1	150	70	77,000
5	1	125	55	56,000

3.3 Statistical analysis

The central composite design, is imported in Statgraphics Plus Version 5.1, a statistical program that is capable of creating and analysing factorial designs.

The statistical method ANOVA is based on the F-test. The F-test examines the significance of the experimental results and involves subdividing of the total variation of the outcome into different parts. Because of this, the ANOVA-table is able to tell which factor or interaction factor is significant and which is not. Within ANOVA, the F-ratio is the ratio of the mean square to its error. The larger it is, the more likely that the variance is significantly larger than random error. In other words: a certain effect is probably caused by a factor and not by chance. The P-value is quite the opposite of the F-ratio: the smaller the P-value, the lesser the chance of no correlation. The low P-value of 0.0186 at 'A: Coating quantity' indicates there is significant correlation with the response variable.

Important for ANOVA is the percentage of R-squared: the coefficient of multiple determinations. The more it inclines to 100 %, the better the relation between the response variable and the combined experimental factors or independent variables. R-squared is 83.7 %, but adjusted for the nine degrees of freedom, R-squared is only 59.2 %. The standard error of estimation is the estimated standard deviation of the error.

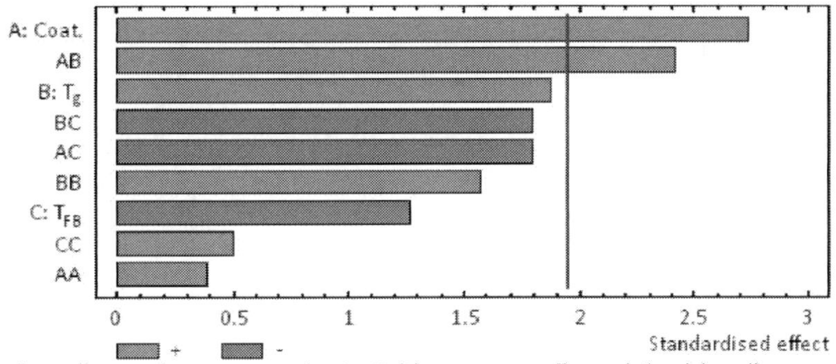

Figure 9. Pareto chart for resistance against attrition: exceeding of the blue line equals significance.

Fig. 9 shows significance of all the variables used in the model. The more the positive (pink-coloured) bars or the negative (red-coloured) bars exceed the vertical blue line of the selected confidence level, the more significance of the effects. Even when the confidence level is lowered to 90 % instead of the default 95 %, taking into account the noise of the experiments, only two effects are significant: A and AB.
Effects are observed by changes in average response, caused by factors or interaction factors when they go from their low level to their high level. A-squared (AA or A^2), C-squared (CC), and C: the fluid bed temperature T_{FB}, are not at all significant. By removing them, a lot of noise disappears and the degree of freedom rises from 6 to 9 degrees of freedom. With augmented degrees of freedom, estimation is more accurate, because the number of independent comparisons increases. After removal of insignificant effects, the outcome of the repeated impact test results into the standardised Pareto chart shown in Fig. 10, in which all main and side effects are significant at the 90 % confidence level.

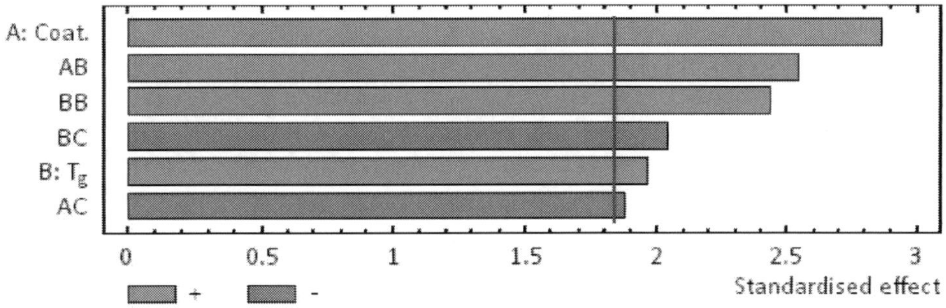

Figure 10. Adjusted Pareto chart for resistance against attrition.

3.3.1 Effects on resistance to attrition

In accordance with the Pareto chart of Fig. 10, six effects are significant: 1st order as well as 2nd order effects are now mentioned in order of significance.

1) The design contains three factors, so there can be three main, 1st order effects. The coating quantity has a predominant (Fig. 11), positive effect on the resistance to attrition of coated particles. Because the effect is positive, the response variable increases along at increasing coating quantity. The steepness of the graph is inherent to the measure of effect on resistance to attrition.

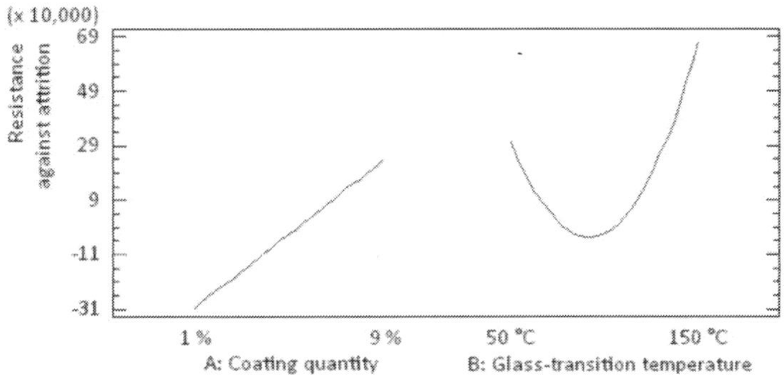

Figure 11. Main effects plots: 1st order effects of the only two significant factors.

2) Also some combined factors influence the results significantly. The major effect is caused by interaction AB between the coating quantity and the glass-transition temperature T_g, illustrated in the interaction plot of AB in Fig. 12. When applying coated material with a high value of T_g like 150 °C, the resistance to attrition rises, as at the same time the coating quantity increases. This linear relationship has a large slope, so the effect is also large. On the other hand, when the T_g is low, hardly any effect is determined of the coating quantity on the resistance to attrition.

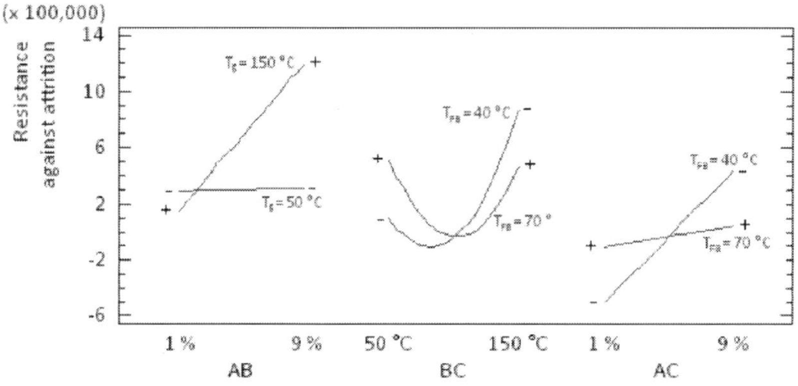

Figure 12. Interaction plots: 2nd order effects of combined factors.

3) BB is a second order effect of the glass-transition temperature T_g. The relation of T_g to the resistance to attrition appears to be quadratic. At this a minimum is noticeable at a T_g of approximately 90°C, which is outlined in Fig. 12.

4) BC stands for the effect of the fluid bed temperature T_{FB} on the influence of the glass-transition temperature T_g, on the response variable resistance to attrition, illustrated by the interaction plot of BC in Fig. 12. At the relation of the glass-transition temperature to the resistance to attrition, the minimum moves to higher glass-transition temperatures at increasing fluid bed temperatures.

5) The fifth effect in the line of significance is B, the main effect of glass-transition temperature T_g, Fig. 12. Alteration of the T_g has a certain influence on the resistance to attrition.

6) The smallest effect that is still significant at a confidence level of 90% is the effect of the fluid bed temperature T_{FB} on the influence of the coating quantity on the response variable, shown in the interaction plot AC, Fig. 12. Increasing of the resistance to attrition at a rising coating quantity is higher at a low value of the fluid bed temperature. This would imply that it is recommended to choose a high value of the fluid bed temperature, when obtaining a strong coating of 1% to 5% of the coated particle. On the other hand, when the coating quantity is more than 5% of the final particle, it is advisable to keep the fluid bed temperature at a low value (40 °C).

3.3.2 Response surface

The most significant effects described so far are summarised in an estimated response surface, illustrated in Fig. 13. It counts for T_{FB} = 55 °C. The factor fluid bed temperature is not represented in one of the axes, because of the low significance of its effect. Under normal conditions, the curvature of the response surface can be used for the prediction of the optimum value of the variables, inside or outside the research area.

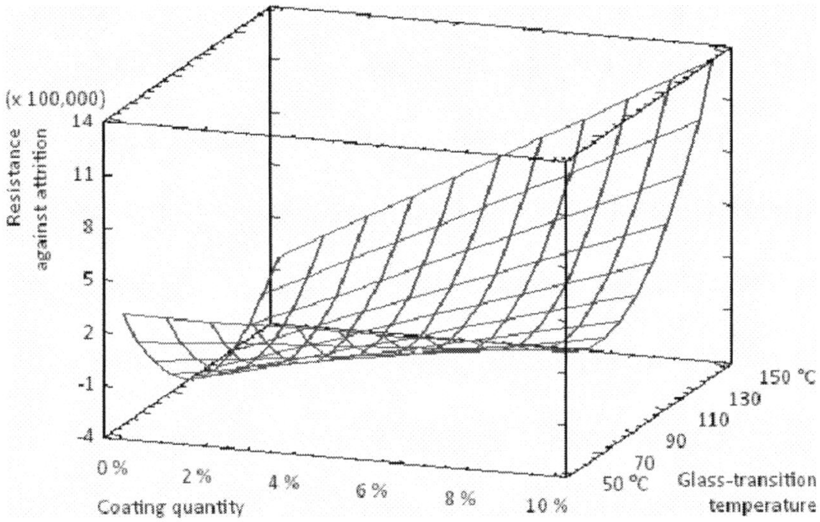

Figure 13. Response surface: the compilation of the main effects and interaction plots.

4. Discussion

4.1 The model for design of experiments

Fig. 14 illustrates the contours of the estimated response surface. The contour plot could be useful for determination of the region outside of the investigated area, where the most optimisation concerning strength of coated particles is suspected. The plot shows a descending ridge to the left in the direction of lesser coating quantity, and a slope towards the right corner at the top. The region of the next study can be determined by using the method of the steepest ascent. In brief, this method is found on an arrow into the surrounding space, which travels at right angles to the altitude lines of the contour plot. Nevertheless, by certain circumstances this is not the case. Due to the mixing of process variables with product characteristics as a factor, the steepest ascent method is inaccurate. The response surface methodology only works properly with quantitative factors, which go continuously from level to level.

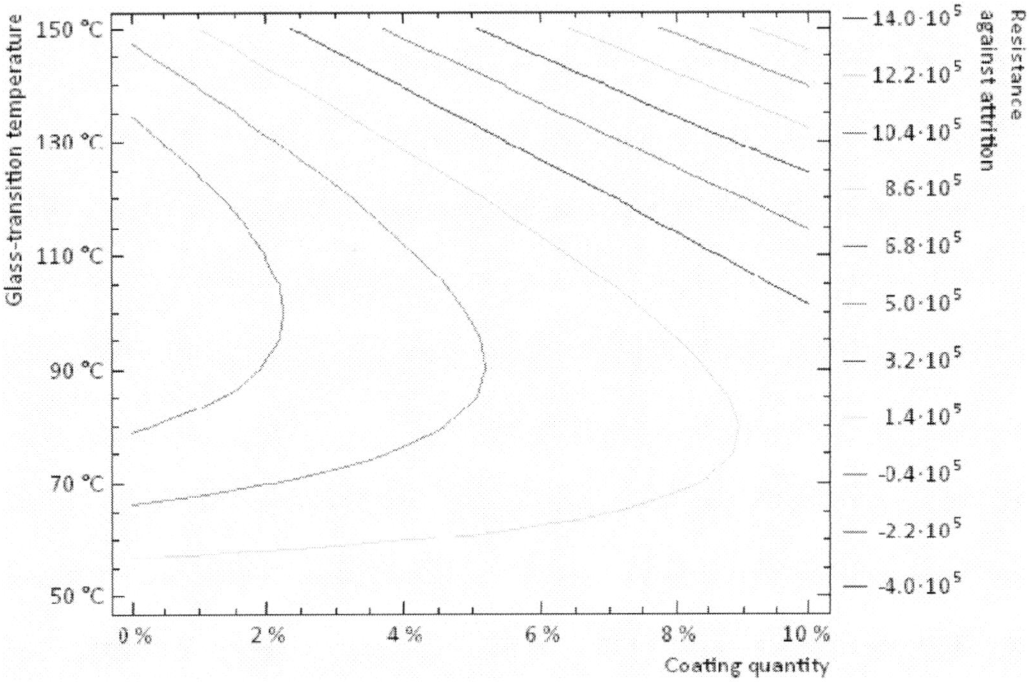

Figure 14. Contour plot of the response surface; T_g as function of the coating quantity. The legend on the right displays the lines, corresponding to resistance against attrition.

Qualitative values like material characteristics only fit in a screening or mixed-level model. In the screening model, the factors are represented by only two levels instead of three. Fractional factorial design is possible within this model in order to reduce the number of experiments. This is in contrast to the mixed-level model, in which the choice of qualitative values and number of levels is free, but all combinations of factorial levels have to be performed: at a full-factorial design, no experiment is left out. Concerning this study, the result is that the main effects and interactions are meaningful as long as the outcome is significant, but the model is still restricted: more profound analyses have no eloquence. All this, is fortified by the absence of an exact centrepoint, which on its turn fortifies the noise of the model. This flaw in the design is due to the lack of a desirable coating material with the right average glass-transition temperature.

4.2 Influence of fluid bed temperature

It turns out, that no direct effect is observed of the fluid bed temperature T_{FB} on the strength of coated particles (at least with T_{FB} in range of 40 – 70 °C), in contrast to the other two main effects of the coating quantity and the glass-transition temperature on the resistance to attrition.
However, there is some influence of the T_{FB} on the effect of T_g, but considerably lesser than suspected. The study is based on the hypothesis that coating with material in the glassy state, as against material in the rubbery state, would lead to different results. On the other hand the effect of the coating quantity on the resistance to attrition is dominant to such an extent, that perhaps consequently the influence of fluid bed temperature is insignificant in this study. More deep studies must be done in order to understand the reason behind it. In particular fluid bed temperature T_{FB} should be related to droplet impacts and spreading on the particle surface which is considered to be the key micro-level process (Werner et al., 2007[b]). In fact within this process, surface tension of the droplets, viscosity of the coating solution and fluid bed temperature are also expected to be the most important parameters. Despite the lack of influence of fluid bed temperature in the present study, it is still considered to be an important factor at fluid bed coating. For economical reasons it is favourable to use as less coating material as possible. When comparing parameters in a study with coating quantity as a constant factor, there might be other significant effects of the independent variables on attrition strength observed.

4.3 Noise of the experiments

Noise in the experiments due to the performance and conditions of the fluidised bed coating needs also to be taken into account and discussed. Possible causes of noise are:

- wall-adhesion;
- inter-particle agglomeration;
- electrostatic effects;
- interruptions of the coating process;
- varying droplet size;
- loss of materials;
- randomisation.

Eventually the noise, together with the non-continuous variables and the reduction of the number of experiments, provides negative values of the estimated response variables (Fig. 15). This is obviously impossible because this variable is the number of collisions necessary to achieve a mass loss of 4 %. Occurrence of this error can be explained by Fig. 4, the projection of the experimental model. The different experiments are represented by corners, axial points and the centrepoint, and when the points are translated to a response surface, there are always data missing. By extending the number of experiments this inaccuracy will be reduced, but can not be prevented. Again, disposition of an exact centrepoint and continuous parameters would also diminish the appearance of negative values in the response surface.

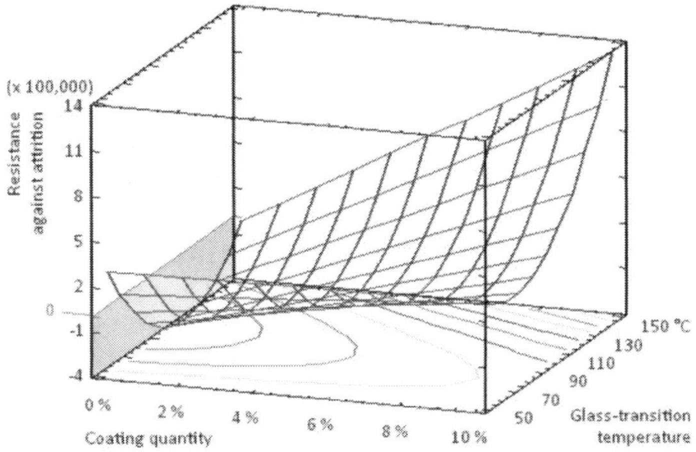

Figure 15. The estimated response surface with contour plot shows, that the response variables have negative values of resistance to attrition, which is in fact impossible.

5. Conclusion

The present study has led to a number of conclusions, which can be divided into the conclusion of the statistical analysis, of the suitability of the statistical model itself and the noise of the experiments.

The experimental factors are assessed on significance of the main effects.
- It appears that the variable quantity of the applied coating has a significant effect on the resistance to attrition of the coated particle. This relationship is linear.
- The relation of glass-transition temperature to the resistance to attrition appears to be quadratic. The main effect of glass-transition temperature is hardly significant.
- No direct effect on the coating strength is detected of the fluid bed temperature.

Three second order effects of combined factors are slightly significant:
- At a low glass-transition temperature (T_g = 50 °C), the coating quantity has little if any effect. At a high glass-transition temperature (T_g = 150 °C), the quantity of the coating is a very important factor.
- The fluid bed temperature affects the effect of the glass-transition temperature on the resistance to attrition.
- The fluid bed temperature has a certain influence on the effect of the coating quantity on the resistance to attrition.

The application of factorial design of experiments is frequently chosen because of the possibility to reduce the number of tests, while the outcome is still significant and useful. Basically, there are two different kinds of models, suitable for the reduction of experiments: models of the screening method and models of the response surface method. It is of utmost importance to make a clear choice of variables to be included. Fractional factorial designs by the screening model allow only two levels per variable, while central composite designs of the response surface methodology may contain more than two levels, but they are restricted to only three quantitative factors. Within this study, the requirements do not fit with the chosen model: the face-centred central composite design.

References

- Antonyuk, S., J. Tomas, S. Heinrich, L. Mörl, 2005. Breakage behaviour of spherical granulates by compression. *Chemical Engineering Science, 60,* p 4031-4044
- Arnold, R.E., N.T. Becker, M.G. Boston, A. Mansikkamaki, M. Simpson, D.J. Wendt, June 1994. *Patent US 5,324,649: Enzyme-containing granules coated with hydrolyzed polyvinyl alcohol of copolymer thereof,* United States of America
- Beekman, W.J., 2000. *Measurement of the mechanical strength of granules and agglomerates.* Ph.D. Thesis, Delft University of Technology, The Netherlands
- Beekman, W.J., G.M.H. Meesters, B. Scarlett, T. Becker, 2002. Measurement of granule attrition and fatigue in a vibrating box. *Particle & Particle Systems Characterization, 19,* p 5-11
- Chan, L.W., E.S.K. Tang, P.W.S. Heng, 2006. Comparative study of the fluid dynamics of bottom spray fluid bed coaters. *AAPS PharmSciTech, 7.2,* p 1-9

- Dale, D.A., A.L. Gaertner, G. Park, N.T. Becker, 1999. *Patent US 5,879,920: Coated enzyme-containing granule,* United States of America
- Dewettinck, K. and A. Huyghebaert, 1998. Top-spray fluidized bed coating: effect of process variables on coating efficiency. *Lebensmittel-Wissenschaft und-Technologie – Food Science and Technology, 31,* p 568-575
- Dewettinck, K., W. Messens, L. Deroo, A. Huyghebaert, 1999[a]. Agglomeration Tendency during top-spray fluidized bed coating with gums. *Lebensmittel-Wissenschaft und-Technologie – Food Science and Technology, 31,* p 576-584
- Dewettinck, K., W. Messens, L. Deroo, A. Huyghebaert, 1999[b]. Agglomeration Tendency during top-spray fluidized bed coating with gelatine and starch hydrolysate. *Lebensmittel-Wissenschaft und-Technologie – Food Science and Technology, 32,* p 102-106
- Dewettinck, K. and A. Huyghebaert, 1999[c]. Fluidized bed coating in food technology. *Trends in Food Science and Technology, 10,* p 163-168
- Dogan, N. and T.H. McHugh, 2007. Effects of microcrystalline cellulose on functional properties of hydroxypropylmethylcellulose microcomposite films. *Journal of Food Science, 72.1,* p 16-22
- Elversson, J. and A. Millqvist-Fureby, 2006. *In situ* coating – An approach for particle modification and encapsulation of proteins during spray-drying. *International Journal of Pharmaceutics, 323,* p 52-63
- Ford, J.L., 1999. Thermal analysis of hydroxypropylmethylcellulose and methylcellulose: powders, gels and matrix tablets. *International Journal of Pharmaceutics, 179,* p 209-228
- Guignon, B., A. Duquenoy, E.D. Dumoulin, 2002. Fluid bed encapsulation of particles: principles and practice. *Drying Technology, 20.2,* p 419-447
- Inoue, T., M.T. Cicerone, M.D. Ediger, 1995. Molecular motions and viscoelasticity of amorphous polymers near T_g. *Macromolecules, 28,* p 3425-3433
- Jones, D.M., 1985. Factors to consider in fluid-bed processing. *Pharmaceutical Technology, 9,* p 50-62
- Jones, D.M., 1988. Air-suspension coating. In: Swarbrick, J. and J.C. Boylan, *Encyclopaedia of Pharmaceutical Technology,* vol 1, Marcel Dekker Inc., New York, USA, p 189-216
- Laarhoven, B. van, S.C.A. Wiers, S.H. Schaafsma, G.M.H. Meesters, 2008. Attrition strength of different coated agglomerates. *Chemical Engineering Science, 63,* p 1361-1369
- Maa, Y.-F., P.A. Nguyen, C.C. Hsu, 1996. Spray-coating of rhDNase on lactose: effect of system design, operational parameters and protein formulation. *International Journal of Pharmaceutics, 144,* p 47-59
- Meesters, G.M.H., 2002. Powder technology at DSM Food Applications: powders in applications, *Proceedings: Powder Research to Promote Competitive Manufacture of Added-Value Food Ingredients,* Brussels, Belgium
- Meesters, G.M.H., Principle Scientist DSM. *Personal note,* Delft [February 15, 2008]
- Mehta, A.M., M.J. Valazza, S.E. Abele, 1986. Evaluation of fluid-bed processing for enteric coating systems. *Pharmaceutical Technology, 10,* p 46-56
- Montgomery, D.C. 2005. *Design and analysis of experiments,* 6th edition, John Wiley & Sons Inc., Hoboken, United States of America, p 405-444
- Myers, R.H. and D.C. Montgomery, 2002. *Response surface methodology: process and product optimization using designed experiments,* 2nd edition, John Wiley & Sons Inc., New York, United States of America, p 156, 321-342
- Obara, S., J.W. McGinity, 1995. Influence of processing variables on the properties of free films prepared from aqueous polymeric dispersions by a spray technique. *International Journal of Pharmaceutics, 126,* p 1-10
- Picker, K.M., 2003. The relevance of glass transition temperature for the process of tablet formation. *Journal of Thermal Analysis and Calorimetry, 73,* p 597-605
- Pitchumani, R., 2003. *Breakage characteristics of particles and granules.* Ph.D. Thesis, Delft University of Technology, The Netherlands, p 1-37
- Pitchumani, R., S.A. Strien, G.M.H. Meesters, S.H. Schaafsma, B. Scarlett, 2004. Breakage of sodium benzoate granules under repeated impact conditions. *Powder Technology, 140,* p 240-247
- Schmedlen, R.H., K.S. Masters, J.L. West, 2002. Photocrosslinkable polyvinyl alcohol hydrogels that can be modified with cell adhesion peptides for use in tissue engineering. *Biomaterials, 23,* p 4325-4332
- Siepmann, J. and N.A. Peppas, 2001. Modeling of drug release from delivery systems based on hydroxypropyl methylcellulose (HPMC). *Advanced Drug Delivery Reviews, 48,* p 139-157
- Soltani, S. H. Asempour, H. Jamshidi, 2007. Investigation of reaction conditions for preparation of medium molecular weight poly(vinyl alcohol) as emulsifier. *Iranian Polymer Journal, 16.7,* p 439-447
- Surutzidis, A., B.M. Heist, M.J. LeBlanc, April 2001. *Patent US 6,221,826 B1: Laundry additive particle having multiple surface coatings,* United States of America
- Tanaka, S., C.C. Pin, K. Uematsu, 2006. Effect of organic binder segregation on sintered strength of dry-pressed alumina. *Journal of the American Ceramic Society, 89.6,* p 1903-1907
- Teunou, E. and D. Poncelet, 2002. Batch and continuous fluid bed coating – review and state of the art. *Journal of Food Engineering, 53,* p 325-340

▪ The MathWorks, 2008. *Statistics Toolbox,* http://www.mathworks.com/access/helpdesk/help/toolbox/stats/f56635.html [March 11, 2008]
▪ The United States Pharmacopeial Convention, 2008. *USP-NF,* http://www.usp.org/USPNF/ [February 26, 2008]
▪ Turton, R. and X.X. Cheng, 2005. The scale-up of spray processes for granular solids and tablets. *Powder Technology, 150,* p 78-85
▪ Werner, S.R.L., J.R. Jones, A.H.J. Paterson, R.H. Archer, D.L. Pearce, 2007[a]. Air-suspension particle coating in the food industry: Part I –state of the art. *Powder Technology, 171,* p 25-33
▪ Werner, S.R.L., J.R. Jones, A.H.J. Paterson, R.H. Archer, D.L. Pearce, 2007[b]. Air-suspension particle coating in the food industry: Part II –micro-level process approach. *Powder Technology, 171,* p 34-45

178d Use of Cfd to Identify and Resolve Cell Culture Issues Observed at Production Scale during Technology Transfer

Zhiwu Fang, Systems Informatics, Amgen Inc, One Amgen center Dr, Thousand Oaks, CA 91320

Substantial experimental work has been dedicated to correlating animal cell culture performance to cell culture process conditions. In this work we present a numerical case study on the environment to which cells are exposed in mechanically agitated and sparged bioreactors. For two modeled production scale bioreactors with different designs and process conditions, we compare mechanical shearing force, Reynolds stress, power input per volume, turbulence energy dissipation rate, detrimental Kolmogorov length scale distribution, as well as bubble residence time in impeller zones and tanks. Detailed and complete information was helpful to understand the culture process performance issues encountered during technology transfer, and to consequently identify an optimal operating parameter design space.

178e Modeling Hot-Melt Extrusion Pelletizers

Stefan Radl[1], Thomas Tritthart[2], and Johannes G. Khinast[1]. (1) Institute for Process Engineering, Graz University of Technology, Inffeldgasse 21a, Graz, 8010, Austria, (2) μnadis Melt Extrusion Technologies, Leoben, Austria

Hot-melt extrusion (HME) is one of the key processes for a continuous production of high-quality pharmaceuticals. Products manufactured by means of HME have significant advantages over conventional drug delivery systems. HME has the benefit to be a solvent free, environmental friendly and cost efficient technology. Furthermore, HME dosage forms have superior drug delivery properties due to the formation of solid dispersions and solid solutions. This is relevant for poorly-soluble pharmaceutically active substances, frequently encountered among novel drugs. Due to this advantages HME technology has been applied to a wide range of dosage forms ranging from granules over transdermal systems to implants.[1-3] However, reliable scale-up of HME technology is still a major challenge and a breakthrough has not been achieved yet. This is mainly due to a limited insight into the process.

In our work we explore the scale-up of hot melt extrusion (HME) devices for direct pelletization of pharmaceuticals by means of computational fluid dynamics (CFD). HME with direct pelletization is a novel concept where the extruded material is cut into small pellets directly after exiting the extrusion die. We investigate non-isothermal, non-Newtonian flow of the melt through different die geometries via a detailed CFD model. To get a refined picture of melt flow through the device, we also take the temperature distribution in the die material into account. In our work we explore different possibilities for stabilizing melt flow including rotating internals.

The results show that a tight temperature control of the die material is necessary to guarantee continuous melt flow. Temperature inhomogeneities using a conventional die design have been identified via simulations. These inhomogeneities act as a trigger to destabilize melt flow and hence lead to a shutdown of the process. We show that a significant improvement in process reliability can be obtained when using an optimized die design. Finally, we highlight some aspects of pellet formation in this novel HME pelletization process.

(1) Breitenbach J. Melt extrusion: from process to drug delivery technology. European Journal of Pharmaceutics and Biopharmaceutics. 2002;54:107-117.

(2) Crowley MM, Zhang F, Repka MA, Thumma S, Upadhye SB, Battu SK et al. Pharmaceutical applications of hot-melt extrusion: Part I. Drug Development and Industrial Pharmacy. 2007;33:909-926.

(3) Miller DA, McConville JT, Yang W, Williams RO, McGinity JW. Hot-melt extrusion for enhanced delivery of drug particles. Journal of Pharmaceutical Sciences. 2007;96:361-376.

178f Formulation and Process Development of Atripla® Tablets

Mark Menning, Gilead Sciences, Inc., 333 Lakeside Dr., Foster City, CA 94404

ATRIPLA® (efavirenz, 600 mg, emtricitabine, 200 mg and tenofovir disoproxil fumarate, 300 mg) is a fixed-dose combination of Sustiva® (efavirenz), Emtriva® (emtricitabine), and Viread® (tenofovir disoproxil fumarate). ATRIPLA is the first single tablet, dosed once daily for the treatment of HIV-1 infection in adults. This presentation will provide a perspective on the challenges associated with the formulation and process development of ATRIPLA tablets under an aggressive timeline. The main points of discussion will be: importance of active ingredient physicochemical properties, selection of formulation composition, manufacturing process and dosage form, and process optimization and scale-up.

178g Expedite Development and Commercialization of Pharmaceuticals with Process Modeling and Scheduling Tools

Demetri Petrides and Charles Siletti. INTELLIGEN, INC., 2326 Morse Avenue, Scotch Plains, NJ 07076

The successful development, scale-up, and commercialization of active pharmaceutical ingredients (APIs) and final pharmaceutical products is a challenging task that requires collaboration of professionals from many disciplines. Process simulation and scheduling tools can facilitate this task by introducing a common language of communication among the various teams involved in process development and product commercialization. They can facilitate answers to the following questions: What is the impact of alternative synthesis routes on the economics and environmental impact of a process? What are the cost sensitive steps of a process that need further development and optimization? What process changes can reduce cycles times and lower manufacturing cost? What changes are required in an existing multi-purpose facility to accommodate the manufacturing of a new product? Our experience in addressing the above and other pertinent questions will be presented using examples from the pharmaceutical and biopharmaceutical industries. The examples will cover production of bulk intermediates as well as manufacturing of final pharmaceutical products.

Web Page: www.intelligen.com/literature

Increasing Productivity in Influenza Virus Production – Virus Yields in High Cell Density Cultivations

A. Bock, A. Wahl, J. Schwarzer, U. Reichl

Max Planck Institute for Dynamics of Complex Technical Systems, Sandtorstr. 1, 39106 Magdeburg, Germany. e-mail: ureichl@mpi-magdeburg.mpg.de

Mammalian cell culture for manufacturing of influenza vaccines plays an increasing role in the prevention and the control of seasonal but also pandemic outbreaks. Due to the enormous socio-economic impact of influenza virus disease not only the development of new vaccination approaches but also the optimization of cell culture-based manufacturing technologies currently being established by various companies is of crucial importance.

The main goal of our group is to develop integrated concepts to optimize cell culture-derived vaccine production processes covering a wide range of unit operations in upstream and downstream processing. As an example we investigate influenza virus replication in various anchorage-dependent and suspension cell lines in stirred tanks and wave bioreactors.

Here, results are presented for cultivation of adherent Madin Darby canine kidney (MDCK) cells in high cell density microcarrier cultures[1]. Maximum cell numbers of about 10^7 cells/mL were obtained by various medium feeding strategies and optimization of incoulum and microcarrier concentrations. Use of serum free medium avoided washing steps before infection[2]. With a cell specific virus yield similar to low density cultivations, productivity of cultivations was increased at least 4 to 6 fold. Level of contaminating proteins of about 220 µg/mL at time of harvest was proportional to cell concentration while concentration of host cell DNA was in the range 8 – 16 µg/mL depending on virus strain and cell lysis. Glycosylation profiles of influenza A virus hemagglutinin showed no significant differences compared to control cultivations performed at lower cell densities[3]. Therefore, no significant problems due to these process modifications are anticipated for use of conventional downstream processing strategies. Quantitative mathematical models for mammalian cell metabolism[4] clearly indicated that precursor supply (amino acids, nucleotides) or consumption of cellular energy (ATP) and redox equivalents (NAD(P)H) is no bottleneck for intracellular virus replication. In contrast, it seems that virus-induced host cell defense and apoptosis are crucial factors to be considered for further increase in cell specific virus yields.

1) Genzel, Y., Reichl, U. (2007): *Vaccine production – state of the art and future needs in upstream processing*, In: Animal Cell Biotechnology – Methods and Protocols, Pörtner, R., Humana Press Inc., Totowa, New Jersey, 457-473.

2) Genzel, Y., Fischer, M., Reichl, U. (2005): *Serum-free influenza virus production without washing steps & medium exchange in large-scale microcarrier culture*, Vaccine, **24**, 3261-3272.

3) Schwarzer, J., Rapp, E., Reichl, U. (2007): *Glycosylation of Influenza A Virus Hemagglutinin*, Proceedings of the 20th ESACT Meeting, Dresden, Germany, June 17 -20, 2007

4) Wahl, A., Sidorenko, Y, Dauner, M., Genzel, Y., Reichl, U. (2007): *Metabolic model for a MDCK cell line: Minimum Substrate Flux Distribution and Phases of Stationarity*, Biotechnology and Bioengineering, in press.

Reducing Batch-to-Batch Variability of *Agrobacterium*-Mediated Transient Protein Expression In Plant Tissue Culture

Jeffrey S. Larsen and Wayne R. Curtis, Chemical Engineering, The Pennsylvania State University, 133 Fenske Laboratory, University Park, PA 16802

Batch-to-batch variations in protein expression levels pose an intractable and often unreported technical hurdle in the development of rapid plant-based heterologous protein expression platforms. Our group is developing protein expression in non-transgenic plant tissue culture which is grown and then co-cultured with *Agrobacterium tumefaciens* to rapidly deliver heterologous transfer DNA (T-DNA) to the nucleus of the plant cells (Curtis 2004). The *Agrobacterium* strain we utilize is an auxotroph that prevents overgrowth within the tissue culture system due to its inability to grow in the absence of a supplied amino acid (Collens, 2004). Since T-DNA transcripts can be rapidly cloned into binary vectors, it is possible to produce kilogram quantities of plant tissue transiently expressing the gene of interest in a matter of days. We have successfully demonstrated this technology at the 50-L pilot scale (O'Neill, 2008) with productivity superior to shake-flask controls. However, even after systematically improving reproducibility via rigor in culture maintenance and synchronization, the concentration of the reporter gene product β-glucuronidase has been observed to vary by orders of magnitude from one experiment to the next. Similar large variations have been observed in plant leaves infiltrated with *Agrobacterium*.

Plant tissue culture-based transient expression provides an ideal platform for controlling variability once the required environmental conditions are identified. We have explored numerous possible physical and physiological sources of batch-to-batch variation in high-level transient protein expression. Some of the important parameters that have been identified and studied as potential contributors to this variation include the organic and inorganic nutrient status of the culture, culture maintenance schedule, gas exchange in shake-flasks, incubator temperature control, and the timing of the nutrient-controlled 'synchronization by starvation' relative to the co-culture of the plant tissues with *A. tumefaciens*.

Achieving reproducible product titers is crucial for developing this protein expression technology for applications ranging from high-throughput screening to commercial-scale production of therapeutic proteins. For large-scale production, predictable and high-level productivity is needed to minimize capital investment as well as develop downstream processing and purification.

Looking forward, we will capitalize on the improved process reliability by utilizing newly developed T-DNA constructs that provide for higher levels of heterologous gene expression (Collens, 2007) and operational strategies that permit implementation of this technology with full gene containment, at a fraction of the cost of traditional pharmaceutical processing. We have developed and patented a low cost (plastic-lined) bioreactor and utilized feed-forward control to grow plant tissues to extremely high density (~40g dry weight / L) at greater than 100 L without the use of expensive pressure-rated vessels, probes or controllers (Curtis, 2004a; 2004b).

References

Curtis WR, Quantitative transient protein expression in plant tissue culture, US #6,740,526, May 25, 2004a.

Curtis WR, Growing cells in a reservoir formed of a flexible sterile plastic liner, US #6,709,862, March 23, 2004b.

Curtis WR, Method and apparatus for aseptic growth or processing of biomass, US #6,245,555, June 12, 2001.

Collens JI, Lee DR, Seeman AM, Curtis WR (2004) The Development of auxotrophic *Agrobacterium tumefaciens* for gene transfer in plant tissue culture. Biotechnol. Prog. 20(3); 890-896.

Collens JI, Mason HS, Curtis WR (2007) Agrobacterium-mediated viral vector-amplified transient gene expression in *Nicotiana glutinosa* plant tissue culture. Biotechnol. Prog. 23(3); 570-576.

O'Neill KM, Larsen JS, Curtis WR (2008) Scale-up of *Agrobacterium*-mediated transient protein expression in bioreactor-grown *Nicotiana glutinosa* plant cell suspension culture. Biotechnol. Prog. 24(2); 372-376.

194c Independent Development of Hydrolysate-Containing and Hydrolysate-Free Cho Cell Culture Processes Yields Different Optimal Conditions

Diane M. Wuest, Polina Rapoport, and Deborah Pascoe. Late Stage Cell Culture BioProcess Development, Genentech, 1 DNA Way, S. San Francisco, CA 94080

In the biotechnology industry, process optimization and performance robustness studies are key to achieving efficient, reliable, and high yield cell culture processes. The use of undefined raw materials, specifically hydrolysates, often enhances the productivity of a culture, but it may also contribute to variability. It is therefore advantageous to remove these undefined materials and develop a capable process without them. Two versions of a fed-batch CHO cell culture process for commercial manufacture of a monoclonal antibody were developed, one with hydrolysates and one without, with a focus on maintaining appropriate product quality relative to the Phase I and II clinical material.

Process development was performed through sequential experiments studying the effects of various process parameters and medium components. Both processes were evaluated in 2L bioreactors and 250mL shake flasks. By performing separate optimization experiments for each process type, different production conditions were defined, demonstrating that the development of a hydrolysate-free process should be performed separately from experiments for a hydrolysate-containing process, and that removal of hydrolysate from a hydrolysate-containing process will not always provide a reliable measure of the potential hydrolysate-free performance of a given cell line.

194d Development of An Animal Component-Free Cell Culture Process

Wen-Lin Tsai, Jennifer L. Autsen, Terrence Allotta, and Judy H. Chou. Oceanside Process Research & Development, Genentech, Inc, 1 Antibody Way, Oceanside, CA 92056

Use of animal component-free cell culture media for the production of recombinant proteins would address regulatory and safety concerns associated with the presence of complex raw materials in cell culture processes. Various approaches were taken through process development efforts to remove an animal-derived peptone, Peptone A, from the cell culture process for a specific antibody. Process development methods included substitution of Peptone A with alternative animal-component free hydrolysates, as well as development of a chemically defined process. Changes were made to media components and process conditions in order to achieve cell culture performance and titer comparable to the peptone-containing process for this antibody. Performance parameters such as cell growth, viability, metabolites, titer, cell cycle and apoptosis levels were monitored throughout production. Spent media was also analyzed for the various conditions to screen for potential limiting components of the peptone-free conditions. The various approaches taken for cell culture process development of an animal component-free process will be discussed.

194e Proteomics and Genome Sequencing to Enhance Chinese Hamster Ovary (CHO) Cell Productivity

Jeffrey C. Swanberg and Kelvin H. Lee. Department of Chemical Engineering, University of Delaware, 150 Academy Street, Colburn Laboratory, Newark, DE 19716

Methotrexate amplification was previously used to increase recombinant secreted alkaline phosphatase (SEAP) production in CHO cells. 2D gel electrophoresis and mass spectrometry were used to identify 21 protein spots with consistently altered expression during amplification. One protein with increased expression, CapZ, inhibits growth of actin filaments. To mimic the effect of CapZ a small molecule drug and actin filament inhibitor, Cytochalasin D (CD), was used. CD was found to destabilize actin and increase SEAP productivity. Here we review the effects of CD on SEAP producing cells in addition to observing the effects two other actin inhibitors, Cytochalasin E and Latrunculin B. Elongation factor 2, which aids translation, and PEST containing protein, associated with protein degradation, were also expressed at increased levels during amplification. We present our observations on the expression of these two proteins on SEAP productivity in CHO cells.

We also discuss the role of non-coding RNA (ncRNA) in CHO cells. Recent studies have shown that over 90% of transcripts in eukaryotic cells are not translated. These transcripts carry out a large variety of regulatory functions. We examine the usefulness of next generation short read sequencing technology as it relates to both ncRNA and genomic DNA in CHO cells. We present low coverage genomic Chinese hamster DNA sequence and its alignment to the mouse and rat genomes.

194f Effects of Cell Culture Conditions on Antibody N-Glycosylation - What Causes High Mannose 5 Glycoform

Efren Pacis, Feng Li, Darren Brown, and Robert Bayer. Process Research & Development, Genentech, 1 Antibody Way, Mail Stop NIMO, Oceanside, CA 92056

The glycosylation of recombinant antibody secreted by mammalian cells affects biological activity in vivo. However, the antibody glycosylation structures are highly variable with cell culture processes. Parameters associated with the cell line, the culture media, and process conditions are known to affect glycosylation. These effects should be better understood in order to consistently make the product with desired quality attributes at different scales.

Antibodies with high mannose glycoforms (Man 5-9) are typically present at low levels, comparing to main antibody glycoforms G0, G1, and G2. Some recent publications indicate that antibodies bearing predominantly the Man5 glycoform have some unique properties in terms of potency, immunogenicity, and plasma clearance rate.

We have studied the effects of cell culture process conditions on antibody Man 5 levels. Multiple process parameters were tested to have better understanding of how to control the product quality in the cell culture process, and improve the consistency of glycosylation profiles. Results of our studies will be presented.

194g Consistent Production of Immunoglycoproteins of Higher Adcc Activity by Controlling Fucosylation In Cho Cells

Vijay Yabannavar, Brian Albarran, Jackie McGourty, and Richard Lowe. Trubion Pharmaceuticals, 2401 4th Avenue, Suite 1050, Seattle, WA 98121

In many immunoglycoproteins (including antibodies) used as biopharmaceuticals, increased ADCC activity is highly desirable. The extent of fucosylation in N-linked glycosylation has a major influence on ADCC activity. Unfortunately, it is reported that the extent of fucosylation varies significantly in a typical cell culture process. Here we report our discovery of the factors that influence the fucosylation in the production of immunoglycoproteins. Using this knowledge, we demonstrate a new system to make more potent immunoglycoproteins consistently. With the use of LC-MS and ADCC (and CD16 binding) assays, we have correlated the structure-ADCC activity relationships for the immunoglycoproteins.

197a Identification of Potential Target Pathways of Recombinant Yeast for Increasing Bioethanol Productivity: In Silico Analysis Using Cybernetic Models

Hyun-Seob Song, John A Morgan, Nancy W. Y. Ho, Arvind Varma, and Doraiswami Ramkrishna. School of Chemical Engineering, Purdue University, Forney Hall of Chemical Engineering, 480 Stadium Mall Drive, West Lafayette, IN 47907

Continuing efforts are in progress to design effective recombinant yeasts converting pentose sugars as well as hexose sugars to ethanol for the development of an economically viable B2B (biomass-to-biofuel) process. While significant advances have been made in the past in the metabolic engineering of fermenting yeasts, further increase of bioethanol productivity is desirable. In this regard, mathematical models are needed for a more systematic approach to genetic modification.

A hybrid cybernetic model has been developed for the Ho-Purdue strain 1400 (pLNH33) by combining cybernetic modeling concept with elementary mode analysis based on the steady-state postulate on intracelluar metabolites. The hybrid cybernetic model is characterized by the dynamical description of metabolic processes as a combination of elementary mode fluxes. The selection of the modes is modulated by cybernetic control laws towards maximizing a predetermined metabolic fitness function (e. g., biomass growth rate or carbon uptake rate). Using this model, we have investigated the effect of various pathway modifications on ethanol productivity of yeast and fermenters. Comprehensive dynamic simulations have been conducted together with metabolic control analysis to identify the metabolic pathway which can lead to the substantial increase of ethanol productivity. While as such, the hybrid model provides critical information on the pathway modification for increasing ethanol productivity, an advanced model based on detailed metabolic description might be required to narrow down the range of genetic modification to a few specific target genes. For this purpose, a more detailed cybernetic model based on Young's modeling framework (Young 2005; Young et al. 2008) is being formulated and will be applied for the development of an improved yeast strain.

References

Young JD. 2005. A system-level mathematical description of metabolic regulation combining aspects of elementary mode analysis with cybernetic control laws. PhD Dissertation, Purdue University

Young JD, Henne KL, Morgan JA., Konopka AE, Ramkrishna D. 2008. Integrating cybernetic modeling with pathway analysis provides a dynamic, systems-level description of metabolic control. Biotechnology and Bioengineering 100(3):542-559

197b Computational Analysis of Yeast Xylose Metabolism

Ljubisa Miskovic and Vassily Hatzimanikatis. Laboratory of Computational Systems Biotechnology, Ecole Polytechnique Federale de Lausanne (EPFL), BCH 3111 (Bâtiment de chimie UNIL), CH - 1015, Lausanne, Switzerland

Xylose fermentation from lignocellulosic material using recombinant yeast cells is a promising resource for efficient ethanol production as a future sustainable energy supply. Experimental approaches emphasize the engineering of xylose-utilizing recombinant Saccharomyces cerevisiae strains to overcome the potential bottlenecks of xylose uptake and cofactor imbalance and to minimize the xylitol accumulation. In this work, we apply systems engineering methodology in order to identify metabolic engineering targets for the optimization of ethanol production by Saccharomyces cerevisiae in the continuous culture. We employ a computational framework we have recently developed for Metabolic Control Analysis, i.e., the quantification of the responses of metabolic properties to changes in genetic manipulations and environmental (process) conditions. We find that redistribution of the metabolic flows around pyruvate among carboxylation, decarboxylation, and transport has significant potential in improving total utilization of hexose and pentose sugar as well as its conversion into ethanol. We also find such objective may be achieved with a more efficient ATP molecule dissipation through maintenance demand and respiration.

197c The Dynamic Biomass Constituting Equation for the Genome-Scale Model of the Biofuels Producer *Clostridium Acetobutylicum*

Ryan S. Senger, Department of Chemical Engineering, University of Delaware, Delaware Biotechnology Institute, 15 Innovation Way, Newark, DE 19711 and Eleftherios T. Papoutsakis, Dept. of Chemical Engineering, Delaware Biotechnology Institute, 15 Innovation Way, Newark, DE 19711.

It has become well-established that genome-scale models using metabolic flux analysis provide the mathematical link between a cellular genotype and its expressed phenotype. However, this is generally done by calculating the metabolic capacity or phenotypic phase plane, which consists of families of solutions. When considering cell growth in a genome-scale model, empirical relations describing the compositions of the biomass building-blocks: DNA, RNA, protein, lipids, cell wall, and solute pools are combined with energetic requirements to form the biomass constituting (or cell growth) equation. Here, we describe the dynamic nature of the biomass constituting equation necessary for predicting vegetative growth and metabolism of *Clostridium acetobutylicum*, an endospore forming, strict anaerobe which is the type strain for the industrial acetone, butanol, and ethanol (ABE) fermentation. The cellular maintenance energy component of the biomass constituting equation was found to be about 4-times greater in the lag and late exponential growth phases than what was calculated for the maximum rate during the vegetative growth phase. Considering that *C. acetobutylicum* takes on energy-intensive processes such as spore germination during the early lag phase, the initiation of sporulation events and the development of solvent tolerance mechanisms in the late exponential and early stationary growth phases, these calculations enabled modeling of the full exponential growth phase consistent with experimental data. Our method was also used to analyze and optimize the lipid component of the biomass constituting equation, a component which is well-known to change in clostridia cultures in response to the accumulation of weak acids and solvents in the extracellular environment.

197d Elucidating Mechanisms of Acetate Tolerance in E. Coli Using Scales

Nicholas R. Sandoval and Dr. Ryan T. Gill. Department of Chemical and Biological Engineering, University of Colorado, ECCH 111, Campus Box 424, Boulder, CO 80309

Creating biofuels from hemicellulosic and cellulosic biomass is an important part in transitioning away from the petroleum-based transportation fuel economy. Acetate is a major toxic side product of the requisite pretreatment steps of the feedstock to the microorganisms producing biofuels. Conferring tolerance of acetate upon the microorganism will increase the viability of the biofuels economy.

The SCalar Analysis of Library Enrichments (SCALEs) method, created previously by the Gill lab at the University of Colorado, combines a traditional genomic library selection with DNA microarray technology providing genotypic data for selected clones. The method employs multiple libraries of different, but defined, insert sizes to allow greater resolution of beneficial genomic regions. The objective is to find mechanisms of acetate inhibition on E. coli growth. These mechanisms can be elucidated using SCALEs. Sections of the genome that confer acetate tolerance, when over-expressed, will be found in a SCALEs selection.

We report here two selections using an E. coli K12 library with a 1.75 g/L or a 2.5 g/L acetate culture titrated to a pH of ~7.0 with potassium hydroxide. The cultures were monitored over a three day period with serial transfers to ensure cultures were not in stationary phase. Top selected clones show an increase in growth rate between 60% to over 100% against control.

Microarray studies were performed and analyzed using the SCALEs method. Regions of the genome were found that conferred tolerance the most. Validation growth rate studies were done via subcloning. These top performing regions contained genes that were important in amino acid transport and metabolism and cell envelope biogenesis. SCALEs data underwent a pathway analysis and those pathways most enriched included amino acid metabolism pathways and central metabolism pathways.

197e Engineered Biosynthesis of Complex Aromatic Polyketides Using Fungal Polyketide Synthase In Escherichia Coli

Wenjun Zhang, Chemical and Biomolecular Engineering, University of California at Los Angeles, 420 Westwood Plaza, 5667 Boelter Hall, Los angeles, CA 90095 and Yi Tang, Chemical and Biomolecular Engineering, University of California, Los Angeles, 420 Westwood Plaza, Los Angeles, CA 90095.

Bacterial aromatic polyketides are a family of structurally diverse natural products with important pharmaceutical and agricultural applications, including antibiotics tetracyclines and anticancer agents doxorubicin, etc. Since most natural polyketide-producing organisms are slow-growing and/or genetically intractable, it is desirable to transfer polyketide biosynthetic pathways to a more robust host such as Escherichia coli (E. coli) for pathway engineering and product isolation. While E. coli has been recently metabolically engineered to produce simple monocyclic or bicyclic compounds, complex aromatic polyketides have not been synthesized by E. coli largely due to the fact that the responsible type II minimal polyketide synthase (PKS) cannot be functionally expressed in E. coli. In this work, the biosynthetic potential of fungal PKS4 from Gibberella fujikuroi will be explored for combinatorial biosynthesis of bacterial aromatic polyketides in E. coli. Using a catalytic platform that employs a reengineered PKS4 and tailoring enzymes, we demonstrate the biosynthesis of complex aromatic polyketides in E. coli for the first time.

197f Deciphering the Late Steps In FR900098 Biosynthesis

Matthew A. DeSieno, Tyler W. Johannes, and Huimin Zhao. Chemical & Biomolecular Engineering, University of Illinois at Urbana-Champaign, 600 S. Mathews Ave., 221 RAL, Box C-3, MC-712, Urbana, IL 61801

Antibiotics containing C-P bonds, phosphonates and phosphinates, are a potent group of bioactive compounds. The phosphonic acid FR-900098 is a novel chemotherapeutic agent for the treatment of malaria and has been shown to be effective in humans and other animals. This compound inhibits 1-deoxy-D-xylulose 5-phosphate (DXP) reductoisomerase, the first committed enzyme in the nonmevalonate pathway for isoprenoid biosynthesis in many bacteria as well as *Plasmodium falciparum*, the parasite responsible for the most virulent form of the disease. The biosynthetic cluster for FR-900098 has recently been cloned from *Streptomyces rubellomurinus* and heterologously expressed in *Streptomyces lividans*. The gene cluster was also successfully reconstituted in *E. coli*, resulting in 6 mg/mL FR-900098.

The initial step in FR-900098 biosynthesis is similar to almost all other known phosphonate biosyntheses, a phosphoenolpyruvate (PEP) mutase catalyzed reaction of PEP to phosphonopyruvate. The subsequent steps parallel the tricarboxylic (TCA) cycle, resulting in the formation of 2-oxo-4-phosphonobutyrate, an analog of α-ketoglutarate. The final steps were proposed based only on sequence homology of the remaining genes, without any definitive assignment of the order or function. Here we report the deciphering of the last steps in FR-900098 biosynthesis, verified through whole cell feeding experiments and *in vitro* assays with purified enzymes. Two novel enzymes were revealed in the process; one a bifunctional enzyme capable of performing a nucleotide transfer and decarboxylation and the other an amide hydroxylase. A complete understanding of the FR-900098 biosynthetic pathway now opens the possibilities for metabolic engineering in the native or heterologous *Streptomyces* strain or in *E. coli* to increase production of the antibiotic and combinatorial biosynthesis to generate novel derivatives of FR-900098.

197g Metabolic Network Tuning to Increasing Malonyl-CoA Biosynthesis In *Escherichia Coli* and Its Application to Flavonoid Production

Zachary Fowler and Mattheos Koffas. Chemical and Biological Engineering, State University of New York at Buffalo, 914 Furnas Hall, Buffalo, NY 14260

Constraint-based modeling is an emerging tool for analyzing and understanding biological systems, specifically the connectivity of the metabolic network. As such, we have developed a Cipher of Evolutionary Design (CiED) to investigate the impact of gene deletions and other network modifications on the metabolite profile of microorganisms. By incorporating constraint-based modeling into an evolutionary algorithm we have created a computational platform where, like other evolutionary algorithms, the process of natural selection is used to find optimal phenotypes for the production of end products, such as recombinant natural products, while maintaining high levels of cellular biomass. Here we report the use of CiED to investigate the metabolic potential of *Escherichia coli* to channel carbon towards malonyl-CoA in an effort to generate recombinant strains with elevated flavonoid production capacity. Flavonoids, along with several of their substituted unnatural analogues, have potential therapeutic value in the treatment of various chronic diseases such as cancer, obesity, type II diabetes, and heart disease. Our engineered *E. coli* strains were first modified by the targeted deletion of native genes controlling competing enzymatic processes predicted by CiED and secondly through the incorporation of selected enzyme overexpressions. These included the coexpression of the plant-derived flavanone genes as well as acetate assimilation, acetyl-CoA carboxylase and the biosynthesis of Coenzyme A. As a result, the specific flavanone production from our optimally engineered strains was increased by over 660% for naringenin (15 to 100 mg/L/OD) and by over 420% for eriodictyol (13 to 55 mg/L/OD). These efforts demonstrate the utility of an evolutionary model based solely on stoichiometry in predicting improved microbial phenotypes for natural product production.

210a Parsing Contact-Inhibition of Motility and Its Role In Multicellular Morphodynamics

Anand R. Asthagiri, Chemical Engineering, California Institute of Technology, 1200 E. California Blvd MC 210-41, Pasadena, CA 91125, Melissa Pope, Bioengineering, California Institute of Technology, 1200 E. California Blvd MC 210-41, Pasadena, CA 91125, and Keiichiro Kushiro, Biochemistry and Molecular Biology, California Institute of Technology, 1200 E. California Blvd MC 210-41, Pasadena, CA 91125.

Cell motility plays a central role in the formation and reshaping of multicellular structures. In normal epithelial tissues, cell-cell contact constrains cell motility. Mutations that relieve contact-inhibition of motility allow epithelial cells to escape from their primary tissue, altering epithelial tissue morphology and contributing to metastasis. Our lab is developing quantitative approaches to better understand the mechanisms that regulate contact-inhibition of motility and its role in multicellular morphogenesis. In this talk, I will describe how we are applying automated image analysis, micropatterned materials and mathematical modeling to quantify how cell-cell interactions affect motility and how this cellular-scale phenomenon alters multicellular phenotypes, such as cell scatter and aggregation.

210b Tissue Development Is Regulated by Endogenous Patterns of Mechanical Stress

Nikolce Gjorevski and Celeste M. Nelson. Chemical Engineering, Princeton University, Department of Chemical Engineering, A321 Engineering Quadrangle, Olden Street, Princeton, NJ 08544

The branched patterns of epithelial organs are generated by spatial differences in the behaviors of individual cells. In addition to biochemical gradients, cells within branching tissues are possibly exposed to mechanical gradients that originate both internally and externally. We have developed lithography-based three-dimensional culture models that recapitulate the architecture of mammary epithelial trees; these engineered tissues undergo a morphogenesis process similar to that of the mammary gland in vivo. Here we combine these models with computational approaches to investigate the role of external and internal mechanical stress gradients during tissue development. We characterize the effects of cell contractility, intercellular adhesion and extracellular matrix stiffness on the emergent pattern and extent of branching in mammary epithelium. Our data provide physical understanding of the morphogenetic processes that drive normal tissue development, and can help elucidate the events that lead to pathologies such as fibrosis and tumorigenesis.

210c A Quantitative Multi-Scale Approach to Study Cell Migration In 3D Matrices

Muhammad H. Zaman[1], Tianyi Yang[2], and Erin Baker[1]. (1) Biomedical Engineering, University of Texas at Austin, ENS 610; C0800; 1 Univ. Station, Austin, TX 78712, (2) Physics, University of Texas at Austin, Austin, TX 78712

Cell migration and cell-matrix interactions have traditionally been studied in the context of 2D environments, where cells are cultured on artificial two-dimensional surfaces. While a number of key cellular and molecular parameters can be studied and quantified accurately with these classical approaches, a comprehensive and systems level understanding of cell migration requires quantification forces, signals and molecular interactions in native like 3D environments. In particular, the mechano-chemical effects of the matrix, the proteolytic pathways and surface receptor dynamics on a 3D surface have all shown to be critical in invasion and tumor metastasis, and can not be fully studied in a 2D environment.

In order to overcome the limited powers of observation in 2D, we utilize a combination of high resolution and high throughput confocal microscopy, bulk and micro-rheological measurements and multi-scale simulations rooted in statistical and continuum mechanics to understand the mechanical and chemical roles of the matrix in regulating cell motility. Our results show that in addition to adhesion and tractile forces, matrix stiffness is a key factor that influences cell movement in 3D in a number of fibrosarcoma, breast and prostate cancer cell lines. We also observe that 3D environments play a far more critical role in modulating cellular viscoelasticity than 2D substrates. Cellular response to minor mechanical changes in its extra-cellular environment is also amplified in 3D than in 2D environments. Our experimental results are complemented by multi-scale simulations, that predict and quantify the synergestic role of receptor-ligand interactions, matrix mechanics and adhesive forces in regulating motility. Our hybrid approach, combining high-resolution experimental and computational techniques demonstrates how a balance of cellular parameters (e.g. integrin expression and MMP activity) co-operate with matrix properties (e.g. composition, stiffness and porosity) to regulate invasion and motility of tumor cells in 3D environments.

210d Systems-Level Analysis of Epithelial Patterning and Morphogenesis: The Control of Cytoskeleton and Adhesion Proteins by Signaling Pathways

Jeremiah Zartman, Nir Yakoby, Jitendra S Kanodia, and Stanislav Y Shvartsman. Department of Chemical Engineering and Lewis-Sigler Institute for Integrative Genomics, Princeton University, Engineering Quadrangle, Princeton, NJ 08544

The morphogenesis of epithelial tissues can be characterized by the sequential execution of a core set of programmed stereotypical shape changes and cellular movements, which result from the differential expression of cytoskeleton and adhesion proteins. The patterned expression of such effector molecules depends on the integration of multiple signaling pathways. Understanding the logic and cellular mechanisms of morphogenesis requires a systems-level analysis of the epithelial patterning and a functional characterization of differentially expressed adhesion and cytoskeleton proteins. The follicular epithelium in Drosophila egg development (oogenesis) serves as an established model for studying epithelial patterning and morphogenesis, amenable to live imaging and sophisticated genetic perturbations. During Drosophila oogenesis, a two-dimensional follicular epithelium encapsulating the developing egg gives rise to an elaborate three-dimensional eggshell, which includes tubular structures called dorsal appendages that project out from the main body, and act as respirator tubes for the developing embryo.

Egg morphogenesis and dorsal appendage formation have been studied extensively, but few connections have been made between signaling pathways, pattern formation, and the physical implementation of the morphogenetic program in this tissue. In an effort to address the knowledge gap between pattern formation and cell mechanics, we have identified the expression pattern of a set of adhesion and cytoskeleton genes, which are required for proper organogenesis. Further, we have characterized the regulation and function of one of these dynamically regulated adhesion genes, the non-classical cadherin, Cad74A, which is expressed starting in mid-oogenesis in all the columnar follicle cells contacting the oocyte except for two dorsolateral patches. The dorsolateral patches correspond to the cells that form the roof of the dorsal appendages by undergoing apical constriction, cell intercalation, and cell migration. Using an antibody to Cad74A, we show that Cad74A repression in the dorsolateral patches is mediated by high levels of the transcription factor Br, which in turn is regulated by the integration of EGFR and Dpp signaling, the two pathways that are essential for patterning the follicle epithelium. We also report the functional analysis of Cad74A which disrupt cell morphology and robust dorsal appendage formation when expression levels are perturbed. Based on the results of the Cad74A case study, we propose a model for the regulation and integrated function of a group of differentially expressed adhesion and cytoskeleton genes during egg development.

210e Egf-Induced Contraction Temporally Separates Traction Generation from De-Adhesion

Ian Schneider[1], Cristen Kern[1], and Clare Waterman[2]. (1) Cell Biology, The Scripps Research Institute, 10550 N. Torrey Pines Road, CB167, La Jolla, CA 92037, (2) National, Heart, Lung and Blood Institute, National Institutes of Health, Bethesda, MD 20892

Cell migration involves coordination of protrusion, adhesion and contraction. Central to this process is the spatiotemporal regulation of focal adhesion (FA) assembly, maturation and disassembly. Myosin II-mediated contraction in the actin cytoskeleton drives FA maturation, which is characterized by FA growth and changes in FA protein composition and phosphorylation state. Epidermal growth factor (EGF), a physiological stimulus for cell migration, is well-studied for its effects on cell protrusion, but how it temporally controls contraction leading to traction generation and FA maturation and disassembly is less clear.

We observed that epidermal keratinocytes stimulated with EGF show a protrusion phase followed by a temporally delayed retraction phase. Using microscopic and biochemical assays, we quantified the temporal ordering of biochemical and mechanical processes with respect to these morphological changes. We found that myosin-II activation occurs very quickly after stimulation. Lagging behind this response was protrusion, traction generation and the formation of small FAs rich in phosphorylated paxillin. Much later, paxillin was dephosphorylated, marking the maturation of FAs, and retraction occurred. Inhibition of myosin-II not only blocked the EGF-stimulated contraction leading to cell retraction; it also eliminated the delay in paxillin dephosphorylation and FA maturation. This suggests that EGF-mediated activation of myosin-II acts as a mechanical signal to promote a period of adhesion maturation that delays paxillin dephosphorylation and adhesion weakening relative to increases in traction generation. This may serve as a mechanism by which the cell temporally segregates protrusion, adhesion and traction generation from retraction during EGF-stimulated cell migration.

210f Investigating Concentration Gradients within a Novel 3D Angiogenesis Tissue Model

Heather Gappa-Fahlenkamp and Subuola Sofolahan. Chemical Engineering, Oklahoma State University, 423 Engineering North, Stillwater, OK 74078

Angiogenesis the growth of new capillaries from pre-existing blood vessels, requiring growth factor driven recruitment, migration, proliferation, and differentiation of endothelial cells (ECs). The study of angiogenesis has many clinical applications in numerous fields, including peripheral and coronary vascular disease, oncology, hematology, wound healing, dermatology, and ophthalmology. In 1980, bovine capillary ECs were found to spontaneously form tubes when cultured in gelatin in vitro. Since then, many in vitro models have been used to recapitulate the basic steps of the in vivo process. Such in vitro models of angiogenesis offer many possibilities: the clinical testing of potential drug therapies; the modeling of pathological conditions; the study of the processes of endothelial cell differentiation, lumen formation, and vascular inoculation; and the investigation of the molecular mechanisms associated with angiogenesis. Endothelial cell differentiation, the lumen or tube formation, can be studied in vitro both in two dimensions and in three dimensions (3D); however, to study the other cellular mechanisms involved, a more complex 3D model that includes cell and extracellular environment interactions is required. Such 3D models involve seeding ECs either on or in gels. Studies involving 3D angiogenesis models have investigated the role of some key design factors on the migration, proliferation, and differentiation of ECs, including endothelial cell type, type of matrix, the concentration and the biochemical conditions of the matrix polymerization that can affect the density and the mechanical properties of the substrate, the culture media composition, and the bioavailability of angiogenic factors. The effect of the key design variables on concentration gradients of oxygen, nutrients, and angiogenic factors within the 3D models has not been fully explored. We have designed a 3D angiogenesis model that can be used to investigate such concentration gradients on the migration, proliferation, and differentiation of ECs. Such a model is comprised of ECs grown on a bovine type I collagen gel formed within a Transwell® membrane plate. This system allows access to compartments both above and below the gel that can be used for the delivery of factors or for sampling. Holding all other variables constant, we saw a significant increase in both the kinetics and the amount of migration, proliferation, and differentiation of ECs within the collagen gel for the samples in the Transwell® membrane plate compared to samples in a typical solid-bottom well plate, which only allows access to a compartment above the gel. We also tested the effect of adding growth factors to either the top or the bottom compartment of the Transwell® membrane plate on both the kinetics and the amount of migration, proliferation, and differentiation of ECs. The results show that both the kinetics and the amount of migration, proliferation, and differentiation of ECs within the collagen was significantly greater for the samples with the growth factors added to the bottom chamber only, showing a clear influence of the concentration gradient on the cellular response. We are continuing to use this system to show that how angiogenic factors are delivered to a 3D system, by taking into account concentration gradients, are just as important as to what angiogenic factors are being delivered.

210g Numerical Modeling of 3-D Single Breast Cancer Cell Migration towards An Egf Concentration Gradient

Doh-Hyoung Lee[1], Temitope R. Sodunke[1], Mauricio J. Reginato[2], and Hongseok Noh[1]. (1) Mechanical Engineering and Mechanics, Drexel University, 3141 Chestnut Street, Philadelphia, PA 19104, (2) Biochemistry and Molecular Biology, Drexel University, 3141 Chestnut Street, Philadelphia, PA 19104

An important aspect of cancer biology is cell migration, which is often mediated by chemical gradients in the environment. Three-dimensional (3D) epithelial culture models utilizing biological based matrices such as Matrigel (a popular biomatrix hydrogel) provide a more physiologically relevant setting for studying cancer cell migration. In this paper, we consider a mathematical model of cancer cell migration on a tissue like matrix, which focuses on the response of single MDA-MB-231 (a breast adenocarcinoma cell line) to a concentration gradient of epithelial growth factor (EGF). The domain of interest consists of a water-like medium layer on top of a Matrigel layer. The diffusion of EGF concentration throughout the domain was solved by a finite element method. The diffusion coefficient of EGF in medium was calculated via a previously developed empirical formula while that in Matrigel layer was estimated after a careful study on the characteristics of thin two-layer diffusion problem. An empirical relationship between cell migration rate and the growth factor gradient was developed based on the previous experimental results obtained by our group. The Lagrangian cell migration was then numerically calculated from the modeled cell migration rate. The cell migration results obtained from numerical computations shows an agreement with experimental results.

214a Microneedle-Based Immunization against Influenza

Vladimir G. Zarnitsyn[1], Sean Sullivan[2], Ioanna Skountzou[3], Dimitrios Koutsonanos[3], Chinglai Yang[4], Ling Ye[3], Richard Compans[4], and Mark R. Prausnitz[1]. (1) School of Chemical and Biomolecular Engineering, Georgia Institute of Technology, 311 Ferst Drive, Atlanta, GA 30332-0100, (2) Wallace H. Coulter Department of Biomedical Engineering, Georgia Institute of Technology, 311 Ferst Drive, Atlanta, GA 30332, (3) Department of Microbiology and Immunology, Emory University, 1510 Clifton Road, Rollins Research Center, Rm 3108, Atlanta, GA 30322, (4) Microbiology and Immunology Department, Emory University, 1510 Clifton Road, Rollins Research Center, Rm 3108, Atlanta, GA 30322

Influenza virus is one of the common causes of serious respiratory illness. Typically, the virus causes seasonal epidemics of disease; during a typical season 36,000 deaths and 200,000 hospitalizations are observed, with up to 1.5 million deaths worldwide, which make it the largest cause of death in the US from vaccine-preventable disease. The vaccines do not prevent infection, but result in a reduction of disease severity and transmission.

The composition of the influenza vaccine is determined annually based on the strains of virus that are expected to be circulating in the human population just a few months before the vaccination season. As a result, the production, storage and distribution of the necessary amount of vaccine puts a lot of pressure on the pharmaceutical industry. Improved vaccine coverage is needed, which can be facilitated by better delivery methods.

In this study we investigated a novel way to facilitate and enhance influenza vaccination, based on arrays of microscopic needles (microneedles) coated with inactivated influenza vaccine. Using dozens of microneedles coated with influenza vaccine, the vaccine will be delivered in a painless yet simple patch-like manner. An additional advantage of intradermal delivery of vaccine is that the upper layers of skin are rich with antigen presenting immune cells which makes vaccine delivery to these areas more efficient. Intradermally delivered vaccine indeed showed better efficacy in elder population. Because vaccine is stored and administered in the solid state, vaccine stability may be sufficient for storage and distribution at room temperature. In addition, this approach may also address other limitations of the current injectable vaccine, including its limited breadth and duration of immunity, and its lack of effective mucosal immune responses.

In this study, microneedles were coated by a dip coating method. In order to achieve an adequate coating and good wetting of microneedle tips, thickening and surfactant additives were used during the process. At the final stage the coating solution was dried on microneedle tips at room temperature and humidity. Drying of vaccine during coating may potentially damage the immunogenicity of vaccine. By varying thickening agents, buffer composition and protective additives we optimized the coating composition and condition in order to achieve optimal amount of coated vaccine and its mechanical and biological stability. Under current conditions it is possible to coat up to 2 micrograms of vaccine per individual needle.

In order to address the vaccine activity we measured the hemagglutination activity (HA) of vaccine after drying on microneedles. This test gives a quick measurement of the activity of surface receptors of influenza with the only downside of possible false-negative outcome to predict antigenicity in vivo. Optimized formulations were found to retain more than 25% of its original activity.

The immunogenic activity of vaccine was tested by vaccination of groups of Balb/c female mice with 10 micrograms of influenza vaccine. The groups of mice were vaccinated by conventional IM injections or by coated microneedles containing H1N1 or H3N2 vaccines. The efficiency of immunization was measured by antibody-specific IgG titer and hemagglutination inhibition (HAI) in plasma. These measurements confirmed that microneedle delivered vaccine is able to increase the level of protective antibodies in blood compared to naïve mice.

At the final step the protection caused by induction of cellular and humoral immune systems was measured by challenging groups of mice by lethal doze of live influenza virus (5 x LD50). Both microneedle vaccine and the positive control IM groups showed good protection against this challenge.

In conclusion, our results showed that vaccine coated on microneedles is easy to administer and provides sufficient protection to mice even after a single vaccination (priming) and even better protection after double vaccination (priming and boosting).

214b Effect of Freezing Rate and Geometry on Protein Particle Formation and Storage Stability

Miguel A. Rodrigues, Maria Andrea Miller, and Keith P. Johnston. University of Texas at Austin, Austin, TX 78712

A variety of freezing processes including spray freeze drying, thin film freezing and bulk freezing are utilized for producing protein particles for drug delivery. Thin film freezing (TFF) enables freezing at considerably fast cooling rates (of ~10^2 K/s), without exposing the proteins to large gas-liquid interfaces which can otherwise lead to unfolding and aggregation. The cooling rates experienced in the TFF process permit the formation of an ice template, where thin liquid channels formed between the ice domains produce high freezing rates. Upon lyophilization of such ice structures, fine protein particles with sizes on the order of a few 100 nm have been harvested. The objective of this work was to relate the protein morphology to the freezing rates for solutions with varying heats of solution and freezing points. The particle morphology was analyzed by scanning electron microscopy, and the size distribution was measured by laser light scattering and aerodynamic time of flight.

Freezing processes are also of interest in storage stability. Freezing processes can lead to losses in protein stability as a result of exposure to air-water and ice-water interfaces and cold denaturation as well as freeze-concentration. We report how the freezing rate and geometry influence ice formation, freeze-concentration and protein stability in the frozen state. Freezing temperature profiles and ice front velocities were measure to understand the stresses experienced by the protein and how they influence protein stability.

214c Glutathione Sensitive PAMAM Dendrimer Conjugates Tailored for Intracellular N-Acetyl-Cysteine Release

Yunus Emre Kurtoglu[1], Raghavendra Navath[2], Bing Wang[3], Sujatha Kannan[3], Rangaramanujam M. Kannan[4], and Roberto Romero[5]. (1) Chemical Engineering, Wayne State University, 5050 Anthony Wayne Drive, Detroit, MI 48202, (2) Chemical Engineering/NICHD PRB, Wayne State University, 5050 Anthony Wayne Drive, Detroit, MI 48202, (3) Pediatrics, NICHD PRB, Wayne State University, 5050 Anthony Wayne Drive, Detroit, MI 48202, (4) Chemical Engineering, NICHD PRB, Wayne State University, 5050 Anthony Wayne Drive, Detroit, MI 48202, (5) NIH NICHD Prb, Wayne State University, Detroit, MI 48202

N-Acetyl-L-Cysteine (NAC) is a drug frequently used as a mucolytic agent as well as acetaminophen overdose cases. The determination of its antioxidant, radical-scavenging properties, gave rise to investigation of its other therapeutic applications. Even though NAC proved to be a very therapeutic drug, delivery problems remain. The stability of NAC is low which increases the effective dose required. The high protein binding is another problem associated with NAC therapies since it affects activity and elimination of the drug. We proposed that the efficacy of NAC could be increased by protecting the drug from degradation, preventing protein binding and enhancing the cellular uptake, in addition to targeting it to tissue of interest in the body. For this purpose, we designed Polyamidoamine (PAMAM) dendrimer NAC conjugates as delivery devices that can address these issues. Evaluation of the nanodevices indicates that we successfully engineered Glutathione sensitive PAMAM dendrimer conjugates with high payload that release NAC specifically inside the cells and have higher in vitro efficacy compared to free NAC.

PAMAM dendrimers are highly hyper-branched synthetic polymers with well defined spherical structure and size. They have shown prospect as drug delivery vehicles with active and passive targeting capabilities, drug solubility and stability enhancement and vast opportunities for innovative designs at molecular level. In this study, we developed PAMAM dendrimer conjugates containing disulfide linkages. The choice of disulfide linkage was aimed at enhancing the intracellular availability of NAC by protecting it from degradation and protein binding and finally selectively releasing it intracellularly employing differences in Glutathione (GSH) levels. GSH is the most abundant thiol species in the cytoplasm and the major reducing agent in biochemical processes. The intracellular GSH concentration (1-10 mM) is substantially higher than extracellular levels (2uM in plasma), thus provides opportunities for intracellular cleavage of disulfide linkages.

In the present investigation, we have successfully synthesized two conjugates of NAC with G4-NH2 and G3.5-COOH PAMAM dendrimers. Additionally Fluorescein Isothiocyanate (FITC) was also conjugated for imaging and cellular uptake studies. The cell studies were carried out on Mouse microglial cells (BV-2). G4-NH2 and G3.5-COOH conjugates contained 16 and 18 NAC molecules per dendrimer respectively as determined by 1H-NMR and MALDI-TOF analysis. The drug release kinetics of the conjugates was evaluated at intracellular and extracellular GSH concentrations. The release mechanism was determined to be via disulfide exchange reactions with the thiol group of GSH. The conjugates released ~50% of their NAC payload within 1 hour at intracellular GSH concentrations whereas they were quite stable at extracellular GSH levels and did not release significant levels of NAC. The release rates indicate that the disulfide bond was easily and rapidly cleaved by GSH to a high extent within a short amount of time. These results have significant implications in achieving controlled release in dendrimer-based delivery systems as well as demonstrating that GSH can be used as a reliable in vivo releasing agent.

Cellular entry of FITC labeled G4-NH2-NAC conjugate was evaluated using flow cytometry. It was determined that most of the G4-NH2-NAC conjugate entered BV-2 cells within 15 minutes after incubation and uptake continued slowly for another 45 minutes. The cellular entry was also visualized by using Confocal Laser Scanning Microscopy. It is evident that the FITC labeled G4-NH2-NAC conjugate localized mostly in the cytoplasm and lysosomes, while the nucleus appears to be relatively free of the presence of any fluorescence two hours after incubation. The results indicate that PAMAM dendrimer conjugates are transported inside the cells efficiently by endocytosis mechanism.

Efficacy of conjugates were evaluated by measuring the Nitrite in BV-2 cell culture medium which indicates the level of anti-oxidation. Free NAC inhibited nitrite release in dose-dependent manner after 72 hours incubation, while only high concentration of free NAC (8mM) inhibited nitrite release significantly after 24 hour incubation. On the other hand both G4-NH2 and G3.5-COOH conjugates had more significant inhibition both at 24 and 72 hours compared to the same concentration of free NAC. The results indicated both dendrimer-NAC conjugates are more effective anti-oxidants compared to equivalent free NAC, especially at lower NAC doses. The results indicate that PAMAM dendrimer conjugates produce a higher local NAC concentration inside the cells by enhancing cellular uptake and blocking degradation of the drug.

ABSTRACT

Statistical figures outline the five-year survival rate for all cancers diagnosed between 1996 and 2002 as 66%, which depicts a marked rise from the 51% that survived in 1975-1977[1]. However, cancer still remains the second leading cause of death in the United States, following heart disease. An American Cancer Society report estimated that in 2007, there will be over 1.4 million new cancer cases and over half a million cancer deaths in the United States[1]. Although significant oncology drug discoveries have been made during the past 30 years, conventional chemotherapeutic agents exhibit poor specificity in reaching the tumor site and are often restricted by toxicity factors. The lack of a uniform biodistribution leads to harmful side-effects to healthy tissues and the need for administration of a larger than necessary drug dosage with a higher repetitive rate so as to elicit a satisfactory pharmacological response.

Wide interest in cancer nanotherapy has led to the development of nanoparticle based "smart drugs" that have not only improved pharmacological and therapeutic properties of anticancer drugs, but also offer a less invasive alternative enhancing the patient's life expectancy and quality of life as well. Dendrimers, due to their unique architecture and macromolecular characteristics are currently used extensively in research of nanoparticles for targeted and controlled drug delivery. The research objective was to design, synthesize and characterize a novel nanoparticle based "PAINT-BRUSH" like multi-hydroxyl capped poly (ethylene glycol) (PEG) conjugate

(Figure 1) using the dendron – bishomotris that may have a potential use in targeted cancer nanotherapy.

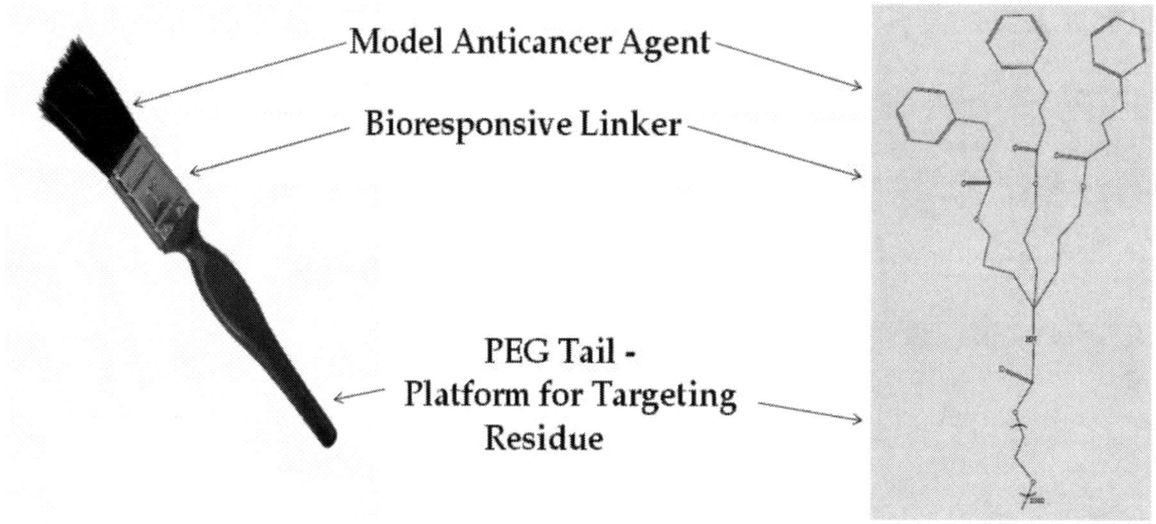

Figure 1) The "PAINT-BRUSH" Conjugate

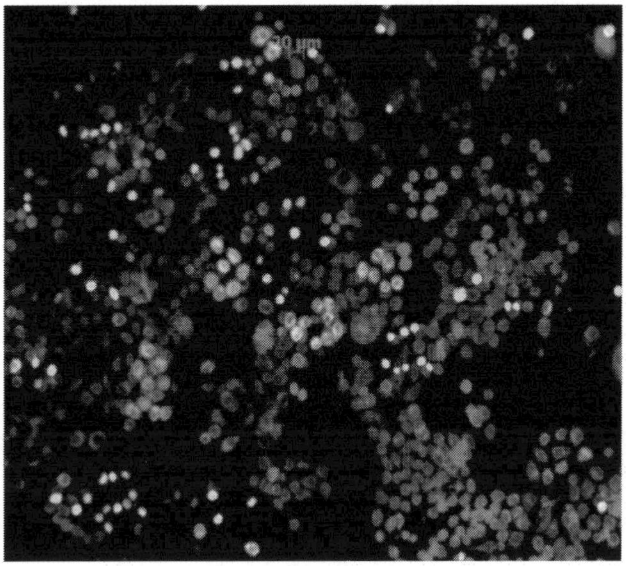

(A)

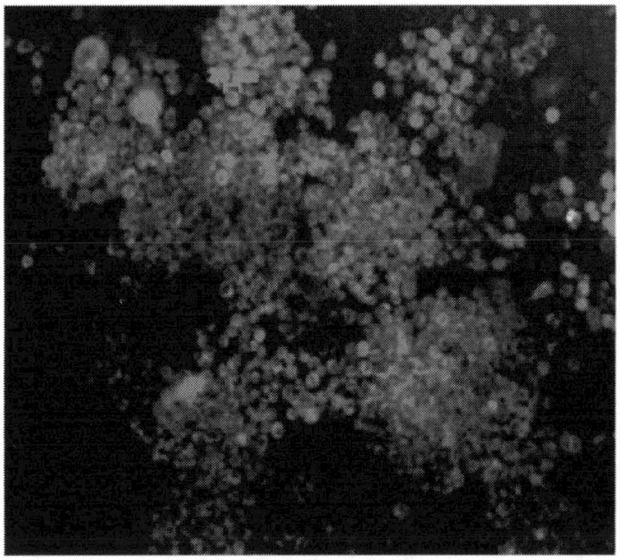

(B)

Figure 2) 12 h fluorescence images of HeLa cells treated with BH (1mg/mL) – (A), PEG-BH (1mg/mL) – (B)

Characterization of the conjugates suggested that the synthesis was successful; resulting in the formation of nanoparticle "PAINT-BRUSH" conjugates. It was also found that these conjugates remain stable under normal physiological conditions but would activate in response to an acidic pH (*a characteristic trait of target cancer cells*) so as to release the anticancer drug. The research also presents the relative cell viability of the human epithelial carcinoma cell line (*HeLa S3*) with respect to *in-vitro* characterization of the conjugate to determine its suitability as a drug delivery vector. The study demonstrated that bishomotris was cytotoxic in nature evidently due to the interaction of positively charged amine group with the surface of HeLa cells. It was also evident that surface modification by PEGylation has led to a distinct reduction in toxicity levels thus laying foundations for further research to realize a promising new scaffold for cancer nanotherapy (Figure 2).

Reference:

1) *Cancer Facts and Figures 2007,* American Cancer Society, Atlanta, 2007

214e Dendrimer-Based Nanodevices for Sustained Intravitreal Drug Delivery

Bharath Rajaguru[1], Rangaramanujam M. Kannan, and Raymond Iezzi[2]. (1) Chemical Engineering, Wayne State University, Detroit, MI 48202, (2) Ophthalmology and Kresge Eye Institute, Wayne State University, Detroit, MI 48202

Neurodegenerative diseases such as retinitis pigmentosa and age related macular degeneration are due to degeneration of photoreceptor cells of retina. Though some of the neuroprotective drugs like fluocinolone acetonide (FA) are available for treatment, it is difficult to deliver therapeutic amount of drug to the diseased site for a sustained period. We have developed polyamidoamine (PAMAM) dendrimer based nanodevices (size °Ö5 nm) for sustained delivery of neuroprotectants. A single intravitreal injection of the nanodevice appears to prevent the neurodegeneration up to one month in RCS rats, in vivo model used for this project. These were based on both electroretinography and cell count measurements. In vivo biodistribution measurements of FITC-labeled PAMAM dendrimers suggested that dendrimers localized in the activated microglial cells and astrocytes. To understand the release of drug from the dendrimer nanodevices, we conducted in vitro release of drug from the nanodevice at different pH conditions.

To enable sustained delivery for 6 months from a single intravitreal injection, we encapsulated these nanodevices in to a microparticle made up of FDA approved biomaterial which can release drug-dendrimer nanodevice for longer period. In vitro release profile of the nanodevices from the microparticles at different pH conditions have been found out and the stability of the drug-dendrimer nanodevices inside the microparticles have been analyzed. We will present results of in vivo efficacy of drug-dendrimer nanodevices, biodistribution of fluorescently-labeled dendrimers in diseased rats and healthy rats, in vitro release profile of drug from dendrimer nanodevice and in vitro release profile of drug-dendrimer nanodevice from the microparticle.

214f Microbubbles for Intravenous Oxygenation

Mark A. Borden[1], John N. Kheir[2], and Francis X. McGowan[2]. (1) Chemical Engineering, Columbia University, 500 West 120th Street, New York, NY 10027, (2) Anesthesiology, Children's Hospital Boston and Harvard Medical School, 300 Longwood Ave., Farley 517, Boston, MA 02115

Gases are a class biologically active species that are normally transported to the blood by the lungs. Examples include oxygen and other metabolites, as well as nitrous oxide and other anesthetics. But what happens when the lungs are not functioning adequately? Respiratory distress syndrome and other forms of lung injury can result in acute hypoxia, which can rapidly lead to severe morbidity and mortality. Extracorporeal membrane oxygenation (ECMO) currently is the standard clinical treatment for acute hypoxia, but the technique is cumbersome and problematic for certain patient populations, including infants. A means of directly injecting oxygen into the bloodstream, thus precluding the need for an extracorporeal loop, is highly desired in the critical care setting. Such a technology could augment or even replace ECMO for some cases. The approach we have taken is to use oxygen-filled microbubbles for direct intravenous oxygenation. Microbubbles are 0.1 to 10 micron diameter gas spheres, coated and stabilized by a lipid monolayer shell, and surrounded by an aqueous medium. Results will be presented which show that oxygen microbubbles can be reproducibly generated and concentrated to 70 vol%. A dose-response relationship was measured for microbubble suspensions combined in vitro with venous blood. The oxygen carrying and release capability is strongly dependent on lipid composition. Lipids above their main phase transition temperatures do not encapsulate oxygen for meaningful time scales. Those in the gel phase encapsulate and release oxygen efficiently, and an optimum composition was found. Phenomena governing microbubble dissolution in blood will be discussed, including modeling of gas exchange kinetics and excess lipid collapse and shedding. Preliminary results are promising and suggest improvements for future testing in vivo.

214g Novel Biodegradable Drug Carriers for Oral Protein Delivery

Erik C. Brewer and Anthony M. Lowman. Chemical and Biological Engineering, Drexel University, 3141 Chestnut St., Philadelphia, PA 19104

Despite advancements in the synthesis of protein therapeutics, the standard administration method has remained by injection. This delivery route has frequently resulted in poor patient compliance, especially among children. Considerable effort has thus been placed on developing alternative and non-invasive delivery routes, in particular through oral administration. Though the oral route is regarded as the most convenient and practical administration route, two major obstacles limit its viability: enzymatic degradation in the gastrointestinal tract (GI) and physical transport across the epithelial layer into systemic circulation. These barriers limit the bioavailability of most proteins to well below 1%.

The proposed delivery mechanism involves encapsulating proteins within a biodegradable polymer shell, which protects proteins from enzyme digestion. In addition, the polymer shell releases proteins in a sustained release manner rather than all at once, maintaining blood levels for longer periods. However, these polymer vehicles typically exhibit a burst release where 10-80% of the encapsulated drug is released within the first 24 hours, raising toxicity concerns. The burst is attributed to the absorption of proteins within the polymer shell rather than the core during the encapsulation process, allowing them to readily diffuse out when delivered. To counteract this initial release, the proposed delivery vehicle will contain protein conjugated to the shell to provide an additional diffusion resistance. To do this, a biodegradable polymer based on PLGA will be engineered that contains functional groups along the repeating unit capable of protein conjugating.

To increase physical uptake across the epithelial layer, the polymer vehicle will be conjugated to transferrin (Tf), an active carrier protein involved in iron transport. Transferrin is highly expressed in the GI tract, and thus been investigated as an active carrier for orally delivered drugs. The nanoparticle-Tf conjugate will take advantage of a receptor-mediated transport pathway that actively transports the drug carrier across the cell rather than relying on passive diffusion. These polymer carrier-Tf conjugates have been shown to increase uptake by up to 400% in tumor cells, where Tf-receptors are also high expressed.

The resulting polymer carrier will reduce the burst release typically seen in these delivery systems, thus creating a sustained-release formulation that protects from enzyme degradation, making this a viable delivery method. Utilizing an active transport pathway to increase epithelial uptake will increase the bioavailability of the orally delivered protein, potentially creating a practical alternative to injected protein therapeutics.

242a Affinity Maturation of Antibody Fragments Containing Non-Canonical Amino Acids

James A. Van Deventer, Chemical Engineering, California Institute of Technology, MC 210-41, 1200 E. California Blvd., Pasadena, CA 91125, Tae Hyeon Yoo, Division of Chemistry and Chemical Engineering, California Institute of Technology, 1200 E. California Blvd., MC 210-41, Pasadena, CA 91125, and David A. Tirrell, Division of Chemistry and Chemical Engineering, California Institute of Technology and Joseph J. Jacobs Institute for Molecular Engineering for Medicine, 1200 E. California Blvd., Pasadena, CA 91125.

The incorporation of non-canonical amino acids into proteins enables us to alter the chemical and/or physical properties of proteins. In this work, we seek to develop high-affinity antibody single-chain variable fragments (scFvs) with altered chemical compositions by replacing the methionine residues in the fragments with one of several non-canonical amino acids that are close structural analogs of methionine. Two amino acids of interest to us, homopropargylglycine (Hpg) and azidohomoalanine (Aha), allow for the incorporation of chemical functionalities useful for future modification of antibody fragments using click chemistry techniques. A third amino acid of interest, norleucine (Nrl), lacks the easily-oxidizable thioether present in methionine side chains and may enable the development of proteins more resistant to oxidative damage. All of these amino acids can nearly quantitatively replace methionine under the appropriate expression conditions. Using a model anti-digoxigenin scFv displayed on the E. coli cell surface with an Lpp-OmpA' display anchor, we can estimate the display levels and dissociation rate constants of the scFvs using flow cytometry. Replacing methionine with any of the non-canonical amino acids described above results in decreased binding activity and/or display levels of the compositionally altered scFvs. We have used error-prone PCR and fluorescence-activated cell sorting (FACS) in an attempt to improve the properties of these proteins. After construction of a library based on the initial scFv construct (Antidig Lib 1), we screened the library for cells capable of increased binding to high amounts of digoxigenin after expression in medium containing Hpg. After three rounds of sorting, we constructed a second error-prone PCR library (Antidig Lib 2) based on the thrice-enriched population isolated from the first library. We screened Antidig Lib 2 using increasing levels of stringency under three different sets of expression conditions. Four rounds of sorting were completed after expression in media supplemented with each of the amino acids Hpg, Aha, and Nrl. Using kinetic competition methods, we have isolated clones possessing low apparent dissociation rate constants with altered compositions at the positions that normally contain methionine. Sequence analysis of clones isolated after three or more rounds of screening of Antidig Lib 2 expressed in Hpg-, Aha-, or Nrl-containing medium indicates that a number of commonly occurring mutations are shared amongst the various isolated populations despite the distinct amino acid contexts employed during the course of screening. Important differences in mutational frequencies between the populations are also observed. We are interested in the contributions of individual mutations to the affinity of the scFv for its cognate antigen and how these contributions change depending on the identity of the amino acids incorporated at positions normally occupied by methionine. Understanding these contributions may enable us to better understand fundamental aspects of protein folding and stability and how subtle changes in amino acid side chain identities affect these properties. From an engineering perspective, performing directed evolution experiments with proteins containing non-canonical amino acids may enable the development of therapeutically relevant proteins with properties that can only be introduced by incorporating non-canonical amino acids.

242b a Comparison of Antibodies Isolated from Immune Libraries by Two Different Anchored Periplasmic Expression (APEx) Systems

Thomas J. Van Blarcom, Sean Carroll, Yariv Mazor, and George Georgiou. Chemical Engineering, University of Texas at Austin, Molecular Biology Building, 2500 Speedway Room 3.310, Austin, TX 78712

High affinity antibodies to the Protective Antigen (PA) component of the Bacillus anthracis toxin are in late stage testing for prophylaxis and therapy against anthrax intoxication. However, recent studies (Nowakowski, A et al.) indicate that treatment with multiple antibodies simultaneously can confer a synergistic protection to toxin challenge. To explore multi-antibody therapeutic approaches, antibodies toward the protective antigen (PA) component of the B. anthracis exotoxin were isolated from immune libraries using two variations of the E. coli based protein library screening technology, Anchored Periplasmic Expression (APEx) (Harvey, B.R. et al., Mazor, Y. et al.). Full length antibodies and scFv antibodies with nanomolar affinities were isolated as early as the second round of flow cytometric screening. While all of the scFvs were easily converted to functional full-length antibodies, the converse was not true. These results indicate the E¬-clonal system is better suited than the original APEx system for the isolation of a diverse array of antibodies with a wide affinity range. Furthermore, all antibodies isolated are currently being evaluated for multi-antibody therapy.

Nowakowski, A. et al. (2002) Proceedings of the National Academy of Sciences of USA 99(17): 11346

Harvey, B.R. et al. (2004) Proceedings of the National Academy of Sciences of USA 101(25): 9193

Mazor, Y, et al. (2007) Nature Biotechnology 25(5): 563

242c Synthetic Library Design Using the Fibronectin Domain as a Model System

Benjamin J. Hackel and K. Dane Wittrup. Chemical Engineering, Massachusetts Institute of Technology, 50 Ames Street, E19-563, Cambridge, MA 02139

The ability to engineer a protein for specific, high-affinity binding to a target is a valuable tool in drug development, biotechnology, and scientific research. Combinatorial library selection, both in vivo through immunization or in vitro, is the most effective method for identification of binders. While immune and naïve repertoires can serve as an effective source of immunoglobulin sequences, synthetic libraries represent an effective alternative in antibody engineering and a requisite source of diversity for alternative scaffolds. The immensity of protein sequence space and the limited capacity of laboratory selection methods necessitates an efficient library design in which the diversities at each position combine to yield a pool of clones that maintain structural integrity while imparting a wide array of binding specificities.

In the current work, we investigate the impact of two library design elements: tailored diversity at putative paratope positions and partial wild-type conservation at structurally important positions within the paratope region. We hypothesize that the amino acid distribution observed in expressed immunoglobulin complementarity-determining regions is superior to both an equal distribution and focused tyrosine/serine degeneracy. We have identified a set of skewed nucleotide mixtures that yields codons that approximately match this antibody-inspired distribution while enabling library synthesis at significantly less expense than triphosphoramidite-based construction. A second design hypothesis is that wild-type bias at structurally critical positions within the paratope region can improve library design by increasing the functional diversity of the library, reducing the entropic cost of binding, and focusing diversity on positions with higher likelihood for antigen contact.

We investigated these hypotheses in the context of the tenth type III domain of human fibronectin (Fn3). Fn3 is a small, stable protein with a beta-sandwich fold and three solvent exposed loops, which is structurally similar to an antibody variable domain. The two new design elements were incorporated into a new Fn3 library. Stability, structural, and genetic analyses were conducted to determine the extent of wild-type bias, if any, desired at each position. Diversity was achieved using skewed nucleotides to yield the antibody-inspired distribution. Direct competition of the new library versus two previous library designs yielded preferential selection of binders from the new library. Sequence analysis reveals the relative importance of the multiple design elements. The impact of wild-type bias was further characterized by stability analysis of library clones providing insight into the structural tolerance of the domain as well as potential refinement of library design. The library design principles and methodology should be directly applicable to other protein engineering efforts both in molecular recognition and elsewhere.

242d Developing Therapeutic Proteins by Engineering Ligand-Receptor Interactions

Douglas S. Jones, Stayce E. Beck, Stephen S. Lee, and **Jennifer R. Cochran**. Bioengineering, Stanford University Schools of Engineering and Medicine, 318 Campus Dr. West, MC5439, James H. Clark Center, W250, Stanford, CA 94305-5439

Ligand–receptor interactions govern a myriad of cell signaling pathways that regulate homeostasis and ensure that cells respond properly to stimuli. Growth factors use these interactions to stimulate cell responses, including proliferation, migration, angiogenesis, and cell death. Proteins that inhibit these processes have potential as therapeutics for cancer, whereas proteins that stimulate these processes offer promise in regenerative medicine. Although much of the focus in this area over the past decade has been on antibodies, recently there has been increased interest in using non-antibody proteins as therapeutic agents. We will discuss our recent successes using directed evolution through yeast surface display to engineer epidermal growth factor (EGF) or hepatocyte growth factor (HGF) ligands with therapeutic potential.

Epidermal growth factor (EGF) is one of the major factors involved in the re-epithelialization and remodeling phase of wound healing. Local and sustained presence of EGF has been shown to accelerate wound healing. However, attempts to develop EGF for clinical applications have been limited by its short circulating half-life. We hypothesized that increasing the affinity of human EGF for its receptor (EGFR) would result in enhanced biological efficacy. We show that EGF mutants engineered for increased receptor binding affinity by yeast surface display elicit increased directional and chemotactic migration of fibroblast cells compared to wild-type EGF. Increased biological activity of the EGF mutants was due to their ability to bind much higher levels of cell surface receptors compared to wild-type EGF after just minutes of incubation. To move towards pre-clinical testing in murine models of wound healing, we demonstrated that EGF mutants also elicited enhanced chemotactic migration over wild-type EGF on fibroblast cells expressing murine EGFR. In full-thickness mouse wounds, one mutant in particular was shown to accelerate wound closure over wild-type EGF, demonstrating promise for clinical translation.

Met (also known as the hepatocyte growth factor receptor) is a receptor tyrosine kinase that plays a central role in controlled cell proliferation and invasion during embryonic development. Conversely, dysregulated Met activation results in a phenotype of invasion and metastasis in many human tumors. Despite its strong role in cancer progression, there are currently no FDA-approved therapeutics targeting the Met receptor. Previously, fragments of the hepatocyte growth factor were rationally engineered as Met receptor agonists and antagonists; however, these ligands bind too weakly to Met (micromolar binding affinity) to be therapeutically useful. To improve the clinical potential of these proteins, we used yeast surface display to engineer mutants with high-affinity binding to the Met receptor. Libraries of yeast-displayed mutants were screened with soluble Met receptor extracellular domain to isolate clones with high affinity binding. The soluble expression, receptor binding, and biological effects of these high-affinity Met-binding mutants will be discussed.

Funded by NIH/NCI 1R21 CA131706, the Wallace H. Coulter Foundation, and fellowship support from NIH 5T32 GM008412-15S1 (DSJ), NIH T90 DK070103 (SEB), and NSF (SSL).

242e Novel Enzyme-Prodrug Therapy for Cancer

Roger G. Harrison[1], Yahya A. Lazrak[1], Luis F.F. Neves[1], Peter S. McFetridge[1], and Arafat Tfayli[2]. (1) Chemical, Biological and Materials Engineering, University of Oklahoma, 100 E. Boyd St., T-335, Norman, OK 73019, (2) Hematology-Oncology, University of Oklahoma Health Sciences Center, 920 Stanton L. Young Blvd., WP 2040, Oklahoma City, OK 73104

We are developing a new type of enzyme-prodrug therapy designed to selectively clot the vasculature of tumors and kill cancer cells. It has been shown that the anionic phospholipid phosphatidylserine (PS) is exposed almost exclusively on the external membrane layer of cancer cells and endothelial cells of the tumor vasculature. It is also known that the human protein annexin V binds specifically and strongly to PS. We have engineered a fusion protein (FP) containing annexin V linked to the enzyme methioninase, which has the ability to convert a biologically safe compound, selenomethionine, to a toxic compound, methylselenol. The annexin V part of the FP will specifically bind to the tumor vasculature, while the methioninase part will convert the selenomethionine prodrug to the toxic methylselenol leading to two mechanisms to kill the tumor: 1) death of the endothelial cells causing cloture of tumor vasculature, thus cutting off the oxygen supply of the cancer cells; and 2) methylselenol will be transported to tumor cells by diffusion and fluid permeation resulting in their death.

We have studied the binding of purified FP to PS immobilized on microtiter plates and to human endothelial cells in which PS has been induced to the on the cell surface by the addition of a low level of hydrogen peroxide. Data has also been obtained for the effect of adding seleonmethionine to endothelial cells in vitro in which the FP is bound on the cell surface. The cell viability is evaluated by performing the Alamar Blue assay, cell counting, and cell observation with a microscope.

242f Fabrication of Engineered Collagen-Like Biopolymers

Sam Wei Polly Chan[1], She-pin Hung[2], Richard Lathrop[3], Nancy A. Da Silva[1], and **Szu-Wen Wang**[1]. (1) Department of Chemical Engineering and Materials Science, University of California, Irvine, 916 Engineering Tower, Irvine, CA 92697-2575, (2) CODA Genomics, Inc., Laguna Hills, CA 92653, (3) Department of Information and Computer Science, University of California, Irvine, CA 92697

Collagen, the most abundant extracellular matrix protein, and collagen-like materials have been widely used in tissue scaffold and drug delivery applications. Recombinant human collagen using the native gene can be expressed in mammalian and yeast cells. However, the characteristic Gly-X-Y protein sequence repetition of the triple-helical domain adds an extra degree of complexity to the *de novo* gene synthesis that results in incorrect oligonuclucleotide hybridizations. This has, up until now, prevented the synthesis of full-length collagen-like biopolymers. Using a computational algorithm, we have successfully created a gene encoding native human collagen III. This gene has been designed to enable modular modifications and facilitate the fabrication of collagen-like variants. We have expressed these biopolymers in *Saccharomyces cerevisiae* using an integration/plasmid system, and these protein products have been purified and characterized. The strategy we have developed sets up the foundation for creating variants of collagen for tailored applications.

242g Non-Associating Heterodimeric DNA Methyltransferases as a Platform for Developing Site-Specific Methyltransferases

Glenna Meister[1], Srinivasan Chandrasegaran[2], and Marc Ostermeier[1]. (1) Chemical and Biomolecular Engineering, Johns Hopkins University, 3400 N. Charles St., Baltimore, MD 21218, (2) Division of Physiology, Johns Hopkins University, Bloomberg School of Public Health, Baltimore, MD 21218

The ability to site-specifically methylate DNA has numerous potential uses including 1) a tool for the study of DNA methylation patterns, 2) as a tool to silence gene of interest, and 3) as a potential gene therapy device to correct conditions caused by hypomethylation. Current approaches include linking methyltransferases to DNA binding domains to localize enzymes next to a target site. This approach has achieved site-biased methylation, however the engineered methyltransferases are still active in the absence of binding their intended target and methylate non-target sites. What is necessary to make a truly site-specific methyltransferase is to require binding to the target site for catalytic activity. We propose to use non-associating fragments of natural heterodimeric methyltransferases as a platform for building such enzymes, with DNA at the target site acting as a template for enzyme assembly. The naturally occurring C5-methylcytosine methyltransferases M. AquI and M. EcoHK31I each have alpha and beta peptide chains that associate to create a functional enzyme. We have made non-associating fragments by creating truncated versions of the beta fragments. We have evaluated these fragments as a platform for creating site-specific methyltransferases by creating small combinatorial libraries of fusions with zinc finger proteins and screening these libraries for site-specific methylation activity.

255a FDA Update on Implementation of Quality by Design

Christine M.V. Moore, Fda/cder/ondqa, Food and Drug Administration, W022 RM2418, 10903 New Hampshire Ave., Silver Spring, MD 20993

Quality by Design is a systematic approach to pharmaceutical development that begins with predefined objectives and emphasizes product and process understanding, and process control, based on sound science and quality risk management. Much progress has been made in the past few years to define QbD implementation approaches for pharmaceutical development and manufacturing. Recent harmonized guidelines have been published to aid in the implementation of QbD approaches including: ICH Q8 - Pharmaceutical Development and more recently ICH Q8R (draft), which discusses the fundamentals of QbD in pharmaceutical development. Additionally, ICH Q9 – Quality Risk Management and ICH Q10 – Pharmaceutical Quality System discuss other essential elements of implementing a QbD approach. In 2005, FDA's Office of New Drug Quality Assessment (ONDQA) began a CMC Pilot Program to explore ways of implementing QbD concepts in regulatory submission and review. This presentation will provide a brief overview of QbD concepts and recent ICH guidelines, and discuss experience gained from recent QbD submissions, including those in the CMC Pilot Program.

255b Novartis/fda Crada Qbd Project – Real Life Case Studies of Qbd Applied to Active Pharmaceutical Ingredient and Drug Product Manufacturing Processes

Marc Goeller, Global PAT, Novartis, East Hanover, NJ 07936-1080

Since 2006, teams from Novartis Pharmaceuticals Corp. and the FDA's Division of Product Quality Research have been collaborating through a Cooperative Research and Development Agreement (CRADA) on the implementation of Quality by Design principles to the API and Drug Product manufacturing processes of a commercial product. This presentation will therefore consist of case studies extracted from this pilot project and will also highlight the key lessons learned.

255c Particle Technology Gaps of Qbd

Fernando J. Muzzio, Chemical and Biochemical Engineering, Rutgers University, 98 Brett Road, Piscataway, NJ 08854

This talk discusses the QbD initiative from a particle technology perspective. The central technical requirement of a sound QbD strategy is to determine how ingredient properties and processing conditions determine product structure and performance. Since particles (both granules and powders) dominate the manufacturing of almost all pharmaceutical products, particle science and technology are therefore critical foundational components of the knowledge base needed for modern product and process design.

While the task ahead is conceptully clear, and much progress is under way, a number of gaps need to be addressed, including:

- Incomplete understanding of raw material properties

- Insufficient knowledge of effect of process variables across scales

- Lack of methods for characterizing structure of products and in-process materials

- Underdeveloped process monitoring tools

- Lack of control methodologies for powder-based processes

- Over-reliance on batch manufacturing methods

In this talk, these gaps are discussed, recent progress in addressing themis presented, and a way forward is outlined.

Quality-by-Design (QbD) Case Study: Powder Blending Process Kinetics Evaluation

Huiquan Wu and Mansoor A. Khan

Division of Product Quality Research (DPQR, HFD-940)

OTR/OPS/CDER/FDA

FDA White Oak Campus Life Science Building 64

10903 New Hampshire Ave

Silver Spring, MD 20993

Disclaimer: The views and opinions expressed in this article are only of the authors, and do not necessarily reflect the views or policies of the FDA.

Abstract: The objective of this project was to develop a Quality-by-Design (QbD) case study for the evaluation of powder blending process kinetics. A mixture design was created to include 26 powder formulations consisting of ibuprofen as the model drug and three excipient components (HPMC, MCC, and Eudragit L100-55). The mixer was stopped at various time points to enable NIR scan of the powder mixture for obtaining the time course of the blending process for each formulation. Previous works demonstrated that NIR spectra of pharmaceutical dosage form are information rich, and may contain physical, chemical, and process information of the formulation components and unit operations. The focus of this work was to develop data analysis and modeling approaches to extract relevant process information, generate process knowledge, and evaluate powder blending process kinetics. Three quantitative approaches were used: (1) Pure component spectra linear superposition method; (2) Characteristic peak method; (3) Moving block standard deviation method. Our study revealed that the blending process experiences three distinct stages: (1) an initial rapid process to reach a quasi- end point within the first a few minutes; (2) demixing; and (3) a real blending end-point as characterized by an inflection point. ANOVA shows that the main components' compositions (Ibuprofen and MCC) are the most statistically significant variables (critical formulation/process variables) that impact the time required to reach the blending end-point. This work as a QbD case study highlighted the critical importance of integration of Design of Experiments (DOE), Near infrared (NIR) process spectroscopy, and chemometrics to extract critical process information and generate essential process knowledge to enable real-time release of the blending process.

Key words: Quality-by-Design (QbD), Process Analytical Technology (PAT), design of experiments, multivariate statistical data analysis, powder blending, process monitoring, blending end-point determination, moving block standard deviation, process dynamics.

INTRODUCTION

In studying the mixing of powders, traditionally the determination of powder blending end-point relies on wet chemical assay of the active drug by methods such as HPLC of several samples over the time course of blending, which is not only labor intensive but also time-consuming. The sampling procedure is also prone to experimental error as demonstrated in the literature[1]. Furthermore, there is always a time lag between the time when the real process events take place and the time when the laboratory analytical results become available. To certain degree, this type of practice reflects the paradigm of so-called testing-into-quality and the low manufacturing efficiency in the pharmaceutical sector. Most importantly, this practice of monitoring only the active drug in blend homogeneity is unlikely to provide explanation for out of specification situation of a product where the variability may be due to excipient composition change by inadequent mixing.

Recent pharmaceutical quality regulatory initiatives such as FDA's Process Analytical Technology (PAT)[2] Initiative, ICH Guidance Q8[3] and Q9[4] have provided excellent opportunities to realize the benefits of at/in/on-line process monitoring and on-line process control[5]. Although process analyzers have been used for some time in the pharmaceutical industry for powder blending monitoring, the focus has been largely on the qualitative aspect of process monitoring. Surprisingly, little attention has been paid to the quantitative evaluation of powder blending process kinetics, which is essential to blending process design and blending equipment design. A number of online techniques available among them the Near-Infrared Reflectance Spectroscopy (NIRS)[6-10] was one of frequent choice of techniques. However, none of them have used an integrated approach of combining statistical experimental design, multivariate data analysis, on-line or at-line process monitoring to understand powder blending process kinetics, blending homogeneity of both API and excipients for a 4 component formulation system.

From a PAT process control[5] perspective, key questions for blending operation may include the following:

(1) How do the formulation component compositions impact the time course of blending operation and the blending end-point?

(2) How to quantitatively determine the blending end-point?

(3) How do the formulation parameters and process variables impact the powder blending process behavior and blending process kinetics?

To explore the above technical challenges in concept at the laboratory scale, an extreme vertices design was created to include 26 formulations which consist of a four-component formulation system for a blending process study. The mixer was stopped at various time points to enable NIR scan of the powder mixture for obtaining the time course of the blending process for each formulation. The focus of this work was on developing appropriate data analysis and modeling approaches to evaluate powder blending process kinetics.

EXPERIMENTAL

Materials and Methods

The following pharmaceutical materials were used as-received for this study, without further processing or purification prior to the powder mixing: USP 70 grade Ibuprofen (Albemarle Corp., LA); Hydroxypropyl Methylcellulose (HPMC), Methocel E15 Pemium LV (Dow Chemical, Midland, Michigan); USP/NF Microcrystalline Cellulose (MCC) (JRS Pharma LP, Cedar Rapids, Iowa); Eudragit L 100-55 (Methacrylic Acid-Ethyl Acrylate Copolymer (1:1), Methacrylic Acid Copolymer Type C NF) (Degussa, Germary).

Experimental Design

An extreme vertices design was used to compute the formulation compositions for 26 formulations using JMP 5.1 software (SAS Institute, Cary, NC), with the following four constraints applied to the weight fractions of corresponding formulation components: for ibuprofen, $0.25 \leq$ Wt. fraction≤ 0.75; for HPMC, $0.01\leq$wt. fraction≤ 0.03; for MCC: $0.19 \leq$wt. fraction ≤ 0.57; for Eudragit L 100-55: $0.05\leq$wt. fraction≤ 0.15.

Powder Blending Experiments

After weighting the components using Mettler AE 240 analytical balance (Mettler Instrument Corp, Highstown, NJ), the components of each formulation (in a total of 5 grams) were transferred to a 20 ml scintillation vial for geometric mixing for 5 seconds.

The vials were then placed inside a basket of a Tubula mixer (Willy A. Bachofen AG, Maschinenfabrik, Basel Switzerland). The Tubula mixer was operated at 72 rpm for powder blending. The mixer was stopped at predefined time points (0, 1, 2, 3,4,5,6,8,10,15,20,25,30,45, and 60 minutes). The powder inside the vial was then subjected to NIR scan. When the NIR scan was done for a pre-defined time point, the vials were placed inside the basket of the Tubula mixer again to resume the mixing operation until the next predefined time point was reached.

NIR Spectroscopy

In this work, near infrared (NIR) spectra of blending powders at various time points were acquired with a Luminar™ acoustic-optic tunable-filter (AOTF) based NIR spectrometer (Brimrose Corporation of America, Baltimore, MD), equipped with a transflectance probe. The acquisition parameters for the NIR spectrometer include: the number of spectra average was 50; no background correction; normal scan type; the gain was 4.Certain measures were taken to average out the potential measurement errors. Figure 1 is the NIR spectra for the four formulation components at static state.

Data Analysis Methods

For data analysis, three quantitative approaches were employed to evaluate blending process kinetics and determine blending end-point from the NIR spectra: pure component spectra superposition method, characteristic peak-based multivariate data analysis method, and moving block standard deviation method.

RESULTS

Powder Blending Process Monitoring

Linear Superposition Method

If there is no interaction between various powder formulation components, synthesized NIR spectrum through linear superposition of pure components by formulation compositions would provide an ideal final spectrum for a well-mixed powder formulation. Subtracting this ideal spectrum from the actual NIR spectrum for an in-process powder mixture at various time points, then dividing the actual NIR spectrum (absorbance values at each wavelength), would enable us to assess the difference between the ideal spectrum for a well-blended mixture and the actual spectrum at various time points. Theoretically, if there are no interactions between components, then these two spectra would eventually converged together. However, if the difference is still easily appreciable even after a relative long time of blending, it probably tells that there are some interactions between the formulation components.

Taking formulation 5 as an example shown in Figure 2, when plotting the relative error between the synthesized NIR spectrum and the actual NIR spectrum vs. wavelength for various blending time points, it was noticed that except the wavelength range of [1600nm, 1950nm] where large relative errors (maximum 5%) occur, other wavelengths exhibit relatively small errors (within ±2%). Interesting, a general trend of decreasing relative error as powder blending time increases can be observed for the wavelength range of [1600nm, 1950nm]: (1) during the first 6 minutes, the maximum relative error rapidly decreases from 4.8% to 2.6%; (2) when the blending time increases from 20 minutes to 30 minutes, the maximum relative error reduces from ca. 1.8% to ca. 1.3%, which is pretty comparable to the relative error values of other wavelengths; (3) when the blending time increases from 45 minutes to 60 minutes, the maximum relative error decreases further to ca. 0.5~0.8%, which is very close to the relative error values of other wavelengths. These observations suggest that: (1) there is no significant interaction between the formulation components; (2) the plot of relative error vs. wavelength could be served as a useful means to evaluate whether and when the powder blending process has approached the process end-point.

Characteristic Peak Multivariate Method

A number of potential characteristic peaks for Ibuprofen, MCC, Eudragit L100-55, and HPMC were identified through applying S-G 1st derivative method to the NIR spectral data of pure formulation components, as summarized in Table 1. For each characteristic wavelength listed in Table 1, at each pre-defined time points during the course of powder blending experiments, three multivariate data analysis methods (PCR, PLS, and MLR) were employed to correlate the formulation compositions with the S-G 1st derivative spectral data for all of the 26 formulations studied. These multivariate calibration models were then used to predict the S-G 1st derivative spectral data for an independent sample formulation set. The relative prediction error was calculated for each formulation and was plotted vs. formulation at certain time point for each individual characteristic wavelength. Figures 3 is such an example for blending 15min.

Theoretically, if the powder formulation components are well mixed, for a well-designed and calibrated measurement system (here at-line NIR system), the process NIR spectra of the powder mixture should capture both physical and chemical information of the corresponding formulation and therefore represent the powder mixture well. When a well-calibrated multivariate model is used to predict the NIR S-G 1st derivative spectral data based on the target formulation compositions, we would expect a low relative prediction error. Therefore, for each characteristic wavelength, computing the average of the absolute values of the relative prediction errors for all the samples and plotting against the blending times could essentially provide a soft sensor to monitor the powder blending process, and to determine whether the blending end-point is reached or not. Figure 4 is such an example to illustrate this application of using multivariate calibrated models to predict the NIR S-G 1st derivative spectral data of the powder under blending at wave length of 1770 nm. The plot for PLS1 model is almost identical with that for PCR model. Therefore it is not shown in Figure 4 for better visualization. As we can see from Figure 4, the plots of PCR model and MLR model merged together first at around 3 minutes of blending, then after 15 minutes until 60 minutes of blending. The times when these plots from different models merged together would be considered as primary indicators of blending process end points.

Powder Blending Process Dynamics Characterized by at-line NIR Spectral Data

For each formulation, a plot of the moving standard deviation vs. wave length at various blending time windows was made in order to examine the process trend, such as whether a global or local blending end-point is approaching. One representative figure is shown in Figures 5 for formulation A3.

From the evolution of STDEV over the wave length range studied during the course of blending operation, it shows that within a few minutes of the initial stage of blending, the shape of the plot changes rapidly, which indicates the mixing efficiency is very high. This is not unexpected since thermodynamically, the driving force for blending (the difference between the component concentration at each individual location within the vial and the target component concentration) is large at the initial stage of the blending operation. This large thermodynamic driving force results in a fast dynamic blending process once the blending process is initiated and maintained by a certain rotation speed of the blending vessel. As blending progresses, the STDEV plot rapidly approaches so-called steady state when the thermodynamic driving force is approaching zero. Graphically, the STDEV plot gradually becomes flat and approximately parallels to the wave length axis. Thus, it approaches the blending end-point.

The blending process dynamics could be further visualized and characterized by the plot of the integrated spectra standard deviation vs. blending time. The integrated spectra standard deviation was the standard deviation of the moving block standard deviation. Figure 6 is such a plot for formulation A3. There are 4 distinguishable stages occurred during the blending operation, as shown in Figure 6. First, there is an initial rapid decline of the Y-axis value within the first 5 minutes or so. After that, the Y-axis value surges within the next few minutes. Then, the Y-axis value declines again, until a minimum point occurs within the time frame of 20 to 30 minutes. After that, there is a small surge of the Y-axis value, but it quickly approaches a steady-state value. The Y-axis value surging at the 2nd stage may suggest that some kind of back-mixing or segregation takes place during the course of blending, as reported elsewhere[11].

Powder Blending Process End-point Determination

The time required to reach powder blending end-point can be influenced by a number of factors, such as powder formulation properties and compositions, blending process conditions and type of the blender, and scale of scrutiny, etc. In this work, the components of 26 powder formulations were determined by a mixture design as discussed in the experimental design subsection. Blender type and blending process conditions were fixed as described previously in the experimental section. The scale of scrutiny will be discussed separatedly. Powder blending process end-point could be determined by various measures. Mathematically, it could be determined by (1) the time when the STDEV plot becomes flat and parallels to the wave length axis. In addition, the neighboring STDEV plots almost overlap with each other; (2) the inflection point on the plot of standard deviation vs. blending time; (3) it could be determined by comparing the predicted component concentrations at various time points with the targeted concentrations determined by the powder dispersion ratio. Table 2 list the results for powder blending process end-point determination based on the first two methods.

Critical Formulation Parameters Which Impacts the Time Required to Reach the Blending End-point

In this study, powder formulation compositions were varied according to the design of experiment (DOE). The time required to reach blending homogeneity is an inferred parameter that was estimated by various methods and was listed in Table 2. Here we treat this time as response variable. Analysis of Variance (ANOVA) as an important data analysis tool for DOE has been used to identify critical parameters for both PAT and

QbD applications[12-13]. It was used here to analyze this DOE dataset. The ANOVA results (Table 3) show that the main components (Ibuprofen and MCC) are the most statistical significant variables or critical formulation variables that impact the blending time required to achieve the blending homogeneity or blending end-point. This makes sense because Ibuprofen and MCC are the main formulation components and the variations embarked by these two components are dominant. Therefore, they become the major factors that dictate the time to reach the blending end-point.

CONCLUSIONS

An integrated Quality-by-Design approach was developed for powder blending process kinetics evaluation for a 4-component pharmaceutical powder blending system. 26 powder formulations of 4-component system constructed through an extreme vertices design were used for the development and evaluation of this QbD approach. This study demonstrated that the time required to reach blending process endpoint could be estimated by three methods: (1) the time when the STDEV plot becomes flat and parallels the wave length axis; (2) the inflection point on the plot of standard deviation vs. blending time; (3) by comparing the predicted component concentrations at various time points with the targeted concentrations determined by the powder dispersion ratio. Our study revealed that the blending process experiences three distinct stages: (1) an initial rapid process to reach a quasi- end point within the first a few minutes; (2) demixing; and (3) a real blending end-point as characterized by an inflection point. ANOVA shows that the main components' compositions (Ibuprofen and MCC) are the most statistically significant variables (critical formulation/process variables) that impact the time required to reach the blending end-point. This work as a QbD case study highlighted the critical importance of integration of Design of Experiments (DOE), Near infrared (NIR) process spectroscopy, and chemometrics to extract critical process information and generate essential process knowledge to enable real-time release of the blending process.

ACKNOWLEDGMENTS

Mr. Maury White (summer intern) and Mr. Mobin Tawakkul at the Division of Product Quality Research (DPQR, HFD-940) in FDA are acknowledged for technical assistance on some of the experimental work of this study. Dr. Meiyu Shen in the Division of Biometrics VI in FDA is acknowledged for conducting ANOVA for this study. Mr. Igor Nazarov, former Spectrometer Division Manager at Brimrose Corp. of America (Baltimore, MD) is acknowledged for instrument assistance. Mr. Andrew Chu and Dr. Dongsheng Bu from Camo Technologies are acknowledged for software assistance.

REFERENCES

1. Muzzio, F J., Robinson P., Wightman C., and Brone D., 1997. Sampling practices in powder blending. Int J Pharm 155:153-178.
2. FDA. 2004. Guidance for Industry, PAT — A Framework for Innovative Pharmaceutical Development, Manufacturing, and Quality Assurance. Available at: http://www/fda.gov/cder/guidance/6419fnl.pdf.
3. FDA/ICH. 2006. Guidance for Industry. Q8 Pharmaceutical Development. Available at:http://www.fda.gov/cder/guidance/6746fnl.pdf.
4. FDA/ ICH. 2006. Guidance for Industry. Q9 Quality Risk Management. Available at: http://www.fda.gov/cder/guidance/7153fnl.pdf
5. Wu, H., Hussain, A., and Khan, M., 2007.Process Control Perspective for Process Analytical Technology: Integration of Chemical Engineering Practice into Semiconductor and Pharmaceutical Industries. *Chem. Eng. Comm.*, 194:760-779.
6. MacDonald BF and Prebble KA, 1993. Some applications of near-infrared reflectance analysis in the pharmaceutical industry. J Pharm Biomed Anal 11(11/12): 1077-1085

7. Sekulic SS, Ward HW, Brannegan DR, Stanley ED, Evans CL, Sciavolino ST, Hailey PA, and Aldridge PK, 1996. On-line monitoring of powder blend homogeneity by near-infrared spectroscopy. Anal Chem 68:509-513.

8. Hailey PA, Doherty P, Tapsell P, Oliver T, and Aldridge PK, 1996. Automated system for the on-line monitoring of powder blending process using near-infrared spectroscopy. Part I. System development and control. J Pharm Biomed Anal 14:551-559

9. Sekulic SS, Wakeman J, Doherty P, Hailey PA, 1998. Automated system for the on-line monitoring of powder blending process using near-infrared spectroscopy. Part II. Qualitative approaches to blend evaluation. J Pharm Biomed Anal 17: 1285-1309

10. El-Hagrasy AS, Morris HR, D'amico F, Lodder RA, Drenen III JK, 2001. Near-infrared spectroscopy and imaging for the monitoring of powder blend homogeneity. J Pharm Sci 90(9): 1298-1307

11. Lachman L, Lieberman, HA, Kanig, JL, 1986. The Theory and Practice of Industrial Pharmacy, 3rd ed. Lea & Febiger, Philadelphia, PA.

12. Wu H. and Hussain AS, 2005. Integration of multivariate statistics and design of experiments to identify critical process variables for pharmaceutical process analytical technology (PAT) applications. In: American Statistical Association 2005 Proceedings of Joint Statistical Meetings, Minneapolis, Minnesota, August 7-11, 2005. CD-ROM , Mira Digital Publishing.

13. Xie L., Wu H., Shen M., Augsburger L, Lyon RC, Khan MA, Hussain AS, Hoag SW. Quality-by-Design (QbD): Effects of Testing Parameters and Formulation Variables on the Segregation Tendency of Pharmaceutical Powder Measured by the ASTM D 6940-04 Segregation Tester. J. Pharm. Sci., 2008 (in press).

Table 1. Characteristic NIR wavelengths identified for the pure components in the formulation system

Formulation components	Characteristic NIR wavelength (nm)
Ibuprofen	1126, 1170, 1196, 1376, 1670, 1686, 1698, 1730, 1770, 2122
MCC	1418, 1468, 1604, 2046
HPMC	1508

Table 2. Approximate end-point of powder blending process (minutes) estimated by the STDEV plot method and Integrated spectral STDEV method

Formulation No.	STDEV plot method	Integrated spectral STDEV method
A1	30	30
A2	30	30
A3	25	25
A4	30	30
A5	20	20
A6	30	30
A7	20	20
A8	25	25
A9	10	10
A10	20	20
A11	45	45
A12	25	25
A13	20	20
A14	20	20
A15	20	20
A16	25	25
A17	20	20
A18	15	15
A19	30	30
A20	15	15
A21	20	20
A22	30	30
A23	15	15
A24	8	8
A25	6	6
A26	30	30

Table 3. ANOVA effect testing results showing the most statistically significant variables that impact the blending time required to achieve the blending homogeneity (BTR)

Dependent	Source	Degree of freedom	Sum of squares	F ratio	Prob > F
BTR	API	1	542.30	5.80	0.028
BTR	MCC	1	603.11	6.45	0.022
BTR	EUD	1	8.67	0.09	0.76
BTR	HPMC	1	77.82	0.83	0.38
BTR	API*MCC	1	48.55	0.52	0.48
BTR	API*EUD	1	18.67	0.20	0.66
BTR	MCC*EUD	1	7.55	0.08	0.78
BTR	API*HPMC	1	80.85	0.86	0.37
BTR	MCC*HPMC	1	74.95	0.80	0.38
BTR	EUD*HPMC	1	75.64	0.81	0.38

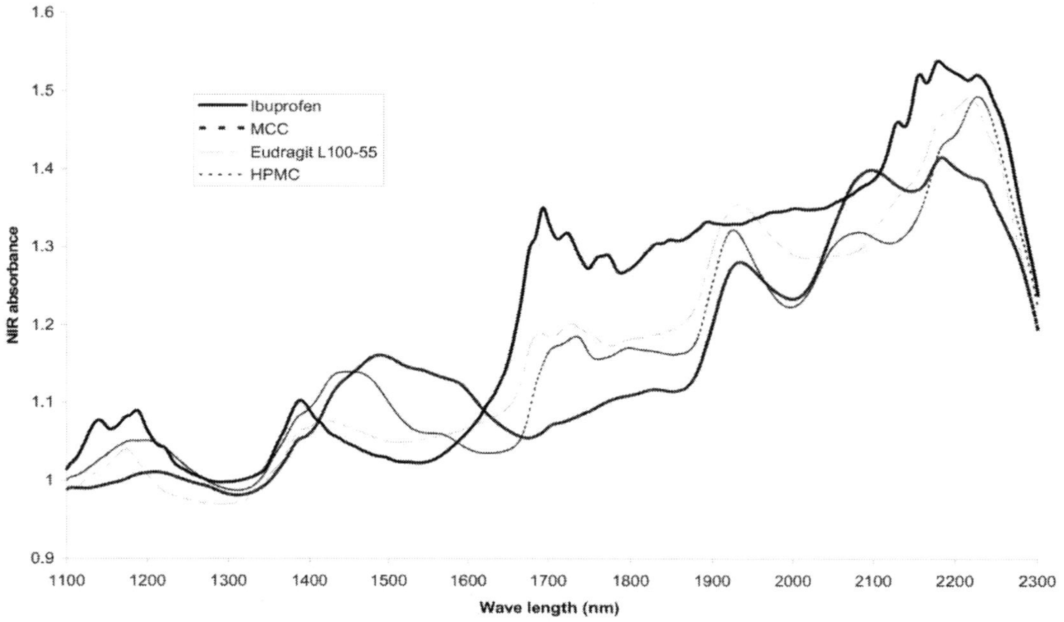

Figure 1 the NIR spectra for the four pure components at static state

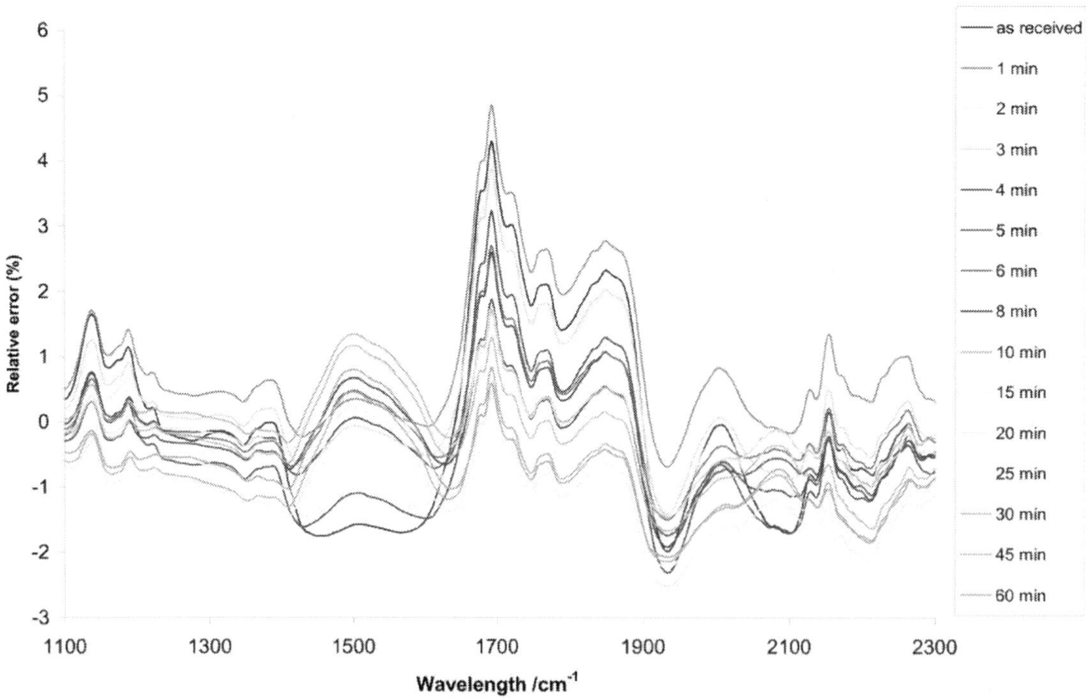

Figure 2. Plot of relative error between the actual NIR spectrum of formulation 5 at certain blending time point and the ideal NIR spectrum of a well-blended powder mixture vs. wavelength.

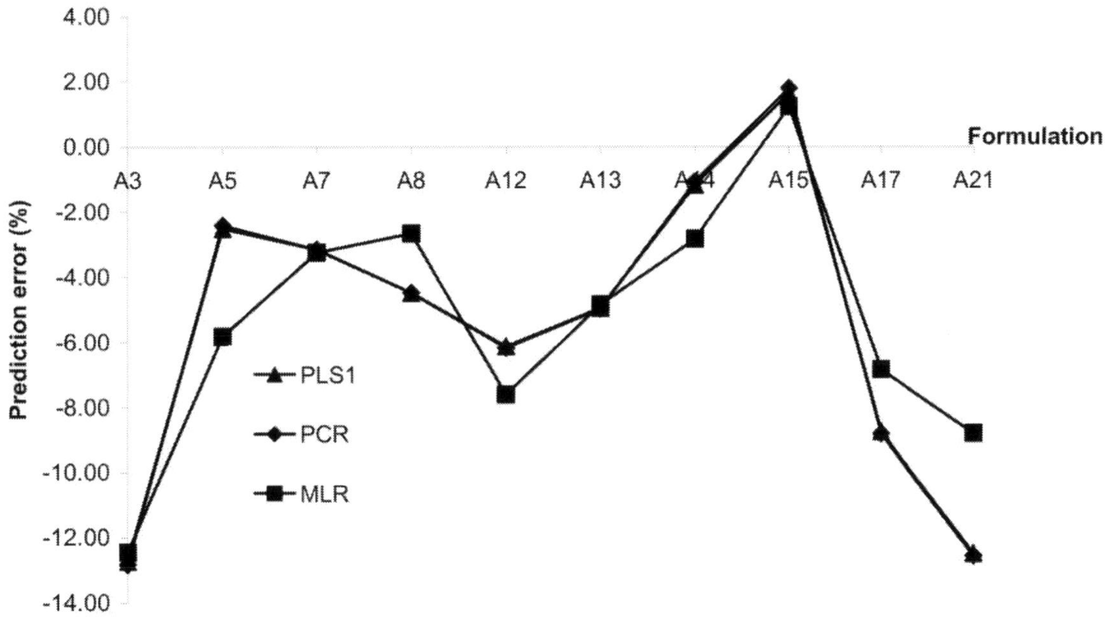

Figure 3 The plot of relative prediction error for HPMC vs. formulation at 15 minutes blending time point at HPMC's characteristic wavelength of 1508 nm.

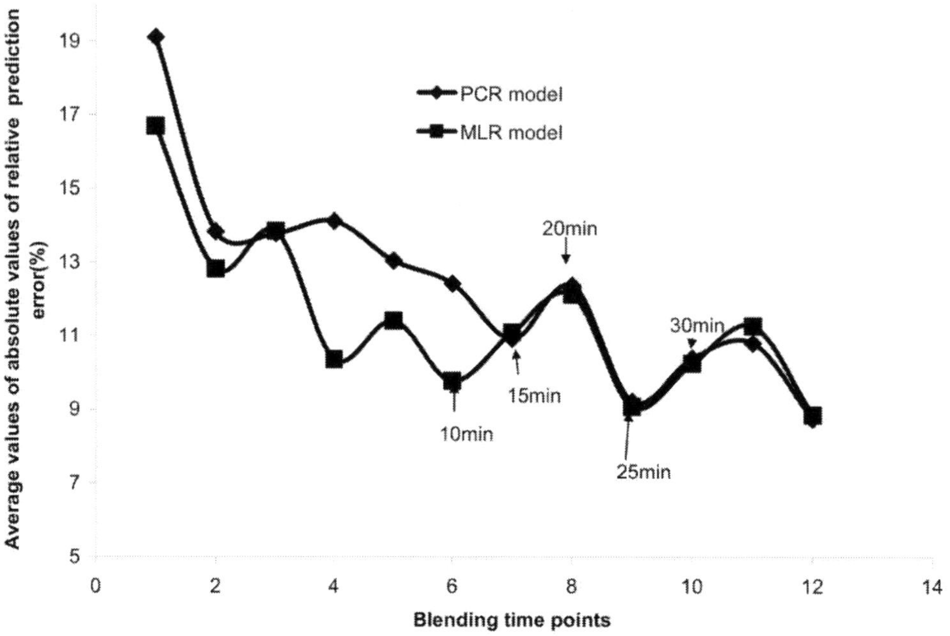

Figure 4 The plot of average of absolute values of relative prediction error (%) for the S-G 1st derivative at wave length of 1770 nm over 9 independent formulation batches vs. blending time point.

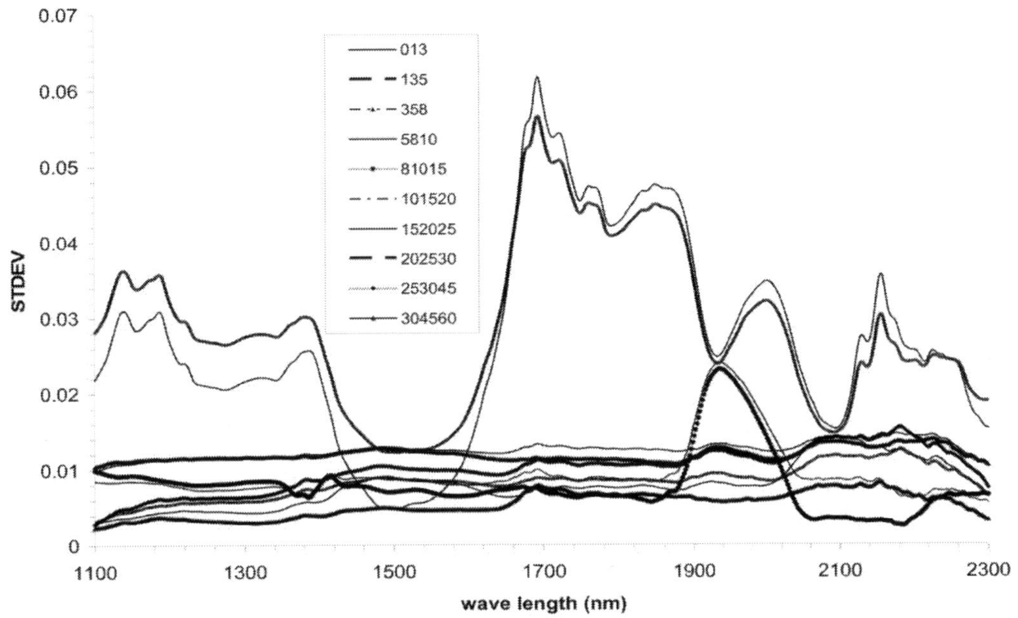

Figure 5 Plot of the moving standard deviation vs. wave length at various blending time windows for formulation A3.

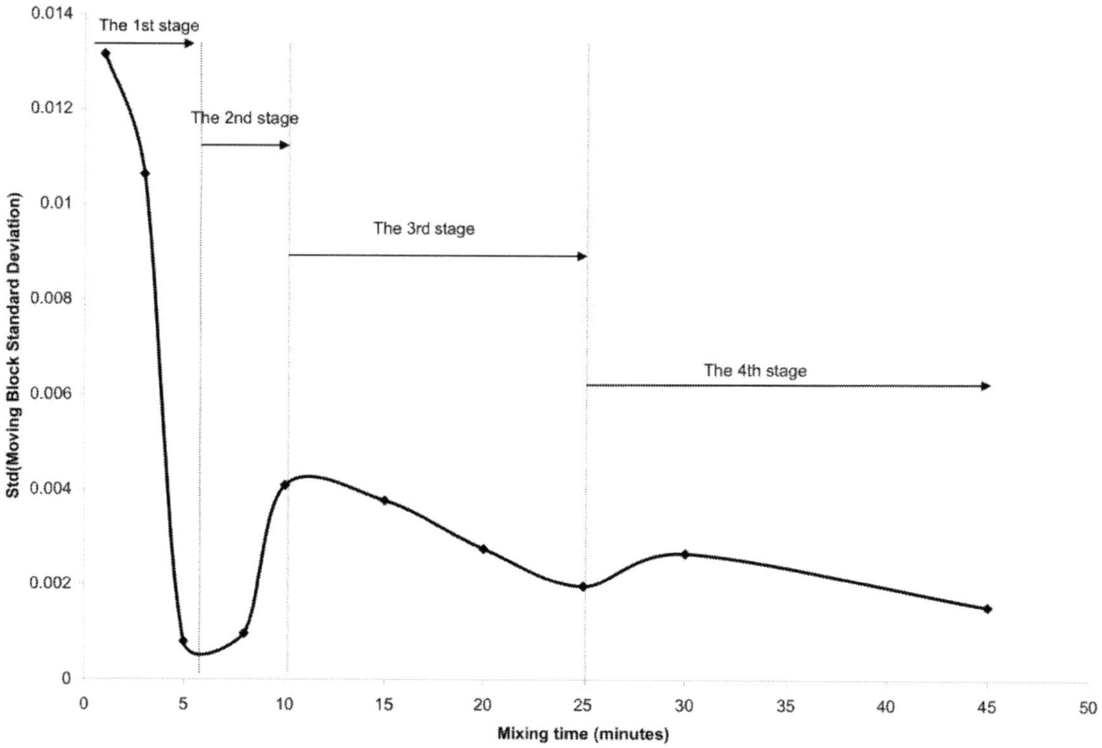

Figure 6 The plot of the integrated spectra standard deviation vs. blending time for formulation A3.

255e Move the Boundary and Raise the Bar

San Kiang and Shih-Ying Chang. Pharmaceutical Research Institute, Bristol-Myers Squibb, One Squibb Drive, New Brunswick, NJ 08903

In the pharmaceutical industry, it has been a tradition to deal with the active ingredient (API) and the formulation (drug product, DP) separately.

As a consequence, large amounts of efforts have been devoted to designing in the desired properties of the API through particle engineering techniques.In addition, the linkage of API properties to the performance of the DP has been a holy grail for many of us in the industry. However, some of the desired properties of the API can be obtained by simply moving the boundary between API and DP. Some of the approaches that can improve density, material flow, solubility, stability and other properties will be reviewed. We hope to provoke some discussion of perceived obstacles to this approach.

255f The Role of Chemical Engineering in Quality by Design

Daniel R. Pilipauskas, Global Manufacturing, Pfizer, 235 East 42nd Street, New York, NY 10017-5755 and David J. am Ende, Chemical R&D, Mail Stop 4016 Bldg 118E room 303L, Pfizer, Inc, Eastern Point Road, Groton, CT 06340.

As we apply the principles embodied by quality by design to gain a deeper understanding of our chemical manufacturing processes for active pharmaceutical ingredients (API), we are challenged with effectively describing the design space in terms of chemical and engineering principles (i.e. via a science based approach). One approach is to combine the appropriate scientific principles, process modeling tools, and multivariant designs to capture and describe the process outputs . In addition we are challenged with understanding and addressing the subtle but important scale up effects, the limitations of equipment to produce optimum process conditions for reactions and isolations, and process control. Developing and leveraging the expertise to perform these activities has potentially immense benefits in process understanding and process design. In this presentation, we present examples that illustrate how chemical engineering is critical toward fully realizing the benefits of quality by design

261a Control of API Powder Properties Via Cake Washing Protocol and Agitated Drying

Jessica L. Defreese, Derek Berglund, Prashant P. Deshpande, and Mauricio Futran. Process Research and Development, Bristol-Myers Squibb Company, One Squibb Drive, New Brunswick, NJ 08903

An API candidate has historically shown a propensity to form hard, fused aggregates during agitated drying. Prior work focused on minimizing aggregate formation through control of drying conditions, and investigated various milling strategies for management of the final powder properties. However, milling is a high-risk operation for this API. This work demonstrates how modifications to the cake-washing protocol mitigated the formation of aggregates and hard lumps during agitated drying. This process modification allowed for the elimination of API milling operations while retaining the desired formulation performance.

261b Flocculated Amorphous Nanoparticles for Enhanced Bioavailability of Poorly-Water Soluble Drugs: In-Vitro and in-Vivo Studies

Maria Andrea Miller[1], Michal E. Matteucci[2], Robert O. Williams III[3], and Keith P. Johnston[1]. (1) Chemical Engineering, The University of Texas at Austin, 1 University Station C0400, Austin, TX 78712, (2) The Dow Chemical Company, Building 1712, Midland, MI 48674, (3) College of Pharmacy, The University of Texas at Austin, 1 University Station, Austin, TX 78712

For a poorly-water soluble drug, a key challenge in oral delivery is to create a high enough bioavailability to be therapeutically effective. Particle engineering in this area has primarily emphasized creating individual amorphous nanoparticles to maximize the surface area not only to cause rapid dissolution, but also to achieve high supersaturation of the drug. Nanoparticle dispersions are formed by mixing an organic solution containing the dissolved drug with water to precipitate the drug in the presence of a stabilizer. However, these particles are difficult to harvest. For this work, we harvest the nanoparticles by purposely flocculating and filtering them to produce a powder highly favorable for further processing. The objectives of this study were two fold; 1) to flocculate the nanoparticles and ensure that they retain their amorphous qualities after filtering and drying and 2) to examine these particles for in vitro - in vivo correlation to determine the dominant effects leading to higher bioavailability for these particles in comparison to the current marketed itraconazole formulation, Sporanox®.

Nanoparticle dispersions stabilized with hydroxypropylmethylcellulose (HPMC) were flocculated and filtered by adding sodium sulfate to recover micron sized aggregates. By adding a tri-block copolymer of poly(ethylene oxide-b-propylene oxide-b-ethylene oxide) (P407), additional steric stabilization allowed the particles to redisperse to their original nanoparticle size. A second set of particles, stabilized with a random copolymer of methyl methacrylate and methacrylic acid, Eudragit®L100-55, were flocculated by lowering the pH to 2.5 to neutralize the acid groups. In vitro testing to simulate the particles through the highly acidic stomach conditions and neutral pH of the intestines correlated well with the in vivo bioavailability measured. This demonstrates the importance of not only achieving a high degree of supersaturation but also of maintaining the supersaturation through the varying conditions in the digestive tract.

261c Preserving API Hemihydrate during Vacuum Drying

Erwin Irdam, Chemical Development, GlaxoSmithKline, Five Moore Drive, Research Triangle Park, NC 27709

Control strategy for vacuum drying of an API hemihydrate to achieve optimum cycle time and to preserve the water content will be presented. This strategy was established based on understanding of drying kinetics and thermodynamics of hemihydrate derived from experiments in a small Ø 81mm x 50 mm agitated vacuum dryer utilizing online analytical techniques such as diffuse reflectance near infrared (NIR) and headspace mass spectrometry (MS). Comparison of drying results obtained from large scale dryer will also be illustrated.

261d Use of In-Line FTIR for the Kinetic Study of the Reaction and Degradation of Propylene Oxide In the Synthesis of a Pharmaceutical Intermediate

Thomas Laporte, Zhongmin Xu, Fernando Quiroz, Lori Spangler, Chiajen Lai, and Jean Tom. Bristol-Myers Squibb Company, One Squibb Drive, New Brunswick, NJ 08903

The process for the manufacture of a final intermediate of an active pharmaceutical ingredient (API) is a reactive crystallization in which the product crystallizes upon formation. The conditions driving that crystallization involves reaction of R-propylene oxide with an anion of the input intermediate. During normal processing, degradation of propylene oxide competes with the desired reaction. For example, elevated water content under basic conditions promotes propylene oxide degradation, primarily to propylene glycol. In-line FTIR was used to measure the conversion of the starting material and propylene oxide along with the degradation of propylene oxide under both model and actual reaction conditions. The network of propylene oxide reactions - degradation and product formation along with mass transfer of the product (crystallization) – was studied and modeled to increase process knowledge and to afford predictive simulation capability under various reaction conditions (temperature, stoichiometry, etc.). The process has been successfully scaled up to the pilot plant in two separate campaigns. Model results and data from both laboratory studies and pilot plant runs will be presented in this talk.

261e Control of Particle Sizes In An API Crystallization Process Via Thermal Cycling

Yao-En (David) Li, Shaila Perez, Warrant Holmes, Yanhui Hu, Gerrie Liaw, and Yubo Yang. Global Pharmaceutical Operations, Manufacturing Science and Technology, Dept 045B, Building R13-2,, Abbott Laboratories, 1401 Sheridan Road, North Chicago, IL 60064

In pharmaceutical industry, most active pharmaceutical ingredients (API) are purified via crystallization process. A good crystallization process should be able to reliably reject impurities to suitably low levels and provide the greatest possible yield. From operational point of view, large crystal sizes would enable to faster liquid-solid separation, which include filtration and centrifugation. Crystallized slurry with large crystals and small amount of fines will result in a faster filtration process.

In this talk, the authors will discuss a thermal cycling crystallization process developed for an actual API manufacturing process in Abbott. The thermal cycling process is based upon the Oswald ripening principle and the solid dissolution kinetics: it promotes the crystal growth during cooling while enhancing the small crystal dissolution rate during heating. Results from in-situ FBRM and PVM during the thermal cycling will be discussed and compared with the resulting wetcake resistance and permeability.

261f Poly(Ethylene) Glycol-Drug Eutectic Compositions: Complex Morphologies and Crystallization Kinetics

Phong Huynh and **Paul Takhistov**. Rutgers University, 65 Dudley Rd., New Brunswick, NJ 08901

Hot-melt extruded drug loading in drug-polymer eutectic formulations of ibuprofen in polyethylene glycol (PEG) have been investigated. The crystallization rate and morphology of solid dispersion of the drugs were found to be depended on carrier molecular weight and its weight fraction. It was experimentally determined that PEG crystallization forms 2D spherulitic structures due to temperature gradient and rapid surface cooling. It was shown by NIR-Raman spectroscopy that the regime of cooling impacts an interaction between ibuprofen and PEG that was a cause for observed complex morphologies solidified melts. Observed phase separation in studied solid dispersions has been compared with the developed mathematical model of the crystallization process including rate of crystallization and spherulites' interaction.

263a Bioactive Microgels for Large-Scale Eukaryotic Cell Culture

John J. Schmidt, Chemical and Biomolecular Engineering, University of Illinois Urbana-Champaign, 1206 W. Gregory Dr, Room #2200, Urbana, IL 61801 and Hyun Joon Kong, Department of Chemical and Biomolecular Engineering, University of Illinois Urbana-Champaign, 118 Roger Adams Laboratory, 600 South Mathews Avenue, Urbana, IL 61801.

One of the critical components for successfully using stem and progenitor cells in clinical treatments is a large-scale cell culture technology, which regulates the diverse cellular activities (i.e. growth, death, and differentiation) in a sophisticated manner. Currently, dextran and collagen-coated dextran microcarriers are used as the cell adhesion substrates on an industrial scale, because they significantly increase the ratio of the surface area for cell adhesion to the volume of bioreactors. However, currently commercialized microcarriers present several limitations, such as the use of organic solvents in preparation of microcarriers, limited control of biological function, and difficulty in the cell isolation from the microcarriers following the completion of cell culture[1]. Previously, we have demonstrated that chemical and mechanical properties of calcium cross-linked polysaccharide hydrogel modified with oligopeptides containing Arg-Gly-Asp sequence (RGD peptides) is able to regulate proliferation rate and differentiation level of cells adhered to the hydrogel disk surface [2,3]. We hypothesized that processing of these calcium cross-linked alginate hydrogels as microgels would allow us to control the cell proliferation and differentiation on a large scale and readily collect the cells from the microgels through the dissolution of gel matrices in aqueous environment. This hypothesis was examined with RGD peptides-presenting calcium cross-linked alginate microgels with diameters ranging from 300 to 700 mm. The density of RGD peptides was varied by mixing unmodified alginate molecules and RGD peptides-presenting alginate molecules at different ratios. The mechanical stiffness of the gel was controlled with the cross-linking density. We tested both bone precursor cells (MC3T3-E1 presoteoblasts) and bone marrow-derived mesenchymal stem cells. We found that the proliferation rate, shown in Figure 1, and the differentiation level of the cells were significantly controlled with these material variables on a large scale. In addition, a process to collect the expanded cells through the dissolution of microgel matrices with citric acid did not affect cell viability. Overall, the results of this study present a unique cell culture device, which allows one to control the proliferation rate, differentiation level and cell recovery on a large scale. The results of this study will be widely useful in culturing a broad array of stem and progenitor cells within a bioreactor with limited volume and thus greatly benefit the use of cells in various therapeutic interventions.

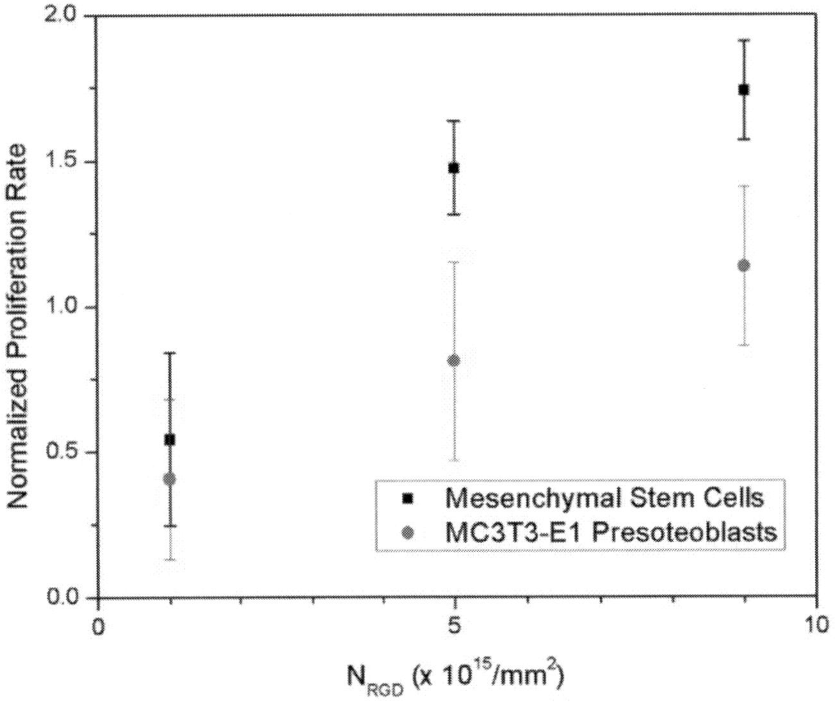

Figure 1: Cell proliferation was regulated by the overall density of RGD peptides on the alginate microcarriers. The cell proliferation rate was hastened with the increase in the number of RGD peptides (N_{RGD}). Values represent the mean and standard deviation from four independent measurements.

1. Malda, J., & Frondoza, C.G. Microcarriers in the engineering of cartilage and bone. *TRENDS in Biotechnology* 24:299-304 (2006)

2. **Kong, H. J.**, Boontheekul, T., & Mooney, D. J. Quantifying the relation between adhesion ligand-receptor bond formation and cell phenotype. *Proceedings of the National Academy Sciences (USA)* 103:18534-18539 (2006).

3. **Kong, H. J.**, Liu, J., Riddle, K., Matsumoto, T., Leach, K., & Mooney D.J. Gene delivery regulated by substrate rigidity *Nature Materials* 4:460 (2005)

263b Aggregation Dynamics In Plant Cell Culture

Martin E. Kolewe, Michael A. Henson, and Susan C. Roberts. Chemical Engineering, University of Massachusetts, 686 North Pleasant Street, Amherst, MA 01002

Plant cell culture is an alternative supply source for plant derived pharmaceuticals which may have no other feasible method of commercial production. In comparison with microbial or mammalian cell culture, plant cell culture presents challenges including low and variable product yields and slow growth rates. One defining characteristic of plant cell suspension cultures is the tendency of cells to grow in aggregates. Plant cells remain connected after cell division, and as a result, aggregates ranging from two to several hundred cells (50 to 2000+ μ m) are found in suspension, which can affect culture dynamics in several ways, including secondary metabolite accumulation and growth rate.

Taxus sp. cell culture for the production of the anti-cancer agent paclitaxel was used as a model system to investigate the effects of aggregate size on secondary metabolite accumulation. We developed a method to separate and subsequently screen different aggregate size fractions for paclitaxel and related taxanes using high performance liquid chromatography (HPLC). Aggregation dynamics in the culture system were phenomenologically described using a mathematical simulation correlated with experimental data. We used electrical zone sensing in conjunction with image analysis to collect aggregate size distribution data, and tracked the evolution of this distribution over the batch culture period. The model was formulated in a population balance equation framework, which could be directly compared to the distribution data collected. The comparison of the data and the model was used as meaningful criteria to evaluate the basic assumptions regarding aggregation phenomena that were used in structuring our model. This enhanced understanding of culture behavior can eventually be used as a tool to guide culture conditions to increase product accumulation.

263c Plant Cell Culture in Shake Flask and Their Relation with Oxygen Transfer Rate

Fernando Orozco-Sánchez[1], Gabriela Sepúlveda - Jiménez[2], Gabriela Trejo-Tapia[2], Alejandro Zamilpa – Alvarez[3], and Mario Rodríguez-Monroy[2]. (1) Escuela de Química, Universidad Nacional de Colombia and CeProBi - IPN, Medellín, Colombia, (2) Departamento de Biotecnología - CeProBi, Instituto Politécnico Nacional, Yautepec, Morelos, Mexico, (3) Centro de Investigaciones Biomédicas del Sur. Instituto Mexicano del Seguro Social, Xochitepec, Morelos, Mexico

Plant cell culture represents a biotechnology alternative to satisfy the demand of many compounds for medicinal, alimentary and agricultural industry. The studies in shake flasks are necessary before to culture the cells in a bioreactor. Cell suspension culture is a better option than organ culture to scale up process from shake flask to bioreactor (Doran, 1999). However, cell suspension cultures of some plant species are unstable in shake flasks, cells decrease its viability without a known reason. In this work, it was analyzed the possibility that oxygen transfer rate (OTR) in shake flasks may be insufficient to satisfy the cell oxygen demand.

Results obtained in our laboratory showed that cell lines of Beta vulgaris (Rodríguez-Monroy and Galindo, 1999) and Uncaria tomentosa (Trejo-Tapia et al, 2007) may be cultivated in shake flasks plugged with aluminum foil. In contrast, the culture of A. indica continuously loses its viability and reduces its pH in these conditions, and it is necessary regenerate periodically new suspension cultures from callus.

In order to know the oxygen demand by each specie, it was used a stirred tank bioreactor stirring with a 45° inclined - 6 blades impeller (400 rpm) and 0.08 vvm of aeration. Oxygen was measured using a polarographic electrode. Cell suspension cultures produced in shake flasks were used to inoculate the biorreactor. OUR was determined by the dynamic method (Doran, 1995). The measurements were done in the first hour after inoculation and the cells were subjected to conditions without oxygen limitation (oxygen dissolved higher than 30 %). OUR was calculated considering the variation of oxygen dissolved with the time, and the biomass concentration present in the bioreactor. The values of OUR obtained were: A. indica 0.100, B. vulgaris 0.027 and Uncaria tomentosa 0.026 kg O2/(kg dry cells.day). Those results indicate A. indica cells consume 4.5 times more oxygen than B. vulgaris or U. tomentosa.

On the other hand, OTR of shake flasks was determined using different plugs, aluminum foil and cotton. So, the overall oxygen mass transfer coefficient (Ko) and the maximum OTR (OTRmax) of the flasks were measured in 500 mL shake flasks at 120 rpm with 100 mL of culture medium, using the dynamic gassing out method. It can be demonstrate that Ko = kplug.kLa.H/(kplug.H + kLa), being kplug the oxygen transfer coefficient through the plug, kLa the oxygen transfer coefficient through the gas-liquid interface and H, the constant of Henry law. Besides, OTRmax = Ko.CO2*, being CO2* the oxygen concentration in the liquid phase at equilibrium with air at the system temperature.

Ko was 0.42 and 3.51 h-1 and OTRmax was 0.07 and 0.58 kg O2/(m3.day) for aluminum and cotton closure respectively. The values of kLa and kplug showed that the main resistance to oxygen flow was in the plug and not in the interface liquid gas, using both kinds of closure. With the values of OTRmax, it is possible to estimate the maximum OUR for a culture with 10 kg dry cells/m3 in shake flasks. The cultures could have a maximum OUR in shake flask of 0.007 and 0.058 kg O2/(kg dry cells.day) for aluminum and cotton closure respectively. Comparing the theoretical OUR obtained in shake flask with those measured in bioreactor for the three species (0.026-0.100 kg O2/(kg dry cells.day)), it is possible to observe that the measured values were higher than some values of maximum OUR in shake flask. This result indicated that the cultures of the three species cultures in shake flasks could have oxygen limitation.

Using oxygen balance is possible to show that A. indica cells have to reduce their OUR below 0.100 kg O2/(kg dry cells.day) when they reach a biomass concentration of 0.7 and 5.7 g DW/L in shake flasks plugged with aluminum and cotton respectively. These results suggests that the cell culture consuming an high quantity of oxygen (A. indica) could no adapt to oxygen limiting condition with foil aluminum, and the products of fermentation under anaerobic conditions such as lactic acid and ethanol, could be a cause of a reduction in pH. On the other hand, U. tomentosa cells have to reduce their OUR below 0.026 kg O2/(kg dry cells.day) when they reache a biomass concentration of 2.7 and 22.1 g DW/L in shake flasks plugged with aluminum and cotton respectively

With the purpose of studying the changes of pH and viability during the subcultures of A. indica, cells of this specie were grown in shake flasks plugged with aluminum foil and cotton. Cell viability and pH were evaluated during 6 continuous subcultures. Results obtained with shake flasks plugged with aluminum foil showed that pH decreased from 5.1 to values below 4.0, while in flasks with cotton, pH was above 5.0. Cell viability decreased from 78 % to 16 % in flasks with aluminum, while the cultures in flasks plugged with cotton had viability above 70 %. Cell suspension cultures with pH below 4.0 stopped their cell growth. A. indica cultures obtained under these conditions presented evident differences in their morphology. Cells cultured in flasks with aluminum foil are composed by aggregates with a diameter lower than 1 mm, while the agglomerates growing in flasks plugged with cotton had a diameter between 3 and 10 mm. Possibly a smaller size of agglomerates facilitates a better oxygen transfer under conditions with too little oxygen.

The results obtained in this work are in agree with those reported by Geigenberger (2003), who showed that plant tissues can reduce the oxygen consume when external concentration oxygen falls o when internal oxygen concentration falls into the tissue even in well oxygenated atmospheres. The plants can reduce the respiration, biosynthetic process, protein degradation or modify their morphology to avoid internal anoxia, and only with oxygen concentration is near to zero, they activate the lactate dehydrogenase and alcohol dehydrogenase for production of anaerobic ATP.

In conclusion, plant cells cultured in shake flask can have an oxygen limitation, so in flasks closured with aluminum foil, as well flasks closured with cotton. If the OTR offered by the flask is too low, the cells can reduce its viability, especially if the plant specie has a higher oxygen uptake rate, as was the case of A. indica. It is necessary to analyze the oxygen consumption of the different plant species in particular way. So, if the plant cell culture has a high oxygen consumption should be studied the effect of one low OTR in primary and secondary metabolism and its morphology. For increasing the OTR in shake flasks could be necessary modify the closure or others variables affecting the OTR (e.g. volume of culture medium in the flask or agitation speed).

Acknowledgements

F. Orozco – Sánchez thanks to Secretaría de Relaciones Exteriores of México for the doctoral fellowship awarded.

References

Geigenberger, P. Response of plant metabolism to too little oxygen. Curr. Opin. Plant Biol. 6: 247–256

Doran, P. 1999. Design of mixing systems for plant cell suspensions in stirred reactors. Biotechnol. Prog. 15: 319-335.

Doran, P. 1995. Bioprocess Engineering Principles. Academic Press. London. U.K.

Rodríguez-Monroy, M. and E. Galindo. 1999. Broth rheology, growth and metabolite production of Beta vulgaris suspension culture: a comparative study between cultures grown in shake flasks and in a stirred tank. Enzyme Microb. Technol. 24 (10): 687-693.

Trejo-Tapia, G, Sepúlveda-Jiménez, G., Trejo-Espino, JL., García-Rojas, C. de la Torre-Martínez, M., Rodríguez-Monroy, M. and Ramos-Valdivia, A. 2007. Hydrodynamic stress induces monoterpenoid oxindole alkaloid accumulation by Uncaria tomentosa (Willd) D. C. cell suspension cultures via oxidative burst. Biotechnol. Bioeng. 98 (1): 230-238.

Sparger and Surface Gas Transfer for Cell Culture Bioreactors
John S. Bowers
Schering-Plough Research Institute
Biotechnology Development
1011 Morris Ave, U-14-2-20
Union NJ 08536

Summary

The impact of bioreactor and sparger scale-up on gas transfer properties was investigated using four geometrically similar bioreactors ranging in size from 15-L to 2500-L total volume with the same sintered metal frit (microporous) sparger design. The mass transfer coefficients for both oxygen and carbon dioxide were measured at each scale at various flow rates. Scale did not directly affect gas transfer properties. Of the four bioreactors investigated, only the 200-L bioreactor had significantly different (higher) values for the oxygen and carbon dioxide sparger mass transfer coefficients.

The rate of carbon dioxide stripping from sparging in all vessels was much lower that expected based from film diffusion theories. The rate was limited by the saturation of the sparge gas by the carbon dioxide dissolved in the medium. The stripping rate was so low that sparging to provide oxygen would not remove enough carbon dioxide to balance cellular respiration.

The measured surface transfer of carbon dioxide and oxygen decreased directly with available surface area per unit volume. In agreement with observations, cultures that exhibit controlled carbon dioxide levels at small scale (15 L and less) may not have stable carbon dioxide levels at larger scale.

Introduction

To understand the effect of scale-up on dissolved gas concentrations in bioreactors, we investigated the impact of bioreactor and sparger scale-up on gas transfer properties with particular emphasis on carbon dioxide removal. Aunins (Aunins 1993) divided spargers into two categories, microporous and orifice. Microporous spargers are typically frits with holes less than 0.1 mm in diameter, whereas orifice spargers have drilled holes that are greater than 0.1 mm in diameter. Previous work has demonstrated that sintered metal frit spargers will provide high gas transfer coefficients by creating very small bubbles. The high oxygen transfer is several times what is achieved for orifice spargers, which create larger bubbles.(Aunins 1993), (Bovonsombut 1987), (Zhang 1992), (Zhang 1993), (Chisti 1993)

Authors have associated spargers generating very small bubbles with poor carbon dioxide removal(Aunins 1993). Gray (Gray 1996) attributed poor carbon dioxide removal to the shrinking of pure oxygen bubbles as oxygen is transferred to the media. Poor carbon dioxide removal can lead to difficulties in controlling pH (Aunins 1993), and accumulation of carbon dioxide can affect cell culture behavior, product titer and desired product quality(Zhangi 1999), (Kunkel 2000),(Jardon 2003),(Mostafa 2003). Mostafa and Gu (Mostafa 2003) found that replacing a frit sparger with an open pipe sparger, and adjusting the impeller position enhanced carbon dioxide removal in a large scale (1000-L) cell culture bioreactor, and resulted in increased titers.

To assess how scale affects the performance of these spargers, we studied four bioreactors ranging in size from 15-L to 2500-L total volume, with the same sintered metal frit sparger design. We measured mass transfer coefficients for oxygen and carbon dioxide for the sparger and liquid surface.

Material and Methods

Four Bioreactors were included in the study with nominal total volumes of 15-L, 200-L, 500-L and 2500-L. The physical dimensions are provided in Table 1. Marine impellers were used in all bioreactors. The bioreactors were filled to a liquid height to vessel diameter of 1:1. Microporous (10 to 20 μm) frit spargers were installed in all vessels, and where feasible, spargers were scaled to provide the same sparger frit area per unit volume. Sparger dimensions are provided in Table 2. Spargers for the 15-L and 200-L bioreactors were provided by BioTech Services Inc. (Halesite, NY), and spargers for the 500-L and 2500-L bioreactors by the bioreactor manufacturer (Bioengineering, Wald, Switzerland).

Table 1: Bioreactor Dimensions

Total Volume	Fill Volume	Vessel Diameter	Impeller Diameter
15-L	8 L	20 cm	10 cm
200-L	160 L	61 cm	40 cm
500-L	420 L	80 cm	50 cm
2,500-L	2,000 L	150 cm	90 cm

Table 2: Sparger Dimensions

Bioreactor	(Number x) Length, Diameter
15 L	(1x) 5 cm, d = 1.0 cm
200 L	(4x) 25 cm, d = 1.0 cm
500 L	(2x) 14cm + (2x) 23 cm, d = 1.4 cm
2,500 L	(2x) 14cm + (2x) 23 cm, d=3.8 cm

Dissolved oxygen and pH were monitored continuously with probes installed through 25 mm Ingold ports. pH probes (model GT-DJ-PG13.5-DTCH, ISI) were calibrated prior to insertion into the vessel. Each vessel was sterilized by steam-in-place, in accordance with established procedures. The pH readings were periodically confirmed with at-line measurements of samples from the bioreactor. Dissolved oxygen probes (model 7803-31535, Mettler Ingold) were calibrated to 100% after the media was equilibrated with air (typically overnight). All vessels were filled with purchased DMEM (HyClone) media with 10% Fetal Calf Serum (HyClone) to a liquid height equal to the diameter of the vessel. The agitation rate for each bioreactor was selected to provide a power per unit volume of approximately 2.5 mW/m^3, i.e. 90, 30, 27 and 17 rpm for the 15-L, 200-L, 500-L and 2,500-L bioreactors respectively. For all experiments, gas was sent to the head space of the bioreactor at a flowrate of approximately 0.05 liters per minute per filled reactor volume (0.05 vvm).

Determination of sparger oxygen mass transfer coefficient $k_{L,Ox}a_B$

Oxygen transfer in cell culture media has been extensively studied at small scale for various sparger designs (Aunin 1993), (Sambanis 1993), (Zhang 1992), (Zhang 1993), (Emery 1995), (Fenge 1993). Oxygen transfer can be described by the mass transfer model for a sparingly soluble gas into a liquid, where the driving force for oxygen transfer is the oxygen concentration that would be in equilibrium with the sparger gas, $[O_2]^*$, minus the dissolved oxygen concentration, $[O_2]$. The oxygen transfer rate, $d[O_2]/dt$ is the oxygen mass transfer coefficient for the bulk ($k_{L,Ox}a_B$) times the driving force:

Equation 1

$$d[O_2]/dt = k_{L,Ox}a([O_2]- [O_2]^*)$$

A similar equation can be written for oxygen surface transfer, using the oxygen mass transfer coefficient for the surface ($k_{L,Ox}a_S$) and the driving force from the headspace gas $([O_2]^*-[O_2])_S$. For larger bioreactors, the surface oxygen transfer is much smaller than the sparger oxygen transfer, and can be neglected during sparging.

To determine the sparger oxygen mass transfer coefficient ($k_{L,Ox}a_B$) for each bioreactor, nitrogen was sparged to reduce the dissolved oxygen to less than 10%, based on 100% is in equilibrium with air. Air or oxygen was then sparged until the dissolved oxygen was $k_{L,Ox}a_B$ over 90%. During the sparging of nitrogen, air or oxygen, the $k_{L,Ox}a_B$ was determined by a linear fit of ln $/[O_2]^*- [O_2]/$ vs time.

Determination of sparger carbon dioxide mass transfer coefficient $k_{L,CO}a_B$

In many media commonly used for cell culture, the pH is maintained by a carbon dioxide – bicarbonate buffer. Several reactions are required to fully describe the behavior of this buffer system, but in the pH range 6.5 to 8.0, the reactions can be summarized in one combined reaction: (Ho, 1986)

Equation 2

$$CO_2 + H_2O \Leftrightarrow H^+ + HCO_3^-$$

The carbonic acid (H_2CO_3) and carbonate ions (CO_3^{-2}) concentrations are not significant compared to bicarbonate and carbon dioxide concentrations in the pH range from 6.5 to 8.0(Stumm, 1981), (Gray 1996). The equilibrium constant for the above reaction in cell culture media (Gray 1996) is about $8x10^{-7}$. Using a mass balance for the species, H^+, CO_2 and HCO_3^- and eliminating insignificant terms, we can obtain the carbon dioxide concentration $[CO_2]$ during stripping at any time relative to an initial value $[CO_2]_0$ as a function of pH:

Equation 3

$$[CO_2]=10^{pH0-pH}[CO_2]_0$$

The rate of carbon dioxide stripping, $d[CO_2]/dt$ is describe by a differential mass balance similar to that for oxygen transfer, except it is based on the carbon dioxide mass transfer coefficient for the bulk ($k_{L,CO}a_B$) and carbon dioxide is not present in the sparger gas ($[CO_2]^*=0$):

Equation 4

$$d[CO_2]/dt=-k_{L,CO}a_B[CO_2]$$

Combining the equilibrium relationship with a mass balance for carbon dioxide, we can show that without the addition of other acids or bases, the bicarbonate level will remain constant, i. e. $[HCO_3^-]=[HCO_3^-]_0$, and the mass transfer coefficients will be directly proportional to the rate of change of the pH, independent of the total inorganic carbon concentration.

Equation 5

$$k_{L,CO}a_B=2.303\ dpH/dt$$

A similar relation can be written for surface CO_2 removal using the carbon dioxide mass transfer coefficient for the surface, $k_{L,CO}a_S$.

To determine the sparger carbon dioxide mass transfer coefficient ($k_{L,CO}a_B$) for each bioreactor, the pH was reduced to between 7.00 and 7.20 by sparging with CO_2. These sparges were not used to determine the $k_{L,CO}a_B$, since the changes in pH with sparging CO2 may not represent the equilibrium of CO_2 related species in water. The pH was allowed to stabilize for approximately 15 minutes before measuring the stripping rate, d(pH)/dt. Nitrogen, air or oxygen was then sparged until a pH change of at least 0.04 units was observed. The stripping rate, d(pH)/dt, was determined by a linear fit to pH vs time and the mass transfer coefficient was then calculated directly from $k_{L,CO}a_B = 2.303$ d(pH)/dt.

Determination of surface oxygen mass transfer coefficient $k_{L,Ox}a_S$ *and surface carbon dioxide mass transfer coefficient* $k_{L,CO}a_s$

To determine the surface mass transfer coefficients for each reactor, nitrogen was first sparged to reduce the dissolved oxygen to less than 10% and the pH was reduced to between 7.00 and 7.20 by sparging with CO_2. Air was sent to the headspace at 0.05 vvm, and the $[O_2]$ and pH was monitored for several hours, typically overnight. After an initial transitional period associated with the dissipation of the foam on the surface, the $k_{L,Ox}a_B$ was determined by the slope of the least squares fit of ln $/[O_2]^* - [O_2]/$ vs time, and stripping rate, d(pH)/dt, was determined by a linear fit to pH vs time. The surface carbon dioxide mass transfer coefficient was calculated directly from $k_{L,CO}a_S = 2.303$ d(pH)/dt.

Results and Discussion

No statistically significant differences were found between determinations made by sparging with air, nitrogen or oxygen. Results for all three gases were therefore combined in the analysis. As can be seen in Table 3, there is no trend with bioreactor scale in the oxygen or carbon dioxide sparger mass transfer coefficients ($k_{L,Ox}a_B$ and $k_{L,CO}a_B$) under typical operating conditions. The mass transfer coefficients of the 15-L, 500-L, and 2,500-L bioreactors are not significantly different from each other.

Table 3: Gas transfer coefficients at 0.005 vvm with scale. Ranges are 95% confidence limits.

Bioreactor	$k_{L,Ox}a_B$ (1/hr)	$k_{L,CO}a_B$ (1/hr)
15-L	3.4±1.2	0.46±0.09
200-L	9.3±1.0	0.65±0.25
500-L	3.0±1.0	0.28±0.05
2,500-L	3.7±1.1	0.42±0.08

The observed ratio of the carbon dioxide to the oxygen mass transfer coefficient ($k_{L,CO}a_B$ / $k_{L,Ox}a_B$), about 0.1, is much lower than would be expected based on film diffusion based theories. The ratio of the mass transfer coefficients should be between 0.8 and 0.9. (Aunins 1993), (Ho 1986), (Ho 1988) Gray proposed that for small oxygen bubbles, the shrinking ofbubbles as they traveled to the surface would reduce the stripping of CO_2.(Gray 1996) However, we observed no significant differences in the CO_2 stripping rate for nitrogen, air or oxygen, so this mechanism does not explain these results.

An explanation that fits our results well is that the sparge gas is saturating with CO2 prior to reaching the headspace. When this occurs, the total amount of stripping can only be the gas volume sparged times the equilibrated gas phase concentration. Using Henry's law, the ideal gas law, and the volume of sparger gas per bioreactor volume per minute (vvm), we can derive an apparent maximum gas transfer coefficient, $(k_{L,CO}a)_{SAT}$.

Equation

$$(k_{L,CO}a)_{SAT} = (H/RT)(60)(vvm)$$

For CO_2 in media at 37 °C, the Henry's Law constant, H, is ~40 atm-L/mol (Onda 1970), (Yagi 1977) and $(k_{L,CO}a)$SAT will be ~94 vvm or 0.47 at 0.005 vvm. The saturation value is consistent with the measured values in table 3. This is so low, sparging with pure oxygen will not remove as much carbon dioxide as is formed by respiration. For typical cell culture conditions, i.e. 10% dissolved CO_2 ($[CO_2]$~2.5 mM) and sparging with pure O_2 ($[O_2]^*$- $[O_2]$)~0.9 mM), the amount of carbon dioxide stripped will be less than half of the carbon dioxide produced by respiration.

To avoid accumulation, CO_2 must be removed by another mechanism such as surface stripping. The surface gas mass transfer coefficients were proportional to the available surface area to volume ratio with scale (Table 4). The ratio $k_{L,Ox}a_s$ / $k_{L,CO}a_s$ for surface stripping agreed better with film diffusion theory, consistent with the headspace gas, having 10 times the flowrate, does not saturate with carbon dioxide.

Table 4: Surface Gas transfer coefficients with scale.

Bioreactor	$k_{L,Ox}a_s$ (1/hr)	$k_{L,CO}a_s$ (1/hr)	Ratio
15 L	0.25	0.15	0.6
200 L	0.10	0.08	0.8
500 L	0.06	0.04	0.7
2500 L	0.05	0.03	0.6

These results also are consistent the problem of carbon dioxide accumulation with scale-up. For the smallest bioreactor, the $k_{L,CO}a_B$ was about 0.15/hr. For typical cell culture conditions (10% dissolved CO_2), the rate of carbon dioxide removal through the surface would be 0.38 mmoles/L-hr, or equivalent to the carbon dioxide produced by cellular respiration when the OUR is about 2.5% per minute. For the largest reactor, the $k_{L,CO}a_B$ is about 0.03/hr and could only remove the carbon dioxide produced by cellular respiration when the OUR is about 0.5% per minute. A process with cellular OUR in the range of 2% to 5% per minute would not have carbon dioxide accumulate in the small reactor, but would have carbon dioxide accumulate in the largest reactor.

Conclusions

The oxygen gas transfer coefficient of the sparger, $k_{L,Ox}a_B$, was not dependent on scale. Of the four bioreactors investigated, only the 200 L had statistically significantly different (higher) values for the oxygen transfer coefficient at the operating condition of 0.005 vvm. As the gas flowrate was increased, the oxygen mass transfer coefficients increased until reaching a maximum. Because of the non-linearity of the relationship between oxygen mass transfer coefficient and flow rate, lowering the flowrate may lead to less gas sparged, and less foam formed in the bioreactor.

The carbon dioxide transfer coefficient, $k_{L,CO}a_B$ was nearly the same for the four bioreactors and was not dependent on sparge gas. The measured values were much lower than what would be expected based on film diffusion theories. The values were consistent with the saturation of sparge gas with carbon dioxide. Because of the low values this sparger design may not remove enough carbon dioxide to keep the levels balanced. As expected, the surface stripping of carbon dioxide decreased directly with the available surface area per unit volume. With this sparger design, cultures that exhibit balanced carbon dioxide levels at small scale (15-L and less) may not have stable CO_2 levels at larger scale.

Acknowledgements
Many individuals effort contributed to this work. Significant time and effort was required to acquire and interpret the data. Individuals who contributed included:

- Anita Chaudry
- Balrina Gupta
- Rajesh Krishnan
- Jason Polite
- Stacey Merz
- Sandeepa Sandadi
- Marcus Tanner

References

(Aunins 1993) Aunins, J. G.; Henzler H.-J. "Aeration in cell culture bioreactors." In *Biotechnology: A Multi-Volume Comprehensive Treatise*; Stephanopoulos, G., Rehm, H.-J., Reed, G., Puhier, A., Stadler, P. J. W., Eds.; VCH Verlag: Weinheim, **1993**; Vol. 3.

(Bauer 2000)Bauer, P., Hancock, L., Rathman, J. and Chalmers, J., Cell-Microcarrier Adhesion to Gas-Liquid Interfaces and Foam, *Biotechnol. Prog.*, **2000,** *16*, 125-132.

(Bavarian 1991)Bavarian, F., Fan, L. S.,and Chalmers, J. J., Microscopic Visualization of Insect Cell-Bubble Interactions. I: Rising Bubbles, Air-Medium Interface, and the Foam Layer, *Biotechnol. Prog.* **1991,** *7,* 140- 150.

(Chalmers 1991)Chalmers, J. J., and Bavarian, F., Microscopic Visualization of Insect Cell-Bubble Interactions. II: TheBubble Film and Bubble Rupture, *Biotechnol. Prog.* **1991,** *7,* 151- 158.

(Bonarius 1995)Bonarius, H., da Gooljer, C.; Tramper, J.; and Schmid, G.; Determination of the Respiration Quotient in Mammalian Cell Culture in Bicarbonate Buffered Media, *Biotech Bioeng*, **1995**, *45*, 624—535.

(Bovonsombut 1987)Bovonsombut, S.; Wilhelm A.-M..; Riba, J.-P.; Influence of Gas Distributor Design on the Oxygen Transfer Characteristics of an Airlift Fermenter, *J. Chem. Tech. Biotech.* **1987** *40*, 167-176,

(Cherry 1986)Cherry, R.S., and Papoutsakis, E. T. "Hydrodynamic Effects on Cells in Agitated Tissue Culture Reactors" *Bioprocess Eng.*, **1986**, *1*, 29.

(Chisti 2000) Chisti, Y. "Animal-cell Damage in Sparged Bioreactors" *Trends in Biotech,* **2000,** *18*, 420-431.

(Chisti 1993) Chisti, Y,. "Animal Cell Culture in Stirred Bioreactors: Observations on Scale-up" *Process Biochemistry*, **1993**, *28*, 511-517

(Dorresteijn 1994)Dorresteijn, R., de Gooijer, C., Tramper, J., Beuvery, E., "A Simple Dynamic Method For On-Line and Off-Line Determination of Kla During Cultivation of Animal Cells" *Biotech Tech,* **1994***, 8,* 675-680.

(Emery 1995)Emery, A. N., Jan, D. C.-H., Al-Rubeai , M., Oxygenation of intensive cell-culture system, *Appl Microbiol Biotechnol,* **1995,** *43*, 1028—1033

(Fenge 1993)Fenge, C., Klein, C., Heuer, C., Siegel, U., and Fraune, E., "Agitation, Aeration and Perfusion Modules for Cell Culture Bioreactors" *Cytotechnology* **1993,** *11*, 233-244.

(Gray 1996)Gray, D,; Chen, S.;, Howarth, W.; Inlow, D.; Malorella, B.; CO_2 in large-scale and high-density CHO cell perfusion culture, *Cytotech.*, **1996**, *22,* 65-78.

(Handa-Corrigan, 1989) Handa-Corrigan, A.,. Emery*, A. N., and Spier, R. E; "Effect of Gas-Liquid Interfaces on The Growth of Suspended Mammalian Cells: Mechanisms of Cell Damage By Bubbles" *Enzyme Microb. Technol.*, **1989**, *11*, 230-235.

(Ho 1986)Ho, C.S. and Shanahan, J. F., "Carbon Dioxide Transfer in Bioreactors" *CRC Rev. in Biotech.*, **1986**, *4*, 185-252.

(Ho 1988)Ho C.S., Ju, L. "Effects of Microorganisms on Effective Oxygen Diffusion Coefficients and Solubilities in Fermentation Media" *Biotechnol. Bioeng.* **1988**, *32*, 313-325.

(Jardon 2003)Jardon, M. and Garnier, A. "pH, pCO and Temperature Effect on R-Adenovirus Production" *Biotechnol. Prog*, **2003**, *19*, 202-208.

(Kunkel 2000)Kunkel, J., Jan, D., Butler, M., Jamieson, J. Comparisons of the Glycosylation of a Monoclonal Antibody Produced under Nominally Identical Cell Culture Conditions in Two Different Bioreactors" *Biotechnol. Prog.*, **2000**, *16*, 462-470.
(Mostafa 2003) Mostafa, S., Gu, X. "Strategies for Improved dCO Removal in Large-Scale Fed-Batch Cultures", *Biotechnol. Prog.* **2003**, *19*, 45—51

(Martens 1996)Martens, D. E.; Nollen, E. A. A.; Hardeveld, M.; van der Veldende-Groot, C. A. M.; de Gooijer, C. D.; Beuvery, E. C.; Tramper, J., "Death Rate in a Small Air-Lift Loop Reactor of VERO Cells Grown on Solid Microcarriers and in Macroporous Microcarriers", *Cytotechnol.* **1996**, *21*, 45-59.

(Meier 1999)Meier, S. J.; Hatton, T. A.; Wang, D. I. C. "Cell Death from Bursting Bubbles: Role of Cell Attachement to Rising Bubbles in Sparged Bioreactors". *Biotechnol. Bioeng.* **1999**, *62*, 468-478.

(Nyiri 1968)Nyiri, L., Lengyel, Z. "Studies on Ventilation of Culture Broths. I. Behavior of CO in Model Systems" *Biotech BioEng.* **1968**, *10*, 133-150.

Ozturk, S. "Engineering challenges in high density cell culture systems" 1996

(Sambanis 1993)Sambanis, A., Hu, W.-S, "Cell Culture Bioreactors", **1993**, *Biotechnology*, *3*, 105-126

Schumpe A.. The estimation of gas solubilities in salt solutions. Chem Eng Sci **1993**, *48*, 153–158.

(Stumm 1981)Stumm, W. and Morgan. J.J. *Aquatic Chemistry*. Chapter 4. **1981** John Wiley & Sons. NY, New York

(Zhang 1993)Zhang, S., Handa, A,. and Spier, R. E., A Comparison of Oxygenation Methods for High-Density Perfusion Cultures of Animal Cells, *Biotech BioEng.* **1993**, *41*, 685-692.

(Zhang 1992)Zhang, S., Handa, A,. and Spier, R. E., Oxygen Transfer Properties of Bubbles in Animal Cell Culture Media, *Biotech BioEng.* **1992**, *40,* 252-259.

(Zanghi 1999)Zanghi, J., Schmelzer, A., Mendoza, T., Knop, R., and Millert, W. "Bicarbonate Concentration and Osmolality Are Key Determinants in the Inhibition of CHO Cell Polysialylation Under Elevated pCO or pH" *Biotech Bioeng,* **1999**, *65*, 182-191.

263e Evaluation of a Novel 2-D Rocking Motion Bio-Reactor System with Superior Gas Transfer Rates and Mixing Capabilities Suited for a Process Development Environment

Kapil Kothary[1], Terry Hudson[1], Ekta Mahajan[2], Jemelle Dizon-Maspat[1], Gayle E. Derfus[1], and Ashraf Amanullah[1]. (1) Oceanside Process R&D, Genentech Inc., 1 Antibody way, Oceanside, CA 92056, (2) Process Development Engineering, Genentech Inc., Mail Box 75B, 1 DNA Way, So San Francisco, CA 94080

Disposable bioreactors are being increasingly used in process development due to improved simplicity of operation and for the economic advantages they offer over the conventional stirred tanks. Rocking motion reactors have proved to be an efficient alternative to produce antibodies. Several rocking bioreactor systems are currently available and are equipped with various probes/sensors and control systems. Recently, a novel 2D rocking motion bioreactor has been introduced in the disposables market by Cell-ution Biotech. The vendor claims that the system is suitable for growing high cell density cultures due to its superior gas transfer rates and mixing capabilities over its peers. In this study we characterized oxygen mass transfer and mixing. The performance of this system was also evaluated by comparing cell growth data, metabolic profiles and controls with similar bioreactor systems.

263f Quantitative Study of Physiological Responses of Cho Cells to Repetitive Hydrodynamic Stress

Ruben Godoy-Silva[1], Jeffrey Chalmers[2], Susan Casnocha[3], and Ningning Ma[3]. (1) Chemical Engineering, Universidad Nacional de Columbia, Bogota, Colombia, (2) Chemical and Biomolecular Engineering, Ohio State University, 140 W. 19th Avenue, Columbus, OH 43210, (3) Bioprocess R&D, Global Biologics, Pfizer, Inc., 700 Chesterfield Parkway, West Chesterfield, MO 63017

Investigations on hydrodynamic sensitivity of mammalian cells have focused mainly on the readout of lethal effect as determined by cell death or lysis. In this talk, we will present the first systematic investigation of sub-lethal responses of biopharmaceutical producing mammalian cells to hydrodynamic stress. In large-scale agitated bioreactors, cells constantly circulate between high hydrodynamic stress zone around the impeller and low hydrodynamic stress zone away from impellers. A novel scale-down model, which reproduces this repetitive stress was used to test GS-CHO cells in a fed-batch process. Lethal effects were monitored through measurement of viability and lactate dehydrogenase (LDH) release. As a focus of this study, a broad array of assays and measurements were used to detect sub-lethal effects, which could affect important process and product quality attributes. These sub-lethal effects include cell growth rate, peak cell density, monoclonal antibody (mAb) production rate, glucose metabolism, post-translational modifications, and product related impurities.

In general, GS-CHO cells showed strong resistance to hydrodynamic stress, even under ten days' repetitive exposure. Cell growth rate and peak cell density were not affected at the highest stress level tested. Reduction in productivity and glucose utilization was observed, but the impact was minimal. However, significant non-lethal and physiological responses were indeed observed at hydrodynamic stress level much lower than that needed to elicit cell death or productivity drop. Cell size became smaller under hydrodynamic stress at which growth rate was not affected and production rate was only minimally reduced. Product quality was found to be the most sensitive parameter to hydrodynamic force. mAb charge profile shifted where the percentage of acidic species increased and that of basic species decreased. A significant shift of glycosylation pattern was also observed. Detailed results, including process performance, product quality, and the threshold hydrodynamic stress that triggers each sublethal effects, will be presented. The implication of this study on large-scale bioreactor design and operation will also be discussed.

263g Effect of Hydrodynamic Conditions on Cell Cycle, Stress Proteins and Recombinant Protein Productivity

Claudia Berdugo, The Ohio State University, 125 Koffolt Labs, 140 W. 19th Ave., Columbus, OH 43210 and Jeffrey J. Chalmers, Chemical and Biomolecular Engineering, Ohio State University, 140 W. 19th Avenue, Columbus, OH 43210.

Cells respond against adverse conditions by expressing a set of proteins, some of which are historically referred to as Heat Shock Proteins. While the name originated from a cell's expression response to a sudden increase in temperature, from a physiological point of view the expression of these proteins are a defense mechanism and is associated with the synthesis of proteins since they act as chaperons.

In this work, we have evaluated the expression of stress proteins under different hydrodynamic conditions and we analyzed the relationship between the expression of stress proteins and culture performance. Also, the expression of stress proteins in highly sensitive cell lines will be discussed. Stress proteins investigated have been studied in the context of heat and nutritional stress but it has not been determined the expression of stress proteins in response to hydrodynamic conditions.

The results strongly suggest that the expression of stress proteins follows a characteristic profile that is related to the state of the culture.

275a Ethanol-Induced Changes In Neutrophil Membrane Tether Mechanics and Adhesion Molecule Expression Depend on Cellular Cholesterol Levels

Hana Oh and Scott L. Diamond. Chemical and Biomolecular Engineering, University of Pennsylvania, 3340 Smith Walk, 1150 Vagelos Labs, Philadelphia, PA 19104

Clinical studies have shown that moderate alcohol consumption protects from atherosclerosis by altering cholesterol levels, but the mechanisms of ethanol-cholesterol effects on the membrane remain unknown. To study the effect of ethanol on neutrophils with low or high cholesterol content, we treated neutrophils with MßCD for 35% depletion or with MßCD-cholesterol complexes for 40% enrichment, which decreased or increased membrane fluidity, respectively. Neutrophils were incubated with physiologically relevant ethanol concentrations achieved in humans by alcohol consumption (<0.5% ethanol by vol.), washed and perfused over parallel-plate flow chambers with P-selectin-coated 1-µm-diameter beads or P-selectin surfaces at a venous shear rate of 100 s-1. Cholesterol enrichment increased the tethering fraction of neutrophils from 0.63 to 0.74, while cholesterol depletion decreased it to 0.45. Tether length and lifetimes also increased from 2.50 to 5.30 µm and 0.53 to 0.62 sec with cholesterol enrichment, while cholesterol depletion reduced them to 0.67 µm and 0.38 sec. Tether growth velocity doubled with cholesterol enrichment. The fraction of firm arrest increased from 0.05 to 0.22, and cholesterol-treated cells were 13% larger in size and more deformable, suggesting that high-cholesterol cells may be more susceptible to activation. Treatment with ethanol mitigated these cholesterol-induced effects, but the effect of ethanol on cholesterol-treated cells was significantly smaller than its effect on neutrophils with basal cholesterol level. Ethanol also reduced CD11b expression and caused PSGL-1 redistribution, which may contribute to the change in neutrophil adhesive behavior. Our results indicate that both ethanol and cholesterol affect bond mechanics, membrane fluidity, and surface molecules to change neutrophil response during receptor-mediated adhesion.

275b Enrichment of CD34+ Hematopoietic Stem and Progenitor Cells from Bone Marrow and Peripheral Blood Using a P-Selectin-Coated Microtube

Srinivas Narasipura[1], Joel Wojciechowski[2], Nichola Charles[1], Jane Liesveld[3], and Michael King[1]. (1) Biomedical Engineering, University of Rochester, Rochester, NY 14627, (2) CellTraffix inc, Rochester, NY 14623, (3) Hematology and Oncology, University of Rochester, Rochester, NY 14627

Enrichment and purification of hematopoietic stem and progenitor cells (HSPCs) is important in transplantation therapies for hematological disorders and for basic stem cell research. Primitive CD34+ HSPCs have demonstrated stronger rolling adhesion than mature CD34- mononuclear cells on selectins. We have exploited this differential rolling behavior to capture and purify HSPCs from human bone marrow and rat peripheral blood by perfusing samples through selectin-coated microtubes.

Bone marrow- mononuclear cells were perfused through the cell capture microtubes coated with adhesion molecules. The device lumen was washed and captured cells were visualized and estimated by video microscopy. Adherent cells were eluted by high shear, calcium free buffer and air embolism. Immunofluorescence staining followed by flow cytometry was used to analyze CD34+ HSPCs. CD34+ HSPC purity of cells captured in adhesion molecule-coated microtubes was significantly higher than the fraction of CD34+ cells found in bone marrow- mononuclear cells (2.5 °" 0.8%). P-selectin coated surfaces yielded 16-20% CD34+ cell purity, while anti-CD34 antibody coated surfaces yielded 12-18%. Although the CD34+ cell purities were comparable between selectin and antibody surfaces, the total number of CD34+ HSPCs captured was significantly higher in P-selectin devices (~5.7-7.1 x 104) when compared to the antibody device (~1.74-2.61 x 104). Furthermore, analysis for cells positive for CD133, a surface marker for more primitive HSPCs, depicted approximately 10-14 fold enrichment in P-selectin samples over control bone marrow mononuclear cells. The captured cells were viable and exhibited in vitro colony forming capabilities.

In a parallel study we aimed at capturing HSPCs directly from circulating blood in vivo. Vascular shunt prototypes, coated internally with P-selectin, were inserted into the femoral artery GCSF mobilized rats. After one-hour blood perfusion, a successful capture of mononuclear cells was observed and analysis of such cells revealed a significant enrichment for CD34+ HSPCs. Blood flow through the cell capture device resulted in a wall shear stress of 4-6 dynes/cm2. Purity of captured CD34+ cells showed 7-fold enrichment (average purity = 28 „b 4%) over levels found in the control mobilized peripheral blood mononuclear cells (average purity = 3 „b 0.5%). Robust cell capture and HSPC enrichment were also observed in microtubes that were implanted in a closed-loop arterio-venous shunt conformation for two hours. Adherent cells were viable and were able to differentiate into burst forming units in culture.

Thus, P-selectin can be used in a compact flow device to capture and enrich HSPCs. This study supports the hypothesis that flow-based adhesion molecule-mediated capture may be a viable physiologic approach to the capture and purification of HSPCs.

References:

1. Kansas GS. Selectins and their ligands: current concepts and controversies. Blood 1996;88:3259-3287.

2. Narasipura SD, Wojciechowski JC, Charles N, et al. P-selectin- coated microtube for enrichment of CD34+ hematopoietic stem and progenitor cells from human bone marrow. Clin Chem 2008;54:77-85.

3. Charles N, Liesveld JL, King MR. Investigating the feasibility of stem cell enrichment mediated by immobilized selectins. Biotechnol Prog 2007;23:1463-72.

4. Wojciechowski JC, Narasipura SD, Charles N, et al. Capture and enrichment of CD34-positive haematopoietic stem and progenitor cells from blood circulation using P-selectin in an implantable device. Br J Haematol 2008;140:673-81.

275c Effect of Fluid Induced Shear on Biofilm Accumulation, Erosion and Cell Adhesion of *Staphylococcus Aureus*

Patrick Ymele-Leki and Julia M. Ross. Chemical and Biochemical Engineering, University of Maryland, Baltimore County, 1000 Hilltop Circle, Baltimore, MD 21250

S. aureus infection is a leading public health threat in community and healthcare settings. Its virulence correlates in part with the range of adhesins present on the cell surface, which are known to mediate adhesion to exposed ECM proteins and immobilized platelets in the bloodstream. The ability to form a biofilm then enables the pathogen to evade the host immune system and antimicrobial measures. This study investigates *in vitro* the potential role of adhesins in the establishment of early *S. aureus* biofilms under physiologically relevant conditions. We demonstrate with immunocytometry and kinetic studies that eroding planktonic cells and fixed sessile cells retain significant adhesion properties with respect to ECM proteins over 24 hours of biofilm growth. Our data reveal a potential mechanism for the promotion of secondary sites of infections via adhesin-mediated binding events. Immunofluorescent confocal microscopy is used to characterize the architectural attributes of *S. aureus* biofilms *in situ*. Results show that volumetric and areal parameters such as biovolume and cluster perimeter level off with increased shear while textural parameters and others including fractal dimension and porosity tend to be shear-dependent over the fluid shear range investigated. Taken together, our findings reveal important biophysical properties of staphylococcal biofilms grown under fluid shear stress and suggest potential therapeutic approaches for the treatment of staphylococcal infections.

Web Page: www.umbc.edu/cbe/ross/

275d Quantitative Analysis of Fibroblast Focal Adhesions

Ehsan Jabbarzadeh and Anand R. Asthagiri. Chemical Engineering, California Institute of Technology, 1200 E. California Blvd., Pasadena, CA 91125

Focal adhesions provide a linkage between an anchorage dependant cell and extracellular matrix (ECM) and mediate cell functions (eg. migration). Our understanding of the focal adhesion regulation of cell behavior, however, has been limited due to the involvement of multiple proteins and their complex network of interactions. The present study is an effort to unfold this complexity by mapping the dynamic organization of focal adhesion proteins (eg. vinculin and paxillin) in fibroblasts using immunofluorescent imaging techniques. Specifically, we explored two main questions: 1) how do substrate properties (eg. ligand density) regulate the spatial and temporal dynamics of focal adhesions? and 2) what are the potential mechanisms of the focal adhesion regulation of cell spreading, shape and motility? To answer these questions, we developed metrics to quantify focal adhesions and cell spreading. We demonstrate that these metrics delineate the effects of cell-ECM interactions on focal adhesion formation.

275e Computational Investigation of How Integrin Clustering Affects Cell Adhesion and Migration

Erik S. Welf, Chemical Engineering, University of Delaware, 150 Academy Street, Colburn Lab, Newark, DE 19716, Ulhas P. Naik, Biological Science, University of Delaware, 150 Academy St., Newark, DE 19716, and Babatunde A. Ogunnaike, Department of Chemical Engineering, University of Delaware, Newark, DE 19716.

Integrins are transmembrane proteins that bind extra-cellular matrix (ECM) proteins to facilitate cell adhesion and migration, and a variety of pathologies have been linked to problems with integrin function [1, 2]. Because integrins function by clustering at points of contact between cells and the ECM, integrin clustering potentially influences many aspects of cell function [3]; however, we currently lack a framework for systematically characterizing the effects of integrin activation states and ECM properties on integrin cluster formation.

In this work, we describe a reaction-diffusion model for integrin clustering, and discuss how it may be used to explore ways in which integrins modulate integrin cluster properties such as cluster shape, size, and formation dynamics. Our model represents integrins by three model states: unbound, bound, and clustered, and their reactivity is described by dimensionless parameters representing integrin activity characteristics, such as affinity and avidity, allowing us to infer how integrin activity directly affects integrin clustering. Because the integrin cluster properties described by the model are directly related to adhesion strength [4] and cell migration speed, we can use the model to understand how different integrin properties affect cell behavior, and how we may manipulate integrin and ECM properties to achieve desired cell adhesion and migration responses. Our results indicate that different combinations of integrin properties and ECM properties lead to integrin clusters with varying capacity for force transduction, given the same extent of integrin binding. Model simulations indicate that patterned ECM surfaces and fast integrin clustering result in integrin clusters that may be capable of transferring adhesive forces more effectively than non-patterned surfaces or slower-clustering integrins. Our results also suggest that ECM patterns and integrin properties can affect cell migration by controlling integrin cluster turnover, which affects cell migration speed. This insight into how integrin and ECM properties control integrin clustering may be used to control cell adhesion and migration on biomaterials by creating material design criteria based on optimum conditions for different cell types as a function of their integrin properties.

[1] Wozniak MA et al. Focal adhesion regulation of cell behavior. (2004) Biochim Biophys Acta 1692, 103-119.

[2] Wehrle-Haller B and Imhof BA, Integrin-dependent pathologies. (2003) J Pathol 200, 481-487.

[3] Miyamoto S et al. Integrin function: molecular hierarchies of cytoskeletal and signaling molecules. (1995) J Cell Biol 131, 791-805.

[4] Ward, M.D., and D.A. Hammer (1993) A theoretical analysis for the effect of focal contact formation on cell-substrate attachment strength. Biophys J. 64, 936-959.

275f Dynamics of Contractile Stress Fiber Turnover Inside Living Cells

Richard B. Dickinson, Robert Russell, Kristen Lee, and Tanmay Lele. Chemical Engineering, University of Florida, PO Box 116005, 221 CHE Bldg., Gainesville, FL 32611-6005

Cells adhere to and spread on substrates by assembling adhesion complexes called focal adhesions. Focal adhesions are physically linked to tension-generating actin stress fibers inside the cell, allowing transfer of intracellular force to the substrate. How stress fibers are assembled by cells is not well-understood. We have developed and experimentally evaluated a mathematical model for the reaction/diffusion/convection processes involved in actin turnover in stress fibers. Model predictions are compared to measured F-actin turnover inside living cells using live cell imaging and FRAP. Our results suggest that new actin adds at sarcomeric dense bodies along the length of stress fibers in diffusion-rate limited manner, as well as at the stress fibers end located at focal adhesions. Preliminary experiments to uncover the relationship between these processes and tension generated by myosin will also be presented.

275g A Cell Migration Model Based on Cell Adhesion and Elasticity

Mahnaz Eskandari[1], Eric M. Brey[2], Megan E. Francis[2], Arsun Artel[3], and Ali Cinar[1]. (1) Chemical and Biological Engineering, Illinois Institute of Technology, Suite 214A, 10 W. 33rd street, Chicago, IL 60616, (2) Biomedical Engineering, Illinois Institute of Technology, Engineering 1 Building, Room 116, 10 W. 32nd Street, Chicago, IL 60616, (3) Department of Chemical and Biological Engineering, Illinois Institute of Technology, 10 W 33rd Street, Chicago, IL 60616

A cell migration model is developed based on the mechanical and physicochemical properties of the cell and its extracellular environment. The model includes four adjustable parameters; average number of cell receptors, average distance of receptors from the cell mass center, average linear elastic modulus of the cell, and average strength of receptor adhesion to substrate. It is assumed that available levels of nutrient(s), oxygen, and growth factor(s) are sufficient for cell migration but not for proliferation.

A cell's movement is broken into two steps, stretching and retraction. A cell stretches to create new bounded receptors (stored elastic energy). In the next step, the stored elastic energy is minimized by reducing the number of bounded receptors. Changes in the positions of bounded receptors are used to determine cell movement (migration). The speed of this movement, fluctuations in movement length, fluctuations in stored elastic energy, and total migration length are variables that are functions of input and adjustable parameters. These variables are studied for various elastic properties of a cell (constant linear elasticity, static non-linear elasticity, and dynamic non-linear elasticity) and physical and mechanical properties of the substrate (ideal cross-linked network, entangled cross-linked network, entangled cross-linked network with static dangling ends, and entangled cross-linked network with dynamic dangling ends).

Comparison of migration speed and total migration length for different cases and to experimental results indicates a strong correlation between the elastic properties of a cell and number of bounded receptors and therefore migration speed and length.

283a Surfactant Effects on the Size, Stability, and Flexibility of Lipid Vesicles

David S. Chen[1], Linh-Yen Do[1], Benjamin Tylka[1], **Kevin Quinn**[1], Nicole Chait[1], Oluwatosin Ogunsola[2], Srinivasa R. Raghavan[2], Keran Lu[2], and Nam Sun Wang[2]. (1) College of Chemical and Life Sciences, University of Maryland, College Park, MD 20740, (2) Department of Chemical & Biomolecular Engineering, University of Maryland, College Park, MD 20742

Flexible lipid vesicles have shown promise as a means of transdermal drug delivery, potentially complementing or even ultimately replacing current needle injection methods. Vesicles are made flexible by incorporation of a surfactant at moderate amounts into the formulation. While surfactant addition increases the apparent flexibility of vesicles, it may also alter vesicle stability. We are studying changes in the size and apparent flexibility of lipid vesicles with varying surfactant concentration over a six-week period. The effects of these changes on vesicle permeation of a skin-like barrier are also being tested. Flexible lipid vesicles are prepared using the lipid, egg phosphatidylcholine (egg-PC) and the nonionic detergent, Tween 80. An agarose gel with uniform pore size is used to simulate the physical barrier presented by the outer layer of skin. Apparent flexibility of the vesicles is measured indirectly via the permeation of the vesicles through a porous membrane. This talk will present results from our studies and will offer a mechanistic explanation for these results.

283b Effect of Particle Size and Shape In Prescribing the Efficacy of Vascular –Targeted Drug Delivery Vehicles

Omolola Eniola-Adefeso and Phapanin Charoenphol. Chemical Engineering, University of Michigan, 2300 Hayward Street, 3074 H. H. Dow, Ann Arbor, MI 48109

Overall, localized delivery of therapeutics offer the possibility of increased drug effectiveness while minimizing side effects often associated with systemic drug administration. Factors that affect the possibility of targeting therapeutics to reach the vascular wall include the ability to 1) identify disease-specific target epitope expressed by the vascular cells; 2) identify optimum drug carrier type, shape and size for efficient interaction with the vascular wall; and 3) identify a drug-carrier combination that allow for the effective release of therapeutics at the intended site. Existing literature have focused mainly on identifying target epitopes and the degradation/drug release characteristics of a wide range of drug-carrier formulations. Absent in the literature, are work focused on the potential roles of particle shape and size on the ability of vascular-targeted drug carriers to interact with the vessels – an important consideration that will control the effectiveness of drug targeting regardless of the targeted disease or delivered therapeutic. Submicron to nanometer-sized spherical particles are widely proposed for use in vascular drug delivery, yet very little evidence has been presented in the literature as to their optimal use for vascular-targeted drug delivery. This talk will discuss how particle size and shape along with blood flow dynamics, vessel size and hematocrit dictates the efficiency of drug carriers to interact with inflamed endothelium in vitro. The presented data will suggest the optimum drug carrier size(s) and/or shape for vascular-targeted drug delivery applications in chronic inflammation and cardiovascular disease.

283c Sterically Stabilized pH-Sensitive Liposomes Modified with a Fibronectin-Mimetic Peptide Targeted to Colon Cancer Cells: Controlled Release and Intracellular Delivery

Ashish Garg[1], Jeremy Lois[2], and Efrosini Kokkoli[1]. (1) Department of Chemical Engineering and Materials Science, University of Minnesota, 421 Washington Ave. SE, Minneapolis, MN 55455, (2) Department of Chemistry, Macalester College, 1600 Grand Avenue, St Paul, MN 55105

pH-sensitive liposomes undergo rapid destabilization under mild acidic conditions found in endocytotic vesicles. Their application is however limited due to their rapid clearance from circulation by the reticulo-endothelial system and researchers have therefore used pH sensitive liposomes sterically stabilized by polyethylene glycol (PEG) molecules on the surface of liposomes. The goal of this study is to bring bio-functionality to pH-sensitive stealth liposomes in order to facilitate their potential use as a targeted drug delivery agent. To improve the selectivity of these nano-particles we included a targeting moiety, PR_b which specifically recognizes and binds to integrin $\alpha_5\beta_1$ expressing cells. PR_b is a novel fibronectin mimetic peptide sequence, designed in our lab, that mimics the cell adhesion domain of fibronectin, and includes both RGDSP (the primary recognition site for $\alpha_5\beta_1$) and PHSRN (the synergy site for $\alpha_5\beta_1$) connected with a $(SG)_5$ linker. Integrin $\alpha_5\beta_1$ is expressed on several types of cancer cells, including colon cancer, and plays an important role in tumor growth and metastasis. Different liposomal formulation encapsulating calcein at self-quenching concentration have been evaluated. Release profiles will be discussed as a function of lipid composition, peptide concentration, pH, and time. Our data has been normalized to account for the changes in fluorescent properties of calcein due to pH and time. The pH-sensitive peptide-functionalized stealth liposomes have further been characterized in-vitro with CT26 colon cancer cell, and flow cytometry and confocal imaging data will be presented. Our studies demonstrate that peptide-functionalized pH-sensitive targeted delivery systems have the potential to deliver water soluble chemo-therapeutic agents directly to cancer cells, in an efficient and specific manner.

283d Towards the Rational Design of Controlled Release Therapeutics

Sam N. Rothstein, William Federspiel, and Steven Little. Department of Chemical Engineering, University of Pittsburgh, 1237 Benedum Hall, Pittsburgh, PA 15261

Controlled release technology presents solutions to the problems of patient compliance and sub-optimal dosing that often plague modern pharmaceuticals. This technology is commonly manifested as biodegradable matrices that gradually release drug as the structure degrades and dissolves, obviating the need for removal. As therapeutics, these matrices, which form the basis of a $110 billion dollar industry, have simplified dosing schedules from daily to monthly or even biannual treatments. However, the design of controlled release therapeutics remains largely heuristic despite nearly 30 years of research, resulting in substantial development costs in the form of exploratory, in vitro release studies. At present, thousands of medications that could offer improved performance as controlled release formulations await development.

We have recently reported the first, broadly applicable, predictive model for biodegradable matrices composed of the most widely used polymers (e.g. polyesters and polyanhydrides). Unlike prior attempts to predict release, the published model uses readily measured design parameters to calculate the in vitro release profiles for a wide variety of agents, including small molecules, peptides and proteins without regression. As all these parameters can be controlled or set with standard microparticle fabrication methods (single/double emulsion processing, melt casting, ect.), it should be possible to produce formulations to match particular release profiles.

As one example, the model suggests that it is possible to precisely control the magnitude and timing of subsequent doses in a burst-lag-burst release profile (e.g. the duration of the lag phase). Such a profile has significant potential to replicate proven vaccine dosing schedules without repeat injections. To date, formulations where the timing of controlled release bursts match the desired vaccine dosing schedule have proven elusive. Using the published model, design parameters were computed for microspheres based on a common murine vaccination schedule. When fabricated to these specifications, release assays showed that dosing from these microspheres did indeed match the desired schedule within the bounds of experimental error.

As another example, theoretical predictions from this model also suggest that a sustained, constant rate of drug delivery may be obtained from bulk eroding polyester microspheres. This release behavior was previously thought to only be obtainable using surface eroding systems. Efforts are currently underway to experimentally confirm such precise specifications for microparticle formulations.

In conclusion, initial experimental evidence shows that our newly published model can serve as a powerful tool for the inexpensive and rapid design of select therapeutics. Using this model to guide matrix fabrication, it may be possible to both improve upon current formulations and also enable the design of new formulations to address clinical needs in the area of controlled release.

283e pH-Sensitive Poly (N,N-Dimethylaminoethyl Methacrylate / 2-Hydroxyethyl Methacrylate Nanoparticles) as Efficient Drug Carriers

Jin-Oh You and Debra T. Auguste. School of Engineering and Applied Sciences, Harvard University, 40 Oxford St. ESL 224, Cambridge, MA 02138

For the triggered release of paclitaxel within a tomor microenvironment, pH-sensitive poly (N,N-dimethylaminoethyl methacrylate (DMAEMA) / 2-hydroxyethyl methacrylate (HEMA)) nanoparticles were prepared. Tumors exhibit a lower exracellular pH than normal tissues. We show that paclitaxel release from DMAEMA/HEMA particles can be actively triggered by small, physiological changes in pH (within 0.2-0.6 pH units). Monodispersed nanoparticles were synthesized by forming an O/W emulsion followed by photopolymerization. Particles were characterized by transmission electron microscopy, dynamic light scattering, electrophoresis, and cytotoxicity. High release rates and swelling ratios are achieved at low pH, low crosslinking density, and high content of DMAEMA. Paclitaxel release is limited to 9% of the payload at pH 7.4 after a 2 hour incubation at 37 ºC. After adjusting to pH 6.8, 25% of the payload is released within 2 hours. Cell viability studies indicate that pH-sensitive DMAEMA/HEMA nanoparticles are biocompatible and may be used as an efficient, feedback-regulated drug delivery carrier.

283f Nanocomposite Biodegradable Hydrogels: Preparation and the Analysis of Heating, Degradation, and Drug Release

Ashley M. Hawkins, Chemical and Materials Engineering, University of Kentucky, Lexington, KY 40506 and J. Zach Hilt, Chemical & Materials Engineering, University of Kentucky, 177 F. Paul Anderson Tower, Lexington, KY 40506.

Biodegradable polymers are extremely versatile materials that can be used in many applications, but current materials exhibit degradation profiles that are pre-programmed. Thus, there is limited or no ability to modulate the degradation once it is applied (e.g., implant, subcutaneous injection, etc.). This research involves the incorporation of nanoscale heat sources into a biodegradable polymer system. It is hypothesized that the heating from the particles can affect the degradation rate and therefore the drug release from the film. For this hypothesis to hold, several requirements must be met. First, the heating obtained from the nanoparticles must be significant. Secondly, the polymer system must have temperature dependent degradation. Finally, the drug must be released through a degradation process as opposed to dissolution.

Nanocomposite biodegradable polymers have been designed and developed that exhibit degradation profiles that can be remotely controlled by an alternating magnetic field. In particular, biodegradable hydrogels have been prepared from macromers synthesized through a condensation reaction between a diacrylate and amine. The incorporation of magnetic nanoparticles allows heating of the films, which can therefore affect many of the other properties of the polymer (e.g., degradation rate). Thus, the drug release can be controlled by altering the degradation rate through remote heating of the nanoparticles. Degradable hydrogels with and without iron oxide nanoparticles (and drugs) were successfully prepared. Characterization of the gels was done through viscometer readings, FTIR, and gel permeation chromatography methods. Heating studies were done using an infrared camera and induction heating instrument and the gels were shown to have temperature dependent degradation based on gravimetric analysis methods. Remote control drug release was also demonstrated by subjecting the drug-loaded gels to magnetic field exposure.

283g Biohybrid Nanoparticles and Hydrogel Platforms with on-Demand Release of Therapeutics by the Application of Nucleic Acid Triggers

Siddarth Venkatesh[1], Jacek Wower[2], and **Mark E. Byrne**[1]. (1) Biomimetic & Biohybrid Materials, Biomedical Devices, and Drug Delivery Laboratories, Department of Chemical Engineering, Auburn University, Auburn, AL 36849-5127, (2) Department of Animal Sciences, Auburn University, Auburn, AL 36849

The incorporation of nucleic acid constructs into hydrogels and nanoparticles can produce novel biomaterials with programmable on-demand switches or modulatory mechanisms, with unprecedented control and sensitivity. Here we demonstrate the controlled release of nucleic acid therapeutics loaded into hydrogels and nanoparticles via enzymatic and physical triggers. These novel biomaterials, designed using the principles of molecular biology, are expected to profoundly impact gene therapy regimes. Substrate non-specific (DNase I) enzymes were used to trigger the release of fluorescent neomycin immobilized onto gold nanoparticles. The neomycin molecules were bound by an unstructured thiolated oligonucleotide hybridized to the neomycin aptamer, which was isolated by affinity chromatography. Surface coverage densities of the DNA were quantified by using oligonucleotides with fluorescent dyes at the 3' end, and were optimized by varying buffer ionic strengths, reaction times and sonication. Release profiles of fluorescent neomycin were further tuned by creating an affinity spectrum of neomycin aptamers by isolating aptamers of varying binding affinities. Such novel schemes could be used for the intracellular delivery of multiple drugs from the same platform, as injectable systems targeted to specific cells using functionalized ligands such as folic acid.

Custom oligonucleotides, which were partially complementary to each other and had a programmed recognition site for the restriction endonuclease BamHI, were modified with a polymerizable acrylate functionality and 32P-labeled using T4 polynucleotide kinase. In-vitro hybridization and restriction enzyme digests were optimized in various concentrations of Tris buffer and water, and confirmed using polyacrylamide gel electrophoresis. Magnesium was determined to be critical for hybridization of the construct. About 10% of the complementary oligonucleotide self-annealed. The functional monomers used for hydrogel synthesis had no effect on hybridization efficiency. Digestion of the DNA helices was carried out by incubating the DNA helices with restriction enzyme BamHI and BamHI buffer. Incubation of the DNA double helix (without BamHI) on ice and at 37°C had no effect on the duplex. BamHI buffer did not disrupt the annealed duplex. Poly(acrylamide–co-N,N' methylene bisacrylamide-co-acrylated DNA) hydrogels were synthesized via redox polymerization and the unincorporated DNA was eluted by electrophoresis. The labeled complementary oligonucleotide, when polymerized into the hydrogel without the anchoring acrylated oligonucleotide, eluted readily from the hydrogel during electrophoresis whereas acrylated DNA duplex was readily incorporated into the network. Quantification of the polymerization process via electrophoresis and phosphoimaging indicated 70% capture of the acrylated DNA, 25 % unincorporated acrylated DNA, and 5% 32P-labeled oligonucleotide. A labeled oligonucleotide bearing a non-complementary sequence to the acrylated oligonucleotide did not get incorporated as annealing did not occur between the two strands. Release studies of 32P-labeled loaded DNA were conducted by incubating the DNA-loaded gels under physiological conditions in the presence of BamHI. The release of DNA due to the penetration of restriction enzyme was shown to be highly specific, with no release in the absence of BamHI, and the presence of another endonuclease, EcoRI. Temperature was used to release the 32P-labeled oligonucleotide as an alternative physical trigger. Temperature responsive release characteristics corresponded to the theoretical melting temperature of the helix (58°C). Poly(2-hydroxyethyl methacrylate-co-polyethylene glycol 600 dimethacrylate-co-acrylated DNA) hydrogels of varying crosslinking densities were triggered to release DNA by the non-specific endonuclease DNase I. Release rate of DNA from biocompatible hydrogels varied inversely with the crosslinking densities, and consequently mesh size.

The physiological significance of this platform was demonstrated by delivering a deoxyribozyme, which bore a catalytic 10-23 motif and was specific to a HIV Tat/Rev mRNA. The HIV-1 Tat/Rev RNA was synthesized by in vitro transcription and labeled using [5'-32P]pCp and T4 RNA ligase. Incubation of the deoxyribozyme-loaded hydrogel and the HIV-1 Tat/Rev RNA resulted in down regulation of the gene. BamHI action on a hydrogel lacking the deoxyribozyme construct had no effect on the HIV-1 Tat/Rev RNA.

300a A Novel Control-Theoretic Approach for Modeling Metabolic Networks and Inferring Pathway Regulation

Jamey D. Young[1], Kristene L. Henne[2], John A. Morgan[3], Allan E. Konopka[4], and Doraiswami Ramkrishna[3]. (1) Department of Chemical Engineering, Massachusetts Institute of Technology, 77 Massachusetts Ave., Building 56-439, Cambridge, MA 02139, (2) Department of Biological Sciences, Purdue University, 915 West State St., West Lafayette, IN 47907, (3) School of Chemical Engineering, Purdue University, 915 West State St., West Lafayette, IN 47907, (4) Biological Sciences Division, Pacific Northwest National Laboratory, P.O. Box 999, MSIN: P7-50, Richland, WA 99352

Cybernetic modeling involves the application of control-theoretic ideas to the modeling and analysis of biological systems. It relies upon optimal control heuristics that describe the system-wide aims and functions of metabolic regulation, and thus provides a teleonomic approach for dynamic modeling of metabolic pathways. Combining cybernetic control laws with concepts from metabolic pathway analysis has culminated in a comprehensive cybernetic modeling strategy, which has been previously lacking. The newly devised framework relies upon the simultaneous application of local controls that maximize the net flux through each elementary flux mode and global controls that modulate the activities of these modes to optimize the overall nutritional state of the cell. These modeling concepts are illustrated using a simple linear pathway and a larger network representing anaerobic *E. coli* central metabolism. The *E. coli* model successfully describes the metabolic shift that occurs upon deleting the *pta-ackA* operon that is responsible for fermentative acetate production. The model also furnishes predictions that are consistent with experimental results obtained from additional knockout strains as well as strains expressing heterologous genes. Because of the stabilizing influence of the included control variables, the resulting cybernetic models are more robust and accurate than their predecessors in simulating the network response to imposed genetic and environmental perturbations.

300b Modeling Glycolysis and Mitochondrial Metabolic Processes In Pancreatic β-Cells

Pradeep Nagaraja, Chemical and Biological Engineering, State University of New York at Buffalo, Buffalo, NY 14260 and Emmanuel (Manolis) S. Tzanakakis, Department of Chemical and Biological Engineering, State University of New York at Buffalo, 907 Furnas Hall, Buffalo, NY 14260.

Diabetes affects more than 150 million people worldwide and its incidence is increasing. Glucose-stimulated insulin secretion (GSIS) is an essential homeostatic process for the control of blood glucose levels and takes place in pancreatic β-cells through interplay between glucose metabolism and various signaling cascades leading to a regulated insulin production and secretion. Many details of this interplay are still elusive and pure experimental approaches aiming to unveil such details are inadequate given the complexity of the system. Mathematical modeling can facilitate our understanding of how GSIS works given that pancreatic β-cell dysfunction in one of the leading causes of diabetes. Although many computational dynamic models for GSIS have been constructed, a complete, detailed kinetic model with a large number of enzyme rate equations mimicking the dynamics from the plasma glucose input and leading to the generation of ATP production based on experimental β-cell data has not been developed so far.

In this study, we have constructed a systems level model consisting of the following detailed subsystems: glycolysis, TCA cycle, oxidative phosphorylation and mitochondrial shuttles. The model is constructed based on detailed kinetic and thermodynamic reaction schemes for the biochemical intermediates involved in the pathway. Parameterization and validation of the model involved utilizing a large amount of the available β-cell experimental data. The model provides a relation between blood glucose levels and the amount of ATP generated. Most of the intermediates are minimal at baseline (low) glucose conditions. However, we observe that increasing extracellular glucose concentration with time in a similar fashion to postprandial increase in blood glucose, causes the flux through the glycolytic pathway to increase resulting in matching the β-cell experimental data of the glycolytic intermediates. The model also shows expected behavior in its outputs, including the response of ATP production to starting glucose concentration and the induction of oscillations in the concentration of metabolites in the glycolytic pathway as well as in the concentrations of ATP and ADP.

We further examined the effect of β-cell acidification on insulin secretion as the role of intracellular pH in GSIS is not clearly understood. Also, we investigated in silico the 'pyruvate paradox' reported in β-cells. The relative effect of the TCA cycle and glycolysis on putative signals (e.g. ATP) was also explored. Results from these studies will be presented. Ongoing efforts focusing on the integration of calcium signaling and insulin exocytosis modules will be discussed. The outcome of this study will facilitate the discovery of novel targets for therapeutic intervention and the identification of new biomarkers of abnormal β-cell function for diagnostic purposes.

300c Unraveling the Role of Central Metabolism In Hepatic Lipoapoptosis Using ^{13}C Flux Analysis and Comprehensive Metabolic Profiling

Jamey D. Young, Yasushi Noguchi, Jose O. Aleman, Michael E. Hansen, Joanne K. Kelleher, and Gregory Stephanopoulos. Department of Chemical Engineering, Massachusetts Institute of Technology, 77 Massachusetts Ave., Building 56-439, Cambridge, MA 02139

Cell dysfunction and death induced by elevated serum free fatty acid (FFA) concentrations is a key contributor to the pathogenesis of obesity and type 2 diabetes. In particular, fat accumulation in the liver can lead to a condition known as non-alcoholic steatohepatitis (NASH), which involves tissue inflammation and damage that may culminate in cirrhosis. Hepatic apoptosis is a prominent feature of NASH and correlates with disease severity. Previous *in vitro* studies have demonstrated that saturated fatty acids (SFAs; e.g., palmitate) induce reactive oxygen species (ROS) generation and apoptosis, while mono-unsaturated fatty acids (e.g., oleate) lead to lipid accumulation but do not trigger apoptosis. Ceramide accumulation has been considered a primary factor responsible for these effects, because ceramide is synthesized *de novo* from palmitate and serine and also has been shown to activate apoptotic signaling. Recent studies, however, have reported that SFAs can induce apoptosis through ROS formation and endoplasmic reticulum stress without altering intracellular ceramide levels. Thus, a consensus mechanism linking SFA-induced metabolic alterations to apoptosis has been difficult to establish.

To identify metabolic pathways causing hepatic lipoapoptosis, we applied metabolic flux analysis (MFA) using [U-$^{13}C_5$]-glutamine as an isotopic tracer to quantify phenotypic changes in H4IIEC3 hepatoma cells treated with either palmitate alone (PA-cells) or both palmitate and oleate in combination (PA/OA-cells). Our results indicate that palmitate inhibited glycolysis and lactate dehydrogenase fluxes while activating TCA cycle flux and glutamine uptake. This decoupling of glycolysis and TCA cycle fluxes occured during the period following palmitate exposure but preceding the onset of apoptosis. Oleate co-treatment restored most fluxes to their control levels, resulting in increased lipid accumulation while preventing apoptosis. In addition, palmitate strongly increased the cytosolic NAD^+/NADH ratio, while oleate co-treatment had the opposite effect on cellular redox.

We next examined the influence of amino acids on these FFA-induced phenotypic changes. Increased medium amino acids enhanced ROS generation and apoptosis in PA-cells, but these effects were not observed in PA/OA-cells. Overloading the medium with non-essential amino acids induced apoptosis, but essential amino acid overloading partially ameliorated apoptosis. Glutamate was the most effective single amino acid in promoting ROS. Amino acid overloading also increased cellular palmitoyl-ceramide; however, ceramide synthesis inhibitors had no effect on measurable indicators of apoptosis. Our results therefore suggest that FFA-induced ROS generation and apoptosis are accompanied by the decoupling of glycolysis and TCA cycle fluxes leading to abnormal cytosolic redox states. Amino acids play a modulatory role in these processes, not through ceramide, but possibly through the control of TCA cycle flux.

300d Modulating the Lipoprotein-Dependent HCV Secretion

Yaakov Nahmias[1], Jonathan Goldwasser[2], Monica Casali, Raymond Chung[3], and Martin Yarmush[4]. (1) Center for Engineering in Medicine, Massachusetts General Hospital, Harvard Medical School, 51 Blossom Street, Boston, MA 02114, (2) Harvard-MIT Division of Health Sciences and Technology, Center for Engineering in Medicine, 51 Blossom Street, Boston, MA 02114, (3) Massachusetts General Hospital, 55 Fruit St., Boston, MA 02104, (4) Research, Center for Engineering in Medicine, 51 Blossom Street, Boston, MA 02114

The hepatitis C virus (HCV) infects over 3% of the world population and is the leading cause of chronic liver disease world-wide. HCV has long been known to associate with circulating lipoproteins and its interactions with the cholesterol and lipid pathways have been recently described. In this work we demonstrate that HCV is actively secreted by infected cells through a Golgi-dependent mechanism, while bound to very low density lipoprotein (vLDL). Silencing apolipoprotein B (ApoB) mRNA in infected cells causes a 70% reduction in the secretion of both ApoB-100 and HCV. More importantly, we demonstrate that the grapefruit flavonoid naringenin, previously shown to inhibit vLDL secretion both in vivo and in vitro, inhibits the microsomal triglyceride transfer protein (MTP) activity as well as the transcription of HMG-Co Reductase and ACAT2 in infected cells. Stimulation with naringenin reduces HCV secretion in infected cells by 80%. Moreover, we find that naringenin is effective at concentrations that are an order of magnitude below the toxic threshold in primary human hepatocytes and in vivo in mice. These results suggest a novel therapeutic approach for the treatment of HCV infection

Web Page: www.forbes.com/forbeslife/health/feeds/hscout/2008/05/06/hscout615158.html

300e Effects of Hormone and Amino Acid Supplementation In Optimal Liver Specific Functions of Hepatocytes

Hong Yang, Department of Chemical and Biochemical Engineering, Rutgers, the State University of New Jersey, 98 Brett Road, Piscataway, NJ 08854, Charles M. Roth, Rutgers University, 599 Taylor Road, Piscataway, NJ 08854, and M.G. Ierapetritou, Chemical & Biochemical Engineering Department, Rutgers University, the State University of New Jersey, 98 Brett Road, Piscataway, NJ 08854.

The metabolic capabilities and constraints of cultured hepatocytes are of great interest for the development of bioartificial liver (BAL) devices. Metabolic flux analysis (MFA) has been to provide a comprehensive view of intracellular metabolism in hepatocyte cultures under various conditions to quantify the intracellular fluxes and to investigate the effects of insulin and amino acids supplementation of primary rat hepatocytes in culture medium following plasma exposure [1]. Our work presents an optimal framework for optimal supplementation by a comprehensive analysis of culture hepatocytes' metabolism in order to achieve the best performance of liver-specific functions of hepatocytes during plasma exposure. And results calculated by mathematic models are further validated using experimental data.

First, a flux balance analysis (FBA) is developed to determine the optimal flux distribution of hepatic metabolism and to design an amino acid supplementation protocol given the flux bounds encompassing a range of physiological states [1]. The effects of several factors governing the specific liver functions, such as insulin level in the precondition of culture, amino acid supplementation and hormone level during plasma exposure, are then investigated for cryopreserved hepatocytes. Hepatocyte morphology, specific cell functions (urea and albumin production), glucose production/consumption and lipid accumulation are studied as markers to validate our hypothesis that liver specific functions are improved and hepatic pathway may change by modulation of hepatic network using amino acids supplementation. Based on the analysis of variance and a factorial design of experiments, amino acid supplementation and hormone level are the most significant factors under the experimental ranges examined during plasma exposure. Both the computationally predicted values and experimental measurements of optimal level of urea production were found to be increased by two fold compared to those that have been observed in the literature [1-3]. Metabolic Flux Analysis is used to calculate internal fluxes given the measured external fluxes in order to validate that higher liver-specific function is achieved simultaneously with an increase of fluxes of gluconeogenesis/glycolysis, TCA cycle, urea cycle fluxes, pentose phosphate pathway and uptake of some amino acids.

The obtained result helps our understanding of hepatocyte metabolism that can be used to develop liver models to test hepatotoxicity of various substances and to improve an extracorporeal bioartificial liver device during different patient treatments.

Reference:

[1] Chan, C., Hwang, D., Stephanopoulos, G. N., Yarmush, M. L. and Stephanopoulos, G., 2003c. Application of multivariate analysis to optimize function of cultured hepatocytes. Biotechnol Prog 19(2), 580-598.

[2] Chan, C., Berthiaume, F., Lee, K. and Yarmush, M. L., 2003. Metabolic flux analysis of hepatocyte function in hormone- and amino acid-supplemented plasma. Metab Eng 5(1), 1-15.

[3] Washizu, J., Berthiaume, F., Chan, C., Tompkins, R. G., Toner, M. and Yarmush, M. L., 2000. Optimization of rat hepatocyte culture in citrated human plasma. Journal of Surgical Research 93(2), 237-246.

300f In Vivo Metabolic Fluxes In Rat Livers: Effect of Burn Injury

Nripen S. Sharma[1], Maria-Louisa Izamis[2], Korkut Uygun[3], Francois Berthiaume[1], and Martin Yarmush[4]. (1) Center for Engineering in Medicine, Massachusetts General Hospital, Harvard Medical School, 51 Blossom Street, Boston, MA 02114, (2) Memp/hst, Harvard Med School/ MIT, 51 Blossom Street, Boston, MA 02114, (3) Center for Engineering in Medicine, Harvard Med School, 51 Blossom Street, Boston, MA 02114, (4) Research, Center for Engineering in Medicine, 51 Blossom Street, Boston, MA 02114

Available animal models provide a limited understanding of the complex, integrated effects of burn injury systemically and on the liver; in vivo studies commonly lack a thorough analysis of metabolic pathways, whereas ex vivo studies such as isolated perfusion carry an increased potential risk of experimental artifacts. The purpose of this study is to develop an in vivo animal model for the comprehensive investigation of the effects of burn injury. The data is presented both as concentrations of metabolites and their fluxes to enable the separation of systemic changes from changes brought about by the liver. A comprehensive representation of the metabolic pathways implicated in in vivo normal vs. burn liver can help identify differences in intracellular pathways for metabolic intervention using therapeutic targets to reduce complications associated with thermal burn injury.

Three groups of Sprague Dawley rats received a sham, 20% or 40% TBSA (Total Body Surface Area) burn and on the fourth day post-injury, following an 18-24 hour fast, flow rates of the portal vein, hepatic artery and suprahepatic vena cava were measured per rat followed by blood sampling of the suprahepatic vena cava, portal vein and hepatic artery. The liver was then excised and weighed for the purposes of normalizing all the data obtained. Blood samples were processed for blood gases and multiple metabolites as well as amino acids were measured with HPLC.

The results of this study verify the use of a 20 % and 40% TBSA burn in 6 week old Sprague Dawley rats as a clinically relevant in vivo animal model to study the effects of this kind of trauma. The data shows significant metabolic alterations as a consequence of burn injury and clearly demonstrates the role the liver plays in facilitating hypermetabolism. Metabolic Flux analysis (MFA) shows an increase in pyruvate synthesis, intracellular glutamate synthesis from histidine and increased tyrosine and phenylalanine uptake and metabolism for the 20 % burn injury conditions as compared to the sham control condition. For the 40 % burn condition, a significant increase in urea cycle, gluconeogenic, TCA cycle, acetyl-CoA synthesis, oxygen uptake and mitochondrial metabolism is observed as compared to the sham condition.

In addition to comparison of in vivo conditions, we performed a comparison of in vitro perfusion vs. in vivo conditions for the available data that revealed striking differences. In vitro perfusion data shows an increase in urea cycle fluxes, glutamine to glutamate conversion, asparagine to aspartate conversion, triacylglycerol conversion to beta-hydroxybutyrate and acetoacetate as compared to the in vivo perfusion data. For the 20 % burn condition, in vitro burn perfusion shows an increase in few hepatic fluxes viz. glycine, histidine, glutamine, arginine, threonine uptake with decrease in alanine, phenylalanine, asparagine and lactate uptake rate. Intracellularly, there is an increase in glutamine and histidine to glutamate metabolism and acetyl-CoA to â-hydroxybutyrate conversion. Overall, a similarity in PPP, TCA cycle and gluconeogenic fluxes between in vivo and in vitro burn injury models might imply application of appropriate metabolic intervention previously applied in the context of therapeutic efficacy of antioxidants in vitro to revert the hepatic hypermetabolic state post-burn injury in vivo.

300g Endothelial Cells Augment Hepatocytes Function In An Organotypical Model of Liver

Rohit Jindal[1], Yaakov Nahmias[1], Arno W. Tilles[1], Francois Berthiaume[1], and Martin Yarmush[2]. (1) Center for Engineering in Medicine, Massachusetts General Hospital, Harvard Medical School, 51 Blossom Street, Boston, MA 02114, (2) Research, Center for Engineering in Medicine, 51 Blossom Street, Boston, MA 02114

In vitro models of liver tissue that mimics the in vivo arrangement of different cell types offer the possibility of capturing complex signaling environment found in vivo. In the liver, hepatocytes and endothelial cells are closely associated in their native environment separated by the extracellular matrix (the space of Disse). This unique configuration was mimicked by embedding primary hepatocytes in collagen gel and overlaying endothelial cells on top of the gel. We found that albumin and fibrinogen secretion was doubled in the co-cultures compared to hepatocytes cultured alone during first week after seeding. However during second week, albumin and fibrinogen secretion profile became similar in co-culture and monoculture indicating the importance of endothelial cells derived factors towards early recovery of hepatocytes after isolation. Endothelial cell-conditioned medium reproduced the effect of endothelial cell co-culture, suggesting a major contribution of soluble factors secreted by endothelial cells. Conditioned medium also increased mRNA levels of various acute phase proteins such as albumin, fibrinogen, transferrin, and á-macroglobulin. The effect of endothelial cell-conditioned medium did not occur beyond 4 days of culture if hepatocytes were not embedded in collagen. This underscores the importance of synergism between the architecture of the culture and soluble factor secreted by endothelial cells. Currently work is in progress to identify the soluble factor responsible for enhanced function.

313a De Novo Protein Design of Complement 5a

Ho Ki Fung and Christodoulos A. Floudas. Department of Chemical Engineering, Princeton University, Engineering Quadrangle, Princeton, NJ 08544

Complement 5a is a 74-residue peptide cleaved from the N-terminal region of the α-chain of complement protein C5. Upon binding of C5a to the C5a receptor, a G protein-coupled receptor, most of the proinflammatory signaling is induced. The level of C5a is regulated by carboxypeptidase, which cleaves the Arg at position 74 and renders the peptide inactive. If the regulation is compromised, excessive prolonged production of C5a will occur which can result in a number of diseases including adult respiratory distress syndrome (ARDS), Alzheimer's disease, cystic fibrosis, rheumatoid arthritis, ischemic heart disease, etc. In view of the crucial role of C5a in regulating the complement cascade, efforts to treat inflammatory diseases have been focused on developing antagonists to the C5a receptor. They are reviewed in refs. [1] and [2] and are based on the C-terminal region of C5a as the design template for designing peptides with variable lengths. So far two potent C5a receptor antagonists have been reported: Me-FKPdChaWr by Merck and a hexapeptidic macrocycle Ac-F[OPdChaWR] by Finch et al.[3]. They have IC_{50}'s of 100 nM and 20 nM respectively against a maximum concentration of 100 nM of human polymorphonuclear leukocyte C5a receptors [3].

In this work, we use our de novo protein design framework to design antagonists to C5a receptor. Our framework consists of two stages: (i) in silico sequence selection, and (ii) fold specificity calculation. It has been successfully applied to the full-sequence design of human beta-defensin-2 [4] and the design of the oligomerization interface of G protein-coupled receptors [5]. We split our study of C5a into three different cases. The first case involves the redesign of the cyclic antagonist Ac-F[OPdChaWR] based on its ten NMR structures as the flexible design templates [6]. In the second case we design C-terminal hexapeptides and decapeptides based on the templates of the twenty NMR structures of C5a elucidated by Zhang et al. [7]. The third case deals with the redesign the three binding sites of C5a to C5a receptor proposed by Huber-Lang et al. [8]. Since all design backbone templates have multiple structures, we will employ both the weighed average and distance bin sequence selection models in our design [9], which are developed explicitly to handle backbone flexibility. As far as fold validation is concerned, we are applying a highly computationally efficient method, CYANA, on a full-atomistic scale driven by the AMBER forcefield, together with a local energy minimization package TINKER, for calculating fold specificities of the new sequences to the native fold.

[1] D. Morikis and J.D. Lambris. " Structural Aspects and Design of Low-Molecular-Mass Complement Inhibitors." *Biochem. Soc. Trans.* 30 (2002): 1026-1036.

[2] D. Ricklin and J.D. Lambris. "Complement-Targeted Therapeutics." *Nature Biotechnology* 25 (2007): 1265-1275.

[3] A.M. Finch and A.K. Wong and N.J. Paczkowski and S.K. Wadi and D.J. Craik and D.P. Fairlie and

S.M. Taylor. "Low-Molecular-Weight Peptidic and Cyclic Antagonists of the Receptor for the Complement Factor C5a." *J. Med. Chem.* 42 (1999): 1965-1974.

[4] H.K. Fung and C.A. Floudas and M.S. Taylor and L. Zhang and D. Morikis. "Toward Full-Sequence De Novo Protein Design with Flexible Templates for Human Beta-Defensin-2." *Biophys. J.* 94 (2008): 584-599.

[5] M.S. Taylor and H.K. Fung and R. Rajgaria and M. Filizola and H. Weinstein and C.A. Floudas. "Mutations Affecting the Oligomerization Interface of G-Protein-Coupled Receptors Revealed by a Novel De Novo Protein Design Framework." *Biophys. J.* 94 (2008): 2470-2481.

[6] L. Zhang and B. Mallik and D. Morikis. "Structural Study of Ac-Phe-[Orn-Pro-dCha-Trp-Arg], a Potent C5a Receptor Antagonist, by NMR." unpublished work (2007).

[7] X. Zhang and W. Boyar and M.J. Toth and L. Wennogle and N.C. Gonnella. "Structural Definition of the C5a C Terminus by Two-Dimensional Nuclear Magnetic Resonance Spectroscopy." *Proteins* 28 (1997): 261-267.

[8] M.S. Huber-Lang and J.V. Sarma and S.R. McGuire and K.T. Lu and V.A. Padgaonkar and E.M. Younkin and R.F. Guo and C.H. Weber and E.R. Zuiderweg and F.S. Zetoune and P.A. Ward. "Structure-Function Relationships of Human C5a and C5aR." *J. Immunol.* 170 (2003): 6115-6124.

[9] H.K. Fung and M.S. Taylor and C.A. Floudas. "Novel Formulations for the Sequence Selection Problem in De Novo Protein Design with Flexible Templates." *Optim. Methods & Software* 22 (2007): 51-71.

313b Design and Characterization of Peptides That Target Transmembrane Helices

Bryan Berger[1], Hubert Yin[1], Joanna Slusky[1], Greg Caputo[1], Rustem Litvinov[2], James Lear[1], Joel Bennett[3], and William DeGrado[1]. (1) Biochemistry and Biophysics, University of Pennsylvania School of Medicine, 1009 Stellar-Chance Laboratory, 36th & Hamilton Walk, Philadelphia, PA 19104, (2) Cell and Developmental Biology, University of Pennsylvania School of Medicine, BRB II/III, Philadelphia, PA 19104, (3) Hematology/Oncology Division, University of Pennsylvania School of Medicine, 924 BRB II/III, 421 Curie Drive, Philadelphia, PA 19104

Transmembrane (TM) helices play essential roles in biological processes such as signal transduction, solute transport and membrane protein folding and stability. In order to examine these processes in greater detail, we developed a computational method, CHAMP (computed anti-helical membrane protein), to design peptides which target the TM domains of membrane proteins in a sequence-specific manner. In particular, we designed peptides that target the TM domains of integrins α_{IIb} and α_v and demonstrate their specificity in bacterial membrane, platelets and micelles. Furthermore, using genetic selection methods, we are able to examine the specificity and affinity of interactions between the designed and target TM helices. Overall, our results indicate that TM peptides can be designed to selectively activate a given integrin TM domain as well as how a CHAMP methodology can be generalized to examine a wide range of TM-mediated processes.

313c High-Resolution Homology Modeling of Antibody Fv Regions and Application to Antibody-Antigen Docking

Arvind Sivasubramanian[1], **Aroop Sircar**[1], Sidhartha Chaudhury[2], and Jeffrey J. Gray[1]. (1) Chemical & Biomolecular Engineering, Johns Hopkins University, 3400 N Charles Street, 221 Maryland Hall, Baltimore, MD 21218, (2) Program in Molecular & Computational Biophysics, Johns Hopkins University, 3400 N Charles Street, 221 Maryland Hall, Baltimore, MD 21218

High-resolution homology models are useful in structure-based protein engineering applications when a crystallographic structure is unavailable. Therapeutic antibodies are powerful tools for the prevention and treatment of human and infectious disease because of their high affinity and specificity for target antigens. Furthermore, the existence of many known antibody structures provides a testing platform for high resolution homology modeling tools. Here, we report the development and implementation of RosettaAntibody, a protocol for homology modeling of antibody variable regions. The protocol combines comparative modeling of canonical complementarity determining region (CDR) loop conformations and de novo loop modeling of CDR H3 conformation with simultaneous optimization of VL-VH rigid-body orientation and CDR backbone and side-chain conformations. The protocol was tested on a benchmark of 34 antibody crystal structures. The median root-mean-square-deviation (rmsd) of the antigen binding pocket comprised of all the CDR residues was 1.3 Å with 80% of the targets having an rmsd lower than 2.0 Å. The median rmsd of the final CDR H3 loop prediction was 2.0 Å, and 70% of the loop predictions had an rmsd lower than 2.7 Å. When the set of ten top-scoring antibody homology models are used in local ensemble docking to antigen, a moderate to high accuracy docking prediction was achieved in eight of fifteen targets. This large-scale antibody-antigen docking study using homology models provides insight into the "functional accuracy" of these structural models towards protein engineering applications. The models promise to widen the applicable targets for practical structure-based antibody engineering tasks such as predicting structural changes due to antigen binding, rational design of CDR loops, predicting the structural consequences of antibody humanization including those related to the change in the VL-VH orientation, dissecting structural aspects relevant to antibody cross-reactivity and specificity, and improving antibody affinity. Figure: The antigen binding surface of the ten top-scoring homology models superposed on the antibody crystal structure for D3H44 Fab (1jpt). The native conformation is colored yellow-green and the low-rmsd model is colored salmon. Side chains are shown for the native and low-rmsd structures for CDR residues that contact the antigen in the co-crystal structure.

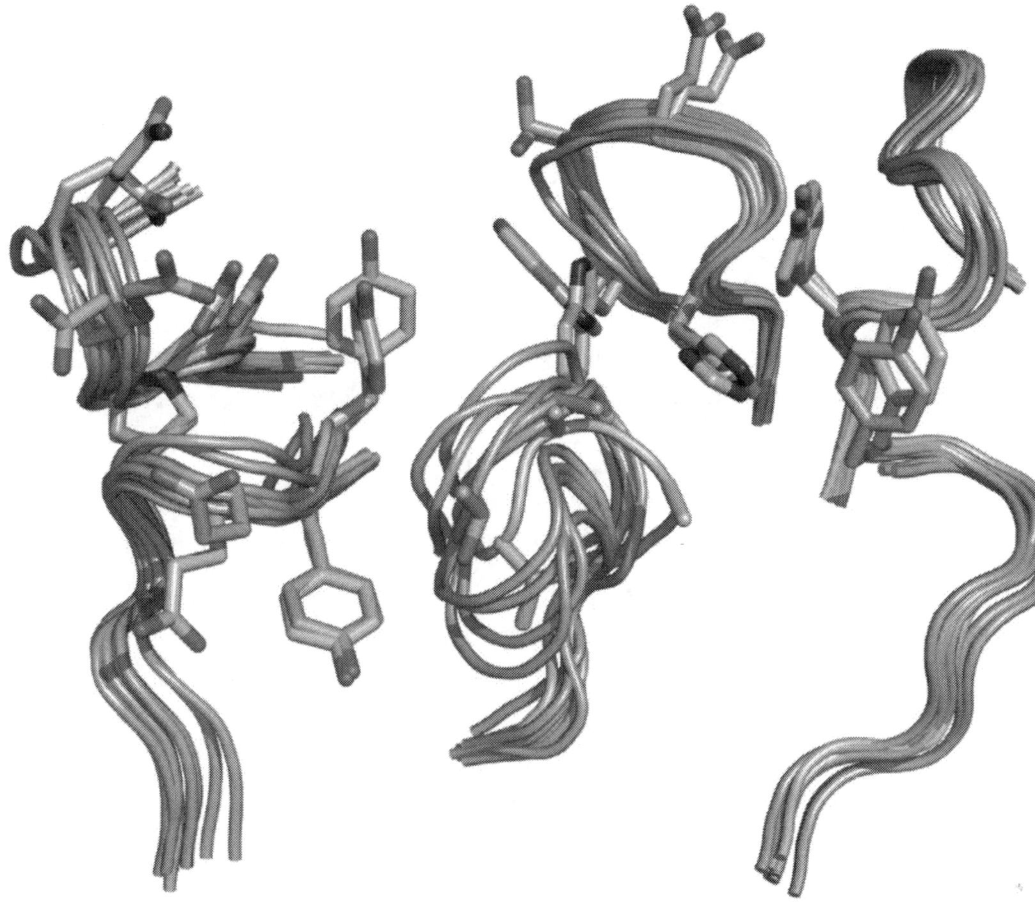

Web Page: antibody.graylab.jhu.edu/

313d Optimal Protein Library Design Using Recombination or Point Mutations Based on Sequence Based Scoring Functions

Robert J. Pantazes, Chemical Engineering, The Pennsylvania State University, 147A Fenske Lab, University Park, PA 16802 and Costas D. Maranas, Department of Chemical Engineering, The Pennsylvania State University, University Park, PA 16802.

In this talk, we introduce and test two new sequence based protein scoring systems (i.e., S1, S2) for assessing the likelihood that a given protein hybrid will be functional. By binning together amino acids with similar properties (i.e., volume, hydrophobicity and charge) the S1 and S2 scoring systems allow for the quantification of the severity of mismatched interactions in the hybrids. The S2 scoring system is found to be able to significantly functionally enrich a P450 library. Given this scoring base, we subsequently constructed two separate optimization formulations (i.e., OPTCOMB and OPTOLIGO) for optimally designing protein combinatorial libraries involving recombination or mutations respectively. Notably, two separate versions of OPTCOMB are generated (i.e., model M1, M2) with the latter allowing for position dependent parental fragment skipping. Computational benchmarking results demonstrate the efficacy of models OPTCOMB and OPTOLIGO to generate high scoring libraries of a pre-specified size. Specifically, we find that the optimal recombination or mutation patterns are complex and difficult to a priori identify.

313e Altering Effector Specificity in An Engineered Protein Switch by a Combination of Computational Design and Directed Evolution

Richard A. Heins[1], Jin Ryoun Kim[2], Loren L. Looger[3], Takayuki Soka[1], and Marc Ostermeier[1]. (1) Chemical and Biomolecular Engineering, Johns Hopkins University, 3400 North Charles Street, Baltimore, MD 21218, (2) Chemical and Biological Engineering, Polytechnic University, Brooklyn, NY 11201, (3) Janelia Farm Research Campus, Howard Hughes Medical Institute, 19700 Helix Drive, Ashburn, VA 20147

We have previously engineered a family of enzyme switches by the in-vitro recombination of two non-homologous genes. These switches were created using the genes encoding TEM1 beta-lactamase (BLA) and maltose binding protein (MBP) and exhibited maltose-dependent beta-lactamase activity. Another approach to creating protein switches is to modify existing switches so that they respond to new effectors. Increasing the affinity for the target ligand while simultaneously decreasing the affinity for the original ligand has proven difficult. We have attempted to convert our maltose-activated switch into one that is activated by sucrose. Our best sucrose-activated switch to date was derived from a cassette mutagenesis library in which five residues in the maltose-binding pocket were varied to all possible amino acids. This switch had a >15,000-fold increase in affinity for sucrose but only a 70-fold reduction in maltose affinity. We have created a second, computationally designed library in which 11 MBP residues are varied and subjected this library to a two-tiered genetic selection designed to identify switches that are specifically activated by sucrose. A comparison of the successes of both libraries will be presented.

313f Re-Wiring Protein Structures

Philippa J. Reeder, Chemical and Biological Engineering, Rensselaer Polytechnic Institute, 110 8th St, Troy, NY 12180, Jonathan Dordick, Chemical & Biological Engineering, Rensselaer Polytechnic Institute, Troy, NY 12180, and Christopher Bystroff, Biology, Rensselaer Polytechnic Institute, 110 8th St, Troy, NY 12180.

This work focuses on design and improvement of functional protein properties, primarily protein stability, by a novel method: protein structure re-wiring. A structure-based protein redesign method, it is focused on direct improvement of protein kinetic stability with minimal effect on protein activity. We hypothesize in this work that enzyme function is primarily conferred by intramolecular contacts between secondary structural units. By maintaining such interactions (and therefore the majority of residue-residue contacts) and intelligently changing the way in which those structural units are connected (i.e., re-wiring it), we demonstrate that it is possible to change the rate of unfolding of a model protein, GFP, without effecting its activity. This method may be universal and therefore applied to any protein with multiple structural units.

Web Page: www.rpi.edu/~reedep/work.html

313g A Rapid, General Platform to Identify Structure/function Relationships of a Viral Protein

James T. Koerber, Department of Chemical Engineering, University of California, Berkeley, 278 Stanley Hall, Berkeley, CA 94720 and David V. Schaffer, Chemical Engineering, University of California, Berkeley, 278 Stanley Hall, Berkeley, CA 94720.

Protein engineering often requires detailed mechanistic knowledge of the essential structure/function relationships for a given protein. Furthermore, obtaining detailed structure/function knowledge for proteins requires extensive structural analysis and biochemical characterization. Here, we have developed a high-throughput platform for the rapid identification of key functional regions of a viral protein shell or capsid. Adeno-associated viral (AAV) vectors have demonstrated considerable potential as gene delivery vectors to treat a broad range of diseases, including hemophilia and Parkinson's. However, both cellular (e.g. attachment and trafficking) and systemic (e.g. neutralizing antibody responses and tissue transport) mass transfer barriers limit efficient gene delivery of AAV vectors. A greater understanding of viral structure/function relationships would benefit both rational design and directed evolution approaches to engineer enhanced vectors. To date, more than 100 different natural variants of AAV have been isolated from both human and non-human tissues. The sequence variation in the viral protein capsid of these variants confers a broad range of gene delivery properties including binding to a variety of cell surface receptors such as heparan sulfate proteoglycan (HSPG), sialic acid, fibroblast growth factor receptor (FGFR), and platelet derived growth factor receptor (PDGFR). Furthermore, exhaustive rational mutagenesis techniques have successfully mapped regions of the AAV2 capsid which are involved in HSPG binding and putatively, FGFR binding. However, the lack of extensive structure/function knowledge of other AAV variants has hindered our ability to engineer more effective variants and understanding of the basic biology of AAV infection.

Accordingly, we have developed a novel, rapid, high-throughput platform to identify regions of the capsid of any AAV variant that are involved in a specific viral function. Specifically, we have generated highly diverse AAV libraries based on AAV5 and AAV6 through random mutagenesis of the cap gene and selected for mutants that are defective in specific steps of viral infection. For example, to map key residues involved in receptor binding, we selected the libraries for variants with decreased cell binding in iterative rounds of binding to CHO cells. Characterization of clones from the selected libraries yielded >15 novel variants of both AAV5 and AAV6 that exhibit decreased affinity for CHO cells. Interesting, a majority of these residues cluster to distinct regions of the capsid for AAV5 as compared to AAV6. Furthermore, several of these variants demonstrate altered gene delivery efficiencies to mutant CHO cell lines lacking sialic acid or HSPG, suggesting some of these residues are necessary for viral binding to sialic acid and other cellular receptors. This work demonstrates that our novel forward genetics platform is an efficient and effective approach to map functional regions of the AAV capsid and potentially other proteins, further our knowledge of basic AAV biology, and enhance efforts to engineer viruses and other proteins with customized properties.

342a Low Viscosity Highly Concentrated Suspensions of Protein Particles for Subcutaneous Delivery

Maria Andrea Miller[1], Joshua D. Engstrom[2], and Keith P. Johnston[1]. (1) Chemical Engineering, The University of Texas at Austin, 1 University Station C0400, Austin, TX 78712, (2) Engineering Technology, Bristol-Myers Squibb, 1 Squibb Drive, New Brunswick, NJ 08903

Delivery of protein therapeutics has been largely limited to large volume intravenous injections due to physical and chemical instabilities of proteins at high concentrations. In addition, monoclonal antibody delivery has been limited to low concentrations due to the dramatic increase in viscosity of a solution at a concentration greater than approximately 125 mg/mL. For subcutaneous delivery, the maximum volume is about 1.5 mL, corresponding to a dose of less than 200 mg for a concentration of 125 mg/mL. The main objective of this study was to formulate a protein suspension that is syringeable through a 27 gauge needle and at a concentration greater than 100 mg/mL in order to deliver a larger dose for a given injection volume.

As the volume fraction of particles increases above 100 mg/mL, the viscosity of the formulated suspension remains low enough to allow syringeability. Here, the low volume fraction approximation of Einstein is no longer accurate, and the viscosities were correlated with the Krieger-Dougherty Equation for concentrated suspensions. The experimental settling rates are correlated with a modified Stokes settling model to account for the high particle volume fraction. The protein particle size and shape are designed with milling and precipitation based processes. In addition, the protein particles must be designed to prevent particle and protein aggregation. Even with concentrations above 100 mg/mL, samples were shown to be syringeable through a 27 gauge needle. In addition, these samples were redispersible by manual shaking after exposure for one year and did not show evidence of particle growth or aggregation.

342b Understanding the Effects of Protein Conjugation with Poly(ethylene glycol) on Protein Adsorption for Therapeutic Protein Drug Delivery

Sheetal S. Pai[1], Todd M. Przybycien[2], and Robert D. Tilton[2]. (1) Chemical Engineering, Carnegie Mellon University, 700 Technology Drive, Room 4212, Pittsburgh, PA 15219, (2) Biomedical Engineering and Chemical Engineering, Carnegie Mellon University, 5000 Forbes Avenue, Pittsburgh, PA 15213

Over the last two decades, proteins have become an important class of drug molecules. Most of these protein drugs are used for life-threatening and seriously debilitating diseases such as diabetes, cancer, and hepatitis. The high activity and specificity of proteins compared with the more conventional, low molecular weight drugs often allows for better treatment of these diseases [1]. Sustained concentrations of the protein may be achieved by incorporating the drug into an implantable or injectable, biodegradable delivery system, such as FDA approved, biodegradable poly(lactide-co-glycolide) (PLG) microspheres. These polymers have an excellent record of biocompatibility and are known to degrade to toxicologically acceptable metabolites. Water soluble protein molecules have been encapsulated within PLG microspheres using a method based on a (water-in-oil)-in-water, (W/O)/W, multiple emulsion technique [2].

However, protein delivery from PLG microspheres has not been fully realized due to the difficulty in maintaining bioactivity when proteins are encapsulated. Protein conformation is very sensitive to local environment, particularly to the damaging effects of protein-surface interactions. Protein adsorption to interfaces limits the performance of sustained protein release from biodegradable depots [3]. During the encapsulation process, protein is exposed to an oil/water interface, where adsorption may cause denaturation or aggregation. During release, protein is exposed to a solid PLG/liquid interface as the microspheres undergo bulk erosion. Protein adsorption to this interface has been implicated as a critical factor leading to incomplete protein release and loss in bioactivity [2]. In order to achieve the full promise of sustained protein release from biodegradable microspheres, adsorption to these interfaces must be understood and minimized.

The covalent attachment of poly(ethylene glycol) (PEG) to proteins, PEGylation, has been shown to increase the in vivo half life of circulating proteins and decrease immune response [4,5]. Our research investigates whether protein PEGylation can also reduce the damaging effects of protein adsorption and thereby increase the extent of protein release from PLG microspheres in an active form. Such knowledge can aid in the design of therapeutic protein drug delivery devices. Furthermore, protein adsorption studies are necessary in order to establish a basis for the prediction, generalization, and control of protein behavior on surfaces.

We employed a total internal reflection fluorescence technique to explore the extent of lysozyme adsorption at the solid PLG/liquid interface and, in particular, to determine how covalent grafting of poly (ethylene glycol) (PEG) to lysozyme affects adsorption. Adsorption results showed that PEGylation significantly decreased the amount of lysozyme adsorption onto PLG surfaces. Additionally, the surface coverage values led to a possible explanation of the orientation of the PEGylated lysozyme versus unmodified lysozyme upon adsorption onto PLG.

References

[1] Van de Weert, M.; Jorgensen, L.; Moeller, E. H.; Frokjaer, S. Expert Opinion on Drug Delivery 2005, 2, 1029.

[2] Arai, T; Norde, W. Colloids and Surfaces 1990, 51, 1.

[3] Crotts, G.; Sah, H.; Park, T. G. Journal of Controlled Release 1997, 47, 101-111.

[4] Harris, J. M.; Martin, N. E.; Modi, M. Clinical Pharmacokinetics 2001, 40, 539.

[5] Michaelis, M.; Cinatl, J.; et al. Anti-Cancer Drugs 2002, 13, 149-154.

342c Stable Suspensions of Protein and Poorly Water Soluble Drug Particles for High Dosages with Pressurized Metered Dose Inhalers

Joshua D. Engstrom[1], Jasmine Tam[2], Maria Andrea Miller[2], Robert O. Williams III[3], and Keith P. Johnston[4]. (1) Engineering Technology, Bristol-Myers Squibb, 1 Squibb Drive, New Brunswick, NJ 08903, (2) Chemical Engineering, The University of Texas at Austin, 1 University Station C0400, Austin, TX 78712, (3) College of Pharmacy, The University of Texas at Austin, 1 University Station, Austin, TX 78712, (4) Department of Chemical Engineering, The University of Texas at Austin, 1 University Station C0400, Austin, TX 78712

A novel concept is presented for the formation of concentrated (up to 50 mg/mL), stable suspensions composed of low density flocs of anisotropic drug particles in hydrofluoroalkane (HFA) propellants without surfactants to achieve high fine particle fractions in pulmonary delivery with a pressurized meter dose inhaler (pMDI). Bovine serum albumin (BSA) and itraconazole (Itz) particles were prepared by thin film freezing (TFF). BSA (0.5% w/v) was dissolved in 10 mM potassium phosphate buffer (pH 7.4). The protein solution was passed though a stainless steel needle and dripped onto a rotating stainless steel drum, which was filled with dry ice to maintain a surface temperature of 223K. On impact, the droplets deformed into thin films and froze. The frozen films were collected and lyophilized to yield a dry powder. Itz particles were made using the same method, where the feed solution consisted of Itz (0.4% w/v) dissolved in 1,4 dioxane. The particles were analyzed with static and dynamic light scattering and scanning electron microscopy (SEM). Aerodynamic properties of the aerosolized particles were determined by an Andersen Cascade Impactor. Anisotropic BSA and Itz particles, in the shape of rods or thin plates, formed highly open, space-filling flocs in HFA, and were stable against settling for one year at drug concentrations up to 50 mg/mL. In contrast, spherical particles produced by milling formed dense aggregates that settled within 60 seconds in HFA. The atomized HFA droplets were shown to break apart and template the highly open flocs. Upon evaporation of HFA, capillary forces shrink the templated flocs to produce porous particles with optimal mass median aerodynamic diameters (MMAD) for deep lung delivery. The pMDI suspensions composed of anisotropic particles produced aerosol particles with MMADs between 3-4 ìm, high fine particle fractions (47-56%), and a delivered dose up to 3.7 mg/actuation. The concept of forming open flocs composed of anisotropic particles, that are stable against settling without surfactants, and templating the flocs during pMDI actuation to achieve optimal aerodynamic diameters and high fine particle fractions is of practical interest for wide classes of low and high molecular weight pharmaceuticals.

342d Synergistic Enhancement of Transdermal Delivery Using a Pore-Forming Peptide and Chemical Enhancer

Yeu-Chun Kim[1], Sameer Late[2], Ajay K. Banga[2], Peter J Ludovice[1], and Mark R. Prausnitz[1]. (1) School of Chemical and Biomolecular Engineering, Georgia Institute of Technology, 311 Ferst Drive, Atlanta, GA 30332-0100, (2) College of Pharmacy & Health Sciences, Mercer University, Atlanta, 30341

1. Introduction

The world-wide transdermal drug delivery market is quite large, but only a small number of agents have FDA approval. The primary reason for such limited development of this market is the difficulty in permeating the *stratum corneum* layer of human skin. Compressed lipid bilayers comprise the continuum portion of this layer, which is the primary barrier in skin. To overcome the barrier function of *stratum corneum,* various physical and chemical methods have been tested to increase the permeability of the *stratum corneum* to drugs. However, few have succeeded in delivering relevant agents at the appropriate flux levels without causing notable skin irritation or damage [1].

In our study, we developed a novel percutaneous delivery enhancing approach. Magainin peptide, our novel candidate of enhancer, is a 23-residue helical peptide that exhibits broad spectrum of antimicrobial activity properties. It has a net +4 charge and binds to negatively charged phospholipid membranes with the aid of electrostatic interactions, forming an amphiphilic helix and permeabilizing the bilayers [2]. Magainin peptide was previously shown to disrupt vesicles which are made from lipid bilayer components representative of those found in human *stratum corneum* and this ability of magainin allows us to propose that magainin can increase skin permeability [3]. Therefore, we tested the hypothesis that magainin, a peptide believed to form pores in bacterial cell membranes, can increase skin permeability by disrupting *stratum corneum* lipid structure. We further hypothesized that magainin's enhancement requires co-administration of a surfactant chemical enhancer to increase magainin penetration into the skin. In support of these hypotheses, synergistic enhancement of transdermal permeation can be observed with magainin peptide in combination of N-lauroyl sarcosine (NLS) in 50% ethanol-PBS solution.

2. Experimental methods

Intact epidermis was isolated from dermis using the heat separation method. A piece of excised skin was placed in a vertical Franz diffusion glass cell and the 2 wt% of NLS solution(N-Lauroyl Sarcosine in 50% ethanol solution) or NLS and 100 μ M magainin peptide mixture solution is placed in the donor chamber and the Franz cell is kept in the refrigerator at 4 °C for 12 h. A fluorescent compound (1 mM fluorescein) or real drug model (1mM Granisetron) in PBS solution (pH from 5 to 11) is applied to the *stratum corneum* side of the skin and the recovery of fluorescein or Granisetron every 1hr for 5hr in the receiver fluid on the epidermal side of the skin sample is then measured by spectrofluorimetry or HPLC.

3. Results

The exposure to a known surfactant chemical enhancer, NLS, in 50% ethanol solution increased *in vitro* skin permeability to fluorescein 15 fold and the combination of magainin and NLS-ethanol synergistically increased skin permeability 47 fold. In contrast, skin permeability was unaffected by exposure to magainin without co-enhancement by NLS-ethanol. (Figure 1.C)

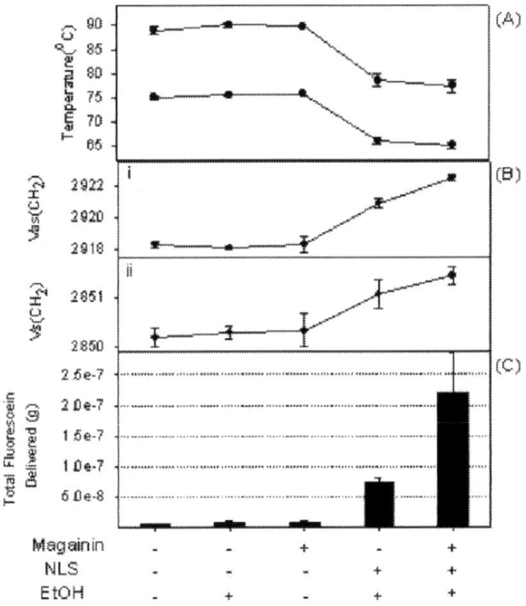

Figure 1. (A) The change of transition midpoint temperature(T_m) measured by DSC (B) CH_2 (i) asymmetric and (ii) symmetric stretching frequency measured by FT-IR (C) transdermal penetration amount of fluorescein with NLS and Magainin peptide in 50% ethanol solution

The exact mechanism of this synergistic effect has not been elucidated, so several characterization methods were applied to investigate the mechanism of synergistic enhancement. Two endothermic transition temperatures in DSC related to the lipid packing were decreased by about 10°C, which shows that NLS and magainin peptide disrupt and increase the fluidity of the *stratum corneum* lipid. (Fig.1a) The exposure of *stratum corneum* to NLS and magainin caused an increase in lipid related carbon-hydrogen stretching frequencies, which suggest an increase in the lipid chain disorder and fluidity. (Fig.1b) From wide angle XRD (Fig 2A), we observed that NLS and magainin peptide reduced the crystalline cholesterol peak and from small angle XRD (Fig 2B), ceramide peak (d=6.13nm) and cholesterol peak (d=3.38nm) is reduced by NLS and magainin peptide. These analyses showed that NLS-ethanol disrupted *stratum corneum* lipid structure and that the combination of magainin and NLS-ethanol disrupted *stratum corneum* lipids even further.

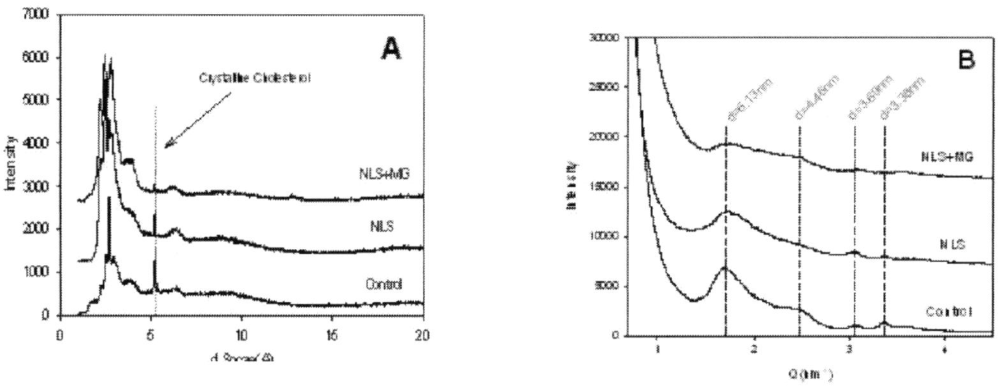

Figure 2. (a) Wide angle X-ray scattergram and (b) small angle X-ray scattergram of human *stratum corneum* treated with NLS or NLS/magainin peptide combination relative to the control

Finally, skin permeability was enhanced by changing the charge of magainin peptide via pH change. We modulated pH from 5 to 11 to change the magainin charge from positive to neutral, which decreased skin permeability to a negatively charged model drug (fluorescein) and increased skin permeability to a positively charged drug (granisetron). This suggests that an attractive interaction between the drug and magainin peptide improves transdermal flux.

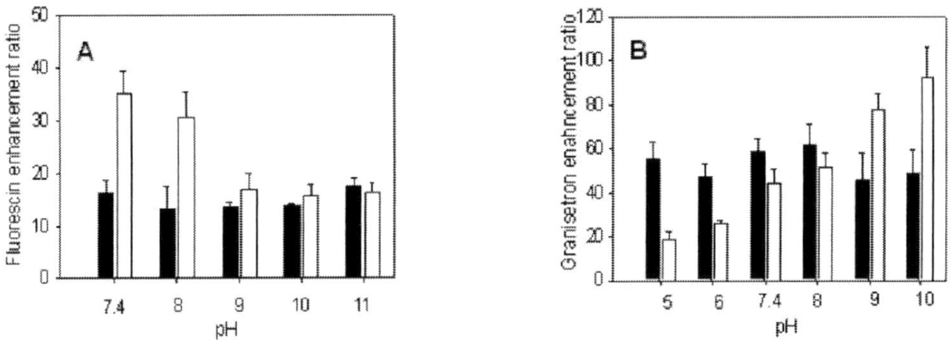

Figure 3. Enhancement of transdermal (A) fluorescein and (B) granisetron delivery as a function of pH with NLS (¢) or magainin + NLS (≤) in 50% ethanol.

4. Conclusion

This is the first study to use a pore-forming peptide as a skin penetration enhancer and the novel approach to use one percutaneous enhancer to increase the penetration of another percutaneous enhancer into the skin. It is suggested that antimicrobial peptides can be used as efficient skin permeability enhancers in conjunction with other enhancers. By changing the pH of the drug solution, we can also control the drug delivery amount.

References

1. M.R. Prausnitz, S. Mitragotri, R. Langer, *Current status and future potential of transdermal drug delivery*, Nature Reviews Drug Discovery (2004) 3, 115-124.

2. K. Matsuzaki, O. Murase, N. Fujii, K. Miyajima, *Translocation of a channel-forming antimicrobial peptide, magainin-2, across lipid bilayers by forming a pore*, Biochemistry (1995) 34, 6521-6526

3. S. Kaushik, A. Krishnan, M. Prausnitz, P. Ludovice, *Magainin-mediated disruption of stratum corneum lipid vesicles*, Pharmaceutical Research (2001)18, 894-896.

342e Hydrogel Filled Micro-Engineered Particles for Oral Delivery of Chemotherapeutic In Shear Conditions

Kristy M. Ainslie, Department of Bioengineering and Therapeutic Sciences, UCSF, 1700 4th Street, Byers Hall Room 203; Box 2520, San Francscio, CA 94158, Casey M. Kraning, Chemistry, Butler University, Indianapolis, IN 46208, and Tejal Desai, Department of Bioengineering and Therapeutic Sciences, University of California, San Francisco, 1700-4th Street, Box 2520, QB3 203, Mission Bay Campus, San Francisco, CA 94143-2520.

Hydrogel laden micro-engineered particles were used to create a novel oral drug delivery device to control the release of the chemotherapeutics. Delivery via these microdevices is advantageous because they deliver drug only in the direction of the cell layer, in contrast to spherical particles that deliver drug radially. The devices are fabricated through multi-layered photolithography. The attachment and release of these devices in vitro under physiologically relevant shear conditions was observed.

Cancer is the second leading cause of death in the United States. Chemotherapeutic delivery is challenged due to drug hydrophobicity, therapeutic degradation and/or lethal toxicity affecting healthy cells. A well engineered chemotherapeutic oral delivery device can help overcome these obstacles. Additionally, oral delivery is preferred over other invasive methods (e.g. intravenous) because it can be self-administered, leading to reduced delivery costs and increased patience compliance. Our laboratory has developed a micro-engineered oral drug delivery device (termed microdevice) that incorporates flat geometry allowing two-dimensional drug release, planar attachment to the intestinal wall and minimal resistance to mechanical forces generated by peristalsis in the intestine. The microdevice consists of a 150x150x8 micron-cubed flat polymer with a walled reservoir.

The microdevice is microfabricated with SU-8 through multi-layered photolithography steps. A poly (ethylene glycol) dimethacrylate (PEGDA) based hydrogel was added into the reservoir through additional photolithography. Acrylate functional groups have been shown to covalently bind to the free-radical groups present on SU-8 and we have shown long-scale binding of the hydrogels to the devices in physiological relevant conditions. Proteins and chemotherapeutics can easily be incorporated into the hydrogel, for controlled release of therapeutics from the device over the course of hours, as is relevant for intestinal delivery. Furthermore, layers with unique therapeutics incorporated in each can be fabricated into the device, allowing for incorporation of combinatorial therapy, such as applied in traditional cancer treatment. The bioactivity of chemotherapeutics released from the hydrogel was measured in vitro. Permeation of drug through intestinal epithelial cells was increased over 400% with the microdevices, compared to bare hydrogels. Chemotherapeutic loaded microdevices placed on a monolayer superior to a culture with colon cancer cells showed a decrease in cancer cell proliferation and viability while not adversely affecting the epithelial cell monolayer.

For specific attachment of the device to the intestinal epithelium, tomato lectin was covalently attached to the hydrogel layer. An acrylate- N-hydroxysuccinimide (Acr-NHS) cross-linker functionalized with avidin was introduced into the hydrogel. The addition of biotinalyted tomato lectin resulted in increased face-down attachment to an in vitro intestinal cell epithelial cell layer under physiologically relevant shear conditions. The percentage of adherent microdevices was shown to be dependent on the synergistic effect of tomato lectin and acrylate-avidin concentration. Drug release and diffusion through an intestinal monolayer was also measured in shear conditions. This work characterizes the further development of a novel micro-engineered oral drug delivery device.

342f Sequential Delivery of Dexamethasone and VEGF to Control Local Tissue Response for Carbon Nanotube Fluorescence Based Micro-Capillary Implantable Sensors

Jaeyun Sung, Chemical and Biomolecular Engineering, University of Illinois at Urbana-Champaign, 114 Roger Adams Laboratory, MC 712, 600 South Mathews Avenue, Urbana, IL 61801, Paul W. Barone, Chemical Engineereing, Massachusetts Institute of Technology, Building 66, Cambridge, MA 02139, Hyun Joon Kong, Department of Chemical and Biomolecular Engineering, University of Illinois Urbana-Champaign, 118 Roger Adams Laboratory, 600 South Mathews Avenue, Urbana, IL 61801, and Michael S. Strano, 66-566 Department of Chemical Engineering, Massachusetts Institute of Technology, 77 Massachusetts Avenue, Cambridge, MA 02139.

In this study, we examined the in vivo pharmacological effects of the sequential delivery of dexamethasone (DX) followed by vascular endothelial growth factor (VEGF) on the immune response and localized vascular network formation around a hydrogel-coated, micro-capillary implant for single walled carbon nanotube (SWNT) based fluorescence sensors. We demonstrate, for the first time, imaging of a SWNT near infrared fluorescent device implanted subcutaneously in a rat model. For tissue response studies, the chick embryo chorioallantoic membrane (CAM) was used as a tissue-model for an 8-day implantation period. The average vascular density of the tissue surrounding a micro-dialysis capillary sensor was evaluated for various types of surface coatings of the capillary designed to minimize protein adsorption and deliver therapeutics designed to reduce inflammation and promote vascularization. A PEG hydrogel loaded either with DX and VEGF for simultaneous delivery, resulted in an average of 1.24±0.35 x 10-3 vessels / μ m2 after the 8 day incubation. VEGF alone with a bolus addition of DX, for sequential delivery, resulted in a comparable 1.15±0.30 x 10-3 vessels / μ m2. The control where no therapeutics were utilized resulted in 0.71±0.20 x 10-3 vessels / μ m2. We introduce the concept of a therapeutic index for implantable sensors defined as the ratio of average vasculature density to average inflammatory cell density from histological analyses. This index reflects the promotion of angiogenesis versus the host immune response. The result was that the index for sequential DX/VEGF delivery was 60.3% and 139.3% higher than that of VEGF and DX release alone, respectively, and was also 32.1 % higher when compared to simultaneous administration, proving to be a more effective strategy in utilizing the pharmacological impact of DX and VEGF around the biosensor-model implant.

342g Metal Core-Shell Nanomagnets as a Carrier for Magnetically Tagged Drugs

Inge K. Herrmann, Robert N. Grass, and Wendelin J. Stark. Chemistry and Applied Biosciences, ETH Zurich, Wolfgang-Pauli-Str. 10, ETH Hönggerberg, Zurich, 8093, Switzerland

Currently, most drugs are applied in a non-targeted delivery mode. The human body is flushed with high doses of active substance in order to achieve the required tissue concentrations in the relevant organ. This application mode is likely to involve adverse effects caused by uncontrolled and irreversible invasion of the active drug in all tissues available [1]. Studies have shown that today's drug delivery systems are less than perfect [2].

Agile and controllable drug carriers are highly demanded to ensure quick action and avoid high concentrations in the body. According to recent findings, nanomaterials seem to meet exactly these requirements [3]. However, the toxic potential of nanomaterials is not yet fully investigated [4]. Therefore, the carrier particle should be either self-destroyed by dissolution or removed completely from the human body after application. Magnetic particles open an easy way to direct the drug to the site of action and to separate it from the body when it is not needed any longer [5]. However, most currently used oxide nanoparticles are only weakly magnetic.

Recently, covalent surface functionalization of carbon-coated magnetic metal nanoparticles has proven to be successful and provide a much stronger magnetic carrier [6]. This provides a powerful linkage between the multi-facetted properties of magnetic core-shell nanoparticles and organic chemistry. Linker molecules connect the magnetic particles with e.g. proteins, such as therapeutic antibodies (Figure 1). The magnetically tagged antibody is applied like in a conventional therapy, but magnetic fields can guide the active substance to the desired target and finally remove the antibody-antigen complex efficiently.

[1] V. P. Torchilin, Europ. J. Pharm. Sci., 2000, 11(2), 81–91.

[2] L. J. Samoy, P. J. Zed, K. Wilbur, et al., Pharmacotherapy, 2006, 26(11), 1578-1586.

[3] K. K. Jain, Med. Princ. Pract., 2008,17, 89-101.

[4] L. K. Limbach, P. Wick, P. Manser, R. N. Grass, A. Bruinink, W. J. Stark, Environ. Sci. Technol., 2007, 41, 4084-4089.

[5] A. Ito, M. Shinkai, H. Honda, T. Kobayashi, J. Biosci. Bioeng., 2005, 100 (1), 1-11.

[6] R. N. Grass, E. K. Athanassiou, W. J. Stark, Angew. Chem. Int. Ed., 2007, 46, 4909-4912.

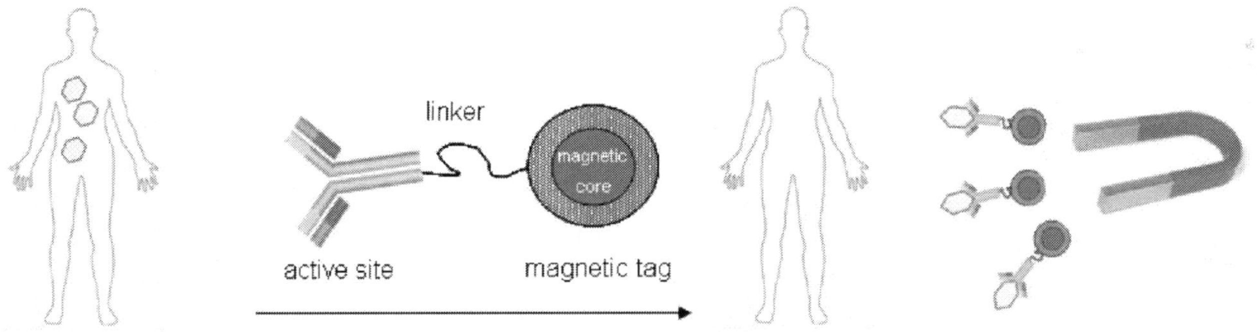

350a Proteomic Analysis of Cardiomyocytes Following Treatment with Soy-Derived Phytoestrogens

Zeyu Sun[1], Karyn L. Hamilton[2], and Kenneth F. Reardon[1]. (1) Department of Chemical and Biological Engineering, Colorado State University, 1370 Campus Delivery, Glover 100, Fort Collins, CO 80523, (2) Department of Health and Exercise Science, Colorado State University, 1370 Campus Delivery, 205A Moby, Fort Collins, CO 80523

Epidemiological observations comparing populations from different ethnic groups and countries suggest that higher consumption of soy-containing foods contributes to the lower incidence of cardiovascular disease in the Asian population than in the American population. Two major phytoestrogens, genistein and daidzein, which constitute the major part of the total isoflavones in soy protein respectively, are thought to have a potential role in improving risk factors for cardiovascular disease. Although phytoestrogens are generally accepted as beneficial for the cardiovascular system, scientific certainty about the relationship between soy and cardiovascular diseases is still lacking. Several studies have suggested that soy-derived phytoestrogens may potentially exert their effects in cardiomyocytes via many means including estrogen receptor-dependent and independent mechanisms. Phytoestrogens were also reported to have other cellular effects including inhibition of cell cycle progression, inhibition of angiogenesis, attenuation of oxidant damage, kinase modulation, and regulation of growth factors. Moreover, the biological activity of soy-derived phytoestrogens is likely to vary substantially in dose- and cell type-specific manners. Due to this complex nature of cellular responses to phytoestrogens, a more powerful approach to evaluate the mechanism systematically is needed.

The overall goal of this research is to investigate changes in the proteome of heart tissue following exposure with genistein and daidzein (10-100 μM). Our first study has focused on cultured rat cardiomyocytes as a model system. 2D electrophoresis has been employed to compare the protein profiles of treated and control samples, and differentially expressed proteins have been analyzed by mass spectrometry. More than fifty proteins were shown to be differentially expressed when cardiomyocytes were treated with 10μM or 50μM of genistein, including voltage-dependent anion-selective channel protein 2 and heat shock proteins. These and other results will be discussed, along with experiments in which HILIC-RP LC-MS based shotgun approach is used to target membrane-associated proteins.

350b Cys-Shotgun Labeling of the Proteome In Intact Cells to Probe Structural Changes

Christine I. Carag and Dennis E. Discher. Chemical and Biomolecular Engineering, University of Pennsylvania, Towne Bldg. Rm 311A, 220 S. 33rd St., Philadelphia, PA 19104

Text Not Available.

350c Glycosyltransferase Assays for Leukocytes, Cancer Cells and Stem Cells

Shilpa A. Patil[1], Dhananjay D. Marathe[1], E. V. Chandrasekaran[2], Khushi Matta[2], and Sriram Neelamegham[1]. (1) Chemical and Biological Engineering, State University of New York at Buffalo, 303 Furnas Hall, Buffalo, NY 14260, (2) Department of Cancer Biology, Roswell Park Cancer Institute, Elm and Carlton Streets, Buffalo, NY 14263

Glycomics is an emerging research area. It aims to integrate large scale experimental data sets and computational techniques to better understand the structure, chemistry and roles of carbohydrates in guiding glycan formation. This in turn regulates diverse processes including cell adhesion, bacterial/viral interaction with host cells and cancer metastasis. The glycan structures are formed by catalytic action of glycosyltransferases enzymes on protein, lipid and small molecule scaffolds. The levels of these enzymes in the Golgi are a key feature regulating glycan structures. Abnormal glycosyltransferase activity is also associated with cancer. In the current paper, we present the development of a methodology to quantify glycosyltransferase activity in nanoliter scale reaction volumes in microarray format. Particular emphasis is placed on studies of enzymes mediating the O-linked glycosylation pathway in human cells. The method involves performing chemical/enzymatic reactions in small volumes for extended durations of time (30m-4h) with an array of carbohydrate substrates, followed by the rapid, tandem separation of product from unreacted radioactive sugar-nucleotide. Quantitative phosphorimaging provides estimation of reaction rate constants. Data are presented with cloned enzymes and with extracts from various cell systems consisting of human leukocytes, metastatic cell lines and stem cells. The activities of 4 families of enzymes, sialyltransferase, fucosyltransferase, galactosyltransferase and N-acetylglucosaminyltransferase, are monitored in these assays. The development of this method, will overcome complex and time-consuming synthesis and separation steps that typically limit the number of runs performed in traditional glycosyltransferase assays. This then allows the comparison of different enzyme families within a single cell system under uniform conditions using a diverse panel of substrates. The use of our novel assay to determine rate limiting steps in glycosylation reaction networks, and insight related to cellular development process will be discussed.

350e Synergy In Cell Proliferation from Combinatorial Gene Expression

Clifford L. Wang, Chemical Engineering, Stanford University, 381 NS Mall, Stauffer 3, Rm 113, Stanford, CA 94305

A common experiment is to over-express or inactivate a gene, and then observe the resulting phenotype. For example, researchers often over-express or inactivate a single proto-oncogene or tumor suppressor gene, and then record the frequency of tumors and the time it took for those tumors to develop. Yet it is generally understood that multiple genes are needed to transform a normal cell into a tumor cell. Secondly, these genes often synergize or cooperate, i.e., the genes acting alone have minimal effect, but together they cause a phenotypic change greater than the sum of their contributions. Furthermore, there is strong evidence that there are maximal levels of gene expression that drive cell immortalization and transformation—the highest expression of genes does not necessarily lead to the greatest phenotypic change. We are developing an experimental platform to evaluate a range of gene expression from combinations of genes in mammalian cells. To achieve this, we have borrowed from the metabolic engineer's toolkit, employing mutagenesis and inducible gene expression methods. Here we report the use of our gene expression platform to quantify synergy between genes in cell proliferation.

350f Transcriptional Regulatory Network Dynamics In the Early Regenerative Response of Rat Liver

Rajanikanth Vadigepalli[1], Egle Juskeviciute[2], and Joannes Hoek[2]. (1) Daniel Baugh Institute for Functional Genomics and Computational Biology, Department of Pathology, Anatomy and Cell Biology, Thomas Jefferson University, 1020 Locust St Room 381, Philadelphia, PA 19107, (2) Department of Pathology, Anatomy and Cell Biology, Thomas Jefferson University, Philadelphia, PA 19107

Liver regeneration is a complex, yet remarkably coordinated, process characterized by system-wide gene expression changes underlying the response of the liver tissue to acute damage through mechanical, chemical or viral mechanisms (Taub, 2004 review). This response involves a variety of growth factors, cytokines, hormones, matrix components and other factors. Following the signals that mark the recognition of tissue damage after partial hepatectomy (PHx) and the onset of regeneration, hepatocytes are primed to enter the pre-replicative phase of the cell cycle (G1) and respond to the mitogenic effect of growth factors. The priming phase is controlled by several cytokines that activate a variety of normally latent transcription factors (TFs) important during the initial stages of liver regeneration before the onset of de novo protein synthesis and entry into the cell cycle. In recent years, the use of microarray technology has considerably expanded the list of identified changes in early gene expression and started bringing new insight into regulatory processes in the course of liver regeneration. A notable observation in these studies is the broad range of cellular processes that appears to be represented among these genes. However, our understanding of the temporal patterns of gene expression and of the upstream regulatory signals responsible for these patterns is still limited. Our goal is to obtain a detailed perspective on the time course of early responses and to gain insight into the network of intracellular signaling and transcriptional/translational events that drive the functional response of the liver tissue.

In this study we used cDNA microarrays to monitor changes in gene expression at 1, 2, 4, 6 h after PHx in remnant livers in the rat. We adopted a novel approach to analyze the microarray data that extends beyond the list of differentially expressed genes and focuses on the characterization of their transcriptional regulation. Candidate TFs responsible for differential expression profiles of the immediate early genes were characterized using the Promoter Analysis and Interaction Network Toolset (PAINT) software (Vadigepalli et al., 2003). Analysis of the gene expression time series data using ANOVA resulted in a total of 309 genes significantly up- or down-regulated at any of the four time points at ~20% false discovery rate (FDR) threshold. The 309 differentially regulated genes were clustered according their expression profiles. Six clusters provided maximum information on distinct temporal patterns and were well distinguishable from randomized data partitioning. We validated these results using qRT-PCR on a total of 17 genes that represent the response profiles in the different clusters.

PAINT analysis identified 22 TF binding sites enriched (FDR<30%) in individual clusters with distinct temporal gene expression patterns. Some of these TFs, e.g., NFκ B, HNF1, C/EBP, CREB, ATF,

GATA, are known to be involved in the early phase of liver regeneration from previous studies, whereas others (AP2α, LEF1, PAX6) are novel predictions. and are known to contribute to the regulation of cellular processes related to proliferation and differentiation in many tissues. In order to corroborate the PAINT analysis, we obtained time series data on the DNA binding activity detected in nuclear extracts from remnant livers for several of the transcription factors implicated by our PAINT analysis. The regulatory dynamics we observe is generally consistent with the differential gene expression pattern between 1h and 6h post PHx. In particular, the temporal patterns of NFκB and GATA-1 activity are consistent with the expression pattern of Cluster 4 in which the NFκB and GATA binding sites were found to be enriched in our computational analysis. The early changes in PAX-6 activity also confirmed the potential role of this TF in regulation of some of the immediate-early gene expression (Cluster 2 genes). It is notable that for many of the TFs analyzed that show an early transient increase, we observed a significant resurgence in the DNA-binding activity between 4h and 6h. The resurgence phase coincides with the gene expression profile that suggests a broad range of functional changes occurring between 4 and 6 hrs after PHx.

Our study highlights significant candidate mechanisms for transcriptional control of specific genes and gene clusters in the onset of liver regeneration. These findings also emphasize the fact that the study of any individual factor will not capture the systemic nature of the regulatory machinery that drives the regenerative response of the liver to PHx. We are actively pursuing chromatin immunoprecipitation and related methodologies to identify the functional role of each TF in individual gene responses. Our study points not only to the complexity of the transcriptional control of the early response to PHx, but also suggests that there is a clear underlying organization to the temporal response of genes in different functional categories that is driven by transcriptional regulation. The data reported here should provide a basis for a more detailed analysis of the role of each of these TFs to the regulation of individual genes and gene categories.

REFERENCES:

1. Taub R. (2004). Liver regeneration: from myth to mechanism. NatRev 5: 836–847.
2. Vadigepalli R, Chakravarthula P, Zak DE, Schwaber JS,Gonye GE (2003) PAINT: a promoter analysis and interaction network generation tool for gene regulatory network identification. Omics 7:235-252.

350g Effect of Frequency of Pulsatile Fluid Flow on Mechanotransductive Signaling of Osteoblasts

Lindsay A. Sharp[1], Yong W. Lee[2], and Aaron S. Goldstein[1]. (1) Chemical Engineering, Virginia Polytechnic Institute and State University, Blacksburg, VA 24061, (2) Biomedical Engineering and Sciences, Virginia Polytechnic Institute and State University, Blacksburg, VA 24061

A promising strategy for the development of tissue engineered bone involves culturing osteoprogenitor cells within porous biomaterial scaffolds under conditions that induce the deposition a bioactive extracellular matrix (ECM). One such condition is perfusion culture, which is known to stimulate the expression of numerous growth and differentiation factors, including bone morphogenetic protein (BMP)-2, vascular endothelial growth factor (VEGF)-A, and transforming growth factor (TGF)-β. Our working theory is that these proteins act back on osteoblastic cells to induce the expression of bone-specific ECM proteins, such as osteocalcin (OC) and bone sialoprotein (BSP). However, the molecular signaling mechanisms by which fluid flow induces the expression of these growth and differentiation factors are not understood.

The objective of this study is to probe the mechanism by which fluid flow stimulates osteoblastic differentiation through the induction of growth and differentiation factors. To accomplish this, MC3T3-E1 cells were seeded onto glass slides, loaded into parallel plate flow chambers (PPFCs), and exposed to various regimens of pulsatile flow (square wave form pattern) to modulate mechanotransductive signaling and gene expression. Experimental conditions included continuous flow, 0.074 Hz pulsatile flow, 0.044 Hz pulsatile flow, 0.015 Hz pulsatile flow, and no flow. The time-average shear stress for all flow conditions was 2.8 dyn/cm^2. To probe the signaling mechanisms we characterized the phosphorylation of extracellular signal-regulated kinase (ERK) by western blot, the activation of transcription factors activator protein (AP)-1 and Osterix by electrophoretic mobility shift assay, and the mRNA expression of BMP-2, BMP-4, BMP-7, and VEGF-A by quantitative real time PCR. Finally, we characterized the induction of OC and BSP mRNA and correlated this to the expression of growth and differentiation factors. Our preliminary data suggests that expression of ECM proteins OC and BSP are correlated to BMP-2 induction and enhanced by dynamic flow regimens.

360a Experimental and Computational Investigation of the Hydrodynamics of Partially Baffled and Unbaffled Stirred Tank Reactor Systems Equipped with a Retreat-Blade Impeller

Giuseppe Di Benedetto, Deepak R. Madhrani, and Piero M. Armenante. Otto H. York Department of Chemical Engineering, New Jersey Institute of Technology, 323 King Blvd., Newark, NJ 07102-1982

In this work, the velocity distribution inside the typical glass-lined vessel/impeller system was experimentally and computationally quantified using Laser Doppler Velocimetry (LDV) and Computational Fluid Dynamics (CFD). Two different reactor configurations were investigated, i.e., a flat-bottom tank and a hemispherical-bottom tank. In each case, two baffling configurations were studied, i.e., a partially baffled tank with a single beaver-tail baffle (the most common baffled configuration used in the pharmaceutical industry), and an unbaffled system. The three velocity components (tangential, axial, and radial) at 8 to 13 radial locations on 7 horizontal planes for the flat-bottom tank case and 5 horizontal planes for the hemispherical-bottom tank case were experimentally determined by LDV for both the partially baffled and unbaffled configuration. Numerical simulations of the velocity distribution inside each vessel were conducted using a commercial mesh generator (Gambit) coupled with a computational fluid dynamic (CFD) package (Fluent). The full 360°-tank geometry was incorporated in the simulations. Agreement between the experimental and simulation results was found to be favorable.

In the unbaffled flat-bottom reactor case, the tangential component of the velocity appears to dominate over the other velocity components at nearly every location, with tangential velocity typically on the order of 40% to 50% of the impeller tip speed. The radial and axial velocities, especially in the region just below the impeller, were found to be very small, with magnitudes typically smaller than 15% for the axial component and 5% to 10% for the radial component. In general, the presence of a hemispherical bottom did not alter significantly the magnitude of the velocity components except in the lower portion of the tank, where the hemispherical bottom generated a stronger axial and radial recirculation pattern. The velocity distribution in the single-baffle case was found to be only partially different from the unbaffled case, and primarily in the upper portion of the tank, where the baffle is. The velocity distribution in the lower portion of these vessels was not significantly affected by the presence of the baffle.

The dominance of the tangential velocity and the small value of the radial and especially axial velocity in all the system investigated here indicate a poor vertical recirculation of the fluid inside the tank and imply that these systems may have a limited mixing performance.

2008 AIChE Annual Meeting
Philadelphia, PA
November 16 – 21, 2008

Low Shear Mixing With a Novel Impeller Design

Chris Hibshman[1], Binxin Wu[1] and Wojtek Wyczalkowski[1]

[1]Philadelphia Mixing Solutions, Ltd., 1221 E. Main Street, Palmyra, PA 17078, USA

ABSTRACT

Hydrofoil impellers are considered to be low shear devices. Recent research indicates that hydrofoils may produce very high local energy dissipation that may lead to particle attrition, cell damage, or product degradation. This paper presents a novel approach to low shear mixing for shear sensitive mixing applications. Lab studies were performed on shear sensitive solids with the hydrofoil impeller and with the new low shear impeller at uniform solids suspension in both cases. The long term mixing indicated no particle attrition while using the Low Shear Impeller and significant damage to solids with the hydrofoil impeller.

INTRODUCTION

The scope of this work was to define an impeller design as a viable alternative to the low shear hydrofoil that provides desired results in applications requiring uniform suspension of shear sensitive solids. The experiment also provides scale up data for a pilot facility and full scale installation.

A unique blade configuration was utilized in the design of an impeller for mixing shear sensitive materials. The wide, large diameter blades promote high flow and uniform distribution of shear. The impeller provides mixed radial and axial flow action, creating greater efficiency than radial action alone. It has a high power number, thus low RPM. With smooth surface operation and easy clean-ability, it lends itself to clean-in-place (CIP) systems with no horizontal surfaces and broad, low-angle vertical projections. The design does have size limitations and for applications that require high polishing, it may have high cost. This design has been dubbed "Super Low Shear (SLS) Impeller".

Hoefken at al presented data on measurements of local energy dissipation for three types of impellers and compared the ratio of local to global dissipation. Hydrofoil impeller, due to high tip speed, has the highest ratio.

Hydrofoil Impeller = 104
Pitched Blade Impeller = 31
Rushton Turbine = 52

Additional calculations made for 19 m^3 vessel using Hoefken data show identical trend. For 60 s blend time the Hydrofoil impeller produce local dissipation of 3.4 W/kg while PBT will produce only 1.8 W/kg.

DESIGN of EXPERIMENT

Experimental and CFD models were setup to provide information on suspension uniformity and particle attrition. The purpose of the experiment was to test impeller performance and to provide scale-up data for full scale installation. Local Dissipation Rates were verified with PIV measurement.

Two different impeller designs were tested: a 3 bladed MHS impeller (wide blade design frequently used for solids suspension) and the Super Low Shear impeller.

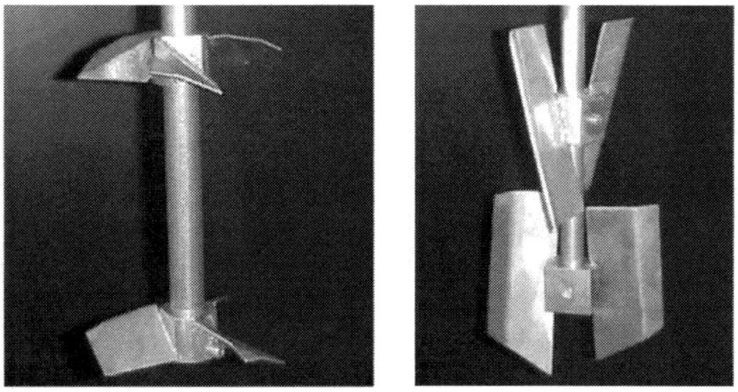

MHS Impeller

Fig 1. Impeller types used in testing

The test vessel configuration had a conical bottom and 5 sampling ports. Dimensions are shown in the diagram below. Vessel volume was 18 liters (4.77 gallons) and vessel diameter was 304mm (12").

Fig 2. Test Vessel

RESULTS

Uniform suspension was determined by measurements of the solids concentration at 5 different fixed levels of the vessel at four different RPM levels. A minimum RPM for uniform solids suspension was established from graphing RPM vs. concentration.
For MHS impeller, uniform suspension level was achieved at 432 RPM. For SLS impeller uniform suspension level was achieved at 132 RPM.

Uniform Suspension in 18 liter vessel	SLS	MHS
Impeller RPM	132	432
Mixer Power (Hp)	0.0098	0.0088
Mixer Power (W)	7.34	6.58
Global Dissipation (Hp/1000 gal)	3.61	2.85
Global Dissipation (W/kg)	0.41	0.37
Unit Torque (in-lb/gal)	1.00	0.27
Unit Torque (N-m/kg)	0.026	0.0078

Fig 3. Experimental data summary

Particle attrition is a function of dissipation level and duration (time) of mixing. It can be quantified by measuring shifts in particle size distribution. Attrition tests were run at three different RPM levels. Dissipation is responsible for particle damage. Measurements of particle size distribution was done at various time intervals during mixing

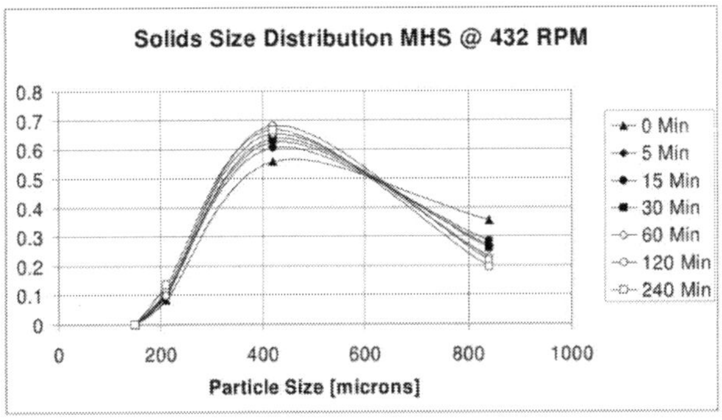

Fig 4. Particle Size Distribution for mixing with MHS impeller

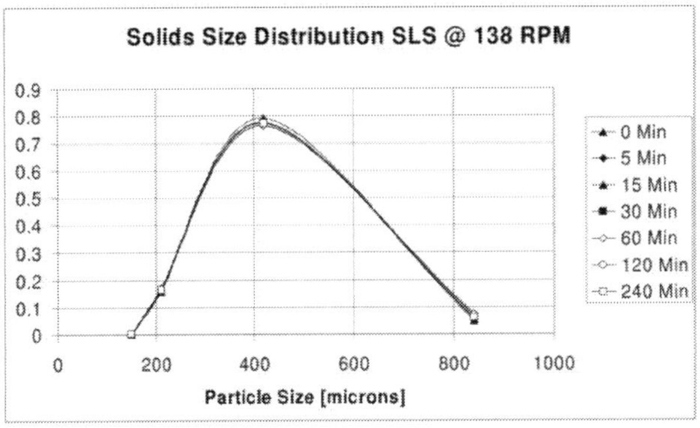

Fig 5. Particle Size Distribution for mixing with SLS impeller

Fig 6. Full scale impeller prototype 0.76 m diameter

CONCLUSION

Both experimental findings and CFD modeling demonstrate the difference in performance between a traditional hydrofoil and the Philadelphia Mixing Solutions low shear impeller.

Particle attrition for uniform suspension was proven to be much less with the SLS impeller than with the MSH hydrofoil impeller.

The experimental data validates the CFD model predicting lower strain rates for the SLS impeller than the MHS hydrofoil impeller.

The SLS impeller can be effective for application where low shear and high flow are required.

REFERENCES

Hoefken, M., Steidl W., Shaefer M., "Micromixing and Trailing Vortices" NAMF Conference – Mixing XVIII Pocono Manor, USA, June 24-28, 2001

Kresta S., Zhou G., "Impact of Tank Geometry on the Maximum Turbulence Energy Dissipation Rate for Impellers" AIChE Journal, September 1996 Vol.42 No 9

Baldi S., Yannieskis M., " On the Quantification of Energy Dissipation in the Impeller Stream of a Stirred Vessel from Fluctuating Velocity Gradient Measurements" Chem. Eng. Sc. 59 (2004) 2659-2671

360c Predictive Modeling for Mixing and Feeding Processes Using Kriging Apporach

Zhenya Jia[1], Eddie Davis[1], Fernando J. Muzzio[2], and M.G. Ierapetritou[3]. (1) Dept. of Chemical and Biochemical Engineering, Rutgers University, 98 Brett Rd., Piscataway, NJ 08854, (2) Chemical and Biochemical Engineering, Rutgers University, 98 Brett Road, Piscataway, NJ 08854, (3) Chemical & Biochemical Engineering Department, Rutgers University, the State University of New Jersey, 98 Brett Road, Piscataway, NJ 08854

Powder mixing is a crucial process in many industries, such as chemical manufacturing, foods, plastics and drugs. A slight change in the ingredient properties or the operating conditions during the manufacturing of a drug can have great impact on its specifications. The performance of a mixer is also dependent on accurate feeding system. Ingredient agglomeration and intermittent flow are known to depend on feeder design, size, and operating conditions as well as ingredient particle size and cohesion.

However, there is lack of a predictive understanding of the relationships between the operating and design parameters and the performance of the equipments, so that it can be used to find an optimal design. Various approaches have been developed to model powder flow and improve the characterization of powder processes, and can be mainly categorized into the following groups: Monte Carlo simulations, particle-dynamic simulations, heuristic models, and models based on kinetic theory. Discrete-element method (DEM) was initially developed by Cundall and Strack (1979). It simulates the interactions and movements of individual particles based on physical laws. However DEM is computationally very expensive since the Newton's equation of motion for each particle must be solved at each time step.

One of the widely used predictive data-driven models is response surface method (RSM), which was first introduced by Box and Wilson in 1951. It is an input-output mapping by fitting data using a set of basis functions. As the first step of RSM is specification of a sampling set within the local region, design of experiment (DOE) tools are usually required. Hence the prediction given by RSM relies greatly on the sampling points. A set of randomly selected sampling points very likely will lead to a poor predictive function. In this work, a predictive model is proposed that can efficiently characterize the performance of feeders and mixers with respect to operating parameters, and thus can be used to determine optimal process design variables and operating conditions. The Kriging method (Issaks and Srivistava, 1989; Gressie, 1993), which is an optimal prediction method designed for geophysical systems, has been utilized to predict the flow variability of feeders and regional concentration of mixers under different operating conditions. Without building an accurate mathematical model, Kriging method can provide a global predictive model starting with a small number of random sampling points and show where additional samplings should be done so as to iteratively improve the prediction. At each iteration, the first step is to determine semivariance of the each pair of the sampling points. After obtaining the set of semivariance scatterpoints, variogram model coefficients can be obtained using regression. Finally a set of equations are solved to get the kriging weights at each testing point, and the prediction value can be calculated as the weighted sum of the observed value of sampling points.

Two case studies of feeding and mixing processes are provided to illustrate the effectiveness of the proposed approach. The predicted flow variability is compared with the experimental data of loss-in-weight feeders, while for the mixing case the predicted concentrations describing mixture homogeneity are compared with the data obtained using a detailed DEM model.

Reference:

[1] Cundall P.A., Strack O.D.L. A discrete numerical model for granular assemblies. Geotechnique.1979, 29: 47–65.

[2] Box, G. and Wilson, K. On the experimental attainment of optimum conditions. J. Royal Stat. Soc. Series B. 1951, 13:1-45.

[3] Isaaks E. and Srivistava R. Applied geostatistics. New York, NY: Oxford Univ. Press, 1989

[4] Cressie N. Statistics for spatial data. New York, NY: Wiley, 1993.

360d Enhancement of Shelf-Life and Handling Properties of Drug Nanoparticles: Nanoscale Mixing of Itraconazole with Silica

Ganesh Sanganwar and Ram B. Gupta. Chemical Engineering, Auburn University, Auburn, AL 36849-5127

Nanoflakes of itraconazole, a poorly-water-soluble antifungal drug, are produced by SAS-EM supercritical antisolvent solvent with enhanced mass transfer (SAS-EM) method. The nanoflakes show poor flowability and shelf-life; which are improved by mixing with silica nanoparticles. To intimately mix at nanoscale, macro mixture of itraconazole nanoflakes and silica nanoparticles is first pressurized with supercritical carbon dioxide, and then rapidly depressurized through a nozzle. The rapid depressurization of supercritical suspension (RDSS) causes deagglomeration and mixing, due to a high CO2 expansion velocity. Upon mixing, due to the presence of silica nanoparticles between itraconazole nanoflakes, growth of the drug particles during storage is avoided or reduced, and particle flow properties are significantly improved. The handling properties are characterized using angle of repose, compressibility index, and Hausner ratio. Physical stability (shelf-life) of drug and drug/silica mixture is tested by storing the samples for 25 days at 90 °C. Agglomeration and growth of pure drug nanoflakes during storage leads to substantial decrease in the dissolution rate. Whereas when the drug nanoflakes are mixed with silica nanoparticles, the dissolution rate almost remains constant during storage. Thus, effective deagglomeration and nanoscale mixing lead to increase in the shelf-life.

Web Page: www.eng.auburn.edu/users/gupta/

360e A Hydrodynamic Method for the Production of Nanoparticles

Eric Jayjock and Fernando J. Muzzio. Chemical and Biochemical Engineering, Rutgers University, PIscataway, NJ 08817

Recent advances in the ability to structure material on the nano-meter size scale offer the pharmaceutical industry a new tool set: (i) high potency drugs can be made into small controlled release spheres which can survive in the body for a long period and slowly release there payload keeping a steady safe concentration in the body (ii) low solubility (lipophilic) compounds can be made into nano-sized spheres which will have a very high surface area to volume ratio for improved solubility (iii) nano-sized drug carriers can be surface modified with specific proteins to target specific areas of uptake and greatly reduce the occurrence of side effects. While creating these functionalities is already a difficult task at the laboratory scale, scaling-up these techniques to the manufacturing scale poses a whole new set of challenges.

In this talk we focus on a method for creating nano sized particles at a rate suitable for full-scale manufacturing. The basic mechanism of the Hydrodynamic Formation pathway is the creation of a nano-emulsion of which the droplets can be hardened to form nano-particles. The droplets of the emulsion are either the therapeutic agent themselves or the therapeutic agent and an encapsulation material. Initially the emulsion is rough and the droplets are on the micron size scale. The emulsion is passed through a static mixer at a high rate. The shear inside the mixer causes the micron-sized drops to be reduced down into the 50-1000 nano-meter range. After the size reduction the droplets are hardened by a phase shift brought about by a rapid change in temperature or solubility conditions, or by extracting a solvent.

This work focuses on identifying the critical parameters for control the width and size of the particles produced by the high intensity process.

360f Coupled Effects Between Process and Formulation Variables on Rheological Properties of Highly Concentrated Water-In-Oil Emulsions

Oscar A. Alvarez[1], Lionel Choplin[2], Veronique Sadtler[2], Marie José Stébé[3], and Philippe Marchal[2]. (1) Chemical Engineering Department, Universidad de Los Andes, Bogotá, Colombia, (2) Gemico, Institut National Polytechnique de Lorraine - Ecole Nationale Supérieure des Industries Chimiques, 1, Rue Grandville - BP 451, Nancy, France, (3) Equipe Physico-chimie des Colloïdes, Université Henry - Poincaré - Faculté des Sciences, BP 239, Vandoeuvre les Nancy cedex, France

Highly concentrated emulsions are mixtures of two immiscible liquids (water and oil generally) in which the volume fraction (Fv) of dispersed phase is higher than 0.90. These emulsions display a mechanical response that strongly depends on intensity of shear stress applied during process preparation. In this work, coupled effects between process and formulation variables on rheological characteristics of highly concentrated water-in-oil emulsions were determined.

Semi-batch process is mainly used for obtaining this kind of emulsions and it consists of two steps, the first one is aqueous phase incorporation into the surfactant-oil mixture phase under given mixing conditions, and the second one is the homogenization step in the same equipment (with possible modifications in the mixing conditions). The experimental set-up consists in a mixing vessel equipped with a 45° pitched blade turbine, whose diameter is 71% of the mixing vessel, and the turbine positioned at the oil phase free surface (before the aqueous phase incorporation). Agitation speed (N) was analyzed during the emulsification process.

The rheological behavior of highly concentrated water-in-oil emulsions was evaluated one hour after their preparation through mechanical spectrometry in the linear viscoelastic domain. The emulsions were prepared with different hydrocarbons oils as dispersed phase (each time only one type of oil was used), deionized water as aqueous phase and sorbitan monooleate (Span 80) as emulsifying agent. The studied formulation variables were volume fraction (Fv) of the dispersed phase, the surfactant/oil weight ratio, Hydrocarbon oil structure, and NaCl concentration in the dispersed phase.

We found a relationship between the rheological characteristics of highly concentrated water-in-oil emulsions (elastic modulus (G')) and energy consumption during emulsification process (E). This relationship can be described by G' is proportional to $E^{0.6}$ for all the formulations prepared with different hydrocarbons oils. Moreover, we established an influence of the hydrocarbon oil structure (carbons number of the chain) on the G' values. Finally we observed that the effect of NaCl on the G' value depends on process variables, specially the agitation speed (N). If N is higher, higher is the NaCl effect.

360g Study of Tablet Coating Thickness Uniformity Using Libs, Dem and Compartment Modeling

Atul Dubey, Patricia M. Portillo, and **Fernando J. Muzzio**. Chemical and Biochemical Engineering, Rutgers University, 98 Brett Road, Piscataway, NJ 08854

The coating uniformity of tablets is an extremely important feature in pharmaceutical manufacturing. Many coating devices have been used, studied and analyzed using experimental as well as computational methods such as Discrete Element Modeling (DEM). This study attempts to explore compartment modeling as a faster and accurate supplement to these methods. Compartment modeling has been utilized to study mixing processes in large scale aerated and stirred reactors, and more recently to study granular mixing in continuous blenders. Here, the approach is used to examine the coating distribution within each subsection. It involves spatially discretizing the system into a number of subsections. In this study we examine two subsections one compartment represents the coating area (spray zone) whereas the other compartment is not exposed to coating. Within each subsection a number of entities exist, the ratio of entities between each subsection has been studied. Another parameter studied is the rate entities are exchanged between subsections.

This study approaches the tablet coating mixing problem from three directions – a) experimental analysis using Laser Induced Breakdown Spectroscopy (LIBS) which shows the content and thickness uniformity of tablets, b) computational analysis using DEM modeling, and c) compartment modeling. The results show good agreement when comparing residence time in the spray zone which is an indicator of the coating thickness of the tablets.

375a A Novel Genetically Engineered Protein Module for a Qd-Based FRET Biosensor

Seung Hyun Kang[1], **Divya Sivaraman**[1], Ashok Mulchandani[2], and Wilfred Chen[1]. (1) Chemical and Environmental Engineering, University of California Riverside, Riverside, CA 92521, (2) Department of Chemical and Environmental Engineering, University of California, Riverside, CA 92521

Proteases are involved in many essential cellular processes such as blood coagulation, fibrinolysis, hormone maturation, and apoptosis. They are also used as the key virulence factors for pathogenic infection. These properties make proteases a prime target for detailed investigation to better understand the disease development process and to identify targets for drug treatment.

One of the most promising methods for probing protease activity is based on the principle called fluorescence resonance energy (FRET) containing two fluorophores located less than 100 • apart. In this study, we develop a genetically engineered protein module that is designed with 1) a quantum dot (QD)-binding moiety containing poly-histidine, 2) a specific site for fluorescent dye incorporation based on unnatural amino acid, 3) a protease cleavage site, 4) a flanking TAT peptide sequence for cell penetration and 5) an elastin-like protein (ELP) domain for thermal purification. We will investigate the use of this new genetically programmable protein module as QD-based FRET substrates for protease inhibitor discovery. The advantages of this approach include the significant reduction of cost on peptide synthesis, easy modulation of the placement of unnatural amino acid and intracellular monitoring of protease activity.

375b Hyperthermophilic Protein Scaffolds for Engineering Molecular Recognition

Balaji Rao[1], Nimish Gera[2], Mahmud Hussain[2], and Robert Clay Wright[2]. (1) Department of Chemical and Biomolecular Engineering, North Carolina State University, Box 7905, Engineering Bldg. I, 911 Partners Way, Raleigh, NC 27695, (2) Chemical and Biomolecular Engineering, North Carolina State University, Box 7905, Engineering Bldg. I, 911 Partners Way, Raleigh, NC 27695

Antibodies are the most commonly used binding molecules in biomolecular measurement. However, antibodies have several disadvantages such as low thermodynamic stability, a large multi-domain structure stabilized by disulfide bonds and high cost of production. As an alternative to antibodies, we are investigating the use of a set of small protein domains from extremophilic organisms as novel scaffolds for engineering molecular recognition, i.e. a "template" for generating protein variants that bind with high affinity and specificity to target species.

As part of this effort, we have identified and cloned multiple small protein domains from the hyperthermophilic organisms Thermotoga maritima, Sulfolobus solfataricus and Pyrococcus furiosus. The characteristic features of these domains are: (i) high thermal stability (ii) lack of cysteines; and (iii) low molecular weight (< ~10-12 kDa). Using a combination of yeast surface display and mRNA display, we are currently screening large libraries of these scaffold proteins to select variants that bind to model targets. Our goal is to create a versatile collection of validated hyperthermophilic protein scaffolds for molecular recognition in biomolecular measurement applications. Collectively, these protein scaffolds will provide a diverse set of binding interface topologies suited for various target species.

375c Enzymatic Hydrogels from Proteinaceous Bifunctional Building Blocks

Ian R. Wheeldon and Scott Banta. Chemical Engineering, Columbia University, 500 West 120th Street, New York, NY 10027

Hydrogels find use in a myriad of applications including biocatalysis, tissue engineering, and drug delivery. There are many examples of protein-engineering efforts that address some of the more technologically challenging aspects of these applications, including stimuli-responsiveness, bioactivity and catalytic activity. Here we present a general method of enzymatic hydrogel design through the development of bifunctional protein building blocks. We present three examples of chimeric proteins, differing in tertiary and quaternary structures, that self-assemble into enzymatic supramolecular hydrogels. Self-assembly functionality is achieved by appending previously designed α-helical leucine zipper domains to the termini of the desired enzyme. Compatible helices allow for the co-assembly of enzymatic and inert building blocks in mixed supramolecular hydrogels. The physical properties and enzymatic activity of the hydrogels are dependent on the identity, amount and ratio of each building block thus allowing for the independent tuning of these characteristics.

An oxidoreductase, SLAC, from *Streptomyces coelicolor* is the basis of a hydrogel that catalyzes the reduction of dioxygen to water; an alcohol dehydrogenase, AdhD, from *Pyrococcus furiosus* is the basis of a thermostable hydrogel that catalyzes the oxidation of secondary alcohols; a phosphotriesterase, OPH, from *Pseudomonas diminuta* is the basis of an hydrogel that catalyzes the hydrolysis of acetylcholinesterase inhibitors. The structure of the appended leucine zipper domains in all building blocks is confirmed by circular dichroism spectroscopy. Physical cross-linking functionality of the helices is confirmed by small amplitude oscillatory shear experiments. The impact of the N- and C-terminal modifications on enzymatic activity is determined by evaluating the kinetic parameters through dilute solution activity assays. We also present data on the rate of hydrogel erosion in open buffer solution and demonstrate the independent tuning of hydrogel strength (in term of the storage modulus) and catalytic activity.

Preliminary data is presented on the use of the enzymatic hydrogels as electrode surface modifications; modified SLAC as a cathode for a biofuel cell, AdhD for the anodic oxidation of alcohols and OPH as a biosensor for nerve agents. These examples, in conjunction with previously work on fluorescent protein building blocks, demonstrate the broad utility of our protein engineering approach to advanced hydrogel design. The bifunctional protein building blocks exhibit dual roles, hydrogel formation and enzymatic activity, a scheme that will prove useful in not only biocatalysis and biosensing but also tissue engineering and drug delivery applications.

375d Engineering Adhesive Ligands Based on Integrin I Domain

Lauren R. Pepper, Department of Bioengineering, University of Pennsylvania, 210 S. 33rd St, 240 Skirkanich Hall, Philadelphia, PA 19104, Daniel A. Hammer, Bioengineering, University of Pennsylvania, 240 Skirkanich Hall, 210 South 33rd Street, Philadelphia, PA 19104, and Eric T. Boder, Department of Chemical and Biomolecular Engineering, University of Tennessee, 1512 Middle Drive, 437 Dougherty Engineering Bldg., Knoxville, TN 37996-2200.

The inserted (I) domain of αLβ2 integrin (LFA-1) contains the entire binding site of the molecule. It mediates both rolling and firm adhesion of leukocytes at sites of inflammation depending on the activation state of the integrin molecule. The affinity change of the entire integrin can be mimicked by the I domain alone through mutations that affect the conformation of the molecule. High affinity mutants of the I domain have been discovered previously using both rational design and directed evolution.

We have developed a novel sorting method to separate firmly adherent I domain molecules out of a library of mutants under shear flow. Using conditions analogous to those experienced by cells along the endothelium, we have identified mutations that support firm adhesion not found using typical directed evolution techniques that select for binding to soluble ligands. Characterization of these mutants is underway to support a greater understanding of the structure-function relationship of the αL I domain.

We have also engineered the I domain as an allosteric switch regulated by ligand binding to secondary domains that flank the I domain. Yeast displaying this I domain fusion on the surface exhibit rolling adhesion along ICAM-1 surfaces under shear flow, and the rolling velocity of these yeast decreases upon addition of the soluble secondary ligand. Work is ongoing to refine this switchable adhesive interaction by directed evolution to support rolling adhesion in the off state, and firm adhesion in the on state.

375e Selecting for Folded Proteins and Their Interactions In the Bacterial Periplasm

Thomas J. Mansell[1], Stephen W. Linderman[2], Adam C. Fisher[3], and Matthew Delisa[1]. (1) School of Chemical and Biomolecular Engineering, Cornell University, Ithaca, NY 14853, (2) Department of Biological and Environmental Engineering, Cornell University, Ithaca, NY 14853, (3) Cornell University, 304 Olin Hall, Ithaca, NY 14853

Reporters of protein folding and solubility deliver valuable insight on the *in vivo* protein folding process and allow for the engineering of proteins with improved solubility. Currently, many folding reporters exist that report folding behavior in the *E. coli* cytoplasm, but to our knowledge, no such systems have been shown to report folding of proteins in the periplasm. The periplasm is an important compartment of gram-negative bacteria where several unique post-translational processes contribute to protein solubility including disulfide bond formation and, more recently, engineered protein glycosylation. Using the recently discovered SRP transport pathway and a C-terminal fusion to TEM-1 β-lactamase, we have developed a reliable reporter of (1) intrinsic protein folding properties and (2) the effects of chaperone activity, disulfide bond formation, and glycosylation on protein folding and solubility. In addition to reporting properties of proteins and their folding environment, we show that this technique can be used to engineer proteins with improved solubility via directed evolution. Furthermore, we have expanded this technique to develop an assay for protein-protein interactions in the periplasm. Using a split β-lactamase reporter system, we have successfully reported interactions of several known interacting domains. Currently, we are using this technique to select for novel interactions *in vivo* from naïve directed evolution libraries. The combination of these two powerful methodologies allows for a broad expansion of the protein engineer's toolbox for periplasmic expression and ultimately works towards a reliable reporter of protein glycosylation.

375f Directed Evolution of Rhli for Introducing a Novel Activity

Pavan Kumar Reddy Kambam[1], Dawn Eriksen[1], Jason Lajoie[1], and **Lianhong Sun**[2]. (1) Department of Chemical Engineering, University of Massachusetts Amherst, 159 Goessmann Lab, 686 North Pleasant St, Amherst, MA 01003, (2) Chemical Engineering, University of Massachusetts Amherst, 159 Goessmann Lab, 686 North Pleasant St, Amherst, MA 01003

LuxI and its homologs catalyze the synthesis of a variety of signaling molecules essential for a cell-cell communication mechanism, quorum sensing. These signaling molecules vary in the length of the side chain and the oxidation states, and these differences can be distinguished by corresponding cognate transcription factors. Such specificity allows the presence of multiple quorum sensing systems simultaneously without interfering with each other's functions. LuxI homologs exhibit high sequence similarity, and highly conserved amino acid residues across the family have been identified. In spite of the intensive biochemical characterizations of the LuxI enzymes, the molecular mechanisms of the LuxI enzyme substrate specificity, amino acid residues important for the substrate specificity in particular, remain unknown. Because quorum sensing is involved in a variety of infectious diseases, illuminating the mechanisms will facilitate the development of novel antibiotics specifically targeting quorum sensing to eliminate bacterial infections, a method regarded as a superior strategy to common antibiotics with the potential of not introducing bacterial drug resistance.

We believe engineering the substrate specificity of LuxI homologs will identify amino acids important for the substrate specificity. As a homolog of LuxI, RhlI catalyzes the synthesis of C4HSL (N-butanoyl homoserine lactone) and C6HSL (N-hexanoyl homoserine lactone) but not OHHL (3-oxo-hexanoyl homoserine lactone), which is synthesized by LuxI. Using a genetic selection we established for the engineering of LuxI, we performed directed evolution on RhlI to introduce a novel OHHL synthesis activity. After three rounds of directed evolution experiments, a mutant with a significant OHHL activity was identified, and the OHHL synthesized by the mutant was detected from the cell culture. In addition to the novel OHHL synthesis activity, this mutant shows a more than 10-fold increase in the C6HSL synthesis activity, an activity which is comparable to the C4HSL synthesis activity. To further increase OHHL synthesis activity, we have developed a strategy allowing for selecting such an activity. In this presentation, we will discuss our efforts in using both directed evolution and rational design in introducing and improving the OHHL synthesis activity.

375g Insights for the Engineering of More Efficient Inteins

Brian Pereira[1], Philip Shemella[2], Shekhar Garde[3], Georges Belfort[2], Saroj Nayak[2], and Marlene Belfort[4]. (1) Chemical and Biological Engineering, Rensselaer Polytechnic Institute, 110 8th ST, Troy, NY 12180, (2) Rensselaer Polytechnic Institute, 110 8th ST, Troy, NY 12180, (3) Department of Chemical and Biological Engineering, Rensselaer Polytechnic Institute, 110 8th Street, Troy, NY 12180, (4) Wadsworth Center, NYS Department of Health, 150 New Scotland Avenue, Albany, NY 12208

Inteins are intervening protein segments that mediate the splicing of flanking protein sequences, termed exteins. This protein splicing mechanism consists of four reactions which together results in the cleavage of two peptide bonds at the intein's N and C termini and the formation of a peptide bond between the exteins. However, cleavage at only one terminus may occur, yielding side products and inefficient splicing. In some applications, the N- or C-terminal cleavage is desired and achieved by altering the splicing reaction mechanism through mutagenesis. Through experimental methods such as site-directed mutagenesis of conserved residues and subsequent assays for intein activity, as well as computational methods including first principles density functional theory and combined quantum mechanics and classical molecular mechanics (QM/MM), we have investigated the reactions of protein splicing. These results indicate that certain mutations either inhibit or enhance specific reaction steps of the overall splicing mechanism, providing insight into how inteins can be engineered for more efficient activity.

376a Integration of PAT for the Development and Scale-up of a Hydrogenolysis Reaction

Jale Müslehiddinoglu[1], Paul Lobben[1], Thomas Laporte[1], Lori Spangler[1], Daniel Fichana[2], Shih-Ying Chang[1], Donald Kientzler[1], Sushil Srivastava[1], John Venit[1], Robert Wethman[3], and San Kiang[1]. (1) Process R&D, Bristol-Myers Squibb Company, One Squibb Drive, New Brunswick, NJ 08903, (2) Csm, Bristol-Myers Squibb Company, One Squibb Drive, New Brunswick, NJ 08903, (3) Analytical R&D, Bristol-Myers Squibb Company, One Squibb Drive, New Brunswick, NJ 08903

Although the Cbz deprotection reaction is a common organic reaction in the pharmaceutical industry, challenges were faced during the development and scale-up of this reaction at the API step due to the instability of the product in the reaction mixture. This led us to develop a control strategy as a part of quality by design approach. Consequently, a PAT (Process Analytical Technology) was used for continuous monitoring and control in order to ensure the quality of the product. This talk will describe the development of three in-process controls, using kinetics information from FTIR monitoring.

376b A Rational Approach to Solvent Choice in the Development of Homogeneous Catalytic Reaction Systems

Claire S. Adjiman[1], Adam J. Clarke[2], Gregory Cooper[2], and **Paul C. Taylor**[2]. (1) Department of Chemical Engineering, Imperial College London, Centre for Process Systems Engineering, South Kensington Campus, London, SW7 2AZ, United Kingdom, (2) Department of Chemistry, University of Warwick, Gibbet Hill Road, Coventry, CV4 7AL, United Kingdom

The choice of solvent or solvent mixture in which to carry out a reaction is usually made in the early stages of process development. This decision can have a huge effect on process performance, by affecting reaction rates and selectivity, by placing demands on downstream purification units and by contributing to the environmental impact of the process. In this work, we present a systematic approach to solvent selection, which greatly widens the options considered during process design, and can lead to significant improvements in reaction performance. The approach was first developed for model single-step reaction [1] and is extended here to more complex reaction schemes. It involves (i) a targeted set of experiments to quantify reaction kinetics and solvent influences on rate (ii) a computer-aided solvent design step in which potentially better solvents are identified, and (iii) synthetic verification of the results.

The approach is presented through application to the ring-closing metathesis (RCM) of a diene catalyzed by the 1,3-dimesityl-4,5-dihydroimidazol-2-ylidene ruthenium complex (Grubbs II catalyst, [2]). The Grubbs II catalyst is in many ways promising for RCM in an industrial context, but the amount of catalyst that is typically used to achieve good productivity makes its adoption expensive. It is difficult to postulate what solvent properties best promote the reaction, as the conversion is the result of complex balance between the rates of catalyst activation, the metathesis reaction itself, and catalyst deactivation. We demonstrate how the proposed methodology has led to the discovery of better solvents for this reaction, in which complete conversion is achieved quickly with 10 times less catalyst than is required for the usual solvent, dichloromethane [3]. Through this approach, we have also been able to gain a better understanding of the kinetics of the key reactions, providing essential information for process design and scale-up.

References

[1] M. Folic, C.S. Adjiman, E.N. Pistikopoulos, AIChE Journal, 53(5):1240, 2007; M. Folic, C.S. Adjiman, E.N. Pistikopoulos, "Computer-aided solvent design: Maximizing product formation", in press, Industrial & Engineering Chemistry Research (2008).

[2] M. Scholl, S. Ding, C.W. Lee, R.H. Grubbs, Org. Lett,, 1: 953 , 1999.

[3] C.S. Adjiman, A.J. Clarke, G. Cooper, P.C. Taylor, Chem Commun., DOI:10.1039:b802921k, 2008.

376c A New Small-Scale Reaction Calorimeter for Data-Oriented Process Development and Its Application to Fast Online Heat Capacity Determination

Gilles Richner, Yorck-Michael Neuhold, and Konrad Hungerbühler. Institute for Chemical and Bioengineering, Swiss Federal Institute of Technology, ETH Zürich, Wolfgang-Pauli-Str.10, Zürich, 8093, Switzerland

During early stages of process development chemical industry needs flexible and versatile tools to assess relevant information about chemical reaction systems. Over the years, reaction calorimeters have become essential devices for data oriented research and development in chemical engineering. They are commonly used for the optimization of a chemical process considering economic factors and safety assessment. Their applications also imply the elucidation of reaction mechanisms and associated parameters (activation energies, rate constants, and heat of reaction).

For the past few years we have been developing a new fully automated small-scale reaction calorimeter. A power-compensation heater combined with a thermoelectrically regulated metal surrounding makes the reactor well suited for fast and highly exothermic reactions. As the baseline for the heat signal is measured online, time-consuming calibrations of the global heat-transfer coefficient are not required. With a working volume from 25 to 45ml the device is particularly useful for the fine and pharmaceutical chemical industry where only small amounts of test substances are available in research and development. Despite its small volume, an ATR-FTIR probe is mounted at the bottom of the reaction vessel. Under strictly isothermal conditions the calorimeter has already demonstrated its capability to investigate fast reactions in the liquid phase such as the hydrolysis of acetic anhydride [1] or the three phases palladium-catalyzed hydrogenation of nitro compounds [2].

Knowledge of the heat capacity is crucial for process development in chemical and pharmaceutical industry. This includes scale-up, thermal safety, and design of the reactor cooling system. Here, we present a method for rapid determination of heat capacity under quasi-isothermal conditions, which is perfectly suited for our small-scale calorimeter. The heat capacity of the reactor content is calculated from amplitudes and phase shifts of the reactor temperature, of the power of the electrical heater and of the cooling rate when defined temperature oscillations are enforced. The heat capacities of eight solvents (water, ethanol, methanol, acetone, 1-octanol, diethylenglycol, toluene, and 1-butanol) covering a wide range of viscosities are calculated for various experimental conditions, including filling volume and stirrer speed. Systematic deviations are detected when compared to the corresponding literature values. Straight-line calibration with the total heat transfer coefficient and a multivariate calibration technique (Partial Least Squares) are applied to correct for these deviations. The two different calibrations show similar precision and allow for an online determination of the heat capacity with an accuracy comparable to other published methods. Successful applications include the determination of the heat capacities for n-heptane, for various homogenous ethanol/water mixtures, and during the course of the hydrolysis of concentrated sulfuric acid.

A heat-flow model of the calorimeter based on the Finite Element Method allows an improved calculation of the heat-balance, particularly when the reactor is to be used under dynamic conditions. Simulation results will be presented that allow a deeper understanding of the above-mentioned systematic deviations when the heat capacity of the reactor content is determined by the temperature oscillation method.

[1] Visentin et al, (2004) Org. Proc. Res. & Dev., 8 (5), 725-737

[2] Visentin et al, (2006) Ind. & Eng. Chem. Res., 45 (13), 4544-4553

Web Page: www.sust-chem.ethz.ch/research/reaction/reaction.html

376d Understanding the Impact of Magnesium to Grignard Reaction Kinetic through on-Line Monitoring

Sze Wing Wong, Utpal Singh, Mark A. LaPack, and Micheal E. Kopach. Chemical Product Research and Development, Eli Lilly and Company, Lilly Corporation Center, Indanapolis, IN 46285

Safety aspects of Grignard reactions are well documented; and PAT (Process Analytical Technology) has been utilized extensively for industrial processing. The presentation reports lab and pilot plant development efforts with a line of sight to manufacturing installation. Results using the Attenuated Total Reflection Fourier Transform Infrared (ATR-FTIR), NIR, and Raman spectroscopy are described along with the challenges encountered with each technology during scale-up. Additionally, the effect of particle size, mixing, and magnesium grade on reaction kinetics were measured on both the laboratry- and pilot plant scale.

376e Towards the Kinetic Analysis of Solid/liquid Reaction Systems from Focused Beam Laser Reflectance Measurements (FBRM) In a High Performance Small Scale Reaction Calorimeter

Sebastien Cap, Yorck-Michael Neuhold, and Konrad Hungerbühler. Institute for Chemical and Bioengineering, Swiss Federal Institute of Technology (ETH) Zürich, Wolfgang-Pauli-Str.10, Zurich, 8093, Switzerland

A wide variety of industrially or biologically relevant chemical processes, e.g. Grignard chemistry or the respiration of living organisms involve multiphase reactions systems. There is no doubt that the understanding of these systems is of considerable importance.

Consider the dissolution of a solid powder into a chemically reactive liquid. Compared to homogeneous reactions in solution the kinetics of solid/liquid biphasic systems are more complex due to the mass transfer of the solid into the reactive liquid. This mass transfer is known to be dependent on the particle size [1] and it is desirable to measure the time resolved particle size distribution (PSD) during the course of a reaction.

Over the past few years, we have developed high performance small scale reaction calorimeters (50mL, -70 to 180°C, up to 70bar) based on the combined principles of heat balance and power compensation [2,3]. The reactor comprises a well defined environment for the kinetic analysis of chemical reactions via the reaction heat and various analytical hyphenated *in situ* devices such as mid-IR and UV-Vis ATR spectroscopy. Importantly, our latest generation reactor was designed to perfectly accommodate a probe for focused beam laser reflectance measurements (FBRM). FBRM provide information on the particle size distribution (PSD) via the measured chord length distribution (CLD). The combination of these analytical devices is ideally suited to study the reaction kinetics from the perspective of both the solid and the liquid phase.

While Beer's law defines a simple linear relationship between spectroscopic absorbance measurements and concentrations in solution there is unfortunately no physical model that extracts quantitative information on the solid particles from FBRM. Few empirical methods have been proposed [4].

For our well defined reactor environment, we investigate the possibilities to quantitatively use the FBRM device. For this, the device is studied under various experimental conditions, such as hydrodynamics, solid particle types, device settings, etc. Results are compared to those obtained from the manufacture's reference setup and/or available literature data. Important issues to be considered include for example reproducibility and sensitivity. From this, we propose a calibration method most suitable for our reactor environment relating the CLD to the amount of mass depending on the mean particle size.

On our path towards a kinetic analysis of solid/liquid biphasic reactive systems, we ultimately intend to incorporate the additional analytical signals from FBRM into our existing established methods for the multivariate non-linear optimisation of kinetic data from time resolved absorbance spectroscopy and calorimetry [5,6].

This project is funded by the Swiss National Science Foundation (SNF) no. 200021-113473.

[1] LeBlanc S.E., Fogler H.S. Population balance modelling of the dissolution of polydisperse solids: rate limiting regimes. *AIChE* (1987) 33: 54-63

[2] Visentin F., Gianoli S.I., Zogg A., Kut O.M., Hungerbühler K. A pressure-resistant small-scale reaction calorimeter that combines the principles of power compensation and heat balance (CRC.v4). *Organic Process Research & Development* (2004) 8: 725-737.

[3] Zogg A., Fischer U., Hungerbühler K. A new small scale reaction calorimeter that combines the principles of power compensation and heat balance. *Industrial and Engineering Chemistry Research* (2003) 42: 767-776

[4] Clarke M.A., Bishnoi P.R. Determination of the intrinsic rate constant and activation energy of CO_2 gas hydrate decomposition using in-situ particle size analysis. *Chemical Engineering Science* (2004) 59: 2983–2993

[5] Puxty G., Fischer U., Jecklin M., Hungerbühler K. Data-oriented process development: determination of reaction parameters by small-scale calorimetry with in situ spectrospcopy. *Chimia* (2006) 60: 605–610

[6] Maeder M. and Neuhold Y.M. *Practical Data Analysis in Chemistry*. Elsevier, Amsterdam 2007.

376f Generalized Tendency Kinetic Modeling for the Rapid Modeling and Optimization of Pharmaceutical Processes - the Case of Asymmetric Catalytic Hydrogenation

Foteini Makrydaki[1], Christos Georgakis[1], and Kostas Saranteas[2]. (1) Chemical & Biological Engineering, Tufts University, 4 Colby Str., Medford, MA 02155, (2) Chemical Process Research and Development, Sepracor Inc., 84 Waterford Drive, Marlborough, MA 01752

This paper aims to demonstrate the application of the generalized Tendency Kinetic Modeling to the systematic modeling and optimization of Asymmetric Catalytic Hydrogenation Reactions. The developed methodology achieves the rapid identification of all plausible model structures along with the identification of their respective parameters, based on statistically informative experimental data. The informative data are acquired by online monitoring of the reactions. Experimental design methods (DoE) are used to plan the necessary experiments.

The industrial example of a catalytic asymmetric hydrogenation reaction is the system on focus, which is very important and frequently encountered in the production of pharmaceuticals. The performance of such process is sensitive to the selection of the proper conditions that favor the maximum yield to the desired product. The asymmetric catalytic hydrogenation of a pharmaceutical compound is the specific experimental system selected. This process will be optimized by directing the selectivity of the asymmetric reaction towards the desired product and maximizing the conversion of the reactant. This is achieved by selecting the optimal reaction conditions, such as temperature, pressure, catalyst loading and initial reactant concentration. The experiments are performed both in the ChemSCAN, a small scale reactor system with 8 parallel reactors of 20mL each, and in a larger scale reactor vessel of 1000mL controlled by the Büchi gas press-flow controller (bpc). The reaction is monitored using Raman spectroscopy. Appropriate chemometric models are developed and they are used on-line to estimate the reactant and product concentrations versus time. The experimental work for this project is performed at the laboratories of Sepracor, Inc.

378a Directed Differentiation of Embryonic Stem Cells towards Insulin-Producing Pancreatic Cells

Ipsita Banerjee, Massachusetts General Hospital, Harvard Medical School, 51 Blossom Street, Boston, MA 02114 and Martin Yarmush, Research, Center for Engineering in Medicine, 51 Blossom Street, Boston, MA 02114.

A promising direction in cell therapeutic treatment of diabetes is the production of sufficient numbers of pancreatic endocrine cells that are functionally identical to primary islets. Mouse embryonic stem (ES) cells, having virtually unlimited replicative capacity and the potential to produce most of the differentiated cell types, have the promise of providing a viable solution. Many studies have reported the differentiation of insulin-producing cells from mouse ES cells and, more recently, from human ES cells. However, inefficiency of differentiation, low insulin content of the insulin-producing cells, a reliance on non-directed ES cell differentiation limits most of these methods. The current promising approach, which is also the focus of this work, is to direct the ES cell differentiation through a process mimicking normal pancreatic development.

The mouse ES cells are cultured on a monolayer, and the pancreatic differentiation path is traced by first inducing endoderm formation. The upregulation of the Pdx-1 gene marks an initial step in pancreatic commitment, which was induced by replating the endoderms on a thin layer of Matrigel. Retinoic Acid (RA), an inducer of pancreas development, when added in the in-vitro culture system in the early differentiation stage was found to enhance Pdx1 expression. Furthermore, inhibition of sonic hedgehog gene was found to have a synergistic effect in the upregulation of Pdx1 expression. The next phase of pancreatic development is endocrine cell specification, which is initiated by the expression of the pro-endocrine gene Ngn3. The Pdx1 positive progenitors, on further maturation, were found to strongly express Ngn3, hence confirming their endocrine commitment. The Ngn3 gene, which is expressed in all endocrine progenitors, initiates a cascade of transcription factors, including Nkx2-2, NeuroD, Nkx6-1, Pax4 and Isl1. All these genes were positively expressed in the ES cell derived endocrine progenitor population. The final stage of the maturation of pancreatic cells is in the formation of islets of Langerhans, consisting of α, β, δ and PP cells, which produce the hormones glucagons, insulin, somatostatin and pancreatic polypeptide, respectively. With further maturation of the ES cell derived endocrine progenitor cells in an appropriate media, gene expression of all these hormones were detected, with the exception of insulin. Strong insulin expression was induced by coculturing the pancreatic progenitor cells with endothelial cells. Although there has been some evidence of the role of endothelial cells during the in-vivo maturation of pancreatic islets, this is the first report demonstrating the effect of endothelial cells in in-vitro differentiating pancreatic β cells. This culture system will be useful in obtaining a renewable source of pancreatic β cells for diabetes treatment.

378c Direct Differentiation of Human Hair Follicle Stem Cells into Vascular Smooth Muscle Lineage for Cardiovascular Therapy

Jin Yu Liu[1], Hao Fan Peng[2], Jun Tian[2], Siddihita Goppinath[2], and Stelios T. Andreadis[3]. (1) Chemical and Biological Engineering, SUNY at Buffalo, 916 Furnas Hall, Amherst, NY 14260, Amherst, NY 14260, (2) Chemical and Biological Engineering, SUNY at Buffalo, 916 Furnas Hall, Amherst, NY 14260, (3) Chemical and Biological Engineering, SUNY-Buffalo, 908 Furnas Hall, Buffalo, NY 14260-4200

Adult stem cells that were obtained from human hair follicles (HF-SC) exhibited high proliferation potential and expressed surface markers characteristic of mesenchymal stem cells such as CD44, CD49b, CD73, CD90 and CD105. Notably, HF-SC could differentiate towards the adipogenic, chondrogenic and osteogenic lineages suggesting that these cells are similar to mesenchymal stem cells from bone marrow. In addition, TGF-beta promoted HF-SC differentiation towards the smooth muscle cell (SMC) lineage. Differentiation towards SMCs was further explored using tissue specific promoters, namely smooth muscle alpha-actin (SMa) and myosin heavy chain (MHC) promoters driving expression of a green fluorescence protein variant (ZsGreen) or red fluorescence protein (RFP), respectively. Using flow cytometry we found that the expression level of ZsGreen or RFP decreased significantly by bFGF and increased by TGF-beta. In addition, bFGF was found to maintain the potential of HF-SC for multilineage differentiation even after long time in culture.

We also prepared tissue engineered blood vessel constructs (TEV) using smooth muscle cells (SMC) derived from HF-SC and designated as HF-SMC. Notably, TEVs from HF-SMC exhibited receptor and non-receptor medicated contractility in response to vasoactive agonists such as KCl, endothelin and the thromboxane mimetic, U46619. In conclusion, HF-SC exhibit high proliferation ability and multilineage differentiation potential and therefore, they provide an easily accessible source of stem cells for tissue engineering and regenerative medicine.

378d Differential Regulation of p21 and p27 In cAMP Induced Differentiation and Apoptosis In Mesenchymal Stem Cells

Linxia Zhang[1], Linsey Seitz[2], and Christina Chan[1]. (1) Chemical Engineering and Material Science, Michigan State University, 2527 Engineering Building, East Lansing, MI 48824, (2) Chemical Engineering and Materials Science, Michigan State University, 2527 Engineering Building, East Lansing, MI 48824

Since their first identification by Friedenstein and Petrakova in 1966, mesenchymal stem cells (MSCs) have gained increasing interest due to their potential use in tissue engineering, drug delivery, etc. As adult stem cells, MSCs are able to undergo multi-lineage differentiation into osteoblasts, adipocytes, chondrocytes and other lineages such as neural cells. The small molecule cAMP (cyclic adenosine monophosphate) has been reported to induce MSCs differentiation into neural cells, but the mechanism is unclear. In our lab, we have found cAMP caused G1/S cell cycle arrest in MSCs when a subpopulation of cells underwent differentiation. The G1/S cell cycle arrest is caused by down-regulation of positive cell cycle regulator such as cyclin D1 and up-regulation of negative cell cycle regulator such as p27. Among the negative cell cycle regulators, two CKI (CDK inhibitor) family proteins, p21 and p27, were oppositely regulated. While both p21 and p27 were transcriptionally upregulated, the total protein level of p21 was significantly downregulated whereas p27 total protein level was greatly enhanced. Further experiments indicated that the increase in the p27 protein level mainly resulted from enhanced transcription, while reduction in p21 protein level is largely due to enhanced protein degradation. In addition to being cell cycle regulators, p21 and p27 play other important roles in cellular processes such as differentiation and apoptosis. p21 is reported to play a protective role in many different cells. Interestingly, apoptosis was observed in the cAMP induced MSC neural differentiation, which is likely related to p21 down-regulation. The cells with neural like morphology upon cAMP elevation had re-organized cytoskeleton which may be attributed to p27 modulation. p27 interferes with actin filament dynamics when it translocates to the cytoplasm. Further studies are underway to uncover whether p21 and p27 are the major players in cAMP induced MSCs neural differentiation and concurrent apoptosis.

378e Expansion and Directed Differentiation of Human Embryonic Stem Cells In a Microcarrier Suspension Culture System

Lye T. Lock, Chemical and Biological Engineering, State University of New York at Buffalo, 1015 Furnas Hall, Buffalo, NY 14260 and Emmanuel S. Tzanakakis, Department of Chemical and Biological Engineering, State University of New York at Buffalo, 907 Furnas Hall, Buffalo, NY 14260.

Human embryonic stem cells (hESCs) hold great promises as sources of therapeutics due to their self-renewal and multilineage differentiation. Human ESCs can serve as a source of insulin-producing cells for diabetes patients. Current protocols for hESC differentiation to insulin-producing cells are characterized by low efficiency. This implicates that processes for the expansion and directed differentiation of hESCs in large quantities is highly desirable.

Here, we explored the use of microcarrier suspension culture for the propagation of hESCs and their subsequent differentiation towards islet cells. Microcarriers provide a high surface area-to-volume ratio, which is easily adjustable. Pluripotent hESCs seeded on microcarriers were able to attach and grow on the beads with viability of at least 90% for 10 days. A doubling time of ~48 hrs was calculated for hESCs in suspension culture compared to ~32 hrs for hESCs cultured on dishes. The proliferation of hESCs on microbeads depended on culture conditions including the agitation rate. In addition, the cells maintained high expression levels of pluripotency markers OCT4, NANOG, REX1, TRA-1-81 and SSEA4 as assessed by quantitative PCR, immunostaining and flow cytometry. These results support the use of microcarrier suspension systems for hESC expansion.

The observed high expression of pluripotency markers and low or absent expression of germ layer genes suggests that hESCs propagated on microbeads may retain their potential for further multi-lineage differentiation. Current work focuses on directing hESCs cultured in suspension towards endoderm progeny, especially pancreatic islet cells.

378f A Pulsatile Flow Bioreactor and the Effect of Frequency on the Differentiation of Mesenchymal Stem Cells

Katherine D. Kavlock, School of Biomedical Engineering and Sciences, Virginia Tech, Blacksburg, VA 24061 and Aaron S. Goldstein, Department of Chemical Engineering, 0211, Virginia Tech, Blacksburg, VA 24061-0211.

An engineered bone substitute consisting of a bone-like extracellular matrix deposited on the internal pores of a resorbable biomaterial scaffold is postulated to stimulate integration, vascular infiltration, and normal bone remodeling when implanted *in vivo*. Bone marrow derived mesenchymal stem cells (MSCs) have the potential to synthesize a bone-like extracellular matrix *in vitro* but culture strategies are needed to facilitate the development of clinically applicable engineered bone graft substitutes. Perfusion bioreactors have been investigated as a component of engineered bone strategies because they supply oxygen and nutrients to cells seeded within a scaffold while applying a mechanical stimulus to the cells. Evidence with planar cell cultures has shown that dynamic flow regimens elicit an enhanced cellular response over steady flow regimens but dynamic perfusion strategies have not yet been translated to 3D scaffold architectures. The objective of this research is to evaluate the effect of frequency of pulsatile flow patterns on the osteogenic differentiation of mesenchymal stem cells seeded in porous scaffolds. We hypothesize that pulsatile flow may be more effective at stimulating osteogenic differentiation of MSCs than continuous perfusion alone and that the cellular response to pulsatile flow may be frequency dependent.

To test the effect of pulsatile flow on osteogenic differentiation we have cultured MSCs in polyurethane foam scaffolds. The scaffolds were either kept under static conditions or subjected to continuous flow or pulsatile flow at frequencies of 0.083, 0.050, and 0.018 Hz. Scaffolds were collected after 14 days of flow to measure proliferation and osteoblastic differentiation of MSCs as indicated by levels of alkaline phosphatase activity and bone extracellular matrix proteins. Preliminary results indicate that all flow conditions enhanced alkaline phosphatase activity and osteopontin expression. Additionally, these markers are preferentially enhanced by pulsatile flow over continuous flow. A trend of increasing alkaline phosphatase activity with decreasing pulse frequency suggests that cells are sensitive to frequency of pulsatile flow. These results indicate that dynamic perfusion may be a useful component of the engineered bone tissue strategy.

378g Regulation of Smad2 Signaling In Self-Renewing Human Embryonic Stem Cells

Balaji Rao, Department of Chemical and Biomolecular Engineering, North Carolina State University, Box 7905, Engineering Bldg. I, 911 Partners Way, Raleigh, NC 27695 and Prasenjit Sarkar, Chemical and Biomolecular Engineering, North Carolina State University, Box 7905, Engineering Bldg. I, 911 Partners Way, Raleigh, NC 27695.

Human embryonic stem cells (HESCs) are pluripotent cells that are derived from the inner cell mass of blastocyst stage embryos. HESCs can be propagated extensively in culture while retaining the ability to differentiate into different somatic cell types. The pluripotent HESC state is maintained by the interactions between (i) extracellular factors added to cell culture and (ii) endogenous factors secreted by HESCs and differentiated cells in cell culture. Specifically, signaling mediated by fibroblast growth factors (FGFs), Wnts and members of the transforming growth factor - beta (TGF-beta) superfamily have emerged as important regulators of HESC self renewal.

We have previously demonstrated that the effect of endogenous factors and the resultant spatial gradients in signaling can be controlled by manipulating the spatial organization of HESCs in culture. This is achieved through micropatterning of HESC colonies with defined diameters and separation. Control of HESC colony size minimizes the variability arising from endogenous signaling in conventional HESC cultures. Using experimental data generated in our system and simple mechanistic modeling, here we show that the level of Smad2 activity is maintained in a specified range in self-renewing HESCs. Smad2 signaling activity is regulated through intracellular and extracellular control mechanisms. Specifically, we focus on extracellular interactions between Nodal, Lefty, Cripto, Activin and the Activin Receptors and intracellular crosstalk between the PI3 kinase, Wnt and Smad2 signaling pathways.

Development of the Photoheterotroph Rhodobacter for Functional Membrane Protein Expression

Brandon S. Curtis[1], Philip D. Laible[2], Deborah K. Hanson[2] and Wayne Curtis[1], (1) Chemical Engineering Department, Pennsylvania State University, Fenske Building, University Park, PA 16802, (2) Biosciences Division, Argonne National Laboratory, Argonne, IL 60439

Abstract:

This paper presents progress on the development of *Rhodobacter sphaeroides* as a platform for high-level expression of functional membrane proteins. The high capacity of *Rhodobacter* for membrane protein expression stems from the dense intracellular membranes synthesized by this organism under anaerobic photoheterotrophic conditions. Using plasmids with promoters from the light-harvesting pigments, inducible heterologous expression is accomplished simultaneously with membrane formation when the organism experiences reduced oxygen tension and low-light conditions. Unique growth characteristics of this organism are presented: although anaerobic growth ceases in the dark, anaerobic photoheterotrophic growth is as fast as aerobic growth and results in nearly 3-times greater biomass yield as a result of photosynthetic ATP production. Growth yield (on carbon basis) is also several-fold higher on defined media as compared to complex media – which is of particular interest for the application of isotopic labeling for membrane protein structural studies. Although transformed lines display dramatically reduced growth under low light conditions, this can be overcome with high light, even with 'knockout' strains where the dominant light harvesting protein (LH2) has been deleted to facilitate greater capacity for functional heterologous expression. Preliminary growth testing in a thin film photobioreactor combined with mass-balance rationalized fed-batch operation are shown to facilitate photoheterotrophic growth to nearly 10 grams DW/L (an order of magnitude higher than current practice in flask culture). Preliminary results demonstrate high-level membrane protein expression.

Background:

Rhodobacter is an extremely versatile organism, capable of growth under a myriad of unique culture conditions. Much of the preliminary work focused on characterizing growth behavior, since the expression system couples expression to promoters involved in the transition from aerobic heterotrophic growth to anaerobic photoheterotrophic growth. Despite extensive study of this versatility in growth behavior, there are very few reports of even simple growth kinetics; its inconvenient growth rate requiring sampling every few hours for 2-4 days, invariably contributed to the paucity of data. An important auxiliary objective was to achieve photoheterotrophic growth, and growth using a defined medium toward the objective of using this system for isotopic labeling of membrane proteins.

Results

Figure 1 shows growth curves executed in Klett flasks (shake flasks with side-arm for insitu optical density measurement) where OD @ 660nm were measured during aerobic, anaerobic and photoheterotrophic growth. In this experiment, nitrogen overlay was used to remove oxygen at an OD~0.2 This experiment displays three important characteristics of

<u>*Rhodobacter*</u>: **(1)** Anaerobic photo-heteterotrophic growth results in growth as fast as aerobic growth, **(2)** Anaerobic photoheterotrophic growth provides much higher yield (as much as 3x greater yield based on dry weight measurements – data not shown); **(3)** *Rhodobacter* does not display significant anaerobic growth in the dark, though it does retain high viability as demonstrated by dilution plating (data not shown). Other useful characteristic of *Rhodobacter* include the distinct coloration patterns of colonies and growth rate, which allow for facile contamination testing.

While Klett flasks provide an excellent experimental system for volumes of material required for extensive protein analysis, they are cumbersome and time consuming to manipulate. A highly multiplexed means of growing Rhodobacter in anaerobic culture tubes will be presented that allows much more detailed examination of growth kinetics. A variety of *Rhodobacter* strains developed with and without plasmids (and blank expression vectors) have been characterized.

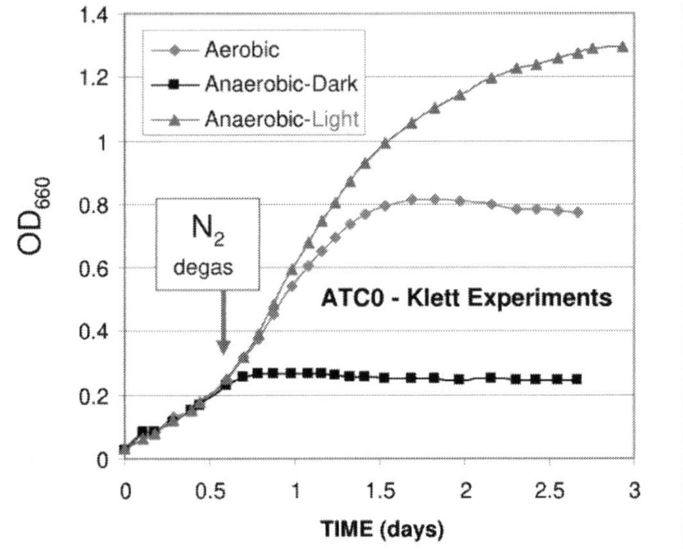

<u>Figure 1</u>: Klett flask growth curves establish basic growth kinetics. Experiments could be degassed to provide anaerobic photoheterotrophic growth which displayed greater yield and comparable growth rates to aerobic growth. Dilution plating confirmed long-term viability of anaerobic cultures despite lack of growth.

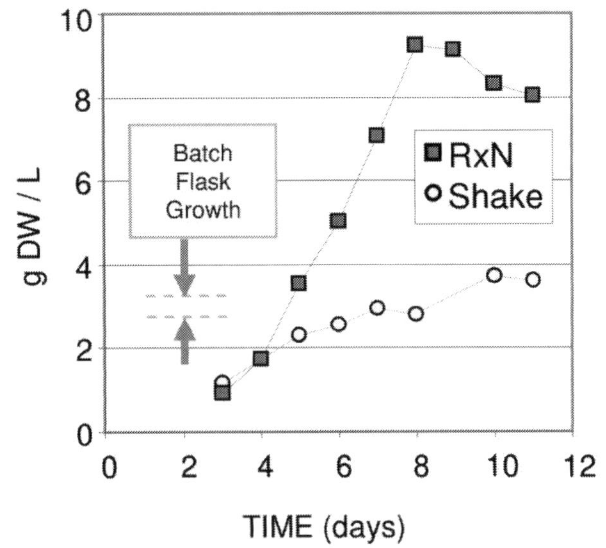

<u>Figure 2</u>: Fed-batch growth in a anaerobic photobioreactor using defined media to achieve cell densities > 10-times higher than Argonne's baseline flask culture growth (indicated in green).

An initial study of a novel bioreactor design was undertaken at a 1-L liquid volume scale as shown in **Figure 2**. This fed-batch strategy utilized mass balances and pH control to allow for accumulation of biomass to more than 10-times the standard batch flask culture.

<u>References</u>:

Laible PD, Scott HN, Henry L, Hanson DK. Towards higher-throughput membrane protein production for structural genomics initiatives. J Struct Funct Genomics. 2004;5(1-2):167-72.

Laible PD, Mielke DL, Hanson DK (2008). <u>The Purple Phototrophic Bacteria</u>. Series: Advances in Photosynthesis and Respiration , Vol. 28 Hunter, C.N.; Daldal, F.; Thurnauer, M.C.; Beatty, J.Th. (Eds.) ISBN: 978-1-4020-8814-8; In press, Available: November 3, 2008.

392b Engineering *Saccharomyces Cerevisiae* for the Heterologous Expression of Mammalian G-Protein Coupled Receptors

Michelle O'Malley, J. Dominic Mancini, Carissa L. Young, David Raden, and Anne Skaja Robinson. Chemical Engineering, University of Delaware, 150 Academy Street, Newark, DE 19716

The G-protein coupled receptors (GPCRs) constitute the largest family of mammalian membrane proteins, and are targets for nearly half of all pharmaceuticals on the market. However, the development of improved, structure-based therapeutics to target these proteins requires milligram amounts of properly folded, purified receptors, which are generally not achievable from native host systems. Thus, over-expression of mammalian G-protein coupled receptors is a necessary first step towards high-resolution structure determination for these important pharmaceutical targets. Though bacterial, yeast, and insect expression systems have been employed to express mammalian GPCRs at high levels, many receptors are improperly processed or otherwise mis-folded in these systems. Even with several host systems available, heterologous expression has not yet systematically allowed for large-scale over-expression of GPCRs, and strategies used to over-express these proteins have typically been empirical.

In this work, we expressed a panel of 12 representative GPCRs from the rhodopsin family of receptors in the yeast *Saccharomyces cerevisiae*, and systematically compared cellular expression patterns in order to investigate limitations associated with GPCR over-expression in this system. Our findings indicate that only the adenosine A_2a receptor can be properly over-expressed in a vacuolar-deficient cell strain, whereas other GPCRs are mainly mis-localized within the peripheral or nuclear endoplasmic reticulum (ER). GPCRs which fail to reach the plasma membrane activate the unfolded protein response pathway within the ER, and many trigger the cellular heat shock response, whereas expression of A_2aR escapes both these quality control checkpoints. Furthermore, mis-localized receptors associate with BiP, an ER-resident chaperone, which indicates probable mis-folding of these receptors. N-terminal sequencing has suggested that these problems may stem from translocation failure across the ER membrane. However, expression levels and protein yields for almost all mis-localized receptors are significant (~ 2-4 mg/L of culture), and some are comparable to that of A_2aR (~10 mg/L of culture). Overall, our findings suggest that the main limitation associated with heterolgous GPCR expression in *S. cerevisiae* is associated with a folding problem within the ER.

392c Enhancing Protein Secretion In *E. Coli* by Codon Engineering

Prateek Gupta, Chemical and Biomolecular Engineering, Cornell University, 120 Olin Hall, Ithaca, NY 14853-5201 and Kelvin H. Lee, Department of Chemical Engineering, University of Delaware, 150 Academy Street, Colburn Laboratory, Newark, DE 19716.

Secretion of recombinant proteins to the extracellular media has the potential to reduce protein aggregation and to simplify downstream purification. Here we have studied the effect of synonymous codon changes on recombinant protein secretion via Type-I pathway in *E. coli*, based on our hypothesis that translation rate of heterologous protein may be related to the ability to secrete higher levels of product via this pathway. We observed that certain synonymous changes in codon usage in specific parts of the gene of interest can result in nearly 10 fold increase in secretion of active protein via Type-I pathway. We also investigated the effects of both the position of synonymous codon clusters as well as the types of codon changes on the intracellular and extracellular expression. The effect of synonymous codon changes on the folding of the polypeptide destined for secretion, and its intracellular fate was explored. We also tested the strategy to improve secretion via multiple secretion pathways and observed similar useful effects of codon engineering.

392d Expression and Reconstitution of Fungal Polyketide Synthases

Suzanne M. Ma[1], Jin Wook Choi[2], Yi Tang[1], and Nancy A. Da Silva[3]. (1) Chemical and BioMolecular Engineering, University of California, Los Angeles, 420 Westwood Plaza, Los Angeles, CA 90095, (2) Chemical Engineering, University of California, Irvine, The Henry Samueli School of Engineering, University of California, Irvine, Irvine, CA 92697, (3) Chemical Engineering and Materials Science, University of California, Irvine, ET 936 The Henry Samueli School of Engineering, University of California, Irvine, Irvine, CA 92697

Hypercholesterolemia is the primary risk factor for coronary heart diseases. By inhibiting 3-hydroxy-3-methylglytaryl coenzyme A (HMG-CoA) reductase, the enzyme responsible for catalyzing the committed step in cholesterol biosynthesis, serum levels of cholesterol can be effectively lowered, thus lowering plaque accumulation and decreasing the risk for coronary heart disease. The cholesterol lowering statin natural products, such as the prototypical lovastatin, are the drug therapy of choice in effectively treating hypercholesterolemia. Lovastatin is biosynthesized in its natural host, Aspergillus terreus by an iterative type I fungal polyketide (IPKS). Using the genetically amendable host Saccharomyces cerevisiae, we have cloned and expressed two central enzymes involved in lovastatin biosynthesis: the lovastatin nonaketide synthase LovB (LNKS, 335kDa) and the accessory enoylreductase LovC (40kDa). We demonstrate for the first time successful verification of the activities of LovB and LovC in vitro, essentially reconstituting the upstream biosynthetic pathway in lovastatin biosynthesis. We verified 1) the recombinant LovB retained complete biosynthetic capabilities of all their catalytic domains; 2) LovB megasynthase alone afforded a series of truncated, elongated polyketide products derived from malonyl-CoA; and 3) LovB and LovC can interface in vitro towards the coordinated catalysis of the lovastatin precursor dihydramonacolin L. Our work paves the way for detailed mechanistic analysis of the programming rules associated with this large, yet cryptic family of fungal megasynthases. The successful reconstitution study also enables combinatorial biosynthesis of the next generation of therapeutically important statin molecules.

392e Investigating the Status of Oxidative Protein Folding in Escherichia Coli Missing Components of the Thiol-Redox Pathways

Karthik Veeravalli[1], Melinda Faulkner[2], Constantine Chrysostomou[1], Jon Beckwith[2], and George Georgiou[1]. (1) Chemical Engineering, University of Texas at Austin, 1 University station C0400, Austin, TX 78712, (2) Department of Microbiology and Moleculae Genetics, Harvard Medical School, Boston, MA 02115

The thioredoxin and glutaredoxin pathways in the cytoplasm of Escherichia coli are responsible for the maintenance of cytoplasm in a reduced state, thus strongly disfavoring the formation of structural disulfide bonds in cytosolic proteins. E. coli mutants missing components of both these pathways are not viable as essential enzymes like ribonucleotide reductase depends on atleast one of these pathways for the reduction of disulfide bond formed during its catalytic cycle. Such mutants having non functional thioredoxin and glutaredoxin pathways accumulate suppressor mutations in the gene ahpC encoding a peroxiredoxin, which rescues the growth defect. In this study, we explored oxidative protein folding in cytoplasm of E. coli mutants, missing components of both the thioredoxin and glutaredoxin pathways, to determine the optimal conditions for maximizing the expression yield of multiple disulfide-bonded proteins. Our results indicate that the correct folding of proteins with complex patterns of disulfide bonds like tissue plasminogen activator depends on whether the strains can synthesize glutathione or not as well as the nature of the suppressor allele in the peroxiredoxin AhpC to a certain extent. Also, we found that the strains missing thioredoxin reductase and glutathione accumulate high yields of proteins with fewer disulfide bonds like alkaline phosphatase, and antibody fragments. Consequently, the latter strains provide an exciting platform for high level expression of antibody fragments and as a result could be a useful expression host for biotechnological purposes. To understand some of the differences observed among the thiol-redox pathway mutants in supporting disulfide bond formation, a DNA microarray based global transcriptional profiling was performed. While the genes involved in chemotactic responses and flagellar biosynthesis seem to be greatly repressed in the thiol-redox pathway mutant strains, genes involved in oxidative stress reponses, cold shock and envelope stresses are highly upregulated. The results from the transcriptional analysis data are being employed to construct strains that exhibit more extensive disulfide bond formation in the cytoplasm.

392f Incomplete Quality Control of Yeast-Secreted Multidomain Proteins

Andrew W. Nields, Department of Chemical and Biomolecular Engineering, University of Pennsylvania, 220 South 33rd St., Philadelphia, PA 19104 and **Eric T. Boder**, Department of Chemical and Biomolecular Engineering, University of Tennessee, 1512 Middle Drive, 437 Dougherty Engineering Bldg., Knoxville, TN 37996-2200.

Saccharomyces cerevisiae is an advantageous host organism for expression of a number of proteins, particularly those requiring post-translational processing such as disulfide bond formation. Proteins bound for secretion from yeast are subject to quality control mechanisms in the endoplasmic reticulum believed to ensure only fully folded, native proteins are exported from the ER and secreted from the cell. This quality control mechanism has been exploited to engineer proteins for increased thermal stability and secretion level using yeast surface display. We have recently probed the limits of yeast's ability to discriminate two-domain fusion proteins based on differences in thermostability. Secretion products of model proteins demonstrating moderate to high stability were characterized by flow cytometry and Western blotting of surface-expressed fusions. Results indicate that proteins with domains of moderate stability can be secreted in a heterogeneously truncated form, with truncations proceeding from the C-terminus. Inhibitor studies suggest that truncation results from partial proteasomal degradation, suggesting that targeting of incompletely folded secretory proteins to the endoplasmic reticulum-associated degradation pathway in yeast is not irreversible.

392g Triple Light Chain Antibodies in Thiomabs

Natalia Gomez, Abigail Vinson, Jun Ouyang, and Inn Yuk. Genentech, South San Francisco, CA 94080

ThioMabs are recombinant monoclonal antibodies that have been engineered to have a point mutation by replacing one amino acid residue with cysteine. Antibody-drug conjugates (ADC) for targeted chemotherapeutics can be formed by attaching cytotoxic drugs of interest via a maleimide functional group to these residues. Chinese Hamster Ovary (CHO) stable cell lines have been developed for the production of ThioMabs in suspension cell culture.

During cell culture, the additional ThioMab cysteine residues are usually capped with cysteine or glutathione from the media. However, in some cases ThioMabs composed of two heavy chains (HC) and three light chains (LC) have also been found. These antibodies, known as Triple Light Chain Antibodies (3LC), are formed due to an improperly capped engineered cysteine that disulfide bonds to an extra LC. In this work, the underlying causes for the formation of these 3LC antibodies in CHO cell lines were studied.

Stable cell lines expressing several types of ThioMabs were cultured in suspension in shake flasks with a temperature shift after 24-72 hours, and harvested after 14 days. The Harvest Cell Culture Fluid (HCCF) was analyzed for titer and percentage of 3LC antibodies. Additionally, cell lysates were analyzed for mRNA levels and protein expression for light and heavy chains. Correlations between LC/HC mRNA ratio, free LC expression, concentrations of capping agents and 3LC levels were determined.

Methods for decreasing 3LC formation were also investigated. Analyses of the impact of cell culture conditions such as temperature and pH on decreasing amount of 3LC yielded cell-line specific results. Addition of extra cysteine to the cell culture fluid for facilitating the capping process of ThioMabs was evaluated and it was found to decrease 3LC levels in a time-dependent manner.

405a Application of Pharmaceutical Solvates In Drug Development

Yanfeng Zhang, Yaling Wang, and Louis S. Crocker. Merck Research Laboratories, Merck, 126 E Lincoln Ave, RY818-B112, Rahway, NJ 07065

The presentation will focus on physicochemical characterization of pharmaceutical solvates and their application in drug development. Case studies from Merck will be presented: (a) a solvate of an active pharmaceutical ingredient (API) was selected for further development; (b) a solvate was used to obtain a highly crystalline API hydrate; (c) a solvate was applied in process optimization to reject an impurity and improve product yield. Emphasis will be placed on how a good understanding of the characteristics of pharmaceutical solvates leads to successful applications of the solvates in drug development.

405b Phase Diagram Visualization Via Continuously-Fed Crystallization: Experiments and Model

Masano Sugiyama, Department of Chemical Engineering and Materials Science, University of Minnesota, 151 Amundson Hall, 421 Washington Avenue SE, Minneapolis, MN 55455 and Victor H. Barocas, Department of Biomedical Engineering, University of Minnesota, 312 Church St SE, 7-115 Hasselmo Hall, Minneapolis, MN 55455.

X-ray diffraction is the most common way to determine structural information of proteins at an atomic level. Elucidation of these structures has a dynamic impact on biotechnology and pharmacology. To determine the protein structure, a high-quality crystal of sufficient size is required. However, the production of such crystals is a current bottleneck. Current technology involves hundreds of conditions in this multi-parametric process to be tested by either batch or vapor diffusion crystallization techniques. Knowledge of the phase diagram allows experimenters to better design experiments to grow large high quality crystals.

A continuous-feed crystallization chamber is manufactured to allow phase diagram visualization to predict the phase diagram of a protein in a single experiment. This microfluidic system allows the experimenter to screen a large range of salt and protein concentrations by controlling the convection and diffusion of both protein and salt. A continuous-feed crystallization chamber has been successfully fabricated and characterized in terms of its flow profile. The protein and salt concentration profiles are determined using a computational model. Experimental results determine locations where crystals form and in combination allow determination of the phase diagram. This device has successfully predicted the well-known phase diagram for tetragonal lysozyme and has been utilized on triclinic lysozyme, a protein crystal for which the full phase diagram is unknown.

405c Separation and Purification of (–)Schisandrin B from Schisandrin B Stereoisomers

Ka F. Luk[1], Robert Kam M. Ko[2], and Ka M. Ng[1]. (1) Department of Chemical Engineering, Hong Kong University of Science and Technology, Clear Water Bay, Kowloon, Hong Kong, China, (2) Department of Biochemistry, Hong Kong University of Science and Technology, Clear Water Bay, Kowloon, Hong Kong, China

Schisandrin B (Sch B), consisting of a mixture of its stereoisomers, namely (−)Sch B and (±)γ-schisandrin, is the most abundant and biologically active dibenzocyclooctadiene lignan present in *Fructus Schisandrae* (FS). Among the stereoisomers of Sch B in FS, (−)Sch B was found to produce more potent enhancing effect on cellular glutathione and protection against oxidative injury in both cultured cardiomyocytes and hepatocytes. Therefore, it is desirable to recover (−)Sch B in a purified form to achieve the highest bioactivities.

The objective of this study is to develop a process for large-scale separation and purification of (−)Sch B from the Sch B stereoisomeric mixture. To this end, a crystallization-based separation and purification process has been conceptualized. Bench-scale crystallization experiments guided by experimental solid-liquid equilibrium phase diagrams were performed to verify process feasibility. A (−)Sch B product with a purity of 98.5 wt% and a (±)γ-schisandrin-enriched product with a purity of 65.0 wt% were obtained. The (−)Sch B product caused a 32% increase in cellular glutathione level and the (±)γ-schisandrin-enriched product a 26% increase, indicating a potentially more efficacious pharmaceutical preparation.

405d Solvent Effects on the Molecular Self-Assembly of Tetrolic Acid In Solution and Its Implications of the Polymorph Formed from Crystallization

Jie Chen and Bernhardt L. Trout. Department of Chemical Engineering, Massachusetts Institute of Technology, 77 Massachusetts Avenue, E19-528, Cambridge, MA 02139

Solution crystallization is a commonly used technique to separate and purify APIs (Active Pharmaceutical Ingredients) in the pharmaceutical industry. The selection of the solvent is of great importance in solution crystallization. Solvent affects many properties of the crystal product, including polymorph form, morphology, size distribution, flowability, and compressibility. It therefore has consequences for the dissolution rate and bioavailability of the final drug product. Current solvent selection techniques for solution crystallization remain ad hoc, and typically do not have a theoretical underpinning. In this study, computer simulations are applied to elucidate the interactions between solvent and solute molecules at molecular level and also the mechanism underlying the solvent effects on the properties of the crystal product. By doing this, we want to pave the way to the rational selection of solvent for solution crystallization.

Molecular dynamics was applied to study the solvent effects on the molecular self-assembly of a model compound tetrolic acid (TTA) in four different solvents and implications on the polymorph formed during crystallization. TTA has two polymorphic forms (form I and form II) and a solvate form, which are known to form from chloroform, ethanol and dioxane respectively. A hypothesis in literature states that there is a link between the solution chemistry and the solid state polymorphic outcome: the most stable motif (growth synthon) in the solution has the highest probability of crystallizing into the crystal as the structural synthon. This link hypothesis was investigated using molecular simulations in our study. Our results suggest that the strong interactions between TTA and solvent molecules (ethanol or dioxane) prevent the formation of a carboxylic acid dimer in solution and thus promote the crystallization of TTA in a catemer based form or a solvate form. However weak interactions between TTA and solvent molecules (carbon tetrachloride or chloroform) facilitate the formation of carboxylic acid dimers in solution and thus promote the crystallization of a dimer based crystal. Detailed solvent structure plays an important role in determining the relative stability of various growth synthons in solution and also the barriers along the pathway connecting them. Moreover, a procedure to study the solvent-solute interactions in solution and the initial solvent screening for the desired polymorph is proposed, a procedure which can be extended easily to other polymorphic systems.

405e Biological Macromolecule Crystallization for Structure Based Drug Design: Finding the Keys for Crystallization Success

Russell A. Judge, Structural Biology, Abbott Laboratories, GPRD, R46Y, AP10, 100 Abbott Park Road, Abbott Park, IL 60064

The challenge in supporting biological macromolecule crystallization projects for structure based drug design is to develop robust crystallization systems, from which high-resolution crystal structures for a variety of ligand complexes can be obtained. Each biological macromolecule is unique, often requiring the testing of binding partners, additives, purification tag placement and crystallization methods, in addition to solution condition screening. It is often a combination of a number of these factors that lead to the most robust crystallization system. In this presentation we will discuss case studies illustrating how the use of additives such as ionic liquids, second site binders, crystallization method, tag placement and purification methods have provided high quality crystals for a number of biological macromolecules.

405f Protein Crystallization: Studies with Gamma D-Crystallin, a Protein Involved In Cataract Formation

Quinta Nwanosike, Athanassios Sambanis, and Ronald W. Rousseau. School of Chemical & Biomolecular Engineering, Georgia Institute of Technology, 311 Ferst Drive, Atlanta, GA 30332-0100

Cataracts are the leading cause of blindness worldwide. At present, the only cure for cataracts is surgical removal of the lens and replacement with a plastic intraocular lens. Post-biosynthetic modifications and chemical stress of the major lens proteins, which are known as crystallins, may lead to formation of crystallin aggregates, precipitates and crystals, ultimately resulting in development of cataracts in the lens.

Gamma D-crystallin is one of the major proteins in the lens. In the present study, it has been expressed using *E. coli* and purified with ion exchange chromatography. Initial crystallization of this protein has been carried out in Eppendorf tubes to determine suitable concentrations for observable crystals. Further crystallization is carried out on an automated stage and with varying solution conditions.

Crystallization of a gamma D-crystallin could be predicted by measuring pre-crystallization or nucleation determinants using light scattering techniques. In this work an effort is made to correlate the crystallization habits of gamma D-crystallin and parameters observed with light scattering.

405g Monoclonal Antibody Interactions and Phase Behavior

Rachael A. Lewus[1], Abraham M. Lenhoff[2], and Stanley I. Sandler[1]. (1) Department of Chemical Engineering, University of Delaware, 150 Academy St, Newark, DE 19716, (2) Department of Chemical Engineering, University of Delaware, 226 Colburn Lab, Newark, DE 19716-3110

Protein phase behavior is implicated in numerous aspects of downstream processing either by design, as in crystallization or precipitation processes, or as an undesired effect, such as aggregation. An improved understanding of protein phase behavior is therefore critical to developing rational design strategies for important process steps. This presentation will explore the phase behavior of monoclonal antibodies (mAbs), which may exhibit a wide range of forms despite the structural similarities at the molecular level. Liquid-liquid separation, aggregation, gel networks, and crystals have been observed. A systematic study of numerous factors, including the effects of solution composition and pH, is conducted in order to explore antibody behavior. Examples of phenomena observed for individual mAbs include a significant dependence of the spinodal boundary on the cation in sulfate salts and non-monotonic trends in pH dependence. Protein-protein interactions are used to interpret these observations. Comparisons of behavior among different mAbs, and possible mechanistic origins, will be discussed.

414a Assessment of *In Vivo* Murine Transduction Efficiency and Biodistribution Profiles with AAV/cationic Lipid Co-Formulations

David E. Fein[1], Maria P. Limberis[2], James M. Wilson[2], and Scott L. Diamond[1]. (1) Chemical and Biomolecular Engineering, University of Pennsylvania, 1150 Vagelos Laboratory, 3340 Smith Walk, Philadelphia, PA 19104, (2) Gene Therapy Program, University of Pennsylvania School of Medicine, 125 South 31st Street, Philadelphia, PA 19104

Successful viral vector-mediated delivery of therapeutic transgenes to the airway epithelium has been limited by both low efficiency and induced immune response to the vector and encoded transgene. Adeno-associated viral (AAV) vectors are attractive due to their non-pathogenic nature and sustained transgene expression; however, neutralizing antibodies are generated against the viral capsids preventing effective readministration. A synthetic cationic glucocorticoid, dexamethasone-spermine (DS), can associate with the AAV capsid altering the biodistribution and immune response following *in vivo* administration of co-formulated vector/lipid complexes. Intranasal administration of 1E+11 genome copies of AAV co-formulated with 6.4 µg (total lipid) of liposomes with the neutral lipid dioleoylphosphatidylethanolamine (DOPE)/mouse, resulted in average 1.1 to 2.2 fold increase in transgene expression for AAV2/6.2 in the lung at Day 7 through Day 31 for two different lipid formulations compared to AAV2/6.2 formulated with vehicle control. Similarly, AAV2/9 co-formulated with lipid resulted in a 1.3 to 3.7 fold increase in transgene expression in the lung at Day 7 through Day 31 for the two lipid formulations compared to AAV2/9 formulated with vehicle control. Histological examination of lung cross-sections indicated that cohorts receiving AAV/lipid co-formulations had moderately elevated transduction of both the alveolar epithelium and conducting airways following homologous vector re-administration. Our results suggest that formulating AAV with DS/DOPE may improve total transduction efficiency and possibly reduce the host immune response to the vector in mouse airway epithelium *in vivo*.

414b Directed Evolution of AAV to Target Tissue-Specific Vascular Endothelium

Inchan Kwon, James T. Koerber, and David V. Schaffer. Department of Chemical Engineering, University of California, Berkeley, 278 Stanley Hall, Berkeley, CA 94720

Gene delivery vectors based upon adeno-associated virus (AAV) are highly promising, due to lack of pathogenicity, efficient infection of both dividing and non-dividing cells, and sustained maintenance of the transgene. While vectors based on AAV serotype 2 (AAV2), the best characterized variant, show broad tissue and cell tropism, AAV2 vectors do not efficiently transduce vascular endothelium. The vascular endothelium is associated with many diseases, including pulmonary hypertension and tumor angiogenesis. We hypothesized that the development of AAV vectors capable of efficient gene delivery targeted to vascular endothelium of a specific tissue would have therapeutic benefits for the tissue-associated diseases including cancer. First, tissue-specific AAV2-vectors can reduce potential side effects caused by the delivery of the transgene into unwanted tissues. Second, the amount of vectors required to provide comparable therapeutic effects can be lowered. In order to target AAV2 to vascular endothelium, three different AAV2 plasmid libraries were generated via error-prone PCR, shuffling of cap genes obtained from several serotypes, or insertion of randomized peptides in a defined capsid location. In vivo selection was then performed by tail vein injection of mice with these libraries followed by tissue harvest at 24 hours post-injection. Recovered cap genes of the AAV2 variants were used to construct next round of selection. After three rounds of in vivo selection for lung, the vascular distribution of AAV variants selected was evaluated. Among them, R4L13 variant, showed substantially enhanced localization to lung when compared with wild-type AAV2. R4L13 variant was further evolved via introduction of random mutations in the capsid protein. Another four rounds of in vivo selection for lung were performed. One of the AAV2 variants selected, EP09, showed further improvement in localization to lung compared to R4L13 variant. Recombinant vectors based upon R4L13 and EP09 are being assessed for targeted delivery of the gene expressing luciferase to lung, relative to recombinant vectors with wild-type AAV2 capsid. Similarly AAV libraries are being selected for other tissue-specific vascular endothelium including brain and spinal cord.

414c Engineering of a Novel AAV Vector In a Human Airway Model System for Cystic Fibrosis Gene Therapy

James T. Koerber[1], Katherine Excoffon[2], Brian Kaspar[3], Joseph Zabner[2], and David V. Schaffer[4]. (1) Department of Chemical Engineering, University of California, Berkeley, 278 Stanley Hall, Berkeley, CA 94720, (2) Department of Internal Medicine, University of Iowa, (3) Center for Gene Therapy, Nationwide Children's Hospital, (4) Chemical Engineering, University of California, Berkeley, 278 Stanley Hall, Berkeley, CA 94720

Cystic Fibrosis is a common autosomal recessive disease caused by mutations in the CFTR gene. Although current treatments have improved the lives of patients, none are directed at the underlying genetic defect, and CF remains a lethal disease. Gene transfer is attractive because it could correct all of the physiologic abnormalities due to the loss of CFTR. However, recent clinical trials with a variety of delivery vehicles, both viral and nonviral, have proven inefficient. In particular, recent clinical trials for CF have shown that adeno-associated virus (AAV2) is extremely safe; however, failed to achieve therapeutic benefits due to a range of barriers to effective delivery, including poor binding to the apical surface – due to low apical levels of receptors – and inefficient intracellular trafficking.

Rational efforts to engineer novel AAV vectors have resulted in moderate success; however, lack of sufficient knowledge regarding the structure/function relationships of AAV2 and other serotypes has hindered further rational engineering methods. We thus hypothesized that a directed evolution approach employing recombination and mutagenesis, akin to natural evolutionary mechanisms, have the potential to engineer novel viral capabilities under artificial, clinically relevant selective pressures not found in nature. Specifically, we employed directed evolution of the AAV capsid to generate novel viral variants with enhanced gene transfer to human pulmonary epithelium. PCR based mutagenesis coupled with high-throughput in vitro recombination generated a diverse library of chimeric cap genes with components from two divergent serotypes that utilize distinct receptors, AAV2 and AAV5. Subsequent selection of this library, consisting of two rounds of evolution and five rounds of selection, in polarized primary human airway epithelia (HAE) identified a single novel AAV chimera with a novel point mutation. Impressively, binding to HAE was increased approximately 100-fold and luciferase gene transfer showed 100-fold improvement suggesting that the changes improved both binding and post-binding steps. Furthermore, the novel AAV virus mediates successful CFTR cDNA-gene transfer to correct the Cl-transport defect in human CF epithelia. In summary, we have engineered a novel AAV variant that is capable of significantly greater airway gene transfer and holds the promise of therapy for CF as well as other pulmonary diseases, and these results highlight the potential of directed evolution approaches to engineer 'designer' gene delivery vectors.

414d JNK Signaling Is Necessary for Lentivirus Gene Transfer

Meng Horng Lee, Chemical Engineering, SUNY Buffalo, 916 Furnas Hall, SUNY Buffalo, Buffalo, NY 14260 and Stelios T. Andreadis, Chemical and Biological Engineering, SUNY-Buffalo, 908 Furnas Hall, Buffalo, NY 14260-4200.

Recombinant lentivirus is widely used as a gene transfer vehicle for gene therapy and tissue engineering due to high transduction efficiency and its ability to infect non-dividing cells and stem cells in vitro and in vivo. However, the interactions between lentivirus and the target cells during transduction process are only partially understood. Here we show that transduction with recombinant lentivirus activated the c-jun-N-terminal kinase (JNK) pathway by phosphorylating JNK in primary epidermal cells and several epithelial cell lines. Blocking JNK phosphorylation with chemical inhibitors and siRNA decreased gene transfer in a dose dependent manner. Similar results were observed with actively growing as well as non-proliferating cells indicating that JNK may affect gene transfer independent of cell cycle status of target cells. In addition, we found that inhibition of JNK affected early events in the gene transfer process. Specifically, activation of JNK was necessary of internalization of viral particles in the cell cytoplasm as evidenced by measurements of internalized viral proteins using western blots and ELISA. Immunostaining results further indicated that JNK affected viral internalization may through regulating filopodia formation of the target cells. These results suggest activation of JNK may be necessary for lentivirus to gain entry into the target cells.

414e Co-Delivery of Cationic Polymers and Adenovirus Enhance Tumor Protection In a Murine Model of Prostate Cancer

Jessica B. Graham, Department of Chemical Engineering, University of Iowa, 115 S Grand Avenue, 201 PHAR, Iowa City, IA 52242, Elisabeth Donahey, Division of Pharmacy, University of Iowa, 201 Pharmacy Building, 115 S Grand Avenue, Iowa City, IA 52242, David M. Lubaroff, Department of Urology; Holden Comprehensive Cancer Center, University of Iowa, 375 Newton Road, 3210 MERF, Iowa City, IA 52242, and Aliasger K. Salem, Department of Chemical Engineering, Division of Pharmaceutics, Holden Comprehensive Cancer Center, University of Iowa, 115 S Grand Avenue, S228 PHAR, Iowa City, IA 52242.

Prostate cancer is the most common non-skin cancer in America, and the most commonly diagnosed cancer among males. When metastatic, the disease is ultimately incurable. Consequently, alternative strategies to current treatments are sought, especially in the area of immunotherapy. Vaccine immunotherapy using a specific antigen can be used to stimulate the body's immune response for targeted destruction of cancer cells and for prevention of tumor recurrence. Adenovirus-based vaccines have been successful in inducing an immune response to certain cancers and limiting tumor growth. To enhance this anti-tumor activity, we propose the combination of a viral system with a non-viral system, in the form of the cationic polymer poly ethylenimine (PEI). PEI complexes with DNA to form nanoparticles that can be used in gene delivery. We show that PEI can upregulate the CD80 and CD86 markers of maturation and activation in dendritic cells, a process necessary for launching a robust antigen-specific immune response in the body. We also show that PEI complexed with an antigen-specific adenovirus provides increased tumor protection in a murine tumor model.

414f DNA Delivery to Neuronal-Like Cells Using Designed Recombinant Fusion Proteins

Shan Gao[1], Melissa J. Simon[2], Barclay Morrison III[2], and Scott A. Banta[1]. (1) Department of Chemical Engineering, Columbia University, 500 W. 120th Street, New York, NY 10027, (2) Biomedical Engineering, Columbia University, 1210 Amsterdam Avenue, New York, NY 10027

A number of methods have been developed for gene delivery including recombinant viral vectors, synthetic polymers, peptides, lipids and targeted vesicles. Recombinant proteins, with their inherent biocompatibility, versatility for incorporating targeting function, low toxicity, and facile preparation, have shown great advantages as vehicles for non-viral gene delivery. In this study, a recombinant protein was developed for delivering DNA oligonlucleotides to PC12 cells. This recombinant protein (p50-GFP-TAT) consists of a DNA binding domain (p50), a cell penetrating domain (TAT), and a fluorescent protein domain (GFP). Two other recombinant proteins, p50-GFP and GFP-TAT, were also prepared as controls. The efficiency of this protein vehicle was evaluated by delivering fluorescently-labeled dsDNA molecules of 30bp and 294bp lengths at varying molar ratios of protein to DNA in a physiological medium containing 17.5% serum. The results show that the three recombinant protein constructs can transfect PC12 cells in a dose dependent pattern but p50-GFP-TAT exhibits the highest transfection efficiency compared to p50-GFP and GFP-TAT. However, among the three proteins, only p50-GFP-TAT can deliver DNA molecules to the PC-12 cells, which indicates both DNA binding domain (p50) and cell penetrating domain (TAT) are necessary for the successful delivery of DNA. The optimal molar ratios of protein to DNA are 11.2 for delivery of the 30bp DNA oligonucleotide and 560 for delivery of 294bp DNA oligonucleotide. Our study has also shown that TAT exhibits inherent transfection specificity to different cell types and even the same cell type under varying conditions. For example the GFP-TAT peptide preferentially penetrates serum-stimulated astrocytes over quiescent astrocytes and neurons grown in a serum-free environment, which indicates a potential application of the construct in targeting activated astrocytes for the treatment of brain injuries. Pilot studies exploring the efficiency of the recombinant protein containing both p50 and TAT domain for the delivery of siRNA will also be presented. Overall, this study will demonstrate the potential of designed recombinant protein constructs as non-viral gene therapy vehicles for the delivery of various DNA and RNA cargoes.

414g Formulation of a Hierarchically Designed Peptide Nucleic Acid Based DNA Delivery Construct

Peter G. Millili[1], Ulhas P. Naik[2], and Millicent O. Sullivan[1]. (1) Chemical Engineering, University of Delaware, 150 Academy St., Newark, DE 19716, (2) Biological Sciences, University of Delaware, 118 Wolf Hall, Newark, DE 19716

Gene delivery biomaterials need to be designed to efficiently achieve nuclear delivery of plasmid DNA. To achieve this goal, various extracellular and intracellular barriers must be overcome. Polycationic polymers have been used to package DNA and other nucleic acids within sub-micron sized particles, offering protection from shear-induced or enzymatic degradation. However, cytotoxicity issues coupled with limited transfection efficiencies minimize the effectiveness of this approach. In an effort to improve upon existing technologies aimed at delivering nucleic acids, we have explored an alternative approach to DNA packaging. Our method involves the utilization of peptide nucleic acids (PNAs) to directly functionalize DNA and allow for the formulation of modified plasmids complexed with lower molecular weight polymers. PNAs are nucleic acid analogs with a peptide backbone that can bind to double stranded DNA to form a stable PNA-DNA-PNA triplex. Polyethylene glycol (PEG) chains can be directly added to a plasmid via this conjugation chemistry, providing a steric layer which prevents opsonization and promotes the colloidal stability of the DNA prior to its complexation with polycations. Peptides with specific amino acid sequences related to physiological targets can also be incorporated onto this hierarchically designed structure, promoting preferential delivery to a particular location. We have created DNA-PNA-peptide conjugates using this approach and have post-complexed the DNA using spermidine and polyethylenimine. We have characterized the size, morphology, and surface chemistry of these conjugates by dynamic light scattering, atomic force microscopy, and zeta potential analyses, respectively. We can create 100 – 300 nm stabilized particles whose diameter depends upon the nature and concentration of the condensing polycations. Proof of principle of our PNA-based design will lay the foundation for their use as tools to probe the kinetics of intracellular transport.

425a Multiscale Modeling of Supported Lipid Bilayers

Roland Faller, Chenyue Xing, and Matthew Hoopes. Chemical Engineering and Material Science, UC Davis, One Shields Ave, Davis, CA 95616

Supported lipid bilayers (SLBs) have been a standard model system for experimental investigations of biomembranes. The difference between supported and unsupported membranes which is crucial for the applicability of these models has, however, not been studied very systematically. In simulations this is even more true as computational studies of supported lipid bilayers are essentially lacking from the literature. We are aiming at filling this gap by a multiscale approach to address the influences various types of supports have on the structure, dynamics and phase behavior of lipid bilayers. We are using various molecular models on different time and length scales in order to address the multitude of effects. Our simulations demonstrate that these simulation models can predict different aspects of the change of the system under the influences of hydrophilic and hydrophobic support. It becomes very clear that the support breaks the symmetry of the membrane and suppresses large amplitude fluctuations. The structural influence on the proximal leaflet is much more strongly pronounced than for the distal leaflet. It turns out that the distal leaflet is a reasonable model for a leaflet in a free bilayer but the proximal leaflet is strongly coupled to the support. We also find that the lateral pressure profile close to the support is strongly changed and this may provide clues why proteins in supported lipid bilayers can loose their function.

425b Multiscale Investigation of Nucleotide Effects In Actin Structures

Jim Pfaendtner[1], Davide Branduardi[2], Michele Parrinello[2], and Gregory A. Voth[3]. (1) The University of Utah and ETH Zurich, Salt Lake City, UT 84112, (2) Department of Chemistry and Applied Biosciences, ETH Zürich, USI-Campus, Via Giuseppe Buffi 13, Lugano, 6900, Switzerland, (3) Center for Biophysical Modeling & Simulation, University of Utah, Department of Chemistry, 315 S. 1400 E. Rm 2020, Salt Lake City, UT 84112-0850

The polymerization of actin monomers into filaments and larger structures, e.g., bundles, is a key process in the machinery of the cell. In the last decade, experimental measurements of actin and actin-related proteins have greatly increased our understanding of both the mechanism of cell motility and the role of cellular components that regulate filament growth, i.e., capping proteins. In spite of these tremendous achievements, there are still numerous fundamental questions that remain unanswered. Chief among these questions is the role of the actin-bound nucleotide in the structure, properties and function of the actin cytoskeleton. This presentation describes an investigation of the folding of the alpha-helical domain in the DNase I-binding loop of actin.

Historically, the main challenges in applying the principles of molecular modeling to systems of biological relevance have been the poor scaling of these techniques and lack of computational power. Of particular relevance to the present study is the issue of sampling rare-events. For example, the folding or unfolding of an alpha helix such as the one described above requires the formation or breaking of many hydrogen bonds along the peptide backbone. Transformation between these two states requires the crossing of a free-energy barrier that is much larger than kBT, and therefore extremely unlikely to be observed during a classical molecular dynamics (MD) simulation, wherein the longest accessible timescales are on the order of 100 ns. To address the issue of sampling rare events, the method of metadynamics was introduced, wherein the simulation is biased along one or more collective variables that offer a reduced description of the transition of interest. Due to the long timescales under consideration, application of such a technique is essential for accurate modeling and simulation of the process of interest.

First, a systematic methodology for using metadynamics to obtain free-energy landscapes for folding alpha helices will be presented. Various strategies for selection of collective variables will be discussed with an emphasis on obtaining an efficient yet general procedure for studying the folding of the alpha helix. Next, free-energy landscapes for the folding of the alpha helix within the DNase I-binding loop will be presented. We have studied this folding process as a function of the bound nucleotide within actin, and differences between the landscapes will be discussed. Finally, the role of folding in the DNase I-binding loop will be discussed in the context of the supramolecular structures F-actin and the Arp2/3 complex.

425c Simulated Mutagenesis and Structural Analysis for Protein Fragment Engineering

Camilo A. Velez Vega[1], Michael K. Fenwick[2], and Fernando A. Escobedo[1]. (1) School of Chemical & Biomolecular Engineering, Cornell University, Ithaca, NY 14853, (2) Department of Molecular Medicine, Cornell University, Ithaca, NY 14853

Mutagenesis analyses were performed using Molecular Dynamics Replica Exchange (REM) as a tool for systematically evaluating the structural variability of protein domains of interest in the area of protein biotechnology. A novel REM scheme, termed Multiple Mutant REM (MMREM), was developed to quickly screen diverse mutants based on their relative propensities for attaining favorable structures.

In our initial implementation, the wildtype and mutated hypervariable regions of an anti-hCG llama VHH antibody were simulated via REM on a constrained scaffold. Llama antibody binding domains such as this (which are smaller than those of conventional antibodies) are of interest as minimalistic models for antibody engineering, but are known to have uncommon structures which are not predictable by bioinformatic tools. The loop structures of the wildtype crystal (Spinelli et al., Nat. Struct. Biol. 3:752-757, 1996) were recovered, starting from an "open" conformation. Furthermore, the structural variability observed for the H1 and H3 loops of the simulated wildtype antibody agrees with that found in the NMR study of this antibody (Renisio et al., Proteins Struct. Funct. Genet. 47:546-555, 2002). In addition to the wildype, seven mutants were simulated via REM with the goal of identifying structural determinants that return the non-canonical H1 loop of the wildtype antibody to the type 1 canonical structure predicted by database methods formulated for conventional antibodies. Two cases with three point mutations yield a stable type 1 H1 structure. Other mutants with fewer mutations were also found to give a significant fraction of such conformations. MMREM was further used to successfully identify additional mutations that stabilize a canonical H1 loop grafted on the llama antibody scaffold. Overall, the mutagenesis results suggest an influence of interloop interactions on the attainment of canonical conformations for this antibody.

In the second implementation of our proposed scheme, wildtype and mutant conformers of the Alzheimer's Aβ42 peptide are studied in order identify "soluble" - stable mutants that are likely to have reduced proclivity towards aggregation. The results are compared to those obtained experimentally for this system via a recently developed mutant screening method (Fisher et al., Protein Sci. 15:449-458, 2006).

The use of MMREM and REM for screening mutants and assessing structural stability may be useful for elucidating key interactions that lead to the appearance of particular structural features in proteins. The latter can ease the identification of sequences that may promote disease prone conditions and facilitate a rational protein design.

425d Simulating Mechanisms for Control of Vesicle Fusion

Peter M. Kasson, Departments of Chemistry and Structural Biology, Stanford University, 318 Campus Drive MC 5447, Stanford, CA 94305-5447 and Vijay S. Pande, Department of Chemistry, Stanford University, 318 Campus Drive MC 5447, Stanford, CA 94305-5447.

Lipid vesicle fusion is an important therapeutic target for control of neurotransmission, viral infection, and endocrine secretion. Synthetic fusion systems also provide an attractive means of drug delivery. One challenge in designing systems for the control of membrane fusion is understanding how changes to the physical properties of lipid membranes affect fusion behavior. Biological mechanisms for controlling fusion behavior are thought to include changes to bilayer curvature, control of lipid composition, and induction of local lipid disorder by fusion proteins. We have used ensemble molecular dynamics simulation to develop physical models for each of these mechanisms and its effects on the fusion process. By analyzing thousands of such simulations, we can derive a more general model for how lipid composition and curvature control fusion kinetics. Here we report simulations where both the composition and the curvature of lipid membranes were varied, with curvature regimes ranging from small 15-nm vesicles to larger vesicles to planar bilayers. We also consider the effects of curvature fluctuations in bilayers and changes to bilayer structure induced by fusion peptides. We find a pronounced effect of curvature on fusion rates in our simulations; in addition, the effects of lipid composition remain relatively independent of curvature over the systems simulated. We then apply these models to interpret the effects of fusion proteins on membrane curvature.

425e Protein Structure Prediction and Refinement Using Folding Mechanism-Informed Replica Exchange Methods

M. Scott Shell, Chemical Engineering, University of California, Mail Code 5080, Santa Barbara, CA 93106 and S. Banu Ozkan, Center for Biological Physics, Arizona State University, PO Box 871504, Tempe, AZ 85287.

Efficient conformational sampling is the foremost bottleneck to the prediction of native structures in all-atom, physics-based simulations of proteins and peptides. The use of replica exchange molecular dynamics (REMD) methods improves sampling efficiency; however, it remains challenging to converge simulations of polypeptides of even 15 residues long in a day's worth of real time using commodity cluster computing, a time comparable to that taken by modern bioinformatics-based structure prediction web servers. Yet it is critical that physics-based methods succeed, as bioinformatics methods fail for many emerging synthetic protein and peptide technologies.

Here, we discuss a strategy for accelerating physics-based REMD simulations of proteins by taking note of the major driving force for folding: the hydrophobic interaction. We show that by detecting and encouraging the formation of strong contacts between hydrophobic residues along the chain (through restraints added to the molecular potential energy function), convergence of REMD simulations is accelerated. We demonstrate this approach in three different case scenarios spanning small peptides to complete protein structures: (1) on a library of small peptides that have stable secondary structures, (2) at the level where secondary structure pieces are assembled into a complete protein, and (3) during refinement and selection of full protein structures predicted by other algorithms, such as bioinformatics web servers. In each of these cases, hydrophobic restraints are selected using a combination of short coarse-grained simulations that generate a large ensemble of candidate protein structures with compact hydrophobic cores, and by examining the free energies of residue-residue contact formation computed on the fly during the REMD simulations. Periodic conformational clustering and re-seeding of the REMD simulations are also used to accelerate convergence. Importantly, any biasing effects of the added hydrophobic restraints (beyond accelerating convergence) can ultimately be removed by either scaling their strength across the replica cascade, or by running a restraint-free final "consensus" REMD run using all of the predicted structures in competition with each other. We demonstrate the success of the approach for a number of model test peptides and small proteins, and for several proteins disseminated during the CASP8 competition.

425f Structural Fluctuations In the Nucleosome Core Particle

Andrew Spakowitz and Bariz Sudhanshu. Chemical Engineering, Stanford University, Stauffer III - Room 113, 381 North South Mall, Stanford, CA 94305

All of the processes necessary for the survival of a eukaryotic cell hinge on the cell's ability to store and read the genetic information encoded in its DNA. The daunting task of packaging the meter-long genome into a micron-sized nucleus is complicated by the necessity of maintaining the accessibility of the DNA to the cell's enzymatic machinery. The fundamental unit of packaged DNA, the nucleosome core particle, contains 146 base pairs of DNA wrapped 1.7 times around a cationic protein complex called the histone octamer. A string of nucleosomes is organized into higher-order structures at several hierarchical levels to form chromatin, a remarkable complex that is compact yet maintains accessibility for gene expression. The interactions between DNA and histones constitute the initial barrier to protein accessibility to chromosomal DNA, thus a quantitative picture of DNA-histone interactions is critical to understand the biological function of chromatin.

We develop a statistical-mechanics model of a nucleosome as a wormlike chain bound to a spool, incorporating fluctuations in the number of bases bound, the spool orientation, and the conformations of the unbound polymer segments. This model is directly compared to single-molecule experiments conducted in Carlos Bustamante's lab; we find good agreement between our theory and the experimental data. We then proceed to consider the role of structural fluctuations in the accessibility of the packaged DNA to regulatory proteins. Nucleosome translocation and transient unwinding are considered as molecular mechanisms for protein accessibility. We discuss the impact of structural fluctuations within the nucleosome core particle on gene regulation.

425g Molecular Dynamics Simulations of the Association of Model Transmembrane Domains of Epidermal Growth Factor Receptors

Anupam Prakash and Manolis Doxastakis. Department of Chemical and Biomolecular Engineering, University of Houston, S222 Engineering Bldg. 1, Houston, TX 77204

Epidermal growth factor receptors have a major impact in cellular signal transduction. Dimerization/ oligomerization of ErbB's has been attributed to ligand binding to the extracellular protein domain. Recent experimental studies suggest that protein transmembrane domains promote dimerization. Interactions between amino acids similar to proposed motifs in the well-studied dimer of Glycophorin-A, could favor the close packing of helix interfaces. We performed detailed atomistic simulations of individual ErbB transmembrane domains in a lipid bilayer and compared to the Glycophorin-A helical transmembrane domain. Using the structure from the single atomistic domains, we parameterized efficient coarse grain models to study the dimerization process. Molecular dynamics simulations, longer than a microsecond, suggest that all transmembrane domains of Glycophorin, ErbB-1, ErbB-2, ErbB-3 and ErbB-4 associate in well defined conformations directed by specific interactions. Our simulations support experimental evidence that transmembrane domains contribute significantly to the dimerization process.

A SCALABLE 30 ml BIOREACTOR FOR BIOPHARMACEUTICAL SCREENING

Bulnes D, G., Aguilar, J.M., Carrillo-Cocoom L.M., Murugappan G., Moreno-Cuevas, J. and Alvarez M.M.

Centro de Biotecnologia. Instituto Tecnológico y de Estudios Superiores de Monterrey. Ave. Eugenio Garza Sada 2501 sur. Monterrey, N.L. México. C.P. 64849.

Abstract.

Glycosylated Biopharmaceutcals are usually produced by suspended mammalian cells cultured in stirred tank bioreactors. However, screening protocols at the lab scale (i.e. to find optimum culture media or initial cell density) mainly rely on Petri dish, bottle or Erlenmeyer flask culture. There is an evident need for small scale fully scalable mammalian cell culture mini-bioreactors for screening processes. Here, we study the mixing performance, mass transfer capabilities, and culture adequacy of 30 ml eccentric stirred tanks. A detailed mixing characterization of the proposed bioreactor is presented. LIF experiments and CFD computations are used to identify operational conditions required for adequate mixing (globally chaotic conditions). Adult stem cells are cultured in this novel bioreactor system. Our results demonstrate good mixing performance, culture convenience, and potential scalability of the proposed screening bioreactor.

To be submitted as extended abstract
Proceedings AIChE Meeting, 2008

Key words: Mammalian cells, mixing, screening, biopharmaceutical, scalable, microbioreactor

INTRODUCTION

Nowadays, most biopharmaceuticals are recombinant glycoproteins secreted by mammalian cells under culture in large scale stirred tank bioreactors (Nienow, 2006), typically in the range from 50 to 20,000 L. Preceding that stage, the development path for a biopharmaceutical molecule implies extensive rounds of screening to select fast growth high producing clones adapted to suspension conditions. Commonly, the first portion of that selection process is done in very small volume systems, in vessel geometries that share nothing in common with industrial stirred tanks. Petri dishes and culture bottles of 20 to 50 ml continue to be the preferred containers to grow and select high producer cells at the lab scale. Afterwards, selected clones are adapted to suspension conditions in Erlenmeyer flasks ranging in volume from 200 to 1000 ml. However, the hydrodynamic conditions in Erlenmeyer flasks are very different with respect to the next scaling stage, typically a stirred tank bioreactor. Recently, new commercially available small scale bioreactors had appeared as options to the use of Erlenmeyer flasks. For example, DASGIP®, a medium size firm specialized in the manufacture of bioreactor systems for culture applications, offers platforms of fully instrumented bioreactors capable of controlling oxygen concentrations, pH, inlet flow rates, etc. to operate in the range of 300 to 1000 ml. Therefore, it is until scales of 300 and up are reached that stirred tank systems can be used to continue development. In consequence, there is an uncovered niche for screening systems, stirred tank-like, that can partially substitute the activities now done in culture bottles and Petri dishes. In this contribution, we examine the performance of a 30 ml stirred tank micro-reactor system potentially useful for mammalian cell culture operations.

Mixing and adequate oxygen mass transfer is not easily achievable in small scales. In larger systems, agitation with marine turbines causes turbulence as the main mixing mechanism. Under turbulent conditions, oxygen dosed through fine spargers is correctly distributed, and typical bubble residence times allow for adequate oxygen mass transfer. As scale reduces, adequate gas dispersion will demand higher agitation intensities. On the other hand, smaller scales will cause shorter residence time for air bubbles. More of them will achieve the surface of the tank, trapping cells at their surface. Both, high shear rates and bubble damage have been studied as possible factors of lethal and sub-lethal damage to mammalian cells, recognized as more fragile to forces than bacteria or yeast (Keane *et al.*, 2003; Amanullah *et al.*, 2004). This framework almost precludes the use of typical stirred tank or impeller designs for very small scales. Here we propose a geometry design, capable of adequately mix in the laminar and transitional regime, as we demonstrate by Laser Induced Fluorescence and CFD analysis. We culture human adult stem cells to demonstrate the potential of the proposed system for screening applications.

MATERIALS AND METHODS

Stirred tank system. The stirred tank geometry analyzed here is depicted in figure 1. A 30 ml glass round bottom vessel was agitated using a disc impeller eccentrically located. The impeller (0.6 cm radius) was positioned at an eccentricity of E=0.21, where eccentricity (E) is defined as the distance between the shaft and the vertical centerline of the tank/tank radius. Agitation was set at 75 RPM. Temperature was maintained at 37°C by wrapping the tank with heated plastic tubing.

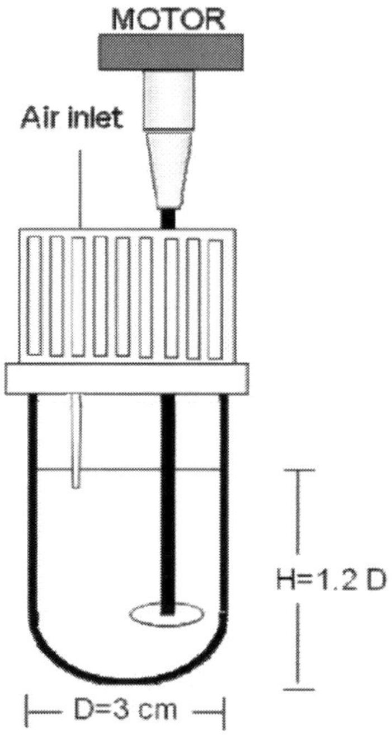

Figure 1. Schematic representation of the bioreactor system.

Flow visualization experiments. All visualization experiments reported here were done using glycerin (viscosity equal to 1200 cP/mPa s at 22°C) as fluid. Mixing 3D patterns within the bioreactor where observed through fluorescein injections (a fluorescent tracer) in a dark room where UV light was the only illumination source. To better resolve the evolution of mixing structure in time, laser induce fluorescence experiments (LIF) were conducted. In an otherwise dark room, a plane of laser light was projected towards the experimental system, to visually expose the mixing structure at a vertical 2D plane (slightly behind the rotation axes) of the bioreactor originated by the injection of a pulse of rhodamine dye. Photographic documentation of the visualization experiments was done using a professional digital camera model Rebel XSi from Canon®. In order to correct for optical aberration, the bioreactor was immersed in a transparent cubic chamber filled with glycerin to also match refractive index.

CFD simulations. Computational Fluid Dynamics (CFD) experiments were done using a xw8600 HP Workstation computer with 2 Intel Xeon Quad Core processors and a 512 Mb. Nvidia Quadro FX3700 graphic card. . Geometries were constructed using the software SolidEdge ® from UGS. Meshes were built using software from ACUSIM®. Velocity field was calculated by solution of the Navier Stokes Equations at 200,000 nodes using ACUSOLVE from ACUSIM® Sofware Inc., Mountview Ca. ACUSIM® is based on a special stabilized Galerkin/Least Squares (GLS) finite element technology formulation of the full transient 3D non-linear equations of fluid dynamics. The fluid was modeled primarily as a newtonian fluid with the nominal properties of water.

Stem cell culture. 25 ml of cell suspension at a concentration of 1.75X10^4 cells/ml was seeded in the tank, such that the ratio of vessel width to liquid height was 1:1.2. DMEM F-12 supplemented culture media [10% FBS (lot), 1% penicillin-streptamycin solution, 10 mM HEPES,] was changed every 24 hours according to a feed batch protocol, where agitation was suspended for 5 minutes to allow cells to settle in the bottom of the tank and 80% of the culture media was removed from the upper portion of the tank using a sterile syringe and replaced by fresh media warmed to 37°C. 1 ml samples of cell suspension were removed from the upper third portion of the tank for counting every 36 hours. Cells were stained with Trypan blue at a concentration of 10% and counted using a Neubauer camera. Here, we report results from daily replacement with pH-controlled culture media. RNA profiles were analyzed by PCR following addition of primers for CD133 and CD34 regions to verify if cells maintained un-differentiation during culture.

RESULTS AND DISCUSSION

The mixing visualization experiments reported here were conducted in glycerin. Glycerin retards the effect of diffusion, allowing for a cleaner observation of the structural effect of convective mixing at very low speed. Figure 2(a) and(b) shows the 3D mixing patterns arising from operation at laminar speed (Re~30, agitation=300 RPM in glycerin). In terms of Reynolds number values, and consequently mixing structure, these conditions will be approximately equivalent to those observed in water at 0.25 RPM.

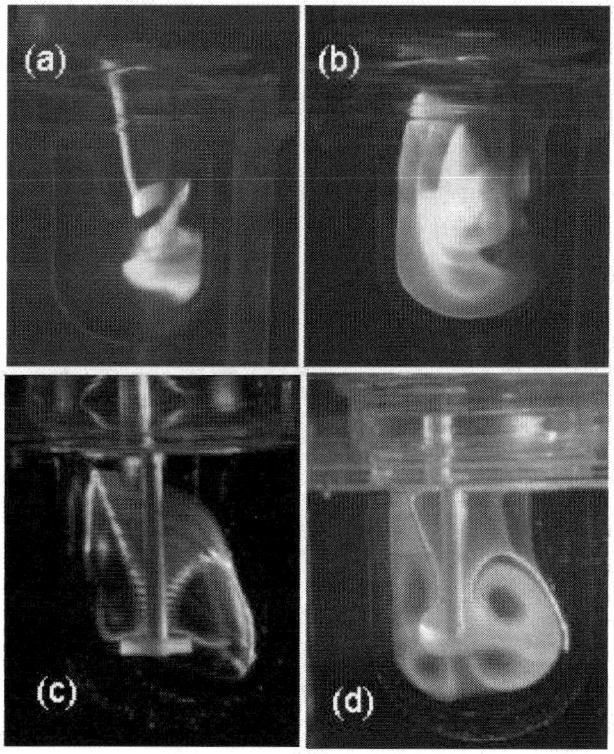

Figure 2. Mixing visualization experiments: (a) laminar flow patterns are revealed through a fluorescent tracer injection under UV light in a eccentric system (E=0.25) 10 seconds after injection, and (b) one

minute after injection; (c) LIF experiments reveal the mixing structure of eccentric stirred tank systems agitated by horizontal flat disc impellers and (d) 45 angle disc impellers at low Re number.

The mixing structure observed in the proposed bioreactor is highly complex. But even at this extremely low agitation values, liquid mixing is achieved in reasonable times (less than 10 minutes). Figure 2(c) and (d) show 2D planes of the mixing structure observed at such low Re values. Small structural features of the flow reveal the coexistence of chaos and order. Chaotic regions of adequate mixing coexist with areas of low mixing intensity and regular behavior, as previously reported (Lamberto *et al.*, 1999; Álvarez, 2002; Álvarez et al., 2002; Cervantes *et al.*, 2006).

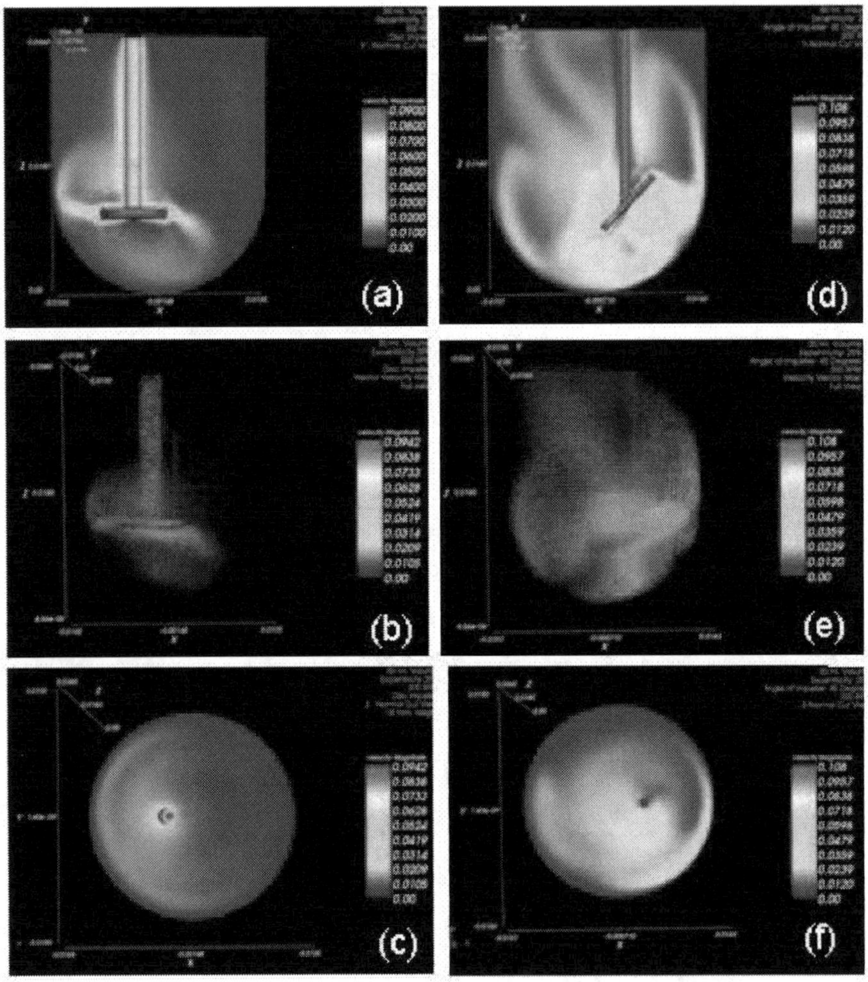

Figure 3. Velocity fields (obtained by CFD calculations) within eccentric stirred tanks (a) agitated by a horizontal disc impeller(vertical velocity contour at the axes level); (b) (vertical velocity field), and (c) (horizontal velocity contour at the impeller level); and (d) agitated by a 45° disc impeller (vertical velocity contour at the axes level); (b) (vertical velocity field), and (c) (horizontal velocity contour at the impeller level).

Figure 3 shows CFD results from simulations done with two different disc impeller configurations (varying in angle) rotating at 150 RPM in a water like liquid (matching average culture media properties) at an eccentric position (E=0.25) in a round bottom flask. A flat, horizontal disc, impeller geometry previously analyzed by Alvarez *et al.* (2002) and Cervantes *et al.* (2006), is compared versus a disc impeller, were the angle of the impeller with respect to its axes is different to zero (for the case reported here, 45 degrees). The effect of the modification in the impeller angle is significant. In general, circulation patterns were dramatically modified, as observed in the velocity field presented in Figure 3. Particularly, axial circulation was greatly improved.

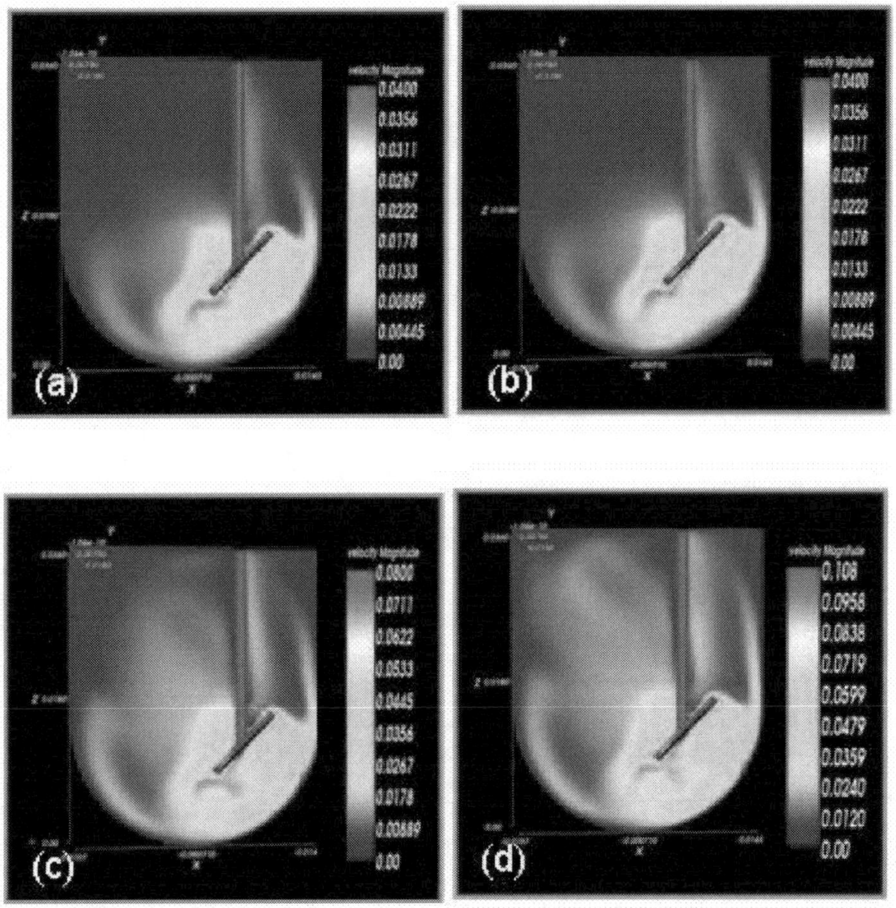

Figure 4. Velocity fields (obtained by CFD calculations) within eccentric stirred tanks agitated by a 45° disc impeller agitated at (a) 75 RPM; (b) 100 RPM; (c) 150 RPM; and (d) 200 RPM.

Figure 4 presents CFD results for 45 degree angle configurations at slightly higher eccentricity values (E=0.33). Different velocity field contours (from 75 to 200 RPM) are compared. Although the flow structure is similar among them, there are significant differences in the axial outreach of the flow field. This type of analysis, complementing experimental information, is useful to take decision like where to inject reagents for fast dispersion, or how low to locate the impeller to achieve adequate suspension. In our experiments with stem cells and CHO cells, agitation was set at 75 RPM and 200 RPM (in water like media) to allow for adequate mixing while minimizing mechanical stress. Although adequate liquid-liquid mixing is achievable at much lower speeds, visualization experiments with neutrally buoyant 20 microns polystyrene particles (results not presented here) demonstrate that higher speeds are needed to maintain cells in suspension. The "just suspension velocity" for polystyrene particles in water was estimated in 200 RPM.

Cells cultured in the eccentric stirred tank bioreactor at an initial concentration of $1.75X10^4$ cells/ml exhibited the growth curve shown in Figure 5(a), with characteristic latency, exponential and stationary phases. Cell doubling time, calculated using a linear regression analysis on the exponential growth phase, was 31.46 hours, 19.18% increase from doubling time in stationary culture in Petri dishes (standard culture protocol for this type of cells). Cell viability in time is presented in Figure 5(b).Cell viability is maintained high (in the range of 80% to 100%). Remarkably, in exponential phase, cell viability is sustained in the range of 95 to 100% for several hours, evidencing that the flow field conditions are acceptable to favor survival and growth of extremely shear sensitive cells. Cells remained non-differentiated (CD 34^+ CD 133 $^+$) during culture, as demonstrated by expression analysis using PCR protocols.

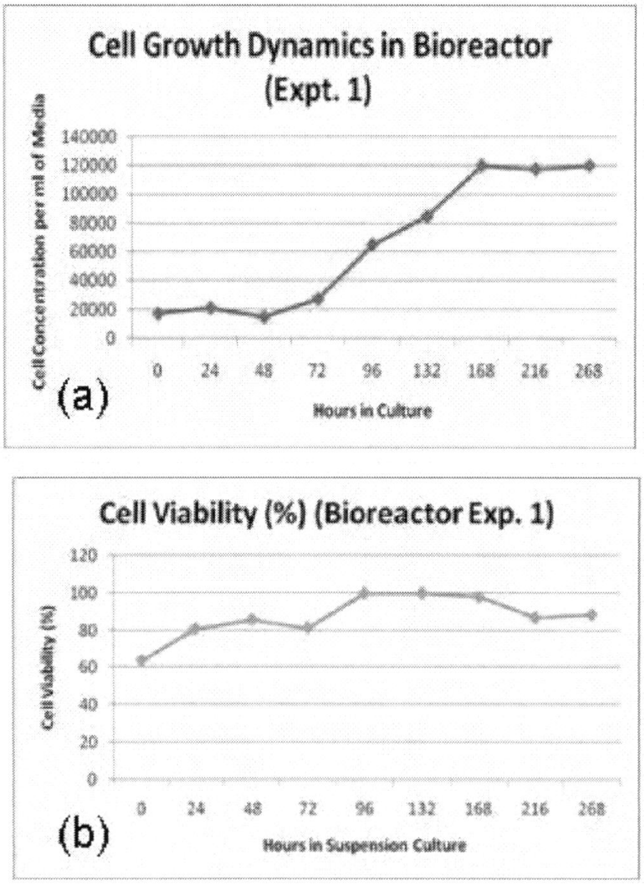

Figure 5. Non-embryonic stem human hematopoietic stem cells growth in an eccentric stirred tank system: (a) cell growth profile in time; (b) culture cell viability in time.

CONCLUSIONS

A 30 ml round bottom vessel stirred by a disc impeller eccentrically located at relative low speed has been proposed for biopharmaceutical screening applications. A mixing analysis was conducted using fluorescent tracer experiments under UV light, LIF visualizations and CFD simulations. Based on this analysis, operational conditions were decided for the culture of human non-embryonic stem cells. The stem cell culture was monitored over a period of 10 days, showing adequate performance in terms of growth rates, cell densities and viability.

REFERENCES

1. Álvarez MM. (2000). Using spatiotemporal asymmetry to enhance mixing in chaotic flows: from maps to stirred tanks. Ph. D. Thesis, Chemical and Biochemical Engineering. Rutgers University, New Brunswick, N.J.
2. Álvarez MM., Arratia PE., Muzzio FJ. (2002). Laminar Mixing in eccentric stirred tank systems. *Canadian Journal of Chemical Engineering, 80, 546.*
3. Amanullah A, Buckland BC, and Nienow AW. Mixing in the Fermentation and Cell Cuture Industries in Handbook of Industrial Mixing, Science and Practice, edited by Paul EL, Atiemo-Obeng, V and Kresta SM. Willey Interscience. United States of America, 2004, pag1071-1170.
4. Cervantes MIS, Lacombe J, Muzzio FJ, and Alvarez MM. 2006. Novel bioreactor design for the culture of suspended mammalian cells. Part 1: Mixing characterization. *Chemical Engineering Science*, 61(24): 8075-8084.
5. Keane JT, Ryan D, and Gray PP. 2003. Effect of shear stress on expression of a recombinant protein by Chinese hamster ovary cells. *Biotechnology And Bioengineering* , 81(2): 211-220.
6. Lamberto DJ, Álvarez, MM, Muzzio FJ. (1999). Experimental and computacional investigation of the laminar flor structure in a stirred tank. *Chemical Engineering Science, 54, 919.*
7. Nienow AW. 2006. Reactor engineering in large scale animal cell culture. *Cytotechnology,* 50(1-3): 9-33.

432b A 24-Well Plate Microbioreactor Array for High-Throughput Cell Culture Process Development

Yuan Wen[1], Ning Liu[2], Xudong Zhang[3], and Shang-Tian Yang[3]. (1) Chemical and Biomolecular Engineering, The Ohio State University, 125A Koffolt Labs, 140 West 19th Ave., Columbus, OH 43210, (2) Ohio State University, 125A Koffolt Lab, 140 West 19th Ave., Columbus, OH 43210, (3) Chemical and Biomolecular Engineering, Ohio State University, 125 Koffolt Laboratories, 140 West 19th Ave., Columbus, OH 43210

We have designed and fabricated a novel plate-based 24-well microbioraector array for high-throughput cell culture process development. Each well of the plate has a set of observation windows at the bottom, where optical detection methods can be used for monitoring cell number, pH and dissolved oxygen with corresponding sensors in a commercially available plate reader. Good mixing was achieved by a center stir pillar in each well under agitated conditions. This plate also features two membranes as bottom and lid which are highly permeable to non-polar gases, such as oxygen and carbon dioxide, critical for high density cell culture. With these membranes, in addition, low evaporation of water from each well can also be achieved. Suspension culture can be grown to a viable cell density as high as over 2 million cells/mL with the microplate. Using experimental design techniques, we also developed and optimized a serum-free medium for CHO cell culture in the microplate. This platform is a useful high-throughput scale-down model for cell culture process development.

432c Modeling the Platform: Leveraging Cross-Project Data to Understand Upstream Bioprocess Scale-up

Brian J. Stamper, Eli Lilly & Company, Indianapolis, IN 46285

A platform approach for mAb and mAb-like therapeutic protein cell culture process development has been established to speed bioprocess development times and reduce complexity associated with technology transfer to manufacturing. To enable this strategy, process understanding must be achieved across multiple projects within the platform. For upstream processes, this entails understanding the first principles such as mass transfer and mixing that dictate the bioreactor performance at lab and manufacturing settings. Furthermore, empirical modeling can be performed to identify the important scale-up factors that affect the upstream bioprocess outcome.

Empirical modeling was completed using several platform-based projects that had been scaled up to the cGMP CT pilot plant. Factors such as dissolved gases, seed train performance, and process controllability were examined while all process setpoints were assumed constant between the scales (i.e. temperature, pH). The resulting models indicated that matching seed train performance and dissolved carbon dioxide levels between the bench and pilot scales would result in equivalent productivities.

These models have been implemented and are being used to focus attention on the most important scale up parameters during technology transfer to increase the probability of success at scale. As a result, the productivities demonstrated by the CT pilot plant have matched that of the bench scale to date.

432d Effect of Manufacturability Considerations In Facilitating Efficient Development and Transfer of An Escherichia Coli Process Producing Fc-Fusion Proteins

Henry Lin, Heideh Barkhordarian, Irene Liu, and Jonathan Lull. Amgen, One Amgen Center Drive, MS 18S-1-A, Thousand Oaks, CA 91320

A commercial fermentation process that produces Fc-fusion proteins as inclusion bodies in E. coli was successfully developed from a clinical process. Various obstacles were encountered as the process performance was further improved. For example, a substantial improvement in titer led to higher oxygen usage and heat removal requirement, which posed a challenge on manufacturing equipment. Subsequently, changes in the fermentation process led to changes in the harvest product profile creating challenges for downstream development. In the end, such obstacles were successfully resolved. Early consideration of manufacturability coupled with clear understanding of equipment design, process parameters and interactions between unit operations were key for an efficient and rapid development of the fermentation process.

432e Production of Growth Hormone Receptor Antagonist In Nicotiana Tabacum Cells: Process Development, Optimization, Scale-up and Technology Transfer

Seshu Tyagarajan and Venkatesh Srinivasan. BioProcess Development, Phyton Biotech, 279 Princeton Hightstown Road, East Windsor, NJ 08520

Production of recombinant Growth Hormone Receptor Antagonist in Nicotiana Tabacum cells serves as a model for production of O-Glycosylated Arabinogalactan proteins in plant cell culture. Data presented indicates efforts involved in process development as well as process optimization, leading to a 10-100 fold increase in protein titers. Experimental design and strategy involved a good understanding and implementation of the pilot scale and large scale manufacturing conditions, such as mixing, dissolved oxygen and air flow. Process development was also carried out with a view to simplifying the downstream processing requirements. As a result of comprehensive understanding and integrated process development efforts, results obtained show identical titer progression at all three levels, namely, lab scale, pilot scale and manufacturing scale.

432f Mammalian Cell Death Pathways Induced by Different Types of Shear Flow

Timm Tanzeglock, Institute for Chemical and Bioengineering, ETH Zurich, Wolfgang-Pauli-Str. 10, Zurich, 8093, Switzerland, Miroslav Soos, Institute for Chemical and Bioengineering, Department of Chemistry and Applied Biosciences, ETH Zurich, Wolfgang-Pauli-Str. 10, Zurich, 8093, Switzerland, Gregory N. Stephanopoulos, Department of Chemical Engineering, Massachusetts Institute of Technology, 77 Massachusetts Ave., 56-439, Cambridge, MA 02139, and Massimo Morbidelli, Institute for Chemical and Bioengineering, Department of Chemistry and Applied Biosciences, ETH Zurich, Wolfgang-Pauli-Str. 10, HCI F 133, Zurich, 8093, Switzerland.

During cultivation of mammalian cells in a bioreactor, cells are constantly subjected to shear stresses arising through agitation and aeration. In this work we investigated whether the type of shear flow, to which cells are exposed, will influence the cellular response. It will be shown that mammalian cells, indeed, distinguish between discrete types of flow and respond differently. Two flow devices were employed to impose accurate hydrodynamic flow fields: uniform steady simple shear flow and oscillating extensional flow. The cell response was evaluated by measuring the release of DNA in the culture supernatant, as indicator for membrane rupture and necrotic death, and by means of fluorescence activated cell sorting, to distinguish between necrotic and apoptotic cell death. Results show that Chinese Hamster Ovaries and Human Embryonic Kidney cells will enter the apoptotic pathway when subjected to low levels of hydrodynamic stress (around 2.0 Pa) in oscillating, extensional flow. In contrast, necrotic death prevails when the cells are exposed to hydrodynamic stresses around 1.0 Pa in simple shear flow or around 500 Pa in extensional flow. These threshold values at which cells will enter the respective death pathway should be avoided when culturing cells for recombinant protein production to enhance culture longevity and productivity.

432g Modeling Shear-Induced Cho Cell Death In Rotary Lobe Pump

Hari Kamaraju, Kenneth Wetzel, and William J. Kelly. Villanova University, 800 Lancaster Ave., Villanova, PA 19085

Rotary lobe pumps are commonly used in the biotechnology industry for a variety of purposes. Shear damage to animal cells within the rotary lobe pump can adversely affect the product yield or purity during, for example, cell concentration via cross-flow filtration. In this research, CHO cells grown in 20L bioreactors where fed to a rotary lobe pump and both single pass and recycle experiments were conducted at different RPMs and back pressures. The results indicate that the slip flow rate more severely impacts the viability of the CHO cells than the pump RPM. Preliminary results also suggest that a low ratio of the net flow rate to the amount of slip has detrimental effects on the cell viability. Hence a mathematical model is presented which recommends the use of a cell death rate constant, K, which accounts for the RPM, net flow rates, amount of slip, back pressures and more importantly the time of exposure of the cells to the shear environment. Based on this model we can attribute a cell death rate constant to a cell line for a given pump and use this constant to compare pump performance from the cell viability point of view. The model also gives a better insight into pump design and selection as it indicates that a smaller pump with higher efficiency and higher RPM might be better than a larger pump with lower efficiency and lower RPM.

446a Stem Cell Differentiation to Muscle Induced by a Suitably Compliant Matrix

Dennis E. Discher, Chemical and Biomolecular Engineering, University of Pennsylvania, Room 129 Towne Building, 220 South 33rd Street, Philadelphia, PA 19104-6393

Stem cells can in principle develop into a wide range of soft tissue cell types, but the importance of the 'softness' of the tissue microenvironment in differentiation is only now becoming clear. In fact, tissue compliance varies considerably between tissues, and muscle is one soft tissue that at rest appears to have a well-defined matricellular elasticity (10-15 kPa) which is readily achieved with polymeric hydrogels. We use a molecular layer of collagen-I on crosslinked polyacrylamide hydrogels with muscle-like elasticity to induce adult human mesenchymal stem cells (MSCs) cultured in 10% serum to become spindle-shaped myoblasts that express key early markers (Pax7, MyoD). When replated at high density, these cells further fuse into multi-nucleated myotubes expressing late skeletal muscle markers. On hydrogels that are too stiff, the MSCs do not differentiate – unless we add a potent glucocorticoid used in the clinic to treat muscular dystrophy. This single glucocorticoid also tends to downregulate adhesive/ cytoskeletal structures involved in substrate attachment and sensing, and it appears to suppress osteogenic markers, consistent with bone loss in treated patients. This replaces typical soluble induction cocktails of multiple glucocorticoids plus other factors, and in such a case, we find muscle induction occurs regardless of matrix elasticity. Nonetheless, the best results at low dose clearly occur in combination with an optimally compliant matrix.

446b N-Isopropylacrylamide (NIPAM)-Based Thermoresponsive Polyelectrolyte Multilayer Films for Human Mesenchymal Stem Cell Expansion

Tianqing Liao[1], Maroun D. Moussallem[2], Joseph B. Schlenoff[2], and Teng Ma[1]. (1) Department of Chemical and Biomedical Engineering, Florida State University, FAMU-FSU College of Engineering, 2525 Pottsdamer Street Suite A131, Tallahassee, FL 32310, (2) Department of Chemistry and Biochemistry, Center for Materials Research and Technology, Florida State University, Tallahassee, FL 32306

Human mesenchymal stem cells (hMSCs) are colony-forming unit fibroblasts (CFU-F) derived from adult bone marrow and have significant potential for many cell-based tissue-engineering applications. Their therapeutic potential, however, is restricted by the diminishing plasticity as they are expanded in culture. In this study, we investigated using N-isopropylacrylamide (NIPAM)-based thermoresponsive polyelectrolyte multilayer films as a culture substrate to support hMSC expansion without compromising their stem cell characteristics. The thermoresponsive surface is hydrophobic at culture temperature (37 °C) and hydrophilic at lower temperature (<4 °C), which enabling cell detachment without enzymatic treatment. The thermal responsive PEMU films were made via layer by layer adsorption of thermoresponsive polymers copolymerized with oppositely charged polyelectrolytes. The coating has been made to both 2-D and 3-D constructs. Initial study found that NIPAM-based surfaces supported efficient cell detachment but at lower growth rate compared to the control surfaces. To improve cell adhesion and growth, NIPAM surfaces were further modified by positively charged Poly (allylamine hydrochloride),PAH and negatively charged Poly(styrene sulfonic acid), PSS and compared to FBS coated surfaces. Surface charges were shown to alter the ECM structure and subsequently regulate hMSC responses including integrin expression, adhesion, proliferation, detachment, and colony forming ability. The positively charged thermal responsive surfaces improved cell adhesion and growth in a range comparable to control surfaces while maintaining significantly higher CFU-F forming ability. Immunostaining and Western blot results indicate that the improved cell adhesion and growth on the positive charge surfaces are resulted from the elevated adhesion of fibronectin on the positively charge surfaces. These results demonstrate that layer-by-layer approach is efficient to form PNIPAM-based thermal responsive surfaces for hMSC growth and removal without enzymatic treatment. The results also show that surface charge regulates ECM adhesion, which in turn influences not only cell adhesion but also CFU-forming ability and their multi-lineage differentiation potential. We have also investigated the effects of NIPAM-based coating of 3-D PET scaffold surfaces on cell growth, thermal removal, and multilineage differentiation potential. We will compare the effects of surface properties on hMSC expansion in 3-D scaffold and determine the condition for efficient cell removal from 3-D scaffold while preserving hMSC property. Overall, NIPAM-based surfaces provide many advantages over the conventional culture substrate for hMSC expansion.

446c Directed Differentiation of Human Embryonic Stem Cells Via 3-D Biofunctionalized Substrates

Jocie F. Cherry[1], Aaron Carlson[1], Rebecca N. Moore[1], Charles Florek[1], Christopher Ricupero[2], Vani Mathur[1], Melitta Schachner[2], Martin L. Grumet[2], Joachim Kohn[3], and Prabhas V. Moghe[4]. (1) Biomedical Engineering, Rutgers University, 599 Taylor Road, Piscataway, NJ 08854, (2) Cell Biology & Neuroscience, Rutgers University, 599 Taylor Road, Piscataway, NJ 08854, (3) Chemistry & Chemical Biology, Rutgers University, 599 Taylor Road, Piscataway, NJ 08854, (4) Chemical and Biochemical Engineering; Biomedical Engineering, Rutgers University, 599 Taylor Road, Piscataway, NJ 08854

Human embryonic stem (hES) cells represent a promising source for cell transplantation because of their high degree of self-renewal and their unique ability to give rise to most somatic cell lineages. Despite the promise of obtaining various lineages, the ultimate goal of expanding and differentiating hESCs in three-dimensional environments remains challenging. We report on the idea that three-dimensional substrates can be designed for specific geometry and molecular matrix presentation to maintain the stemness of hESCs or induce the directed differentiation of isolated hES cells into strategic lineages (e.g. neural) without the use of aggregates, embryoid bodies or neurospheres.

To this end, our study offers a three-pronged strategy to direct the differentiation of hES cells into neural lineage on 2-D substrates and in 3-D scaffold cultures. The first relies on the display of adhesion ligands that can control directed differentiation vs. pluripotency. As a surface molecular engineering approach, we have incorporated 2-D and 3-D substrates with fragments of cadherin, fibronectin or L1 molecules that can be used to more precisely control pluripotency versus lineage commitment fates of hESCs in situ. Second, we have designed 3-D substrates for hES cells fabricated from a subset of combinatorial library of biodegradable polymers, tyrosine-derived polycarbonates. Using electrospinning, various fibrous scaffolds with varying microgeometries and surface composition were compared. Such scaffolds can yield controlled adhesion and outgrowth of individual hESCs, which can be used as another basis to regulate cell-cell vs. cell-matrix interactions and thereby induce expansion while limiting spontaneous differentiation. The third novel approach of the paper will demonstrate the possibility of integrating nanoscale adhesion ligands within the polymeric scaffolds, so as to switch cell fates to specific neuronal differentiation pathways in situ within three-dimensional microenvironments.

446d Effect of Swelling Ratio of Injectable Hydrogel Composites on Chondrogenic Differentiation of Encapsulated Rabbit Marrow Mesenchymal Stem Cells *In Vitro*

Xuan Guo, Department of Chemical and Biomolecular Engineering, Rice University, P.O.Box 1892, MS 362, Houston, TX 77251-1892, Hansoo Park, Bioengineering, Rice University, P.O. Box 1892, MS 142, Houston, TX 77251-1892, Johnna S. Temenoff, Department of Biomedical Engineering, Georgia Institute of Technology, Atlanta, GA, Yasuhiko Tabata, Department of Biomaterials, Kyoto University, Kyoto, Japan, Arnold I. Caplan, Department of Biology, Case Western Reserve University, Cleveland, OH, and Antonios G. Mikos, Department of Bioengineering, Rice University, P.O.Box 1892, MS 142, Houston, TX 77251.

During the past decade, cartilage tissue engineering that involving the use of cells, bioactive molecules and a three dimensional biodegradable scaffold material has received attention as a promising approach to aid regeneration of cartilage defects. In particular, injectable hydrogels hold great promise as a carrier for cells and bioactive molecules since constructs can be implanted minimally invasively and their high water content facilitates exchange of nutrients and waste.

Our laboratory has recently developed an injectable hydrogel composite consisting of oligo(poly (ethylene glycol) fumarate) (OPF) and gelatin microparticles (MPs) as a novel carrier system for cells and growth factors. In this study, hydrogel composites with different swelling ratios were prepared using OPF macromers with poly(ethylene glycol) repeating units of molecular weight 35,000 (OPF 35K), 10,000 (OPF 10K), 3,300 (OPF 3K) and 1,000 (OPF 1K). 250,000 rabbit marrow mesenchymal stem cells (MSCs) and MPs loaded with 25 ng transforming growth factor-β1 (TGF-β1) were encapsulated in each hydrogel composite to examine the effect of the swelling ratio of hydrogel composites on the chondrogenic differentiation of the cells. Hydrogel composites encapsulating rabbit marrow MSCs and blank MPs served as controls. Quantitative reverse transcriptase-polymerase chain reaction (RT-PCR) results showed that rabbit marrow MSCs encapsulated in higher swelling ratio hydrogel composites (OPF 35K and OPF 10K) had a significant increase in the expression of chondrocyte specific genes such as collagen type II and aggrecan at day 28 compared with those encapsulated in low swelling ratio hydrogel composites (OPF3K and OPF 1K), both in the presence and absence of TGF-β1. Specifically, OPF 35K samples with TGF-β1-loaded MPs exhibited a 159 ± 95 fold increase of type II collagen gene expression and a 24 ± 12 fold increase of aggrecan gene expression after 28 days of *in vitro* culture as compared to controls at day 0. The results indicate that chondrogenic differentiation of rabbit marrow MSCs encapsulated within OPF hydrogel composites could be affected by the swelling ratio of the hydrogels, thus suggesting the potential of OPF composite hydrogels as part of a novel strategy for controlling the differentiation of stem cells.

446e Regulating Migration of Bone Marrow Stromal Cells to Injured Bronchial Epithelium

Kim C. OConnor[1], Bonnie Barrilleaux[1], Ben Fischer-Valuck[1], Darwin Prockop[2], and Donald Phinney[2]. (1) Chemical and Biomolecular Engineering Dept., Tulane University, New Orleans, LA 70118, (2) Gene Therapy Center, Tulane University Medical School, New Orleans, LA 70112

Human bone marrow stromal cells (hMSCs) are a promising source of multipotent progenitors for a broad range of stem cell therapies, including the treatment of lung fibrosis and loss of pulmonary function in cancer patients given the chemotherapy bleomycin. An early step in treatment is hMSCs homing via migration to the site of injury. This study investigates factors regulating migration of hMSCs to bleomycin-exposed bronchial epithelium. The hMSCs employed in this study are multipotent in exhibiting adipogenesis, chondrogenesis and osteogenesis, and the bronchial epithelium is regenerated in vitro at an air-liquid interface. Bleomycin exposure decreases cell density, viability, intracellular adhesion and membrane integrity of the epithelium. Pituitary extract in culture medium strongly suppresses hMSC migration. Conditioning of the medium by the bronchial epithelium results in a 10-fold increase in the number of migrating hMSCs. After the epithelium is exposed to bleomycin, hMSC migration diminishes to the level of the negative control. The inhibitory effect of pituitary extract is observed in different donor preparations of hMSCs that range in colony forming efficiency from 22% to 66%. Migration inhibition factor (MIF), a component of pituitary extract, suppresses migration in a dose-dependent manner. Epithelial culture medium employed in this study contains 85.4 +/- 3.5 ng/ml MIF. At this concentration, rMIF inhibits hMSC migration by 45% to 55%. This is the first report of MIF regulating hMSC migration. MIF, which is released into the bloodstream from the pituitary, may impede hMSC treated of bleomycin-exposed lung. Systemic administration of a MIF inhibitor may improve the efficacy of this stem cell therapy. This work is funded with a grant from the National Science Foundation.

446f Synthetic Stem Cell Niche for In Vitro Spermatogonial Stem Cell Culture

Cathy Chu[1], John Schmidt[1], Kay Carnes[2], Zhen Zhang[2], Marie-Claude Hofmann[2], and Hyunjoon Kong[1]. (1) Chemical and Biomolecular Engineering, University of Illinois at Urbana-Champaign, IGB 2200, 1206 W. Gregory Dr., Urbana, IL 61801, (2) Veterinary Biosciences, University of Illinois at Urbana-Champaign, IGB 2200, 1206 W. Gregory Dr., Urbana, IL 61801

Stem cells are increasingly studied for potential use in various therapies including tissue regeneration. Specifically, spermatogonial stem cells (SSCs) are attracting attention due to their potential pluripotency arising from their ability to dedifferentiate into embryonic stem cell-like cells [1]. For their successful therapeutic use, these cells first must be expanded in vitro using an appropriate culture system. We have previously shown that the mechanical stiffness and cell adhesion cues of hydrogel substrates can be engineered to control cell proliferation [2]. Therefore, we hypothesized that a hydrogel with proper biochemical and biomechanical properties may mimic the composition and structure of the native basement membrane onto which SSCs reside, thus allowing us to control SSC proliferation. This hypothesis was examined with a hydrogel containing chemically linked synthetic oligopeptides including the Arg-Gly-Asp sequence (RGD peptides). The RGD peptide density (N_RGD) and the mechanical stiffness of the hydrogel were varied to examine their effects on cell proliferation in both 2D and 3D cultures. Interestingly, cell proliferation in 2D cultures was greatly dependent on N_RGD but minimally influenced by mechanical stiffness. The N_RGD also modulated the number and size of SSC colonies formed in 3D cultures. Overall, the results of this study elucidate an important factor regulating SSC proliferation and also present a bioactive hydrogel that can be used as a 3D synthetic basement membrane. In addition, the results of this study will be broadly useful in controlling the proliferation of various stem cells and expedite the use of stem cells in clinical treatments. References: [1] Kanatsu-Shinohara, M., Inoue, K., Lee, J., Yoshimoto, M., Ogonuki, N., Miki, H., Baba, S., Kato, T., Kazuki, Y., Toyokuni, S., Toyoshima, M., Niwa, O., Oshimura, M., Heike, T., Nakahata, T., Ishino, F., Ogura, A., and Shinohara, T. Generation of pluripotent stem cells from neonatal mouse testis. Cell 119, 1001, 2004. [2] Kong, H.J., Boontheekul, T., and Mooney, D. J. Quantifying the relation between adhesion ligand-receptor bond formation and cell phenotype. Proc Natl Acad Sci USA 103, 18534, 2006.

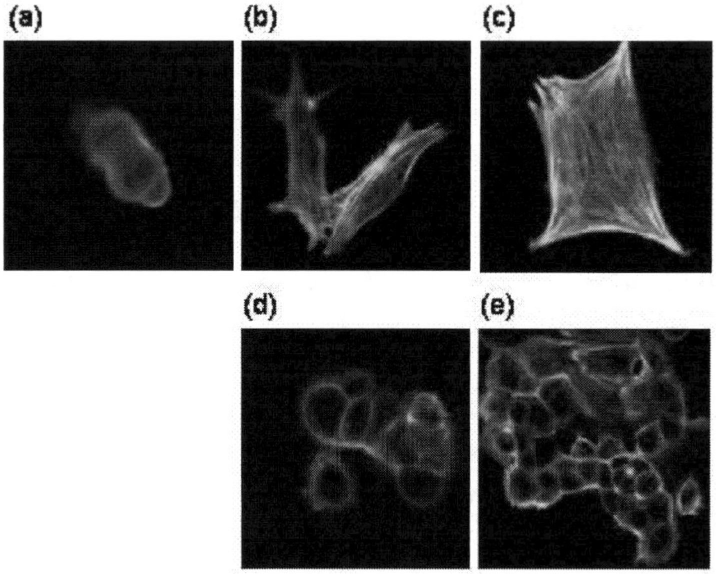

The overall density of RGD peptides immobilized on hydrogel surfaces (N_RGD) influenced cellular adhesion and colony formation.
Intracellular actin stress fiber formation was visibly enhanced as N_RGD was increased. (a), (b), and (c) represent C18-4 cells attached to gels presenting N_RGD of 1, 4.6, and 9×10^{15} RGD/mm^2, respectively. The number of cells per colony also increased as N_RGD was increased from 4.6×10^{15}/mm^2 (d) to 9×10^{15}/mm^2 (e). Green fluorescence in microphotographs represents actin fibers stained with Oregon Green 514 phalloidin. Images were captured after 5 days of culture.

446g Embryonic Stem Cell-Derived Cardiomyocytes Enhance Action Potential Propagation through Cardiac Scaffolds

Elizabeth A. Lipke and Leslie Tung. Department of Biomedical Engineering, Johns Hopkins University, Traylor Rm #703, 720 Rutland Ave., Baltimore, MD 21205

One potential cell source for cardiac tissue engineering is embryonic stem cell-derived cardiomyocytes (ESC-CMs). However, obtaining pure cultures of cardiomyocytes in sufficient numbers to form large-scale tissues from these cells is challenging. In addition, the electrophysiological properties of tissues formed using ESC-CMs are unknown. This study used genetic selection to obtain pure cardiomyocyte cultures from mouse ESCs and examined the large-scale electrophysiological properties of these mESC-CMs monolayers (2 cm diameter). Methods: In order to obtain sufficient numbers of pure cardiomyocytes, mESCs were stably transfected with a DNA construct allowing expression of the neomycin resistance gene under the control of an •-myosin heavy chain promoter (α-MHC, Klug et al., 1996). The mESCs were differentiated through embryoid body formation. Following neomycin selection, cells were dissociated on day 10 post-initiation of differentiation and seeded at confluency to form functional mESC-CM monolayers. On day 16-18 the monolayers were stained with voltage-sensitive dye di-4-ANEPPS and optically mapped at 253 sites in Tyrode's solution (36°C). Results: Differentiation and selection resulted in mESC-CMs that immunostained positively for the cardiac markers α-sarcomeric actinin, α-MHC, and cardiac troponin I. Obtained mESC-CMs beat in culture up to 5 months. Using optical mapping, planar wave propagation was observed across the confluent monolayers of mESC-CMs (n = 6). Planar conduction velocity was slow (2.4±0.9 cm/sec) compared to control neonatal rat cardiomyocytes (16±3 cm/sec), suggesting additional cues will be needed to further direct ESC-CMs toward the adult cardiomyocyte phenotype. This study demonstrates that cardiomyocytes derived from mESCs can be used to create electrophysiologically functional cardiac tissues. After additional electrophysiological characterization and optimization of cardiac phenotype, forming tissues from ESC-CMs may be a viable option for cardiac tissue engineering applications.

473a Crystallization of Pharmaceuticals Via Microfluidic Crystallization

Defne Kayrak-Talay[1], Jared Baird[2], Lynne S. Taylor[2], and James D. Litster[3]. (1) Chemical Engineering, Purdue University, 480 Stadium Mall Drive, West Lafayette, IN 47907, (2) Industrial & Physical Pharmacy, Purdue University, 575 Stadium Mall Drive, West lafayette, IN 47907, (3) Chemical Engineering & Industrial and Physical Pharmacy, Purdue University, 480 Stadium Mall Drive, West Lafayette, IN 47907

Microfluidic crystallization is a promising method to produce active pharmaceutical ingredient (API) crystals of extremely narrow size distribution and controlled polymorphic form. The microfluidic crystallization system consists of fine Teflon capillary tubes (50 to 300 μm I.D.) and a T-junction and were originally designed and used for chromatographic applications. A continuous oil phase and a supersaturated aqueous phase are introduced to a T-junction where the aqueous phase is segmented into monodisperse drops. Nucleation and growth occur in these very small, uniformly sized drops under highly controllable temperature, flow and anti-solvent addition conditions. By confining crystal growth, controlled nucleation and growth without agglomeration is achieved and effects of growth rate dispersion are overcome. The novelty of the process was shown by production of lactose and lysozyme crystals (Dombrowski et al., 2007 and Dombrowski, 2007). The process is now used to produce seed API crystals of a controlled polymorphic form. Flufenamic acid is used as the model API. The key point in microfluidic crystallization is to obtain one crystal per one drop. Using temperature cycles (applying different temperatures for nucleation and growth), the process is optimized to increase overall nucleation rates without producing multiple crystals in a single drop.

References

Dombrowski, R. D, Litster, J. D, Wagner, N. J and He, Y. (2007). Crystallization of alpha-lactose monohydrate in a drop based microfluidic crystallizer, Chem. Eng. Sci., Vol 63, pp 4802-4810.

Dombrowski, R. D. (2007) Crystallization in a Drop Based Microfluidic Crystallizer, Ph.D Thesis, The University of Queensland.

473b Continuous Plug Flow Crystallization of Pharmaceutical Compounds

Alejandro J. Alvarez and Allan S. Myerson. Department of Chemical and Biological Engineering, Illinois Institute of Technology, Perlstein Hall, Suite 127, 10 W. 33rd Street, Chicago, IL 60616

The present study introduces the application of continuous crystallization to control the critical properties of a pharmaceutical product. Most of the active pharmaceutical ingredients (API) of the pharmaceutical products are crystals of small organic molecules. The effectiveness of the API depends on key characteristics of the crystals. One of the relevant characteristics is the Particle Size Distribution, which affect the transport of the API through the circulatory system to the target organs. Particles with small size and narrow size distribution are required for efficient drug delivery. In order to obtain small crystals with narrow Particle Size Distribution, this research work follows an approach based on antisolvent crystallization with high supersaturation levels in a continuous crystallizer operating at plug flow conditions. The process is mathematically modeled using the Population Balance Equations. An optimization problem is solved to estimate the model parameter s. Particle Size Distribution was measured with Laser Diffraction methods, and the crystallization process was monitored measuring concentration of the solute in the liquid phase with Ultraviolet Spectroscopy (UV). As a result of the study, a process configuration is proposed, which is able to produce API crystals with controlled size and narrow size distribution.

473c Screening for Cocrystallization Tendency: The Role of Intermolecular Interactions

Guangwen He, Pui Shan Chow, and Reginald B. H. Tan. Institute of Chemical & Engineering Sciences, 1 Pesek Road, Jurong Island, Singapore, Singapore

Much work has been focused on exploring the crystal engineering and design strategies that facilitate supramolecular synthesis of new crystalline phases (cocrystals) of APIs and cocrystal formers. However, few attempts have been made to understanding the equilibrium phase behavior and phase transition kinetics of the cocrystallizing solutions. Such lack of understanding of essential knowledge about the solution physical chemistry often leads to difficulty in screening for potential molecular pairs for API and cocrystal former that form cocrystals effectively. Early trials on designing and making cocrystals are mostly limited to Edisonian approaches or high throughput surveys.

Given these limitations and the increasing interest in making cocrystals in the pharmaceutical industry, the proposed work here focuses on characterization of intermolecular interactions between API and cocrystal former in solution cocrystallization. The self diffusivities of various model systems in solution are measured using pulsed gradient spin-echo nuclear magnetic resonance (PGSE NMR). The pair contribution of the self diffusivity is used to probe the intermolecular interactions for various solutions. The relative magnitude of the pair contributions of homomeric and heteromeric interactions is then used as a novel screening method in determining whether the formation of cocrystals is feasible.

473d Microfluidic Platforms for (Membrane) Protein Crystallization

Paul J. A. Kenis[1], Sarah L. Perry[1], Griffin W. Roberts[1], Sameer Talreja[1], Joshua D. Tice[1], Charles F. Zukoski[1], and Robert B. Gennis[2]. (1) Chemical & Biomolecular Engineering, University of Illinois at Urbana-Champaign, 600 South Mathews Avenue, Urbana, IL 61801, (2) Biochemistry, University of Illinois at Urbana-Champaign, 600 South Mathews Avenue, Urbana, IL 61820

Despite their critical role in many biological processes, the precise structure of a disproportionate number of membrane proteins is still unknown. Their limited availability and amphiphilic nature have seriously hampered both the identification of suitable crystallization conditions as well as the subsequent growth of X-ray quality crystals. We exploit microfluidics to facilitate the identification of suitable crystallization conditions on-chip using minute quantities of proteins on both the in-surfo and in-meso approach.

In Surfo: Determining Phase Diagrams and Improving Crystal Quality.

We have shown that kinetic control of crystallization by gradual, direct evaporation of the solvent is an effective method to identify and optimize crystal-producing conditions for lysozyme, thaumatin, and RNAse. We created a microfluidic platform comprised of 144 wells for ~20 nL-sized droplets, capable of screening 48 precipitants at 3 different protein/precipitant ratios. We have also added a reverse vapor diffusion component to the protocol to decouple crystal nucleation and growth, thereby reducing the number of nuclei and increasing crystal quality. Using a similar protocol, we can determine the solubility diagram of a protein using less than 20 μL of protein solution. We have done this for example for the membrane protein bacteriorhodopsin.

In Meso: Membrane Protein Crystallization from Lipidic Phases.

Landau & Rosenbush, and later Caffrey and coworkers have developed the crystallization of membrane proteins from a lipidic mesophase. An aqueous protein solution and a viscous lipid are mixed to form a lipidic mesophase with the protein embedded in the bilayers that form a bicontinuous cubic structure. Upon the addition of salt, a lamellar phase forms in which the membrane proteins align into stacks of 2D sheets, thus facilitating crystallization. We have created a microfluidic chip capable of mixing viscous lipids (e.g. monoolein) with a 30x less viscous, protein containing aqueous phase. The required cubic mesophase is formed by pushing the solutions back and forth between a central lipid containing compartment and two adjacent protein solution containing compartments. Upon the addition of salt the solution transitions to a lamellar phase. We have successfully grown crystals of bacteriorhodopsin on chip. Presently we use IR microscopy to confirm that the crystals are indeed protein and not salt. The on-chip, in meso crystallization of several other membrane proteins is in progress as well as the development of chips that allow for on-chip X-ray evaluation of crystal quality.

473e Modeling Transport Processes within a High-Throughput Evaporation Platform

Michael L. Rasche, Chemical and Biomolecular Engineering, University of Illinois at Urbana-Champaign, 600 South Mathews Avenue, 200 Roger Adams Laboratory, Box C-3, Urbana, IL 61801-3602 and Richard D. Braatz, Department of Chemical and Biomolecular Engineering, University of Illinois at Urbana-Champaign, 600 South Mathews Avenue, Box C-3, 293 Roger Adams Laboratory, Urbana, IL 61801-3602.

There has been increasing industrial interest in the crystallization of pharmaceuticals in impinging jet and other types of crystallizers that operate at high supersaturation. A major roadblock in the development of such crystallizers is the difficulty in directly measuring first-principles nucleation kinetics at high supersaturation. The evaporation of hanging drops within microscale crystallization platforms can be used to measure induction times for a wide range of solvents and evaporation rates at high supersaturation. In these systems, solvent evaporates at a controlled rate from a hanging drop which causes the dissolved pharmaceutical to become more concentrated. Relative supersaturations of nearly ten have been achieved in these systems.

This research focuses on modeling the transport phenomena within the hanging drop caused by the evaporation, as this information is required to determine first-principles nucleation kinetics from the measured induction times. The solute concentration increases locally near the liquid-air interface resulting in a diffusive flux toward the center of the hanging drop. A density gradient is associated with the concentration gradient, which induces natural convection within the droplet. A third transport phenomenon is Marangoni (surface tension-driven) convection which results from the dependency of surface tension on local concentration. Although the focus of this presentation is on the pharmaceutical application, the analyses can be applied to the crystallization of other organic molecules such as proteins and amino acids.

Klupsch et al (2003) provided an analytical solution to a pure diffusion problem involving a drop undergoing evaporation by Langmuir kinetics. A computational simulation involving combined diffusion, natural convection, and Marangoni convection was reported by Savino and Monti (1996). The analytical solution for an isothermal system undergoing solely Marangoni convection was presented by Hu and Larson (2005), and Ristenpart et al (2007) indicated that the direction and magnitude of Marangoni flow is related to the ratio of thermal conductivities between the substrate and the droplet.

This presentation describes a mix of exact and approximate analytical solutions for the solution concentration within the drop in the case of diffusion, as well as computational solutions for the cases of diffusion and diffusion combined with natural convection. Analytical expressions are derived for the maximum concentration difference within the drop, which is important for assessing the accuracy of the commonly made assumption of well-mixedness. These expressions characterize crystal-solvent systems and operating conditions for which the drop is not well-mixed. Comparisons are made to the analytical and computational models developed by other researchers who address similar problems.

References:

Hu, Larson, *Langmuir*, **21**, 3972-3980, 2005.

Klupsh, Mühlig, Hilgenfeld, *Colloids & Surface Science A*, **231**, 85-102, 2003.

Ristenpart, Kim, Domingues, Wan, Stone, *Physics Review Letters*, **99**, 234502, 2007.

Savino, Monti, *J Crystal Growth*, **165**, 308-318, 1996.

473f Microfluidic Systems for Continuous Crystallization

Mahmooda Sultana and Klavs F. Jensen. Chemical Engineering, Massachusetts Institute of Technology, 25 Ames Street, Bldg: 66-513, Cambridge, MA 02139

Traditionally, crystallization has been achieved in batch processes that suffer from non-uniform process conditions across the reactors such as temperature, important in temperature-driven crystallization, and concentration, important in evaporation-driven crystallization. Traditional batch processes also suffer from poorly controlled mixing of the reagents, important in precipitation and antisolvent-driven crystallization. Consequently, crystal growth can vary across the reactors, giving rise to polydisperse crystal size distribution (CSD). This reduces reproducibility of the crystallization process and increases difficulty in obtaining accurate kinetics data.

Microfluidic systems offer a unique toolset for studying the growth kinetics of crystal systems because of well-defined laminar flow profiles and online optical access for measurements. The short length scale in microfluidic devices also allows for better control over the process parameters, such as the temperature, and the contact mode of the reagents, creating uniform process conditions across the reactor channel. Thus, these devices have the potential to generate a more uniform size distribution and more accurate kinetics data. In addition, microfluidic systems decrease waste, provide safety advantages, and require only minute amounts of reactants, which is most important when dealing with expensive materials such as pharmaceutical drugs. Some of these advantages have been demonstrated for evaporation-driven batch crystallization of macromolecules in microfluidic devices [1, 2], but so far not for continuous crystallization.

We have developed microfluidic based continuous microcrystallizers for small organic molecules by using soft lithography techniques. A key issue for achieving continuous crystallization in microsystems is to eliminate heterogeneous crystallization – irregular and uncontrolled formation and growth of crystals at the channel surface, and aggregation of crystals, which ultimately clogs the reactor channel. We have designed and fabricated microcrystallizers that introduces the reagents to the reactor channel in a controlled manner preventing heterogeneous nucleation and aggregation. We have used optical microscopy for in situ characterization of the crystal size distribution and measurement of the crystallization kinetics. We have used Glycine as the model system for our studies and have determined the growth kinetics of the different polymorphs of Glycine. In addition, we have also integrated a Raman spectroscopy tool with the microcrystallizer in order to distinguish between the different polymorphs in situ.

1. Hansen, C.L., et al., A robust and scalable microfluidic metering method that allows protein crystal growth by free interface diffusion. Proceedings of the National Academy of Sciences of the United States of America, 2002. 99(26): p. 16531-16536.

2. Zheng, B., et al., A droplet-based, composite PDMS/glass capillary microfluidic system for evaluating protein crystallization conditions by microbatch and vapor-diffusion methods with on-chip X-ray diffraction. Angewandte Chemie-International Edition, 2004. 43(19): p. 2508-2511.

473g Formation of Lamellar Structures from Spherical Particles

Jianzhong Wu, Chemical and Environmental Engineering, University of California, Riverside, Bourns Hall A242, 900 University Avenue, Riverside, CA 92521 and Zhidong Li, Reservoir Engineering Research Institute (RERI), 385 Sherman Avenue, Suite 5, Palo Alto, CA 94306.

We report disorder to lamellar transitions in a system of spherical particles that invoke a short-range attraction and a long-range repulsion. The system provides a simplified model for aqueous dispersions of colloidal particles and globular proteins that may exhibit stable/metastable clusters or microscopic phases. We show that under appropriate conditions, a lamellar phase with alternating condensed and dilute layers of particles is thermodynamically more stable in comparison to a uniform disordered phase and the macroscopic fluid-fluid coexistence can be inhibited by formation of the lamellar structure. While the width of the condensed layers increases with the particle density, the trend is opposite for the dilute layers. A minimal lamellar thickness is obtained when the condensed and dilute layers approximately have the same width.

478a Efficient Polyethylenimine-Mediated Gene Delivery Proceeds Via Caveolae -Mediated Uptake In HeLa Cells

Nathan Gabrielson, Chemical & Biomolecular Engineering, University of Illinois, Box C-3, 600 S. Mathews Ave., Urbana, IL 61801 and Daniel W. Pack, Chemical & Biomolecular Engineering, University of Illinois Urbana-Champaign, 600 S. Mathews, Box C-3 MC 712, Urbana, IL 61801.

Cancer gene therapy will require cell-specific targeting to ensure that diseased cells are treated while healthy cells are untouched. To direct genes to particular cells, peptides, proteins and small molecule ligands have been attached to the surface of gene delivery complexes. Although such targeting methods have met with success in generating particles that are capable of specific cellular interactions, little attention has been given to the effect such ligands can have on subsequent intracellular processing. For example, transferrin and folic acid are common ligands used for drug and gene delivery targeting due to the overexpression of their respective receptors on many cancer cells. Transferrin is internalized by clathrin-coated pits while folic acid is internalized by caveolae – flask-like invaginations of the cell membrane. Endocytosis via clathrin-coated pits follows the "traditional" model of intracellular processing – trafficking through endosomes to lysosomes with concomitant acidification of the endocytic vesicles. Caveolae are believed to be directed towards neutral pH vesicles termed caveosomes. The difference in pH microenvironment may affect cytoplasmic release of the vectors, especially for so-called "proton-sponge" materials.

Folic acid and human transferrin were conjugated to 25-kDa branched polyethylenimine (PEI) to yield folic acid-targeted PEI (PEI-Fol) and transferrin-targeted PEI (PEI-Tf). Caveolae inhibitors (genistein, methyl-β-cyclodextrin) inhibited uptake of PEI-Fol in HeLa cells. Likewise, clathrin inhibitors (chlorpromazine, amantadine) inhibited PEI-Tf uptake. Thus, it appears that the folic acid- and transferrin-targeted complexes are predominantly endocytosed by caveolae and clathrin, respectively. When folic acid- and transferrin-targeted polyplexes were used to transfect cells in the presence of the same drugs, inhibitors of caveolae-mediated endocytosis drastically reduced gene expression while inhibitors of clathrin-mediated endocytosis had no effect. This suggests that successful gene delivery proceeds via a caveolae-mediated process, which avoids acidic environments and lysosomes. In addition, clathrin inhibition by chlorpromazine increased gene delivery by unmodified PEI by 1.8-fold and transferrin-targeted PEI by 2.7-fold. These results suggest that the caveolin-specific pathway, as opposed to the clathrin-pathway, is effective for gene delivery in HeLa cells and suggest that the conventional understanding of PEI's proton-sponge mechanism may be flawed. Most importantly, we demonstrate that the uptake mechanism and subsequent endocytic processing are important design parameters for gene delivery materials.

478b Selective Intracellular Localization of Acid-Degradable and Biocompatible Polyplexes for Controlled Therapeutic Effects

Min Suk Shim, Chemical Engineering and Materials Science, University of California, Irvine, CA 92697 and Young Jik Kwon, Chemical Engineering and Materials Science, and Pharmaceutical Sciences, University of California, Irvine, CA 92697.

Among many processes involved in nonviral gene delivery, intracellular fates of gene carriers are of great interest. Unlike viral vectors, which have been evolved to maximize intracellular delivery processes, inefficient and non-selective intracellular transport of synthetic gene carriers (e.g., polymer/DNA polyplexes) is believed as one of the major reasons for poor gene delivery efficiency. Utilizing various localization signals (e.g., NLS peptides) has been a major approach to overcome this challenge. In this study, targeted localization of synthetic polyplexes in the cytoplasm and the nucleus was explored by tailoring structures of a carrier polymer. It was also surmised that controlled types of therapeutic effects could be achieved by selectively localizing siRNA and plasmid DNA-containing polyplexes in their different intracellular destinations. A commonly used cationic polymer, polyethylenimine (PEI), was ketalized to increase cytosolic release but reduce cytotoxicity of polyplexes, while superior complexation and buffering capacity of the polymer were retained. Ketalized PEI (K-PEI) with different molecular weights at various ketalization ratios was synthesized. Acid-degradability, capability of nucleic acid condensation, and transfection and RNA interference efficiency of K-PEI/nucleic acid polyplexes were evaluated. It was clear that both siRNA and plasmid DNA condensed by K-PEI were efficiently released upon hydrolysis of ketal branches at endosomal pH. Almost completely diminished cytotoxicity of K-PEI/nucleic acid polyplexes was also demonstrated, regardless of molecular weights and N/P ratios. Surprisingly, polyplexes prepared with K-PEI of molecular weight of 0.8kDa transfected cells even more efficiently than 25kDa unketalized PEI, while there was no evidence of tranfection by unketalized 0.8kDa PEI polyplexes. This implies a possibility of using low molecular weight K-PEI for clinical nonviral gene therapy by utilizing its high transfection efficiency, low cytotoxicity, and eventual secretion from the body. More importantly, transfection efficiency of K-PEI polyplexes at the same ketalization ratio (i.e., 35%) was found to be inversely proportional to molecular weights of K-PEI, while higher RNA interference was observed with larger ketalized polymers. Further studies were carried to differentially control intracellular localization of siRNA and plasmid DNA-containing polyplexes by precisely modulating ketalization ratios of K-PEI in the range of 17-96%. Nucleic acid condensation, transfection efficiency, and RNA interference were found to be strongly correlated with ketalization ratios and molecular weights of K-PEI. On the contrary, it was proved that the the minimal ketalization (17%) was sufficient enough to diminish cytotoxicity of the polymer. The presentation will convey 1) logics of polymer design and synthesis, 2) preparation and characterization of K-PEI polyplexes using TEM, dynamic light scattering (DLS) analysis, and gel electrophoresis, 3) cytotoxicity test mainly by MTT assay, 4) results of transfection and RNA interference, and 5) evidence of selective localization of various K-PEI polyplexes by intensive confocal microscopic studies. Finally, implications of the study in achieving potent, biocompatible, and controlled nonviral gene delivery will be discussed.

478c Polymersome Delivery of siRNA and Antisense Oligonucleotides

Dennis E. Discher[1], J. David Pajerowski[2], and Younghoon Kim[1]. (1) Chemical and Biomolecular Engineering, University of Pennsylvania, Room 129 Towne Building, 220 South 33rd Street, Philadelphia, PA 19104-6393, (2) Bioengineering, University of Pennsylvania, Room 129 Towne Building, 220 South 33rd Street, Philadelphia, PA 19104-6393

siRNA and antisense oligonucleotides, AON, have similar size and negative charge and are often packaged for in vitro delivery with cationic lipids or polymers — but positively charged assemblies are problematic in vivo. Here we demonstrate loading and functional delivery of RNAi and AON with neutral, nano-transforming polymersomes. These degradable carriers are taken up passively by cultured cells after which the vesicles transform into micelles that allow endolysosomal escape and delivery of either siRNA into cytosol for mRNA knockdown or else AON into the nucleus for exon skipping within pre-mRNA. Polymersome-mediated knockdown appears as efficient as common cationic-lipid transfection and about half as effective as Lenti-virus after sustained selection. For AON we further show that intramuscular injection into a mouse model of muscular dystrophy leads to the expected protein expression, which occurs along the entire length of muscle. The lack of cationic groups in antisense polymersomes together with initial tests of efficacy suggests broader utility of these non-viral carriers.

478d Liposomal siRNA Delivery to Silence Aromatase for Breast Cancer Treatment

Megan Lynn Cavanaugh[1], L. James Lee[1], Robert J. Lee[2], Yasuro Sugimoto[2], and Robert W. Brueggemeier[2]. (1) Chemical and Biomolecular Engineering, The Ohio State University, 125 Koffolt Labs, 140 W 19th Ave, Columbus, OH 43210, (2) College of Pharmacy, The Ohio State University, 542 Parks Hall, 500 West 12th Avenue, Columbus, OH 43210

Increased expression of certain genes is common in many diseases and inhibiting these over-expressed genes is an option for treating them. RNA interference is one such gene silencing technique and is a potentially very powerful approach to cancer therapy. This collaborative research effort seeks to develop a novel liposomal vector to successfully deliver siRNA that will suppress aromatase activity in tumor cells for breast cancer treatment. Aromatase is an enzyme that produces estrogens and the expression and activity of aromatase are higher in breast cancer cells than in normal breast tissue. Approximately 70% of breast cancers are estrogen receptor positive and the current first line treatment for postmenopausal women with these cancers is aromatase inhibitors. However, current aromatase inhibitors bring about estrogen deprivation in the entire body, causing some side-effects. Using siRNA targeting aromatase mRNA specific to breast cancer cells allows estrogen levels to be decreased in a tissue specific manner leading to diminished side-effects. Additionally, a liposomal drug delivery system will not only help prevent siRNA degradation, but can increase circulation time, target tumor cells, improve cellular uptake, and lower the required effective dose.

478e High Efficiency siRNA-Mediated Knockdown Using High-Molecular-Weight Branched Polyethylenimine In Mammalian Cells

Daniel W. Pack, Chemical & Biomolecular Engineering, University of Illinois Urbana-Champaign, 600 S. Mathews, Box C-3 MC 712, Urbana, IL 61801 and **Lily Wong**, Chemical and Biomolecular Engineering, University of Illinois, Box C-3, 600 S. Mathews Ave, Urbana, IL 61801.

RNA interference has been widely used to down-regulate the expression of many targeted genes and is currently being explored for gene therapy. The main mediator of this phenomenon is short interfering RNA (siRNA). Commercially available high-molecular-weight (HMW) polyethylenimine (PEI) of 25-kDa in the branched-form was used to deliver siRNA in mammalian cells. siRNA polyplexes were characterized with regard to siRNA condensation and determined to be fully condensed at 0.4 μ g PEI/ μ g siRNA by gel retardation and significant quenching of ethidium bromide measured at 0.2 μ g PEI/ μ g siRNA. Sizes of all the polyplexes measured by dynamic light scattering were roughly 120 to 240 nm. Model cell lines, HEK293, NIH/3T3 and HeLa, were transiently co-transfected with luciferase-encoding plasmid (pGL3)-Lipofectamine 2000 lipoplexes and 5-50 nM luciferase-targeting siRNA (siGL3)-PEI polyplexes. The PEI/siGL3 was varied over a wide range from 4:1 to 40:1. The delivery efficiency of siRNA is highly dependent on the amount of PEI applied to the cells and not on the PEI/ siGL3 ratio. At ≥1.33 μ g/mL of PEI, a downregulation >80% is achieved with as little as 5 nM of siRNA (corresponding to PEI/siGL3 = 20:1 w:w). Polyplexes formed at high PEI concentration exhibit better cellular uptake and transfection efficiencies. The uptake of polyplexes formed at 2.67 μ g/mL of PEI is ≥10-fold higher than polyplexes formed at 0.27 μ g/mL of PEI and the knockdown rate improves by ≥30% at all siRNA concentrations.

478f Novel Cationic Steroid Family and Its Application In Gene Delivery and as Antibacterials

Renee A. Randazzo[1], Scott L. Diamond[1], Robert Bucki[2], and Paul A. Janmey[2]. (1) Chemical and Biomolecular Engineering, University of Pennsylvania, 3340 Smith Walk, 1150 Vagelos, Philadelphia, PA 19104, (2) Institute for Medicine and Engineering, University of Pennsylvania, 3340 Smith Walk, Philadelphia, PA 19104

A novel family of cationic steroids has been synthesized and found to display potential both in non-viral gene delivery and as antimicrobials. The compounds were formed by amide couplings of the polyamines spermine, spermidine, and putrescine with cholenic and cholanic acid derivatives and span a hydrophobicity range of 0.64 to 5.63 logD (octanol-water distribution coefficients). On liposomal formulation with the neutral helper lipid 1,2-dioleoyl-*sn*-glycero-3-phosphoethanolamine (DOPE), 5β-cholanyl-spermide-3α-ol and 8-cholenyl-spermide-3α,12α-diol were the most successful as gene delivery vectors achieving transfection efficiencies comparable to those found with Lipofectamine2000™ in bovine aortic endothelial cells. The toxicity to the cells was reduced by nearly half that of Lipofectamine2000™ for both of these compounds. For the more hydrophilic 5β-cholanyl-spermide-3α,7α,12α-triol, transfection efficiency was enhanced by over 50% through incorporating 20-30mol% bis-substituted amide in the liposome formulation without increasing the toxicity. Evaluation of antibacterial activity against gram positive bacteria (*B. Subtilis ATCC 6051*) revealed MIC_{50} (concentration required to inhibit the growth of 50% of the organism) of less than 25 mg/L for the majority of the compounds. When tested against gram negative bacteria (*E. Coli MG 1655*), the MIC_{50} values of 5β-cholanyl-spermide-3α-ol and 8-cholenyl-spermide-3α,12α-diol were less than 25 mg/ml and the effective therapeutic indexes (ratio of hemolytic activity over MIC_{50}) were 37 and 53, respectively.

478g Effect of Hydrophilic Spacer Length on the Complexation of Novel Biodegradable Hyperbranched Polycationic Polymers and Plasmid DNA for Non-Viral Gene Delivery

Sue Anne Chew, Michael C. Hacker, Anita Saraf, and Antonios G. Mikos. Department of Bioengineering, Rice University, 6100 Main Street, MS-142, Houston, TX 77005

Biodegradable polycationic polymers that have high transfection efficiency are much needed to replace current non–biodegradable polycationic polymers for non-viral gene delivery. Although these existing polymers are known to be successful in assisting in the delivery of plasmid DNA (pDNA), the non-degradability of the polymers presents a problem with repeated dosage and long-term usage due to their accumulation in the endosomal compartment or cell nucleus.

In our laboratory, novel biodegradable polycationic polymers were synthesized by Michael addition polymerization of an amine monomer, 1-(2-aminoethyl) piperazine (AEPZ) with triacrylate monomer, trimethylolpropane triacrylate (TMPTA) or different types of trimethylolpropane ethoxylate triacrylates (TMPETA) for non-viral gene delivery. Different polymers were synthesized by reacting AEPZ with TMPTA (P1), TMPETA with 1 ethyleneoxy group in each acrylate chain (P2), TMPETA with 7 ethyleneoxy groups in each monomer (P3) and TMPETA with 14 ethyleneoxy groups in each monomer (P4). Previous work in our group has shown that these polymers are non-cytotoxic and the degradation rate of the polymers can be increased by lengthening the hydrophilic spacers in the triacrylate monomer. Furthermore, the buffering capacity of the polymers can be increased with the decrease in hydrophilic spacer length. In this work, the effect of the hydrophilic spacer length of the polycationic polymers on the complexation of the polycationic polymers and pDNA were investigated at different polymer to pDNA (pCMV-GFP, 4.7 kb) weight ratios. The band retardation of the complexes was evaluated by gel electrophoresis with 1 ìg of pDNA. All the polymers were able to complex with the pDNA and neutralize the charges as evident by the retardation of the migration of the pDNA at a weight ratio of 10 or higher except for P4, which was able to retard the migration at a weight ratio of 20 or higher. Compared to the other polymers, the longer spacer length in P4 resulted in a polymer with the lowest number of protonated amines for a given weight of polymer, thus, has less positive charges to interact with the negative charges on the pDNA. The hydrodynamic radius of the different polycationic polymers were evaluated by dynamic light scattering (DLS) at 25°C. The polycationic polymers were able to partially condense the pDNA and resulted in a hydrodynamic radius as low as 160 nm. The hydrodynamic radius of the complexes formed varied with the different polycationic polymers and was seen to decrease as the weight ratio of polymer to pDNA increased and levelled off at a certain weight ratio. In summary, hydrophilic spacers can be incorporated into polymers to produce non-cytotoxic biodegradable polymers with different characteristics which are important in generating vectors for gene delivery.

480a Liver Tissue Engineering with Inverted Colloidal Crystal Scaffolds for In Vitro Drug Testing Application

Jungwoo Lee, Biomedical Engineering, University of Michigan, 2300 Hayward St., 3074 H. H. Dow Bldg., Ann Arbor, MI 48109-2136 and Nicholas A. Kotov, Chemical Engineering Department, Materials Science Department, Biomedical Engineering Department, University of Michigan, 2300 Hayward, Ann Arbor, MI 48109.

Effective early stage drug screening is important to minimize failures in late clinical stages. 2D cell culture has been predominantly used, but it is becoming apparent that 2D culture is extremely limited when predict a drug's toxicity and efficacy in the body. As a potential solution, we developed a 3D cell culture based in vitro drug testing system representing tissue level physiology in a standardized manner.

Our approach started with preparing hydrogel inverted colloidal crystal (ICC) scaffolds. Primary colloidal crystals are hexagonally packed lattices of microspheres (D=50-300μm). ICC is a similarly organized structure where the spheres are replaced with cavities, while the interstitial spaces are filled with poly (acrylamide) hydrogel. Highly packed, identically sized pores and well interconnected channels promoted homogenous cell seeding in each pore. The ICC geometry created by a cell repulsive hydrogel provides supportive physical and chemical environments for spheroid formation of human hepatocarcinoma (HepG2). After 3-5 days of culture, HepG2 cells formed solid spheroids with a narrow size distribution and filled approximately 70% of the pore volume. The spheroid culture exhibited tissue level morphology and specific liver functions such as albumin secretion. Also, it showed enhanced CYP450 induction and reduced CdTe nanoparticle toxicity compared to the 2D cell culture.

Overall our data show that hydrogel ICC scaffolds would be a promising tool for engineering liver tissue spheroids while retaining a high analytic capability. It was substantiated by tissue level morphology, function and toxicity. This cell-scaffold system will contribute to the pharmaceutical industry as a practical and valuable early stage drug testing tool by reducing the gap between 2D culture and in vivo.

480b Optimization of Chemically-Defined Media for High Cell Density Growth of Lactococcus Lactis Using Highly-Efficient Nonlinear Experimental Design Methods

Guiying Zhang, Department of Food Science & Technology and Department of Viticulture and Enology, University of California, One Shields Avenue, Davis, CA 95616 and David E. Block, Department of Chemical Engineering and Materials Science and Department of Viticulture and Enology, University of California, One Shields Ave, Davis, CA 95616.

Optimization of fermentation media and processes is a difficult task due to the potential for high dimensionality and non-linearity. Two novel and highly efficient experimental design methods were developed and evaluated for experimental fermentation optimization. The first approach is based on using a truncated genetic algorithm with a developing neural network model to choose the best experiments to run for process optimization. The second approach uses information theory, along with Bayesian regularized neural network models, for experiment selection. To evaluate these methods experimentally, we used them to develop new chemically-defined media for high cell density growth of Lactococcus lactis IL1403. A leave-one-out approach was employed to identify the necessity of each of the 57 chemical components used in the medium development. The final fermentation system included 19 defined components or groups of components, along with an optimal temperature and initial pH. The maximum cell growth from the optimal process resulting from each novel method is generally comparable to or higher than that achieved using a traditional statistical experimental design method, but these optima are reached in about half of the experiments (73 to 94 vs. 161, depending on the method). All the optimal chemically-defined media developed in this work can support high cell density growth for L. lactis, generally 2.5 to 3 times higher than the best previously reported synthetic medium (SA) and 72% higher than the commonly-used complex medium (M17) at optimization scale. The best chemically-defined medium developed was evaluated and compared to other media at flask- and fermentor-scales. All optimal chemically-defined media can be used for the growth of eight other L. lactis ssp and other organisms (e.g., enterococci, and streptococci) for which M17 is a common complex growth medium.

480c Mixing Studies with Elephant Ear Impellers: Are They Low Shear Agitators?

Alvin W. Nienow, Sophie Zhu, Waldek Bujalski, and Mark J. H. Simmons. Birmingham University, Department of Chemical Engineering, Edgbaston, Birmingham, B15 2TT, United Kingdom

Ever since the 1980s when animal cells first started to become important for the production of biopharmaceuticals, there has been concern for their 'shear sensitivity'. As a result, initially, there were many attempts to find 'low shear' systems in which to grow them rather than the agitated 'submerged' fermentation vessels used for other more robust organisms. For large-scale production, airlift bioreactors were introduced, as it was perceived that they met this criterion. However, later work soon showed that the stresses associated with bursting bubbles were much more damaging to cells than those due to agitation. Fortunately, it had been shown much earlier that the surfactant, Pluronic F68, was able to prevent such damage whatever the reactor type. As a result, and perhaps also because of the inherent flexibility, the stirred bioreactor has become the bioreactor of choice for large-scale free suspension animal cell culture up to 20,000 L that are now being installed. However, there is still a belief that cells will grow better if agitation is provided by 'low shear' agitators. Recently, a particularly favoured one has been the type colloquially known as an 'elephant ears agitator.' In this presentation, a detailed analysis will be given of the flow characteristics of such an agitator using the advanced fluid dynamic measurement technique, particle imaging velocimetry (PIV). Both unaerated and aerated conditions were used at specific energy dissipation rates up to 0.1 W/kg and air flow rates of 0.01 to 0.05 vvm, which are typical for animal cell culture. In particular, the global flow field and turbulence parameters have been measured, the latter characterising 'shear sensitivity', as well as unaerated and aerated power numbers and air dispersion characteristics. These data are compared with equivalent measurements for a standard pitched blade turbine and with a Lightnin' A-315 hydrofoil agitator, which has found favour relatively recently as a 'low shear' agitator for mycelial and bacterial fermentations. The results indicate that with respect to 'shear sensitivity', all the agitators are essentially the same. A brief review of the use of other agitators for growing animal cells, including the so-called 'high-shear' Ruston turbine, will be presented. It is concluded that though different agitators may offer advantages in other ways, e.g. better bulk blending, aerated power or air dispersion characteristics, they do not do so in relation to 'shear sensitivity'.

480d Evaluation of a Pendulum Agitator In Biopharmaceutical Applications

Lauren Speciner[1], Al Stern[2], Tom Bewley[2], Liza Munda[3], Brad Wolk[1], and Adeyma Arroyo[1]. (1) Process Development Engineering, Genentech, Inc., One DNA Way, South San Francisco, CA 94080, (2) LS Pharmaceutical and Device Development, Genentech, Inc., One DNA Way, South San Francisco, CA 94080, (3) Corporate Engineering, Genentech, Inc., One DNA Way, South San Francisco, CA 94080

The Chemineer Clean Sweep is a pendulum mixer which applies a side to side mixing motion instead of the traditional rotational motion. This design allows for low volume mixing which is an important requirement for many pharmaceutical applications. In addition, the design utilizes a diaphragm seal eliminating the need for mechanical seals which can cause particulate shedding. This paper presents an evaluation of the Clean Sweep mixer. Cleanability and product quality were assessed under different operating conditions (agitation speeds and mixing volumes). The mixer design, mixing efficiency, ease of disassembly, and scalability parameters were also assessed. Results showed acceptable mixing and cleanability with no product quality effects based on size exclusion chromatography (SEC), filtration studies, particle counting, and turbidity measurements.

480e Automated Monitoring of Mammalian Cell Cultures Via An Auto-Sampler and An Integrated Multi-Functional off-Line Analyzer

Gayle E. Derfus[1], Daniel Abramzon[2], Meg Tung[3], Robert Kiss[3], and Ashraf Amanullah[2]. (1) Oceanside Process R&D, Genentech Inc., 1 Antibody way, Oceanside, CA 92056, (2) Oceanside Process R&D, Genentech, Inc., 1 Antibody Way, Oceanside, CA 92056, (3) Late Stage Cell Culture, Process Development, Genentech, Inc., 1 DNA Way, South San Francisco, CA 94080

Mammalian cell-based bioprocesses are used extensively for production of therapeutic proteins. Some bioprocess control parameters, such as temperature, dissolved oxygen and pH, can be monitored on-line using *in situ* sensors. Other key parameters, however, are still typically monitored via manual sampling and several off-line analyzers. Off-line monitoring is labor-intensive, and introduces operator-dependent error into the process. The BioProfile FLEX, manufactured by Nova Biomedical, combines the functionality of three off-line analyzers (cell counter, osmometer, gas/electrolyte/nutrient/metabolite bio-profile analyzer) into one device. An auto-sampler has also been developed to support up to ten reactors with output to the FLEX. In combination, the FLEX and auto-sampler offer the potential for more robust, operator-independent processes, resulting in increased efficiency and process reproducibility. Furthermore, the ability to sample more frequently with less operator effort could lead to improved understanding of bioprocesses.

This work demonstrated the function of the FLEX and auto-sampler for on-line monitoring of mammalian cell cultures, with data output to a DeltaV bioreactor control system via OPC. FLEX measurements were generally equivalent to those obtained using instruments it would replace (Beckman Coulter Vi-Cell, Advanced Instruments Osmometer 3900, Nova BioProfile 400). FLEX measurements correlated particularly well with those of the other instruments for key process decision parameters including glucose concentration, pH, and cell counts. Inter-instrument variability, assessed using three FLEX analyzers, was minimal for most parameters. Measurements of samples provided to the FLEX by the auto-sampler prototype correlated well with those from manual samples, indicating that the auto-sampler did not alter the samples. In addition, bioreactors connected to the auto-sampler remained contamination-free during fifty days of frequent automated sampling. This work suggests that the FLEX and auto-sampler, along with OPC-based communication between the FLEX and a bioreactor control system, have the potential to dramatically reduce the manual labor involved in maintaining mammalian cell bioprocesses.

480f The Design and Synthesis of Novel Excipients for the Suppression of Protein Aggregation

Curtiss P. Schneider and Bernhardt L. Trout. Chemical Engineering, Massachusetts Institute of Technology, 77 Massachusetts Avenue, E19-531, Cambridge, MA 02139

Due to the advancement of recombinant DNA technology and cell biology, the past two decades have seen an explosive growth in protein based therapeutics, e.g. monoclonal antibodies. Over 130 protein based therapeutics have been approved by the FDA for clinical use and well over 200 more are in some phase of development. However, rapid commercialization of these products has not been fully realized due to the presence of many barriers, namely the physical and chemical instabilities of proteins. Of these degradation pathways, protein aggregation is arguably the most common and troubling manifestation of protein instability, occurring in almost all phases of development. In addition to reducing efficacy, if administered to a patient, aggregates can cause adverse reactions, such as immune response, sensitization, or even anaphylactic shock. The structural differences among different proteins are so significant that the application of a universal stabilization strategy has not yet been successful, though the effects of common excipients are generally universal (e.g. sugars increase the free energy of unfolding thus deter the formation of aggregation prone species). The current approach to stabilizing protein drugs against aggregation is by trial-and-error testing of different combinations of cosolutes (e.g. salts, sugars, surfactants, arginine, etc.) using empirically derived heuristics. While ubiquitously used, this approach is inefficient and does not always enable the discovery of stable protein solution formulations. Thus many products must be lyophilized and reconstituted prior to injection, which is highly undesirable.

In response to this major problem, we have developed a theory for how a novel class of excipients, termed "neutral crowders", should inhibit protein association reactions. This new class of excipients has the potential for wide spread application as a universal stabilizer of biopharmaceuticals. The basis for the mechanism of a "neutral crowder" is that a large additive which is neither excluded from nor bound to the surface of the protein will solvate the protein much like water but will tend to deter association due to the unfavorable entropic effect arising when the large additive is forced to be excluded from the gap formed between two associating protein molecules. To explore the possibility of such additives, we have synthesized a wide variety of large compounds with different surface groups known to bind to the surface of the protein in the hopes of developing an additive that has the right balance of attraction and steric exclusion, and thus behave as a "neutral crowder". So far we have had some initial success at developing such excipients and at inhibiting aggregation, allowing us to refine our designs for developing excipients that will significantly slow aggregation.

480g Development of a 'Cycle to Cycle' Control for the Mcsgp-Process for a Monoclonal Antibody Variant Separation

Cristian Grossmann[1], Manfred Morari[1], Massimo Morbidelli[2], and Guido Ströhlein[3]. (1) Automatic Control Laboratory, ETH Zurich, Zurich, Switzerland, (2) Institute for Chemical and Bioengineering, Department of Chemistry and Applied Biosciences, ETH Zurich, Wolfgang-Pauli-Str. 10, HCI F 133, Zurich, 8093, Switzerland, (3) ChromaCon AG, Technoparkstr. 1, Zürich, 8005, Switzerland

The MCSGP process is a new continuous chromatographic process specifically designed for the purification of proteins and peptides. Due to its counter-current purification principle, the performance increase with respect to the batch process can be up to 10x in productivity and more at even higher yields [1]. Various successful applications of the MCSGP-technology have been reported, e.g. the purification of monoclonal antibodies from high-titer supernatants with ion-exchange resins and the purification of polypeptides.

However, optimum MCSGP operation is a challenge and the current practice is to operate the MCSGP units at sub-optimal conditions to guarantee robustness. As a result, control and automation of MCSGP is of great interest in order to exploit the full economic potential of this process. An automatic control algorithm for MCSGP units that guarantees an optimal, robust operation with product purities in specification is a challenging problem because of the complex dynamics involved in this process, i.e. its cyclic and hybrid nature due to the inlet/outlet port switching with strong nonlinearities and delays in the feedback information.

In this work, a control algorithm based on previous work[2] is developed for the MCSGP-technology. The flow rates as well as the salt gradients have been chosen as manipulated variables. The suitability of the controller is proven using the separation of three monoclonal antibody variants as model system for the simulations.

[1] http://www.chromacon.biz/PressRelease_MCSGPpeptides_Novartis.pdf

[2] Grossmann, C., Erdem, G., Morari, M., Amanullah, M., Mazzotti, M., Morbidelli, M., 'Cycle to cycle' optimizing control of simulated moving beds. AICHE J, Volume: 54, Pages: 194-208, Published: 2008

489a How Do Antimicrobial Peptides Work? Simulations of the Pore Structure of Protegrin-1

Allison Langham and Yiannis Kaznessis. Chemical Engineering and Materials Science, University of Minnesota, 117 Pleasant St SE, Walter Library 488, Minneapolis, MN 55455

Due to increasing reports of antibiotic resistance, antimicrobial peptides have been studied over the past few decades as possible substitutes for traditional antibiotic drugs. Their mechanism of action, however, is not fully understood and this hinders efforts to design new peptides for use in human therapies. We previously performed over 150ns of simulation of a protegrin-1 (PG-1) pore in a lipid bilayer composed of plamitoyloleoylphosphatidylethanolamine (POPE) and palmitoyloleoyl phosphatidylglycerol (POPG) lipids meant to mimic the inner membrane of a bacterial cell [1]. The simulation improves on a model of an octameric pore proposed from NMR experiments [2]. Additional experimental data was made available to us that describes the movement of potassium ions from E. coli treated with PG-1. In the experiments, potassium ion leakage, membrane potential, and membrane permeability were monitored with respect to time for samples treated with various PG-1 concentrations for cells in two growth phases. Quick, significant loss of potassium from the cells combined with a decrease in cell volume is observed and for samples treated with higher concentrations of PG-1, this is followed by an increase in the volume and eventual cell lysis. For a more direct comparison to these experimental data, a simulation of the pore was carried out with potassium ions investigate the movement of this ion type through the pore. From the results of the experiments and simulations, we propose a timeline of the events that result in bacterial cell death: after pores form in the membrane potassium leaves the cell due to differences in its intracellular and extracellular concentrations. Other species such as sodium and chloride ions will move into the cell due to concentration gradients and to balance the charge distribution. As this happens, the transmembrane potential is disrupted, and the bacteria are no longer able to control their cellular processes. Water will enter the cell, at rates determined by the number of pores on the cell. For bacteria treated with a high concentration of protegrin, this will result in osmotic lysis of the cell.

This work was supported by a grant from NIH (GM 070989). Computational support from the Minnesota Supercomputing Institute is gratefully acknowledged. This work was also partially supported by National Computational Science Alliance.

1. Langham, A.A., A. Sayyed Ahmad, and Y.N. Kaznessis, On the nature of antimicrobial activity: a model for Protegrin-1. J Am Chem Soc, 2008. 130(13): p. 4338-46.

2. Mani, R., et al., Membrane-bound dimer structure of a beta-hairpin antimicrobial peptide from rotational-echo double-resonance solid-state NMR. Biochemistry, 2006. 45(27): p. 8341-9.

489b Modeling the Open-to-Closed Transition of Adenylate Kinase: All-Atom Molecular Dynamics Simulations and a Double-Well Network Model

Jhih-Wei Chu, Chemical Engineering, University of California, Berkeley, 101A Gilman, UC Berkeley, Berkeley, CA 94720 and Jason Brokaw, Chemistry, University of California, Berkeley, 101A Gilman, UC Berkeley, Berkeley, CA 94720.

An intrinsic property of protein critical to function is the ability of undergoing conformation change upon ligand binding or biochemical modification. In this work, we study Adenylate Kinase (AKE), an important enzyme controlling the balance of ATP in prokaryotic cells. X-ray crystallography indicates that AKE has two distinct conformations, open and closed, depending on whether it is bound with substrates (ATP and AMP). Conformation difference in AKE is determined by the relative position of two separate domains, the lid domain and the NMP binding domain, to the core. In this work, we use all-atom molecular dynamics (MD) simulations and coarse-grained modeling to understand the effects of ligand binding on AKE conformation. Results based on four ~100ns all-atom trajectories indicate that ATP binding induced the closing of lid domain and suggest that the relative population between closed to open structure is increased as a result. The closing of NMP binding domain, however, is found to be more specific and may require a timescale longer than 100ns to close. The mechanical property of a hinge region is found to correlate with lid closing; residues in this region may be mutated to alter the rate of conformational change and hence enzyme catalysis. This prediction agrees well with the results of recent single molecule experiments. Using a double-well network coarse-grained model, multiple pathways of open-to-closed transition can be found. Motions of lid-domain and NMP binding domain are not entirely concerted and may be treated as two distinct events. This picture is different from the result of using elastic network model and agrees better with atomistic simulations. In addition to open-to-closed transition, solvation structures and intrinsic mechanical properties of AKE are also characterized to identify key residues that may control the conformational change of AKE from mechanical perspectives.

489c Computer Simulation of Beta-Amyloid Aggregation with An Intermediate Resolution Model

Victoria Wagoner and Carol K. Hall. Chemical and Biomolecular Engineering, North Carolina State University, College of Engineering 1, Box 7905, 911 Partners Way, Raleigh, NC 27695

Assembly of normally soluble proteins into ordered aggregates, known as amyloid fibrils, is a cause or associated symptom of numerous human disorders, including Alzheimer's and the prion diseases. The fibrils found in Alzheimer's disease are primarily composed of the 39-43 amino acid peptide, beta-amyloid. Recent experimental studies have offered tantalizing clues regarding the fibril structure adopted by beta-amyloid, but our understanding of its assembly is still far from complete. The long term goal of our work is to determine the underlying physical forces responsible for the misfolding and aggregation of proteins. Our objective is to extend the intermediate-resolution protein model, PRIME (Protein Intermediate Resolution Model) that we have already developed for polyalanine to the description of beta-amyloid. PRIME is well suited for modeling protein aggregation because it provides a faithful representation of protein geometry while also capturing the essential features of the forces responsible for protein folding, hydrogen bonding and hydrophobicity.

We are using a two-pronged approach in the modeling of beta-amyloid. One model attempts to capture the basic physical features of beta-amyloid using a simple single-sphere side chain representation and an "HP" model description of the energetics of each side chain. The second model characterizes specific side chain interactions essential to the aggregation of beta-amyloid using either a one-or-two-sphere side chain representation and an "HP" + electrostatic interaction description of the energetics of each side chain. We explore how variations in the temperature, concentration, hydrophobic strength, and quench rate affect the formation and structure of beta-amyloid. Movies will be shown of the aggregation process.

489d Potential of Mean Force of Glycophorin-a Alpha-Helix Dimerization

Lorant Janosi and Manolis Doxastakis. Chemical and Biomolecular Engineering, University of Houston, S222 Engineering Bldg. 1, Houston, TX 77204

Recognition and association plays a crucial role in the assembly and function of transmembrane proteins. Despite significant research in the last years, these processes are only partially understood. We investigate the association of model transmembrane alpha-helices embedded in lipid bilayers by means of uniform sampling along their separation distance and accurate potential of mean force (PMF) calculations. The model system consists of a pair of alpha-helices of Glycophorin A, a system that experimentally exhibits dimerization in a lipid membrane. Using established coarse-grained models, we develop an efficient Monte Carlo methodology to overcome sampling limitations imposed by long characteristic times present in conventional lipid membrane simulations. Designed Monte Carlo moves aim to increase conformational sampling for the helices in the hydrophobic environment. A combination of the Expanded Ensemble Density of States formalism and hybrid molecular dynamics allow for efficient and accurate calculations on the association of the helices. Using our methodology, we evaluate the effect of different lipid composition on the mechanism of the recognition process of protein molecules in a hydrophobic environment.

489e Dynamic Coupling Machinery In DNA Polymerases:

Ravi Radhakrishnan and Ravindra Venkatramani. Department of Bioengineering, University of Pennsylvania, 210 South 33rd Street, 240 Skirkanich Hall, Philadelphia, PA 19104

Polymerases are proteins that replicate DNA not only with remarkable efficiency but also with high accuracy. They are important players in the error-free DNA replication process that is central to the integrity of our genome. Through computer simulations of the dynamics of a polymerase DNA complex at atomic resolution we suggested that an intriguing dynamical coupling between the cooperative motion of polymerase and DNA atoms may play a significant role in aiding and abetting the chemical reaction (chemical step) that leads to nucleotide incorporation during DNA replication. We also suggest that the coupling is disrupted to varying extents in a context-specific fashion, i.e., when the inserted nucleotide is not complementary to the template base (mispair) or when the template base in the DNA is oxidatively damaged. Using an elastic response formalism, we also show that as a direct consequence of the dynamical coupling, the rate of the chemical step is dependent on the applied force on the DNA (template) strand, and that the force-dependence is also context-specific. The motivation for performing computational studies on atomically-detailed systems is to challenge the status quo, i.e. the existing paradigm which attributes the force-response to elasticity changes between single and double-stranded DNA and hence does not predict a context specific force-response. Our finding that the dynamic coupling resulting from the atomic details of the protein architecture is directly responsible for the force-dependence of the rate is in stark contrast to the DNA-elasticity paradigm. Our predictions can be tested directly through single-molecule experiments.

Web Page: www.seas.upenn.edu/~biophys

489f Quantitative Structure-Property Relationship Modeling of Skin Sensitization. A Quantitative Prediction

S. Golla, K. Yerramsetty, B. J. Neely, S. V. Madihally, R. L. Robinson Jr., and K. a. M. Gasem. School of Chemical Engineering, Oklahoma State University, 423 Engineering North, Stillwater, OK 74078

With the discovery of numerous new chemicals for various scientific applications the need for assessing their toxic effects with reasonable accuracy arises. Extensive efforts have been carried out in the literature to identify assessment methods for evaluating the toxicity of various chemical compounds. Of the toxicity related issues, skin sensitization due to exposure of toxic chemical compounds has been a major work-related problem comprising up to 95% of the occupational contact dermatitis cases. Though reliable test procedures for skin sensitization exist, their application to chemical compounds is limited either by the time consumed or cost involved. Hence, it becomes necessary for development of computational techniques that not only reduce time and cost, but also ensure animal welfare. Non-testing procedures such as quantitative structure-property relationships (QSPRs) can be used as effective methods for a priori prediction of physical properties of chemical compounds. QSPR models offer an attractive alternative with the potential to provide reliable property estimates based on information obtained solely from the chemical structure.

In this work, an effort has been made to obtain a QSPR skin sensitization model with wide applicability. An extensive database comprised of test results from three exclusive test procedures were used for the QSPR model development. This work focuses primarily on the following objectives: (a) Develop QSPR models to predict skin sensitization. Since the experimental procedure and end point ranking for local lymph node assay (LLNA), guinea pig maximization test (GPMT) and Federal Institute for Health Protection of Consumers and Veterinary Medicine (BgVV) are different, three exclusive QSPR models were developed, and (b) Improve the predictive capability of the QSPR models using a combination of literature recommended and statistically determined descriptors. The resultant QSPR models are capable of predicting critical properties of the diverse set of molecules considered with an accuracy of 88%, 93% and 90% for LLNA, GPMT and BgVV datasets, respectively.

489g Using 3-D Dense Packing Models to Predict Surface Tension Change Due to Protein Adsorption

Joshua W. Lampe, Emergency Medicine, Hospital of the University of Pennsylvania, 125 South 31st Street, Suite 1200, Philadelphia, PA 19104, Portonovo S. Ayyaswamy, Mechanical Engineering and Applied Mechanics, University of Pennsylvania, 229 Towne Building, 220 South 33rd Street, Philadelphia, PA 19104, and David M. Eckmann, School of Medicine, University of Pennsylvania, 331 John Morgan Building/6112, 3620 Hamilton Walk, Philadelphia, PA 19104.

Recently, work has been published which demonstrates the significant role of the solvent, usually water, in the protein adsorption process. A 'Traube-rule-like' dependence of protein adsorption on molecular weight has been reported. This work has broadly demonstrated that the adsorption of human blood-borne protein molecules of widely varying size and purpose is more similar than different, and a quasi-thermodynamic model has been explored. In this presentation we independently evaluate the efficacy of this model, as well as provide a method to generalize the model for application to non-spherical protein molecules. This generalization allows for the application of this model to molecules such as fibrinogen, of significant importance to thrombogenic processes, as well as more accurate prediction of the tendency of BSA to adsorb into multi-layers at physiological concentrations, while HSA adsorbs into a monolayer. This difference elucidates the importance of molecular shape, as opposed to molecular weight, in the adsorption process, and suggests that this model can be applied to proteins from all species.

504a Synergetic Approach to Nonnative Aggregation by Multi-Assay Design of Experiments, Multivariate Statistics, and Mechanistic Kinetics

Yi Li[1], Babatunde A. Ogunnaike[2], and Christopher J. Roberts[1]. (1) Department of Chemical Engineering, University of Delaware, 150 Academy St., Colburn Laboratory, Newark, DE 19716, (2) Chemical Engineering, University of Delaware, 150 Academy St., Newark, DE 19716

Nonnative protein aggregation is a common challenge during biopharmaceutical product development. Protein aggregates are often observed during protein expression, purification, and fill-finish operations, as well as during product storage prior to use. Aggregates may be soluble or insoluble, and the former are often further categorized as low or high molecular weight (MW). Nonnative aggregates of all types are potentially problematic with respect to regulatory approval of a product, due to purity and safety requirements. To better predict and control aggregation under a variety of conditions it would be beneficial to have a general experimental and modeling approach to efficiently and accurately assess the underlying kinetics and their dependence on a range of experimental variables, such as pH, temperature, protein concentration, and excipient type / concentration. This challenge is made more difficult by limitations of many standard experimental techniques to detect and monitor one or more key intermediates in multi-stage aggregation pathways, and outstanding questions regarding which techniques, if any, are best suited for monitoring aggregation.

The work here uses alpha-chymotrypsinogen A (aCgn) as a model protein that readily undergoes nonnative aggregation under accelerated conditions [1]. aCgn is shown here to display a range of different aggregate morphologies and kinetic behaviors that depend primarily on sample pH and electrolyte concentration. Extensive, multivariate data sets for aggregate formation kinetics, structure, morphology, and (soluble) aggregate size distributions were obtained as function of temperature, pH, [NaCl], [sucrose], and protein concentration. The approach includes: (1) statistical design of experiments and multivariate data analysis; (2) quantitative mechanistic kinetic modeling with a Lumry-Eyring Nucleated Polymerization treatment that was extended here to explicitly account for aggregate-aggregate condensation steps; (3) semi-quantitative characterization of aggregate size, structure, and morphology. Together, (1) to (3) provide means to efficiently interpolate and predict qualitative and quantitative differences in aggregate characteristics and changes in the underlying mechanism of aggregation across a wide range of experimental variables. The results also clearly illustrate the potential power of combining in-line multi-angle static laser light scattering with size-exclusion chromatography for systems with high-MW aggregates. The experimental and modeling framework further facilitates categorizing and discriminating among a variety of existing aggregation models.

[1] Andrews JM, Roberts CJ. Biochemistry 46 7558-7571 (2007); Weiss WF IV, Hodgdon TK, Kaler EW, Lenhoff AM, Roberts CJ. Biophys. J. 93 4392-4403 (2007); Andrews JM, Weiss WF IV, Roberts CJ. Biochemistry 47 2397-2403 (2008).

504b Folding and Aggregation of Polyglutamine Peptides and Proteins

Regina M. Murphy, Robert Walters, Matthew Tobelmann, and Joseph Bernacki. Chemical and Biological Engineering, University of Wisconsin, 1415 Engineering Drive, Madison, WI 53706

Huntington's is the most prevalent example of the "CAG diseases". Each of these neurodegenerative diseases is associated with a unique protein and unique clinical symptoms, but they share two common features: the disease-associated protein has an abnormally expanded polyglutamine (polyQ) domain, and the protein aggregates into nuclear and/or cytoplasmic inclusions. A causal link between aggregation and toxicity is supported by many studies, but the mechanism of toxicity is still unknown. We are interested in developing a quantitative description of the kinetics of aggregation of peptides and proteins with expanded polyQ domains. We believe this information is relevant in determining the size and nature of the polyQ aggregates that are responsible for toxicity, and in developing new therapeutic approaches for inhibiting toxicity. Our first objective is to examine the mechanism of aggregation of synthetic polyQ peptides, and the role of the length of the polyQ domain in moderating aggregation. The dominant hypothesis in the literature is that aggregation proceeds via formation of a monomeric nucleus that is thermodynamically unfavorable. We show, through statistical model discrimination methods, that the published data do not support this conclusion. Further, we present experimental data indicating that aggregation proceeds through formation of soluble oligomers that lack defined secondary structure. Through FRET analysis, we show that shorter, aggregation-resistant polyQ peptides are expanded in physiological buffer whereas longer, aggregation-prone peptides are collapsed, and we propose a mechanism by which the extent of collapse is related to aggregation. Our second objective is to examine the influence of polyQ domains on protein misfolding and aggregation. We report on a strategy for generating "length libraries" of proteins with variable-length polyQ inserts. We have produced recombinant proteins containing polyQ inserts and have characterized how folding, misfolding, and aggregation of these proteins is correlated with polyQ length.

504c Synthetic Prions Shed Light on Conformational Basis of Strain Diversity

David Colby, Institute for Neurodegenerative Diseases, UCSF, Box 0518, San Francisco, CA 94143 and Stanley Prusiner, Ind, UCSF, Box 0518, San Francisco, CA 94143.

Prions are proteins which are able to adopt self-replicating conformations. In both fungi and mammals, multiple prion strains have been observed to arise from a single protein sequence, indicating that distinct protein conformations underlie prion diversity. We are generating novel prion strains synthetically, by refolding recombinant protein into different conformations *in vitro* and introducing these conformations into mice. We are also attempting to develop stem cell models of prion disease and to infect these with both naturally occurring and synthetic prion strains. In one major breakthrough, we have successfully produced novel synthetic prion strains that mimic rare and controversial prions occurring naturally in humans and sheep. These prion strains lack resistance to protease digestion, once believed to be an essential hallmark of mammalian prions. In other work, we have generated an array of synthetic prions with varying degrees of protease resistance. These synthetically produced prions are shedding light on the role of prion conformation in disease pathogenesis.

504d Analysis of Protein Misfolding and Aggregation Using Peptide Microarrays

Moumita Bhattacharya, Jason Lin, Sumitra Sukumaran, Jonathan Dordick, and **Peter M. Tessier**. Dept of Chemical & Biological Engineering, Rensselaer Polytechnic Institute, Troy, NY 12180

The "misfolding" and assembly of proteins into ordered and amorphous aggregates is implicated in a wide range of detrimental and beneficial biological functions. Although much less is known about protein misfolding than protein folding, it appears that the molecular interactions governing these two processes are very different. To further illuminate the molecular determinants of protein aggregation, we have sought to understand the underlying protein-protein interactions. Recently we have demonstrated that peptide microarrays can be used to identify small sequence elements within several amyloid-forming proteins that govern their aggregation behavior (Tessier & Lindquist, Nature, 2007). In this presentation we will discuss our recent work using peptide microarrays to understand the interactions governing aggregation of closely related prions and how these interactions regulate species barriers. Moreover, we will discuss analysis of the peptide interactions driving Abeta aggregation associated with Alzheimer's disease, as well as mechanisms of small molecule inhibition of Abeta aggregation.

Web Page: www.rpi.edu/~tessip

504e Computational Predictions of the Enzymatic Activity of Single Deletion Mutants

Monica Berrondo and Jeffrey J. Gray. Chemical and Biomolecular Engineering, Johns Hopkins University, Baltimore, MD 21218

Single amino acid mutations, (substitutions, insertions, or deletions) in proteins can vary in effect fron negligible to complete loss of function but such effects are not generally predictable. It is generally thought that a deletion is more disruptive to the structure, and thereby the function, of a protein since the gap created must be closed in the folded conformation. Such deletions are widely observed in nature and are often detrimental to the activity and stability of the enzymes, resulting in diseases such as cancer, cystic fibrosis, or osteogenesis imperfecta. Using a set of 50 amino acid deletions from TEM-1 beta-lactamase (Simm et al., FEBS Letters, 581:21, 2007 3904,) we have developed an algorithm to model the perturbed structure of deletion mutants and predict their activity. Preliminary results show that metrics such as distortion of the active site, formation of hydrogen bonding networks, and steric clashes between the ligand and the mutant protein can be used to predict the activity in the mutant proteins. We will further attempt to distinguish mutants which alter substrate concentration required for catalysis (Km) and substrate turnover (kcat). The ability to predict the activity of enzyme mutants has implications for engineering new proteins and understanding the molecular basis of disease. Figure: TEM-1 beta-lactamase with deletion points from the Simm dataset shown in gray. Residues that were deleted are shown in shades of gray representing the amount of activity observed experimentally; no activity (black), wild-type activity (light gray)

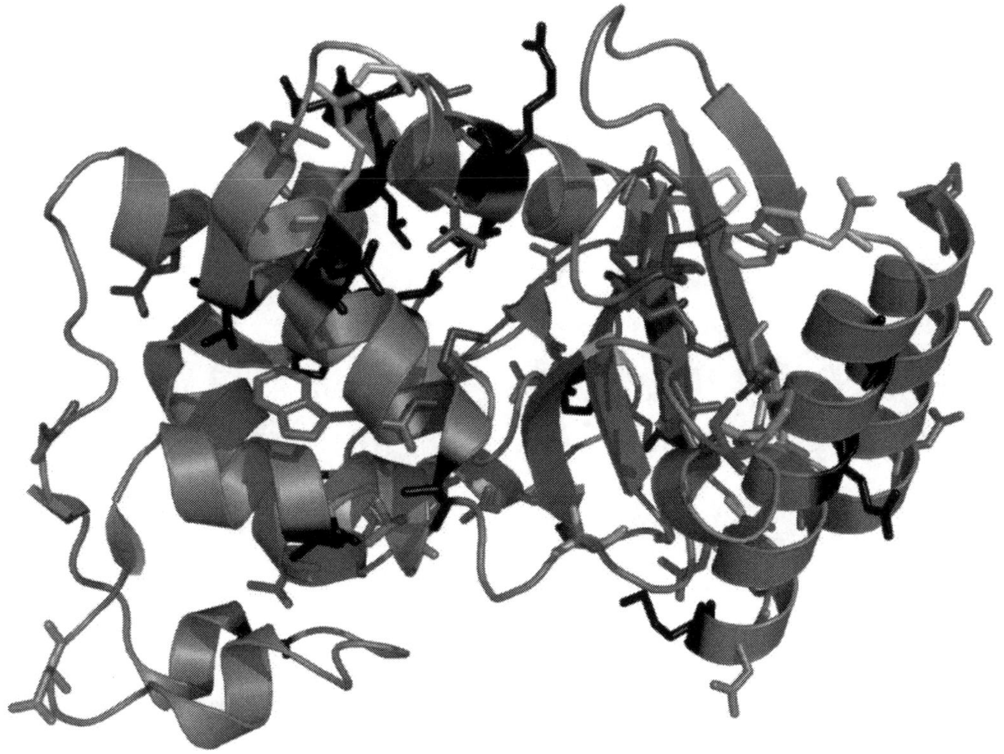

504f Computational Analysis and Design for Altered Cofactor Specificity

George A. Khoury, Hossein Z. Fazelinia, and Costas D. Maranas. Department of Chemical Engineering, Pennsylvania State University-University Park, 112A Fenske Laboratory, University Park, PA 16802

Receptors have previously been computationally designed to bind new cognate ligands selectively to closely related molecules and have been shown to have affinities surpassing their naturally evolved parent proteins. In this work, we compiled a comprehensive dataset of previous protein engineering studies that changed cofactor specificity from NADP(H)→NAD(H), and NAD(H)→NADP(H). We analyzed the metrics of volume, hydrophobicity, and charge, and rationalized their effects on the cofactor affinities for different mutations on a cross-section of that dataset. Calculated cofactor binding energy was also explored and verified as a surrogate for computational cofactor alteration. The implicit solvation models generalized born with molecular volume and generalized born with simple switching were integrated in the Iterative Protein Redesign and Optimization framework to redesign Candida boidinii xylose reductase (CbXR) to use NADH as its cofactor by finding the optimal set of mutations in the CbXR binding pocket. NAD(H) is an order of magnitude less expensive than NADP(H), much more stable, and its concentration is roughly an order of magnitude greater than that of NADP(H). IPRO generated five libraries utilizing various tolerances in the energy cutoff and metropolis criterion. The modified version of IPRO was also used to predict mutations for different design positions of the bacterial transcriptional regulatory protein, AraC to bind different conformations of D-arabinose. This computational framework was conducted in coordination with an experimental study that successfully redesigned the AraC protein to have altered effector selectivity capable of distinguishing between different forms of the arabinose sugars.

504g Molecular Design of Therapeutic Antibodies with Enhanced Stability

Naresh Chennamsetty[1], Vladimir Voynov[1], Veysel Kayser[1], Bernhard Helk[2], and Bernhardt L. Trout[1]. (1) Chemical Engineering, Massachusetts Institute of Technology, 77 Massachusetts Ave, E19-528, Cambridge, MA 02139, (2) Novartis Pharma AG, Basel, Switzerland

Antibodies currently constitute the most rapidly growing class of human therapeutics. One of the major problems encountered in antibody-based therapies is that these antibodies tend to aggregate under high concentration formulations required for disease treatment. The aggregation in turn leads to a decrease in antibody activity and could elicit an immunological response. In this study, we employ molecular simulation tools in collaboration with experimental techniques to understand the mechanism behind antibody aggregation. Based on this mechanistic understanding, we engineer antibodies with enhanced stability. Using detailed atomistic molecular dynamics simulations of the antibody Fab, Fc fragments and that of the full antibody we show that there are specific regions on the surface of antibody that are prone to aggregation. These aggregation prone regions are either always surface exposed, or exposed due to dynamic fluctuations or conformational changes observed in our molecular simulations. Based on these simulations, we develop a parameter that indicates the location and size of these aggregation prone regions. Furthermore, we perform mutations in these specific aggregation prone regions to engineer antibodies with enhanced stability.

Web Page: web.mit.edu/naresh/www/

507a JNK Regulates Adherens Junctions by Phosphorylating Beta-Catenin

Meng Horng Lee, Chemical Engineering, SUNY Buffalo, 916 Furnas Hall, SUNY Buffalo, Buffalo, NY 14260 and Stelios T. Andreadis, Chemical and Biological Engineering, SUNY-Buffalo, 908 Furnas Hall, Buffalo, NY 14260-4200.

Cell-cell adhesion is dynamically regulated under different biological circumstances such as the establishment of epithelial cell polarity, wound healing, cell scattering and tumorigenesis. However, the mechanism of adherens junction regulation is not completely understood. The current work led to the discovery that the well known kinase, c-Jun N-terminal kinase (JNK) is associated with adherens junctions and regulates cell-cell adhesion.

Specifically, we found that blocking JNK led to translocation of E-cadherin and beta-catenin to the cell surface and formation of compact colonies, suggesting a possible role of JNK in cell adhesion. Indeed, immunostaining and co-immunoprecipitated showed that JNK co-localized and bound to adherens junctions (AJs) complexes. In vitro kinase assays and mass spectrometry determined that JNK phosphorylated beta-catenin at serine 33/37 and threonine 41. Surprisingly, these were the same sites phosphorylated by GSK3-beta, which is involved in beta-catenin transcriptional activity. In addition, a combination of chemical inhibitors, dominant negative JNK constructs and siRNA indicated that JNK binds to and phosphorylated beta-catenin and that this phosphorylation leads to dissolution of adherens junctions and cell scattering. Our results showed for the first time that JNK is involved in regulation of adherens junctions by binding to and phosphorylating beta-catenin. These findings contribute to our understanding of cell-cell adhesion and may have wider implications in wound healing, embryonic development, tissue engineering and cancer metastasis.

507b Mechanical Shedding of L-Selectin from the Leukocyte Surface: Theory and Experiment

Dooyoung Lee[1], Kelly E. Caputo[2], Daniel A. Hammer[1], and Michael R. King[3]. (1) Bioengineering, University of Pennsylvania, 240 Skirkanich Hall, 210 South 33rd St, Philadelphia, PA 19104, (2) Chemical and Biomolecular Engineering, University of Pennsylvania, 311A Towne Building, 220 South 33rd St, Philadelphia, PA 19104, (3) Biomedical Engineering, University of Rochester, 218 Goergen Hall, RC 270168, Rochester, NY 14627

L-selectin-mediated leukocyte rolling is a key initial step in the multistep leukocyte adhesion cascade in blood. We have previously shown that constitutive L-selectin is cleaved from the neutrophil surface during rolling on a sialyl Lewis x-coated planar surface under physiological shear flow without the addition of exogenous stimuli. Based on the experimental work, we have extended the Adhesive Dynamics simulation of L-selectin-mediated cell rolling by incorporating the shear-dependent mechanical shedding of L-selectin. Utilizing the Bell model to describe a shedding rate which presumably increases exponentially with force, we were able to recreate the characteristics of L-selectin-mediated neutrophil rolling observed in flow experiments. First, the rolling velocity was found to increase during rolling due to mechanical shedding of L-selectin. The number of L-selectin shed during rolling depends strongly on the shear stress applied. When most of the L-selectin located on the tips of deformable microvilli was cleaved by force exerted on the L-selectin bonds, the cell detached from the reactive plane to join the free stream as observed in experiments. In summary, we showed through detailed computer simulations that the force-dependent shedding of L-selectin could fully explain the rolling behavior of neutrophils mediated by L-selectin in vitro.

507c Carcinoembryonic Antigen Is An E- and L-, but Not P-, Selectin Ligand

Susan N. Thomas[1], Fei Zhu[1], Christina Alves[1], **Matthew Dallas**[1], Ronald Schnaar[2], and Konstantinos Konstantopoulos[1]. (1) Chemical and Biomolecular Engineering, Johns Hopkins University, 3400 North Charles St., Baltimore, MD 21218, (2) Pharmacology and Molecular Sciences, Johns Hopkins University, 3400 North Charles St., Baltimore, MD 21218

Selectin-mediated adhesion of tumor cells to platelets, leukocytes and endothelial cells may regulate their hematogenous dissemination in the microvasculature. We recently identified CD44 variant isoforms (CD44v) as functional P-, but not E- or L-, selectin ligands on colon carcinoma cells. Moreover, a ~180-kDa sialofucosylated glycoprotein(s) mediated selectin binding in CD44-knockdown cells. Using immunoaffinity chromatography and tandem mass spectrometry, we identify this glycoprotein as the carcinoembryonic antigen (CEA). Blot rolling assays and flow-based adhesion assays using microbeads coated with CEA immunopurified from LS174T colon carcinoma cells and selectins as substrate reveal that CEA possesses E- and L-, but not P-, selectin ligand activity. CEA on CD44-knockdown LS174T cells exhibits higher HECA-452 immunoreactivity than CEA on wildtype cells, suggesting that CEA functions as an alternative acceptor for selectin-binding glycans. The enhanced expression of HECA-452 reactive epitopes on CEA from CD44-knockdown cells correlates with the increased CEA avidity for E- but not L-selectin. Through the generation of stable knockdown cell lines, we demonstrate that CEA serves as an auxiliary L-selectin ligand, which stabilizes L-selectin-dependent cell rolling against fluid shear. Moreover, CEA and CD44v cooperate to mediate colon carcinoma cell adhesion to E- and L-selectin at elevated shear stresses. The novel finding that CEA is an E- and L-selectin ligand may explain the enhanced metastatic potential associated with tumor cell CEA overexpression and the supportive role of selectins in metastasis.

507d Fibrin(ogen) Is a Functional CD44 Ligand

Christina S. Alves[1], Sergiy Yakovlev[2], Leonid Medved[2], and Konstantinos Konstantopoulos[3]. (1) Chemical and Biomolecular Engineering, Johns Hopkins University, 3400 North Charles St., Baltimore, MD 21218, (2) Biochemistry and Molecular Biology, University of Maryland, Baltimore, Baltimore, MD 21201, (3) Chemical & Biomolecular Engineering, Johns Hopkins University, 3400 N. Charles Street, Maryland Hall Room 223, Baltimore, MD 21218

Fibrin(ogen) plays a critical role in the hematogenous dissemination of tumor cells, including colon carcinoma, presumably by mediating their sustained adhesion and survival in the high shear environment of secondary target organs. However, the fibrin(ogen) ligand(s) on colon carcinoma cells has yet to be defined along with its relative capacity to bind fibrinogen versus fibrin under flow. Using CD44-knockdown LS174T colon carcinoma cells and microspheres coated with CD44 immunopurified from control LS174T cells, and purified fibrin(ogen) as substrate, we demonstrate that CD44 variant isoforms (CD44v) act as the major fibrin, but not fibrinogen, ligands on colon carcinoma cells. In contrast, the standard form of CD44, CD44s, immunoprecipitated from human myeloid cells is capable of binding to both fibrinogen and fibrin in shear flow. CD44v-fibrin binding is primarily dependent on the presence of O-linked glycans on CD44v, whereas the CD44s-fibrin interaction has an absolute requirement for N-linked glycans. Interestingly, both N- and O-linked glycans on CD44s are necessary for optimal CD44s-fibrinogen binding. Enzymatic assays reveal that chondroitin and dermatan sulfates are also implicated in the CD44-fibrin(ogen) molecular recognition. To characterize the CD44 binding site on the fibrin(ogen) molecule, we perfused CD44-coated microspheres over the major fibrin(ogen) proteolytic fragment D-D and (t)-NDSK, and the recombinant Aα221-610 and $(\beta 15\text{-}66)_2$ fragments. Our flow-based adhesion assays disclose that CD44-bearing microspheres interact avidly and extensively with the $(\beta 15\text{-}66)_2$ and t-NDSK fragments. We verified these results and determined the relative affinity of these fragments for CD44v and CD44s via Surface Plasmon Resonance. Use of the function-blocking monoclonal antibody (mAb) Hermes-1 enabled us to localize the fibrin epitope on the CD44v molecule, which is identical to that of hyaluronic acid. In contrast, this mAb failed to inhibit the adhesive interactions of CD44s-coated microspheres with fibrin, suggesting the existence of an additional fibrin binding epitope on CD44s. Taken together, our data bring together disparate observations about the contributions of fibrin(ogen) and CD44 to hematogenous metastasis in a unifying mechanistic interpretation, and provide a rational basis for the design of novel therapeutic strategies to impede metastasis.

507e Transmembrane Domain Interactions in Regulation of β_1 Integrin Function

Bryan Berger[1], Lisa Span[1], Joel Bennett[2], and William DeGrado[1]. (1) Biochemistry and Biophysics, University of Pennsylvania School of Medicine, 1009 Stellar-Chance Laboratory, 36th & Hamilton Walk, Philadelphia, PA 19104, (2) Heme/Onc Division, University of Pennsylvania School of Medicine, 914 BRB II/III, 421 Curie Drive, Philadelphia, PA 19104

Integrins are a diverse family of adhesion molecules that mediate cell-cell and cell-matrix interactions. Of these, $\alpha_{IIb}\beta_3$ has been most extensively characterized, and provides a model for regulation of integrin function. A central feature of integrin regulation is transmembrane (TM) domain interaction: when integrins are inactive, the TM domains of their α and β subunits interact, but these domains separate when the integrin assumes its active conformation. Previously, we observed that the α_{IIb} and β_3 TM domains undergo both heteromeric and homomeric interactions and there is overlap of the helical interface that mediates these interactions. Less is known about whether other platelet integrin TM domains interact. To address this question, we examined the role of TM domain interactions in β_1 integrins, particularly $\alpha_2\beta_1$ and $\alpha_5\beta_1$. Using a novel ToxR-based assay for TM domain interactions in *E. coli* coupled with scanning mutagenesis, we identified residues important for homo- and heterooligomerization of the TM domains for each α/β chain, particularly small, polar residues that lie along one face of the TM helix. We also find overlap between the homo- and heterooligomeric interfaces for each of the integrin TM domains. We then determined the effect these residues have on the function of the full-length integrins by introducing several of the mutations into Jurkat and HEK293 suspension cell lines. Several of the TM domain mutations identified as important from the ToxR-assay caused activation of the full-length integrin, indicating these residues may interact in the inactive state. Overall, our results point to the essential role of TM regions in regulating integrin function and demonstrate the importance of small, polar residues in maintaining these integrins in their inactive conformations.

507f Formation and Deposition of Leukocyte Microparticles during Rolling

Phillip Coghill, Alexander Christov, and **David W. Schmidtke**. School of Chemical, Biological, and Materials Engineering, University of Oklahoma, 100 East Boyd, Norman, OK 73019

During inflammation and thrombosis, interactions of selectins with cell-surface adhesion molecules allow the tethering and rolling of leukocytes on blood vessel walls. Leukocytes express L-selectin, whereas activated endothelial cells express P- and E-selectin. Leukocyte rolling allows regional sampling of chemokines and other mediators of leukocyte targeting, which leads to integrin-dependent arrest and emigration of leukocytes into the underlying tissues. Rolling requires the rapid formation and rapid dissociation of selectin-ligand bonds that are subjected to tensile forces applied by wall shear stress. We have previously demonstrated the rapid formation of membrane tethers as neutrophils roll on P-selectin. Normally these membrane tethers retract back to the cell body, however we have also observed that membrane tethers can break and deposit membrane microparticle (< 1 &mum) or microfragment (> 1 &mum) on the underlying surface. Breakage of membrane tethers was observed when neutrophils rolled over a wide range of P-selectin site densities (170-1096 sites/&mum2) and shear rates (100-800s-1). We demonstrate that the frequency of membrane tether formation (40-100%) and microparticle deposition (0-40%) both increase as the wall shear stress increases, while there was a small effect of ligand density. We also demonstrate that deposited membrane microfragments may serve as adhesion points for other cells.

507g Design and In Vitro Characterization of Polymer Vesicles with Adhesive Properties That Mimic Leukocytes

Gregory P. Robbins, Chemical and Biomolecular Engineering, University of Pennsylvania, Room 311A Towne Building, 220 South 33rd Street, Philadelphia, PA 19130 and Daniel A. Hammer, Bioengineering, University of Pennsylvania, 240 Skirkanich Hall, 210 South 33rd Street, Philadelphia, PA 19104.

Cell sized polymer vesicles self-assembled from amphiphilic polymers are ideal platforms on which to design in vivo drug delivery, imaging, or cellular therapeutics. Increased bilayer thickness leads to tougher vesicles and larger hydrophobic bilayer cores in which to carry cargo compared to analogous phospholipids vesicles. Additionally, the use of synthetic materials allows for both simple tuning of membrane properties and easy attachment of targeting ligand to the exterior surface of vesicles. In this study we functionalize the surface of polybutadiene-polyethylene oxide based vesicles with anti-ICAM-1 antibody and sialyl Lewisx in order to mimic the adhesive characteristics of activated leukocytes which bind to the targets of these ligands, ICAM-1 integrin and selectins. We chose these targeting ligands because activated leukocytes adhere to sites of inflammation in blood vessel endothelium, and by delivering vesicles to areas of inflammation, we target sites of infection, inflammatory disease, and cancer. We used a modular biotin-avidin system to functionalize the vesicles, and we characterized the surfaces of these vesicles using confocal microscopy and flow cytometry. Using parallel plate flow chamber assay at typical post-capillary venule shear rates and MATLAB-based particle tracking, we show that these vesicles mimic the rolling to firm adhesion trajectories of leukocytes. This study shows that we can mimic the adhesive characteristics of a leukocyte under flow with a cell-free system capable of delivering various types of cargo. These vesicles serve as a basis for the design of a targeted multifunctional vehicle for simultaneous imaging and delivery of small molecules or therapeutic proteins.

The Pharmaceutical Industry is notoriously conservative with respect to new technology. This attitude is understandable, given the real constraints of the healthcare field. Every innovation carries with it a burden: New processes and perhaps new equipment must be developed, along with new test methods and validation packages. New types of clinical trials may be needed. Manufacturing plants may need to be constructed or reconfigured. Various review boards and regulatory authorities must also be convinced of the soundness of any new approach, and human health and safety must always come first. These factors combine to impede implementation of new technology, yet there are equally important drivers for innovation: society continues to demand higher-quality healthcare and innovative medical treatments. Our attitude has evolved from the amazement that greeted new "miracle drugs" in the last century to an expectation that science will provide a cure (or at least a treatment) for every ailment that plagues us. Medical science and the healthcare industry have achieved the successes that led to this expectation, but each new increment of progress seems to come with increasing effort and sharply increasing costs. Both medical innovations and containment of development costs depend on adoption of new technologies.

Technology inventions occur in university laboratories, and in the research labs of private companies large and small. Technology development (i.e. to the point of application to make a product) tends to occur mostly in industry. Out of the hundreds of new technology ideas over the last few decades, which ones have become, or are becoming mainstream practices in the industry? This talk will examine a number of examples of "newer" technologies that are increasingly mainstream in industrial practice. These examples will be divided into two broad categories: (1) Technologies that increase efficiency or productivity, or otherwise offer benefits to the manufacturing process, and (2) technologies that enable new medical treatments, thus benefiting patients directly (e.g. drug delivery), with an emphasis on the second category. The author will also offer a perspective on technologies that have received publicity, but have not yet been widely adopted, and what might be required for them to attain wide application.

536c Real-Time Monitoring of Viral Replication and Anti-Viral Drug Discovery by Molecular FRET Probes

Wilfred Chen, Chemical and Environmental Engineering, University of California, Riverside, Bourns Hall A242, University of California, Riverside, Riverside, CA 92521

The ability to detect infectious viruses is of critical importance in medical diagnostic and environmental/agricultural protection. Current methods to assess the presence of infectious viruses are based on mammalian cell culture and rely on the production of visible cytopathic effects (CPE). Depending on the specific virus type and concentration in the sample, it may take several days or weeks for CPE to appear. In this talk, I will discuss several FRET-based approaches recently developed in my lab for real-time monitoring of viral infection. Quantum dot-based molecular probes were designed to detect the presence of viral RNAs or proteases as an indicator of viral infection. For intracellular delivery, a cell-penetrating TAT peptide was appended to the probes to enable real-time monitoring of viral infection in living cells. Fluorescence microscopy and flow cytometry were used to directly visualize infected cells and to subsequently follow virus spread in situ. As an added benefit, the specific nature of these probes also enables their utility as a sensitive agent for viral inhibitors discovery. Some progress in this area will be discussed.

563d Nanoparticles for Drug Delivery and Biochemical Sensing

Michael V. Pishko, Department of Chemical Engineering, The Pennsylvania State University, 104 Fenske Laboratory, University Park, PA 16802-4400

Text Not Available.

569a A Unified Mathematical Model for Predicting Controlled Release from Surface and Bulk Eroding Matrices without Regression

Sam N. Rothstein, William Federspiel, and Steven Little. Department of Chemical Engineering, University of Pittsburgh, 1237 Benedum Hall, Pittsburgh, PA 15261

The development of polymeric matrices to control the release of small molecules, peptides and proteins has created an industry valued at over $110 billion dollars. Spurred by a desire to explain how agent egress occurs, many attempts have been made to model drug release based on the physical properties of the matrix, drug, and polymer. Over time, these models have progressed from empirical equations that provide descriptive insights into more mechanistic forms capable of predicting release data in specific matrix systems, such as small molecules encapsulated in PLGA microspheres or polyanhydride implants.

We have recently published the first, broadly applicable, predictive model for the most widely used polymers in bulk-eroding biodegradable matrices (e.g. polyester formulations such as Lupron Depot®). We have also recently extended this model to traditionally surface eroding systems, such as polyanhydride implants (e.g. Gliadel wafer®). In its new form, the model considers two phases (solid and aqueous) of the encapsulated agent to account for dissolution kinetics. Accounting for these kinetics is essential, as many polyanhydride matrices, being too small to prolong release via surface erosion, extend release based solely upon solubility limitations. Diffusion reaction equations were also added to account for matrix hydration kinetics, which can have a significant duration over an implant's lifetime. Accounting for the hydration kinetics allows the calculation of the critical length where a system transitions from surface to bulk erosion. Unlike prior methods of calculating this critical length, the current model uses a mechanistically relevant second order degradation rate expression, which also accounts for changes in the polymer's initial molecular weight.

This extended form of the model can now be used to predict release data without regression for both bulk and surface eroding systems. Further, the plasticity of the model's framework allows modeling of systems that noticeably transition from surface to bulk erosion. To test the model regression free predictions were compared to published data from all three types of systems. Theoretical predictions from this model also indicate that properly sized implants, composed of common biodegradable polymers with specific molecular weights, can prolong the duration of release for hydrophilic agents to at least one month by transitioning between erosion schemes.

In conclusion, extending our published model to account for dissolution and hydration kinetics allows it to accurately describe a broader range of systems. Accounting for the hydration kinetics also enabled the calculation of a critical length which hints at a transitioning erosion scheme for many systems originally thought to be dominated by surface erosion. With this increased understanding of biodegradable matrix controlled release, therapeutics design may be improved based on model predictions and new materials may be targeted based on their potential to control release.

569b Numerical Analysis of Multicomponent Flow and Condensation In a Dhp Chamber

Radompon Sungkorn, Pietro Santagati, Stefan Radl, and Johannes Khinast. Institute for Process Engineering, Graz University of Technology, Inffeldgasse 21a, Graz, Austria

Surface bio-decontamination is an important aspect in many sterile pharmaceutical applications. Frequently, the method of choice is the exposure of the surfaces to hydrogen peroxide (H2O2) vapor [1, 2]. This operation is called decontamination by hydrogen peroxide (DHP). The work's objectives are the simulation of the decontamination processes in a closed volume, known as a DHP chamber. The flow in the DHP chamber may be affected by the condensation of water and H2O2 vapor. Depending on the parameters local condensation occurs or is avoided. Thus, a detailed study of the influence of operating conditions on the flow characteristics and condensation is the basis for the design of the DHP chamber.

In this work we have developed a detailed model to study the condensation of water and H2O2 vapor in the presence of air as a non-condensable species in a DHP chamber. The film model based on the Stefan-Maxwell equations for the multicomponent diffusion is adopted. In this work, the model is extended to compute the condensation in complex geometries like DHP chambers and is solved using exact solution of Taylor and Krishna [3]. Effects of buoyancy are also included. The flow fields in the DHP chamber are solved using Reynolds-averaged Navier-Stokes (RANS) equations. The species transport equations are used to compute the distribution of water, H2O2 vapor and air in the chamber. The binary condensation is accounted for by a new wall boundary condition. The activity coefficient of the liquid phase is calculated using the Redlich-Kister approach with temperature-dependent coefficients [4]. Thus, the dew point of the mixture and the condensation phenomenon in the DHP chamber can be predicted accurately. The model is validated by experimental data available in literature and shows good agreement.

References

[1] C. Hultman, A. Hill and G. McDonell, The physical chemistry of decontamination with gaseous hydrogen peroxide, Pharmaceutical Engineering (2004).

[2] M. Park, D. Walting, The relationship between saturated hydrogen peroxide, water vapour and temperature. Pharmaceutical Technology Europe (2004).

[3] R. Taylor, R. Krishna, Multicomponent mass transfer, John Wiley and Sons, New York (1993).

[4] S. Manatt, M. R. Manatt, On the analyses of mixture vapor pressure data: The hydrogen peroxide/ water system and its excess thermodynamic functions, Chemistry-A European Journal (2006).

569c Evaluating Biopharmaceutical Economics and Capacity with Process Simulation Tools

Charles Siletti[1], Victor Papavasileiou[2], and Demetri Petrides[1]. (1) INTELLIGEN, INC., 700 Walton Ave., Mount Laurel, NJ 08054, (2) Intelligen Europe, Rijnsburgerweg 119, Leiden, 2334, Netherlands

The capital investment for new biopharmaceutical manufacturing facilities is around $7000 per liter of bioreactor capacity. Moderate to large scale facilities can cost hundreds of millions of dollars, yet such investments are often made with uncertain information about process performance, media and materials prices, and market conditions. This paper presents a systematic way to evaluate the critical costs and capacity issues in biopharmaceutical plants. Examples will be presented on determining the target product titer, optimizing bioreactor batch size and choosing where or whether to use disposable (single-use) equipment. Since many process decisions must be based on partial or uncertain information, this presentation will also describe how to evaluate the risk associated with process and economic assumptions.

Web Page: www.intelligen.com/literature

569d Analysis and Evaluation of Pharmaceutical Processes Under Uncertainty Using Process Simulation and Risk Analysis Tools

Charles Siletti and Demetri Petrides. INTELLIGEN, INC., 700 Walton Ave., Mount Laurel, NJ 08054

Process simulation tools used for batch process design, cycle time analysis, and cost estimation, typically rely on deterministic models. However most industries, and especially the pharmaceutical and specialty chemical industries, are characterized by uncertainty and variability in crucial process and market parameters that greatly influence the profitability of a project. This paper presents a framework for integrating reliable state of the art process simulation tools, which rely on deterministic models, with risk analysis tools, which allow for stochastic modeling of uncertain variables. A methodology is presented for performing a parametric study of batch process design and analysis under uncertainty, providing precise forecast statistics for the decision variables. The results of such analyses are useful for quantifying risks and facilitating decision making under uncertainty. The risk analysis methodology is illustrated with two case study, the involving the synthesis of a small pharmaceutical molecule, and the second dealing with production of pharmaceutical tablets.

Web Page: www.intelligen.com/literature

569e Facilitating Bioprocess Development and Technology Transfer with Computer Aids

Demetri Petrides and Charles Siletti. INTELLIGEN, INC., 2326 Morse Avenue, Scotch Plains, NJ 07076

The successful scale up and commercialization of biopharmaceuticals is a challenging task that requires collaboration of professionals from many disciplines. Process simulators and other computer aids can facilitate this task by assisting scientists and engineers to answer the following and other related questions: What is the impact of product titer increase on the capacity load of the downstream section, the overall throughput of a plant, and the cost of goods? What changes are required in an existing multi-product facility to accommodate the process of a new product? What is the range of variability that a process can accommodate if it operates under a tight cycle time? What is the impact of single-use equipment on the demand for utilities and the cost of goods? Our experience in addressing the above questions will be presented using industrial examples in which we evaluated alternative technologies for producing therapeutic monoclonal antibodies and vaccines.

Web Page: www.intelligen.com/literature

Synthesis and purification of 2-heptylpropargylamine

Lin Yang, Rongji Dai, Wei Liu and Yulin Deng

School of Life Science and Technology, Beijing Institute of Technology, 5 South Zhongguancun Street, Haidian District, Beijing, China

R-N-(2-heptyl)-N-methylpropargylamine (R-2-HMP) which is analogous to (R)-deprenyl, was developed as a monoamine oxidase type B inhibitor and putative antiapoptotic agent [1]. In the rat, studies of R-2-HMP identified that R-N-2-heptylpropargylamine (R-2-HPA) was the active metabolite of R-2-HMP. Wherever an effect of R-2-HMP has been observed, R-2-HPA has been shown to have the same effect and normally a more potent effect than R-2-HMP. This aliphatic desmethyl compound was shown to be moderate MAO-B inhibitors, but very good neural rescue agent; recently, it has been studied for development in the area of Parkinson's disease and Amyotrophic Lateral Sclerosis (ALS). Therefore, there is an increasing interest focused on the development of methods to synthesis this compound [2,3].

In this report, a new synthesis method of 2-HPA was described via 2-heptylamine (2-HA), which operating procedure was simpler and more convenient; solvent extraction was used to purify the synthesized products, and that the effect of extraction condition on the purification of 2-HPA was done; purity analysis was carried out on GC system.

1. Experimental

1.1 Instruments and reagents

GC (HP 5890, FID) ; Waters 717 HPLC Autosampler; NMR spectra were measured on a Bruker ARX400 spectrometer (Germany); EI-MS data were measured with VG ZAB-HS mass spectrometer; 2-HA and propargyl benzensulfonate (PBS) were purchased from Aldrich Chemical Inc., all other chemicals such as ether, anhydrous sodium carbonate, anhydrous magnesium sulfate as well as sodium hydroxide were of analytical grade.

2.2 Synthesis and purification of 2-HPA

2-HA reacted with PBS in aqueous Na_2CO_3 and diethyl ether to give the desired product 2-HPA and a byproduct N-2-heptyl dipropargylamine (2-HPPA). 2-HPA was then separated and purified by serials solvent extractions.

Synthesis: 2-HA (2.3g, 20mmol), 20 ml diethyl ether and 32 ml 10% aqueous Na_2CO_3 were placed respectively to 100 ml flask, added PBS (4.5g, 22mmol) dropwise to the above vigorously stirred solution through a dropping funnel and the mixture was refluxed (40°C) for 17hr. The reaction mixture was cooled to room temperature and moved to separatory funnel to separate and save the organic layer.

NMR ^1H: δ 3.49 (q, 2H, propargyl, -CH_2-, J=9.37), δ 2.85(q, 1H; NH; J=3.27), δ 2.18 (t, 1H; propargyl-CH-, J=2.34),δ 1.40(m, 9H, -C_4H_8-CH-), δ 1.03(d, 3H, 1-CH_3, J=4.68), δ 0.89(t, 3H, 2-CH_3)。 NMR ^{13}C: δ 82.35(C-9), δ 70.94(C-10), δ 51.34(C-6), δ 35.69(C-8), δ 35.50(C-6), δ 35.50(C-5), δ 31.97(C-3), δ 25.46(C-4), δ 22.58(C-2), δ 19.67(C-7), δ 13.98(C-1).

MS, m/e: 154 (M^++1) .

Purification: Divided the organic layer to several equals and placed in a 100mL separatory funnel, then sequentially extracted with (0.33 mol/L) NaH_2PO_3 buffer solution (pH=3.3-3.4). combine all the buffer extraction to a flask and adjusted appropriate pH, and then washed the buffer extraction with appropriate organic solvent to extract the residual 2-HPPA. The organic layer was quantitatively detected on HPLC-UV, TLC was also performed on the organic. If TLC showed there is no longer 2-HPPA was detected in the organic layer, adjusted the buffer extraction to pH=7 with 15% aqueous NaOH and extracted 2-HPA with diethyl ether, and to pH=11 to extract 2-HA.

Purity analysis: 2-HPA quantitative measurement was carried out on a HP 5890 Gas Capillary Chromatograph equipped with flame ionization detector (FID). Column: 2.6-di-o-pentyl-β-cyclodextrin (CD) was synthesized following the method of Dai.et.al (2004) [6]. Sample treatment: the solvent was evaporated to dryness by rotary evaporator; 0.1μl of this sample was injected into the GC-FID system.

3. Results and discussion

3.1 Synthesis, purity analysis (GC)

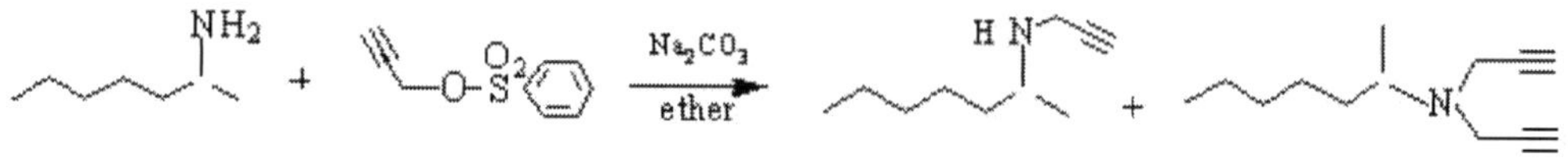

scheme 1 the reaction formula of 2-HPA

From the above-mentioned synthesis procedure, 2-HPA was synthesized; the reaction formula was shown in the scheme 1. In this paper, 2-HA and PBS were chosen as the raw materials to prepare the novel drug 2-HPA through a Nucleophilic Substitutions reaction, benzensulfonate group is a good leaving group, it can be displaced easily by amino group to give the desired product 2-HPA and byproduct 2-HPPA.

In this study, this Chemical reaction required much less time for completion. The reaction condition is mild with a yield of 2-HPA (61.4%), 2-HPPA (25.6%) and unreacted material 2-HA (13%), then the product was pale yellow which didn't need to discolor by boiling with activated carbon. After purification, 2-HPA GC purity: 99.6%.

3.2 Effect of extraction condition on the purification of 2-HPA

Purification of the 2-HPA was carried out through solvent extraction, it included two processes: 1) using NaH_2PO_3 buffer solution extracted the solutes from the reaction solution; 2) using organic extraction solvent to extract the different amines respectively under various PH conditions. In the first process, the desired product 2-HPA, unreacted material 2-HA and part of side-product 2-HPPA would be extracted to the NaH_2PO_3 buffer solution; in the second process we found that different organic extraction solvent and PH values of buffer phase played a vital role in the separation process.

3.2.1 Effect of organic extraction solvent.

During the extraction process, the polarity of organic solvent may have considerable effect in the separation process. Thus, different common organic extraction solvent such as heptanes, ethyl acetate, diethyl ether and petroleum ether were chosen to extract respectively the NaH2PO3 buffer with PH3.4 in the second process, as shown in the Fig.1, Solutes 2-HPA and 2-HPP were extracted by different ratio. These two compounds with varying degrees of N-substitution have the differential pka values and differential solubility in organic solvents.

Table 1 demonstrated the effect of different organic extraction solvent on the distribution coefficient and the corresponding selectivity coefficients of the two compounds. These results indicated that diethyl ether has the best selectivity for separation of the compounds 2-HPA and 2-HPPA. Therefore, diethyl ether is the most suitable among the solvents studied for our purposes.

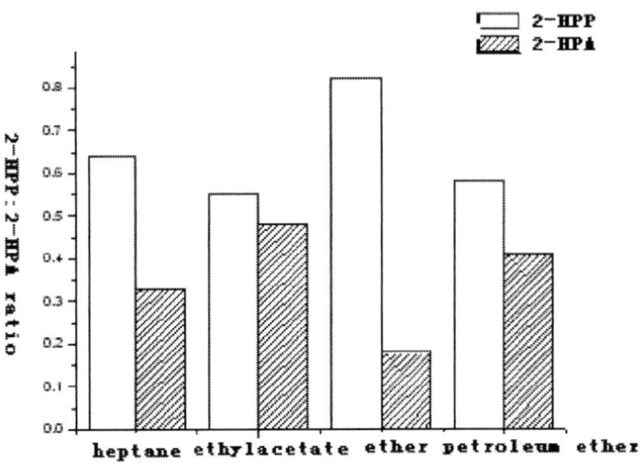

Fig. 1 Quantitative experiments illustrating separation of 2-HPP and 2-HPA in 1.0 N NaH2PO3 buffer at pH 3.4 using various solvents

Table 1 effect of different organic solvents on the distribution coefficient of 2-HPA and 2-HPPA

extraction solvents	Da (2-HPA)	Db (2-HPP)	$\beta = \frac{D_a}{D_b}$
heptanes	0.054	0.112	0.464
ethyl acetate	0.897	1.049	0.451
diethyl ether	0.475	2.164	0.219
petroleum ether	0.094	0.143	0.657

3.2.2 Effect of PH of the buffer solution

During the first process, for 2-HPP has weakly solubility in the NaH2PO3 buffer (PH3.4), thus majority of 2-HPP was separated from the mixture. NaH2PO3 buffer solution could selectively dissolve 2-HPA, 2-HA from the reaction solution, nevertheless, there was still a fraction of 2-HPP dissolved in the buffer, so the buffer should be washed by solvent to discard the residual 2-HPPA.

However, while the amines being extracted, the PH of the NaH2PO3 buffer solution increased pH=3.4~6.1. It was found that unequal amount of 2-HPA would be extracted at the same time while the buffer was washed. Table 2 was shown the extraction ratio of the two compounds extracted under buffer with various PH values and the yield of 2-HPA .Fig.3 illustrated the different extraction distribution ratio (D) of the two compounds extracted under buffer with various PH values.

It is apparent that under acidic conditions (PH3.3-3.4) the solubility of 2-HPP in diethyl ether was higher largely than 2-HPA. As the pH of the buffer increased, the solubility of 2-HPA in diethyl ether increased much faster than 2-HPP. This clearly indicated that the residual 2-HPPA can be selectively separated under pH=3.3-3.4 while losing the minimal 2-HPA.

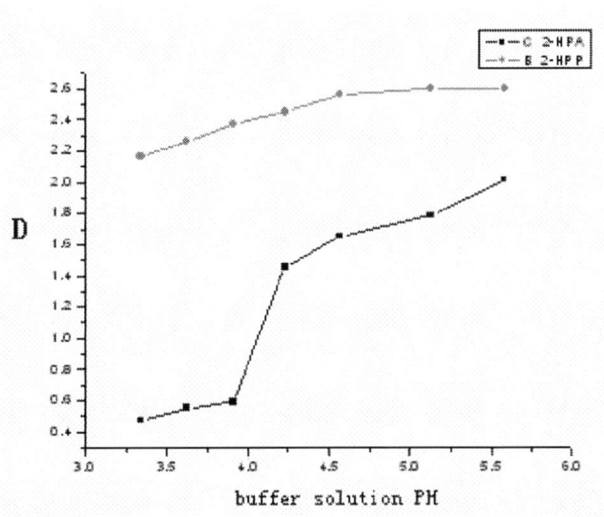

Fig. 2 effect of the various buffer PH values on the extraction distribution ratio of the two compounds

Table 2 quantitative experiment illustrating the different extraction distribution ratio of the two amines under buffer with various PH values

PH	2-HPP:2-HPA Ratio (HPLC)	Yield of 2-HPA (%)
3.34	6.69:1	61.4
3.62	5.67:1	59.7
3.91	5.37:1	57.7
4.23	3.04:1	53.4
4.57	2.81:1	50.6
5.12	1.74:1	47.3
5.58	1.04:1	40.7

These results indicated that, while pH 3.3-3.4 would be ideal to remove residual 2-HPP from the mixture, and then the buffer would be basified to PH 7.0 to extract 2-HPA. Furthermore, during the second process, 2-HA could be extracted unless the buffer solution was adjusted to PH11. Therefore, through the two procedures, 2-HPA with a approving yields by GC purity: 99.6% was obtained.

Conclusion

An cellular rescue drug 2-HPA was synthesized under a mild condition with an acceptable yield of 61% and its structure was identified by NMR and MS spectra; furthermore, we have demonstrated that the mixture of 2-HPA, 2-HPPA and 2-HA can be successfully separated by a buffer-based extraction procedure: while pH 3.3-3.4 for separating tertiary amines 2-HPP from the mixture ; pH=7.0 for extracting the secondary amines 2-HPA and PH 11 for removing the primary amines 2-HA.This process of amine purification has more potential than the time-consuming column chromatography and can routinely be employed in laboratory practice.

Reference

[1] D. Zhang, M.D. Berry, I.A. Paterson. Loss of Mitochondrial Membrane Potential Is Dependent on the Apoptotic Program Activated: Prevention by R-2HMP [J]. Journal of Neuroscience Research. 1999,58: 284-292

[2] Makoto Naoia, Wakako Maruyamab. Future of neuroprotection in Parkinson's disease Parkinsonism and Related Disorders [J]. 2001,8: 139-145

[3] Makoto Naoi, Wakako Maruyama, Apoptosis induced by an endogenous neurotoxin: N-methyl(R) salsolinol [J]. Toxicology. 2000, 153: 123-141.

[4] David A,Durden, Lillian E,Dyck Metabolism and pharmacokinetics, in the rat, of (R-2-(heptyl)methylpropargylamine (R-2HMP), a new potent monoamine oxidize inhibitor. Drug Metabolism and Disposition [J].2000; 28(2):147-154.

[5] Peter H. Yu, Bruce A. Davis, Alan A. Boulton. Aliphatic Propargylamines: Potent, Selective, Irreversible Monoamine Oxidase B-Inhibitors. J. Med. Chem. [J]. 1992; 35:3705-3713.

[6] Dai Rongji, Zhang Shu, Li Fang, Jin Hui, Gu Junling, Fu Ruonong. Synthesis and Characterization of B-Cyclodextrin Derivatives Containing Amino Groups and Carboxyl Groups. Journal of Beijing Institute of Technology [J].1998, 18(2): 159-164.

Preparation of Aliphatic Amines by the Leuckart Reaction

Lin Yang, **Rongji Dai**, **Wei Liu** and **Yulin Deng**, School of Life Science and Technology, Beijing Institute of Technology, 5 South Zhongguancun Street, Haidian District, Beijing, China

The in situ reductive amination of carbonyl compounds has been an important part of the synthetic chemist's repertoire since the introduction of a procedure based on the Leuckart reactions, conversion of certain ketones and aldehydes to the corresponding amines by heating with excess ammonium formate[1,2]. While this reaction has been very extensively investigated in aromatic compounds very little work has been done with aliphatic compounds other than terpenoid ketones. Wallach, representative of many predecessors proposed the mechanism of the reaction were as follows[3,4]:

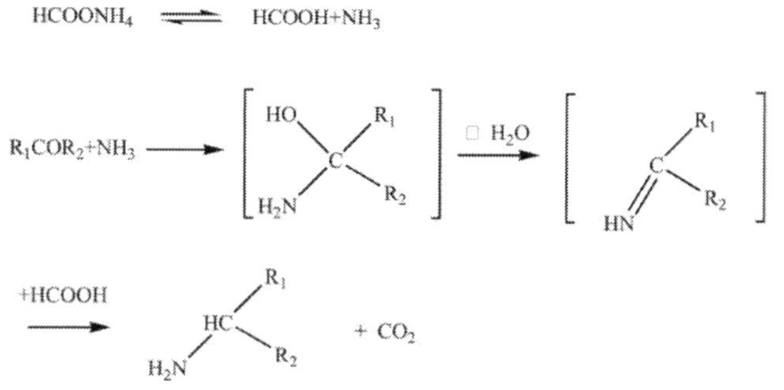

And they thought the impossible intermediate for this reaction in the form of the aliphatic imine ($R_1R_2C=NH$), which only one double bond was present. The proposed intermediate, however, contains a conjugate system of two double bonds at least[5].

We have been studying this reaction as a possible preparative procedure for wholly aliphatic amines and have obtained some rather interesting results. The Leuckart reaction gives generally good yields of products, furthermore, we do propose to offer some novel mechanisms in this report.

1.Experimental

1.1 Apparatus and reagents

A Milestone ATC-300 microwave laboratory system was used to perform the microwave reaction. ^{1}H NMR spectra were measured on the Bruker ARX-300 and ARX-400 spectrometer using $CDCl_3$ as solvent. Mass chromatographic analysis were performed using a Agilent-1100 series LC/MS Trap mass analyser. All the regents used were analytical grade without further purification.

1.2 Procedure

The following procedures are typical example of the experiments reported in the following Table 1. This reaction may be used to prepare the corresponding primary amines in satisfactory yields from a large number of aliphatic ketones with only a slight modification of the experimental conditions.

2-Heptylamine (i). To a 250ml four-necked flask, equipped with a dropping-funnel, thermometer, water segregator and down-directed condenser, was added with care 54g (0.86 mol) of ammonium formate. The temperature was raised to 120°C by melting the solid with stirring, and 19.8 g (0.172 mol, 99%) of 2-oenanthone was added when the ammonium formate melted completely. The temperature was maintained at 130-140° for seven hours and any ketone which distilled was returned to the flask at intervals. The formyl derivative was hydrolyzed in the reaction mixture by refluxing for eight hours with 110°C of concentrated hydrochloric acid (36-38%). After standing overnight, the mixture was diluted with 100 ml of water and filtered the foreign matter. The Filtrate extracted with 50ml ether to remove water-insoluble material. Aqueous solution was made alkaline with 15% NaOH solution and the oil thus produced extracted with 150ml ether. The ether solution was put together and washed three times with 100 ml NaH_2PO_3 buffer (pH 3.3-3.4) , so most of 2-heptylamine was solved in the buffer solution. Adjusted alkalinity (pH 9-10) with 15% NaOH solution again and extracted with ether three times (80ml, 50ml, 30ml) dried with anhydrous calcium chloride and the ether removed by distillation under reduced pressure. The residue gave 11 g (56%) of 2-heptylamine (purity: 99%). ^{1}H NMR ($CDCl_3$, δ ppm): 2.86 (t, 1H, J=11.6Hz, CH), 2.17(d, 2H, J=6.4Hz, NH_2), 1.25~1.31(m, 8H, CH_2), 1.04~1.07(m, 3H, CH_3), 0.84~0.87(m, 3H, CH_3). MS (ESI-MS): 116.1 $[M+H]^+$.

(2) Propylamine (a)

^{1}H NMR ($CDCl_3$, δ ppm): 2.65~2.76 (m, 2H, CH_2), 1.45~1.53 (m, 2H, CH_2), 1.24(s, 2H, NH_2), 0.92 (t, 3H, J=6.8Hz, CH_3). MS (ESI-MS): 60.1 $[M+H]^+$.

(3) Isopropylamine (b)

^{1}H NMR ($CDCl_3$, δ ppm): 2.92~3.00 (m, 1H, CH), 1.17 (t, 2H, J=3.6Hz, NH_2), 0.91 (d, 6H, J=6.4Hz, CH_3). MS (ESI-MS): 60.3 $[M+H]^+$.

(4) Butylamine (c)

^{1}H NMR ($CDCl_3$, δ ppm): 2.68~2.75 (m, 2H, CH_2), 1.77 (s, 2H, NH_2), 1.33~1.66 (m, 4H, CH_2), 0.92 (t, 3H, J=6.4Hz, CH_3). MS (ESI-MS): 74.6 $[M+H]^+$.

(5) 2-Butylamine (d)

^{1}H NMR ($CDCl_3$, δ ppm): 2.79~2.95 (m, 1H, CH), 1.36~1.50 (m, 2H, CH_2), 1.25 (t, 2H, J=4.0Hz, NH_2), 1.05 (d, 3H, J=5.6Hz, CH_3), 0.92 (t, 3H, J=7.2Hz, CH_3). MS (ESI-MS): 74.3 $[M+H]^+$.

(6) Pentylamine (e)

^{1}H NMR ($CDCl_3$, δ ppm): 2.68~2.77 (m, 2H, CH_2), 1.02~1.58 (m, 6H, CH_2), 1.12 (s, 2H, NH_2), 0.91 (t, 3H, J=6.4Hz, CH_3). MS (ESI-MS): 88.2 $[M+H]^+$.

(7) 1-Methylbutylamine (f)

^{1}H NMR ($CDCl_3$, δ ppm): 2.88~3.12 (m, 1H, CH), 1.24~1.36 (m, 4H, CH_2), 1.26 (d, 2H, J=3.6Hz, NH_2), 1.01 (d, 3H, J=6.0Hz, CH_3), 0.88~0.93 (m, 3H, CH_3). MS (ESI-MS): 88.1 $[M+H]^+$.

(8) 1-Ethylpropylamine (g)

^{1}H NMR ($CDCl_3$, δ ppm): 2.55~2.64 (m, 1H, CH), 1.70 (d, 2H, J=5.2Hz, NH_2), 1.46 (m, 2H, CH_2), 1.28 (m, 2H, CH_2), 0.91 (t, 3H, J=8.0Hz, CH_3). MS (ESI-MS): 88.2 $[M+H]^+$.

(9) Cyclopentylamine (h)

^{1}H NMR ($CDCl_3$, δ ppm): 3.31~3.43 (m, 1H, CH), 1.82~1.93 (m, 2H, CH_2), 1.72~1.76 (m, 2H, CH_2), 1.55 (t, 2H, J=5.6Hz, CH_2), 1.37 (d, 2H, J=5.6Hz, NH_2), 1.28~1.32 (m, 2H, CH_2). MS (ESI-MS): 86.1 $[M+H]^+$.

(10) Hexylamine (i)

^{1}H NMR ($CDCl_3$, δ ppm): 2.68~2.88 (m, 2H, CH_2), 1.44~1.63 (m, 2H, CH_2), 1.08~1.41 (m, 6H, CH_2), 1.17 (s, 2H, NH_2), 0.90 (t, 3H, J=6.8Hz, CH_3). MS (ESI-MS): 102.1 $[M+H]^+$.

(11) Cyclohexylamine (j)

^{1}H NMR ($CDCl_3$, δ ppm): 2.62~2.78 (m, 1H, CH), 1.44~1.97 (m, 5H, CH_2), 0.72~1.44 (m, 5H, CH_2), 1.18 (s, 2H, NH_2). MS (ESI-MS): 100.6 $[M+H]^+$.

(12) Heptylamine (k)

^{1}H NMR ($CDCl_3$, δ ppm): 2.63~2.77 (m, 2H, CH_2), 1.41~1.53 (m, 2H, CH_2), 1.40 (t,

2H, J=3.2Hz, NH_2), 1.18~1.39 (m, 8H, CH_2), 0.89 (t, 3H, J=6.8Hz, CH_3). MS (ESI-MS): 116.3 $[M+H]^+$.

(13) Octylamine (m)

^1H NMR ($CDCl_3$, δ ppm): 2.68~2.79 (m, 2H, CH_2), 1.69 (s, 2H, NH_2), 1.28~1.43 (m, 12H, CH_2), 0.89 (t, 3H, J=5.2Hz, CH_3). MS (ESI-MS): 130.7 $[M+H]^+$.

(14) 2-Octylamine (n)

^1H NMR ($CDCl_3$, δ ppm): 2.84 (t, 1H, J=9.6Hz, CH), 1.55 (s, 2H, NH_2), 1.29~1.40 (m, 10H, CH_2), 1.05~1.09 (m, 3H, CH_3), 0.89~0.91 (m, 3H, CH_3). MS (ESI-MS): 130.3 $[M+H]^+$.

concentrated hydrochloric acid (36-38%) as the reducing agent refluxing for eight hours.

2. Results and Discussion

2.1 Effect of different reagents

We found that some water was generated in the water segregator within ammonium formate melting. According to Wallach's view there were not water except for formic acid and ammonia. We suggested formamide was generated by deaquation of ammonium formate, and the reaction system was quite complex including formamide, water, ammonium formate, formic acid and ammonia (the last two were generated by hydrolysis). In order to study each reagent how to effect the the yield of amine, a series of experiments were ran with 2-oenanthone. The reactions were carried out in the reaction to undergo the same steps as above.

entry	reagent	temp(℃)	time (h)	yield (%)
1	formamide	170-180	21	15
2	formamide and water	170-180	21	16
3	ammonium formate	120-130	7	56
4	ammonium formate*	120-130	7	27
5	formamide and ammonium formate	170-180	12	51
6	formamide and ammonium formate *	170-180	12	58
7	formic acid and ammonia	120-130	8	55
8	formamide and formic acid	170-180	20	63

*: removed the water segregator and kept water in reaction system.

Formamide gives a lower yield of product whether the water is removed. When water is removed, a mixture of formamide and formic acid gives the best yields; ammonium formate gives better yields than formamide; a mixture of formamide and ammonium formate is as good as ammonium formate alone but no better. However, when no water is removed, the mixture is better than ammonium formate alone. It will be observed that in every case formamide is initially present or can easily be formed by dehydration of ammonium formate, a reducing agent is required. The examination of Table 2 shows that all of the compositions played respective role in the reaction.

2.2 Possible mechanism

Now that formamide, water, ammonium formate and formic acid are active in this reaction and influenced each other, we must consider all of the facts. A possible general mechanism system is proposed and these reactions were taking place simultaneously.

At first ammonium formate easily dehydrated to formamide under heating

$$HCOONH_4 \longrightarrow HCONH_2 + H_2O$$

And then it is the addition of formamide to the carbonyl carbon, present as the familiar resonance form [$R_1R_2C^+$-O^-].

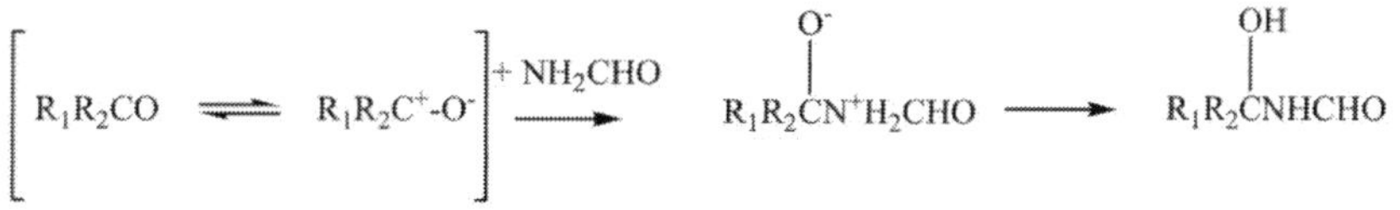

1

When water is split out, a stabile conjugate system is formed.

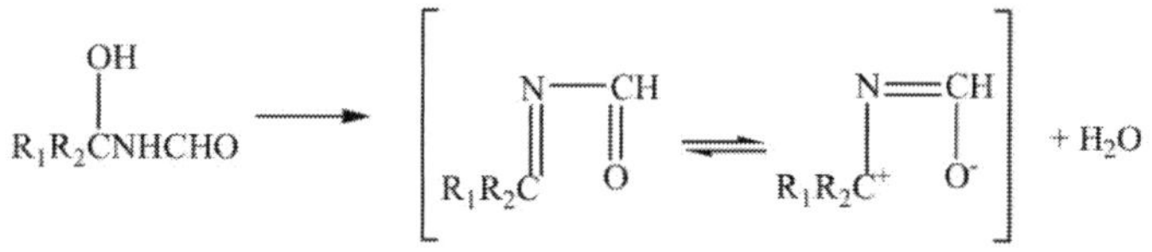

2

Little water still in the vessel under 120-130°C, therefore it can hydrolyze some formamide to give a small concentration of ammonium formate, including some undecomposed ammonium formate, which can serve as the reducing agent which the reaction requires.

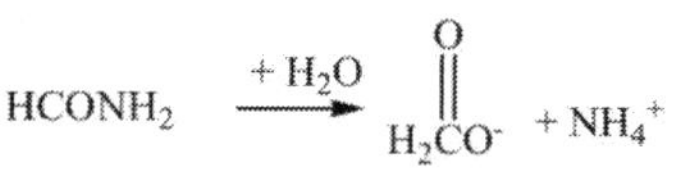

3

In a solution containing formamide (which has a very high dielectric constant) ammonium formate is probably highly ionized. The most reasonable mechanism for the reduction by ammonium formate consists of the addition of the formate ion to the positive center of the formimido compound, followed by **a** hydride ion shift with the subsequent release of carbon dioxide.

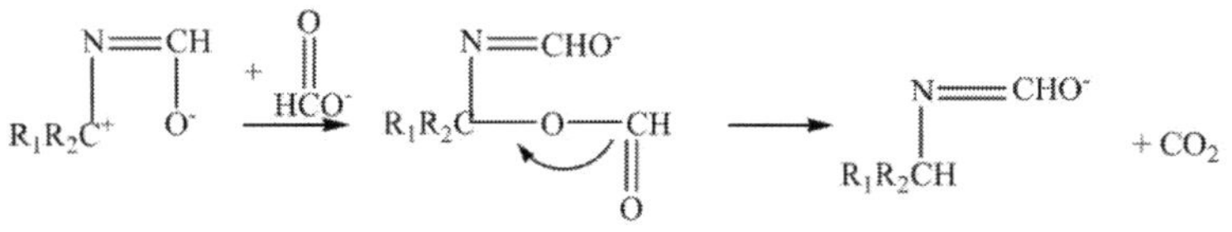

4

The ammonium ion gives up a proton to the intermediate and forms the aliphatic amide.

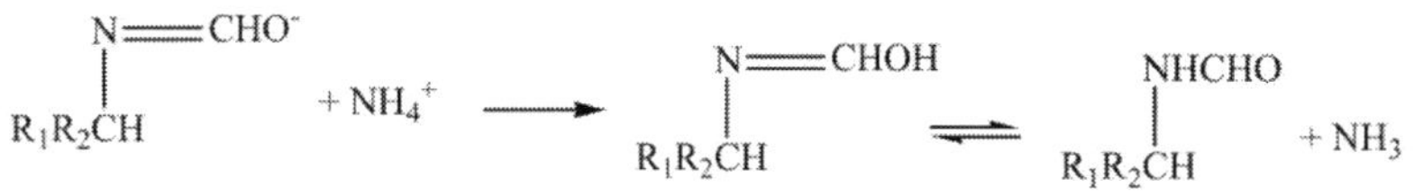

5

At last the aliphatic amide is hydrolyzed to aliphatic amine.

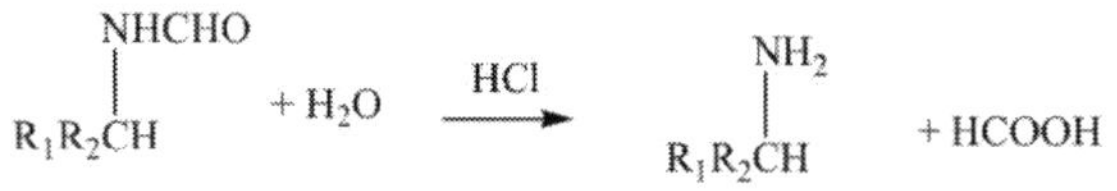

6

Water is a very important factor fellow Table 1 and the mechanism. The removal of water facilitates the formation of formamide from the ammonium formate as well as the formation of the intermediate of step 2. But it is unwise to remove water completely because formamide can't hydrolyze in step 3.

3 conclusion

A good method of obtaining aliphatic amines, based on the Leuckart synthesis, is described. It has been applied to the preparation of several old and new aliphatic amines. Comparative datas on various reagents have been tabulated and a systemic mechanism is advanced for the synthesis of aliphatic amines.

Reference

1. David T. Hill, Bernard Loev. 4-(2-Thienvl)-5-methylpyrimidine. Anomalous Leuckart product. J. Org. Chem., 1973, 38(11): 2102-2103.
2. Donald S. Noyce, Frank W. Bachelor. The Stereochemistry of the Leuckart

Reaction. J. Am. Chem. Soc., 1952, 74(18): 4577-4579.

3. Vincent J. Webers, William F. Bruce. The Leuckart Reaction: A Study of the Mechanism. J. Am. Chem. Soc., 1948, 70(4): 1422-1424.
4. FRANK S. CROSSLEY, MAURICE L. MOORE. Studies on the Leuckart reaction. J. Org. Chem..1944, 9(6): 529-536.
5. Thalji, R. K.; Ahrendt, K. A.; Bergman, R. G.; Ellman, J. A.. Annulation of aromatic imines via Directed C-H bond activation. J. Org. Chem.. 2005, 70(17): 6775-6781.

569h Micro Thermal Ablation for Transdermal Drug Delivery

Jeong-Woo Lee[1], Priya Gadiraju[1], Jung-Hwan Park[2], Mark Allen[3], and Mark Prausnitz[1]. (1) Chemical and Biomolecular Engineering, Georgia Institute of Technology, 311 Ferst Drive, Atlanta, GA 30332, (2) Bionano Technology, Kyungwon University, Seongnam, South Korea, (3) Electrical and Computer Engineering, Georgia Institute of Technology, 311 Ferst Drive, Atlanta, GA 30332

Introduction

The traditional passive transdermal patch systems have been limited to only small molecular weight and moderately lipophilic drugs due to the barrier properties of the skin, which reside in the cornified layer of keratinized epidermal cell, stratum corneum. Many research groups have demonstrated that the removal of stratum corneum could enhance the drug permeability to the skin and designed a number of active transdermal applications to overcome the stratum corneum barrier and deliver drugs efficiently in painless ways.

Although the main function of the skin is the protection of the human body from the external environment, it is composed of mechanically weak biomaterials. Thus, sufficient energy to ablate the skin can be applied to the skin. This has been demonstrated using lasers, RF heating using microelectrode arrays, and conductive heating from ohmic microheaters. Phase 1 and 2 clinical trials are underway using the latter two methods to deliver various drugs into the skin.

The key to success with thermal ablation of skin is to apply extremely short thermal pulses that heat just the surface of the skin, and thereby remove microscopic patches of the rate-limiting stratum corneum layer on the skin surface without damaging deeper, viable tissue. Because heating occurs on the skin surface, the short pulse assures a steep temperature gradient that does not persist long enough for significant thermal diffusion to deeper tissue. Typical ablation temperatures are hundreds of degrees Celsius and stratum corneum thickness is just 10 – 20 microns. Thermal pulses applied using current technology are milliseconds long.

In this study, we sought to apply microsecond thermal pulses of hundreds of degrees Celsius to ablated skin's stratum corneum. For the first time, we developed an arc-based energy release to achieve the extremely rapid heating.

Methods and Materials

A microdevice shown in Figure 1 was designed and fabricated to generate the ablation energy. The system has a 1 μl microchamber having two metal electrodes on each side and a nozzle at top; and a power supply circuit with discharging capacitors providing the electrical energy to the electrodes. The microchamber was filled with PBS serving as the medium through which the arc discharge occurs. The metal mask covered the outer surface of the nozzle and additional masks were used to localize the skin ablation effect.

The ablation energy generated by the system was simulated with the software COMSOL MULTIPHYSICS¢ç and the temperature was measured by using thermal indicating papers coated with heat sensitive polymer film, which turns transparent and shows the color of the background. The thermal paper was placed on the metal mask exposed to hot medium ejected from the system and the temperature of hot medium was computed on the basis of the temperature measurement. The recoil force generated by arc discharge was measured by the force sensor installed at the back side of the microchamber. The force sensor recorded the change of the generated recoil force according to the time after triggering arc discharge phenomenon in the microchamber.

For histological analysis, shaved pig cadaver skin was placed on the metal mask in the same way as when measuring the temperature. Then, a viscous drug solution having a model drug, sulforhodamine, was mounted on the ablated skin surface to visualize the effect of the skin ablation on drug permeation. Tissue samples were imaged by brightfield and fluorescent microscopy, cryo-sectioned, and stained with hematoxylin and eosin to observe the removal of stratum corneum.

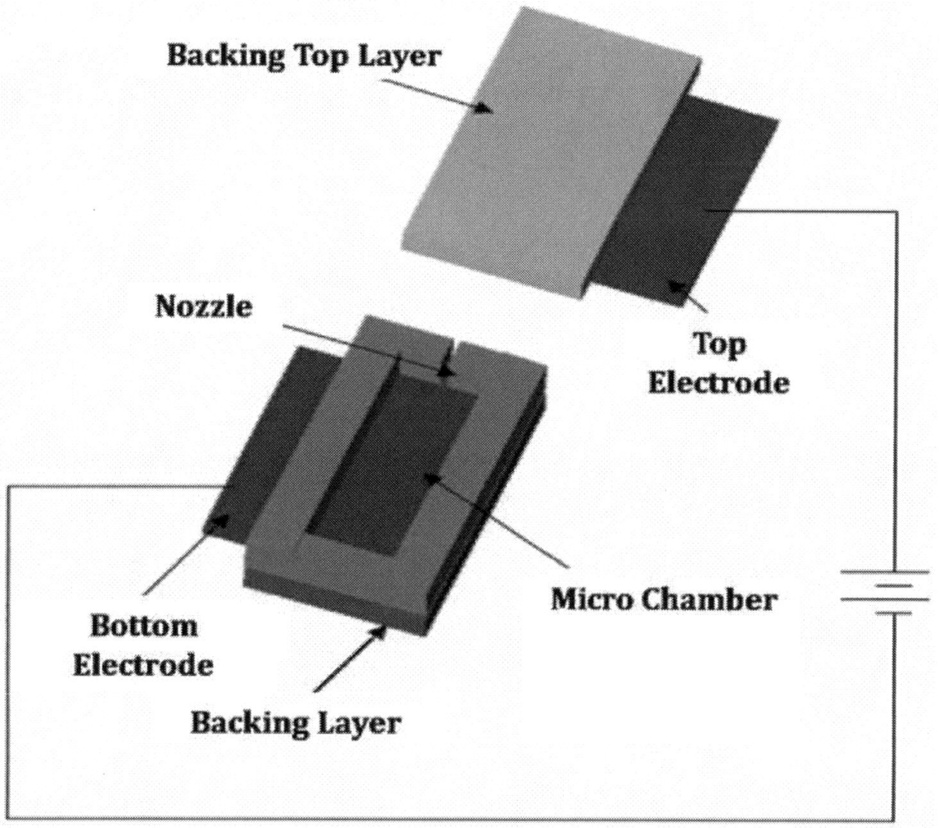

Figure 1.Schematic of micro thermal ablation system.

Results and Discussion

In this study, the ablation system was operated using a voltage of 100-150 V, which was experimentally determined as the threshold voltage range to obtain reliable skin ablation. The amount of voltage required to cause an arc discharge is strongly dependent on the gap between electrodes. The total energy stored in the system while supplying electrical energy was simulated between 4 and 5 J (data not shown) and the temperature of water medium filled in the microchamber was computed up to 600°C with the assumption that the energy generated by arc discharge phenomenon between two electrodes was used to heat the total amount of medium in the microchamber (calculation not shown). Compared to the other ablation techniques using the energy range of 30-50 mJ [1,2], the amount of energy released from this system is higher almost by 2 orders of magnitude.

Because direct measurement of the temperature of the ejected medium was not easy due to the high speed arcing within the microsecond timescale, we measured the temperature of heat energy transferred through the metal mask from the ejected hot medium. The temperature of heat energy on the underside of a 25 μm-thick tungsten mask was measured to be at least 290°C. To determine the temperature of the jet that impacts the top side of the mask, we carried out a simulation, which predicted the temperature of the jet to be at least 1000°C.

The recoil force generated by the system ranged from 1 to 2.5 N over a timescale of less than 100 μs. This force may be sufficient to have mechanical effects on the skin. Our research is continuing to address the relative roles of the high temperature and the high force of the arc-generated jet on the skin.

Guided by this physical characterization of the arc-based jet, we next examined its effect on the skin. After ablating the skin in this way, the selective removal of stratum corneum was achieved. Figure 2A shows the surface of the untreated skin (control) and the ablated skin after applying the system to the skin surface and Figure 2B shows the same sample after 12 h delivery of the model drug, sulforhodamine. Comparing to the control sample, the skin surface around the ablated site showed the radially decreasing intensity of purple staining by sulforhodamine, implying the diffusion of sulforhodamine into the skin. This diffusion was identified with histological examination shown in Figure 3.

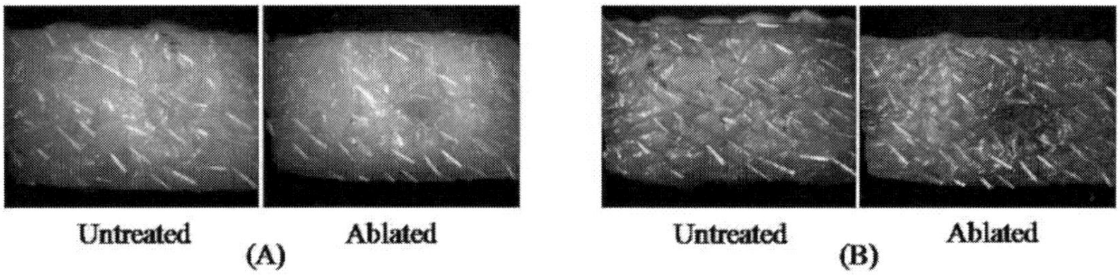

Figure 2. The surface of pig cadaver skin before (A) and after (B) delivering sulforhodamine for 12 h.

While the control sample shows no diffusion of hydrophilic sulforhodamine, the brightfield and fluorescent images (Figure 3A and 3B) of the ablated skin sample show a gradient in sulforhodamine from the ablated top to the deeper tissue and fluorescence due to the diffused sulforhodamine, respectively. Figure 3C shows a cross section of the stained skin sample which had intact stratum corneum and which lost stratum corneum without epidermis damage. Furthermore, an additional mask having the localizing windows enabled the separated and controlled effect of the heat ablation. It is highly probable that the localization of the skin ablation can determine the release rate of drug by controlling the area where the diffusion occurs.

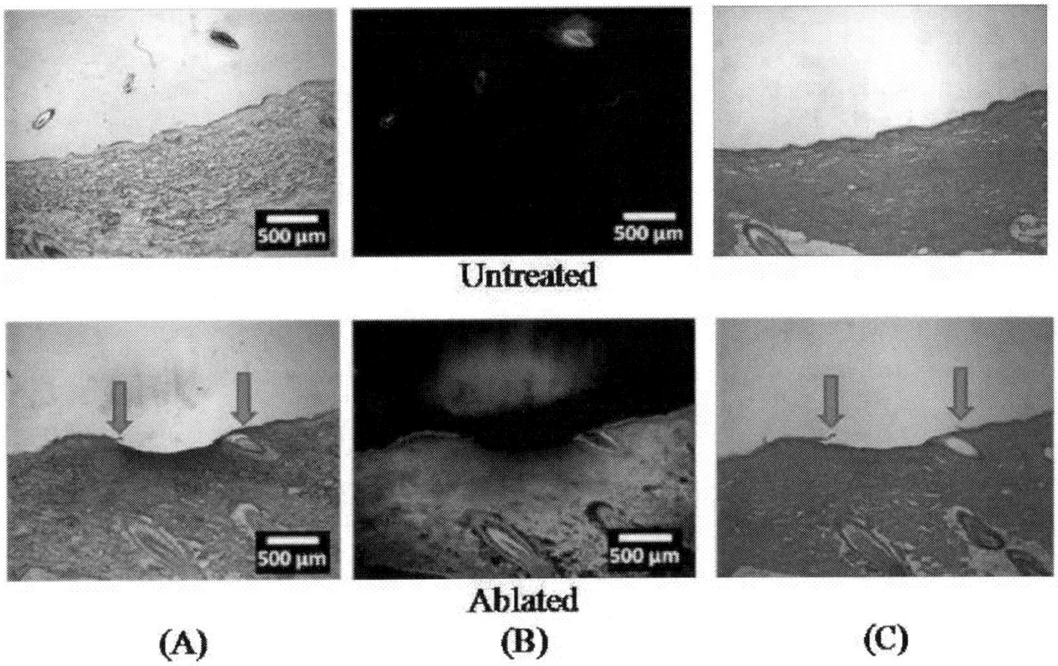

Figure 3. Histological images of untreated (above) and ablated (below) skin samples. Brightfield (A), Fluorescent (B), and H&E Stained (C).

We expect that the selective removal of stratum corneum in this way can enhance the skin permeability of high molecular weight and hydrophilic biomolecule drugs. Moreover, the skin ablation using this system was completed within the microsecond timescale so that this system might be an effective and safe transdermal delivery technique based on heat ablation.

Conclusion

A micro thermal ablation system based on arcing was designed to generate and deliver the heat ablation energy to the skin on the microsecond timescale, which is faster than previous millisecond systems. Arc discharge phenomenon released the high heat energy sufficient to ablate the skin and thereby selectively remove stratum corneum. This was enabled by the microsecond timescale of ablation minimized thermal diffusion to deeper tissue. Through the micro size channel created by the heat ablation, hydrophilic molecule, sulforhodamine, was effectively delivered into the skin, which might enable the transdermal delivery of other hydrophilic drugs, including macromolecules.

Reference

1. Steven L. Jacques, Daniel J. McAuliffe, Irvin H. Blank, John A. Parrish (1987), "Controlled Removal of Human Stratum Corneum" *The journal of investigative dermatology*, Vol. 88, pp. 88-93.

2. Basil M. Hantash, Vikramaditya P. Bedi, Bhumika Kapadia, Zakia Rahman, Kerrie Jiang, Heather Tanner, Kin Foong Chan, Christopher B. Zachary (2007), "In Vivo Histological Evaluation of a Novel Ablative Fractional Resurfacing Device" *Lasers in Surgery and Medicine*, Vol.39, pp.96-107

SUPERCRITICAL IMPREGNATION OF ANTICANCER DRUG INTO BIODEGRADABLE POLYMER

Keisuke Sato[1], Satoshi Yoda[2], Hideko T. Oyama[1],
1 Department of Science, Rikkyo Univ., Tokyo, Japan,
2 National Institute of Advanced Industrial Science and Technology (AIST), Tsukuba, Japan

Introduction

Poly (lactic acid) (PLA) is a useful biodegradable polymer with biodegradability, biocompatibility and bioresorbability. PLA is thus employed as a variety of medical materials including controlled release drug delivery system (DDS). There are two major routes on impregnating drug into PLA matrix; one is immersion using an organic solvent and the other is mechanical mixing. However remaining harmful solvent in the PLA matrix or deterioration of drugs processed at elevated temperature on mixing have been problems on DDS material fabrication.

Recently an alternative route using supercritical carbon dioxide ($scCO_2$) as a media for drug impregnation has been developed. PLA matrix is immersed in the $scCO_2$ solution of drugs under high pressure. $scCO_2$ is an environmental friendly, non-flammable, and non-toxic solvent and is evaporated from the PLA matrix after use. $scCO_2$ is capable of plasticizing and swelling PLA at lower temperature than that in atmosphere, and it is helpful to prevent drugs from deterioration. Thus the $scCO_2$ route is quite beneficial for preparation of DDS materials.

Solubility of drugs into $scCO_2$ is one of the most important factors for successful impregnation process. However, solubility of drugs, which generally have a large molecular weight and /or molecular polarity, is quite low. It is well known that addition of polar solvents such as ethanol can increase solubility of many polar substances into $scCO_2$. However, mainly due to difficulty of measurements, there are limited data of drug solubility in the $scCO_2$-organic solvent mixture. Additionally, there are few literature on PLA swelling or plasticization in the mixture.

We have developed the novel fabrication route of PLA-based DDS material using ethanol (EtOH) and CO_2 mixture. EtOH is believed to be the only one organic solvent acceptable to preparation of medical materials. Here we demonstrate impregnation of an anticancer drug, paclitaxel (PT) into amorphous poly ($_{DL}$-lactic acid) ($P_{DL}LA$) in the EtOH-CO_2 mixture at 313 K and 20 MPa. Solubility measurement of PT and observation of the $P_{DL}LA$ swelling in the EtOH-CO_2 mixture were performed. The solubility of PT to $scCO_2$ have already reported and that to EtOH-CO_2 mixture was mentioned in a literature[1)] but the data region is limited. $P_{DL}LA$ swelling in $scCO_2$ was reported in a few researches[2)3)] but no literature on the swelling in the EtOH-CO_2 mixture. The efficacy of the mixture will be discussed in detail.

Experimental

Materials

The paclitaxel (PT) (purity 99.74 %, Tokyo supply, Co. Ltd) was used without further purification. The $P_{DL}LA$ used in this work (Mitsui Chemicals, Inc.) contains approximately 12 % D-lactide, which has the number average molecular weight (M_n) and the weight average molecular weight (M_w) of 117,000 and 157,000, respectively. This $P_{DL}LA$ denoted as PLA117k in the present work was molded to an adequate shape first by a hot-press at 463 K, 5 min at 10 MPa and then by a cold-press at 273 K. CO_2 gas (99.99%) was dehydrated using a zeolite column and used for all experiments. Conventional EtOH (99.9%) and acetonitrile (analytical grade) for HPLC were used without further purification.

Solubility measurement of PT in EtOH-CO_2 mixture

Solubility of PT in EtOH-CO_2 mixture was measured by direct injection of the solvent into high performance liquid chromatography (HPLC). The details of measurement using HPLC has been described in our previous publication.[4] Schematic diagram of the high pressure flow system is shown in Fig.1. A molar composition of EtOH and CO_2 were controlled by flow rates of pumps (2) and (3). EtOH and CO_2 were mixed in the accumulator (4) and adjusted at 313 K and 10-37 MPa by a thermo static bath (7) and a backpressure regulator (9). PT was filled up into the sample cell (5) with glass beads (0.7-1 mm in diameter). PT was then dissolved to be saturated in the EtOH-CO_2 mixture. The flow rate was carefully adjusted to keep the dissolution equilibrium.

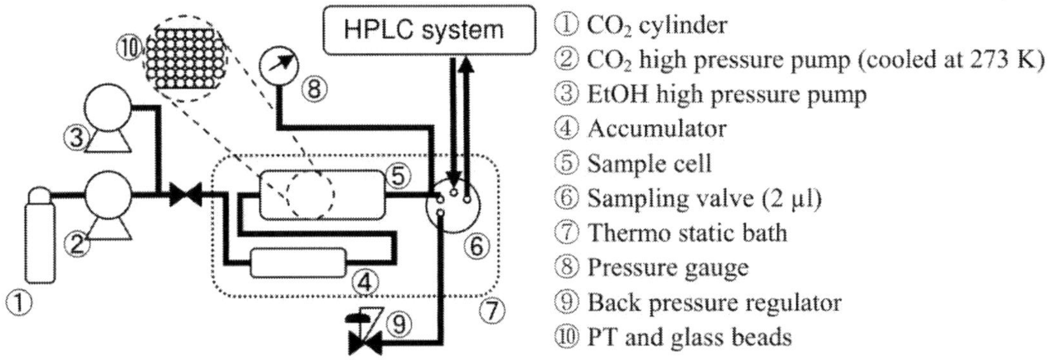

Fig.1 Solubility measurement apparatus

The HPLC analysis involved acetonitrile-water (50: 50 (v/v)) mixture as eluent and a column (MetaChem taxsile, 250 mm (L) x 4.6 mm (ID), VARIAN, Inc.). The absorption of PT at 230 nm was monitored via UV-VIS detector. The PT is detected in HPLC after the EtOH and CO_2 are dissolved in the HPLC eluent. In other words, molar composition of EtOH and CO_2 and the mixture's pressure do not directly affect the PT analysis. A calibration curve of PT dissolved in the acetonitrile-water mixture was established before the analysis. The mole fraction solubility of PT in the mixture, S_{PT} was determined using the following equation:

$$S_{PT} = \frac{n_{PT}}{n_{PT} + n_{CO2} + n_{EtOH}} \tag{i}$$

where n_{PT}, n_{CO2} and n_{EtOH} are the number of moles in the sampling valve(2 μl) of PT, CO_2 and EtOH, respectively. n_{PT} was calculated from the calibration curve and the volume of the valve. Due to the lack of established density data of EtOH-CO_2 mixture, n_{CO2} and n_{EtOH} are defined as follows:

$$n_{CO2}=(\rho_{CO2}/M_{CO2})\,[V_{valve}\,\{FR_{CO2}/(FR_{CO2}+FR_{EtOH})\}] \qquad \text{(ii)}$$

$$n_{EtOH}=(\rho_{EtOH}/M_{EtOH})\,[V_{valve}\,\{FR_{EtOH}/(FR_{CO2}+FR_{EtOH})\}] \qquad \text{(iii)}$$

where ρ_{CO2} is CO_2 density at the measurement temperature and pressure, ρ_{EtOH} is EtOH density at 293 K, M_{CO2} and M_{EtOH} are the molecular weight of CO_2 and EtOH, FR_{CO2} and FR_{EtOH} are the flow rate of pumps (2) and (3), and V_{valve} is the volume of sampling valve(2 μl). The ρ_{CO2} was calculated from the equation of state developed by Span and Wagner[5]. EtOH density at atmospheric pressure was employed here due to the lack of established density data around the experimental condition. Then the solubility of PT is not high ($S_{PT} < 10^{-4}$), we assumed that the volume of PT in the sampling valve was negligible.

Swelling observation of PLA117k in EtOH-CO_2 mixture

PLA117k swelling in EtOH-scCO_2 mixture was observed at 313 K and 6-20 MPa by direct visualization using a high pressure view cell. Fig. 2 shows a schematic structure of the view cell (7). The P_{DL}LA disk (12) (6 mm in diameter and 0.6 mm in the thickness) was fixed between sapphire windows (10) with c-style spacer ring (11) (0.5 mm in thickness). The cell was maintained at 313 K during the experiments. After evacuating the sample cell at least for 24 h, the EtOH-CO_2 mixtures (0-100 mol % EtOH) at 313 K and 6-20 MPa was prepared using pumps (2 and 3) and introduced in the cell (7). Homogenous phase conditions of the EtOH-CO_2 mixture[6] was selected to simplify the experiment.

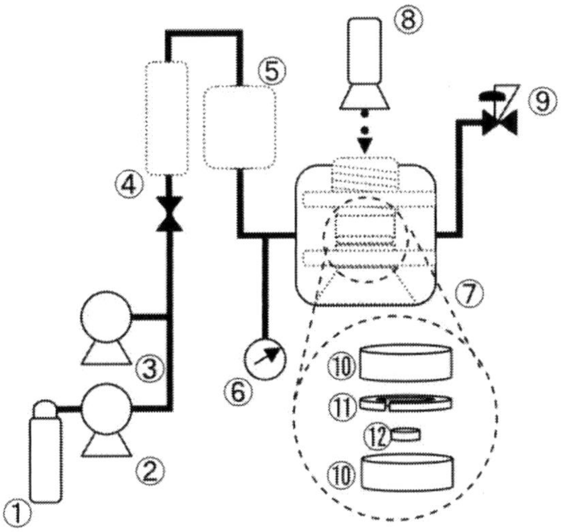

① CO_2 cylinder
② CO_2 high pressure pump (cooled at 278 K)
③ EtOH high pressure pump
④ Accumulator
⑤ Buffer tank
⑥ Pressure gauge
⑦ High pressure view cell with cartridge heater
⑧ Video camera
⑨ Back pressure regulator
⑩ Sapphire glass (view window)
⑪ C style stainless spacer (0.5 mm in thickness)
⑫ PLA117k sample

Fig.2 Direct visualization of polymer swelling apparatus

The appearance, the size and the area of PLA117k were recorded by video camera (8). Since the PLA117k is fixed between the windows to have a constant gap, the vertical swelling can be neglected. Then the swelling degree (s) is defined as the following equation;

$$s = \frac{A - A_0}{A_0} \qquad \text{(iv)}$$

where Ao and A are areas of PLA117k before and after the swelling experiments, respectively.

Impregnation of PT in PLA117k

Apparatus for the impregnation experiments is shown in Fig.3. Ten mg of PT and the PLA117k disk (5 mm in diameter and 1 mm in thickness) were placed separately in a high pressure cell (5). The EtOH-CO_2 mixture (0-100 mol% EtOH) at 313 K and 20 MPa was then introduced into the cell and maintained the condition for 24 h without flowing. After cooling the cell to room temperature, the EtOH-CO_2 was slowly released using the back pressure regulator (8). The PT impregnated sample was evacuated to remove ethanol and used for PT content analysis.

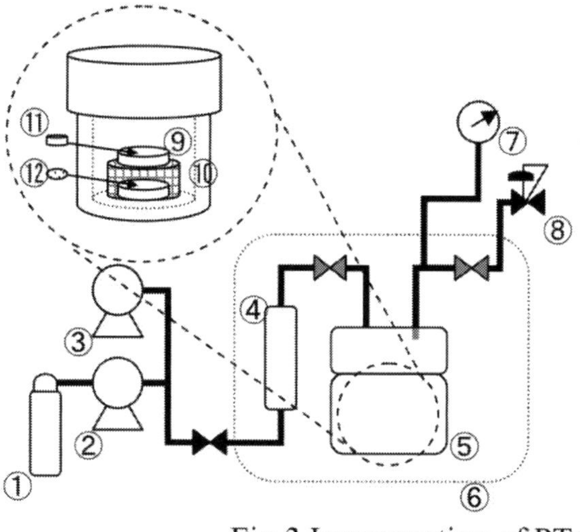

① CO_2 cylinder
② CO_2 high pressure pump (cooled at 273 K)
③ EtOH high pressure pump
④ Accumulator
⑤ High pressure cell
⑥ Thermostatic bath
⑦ Pressure gauge
⑧ Back pressure regulator
⑨ Glass plate
⑩ Stainless mesh
⑪ Polymer sample
⑫ PT

Fig.3 Impregnation of PT into PLA117k apparatus

The amount of impregnated PT in the PLA117k disks were determined using gel permeation chromatography (GPC). The sample was dissolved into chloroform. The GPC analysis was conducted using (TSK-GEL, 300 mm (L) x 7.8 mm (ID), Tosoh corp.) and chloroform as eluent at (313 K). The PT and PLA117k can be detected separately at this condition. Both UV-VIS and fluorescence detectors were employed for PT detection. The UV-VIS absorbance of PT was measured at 250 nm. The fluorescence intensity of PT was monitored at 310 nm with the excitation at 250 nm.

Results and discussions

Solubility of PT in EtOH-CO_2 mixture

Solubility of PT in scCO_2 and EtOH-CO_2 mixture was shown in Fig. 4. The solubility of PT in scCO_2 located between S_{PT}=1.2 to 2.1 x 10^{-6}. Results were in accordance with those in literatures [7-8]. Solubility of PT in 10 mol% EtOH-90 mol% CO_2 mixture was S_{PT}=1.5-1.9 x 10^{-5} at 313 K, 10-15 MPa, which is 9.5-13 times larger than that in the pure scCO_2. Since the solubility of PT in EtOH is S_{PT}=1 x 10^{-3}, the solubility in the mixture is expected to increase with enhancement of the EtOH content. Solubility in the mixture with larger EtOH content is under investigation.

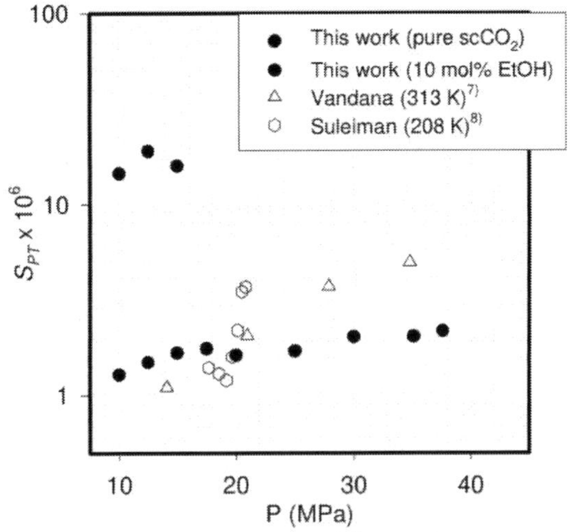

Fig.4 Solubility of PT in the $scCO_2$ and EtOH-CO_2 mixture

Swelling of PLA117k in EtOH-CO_2 system

Fig.5 shows profiles of PLA117k swelling in a variety of EtOH-CO_2 mixture at 313 K. Larger PLA117k swelling was observed in EtOH-CO_2 mixture than that in $scCO_2$. No swelling of PLA117k was observed in EtOH, like crystalline Poly (L-lactic acid) (P_LLA) as mentioned in the literature[9]. The mixture with 25-50 mol% EtOH induced substantial swelling especially at high pressure. Within our experiments, the maximum swelling was observed in the mixture with 40 mol% EtOH at 20 MPa. It is noteworthy that appropriate EtOH-CO_2 mixture was observed to conduce to significant PLA117k swelling, which is larger than that in $scCO_2$.

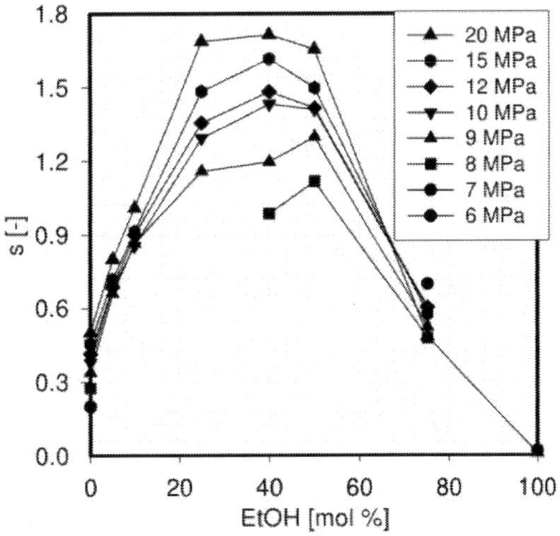

Fig.5 Swelling of PLA117k in EtOH-CO_2 mixture

Impregnation of PT in PLA117k via EtOH-CO_2 systems

Impregnations of PT into PLA117k at 313 K, 20 MPa with a variety of EtOH-CO_2 mixtures (0-100 mol% EtOH) were investigated. The amount of impregnated PT in pure $scCO_2$ was 0.1 μg/ mg (impregnated PT/ PLA), whereas that in EtOH-$scCO_2$ mixtures were 4-10 times larger than that in pure $scCO_2$ (0.4–1.1 μg/ mg). Little amount of PT was impregnated into PLA117k in EtOH solution.

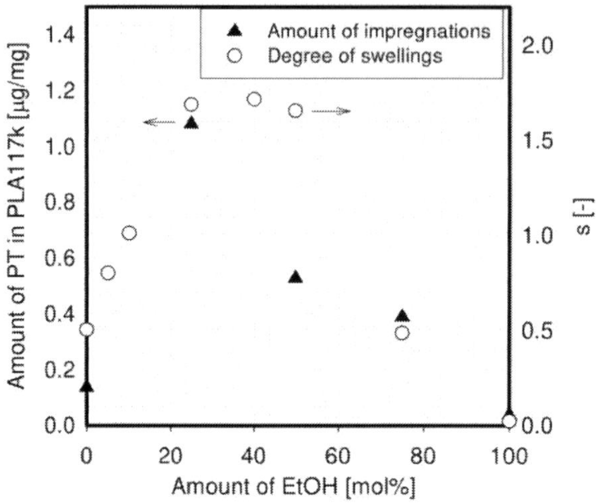

Fig.6 Amount of impregnations and degree of swellings in the mixtures

Fig.6 shows the relationship between the EtOH molar ratio in EtOH-CO_2 mixture at 20 MPa, 313 K and the amount of impregnated PT. The maximum swelling at the same condition was also shown for comparison. The results of PT content are approximately correlated with results of swelling. Even though PT solubility to the mixture expected to be increase significantly with EtOH, the amount of PT inversely decreased with EtOH content. It is presumed that diffusion of impregnating media in PLA117k matrix plays important role in the system. The use of EtOH-CO_2 mixture is more beneficial than the scCO_2 system due to larger PT solubility and swelling, and than the EtOH system having no swelling effects.

The maximum of PT amount was not in accordance with the maximum of swelling. We consider the effects of depressurization step. A part of impregnated PT in the mixture under a high pressure could be removed from the PLA117k matrix during the depressurization process. This process would result in complicated change in both solubility and swelling. Further research on depressurization process should be conducted.

Conclusion

Impregnations of PT into PLA117k in various compositions of the EtOH-CO_2 mixture were investigated at 313 K and 20 MPa. It was proved to be more advantageous to use the EtOH-CO_2 mixture for impregnation of PT than the pure scCO_2. The amount of impregnated PT in EtOH-scCO_2 mixtures (0.39–1.08 μg/ mg) was 3-8 times larger than that in pure scCO_2 (0.14 μg/ mg), and was 10-27 times than that in EtOH (0.04 μg/ mg). The EtOH-CO_2 mixture with 25 mol% EtOH showed the maximum amount of impregnated PT in PLA117k. The increase in the amount of impregnated PT was probably attributed to the solubility of PT into the EtOH-CO_2 mixture and the swelling of PLA117k. The solubility of PT in the mixture with 10 mol% EtOH was 13 times higher than that in scCO_2 at 313 K and 15 MPa. The maximum swelling of PLA117k in the mixture containing 40 mol% EtOH at 313 K and 20 MPa was observed as 1.7 times larger than the initial size. Therefore, it was demonstrated that EtOH-CO_2 system is promising for fabrication of polymer-based DDS materials.

References

1) Vandana,Vishnu and A. S. Teja (1995), "*Supercritical Extraction of Paclitaxel Using CO_2 and CO_2-Ethanol Mixtures*", in *ACS Symp. Ser 608 chapter 29*, Washington, D.C., American chemical society.
2) Fujiwara, Tomoko and T. Yamaoka, Y. Kimura, K. J. Wynne (2005), "Poly(lactide) Swelling and Melting Behavior in Supercritical Carbon Dioxide and Post-Venting Porous Material", *Biomacromolecules*, 6, pp. 2370-2373.
3) Pini, Ronny and G. Storti, M. Mazzotti, H.Y. Tai, K.M. Shakesheff, S.M. Howdle (2007), "Sorption and Swelling of Poly($_{DL}$-lactic acid) and Poly(lactic-co-glycolic acid) in Supercritical CO_2: An Experimental and Modeling Study", *J. Polym. Sci. Part B: Polym. Phys.*, 46, pp. 483-496.
4) Yoda, Satoshi and Y. Mizuno, T. Furuya, Y. Takebayashi, K. Otake, T. Tsuji, T. Hiaki (2008), "Solubility measurements of noble metal acetylacetonates in supercritical carbon dioxide by high performance liquid chromatography (HPLC)", *J. Supercritcal Fluids*, 44, pp. 139-147.
5) Span, R and W. Wagner (1996), "A new equation of state for carbon dioxide covering the fluid region from the triple point temperature to 1100K at pressures up to 800MPa." *J. Phys. Chem. Ref. Data*, 25, pp. 1509-1596.
6) Suzuki, Kazuhiro and H. Sue, M. Ito, R.L. Smith, H. Inomata, K. Arai, S. Saito (1990), "Isothermal Vapor-Liquid Equilibrium Data for Binary Systems at High Pressures: Carbon Dioxide-Methanol, Carbon Dioxide-Ethanol, Carbon Dioxide-I-Propanol, Methane-Ethanol, Methane-I-Propanol, Ethane-Ethanol, and Ethane-I-Propanol Systems", *J. Chem. Eng. Data*, 35, PP. 63-66.
7) Vandana, Vishnu and A. S. Teja (1997), "The solubility of paclitaxel in supercritical CO_2 and N_2O", *Fluid Phase Equilibria*, 135, pp. 83-37.
8) Suleiman, David and L. A. Esteves, J. C. Pulido, J. E. Garcia, C. Mojica (2005), "Solubility of Anti-Inflammatory, Anti-Cancer, and Anti-HIV Drugs in Supercritical Carbon Dioxide",*J. Chem. Eng. Data*, 50, pp. 1234-124.
9) Tsuji, Hideto and K Sumida (2001), "Poly (L-lactide): V. Effects of Storage in Swelling Solvents on Physical Properties and Structure of Poly (L-lactide)", *J. Appl. Polym. Sci.*, 79, pp. 1582-1589.

569j Polymer Films as Carriers for Spatially Dispersed Micro- and Nano-Size Drug Particles

Bo Zhou[1], **Teresa Carvajal**[1], David Nivens[2], and Rodolfo Pinal[1]. (1) Department of Industrial and Physical Pharmacy, Purdue University, 575 Stadium Mall Drive, West Lafayette, IN 47907, (2) Department of Food Science, Purdue University, West Lafayette, IN 47906

Strip films are considered a potential alternative delivery system and also as a mechanism for maintaining small particles in a non-agglomerated form. This work focuses on the formation and characterization of API particle-laden strip films, the impact of film formulation and formation process on the final film properties.

Film strips made from HPMC and sodium alginate containing dispersed griseofulvin (model drug) microparticles were prepared by film casting. The films were characterized by their physical and mechanical properties as well as by their performance in terms of drug release upon dissolution. Sodium alginate films containing 9.3% (w/w) griseofulvin particles, release 95% of the drug in 50 min. This is significantly faster than HPMC, which attain the same level of release after 80 min. Either film matrix results in much faster release than that observed from the dissolution of the unprocessed drug. Sodium alginate films have a greater drug load capacity than HPMC. HPMC on the other hand, offers a less hygroscopic environment for the particles in the film.

The results suggested that particle-laden polymer strip films present an effective configuration for harnessing the potential of increased dissolution rate offered by small drug particles, while eliminating the otherwise unavoidable issue of particle agglomeration (and reduced dissolution rate) typical of fine particles in powdered form. Different polymer matrices result in different drug release rates with the added possibility of controlling exposure of the drug to environmental factors such as moisture.

EVALUATION OF THE FLOW PROPERTIES OF POLLEN SHAPE HYDROXYAPATITE PARTICLES FOR DRY POWDER INHALATION

***Meer Saiful Hassan*[1]**, *Yongsheng Wang*[2], *Rong Xu*[1], *and Raymond Lau*[1]
(1) School of Chemical and Biomedical Engineering, (2) School of Mechanical and Aerospace Engineering, Nanyang Technological University, Singapore

Abstract

The use of pollen shape (spiked sphere) particles for dry powder inhalation is studied. Pollen shape hydroxyapatite (HA) particles are synthesized to cover both size ranges of typical drug and carrier particles for pulmonary drug delivery. The HA particles produced have geometric diameter range from 5-50 μm and effective density range from 0.2-0.72 g/cm^3. The flow properties of the powders are characterized by the Carr's compressibility index (CI). The aerodynamic properties of the particles are studied in vitro by using an eight-staged Anderson Cascade Impactor with a Rotahaler®. HA samples of different sizes are used as single formulation in the experiment. The HA particles are found to have lower CI values than commonly used lactose carrier particles with similar size range. The pollen shape particles also show high emitted dose and fine particle fraction and correlates well with the corresponding mean aerodynamic diameter.

Introduction

Pulmonary drug delivery is attracting interests as an alternative route of drug administration. Aerodynamic diameter is commonly used to characterize aerosol drug particles for inhalation delivery. Researchers showed that particles with an aerodynamic diameter (d_a) lower than 5 μm have the highest possibility to get deep into the lung[1]. However, shape irregularity affects drag force and terminal settling velocity of the particles. Therefore, a shape correction factor is needed for the estimation of their aerodynamic diameter, which eventually demonstrates their distinct aerodynamic behavior[2,3]. Researchers found an increase in respirable fraction by lowering the aerodynamic diameter of the particles through surface treatment and turn the particles into elongated shape[2,4]. Large geometric diameter and low density particles have also been used as a means of improving the delivery of inhaled therapeutics. Particles having large geometric diameter and low-density can have the same aerodynamic diameter as small and high density particles but the particle dispersion is better due to the lower van der Waals forces. Studies also showed that large geometric diameter and low density particles have the advantages of reduced effect of diffusion and reduced macrophage clearance for better bioavailability [5,6].

Another solution to improve pulmonary drug delivery is to mix drug particles with larger sized carrier particles. In that case, the morphology of these carrier particles has important effects on deposition. Zeng et al. [7] showed that increasing the elongation ratio of lactose carrier particles can increase the respirable fraction of drug from dry powder formulation for inhalation. Most researchers [8-10] found that increasing the surface roughness of the carrier particles can enhance the particle respirable fraction while Zeng et al. [7] found otherwise.

The literature results show that shape and surface features of drug/carrier particles can affect the efficiency of dry powder inhalation. However, systematic studies of these effects are limited. Theoretically, a pollen shape structure can attribute both the characteristics of large geometric diameter and low-density. The spiky surface structure will limit the approach of two particles, increase the effective separation and reduce their contact area and van der Waals force[11]. Therefore, pollen shape particles are expected to have good dispersion properties. In this study, pollen shape hydroxyapatite (HA) particles are synthesized with different size range. The particle properties are measured and the dispersion and deposition properties of these particles are studied in vitro using a cascade impactor. .

Experimental

Preparation for HA

The HA particles are prepared according to Wang et al.[12] by hydrothermal reaction at a temperature range of 100-200°C. Pollen shape HA particles are synthesized using KH_2PO_4, (Panreac), $Ca(NO_3)_2{\cdot}4H_2O$ (Sigma-Aldrich), PSS (poly(sodium-4-styrenesulfonate), 30 wt% solution in water, Sigma-Aldrich, MW = 70,000) and urea (Sigma-Aldrich). 30 ml of KH_2PO_4 (0.02 M) solution is mixed with 50 ml of Ca $(NO_3)_2.4H_2O$ (0.02 M) solution. Then PSS is added to the mixture to get a PSS concentration of 30-40 g/liter. Sufficient amount of urea is also added to get different urea concentration. This mixture is then kept for half an hour under gentle stirring. After the urea is dissolved in the liquid, the solution is poured into an autoclave and then kept in an oven for 6-12 hours for hydrothermal reaction. After the reaction, the precipitate is collected and washed few times with de-ionized water to remove unreacted reactants and then dried at 70°C for 24 hours.

Preparation of Lactose

α-Lactose monohydrate (Sigma) is used as the control carrier particles to compare against the performance of pollen shape HA particles. Lactose is separated into three different size ranges by sieving. The ranges are 38-75 μm, 20-38 μm and <20 μm. Sieves (Retsch, Haan, Germany) with a mesh size of 20 μm, 38 μm and 75 μm are used for the separation.

Particle characterization

The size distribution and surface morphology are examined by scanning electron microscopy (SEM) (JEOL JSM-5600) images. The SEM images are taken randomly from different area of the sample. The largest dimension of the particle is considered as the geometric diameter. At least two hundred particles are measured to get the mean diameter of the particles. The surface morphologies of the particles are assessed qualitatively based on the SEM images. The bulk (ρ_{bulk}) and tap (ρ_{tap}) densities of all the samples are measured according to Shi et al.[13]. The flow behavior of the samples are characterized by the Carr's compressibility index (CI) [9,14,15]. CI can be determined by:

$$CI = \frac{\rho_{tap} - \rho_{bulk}}{\rho_{tap}} \times 100\%$$

In Vitro dispersion and deposition study

The in vitro aerosol dispersion is determined using an eight staged Anderson cascade impactor (ACI) with a preseparator (Copley, UK) operating at an airflow rate of 30 liter/min.

The preseparator contained 8 ml of solvent (2% HNO_3) to prevent particle bounce and re-entrainment. The powder mixtures are aerosolized using a dry powder inhalation device (Rotahaler®, GSK, U.K.). 20mg of each sample is loaded into a hard gelatin capsule (Gelatin Embedding Capsules, size 4, 0.25 cc, Polysciences, Inc., Warrington, PA) manually. An actuation time of 20 seconds is allowed for each capsule for complete dispersion of all the particles. Experimental runs are conducted in triplicate. The emitted dose (ED) and fine particle fraction (FPF) are determined.

Results and Discussions

Physical properties of the HA particles

Pollen shape HA particles are prepared by a hydrothermal reaction. The size and morphology can be varied by controlling the PSS concentration, the urea concentration and the reaction temperature. The physical properties of different HA particles are shown in Table 1. The aerodynamic diameter is defined as the diameter of a sphere of unit density that has the same terminal settling velocity as the particle under consideration, $d_a = d_g\sqrt{\frac{\rho_e}{\lambda \rho_s}}$ where ρ_s= 1 g/cm^3, d_g is the particle geometric diameter, ρ_e is the effective particle density and λ is the particle shape factor [1,3,16]. The maximum dimension of a particle is considered as its geometric diameter. The effective density is approximated by the tap density, ρ_{tap}. The shape factor is defined as the ratio of the particle surface area to the surface area of a sphere with same volume as the particle. Since the pollen shape closely resembles a spherical particle, a shape factor of 1 is assumed.

Table 1. Physical properties of HA particles

Sample	d_{mean} (µm)	σ (µm)	ρ_{bulk} (g/cm^3) (n=0)	ρ_{tap} (g/cm^3) (n=2500)	$d_{a,theoretical}$ (µm)	CI (%)
HA-1	5.70	0.53	0.346	0.719	4.83	51.88
HA-2	13.1	1.7	0.217	0.357	8.56	39.16
HA-3	21.1	3.3	0.233	0.411	13.51	43.17
HA-4	48.55	10.21	0.215	0.289	26.64	18.72
HA-5	28.2	4.02	0.105	0.218	13.18	51.95

Table 2. Physical properties lactose samples

Sample	Size range (µm)	ρ_{bulk}(g/cm^3) (n=0)	ρ_{tap} (g/ cm^3) (n=2500)	CI (%)
Lactose1	< 20	0.217	0.645	57.98
Lactose2	20-38	0.316	0.760	58.42
Lactose3	38-75	0.586	0.874	32.95

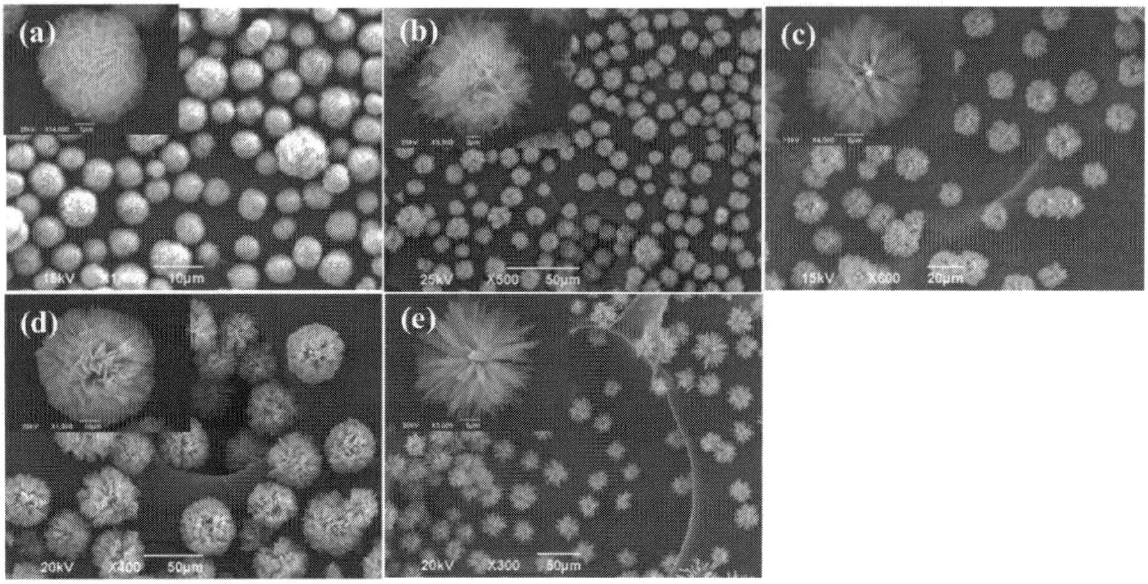

Figure 1. The SEM image of HA particles produced by using (a) PSS-40 g/liter & urea- 7.5M, at 200^0C (HA1); (b) PSS-40 g/liter & urea- 3M, at 200^0C (HA2); (c) PSS-40 g/liter & urea- 0.5M, at 150^0C (HA3); (d) PSS-40 g/liter& urea- 0.5M, at 100^0C (HA4); (e) PSS-30 g/liter & urea- 0.5M, at 200^0C (HA5).

Figure 1 shows the SEM images of the five HA samples synthesized in this study. It can be seen from the SEM images that with PSS concentration of 40 g/liter, particles with denser surface structure are produced. They comprise petal like surface structure whereas the particles produced with PSS 30 g/liter have needle like surface structure (HA5). At a temperature of 100 ^{o}C larger HA particles (HA4) are found. With higher urea concentration of 7.5 M and 3 M at 200 ^{o}C temperature, HA particles with smaller size ranges are generated (HA1, HA2). It can also be seen that the HA particles are well segregated which is very effective in dispersing them into air.

Figure 2. SEM images of lactose monohydrate with a size range of (a) less than 20 μm (lactose1); (b) 20-38 μm (lactose2); and (c) 38-75 μm (lactose3).

The flow behavior of the HA particles are compared with commonly used lactose particles of similar size range. The lactose particles are sieved into three different size ranges and their SEM images are shown in Figure 2. It is to note that sieving alone is not possible to completely separate all the smaller particles from the larger ones. A fraction of fine particles is present in the samples. The fine particles may affect the flow behavior of the samples by forming large aggregates. Their physical properties are stated in Table 2.

Powder Flow

Carr's compressibility index (CI) is commonly used to characterize powder flow. CI value quantifies the powder flow based on the concept that compressible powders are more cohesive and would have poor flow property. Cohesive particles can form aggregates and through tapping, these aggregates may collapse. The overall volume of the particles will be reduced and the powder will appear to be compressible. Non-cohesive particles, on the other hand, will not collapse and the powder would appear to be incompressible. From Figure 3, it can be seen that, particles with lower size range (< 38 μm) have higher CI values. Interparticle forces are dominating in small particles and therefore these particles normally exhibit poor flow property. Nonetheless, HA particles still have lower CI than lactose particles in a similar size range. This can be attributed to the surface structure of the HA particles that the interparticle forces are weakened and less aggregates are formed. It can be seen from the SEM images that the HA samples (HA1, HA2 and HA3) are well segregated despite the small particle size.

A comparison of the different HA particles in a similar size range indicate that a petal like surface structure would give a lower CI than a needle like surface structure. The needle like surface may induce particle interlocking and cause particle aggregation. Therefore, the flow property of needle surface can be expected to be poorer than petal like surface.

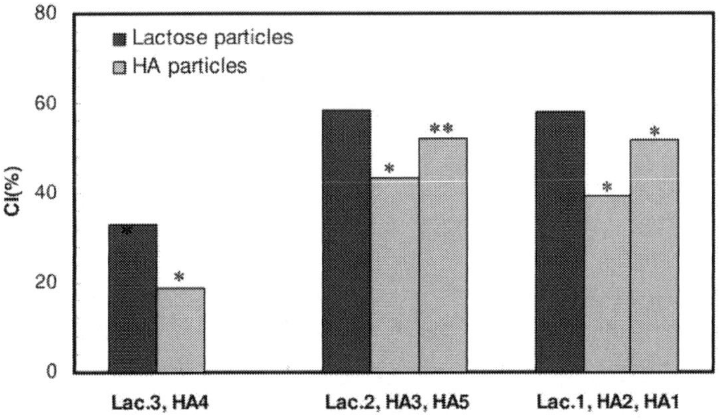

Figure 3. Comparison of the CI index of the pollen shape HA particles with lactose particles of similar size range. * = petal like surface structure, **= needle like surface structure.

In vitro deposition property of HA particles in single formulation

The deposition properties of the pollen shape HA particles are studied in terms of ED and FPF. ED is defined as the mass percentage of particles delivered from the inhaler. FPF is defined as the mass percentage of the particles deposited in stage 2 or lower in the cascade impactor. ED can be used as the index of aerosolization property. It can be seen from Figure 4(a) that the ED can correlates well with the mean aerodynamic diameter of the HA particles. There is limited dispersion and deposition study using single particle formulation in cascade impactor. The ED of the HA particles are compared with angular jet milled and spherical spray dried mannitol particles of various size and shape studied by Louey et al.[14]. The mannitol particles were aerosolized at an air flow rate of 60 liter/min whereas the flow rate used in this study is 30 liter/min. Despite the lower air flow rate, the HA particles show higher ED than the angular and spherical particles as shown in Figure 4(a). An increase in flow rate can substantially increase the total powder emission from an inhaler[17]. It is anticipated that the HA particles would exhibit even higher ED at a flow rate of 60 liter/min. Figure 4(b) shows a good correlation between the FPF and the mean HA aerodynamic diameter. The FPF of the HA particles is also compared against the FPF of the mannitol particles[14]. The improvement in FPF of HA particles over the FPF of the mannitol particles is even more obvious than the ED. It is apparent that pollen shape surface structure is highly effective in improving the ED and FPF of drug particles in inhalation delivery.

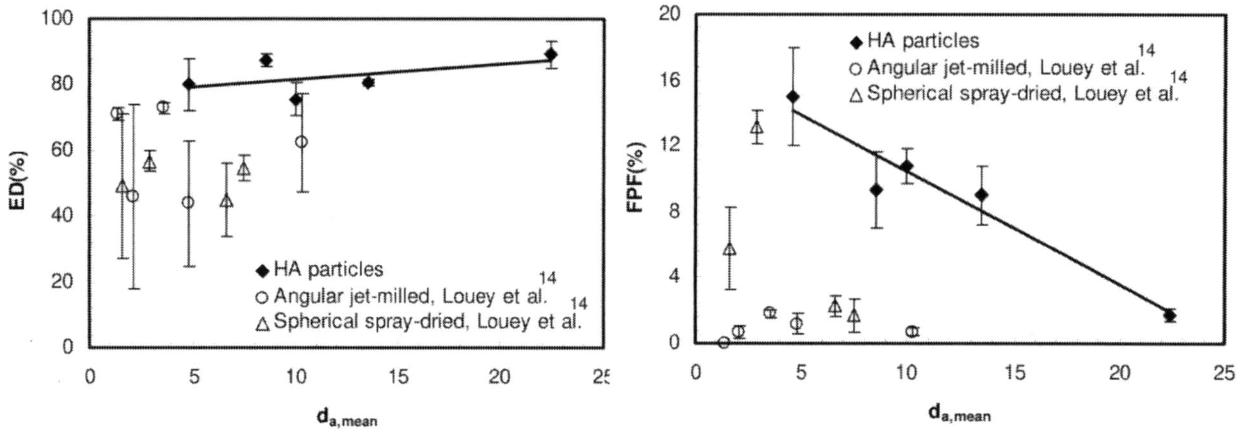

(a) (b)

Figure 4. Comparison of (a) ED and (b) FPF of the HA particles with mean aerodynamic diameter.

Concluding Remarks

Pollen shape HA particles with a geometric diameter range of 5 -50 μm are synthesized. The flow behavior of the particles is assessed with CI index. In comparison, the HA particles show better flow behavior than that of commonly used lactose carrier particles with similar size range. The cascade impactor experiment shows a good correlation of the ED and FPF with mean aerodynamic diameter of the HA particles. A comparison with literature results show that pollen shape surface structure is effective in improving the ED and FPF of drug particles in inhalation delivery.

Acknowledgement

Support by NTU/SUG grant is gratefully acknowledged.

References

1. Hicky AJ. *Pharmaceutical inhalation aerosol technology.* Vol 134. New York:Marcel Dekker, Inc.,2004.
2. Crowder TM, Rosati JA, Schroeter JD, Hickey AJ, Martonen TB. Fundamental effects of particle morphology on lung delivery: Predictions of stokes' law and the particular relevance

to dry powder inhaler formulation and development. *Pharmaceutical Research*. 2002;19: 239-245.

3. Edwards DA. Delivery of biological agents by aerosols. *AIChE Journal*. 2002;48: 2-6.
4. Fults KA, Miller IF, Hickey AJ. Effect of particle morphology on emitted dose of fatty acid-treated disodium cromoglycate powder aerosols. *Pharmaceutical Development and Technology*. 1997;2: 67 - 79.
5. Edwards DA, Hanes J, Caponetti G, Hrkach J, Ben-Jebria A, Eskew ML, Mintzes J, Deaver D, Lotan N, Langer R. Large porous particles for pulmonary drug delivery. *Science*. 1997;276: 1868.
6. Edwards DA, Ben-Jebria A, Langer R. Recent advances in pulmonary drug delivery using large, porous inhaled particles. *Journal of Applied Physiology*. 1998;85: 379-385.
7. Zeng XM, Martin GP, Marriott C, Pritchard J. The influence of carrier morphology on drug delivery by dry powder inhalers. *International Journal of Pharmaceutics*. 2000;200: 93-106.
8. Heng PWS, Chan LW, Lim LT. Quantification of the surface morphologies of lactose carriers and their effect on the in vitro deposition of salbutamol sulphate. *Chemical & pharmaceutical bulletin*. 2000; 48: 393-398.
9. Iida K, Hayakawa Y, Okamoto H, Danjo K, Leuenberger H. Evaluation of flow properties of dry powder inhalation of salbutamol sulfate with lactose carrier. *Chemical & Pharmacutical Bulletin*. 2001;49: 1326.
10. Flament M-P, Leterme P, Gayot A. The influence of carrier roughness on adhesion, content uniformity and the in vitro deposition of terbutaline sulphate from dry powder inhalers. *International Journal of Pharmaceutics*. 2004;275: 201-209.
11. Visser J. Van der waals and other cohesive forces affecting powder fluidization. *Powder Technology*. 1989;58: 1-10.
12. Wang Y, Gunawan P, & Xu R. Polyelectrolyte-mediated formation of hydroxyapatite microspheres: From anisotropic to isotropic growth. *Submitted*. 2008.
13. Shi L, Plumley CJ, Berkland C. Biodegradable nanoparticle flocculates for dry powder aerosol formulation. *Langmuir*. 2007;23: 10897-10901.
14. Louey M, Van Oort M, Hickey A. Aerosol dispersion of respirable particles in narrow size distributions produced by jet-milling and spray-drying techniques. *Pharmaceutical Research*. 2004;21: 1200-1206.
15. Louey M, Van Oort M, Hickey A. Aerosol dispersion of respirable particles in narrow size distributions using drug-alone and lactose-blend formulations. *Pharmaceutical Research*. 2004;21: 1207-1213.
16. Hinds WC. *Aerosol technology: Properties, behavior, and measurement of airborne particles*. New York:John Weily & Sons,1982.
17. French DL, Edwards DA, Niven RW. The influence of formulation on emission, deaggregation and deposition of dry powders for inhalation. *Journal of Aerosol Science*.1996; 27: 769-783.

569m Reverse Aqueous Aggregates In Hydrofluoroalkanes for the Non-Invasive Regional Delivery of Anti-Cancer Therapeutics to the Lungs

Parthiban Selvam[1], Balaji Bharatwaj[2], and Sandro R. P. da Rocha[2]. (1) ChE - Wayne State University, 5050 Anthony Wayne Dr., Detroit, MI 48202, (2) Chemical Engineering and Materials Science, Wayne State University, Detroit, MI 48201

Inhalation chemotherapy has been suggested as an attractive alternative for the local delivery of chemotherapeutics to the lungs due to high therapeutic efficacy of antineoplastic agents and low systemic toxicity. Aqueous reverse aggregates in hydrofluoroalkanes (HFAs) can be used for the systemic delivery of anti cancer drugs to and through the lungs using pressurized metered dose inhalers (pMDIs). There are several potential advantages in using reverse aggregates compared to dispersion-based dispersions. For high potency drugs, where the total surface area of the drug may be comparable to that of the container walls, reverse aggregates may serve to improve dose reliability and control dose concentration. However, the methyl-based FDA approved surfactants for pMDIs have extremely low solubility in HFAs, and cannot stabilize water-in-HFA (W/HFA) microemulsions. Hence understanding the relationship between the chemistry of candidate surfactant tail moieties to their HFA-philicity, and surfactant balance at the HFA-water (HFA|W) interface is of great relevance for the development of such formulations.

In this work we investigate the activity of a series of non-ionic amphiphiles at the HFA134a-water interface using in situ high-pressure tensiometry. An optimal candidate surfactant molecule was identified, and its ability in forming aqueous reverse aggregates in HFA134a was studied by in situ UV-Vis spectroscopy (molecular probe). Small angle neutron scattering was used to investigate the structure of the reverse aggregates in the presence of the anti-cancer drug irinotecan hydrochloride (IrHCl). The aerosol characteristics of the formulations in the presence of IrHCl, and the effect of non-volatiles on the aerosol properties were investigated with an Anderson Cascade impactor. Cytotoxicity of the selected amphiphile was studied on the A549 cells, an alveolar type II epithelial cell line. This work is relevant in that it demonstrates the applicability of reverse-aqueous aggregates in HFAs as potential pMDI formulations for the non-invasive regional delivery of anti-cancer therapeutics to the lungs.

569n A Study of Diffusion and Release of Doxorubicin and Sulforhodamine In Porous Alumina

Bradley R. Gordon[1], Charles E. Luckett[2], Douglas S. English[2], Sheryl H. Ehrman[1], and Daniel D. Lim[2]. (1) Department of Chemical and Biomolecular Engineering, University of Maryland, College Park, MD 20742, (2) Department of Chemistry and Biochemistry, University of Maryland, College Park, MD 20742

A model is presented for the diffusion and release of Doxorubicin (Dox) and Sulforhodamine 101 (SR101) in porous alumina. The diffusion of Dox and SR101 were studied using fluorescence recovery after photobleaching (FRAP) at pH's 5.0 and 7.4 over the temperature range 273K to 298K. Data from these experiments were fitted to first-order kinetics to predict the effects of pH and temperature on the adsorption of Dox and SR101. It was observed that the diffusion of Dox is dependent upon the pH whereas the release is not. It was further found that the diffusion of SR101 is independent of pH and that the release is temperature dependent. These observations suggest that the pH induced charge interactions between the diffusing species and the alumina surface play a significant role in the uptake and release from porous alumina.

Hydrogels and hydrogel composites are currently being extensively studied for various high interest biomedical applications including drug delivery, tissue engineering, and microfluidic valves. The incorporation of nanoparticles into a hydrogel matrix can provide unique properties including remote actuation and also improve properties such as mechanical strength. In this poster, development and characterization of magnetic hydrogel nanocomposites, along with remote controlled (RC) demonstration of pulsatile drug delivery and soft actuator applications, will be highlighted.

Magnetic nanocomposites of temperature responsive poly (N-isopropylacrylamide) (PNIPAAm) hydrogels were developed by incorporation of superparamagnetic Fe_3O_4 nanoparticles. The nanocomposites show negative temperature sensitivity, and the temperature sensitivity and swelling transition temperature can be controlled by composition of PNIPAAm in the hydrogel system. When exposed to external alternating magnetic field (AMF), heating of superparamagnetic Fe_3O_4 particles led to rise in temperature of the nanocomposite system. The rise in temperature can be controlled by particle loadings of the system. Application of the nanocomposites as RC drug delivery systems is of high interest. A schematic of the proposed actuation and controlled delivery is illustrated in Figure 1 (a).

RC drug release was demonstrated from magnetic nanocomposites by application of alternating magnetic field. On application of field, the nanocomposite temperature increased above the LCST resulting into accelerated collapse. The collapse expelled large amounts of imbibed water, resulting in drug release at an increased rate. Magnetic nanocomposites can thus give pulsed release when needed in addition to continuous Fickian release profile. Figure 1(b) shows release profile of vitamin B_{12} to show effective modulation in release by pulse application of AMF.

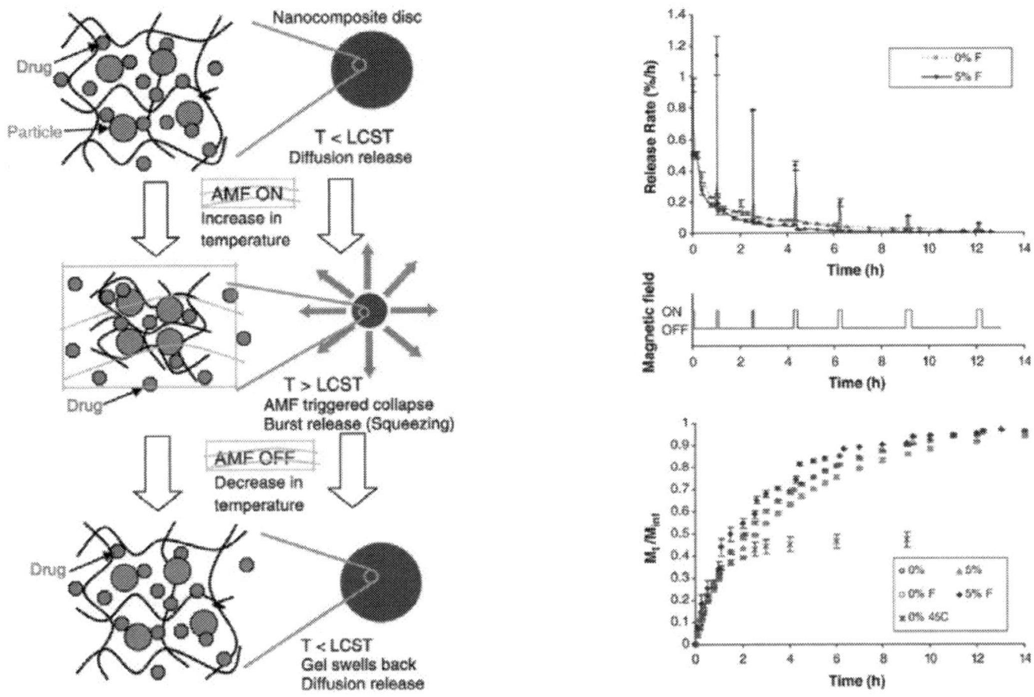

(a) (b)

Fig. 1 (a) Schematic showing the effect of ON-OFF cycles of AMF to the magnetic nanocomposites of PNIPAAm. (b) Vitamin B_{12} release from magnetic nanocomposites of PNIPAAm on pulsed application of AMF. % represents particle loading by weight. F represents application of AMF. M_t represents cumulative mass released at time t. M_{inf} represents cumulative mass released over 48 h.

The rate of hydrogel collapse and recovery are key parameters in design of hydrogel as a component of drug delivery device. The goal of another study was characterization of magnetic PNIPAAm hydrogels with different film thickness. AMF was applied to gels of different thickness, resultant heating and triggered collapse was studied. Heat and mass transfer analysis was conducted to correlate the film thickness with kinetics of AMF triggered heating, collapse, and subsequent recovery.

Present studies are underway to incorporate the hydrogel nanocomposite as a valve in a ceramic microfluidic channel. Application of AMF can heat and collapse the gel, leading to opening of the valve. Thus, flow through the channel can be regulated remotely using AMF.

Wet Granulation Process via Foams and Drops

Melvin X.L. Tan and Karen P. Hapgood

Monash Advanced Particle Engineering Laboratory
Chemical Engineering Department, Monash University, Clayton, VIC, 3800
Email address: melvin.tan@eng.monash.edu.au

ABSTRACT

Wet granulation using foaming binder solution is a new approach of liquid delivery to create granules. This study compares wet granulation via foamed binder addition method and sprayed binder addition method by studying the kinetics of foam and drop penetrations into loosely packed powder beds.

Foam and drop nucleation kinetics were compared by examining specific penetration time (penetration time per unit mass of liquid drop or foam) and nucleation ratio (ratio of mass of nuclei granules to mass of liquid binder).

The mechanisms of foam decay such as liquid drainage and bubble coalescence are critical to the process of foam penetration into the powder pores. Foam penetration rate cannot be completely revealed until the foam characteristics are clearly determined. Nucleation ratio results suggest that foam nucleation method allows more efficient liquid usage and liquid distribution to induce nucleation of powder compared to drop nucleation method.

INTRODUCTION

A new liquid delivery method – foam granulation has been developed as an alternative to the traditional application of atomized liquid spray in pharmaceutical wet granulation processes. Unlike spray granulation, the liquid binder is sent through a foam generator and distributed to the powder bed as aqueous foam to induce granulation. This novel approach has been shown to produce granules and tablets in the pharmaceutical processing. Initial experiments indicate that foam granulation uses less liquid binder; discards the need for nozzles which subsequently eliminates the problems associated with nozzles; and has better liquid dispersion and wetting through the powder mass (Keary and Sheskey, 2004). Scale-up studies has also demonstrated the technical operating potential of this technology (Sheskey, et al., 2007).

Research on foam granulation technology has focused on a series of foam granulation industrial case studies (Keary and Sheskey, 2004). However, there is still lack of understanding on how nucleation and granulation occur when foamed addition method is used. This paper investigates the penetration kinetics of foam by depositing a single foams on a loosely packed powder bed, and it was compared to that of using single drops of the same fluid. The nucleation kinetics of foams and drops are compared by examining the penetration time and nucleation ratio. Both parameters are useful in quantifying the liquid distribution efficiency and the nuclei formation kinetics (Hapgood, et al., 2002, Schaafsma, et al., 2000).

The penetration time is defined as the time required for a drop or foam to completely penetrate into the powder bed. Studying the penetration kinetics of liquid can be useful to understanding nucleation kinetics (Hapgood, et al., 2002). Due to variations of foam mass dispersed, specific penetration ($\hat{t}_p$), defined as the penetration time (t_p) per unit mass of binder (m) is introduced in this study:

$$\hat{t}_p = \frac{t_p}{m} \tag{1}$$

Nucleation ratio is defined as the amount of powder nucleated per unit mass of liquid binder required. This ratio of liquid binder mass to nucleus mass reflects the efficiency of liquid binder distribution through the powder mass to create granule (Schaafsma, et al., 2000). In this paper, nucleation ratio is calculated in terms of mass ratios:

$$K_m = \frac{M_n}{M_b} \tag{2}$$

where M_n and M_b are the nucleus mass and liquid binder mass respectively.

EXPERIMENTAL

Materials

Lactose monohydrate was used in the experiments. A loosely packed powder bed was prepared by sieving the particular powder into a petri dish and the bed surface was scraped smooth using a metal spatula. Several concentrations of hydroxylpropyl cellulose, HPC were used in the experiments. The binders were prepared by dispersing the HPC powders in distilled water according to the required concentration (3% w/w HPC and 6% w/w HPC).

Procedures

Drop and foam penetration experiments are performed with a 6ml syringe or a handheld Airspray 150ml foam dispenser. The amount of binder added can be measured by placing the powder bed on the balance.

Penetration time is measured using a stopwatch. A small amount of food dye was added to the solutions to assist visibility of liquid penetration. Nucleus collection was performed by sieving method. Excavated nuclei were weighed using the balance.

RESULTS AND DISCUSSION

Comparisons of Foam and Drop Specific Penetration Times

The plot of specific penetration time against binder mass for both foam and drop yields similar trends, where decreasing foam and drop mass increase specific penetration time. This similar trend may be attributed to the fact that liquid drains from the foam will also penetrate into the powder mass, which is similar to the drop penetration process.

The results show that 3% HPC foam penetration on lactose powder requires a longer specific penetration time compared to a drop of the same fluid. Longer specific penetration time indicates that foam dispersion will have to be controlled by mechanical mixing during foam granulation process. This may presumably due to the extra time needed for the occurrences of bubble coalescence and liquid drainage to induce liquid penetration into the powder pores. However, the results for 6% HPC binder penetration on 100mesh lactose do not clearly show whether foam or drop has a shorter or longer penetration time. As both liquid binder and foam properties are interactively affecting the penetration behaviour of foam on the powder surface, we cannot reach any general conclusions until the effects of liquid binder properties and foam properties on the penetration process are clearly determined.

Comparisons of Foam and Drop Nucleation Ratios

For lactose powder, foamed addition method has generated a larger nucleation ratio compared to drop addition method. This implies that nucleation via foamed addition method requires less liquid binder; which supports a key finding of Keary and Shesky (2004), who found that foam granulation used lower amount of fluids to achieve similar average granule size. A larger nucleation ratio also indicates that foamed addition method provides better liquid distribution within the non-uniform powder beds to create a lager final nucleus size. Nucleation of powder particles is often restricted due to the presence of macrovoids in lactose powder beds with irregular packing structure (Hapgood, et al., 2002), while foamed addition method offers better liquid distribution efficiency within the powder bed to form nuclei.

CONCLUSIONS

Comparisons were made between foam and drop penetration kinetics by examining the penetration time and nucleation ratio. The properties of foam are critical to the rate at which foam penetrates into the powder pores. This was seen from the microscopy observation that the foam penetration process was largely influenced by the liquid drainage and bubbles movement mechanisms. Comparisons of foam and drop nucleation ratios have shown that foam dispersion provides better liquid distribution efficiency and uses less liquid binder to nucleate the same number of powders.

REFERENCES

Hapgood K. P., Litster J. D., Biggs S. R., Howes T., 2002. Drop penetration into porous powder beds. Journal of Colloid and Interface Science 253 (2), 353-366.

Keary C., Sheskey P., 2004. Preliminary report of the discovery of a new pharmaceutical granulation process using foamed aqueous binders. Drug Development & Industrial Pharmacy 30 (8), 831-845.

Schaafsma S. H., Vonk P., Kossen N. W. F., 2000. Fluid bed agglomeration with a narrow droplet size distribution. International Journal of Pharmaceutics 193 (2), 175-187.

Sheskey P., Keary C., Clark D., Balwinski K., 2007. Scale-up trials of foam-granulation technology--high shear. Pharmaceutical Technology 31 (4), 94-108.

569q Determination of Atorvastatin In Human Plasma with Solid Phase Micro-Extraction Followed by LC Ms/ms Method and Its Application to Pharmacokinetics Study

Pattarawit Rukthong and Korbtham Sathirakul. Pharmaceutics, Mahidol University, Faculty of Pharmacy, Payathai District, Bangkok, Thailand

Determination of drugs in human plasma using the HPLC technique has been widely accepted as an effective method owing to its practical applicability in routine drug analysis in clinical trials. The method sensitivity can be compromised once limited sample volume is available. This problem is typically encountered in pharmacokinetics studies from which small volume of plasma samples either form humans or animals can be obtained. Atorvastatin, an HMG-CoA reductase inhibitor widely prescribed for the treatment of hypercholesterolemia and the prevention of cardiovascular diseases was selected as a model drug in this study. Drug extraction from human plasma was performed with the aid of reversed-phase C18 solid phase micro-extraction. The detection was accomplished by triple quandruple mass spectrometer interfaced operated in electrospray positive ion mode. Quantitation was achieved by using multiple reaction monitoring (MRM) precursor-production transitions at m/z 559.2 and 440.1 for atorvastatin and m/z 412.2 and 224.2 for fluvastatin (internal standard). Linearity was found within the atorvastatin concentration range of 0.2-80 ng/ml. The limit of detection was found to be 0.05 ng/ml. The lower limit of quantification was 0.2 ng/ml with a relative standard deviation (%RSD) of less than 12%. Acceptable precision and accuracy were obtained for the concentrations within the calibration curve range. The average recovery at low, medium and high of atorvastatin concentrations from spiked plasma: quality control samples, were 84.16 ± 7.99%, 96.82 ± 6.14%, and 102.84 ± 6.73%, respectively. The need for less than 250 ml of plasma volume each sample made it possible to decrease sample preparation time. The method was successfully validated and proved appropriate for the analysis of atorvastatin in human plasma and can be applied for pharmacokinetics, bioavailability and bioequivalence studies.

569r Analyzing the Data from An Anderson Cascade Impactor. A Computational Perspective

Yuriy Gulak[1], **Eric Jayjock**[2], and Fernando J. Muzzio[2]. (1) Rutgers, The State University of New Jersey, Piscataway, NJ 08817, (2) Chemical and Biochemical Engineering, Rutgers University, PIscataway, NJ 08817

The Delivery of active ingredients through the pulmonary pathway has received a considerable amount of attention in the past 10 years. The large amount of surface area and relatively quick uptake time make the lungs a good pathway for drug delivery. In recent years researchers have focused on delivering dry powder aerosol particles to the lungs. It is well understood that in this mode of drug delivery the particle size distribution of the active aerosol determines the success or failure of the drug delivery. This creates the need for tools to accurately and quickly test the particle size distribution being emitted from a dosation device. Currently the most commonly used device to try to determine aerosol size distribution is an Anderson Cascade Impactor.

Anderson Cascade Impactors is designed to impinge air onto a metal collection plate at increasing velocities as the air travels through the impactor. In the initial stage of the impactor, where the velocity is the lowest only larger particles with more thermal momentum impact on the collection plates. As the velocity of the air impinging on the surface increases smaller and smaller particles are collected. The plates can then be weighed for mass of impacted particle or sent for HPL analysis.

Inspite of these impactors serving a relatively important roll in dry powder delivery research little work has been done to evaluate the ability of these impactors to differentiate between different particle size distributions of similar size. The goal of this work is utilize currently available computational tools to determine how much information can be be taken from the ACI devise. Special focus will be placed on what practical information can be derived from the output of the ACI test (mass on each stage).

In order to achieve this a computational model has been developed that closely matches real world ACI performance. Variable realistic PSD have been used as inputs and mass impaction is the model output. Various methods for inverting the output of the model back to an approximate PSD. These estimated PSD are then compared to the original input PSD and the method of inversion can be characterized for its accuracy.

569s A Quantitative Study of the Effects of Humidity on the Adhesion of Pharmaceutical Particles to Stainless Steel

Kyung Min Lee[1], Dave Balachandran[1], M. Teresa Carvajal[2], and Stephen P. Beaudoin[1]. (1) School of Chemical Engineering, Purdue University, 480 Stadium Mall Drive, West Lafayette, IN 47907-2100, (2) Department of Industrial and Physical Pharmacy, School of Pharmacy, Purdue University, 575 Stadium Mall Drive, West Lafayette, IN 47907

Pharmaceutical particle interactions with conveyers, tanks, other particles, and other surfaces are important in the development of processes for drug formulation and production. Relative humidity (RH) has a critical effect on these adhesion phenomena. Model particles, including griseofulvin (model active pharmaceutical ingredients, API) and lactose (model excipient), were studied to investigate their adhesion forces against stainless steel and perspex. Stainless steel, a common conveyor belt material in API production, and Perspex, a common substrate in particle synthesis reactors and mixers, were studied. Relative humidity was controlled in the range of 15 % to 65 %. Particle adhesion forces were generated using colloidal probe microscopy (CPM) with a Veeco atomic force microscope (AFM). Images of the particle were obtained before and after their use in CPM using environmental scanning electron microscopy (ESEM). The images allowed an existing in-house simulation code to be applied to predict adhesion forces as a function of particle size, shape, roughness, composition, and elastic properties, as well as substrate composition and elastic properties. Experimental and modeling results were obtained which show that the primary role of increasing relative humidity was to cause surface annealing of the particles, effectively reducing their roughness, and to reduce their elastic modulus, allowing them to deform more extensively in contact.

569t Finite Element Analysis of Tablet Compaction Using a Modified Material Model

Tuhin Sinha[1], Jennifer S. Curtis[2], Bruno C. Hancock[3], and Carl R. Wassgren[1]. (1) School of Mechanical Engineering, Purdue University, 585 Purdue Mall, Purdue University, West Lafayette, IN 47907-2088, (2) University of Florida, Chemical Engineering Department, P.O. Box 116005, Gainesville, FL 32611-6005, (3) Pfizer Inc., Groton, CT 06340

In this poster, finite element methods are used to investigate the compaction process of pharmaceutical tablets. The pharmaceutical powder is treated as a continuum with its elastic-plastic constitutive behavior described by the Drucker-Prager Cap (DPC) model. The model captures the shearing and densification of the powder during compaction and predicts the localized stress and relative density distributions. The DPC model was implemented for a cylindrical tablet geometry using the commercial software, Abaqus, for micro-crystalline cellulose, with the material parameters taken from the literature. The elastic-plastic material parameters of the constitutive model are strongly dependent on the relative density of the powder and evolve during the tablet compaction process. In the current study, a modified material model is used in which the DPC parameters are dependent on the current relative density of the powder and, hence, provide more accurate macroscopic property distributions within the tablet. The results show that significant differences in the final tablet stress and relative density distributions occur when constant material properties and relative density dependent properties are used.

569u Ultrasonic Sound Propagation In Porous Materials

Omar L. Sprockel, Drug Product Process & Package Development, Bristol-Myers Squibb, One Squibb Drive, New Brunswick, NJ 08903, Erik Yang, School of Pharmacy and Pharmaceutical Sciences, Purdue University, 1102 nooning tree dr, Chesterfield, MO 63017, Dimuthu A. Jayawickrama, Analytical R&D, Bristol-Myers Squibb, One Squibb Drive, New Brunswick, NJ 08903, Shih-Ying Chang, Process R&D Engineering Technologies, Bristol-Myers Squibb, One Squibb Drive, New Brunswick, NJ 08903, and Lei Li, Baby Care, Proctor & Gamble, 6280 Center Hill, Cincinatti.

Objective

The objective of this study was to develop the theoretical basis for the dependency of sound velocity through porous materials on density so that active ultrasonics can be used to determine the density of ribbons that are intermediates in the roller compaction manufacturing process.

Theoretical

Equation 1 was developed and predicts that sound velocity depends on density for a porous material above the percolation threshold, where V is the sound velocity, E^{*}_{Max} is the modified Young's elastic modulus, S is a scaling factor, ρ is the ribbon density, and ρ_c is the percolation threshold density. According to percolation theory, the relationship between sound velocity and density of porous materials holds above the percolation threshold density, where the properties of the system would not change abruptly.

$$V = \sqrt{E^{*}_{max} \cdot S} \bullet \frac{(\rho - \rho_c)}{\sqrt{\rho}}$$

Equation 1

Experimental

Sound velocity through a sample was determined by measuring the time of travel of an ultrasound pulse though a sample of a given thickness. The sample was placed between two sensors. Aqualene™, an elastomer with a density close to water, was used as an ultrasonic buffer conduit. Aqualene™ was placed between the ribbon sample and the sensors. A constant force is applied between the sensors. Ultrasonic waves were passed through the sample between the two sensors. The waveform data was acquired and the time of travel between the two sensors was determined, based on a threshold value set at three times the background ultrasound noise level. The sound velocity was calculated from the distance of travel and the time of travel. The ribbon densities were determined using the weight and volume of the sample.

Results

Five sets of ribbons (n=3) were used to estimate the value of E^{*}_{Max}•S from the ribbon density and sound velocity data set. The sound velocity through twelve additional ribbon samples was determined and their density predicted. The average difference between the predicted and determined ribbon density values was ~ 1.76% (0.00 – 5.25%) and the slope of the predicted versus experimental density was ~1.07.

Conclusion

The theoretical basis for dependency of sound velocity on the density of porous ribbons was developed and the application of active ultrasonics as a method to predict ribbon density with a reasonable degree of accuracy was demonstrated.

569v Powder Flow Characterization from Avalanche Behavior

Jose E. Tabora, **Ryan Cullen**, Steven Chan, and Steve Wang. Bristol-Myers Squibb, One Squibb Drive, New Brunswick, NJ 08901

Characterization of powder flow is an important aspect in determining the manufacturing process of active pharmaceutical ingredients (API). The Aeroflow® is a commercial flowability analyzer which measures the avalanche behavior of a powder in a constantly rotating disc. Generally the mean time to avalanche (MTA) and the scatter is measured and used as a comparative value of the flow properties of the powder. Previous work [1] suggests that the underlying dynamics from this process are chaotic. In this work the raw data is converted to a spike train denoting the time and intensity of the avalanche. Non-linear statistical analysis is performed on the time series to characterize the underlying non-linear process involved. This methodology was applied to a series of lots of API's manufactured at pilot plant scale under different processing conditions. The results are compared to conventional powder characterization.

[1] B.H.Kaye, J. Gratton Liimatainen, N. Faddis, Particle System Characterization, 1995, 12, 232-236

569w Selection of An Appropriate Auger and Funnel Combination Using Statistical Analysis of in Process Data to Achieve a Stable and Consistent Sachet Filling Operation

Ganeshkumar A. Subramanian, Admassu Abebe, Robert Jerzewski, James Bergum, Faranak Nikfar, Sushmita Bhattacharjya, and Xin Bu. Biopharmaceutics, Bristol Myers Squibb, 1 Squibb Drive, New Brunswick, NJ 08903

Purpose: A powder for oral use formulation was filled into sachets. The low % of active ingredient (< 4%) with a low fill weight of 1.5 g/sachet, combined with the cohesive and non flowing nature of the blend makes the filling operation critical to ensure appropriate weight control to achieve the correct dosing. The objective of this poster is to explain the filling process, selection of appropriate auger/funnel combination with the critical controls derived using statistical analysis of the in process weight data.

Method:

A sachet filling machine works on the principle of delivering a measured amount of powder using an auger and funnel set up. With no feed back loop to automatically adjust the amount delivered, the selection of the auger and funnel combination is critical to the accuracy of the powder fill weight. Trails with several different size augers namely, 3/8 inch, 1/2 inch, 9/16 inch and 5/8 inch were carried out to select an appropriate auger. In process weight checks were used to monitor the filling operation and this data was statistically analyzed and correlated to potency of the filled sachets.

Results:

The smaller augers 3/8 inch and 1/2 inch gave better accuracy of fill, but resulted in a process that was less stable. The 5/8 inch auger was less accurate and in process weight variations were high. The 9/16 inch auger gave the right combination of accuracy and consistency with a stable filling process.

Conclusion:

Based on statistical analysis of in process data and confirmed by analytical testing for potency, a 9/16 inch auger was selected for filling this powder for oral use formulation into sachets.

569y Development of a Process for Crystallizing the Desired Crystal Form of An API

Christianto Wibowo, ClearWaterBay Technology, 4000 W. Valley Blvd. Suite 100, Pomona, CA 91789

During scale up of a generic drug preparation process that involves crystallization of an API from a solvent, some of the production batches had exceptionally small particle size and high residual solvent levels after drying. The problem was traced to the formation of solvate crystals, which lose the solvent during drying and rapidly transform to very small crystals of the anhydrous form. Therefore, it is desirable to define the best operational parameters, such as temperature, cooling rate, and solvent ratio to directly obtain anhydrous crystals by crystallization from the solvent, without the contamination of other polymorphs and/or impurities.

Based on the notion that understanding the fundamentals is the key in developing a stable and reproducible process, an integrative approach that combines experimental observations, physical insights, and modeling has been applied to analyze the available data and suggest a suitable recipe for crystallizing the desirable anhydrous form of the API. The approach includes consideration of both solid-liquid equilibrium phase behavior of the system and kinetic effects. The relevant SLE phase behavior is studied, and used to explain the findings of the previous experiments. An initial recipe for obtaining the desired product form is developed and validated in a trial run.

AUTHOR INDEX

AUTHOR INDEX

AUTHOR INDEX

AUTHOR INDEX

AUTHOR INDEX

AUTHOR INDEX

AUTHOR INDEX

AUTHOR INDEX

AUTHOR INDEX

AUTHOR INDEX

AUTHOR INDEX

AUTHOR INDEX

AUTHOR INDEX

AUTHOR INDEX

AUTHOR INDEX

AUTHOR INDEX

AUTHOR INDEX

AUTHOR INDEX

AUTHOR INDEX

AUTHOR INDEX

AUTHOR INDEX

AUTHOR INDEX

AUTHOR INDEX

AUTHOR INDEX

AUTHOR INDEX

AUTHOR INDEX

American Institute of Chemical Engineers
3 Park Avenue
New York, New York 10016-5991

CURRAN ASSOCIATES INC
proceedings.com

Food, Pharmaceutical & Bioengineering Division

Presentations at the 2008 AIChE Annual Meeting

Philadelphia, Pennsylvania
16 - 21 November 2008

Volume 2 of 2

ISBN: 978-1-61567-228-8

AIChE

Food, Pharmaceutical & Bioengineering Division

Presentations at the 2008 AIChE Annual Meeting

Philadelphia, Pennsylvania
16 - 21 November 2008

Volume 2 of 2

ISBN: 978-1-61567-228-8

Printed from e-media with permission by:

Curran Associates, Inc.
57 Morehouse Lane
Red Hook, NY 12571

Printed by Curran Associates, Inc. (2009)

For permission requests, please contact AIChE
at the address below.

AIChE
3 Park Avenue
New York, NY 10016-5991

Phone: (203) 702-7660
Fax: (203) 775-5177

www.aiche.org

Additional copies of this publication are available from:

Curran Associates, Inc.
57 Morehouse Lane
Red Hook, NY 12571 USA
Phone: 845-758-0400
Fax: 845-758-2634
Email: curran@proceedings.com
Web: www.proceedings.com

TABLE OF CONTENTS

VOLUME 1

VOLUME 2

Protocols for the Identification of Membrane Protein Interaction Networks

Chiann-Tso Lin[1], Kristin D. Victry[1], Baowei Chen[1], Dale A. Pelletier[2], H. Steven Wiley[1] and Brian S. Hooker[1,*]

[1]Pacific Northwest National Laboratory, Richland, Washington, USA
[2]Oak Ridge National Laboratory, Oak Ridge, Tennessee, USA
*E-mail: brian.hooker@pnl.gov

Nearly one-third of all genes in various organisms encode membrane associated proteins that participate in numerous protein-protein interactions important to the processes of life. Various methods have been developed to identify and characterize these interactions including pull-down, yeast two-hybrid and colocalization methods, among others. However, membrane protein complexes pose significant challenges due to the need to solubilize membranes without disrupting protein-protein interactions. Aiming at development of protocols for the identification and characterization of membrane protein interactions, we have tested the feasibility of isolating and identifying the several protein complexes in *Rhodopseudomonas palustris*. *R. palustris* is a purple, non-sulfur phototrophic bacterium commonly existing in soils and water. Towards this purpose, subunits of several complexes were independently cloned with both a 6-histidine tag and a V5 epitope tag at the protein C-terminal end and used as baits for isolation of complexes using endogenous pull-down methods as well as protein correlation profiling chromatographic methods. Cells expressing individual tandem tag fused proteins were then cultured under illumination at 30°C under anaerobic conditions.

Cells solubilized with non-ionic or zwitterionic detergents were subjected to affinity "pull-down" experiments, using either Ni-NTA gels alone or a two-step tandem procedure, first with Ni-NTA, and then using a V5 epitope specific antibody. Tryptic digested peptide fragments of the pull-down products were then subjected to LC-MS/MS analysis.

Our results show that independent of the subunits selected (RPA0176 [β, 51 kD], RPA0844 [F_0-b, 19.2 kD], RPA0177 [γ, 32 kD], or RPA0178 [F_1-α, 55.17 kD]) as baits for pull-down experiments, the vast majority of the ATP synthase subunits were found in the eluted fractions from the Ni-NTA column. The electrophoretic pattern of proteins also confirms the complex similarity when using different baits. Further evidence for the feasibility to isolate the ATP synthase complex with the Ni-NTA gel in the presence of detergents has been solidified with the tandem pull-down using V5 antibody.

An example of a pull-down experiment for the ATP synthase is shown in Figure 1. In this instance, different detergent concentrations above and below the critical micelle concentration for octyl-β-D-glucopyranoside (OG) were employed in tandem Ni-NTA/V5 antibody pull-downs using the γ chain and the F_0-b chain as baits. Out of the four conditions tested, 0.75% OG, 2% OG and no detergent controls performed comparably, yielding 8-9 known prey proteins. In contrast, 0.25% OG yielded just 4 known interactions between the two bait proteins. Furthermore, beside the known subunits of the ATP synthase complex, more proteins could also concomitantly detected, for example, 105, 92 and 103 proteins were identified for RPA0844, RPA0177 and RPA0178, respectively suggesting that this method may be used as a discovery tool to identify new interactions.

In addition to endogenous pull-down experiments, protein complexes have been elucidated using protein correlation profiling completed employing tagged "bait" protein cell lines. Using this method, protein complexes are identified and characterized by matching elution profiles from multidimensional separation techniques. The keys to success of such an approach are that the combination of various separation strategies provides sufficient aggregate discrimination to achieve a fractionation pattern that defines a single complex.

To complete this method, cell lysate was first detergent solubilized and pre-fractionated using differential ammonium sulfate precipitation under gentle conditions to maintain intact protein complexes. Fractions containing the bait protein (as detected via Western blot assay) were each further separated using anion exchange chromatography. Assuming the complex is stable under these separation conditions, proteins that constitute a complex should co-fractionate. Complexes were identified by following the elution profile of the tagged protein via Western blot analysis in conjunction with LC-MS/MS of the various elution fractions. Further resolution of complexes was completed using Blue Native PAGE of elution fractions followed by in-gel digestion of bands and LC-MS/MS analysis.

Pull-down assays and protein correlation profiling may be used as orthogonal methods to characterize the same complex in order to increase the confidence of protein-protein interaction identifications. Results using these methods will be presented for several different protein complexes.

571b A Novel Approach for Residue Contact Prediction in Beta and Alpha/beta Proteins

Rohit Rajgaria[1], Scott R. McAllister[1], and Christodoulos A. Floudas[2]. (1) Chemical Engineering, Princeton University, Dept of Chemical Engineering; A215 Engineering Quadrangle, Princeton, NJ 08544, (2) Department of Chemical Engineering, Princeton University, Dept of Chemical Engineering; A215 Engineering Quadrangle, Princeton, NJ 08544

Residue contact prediction is an important component of protein structure prediction and it is used to predict contacts between residue pairs which are far apart in the primary sequence of a protein but proximal in their three dimensional structure. This method can be particularly useful in first principles based approaches where no database information is used for the secondary and tertiary structure prediction. These residue contacts are highly useful in guiding the search and help a structure prediction algorithm identify better quality structures. A number of different methods have been developed to predict these non-local contacts. In a very broad sense, these techniques either use correlated mutations analysis or machine learning approach (hidden Markov models, neural networks etc.) [1-5]. Recently, an integer-linear programming based framework was proposed to predict hydrophobic contacts in alpha-helical proteins [6].

In this presentation, a novel optimization based method framework is introduced to predict contacts in beta, alpha+beta and alpha/beta proteins. In this method, a Calpha-Calpha distance dependent force field [7] is used to assign contact energy for a particular contact based on the identity of the amino acids [8]. This problem has been formulated as an integer linear programming problem where the objective function is to minimize the contact energy. A set of constraints is also included in the model to produce physically possible contacts and topologies for alpha, beta, alpha+beta and alpha/beta proteins. This formulation not only offers the advantage of finding the residue contacts corresponding to the global minima, but it can also produce a rank-ordered list of residue contacts. A rank-ordered list like this helps in finding the most frequent contacts. This model also offers the flexibility of incorporating additional constraints where a user can add unique and problem specific constraints to the model. The presented method was tested on few test proteins and produced an average accuracy of ~60%.

[1] Klepeis, J., and Floudas, C. A., 2003a, Prediction of Beta-Sheet Topology and Disulfide Bridges in Polypeptides. Journal of Computational Chemistry, 24: 191-208.

[2] Vicatos, S., Reddy, V. B., and Kaznessis, Y., 2005, Prediction of Distant Residue Contacts With the Use of Evolutionary Information. Proteins: Structure, Function, and Bioinformatics, 58: 935-949.

[3] Fariselli, P., Olema, O., Valencia, A., and Casadio. R, 2001, Progress in Predicting Inter-Residue Contacts of Proteins With Neural Networks and Correlated Mutations. Proteins: Structure, Function, and Genetics, 5: 157-162.

[4] Hamilton, N., Burrage, K., Ragan, M. A., and Huber, T., 2004, Protein Contact Prediction Using Patterns of Correlation. Proteins: Structure, Function, and Genetics, 56: 679-684.

[5] Cheng, J., and Baldi P., Improved Residue Contact Prediction Using Support Vector Machines and A Large Feature Set. BMC Bioinformatics, 56: 679-684.

[6] Rajgaria, R., McAllister, S. R., and Floudas, C.A., 2007, Towards Accurate Residue-Residue Hydrophobic Contact Prediction for Alpha-Helical Proteins Via Integer Linear Optimization , Proteins, Submitted.

[7] Rajgaria, R., McAllister, S. R., and Floudas, C.A., 2007, Development of A Novel High Resolution Calpha- C-alpha Distance Dependent Force Field Using A High Quality Decoy Set. Proteins: Structure, Function, and Bioinformatics, 65: 726-742.

[8] Rajgaria, R., and Floudas, C.A., 2008, In Preparation.

571c Improving Microcin J25 Production In E. Coli

Si Jia Pan and **A. James Link**. Chemical Engineering, Princeton University, A208 Engineering Quadrangle, Princeton, NJ 08544

The microcins are a class of antimicrobial peptides produced by Gram-negative bacteria that putatively function to give the producing cell a selective advantage during periods of stress or nutrient depletion. One of these peptides, microcin J25 (MccJ25) has received attention due to its unusual lasso structure which imbues it with extraordinary thermal stability. Consistent with its ecological role, however, the production of MccJ25 is initiated only during stationary phase, limiting the yield of active peptide to only several milligrams per liter of bacterial culture. To improve the yield of MccJ25, we have reengineered the gene cluster responsible for the expression and maturation of the peptide. Preliminary data indicate significantly improved yields of MccJ25 in our engineered system relative to natural producer cells. This improved system for the production of MccJ25 also serves as a platform for the engineering of the peptide for improved and novel functions.

571d Association Between Human (pro)Renin Receptor Lacking Putative Transmembrane and Cytoplasmic Domains with Autographa Californica Multiple Nucleopolyhedrovirus Lacking Other Transmembrane Domains

Tatsuya Kato[1], Dongning Du[2], Fumiaki Suzuki[3], and Enoch Y. Park[2]. (1) Department of Applied Biological Chemistry, Shizuoka University, 836 Ohya Suruga-ku, Shizuoka, 422-8529, Japan, (2) Laboratory of Biotechnology, Graduate School of Science and Technology, Shizuoka University, 836 Ohya, Suruga-ku, Shizuoka, 422-8529, Japan, (3) Laboratory of Animal Biochemistry, Faculty of Applied Biological Sciences, Gifu University, 1-1 Yanagido, Gifu, 501-1193, Japan

Human (pro)renin receptors (hPRR) with native putative transmembrane and cytoplasmic domains (hPRR-wTM) were purified from the microsomal fraction of baculovirus-infected Tn-5B1-4 cells. However, hPRR produced with deleted transmembrane and cytoplasmic domains (hPRR-w/oTM) could not be adequately purified. In this report, the expression of hPRR-wTM and hPRR-w/oTM was analysed. hPRR-w/oTM was detected in budded baculovirus (BV) fractions isolated from culture supernatant, whilst hPRR-wTM was not. This suggests that hPRR-w/oTM might be associated with BV. hPRR was expressed in the peripheral domains of the nucleus in infected Sf-9 cells. In contrast, hPRR-w/oTM was observed in punctate domains in the cytoplasm, indicating that hPRR-wTM and hPRR-w/oTM localized in the endoplasmic reticulum (ER) and cytoplasmic organelles, respectively. Moreover, intracellular hPRR-w/oTM did not co-localize in the Golgi apparatus and lysosomes. hPRR-wTM could not be extracted from infected Sf-9 and Tn-5B1-4 cells by Triton X-100 treatment at 4°C. It is suggested that hPRR-wTM might not reside in the plasma membrane and that localization of hPRR-wTM in the ER might result in a lack of association with the BV. hPRR-w/oTM was detected in the BV fraction despite lacking a transmembrane domain, but was not detected in the virus envelope fraction, suggesting that hPRR-w/oTM might be associated with other areas in the BV.

571e Single Sequence Secondary Structure Prediction for Globular Proteins

Ashwin Subramani and Christodoulos A. Floudas. Department of Chemical Engineering, Princeton University, Engineering Quadrangle, Olden Street, Princeton, NJ 08544

The secondary structure prediction of a protein is an important intermediate step in the three dimensional structure prediction of a protein. Its importance comes particularly in focus in first principles based approaches to protein structure prediction, which do not use any database information in the form of homology information [1]. The most common techniques for secondary structure prediction feature a 3 – class secondary structure state for each residue of a protein. A number of methods are available in literature for secondary structure prediction. The most common methods use multiple sequence alignment techniques like BLAST and PSI-BLAST [2] to derive profile information for the target protein, which is fed as input to machine learning techniques like neural networks [3] and SVMs [4] .

We have developed an optimization – based method to predict the secondary structure of a target protein without the use of profile information. Hence, this method can be applied to proteins which do not produce reliable profile information using sequence alignment tools. The model combines two models, an α – helix prediction model HELIOS (HELical prediction using Integer Optimization approacheS) and a β – strand prediction prediction model BEST-PRED (BEta STrand PREDiction). For α – helix prediction, a two – stage infeasibility minimization problem has been introduced. The first stage is a linear programming (LP) model for parameter estimation, while the second stage is an integer programming (ILP) model for helix prediction. The residues of a target protein are divided into 4 regions depending on their putative proximity to the helix termini, and propensity to be in helices is compared to a pre – evaluated residue – dependent threshold propensity, using overlapping nonapeptides surrounding the central residue. BEST-PRED for β – strand prediction has been introduced as an integer programming (ILP) model, which maximizes a residue's propensity to be in a β – strand. The protein is divided into overlapping pentapeptides. The β – strand propensity weight for the central residue is evaluated by implementing a novel combination of Naïve – Bayesian and first order Markov models, which represent the physical nature of a β – strand. In both models, important mathematical constraints are introduced to ensure that biologically meaningful results are presented. These constraints refer to the physical nature of the residues [5], along with the minimum and maximum secondary structure content [6]. A formulation of this kind not only provides the secondary structure prediction corresponding to the evaluated global minima, but also has the ability to provide a rank – ordered list of best solutions. Such a rank – ordered list can help in finding the most frequent predictions in a particular class for a given residue. Further, the formulation allows the user to add any form of prior knowledge about the secondary structure easily. This method was tested on a set of α, β and mixed α- β proteins, and the preliminary results are very encouraging. A Q_α accuracy of 82% was obtained for purely α – helical proteins using HELIOS, while a Q_β accuracy of 78.9% was obtained for purely β – proteins using BEST-PRED. These results compare very favorably with some of the standard secondary structure prediction servers.

[1] Klepeis J.L. and Floudas C.A., 2003, ASTRO – FOLD: a combinatorial and global optimization framework for ab initio prediction of three – dimensional structures of proteins from the amino acid sequence, Biophysical Journal, 85, 2119 – 2146.

[2] Altschul SF, Gish W., Miller W., Myers E.W. and Lipman D.J., 1997, Gapped BLAST and PSI – BLAST: a new generation of protein database search programs, Nucleic Acids Research, 25, 3389 – 3402.

[3] McGuffin L.J., Bryson K and Jones D.T., 2000, The PSIPRED protein structure prediction server, Bioinformatics, 16, 404 – 405.

[4] Gassend B., O'Donnell C.W., Thiel W., Lee A., van Dijk M. and Devadas S., 2007, BMC Bioinformatics, 8, S3.

[5] Aurora R. and Rose G.D., 1998, Helix Capping, Protein Science, 7, 21 – 38.

[6] Homaeian L., Kurgan L. A., Ruan J., Cios K. J. and Chen K., 2007, Prediction of protein secondary structure content for twilight zone sequences, Proteins, 69, 486 – 498.

571f Reconstruction of a Genome-Scale Model of B. Subtilis to Facilitate Development of a Minimal Genome

Christopher Henry[1], Jenifer Zinner[1], Matthew Cohoon[1], Kosei Tanaka[2], Edmond Jolivet[2], Sabine Autret[2], Philippe Noirot[2], and Rick Stevens[1]. (1) Mathematics and Computer Science, Argonne National Laboratory, 9700 S. Cass Avenue, Argonne, IL 60439, (2) Microbial Genomics, INRA, 147 rue de l'Université, Paris, France

Bacillus subtilis is a gram positive bacteria often utilized in industry as a producer of high quality enzymes and proteins [1]. One of the primary challenges involved in the use of *B. subtilis* in industry is the extensive regulatory pathways in the cell, making the flux through the metabolic reactions of the cell extremely resistant to alteration by genetic manipulation [2]. It has already been demonstrated that removing a portion of the regulatory genes in *B. subtilis* results in significantly enhanced protein production by the cell [3]. Now, we are endeavoring to produce a minimal strain of the *B. subtilis* genome. This minimal strain will lack every dispensable alternative metabolic pathway and every dispensable regulatory gene, making the strain much more amenable to alteration for industrial use. *B. subtilis* was selected as the platform organism for the construction of this minimal strain because the natural competence of *B. subtilis* allows for rapid knockouts of single genes and intervals of genes [4]. Additionally, the extensive information available about *B. subtilis* will allow for a systematic and planned approach to be used during the construction of the minimal strain. To facilitate the development of our minimal organism, we have constructed a new genome-scale metabolic model of *B. subtilis* based on the annotations available in the SEED subsystems-based annotation environment [5] and supplemented by data included in two previously developed *B. subtilis* models [6, 7]. The new model includes elements of the *B. subtilis* regulation when necessary to properly predict the effect of gene knockouts. The thermodynamic properties of the model reactions were estimated to predict the reversibility of model reactions [8]. The new model also includes numerous genes and pathways not found in any previous models. The new model was validated against a variety of experimental observations including biolog phenotyping array results [7], gene essentiality data [9], 60 published gene interval knockout experiments [3], and over 300 new gene-interval knockout experiments. When model predictions did not match experimental observations, a variety of methods were developed and applied to improve the model accuracy. As a result of these efforts and the inclusion of features from currently published models, this is the most complete and accurate metabolic model of *B. subtilis* constructed to date. We will discuss the methods used to assemble and correct this new genome-scale model; we will compare the accuracy and content of the new model with previously published models; and we will discuss the application of the model to the design of a systematic and optimized strategy for combining gene interval knockouts to produce a minimal strain of *B. subtilis*.

1. Zweers, J.C., et al., Towards the development of *Bacillus subtilis* as a cell factory for membrane proteins and protein complexes. Microb Cell Fact, 2008. 7: p. 10.

2. Fischer, E. and U. Sauer, Large-scale in vivo flux analysis shows rigidity and suboptimal performance

of *Bacillus subtilis* metabolism. Nat Genet, 2005. 37(6): p. 636-40.

3. Morimoto, T., et al., Enhanced Recombinant Protein Productivity by Genome Reduction in *Bacillus subtilis*. DNA Res, 2008. 15(2): p. 73-81.

4. Fabret, C., S.D. Ehrlich, and P. Noirot, A new mutation delivery system for genome-scale approaches in *Bacillus subtilis*. Mol Microbiol, 2002. 46(1): p. 25-36.

5. Aziz, R.K., et al., The RAST Server: rapid annotations using subsystems technology. BMC Genomics, 2008. 9: p. 75.

6. Goelzer, A., et al., Reconstruction and analysis of the genetic and metabolic regulatory networks of the central metabolism of *Bacillus subtilis*. BMC Syst Biol, 2008. 2: p. 20.

7. Oh, Y.K., et al., Genome-scale reconstruction of metabolic network in *Bacillus subtilis* based on high-throughput phenotyping and gene essentiality data. J Biol Chem, 2007. 282(39): p. 28791-9.

8. Jankowski, M.D., et al., Group contribution method for thermodynamic analysis of complex metabolic networks. Biophysical Journal, 2008. Accepted.

9. Kobayashi, K., et al., Essential *Bacillus subtilis* genes. Proc Natl Acad Sci U S A, 2003. 100(8): p. 4678-83.

571g Genetic Engineering of Escherichia Coli for High-Level Expression of Integral Membrane Proteins

Georgios Skretas and George Georgiou. Chemical Engineering, University of Texas at Austin, 1 University Station C0400, Austin, TX 78712

Biochemical and structural studies of eukaryotic membrane proteins (MPs) are hindered by the very low abundance of this type of proteins in their natural tissues. This problem can be overcome by heterologous overexpression of MPs in simple microbial organisms, such as the bacterium Escherichia coli. However, MP overexpression in bacteria typically results in very small amounts of membrane-integrated protein or in accumulation of the target MP in inclusion bodies. We have employed genetic strategies to engineer the E. coli cell machinery to be able to achieve high-level expression of eukaryotic MPs. We have constructed MP fusions with the green fluorescent protein (GFP) and used fluorescence-activated cell sorting (FACS) to screen libraries of E. coli cells carrying different genetic modifications and isolate mutants that exhibit large increases in cell fluorescence due to enhanced MP production. In this manner, we have identified E. coli variants with the ability to produce markedly enhanced amounts of membrane-embedded protein for a variety of different MPs. Using these engineered bacterial strains, we have produced sufficient amounts of high-quality MPs which are currently under structural investigations.

571h Cloning and Expression of Antimicrobial Peptides In Yeast

Saadet Albayrak, Jean-Marie Rouillard, and Erdogan Gulari. Chemical Engineering, University of Michigan, Ann Arbor, MI 48109

The inevitable emergence of the bacterial resistance to conventional antibiotics is the driving force for the development of novel class of anti-infectives, such as Antimicrobial Peptides (AMPs). These peptides are considered as promising drug candidates to replace and/or supplement the existing antibiotics. The rapid increasing need for the potential AMPs can be met by combining the high-throughput parallel oligonucleotide synthesis with eukaryotic protein expression system to discover novel antimicrobial peptides. The main goals of this part of the study are to show that oligonucleotides coding for AMPs can be cloned in yeast by homologous recombination, expressed and secreted by yeast cells, and screened for their activities against microbial cells. We have constructed two different expression systems in *Saccharomyces cerevisiae* and *Pichia pastoris* to investigate the expression and secretion levels of 13 and 44-amino acid long two control AMPs. Our initial results with *S. cerevisiae* containing GAL10-CYC1 promoter and alpha-MF secretion signal showed that while the longer peptide was expressed at high levels, the production of the 13aa peptide was quite low. The expression level of the shorter peptide was enhanced in *P. pastoris* when the AMP coding genes were fused with SUMO (small ubiquitin-like modifier) gene tag. We also observed that both peptides maintained their antimicrobial activities in yeast culture when tested against Gram negative and Gram positive bacteria.

571i Intensified Disulfide Bond Shuffling for Protein Refolding In Vitro: Molecular Dynamics and Experimental Observations

Zheng Liu and Diannan Lu. Department of Chemical Engineering, Tsinghua University, Beijing, 100084, China

One challenge to protein refolding is to dissociate the non-native disulfide bonds and promote the formation of the native ones. Here we proposed a coarse-grained off-lattice model with disulfide bonds and simulated the shuffling of the disulfide bonds. It was shown that the introduction of disulfide bonds enhances the conformational stability while reduces the foldability of the model protein, as compared to its counterpart without the disulfide bonds. The shuffling of disulfide bonds occurs at a suitable redox environment, in which the non-native disulfide bonds can rearrange to the native ones. The folding trajectory suggested that disulfide bonds located in the hydrophobic core form before protein collapsing while those at protein surface form after protein rearrangement. A reductive environment at an early stage of folding favors the shuffling that gives the native disulfide bonds in the hydrophobic core. For those disulfide bonds at protein surface, however, an oxidative environment at late stage of folding is requested. Thus a dynamic redox environment, i.e., reductive environment to oxidative one, can intensify the shuffling of disulfide bonds and result in an improved yield of native protein. A dynamic redox environment can facilitate disulfide shuffling to native state and thus improve the folding yield. To validate the simulation results, we have also conducted protein refolding experiments using lysozyme as model protein. It is shown that there is optimal ratio of GSSG to GSH, which corresponds to optimal redox environment, for the effective protein folding. We have also conducted lysozyme refolding under a dynamic redox environment, i.e., form reductive environment at early folding stage to oxidative one at late folding stage. It is shown that the folding yield under this dynamic redox environment can effectively improve the folding yield than those at constant redox environment, agreeing well with above predictions by molecular dynamic simulations.

571j Data Mining PubChem with Signature: Prediction of Biological Activity for Small Molecules

Donald P. Visco[1], **Derick C. Weis**[1], and Jean-Loup Faulon[2]. (1) Chemical Engineering, Tennessee Technological University, Department of Chemical Engineering, P.O. Box 5013, Cookeville, TN 38505, (2) Computational Biology Department, Sandia National Laboratories, Albuquerque, NM 87185

High-throughput screening (HTS) is a technique to discover new lead compounds by physically screening a large library against a specified biological target. HTS was primarily available only to the pharmaceutical industry in the past. Because of the Molecular Libraries Initiative, part of the NIH Roadmap for Medical Research, HTS is now accessible to academic researchers where the data collected is deposited in a public database called PubChem. The results from more than 1,000 different HTS experiments are currently readily available in PubChem to download. Cheminformatic tools are crucial to effectively interpret and utilize this vast amount of data.

In this work, we demonstrate a method to create a model from existing HTS data in PubChem, and predict new compounds likely to be active for additional screening.

PubChem bioassay 846 [1] screened for potential anticoagulant therapeutics by identifying inhibitors of factor XIa, which is involved in the blood coagulation mechanism. A classification model with 89% accuracy was created using a support vector machine (SVM) with the Signature molecular descriptor. Approximately 12 million compounds deposited in PubChem, but not present in the factor XIa assay, were virtually screened by the SVM. Based on metrics associated with SVM magnitudes and molecular descriptor overlap between candidate molecules with those from bioassay 846, we identified 296 compounds (from the 12 million not previously tested) as active. Docking studies using the crystal structure of factor XIa were performed on known actives and on these 296 predicted actives, with the predicted actives all showing binding energies consistent with the known actives.

A new feature of our approach (relative to current methods employed) [2] is the use of a wrapper method (in contrast to a filter method) that allows for recursive cluster inclusion in order to arrive at an improved SVM model. We compare these approaches in this work.

In conclusion, a primary HTS can identify new lead compounds, but generally has a very low success rate. The focused database of 296 predicted actives from this work could significantly add to existing knowledge, and increase the chance of finding active compounds by building on previous research. The data mining technique described here is not unique to factor XIa, and could be applied to any bioassay in the ever expanding PubChem database.

[1] Factor XIa 1536 pubchem.ncbi.nlm.nih.gov/assay/assay.cgi?aid=846

[2] Yousef, M., et. al., "Recursive cluster elimination (RCE) for classification and feature selection from gene expression data", BMC Bioinformatics, 2007, 8, 14.

571k Early Events In Atherosclerosis: Modeling of Transmural Water Flow and the Effect of Transmural Pressure on It

Shripad D. Joshi[1], David Rumschitzki[1], and Kung-Ming Jan[2]. (1) Chemical Engineering, City College and Graduate Center of City University of New York, 140th Street, Convent Avenue, Steinman Hall, New York, NY 10031, (2) College of Physician and Surgeons, Columbia University, 3675 Riverdale Ave. Suite 5, Bronx, NY 10463

The macromolecular transport of lipoproteins into the artery wall and their spread in the subendothelial intima are believed to be critical prelesion events in atherogenesis. Large molecules (e.g., lipoprotein cholesterol) cross the arterial endothelium via very rare (~1 in 2000-6000) endothelial cells, many associated with widened junctions around endothelial cells that are either dividing or dying[2,5,6,11]. Our group has established that[2,3] this focal transport is due to the advection of macromolecules by transmural pressure-driven water transport through these leaky junctions. In contrast, water is known to cross the normal endothelium uniformly e.g., through normal tight junctions. Thus water transport, together with the diffusion, can further transport lipoproteins that have already entered the vessel's intima. Hence, the total transmural water transport, and not just the portion through leaky junctions, appears to play a central role in delivering low-density lipoprotein (LDL) cholesterol to the subendothelial space. The basic aim behind this work is to develop a theory, that explains the transcellular pathway for water transport, alongside the generally accepted paracellular route, and to quantify its percentage contribution to the vessel endothelium's hydraulic conductivity Lp, the ratio of the tranmural water flux to the driving pressure difference, as a function of transmural pressure. First some background.

Experimental results of Tedgui & Lever[10] and Baldwin & Wilson[1] show a higher value of Lp at low pressures (60-70 mmHg) which drops by ~40% at higher pressures and remains pressure-insensitive beyond, up to ~180mmHg. Endothelial removal doubles Lp and renders it pressure-insensitive. Huang *et al* [4] explained this behavior by postulating that intimal compression under transmural pressure loading causes the endothelium to partially block internal elastic lamina fenestrae, thereby causing the observed Lp lowering with ΔP. Endothelial removal eliminates both a layer of resistance and thus fenestral blocking potential, thereby increasing Lp and explaining the denuded vessel Lp's pressure insensitivity. Jimmy Toussaint of our group has established that rat aortic endothelial cells avidly express Aquaporin-1 (AQP), a membrane protein that acts as a specific water channel. These highly selective water channels allow high throughputs of water (~3×10^9 molecules/sec) in response to, e.g., osmotic gradients, at little or no cost in ATP[7]. Tieuvi Nguyen of our group has shown experimentally[9], by measuring the hydraulic conductivity of an excised vessel *ex vivo* as a function of pressure, first with functioning and then with chemically blocked AQPs (using $HgCl_2$ as a blocker [8]), on the same vessel, that these AQPs play a significant functional role in transmural water transport in rat aortas.

New theory to determine AQP contribution to endothelial Lp vs ΔP: Nguyen's experiments showed significant decreases of ~36%, 13% & 8% in intact vessel's total hydraulic conductivity, Lp_t ($1/Lp_t = 1/$

Lp_{e+i} + 1/ Lp_{m+I}, where Lp_{e+i} is the hydraulic conductivity of endothelium plus subendothelial intima and Lp_{m+I} is the hydraulic conductivity of the denuded vessel, representing media and internal elastic lamina contributions) at 60,100,140 mmHg respectively, when $HgCl_2$ is used to chemically block AQP. The variation in these numbers clearly shows that more is going on than a simple blocking of AQP. Recall Huang *et al's* [4] intimal compression theory (without blocker) argued that above 80 mmHg the intima remains fully compressed and therefore Lp is ΔP-independent. Note that in case of blocked AQPs, Nguyen showed nearly ΔP-independent values of Lp that extend to her lowest value of 60 mmHg, where the unblocked intima is not compressed. This suggests that AQP-blocking may lead to intima compaction at a lower overall ΔP. We propose a theory to explain this Hg++ dependence of Lp vs ΔP. We shall also extract the percentage of intrinsic Lp due to AQP, by extending the filtration model given by Huang *et al* [4] . The idea behind the new model is that blocking the endothelial AQPs decreases the intima's intrinsic Lp, which decreases intimal pressure (P_i) at a fixed ΔP. Thus there is a larger endothelial force per unit area (P_L-P_i),where P_L is lumen pressure,which can compress the intima at lower total ΔP. Specifically, we extend the transport model of Huang *et al* [4] by including transcellular flow characterized by Lp_{EC} (hydraulic conductivity of endothelial cell), which is the fraction of Lp_e (hydraulic conductivity of endothelium) due to the AQPs. From this model, we calculate the experimentally measured Lp_{e+i}, which depends on the thickness of intima, and compare with Nguyen's corresponding experimental values.

References:

(1) Baldwin A. L., Wilson L.M.,Simon B. R., "Effect of pressure on hydraulic conductance", Arteriosclerosis and Thrombosis,12,163-171,1992.

(2) Chuang P.T, "Macromolecular transport across arterial and venous endothelium in rats: Studies with Evans blue-albumen and horseradish peroxidase", Arteriosclerosis, 10, 188-197,1990.

(3) Huang Y., Rumschitzki D.,Chein S.,Weinbaum S., "A fiber matrix model for the growth of macromolecular leakage spots in the arterial intima", ASME Journal of Biomechanical Engineering,116,430-445,1995.

(4) Huang Y., Rumschitzki D.,Chein S.,Weinbaum S., "A fiber matrix model for the filtration through fenestral pores in a compressible arterial intima", American Journal of Physiology,272,H2023-2039,1997.

(5) Lin S.L., Jan K.M., Schuessler G.B., Weinbaum S., Chein S., "Transensendothelial transport of LDL in association with cell mitosis in rat aorta", Arteriosclerosis, 9,230-236,1989.

(6) Lin S L., Jan K M., Schuessler G B., Weinbaum S., Chein S., " Enhanced macromolecular

permeability of aortic endothelial cells in association with mitosis", Atherosclerosis, 73,223-232,1988.

(7) Murata K., " Structural determinants of water permeation through aquaporin-1", Nature, 407,599-605,2000.

(8) Preston G. M., Jung J. S., Guggino W. B., Agre P., " The mercury sensitive residue at cystine 189 in the CHIP28 water channel", Journal of Biological Chemistry, 268,17-20,1993.

(9) Shou Y., Jan K.M., Rumschitzki D., "Transport in rat vessel wallsI: The hydraulic conductivities of aorta, pulmonary artery and inferior vena cava with intact and denude endothelia", American Journal of Physiology, 291, H2758-2771, 2006.

(10) Tedgui A., Lever M J., "Filtration through damaged and undamaged rabbit thoracic aorta", American Journal of Physiology, 247, H784-791,1984.

(11) Weinbaum S., Tzeghai G, Ganatos P, Pfeffer R,Chein S, "Effect of cell turnover and leaky junctions on arterial macromolecular transport", American Journal of Physiology, 248, H945-960,1985.

The Beta Roll Peptide as a Reversible, Calcium Sensitive, and Modular Scaffold for the Engineering of Biomolecular Recognition

Mark A. Blenner[1], Geza R. Szilvay[1], Oren Shur[1], Donald Cropek[2], Scott A. Banta[1]

Beta roll peptide motifs are composed of repetitive calcium binding nonamers. These motifs are found in several proteases, lipases and haemolysins, and are defined by the consensus: GGXGXDXUX, where U is an aliphatic amino acid and X is any amino acid. Repeating peptide scaffolds have been increasingly utilized as a stable and modular interface for biomolecular recognition of target molecules. These repeat peptides can be concatenated to increase the area available to design biomolecular interaction, making possible the design of proteins that can bind targets of varying size. Calcium binding triggers the folding of the beta roll repeats such that two variable amino acids of every other nonamer form an interface on one side of the beta roll. Thus, the binding of calcium controls the assembly of the interface designed to bind its target, and in the absence of calcium, the target can be released from the beta roll.

Here we present the engineering and characterization of beta roll motifs from a Serralysin and adenylate cyclase. The functional folding of the beta roll peptide was determined using circular dichroism (CD) spectroscopy, dynamic light scattering (DLS) and fluorescence resonance energy transfer (FRET). CD spectroscopy indicated that unstructured peptide formed calcium induced beta sheet secondary structure for N- and C- terminally capped beta roll constructs. Similarly, DLS experiments indicated a decrease in the hydrodynamic diameter of beta roll peptides upon calcium binding. FRET experiments similarly showed a calcium dependent decrease in fluorophore separation. In an effort to understand the roles of regions adjacent to the beta-roll, both the N- and C- terminal capping groups were truncated in several logical positions, based on secondary structure prediction models. We found that either N-terminal or C-terminal capping was sufficient for calcium induced beta roll folding. This is consistent with studies on leucine-rich repeat proteins, where capping is necessary to stabilize the hydrophobic core.

571m A Computational and Experimental Investigation of Mechanisms for Caveolae-Mediated Endocytosis

Belinda S. Akpa[1], Ludwig C. Nitsche[1], Lewis E. Wedgewood[1], and Richard D. Minshall[2]. (1) Department of Chemical Engineering, University of Illinois at Chicago, 810 S Clinton Street, MC-110, Chicago, IL 60607, (2) Departments of Anesthesiology and Pharmacology, University of Illinois at Chicago, 835 S. Wolcott Ave., MC-868, Chicago, IL 60612-3748

In this work, we report the development of a computational model which can address how changes in caveolin-1 oligomerization and phosphorylation impact upon membrane invagination, budding, and internalization by (i) controlling the stability of caveolin polymers in the plasma membrane and (ii) their ability to facilitate changes in membrane curvature. These studies will assess the role of *Src*-mediated phosphorylation of caveolin-1 N-terminal tyrosine residue Y14 in controlling the stability of caveolin-1 oligomers in endothelial cells and thereby in the mechanism of caveolae-mediated endocytosis. Central to this mechanism is the interaction between (i) Brownian motion of the individual caveolin chains in the heptameric structures, and (ii) repulsion between the charged groups resultant from phosphorylation – to produce asymmetric forces leading to deformation and invagination.

Biophysical background of the modeling hypothesis: Caveolae, small, flask-shaped invaginations of mammalian plasma membranes, are ubiquitous features of endothelial cells. They comprise about 15% of the cell volume[1] and account for >95% of the plasmalemmal vesicles within the cells[2]. Assessment of caveolar internalization using cholera toxin subunit B[3], a caveolae-specific tracer[4], showed that plasma membrane-bound caveolae are released into the cytosol[2, 5]. Using methyl-β-cyclodextrin (MβC), a specific cholesterol-binding agent that flattens caveolae[6], we observed a reduction in [^{125}I]-albumin endocytosis dependent on the concentration of MβC[5]. Thus, caveolae are implicated as essential vesicle carriers mediating endocytosis in endothelial cells[7-11].

Caveolin-1, the 22 kDa protein that coats the cytoplasmic surface of caveolae, is the defining protein constituent of caveolae[6, 12]. The characteristic invaginations were absent in endothelial cells from caveolin-1 knockout mice[7, 8, 10, 11] – indicating the importance of caveolin-1 in determining the caveolar structure. An important property of caveolin-1 is its ability to form oligomers in the cytoplasmic domain of the caveolar membrane[12]. Caveolin-1 oligomerization is crucial in regulating the invagination of the membrane[6, 13, 14]. Moreover, the stability of caveolin oligomers may be an important determinant of the release of caveolae from the plasma membrane[15, 16]; that is, destabilization of caveolin-1 oligomers may facilitate the release of caveolae.

Caveolin-1 is a well-described substrate of *Src*[17-20]. *In vitro* phosphorylation of caveolin-1-derived synthetic peptides and site-directed mutations showed that Tyr^{14} is the primary *Src* phosphorylation site[17]. Aoki and coworkers[20] demonstrated that phosphorylation of caveolin-1 Tyr^{14} in rat tissue occurred primarily in endothelial cells of capillaries and venules. Importantly, this study suggested that phosphorylation of caveolin-1 at Tyr[14] induced caveolae endocytosis; these investigators observed fewer plasma membrane-attached and invaginated caveolae and more free cytoplasmic vesicles in endothelial cells treated with the tyrosine phosphatase inhibitor pervanadate. Although studies from the Fujimoto lab[19, 20] have implicated an important role of caveolin-1 phosphorylation in regulating caveolae internalization, the mechanism by which phosphorylation mediates this process has not been described.

Features of the model: A coarse-grained atomistic (CGA) approach is developed to simulate key mechanistic features of the budding of vesicles, thereby illuminating specific biochemical hypotheses. It is postulated that Coulombic inter-chain repulsion resulting from phosphorylation-induced charge creates asymmetric forces that cause the formation of curvature on the cell membrane. A Brownian dynamics simulation adapted from polymer kinetic theory represents the protein chains as bead-spring assemblies, and tracks their stochastic motion as forces are transmitted to the caveola. A novel coarse-grained, particulate model is developed for the lipid bilayer, in which local directionality (surface normal vector) is extracted from the relative arrangement of isotropic particles rather than being embedded in the more complex (i.e., rod-like or multi-bead) internal structure of lipid particles used thus far in CGA simulations. An anisotropic inter-particle force law results in effective bending elasticity of the sheet-like aggregate of particles, and preserves two-dimensional Lennard-Jones liquid behavior in the tangential direction. By combining Brownian dynamics of heptamers anchored to the membrane with the elastic membrane response, a model of caveolin-mediated curvature is obtained.

Future extensions of the model are planned to incorporate the stabilizing influence of the adjoining cytoskeleton (using a bead-spring girder model) and also viscous damping by the surrounding cytoplasm and extracellular fluid (by superposing Stokes hydrodynamic kernels).

References:

1. D. Predescu and G. E. Palad. Am. J. Physiol. 265, H725-H733 (1993)

2. D. N. Predescu, S. A. Predescu, *et al.* Mol. Biol. Cell. 14, 4997-5010 (2003)

3. A. Gilbert, J. P. Paccaud, *et al.* J. Cell Sci. 112, 1101-1110 (1999)

4. R. G. Parton, B. Joggerst, *et al.* J. Cell Biol. 127, 1199-1215 (1994)

5. T. A. John, S. M. Vogel, *et al.* Am. J. Physiol. Lung Cell. Mol. Physiol. 284, L187-L196 (2003)

6. K. G. Rothberg, J. E. Heuser, *et al.* Cell. 68, 673-682 (1992)

7. M. P. Drab, P. Verkade, *et al.* Science. 293, 2449-2452 (2001)

8. B. Razani, J. A. Engelman, *et al.* J. Biol. Chem. 276, 38121-38138 (2001)

9. W. Schubert, P. G. Frank, *et al.* J. Biol. Chem. 276, 48619-48622 (2001)

10. Y. Y. Zhao, Y. Liu, *et al.* Proc. Natl. Acad. Sci. USA. 99, 11375-11380 (2002)

11. D. Predescu, S. M. Vogel, *et al.* Am. J. Physiol. Lung Cell. Mol. Physiol. L895-901 (2004)

12. T. Kurzchalia, P. Dupree, *et al.* J. Cell Biol. 118, 1003-1014 (1992)

13. S. Monier, R. G. Parton, *et al.* Mol. Biol. Cell. 6, 911-927 (1995)

14. R. G. Anderson. Annu. Rev. Biochem. 67, 199-225 (1998)

15. B. van Deurs, K. Roepstorff, *et al.* Trends Cell Biol. 13, 92-100 (2003)

16. I. R. Nabi and P. U. Le. J. Cell Biol. 161, 673-677 (2003)

17. S. Li, J. Couet, *et al.* J. Biol. Chem. 271, 29182-29190 (1996)

18. C. Tiruppathi, W. Song, *et al.* J. Biol. Chem. 272, 25968-25975 (1997)

19. R. Nomura and T. Fujimoto. Mol. Biol. Cell. 10, 975-986 (1999)

20. T. Aoki, R. Nomura, *et al.* Exp. Cell Res. 253, 629-636 (1999)

571n Reconstitution of Neutrophil Chemokine Receptor In Yeast

Hyungjoon Cho, Inseong Hwang, and **Sheldon Park**. Chemical and Biological Engineering, University at Buffalo, SUNY, 905 Furnas Hall, Buffalo, NY 14260

Chemokines are detected by cell surface receptors that are members of the G protein coupled receptor (GPCR) family. GPCR proteins serve a wide range of functions, making them attractive targets for therapeutic intervention. Most commonly, the mechanism of ligand binding and receptor activation is studied in model cellular systems through transient expression of the proteins. However, an increasingly large number of mammalian GPCRs are being investigated using yeast as an alternative system. Yeast is an attractive expression host for studying mammalian GPCRs because the signaling pathway is conserved in both organisms, which allows heterologously expressed mammalian GPCR proteins to be coupled to the endogenous pheromone pathway. Yeast offers other advantages over cellular systems, including a faster growth rate, lower experimental cost, and ease of handling. The molecular biology tools are also well developed for yeast, which further facilitates the development of highly sensitive reporter assays in yeast that may be used to measure GPCR activity in high throughput.

In this study, we report the construction of a yeast-based screening system for the neutrophil chemokine receptors. Neutrophils are recruited to a site of injury by the various chemokines secreted by the tissue cells and other leukocytes, yet excessive activation of neutrophil causes a variety of inflammatory disorders, including Crohn's disease and psoriasis. Thus, an efficient screening system to identify novel inhibitors of neutrophil activation would be therapeutically useful. We have expressed the human neutrophil chemokine receptor, CXCR1, in engineered yeast. In our system, the activation of CXCR1 is coupled to the synthesis of various reporter gene products, including LacZ and URA3. This system forms the foundation of high throughput screens for small molecule CXCR1 agonists and antagonists using high density plates. One potential readout system includes measurement of OD600 to monitor cell survival, which allows easy identification of active compounds. Since the paradigm of the assay is invariant with respect to the exact identity of the receptor, similar assays may be developed to find active compounds against other therapeutic GPCRs, as well as to study receptor-ligand interaction in high throughput.

571o A Novel Synthetic Riboswitch Triggered by Small Chemicals

Joonjung Jo and **Jongshik Shin**. Department of Biotechnology, Yonsei University, B509, Engineering Building 2, Shinchon-Dong 134, Seodaemun-Gu, Seoul, 120-749, South Korea

Synthetic biology has emergied as a promising research area that enables researchers to simplify complicated biological systems into managable pieces and to redesign brandnew biological circuits by assembling the simplied biological modules recruited from different living organisms. Several approaches have been investigated to provide logical connectivity between well-defined binding events among relevant biomolecules and the way how the biomolecualr interactions lead to programmable circuit behaviors. These include engineering of biological circuits like gene regulatory systems, metabolic fluxes and signal transduction pathways. Here we present a novel synthetic riboregulatory system that is switchable by small chamicals. The riboswitch shows more than ten-fold decrease in translational efficiency of a reporter gene by addition of small chemicals. Sequence design of the riboswitch is easily achieved to fine-tune mRNA secondary structures and the consequent switch behaviors including target specificity and sensitivity. Owing to the good sequence designability, the riboswitch seems promising for biomedical applications to high-throughput screening systems, biosensors and in vivo bioimaging.

571p Assessment of the Stability and Biological Activity of Heme-Proteins Encapsulated In Hydrogel Membranes

Dr. Madeline Torres-Lugo[1], Celimar Valentín-Rodriguez[1], and Dr. Juan López-Garriga[2]. (1) Chemical Engineering, University of Puerto Rico - Mayaguez, PO Box 9046, Mayaguez, PR 00681-9046, (2) Chemistry, University of Puerto Rico - Mayaguez, PO Box 9046, Mayaguez, PR 00681-9046

Proteins are biological macromolecules which have a unique spatial conformation. This spatial conformation can be affected by extremes in pH, temperature, and organic solvents. Once this 3D spatial conformation is compromised, the protein's biological stability and activity can be severely limited. This investigation focuses on the effects of pre-polymeric solution components on the behavior of proteins to be encapsulated by the entrapment technique in anionic, cationic, and neutral hydrogel membranes. Four proteins were utilized in this investigation: Equine skeletal muscle and heart myoglobin and Bovine and Porcine hemoglobin. Three hydrogel morphologies were examined: methacrylic acid-poly(ethylene glycol) dimethacrylate (n=1000), dimethylaminoethyl methacrylate-poly(ethylene glycol) dimethacrylate (n=1000), and poly(ethylene glycol) (200) monomethyl ether-methacrylate-poly(ethylene glycol) dimethacrylate (n=1000). The heme-proteins were put in contact with different ratios of the pre-polymeric solution components. UV-vis spectroscopy was utilized to monitor displacements in the Soret Bands to determine any changes in biological stability of the proteins in their metaquo, deoxy, and carboxy states. Optimized morphologies were then synthesized. The Soret bands of the proteins in all the pre-polymeric solutions were located at the expected wavelengths, 408nm and 406nm for myoglobins and hemoglobins, respectively. Upon polymerization, the Soret bands of the encapsulated proteins in ionic morphologies suffered blue shifts of 14nm, on average. Soret bands of the deoxy, and carboxy states were all blue-shifted by 16 and 18nm, respectively. The metaquo Soret bands of the four proteins encapsulated in the neutral morphology experienced a red shift of 4nm, whilst the deoxy and carboxy Soret bands presented a blue shift of 12nm and 5nm, respectively. These results suggest that the proteins undergo structural changes, most likely an iron-histidine bond breakage, due to the electrostatic interactions between the polymer and proteins. It appears that the neutral morphology did not cause such a remarkable alteration of the heme-proteins' spatial arrangement. Examination of the results also revealed a possible coexistence of metaquo/oxy and carboxy/oxy states of the proteins within the polymerized membranes.

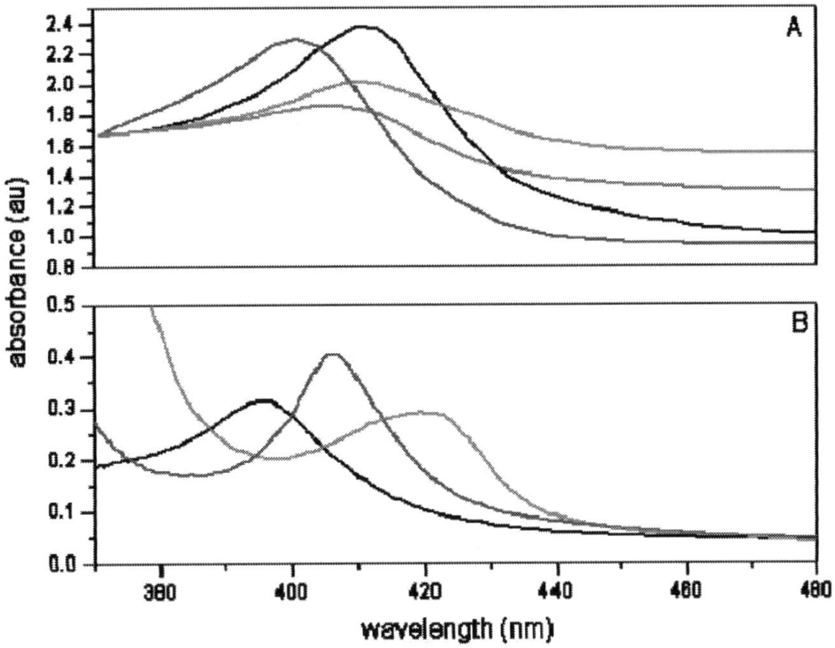

(A) Effect of MAA:PEGDMA1000 Pre-Polymeric Solution Components on the UV-vis Spectrum of mMb. —— MAA+NaOH (λ_{max}=410); —— MAA+PEGDMA1000 (λ_{max}=401); —— Pre-Polymeric Solution (λ_{max}=407); —— EtOH (30%) (λ_{max}=410) (B) Activity of mMb in MAA:PEGDMA1000 hydrogel membranes. —— Metaquo state (λ_{max}=395); —— Deoxy state (λ_{max}=422); —— Carboxy state (λ_{max}=406)

571q Harnessing the Power of Statistical Design and the Efficiency of a High Throughput System for Cho Media Development

Sonali Nigam, Robb Shawley, Martin Gawlitzek, Louis Cheung, and Christina Bevilacqua. Late Stage Cell Culture, Process Research and Development, Genentech, Inc., South San Francisco, CA 94080

A challenge in CHO cell culture process development lies in formulating optimal media to fulfill the various nutritional requirements of cell cultures. Due to the potential for diversity between CHO cell lines, media development is a continuous process of optimization which involves component adjustments, blending and testing new formulations, and analyzing the metabolic pathways of the cells. Given DOE tools and a large set of components and formulations, the process development engineer is often faced with the dilemma of choosing between testing a small subset at many different levels vs. a larger grouping at fewer levels. Additionally, as the size of the design space increases, component additions become a labor intensive and error prone task.

To utilize powerful statistical designs and obtain meaningful results, we have designed and evaluated a convenient and efficient media development format by utilizing a high-throughput cell culture system. Data will be shared demonstrating the ability of this system to increase the coverage of the media design studies in delivering a more detailed picture of the optimal media performance space.

571r Understanding the Role of Secondary Structure In Protein Aggregation

Suwimol Wongsakulphasatch and Robin Curtis. University of Manchester, PO Box 88, Sackville Street, Manchester, M 60 1QD, United Kingdom

Understanding protein aggregation is key to solving problems both in bioprocessing where aggregation of partially folded proteins leads to reduced yields and in biomedicine where many diseases have been linked to the formation of fibrillar protein aggregates. However, studying the early steps in the aggregation is difficult as the process occur quickly and irreversibly. Consequently, computer simulation provides an attractive alternative or complementary method to experimental studies. However, accessing the time and length scales of interest in the aggregation requires using coarse grained models. One of the key ingredients of these models is that they are able to capture secondary structure formation. In this work we study a simple beta-sheet forming model which consists of overlapping beads with fixed bond lengths and angles interacting through directional bonds, a model which is based on the Kemp and Chen helical forming model.1 The directional bonds are meant to mimic the hydrogen bonding interactions found in beta sheets. As such we first find the directionality which is required so that only one hydrogen bond can be formed per bead and determine the phase behaviour for this model. We then compare the thermodynamics of the sheet-globule transition to that of the helix-globule elucidating the role of secondary structure in aggregation.

1. J. P. Kemp and J. Z. Y. Chen, "Helical structures in proteins", Biomacromolecules 2, 389 (2001).

571s In Silico Simulation of Corticosteroids Effect on a Nuclear Factor - Kb Dependent Indirect Response Model of Systemic Inflammation In Peripheral Human Blood Leukocytes

P.T. Foteinou, Biomedical Engineering, Rutgers University, Piscataway, NJ 08854, S.E. Calvano, Department of Surgery, UMDNJ-Robert Wood Johnson Medical School, New Brunswick, NJ 08901, S. F. Lowry, Surgery, UMDNJ-Robert Wood Johnson Medical School, New Brunswick, NJ 08901, and Ioannis P. Androulakis, Biomedical and Chemical & Biochemical Engineering, Rutgers - The State University of New Jersey, 599 Taylor Road, Piscataway, NJ 08854.

Background: The systemic inflammatory response syndrome often accompanies critical illness and can be an important cause of morbidity and mortality (Nystrom 1998). In any physiological system, signaling pathways are characterized by the presence of adequate controlling steps (Chung and Adcock 2005). In this study, we develop an indirect response model of a human endotoxin-induced inflammation that integrates the opposing effect of two signaling pathways, one associated with the activation of NF-kB that drives downstream transcriptional activation of inflammatory mediators and one related to the signaling of exogenous corticosteroids as the putative controllers of inflammation (Hoffmann, Levchenko et al. 2002; Barken, Wang et al. 2005; Covert, Leung et al. 2005; Rhen and Cidlowski 2005).

Materials: The experimental data analyzed in this study were generated as part of the Inflammation and Host Response to Injury Large Scale Collaborative Project funded by the USPHS, U54 GM62119 (Calvano, Xiao et al. 2005; Storey, Xiao et al. 2005). Human subjects were injected intravenously with endotoxin (CC-RE, lot 2) at a dose of 2-ng/kg body weight or 0.9% sodium chloride. The transcript abundance of leukocytes was measured by microarray before endotoxin administration (0hr) and 2, 4, 6, 9 and 24 hrs after endotoxin administration.

Methods: Based on our prior work, we developed an indirect response model of endotoxin-induced inflammation that integrates the extracellular signal (endotoxin, LPS) with the activation of the essential transcriptional responses. One of the key aspects of this model is the systematic identification of an elementary set of temporal responses that describe the trajectory of systemic inflammation in human blood leukocytes when exposed to endotoxin stimulus (Foteinou, Calvano et al. 2007). Therefore such responses define the state space of the probed system and we effectively integrated these responses into a mathematical model using the basic principles of an Indirect Response Model (IDR) (Krzyzanski and Jusko 1997). However, in this study we propose to improve the structure of the inflammatory model taking regulation into consideration. We suggest the association of the activity of NF-kappaB with the up-regulation of inflammatory mediators. We aim to simulate the dynamic interactions of a minimal model of NF-kappaB that consists of a basis set of signaling molecules that affect maximally the dynamic behavior of NF-kappaB. In addition to this, we prompt to model the pharmacodynamic effect of exogenous corticosteroids against inflammation based on the "Fifth Generation Model for Corticosteroid Pharmacodynamics" (Ramakrishnan, DuBois et al. 2002). The resulting model is described by a set of ordinary differential equations that contain the NF-kappaB dependent propagation of LPS signaling coupled with the opposing signaling effect of corticosteroids. We demonstrate the potential of the proposed in silico model evaluating a series of biological relevant scenarios. Such a model is coupled with the corticosteroid intervention envelope and results show the beneficial effect of an early intervention strategy. Regardless of the implications of high LPS, pre-exposuring the system into hypercortisolemia “reprograms” the dynamics of the system towards a balanced immune response. The development of such a mechanistic modeling approach allows us to gain insight about how the system responds to a multitude of external signals through the dynamic interaction of signaling modules. The proposed model opens the research avenues to evaluate the effectiveness of corticosteroids under various treatment schedules establishing an in silico zone of therapeutic opportunity thereby facilitating the clinical making decision

Results: The potential of the proposed in silico model evaluating a series of biologically relevant scenarios is demonstrated. Such a model is coupled with the corticosteroid intervention envelope and the results show the beneficial effect of an early intervention strategy. Regardless of the implications of high LPS, pre-exposing the system to hypercortisolemia “reprograms” the dynamics of the system towards a more balanced immune response.

Conclusions: We have developed a mechanistic modeling approach that allows us to gain insight about how the system responds to multiple external signals through the dynamic interaction of signaling modules. The proposed model opens research avenues to evaluate the effectiveness of corticosteroids under various treatment schedules establishing an in silico zone of therapeutic opportunity which may facilitate clinical decision making.

Barken, D., C. J. Wang, et al. (2005). "Comment on "Oscillations in NF-kappaB signaling control the dynamics of gene expression"." Science 308(5718): 52; author reply 52.

Calvano, S. E., W. Xiao, et al. (2005). "A network-based analysis of systemic inflammation in humans." Nature 437(7061): 1032-7.

Chung, K. F. and I. M. Adcock (2005). "Signalling and transcriptional regulation in inflammatory and immune cells: importance in lung biology and disease." Eur Respir J 26(5): 762-3.

Covert, M. W., T. H. Leung, et al. (2005). "Achieving stability of lipopolysaccharide-induced NF-kappaB activation." Science 309(5742): 1854-7.

Foteinou, P., S. Calvano, et al. (2007). "An indirect response model of endotoxin-induced systemic inflammation." Journal of Critical Care 22(4): 337-338.

Hoffmann, A., A. Levchenko, et al. (2002). "The IkappaB-NF-kappaB signaling module: temporal control and selective gene activation." Science 298(5596): 1241-5.

Krzyzanski, W. and W. J. Jusko (1997). "Mathematical formalism for the properties of four basic models of indirect pharmacodynamic responses." J Pharmacokinet Biopharm 25(1): 107-23.

Nystrom, P. O. (1998). "The systemic inflammatory response syndrome: definitions and aetiology." J Antimicrob Chemother 41 Suppl A: 1-7.

Ramakrishnan, R., D. C. DuBois, et al. (2002). "Fifth-generation model for corticosteroid pharmacodynamics: application to steady-state receptor down-regulation and enzyme induction patterns during seven-day continuous infusion of methylprednisolone in rats." J Pharmacokinet Pharmacodyn 29 (1): 1-24.

Rhen, T. and J. A. Cidlowski (2005). "Antiinflammatory action of glucocorticoids--new mechanisms for old drugs." N Engl J Med 353(16): 1711-23.

Storey, J. D., W. Xiao, et al. (2005). "Significance analysis of time course microarray experiments." Proc Natl Acad Sci U S A 102(36): 12837-42.

571t Small-Scale Method for Determination of Cell Lysis in Harvested Cell Culture Fluid

Yael Hirsch, Jason Chiu, Brian Horvath, and Mike Laird. Late Stage Cell Culture, Process Research and Development,, Genentech Inc, South San Francisco, CA 94080

Platform cell culture processes can show variable characteristics in the harvested cell culture fluid with respect to cell lysis, as determined by LDH release. An increase in cell lysis during harvest has the potential to introduce challenges for downstream processing and the resulting product quality. As a risk mitigation effort, a small-scale laboratory method was developed to help predict a cell line's susceptibility to lysis early on in its development. This method could also be used to screen harvest conditions and reagent additions that may reduce cell lysis. Using this small-scale method, we have generated data to show the effects of varying culture conditions on cell lysis as well as comparative data among cell lines that may be used to predict future efforts related to lysis effects on product quality.

571u Microfluidic Thrombosis Models Using Collagen and Immobilized Relipidated Tissue Factor

Sean F. Maloney, Keith B. Neeves, and Scott L. Diamond. Chemical and Biomolecular Engineering, University of Pennsylvania, 1150 Vagelos, 3340 Smith Walk, Philadelphia, PA 19104

Background: The aim of this research is to develop an in vitro system with spatially controlled presentation of collagen and tissue factor. A technical challenge is to vary the tissue factor concentration in a controlled manner. We have developed a bead-based protocol for presenting tissue factor in conjunction with micropatterned collagen into a microfluidic thrombosis model.

Methods: Biotin-streptavidin chemistry is used to link relipidated recombinant human tissue factor to 1μm-diameter latex beads. Beads are delivered by acoustic dispensing to the surface of glass slides with high spatial resolution. Standard soft-lithography methods are utilized to cast PDMS devices which are bonded reversibly to glass for protein deposition and blood presentation. The accumulation of platelets and fibrin is monitored in real time by epifluorescence.

Results: The activity of tissue factor was measured and confirmed at each level of synthesis, and two modes of modulation allows for precise presentation of active tissue factor at a specific area: the tissue factor concentration in the liposomes (5 to 20×10^3 lipid molecules per tissue factor molecule) and the number of particles on the surface (0.3 to 2.6 particles/μm^2). Microfluidic devices allow for perfusion of less than 100μL of human or murine blood over tissue factor and collagen, which was performed to investigate the relative roles of the two main receptors ($\alpha 2\beta 1$ and GPVI) in knock-out mice. The data imply independent roles of firm adhesion and activation of murine platelets to collagen at an average wall shear rate of $400 s^{-1}$.

571v Systematic Reconstruction of Expanded Metabolic Networks

Goncalo Maia, University of Maryland Baltimore County, 1000 Hilltop Circle, Baltimore, MD 21227 and Mariajose Castellanos, Chemical and Biochemical Engineering, UMBC, 1000 Hilltop Circle, Bldg. ECS Rm 314, Baltimore, MD 21250.

The complexity of the metabolic pathway interaction leads to the accepted truth that it is impossible to separately analyze a single pathway and proteomic information has shown that genetic modifications pertaining a specific pathway also affect other metabolic processes. Genome-scale models have the whole biochemical network in perspective and can predict the effects of genetic modifications in microbial metabolism, such as uptake/secretion rates, exceedingly better than single pathway analysis. Constructing genome scale models is an extremely laborious task. Also, in every microorganism the genome is not yet fully annotated or has errors which will require some degree of manual correction. In our work, we suggest that the development time for metabolic model building can be decreased with usage of automated import procedures from public databases such as KEGG and by eliminating or correcting, *a priori*, biochemical reactions that will result in metabolic dead ends. These consist in specific compounds which lack known metabolic routes that prevent them from accumulating in the cell. Our algorithms have imported and reconstructed the networks from 619 microbes available in KEGG as well as a hyper network consisting of every known metabolic reaction (7353 in total). Using the hyper network, we also successfully identified 2817 reactions (38%) responsible for 2315 general metabolic dead ends. We have shown that metabolic networks and proteomic gel analysis can be mutually enhanced. This combination provides us with additional model constraints and increases the gel resolution. For this project, we reconstructed the network of *Aspergillus niger* and analyzed the proteomic gels of the wild strain and single gene mutant.

571w Semi-Degradable Nanofiber Composites for Delivery of Paclitaxel

Vanessa Vardon, Chemical and Biological Engineering, Drexel University, 3141 Chestnut St., Philadelphia, PA 19104, Anthony Lowman, Chemical Engineering, Drexel University, 3141 Chestnut St., Philadelphia, PA 19104, and James Schuster, Neurosurgery, University of Pennsylvania, Philadelphia, PA 19104.

Chemotherapeutic engineering is a term used to describe the application of chemical engineering principles such as process and system design to enhance efficacy of chemotherapy treatment for cancer and other diseases. Significant research is being done to develop new treatments such as targeted chemotherapy using cellular markers and circulating nanoparticles. Another important aspect of chemotherapeutic engineering is improving delivery methods for currently available treatments.

Paclitaxel is often referred to as one of the most important advances in chemotherapy in years. The extremely hydrophobic nature of paclitaxel however poses significant problems to traditional delivery methods. The only currently available dosage form utilizes the adjuvant Cremophor EL (a polyoxyethylated castor oil and dehydrated alcohol blend) which itself has been shown itself to be detrimental, with side effects including but not limited to neurotoxicity, nephrotoxicity, cardiotoxicity, and hypersensitivity. Cremophor EL has even been shown to inhibit the action of paclitaxel at certain concentrations.

To achieve controlled local delivery of paclitaxel a semi-degradable composite utilizing degradable electrospun nanofibers and microspheres of poly(lactic-co-glycolic acid) embedded within a macroporous poly(vinyl alcohol) hydrogel has been proposed. Through a combination of degradation and diffusion local paclitaxel delivery can be achieved over the course of 36 days.

A family of paclitaxel loaded PLGA fibers and microspheres were created utilizing traditional electrospraying methods. By varying polymer and concentrations, fibers and particles of varying diameters were achieved. These fibers were analyzed for loading and release kinetics using high performance liquid chromatography, HPLC. Fibers were then loaded into PVA hydrogels using suspension and freeze/thaw cycles to induce physical cross-linking. These composite systems were evaluated for mechanical properties and surface characteristics to determine distance between cross-links and porosity. The systems were then incubated in PBS maintained at 37C for various time points from 1 day to 6 weeks. The release media was sampled and analyzed using HPLC for release kinetics. The entire release media was then incubated with A549 (human lung carcinoma) and MCF-7 (human breast carcinoma) cell cultures. Following two days incubation cells were analyzed for apoptosis using acridine orange/ethidium bromide double staining.

571x Production of Bioactive Human Beta-Defensin 5 and 6 In Escherichia Coli by Soluble Fusion Expression

Lei Huang, Chi Bun Ching, **Rongrong Jiang**, and Susanna Su Jan Leong. School of Chemical & Biomedical Engineering, Nanyang Technological University, 62 Nanyang Drive, Singapore, 637459, Singapore

This work reports the first successful recombinant expression and purification of human beta-defensin 5 (HBD5) and human beta-defensin 6 (HBD6) in E. coli. HBD5 and HBD6 are cationic antimicrobial peptides with three conserved cysteine disulfide bonds. Two codon-optimized sequences coding the HBD5 gene (sHBD5) and HBD6 gene (sHBD6), respectively, were synthesized, and each gene fused with thioredoxin A (TrxA) to construct the expression vectors. The plasmids were transformed into E. coli BL21 (DE3) strains and cultured in MBL medium, which gave high volumetric productivity of HBD5 and HBD6 fusion proteins of up to 1.49 g l-1 and 1.57 g l-1, respectively. Soluble HBD5 and HBD6 fusion proteins account for 95.2% and 97.6% of the total fusion proteins, respectively. After cell disruption, the soluble fusion proteins were recovered by affinity chromatography and cleaved by enterokinase. Pure HBD5 and HBD6 were recovered using cationic exchange chromatography. The overall recoveries of HBD5 and HBD6 were 38% and 35%, respectively. Importantly, both HBD5 and HBD6 products showed antimicrobial activity against E. coli but not S. aureus. Antimicrobial activity against E. coli of both HBD5 and HBD6 were suppressed by NaCl.

571z Analyzing and Overcoming of Baculovirus Mutants Obtained from Passaging In Insect Cell Culture. A Potential Approach to Low-Cost and Large-Scale Baculovirus Production

Lopamudra Giri[1], Huorang Li[2], David Sandgren[2], David W. Murhammer[1], Bryony Bonning[2], Mike Feiss[3], and Richard Roller[3]. (1) Chemical and Biochemical Engineering, University of Iowa, 302 Finkbine Lane, # 6, Iowa City, IA 52246, (2) Entomology, Iowa State University, Insectary Building, Iowa State University, Ames, IA 50011, (3) Microbiology, University of Iowa

Introduction: Biopesticides/biological controls are extremely useful in overcoming the significant disadvantages of chemical pesticides. However, the current production cost of a potential biopesticide, such as baculovirus (which kills only targeted insects and is benign to human health and environment), is much higher than that of chemical pesticides. Moreover, the baculovirus insect-cell expression system is widely used for the expression of recombinant proteins in eukaryotic environment. Specifically, the proteins expressed in this system often show biological activities similar to eukaryotic species due to the proper folding and post-translational modification. Hence, the establishment of large scale continuous production of baculovirus-expressed proteins can be of great importance in the area of pharmaceuticals, vaccines (subunit) and diagnostics. But a low cost, large-scale production of the baculovirus *Autographa californica* multiple nucleopolyhedrovirus (AcMNPV) using a continuous insect cell culture is greatly hindered by undesirable mutations in the baculovirus genome, such as few polyhedra (FP) and defective interfering particle (DIP) mutants. Accumulations of these mutants affect the normal functioning of the virus and thereby the quality and quantity of the product virus and recombinant protein is greatly compromised. Overcoming such mutations is an extremely important step in enabling a cheaper large-scale production of baculovirus in insect cell culture. In this context, our primary goal is to overcome the FP mutation by genetic modification of the baculovirus or the host cell. In addition, we will examine whether FP mutants are a necessary precursor of the deletion mutants, namely, defective interfering particle (DIP) mutants by characterization of the mutant formation during continuous passaging.

Background on budded virus and polyhedra: Baculoviruses are a family of invertebrate viruses having a double-stranded circular DNA genome. They have two morphological forms: the budded virus (BV) and the occluded virus (OV). The BV's are encased in a membrane and bud through the cellular membrane to multiply within an infected insect. On the other hand, OV's are enveloped singly or multiply in an occlusion body within the cell's nucleus; OV's are embedded in polyhedrin protein matrix and called "polyhedra" which can be used as a pesticide due to its host specificity.

Background on FP mutation and DIP mutation: The FP mutation generally results from the insertion of host insect cell DNA sequences, known as transposons, into the baculovirus *fp25k* gene at TTAA target sites (Fraser et al., 1983). FP mutation/FP phenotype in baculovirus is characterized by reduced virion occlusion, fewer polyhedra/cell, fewer or no viruses/polyhedra and increased budded virus production and altered intranuclear envelopment. On the other hand, the DIP mutants result from deletion of almost 40% genome from the WT virus and it causes significant reduction in viral infectivity as well as the level of recombinant protein production (Kool et al., 1991).

Genetically engineered "stabilized virus" and "passaging experiment": In order to overcome the FP mutation resulting from the insertion of transposons in the *fp25k* gene, the wild type (WT) AcMNPV is modified by removing the TTAA sites from *fp25k* gene using template directed ligation and PCR. This new engineered virus is expected to be more stable during passaging in cell culture and named as "stabilized virus". A "passaging experiment" was performed where three types of viruses, WT, stabilized and delta (fp25k gene deleted, used as negative control) AcMNPV, were passaged upto the 32 times in *Spodoptera frugiperda* (Sf-21) cells. As the quality of the product virus in cell culture is dependent on the "passaging event", a detailed analysis of the mutant formation was performed with the passaged viruses to evaluate the effectiveness of the stabilized virus during passaging.

Immunofluorescent assay and western blot (Fig 1): An immunofluorescent assay was designed to detect the infected cells expressing FP25K in a qualitative and quantitative manner. It was found that the fraction of cells expressing FP25K protein was higher for the stabilized virus upto passage 16. For further verification, western blotting was performed with the infected cell lysates using FP25K antibody. It was shown that with passaging, the degradation of functional FP25K protein occurs at a lower rate in case of the stabilized virus.

Cell growth/infection pattern: The host cell growth curve and infection pattern was compared for the two types of viruses (passage 1, 16 and 32) by monitoring cell density, viability, viral infectivity, polyhedra production at 48,72,96 hrs post infection. Stabilized virus gave a better productihvity until late passages (passage 11-15) but did not sustain it to later passages.

Infectivity/DNA: A novel assay was designed to quantify the total viral DNA in a high throughput format by fluorescent staining of DNA and measuring the fluorescence in a fluorometer with plate reader. Using this assay, the specific infectivity (infectivity/DNA) of the passaged virus was measured to observe the effect of passaging on virulence. Very late passaged virus was found to have almost 20 times less infectivity/DNA than the early passaged virus for both the WT and stabilized viruses.

PCR analysis: Polymerase chain reaction (PCR) was performed to amplify the *fp25k* region to detect the host DNA insertions. No evidence of insertion was found in the *fp25k* gene at 11th, 15th, 26th and 32nd passage for both the WT and stabilized virus. Further analysis is underway to check for the mutations through selection of FP mutants and sequencing of the *fp25k* gene.

Transmission electron microscopy (TEM) of budded virus: The budded virus was precipitated from early and late passages and a size analysis was performed using a transmission electron microscope. A dramatic change was observed in the geometric size distribution of WT and stabilized virus from very late passage compared to that from early passage.

Pulse field gel electrophoresis of viral DNA and southern blotting (Fig 2): Pulse field gel electrophoresis was performed to identify the DIP formation and characterize the genomic size distribution of the early and late passaged viruses. It was found that there are four distinct genome-sizes present other than the WT genome-size all having deleted sequences. Southern blotting of these DNA (WT virus DNA as the probe) revealed that almost 50% of the viruses had deletion of large amount of DNA from their genome during passaging. For further confirmation of DIP formation, a restriction enzyme (EcoNI, which cuts the viral genome three times) analysis was performed with the passaged viral DNA.

Conclusion: We demonstrated that the removal of the transposon target sites from the wild type baculovirus *fp25k* gene (stabilized virus) reduced the incidence of the FP phenotype in late passages (passage 15), but did not eliminate the FP phenotype in very late passages (passage 30). Further, Genotypic and phenotypic analysis of late passaged virus showed that deletion of genomic sequences also contributed to the reduced infectivity and productivity. Production of DIP's with deleted sequences was shown for both WT and stabilized virus by characterization of the geometric size distribution and DNA size distribution of early and late passaged virus. Further, we corrected a commonly-held assumption (Kool et al.,1991) that FP mutations are precursor to DIP mutations, and showed that DIP mutations sometimes arise on their own. Hence, removal of transposon target sites from AcMNPV *fp25k* delayed incidence of the FP phenotype but had no impact on DIP production in cell culture.

Future plan: A two-stage continuous bioreactor (1.5L) set up is underway with controlled pH, oxygen concentration and temperature to evaluate the effectiveness of the stabilized virus. The quality and mutation (FP and DIP) pattern of the passaged stabilized virus with passaged WT virus. Specifically, The FP mutation will be identified by western blotting and Immunofluoroscence assay and DIP mutation will be identified by TEM and pulse field gel electrophoresis.

References:

Fraser, M.J.; Smith, G.E.; Summers, M.D. (1983). *J Virol.* 47(2): 287-300.

Kool, M.; Voncken, J.W.; Van Lier, F.L.J.; Tramper, J.; Vlak, J.M. (1991). *Virology* 183: 739-746.

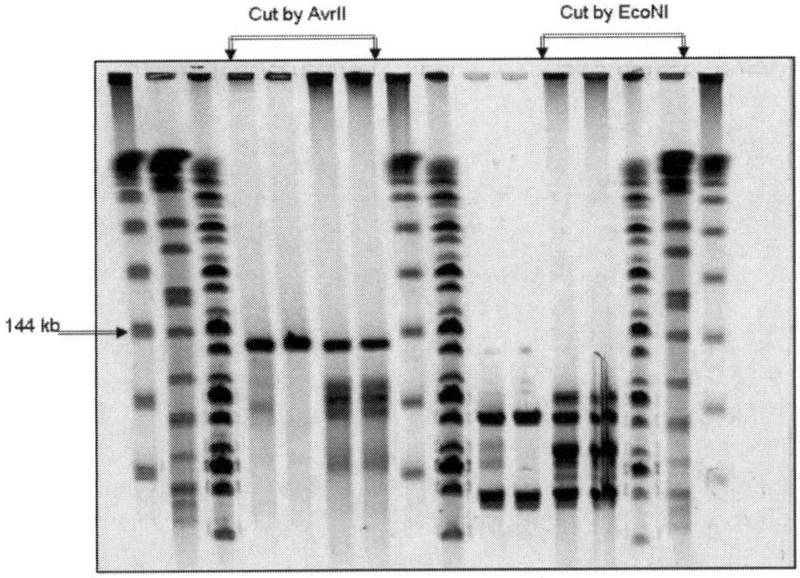

Figure 2: Analysis of AcMNPV genomic DNA by pulse field gel electrophoresis: Lane 4,10: WT AcMNPV passage 1; 5,11: Stabilized AcMNPV Passage 1; 6,12: WT AcMNPV Passage 32; Stabilized AcMNV: passage 32.

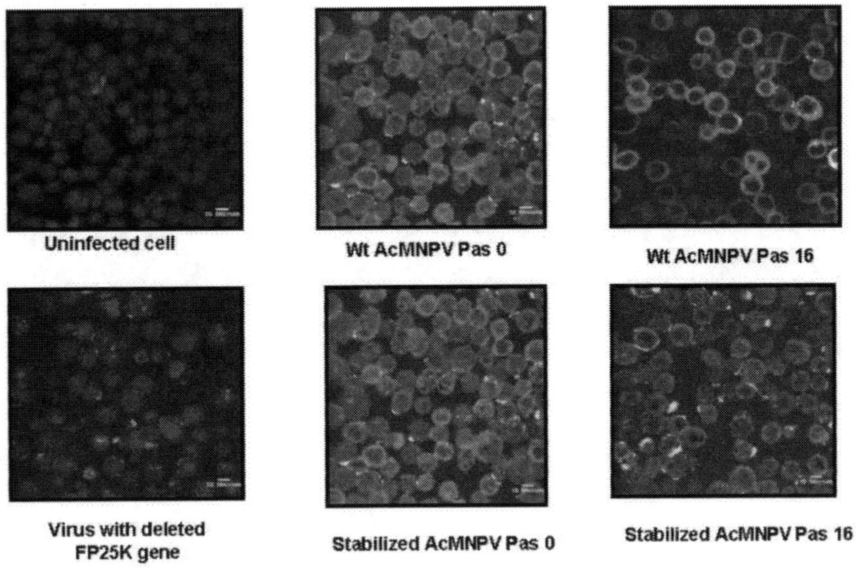

Figure 1: Number of cells expressing FP25K is more in case of stabilized virus: Green = gp64 stain; red = fp25k stain ; blue = nuclear DNA

571aa Protein Dissection and Shuffling for Generating Recombined Proteins

Yong Chen[1], Tingjian Chen[1], Shuang Li[2], Hui Hua[1], and Zhanglin Lin[1]. (1) Department of Chemical Engineering, Tsinghua University, 1 Tsinghua Garden Road, Beijing, 100084, China, (2) School of Bioscience and Bioengineering, South China University of Technology, Guangzhou, 510006, China

A simple method that allows for the facile recombination of distantly related proteins at multiple discrete sites was developed. The approach relies on protein random dissection to search for possible breakage points, which involves the use of a folding reporter to identify soluble shorter peptides independently of protein function. These dissection points are then used as candidate sites for protein recombination. To evaluate this method, two variants of aminoglycoside-3'-phosphotransferase (APHI and APHII) were chosen as model proteins, which share 31.4% amino acid identity. APH I and APH II were recombined at three sites, the library of 16 (2^4) chimeras was screened, among which 4/16 were found to show resistance to kanamycin, albeit at lower levels compared with the wild-type APH. We further applied random mutation to improve the activity of these chimeras. The best mutant showed a MIC of 800 μg/ml, which was an improvement of about 8-fold after one round of evolution. An interesting chimera which had a large extra insertion of about 100 amino acids into APH I at the first recombination site also showed a high catalytic efficiency with a MIC of 800 μg/ml. These results demonstrated that the protein shuffling method should be useful for creating artificial biocatalysts from different templates with low homology.

571ab Identification of Key Residues for Metal Chelate Complex Recognition by Antibody Simulation

Amit S. Jain, Department of Chemical and Biomolecular Engineering, Tulane University, 300 Lindy Boggs Center, 6823, St. Charles Avenue, New Orleans, LA 70118, Hank Ashbaugh, Chemical and Biomolecular Engineering, Tulane University, 327 Lindy Boggs Building Tulane University, New Orleans, LA 70118, and Thomas C. Bishop, Center for Computational Sciences, Tulane University, 6823, St Charles Avenue,, Lindy Boggs center, Suite 500, New Orleans, LA 70118.

The intelligent engineering of antibodies for the specific recognition of heavy metal complex antigens via single site mutations is a key step in the development of antibody based biosensors for the testing of contaminated soils and water streams. The recognition subunit of the antibody is composed of a heavy and light chain sequence, which come together to form the binding pocket. The pocket region is surrounded by six highly variable loops (3 heavy chain and 3 light chain loops) which confer binding specificity to the antibody. Using molecular simulations, we characterize the loop regions to determine which amino acid side chains most likely contribute to heavy metal complex binding for antibody 5B2 described in Delehanty et al., Biochemistry 2003, 42. The loop region is known to be relatively inflexible over the time scales (~10 ns) of brute force molecular dynamics calculations. We therefore employ advanced Replica Exchange techniques to more exhaustively explore the loop conformations. Harvesting configurations generated from these simulations, we minimize their potential energies to characterize the underlying inherent structures of the loops. These calculations yield insight into the residues closest to the binding pocket and allow us, in collaboration with our experimental colleagues, to suggest those amino acids which can be manipulated to more time efficiently optimize the binding affinity of Pb(II)-DTPA complexes to 5B2.

571ac A Design Space Approach towards the Development of the Viral Filtration and Tangential Flow Ultrafiltration Unit Operations for the Commercial Purification Process of a Monoclonal Antibody

Yik Lam, Ryan Haverstock, and Matthew Westoby. Biopharmaceutical Development, Biogen Idec, 5200 Research Place, San Diego, CA 92122

A design space approach was implemented towards the development of the viral filtration and ultrafiltration / diafiltration (UF/DF) unit operations within the manufacturing scale purification process for a monoclonal antibody. Traditionally, development of these and other purification unit operations have been centered around assuring that process performance and product quality attributes met acceptance criteria at and around target manufacturing operating conditions. The design space approach enhances conventional development by exploring process parameters and parameter ranges outside those traditionally explored, to determine the multidimensional combination and interaction of process parameters that assure product quality and process consistency.

The implementation of a design space approach towards development of purification processes has the potential to result in expanded manufacturing operating ranges and process flexibility as well as a better understanding of factor effects on process and product quality responses. Expanded operating ranges decrease the potential of deviations at manufacturing scale, and in the case a deviation does occur, the increased process knowledge gained during development can aid in determining the root cause of a deviation. Additionally, process changes within the design space should decrease future development and reduce regulatory filings associated with such changes.

The Viral filtration and UF/DF unit operations are ideal candidates for the implementation of a design space approach towards development because of their place in the purification process and the number of input parameters associated with each unit operation. These two unit operations are the last two unit operations in the purification process prior to bottling. Therefore, assurance that operating ranges encountered at manufacturing scale do not adversely affect product quality is essential. Additionally, the statistical analysis and modeling implemented in a design space approach are ideal for finding significant parameters from the myriad of factors associated with the viral filtration and UF/DF unit operations.

The development performed resulted in statistical process models that were utilized to set VF and UF/DF operating parameter ranges at manufacturing scale.

571ad Improvement of Artificial Genetic Circuits Using Mutant Libraries of the Luxr Transcriptional Activator

Daniel J. Sayut and Lianhong Sun. Chemical Engineering, University of Massachusetts, 159 Goessmann Laboratory, 686 North Pleasant St, Amherst, MA 01003

The use of biological components for the construction of artificial genetic circuits has the potential for creating novel behaviors in host organisms, and developing a deeper understanding of the design principles underlying natural systems. Natural components, however, have been evolutionarily optimized to function in their specific cellular contexts and commonly do not meet the design requirements for a desired function when moved to artificial systems. This problem is made worse by uncertainties in the design properties that govern natural systems, and in the properties of the components, both resulting from the inability to accurately measure biological parameters. The directed evolution of biological components has been suggested as a way to overcome this problem, as it can be used to generate mutant libraries that have diverse properties while retaining the original function of the mutated protein. Following this approach we have generated a library of mutant LuxR transcriptional activators that have a ten-fold range of increased sensitivities to the 3-oxo-C6 homoserine lactone signaling molecule. Previously, we used this library to improve the properties of artificial positive feedback loops. As an extension of this work, we will show how the LuxR mutant library can also be used to improve the logical outputs of a genetic AND gate where circuit responses are dependent on interactions between LuxR and the LacI repressor. The increased logic of the AND gate was related to the increased sensitivities of the LuxR mutants and a maximum logical improvement of ~1.5-fold was obtained. Finally, we will discuss recent progress on the use of the mutant library to alter the dynamic outputs of a minimal LuxR-LacI based relaxation oscillator.

571ae 1H Spectroscopic Analysis of Hmsc Microenvironment In a 3D Perfusion Bioreactor

Jason J. Crowe[1], Teng Ma[2], Samuel C. Grant[3], and Timothy Logan[1]. (1) Department of Molecular Biophysics, Florida State University, FSU - IMB/KLB Lab Building Code 4380, Mail Code 4380, Tallahassee, FL 32306-4380, (2) Chemical and Biomedical Engineering, Florida State University, 2525 Pottsdamer Street, Tallahassee, FL 32310, (3) Department of Chemical and Biomedical Engineering, Florida State University/FAMU College of Engineering, 2525 Pottsdamer Rd., Tallahassee, FL 32310

Mesenchymal stem cells (MSCs) are a multipotent population of cells which reside in the bone marrow and have significant therapeutic potential due to their plasticity and easy accessibility. In vivo, MSCs have the ability to differentiate into tissues of a variety of lineages and are a significant source of trophic factors. These abilities are mediated via the complex in vivo microenvironment including local physiological conditions. An important approach to tissue regeneration using MSCs is to develop functional cellular constructs by combining MSCs with 3D scaffolds through in vitro cultivation. The success of this approach depends on the innate ability of MSCs to proliferate and differentiate in response to environmental cues that are provided by the 3D construct. To optimize this approach, the in vivo microenvironment of hMSCs must be closely monitored and mimicked in vitro.

We have developed a novel NMR compatible perfusion bioreactor that integrates cell seeding and growth in a 3D construct into a single device that has the flexibility to provide continuous monitoring of the local hydrodynamic and metabolic microenvironment over a period of 20 days under controlled physiological conditions. The bioreactor is capable of supporting long-term construct development and the non-invasive monitoring of metabolic, hydrodynamic, and cellular microenvironment under perfusion flow using both nuclear magnetic resonance (NMR) and magnetic resonance spectroscopy (MRS) imaging technologies. We were able to achieve a uniform cell distribution within the growth construct and demonstrate high-density hMSC growth for a period of 14-21 days. The reactor system also provides the flexibility to acquire spectroscopic data from various parts of the growth construct using MRS imaging. The specific voxels within the construct were targeted to study the local metabolic state. 1H NMR spectroscopy verified the viability of the construct by acquisition of choline and lactate spectral peaks from within the growth construct. Spectroscopy also verified an even distribution of hMSCs within the construct, with no spatial variation in metabolic activities. We also investigated the metabolic states under perfusion as the reactor was perfused within the magnet for a period of 24 hours. We were able to measure the apparent diffusion coefficient (ADC) within the 3D growth construct and determine the localized ADC. The ADC within the 3D growth construct was significantly lower than the surrounding media, indicating limited diffusion due to the presence of cellular membranes. The construct to surrounding media ADC ratio was 0.8 within the experimental reactors, compared to 0.97 in the control reactor. Using the above results we have discovered the baseline hydrodynamic environment within the perfused reactor as well as a baseline metabolic profile of the bioreactor growth construct.

Current and future experiments involve expanding on the aforementioned research. Recent studies have shown that hMSC are highly responsive to their hydrodynamic microenvironment and the presence of media flow regulates hMSC proliferation and progenicity. Our reactor can support both parallel flow around the construct and transverse flow across the construct, thus giving us the ability to study the differences in diffusive and convective transport of macromolecules. Using the bioreactor and 1H MRS and ADC measurements, we will study the effects of varying the flow mode has on the osteogenic differentiation of hMSCs within the 3D growth construct. Differing the flow conditions within the reactor will allow for determination of the optimal method of culturing hMSCs within a 3D construct so that they may have the most beneficial effect on growing a therapeutically useable tissue construct.

571af Bacterial Adhesion on Surfaces with Gradients of Bio-Inertness and Roughness

Shuyu Hou[1], Erik A. Burton[2], Karen A. Simon[2], Yan-Yeung Luk[2], and Dacheng Ren[1]. (1) Department of Biomedical and Chemical Engineering, Syracuse University, 121 Link Hall, Syracuse, NY 13244, (2) Department of Chemistry, Syracuse University, 121 Link Hall, Syracuse, NY 13244

Surface chemistry and roughness play important roles in bacterial cell adhesion and biofilm formation. We recently reported that self-assembled monolayers (SAMs) presenting tri(ethylene) glycol (TEG) are highly resistant to adhesion and biofilm formation of *Escherichia coli*, while SAMs presenting methyl groups are non-resistant (Hou et al., Applied and Environmental Microbiology, 2007. 73: 4300-4307). To further study the response of cell adhesion and biofilm formation to surface conditions, SAM surfaces with molecularly well-defined chemical and roughness gradients were formed on gold films. The surfaces with chemical gradients were prepared with a mixture of two SAMs to obtain the gradient in bio-inertness of TEG-SAM, with the other one being either OH-SAM or CH_3-SAM. *Escherichia coli* RP437 was labeled with a plasmid constitutively expressing DsRed-Express to form biofilms, which were analyzed with confocal laser scanning microscopy. The three dimensional structure of biofilms was analyzed with the COMSTAT software to quantify biomass, surface coverage, and thickness. Measurements of surface coverage indicate that the amount of biofilm formed on surfaces is linearly proportional to surface adhesiveness. In addition to chemical gradient, a gradient with the nanometer-scale topography was introduced in the gold films by creating a continuous increase in the angle of incidence of gold atoms during vapor deposition. The strong correlation between cell adhesion and surface roughness was observed. The results obtained in this study will be useful for understanding bacteria-surface interactions and for developing effective control methods.

Keywords: molecular gradient, bio-inert surface, self-assembled monolayer, biofilm, cell adhesion.

Quantitative Evaluation of Oligonucleotide Surface Concentrations Using Polymerization-Based Amplification

Ryan R. Hansen, Heather J. Avens, Raveesh Shenoy and Christopher N. Bowman, Department of Chemical and Biological Engineering, University of Colorado, ECCH 111, Campus Box 424, Boulder, CO 80309

Quantitative evaluation of biomolecule concentrations is becoming increasingly necessary due to the expanding discovery and characterization of potential disease markers and their corresponding applications in molecular diagnostics. Quantitative, sequence-specific analysis of polynucleotide biomarkers potentially allows for the monitoring of bacterial or viral loads, drug efficacy evaluation based on monitoring viral burden, elucidation of various disease states and progressions, and may provide clinical validation of a diagnosis. Specifically with application to human cancers, several types of nucleic acid biomarkers require quantitative analysis, such as evaluation of over or under expressed RNA transcripts, micro RNAs, or analysis of genomic amplifications or deletions. Mutation detection, traditionally performed on a qualitative basis, can also benefit from quantitative analysis through determination of wild type to mutant type ratios enabling distinction between stochastic and pathogenic mutations.

Polymerization-based amplification has been demonstrated to offer rapid, sensitive, and visual detection of the presence of labeled DNA in a multiplex biochip format. To access the feasibility of this method as a quantitative tool, the amount of polymer growth achieved from amplification of such Bio-recognition events as a function of surface-bound, labeled DNA concentration is evaluated. The approach taken involves fabricating aminosilated glass surfaces with oligonucleotide spots containing controlled concentrations of immobilized, biotinylated DNA targets capable of capturing amounts of Streptavidin (coupled to the initiator, Eosin) followed by contact with a monomer solution and a discrete light exposure. The thickness profile that results shows a dynamic (50 to 4800 biotin markers/micrometer^2) and a saturated region (6,000 to 15,000 biotin markers/micrometer^2). Efforts made to characterize the saturated region by shifting it to a dynamic and quantifiable response resulted in broadening of the dynamic range from 50 to 18,000 biotins/micrometer^2.

Eosin is a weak fluorophore and the fluorescent signal obtained after fabrication of slides and before polymerization-based amplification, can be used to quantify the binding events between the biotinylated oligonucleotide and the Streptavidin-initiator molecule. Incorporation of an amplified fluorescent, radiological, or chemiluminescent response with polymer film growth is desirable due to the numerous types of inexpensive detection instrumentation consistent with this method when signals are large. In an effort to develop this amplification method towards compatibility with fluorescent-based methods of quantification, incorporation of fluorescent nanoparticles into the polymer films is also evaluated. Polystyrene microspheres that encapsulate multiple hydrophobic fluorophores per particle, which prevent non specific initiation as a result of elimination of contact with the surrounding environment, are used. The resulting amplified fluorescent signal enables detection and quantification using inexpensive CMOS-based fluorescent readers as opposed to expensive laser-based microarray scanners employing PMT detectors.

As a concluding remark, the photoinitiator density is shown to be a crucial parameter in determining the amount of polymer growth observed from the surface, as this parameter can be manipulated depending on binding conditions to tune the amount of amplification generated from a given number of biotinylated targets localized on a biosensor surface. The visible light system is amenable to obtaining an amplified fluorescent response correlating with film thicknesses, enabling evaluation of biomarker quantities with inexpensive fluorescent instrumentation. The large, dynamic variation in film thicknesses and amplified fluorescent signal obtained from polymerization-based amplification opens up many potential uses for this assay that were previously unachievable from only a positive/negative response. For example, background polymer growth arising from non-specific biomolecule interactions can be characterized such that only an increase over this background becomes significant. This gives polymerization-based amplification potential use in applications where both high specificity and sensitivity are critical, such as in a high-throughput SNP analysis. Similarly, markers that become significant only when deviating from characterized background levels can also be analyzed, which is the case for the majority of proteomic biomarkers.

571ah Spatial and Temporal Resolution of Shear In Orbiting Dishes

Jonathan Michael D. Thomas[1], Amlan Chakraborty[1], M. Keith Sharp[2], and **R. Eric Berson**[1]. (1) Chemical Engineering, University of Louisville, Louisville, KY 40292, (2) Mechanical Engineering, University of Louisville, Louisville, KY 40292

It is well documented that physiological and morphological properties of anchored cells in culture are influenced by fluid shear stress. Common orbital shakers are prevalent throughout the cell culture industry because of their simplicity and they provide a means for simultaneously running tens to hundreds of tests by loading the shaker with several dishes. However, the complex flow in orbiting dishes is amenable to analytical solution for resolving shear created by the fluid motion only for simplified conditions. The only existing quantification of shear in this flow is an equation that estimates a constant scalar value of shear for the entire surface of the dish. In practice, wall shear stress values will be oscillatory rather than steady due to the traveling waveform and will vary across the surface of the dish at any instant in time. A computational model is presented here that provides complete spatial and temporal resolution of wall shear stress over the bottom surface of a dish throughout the orbital cycle. The model is reasonably well validated by reported experimental WSS values obtained using optical velocimetry at discreet locations on the bottom of an orbiting dish and by comparison with an analytical solution. The model significantly enhances the usefulness of simple shaker apparatuses in the study of hemodynamic effects on endothelial and other anchored cell cultures by providing WSS as a function of time over all dish locations.

571ai Cis and Trans Interactions Can Compete In Heterotypic Adhesive Signaling

Richard Tsai, University of Pennsylvania, Room 129 Towne Building, 220 South 33rd Street, Philadelphia, PA 19104-6393, Shyamsundar Subramanian, Fermentation & Cell Culture, Merck & Co., Inc, 770 Sumneytown Pike, Mailstop WP17-201, West Point, PA 19486-0004, Pia Rodriquez, Chemical Engineering, University of Pennsylvania, Room 129 Towne Building, 220 South 33rd Street, Philadelphia, PA 19104-6393, and Dennis E. Discher, Chemical and Biomolecular Engineering, University of Pennsylvania, Room 129 Towne Building, 220 South 33rd Street, Philadelphia, PA 19104-6393.

Membrane protein-A on cell-A interacting with membrane protein-B on cell-B is the usual 'trans' interaction in cell-cell adhesion, but if cell-A also expresses protein-B, then 'cis' interactions might also compete. The Marker of Self adhesion protein CD47 is found on all cell types and binds SIRPa on Macrophages in trans at sub-microMolar affinities, but human Macrophages also express CD47 and competition in cis proves significant. Binding of CD47 in trans inhibits phagocytosis of a CD47-displaying target [Tsai and Discher, JCB 2008], and we find here that knockdown of CD47 on Macrophages is shown to lead to greater binding of soluble CD47 and, separately, enhanced phagocytosis of opsonized targets. The effective Kd for binding in trans depends linearly on cis CD47 on the macrophage, consistent with the simplest competition model. Additionally, CHO cells engineered to express CD47-GFP bind more soluble SIRPa than CHO cells displaying both CD47 and SIRPa, although high affinity antibodies can out-compete cis interactions. Both Hematopoietic and Mesenchymal Stem Cells also express both CD47 and SIRPa, and inhibitory cis interactions on these cells are also clear. The fact that Macrophages need the Marker of Self CD47 seems to make cis interactions unavoidable, and any other cell type that expresses both proteins will require accounting for this intrinsic cis competition.

571aj Efficient Expression and Purification of α2,6-Sialyltransferase in the Hemolymph of *Bombyx Mori* (silkworm) Larvae Using Bmnpv Bacmid and Its Application on Glycotechnology

Makoto Ogata[1], Makoto Nakajima[1], Takeomi Murata[2], Taichi Usui[2], and Enoch Y. Park[1]. (1) Laboratory of Biotecnology, Faculty of Agriculture, Shizuoka University, 836 Ohya, Suruga-ku, Shizuoka, 422-8017, Japan, (2) Department of Applied Biological Chemistry, Faculty of Agriculture, Shizuoka University, 836 Ohya, Suruga-ku, Shizuoka, 422-8017, Japan

In previous study we demonstrated, *Bombyx mori* nuclear polyhedrosis virus (BmNPV) bacmid system provides the rapid protein production in silkworm as long as 10 days, is free from biohazard, thus will be a powerful tool for the future production factory of recombinant eukaryotic proteins (T. Motohashi *et al.*, Biochem. Biophis. Res. Commun., **326** (1), 2005, 564-569). In this study, we performed expression and purification of α2,6-sialyltransferase (α2,6-SiaT) using *B. mori* and also its application on the synthesis of human type influenza virus infection inhibitor.

First, fusion gene encoding rat α2,6-SiaT containing bombyxin (bx) signal peptide, histidine (His) tag and enterokinase recognition site was constructed, and then we created BmNPV bacmid containing bx-His-α2,6-SiaT fusion gene. Moreover using bacmid system, the α2,6-SiaT was expressed in the hemolymph of silkworm larvae. α2,6-SiaT activity was 0.93 U/ml, and possessed strong sialic acid transfer activity. In addition, produced α2,6-SiaT was partial purified about 100-fold to homogeneity from crude extract of the *B. mori* silkworm larvae with a yield of 30% using a cation exchange chromatography.

This enzyme was shown to be useful for a service tool in the synthesis of human type influenza virus infection inhibitors (M. Ogata *et al.*, Bioorg. Med. Chem., **15**, 2007, 1383-1393).

571ak A Dynamic Light Scattering Study of Antibody Aggregation

Andrea L. Cascio, Chemical Engineering, Missouri University of Science and Technology, 1870 miner circle, cbe-umr, Roll, MO 65409 and Daniel Forciniti, Cbe, Missouri University of Science and Technology, 1870 miner circle, cbe-umr, Roll, MO 65409.

Glycosylation is a post-translation modification of proteins, which is associated with protein stability and function. Glycosylation patterns are different in different types of cells and some are unable to add glycosyl groups. For example, prokaryotic cells do not have the necessary machinery to add glycosyl groups to their proteins; therefore, human proteins expressed in bacteria are not glycosylated. Mammalian cells and plant cells can both glycosylate proteins but their glycosylation patterns are different. This is important because, recently, therapeutic human proteins have been expressed in plants. Although the expression of human antibodies in plants like tobacco or corn has been quite successful, the plant glycosylation machinery is often turned off because of differences in the glycosylation pattern between the human and the plant variety. Unfortunately, the non-glycosylated antibody is less stable and thus has a shorter shelf life than the native one. The decrease in stability is manifested by the formation of irreversible aggregates, which decreases the solubility of the preparation impeding their normal function.

The objective of this work is to investigate the effect of various stresses found during processing on the stability of fully glycosylated and non-glycosylated human antibodies. The stresses that we have chosen to investigate are: exposure to high temperatures, exposure to shear rate, exposure to freeze and thaw cycles and the effect of freeze drying. Experiments were performed in the presence and absence of various amounts of sucrose.

All the preparations were inspected by second derivative UV spectroscopy to detect coarse changes in the secondary structure. No major changes in protein structure were detected upon the applied stresses. Samples were run through a GPC column and monitor with a UV detector in tandem with a multiple angle light scattering detector. All the chromatograms show the presence of three species, the monomer (MW 130,000), the dimer (MW 260,000) and higher aggregates of molecular weights 1 million or higher. These relatively large aggregates consist of approximately ten antibody molecules. Assuming spherical shapes and closed packing (a density of 0.74), these aggregates are roughly 20 nm in diameter. The glycosylated antibody has a higher percent monomer than the non-glycosylated one (~92% vs. 88%). There are only minor differences between the ratio of the different species under various stresses with the exception of experiments performed at 50 C which seem to have less monomer.

Dynamic light scattering was used in this work to study the effect of various stresses on the formation of large aggregates (> 100 nm) of both antibody preparations. The experiments were performed with a backscattered instrument (Brookhaven) which is not very sensitive to the overall quality of the sample. Stirring (at either 20 C or 50 C) as well as freeze drying using an ethanol bath have the largest effects on both preparations. There is a minor effect by freeze and thaw (either one or two cycles) but a pronounced effect by freeze drying using both freezing methods (ethanol or liquid nitrogen). In the presence of sucrose large aggregates (larger than 100 nm) are absent in the control preparation but the protective effect of sucrose is more pronounced with the non-glycosylated preparation.

A few selected samples were inspected using a research goniometer equipped with a 2W laser and a digital correlator (Brookhaven). These experiments revealed details of the shape of the aggregates. Not only the non-glycosylated preparation has a larger number of aggregates but also they are more asymmetric than those of the glycosylated one. The presence of sucrose seems to have an "equalizer" effect for both preparations.

571al Titanium Implant Coated with Hydroxyapatite Microcapsules and Poly (L-lactic acid) Releasing An Antimicrobial Agent

Isao Kimura[1], Takanori Kawaguchi[1], Takaaki Tanaka[2], Masayuki Taniguchi[2], and Mitsugu Kanatani[3]. (1) Graduate School of Science and Technology, Niigata University, 8050, Ikarashi 2-no-cho, Nishi-ku, Niigata-shi, Japan, (2) Department of Materials Science and Technology, Niigata University, 8050, Ikarashi 2-no-cho, Nishi-ku, Niigata-shi, Japan, (3) Graduate School of Medical and Dental Sciences, Niigata University, 2-5274, Gakko-cho, Chuo-ku, Niigata-shi, Japan

A titanium implant coated with hydroxyapatite (HAp) is expected to improve the osseointegration. When HAp is provided as microcapsules containing an antimicrobial agent, it must be effective in preventing infection after the implanting operation as well. This study was carried out as a fundamental study on preparing such an implant. A screw made of pure titanium of Japanese Industrial Standards (JIS) type 2 was used as a model of implants. Hollow HAp microspheres were prepared by the interfacial reaction of calcium nitrate with potassium hydrogen phosphate in a multiple emulsion at 323 K. The titanium screw was dipped into hydrofluoric acid (HF) to alter the roughness. The HAp microspheres were suspended in Hank's balanced salt solution (HBSS), and the surface-treated screw was immersed in it to be coated with the HAp microspheres. The adhering state of HAp microspheres was observed by scanning electron microscopy. The amount of HAp adhering on the titanium surface was evaluated by inductive coupled plasma emission spectroscopy. An ampicillin sodium salt (Amp) solution as an antimicrobial agent was loaded into the HAp microspheres on the titanium screw, and then the screw was covered with poly(L-lactic acid). The release profile of Amp in HBSS for several days was characterized by ultraviolet spectroscopy. The antimicrobial activity of released Amp was evaluated with *Escherichia coli* as an indicator strain.

571am Economical Protein Purification In *Pichia Pastoris*

Baley A. Fong and David W. Wood. Chemical Engineering, Princeton University, Princeton, NJ 08544

Pichia pastoris is a cheaper alternative to CHO for the expression of glycosylated proteins. A method has been developed and implemented in *Pichia* for large scale, economical purification of recombinant proteins. This method utilizes self-cleaving inteins to link purification tags and target proteins, thereby eliminating the need for proteolytic tag removal and making the method economically feasible on larger scales. Both chromatographic and non-chromatographic purification tags were investigated in shake flask and fermentation cultures. A chitin-binding domain tag was successfully used to purify various proteins on chitin affinity resin. In addition, a non-chromatographic elastin-like polypeptide (ELP) purification tag was investigated. In the latter purification scheme, mild temperature shifts induce ELP aggregation and dissociation. This reversible precipitation allows for the separation of the ELP-bound target protein using a series of temperature shifts and centrifugation, ultimately resulting in a simple, economical purification system.

571an Efficient Expression and Purification of Human Membrane Protein in Fat Body of Silkworm Larvae Using Bmnpv Bacmid and Characterization of Its Biological Property

Dongning Du[1], Tatsuya Kato[2], Fumiaki Suzuki[3], and Enoch Y. Park[1]. (1) Laboratory of Biotechnology, Graduate School of Science and Technology, Shizuoka University, 836 Ohya, Suruga-ku, Shizuoka, Japan, (2) Faculty of Agriculture, Shizuoka university, Shizuoka, Japan, (3) Faculty of Applied Biological Sciences, Gifu, Japan

Membrane proteins play important roles in biological system, and are targeted in the screening new medicines. We tried the production of membrane protein in the fat body of silkworm using BmNPV bacmid1). Full length human (pro)renin receptor (hPRR-wTM) and only extracellular domain of human (pro)renin receptor (hPRR-w/oTM) lacking the transmembrane and cytoplasmic domain were used as membrane protein in this work, and was successfully expressed in the fat body of silkworm larvae. To investigate the intrinsic structural and functional property of (pro)renin receptor, this membrane protein was solubilized from the membrane with dodecyl-â-D-maltoside, and get the protein with yield of more than 60 %. And both purified full-length hPRR and its extracellular domain were comparatively characterized in binding assays by real-time monitored surface plasmon resonance (SPR) assay with immobilized human renin on the cuvette. Our results demonstrate that the full-length hPRR showed the biological function of (pro)renin receptor judging by SPR assay. On the other hand, hPRR-w/oTM didn't show any binding property to human renin. This indicates that transmembrane domain of hPRR is indispensable for biological function of human (pro)renin receptor. And an unidentified factor in the fat body was presumed which may be plays an important role in the mechanism of binding of hPRR with human renin. This study is the first findings describing the interaction of full-length hPRR and its extracellular domain with human renin and may provide an important contribution to the study of hPRR.

1) Motohashi et al, Biochem Biophys Res Commun. 326 (2005) 564-569.

571ao Biosurfactants: Efficient Production and Applications

James Winterburn[1], Simon Baker[2], and Peter Martin[1]. (1) School of Chemical Engineering and Analytical Science, The University of Manchester, The Mill, Sackville Street, Manchester, M60 1QD, United Kingdom, (2) School of Life Sciences, Oxford Brookes University, Headington Campus, Gipsy Lane, Oxford, OX3 0BP, United Kingdom

The proposed poster will set innovative methods for the recovery and purification of biosurfactants in the broader context of the need for sustainable alternatives to petrochemical based commercial surfactants.

The global market for surfactants is approximately 8 M tonnes per annum with a total value of $ 13.9b[1]. The bulk of these surfactants are obtained from non-renewable oil-based feedstocks. Our rate of consumption of petrochemicals has reached a point that many believe to be unsustainable. An alternative route of surfactant production exists in nature; "Carbohydrate-based surfactants, being based on plant-derived chemicals, use renewable resources, are readily biodegradable, non toxic and do not add to the Earth's CO_2 burden."[1] Biosurfactants can be produced via fermentation, for example the heptapeptide surfactin, which has antibiotic properties, is a metabolite of *Bacillus subtilis* BBK006. Using genetic modification to insert a specific protein-expressing gene into a suitable host gives control over the production of biosurfactants during a fermentation.

A major obstacle to the economic production of biosurfactants, and hence their use instead of petrochemical based surfactants, is the downstream processing required. Separation techniques such as microfiltration are costly when applied to typical bioproduct process streams, i.e. large volumes, with a low product concentration. An alternative method for biosurfactant recovery is foam fractionation. Foam Fractionation is an adsorptive bubble separation method[2], which is used to enrich and partially separate solutions of surface active species[3]. Although foam fractionation dates from the 1960's its application to the recovery of biosurfactants from fermentations is relatively new[4].

[1] Hargreaves, T. (2003) Surfactants: the ubiquitous amphiphiles. Chemistry World. Accessed online, 30/3/08 http://www.rsc.org/chemistryworld/Issues/2003/July/amphiphiles.asp

[2] Lemlich, R. (1968) Adsorptive bubble separation methods- Foam fractionation and allied techniques. *Industrial and Engineering Chemistry*. **60**, 16-29.

[3] Jashnanil, I.L. and Lemlich, R. (1973) Transfer units in foam fractionation. *Ind. Eng. Chem. Process Des. Develop*. **12**, 312-321.

[4] Chen, C., Baker, S.C. and Darton, R.C. (2006). Continuous production of biosurfactant with foam fractionation. *J. Chem. Technol. Biotechnol*. **81**, 1915–1922.

571ap A New Turbidimetric Method to Measure the Activity of Lipase by Using Polyester Urethane as a Solid Substrate

Rajeeb Gautam, Department of Chemical and Biochemical Engineering, The University of Western Ontario, 1151 Richmond Street, London, ON N6A5B9, Canada, Amarjeet S. Bassi, Chemical & Biochemical Engineering, The University of Western Ontario, 476, Thomson Engineering Bldg., London, ON N6A5B9, Canada, and Ernest K. Yanful, Department of Civil and Environmental Engineering, The University of Western Ontario, 1151 Richmond Street, Loncon, ON N6A5B9, Canada.

Lipases belong to a class of enzyme that can catalyze the heterogeneous reactions involving insoluble substrates. Lipases hydrolyze ester bonds in lipids and solubilize them into other products. Lipase plays an important role in lipid metabolism and therefore it is one of the very important enzymes in nature. Due to the heterogeneous characteristic of the enzymatic reactions involving lipase, it is difficult to accurately measure its activity in a system containing insoluble substrates. One way of measuring its activity is by the turbidimetric method.

Synthetic polyester urethane (PUR) is a water insoluble solid polymer that contains ester bonds in its polymer backbone. Hydrolysis of PUR by using esterase enzymes has been well documented in the literature. Therefore, solid PUR particles can provide a unique opportunity to measure the activity of lipases/esterases in a heterogeneous system.

In this research, the activity of Candida rugosa lipase (type VII, 59 kDa) was measured by using solid PUR particles in aqueous suspensions. Commercially available aqueous suspension of PUR particles (trade name Impranil DLN) by Bayer USA was used as a substrate. Impranil DLN is a milky white, opaque, 40% (w/w) aqueous suspension of PUR with average particle diameter of 0.2 micro meter and specific gravity of 1.0485. The decrease in optical density of the PUR suspension at 600 nm was monitored by using UV-Visible spectrophotometer under various temperature, pH, enzyme and substrate concentrations. This research shows that aqueous suspensions of PUR particles can be used to estimate the activity of lipase by turbidimetric method.

Key words: Turbidimetry, activity assay, Candida rugosa, lipase, polyester urethane, hydrolysis

571aq Msc Differentiation on Nanostructured Polysaccharide-Based Surfaces

Jorge Almodovar and Matt J. Kipper. Chemical and Biological Engineering, Colorado State University, 1370 Campus Delivery, Colorado State University, Fort Collins, CO 80523

Mesenchymal stem cells (MSC) are pluripotent cells that are able to differentiate into chondrocytes and/ or osteoblasts. MSC differentiation is driven by many factors including surface topography and chemistry, and the presence of growth factors. We are studying the performance of nanostructured, polysaccharide-based surfaces with respect to MSC response. To accomplish this, primary sheep MSC, isolated from bone marrow are seeded on nanostructured surfaces containing biologically functional polysaccharides. The MSC response is assessed qualitatively by microscopic imaging, and quantitatively by functional assays. We are also studying the effects of stabilized growth factors (FGF-2 or TGF-β1) in the presence and absence of nanoscale surface features. In our previous work we have demonstrated the stability of model proteins on these surfaces, and thus we are interested in studying the growth factor stability on these surfaces and their effect in MSC response.

571ar Characterization and Optimization of Lovd towards Enzymatic Biosynthesis of Simvastatin

Xinkai Xie, Xue Gao, and Yi Tang. Chemical and Biomolecular Engineering, University of California at Los Angeles, 420 Westwood Plaza, 5667 Boelter Hall, Los Angeles, CA 90095

Simvastatin (Zocor) is a pharmaceutically important semisynthetic compound due to its potent inhibitory activity towards hydroxymethylglutaryl coenzyme A reductase (HMGR), the rate-limiting step of cholesterol biosynthesis. We recently discovered an acyl-transferase, LovD (46 kDa), which is involved in the lovastatin biosynthetic pathway. LovD catalyzes the acyltransfer of the side chain α-methylbutyrate synthesized by LovF to the C-8 hydroxyl of the direct biosynthetic precursor monacolin J sodium salt (MJSS) to yield lovastatin sodium salt. We demonstrated that LovD has broad substrate specificity towards: 1) the acyl group, 2) the thioester acyl carrier, and 3) the decalin core. We showed an Escherichia coli based whole-cell platform can readily convert MJSS to simvastatin sodium salt using the cost-effective •-dimethylbutyryl-S-methyl-3-mercaptopropionate (DMB-S-MMP) as the thioester acyl donor. Further optimization of the E. coli expression host through deletion of a competing esterase bioH led to significant improvement in the kinetics of the biocatalyst. In addition, we rationally engineered LovD by using a combination of homology structural prediction and site-directed mutagenesis. Through systematic cysteine replacement with other amino acids, we created a double mutant C40A/C60N which displayed > 50% improvement in protein solubility and whole cell activity. An optimized fed-batch fermentation process that combines the above improvements was developed to quantitatively convert > 18 g/L MJSS to simvastatin in 18 hours. Taken together, the whole process can be economically competitive with the optimized synthetic methods currently used to manufacture simvastatin.

571as Systems Analysis of Pentose Transport Regulation

Tasha Desai, Chemical and Biomolecular Engineering, University of Illinois, 600 South Matthews Ave., 210 Roger Adams Laboratory, Box C-3, Urbana, IL 61801 and Christopher V. Rao, Department of Chemical Engineering, University of Illinois, 600 South Matthews Ave., 211 Roger Adams Laboratory, Box C-3, Urbana, IL 61801.

The Escherichia coli genome encodes for a number of pentose transporters. Each pentose regulon encodes for at least one transporter, in a few instances two. While the expression of specific transporters is conditional on the presence of its cognate sugar (e.g. xylose transporters are expressed in the presence of xylose), significant regulatory crosstalk exists between different pentose regulons. Furthermore, the transporters are not specific and therefore capable of transporting a broad spectrum of sugars.

By profiling gene expression in different transporter knockouts and regulatory mutants, we have developed a model for pentose transport regulation in the presence of multiple sugars. Our model suggests that this regulation induces a hierarchy in pentose utilization. As transport is a rate-limiting step in pentose metabolism, our results and model may aid in the metabolic engineering of strains using pentose sugars as feedstock, such as in cellulosic biofuel production and natural product synthesis.

Anaerobic Treatment of Cattle Manure for Biogas Production

R. Omar, R.M. Harun, T.I. Mohd. Ghazi, W.A.K.G Wan Azlina, A. Idris and R. Yunus
Department of Chemical and Environmental Engineering, Faculty of Engineering, Universiti Putra Malaysia, Selangor, Malaysia.

Abstract

Feedlot farming with abattoir integration in Malaysia is relatively new; therefore its waste management needs to be carefully administered. Although the cattle are only stored for 3 months cycle to be fatten up before being slaughtered, the manure and waste produced from abattoir can be as much as 200-300 tonnes per day. Batch anaerobic digestion of cattle manure was studied in a 10 L bioreactor at 53°C and pH controlled at 6.95. Palm oil mill effluent activated sludge at ratio of 1 to 5 was used as seed sludge. The initial COD of the mixture was 42.5 g/L and it reduced by 51%. The biogas produced is 207 L/kg VS at hydraulic retention time of 17 days with ultimate methane yield of 184 L/kg VS. The biogas production dipped slightly on day 10 due to the availability of nutrient that has decreased tremendously. Increased N-NH_4^+ concentration above 1000 mg/L may also have inhibitory effect on the biogas production. Methane content obtained is between 58-83% throughout the culture. The H_2S concentration is well below than 0.26 ppm, therefore no further treatment is needed. Effluent on the final days was found to be high in COD (8.75 mg/L) and N-NH_4^+ (260 ppm) which are unsuitable for irrigation according to the Malaysian Water Standard. Further treatment such as post-digestion or chemical/physical treatment is needed.

Introduction

The application of anaerobic digestion (AD) technology to livestock treatment has received considerable attention because it can be used to produce valuable by-products such as biogas. The technology of biogas has been developed due to the advantage of producing energy as well as generating odor-free residues rich nutrients, which can be used as fertilizers (Karim et al., 2005). This would encourage sustainable agricultural practices in mitigating possible manure pollution problems, thereby sustaining development while maintaining environmental quality. Moreover, rural economic development would benefit from the implicit multiplier effect resulting from jobs created by implementing digester systems. Therefore, promising future waste-to-profit may add economic performance of an AD.

Although large scale production and advanced AD systems are well developed and in operation in countries like Europe, United States, Japan, China and Taiwan (Rodriguez et al., 1997), no known AD of cattle manure is found in Malaysia. Nonetheless, there are a few guidelines for cattle and poultry farming which suggest the integration of an anaerobic digester for waste management (Jabatan Perkhidmatan Haiwan, 2003). However, this system is not favored by small farmers due to high capital cost and lack of environmental consciousness. With a population of more than 300,000 pigs and cattle recorded in Penang alone, there is an urgent need to set up this technology (Jabatan Perkhidmatan Veterinar Pulau Pinang, 2001).

An estimation of the potential to produce methane of animal manures is essential. Parameters such as nutrients content, pH, temperature, organic loadings, solid content, mixture ratio and hydraulic retention time (HRT) are detrimental to biogas quality and quantity. These parameters would also determine the size of digester, auxiliary equipment needed, as well as the sizes and specifications of biogas treatment facilities. Maximum methane yield requires adequate and efficient nutrient supply for micro-organisms in the digester. Methane production from organic substrates mainly depends on their content of substances that can be degraded to CH_4 and CO_2. Composition and biodegradability are key factors for the methane yield from energy crops and animal manures (Balsari et al., 1983).

In well-established digesters, pH of the fermenting mass are buffered between 6.8 and 7.4. Bacteria have limited range of temperature, in which they are active. Methanogens, in particular, are very sensitive to temperature changes. In general, mesophilic anaerobic digestion of organic sludge is more widely used compared to thermophilic digestion, mainly because of the lower energy requirements and higher stability of the process. Thermophilic digestion, however, is more efficient in terms of organic matter removal and methane production (Buhr and Andrews, 1977) and (Ahring, 2001). Moreover, it enhances the destruction of pathogens, weed seeds and insect eggs; thus enabling effluent hygienization (Zábranská, 2000) which might be required in the short term for land application. During fermentation of organic wastes, acetic acid is usually the main product. The excess production of volatile fatty acids (VFA) may result in an inhibitory effect on the fermentation of organic wastes (Noike et al., 2000).

The most potential biochemical methane production yield from animal waste and water can be as 0.2 m^3/kg of VS added (Chynoweth et al., 1993). Furthermore, the contents of nitrogen and phosphorus in the animal waste are sufficient to satisfy the cell growth requirements during the biogas production. The others elements, such as sodium, potassium, calcium, magnesium and iron are present in low concentrations. However, they may exhibit inhibitory effects at higher concentrations (Shyam and Sharma, 1994). On the contrary, there was insignificant effect of the retention time on biogas production. Van Der Vlugt and Rulkens (1985) performed the biogas production in 20 liter laboratory digesters and found that the biogas produced was not affected when reducing the hydraulic retention time (HRT) from 40 to 20 days. In contrast, in a 200 L farm scale digester, Kiely (1985) found that cattle slurry of higher VS produced higher amount of biogas over the total digestion period of 66 days.

Bioconversion of animal waste is a non-polluting and environmentally feasible process. The main aim of this present study is to evaluate the potential biogas production from cattle manure. The influence of anaerobic digestion parameters on the manure treatment and methane yield from cattle manure was investigated.

Materials and Method

Feedstock

Cattle manure was collected from the Alifah's farm situated at Kluang, Johore, Malaysia. The samples were kept in tightly closed polycarbonate bottle and stored at 4°C until further use. Seed sludge was obtained from the palm oil mill effluent (POME) aerobic sludge and used without further purification.

Apparatus and Operation

A 5 liter volume of dilute cow manure (1:1) was mixed with 1 L of POME sludge and digested in a 10 L jacketed fermenter. The bioreactor system (Biostat B, Germany) is equipped with pH, mixer and temperature controller. The sample in the bioreactor was flushed with N_2 gas for 10 minutes at the beginning of experiment ensure anaerobic condition. The temperature was controlled at 53°C and homogenous mixing was aided by a 150 rpm stirrer. The pH was controlled at 6.95 via the addition of 1 N HCl and NaOH. Hydraulic retention time (HRT) of 17 days was used in this study. The produced biogas was collected using gas sampling bag.

Analytical methods

Total solid, volatile solid, biological oxygen demand (BOD) and chemical oxygen demand (COD) were measured by following the procedures listed in Standard Methods (APHA, 1998). Total nitrogen content was determined via Kjeldal method while ammonia nitrogen was determined via Autoanalyzer (Lachat QC 800, USA). The metal content was analyzed via ICP-AES using the EPA3050B extraction method. While elemental content was measured using elemental analyzer (LECO CHNS932). Methane and carbon dioxide concentrations in the biogas were determined with a gas chromatograph (Shimadzu, Japan) equipped with a thermal conductivity detector (TCD) and Carboxen-1010 plot column 30m x 0:53mm ID (Supelco, USA). The injector, detector and oven temperatures were 150, 200 and 120°C, respectively. Argon served as the carrier gas.

Results and Discussion

Characteristics of feedstock

The fresh cow manure were analyzed with respect to total solids and volatile solid content, total nitrogen and ammonia nitrogen content, metal content such as phosphorus, potassium, calcium, magnesium, iron, sodium and aluminium, elemental content and pH. The results are tabulated in Table 1. The total solid content was not very high therefore minor dilution is needed as most existing on-farm digesters are operating at total solid between 8 and 12% (Lo et al., 1983). Furthermore, the addition of seed sludge increases the solid content at the start of fermentation. Theoretically, high volatile solid will produce high yield of biogas, given that optimum condition applies in the digester. High content of 9.3 g/L of the total nitrogen could be attributed by organic nitrogen with only 800 mg/L exist as ammonia nitrogen form. Caution must be taken when digesting high nitrogen containing biomass because its conversion to free ammonia might cause toxicity to the bacteria. Free ammonia is known to be able to pass through the microbes' cell membrane, causing proton imbalance and potassium deficiency (Chen et al., 2008).

Table 1: Composition of fresh undiluted manure

Types of Analysis	**Values**
Total Solids (% wet)	16.7
Volatile Solids (mg/L)	1187.9
Total Nitrogen (mg/L)	9260
Nitrogen Ammonium (mg/L)	800 mg/L
Metals (ppm)	
Phosphorus	24.7
Potassium	44.2
Calcium	65.9
Magnesium	14.7
Ferum	10.3
Sodium	3.9
Aluminium	2.6
Elemental (wt %, dry basis)	
Carbon, C	43.5
Hydrogen, H	5.47
Nitrogen, N	1.84
Oxygen, O	49.2
pH	5.33

Cationic elements such as sodium, potassium and others are required for microbial growth in anaerobic digestion of waste, but can be inhibitory to microbial activity if present in high concentrations (Appels et al., 2008). The metal contents are however low in this case therefore, their presence effects the microbial growth positively. Elemental analysis of manure shows high content of oxygenated compound, probably due to the nature of the food given to the cattle. The food comprises of palm kernel cake (PKC), spent brewer grain (wheat and barley) and soy residue. PKC is a by-product of palm kernel oil milling containing 14-16% of crude protein (Chin, 2001) while soy residue contains around 16% protein (Hsieh and Yang, 2004) and spent brewer grain contains of 28% protein (Lim, 1967). PKC also contains oil between 5 and 12%. Though the oil content is not significant, this may give rise to the lipid content as in the manure. The carbohydrates in the feed may not be fully digested in the cow stomach therefore the remaining carbohydrates may have given substantial value of oxygen content.

Biogas and Methane Production

The cumulative biogas production and its respective methane content are shown in Figure 1. The first three days of the digestion indicates slow production of biogas concurring with the first phase of biomass decomposition via acetogenesis process. From observation, automatic addition of NaOH was intensive during this time. The production thereafter rises rapidly for 2 days and then less rapidly up to day 12. This rapid increment of biogas production is a result of acclimatized methane forming bacteria activities and high temperature used in the experiment. Only small increment was observed after day 12 until finishing run at day 17. The bacteria activities began to cease during this time possibly due to toxicity of high ammonia nitrogen content above 1 g/L especially for thermophilic digestion (Hansen et al., 1997). However, the methane content varied between 58 and 68% but exceptionally high on the final day at 83%. The average methane content is calculated to be 65% which is pretty decent when optimisation of the parameters is yet to be attempted. The composition of biogas was also analyzed to yield around 14 to 19% of CO_2 and maximum of 0.26 ppm H_2S. Low H_2S content in this case is very encouraging because post-treatment for the biogas is not needed.

Figure 2 represents the total solid and volatile solid concentration during the anaerobic digestion period. For this batch digestion, initial volatile solid was analyzed to be 290 mg/L giving rise to biogas production of 207 L/kg VS. The elemental composition of the raw manure was analyzed to calculate the theoretical methane yield, B_u using Bushwell's formula (Symons and Bushwell, 1933). The theoretical methane yield was calculated to be 393 L/kg VS as opposed to the ultimate methane yield of 184 L/kg VS, less than half methane was produced than expected. The cessation of biogas production on day 12 is obviously the reason. Volatile solid content on this day is still high at 265 mg/L proving food is still sufficient at this point. Toxicity of ammonia still stands as the main reason of the cessation and will be discussed in detail later. The total solid decreased insignificantly for the first seven days but increased quite significantly thereafter. This is most probably due to sampling difficulties. The same reasoning goes to the volatile solid.

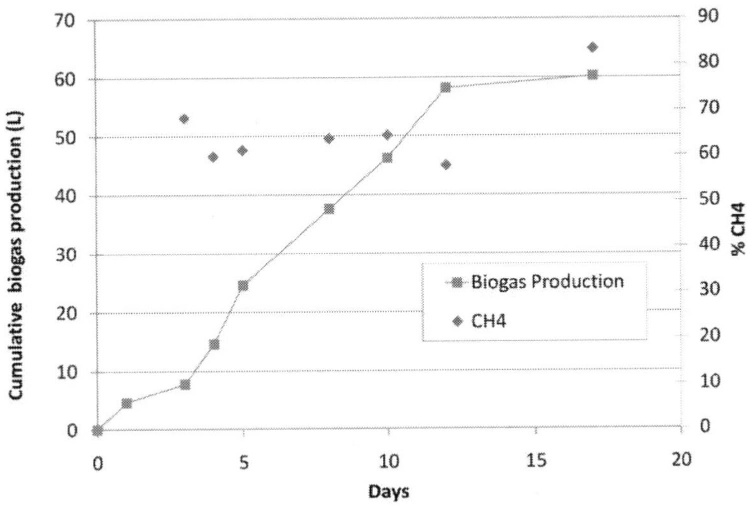

Fig ure 1: Cumulative biogas production and respective methane content for batch anaerobic digestion of cow manure.

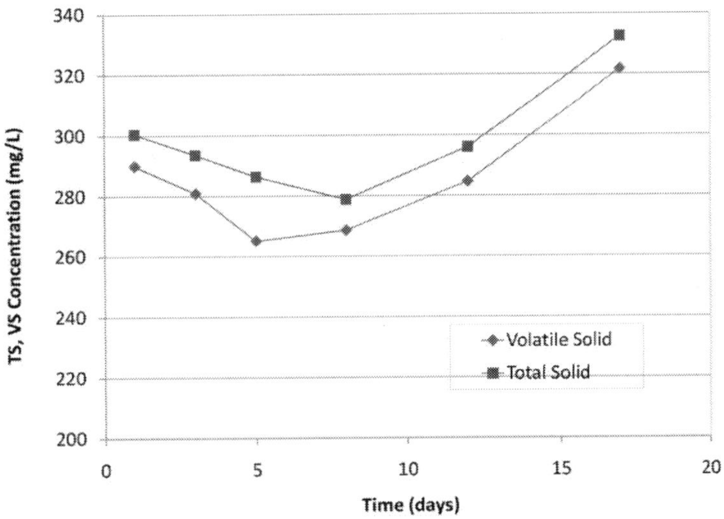

Figure 2: Total solid and volatile solid concentration with respect to the biogas methane yield during the digestion period.

COD removal and Ammonia

The initial COD of undiluted cattle manure is 17.4 g/L however, addition of seed material, the palm oil mill effluent (POME) digested sludge has increased the COD to 42.5 g/L. POME is well known for its high COD value due to its fat contents. Nonetheless, this high initial COD did not restrain the bacterial activity for at least the first 12 days of this batch culture. By the 12th day, the COD value has decreased to 21.7 g/L, half of that initial value. Thereafter, the value started to stabilize around 21 g/L. It can be concluded that with little bacterial activity at this point, no further treatment of the waste took place.

Ammonia content was stable at the beginning of the culture at 400 mg/L until day 12 when the value shot up to 1.2 g/L. The organic nitrogen in the form of protein and lipids contained in the manure after degasification was converted to inorganic nitrogen in the form of ammonium during the mineralization process. It was found that inorganic nitrogen increased by 32% and organic nitrogen was reduced by 50% in the digested manure (Moller, 2006). Inhibition of ammonia has been found at concentrations as low as 0.7 g/L up to 1.1 g/L for adapted anaerobic digestion of manures (Hansen et al., 1998). The limit of which inhibition of methanogenesis is mainly depends on operating temperature and pH. At higher temperature, i.e. thermophilic conditions as in this study, although encompassing advantages such as high degradation rates and better sanitation effects, the production of ammonia was higher as compared to mesophilic conditions (Angelidaki and Ellegaard, 2003). Sung and Liu (2003) found that 50% inhibition of methanogenesis at pH 7.6 was under thermophilic condition even at 0.57 g/L of ammonia. Higher tolerance of the current study as compared to Sung and Liu's possibly attributed to differences in substrates, inocula, pH and acclimatization period. The pH in this study was controlled at 6.9 and from observation, as the digestion progressed, more acid was automatically added. This may have prevented the pH to increase further and prohibit growth. Nonetheless, the ammonia produced due to high temperature is adequate to cease the microbial growth. Figure 3 represents the amount of COD removed and ammonia-nitrogen being produced during the digestion period.

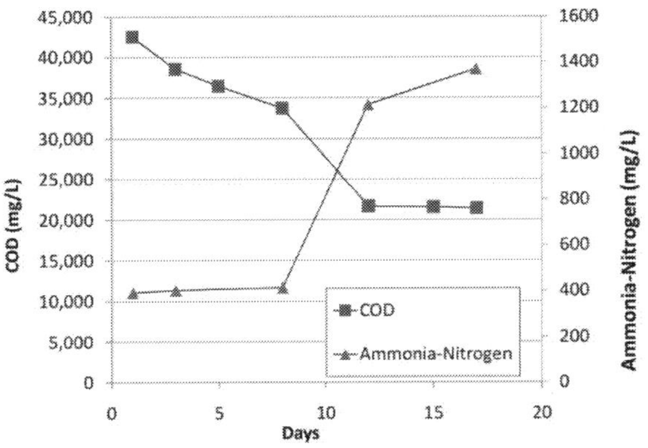

Figure 3: COD removal and ammonia production during the anaerobic digestion of cow manure

Conclusions

There is no doubt that methane and power produced in anaerobic digestion facilities are becoming favorably utilized to replace energy derived from fossil fuels, and hence reduce emissions of greenhouse gases. In this study, biogas methane was produced via anaerobic digestion of cow manure in a 10 L stirred bioreactor. The digestion was carried out at thermophilic temperature of 53°C and a controlled pH at 6.9. Findings suggested that the ammonia produced led to the prohibition of microbial growth. The COD was removed by 51% from 42.5 g/L to 21 g/L, indicating no further treatment of the waste took place due to minimal bacterial activity. It was found that the inhibitory effect caused by the increasing ammonia concentration led to reduction of the biogas methane content during the digestion period. However, the average biogas methane produced is about 65% with only traces of H_2S concentration. This would give a promising evidence of the anaerobic digestion as a viable and feasible treatment technology of organic waste, in particular, the animal manure, simultaneously, generating biogas as the renewable energy. Further investigation is yet to be carried out to finalize the optimization of the parameters involved in this study.

References

Ahring, B.K., Mladenovska, Z., Iranpour, R. and Westerman, P. (2001). State of the art and future perspectives of thermophilic anaerobic digestion, *Proceedings of the 9th World Congress on Anaerobic Digestion, vol. 1* Antwerpen, Belgium, pp. 455–460.

Angelidaki, I. and Ellegaard, L. (2003). Codigestion of manure and organic wastes in centralized biogas plants, *Applied Biochemistry and Biotechnology*, 109: 95-105.

American Public Health Association (1998). Standard Methods for the Examination of Water and Wastewater 15th Ed. Washington, APHA.

Appels, L., Baeyens, J., Degreve, J. and Dewil, R. (2008). Principles and potential of the anaerobic digestion of waste-activated sludge, *Progress in Energy and Combustion Science*, *In press*.

Balsari, P., Bonfanti, P., Bozza, E. and Sangiorgi, F. (2008). Evaluation of the influence of animal feeding on the performances of a biogas installation (mathematical model), In: Third International Symposium on Anaerobic Digestion. Boston, MA, USA, 14–20 August 1983, A 20, pg. 7.

Buhr, H.O. and Andrews, J.F. (1977). The thermophilic anaerobic digestion process, *Water Resources,* 11:129–143.

Chen, Y., Cheng, J. J., Creamer, K. S. (2008). Inhibition of anaerobic digestion process: a review, *Bioresource Technology*, 99: 4044-4064.

Chin, F. Y. (2001). Palm kernel cake (PKC) as a supplement for fattening and dairy cattle in Malaysia, *Report of the 26th Session of the Animal Production and Health Commission for Asia and the Pacific (APHCA),* 24-26 August 2002. Subang Jaya, Malaysia.

Chynoweth, J.N. Owen (1993). Biochemical methane potential of municipal solid waste components, Water Sci. Technol. 27, 1–14.

Hansen, H. H, Angelidaki, I. and Ahring, B. K (1998). Anaerobic digestion of swine manure: Inhibition by ammonia, Water Resource. 33 (8), 1805-1810.

Hsieh, C. and Yang, F. C. (2004). Reusing soy residue for the solid-state fermentation of *Ganoderma lucidum*, *Bioresource Technology*, 91(1): 105-109.

Jabatan Perkhidmatan Haiwan (2003). *Manual Perternakan Lembu Fidlot.* Malaysia Ministry of Agriculture.

Jabatan Perkhidmatan Veterinar Negeri Pulau Pinang (2001). *Garis Panduan Penternakan Khinzir Negeri Pulau Pinang.* Penang State Government. Malaysia.

Karim, K., Hoffmann, R., Klasson, K. T. and Dahhan, M. H. (2005). Anaerobic digestion of animal waste: effect of mode of mixing, Water Research 39, 3597-3606.

Kiely, P.V. (1985). Anaerobic digestion of slurry from cattle fed high roughage diet, Proceedings of the Symposium under EEC programme on recycling urban and industrial waste, 251-258.

Lim, H. K. (1967). Animal feeding stuffs compositional data of feeds and concentrates, *The Malaysian Agricultural Journal*, 46(1): 63-70.

Lo, K.V., Bulley, N.R., Liao, P.H. and Whitehead, A.J. (1983). The effect of solid-separation pretreatment on biogas production from dairy manure. *Agricultural Wastes,* 8: 155-165.

Noike, T., Takabatake, H., Mizuno, O. and Ohba, M. (2002). Inhibition of hydrogen fermentation of organic wastes by lactic acid bacteria, *Int. J. Hydrogen Energy*, 27: 1367-1371.

Rodriguez, L. and An, B.X. (1997). Installation and performance of low cost polyethylene tube digesters on small scale farms. *World Animal Review,* 88: 38 – 47.

Shyam, M. and Sharma, P.K. (1994). Solid state anaerobic digestion of cattle dung and agroresidues in small capacity field digester, *J. Biores. Technol.*, 48:203–207.

Sung, S. and Liu, T. (2003), Ammonia inhibition on thermophilic anaerobic digestion. *Chemosphere*, 53: 43-52.

Symons, G. E. and Bushwell, A. M. (1933). The methane fermentation of carbohydrate, *Journal of the American Chemical Society*, 55: 2028-2039.

Van Der Vlugt, A. J., and Rulkens, W. H. (1985). Biogas production from a domestic waste fraction, Proceedings of the Symposium under EEC programme on recycling urban and industrial waste, 245-250.

Zábranská, J., Dohányos, M., Jeníček, P. and Kutil, J. (2000). Thermophilic process and enhancement of excess activated sludge degradability—two ways of intensification sludge treatment in Prague central wastewater treatment plant, *Water Sci. Technol.* 41: 265–272.

Pyruvic acid is a valuable chemical intermediate in the production of fine chemicals used by agrochemical, pharmaceutical, and food industries. In 2005, world-wide production of pyruvic acid was estimated at 11 million metric tons at a bulk price of $9-11/Kg. Current technology for production of pyruvate is based on conversion from tartaric acid and results in undesirable byproducts. An enzymatic approach to production of pyruvate was developed by cloning the glycolate oxidase (GO) gene from spinach into *Pichia pastoris* (1). GO is a flavoprotein (FMN dependent) which catalyzes the oxidation of lactate to pyruvate with a stoichiometric production of hydrogen peroxide. Hydrogen peroxide can lower GO activity and create non-catalytic byproducts; hence, catalase was also cloned into *P. pastoris* to create a double-transformant designated NRRL Y-21001.

Process development work was performed at the University of Iowa's Center for Biocatalysis and Bioprocessing. High density, double recombinant, *P. pastoris* fermentation (100 g cells/L) was achieved at the 30 L scale. After fermentation, these cells were treated with a proprietary process (2) to enable whole-cell biocatalysis and increase enzyme activity (85 U/g cells‡ for GO, 200 U/g cells for catalase). Current efforts have focused on optimizing these catalytically competent cells in a reaction to generate pyruvate at a high enzyme to product ratio.

Reaction optimization was done at the one-liter scale with a DASGIP Fed-batch Pro system. The DASGIP system contains four parallel reactors with control of temperature, pH, and dissolved oxygen. Oxygen enrichment at large scale would be costly, so air was bubbled through the reaction medium at 1 L/min under maximum stirring. Operating temperatures of 18 °C and 25 °C were tested along with pH values ranging from 5 to 8. Other key variables included: substrate loading, conducting the reaction in water, minimizing enzyme concentration, and maximizing number of recycles. Lactate loading ranged from 0.5 to 4 M, and significant substrate inhibition was observed above 1 M. Based on the above studies, optimum performance was achieved in water at pH 7 with an operating temperature of 18 °C and 1 M substrate loading. Enzyme loading was at 15 g/L for the first 3 cycles. Subsequently, 2 g/L additional enzyme was added every alternate cycle to achieve 10 cycles. Under the above conditions, >90% conversion of lactate to pyruvate was achieved in approximately 11 hours. Based on these parameters, an enzyme to product ratio of 1:38 was achieved.

‡ Enzyme activity needed to catalyze transformation of 1 μmol of substrate per minute.

1. Payne, M. S., Petrillo, K. L., Gavagan, J. E., DiCosimo, R. Wagner, L. W., Anton, D. L., Engineering *Pichia pastoris* for biocatalysis: co-production of two active enzymes, Gene. **194**, 179 (1997).
2. Subramanian, V., Das, S., Glenn, J. H., Patent Pending.

571av Production of Three Tyrosinase Inhibitors, Anthocyanin, Pro-Anthocyanin and Ellagic Acid, In Grape CELL Culture

Chien-Kuo Wang[1], C.K. Hsu[2], M.C-. Mong[2], Y.C. Chung[3], and M.C. Hsu[1]. (1) Department of Biotechnology, Asia University, 500 Liu-Fong Rd, Wu-Fong Shiang, Taichung, Taiwan, (2) Department of Health and Nutrition Biotechnology, Asia University, 500 Liu-Fong Rd, Wu-Fong Shiang, Taichung, Taiwan, (3) Department of Food and Nutrition, Providence University, 200 Chung Chi Rd, Taichung County, Taiwan

It has been shown that grape exhibits antioxidant activity, anti-inflammitory activity and immune modulation function. We are conducting a research project funded by the National Science Council (NSC), Taiwan to determine the biological functions of a grape cell culture, FC-01. Currently, the tested functions focus on the inhibitory effect toward tyrosinase activity.

The FC-01 grape cell culture was incubated under light in GB-5 medium with 30 g/l sucrose. NAA (0.1 mg/l) and Kinetin (0.2 mg/l) were used as growth regulators.

The extract of FC-01 showed inhibitory effect toward the activity of mushroom tyrosinase. When compared to the ethanolic extract of grape, the extract of FC-01 showed higher inhibitory effect than the extracts from the peel, seed and flesh of grape. The concentration of three tyrosin inhibitor, proanthocyanin, and ellagic acid in FC-01cell culture were 5.7 ug/ml, 10.3 ug/ml and 12.1 ug/ml respectively. For the future task, increasing the production of these three compounds and resveratrol, another important tyrosinase inhibitor, in FC-01cell culture by changing medium contents and environmental factors and the interaction between the active compounds and tyrosinase in the cell need to be evaluated.

571aw Regulation of Intrinsic Multiple Antibiotic Resistance In *Escherichia Coli* and *Salmonella*

Lon Chubiz, Chemical and Biomolecular Engineering, University of Illinois, 600 South Matthews Ave., 211 Roger Adams Laboratory, Box C-3, Urbana, IL 61801 and Christopher V. Rao, Department of Chemical Engineering, University of Illinois, 600 South Matthews Ave., 211 Roger Adams Laboratory, Box C-3, Urbana, IL 61801.

Bacterial resistance to antibiotics has been a persistent problem in clinical and public health situations for decades. While most bacteria acquire antibiotic resistance via genes encoded in mobile genetic elements such as plasmids and transposons, many bacterial species possess an intrinsic mechanism for resistance to antibiotics, organic solvents, oxidative stressors, and household disinfectants. In the enteric bacteria Escherichia coli and other closely related bacterial species, such as pathogenic forms of E. coli and Salmonella enterica, this resistance is mediated by the activation of chromosomally encoded efflux systems, cytoplasmic reducing enzymes, and metabolic enzymes. Governing the response to these stresses are transcriptional regulatory proteins of the multiple antibiotic resistance (mar) and superoxide (sox) regulons. Although these networks respond to different environmental queues, the mar and sox systems are known to have interlocked positive and negative feedback loops that allow for coordinated, rapid response to a variety of antibiotics. In this work, we investigate how these feedback loops coordinate antibiotic resistance. Through the combined use of experimental and computational approaches, we demonstrate that the network topology of the interlocked regulons provides a robust mechanism for antibiotic resistance over a wide range of concentrations and in response to dynamic fluctuations. These results may aid in the development of new therapeutic approaches for dealing with multidrug resistance in bacterial pathogens.

571ax Optimization of Monoclonal Antibody Production Using Process Simulation and Production Scheduling Tools

Victor Papavasileiou[1], Charles Siletti[2], and Demetri Petrides[2]. (1) Intelligen Europe, Rijnsburgerweg 119, Leiden, 2334, Netherlands, (2) INTELLIGEN, INC., 700 Walton Ave., Mount Laurel, NJ 08054

The new generation of monoclonal antibodies that are required in large quantities (hundreds of kilograms per year) present a number of manufacturing challenges to the biotech industry. Such products typically require facilities with multiple upstream and downstream suites. Furthermore, such biopharmaceuticals are often produced in multi-product manufacturing facilities. In such environments, the multiple production lines interact with each other through sharing of labor, utilities (e.g., steam, purified water), auxiliary equipment (e.g., CIP skids, transfer panels), buffer preparation tanks, etc. Occasionally, they also share rooms and other work areas. Sharing of resources, however, leads to scheduling bottlenecks and headaches that limit the throughput of the entire facility. Our experience in using process modeling and scheduling tools to improve the design of new biopharmaceutical facilities and the operation of existing ones will be presented using an industrial case study.

Web Page: www.intelligen.com/literature

571ay Urea-Induced Dissociation of Non-Native Aggregates of α-Chymotrypsinogen: Kinetics, Thermodynamics, and Competing Pathways

Rebecca K. Brummitt and Christopher J. Roberts. Department of Chemical Engineering, University of Delaware, 150 Academy Street, Newark, DE 19716

Non-native aggregation is problematic during the development of protein-based pharmaceuticals, and is commonly implicated in several neurodegenerative diseases. A comprehensive understanding of the specific interactions that lead to the stabilization or destabilization of protein aggregates relative to free monomers is important to help alleviate these issues. This work focuses on the effects of urea on the dissociation of pre-formed aggregates of α-Chymotrypsinogen A. The primary experimental results are the kinetics of monomer recovery and changes in aggregate structure, size, and morphology during urea-promoted dissociation as a function of initial aggregate size, overall protein concentration, urea concentration, and salt concentration. Biphasic kinetics are observed where extent of monomer recovery and aggregate dissociation is primarily controlled by the final urea concentration and the initial size of the aggregates. Final monomer recovery is independent of the overall protein concentration, indicating kinetic control of the process. Light scattering measurements indicate aggregate weight-average molecular weight decreases, morphology is conserved, and polydispersity increases during dissociation. A number of candidate models are presented and evaluated to quantitatively and qualitatively capture all of the experimental results.

Engineering hyper-crystallizable single chain antibodies for crystallization scaffold

Jennifer C. Pai[1] and Jennifer Maynard[1], (1)Department of Chemical Engineering, University of Texas at Austin, Austin, TX, 78752

Crystallization of protein has been limited by expression and purification of the protein as well as the identification of conditions that promote crystal growth. Using protein engineering, the protein can become the variable in crystallography thereby offering a promising new pathway to the production of high-quality crystals. Varying amino acids result in altering the biophysical parameters of the protein which include, but are not limited to, expression level, ligand binding affinity, stability and immunogenicity. Manipulation of amino acid sequence was done with both directed and random evolution with the purpose of creating a hyper-crystallizable single chain antibody.

Genes for a hyper-crystallizable single chain antibody (scFv/scAb), were cloned and expressed in E. coli bacteria. The antibody was engineered with the capability of crystallizing in multiple space groups and lattice structures and intended for the development of a crystallization scaffold. The utility of this protein is its specificity and binding for a peptide tag (EE tag; amino acid sequence EYMPME), which can be appended to any protein of interest. CDRs with this EE-tag specificity have been grafted onto frameworks of previously crystallized antibodies. Biophysical properties of these constructs were characterized and recorded. The antibodies were evolved using mutator cells with multiple rounds of growth, selecting for variants with improved biophysical characteristics correlating with crystallizability. Directed evolution was performed to preserve amino acids that were significant in maintaining crystal lattice contacts. *CCP4 program suite for protein crystallography* was used to derive locations for directed evolution. These locations were selected based on surface area energetic and crystallographic work on the original antibodies from which the grafts derived their framework. Variants selected by phage display had significant EE-tag binding affinity, soluble expression and high stability. The engineered single chain antibodies produced were tested for their ability to crystallize independently using the Phoenix robot at the University of Texas.

The generation of a hypercrystallizable antibody will allow for a readily formed complex with any second protein of interest by simply attaching the EE-tag. Its use can then be extended to create scaffolds aiding the crystallization of proteins that are otherwise difficult to crystallize, producing a general engineering approach to creating crystalline proteins. This work allowed better understanding of the correlates for crystallization from biophysical parameters and energetics of proteins. The ability to crystallize biomolecular structures will expand our ultimate goal of understanding and manipulating proteins for improved properties and therapeutic treatments. Long term goals are the development of crystal based separation processes; potentially allowing us to separate out distinct proteins of interest regardless of its size.

571ba Development and Application of High-Throughput Flow-Cytometric Techniques for Analyzing Prokaryotic Cell Cultures

Bryan P. Tracy, Chemical and Biological Engineering, Northwestern University, 2145 Sheridan Road, Evanston, IL 60208 and Eleftherios Terry Papoutsakis, Dept. of Chemical Engineering & Delaware Biotechnology Institute, University of Delaware, 15 Innovation Way, Newark, DE 19711.

The improvement of microorganisms for specialty chemical production, bioremediation, and biofuel generation, remains an immediate scientific and industrial goal. Particularly for bacteria, the goal of converting low value biomass into transportation fuels (ethanol and butanol) has motivated a tremendous amount of bacteria strain development. Subsequently, forward genetics and genome scale approaches are gaining considerable popularity for accelerating the discovery phase of strain development. A popular approach is to generate mutant, plasmid, or species libraries, and then screen for desirable phenotypes via environmental stress enrichment. Although often successful, these approaches are flawed by single parameter detection (fitness), and are typically not high-throughput. Flow cytometry epitomizes high-throughput and is ideal for multi-parametric detection of individual cells. Unfortunately, flow cytometric analysis of prokaryotic organisms is not as developed as for eukaryotes.

Thus, we developed and adapted multiple flow cytometry assays for the analysis of bacterial cell cultures. To accomplish this task, we employed the bacterium, *Clostridium acetobutylicum. C. acetobutylicum* exhibits a well known cycle of cellular differentiation consisting of at least four unique morphologies. Our goal was to develop an assay capable of identifying all four morphologies. Through light scattering and a modified *Bac*Light™ (Invitrogen) DNA staining, we definitively accomplish this task and also demonstrate the efficacy of an alternative approach via pencillin-binding-protein (PBP) detection. In general, we demonstrate the high-throughput and multi-parametric capabilities of these assays. Additionally, we demonstrate their application in fluorescence assisted cell sorting, generating enriched populations for electron microscopy analysis, and analyzing cultures in a manner previously not possible. In future work we hope to utilize these and other assays in screening and sorting mutant libraries for desirable phenotypes, for which environmental stress alone could not enrich.

Web Page: www.papoutsakis.org/page.php?id=63

571bb Construction and Analysis of Cell Population Heterogeneity of Negative Feedback Gene Auto-Regulatory Systems In Escherichia Coli

Mai Li[1], Jiangfeng Zhu[1], Rebecca Simonette[2], Mary L. Harrison[2], Stephanie Portle[3], George N. Bennett[2], and Ka Yiu San[4]. (1) Department of Bioengineering, Rice University, Houston, TX 77005, (2) Biochemistry and Cell Biology, Rice University, Houston, TX 77005, (3) Rice University, Chemical and Biomolecular Engineering Department, Houston, TX 77005, (4) Department of Bioengineering, Department of Chemical and Biomolecular Engineering, Rice University, Houston, TX 77005

The accumulation of genetic information and the development in molecular biologic techniques enable us to design and construct the artificial gene regulatory networks. Since there is a tight relationship between the regulation of gene expression and cell population heterogeneity, analysis of cell population heterogeneity of well-designed artificial gene regulatory networks can provide insights into the underlying mechanisms of gene regulation mechanisms. In the current work we are interested in examining the role of gene network structure of modified lacP/O systems on gene expression patterns, particularly the degree of heterogeneity within a population as indicated by a reporter protein. We are designing and constructing negative feedback gene regulatory systems by placing the gene encoding the green fluorescent protein (gfp), which serves as a reporter, and the gene encoding the LacI repressor protein under the control of the same or different lac promoter systems in Escherichia coli. We are also investigating the difference of cell population heterogeneity between the systems with one or two lac operators by monitoring the GFP expression at the single cell level using a flow cytometer. Results from chemostat experiments of strains carrying various genetic constructs will be presented and discussed.

571bc Three-Dimensional Tubular Porous Materials for Directed Cellular Growth

Jung Hun Song, Department of Chemical Engineering, Graduate Center and City College of City University of New York, 140th St. at Convent Ave., New York, NY 10031, Raymond Tu, Chemical Engineering, City College of New York (of CUNY), T313 Steinman Hall, 140th St at Convent Ave, New York, NY 10031, and Ilona Kretzschmar, Department of Chemical Engineering, The City College of City University of New York, 140th St. at Convent Ave., New York, NY 10031.

Scaffolds have been of great interest to the field of tissue engineering for their applicability in cell culturing and cell delivery. Especially, the topographical features of 3D scaffolds actively regulate the growth of many types of cells and promote organization of cell clusters. Biodegradable and synthetic porous polymeric materials used in biological scaffolding offer superior flexibility and compatibility for directed cellular growth. Our work is concerned with the development and study of cell attachment and propagation in tubular polymeric scaffolds. The effect of a highly periodic pore structure and the size of the pores are the focus of this study.

We have assembled 10 and 16 μ m sized sulfate-polystyrene (PS) colloids in 105 and 360 μ m inner-diameter polymethylmethacrylate (PMMA) capillary templates. Fabrication of a porous material from this template is achieved by infiltrating biodegradable and synthetic polymeric materials into the interstitial spaces of the close packed colloidal crystal. Each polymeric solution is subsequently polymerized and cross-linked leading to a colloid-polymer matrix. Matrix is treated with a sequence of solvents to remove both the PMMA capillary and the PS colloids forming tubular porous 3D materials.

We have investigated the structures of the 3D porous materials using scanning electron microscopy. Additionally, initial compatibility tests with various types of cells are presented.

571bd Effects of Nitric Oxide on Growth and Terpenoid Indole Alkaloid Production In Catharanthus Roseus Hairy Root Cultures

Mai Li[1], Christie A.M. Peebles, Jacqueline V. Shanks[2], and Ka Yiu San[3]. (1) Department of Bioengineering, Rice University, Houston, TX 77005, (2) Chemical and Biological Engineering, Iowa State University, 2114 Sweeney Hall, Ames, IA 50011, (3) Department of Bioengineering, Department of Chemical and Biomolecular Engineering, Rice University, Houston, TX 77005

Nitric oxide is known as a signal molecule involved in elicitor-induced defense responses of plants. Sodium nitroprusside (SNP) can be used as the donor of nitric oxide. It has been reported SNP stimulated catharanthine formation in Catharanthus roseus cells. Two important terpenoid indole alkaloids produced in small quantities within C. roseus are vinblastine and vincristine which are being used clinically as anticancer drugs. We are interested in engineering C. roseus hairy roots to increase the production of the TIAs. In the present study, we explored the effects of feeding the hairy root cultures of different concentration of SNP. The alkaloid levels were analyzed by sampling 9, 14, 17, 20, 23, 26 and 30 days after feeding of SNP on day 0, and mRNA expression levels of some important genes were investigated. We also studied the transient effects of the feeding of SNP during the exponential phase of the C.roseus hairy roots.

571be Optimizing An RF Plasma Reactor for Biocompatible "Smart" Surfaces

Adrianne E. Lucero and Heather E. Canavan. Chemical and Nuclear Engineering, University of New Mexico, 1 University of New Mexico, MSC 01 1120, Albuquerque, NM 87131

Poly(N-isopropyl acrylamide) (pNIPAM) undergoes a sharp property change in response to a moderate thermal stimulus at physiological temperatures. This has generated great interest in the biomaterials community, and pNIPAM is being investigated as a "smart" biofouling and release coating to harvest intact cell monolayers. Currently, many techniques are used to deposit pNIPAM, including electron beam irradiation and solution deposition (e.g., silanes and self-assembled monomers). Recently, we constructed a radio frequency (rf) plasma reactor for plasma polymerization of NIPAM from the vapor phase based on a previous design. Plasma polymerization is a sterile, solvent-free, and compatible with surfaces of any geometry or chemistry. These factors make plasma polymerization extremely useful for cell and tissue culture, which often rely on plastic tissue culture plates. Due to the inherently energetic conditions of the plasma, it is extremely important to characterize the impact that parameters such as maximum rf wattage, pressure of deposition, and location/position of the samples in the chamber have on the resulting films. In this work, pNIPAM films resulting from those varying conditions are characterized using X-ray photoelectron spectroscopy (XPS) for film composition, interferometry for film thickness, contact angles for thermoresponse, and cell detachment for cell releasing properties.

571bf Characterizing Deposition and Culture Conditions for Optimal Cell Release Kinetics from Smart Polymer

Jamie A. Reed and Heather E. Canavan. Chemical and Nuclear Engineering, University of New Mexico, Albuquerque, NM 87131

Methods traditionally used to detach cells from surfaces destroy the integrity of the extracellular matrix (ECM). This, in turn, damages proteins and receptors on the cell. Poly(N-isopropyl acrylamide), or pNIPAM, is a smart polymer that has become of great interested in the bioengineering community because it has been investigated as a non-destructive cell release method. Above its lower critical solution temperature (LCST), 32C, the polymer is relatively hydrophobic, allowing cells to grow and proliferate just like they do on tissue culture polystyrene (TCPS). Below the LCST the polymer physically changes, becoming hydrophilic and swelling causing cells to spontaneously release. In this work, we investigate the optimization of pNIPAM films for rapid cell release. The pNIPAM films were deposited using a novel solution deposition method (spNIPAM) as well as plasma deposition (ppNIPAM). Characterization of the pNIPAM substrates was done utilizing traditional surface chemistry techniques (e.g., X-ray photoelectron spectroscopy, XPS; interferometry; and contact angle measurements). Bovine aortic endothelial cells (BAECs) were then seeded onto the coated surfaces and allowed to grow to confluence. At confluence, the media used to culture cells was exchanged for a solution at 4, 25, or 37C to determine which temperature and solution combination demonstrated the most rapid cell release. The cells were then observed using light microscopy and the time needed for 100% cell release was recorded. From these results, we conclude that the fastest, most reliable release of BAECs occurred using 4C serum free media (SFM).

571bg Use of SBA-15 to Enhance the Response Range of Microparticle Based Glucose Sensors

Mike McShane and **Saurabh Singh**. Biomedical Engineering, TAMU, Zachry Engineering Center, College Station, TX 77843

Microparticle-based glucose sensors with tunable response properties have been developed, which use an enzymatic reaction scheme and oxygen sensitive lumophores to transduce glucose levels to phosphorescence intensity. The sensing chemistry is immobilized in alginate-silica ("algilica") microparticles and Layer-by-Layer (LbL) technique is used to form thin nanofilm coatings with precise thickness. The objective is to use these sensors for continuous, non-invasive glucose monitoring following implantation in the dermis. As our sensors are based on enzyme kinetics, their performance depends on reaction kinetics and the transport rates of the analytes (i.e., glucose and oxygen), which can be adjusted by varying the thickness and material properties of the ultrathin film coating. More specifically, the response range has been shown to increase through deposition of thicker films on the microparticles. Nevertheless, a 10-bilayer PAH/PSS film with 0.2 M NaCl (~ 40 nm thick) resulted in a response range of 0 – 200 mg/dL, which does not cover the full clinical range for glucose monitoring (0 – 600 mg/dL). Another route of increasing the response range of such sensors is to increase the rate of oxygen diffusion. This can be accomplished by using mesoporous silica microparticles (SBA-15), for which oxygen diffusion is much higher than algilica. Simulations show that using silica microparticles instead of algilica microparticles can increase the range of these sensors up to 600 mg/dL. To validate the estimates obtained from modeling, experiments will be performed and results will be presented in this paper.

571bh Control of Alpha Synuclein Aggregation Using a Peptidyl Modulator

Michael Hernandez and **Jin Ryoun Kim**. Chemical and Biological Engineering, Polytechnic University, 6 MetroTech Center, Brooklyn, NY 11201

Aggregation of a peptide, alpha synuclein, into beta-sheet structured amyloid has been believed to be implicated in the pathology of Parkinson's disease. A considerable amount of data has identified the soluble alpha synuclein oligomers as potentially significant toxic agents. However, the possibility that the alpha synuclein fibrils may be associated with neurotoxicity cannot be ruled out, since fibrillar aggregates can serve as a pool of soluble intermediate species through a dynamic exchange with monomers or oligomers. Development of compounds controlling alpha synuclein aggregation would be greatly beneficial in not only reducing toxic effects but also developing accurate functional correlations between an aggregation state and toxicity. We have designed a peptidyl aggregation modulator composed of divalent "binding domain" and "conformational domain". The binding domain contains an amino acid sequence responsible for binding toward alpha synuclein. The conformational domain was added to prevent self assembly of binding domains. A peptidyl modulator was tested in the recently developed Escherichia coli expression system, where the effects of aggregation modulators on alpha synuclein amyloidogenesis can be examined in vivo. Our results indicated that the addition of the designed peptidyl modulator caused conversion of insoluble alpha synuclein aggregates to small soluble species in vivo. Examination of alpha synuclein aggregation in vitro in the presence of the peptidyl modulator is underway.

571bi Production of C4 Compounds by Metabolic Engineered *Escherichia Coli*

Irene Martinez[1], Amanda Lee[1], George N. Bennett[2], and Ka Yiu San[3]. (1) Bioengineering, Rice University, 6100 Main Street MS 142, Houston, TX 77005, (2) Biochemistry and Cell Biology, Rice University, 6100 Main st, Houston, TX 77025, (3) Department of Bioengineering, Department of Chemical and Biomolecular Engineering, Rice University, Houston, TX 77005

Several C4 compounds, such as malate, succinate and fumarate, are widely used by the industry to produce synthetic resins, food additives, pharmaceuticals and other chemicals.Currently, their production is mainly accomplished by chemical processes using petroleum derivatives as starting materials, these processes could lead to pollution and to a potential shortage in the future. On the other hand, the bacteria *Escherichia coli* synthesize some of these compounds naturally but with poor product yields. We will present metabolic engineering approaches to modify *E. coli* metabolic pathways to increase product yields with the purpose of developing a fermentation process from renewable feedstocks ("green technology") with the potential to replace the traditional petrochemical processes.

571bj Genetic Engineering Hydrocarbon Production in Algae

Stephen Gabauer[1], Thomas D. Niehaus[2], Waqas Khatri[3], Rachel Arndt[1], Joeseph Chappell[2], and Wayne Curtis[1]. (1) Chemical Engineering, The Pennsylvania State University, 232 Fenske Laboratory, University Park, PA 16802, (2) Department of Plant and Soil Sciences, University of Kentucky, 301B Plant Sciences Building, 1405 Veterans Drive, Lexington, KY 40546-0312, (3) Industrial Engineering, The Pennsylvania State University, 232 Fenske Laboratory, University Park, PA 16802

Algae can be used to capture and reduce CO_2 to hydrocarbons using photosynthetic energy. While our work with the native hydrocarbon producing algae strain *Botryococcus braunii* is presented in a different session (#202d), this poster will focus on the genetic engineering strategy to move this hydrocarbon synthetic pathway to the model algae host *Chlamydomonas reinhardtii*. This approach to introducing heterologous isoprenoid metabolism includes manipulating fluxes in both the cytoplasm and chloroplast. Transformation of *C. reinhardtii* can be accomplished with either brute force 'bead beating' or the plant vector *Agrobacterium tumafaciens* although the latter preliminarily appears to be less efficient. Future work will compare these alternative transformation approaches using our Agrobacterium auxotroph system developed for gene delivery to plants (as described in a different talk at this meeting; #194b). This work will ultimately allow for a productivity comparison of transgenic lines to natural isolates for hydrocarbon productivity.

571bk Genetic Analysis of G Protein-Coupled Receptor Expression In Escherichia Coli

Georgios Skretas and George Georgiou. Chemical Engineering, University of Texas at Austin, 1 University Station C0400, Austin, TX 78712

The overexpression of G Protein-Coupled Receptors (GPCRs) and of many other heterologous membrane proteins in simple microbial hosts, such as the bacterium Escherichia coli, often results in protein mistargeting, aggregation into inclusion bodies or cytoplasmic degradation. Furthermore, membrane protein production is very frequently accompanied by severe cell toxicity. In this work, we have employed a genetic strategy to isolate E. coli mutants that produce markedly increased amounts of the human central cannabinoid receptor (CB1), a pharmacologically significant GPCR that expresses very poorly in wild-type E. coli. By utilizing a CB1 fusion with the green fluorescent protein (GFP) and fluorescence-activated cell sorting (FACS), we screened an E. coli transposon library and identified an insertion in dnaJ that resulted in a large increase in CB1-GFP fluorescence and a dramatic enhancement in bacterial production of membrane-integrated CB1. Furthermore, the dnaJ::Tn5 inactivation suppressed the severe cytotoxicity associated with CB1 production. This revealed an unexpected inhibitory role of the chaperone/co-chaperone DnaJ in the protein folding or membrane insertion of bacterially produced CB1. Our strategy can be easily adapted to identify expression bottlenecks for different GPCRs or any other integral membrane protein, provide useful and unanticipated mechanistic insights, and assist in the construction of genetically engineered E. coli strains for efficient heterologous membrane protein production.

571bl Metabolic Engineering of Rhizopus Oryzae for Enhanced Production of Fumaric Acid from Sugars

Baohua Zhang and Shang-Tian Yang. Department of Chemical and Biomolecular Engineering, The Ohio State University, 140 West 19th Avenue, Columbus, OH 43210

Fumaric acid is a dicarboxylic acid used extensively in resins, food acidulants, and other applications including oil field fluids, esters etc. Some strains of Rhizopus oryzae, which is a filamentous fungus, can produce fumaric acid from sugars under aerobic conditions but the fumarate yield is relatively low due to the co-production of ethanol and other fermentation byproducts. In this work, metabolic engineering technique is applied to optimize the fermentation for enhanced fumaric acid production. More fumaric acid can be produced from sugars when the alcohol dehydrogenase and pyruvate decarboxylase genes are inactivated. Fumaric acid production can also be increased by overexpressing the enzymes, such as fumarase and pyruvate carboxylase, in the fumaric acid production pathway. These genes were obtained from Rhizopus oryzae genome by using PCR, and then cloned into the fungal cells by biolistic transformation. The effects of transformation on the mutants and fumaric acid fermentation kinetics were studied and will be reported in this paper.

571bm Engineering of n-Butanol Production in Bacteria

David R. Nielsen[1], Effendi Leonard[2], Sang-Hwal Yoon[1], Hsien-Chung Tseng[3], and Kristala L. Jones Prather[4]. (1) Department of Chemical Engineering, Massachusetts Institute of Technology, 77 Massachusetts Ave., 66-425, Cambridge, MA 02139, (2) Chemical Engineering, Massachusetts Institute of Technology (MIT), 77 Massachusetts Ave, Cambridge, MA 02139, (3) Chemical Engineering, Massachusetts Institute of Technology, 77 Massachusetts Ave., 66-425, Cambridge, MA 02139, (4) Chemical Engineering, MIT, 77 Massachusetts Ave, Cambridge, MA 02139

Amid rising fuel costs, emergent environmental issues, and mounting international security concerns, considerable attention has recently returned to reviving biologically derived n-butanol as a renewable transportation fuel alternative. n-Butanol boasts physical and thermodynamic properties which are akin to gasoline and are in many ways superior to those of biologically derived ethanol. The natural n-butanol fermentation pathway in *Clostridium* competes with several other native pathways for metabolite fluxes which can thereby reduce its final yield. Furthermore, the cytotoxic effects of n-butanol contribute to low titers in natural organisms due to feedback inhibition. Accordingly, there is much interest in designing robust, alternative microorganisms capable of producing n-butanol. Through metabolic engineering efforts we have explored the potential for n-butanol production in a variety of bacterial hosts, including *E. coli, P. putida,* and *B. subtilis*. Within these hosts, we have explored both the reconstruction of the natural *Clostridium* pathway as well as the use of homologous genes from a variety of source organisms. The development of alternative hosts for the heterologous n-butanol pathway reveals important implications for both production and product tolerance.

Optoelectronic enzymatic biosensors (OEB) were constructed with immobilized whole cells of *E. coli* pGELAF+ and were used to sense dichloroethane (DCA) in concentrations of parts per billion. The sensor physical characteristics are well understood to be mass transfer limited; therefore OEBs of different volumes of immobilized matrices (0.5μL and 6μL) were constructed and tested. This work is the first step of a complete evaluation and subsequent formulation of an iterative experimental modeling approach for the sensing system used in the Reardon laboratory. Prior model formulation suggests mathematical forms for reaction and diffusion of substrate through the immobilized matrix and a pH dependant quenching mechanism of the immobilized fluorophore. Two modeling strategies are evaluated as an extension to prior modeling efforts; 1) mixed regression (i.e. predictor-response methodology) upon sensor response and 2) a mechanistic approach cast in a symmetric reaction-diffusion equation form. Mixed regression was performed with; optode sensitivity, matrix size, days after biosensor construction, and photomultiplier setting during sensor testing. The following sensing mechanism was formulated; diffusion through a stagnant boundary layer about the cell matrix (I), diffusion through the immobilization matrix to a dehalogenase active site (II), analyte reaction with the dehalogenase (III), diffusion of resultant protons from the dehalogenase to the immobilized fluorophore (IV) and fluorophore quenching (V). Mathematical analysis techniques were developed in Polymath and MATLAB (with the statistics and curve fitting toolboxes).

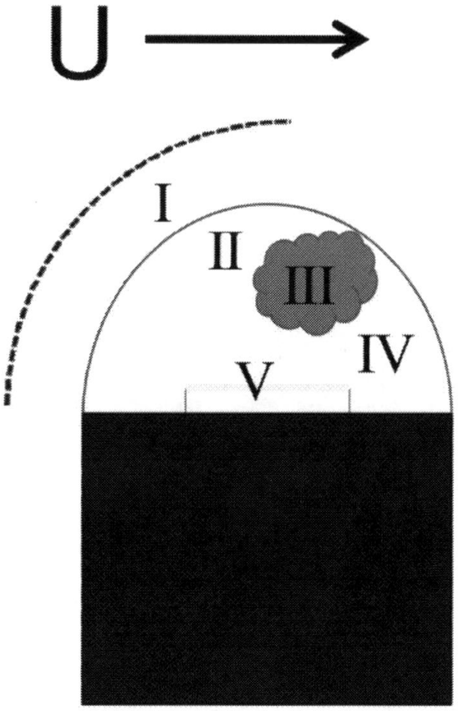

571bo Chemically Selective Displacement Chromatography: Development and Application for Bioseparations

Christopher J. Morrison[1], Sun Kyu Park[2], Scott McCallum[3], Rahul Godawat[3], Shekhar Garde[4], James Moore[2], and Steven M. Cramer[5]. (1) Chemical and Biological Engineering, Rensselaer Polytechnic Institute, 110 8th St., Ricketts Building, Troy, NY 12180, (2) Chemistry and Chemical Biology, Rensselaer Polytechnic Institute, 110 8th St., Troy, NY 12180, (3) Rensselaer Polytechnic Institute, 110 8th Street, Troy, NY 12180, (4) Department of Chemical and Biological Engineering, Rensselaer Polytechnic Institute, Ricketts Building, 110 8th St, Troy, NY 12180, (5) Department of Chemical and Biological Engineering, Rensselaer Polytechnic Institute, Ricketts Bldg., Troy, NY 12180

Previous studies have demonstrated that chemically selective displacement chromatography can be employed to add additional dimensions of selectivity to preparative ion exchange separations. Robotic high throughput screening was employed in concert with several analytical techniques (STD-NMR, SPR and MD Simulations) to identify and evaluate the behavior of chemically selective displacers for protein purification. These results show that selective displacement is achieved by incorporating a binding affinity for the targeted molecule into the design of the chemically selective displacer while also retaining the displacer's affinity for the resin. From these results fluorescent displacers were rationally designed and tested in order to facilitate online detection. Finally, a novel approach was developed for applying selective displacement for the removal of key industrial impurities from product streams. This work demonstrates the ability of chemically selective displacement chromatography to provide high resolution separations while simultaneously removing a targeted molecule, thus eliminating further downstream processing steps.

Web Page: www.rpi.edu/dept/chem-eng/WWW/faculty/cramer/

571bp a Novel Yeast-Based Mutagenicity Screen

Michael Benton, Cain Department of Chemical Engineering, Louisiana State University, South Stadium Drive, Rm 110, Baton Rouge, LA 70803, Nathaniel R. Glasser, Chemical and Biological Engineering, University of Wisconsin - Madison, 1415 Engineering Drive, Madison, WI 53706, and Sean P. Palecek, Chemical and Biological Engineering, University of Wisconsin-Madison, 1415 Engineering Drive, Madison, WI 53706.

Many industrially relevant chemicals have incomplete mutagenicity or cytotoxicity data available. Current testing methods are often expensive and tedious. As an alternative to traditional mutagenicity assays such as the Ames test or the SOS Chromotest, we have developed a yeast-based mutagenicity screen. We have genetically engineered S. cerevisiae cells such that the promoter of a DNA-damage specific gene is regulating expression of green fluorescent protein (GFP). When these recombinant strains are exposed to mutagenic conditions, GFP is produced and the cells fluoresce. Fluorescence can be observed qualitatively or measured quantitatively with flow cytometry or a fluorescent plate reader. We are able to detect multiple types of DNA damage including alkylating agents, oxidative agents, and ionizing radiation. Optimum results are available in 4-6 hours, a great improvement over other assays. Additionally we are able to detect lower mutagenic concentrations than traditional assays.

Engineering of Aglycosylated Antibody Fc for Selective Binding to Fc Gamma Receptors and Effector Function

Sang Taek Jung[1], Tae Hyun Kang[2], Inger Sandlie[3], Philip W. Tucker[2], and George Georgiou[1]. (1) Department of Chemical Engineering, University of Texas at Austin, University Station C0400, Austin, TX 78712, (2) Institute for Cellular and Molecular Biology, University of Texas at Austin, University Station C0400, Austin, TX 78712, (3) Department of Molecular Biosciences and Centre for Immune Regulation, University of Oslo, Oslo, Norway

In IgG molecules removal of N-linked glycan results in a conformational change that abolishes binding to Fc gamma receptor proteins on the surface of leukocytes and clearance of abnormal target cells. We have developed a robust flow cytometric screening system for the isolation of aglycosylated Fc mutants exhibiting high affinity to desired Fc*f*× receptors despite the lack of glycosylation. A series of complex libraries (107-x109 clones) were generated by random or insertional mutagenesis of the Fc domain and screened for binding to purified Fc gamma receptors. Mutant aglycosylated Fcs that bind with high affinity and specificity were isolated and characterized in vitro. The aglycosylated Fc gamma receptor binding mutant proteins contain between 1-11 amino acid substitutions. We found that the mutations allowed highly selective binding of an aglycosylated, anti-Her2 (trastuzumab) to the Fc*f*×RI receptor, with desired affinity indistinguishable from that of CHO-derived, glycosylated trastuzumab (Herceptin). New screening using this methodology and judiciously constructed new mutant library, ultra-high affinity aglycosylated antibody Fc mutant exhibiting much higher affinity to Fc*f*×RI than commercial glycosylated trastuzumab have been also isolated. For the isolated aglycosylated antibody Fc mutants, antibody dependent cell mediated cytotoxicity (ADCC) and antibody dependent cell mediated phagocytosis (ADCP) are being investigated.

572a Microarray Biochips for High-Throughput Drug Screening

Prashanth Asuri[1], Moo-Yeal Lee[1], Anand K. Ramasubramanian[2], Michael Hogg[1], Bilge Eker[1], Sana Butt[1], Jessica Ryan[1], Sumitra Sukumaran[3], Douglas S. Clark[4], and Jonathan S. Dordick[5]. (1) Solidus Biosciences, Inc., 1223 Peoples Avenue, Troy, NY 12180, (2) UC - Berkeley, Dept of Chemical Engineering, Berkeley, CA 94720 - 1462, (3) Chemical & Biological Engineering, Rensselaer Polytechnic Institute, Troy, NY 12180, (4) Chemical Engineering, University of California, Berkeley, Berkeley, CA 94720, (5) Chemical and Biological Engineering, Rensselaer Polytechnic Institute, 4005 Center for Biotechnology, Troy, NY 12180

Combinatorial chemistry and advances in genomics and proteomics has resulted in a dramatic increase in the number of new chemical entities (NCEs) and screenable drug targets. Nevertheless, these advances have not translated into an increased number of new drug approvals, in part because of the high failure rate due to toxicity of the NCE or its metabolite(s). Thus, Solidus Biosciences, Inc. has developed microarray biochip platforms that can adequately mimic human metabolism and assess cell-specific toxicity of NCEs and their metabolites. Initial work has focused on the development of the Metabolizing Enzyme Toxicology Assay Chip (MetaChip) suitable for high-throughput analysis of drug metabolism and toxicity. Human cytochromes P450 are encapsulated in an array of alginate-gel spots on (MTMOS) treated glass slides, followed by the addition of a drug or prodrug to be tested, and the metabolites of the P450 reactions are screened in situ for cytotoxicity on a cell monolayer. To expand our 2D cell monolayer culture for cytotoxicity assays to a more physiologically relevant and robust three-dimensional (3D) cell culture for both cytotoxicity and growth inhibition assays, we have developed a miniaturized 3D cellular array chip (Datachip) that can be used for in vitro toxicological assessment of NCEs and their P450-generated metabolites. Various human cell lines were encapsulated and cultured in an array of alginate-gel drops spotted on poly(styrene-co-maleic anhydride) (PSMA) treated glass slides. We envisage that these novel technologies represents a promising, high-throughput microscale alternative to conventional in vitro multi-well plate platforms and creates new opportunities for rapid and inexpensive assessment of ADME/Tox at very early phases of drug development, thereby enabling unsuitable candidates to be eliminated from consideration much earlier in the drug discovery process.

572b Systems Engineered for Cancer Research

Zoe N. Demou, Northwestern University, Children's Memorial Research Center, 2300 Children's Plaza, Box 204, Chicago, IL 60614

The study of cellular responses to endogenous mechanochemical stimuli in situ is often stymied by technical challenges. However, bioengineered in vitro models have simulated in vivo conditions and produced significant breakthroughs in our understanding of cellular and molecular processes. This presentation focuses on custom-made hardware and software designed to model specific aspects of tumor biology. These systems emphasize three-dimensional substrates, physiologic conditions for long-term cell culture, automation, and compatibility with time-lapse imaging. A fully automated system, which tracks 3D cellular motion in combination with statistical analysis of the cell trajectories, has enabled measurements of the migratory and invasive potential for various types of tumor cells. This system can be used as a screening tool for cancer motility and potentially therapeutic drugs. Time-lapse analysis also revealed that defects in collagen polymerization differentially affect the 3D migration of mammary carcinoma cells by enhancing their morphodynamic activity and promoting motility and invasion. Another model system, the 3D cell-pressor, was developed to simulate the stress built-up in a growing tumor and allowed quantification of the effects of mechanical compression on cancer cell morphology, metastatic potential, and in vivo aggressiveness. Another system, the time-lapse analysis chamber for laser capture microdissection (LCM), enables time-lapse topography of differentiating cell cultures and pre-selection of cell targets thus minimizing LCM time and optimizing the quality of isolated biomolecules for genomics and proteomics analyses. The first ever application of LCM in living 3D cultures, led to identification of molecules that could enable autopoiesis of vascular-like networks by melanoma cells with potentially significant clinical implications, such as response to anti-angiogenic treatments. Overall these systems, used independently or in tandem, provide discovery engines that enable novel insights into the interplay and the individual roles of tumor cells and mechanochemical components of the microenvironment in the process of cancer metastasis.

572c Trbp Binding Preferences for siRNA

Joseph A. Gredell[1], Michael J. Dittmer[1], and S. Patrick Walton[2]. (1) Department of Chemical Engineering and Materials Science, Michigan State University, 2527 Engineering Building, East Lansing, MI 48824, (2) Chemical Engineering and Materials Science, Michigan State University, East Lansing, MI 48824

RNA interference (RNAi) is a natural phenomenon resulting in potent and primarily specific gene silencing that occurs in most eukaryotes. It is initiated in cells by the presence of short interfering RNAs (siRNAs), which are double-stranded, non-coding RNA molecules ~21 nucleotides long. siRNAs guide the active enzyme complex of RNAi, the RNA induced silencing complex (RISC), to mRNAs possessing regions of sequence complementarity, leading to degradation of the target message. TAR RNA binding protein (TRBP), historically associated with binding the TAR structure present on all HIV-1 transcripts, was recently discovered to be a component of RISC. It has been proposed that TRBP facilitates the production of siRNAs by sensing the thermodynamic asymmetry of the siRNA, thus leading to incorporation of the guide strand into RISC, analogous to the role of R2D2 in the Drosophila silencing pathway. We will present our results detailing the precise structural and sequence requirements for TRBP recognition of siRNAs. The contributions of these features to the sensing of duplex asymmetry will also be described.

572d The Coordinated Control of EGFR Signaling by Multiple Feedback Loops: An Experimental and Computational Study In a Developing Tissue

Jeremiah J. Zartman and Stanislav Y. Shvartsman. Department of Chemical Engineering and Lewis-Sigler Institute for Integrative Genomics, Princeton University, Engineering Quadrangle, Princeton, NJ 08544

Hyperactivation of the EGFR pathway has been implicated in the majority of human epithelial cancers (1). Studying the role of EGFR signaling in developing tissues has provided a natural paradigm for controlling complex regulatory networks, and has provided new tools for the therapeutic treatment of cancer in the form of EGFR signaling antagonists (2). In particular, Drosophila oogenesis is a well established system for studying EGFR signaling feedback and is amenable to sophisticated genetic and imaging tools, as well as computational analysis (3). During Drosophila oogenesis, a two-dimensional monolayer follicular epithelium, encapsulating the developing egg, is patterned based on the integration of the Epidermal Growth Factor Receptor (EGFR) and the BMP (Bone Morphogenetic Protein) signaling pathways. The resulting differential expression of adhesion and cytoskeleton genes in the follicle cells (FCs) drives the morphogenesis of an elaborate three-dimensional eggshell, which includes a pair of tubular structures called dorsal appendages that project out from the main body of the egg. Previous research has made significant progress in elucidating the signaling and patterning events that specify the two dorsolateral patches of follicle cells that later form the roof of the dorsal appendages (roof cells). However, a qualitative and quantitative understanding of the role of EGFR inhibitory feedback in FC patterning is still lacking, and the regulatory network that specifies the dorsolateral roof cells as well as the dorsal midline cells separating the two patches of roof cells is incomplete.

We have taken a combined computational and experimental approach to identify the various roles of multiple EGFR inhibitory feedback loops in modulating the spatially distributed EGFR input (4). EGFR feedback is involved in specifying the size, shape, and relative location in the epithelium of the two dorsolateral patches of roof cells. Based on the expression and loss-of-function analysis, we show that, contrary to previously reports, the secreted EGF inhibitor, argos (aos) (5), does not play a significant role in specifying the specification of two patches of roof cells from a single peak of EGFR signaling. Instead, we find that two other EGFR inhibitors, sprouty (sty) (6) and kekkon (kek) (7) are important for modulating EGFR signaling and shaping the patches of roof cells. We present a computational model that incorporates inhibitory feedback loops, which is used to make predictions on the changes in the shape of the roof cell patches due to perturbations in the connectivity of the regulatory network. We then test the computational predictions using the sophisticated genetic tools available in Drosophila to quantitatively delineate the spatial patterning of the roof cells by sty and kek. This study provides quantitative insight into the role of inhibitory feedback for the proper patterning of a model epithelial tissue and the control of a medically important signaling pathway.

References

1. Ciardiello F, Tortora G. EGFR antagonists in cancer treatment. N Engl J Med. 2008 Mar 13;358 (11):1160-74.

2. Yarden Y. The EGFR family and its ligands in human cancer. signalling mechanisms and therapeutic opportunities. Eur J Cancer. 2001 Sep;37 Suppl 4:S3-8.

3. Shvartsman SY, Muratov CB, Lauffenburger DA. Modeling and computational analysis of EGF receptor-mediated cell communication in Drosophila oogenesis. Development. 2002 Jun;129(11):2577-89.

4. Goentoro LA, Reeves GT, Kowal CP, Martinelli L, Schupbach T, Shvartsman SY. Quantifying the Gurken morphogen gradient in Drosophila oogenesis. Dev Cell. 2006 Aug;11(2):263-72.

5. Klein DE, Nappi VM, Reeves GT, Shvartsman SY, Lemmon MA. Argos inhibits epidermal growth factor receptor signalling by ligand sequestration. Nature. 2004 Aug 26;430(7003):1040-4.

6. Reich A, Sapir A, Shilo B. Sprouty is a general inhibitor of receptor tyrosine kinase signaling. Development. 1999 Sep;126(18):4139-47.

7. Ghiglione C, Carraway KL, 3rd, Amundadottir LT, Boswell RE, Perrimon N, Duffy JB. The transmembrane molecule kekkon 1 acts in a feedback loop to negatively regulate the activity of the Drosophila EGF receptor during oogenesis. Cell. 1999 Mar 19;96(6):847-56.

572e Beta Amyloid Channel. A Mathematical Model

Mariajose Castellanos and Bradford Gates. Chemical and Biochemical Engineering, UMBC, 1000 Hilltop Circle, Baltimore, MD 21250

There has been research into beta amyloid (Aβ - one of the primary constituents in Alzheimer's disease) creating an ion channel inside the cell membrane. The proposed beta amyloid channel shares many qualities with other ion channels. It has both an open and closed state, and, while in the open state, ions travel through it via diffusion. The channel is voltage gated. It can be inhibited by various substances but no model of this channel yet exists. Using kinetic and equilibrium data from literature, a prototype model was created and implemented in a neuron model. This model channel showed strong parallels with experimental data. For example: the presence of beta amyloid is known to increase the excitability of a neuron's membrane. Running the model with an increase in the concentration of beta amyloid causes an increase in the frequency of spikes in the voltage profile. This is a clear indication of increased membrane excitability. While the actual frequency of voltage spikes is likely higher than would happen experimentally, the trend of increasing excitability is correct. The prototype model shows a strong correlation with all of the trends expected when Aβ interacts with neurons. This shows that the Aβ channel has a distinct possibility of being a prime part of Aβ-neuron interactions.

572f Receptor Mobility Increases Internalization of Vasopressin In Injured Endothelial Cells

Rico C. Gunawan and Debra T. Auguste. School of Engineering and Applied Sciences, Harvard University, 40 Oxford St., ESL Rm. 224, Cambridge, MA 02138

Targeted drug delivery to inflamed vascular endothelial cells will provide a precise and effective treatment for a range of therapeutic applications. We have explored the use of liposomes to target the delivery of vasopressin to injured endothelial cells during hemorrhagic shock. Liposomes are modified with different ratios of two antibodies against intercellular adhesion molecule-1 (ICAM-1) and endothelial leukocyte adhesion molecule-1 (ELAM-1), which are upregulated in inflamed endothelial cells. Liposome uptake by human umbilical vein endothelial cells (HUVEC), exposed with interleukin-1 to induce inflammatory events, reaches a maximum at an equal molar ratio of ICAM-1 to ELAM-1. In comparison, the uptake of poly(lactic-co-glycolic acid) (PLGA) tagged with similar molar ICAM-1/ ELAM-1 ratios is significantly lower and relatively equal at all ICAM-1/ELAM-1 ratios. This result suggests that receptor mobility plays a key role in increasing cellular uptake. We have created an in vitro coculture model to show that site-specific delivery of vasopressin to injured endothelial cells induces aligned smooth muscle cells to contract simulating blood vessel constriction.

572g Parameter Estimation of Oscillatory Systems

Kok Siong Ang and **Rudiyanto Gunawan**. Department of Chemical and Biomolecular Engineering, National University of Singapore, Singapore, Singapore

Introduction:

Oscillatory behavior is common in biology with examples from circadian rhythms [1] to p53-mdm2 DNA damage response [2] to NF-κB oscillations during an immune response [3]. These dynamical behaviors are important and integral in normal biological functions. While there had been considerable modeling and analysis efforts on such systems [4-6], little effort has been spent in reconciling the experimental data with the model output through parameter estimation. Instead, the models were usually built with parameters chosen on an ad-hoc basis (i.e., those that produce oscillations). The purpose of this work is to establish a framework to facilitate data reconciliation between these models and experiments.

Theory and problem formulation:

Two key characteristics of oscillatory behavior are its periodicity and shape. Period is the time interval for one complete oscillation cycle, while shape describes the state trajectory over one cycle. Both contribute to the error between experimental data and model prediction, which is in contrast to standard non-oscillatory systems that only possess shape characteristic and no period. Additional difficulty can arise in oscillatory systems for which the parametric sensitivities and consequently the gradient of objective function in the standard parameter estimation (sum of squares of error), grow unbounded over time [7]. This problem mainly stems from the accumulation of phase differences due to periodicity mismatch.

In this work, we convert time series data to phase-based data by normalization with the periodicity. This serves to decouple the period effect from the shape (contained in the phase-based data). In addition, mismatch in the initial condition between the (noisy) data and model simulation can also contribute to a phase difference. Hence, the initial condition will also be fitted to experimental data.

Methodology:

Based on Maximum Likelihood Estimation [8] under Gaussian error assumption, the objective function is constructed from 2 contributions: period mismatch and shape error. In the parameter estimation, the solution space contains both oscillatory and non-oscillatory solutions. Among the oscillating solutions, only sustained oscillations (i.e., limit cycles) are feasible. Hence, a finite Fourier transform is first used to separate out oscillating solutions when calculating the objective function. Thereafter, oscillation crest comparisons are used to pick out solutions with sustained oscillations. Finally, periods are estimated using the time difference between points of the same phase on each oscillating cycle. These points are chosen to have the steepest gradient to maximize accuracy. Finally, a stochastic optimizer Differential Evolution [9] is used to search for the optimum solution with the lowest objective function value.

Case Studies:

To illustrate this framework's effectiveness, *in-silico* data corrupted with 10-20% white noise is generated using published circadian rhythm models for parameter estimation. The use of *in-silico* data eliminates the issue of plant-model structure mismatch, which allows us to judge the efficacy of the framework in reducing parametric uncertainty. Although the models used are limit cycle models of circadian rhythms [4, 10], the methodology described above has general applicability to any oscillatory systems. The application to both case studies gave solutions with good agreement to experimental data. Figure 1 shows the data and model simulation from a case study [10]. However, some parameter estimates deviate from their true values by as much as 50%. Calculation of the Fisher Information Matrix [8] showed that these parameters have large variances and were practically unidentifiable based on 95% confidence interval. In other words, these calculations imply that the oscillatory behavior is not sensitive to these parameters, which is the main reason for their poor identifiability from experimental data.

Figure 1

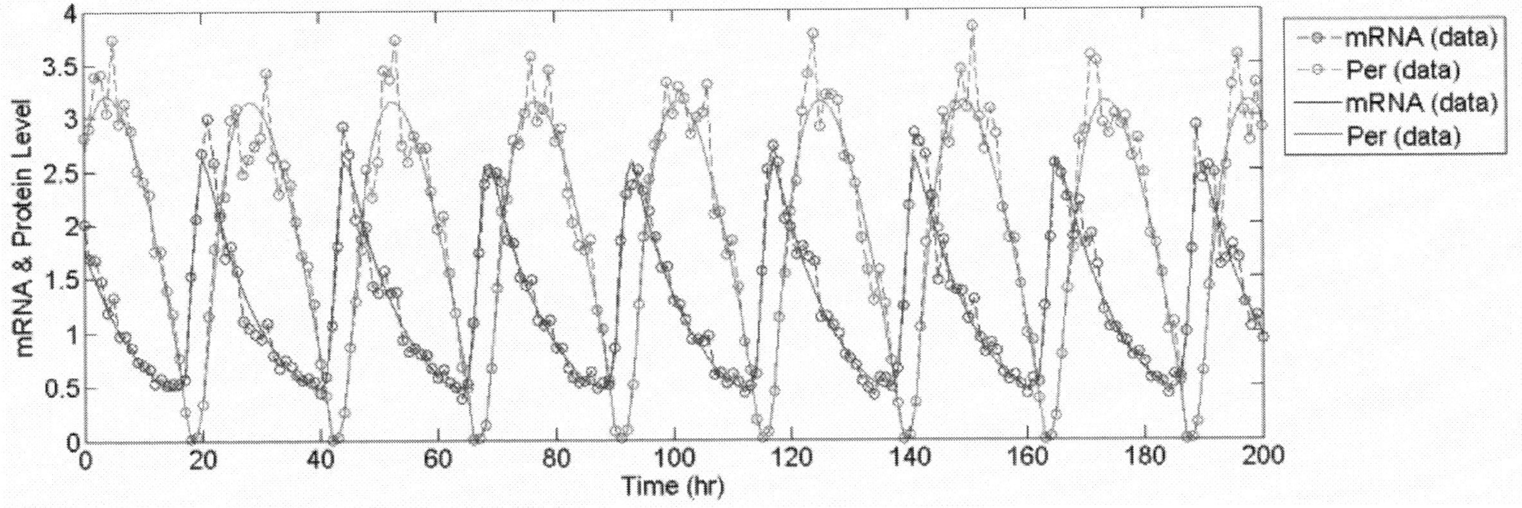

References:

[1] J. C. Dunlap. Molecular bases for circadian clocks. Cell, 96:271-290, 1999.

[2] N. Geva-Zatorsky, N. Rosenfeld, S. Itzkovitz, R. Milo, A. Sigal, E. Dekel, T. Yarnitzky, Y. Liron, P. Polak, G. Lahav and U. Alon. Oscillations and variability in the p53 system. Mocular Systems Biology, 2006.0033, 2006.

[3] A. Hoffmann, A. Levchenko, M. L. Scott and D. Baltimore The IkB-NF-kB signaling module, temporal control and selective gene activation. Science, 298:1241-1245, 2002.

[4] A. Goldbeter. A model for circadian oscillations in the Drosophila period protein (PER). Proc. R. Soc. Lond. B, 261:319-324, 1995.

[5] R. L. Bar-or, R. Maya, L. A. Segel, U. Alon, A. J. Levine and M. Oren. Generation of oscillations by the p53-Mdm2 feedback loop: A theoretical and experimental study. Proc. Nat. Acad. Sci., 97:21:11250-11255, 2000.

[6] N. A. M. Monk. Oscillatory expression of Hes1, p53, and NF-kB driven by transcriptional time delays. Current Biology, 13:1409:1413, 2003.

[7] R. Gunawan and F. J. Doyle III. Isochron-based phase response analysis of circadian rhythm. Biophysical Journal, 91:2131-2141, 2006.

[8] Y. Bard. Nonlinear parameter estimation. Academic Press, New York. 1974.

[9] K. V. Price, R. M. Storn and J. A. Lampinen. Differential Evolution: A practical approach to global optimization. Springer, Amsterdam. 2005.

[10] J. J. Tyson, C. I. Hong, C. D. Thron and B. Novak. A simple model of circadian rhythm based on dimerization and proteolysis of PER and TIM. Biophysical Journal, 77:5:2411-2417, 1999.

572h Modeling Drug Delivery for Design of PLGA Microparticles

Ashlee N. Ford, Chemical & Biomolecular Engineering, University of Illinois, 600 South Mathews Avenue, 201 Roger Adams Laboratory, Box C-3, Urbana, IL 61801-3602, Daniel W. Pack, Chemical & Biomolecular Engineering, University of Illinois Urbana-Champaign, 600 S. Mathews, Box C-3 MC 712, Urbana, IL 61801, and Richard D. Braatz, Department of Chemical and Biomolecular Engineering, University of Illinois at Urbana-Champaign, 600 South Mathews Avenue, Box C-3, 293 Roger Adams Laboratory, Urbana, IL 61801-3602.

The objective of this project is to develop a mechanistic model that is sufficiently accurate for predicting the drug release behavior that the model can be used for the optimal design of controlled-release drug delivery devices. A model is described for capturing the heterogeneity of autocatalytic degradation of poly(lactic-co-glycolic acid) (PLGA) polymer microspheres. The model is extended to apply to core-shell microparticles and microcapsules, which are important options for encapsulating drugs for delivery in a multi-stage pulsatile release fashion or for protecting proteins from being deactivated by suspension in an aqueous core for time-delayed delivery after polymer microcapsule degradation. The predictions for the core-shell microparticles and microcapsules are compared to experimental results. These mechanistic models contain no empirical parameters such as time-varying effective diffusivities--all model parameters are true physicochemical parameters taken from calculations of transport through pores or direct observation of depolymerization in well-characterized well-mixed reactors.

This poster describes the use of the mechanistic model to optimally design biodegradable polymeric microparticles for controlled release, which can be made reproducibly by the precision particle fabrication technique that yields highly uniform distributions with tight control of the specific sizes and thicknesses of core-shell microparticles and microcapsules. Changing the design variables of core diameter and shell thickness along with the distribution of molecular weights and pore sizes enables the design of microparticles to produce a large spectrum of obtainable release profiles. These profiles include zeroth-order release and pulsatile release with a range of shapes for the individual pulses. The model also determines the pH as a function of position within the microparticle, which can be used to design microparticles that limit the pH to ranges in which the released molecule is stable. The model can also be used to compute an optimal distribution of microparticles, which can be relevant when restrictions are placed on the microenvironment within the microparticles, such as limits to the pH.

572i

Modeling of Genetic Regulatory Networks in the Differentiation of Neural Crest Stem Cells to Sensory Neurons by Means of Boolean Networks

Jorge Marcelo Araus Patiño[1], Helena Groot[2], and Andres Fernando Gonzalez Barrios[1]. (1) Chemical Engineering and Biology, Grupo de Diseño de Productos y procesos, Universidad de los Andes, Carrera 1a este No. 19 A 40 7° piso, Edificio Mario Laserna, Universidad de los Andes, Bogotá, Colombia, (2) Biology, Universidad de los Andes, Cra 1A No. 18A-10, Universidad de los Andes Oficina M201, Bogotá, Colombia

Biological processes such as development can now be molecularly analyzed through genetic regulatory networks (GRN). In the present study we have generated a GRN comprising the process by which neural crest stem cells develops to two types of sensory neurons (Propioceptors and Nocioceptors). We have also been able to find recurrent patterns of regulation (motifs) that act coordinately to control such process. The results indicate that these motifs occur in similar stages found in the development of erythrocytes from hematopoietic stem cells. The GRN developed is made up of transduction signals, general and specific transcription factors and genes that confer the specific phenotype. The GRN is also suitable for modeling with an appropriate algorithm. Regarding its complexity we then used Classic Boolean Networks (CBN) for this purpose. The model showed key components, as well as the dynamic of the network through changes in initial conditions. The results show that these systems reach attractors of length two and have sense molecularly and morphologically. Finally, the motifs were reflected in the model, suggesting insights for further studies.

572j The Role of Biochemical and Mechanical Signals In Epithelial-Mesenchymal Transition

Qike Kyle Chen[1], Celeste M. Nelson[1], and Derek C. Radisky[2]. (1) Chemical Engineering, Princeton University, Princeton, NJ 08544, (2) Mayo Clinic Cancer Center, Jacksonville, FL 32224

Epithelial-mesenchymal-transition (EMT) is a characteristic phenotypic alteration in which epithelial cells detach from their neighbors and become invasive. EMT is essential for proper embryonic development, but the process is co-opted by cells during pathological conditions including fibrosis and neoplastic progression. Matrix metalloproteinase-3 (MMP3) induces EMT, fibrosis, and tumor formation in the mammary gland in vivo and in cultured mammary epithelial cells. We have used micropatterning approaches to investigate how the biochemical and mechanical properties of the microenvironment contribute to EMT. Mouse mammary epithelial cells are able to undergo EMT in response to MMP3 when cultured on fibronectin, but only if the underlying substratum permits cellular extension. Altering the identity of the extracellular matrix (collagen I, laminin-111, etc.) changes the EMT response. Understanding these microenvironmental controls will permit construction of an in vitro model of fibrosis genesis, and could reveal novel therapeutic strategies.

572k Bakers Yeast as a Prototype for Metabolic Syndrome

Goutham Vemuri, Quantitative Systems - Biology Department of Chemical and Biological Engineering, Chalmers University, Kemivagen 10, Goteborg, 412 96, Denmark and Jens Nielsen, Chemical and Biological Engineering, Chalmers University of Technology, Kemivägen 10, Gothenburg, 412 96, Sweden.

Several proteins in the bakers yeast, Saccharomyces cerevisiae, have counterparts in humans with conserved function and regulation. The metabolic pathways in which these proteins participate are usually essential for a continued robust performance by the cells. In humans, a malfunction in the regulation of these pathways triggers metabolic imbalance and eventually leads to disease. For practical and ethical reasons, it is not feasible to study the dynamic processes of disease mechanisms in humans and therefore, model organisms such as mice, fruit flies and even bakers yeast, are employed to understand these processes. Although, multicellular model organisms offer the advantage of evolutionary proximity to humans, it is not possible to rapidly quantify the dynamic phenomena of molecular interactions. We studied the impact of dynamically changing nutrition environments on the cellular energy level and how this controls various aspects of metabolism in S. cerevisiae. Our focus was on the homeostasis of cellular energy by the highly conserved AMP-activated protein kinase and its control of lipid metabolism. Our results indicate that despite being a unicellular organism, the regulation of cellular energy content and lipid metabolism is highly conserved compared to human, and that the yeast system makes a useful model for studying the molecular mechanisms of the metabolic syndrome as well as for screening new drug candidates and discovering new drug targets.

572I Coupling Hemodynamics and Protein Regulation In Cerebral Saccular Aneurysms

Marjan Rafat[1], Martin Heller[2], Howard A. Stone[1], and Debra T. Auguste[1]. (1) School of Engineering and Applied Sciences, Harvard University, 29 Oxford St., Cambridge, MA 02138, (2) Micro- and Nanotechnology, Technical University of Denmark, Building 345 East, DK-2800, Kongens Lyngby, Denmark

The mechanisms behind aneurysm growth and rupture are unknown. Additional analysis of the relationship between hemodynamics, mechanical forces, and biological response may enable aneurysm prediction, improve diagnosis, and allow development of new treatment approaches. We examined the hemodynamic profiles of clinically relevant aneurysm geometries for a range of physiological Reynolds numbers ($100 < Re < 400$) by numerically solving the Navier-Stokes equation. We identified flow regimes where shear stress and mass transfer limitations occur and created an in vitro model that mimics aneurysms using a poly(dimethylsiloxane) (PDMS) mold attached to glass. In order to analyze the cellular response to shear stress and flow profiles, we seeded human aortic endothelial cells (HAEC) onto these constructs and flowed through media at various flow rates (100, 200, and 400 mL/min) and viscosities (3, 4, and 5 cP) that are comparable to physiological conditions. We evaluated the regulation of surface receptors (PECAM-1, ICAM-1, and E-Selectin) as well as extracellular matrix remodeling components (matrix metalloproteinase-2 and 9, collagen types I and IV, elastin, and laminin). Understanding protein regulation in response to physical flows could implicate new strategies for vascular tissue engineering and drug delivery.

572m Histone-Mimetic Gold Nanoparticles as Self-Activating and Tailorable Gene Delivery Scaffolds

Meghan J. Reilly and Millicent O. Sullivan. Chemical Engineering, University of Delaware, 150 Academy St, Newark, DE 19716

The field of gene therapy has garnered significant interest over the past two decades as a method for revolutionizing the treatment of various diseases such as Alzheimer's, Parkinson's, and many types of cancer. In recent years, non-viral methods of delivery have received particular attention due to safety concerns and production limitations associated with viral vectors. However, inefficient DNA release is a common cause of ineffective non-viral DNA delivery. A novel solution to this constraint is the development of biomimetic scaffolds capable of regulating DNA accessibility. The presented study involves the design of histone-mimetic gold nanoparticles (HMGNs) as gene therapy packaging materials. Colloidal gold serves as a scaffold for the incorporation of histone H3 tail peptides trimethylated at lysine 4 (H3K4Me3). H3K4Me3 has a high density of positively charged residues that provide a DNA condensation template and impart protection from nuclease degradation. H3K4Me3 is known to be highly enhanced at the transcription start site for essentially all active genes. In addition, recognition of this trimethylated K4-containing peptide sequence by nucleosome remodeling factors has been implicated in mechanisms for chromatin activation. The purpose of the presented work is to assemble and characterize these HMGNs as well as to investigate the influence of HMGN functionalization on DNA binding, protection, and release. To this end, peptide-functionalized gold particles have been prepared. Their size and morphology have been characterized by methods such as UV-Vis spectroscopy and TEM. In addition, the self-assembly of H3K4Me3 with plasmid DNA has also been investigated by dynamic light scattering, zeta-potential, and gel mobility shift assays. These selected studies are aimed at validating the HMGN approach to gene delivery and will provide the framework for further development of our system.

572o A Continuum Pharmacokinetic Model for Perivascular Drug Transport at the Venous Anastamosis of An Arteriovenous Hemodialysis Graft

Randy Jay Christopherson[1], Robert M. Kirby II[2], Christi M. Terry[3], Alfred K. Cheung[3], and Yan-Ting E. Shiu[1]. (1) Department of Bioengineering, University of Utah, Salt Lake City, UT 84112, (2) Department of Computer Science and the Scientific Computing & Imaging Institute, University of Utah, Salt Lake City, UT 84112, (3) Division of Nephrology, University of Utah, Salt Lake City, UT 84112

Hemodialysis vascular access grafts made of expanded polytetrafluoroethylene (ePTFE) are commonly used in dialysis patients in the U.S. However, they are plagued by stenosis at the venous anastamosis due to the overgrowth of smooth muscle cells and fibroblasts, resulting in neointimal hyperplasia (NH) which causes high failure rates. Previous studies in our lab have investigated the efficacy of preventing stenosis in a porcine model by delivering antiproliferative drugs to the anastamoses of ePTFE grafts from a perivascularly located thermoactive gel (ReGel®). However, large animal studies are inherently time consuming and costly. The goal of this study is to develop a continuum pharmokinetic model to predict the concentration profile of perivascularly delivered drugs, thereby decreasing the number of animals required for experimentation and decreasing overall costs. We have developed a finite elements simulation using an idealized geometry in COMSOL multiphysics' transient diffusion application to model this situation. Diffusion is assumed to be isotropic through vessels, graft, and gel, and convection is assumed to be negligible. The effects of various parameters were evaluated in the simulation. These parameters included the initial drug concentration in the gel, diffusion coefficient and reaction rate of the tissue in the perivascular region, drug reaction rate within the venous tissue, and the shape of the gel. Future work will determine whether the assumed first-order kinetics is an appropriate approximation for reaction rates, and the relative importance of the geometry (i.e. idealized vs. realistic) on the diffusion.

Stabilization and Solubilization of Astaxanthin by Inclusion Complex Formation with Cyclodextrin

Soyoung Kim[1], Eunah Cho[1], Jimin Yoo[1], Man-Jin In[2] and Hee Jeong Chae[1,*]

[1]Department of Food and Biotechnology, and Center for Food Function and Safety, Hoseo University, Asan 336-795, Korea

[2]Department of Human Nutrition and Food Science, Chungwoon University, Hongseong 350-701, Korea

1

Introduction

Astaxanthin

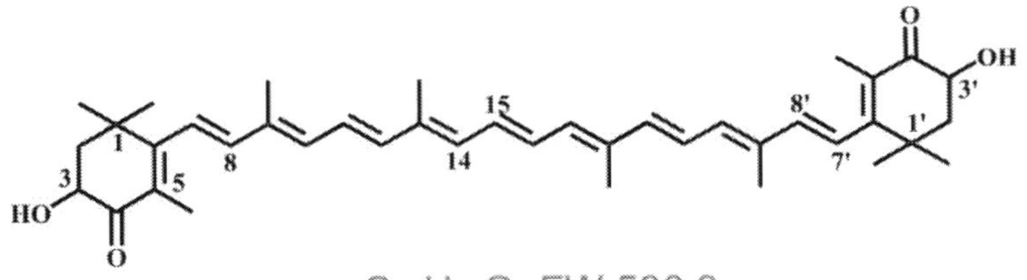

$C_{40}H_{52}O_4$ FW 596.9

- A carotenoid pigment found in certain marine animals and plants such as shrimp and algae
- similar structure to beta-carotene
- protecting DNA, enzyme and cells against oxidative damage
- powerful antioxidant properties

Astaxanthin (550 times) >> Vit E

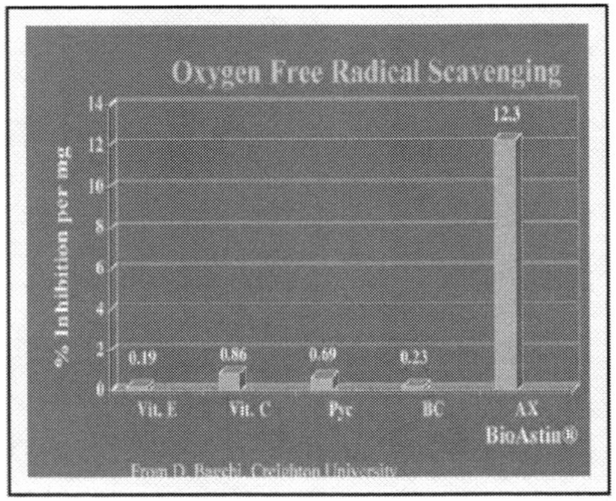

Types of currently available astaxanthin

- Chemical synthetic forms : free
  - Natural forms
    - Krill and crayfish oils (~ 0.15 %), esterified
    - *Phaffia* yeast (~ 0.40 %), free
    - *Hamatococcus* species (1.5-3.0 %), esterified

Dosage of astaxanthin : 2 mg/day (FDA)

Research purpose

The problem of astaxanthin

Poor aqueous solubility	→	limits its application in aqueous phase
Highly unsaturated molecule	→	damaged by heat or light

Water-solubilization
and stabilization of astaxanthin
for functional food & cosmetics

Material and method

■ **Material**

- Astaxanthin
- α-, β-, γ- cyclodextrin(CD) : Wako, Japan
- HPβ-, HPγ-CD : Sigma, USA
- Cyclo B(β-CD) : Corn products, USA

■ **Analysis**

- particle size : Scanning electron microscope(SEM)
- Change of structure : FT-IR

■ **Water-solubility & stablility**

Astaxanthin in dichlo:Ace.

⇩

Cyclodextrin in DW, 50 ℃

⇩

Astaxanthin : CD = 1: 1 - 1: 200

⇩

Stirring (50℃, 7hr)

⇩

cooling

⇩

evaporating

⇩

Inclusion complex

Results

FTIR spectra and SEM : β-CD

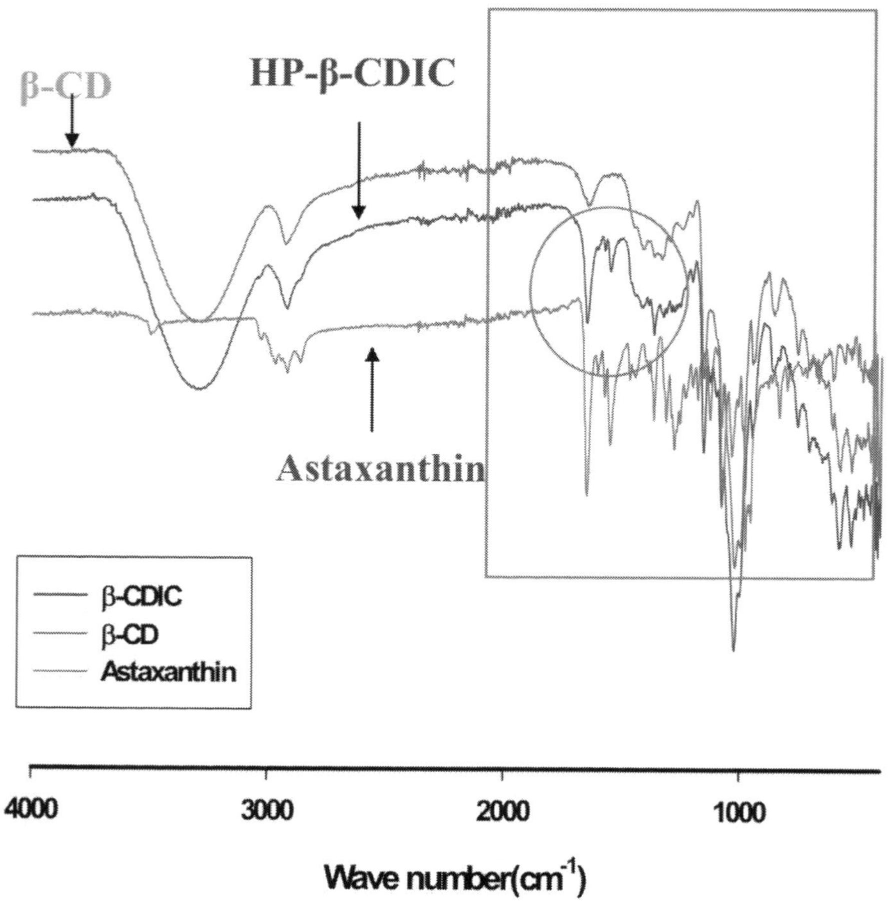

β-CD

β-CD I

FTIR spectra and SEM : HP-β-CD

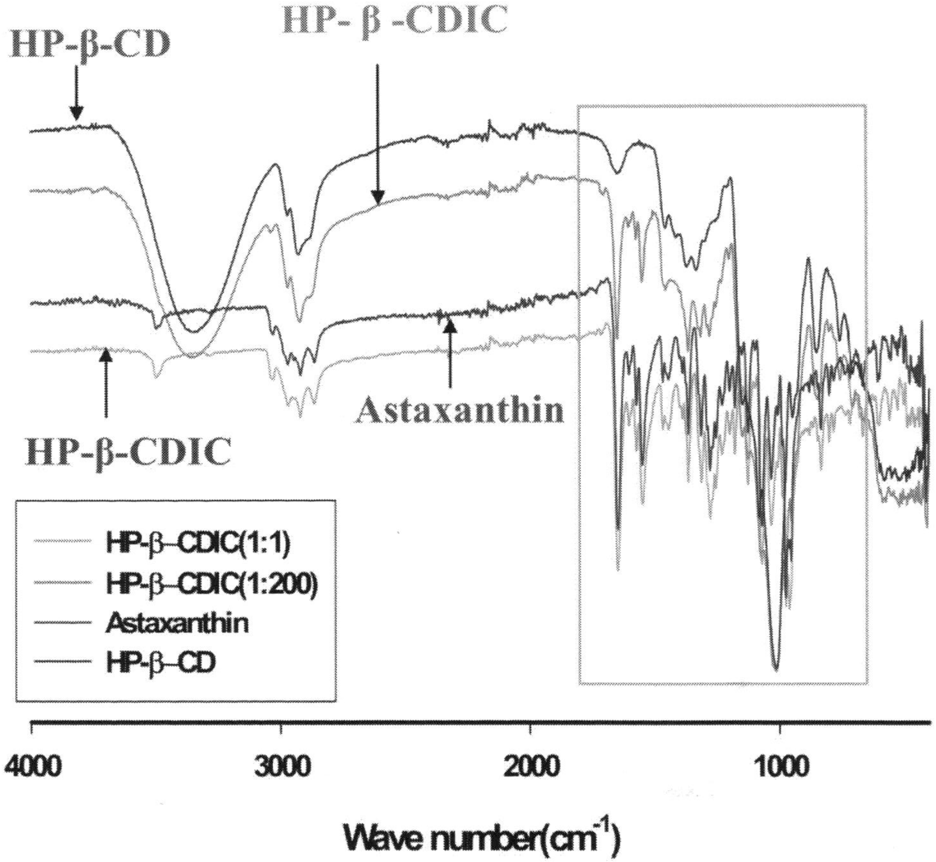

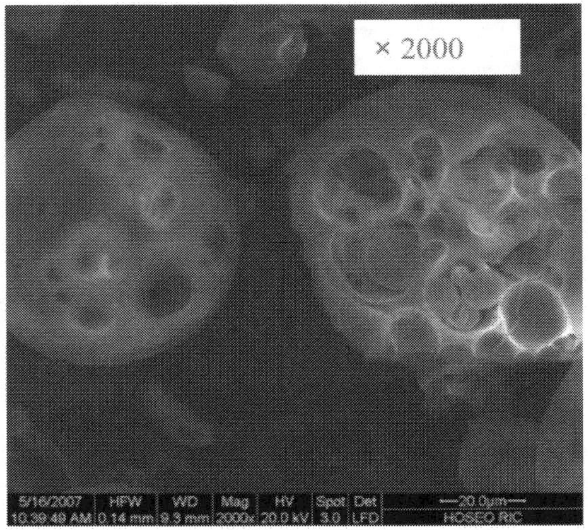

HP-β-CD

HP-β-CDIC

FTIR spectra and SEM : Cyclo B(β-CD)

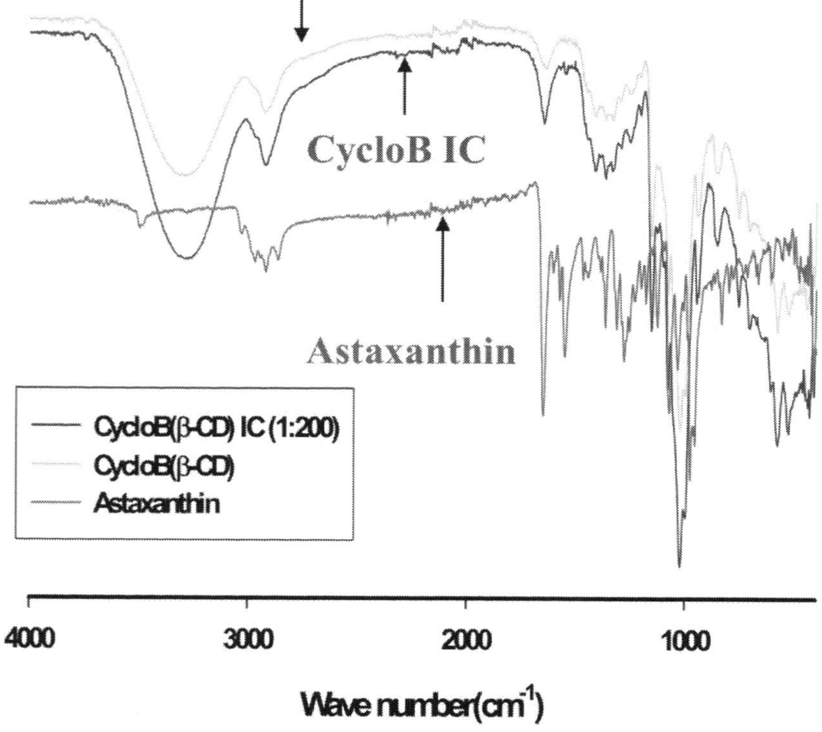

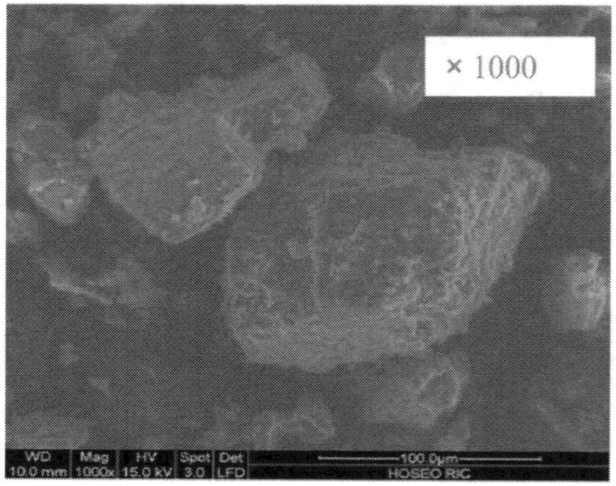

CycloB (β-CD)

CycloB IC (β-CD)

Functional Food & Bioprocess Lab.
biolab. hoseo. ac. kr

Water-solubility of inclusion complex in various pH and temperature

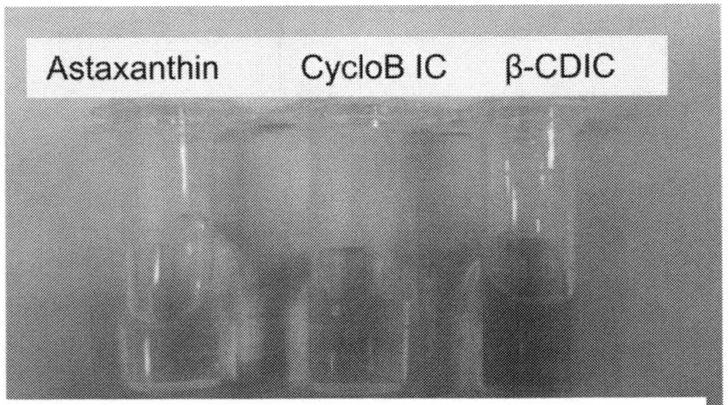

⇒ **β-CDIC is 13.6 times of free astaxanthin at 25℃**

⇒ **β-CDIC is 110 times of free astaxanthin at pH 6.5**

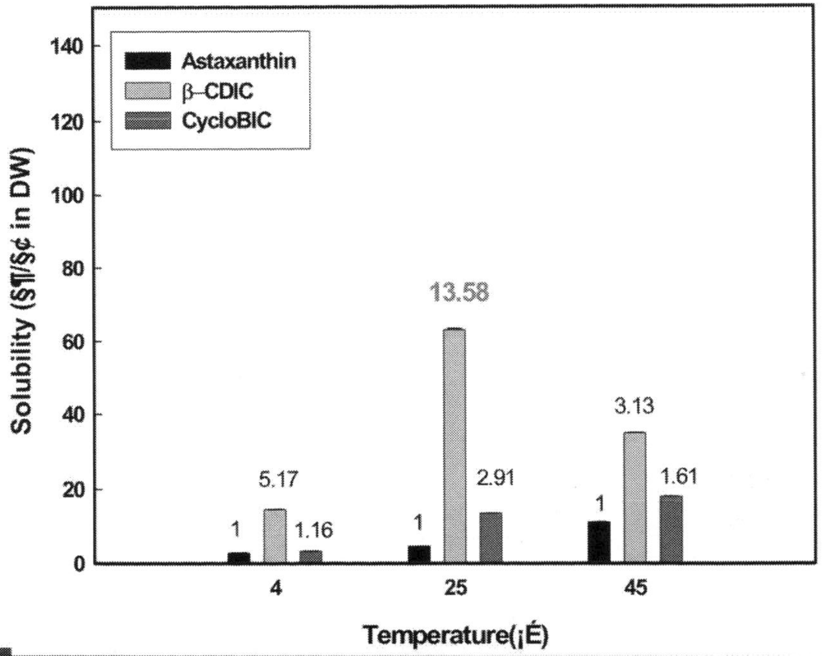

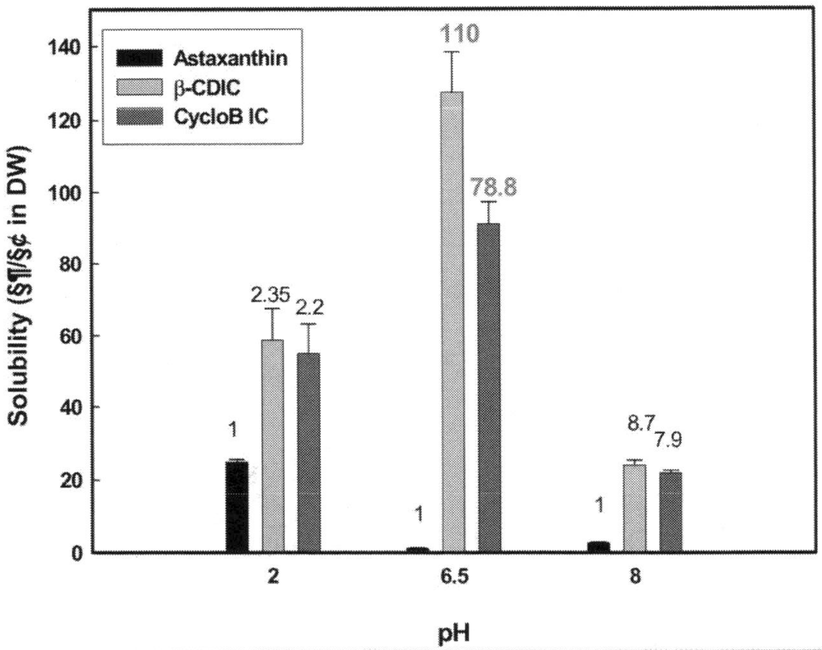

Effects of UV irradiation on the long-term stability of synthetic astaxanthin and inclusion complex

Effect of light

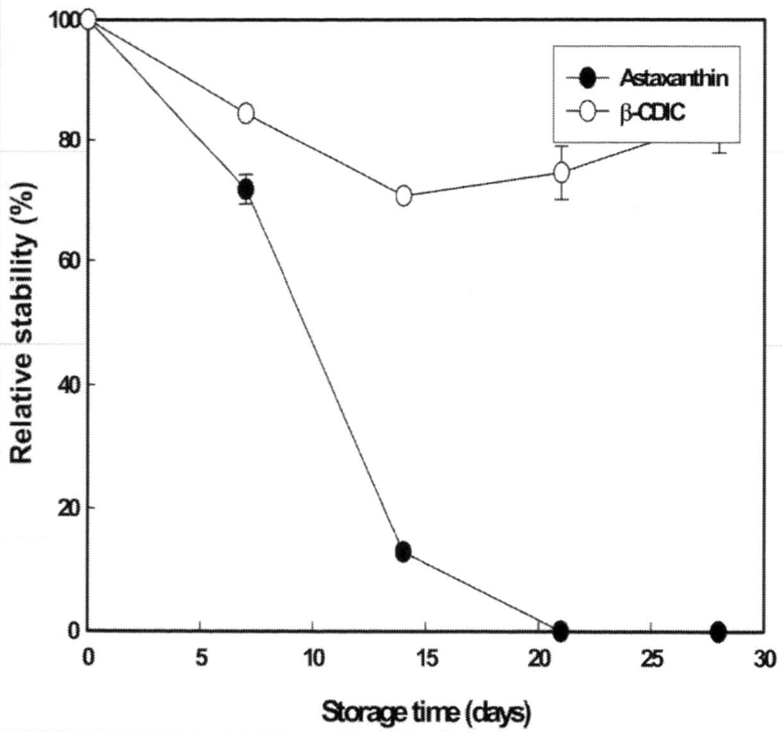

FTIR spectra

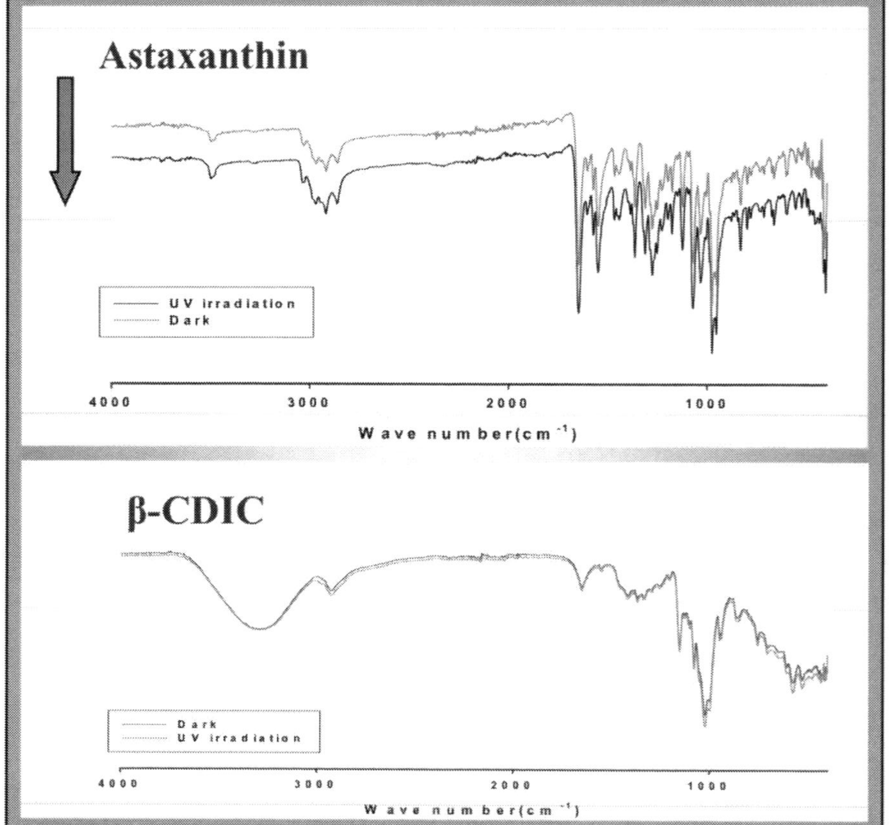

=> β-CDIC is 7 times compared with free astaxanthin(10%) after 14 days

Effects of temperature on the long-term stability of synthetic astaxanthin and inclusion complex

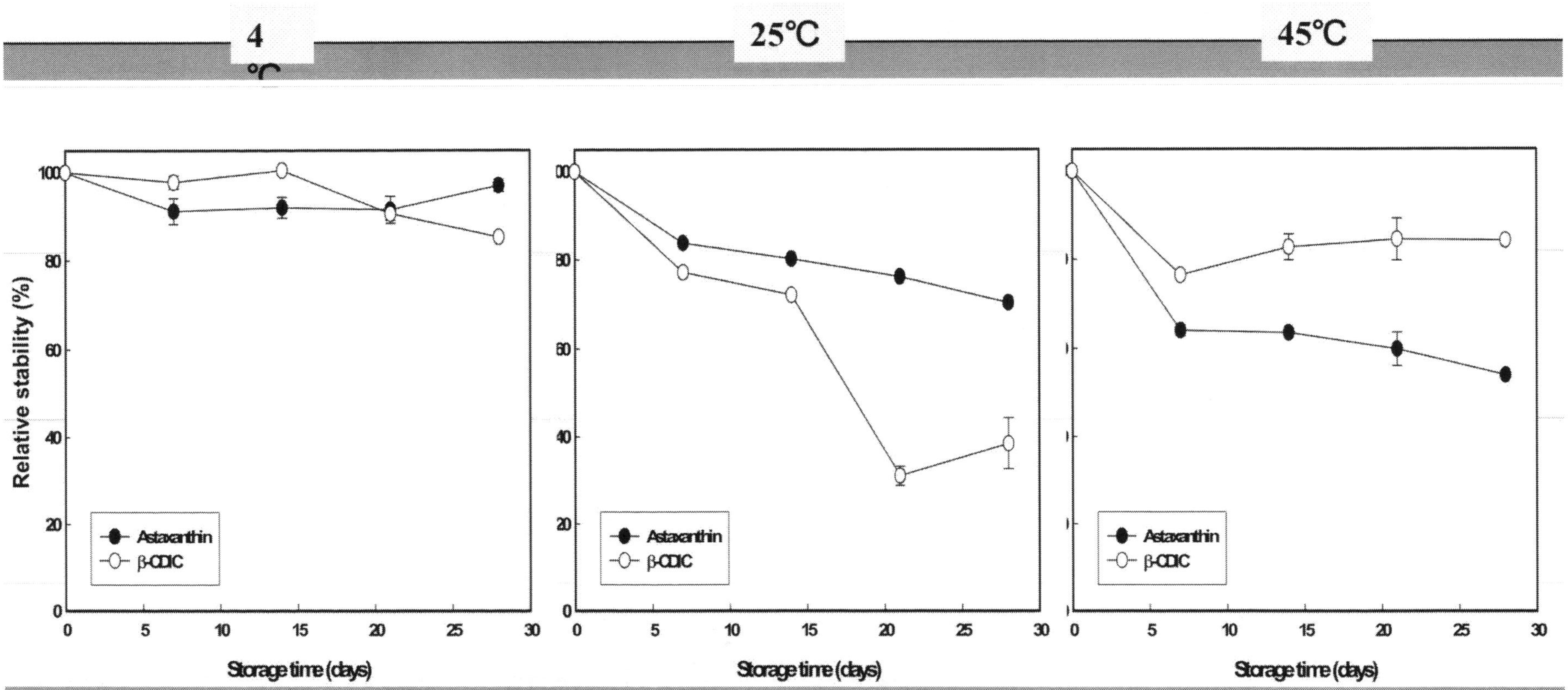

=> β-CDIC is stable than free astaxanthin at 45°C

Effect of heating on the stability of synthetic astaxanthin and inclusion complexes

Effect of heating

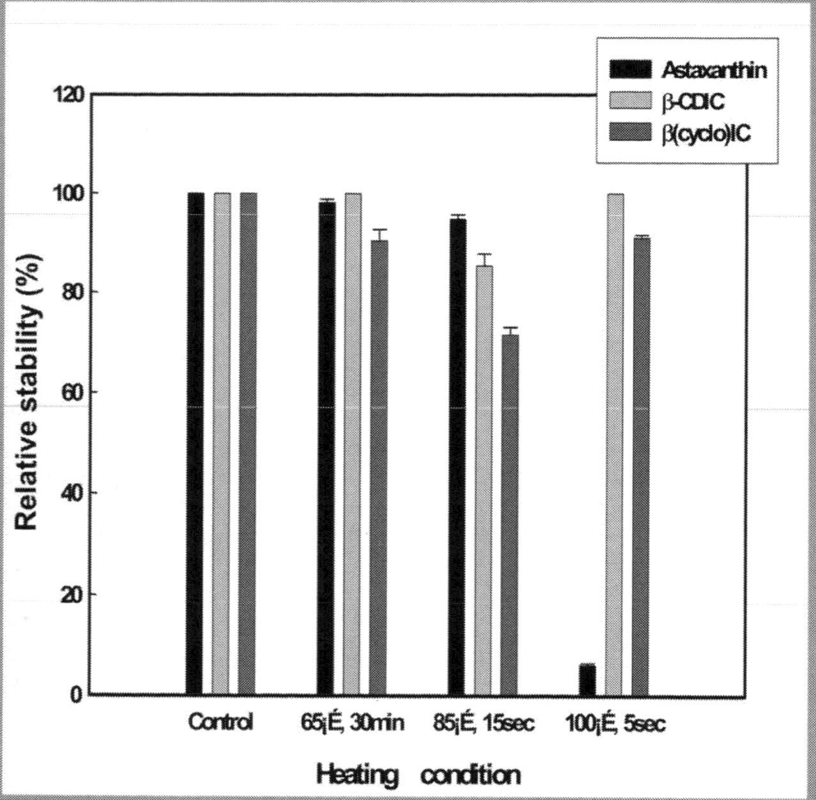

FTIR spectra

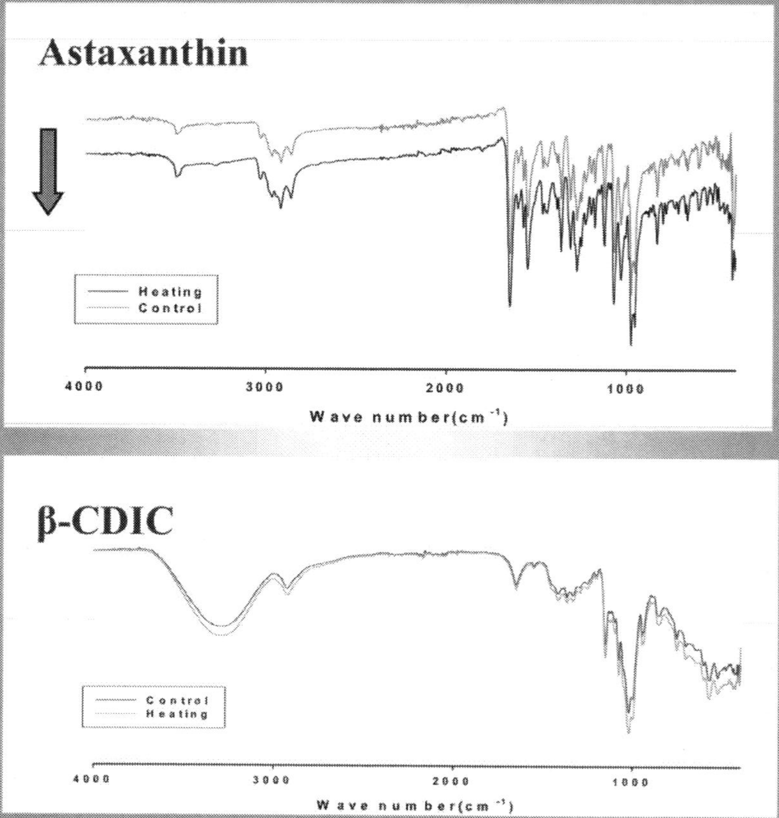

=> β-CDIC is 9 times compared with free astaxanthin after heating(100℃, 5sec)

Effects of oxygen on the long-term stability of synthetic astaxanthin and inclusion complex

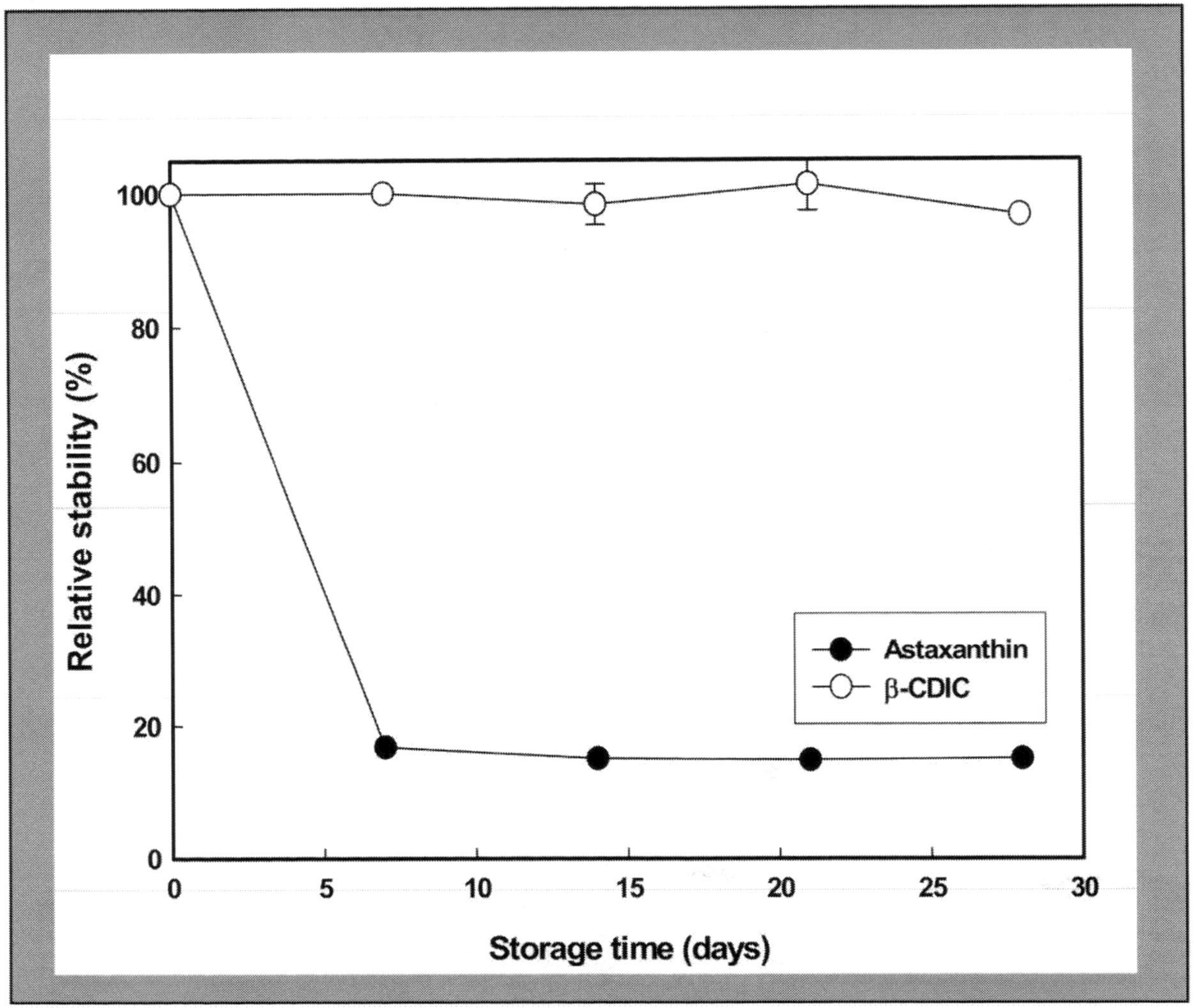

⇒ β-CDIC is very stable than free astaxanthin after 28 days

Effects of pH on the long-term stability of synthetic astaxanthin and inclusion complex

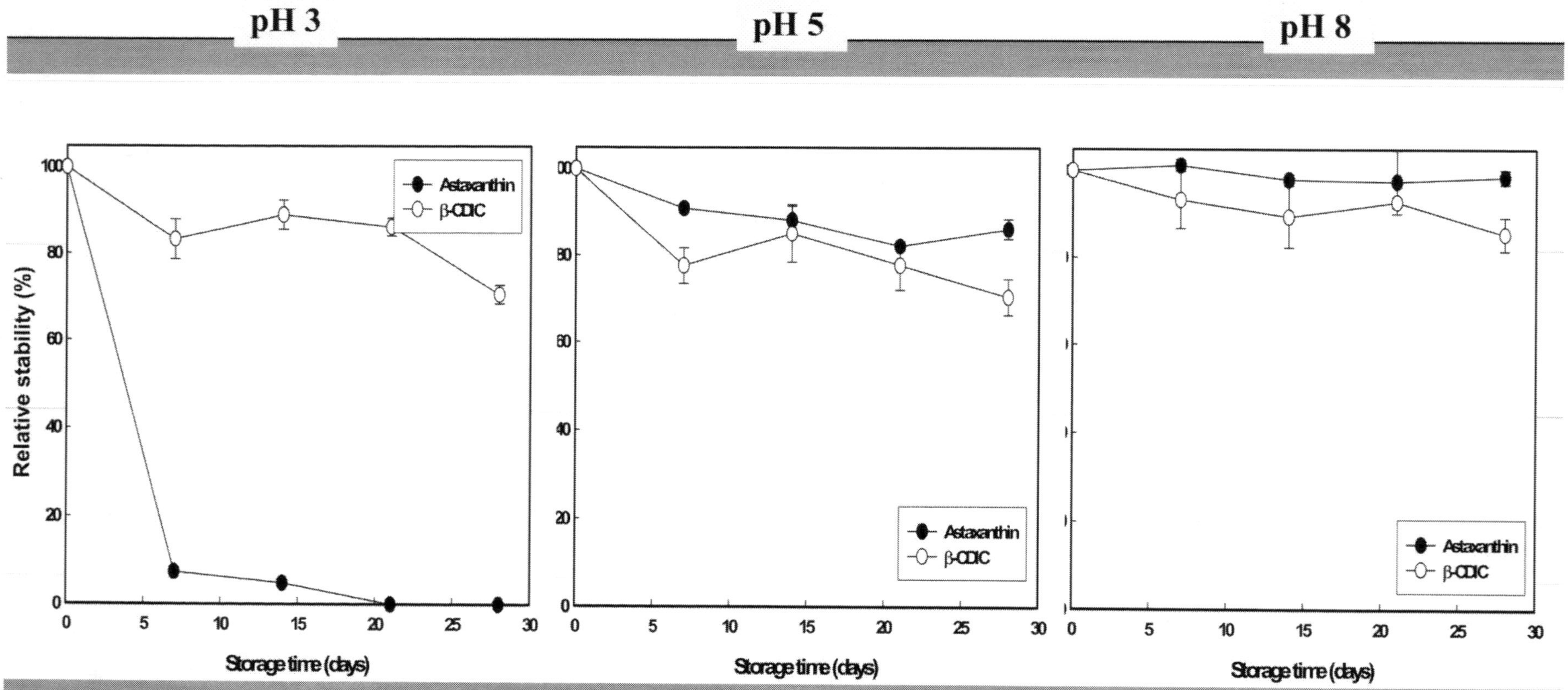

=> β-CDIC is stable than free astaxanthin in acidic condition

Summary

▣ FTIR spectrum, SEM

- Best host material for inclusion of astaxanthin => β-CD
- Reaction ratio between astaxanthin and host molecule ⇒ 1 : 200

▣ Water-solubility of inclusion complex (β-CDIC & CycloB IC)

- at 25℃ : 63.08 & 34.9 ug/ml ⇒ 13.6 & 2.9 times of free astaxanthin
- at pH 6.5 : 127.4 & 91.2 ug/ml ⇒ 110 & 78.7 times of free astaxanthin

▣ Stability of inclusion complex (β-CDIC)

- light : 7 times compared with free astaxanthin(10%) after 14 days
- Oxidation : very stable within 21 days
- heating : 9 times compared with free astaxanthin after heating(100℃, 5sec)
- Storage temperature : stable than free astaxanthin at 45℃

572q Detection and Classification of Multiple Probes *In Vivo* Using Hyperspectral Imaging

Silas J. Leavesley, Chemical Engineering, University of South Alabama, 307 University Blvd. N., EGLB 264, Mobile, AL 36688 and J. Paul Robinson, Biomedical Engineering, Purdue University, Bindley Bioscience Center, Room 122, 1203 W. State St., West Lafayette, IN 47907-2057.

Optical imaging methods have become increasingly popular in biomedical applications, due in part to recent advances in detector technology, increased availability, and decreased pricing. *In vivo* optical imaging techniques are especially promising, as these technologies do not bear the high cost or radiation concerns associated with more traditional imaging methods, such as X-ray, computed tomography (CT), or magnetic resonance imaging (MRI). However, bulk tissue optical properties generally have high scattering and attenuation coefficients that limit the penetration depth, sensitivity, and resolution of optical imaging methods. If fluorescence targeting methods are applied, tissue autofluorescence (background) also serves to limit imaging sensitivity. It is highly important to research methods for increasing the sensitivity to fluorescence signals in tissue, as targeting of specific cellular phenotypes often occurs at pmol or smaller concentrations. This is especially important in detection of rare cellular events, such as some cancers and other pathologies.

In this research, we have used a combination of hyperspectral imaging and mathematical processing techniques to isolate a fluorescence signal from autofluorescence background. We have designed a hyperspectral excitation-scanning imager that has been used to image multiple fluorescence signals in mouse models in the presence of high autofluorescence background. Visualization, analysis, and classification software has also been designed to allow separation of fluorescence signal from autofluorescence background and discrimination amongst multiple fluorescence signals of interest. This presentation will include a summary of the current hyperspectral imager design, optimization and construction as well as a description of the visualization, analysis, and classification methods implemented. Finally, a discussion of current applications and possibilities for translation of hyperspectral technology to the clinical imaging arena will be presented.

Web Page: scitation.aip.org/getabs/servlet/GetabsServlet?prog=normal&id=RSINAK000079000002023707000001&idtype=cvips&gifs=Ye

572r Elucidating the Transcriptional Regulatory Network Underlying the Nts Response to Acute Hypertension

Gregory M. Miller[1], Rajanikanth Vadigepalli[2], James S. Schwaber[2], and Babatunde A. Ogunnaike[1]. (1) Chemical Engineering, University of Delaware, 150 Academy St., Newark, DE 19716, (2) Daniel Baugh Institute for Functional Genomics and Computational Biology, Thomas Jefferson University, 1020 Locust St Room 575, Philadelphia, PA 19107

Hypertension, which affects one in three American adults, can lead to heart disease and stroke (NHLBI, 2008). Individuals with hypertension experience sustained elevated blood pressure because their blood pressure controllers appear to have adapted to a blood pressure set-point higher than that in normotensive individuals, suggesting a critical role for the blood pressure control system in the development and maintenance of hypertension. A major part of this control system is the nucleus tractus solitarius (NTS) in the brain, and NTS function has been shown to be compromised in hypertensive individuals, reflecting differences in NTS neurons between normotensive and hypertensive individuals. To treat hypertension effectively, we must understand the process by which NTS neurons adapt from the healthy state to the hypertensive state. Our overall objective, therefore, is to understand how neurons in the NTS change their cellular state to adapt from the healthy state to the hypertensive state.

To gain insight into how NTS cells are remodeled during the adaptation to the hypertensive state, we subject healthy animals to acute hypertension and then measure the response of their NTS cells at the level of gene and protein activity. Because NTS cells removed from animals subjected to this elevated blood pressure treatment have been found to exhibit a broad, dynamic gene expression response, gene expression is suspected to play a major role in this cellular adaptation (Khan et al, 2008). In this work, we identify the regulatory events orchestrating the gene expression response measured by Khan et al. Because the dominant mechanism by which changes in gene expression occur is through binding of transcription factors to their target gene promoters, we have focused on the network of transcription factors producing the gene expression response in the adaptation of NTS cells to the hypertensive state.

Because genes with AP-1 and CREB binding sites have been found to be significantly differentially expressed in the NTS from animals subjected to elevated blood pressure (Khan et al, 2008), we suspect that AP-1 and CREB are key transcription factors in the adaptation of NTS cells to the hypertensive state. To test this hypothesis, we perform chromatin immunoprecipitation experiments to measure AP-1 and CREB binding to the promoter regions of genes predicted to be regulatory targets of these transcription factors. Our experimental results show AP-1 and CREB bind differentially to genes known to be important for neuronal function in the NTS from animals with elevated blood pressure. These results support the role of AP-1 and CREB in central blood pressure regulation, and may lead to the identification of new pharmaceutical targets for hypertension.

References

Khan R.L. et al. (2008), **Dynamic Transcriptomic Response to Acute Hypertension in the Nucleus Tractus Solitarius**. *Am. J. Physiol. Regul. Integr. Comp. Physiol.*

National Heart Lung and Blood Institute Website, High Blood Pressure (2008). http://www.nhlbi.nih.gov/health/dci/Diseases/Hbp/HBP_WhatIs.html

572s Role of Cancer Cell Phenotype on the Intracellular Trafficking of Nanoparticles

Sutapa Barua and **Kaushal Rege**. Department of Chemical Engineering, Arizona State University, ECG 202, Tempe, AZ 85287-6006

Nanoparticles, including those based on metals, semiconductors, polymers, lipids, and dendrimers are being investigated as therapeutics and imaging agents in advanced cancer disease. Phenotypic changes play an important role in cancer cell survival, progression, and metastasis and changes in intra- and extracellular protein expression patterns and cell polarity contribute to resistance to therapeutics. Investigations into the mechanisms of uptake and intracellular trafficking of nanoparticles in cancer cells are therefore critical for enhancing the efficacy of nanoscale therapeutics and imaging agents. We report that not only are nanoparticles (quantum dots, gold nanoparticles, and polymeric nanoparticles) taken up spontaneously by related human prostate cancer cell lines, but that their intracellular fate is dramatically influenced by the cancer cell phenotype. While nanoparticles were trafficked to vesicles as punctuated structures throughout the cytoplasm in PC3 human prostate cancer cells, they localized at a single, juxtanuclear location ('dot-of-dots') in PC3-PSMA cells which express the biomarker, Prostate-Specific Membrane Antigen (PSMA) on their surfaces. PC3-flu cells demonstrated trafficking characteristics intermediate to those between PC3 and PC3-PSMA cells; nanoparticles localized in punctuated structures throughout the cytoplasm in addition to forming the 'dot-of-dots' structure in some cells. Colocalization analysis, using confocal microscopy revealed that the 'dot-of-dots' formation in PC3-PSMA cells coincided with the localization of transferrin (marker for recycling endosomes), lysotracker (marker for acidic compartments), and the Prostate-Specific Membrane Antigen (PSMA) in these cells. Importantly, the formation of the 'dot-of-dots' was highly dependent on microtubule polymerization. Following clathrin-mediated, lipid-raft independent, endocytosis, nanoparticle-containing vesicles were trafficked along these tubular structures finally leading to their localization in the proximity of the microtubule organizing center (MTOC) and the nucleus. The above approach was expanded to further investigate the effects of nanoparticle size / surface chemistry and intracellular proteins on the uptake and trafficking in these cells. Our results demonstrate that differences in cancer cell phenotype can lead to large differences in intracellular localization of nanoparticles, which in turn, can have significant impact on the efficacy of nanoscale therapeutics and imaging agents.

572t Towards Developing a Reproducible Method for Coaxing Embryonic Stem Cells to Pancreatic Islet Cells

Lye T. Lock, Chemical and Biological Engineering, State University of New York at Buffalo, 1015 Furnas Hall, Buffalo, NY 14260 and Emmanuel (Manolis) S. Tzanakakis, Department of Chemical and Biological Engineering, State University of New York at Buffalo, 907 Furnas Hall, Buffalo, NY 14260.

Diabetes mellitus is the sixth leading cause of death in the US. Current pharmacological treatments and insulin administration can only control the progression of this disease. Islet transplantation has afforded significant benefits including prolonged liberation from frequent insulin injections. Yet, islet transplantation cannot be used widely for diabetes treatment due to the extremely limited availability of donor islets. Embryonic stem cells (ESCs) can be an inexhaustible source of islet cells for transplantation. Previously published protocols have been characterized by low differentiation efficiency and results have been difficult to reproduce.

In this study, we describe an ongoing effort to develop a robust and reproducible method for directing the fate of ESCs towards pancreatic islets. Although we have used mouse ESCs as a model system, our findings translate to hESCs without significant modifications. At the first stage of differentiation, cells are coaxed towards definitive endoderm (DE). To that end, we have used medium containing either low concentration of serum or no serum, along with ligands of the TGF-β and Wnt families. Mouse and hESCs undergo morphological and biochemical changes over 4-6 days in culture. Cells in the resulting population are positive for Sox17, Foxa2 and CXCR4 whereas the expression of pluripotency markers such as Oct3/4 and Nanog is significantly reduced. Marker expression is assessed by quantitative PCR, immunocytochemistry and flow cytometry. A protocol for differentiation of DE cells to posterior foregut and beyond will also be presented. Selection of the islet cells derived from ESCs will be performed via adenoviral delivery of transgenes encoding for fluorescing proteins (e.g. green fluorescent protein; GFP) under an islet-cell specific promoter. To that end, we have generated an adenovirus carrying the red fluorescence protein gene flanked by the insulin gene promoter. We also explore the application of our differentiation scheme in conjunction with large scale expansion of ESCs. Our system will contribute towards developing large-scale bioprocesses for production of islet cells from ESCs for diabetes therapies.

572u Reconstruction of Phosphoproteins Signaling Network In Raw 264.7 Macrophages

Shakti Gupta, Department of Bioengineering, University of California, San Diego, CA 92093, Mano R. Maurya, San Diego Supercomputer Center, University of California San Diego, 9500 Gilman Drive, MC 0505, La Jolla, CA 92093-0505, and Shankar Subramaniam, San Diego Supercomputer Center, Department of Bioengineering, Department of Chemistry & Biochemistry, University of California San Diego, 9500 Giman Drive, La Jolla, CA 92093-0505.

Signaling pathways mediate the response of external stimuli on gene expression. The signaling proteins in these pathways interact with each other and their phosphorylation levels serve as an indicator of the activity in signaling pathways. In most cases, the last protein in signaling pathway is a transcription factor which participates in regulating the transcription of specific genes. Many of these signaling pathways have been well studied. Alliance for Cellular Signaling (AfCS) has measured time-course data in RAW 264.7 macrophage cells on important phosphoproteins, such as the mitogen-activated protein kinases (MAPKs) and signal transducer and activator of transcription (STATs), in single- and double-ligand stimulation experiments for 22 ligands (total 241 experiments). In the present work, we aimed to identify the interactions between signaling pathways using a data-driven approach. We have used early time point data from these experiments to study the phosphoprotein interactions with the assumption that at early time point, degradation effects are negligible. Later time points were discarded because then the degradation effects become prominent and present complications for a linear model based approach. We have used dynamic mapping ($y_{t+1} = f(y_t)$) and calculated the interaction coefficients using a partial least squares (PLS) approach. Significant interaction coefficients were selected based on 99% confidence in T-test by comparing with the coefficient distributions for the corresponding random models (based on random shuffling of the outputs). Using the PLS approach, we have reconstructed the phosphoproteins signaling network and have validated with the interactions already known in the literature. We have shown that this data-driven approach is able to identify most of the known signaling interactions and predicts potentially novel interactions.

572v Finding Your Way. the Drude Initiative for Nanocarriers In Delivery of Small Interfering RNA Therapeutics

Sabrina Pricl[1], Paola Posocco[1], Giovanni Maria Pavan[1], Giulio Scocchi[2], Marek Máły[3], Anastasia Malek[4], Maurizio Fermeglia[1], Andrea Danani[2], and Carlo Catapano[4]. (1) Molecular SImulation Engineering (MOSE) Laboratory - DICAMP, University of Trieste, Piazzale Europa 1, Trieste, 34127, Italy, (2) Icimsi, SUPSI, Galleria 2, Manno, CH-6928, Switzerland, (3) E. Hala Laboratory of Thermodynamics, Institute of Chemical Process Fundamentals, Academy of Sciences of the Czech Republic, 6-Suchdol, Prague, 165 02, Czech Republic, (4) Laboratory of Experimental Oncology, Oncology Institute of Southern Switzerland, Via Vela 6, Bellinzona, CH-6500, Switzerland

The past quarter century of outstanding progress in fundamental cancer biology has not translated into even distantly comparable advances in the clinic. Inadequacies in the ability to administer therapeutic moieties so that they will selectively reach the desired targets with marginal or no collateral damage has largely accounted for the discrepancy. There are two general, synergistic goals that should be striven for to augment the efficacy per dose of any therapeutic formulation: to increase its target selectivity and to endow the agent(s) comprising the therapeutic formulation with means to overcome the biological barriers that prevent it from reaching its target. The realization of such a system faces formidable challenges: nanotechnology, if properly integrated with established cancer research, can provide extraordinary opportunities to meet these challenges.

Intravascularly injectable nanovectors are a major class of nanotechnological devices of interest for use in cancer therapy. Nanovectors in general have at least a bipartite constitution, featuring a core-constituent material and a therapeutical payload. A third function, a biological surface modifier, may be present to enhance the biodistribution and/or tumor targeting of the nanovectors. Nanovector formulations are designed to reduce the clearance time of fast metabolized drugs, provide protection of active agents from enzymatic and environmental degradation, and avoid obstacles to the targeting of active moieties.

RNA interference (RNAi) was originally recognized as an evolutionary conserve defense mechanism in response to double-strand (ds) RNA and represents a powerful, naturally occurring biological strategy for specific gene silencing. It is mediated through 21-25nt double-stranded RNA molecules (small interfering RNA, or siRNAs), which are intracellularly generated from long, endogenous or exogenous ds RNAs, or directly transfected into the cells. Since RNAi is able to provide the relatively easy ablation of the expression of a target gene, it is now commonly used as a powerful tool in biological and medical research. This includes the strategy of siRNA-mediated targeting for functional studies of various genes whose expression is known to be upregulated in a given disease. Since siRNAs play a pivotal role in this process, the critical factors that determine the success of RNAi approaches are (i) the functionality of siRNAs, and (ii) the ability to deliver intact siRNA efficiently to the cells. Chemically unmodified siRNAs, in fact, are rapidly degraded, and mammalian cells do not readily take up naked nucleic acids. Thus, besides retroviruses, adenoviruses, and lentiviruses as vehicles for RNA interference constructs, strategies to deliver siRNA therapeutics to target cells in cell cultures include transduction by physical or chemical transfection by means of nanovectors.

Among these, block and graft co-polymers containing cationic polyelectrolytes, nonionic hydrophilic polymers, and positively charged hyperbranched polymeric structures such as dendrimers have gained significant interest. The task of the cationic moieties is to condense the nucleic acids, whereas the nonionic, hydrophilic parts should increase the solubility, stability and biocompatibility of the resulting supermolecular assembly. Branched polyethylenimine (PEI) has been shown to be one of the most efficient cationic polymers with regard to transfection efficiency. Linear polyethylene glycol (PEG) has been employed in several studies to optimize the biochemical/biophysical properties of transfection complexes. Different combinations of PEI with PEG have different effects on a series of important parameters of the siRNA/PEI complexes, including efficiency of nucleic acid complexation, complex stability in the presence of competing polyanions and nucleases, in vitro toxicity, cellular uptake and siRNA-induced gene silencing. However, the degree of PEGylation and molecular mass of PEG molecules affected the properties of siRNA/PEI complexes in apparently unpredictable ways. Furthermore, current knowledge of the interplay between the molecular structure of PEG-PEI copolymers and the resulting characteristics of siRNA/PEG-PEI complexes is very limited and rather empirical. Utterly analogous considerations can be drawn when the nanocarriers are dendrimeric families, such as the poly(amido amine) or PAMAM series.

Systematic investigation of the structure-activity relationship of the siRNA/nanocarrier assemblies is hindered by the absence of appropriated experimental techniques to study the fine ultra-structure of these nanocarrier-cargo complexes. Molecular modeling can provide a unique approach to simulate the structures of these assemblies and analyze their properties. In the current work, we applied in silico molecular simulation techniques to define the mode of the interaction of siRNAs with copolymeric and/ or dendrimeric delivery agents and identify critical parameters for optimization of carrier/cargo interactions. The combination of computational and experimental approaches will provide essential information for rational design of improved delivery systems for siRNA therapeutics. Optimal delivery would increase the efficacy of these powerful gene-targeted agents and move this strategy toward clinical applications.

572w The Design of Liver-Mimetic Cellular Architectures Using Polyelectrolyte Scaffolds

Yeonhee Kim and Padma Rajagopalan. Chemical Engineering, Virginia Polytechnic Institute and State University, 141 Randolph Hall, Blacksburg, VA 24061

The liver is one of the largest organs in our bodies and performs a multitude of functions such as, metabolism, detoxification, and plays a major role in the body's complex defense mechanisms. The deterioration in any one of the liver's functions can cause serious, life-threatening health problems. The design of tissue-engineered livers and liver-support devices can be accelerated if model hepatic units are available to systematically test their response to a range of stimuli. There is currently no generally applicable methodology to design hepatic culture systems where primary hepatocytes and liver sinusoidal endothelial cells (two of the major cell types found in the liver) are arranged in a stratified structure. We have assembled three-dimensional (3-D) cellular architectures comprised of primary hepatocytes and liver sinusoidal endothelial cells (LSECs) that mimic in several aspects the hepatic structure in vivo. The liver-mimetic architectures are comprised of primary hepatocytes, an intermediate nano-scale biocompatible and biodegradable polyelectrolyte (PE) scaffold and LSECs. Hepatocyte phenotypic function and detoxification profiles are being investigated using these 3D models.

572x Membrane Sandwich Electroporation for in Vitro Gene Delivery to Embryonic Stem Cells and Embryonic Bodies

Zhengzheng Fei[1], Shengnian Wang[1], Brian E. Henslee[1], and Ly James Lee[2]. (1) Department of Chemical and Biomolecular Engineering, The Ohio State University, 140 West 19th Avenue, Columbus, OH 43210, (2) Chemical and Biomolecular Engineering, The Ohio State University, Columbus, OH 43210

Genetic manipulation of cultured Embryonic Stem (ES) cells provides great potential for ex vivo gene therapy. Little progress has been made due to the difficulties involved in successful transfection. Viral transduction of ES cells is very efficient, but safety issues, such as immune and inflammatory responses, have hampered their clinical uses in humans. Electroporation is one of the most popular non-viral gene transfer methods for ES cell transfection. However, cell viability is still an issue due to the high electric voltage used and the nonuniform electric field strength distribution generated during the process. Our previous work demonstrated a new membrane sandwich electroporation (MSE) approach using NIH 3T3 fibroblasts and plasmids GFP and SEAP as model materials. In the MSE method, the focused electric field enhances cell permeabilization at a low electric voltage, leading to high cell viability; more important, the sandwich membrane configuration is able to provide better gene confinement near the cell surface, facilitating gene delivery into the cells. We have tested individual ES cells using our MSE method and got very encouraging results. Our on-going effort is to investigate the transfection of embryonic bodies with controlled size by a modified MSE system. Embryonic Bodies experience very low transfection efficiencies using current electroporation techniques.

572y The Precision of Brain 13C Dynamic Metabolic Modeling Using Different Infusion Protocols

Alexander A. Shestov, Dinesh K. Deelchand, Kamil Ugurbil, and Pierre-Gilles Henry. Center for Magnetic Resonance Research, University of Minnesota, 2021 6th Street SE, Minneapolis, MN 55455

Dynamic metabolic modeling of ^{13}C MRS data is a powerful tool to measure compartmentalized metabolic fluxes in the brain. The goal of the present study was to determine whether the use of ^{13}C labeled glucose or acetate substrates or combination of substrates leads to improved precision on fitted metabolic rates in the model using Monte-Carlo simulations. The metabolic network used was based on a published two compartment neuronal-glial model [1].

Results

The table shows the relative standard deviation on the determination of the neuronal and glial TCA cycle rates (Vtca(n), Vtca(g)) and glutamate-glutamine cycle (Vnt) using two approaches: (i) infusion of [1,6-$^{13}C_2$]glucose and analysis with a positional model (current approach) or (ii) double infusion of [1,6-$^{13}C_2$] glucose and [1,2-$^{13}C_2$]acetate and analysis with an isotopomer model (new approach). The new approach led to great improved precision on all free parameters in the model. The relative SD with the new approach was comprised between 4% and 12%, compared to 7 to 77% with the classic approach with glucose alone.

Conclusion

This study suggests that ^{13}C metabolic modeling using co-infusion of [1,2-$^{13}C_2$]acetate and [1,6-$^{13}C_2$] glucose should provide superior results for the precise determination of metabolic rates using two compartment metabolic modeling in the brain. This double infusion protocol can be expected to advantageously replace more commonly used approaches that utilize ^{13}C-glucose alone.

References

[1] Gruetter et al, AJP, 281, E100, 2001

This work was supported by NIH grants R01NS38672, P41RR0807 and P30NS057091

Gln dilution included in the model?	Metabolic flux	Relative SD with [1,6-$^{13}C_2$]glucose, positional model	Relative SD with [1,6-$^{13}C_2$] glucose + [1,2-$^{13}C_2$]acetate, isotopomer model

NO	Vtca(n)	7%	4%
	Vtca(g)	52%	9%
	Vnt	77%	8%
YES	Vtca(n)	7%	5%
	Vtca(g)	61%	12%
	Vnt	22%	9%

572z Experimental Investigation of the Relationship Between Integrin Clustering and Focal Adhesion Kinase Phosphorylation

Erik S. Welf, Chemical Engineering, University of Delaware, 150 Academy Street, Colburn Lab, Newark, DE 19716, Babatunde A. Ogunnaike, Department of Chemical Engineering, University of Delaware, Newark, DE 19716, and Ulhas P. Naik, Biological Science, University of Delaware, 150 Academy St., Newark, DE 19716.

Integrins are transmembrane proteins that bind extra-cellular matrix (ECM) proteins to facilitate cell adhesion and migration. In this role, the two main functions of integrins are to sense the extra-cellular environment and to regulate cell attachment. Accomplishing these tasks requires that integrins interact dynamically with numerous multi-functional cytoplasmic proteins that alter integrins' ability to bind ECM and transduce signals as a result of integrin binding to ECM. As a consequence of these multi-functional protein-protein interactions that occur where integrins cluster, it is still unknown how integrins and integrin clustering regulate cell adhesion and migration as a function of ECM properties, and how integrins themselves are regulated by the molecules that bind integrins.

To determine how cells use integrins to sense and respond to their environment, we investigate the relationship between integrin clustering and the intracellular protein focal adhesion kinase (FAK) experimentally. FAK senses integrin clustering by phosphorylating itself at locations where integrins cluster [1], and it regulates integrin cluster size by associating with a host of proteins that affect integrin binding and clustering [2]. Because FAK acts as both a sensor and a regulator of integrin clustering, it has been identified as a key protein that facilitates integrin function in both their adhesive role and their signaling role. In this work, we use an in vitro experimental system to investigate how cells use integrin clustering and FAK phosphorylation to sense and respond to ECM protein concentration.

Cells stably expressing the •IIb•3 integrin were allowed to spread on varying concentrations of the ECM molecule fibrinogen (Fg), a specific ligand for •IIb•3. By measuring integrin cluster sizes on different concentrations of Fg, we show that lower concentrations of Fg result in larger integrin clusters. To investigate how cells sense and respond to changes in Fg concentration, we examined the role of FAK in integrin cluster regulation by measuring FAK phosphorylation as a function of Fg concentration. Differences in the magnitude and dynamics of FAK phosphorylation resulting from cell adhesion to different ECM ligand densities suggest bi-directional relationships between FAK activation and integrin clustering. The quantitative relationships between integrin clustering and FAK phosphorylation that we elucidate provide new insight into how integrin clustering and intracellular signaling are coupled to relay important information about the ECM into the cell and simultaneously move the cell across the ECM.

[1] Miyamoto S et al. Integrin function: molecular hierarchies of cytoskeletal and signaling molecules. (1995) J Cell Biol 131, 791-805.

[2] Sieg DA, Hauck CR and Schlaepfer DD (1999) Required role of focal adhesion kinase (FAK) for integrin-stimulated cell migration. J Cell Sci 112, 2677-2691.

572aa Influence of Nanoparticle Composition on siRNA Binding: Effect on Cellular Delivery for RNA Interference

Joseph A. Gredell[1], Erin B. Vogel[2], Christina Chan[1], Gregory L. Baker[2], and S. Patrick Walton[1]. (1) Chemical Engineering and Materials Science, Michigan State University, 2527 Engineering Building, East Lansing, MI 48824, (2) Chemistry, Michigan State University, 534 Chemistry, East Lansing, MI 48824

Delivery is a significant challenge faced during the transition of siRNA initiated gene silencing from an in vitro setting to one of widespread clinical use. Much effort is being expended to develop cationic polymer systems that self-assemble with siRNA via electrostatic interactions with the phosphate backbone. These polymers protect the siRNA from nuclease degradation and facilitate cellular uptake, frequently through the endosomal pathway. Cytoplasmic localization is then commonly accomplished through inclusion of an endosomal escape agent, eventually leading to targeted gene silencing.

Unfortunately, there is not a clear understanding of the effects the various moieties play on the myriad of polymer chemistries currently being pursued for siRNA delivery. In order to better characterize the binding step of polymer/siRNA (nanoparticle; NP) complex formation, we are using a novel polymeric system that allows addition of virtually any moiety. Through our systematic studies, we will show that binding of siRNA by the NP is strongly dependent on the cationic charge density along the backbone and that incorporation of specific functionalities can influence that binding through physical and structural changes. The observed trends also appear to be important for biocompatibility (i.e., toxicity). Together, our results can be used to identify new polymers that are designed to meet the complex requirements for siRNA delivery.

Models of the Aging Cardiovascular System for the Study of Degenerative Diseases

Michael D. VanAuker, John Elliott, Michael J. Carney, and Joel A. Strom
University of South Florida, Tampa, FL, USA

Abstract

Tools to aid in the design and testing of prosthetic hearts valves include cardiac pulse duplicators and computational fluid dynamics simulation. Typically, conditions in the pulse duplicators are well defined and controlled based on established guidelines for prosthetic valve testing. We have created a pulse duplicator to simulate the physiologic effects of the aging and diseased cardiovascular system, including increased aortic impedance (decreased compliance and increased resistance), and cardiac outputs representing heart failure as well as normal states. In this paper we will describe the design and validation of this cardiac pulse duplicator, and its use to study degenerative aortic valve disease. This is a progressive disease with an unpredictable rate of progression characterized by increased leaflet thickness, stiffening and calcification without commissural fusion. These changes increase resistance to cardiac outflow and hence the workload on the left ventricle. This will cause left ventricular hypertrophy and if untreated, heart failure and death. It is believed that "wear and tear", including mechanical stress injury to the endothelium on the valve, mediates the disease process. Risk factors for valve disease are similar for other cardiovascular disease (e.g., age, smoking, hypertension, and hyperlipidemia). Current hemodynamic measurements of the severity of aortic valve disease have little value in predicting progression or onset of symptoms. Early changes in the degenerative process include thickening and stiffening of the leaflets, reflected in altered rates of opening and closing, before becoming restrictive to flow. In this study, we tested the hypothesis that changes in ambient hemodynamic forces associated with aging and related cardiovascular disease (such as hypertension) affect valve mechanics and independently classic measurements of disease severity. For example, hypertension creates increased afterload for the left ventricle, due to increased peripheral resistance, decreased arterial distensibility, and early wave reflections because of increased wave propagation velocity. These contribute to a more rapid initial rise in aortic pressure, a higher systolic peak, and a more rapid fall in pressure during diastole compared to normal. These changes alter forces on the aortic valve and thus stress patterns within it. Data obtained in the pulse duplicator, combined with computational simulation, will be used to develop markers of the properties of the valve itself, independent of ambient hemodynamic state. Such measurements, if validated clinically, would be useful in predicting disease progression and in the design and assessment of new therapeutic strategy.

Introduction

Cardiac pulse duplicators have been developed, primarily with the aim of testing prosthetic devices such as artificial heart valves. These apparati are designed to conform to certain guidance documents or testing standards, but due to differences in design including input waveforms, working fluid, location of measurement sensors for pressure and flow, geometry of testing chamber, and the material properties of the testing chamber, different results can be obtained for a similar valve tested in different systems [1].

In this paper, we present results from an *in vitro* study using a cardiac pulse duplicator that can be used to generate the data outlined in the Food and Drug Administration guidance documents, but also can be used to model hemodynamic conditions associated with cardiovascular pathology and aging. For example, both aging and hypertension create increased afterload for the left ventricle, due to increased peripheral resistance, decreased arterial distensibility, and early wave reflections because of increased wave propagation velocity. These contribute to elevated mean blood pressure, a more rapid initial rise in aortic pressure, a higher and later systolic peak, and a more rapid fall in pressure during diastole compared to normal [2]. Hyperlipidemia, hyperinsulinemia, obesity, diabetes mellitus, and cigarette smoking all affect the vascular system in ways that resemble accelerated aging [3-13].

The experimental apparatus described below allows us to tune these parameters through normal and pathologic ranges. In this way we can test hypotheses regarding the effects of changes in hemodynamics on the performance of the aortic valve, as it relates to degenerative aortic valve disease. Degenerative aortic valve disease is characterized by increased valve leaflet thickness, stiffness and calcification. These changes increase resistance to cardiac outflow and hence the workload on the left ventricle. This will cause left ventricular hypertrophy and, if untreated, heart failure and death. It is an increasingly common disease in the elderly, with degenerative changes seen in up to 75% of patients greater than 85 years old [14]. Degenerative aortic valve disease is progressive, and there is variability in the rate of progression between subjects. Thus, prediction of disease progression in an individual is difficult [15-18].

Below we describe experiments involving aortic valves of different material properties that model the spectrum of aortic valve disease. An echocardiographic method has been proposed to predict an individual's risk for rapid progression, based on the rate of change of the aortic valve area through the cardiac cycle. According to this method, the ratio of the aortic valve area halfway through deceleration to that halfway through acceleration is an indicator of risk [19]. However, changes in ambient hemodynamics, reflected in changes in the pressure and flow waveforms, due to hypertension and aging should affect the distribution of forces on the valve and hence affect its mechanical performance, ***independent*** of changes to the valve itself. Thus, we tested the hypothesis that changes in hemodynamics related to changes in aging and hypertension, such as increased afterload, alter aortic valve mechanics, and that these effects can be measured by observing aortic valve motion during opening and closing phases of the valve and therefore independently influence the proposed echocardiographic method of assessing risk.

Experimental Methods

The pulse duplicator (Figure 1) is essentially a mechanical analog of a modified Windkessel model of the systemic circulation [20, 21]. It consists of an atrial reservoir that feeds a ventricular chamber by gravity. The ventricular chamber houses a compressible bulb. Solenoid valves allow compressed air into the chamber to produce systole, and release the air to produce diastole. Between the model atrium and ventricle is the mitral valve section. Downstream of the ventricular bulb sits the aortic valve section. Compliance and resistance elements are used to "tune" the system to produce physiologic pressure and flow waveforms.

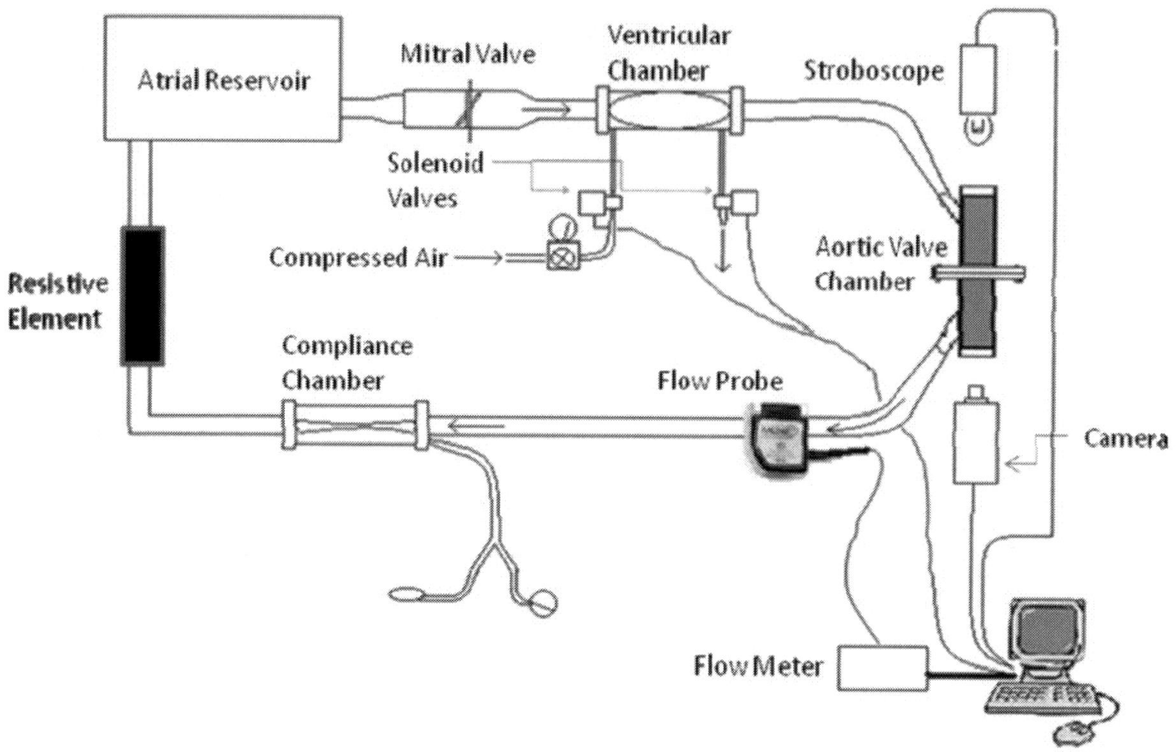

Figure 1. Schematic of the pulse duplicator system (not to scale). Adapted from Ref. 15.

The compliance section consists of an outer chamber connected to a pressurized air reservoir, surrounding an elastic tube through which the fluid travels. Increased afterload can be simulated in the pulse duplicator by increasing the resistance and/or increasing the external pressure in the compliance chamber, rendering the elastic tubing stiffer.

The pulse duplicator is interfaced to a personal computer with LabView™ software and input/output board for data acquisition and display (National Instruments, Houston, TX). Flow measurements are made with an ultrasonic flow meter with cannulating probes (Transonic Systems, Inc., Ithaca, NY). Pressure measurements can be made with fixed-mount pressure transducers at various locations in the flow loop; typical locations include proximal to the aortic valve (P_1), at the vena contracta of the valvular jet (P_2), and downstream at the site of recovered pressure (P_3). High fidelity catheters (Millar Instruments, Inc., Houston, TX) can also be introduced into the system to measure pressure at any location.

Pulse duplicator control is achieved using an external control system. This consists of a stamp, programmable in Visual Basic, to generate signals to control the solenoid valves. In this way, cardiac cycle rate and systolic ejection time can be varied. The pulse duplicator is also equipped with a camera and stroboscope for analysis of the valve motion. A second stamp controls the camera, and the stroboscope is controlled by a temperature compensated crystal oscillator external to the PC. The instantaneous anatomic opening area of the valve can be determined by obtaining a digital photograph of the valve viewed in cross section. A timing signal from the camera and stroboscope is acquired by LabView™ to synchronize the photographs with instantaneous pressure and flow data. Photographs were acquired by oversampling every millisecond through the cardiac cycle and correlated with instantaneous pressure and flow data from that cycle. The instantaneous valve opening area is obtained from planimetry using MATLAB, with functions supplied in the Image Processing Toolbox.

The working fluid was a 41% (by volume) solution of glycerin in phosphate buffered saline (PBS), which at room temperature has viscosity and density (0.037 poise and 1.08 g/cm^3, respectively) similar to the Newtonian properties of whole blood. Blood flow in the heart and great vessels behaves essentially as a Newtonian fluid [22] due to the high shear environment. Because of the sensitivity of the viscosity of the solution to environmental factors such as humidity and temperature, fluid viscosity was repetitively monitored using a Canon-Fenske viscometer and the viscosity of the working fluid was kept at 0.037 ± 0.001 poise throughout the experiments. The cardiac output was varied between 2 and 8 liters per minute, the heart rate was held constant at 70 beats per minute. The systolic ejection period was 35% of the cardiac cycle. Resistance and compliance downstream was varied from a range representing normal blood pressure (120/80 mm Hg, systolic/diastolic) to hypertension (approximately 180/120 mm Hg).

Fresh aortic valves were harvested from pigs (25 to 50 kg) at the Vivarium at USF Health. Approval to use these tissues was obtained from the Institutional Animal Care and Use Committee at the University of South Florida. These were treated either with 2% or 4% glutaraldehyde in PBS or 70% ethanol to produce a range of valve leaflet stiffness. Glutaraldehyde fixation forms crosslinks between carboxyl groups and the lysine groups of the collagen in the valve. This renders a change in leaflet stiffness without a change in valve area. Valves were mounted to sewing rings provided as a generous gift by St. Jude Medical and then set in the aortic chamber of the pulse duplicator (Figure 1).

From the flow and pressure data, peak instantaneous and mean systolic pressure drops, effective orifice area, and valve resistance are calculated. The effective orifice area is calculated by the formula of Gorlin and Gorlin [23]:

$$A_{AV} = \frac{Q}{44.3 \times RS\sqrt{\Delta P_m}}$$

where A_{AV} is the effective aortic valve area (cm^2), Q is the cardiac output (ml/min), R is the heart rate (beats/minute), S is the systolic ejection period (in seconds), and ΔP_m is the mean pressure drop across the valve (mm Hg). Valve resistance is determined by an equation proposed by Ford [24]:

$$R_{AV} = \frac{1333 * \Delta P_m}{Q/60}$$

It is important to note that the anatomic opening area of the valve determined by photography will in general not agree with that calculated by the Gorlin equation from the hemodynamic data. This is due to the fact flow through a stenotic aortic valve forms a turbulent jet (Figure 2). Using the pressure difference in between left ventricle and the *vena contracta* pressure tap will yield an effective area that reflects the coefficient of contraction of the jet, which *in vivo* approaches 0.60 for severely stenotic valves [25]. On the other hand, using the downstream, recovered pressure in the formula more accurately represents the overall energy loss due to the obstruction caused by the valve; therefore, there is an inverse relationship between the amount of pressure recovery and the Gorlin-derived aortic valve area [26]. The result from the Gorlin equation is also flow dependent. Therefore, this data will also be used to elucidate the relationship between anatomic area, calculated area, and hemodynamic state.

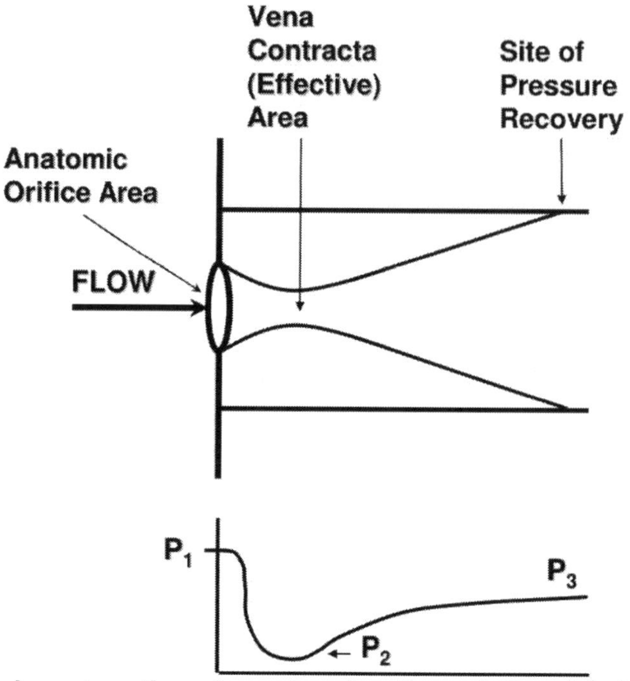

Figure 2. Jet flow through a stenotic valve, showing the relationship between the anatomic and the effective orifice area.

Results

Instantaneous anatomic area was measured from the individual photographs showing the opening of the aortic valve (Figure 3). Plots of anatomic area versus time were then generated (Figure 4). The rate of valve opening was affected by flow rate, whereas valve closing was affected by valve treatment; the stiffer valve would close faster. A higher afterload

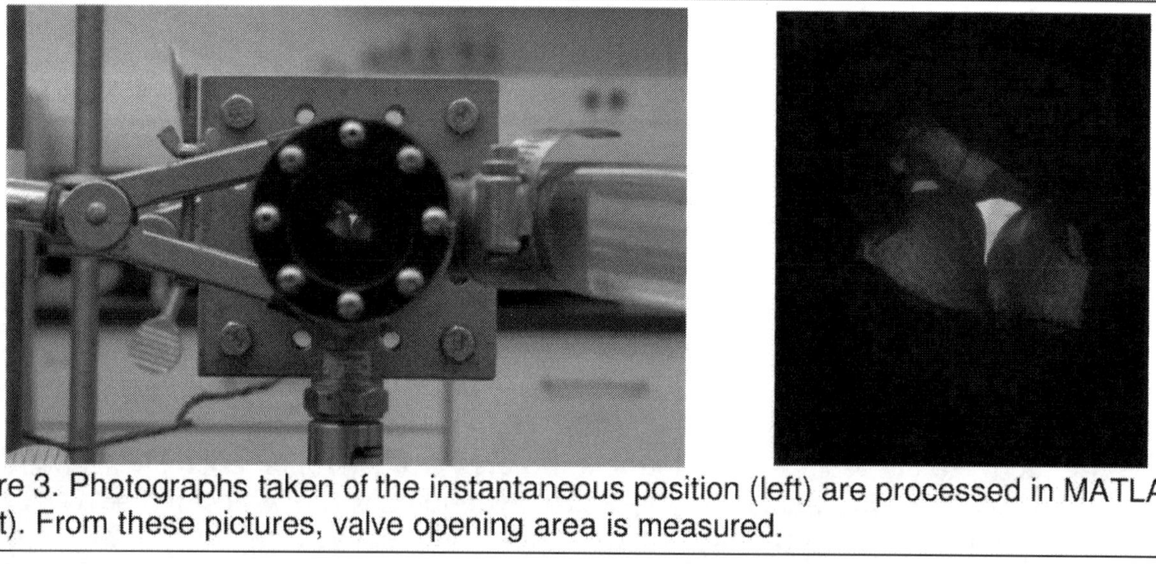

Figure 3. Photographs taken of the instantaneous position (left) are processed in MATLAB (right). From these pictures, valve opening area is measured.

also leads to faster valve closure. Therefore, factors extrinsic to the properties of the valve itself will affect the valve area ratio measured by echocardiography.

Sample pressure and flow traces obtained from are displayed in Figure 5. These data are correlated to the photographic data, and instantaneous flow/pressure/valve area relationships can be studied. The flow dependence of the Gorlin equation is demonstrated with these data; however, the data suggest that valve resistance may be a robust marker of valve stiffness. A marker of the properties of the valve that is independent of other hemodynamic factors may prove valuable in predicting outcome, as well as in evaluating therapeutic strategy designed to slow the progression of the degenerative disease.

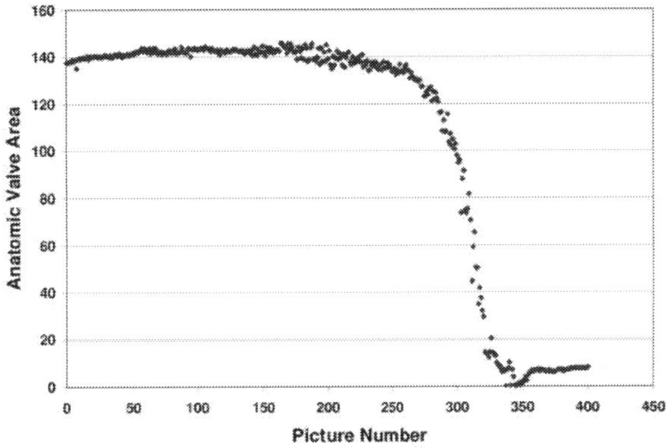

Figure 4. Anatomic area for an ethanol-treated valve at 6 l/min. The closing phase of the cycle is captured. In this graph, Area is in mm^2 and the time between each picture is 1 ms.

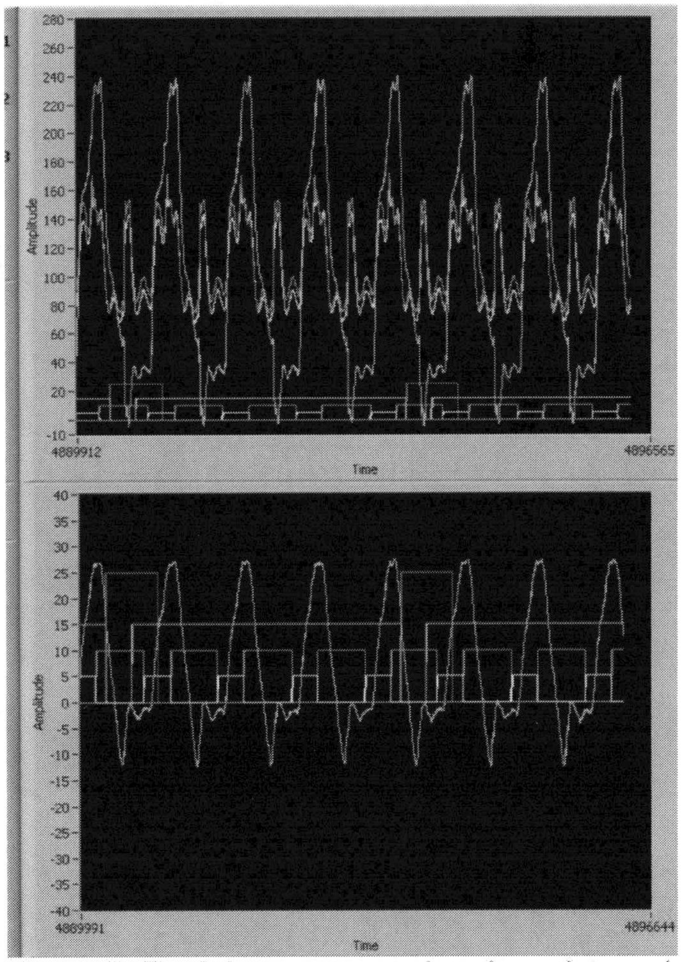

Figure 5. Real-time pressure data from the model ventricle and aorta (top) and flow rate (bottom). Superimposed on the flow waveform iare signals from solenoids, stroboscope and camera shutter.

Acknowledgments

This work was funded in part by an Initial Investigator Grant from the Florida and Puerto Rico Affiliate of the American Heart Association (Grant #0130459B).

References

[1] M. D. VanAuker and J. A. Strom, "Inter-laboratory comparisons: approaching a new standard for prosthetic heart valve testing in vitro," *J Heart Valve Dis,* vol. 8, pp. 384-91, Jul 1999.

[2] W. W. Nichols and M. F. O'Rourke, *McDonald's Blood Flow in Arteries*, 3 ed. Philadelphia, PA: Lea and Febiger, 1990.

[3] J. Levenson, A. C. Simon, F. A. Cambien, and C. Beretti, "Cigarette smoking and hypertension. Factors independently associated with blood hyperviscosity and arterial rigidity," *Arteriosclerosis,* vol. 7, pp. 572-7, Nov-Dec 1987.

[4] R. V. Hogikyan, A. T. Galecki, B. Pitt, J. B. Halter, D. A. Greene, and M. A. Supiano, "Specific impairment of endothelium-dependent vasodilation in subjects with type 2 diabetes independent of obesity," *J Clin Endocrinol Metab,* vol. 83, pp. 1946-52, Jun 1998.

[5] D. S. Celermajer, K. E. Sorensen, D. Georgakopoulos, C. Bull, O. Thomas, J. Robinson, and J. E. Deanfield, "Cigarette smoking is associated with dose-related and potentially reversible impairment of endothelium-dependent dilation in healthy young adults," *Circulation,* vol. 88(5) Pt 1, pp. 2149-55, Nov 1993.

[6] H. R. Black, "New concepts in hypertension: focus on the elderly," *Am Heart J,* vol. 135(2) Pt 2, pp. S2-7, Feb 1998.

[7] J. A. Panza, A. A. Quyyumi, T. S. Callahan, and S. E. Epstein, "Effect of antihypertensive treatment on endothelium-dependent vascular relaxation in patients with essential hypertension," *J Am Coll Cardiol,* vol. 21, pp. 1145-51, Apr 1993.

[8] M. T. Johnstone, S. J. Creager, K. M. Scales, J. A. Cusco, B. K. Lee, and M. A. Creager, "Impaired endothelium-dependent vasodilation in patients with insulin-dependent diabetes mellitus," *Circulation,* vol. 88, pp. 2510-6, Dec 1993.

[9] M. A. Creager, J. P. Cooke, M. E. Mendelsohn, S. J. Gallagher, S. M. Coleman, J. Loscalzo, and V. J. Dzau, "Impaired vasodilation of forearm resistance vessels in hypercholesterolemic humans," *J Clin Invest,* vol. 86, pp. 228-34, Jul 1990.

[10] A. P. Avolio, F. Q. Deng, W. Q. Li, Y. F. Luo, Z. D. Huang, L. F. Xing, and M. F. O'Rourke, "Effects of aging on arterial distensibility in populations with high and low prevalence of hypertension: comparison between urban and rural communities in China," *Circulation,* vol. 71, pp. 202-10, Feb 1985.

[11] K. E. Sorensen, D. S. Celermajer, D. Georgakopoulos, G. Hatcher, D. J. Betteridge, and J. E. Deanfield, "Impairment of endothelium-dependent dilation is an early event in children with familial hypercholesterolemia and is related to the lipoprotein(a) level," *J Clin Invest,* vol. 93, pp. 50-5, Jan 1994.

[12] H. O. Steinberg, H. Chaker, R. Leaming, A. Johnson, G. Brechtel, and A. D. Baron, "Obesity/insulin resistance is associated with endothelial dysfunction. Implications for the syndrome of insulin resistance," *J Clin Invest,* vol. 97, pp. 2601-10, Jun 1 1996.

[13] J. O. Toikka, P. Niemi, M. Ahotupa, H. Niinikoski, J. S. Viikari, T. Rönnemaa, J. J. Hartiala, and O. T. Raitakari, "Large-artery elastic properties in young men : relationships to serum lipoproteins and oxidized low-density lipoproteins," *Arterioscler Thromb Vasc Biol,* vol. 19, pp. 436-41, Feb 1999.

[14] M. Lindroos, M. Kupari, J. Heikkilä, and R. Tilvis, "Prevalence of aortic valve abnormalities in the elderly: an echocardiographic study of a random population sample," *J Am Coll Cardiol,* vol. 21, pp. 1220-5, Apr 1993.

[15] C. M. Otto, I. G. Burwash, M. E. Legget, B. I. Munt, M. Fujioka, N. L. Healy, C. D. Kraft, C. Y. MiyakeHull, and R. G. Schwaegler, "Prospective study of asymptomatic valvular aortic stenosis - Clinical, echocardiographic, and exercise predictors of outcome," *Circulation,* vol. 95, pp. 2262-2270, May 6 1997.

[16] C. M. Otto, A. S. Pearlman, and C. L. Gardner, "Hemodynamic progression of aortic stenosis in adults assessed by Doppler echocardiography," *J Am Coll Cardiol,* vol. 13, pp. 545-50, Mar 1 1989.

[17] M. Peter, A. Hoffmann, C. Parker, T. Lüscher, and D. Burckhardt, "Progression of aortic stenosis. Role of age and concomitant coronary artery disease," *Chest,* vol. 103, pp. 1715-9, Jun 1993.

[18] V. L. Roger, A. J. Tajik, K. R. Bailey, J. K. Oh, C. L. Taylor, and J. B. Seward, "Progression of aortic stenosis in adults: new appraisal using Doppler echocardiography," *Am Heart J,* vol. 119(2) Pt 1, pp. 331-8, Feb 1990.

[19] S. J. Lester, D. B. McElhinney, J. P. Miller, J. T. Lutz, C. M. Otto, and R. F. Redberg, "Rate of change in aortic valve area during a cardiac cycle can predict the rate of hemodynamic progression of aortic stenosis," *Circulation,* vol. 101, pp. 1947-1952, Apr 25 2000.

[20] N. Westerhof, F. Bosman, C. J. De Vries, and A. Noordergraaf, "Analog studies of the human systemic arterial tree," *J Biomech,* vol. 2, pp. 121-43, May 1969.

[21] M. J. Carney, "Literary Review of Aortic Valve Replacement Testing, Production, Problems and Solutions," in *Medicine.* vol. M.S., Biomedical Science Tampa, FL: University of South Florida, 2008.

[22] C. G. Caro, T. J. Pedley, R. C. Schroter, and W. A. Seed, *The Mechanics of the Circulation.* Oxford, UK: Oxford University Press, 1978.

[23] R. Gorlin and S. G. Gorlin, "Hydraulic formula for the calculation of the area of the stenotic mitral valve, other cardiac valves, and central circulatory shunts.," *American Heart Journal,* vol. 41, pp. 1-29, 1951.

[24] L. E. Ford, T. Feldman, Y. C. Chiu, and J. D. Carroll, "Hemodynamic resistance as a measure of functional impairment in aortic valvular stenosis," *Circ Res,* vol. 66, pp. 1-7, Jan 1990.

[25] W. Gotsis, J. S. Meisner, J. Shirani, E. S. Monrad, K. Lerrick, and J. A. Strom, "Measurement of coefficient of velocity for stenotic aortic valves," in *Scientific Sessions of the American Heart Association*, 1995.

[26] W. K. Laskey and W. G. Kussmaul, "Pressure recovery in aortic valve stenosis," *Circulation,* vol. 89, pp. 116-21, Jan 1994.

572ac Formulation and Analysis of An Ultrascale Protein Interaction Network Involved In the Androgen Response of Prostate Cancer Epithelial Cells

Ryan A. Tasseff[1], Satyaprakash Nayak[2], Poorvi Kaushia[3], Noreen Rizvi[4], Saniya Salim[4], and Jeffrey D. Varner[4]. (1) Chemical Engineering, Cornell University, 120 Olin Hall, Ithaca, NY 14850, (2) Department of Chemical and Biomolecular Engineering, Cornell University, Olin Hall, Ithaca, NY 14853, (3) Computational Biology, Cornell University, 120 Olin Hall, Ithaca, NY 14850, (4) Chemical and Biomolecular Engineering, Cornell University, 120 Olin Hall, Ithaca, NY 14850

At a projected 218,000 new incidences in 2007 prostate cancer is the most common cancer found in men in the United States, and the second leading cause of caner related death with 27,000 new deaths projected in 2007. Currently androgen ablation therapy is the primary treatment for metastatic androgen-dependent prostate cancer. The inhibition of androgens initially leads to decreased tumor growth and decreased prostate specific antigen, a common biomarker of prostate cancer. Unfortunately, in nearly all cases such treatment eventually fails due to the development of androgen-independent prostate cancer (AIPC). To the authors knowledge there is currently no available treatment for metastatic AIPC.

One possible mechanism for AIPC action is the ligand independent activation of androgen receptor (AR) via cross talk with other signaling pathways. Experimental observations have, in fact, revealed that proteins in the mitogen activate protein kinase (MAPK) cascade act as kinases for this outlaw AR activation. Specifically, ligand independent activation of the transmembrane growth receptor Her2 has been observed to initiate the MAPK cascade, resulting in AR activation, increased PSA expression and an increased rate of proliferation in several prostate cancer model cell lines. The proportion of active Her2 is strongly dependent on the expression level of the inter cellular form of Prostatic acid phosphatase (PAcP). Not surprisingly, inter cellular PAcP expression negatively correlates with prostate cancer tumor grade.

Currently, mathematical modeling has not played a major role in the study of the molecular basis of disease progression and the development of therapeutics. In this study we explored the working hypothesis that ultrascale mathematical modeling of complex protein interaction networks may be used to identify critical AIPC mechanisms despite model uncertainties. We present the results of a kinetic model for the outlaw activation of AR via MAPK and the regulation of several key proteins. The first generation ordinary differential equation prostate model architecture was composed of a collection of independently verified subsystems, such as MAPK signaling and translation initiation, together with other molecular interactions supported by the experimental literature. In total, our prostate model described the dynamics of 212 species and 384 interactions, where mass action kinetics were assumed for each interaction rate. Due to the limited data, the model parameters were under defined and poorly constrained. To properly account for the effect of unconstrained parameters a Monte Carlo ensemble approach was employed for predictions and for the estimation of prediction errors. Although model parameters in the ensemble varied over several orders of magnitude, the predictions were constrained.

This observation demonstrated the reliability of model predictions despite parametric uncertainty.

The preliminary model captured the androgen response of both the androgen-dependent and androgen-independent clones of the LNCaP prostate cancer cell-line. The model also qualitatively reproduced experimental data pertaining to key proteins, such as Her, PAcP, and the kinases of the primary MAPK cascade. Through a preliminary sensitivity analysis we observed that models of the more aggressive, androgen-independent, clones were notably more robust to perturbation. This result may provide insight into a cancer cell's ability to thrive in an environment where healthy cells would either growth arrest or die. However, our analysis, also, predicted that the relative sensitivity of certain mechanisms pertaining to translation in androgen-independent clones to be elevated compared to androgen-dependent clones. When taken together, these results suggest that the manipulation of the translation subsystem may have a larger impact on cancer cells relative to healthy cells, and therefore, suggests these mechanisms may be low side effect, novel therapeutic targets for the treatment of AIPC.

Web Page: www.varnerlab.org

572ad Effect of Dexamethasone Levels on Osteogenic Differentiation of Wharton's Jelly Stem Cells from the Human Umbilical Cord

Erica Brown and Vassilios I. Sikavitsas. Chemical, Biological, and Materials Engineering, University of Oklahoma, 100 E Boyd Str, RM T-335, Norman, OK 73019

Wharton's Jelly Stem Cells (WJSC) constitutes the mucuoid connective tissue surrounding the perivascular region of the human umbilical cord. These cells have been characterized as a primitive stem cell population as they possess differentiation capabilities similar to that of adult stem cells. The aim of this study was to investigate WJSC for use as a novel allogenic cell source for regenerative bone therapies.

Experiments suggest osteogenic differentiation of Wharton's Jelly Stem Cells (WJSC) in static 2D and 3D cultures as evidenced by mineralization, alkaline phosphatase activity, and other phenotypic changes. Wharton's Jelly Stem cells, cultured with osteogenic media containing 10nm of dexamethasone in 2D static cultures, demonstrated a significant decrease in cell number after 7 days in comparison to control groups. This observation led us to study the influence of dexamethasone levels on osteogenic differentiation and to identify the optimal dexamethasone concentration in the cell culture that maximizes osteogenic differentiation without compromising the proliferative capacity of the cells.

For this reason 2D and 3D static cultures of WJSC were cultured in osteogenic media with dexamethasone concentrations of 10nM, 1nM, 0,1nM, and no dexamethasone for time periods of 4 and 7 days. An additional group was investigated in which WJSC cultures were spiked with osteogenic media containing dexamethasone at 10nM for 1 hour with a continuation of 71 hours in osteogenic media containing no dexamethasone.

572ae A Computational Study of Renal Artery Aneurysms and Possible Arterial Occlusions Leading to Renovascular Hypertension

Linden A. Heflin, Dimitrios Papavassiliou, and Edgar O'Rear. Chemical, Biological and Materials Engineering, University of Oklahoma, 100 E. Boyd St., T-335, Norman, OK 73019

Alterations in renal artery geometry can lead to increased systemic blood pressure. This is observed clinically in cases of renal artery stenosis, where the constriction significantly alters the hemodynamics in the artery. This is thought to be primarily due to unwarranted initiation of the renin-angiotensin-aldosterone system, a renal hormonal feed back loop that responds to changes in blood pressure. The hormonal cascade that leads to vasoconstriction is initiated by baroreceptors in the kidney that sense incoming blood pressure and release renin as needed. Unusual renal artery geometry, such as a stenosis, could lead to increased pressure losses through the artery, causing the pressure at the baroreceptors to be low, and unnecessarily initiating the cascade. However, it is not known precisely what variations in renal artery geometry can induce hypertension. In addition to stenosis, renal artery aneurysms are strongly associated with renin-dependent hypertension. There is much debate over what renal artery aneurysm morphologies indicate surgical intervention, but the conversation focuses mainly on the size of, or symptoms associated with, the aneurysm. The results of multiple pulsatile flow simulations show that for some aneurysm geometries, high pressure within the aneurysm could cause the weakened artery wall to be forced into the blood stream, creating a blockage. This work seeks to investigate a new hypothesis to describe the relationship between renal artery aneurysms and renin-dependent hypertension.

572af Epithelial-Mesenchymal Transition Is Regulated by Cell Shape and Matrix Compliance

Esther W. Gomez[1], Derek C. Radisky[2], and Celeste M. Nelson[1]. (1) Chemical Engineering, Princeton University, A314 Engineering Quadrangle, Princeton, NJ 08544, (2) Mayo Clinic Cancer Center, Jacksonville, FL 32224

Epithelial-mesenchymal transition (EMT), a phenotypic alteration in which epithelial cells detach from their surroundings and become more migratory, is becoming increasingly acknowledged as playing an important role in pathological conditions including fibrosis, tumor progression, and metastatic invasion. Here, we show that under certain conditions cell shape and matrix compliance are critical in determining whether epithelial cells will undergo EMT. EMT can be induced in cultured mouse mammary epithelial cells by matrix metalloproteinase-3 (MMP-3) resulting in expression of Rac1b, alterations in the cytoskeletal structure, and increased spreading of the cells. We find that increased cellular spreading is a requirement for MMP-3-induced EMT in mammary epithelial cells. Our results provide insight into how biophysical properties of the cellular microenvironment contribute to EMT and may aid in future efforts to identify therapeutic targets for blocking MMP-mediated fibrosis and malignancy *in vivo*.

572ag Human Tissue Factor: A Molecular Dynamics Study

Kristin Patterson[1], Coray M. Colina[1], Lalith Perera[2], and Lee G. Pedersen[3]. (1) Department of Materials Science and Engineering, Pennsylvania State University, 121 Steidle Building, University Park, PA 16802, (2) Laboratory of Quantitative and Computational Biology, National Institute of Environmental Health Science, Research Triangle Park, Raleigh, NC 27709, (3) University of North Carolina, Campus Box 3290, Chapel Hill, NC 27599

Most descriptions of the coagulation process state that factor VII (FVII) binds to tissue factor (TF) when an injury occurs and blood leaves the vasculature and encounters TF in the extravascular compartment. Assembly of the FVII(a)/TF complex then leads to initiation of coagulation. The Tissue Factor (TF) pathway (or the extrinsic pathway) is thus believed to play a primary role in initiating blood coagulation during normal hemostasis as well as during many pathologic situations, including arterosclerosis and septicemia [1, 2]. However, there is indirect evidence that this scenario is not accurate [3, 4].

Several studies have also been performed to elucidate the importance of the so-called allosteric disulphide bonds. Human TF consists of 263 residues, the first 219 of which comprise the extracellular region [5]. Extracellular TF (219) consists of two immunoglobulin-like domains connected by a single polypeptide linker (PRO 102 to ASN 107). Each domain is formed by two antiparallel b-sheets. TF has two disulphide bridges at positions 49-57 and 186-209. It has been suggested that the Cys-186-Cys209 disulfide repositions the adjacent F and G beta strands that reorient nearby residues and enable productive binding of the TF substrates.

In this work, we have used Molecular Dynamics (MD) simulations to predict the solution

structure of free TF (fTF) and TF bound to factor VIIa (bTF). A complete refined solution equilibrated model for TF based on available crystal structures (1BOY, 2HFT, 1DAN) using MD simulations in aqueous medium for over 15 ns is presented.

The X-ray crystal structures (1BOY, 2HFT, 1DAN) were used initially as starting points in modeling the unbound and bound structures. The complete models (after missing regions, water molecules, and ions have been added) were subjected to aqueous-phase molecular dynamics (MD) simulations, where unconstrained dynamics were performed using the Cornell et al. force-field (using AMBER9/PMEMD9 programs [6]). The Particle Mesh Ewald (PME) method was used in all aqueous simulations to account for long-range interactions during dynamics. NPT simulations at 300 K, SHAKE algorithm for bond constraints and a 1.0 fs time step were used. The solute, ions/counter-ions and crystal water molecules, were immersed in rectilinear periodic boxes of at least 20Å each side. Standards steps of energy minimization and dynamics were carried out before beginning the "production run". Finally, the post-equilibration NPT dynamics at 300 K, or production run, were completed for at least 15 ns depending of the study case.

We observe significant inter-domain motion of fTF in the present simulations, where the polypeptide linker has the major contribution to the overall motion. From our preliminary molecular dynamic simulations however, we observe only small changes relative to the disulphide bond presence (or not). We have performed these simulations to compare and complement mutational studies. From our long molecular dynamic simulations, the Cys186-Cys209 disulfide bond of TF (the so-called allosteric disulfide), does not appears to play a structural role on FVIIa activation or reorientation of nearby residues in TF. Although these preliminary results seems reasonable based on the antiparallel beta sheets structure of TF, where each beta sheet contain 7 strands, and where the presence of the disulfide bond near the membrane surface does not appear to provide extra strength to the structure, a much more thorough study is needed.

Finally, these equilibrated solution structures for tissue factor should provide insight into the details of how tissue factor enhances the action of FVIIa. The possibility to predict these dynamic conformations could represent a crucial impact for the areas of

protein and drug design. It is also important to keep in mind that the propose carefully constructed models will, since they intrinsically incorporate the experimental data, provide a complete, dynamical molecular understanding of structures and, as such, will be a natural starting point for suggesting mutational experiments, protein engineering and therapeutic design leads, in addition to provide a uniform foundation with which to incorporate new structural information.

References.

1.Bajaj, S. P., Schmidt, A. E., Agah, S., Bajaj, M. S. and K. Padmanabhan, J. Biol. Chem. 281, 24873-24888 (2006)

2.Davie, E. W., Fujikawa, K., and W.Ksie, Biochemistry, 30, 10363-10370 (1991)

3.Hoffman, M., Colina, C. M., Harger, A. G., Arepally, G., Pedersen, L. and D. M. Monroe, J. Thromb. Haesm.

4.Bach, R. R. Arterioscler Thromb, Vasc. Biol. 26, 456-461 (2006).

5.Daubie, V., Pochet, R., Houard, S. and P. Philippart, J. Tissue Eng. Reg. Med. 1, 161-169 (2007)

6.Case, D.A. , Darden, T.A. Cheatham, III, T.E. Simmerling, C.L. Wang, J. Duke, R.E. Luo, R. Merz, K. M. Pearlman, D.A. Crowley, M. Walker, R.C. Zhang, W. Wang, B. Hayik, S. Roitberg, A.,G. Seabra, K. F. Wong, F. Paesani, X. Wu, S. Brozell, V. Tsui, H. Gohlke, L. Yang, C. Tan, J. Mongan, V. Hornak, G. Cui, P. Beroza, D.H. Mathews, C. Schafmeister, W.S. Ross, and P.A. Kollman (2006), AMBER 9, University of California, San Francisco.

572ah Engineering Stem Cell Derived Tissue In a Novel 3D Perfusion Bioreactor

Matthew H. M. Lim[1], I. Gabrielle M. Brons[2], Karel Domansky[3], Linda G. Griffith[3], and Roger A. L. Pedersen[2]. (1) Chemical Engineering, University of Cambridge, Pembroke Street, Cambridge, CB23RA, United Kingdom, (2) Department of Surgery, University of Cambridge, Box 202, Level E9, Addenbrooke's Hospital, Hills Road, Cambridge, cb20qq, United Kingdom, (3) Biological Engineering Division and Biotechnology Process Engineering Center, Massachusetts Institute of Technology, 77 Massachusetts Avenue, Cambridge, MA 02139

Using a three-dimensional perfused flow bioreactor we differentiate murine epiblast stem cells (Brons et al, 2007) into cardiomyocytes. Differentiating the cells under physiologically relevant conditions of flow enables us to promote angiogenesis resulting in vascularised cardiac tissue.

Compared to the other organs in the body the heart has one of the poorest capacities for regeneration. As the adult human heart has only a very limited regenerative capacity, any significant heart cell loss or dysfunction due to ischemia, viral infection, or inflammation may possibly lead to the development of progressive heart failure. Congestive heart failure is currently a growing epidemic that results in significant disability and mortality while placing a heavy burden on health care systems (Cohn et al., 1997). For example, it is estimated that heart failure is responsible for more hospitalizations than all forms of cancer combined. Despite advances in pharmacological, interventional, and surgical therapeutic measures, the prognosis for heart failure patients remains poor. With chronic lack of donors limiting the number of patients who can benefit from heart transplantations, the development of new therapeutic methodologies have become imperative (Capri and Gepstein, 2006).

Zimmerman et al., (2006) stated that non terminally differentiated but electrically and functionally interconnected cardiac myocyte networks have a better chance to survive after implantation in vivo and stress the importance of culturing cells in three-dimensions to effectively engineer tissue in this manner. It has previously been shown that embryonic stem cell–derived cardiomyocytes form stable grafts following transplantation into immune-suppressed animals (Klug et al., 1996). Embryonic stem cell–derived cardiomyocytes have also been shown effectively cultured in large volume using bioreactors (Zandstra et al., 2003). Although embryonic stem cells are highly proliferative and suitable for mass production, an efficient protocol is yet to be established to ensure selective cardiomyocyte induction using these cells.

Recent advances in developmental biology have clarified the involvement of critical factors in cardiomyocyte differentiation, including bone morphogenic protein and Wnt signaling proteins, and such factors have been shown to improve the efficiency of stem cell induction (Fukuda and Yuasa 2006). Current protocols yield cardiomyocytes in sufficient quantities for most basic research needs. However, a limitation of the use of embryonic stem cells for therapeutic purposes is the inefficiency with which cardiomyocytes are generated (typically 1% of a differentiating culture). Strategies based on developmental paradigms have used directed differentiation to increase cardiogenesis (Lev, et al., 2005). These methods have had some success; however, not surprisingly, no one has been able to fully recapitulate the complex mixture of cardiogenic factors and environmental cues that efficiently induce cardiomyogenesis during embryonic development (Murray et al., 2006).

Angiogenic paracrine activity has recently been suggested as the underlying mechanism of improved cardiac function following bone marrow–derived cell transplantation into injured hearts (Kinnaird et al., 2004). The potential of endothelial precursor cells to differentiate into cardiomyocytes has been demonstrated in vitro (Badorff et al., 2003 and Yamashita et al, 2000), but not yet in vivo. Nonetheless, the potential of exogenous endothelial precursor cells to contribute to angiogenesis is well established. The blood vessel wall is inherently subjected to and affected by the pulsatile hemodynamic stimulus of blood flow within the vascular lumen, and biomechanical forces intrinsically present as a result of this hemodynamic flow are believed to play an important role in vascular development, remodeling, and lesion formation (Ghajar et al., 2006). In particular, cells lining the vascular lumen are constantly subjected to shear stress at the apical endothelial surface exerted by blood flow (Ballermann et al., 1998 and Lehoux et al., 2003). Shear stress has been recognized as an important modulator of endothelial phenotype, morphology, gene expression, and, especially, differentiation (Gloe et al., 2002 and Yamamoto et al., 2003).

Huang et al., (2005) studied the impact of shear stress on embryonic stem cells seeded into the walls of a microporous tube. After a 2-day incubation under a weak pulsatile flow they showed that [wall shear stress from -0.98 to 2.2 dyne/cm2; circumferential strain 4.6–9.6 × 104 dyne/cm2] without vascular endothelial growth factor (VEGF), embryonic stem cells in the superficial layer were regularly oriented in the direction of the pulsatial flow. The cells growing into the interstices in the deeper layer showed smooth muscle-like appearance, and were smooth muscle actin (SMA)-positive. Differentiation to two different cell types and segregation of incorporated ES cells may be simultaneously encouraged by the combination of wall shear stress and circumferential strain.

In culturing large bodies of tissue, the maximal size and thickness will critically depend on oxygen and nutrient supply. Perfusion in vitro may improve the metabolic supply and could be achieved by induction of angiogenesis or vasculogenesis (Risau, 1995 and Risau et al., 1997). Although it has been shown that engineered myocardium survives if implanted on healthy and infracted hearts, under these conditions, i.e. in the absence of immediate vascularization, engineered heart muscle must be nurtured exclusively by diffusion, which is unlikely to sufficiently support thick vascular myocardial constructs. Hence, ‘‘prevascularized" heart muscle grafts would be preferable, on the one hand, to allow construction of complex tissues in vitro and, on the other hand, to connect the vascular bed of such grafts to the recipient circulation at the time of implantation (Zimmerman et al. 2006).

In our studies we show that culturing epiblast stem cells in our bioreactor system promotes the formation of tissue expressing VEGF Receptor 1 (Flk-1) shown by both gene expression and immunostaining. Flk-1 is a marker for endothelial cells and has been shown to be a marker for vascular progenitors as well . Thus in carrying out a cardiomyocyte differentiation protocol concurrently during the establishment of the microvessel network allows us to form vascularised cardiac tissue in addition also expressing cardiac actin, troponin and SMA. Culturing vascularised tissue enhances the transport of factors and oxygen to all the cultured tissue mass giving us the ability to culture larger volumes of tissue whilst avoiding hypoxia. This will also provide a more physiologically relevant model for planned transplantation studies.

Badorff C, Brandes RP, Popp R, Rupp S, Urbich C, et al. 2003. Transdifferentiation of blood-derived human adult endothelial progenitor cells into functionally active cardiomyocytes. Circulation 107:1024–32

Ballermann BJ, Dardik A, Eng E, Liu A. Shear stress and the endothelium.Kidney Int Suppl. 1998;67: S100 –S108.

Cohn J.N., Bristow M.R., Chien K.R., Colucci W.S., Frazier O.H., Leinwand L.A., Lorell B.H., Moss A. J., Sonnenblick E.H., Walsh R.A., Mockrin S.C., Reinlib L., Report of the National Heart, Lung, and Blood Institute Special Emphasis Panel on Heart Failure Research, Circulation 95 (1997) 766–770.

Fukuda K., Yuasa S., Stem Cells as a source of regenerative cardiomyocytes, Circulation research 98 (2006) 1002-1013

Gloe T, Sohn HY, Meininger GA, Pohl U. Shear stress-induced release of basic fibroblast growth factor from endothelial cells is mediated by matrix interaction via integrin alpha(v)beta3. J Biol Chem. 2002;277: 23453–23458.

Huang H.Y., Nakayama Y, Qin K.R, Yamamoto K, Ando J, Yamamshita J., Itoh H, Kanda K., Yaku H., Okamoto Y., Neumoto Y., Differention from embryonic stem cells to vascular wall cells under in vitro pulsatile flow loading, Journal of Artificial Organs 8 (2005) 110-118

Kinnaird T, Stabile E, Burnett MS, Shou M, Lee CW, et al. 2004. Local delivery of marrow-derived stromal cells augments collateral perfusion through paracrine mechanisms. Circulation 109:1543–49

Klug M.G., Soonpas M.H., Koh G.Y., Field L.J., Genetically selected cardiomyocytes from differentiating embryonic stem cells for stable intracardiac grafts, Journaol of Clinical Investigation 98, 1 (1996) 216-224

Lehoux S, Tedgui A. Cellular mechanics and gene expression in blood vessels. J Biomech. 2003;36:631– 643.

Risau W. Mechanisms of angiogenesis. Nature 1997;386:671– 4.

Risau W, Flamme I. Vasculogenesis. Annu Rev Cell Dev Biol1995;11:73–91.

Yamamoto K, Takahashi T, Asahara T, Ohura N, Sokabe T, Kamiya A, Ando J. Proliferation, differentiation, and tube formation by endothelial progenitor cells in response to shear stress. J Appl Physiol. 2003;95: 2081–2088

Zandstra PW, Bauwens C, Yin T, Liu Q, Schiller H, et al. 2003. Scalable production of embryonic stem cell-derived cardiomyocytes. Tissue Eng. 9:767–78

Zimmerman W.H., Didie M., Doker S., Melnychenko I, Naito H., Rogge C., Tiburcy M., Eschenhagen T., Heart muscle engineering: An update on cardiac muscle replacement therapy, Cardiovascular Research 71 (2006) 419-429

572ai Systems Biology of Mucosal Injury In Cancer Patients

Douglas E. Peterson[1], **Ranjan Srivastava**[2], Jason R. White[2], Rajesh V. Lalla[1], and Leslie M. Loew[3]. (1) University of Connecticut Health Center, Farmington, CT 06030, (2) University of Connecticut, 191 Auditorium Rd., Unit 3222, Storrs, CT 06269, (3) Center for Cell Analysis and Modeling, University of Connecticut Health Center, Farmington, CT 06030

Background: Mucosal injury (mucositis) caused by high-dose or targeted cancer therapy can result in significant morbidity and financial cost. Understanding the causative complex molecular pathways of this toxicity will enhance design of new therapies. Computational reaction kinetic biological models represent an important strategic advance in this regard. These simulations are cost- and time-efficient and can identify new hypotheses and projected experimental outcomes. The technology also permits integration of complex models based on animal and human data. **Methods:** A mucositis reaction network structure was developed using COX-2 pathway data derived from literature-based animal and human studies as well as pre- and post-transplant oral mucosal biopsies from three hematopoietic cell transplant patients. Virtual Cell software was used to simulate network dynamics. **Results:** The nonlinear network mucositis model incorporated feedback loops and accounted for dynamic response of COX-1, COX-2, TNF-α, PGE-2, IL-1β, IL-6, PGI-2, TXA-2 and NF-κB with nonlinear correlation coefficients of -0.73, -4.27, -89.1, -7.85, 0.02, -2.79, 0.63, 0.86 and -0.02 respectively. For 7 of 8 molecular species, the model was able to predict final steady-state value within experimental error with exception of TNF-α. The ninth species, TXA-2, did not achieve steady-state within the timeframe monitored. Conclusions: Using the COX-2 pathway as a prototype, this model successfully incorporated experimental and computational approaches to elucidate mucositis network pathways. The results identify research areas requiring further study such as TNF-α dynamics, thus highlighting benefit of the model. This technology could also facilitate design of new mucositis research protocols including (i) mechanisms of neuropeptide-mediated pain, (ii) identification of tissue-based risk factors and (iii) integration of genetic, cytomorphologic and functional expression of oral and gastrointestinal mucosal injury.

572aj High-Capacity Assay to Evaluate Colony-Forming Efficiency and Multipotency of Bone Marrow Stromal Cells

Katie Russell[1], Kristin Meyertholen[1], Bonnie Barrilleaux[1], Darwin Prockop[2], Donald Phinney[2], and **Kim OConnor**[1]. (1) Chemical and Biomolecular Engineering Dept., Tulane University, New Orleans, LA 70118, (2) Gene Therapy Center, Tulane University Medical School, New Orleans, LA 70112

Human bone marrow stromal cells (hMSCs) are a promising source of multipotent progenitors for a broad range of stem cell therapies for regenerative medicine. A basic prerequisite to realizing the therapeutic potential of hMSCs is the ability to identify stem-like progenitors in a heterogeneous culture. To overcome current limitations in stem cell identification, we have developed a novel high-capacity assay that will characterize the heterogeneity inherent to hMSCs by (1) classifying cells according to colony-forming efficiency as a measure of proliferation capacity and trilineage potential as a measure of multipotency and (2) preserving a frozen template of these cell populations for future use. Specifically, the assay quantifies the percentage of tripotent progenitors, bipotent progenitors, lineage-committed unipotent progenitors and mature cells in an hMSC culture. The high-capacity format facilitates the analysis of large sample sizes required to obtain statistically significant data on hMSC populations. The assay has numerous clinical and basic research applications, for example, to monitor hMSC preparations in the clinic for consistent content of stem-like progenitors and to evaluate the effect of culture conditions on progenitor content in the laboratory. This work was funded with grants from NSF and NIH.

572ak Design and Characterization of Peptide Amphiphile-Based Immunotherapeutics: Amyloid-â Peptides and Alzheimer's Disease

Amanda Trent, Dimitris Missirils, and Matthew Tirrell. University of California, Santa Barbara, Materials Research Laboratory, MC5121, University of California, Santa Barbara, CA 93106-5121

Peptides play a critical role in determining the magnitude and specificity of cellular and humoral immunity. For this reason, peptide-based antigens capable of stimulating a specific and efficient immune response have long been sought to treat or prevent infectious diseases and other chronic ailments such as cancer. However, strategies for developing peptide-based immunotherapy face significant hurdles, including achieving potent immunogenicity, minimizing toxicity, and overcoming delivery challenges.

One possible alternative for satisfying these requirements is the design of peptide amphiphiles (PA), in which peptide antigens acylated with hydrophobic carbon tails can self-assemble into supra-molecular aggregates. The general properties of PA aggregates, including the multivalent presentation of antigen, the flexibility to control formation of various micellar and vesicular structures for delivery purposes, and the fact that peptide secondary structure is induced upon assembly, make PAs attractive for immunotherapeutic drug design. Moreover, the potential of hydrophobic tails to act as adjuvants may render the use of common toxic adjuvants dispensable.

We propose using amyloid-β peptides (involved in Alzheimer's Disease pathology) as model antigens to demonstrate the proof-of-principle concept of PA-based vaccines. Data revealing aggregate size, shape, composition, and peptide conformation will be presented. Subsequent to this basic physical characterization, in vitro assays as well as animal studies will be used to determine the strength and specificity of the elicited immune response of the amyloid-β constructs.

572al Affect of (*5Z*)-4-Bromo-5-(bromomethylene)-3-Butyl-2(*5H*)-Furanone on *Bacillus Anthracis* Metabolism

Christine Endicott[1], Thomas K. Wood[2], and Ranjan Srivastava[1]. (1) University of Connecticut, 191 Auditorium Rd., Unit 3222, Storrs, CT 06269, (2) Artie McFerrin Department of Chemical Engineering, Texas A&M University, 200 Jack E. Brown Building, MS 3122, College Station, TX 77843-3122

Bacillus anthracis is the etiological agent of anthrax. Due to its potential as a weapon for bioterrorism, it has been classified by the CDC as a Category A pathogen, the highest possible priority rating. Recent studies have shown that (*5Z*)-4-bromo-5-(bromomethylene)-3-butyl-2(*5H*)-furanone (furanone) is capable of inhibiting *B. anthracis* virulence [1]. However, the mechanism by which furanone accomplishes this result, as well as its impact on *B. anthracis* metabolism, are unknown. To begin to address these issues, we monitored the affect on growth rate, glucose uptake rate, and oxygen uptake when *B. anthracis* was grown in the presence of furanone. Growing the Sterne strain of *B. anthracis* in R media in the presence of furanone resulted in a 36-fold increase in glucose uptake rate and an 18-fold increase the oxygen uptake rate as compared to cultures without furanone. Remarkably the growth rate was not affected by the presence of furanone, showing a growth rate of approximately 0.65 hr^{-1} for both experimental and control cultures.

1. Jones, M.B., R. Jani, D. Ren, T.K. Wood, and M.J. Blaser, Inhibition of *Bacillus anthracis* Growth and Virulence-Gene Expression by Inhibitors of Quorum-Sensing. *J Infect Dis*, 2005. 191(11): p. 1881-8.

572am Implementing Miame - Quantification of Oligonucleotide Array Fidelity for Expression Profiling

Ian J. Laurenzi and **Erdem Arslan**. Chemical Engineering, Lehigh University, B330 Iacocca Hall, 111 Research Dr., Bethlehem, PA 18015

Today, scientists use a variety of microarray assays for biological investigations such as expression profiling. However, results of investigations using different platforms may not be the same due to the differences of the assay designs (i.e. oligonucleotide probe sequences). Although the Microarray and Gene Expression Data (MGED) Society recommends that certain information be provided with microarray data to assist in microarray data standardization, implementation of that information is a considerable challenge: quantification of the complex nature of the underlying hybridization reaction network yields either exceptionally coupled population balance equations. In this paper, we evaluate a number of oligonucleotide microarray design features using a computational method that we have developed for simulation of microarray experiments, namely, the number of reporters (probes) per gene, and the inclusion of mismatched (MM) probes for the ostensible development of a cross-hybridization-free differential signal (PM-MM). This method permits the evaluation of the sensitivity and specificity of each probe as well as the specific cross-hybridizing cDNA associated with each. Lastly, we use our simulation technique to evaluate the relative fidelities of probe sets designed using different oligomer design tools. Our results illustrate the utility of the Minimum Information About a Microarray Experiment (MIAME) criteria for engineering and standardization of microarrays and suggest methods for standardization of microarray technology via the robust population and thermodynamic quantification of microarray assays.

The αII spectrin is essential for nuclear envelope integrity

Zhixia Zhong, Kris Noel Dahl and Katherine L. Wilson

Introduction

The nuclear envelope consists of two membranes (INM and ONM), nuclear pore complexes, and an underlying protein structural support, the nuclear lamina network. In mammalian cells, the lamina is composed of A type lamin proterins (Lamin A and C) and B type lamin proteins which recruit and anchor most INM proteins, including emerin, LBR, nesprin-1 α, et al. At least twelve disease syndromes have been linked to mutations in lamin A and lamin-binding proteins. These diseases are termed as laminopathies including Emery-Dreifuss Muscular Dystrophy, Dilated Cardiomyopathy with Conduction System Disease, and Hutchinson-Gilford Progeria Syndrome, etc. The mechanisms of laminopathies have not been determined, but current evidence is consistent with defects in tissue-specific gene expression coupled to mechanical weakness of the nuclear envelope. The nuclear lamins are stiff proteins which make up a majority of the nuclear lamina. Our previous work showed the nuclear lamina is not a stiff network but has 'spring-like' memory, which can swell under osmotic pressure and shrink back to its original size and shape. Spectrins are a family of cytoskeleton proteins characterized with spring-like elastic structure. We hypothesize the sprectrin repeat proteins play important roles in mechanical properties of nuclear envelope, especially for the nuclear lamina network.

In humans, spectrin repeat proteins αII spectrin, nesprin-1α, nesprin-2β and βSpIV∑5 have been identified inside nucleus. Specifically, nuclear αII spectrin is under the study focus due to its nuclear interactions with the proteins associated with Fanconi anemia and its function in recruiting DNA interstrand crosslinks for DNA repair. Recently, the αII spectrin was also revealed to interact with five groups of nuclear proteins *in vitro*, including structural proteins locate at nuclear envelope, DNA repair proteins, chromatin remodeling proteins, Fanconi anemia proteins, transcription and RNA processing proteins. All these suggest that αIIS may be an important nuclear protein involve in both nuclear envelope architecture and nuclear function. However, little is known about the αII spectrin roles in the nuclear envelope and its influence in nuclear lamina and laminopathies. Our present work will focus on the study of nuclear α-II spectrin and our results suggest it is an essential structural component in nuclear envelope. Either overexpression or down-regulation of α-II spectrin will influence the nuclear envelope integrity as well as the cell mitosis.

Results and discusstion

αII spectrin localized in nucleus

In humans, αII spectrin (αIIS) is highly expressed in vertebrate nucleated cells especially brain, muscle and epithelial cells. It is mainly studied as a cytoskeletal protein that lines under the plasma membrane of many cell types, playing an important role in maintenance of plasma membrane integrity and cytoskeletal structure. In most cell types, the nuclear localized αII spectrin is a small portion compared to the cytoplasmic localized αII spectrin. In present work, western blot result reveals that αII spectrin isoforms localized in the nucleus of Hela cell, but little in cytoplasm of Hela (figure to be shown in poster). It suggests that Hela cell is an ideal cell line to analyze the αIIS functions in nucleus, without been largely affected by cytoplasmic αIIS. We also constructed the GFP-tagged cDNA of αII spectrin and transfected into Hela cell. The overexpressed αII spectrin finally localized in nuclear envelope (NE) supported by both time-depend statistical analysis of green signal localization (figure to be shown in poster) and the GFP αII spectrin dynamics tracked by live imaging (Fig. 1).

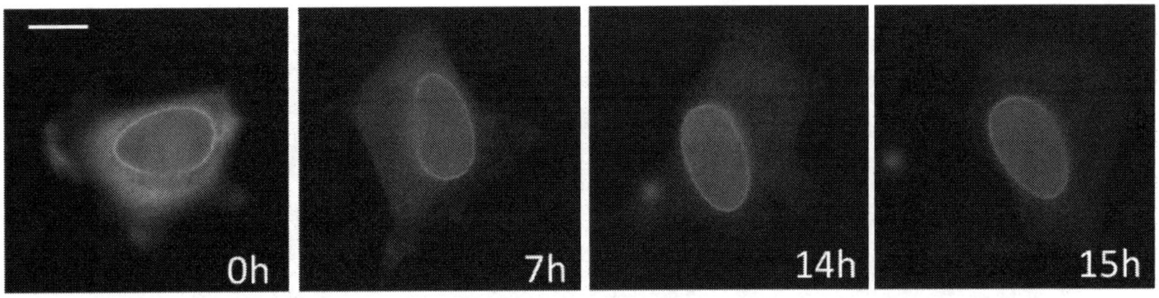

Fig.1 The time-lapse sequence of GFP- αIIS localization tracking. The cytoplasmic GFP- αIIS is transported into NE with time increase. Imaging taking started at three days after transfection labeled as 0h. Bar, 10 μ m.

αII spectrin influence nuclear envelope integrity and lamina network

Nuclear lamina is the major mechanical support of the cell nucleus. We examined the effect of αII spectrin knockdown on lamina network. We labeled lamin A/C and lamin B in αIIS knockdown cells. Most cells had lamina weakening area of the NE, present as both nuclear herniation or even leakage and reduced fluorescence (Figure 2 A, B). This reduced fluorescence signal was true for both lamin A/C and lamin B. The percentage of nuclear herniation and DNA leaking cells as well as the severity increased with the knockdown time increase. Such DNA-leaking cells are rarely found, even in the parallel control cells with lamin A/C knockdown (figure to be shown in poster). Moreover, statistical analysis (n>200) showed that the average size of αII spectrin knockdown nuclei is about 40% larger than normal cells and lamin A/C knockdown cells (figure to be shown in poster). And DNA signal of the large nuclei in αII spectrin knockdown cell reduced comparing to normal size nuclei, suggesting the size increase of αII spectrin knockdown nucleus may be due to nuclear expansion rather than proliferation of chromatin. Supporting this idea, the reduced fluorescence signal of lamin B in nucleus is correlated with the reduced fluorescent

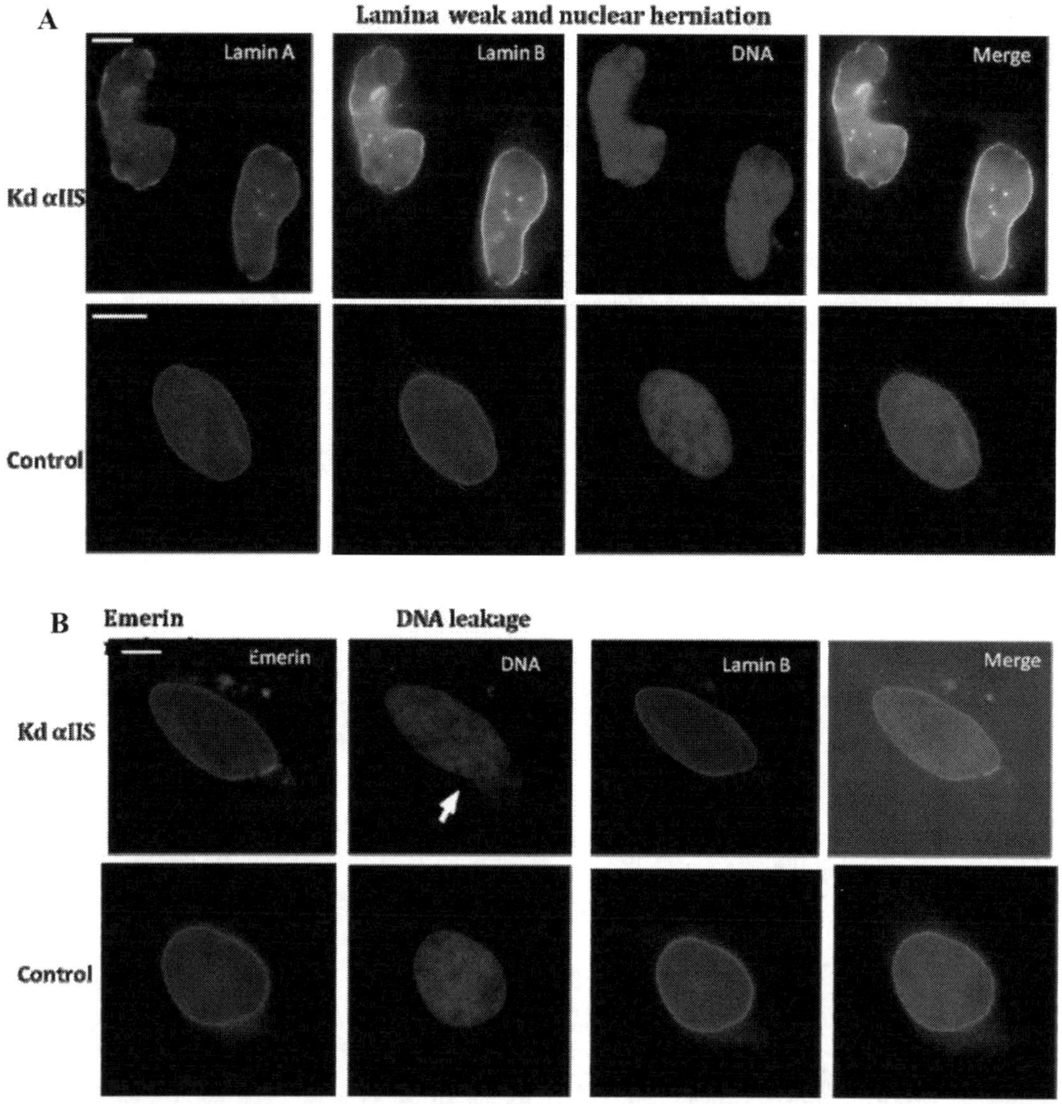

Fig.2 Cellular misbehaviors after knocking down αIIS. A) Nuclear herniation and lamina weakness B) DNA leakage and emerin mislocalization. Bar 10 μ m.

signal of DNA (Figure 3), suggesting loss of 'spring-like' structural protein αII spectrin results in the expansion of both lamina network and DNA. Thus, we suggest that αII spectrin is essential for the integrity of nuclear envelope and may contribute to the 'spring-like' properties of lamina network.

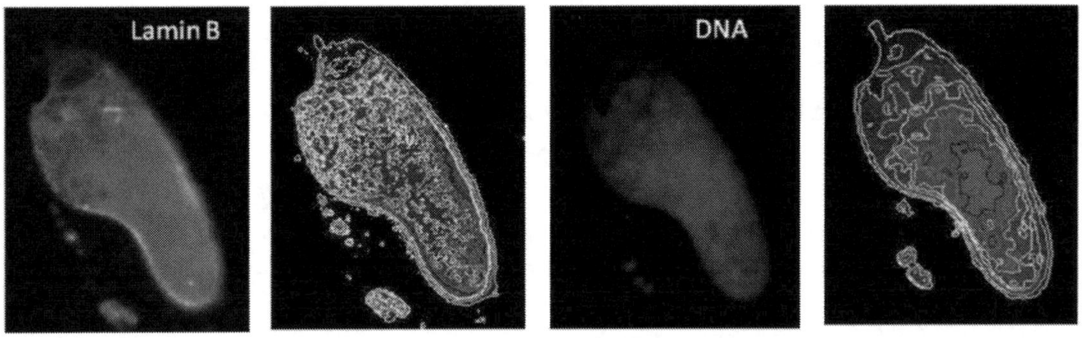

Fig. 3. Intensity analysis shows the reduced signal of local lamina is associated with DNA weakness and the nucleus expansion. From high pixel intensity to low intensity: black, red, orange, yellow, green, blue and white.

αII spectrin interacts with emerin and influence emerin localization

Emerin is an INM protein that can bind to lamin A and has shown direct interactions with αII-spectrin in vitro. Mutations in emerin and lamina are known to cause Emery Dreifuss Muscular Dystrophy, a typical laminopathy. Our results shown that the αII spectrin also play important role in emerin localization. In our previous work, the purified emerin complex from Hela nuclear extract contains lamin A, αII spectrin and other nuclear proteins. After αII spectrin knockdown, most cells showed emerin mislocalization to cytoplasm and abnormal emerin aggregates in cytoplasm (Figure 2B). With increasing time after transfection of the shRNA for knockdown, the percentage of cells with emerin mislocalization increases (figure to be shown in poster). Previous studies have reported that emerin mislocalizes to ER in lamin A deficient cells. Mice lacking the A-type lamins exhibit mislocalization of emerin as well as the appearance of muscular dystrophy. Here, we detected the emerin mislocalization by loss of αII spectrin. Whether αII spectrin plays roles in muscular dystrophy will also be interesting. Supporting the high affinity between αII spectrin and emerin, when we overexpress GFP-αIIS, the mislocalization of GFP-αIIS in cytoplasm is correlated with weak fluorescent signal of nuclear emerin as well as emerin mislocalization to cytoplasm, while the cells with pure NE localized GFP-αIIS shown normal emerin distribution. All these suggest that αII spectrin is an important binding partner of emeirn and may play roles in emerin associated disease.

αII spectrin influence the cell mitosis

The neclear enveope disassembles and reforms during mitosis. Tracking the protein dynamics during the cell mitosis is a good way to analyze the targeting and localization of nuclear envelope protein and its influence in cell poliferation. We tracked the GFP- αIIS expression cell from prometaphase until the two daughter cells come to interphase (movie to be shown in poster). We can see that GFP-αIIS dissemble at the prometaphase, together with the nucleus condensation. The reassembly of GFP-αIIS start at the telophase and nearly all the cytoplasmic GFP-αIIS assembled at NE in the cytokinesis. Interstingly, the overexpression of this nuclear structural protein tend to influence the cell mitosis including elongate the mitosis time and generate DNA tether after cell division. Therefore, we analyze the mitosis index and the percentage of cells with multinuclei and dumbbell-shaped nuclei. The multinuclei may be caused by incomplete division of daughter cells during mitosis and the dumbbell-shaped nuclei are possible to the cells failed in chromatin condensation or division. The results reveal that the GFP- αIIS expression cells shown a slightly higher mitotic index and much higher percentage of multinuclei and dumbbell-shaped nuclei (figure to be shown in poster). The slightly higher mitotic index may be caused by the elongation of mitosis process. And the higher percentage of multinuclei and dumbbell-shaped nuclei is in agreement with the live cell imaging observation that the overexpression of GFP- αIIS will cause mitosis failure or incomplete cell division.

Conclusion

In this work, we analyzed the roles of spectrin repeats protein αII spectrin in nucleus. Our results showed the knockdown of αII spectrin by siRNA causes multiple nuclear abnormalities, including lamin A and lamin B network weakening, high percentage of cells with DNA herniation or even DNA leakage, and mislocalization of emerin. The average size of αII spectrin knocked down nuclei is 40% larger than normal nuclei, suggesting the nuclei expansion due to loss of αII spectrin. The overexpression of GFP-αII spectrin also causes emerin mislocalization especially when GFP-αII spectrin mislocalized to the cytoplasm. The overexpression of GFP-αII spectrin elongated the cell mitosis procedure and influence the cell division. All these suggest that the αII spectrin is an important structure component in nuclear envelope.

572ao Micropatterned Co-Culture of Embryonic Stem Cells with Adult Hepatocytes for Directing Hepatic Differentiation

Ji Youn Lee, Nazgul Tuleuova, Caroline N. Jones, and Alexander Revzin. Department of Biomedical Engineering, University of California at Davis, 451 East Health Sciences Dr., Davis, CA 95616

Embryonic stem cells represent considerable potential as a source of tissue-specific cells for regenerative medicine and tissue engineering. However, realizing this potential requires effective methods for guiding lineage specific differentiation of embryonic stem cells. In this work, mouse embryonic stem cells were co-cultured with adult hepatocytes to guide liver-specific differentiation of stem cells. Robotic microarraying was used to print side-by-side columns of collagen (I) and fibronectin spots on silane-modified glass substrates. When incubated with these substrates, murine embryonic stem cells (mESC) showed preference for fibronectin regions and did not appreciable attach on collagen (I) or silane domains. Addition of model hepatocytes (HepG2 cells) resulted in localization of the cells on collagen (I) domains but not on silane regions of the glass substrate. Therefore, sequential assembly of the two cell types allowed us to juxtapose clusters of hepatic cells and mESC (300 um diameter for both cell types) next to each other with ~100 um edge-to-edge spacing. The co-culture was maintained up to 2 weeks in differentiation media supplemented with insulin and dexamethasone. In order to examine the differentiation, mESC were selectively extracted from the surface using laser microdissection technique and analyzed with real-time RT-PCR. Relative expression level of each gene to housekeeping gene, b-actin, was analyzed. The expression level of marker genes for pluripotency (Oct-4 and nanog), three germ layers (alpha fetoprotein, ACTC, and tubulin), and mature hepatocytes (albumin, transthyretine, and glucose-6-phosphatase) were analyzed throughout the culture period. We observed decrease in expression pluripotency genes while endodermal genes were upregulated. Importantly, mESC cultured with HepG2 cells showed higher expression of liver-specific genes compared to mono-culture. Transthyretin expression increased up to 28.3% of that in model mouse hepatoma cell line (Hepa 1-6) after 12 days of co-culture. Co-culture also lead to significant increases in albumin expression. Albumin was increased by 13.8% compared with Hepa 1-6 while that in mono-culture was only increased by 0.9%. In conclusion, we constructed micropatterned co-cultures of mESC and adult hepatocytes on ECM microarrays. This co-culture format allowed to significantly enhance liver-specific gene expression in mESC likely through the "instructive" and "educational" effect of hepatic cells.

572ap Applying Ligands to B Cell Receptors by Microfluidics

J. Christopher Love[1], Yuan Gong[1], and Eliseo Papa[2]. (1) Chemical Engineering, Massachusetts Institute of Technology, 77 Massachusetts Ave., Bldg. 66-456, Cambridge, MA 02139, (2) Health Sciences and Technology, Massachusetts Institute of Technology, 77 Massachusetts Ave., Cambridge, MA 02139

Microfluidic systems provide a unique set of tools for manipulating the environment of live cells while under continuous observation by optical microscopy. This talk will describe the use of a simple microfluidic system to follow the activation of B cells upon exposure to subsaturating doses of ligands by fluidic switching. The timing and sequence of these pulses makes it possible to observe temporally-distinct populations of receptors. It is possible to track the rearrangement and internalization of surface-expressed B cell receptors (BCRs) on primary naïve B cells. Ratiometric imaging also allows kinetic measurements of the activation following stimulation.

572aq Modeling the Mechanical Behavior of Leukocytes as Elastic Solids In Micropipette Aspiration Indentation Experiments

David J. Gee[1], Elena Lomakina[2], Richard E. Waugh[1], and Michael R. King[3]. (1) Biomedical Engineering, University of Rochester, Goergen Hall, Rochester, NY 14627, (2) Pharmacology & Physiology, University of Rochester, Goergen Hall, Rochester, NY 14627, (3) Biomedical & Chemical Engineering, University of Rochester, Goergen Hall, Rochester, NY 14627

The initial events of the inflammatory cascade in the microcirculation include the formation of transient tether bonds between leukocytes and endothelium. The coordinated association and subsequent dissociation of these molecular bonds is primarily effected by selectins and their corresponding counter-receptors and induces cell rolling motion along the endothelium whereby the leukocyte effectively integrates chemical and molecular signals of inflammation. During this phase of inflammatory response, cell deformation is relatively small and the rolling cell generally maintains its spherical shape. We have developed a computer simulation of the cell as an elastic solid in order to model small-scale cell deformation during this phase. Recently, we published data from cell-bead indentation experiments whereby a rigid bead indents and subsequently induces deformation of a neutrophil during controlled impingement. Additionally, we formulated an analytical description of the interfacial contact area as a function of impingement force and cell-bead geometry. In the current study, the Deformable Multiparticle Adhesive Dynamics (DMAD) simulation was used to compute the interfacial contact area during such impingement and was found to show similar agreement in the force-area relationship in the deformation regime where a linear elastic model is applicable. The resulting model will have great utility in modeling multicellular recruitment during inflammation and the effect of cell deformation in enhancing cell adhesion under flow.

Keywords: leukocyte deformation, micropipette aspiration, computer simulation

572ar In Vitro Stem Cell Analog Development: Image Cytometry Analysis of Hematopoietic Stem and Progenitor Cells In the Bone Marrow

Brendan A. Harley[1], Elena Levantini[2], John E. Mahoney[3], Cesar Nombela Arrieta[4], Daniel G. Tenen[2], Alexei Protopopov[3], and Leslie E. Silberstein[5]. (1) Chemical and Biomolecular Engineering, University of Illinois at Urbana-Champaign, Roger Adams Lab, MC-712, Box C-3, 600 S. Mathews Ave., Urbana, IL 61801, (2) Medicine, Beth Israel Deaconess Medical Center, Harvard Medical School, Harvard Institutes of Medicine, Room 907, 4 Blackfan Circle, Boston, MA 02115, (3) Dana-Farber Cancer Institute, Boston, MA 02115, (4) Laboratory Medicine, Children's Hospital Boston, One Blackfan Circle, Karp 10004D, Boston, MA 02115, (5) Children's Hospital, Karp Family Research Building, 10th Floor, Room 10217, 1 Blackfan Circle, Boston, MA 02115

Hematopoietic stem and progenitor cells (HSPCs) are an ideal platform for studying microenvironmental cues on stem cell behavior and for developing stem cell niche analogs. HSPC self-renewal and lineage differentiation depend on cues from distinct microenvironments, termed niches, in the bone marrow (BM) that are defined by cellular components, soluble regulators, and by the extracellular matrix (ECM) [1]. Here we employ Laser Scanning Cytometry (LSC) to define the anatomical localization of HSPCs within the femoral BM cavity as a first step towards quantifying HSPC-niche cell interactions within the BM microenvironment and then mimicking key interactions ex vivo.

HSPC niches exist within distinct anatomical regions: the femoral metaphyses and diaphysis. However, morphologically and functionally, the potential niches within these regions are significantly different. The diaphyseal BM cavity contains an endosteal region (ER), the region of BM in close association with the endosteal surface of the cortical bone, and the central medullary region (CMR) which contains the BM central sinus; the metaphyses are made up of cancellous bone. All three regions contain distinct populations of hematopoietic cells, BM stromal and reticular cells, adipose tissue, and disparate microvascular organizations. However, conventional fluorescence imaging techniques do not permit visualization of a large enough segment of the BM to definitively identify HSPCs or their niches, because of the rarity of HSPCs in the BM: LT-HSCs account for less than 0.005% of all BM cells [2].

LSC and confocal microscopy was used to quantify the distribution of HSPCs and differentiated hematopoietic cell populations in distinct anatomical locations (diaphyseal ER versus diaphyseal CMR versus the metaphyses) of the extravascular BM compartment. HSPCs were be identified using the bmi-1 gene targeted reporter (GFP) mouse model, where GFP levels are strongest for HSPCs, along with fluorescence immunostaining for antibodies (cKit, Sca-1) characteristic of primitive hematopoietic cell populations [3]. LSC allows complete longitudinal sections of the femoral BM cavity to be scanned and the entire cellular content (order 2.5×10^5 cells/image) mapped with single cell level precision, making objective quantification and statistical comparison of the localization of even extremely rare BM populations possible.

Combinations of antibodies (cKit, B220, IgM) were used to identify the morphological distribution of cell populations during HSPC to B cell development. More multipotent hematopoietic (cKit+) cells are predominantly located within the diaphyseal ER while more mature hematopoietic (IgM+) cells are predominantly within the diaphyseal CMR; B220+ cells, incorporating both precursor and mature hematopoietic cells are evenly distributed across the femoral BM cavity [4]. Further segregation has been observed for GFP+cKit+ cells (putative HSPCs); these cells are observed to be localized primarily within 10 cell diameters of the diaphyseal ER and preliminary results suggest a significant localization towards the metaphyseal regions of the BM microenvironment. These results suggest that cellular gradients exist within the BM, that these gradients may influence HSPC self-renewal and hematopoietic lineage differentiation processes, and that LSC techniques can be applied to quantify the morphological characteristics of HSC niches. Improved definition of HSPC niche(s) in the BM may have important implications for ex vivo expansion of clinically relevant HSPC populations, and will contribute to understanding the etiology of pathological conditions such as leukemogenesis, myelodysplasia and immunodeficiency.

These results will help to further characterize the cellular microenvironment surrounding the HSCs and their differentiated progeny. Further understanding the cell environment of HSCs in vivo provides important information regarding developing an appropriate in vitro analog of the BM microenvironment based on currently available collagen-glycosaminoglycan scaffold technology [5] for inducing maintenance (self-renewal) or lineage differentiation of primitive HSCs.

References:

1. Nagasawa T. Nat Rev Immunol 2006;6(2):107-16.

2. Wilson A, Trumpp A. Nature Revs Immunol 2006;6(2):93-106.

3. Hosen N, Yamane T, Muijtjens M, Pham K, Clarke MF, Weissman IL. Stem Cells 2007;25(7):1635-44.

4. Harley BA, Mahoney JE, Levantini E, Park S-Y, Le Y, Manis JP, Massberg S, von Andrian U, Tenen DG, Protopopov A, Silberstein LE. IABMR Annual Scientific Meeting; Boston, MA; 2007.

5. Harley BAC, Gibson LJ. Chemical Engineering Journal 2007;137(1):102-121.

Keywords: hematopoietic stem cell, niche, self-renewal, lineage development, microenvironment

572as Predicting Patient-Specific Crc Incidence from Polyp Prevalence Using a Mutation Network Model

Eric Sherer[1], Seza Orcun[1], Ann E. Rundell[2], and Doraiswami Ramkrishna[3]. (1) e-Enterprise Center, Purdue University, West Lafayette, IN 47906, (2) Biomedical Engineering, Purdue University, 206 S. Martin Jischke Drive, West Lafayette, IN 47907, (3) Chemical Engineering, Purdue University, 1283 Forney Hall, West Lafayette, IN 47907-2100

Seminal work on colorectal cancer (CRC) incidence modeling argued that the slope of the linear log-log CRC incidence with age implies that 6 or 7 somatic mutations are required for transformation to CRC. Subsequent modeling efforts have built on this theme by using models of linear, sequential transformations for predicting CRC incidence in the population in general and specific demographic groups. These random mutation models rarely explicitly account for the intermediate adenoma stages in the adenoma to carcinoma sequence. By including the colon history of individuals, additional patient-specific information can be used to increase the resolution of the CRC models. This data is also relevant since adenoma dysphasia and numbers have been shown clinically to correlate with the likelihood of subsequent advanced adenomas - with CRC likely following thereafter. Our goals are to (1) include polyp formation in a CRC development model and (2) to account for the heterogeneous nature of polyp genetics by incorporating multiple routes of formation inherent using a mutation network model. It is then shown how the model, tuned to both polyp and CRC prevalence, can be used to refine predictions of CRC incidence based on the results of colonoscopies rather than demographics alone.

572at Molecular Affinity Capture Devices to Enhance Snp Genotyping by MALDI-TOF Mass Spectrometry

Ashish Misra, Chemical and Biochemical Engineering, Rutgers University, 98 Brett Rd, Piscataway, NJ 08854 and **Sobin Kim**, Biomedical Engineering, Rutgers University, 599 Taylor Road, Piscataway, NJ 08854.

We present here approaches for increasing throughput of SNP (single nucleotide polymorphism) genotyping using the previously developed solid phase capture-single base extension (SPC-SBE) method. They involve rapid isolation and cleanup of biotinylated oligonucleotides for MALDI TOF MS (matrix-assisted laser desorption/ionization time-of-flight mass spectrometry) analysis.

SNPs are important genetic markers for disease susceptibility and pharmacogenomics studies and high throughput methods are required for their analysis. SPC-SBE is a MALDI MS based technique that uses primer extension with biotinylated terminators for SNP genotyping. Briefly, it involves design of a library of primers that anneal one base upstream of desired SNP sites and their simultaneous extension using biotinylated terminators in an SBE reaction. Extended primers carry a biotin moiety at their 3' end and are selectively isolated using streptavidin coated magnetic beads followed by washing and release. They are then analyzed by MALDI MS simultaneously to reveal SNP identity based on unique mass of each extension product. Due to the removal of unextended primers and other reaction contaminants by washing, SPC-SBE provides sample cleanup prior to MALDI MS analysis. Besides, it allows analysis of larger number of extension products due to the absence of primers peaks in the spectrum.

However, the use of streptavidin coated magnetic beads for isolating biotinylated extended primers in the SPC step limits use of SPC-SBE for high throughput genotyping. The extremely high affinity between biotin and streptavidin requires harsh denaturing conditions for the release of biotinylated fragments and involves long procedure prior MALDI analysis.

To address these issues, we have fabricated devices with monomeric avidin (MA) coated beads to improve the SPC step and increase process throughput. MA coated beads were trapped in a pipette tip using a nylon mesh at its end to act as a filter. We have demonstrated isolation of biotinylated extended primers selectively from an SBE reaction with direct spotting on the MALDI plate utilizing the pipette tip device. The devices have been used for simultaneous analysis of many samples in combination with a multichannel pipette. We have been able to reuse each device a minimum of 5 times with a simple regeneration protocol thus reducing genotyping cost. This demonstrated feasibility of the device for selective isolation of biotinylated fragments. We used a similar approach to trap MA coated beads in a microchannel device and were able to repeat experiments done with the pipette tip device. This can be extended to an arrangement with multiple microchannels for semi-automated processing of multiple samples and be used in high throughput SNP genotyping studies.

Nanoparticle coatings for enhanced capture of flowing cells in microtubes

B.A. Allio[1*], W.J. Han[1*], D.G. Foster[2], and M.R. King[3]

[1]Department of Biomedical Engineering, University of Rochester, Rochester, New York
[2]Department of Chemical Engineering, University of Rochester, Rochester, New York
[3]Department of Biomedical Engineering, Cornell University, Ithaca, New York

[*]Authors contributed equally to this work.

Abstract

Recently, we demonstrated a flow-based selectin-dependent method for the capture and enrichment of specific types of cells (CD34+ hematopoetic stem and progenitor cells and Human Leukemia HL60) from peripheral blood.[1,2,3] However these devices depend on a monolayer of selectin protein which has been shown to have a maximum binding efficiency as a function of surface area. We have designed a novel surface coating of colloidal silica nanoparticles (average particle dia. 12 nm, 30% by weight SiO_2) that alter surface roughness resulting in increased surface area. The nanoparticles were adhered using either an inorganic titanate resinous coating or an organic polymer of poly-L-lysine. Using Alexa Fluor 647 conjugated P-selectin we found an increased protein adhesion of up to 35% when compared to control. During profusion experiments using P-selectin coated microtubes we observed increased cell capture and greatly decreased rolling velocity at equivalent protein concentration compared to control. Atomic force microscopy showed increased surface roughness consistent with the nanoparticles mean diameter, suggesting a monolayer of particles. Our results show that nanoscale surface topography has significant effects on cell behavior.[4] We believe that this coating has potential to improve existing cell capture implantable devices for a variety of therapeutic and scientific uses.

572av Poly(ethylene glycol)-Protected, pH-Dependent Liposomes Enhance RNAi Delivery

Debra T. Auguste[1], Kay Furman[2], **Andrew D. Wong**[1], Jason Fuller[3], Steven P. Armes[4], Timothy Deming[5], and Robert S. Langer[3]. (1) School of Engineering and Applied Sciences, Harvard University, 29 Oxford St., Cambridge, MA 02138, (2) Department of Material Science and Engineering, Massachusetts Institute of Technology, 77 Massachusetts Avenue, Cambridge, MA 02139, (3) Department of Chemical Engineering, Massachusetts Institute of Technology, 45 Carleton St, E25-342, Cambridge, MA 02142, (4) Department of Chemistry, The University of Sheffield, Dainton Building, Sheffield, United Kingdom, (5) Department of Bioengineering, UCLA, Bolter Hall, Los Angeles, CA 90095

Short interfering RNA (siRNA) has potential therapeutic applications in regulating gene expression; however, its use is limited by inefficient delivery. We can enhance the transfection of siRNA using pH dependent liposomes coated with poly(ethylene glycol) (PEG)-b-polycation polymers. The PEG coating enables protection from immune recognition during circulation (pH 7.4) and triggering the release of the PEG layer within the cellular endosome (pH ~5.5) allows subsequent intracellular delivery of siRNA. Polycationic blocks, poly[2-(dimethylamino) ethyl methacrylate] (DMA) (31 or 62 monomers) or polylysine (K) (21 monomers), act as anchors for a PEG (113 monomer) protective block on the lipsomes. Incorporation of 1,2-dioleoyl-3-dimethylammonium-propane (DAP), a titratable lipid (pKa ~ 6.7), allows the liposome's net charge to increase as the pH shifts from 7.4 in the bloodstream to 5.5 within the endosome. The net increase in liposome cationicity repels the polycationic blocks, which decreases the PEG polymer adsorption, and allows the liposome to fuse with the endosomal membrane; thus, releasing the siRNA into the cytoplasm. Our pH dependent (PD) liposomes encapsulate siRNA that demonstrate green fluorescent protein (GFP) silencing in genetically-modified, GFP expressing HeLa cells and GAPD knockdown in human umbilical vein endothelial cells (HUVEC). The liposomes coated with PEG113–DMA31 and PEG113–DMA62 exhibited a 0.32 ± 0.3 and 0.33 ± 0.03 fraction of GFP knockdown, whereas direct siRNA administration and Oligofectamine complexed siRNA only reduced GFP expression by 0.06 ± 0.02 and 0.14 ± 0.02. Our in vitro data indicates that polymer desorption from PD liposomes may enhance siRNA-mediated gene knockdown, contrary to our membrane fusion results which show greater fusion with uncoated PD liposomes.

572aw Microfluidic Gradient Chamber for Two-Dimensional Mapping of Hepatocyte Drug-Response

Katherine A. Carson, Timothy C. Boire, and Kyongbum Lee. Chemical and Biological Engineering, Tufts University, 4 Colby Street, Medford, MA 02155

The potential for chemical injury to the liver is of great concern in the development of new therapeutic drugs. Currently, approximately 20 % of adverse drug reactions (ADRs) are unpredictable and idiosyncratic. In this poster, we present a microfluidic gradient chamber for in-vitro studies on hepatocyte metabolism and response to drug-induced toxicity. A key feature of the designed reactor is that it affords selective localization of micro-scale collagen gels, which serve as three-dimensional (3D) scaffolds for cell culture. The selective localization is accomplished by fabricating regions of high (convective) transport resistance as cell culture compartments. The culture compartment is flanked by columns of asymmetrically positioned posts. These columns form micro-fabricated "membranes" that separate the culture compartment from medium flow channels. The spacing between the posts affords diffusive exchange of essential nutrients, growth factors and drug molecules with the flow channels, but prevents convective transport. Leveraging the convection-free culture chamber design, stable diffusion gradients can be generated across the culture compartment by introducing parallel media streams with different chemical, i.e. drug, compositions into each flow channel.

To optimize the dimensions of the gradient chamber, a coupled convection-diffusion-kinetic model with oxygen as the limiting substrate was developed. The model was used to describe hepatocyte proliferation and metabolism in the gradient chamber. Kinetic parameters were obtained from the literature for human hepatocarcinoma (HepG2) cells. Simulation results predicted that the oxygen tension could be maintained above the effective Km for a culture compartment volume of up to 4.8x10^-4 cm^3 by distributing the medium inlets along the length of the reactor. The maximum attainable density was 10^5 cells.

Based on the simulation results, gradient chambers were fabricated using standard soft-lithography techniques and PDMS molding. To maintain consistency with the mathematical model, the HepG2 cell line was used as the model cell system. Cultures remained viable and exhibited hepatocyte metabolic function for at least 10 days. On day 7 after seeding, cultures were treated with the test drug troglitozone (TGZ), an insulin sensitizing, anti-diabetic compound first approved for clinical use, but eventually withdrawn due to indications of idiosyncratic ADR. A linear gradient of the drug was applied. Viability and metabolic assessments using fluorescent probes estimated that TGZ causes significant toxicity above 100 uM, in good agreement with conventional well-plate assay results. Taken together, our results to date indicate that the gradient chamber is an efficient platform for in vitro drug toxicity studies.

572ax Chemotaxis of Artificial Protocells

Frantisek Stepanek and Jitka Cejkova. Department of Chemical Engineering, Institute of Chemical Technology, Prague, Technicka 5, Prague 6, 166 28, Czech Republic

The ability of both single-cellular and higher organisms to actively move in the concentration gradient of a signalling molecule – whether to seek a source of nutrient or a partner to mate – is one of the most prominent external signs of life: "it moves, hence it must be alive!". Mimicking chemotaxis in artificial or bio-artificial systems is a challenging task; usually the objective is to devise a bio-mechanic apparatus such as a flagellum in order to ensure active propulsion. However, chemotaxis based on such bio-mechanic principles still remains an elusive task. In this work we will present a different approach to achieving chemotaxis in artificial protocell (i.e. simple, synthetic particles possessing one or more functions of real cells). This approach, which we term 'probabilistic chemotaxis' is based on the combination of random Brownian motion and specific particle-substrate interaction, the strength of which is designed to be a function of the concentration of the desired signalling molecule. Provided that the interaction is reversible, the repetition of adhesion-release events leads to a bias random walk, whereby a swarm of proto-cells gradually migrates in the direction of the concentration gradient. We will present experimental data demonstrating this chemotaxis mechanism, and discuss the potential for its further development, in particular the signal amplification and cooperative behaviour due to cell-cell signalling.

572ay Dynamic Mechanical Properties and Viscoelastic Response Analysis of a Complex Biological Interpenetrating Network: The Temporomandibular Joint Cartilage Disc

Joseph Lomakin, Chemical and Petroleum Engineering, Kansas University, The University of Kansas, 4132 Learned Hall 1530 W. 15th Street, Lawrence, KS 66045, Michael Detamore, Dept. of Chemical & Petroleum Engineering, University of Kansas, 1530 W. 15th St., 4132 Learned Hall, Lawrence, KS 66045, and Stevin H. Gehrke, Departments of Chemical & Petroleum Engineering and Pharmaceutical Chemistry, University of Kansas, 1530 W 15th Street, Lawrence, KS 66045-7609.

The temporomandibular joint (TMJ) disc is a robust cartilaginous biogel separating the mandibular condyle and the fossa-eminence of the temporal bone. It is comprised of a highly hydrated interpenetrating network of collagen Type I and proteoglycans, combined in a ratio of approximately 10 to 1. The discs' structure and exceptional mechanical properties are of both clinical and materials research interest as they empower this complex biocomposite to withstand repeat exposure to high shear, compression and tension forces as part of normal jaw movement associated with chewing and talking. While the mechanical performance of the disc in response to these diverse forces have been characterized, comparison of data published in various venues remains difficult due to the lack of fundamental understanding surrounding the complex multivariable response of this and other biological materials.

In this study, the dynamic mechanical properties of the TMJ disc are determined as a function of initial loading (pre-tension), oscillation frequency and dynamic strain in an effort to reconcile mechanical property data on the disc presented thus far in the literature. Furthermore the results and analysis demonstrate the applicability of viscoelastic interpretation of the data, often applied to synthetic polymer materials, to a complex hierarchical biomaterial such as the temporomandibular joint disc.

The elastic modulus (E') of the TMJ disc measured over two orders of magnitude (0.01 rad/s – 500 rad/s) of frequency and (0.01% - 10%) dynamic strain was found to span 10 to 60 MPa and tan δ to vary between 0.08 and 0.18 depending on the properties of the applied strain wave. With respect to oscillation frequency (ω), the E' exhibited a weak power law increase of a factor of 2 between 0.1 and 100 rad/s, an increase consistent with long time relaxation associated with movement of entangled ends within a loosely cross-linked interpenetrating polymer network. The elastic modulus is found to be independent of dynamic strain below 0.1%. Above this strain the modulus decreases sharply, demonstrating a pronounced power law decrease above 1% strain. The results from tests at various dynamic strain and frequencies are combined to separate the effects of two confounding variables relevant to the biomaterial's mechanical response: strain amplitude and strain rate. Strain-rate frequency superpositioning (SRFS) is used to determine the relative importance of these two parameters on material moduli and to project the dependence of the moduli on dynamic strain and frequency outside of the tested range. The master curve constructed from overlaying constant strain rate measurements demonstrates that strain rate amplitude is an important factor in determining TMJ disc material properties, an effect not typically seen with synthetic materials. This viscoelastic analysis of dynamic mechanical data facilitates understanding of the complex dependence of biomaterial mechanical properties on the applied strain in terms of underlying structural relaxations governed by molecular structure and suggests new avenues for biomimetic material design.

Can the Diagnostic and Prognostic Value of Prostate Specific Antigen be Improved by Harnessing the Specificity of Mass Spectrometry?

R. Brent Dixon[1], *Jack R. Edwards*[2], *Alex Sohn*[3], *Thomas A. Dow*[3] *and David C. Muddiman*[1], *(1) Chemistry, W.M. Keck FT-ICR Mass Spectrometry Laboratory, North Carolina State University, 2620 Yarbrough Drive, Campus Box 8204, Raleigh, NC 27695, (2) Mechanical and Aerospace Engineering, North Carolina State University, Campus Box 7910, Broughton Hall 4223, Raleigh, NC 27695, (3) Precision Engineering Center, North Carolina State University, Box 7918, 1001 Capability Drive, Research Bldg. I, Raleigh, NC 27695*

Background and Significance

In the untiring search to identify and quantify cancer biomarkers, the field of proteomics has emerged with great promise.[1] Liquid chromatography coupled to mass spectrometry (LC-MS) has certainly become a vital tool in these experiments. However, the limits-of-detection (LOD) of existing mass spectrometry platforms mandate the integration of devices such as the air amplifier and other strategies to improve the overall sensitivity in order to become clinically relevant. It is important to note that the specificity attained using mass spectrometry is unparalleled and thus provides the stimulus for the efforts to improve the sensitivity. The concentration of the clinically relevant peptides and proteins in early disease stage tends to be below that detectable with current LC-MS technology. Only when a patient is consumed with cancer will biomarkers be in high enough abundance, when the prognosis for a recovery at this disease stage is markedly decreased. An aerodynamic device described herein as the air amplifier will improve the early detection and quantification of low abundance cancer biomarkers by lowering the LOD of LC-MS/MS methods. The development of a technology, through high precision engineering, enabling a robust analytical strategy as applied to a relevant biological concern (i.e. prostate cancer) is described in a manner that can ultimately improve diagnoses and disease treatment options.

Previous reports of protein cleavage isotope dilution mass spectrometry demonstrate the potential for quantitative studies.[2-4] Prostate specific antigen (PSA) has emerged as a significant biomarker in the diagnosis strategy of prostate cancer.[5-8] It is used in a complementary manner with other tests including digital rectal exams (DRE) and invasive tissue biopsies. However it has been demonstrated that elevated PSA accompanies other prostatic conditions including benign prostatic hyperplasia and prostatitis. The immunologically detectable forms of PSA include complexes with α_1-chymotrypsin and α_2-chymotrypsin in addition to other binding proteins. By utilizing mass spectrometry, different protein forms can be identified in hopes to distinguish various prostatic conditions.

Instrumentation

Mass spectrometry is an extremely sensitive and specific technique that can be used to analyze a broad range of compounds over a wide dynamic range of molecular weight and concentration.[9, 10] These factors have led to increasing interest in coupling liquid separation techniques directly to mass spectrometry for effective analysis of complex mixtures (e.g. serum). Electrospray ionization (ESI) facilitates the direct coupling of liquid separation methods (*e.g.*, HPLC, chromatofocusing, capillary electrophoresis) to the mass analyzer. ESI conditions can be manipulated to enhance detection limits ranging from very low (< 100 Da), to extremely high (>100 MDa)[11], molecular weight compounds. The utilization of mass spectrometry for cancer biomarker analysis extends back to the 1970's[12], and recent reports show great promise for its use as a diagnostic tool.[13-17]

In a conventional electrospray ionization only a minute fraction of the ion generated are sampled by the mass spectrometer. This is inherently due to sample losses occurring in the subsequent stages of ion transmission following sample ionization. Many factors contribute to the low transmission efficiency including the generation of ions by multiple Coulombic explosions that occur at the onset of a droplet on the ESI needle tip. Substantial sample loss in mass spectrometry analyses transpires in the atmospheric pressure region between the electrospray emitter tip and the capillary inlet.[18] It is estimated that only one out of 10^3-10^4 ions generated are sampled by the mass spectrometer.[19-23]

The air amplifier has proven useful in addressing this issue by increasing the observed signal by an order of magnitude.[24-29] The air amplifier's function is accomplished by throttling a thin annular jet of compressed gas into a venturi nozzle passage through a small gap. The jet expands to supersonic conditions upon its exit, creating a low-pressure region that enables air entrainment into the venturi nozzle. The annular jet is typically directed toward the exit of the venturi tube via the Coanda effect. In the Coanda effect, a jet moving over a curved surface can remain attached to the surface as long as a balance between the sub-ambient pressure in the jet sheet and the centrifugal force occurs. The entrained air can be accelerated in the axial and radial directions by contouring the venturi tube. Previous studies have considered air amplification as a means of focusing ions generated through an electrospray process into the inlet of a mass spectrometer. Ion abundance increases of a factor of ten or more and increases in the theoretical mass resolving power were observed in several tests, indicating that the air-amplifier concept is potentially useful as a low-cost alternative to other focusing strategies such as ion funnels. However, significant user experience and case-specific tuning of the air-amplifier parameters (gap width, placement of the emitter and mass-spectrometer inlet) were required to achieve optimal results. Recent work in our laboratory has extended the use of the air amplifier to LC-MS by heating the inlet nitrogen and the body of the air amplifier.

Experimental

Recently the air amplifier has been integrated with liquid chromatography and demonstrated improved ion abundance. This enables increased identifications resulting in greater proteome coverage. In a targeted proteomics experiment, the air amplifier is implemented on a triple quadrupole mass spectrometer. Selected reaction monitoring is routinely used for selective and quantitative tandem MS. These experiments are described in further detail below.

The diagnostic relevance of the air amplifier is studied in a discovery-based proteomics experiment. In the discovery experiment, a digest of prostate specific antigen (PSA) will under go LC-MS analysis. Using this system will allow for quantitative measures including improvements in ionization time, number of MS/MS per unit time in each LC-MS/MS run, and the number of proteins identified using the air amplifier. Concurrently, an analysis of an 80 mg/mL human serum albumin (HSA) digest will be performed to determine the actual improvement in limits of detection. Biological samples are highly complex as the number of peptides within a digested serum sample is ~$10^{5\text{-}6}$ with a concentration range of ~10^9. Current LC-MS sensitivity is a limiting factor of candidate biomarkers that fall below the assay LOD by approximately 1000-fold. With the improved air amplifier design enhancing signal 10-fold, and combining this with other methods for decreasing LOD, clinical diagnostics will hold much better promise for disease diagnosis and treatment options. Limits of detection with the air amplifier will be determined with the HSA digest spiked with decreasing amounts of the PSA digest followed by LC/MS analysis. A calibration curve from lowest to highest concentration of the HSA digest will be used to evaluate the limits of detection as a function of the response across the LC gradient with the air amplifier ON and OFF.

An integral portion of the biomarker discovery experiments will employ selected reaction monitoring (SRM). This method allows both detection and quantification of specific molecules. A mock serum sample (HSA digest) will be spiked with PSA and then digested. Stable-isotope labeled tryptic fragments will be added at known concentrations and serve as the internal standard. As a result, the LOD for targeted proteomics (SRM) can be determined for both the air amplifier ON and OFF and compared with previous results by Muddiman and coworkers that had a detection limit of 4.5 mg/L of PSA in serum.[4] With the improved air amplifier design, we anticipate an improvement in ion abundance of >10 fold.

Conclusions

The methodology developed for the detection and quantification of PSA will be widely applicable to other disease systems (e.g. breast cancer, cardiovascular disease). This will be accomplished by establishing disease-specific biomarkers with their respective internal standards for use in the SRM proteomics analysis. Non-specific point of care testing will continue to have its place in clinical diagnostics however the specific information obtained from the mass spectrometry approach will be highly useful for ongoing treatment plans.

References

1. Muddiman, D. C., "Lewis and Clark" Proteomics. *Journal of Proteome Research* **2006,** 5, (2), 221-222.

2. Barr, J. R.; Maggio, V. L.; Patterson, D. G.; Cooper, G. R.; Henderson, L. O.; Turner, W. E.; Smith, S. J.; Hannon, W. H.; Needham, L. L.; Sampson, E. J., Isotope Dilution-Mass Spectrometric Quantification of Specific Proteins: Model Application with Apolipoprotein A-I. *Clinical Chemistry* **1996,** 42, (10), 1676-82.

3. Barnidge, D. R.; Dratz, E. A.; Martin, T.; Bonilla, L. E.; Moran, L. B.; Lindall, A., Absolute Quantification of the G Protein-Coupled Receptor Rhodopsin by LC/MS/MS Using Proteolysis Product Peptides and Synthetic Peptide Standards. *Anal. Chem.* **2003,** 75, (3), 445-451.

4. Barnidge, D. R.; Goodmanson, M. K.; Klee, G. G.; Muddiman, D. C., Absolute Quantification of the Model Biomarker Prostate-Specific Antigen in Serum by LC-MS/MS Using Protein Cleavage and Isotope Dilution Mass Spectrometry. *Journal of Proteome Research* **2004,** 3, (3), 644-652.

5. Oesterling, J. E., Prostate Specific Antigen: a Critical Assessment of the Most Useful Tumor Marker for Adenocarcinoma of the Prostate. *Journal of Urology* **1991,** 145, (5), 907-923.

6. Labrie, F.; Dupont, A.; Suburu, R.; Cusan, L.; Tremblay, M.; Gomez, J. L.; Emond, J.; Stamey, T. A.; Lange, P. H., Serum Prostate Specific Antigen as Pre-Screening Test for Prostate Cancer. *Journal of Urology* **1992,** 147, (3), 846-852.

7. Harris, R.; Lohr, K. N., Screening for Prostate Cancer: An Update of the Evidence for the US Preventive Services Task Force. *Ann. Intern. Med.* **2002,** 137, (11), 917-929.

8. Force, U. S. P. S. T., Screening for Prostate Cancer: US Preventive Services Task Force Recommendation Statement. *Ann. Intern. Med.* **2008,** 149, (3), 185-191.

9. Fenn, J. B.; Mann, M.; Meng, C. K.; Wong, S. F.; Whitehouse, C. M., Electrospray Ionization for Mass-Spectrometry of Large Biomolecules. *Science* **1989,** 246, (4926), 64-71.

10. Fenn, J. B.; Mann, M.; Meng, C. K.; Wong, S. F.; Whitehouse, C. M., Electrospray Ionization-Principles and Practice. *Mass Spectrom. Rev.* **1990,** 9, (1), 37-70.

11. Chen, R. D.; Cheng, X. H.; Mitchell, D. W.; Hofstadler, S. A.; Wu, Q. Y.; Rockwood, A. L.; Sherman, M. G.; Smith, R. D., Trapping, Detection, and Mass Determination of Coliphage T4 DNA Ions of 10(8) Da by Electrospray-Ionization Fourier-Transform Ion-Cyclotron Resonance Mass-Spectrometry. *Anal. Chem.* **1995,** 67, (7), 1159-1163.

12. Gehrke, C. W.; Waalkes, T. P.; Borek, E.; Swartz, W. F.; Cole, T. F.; Kuo, K. C.; Abeloff, M.; Ettinger, D. S.; Rosenshein, N.; Young, R. C., Quantitative Gas-Liquid-Chromatography of Neutral Sugars in Human-Serum Glycoproteins - Fucose, Mannose, and Galactose as Predictors in Ovarian and Small Cell Lung-Carcinoma. *Journal of Chromatography* **1979,** 162, (4), 507-528.

13. Xiao, Y. J.; Chen, Y. H.; Kennedy, A. W.; Belinson, J.; Xu, Y., Evaluation of Plasma Lysophospholipids for Diagnostic Significance Using Electrospray

Ionization Mass Spectrometry (ESI-MS) Analyses. In *Lysophospholipids and Eicosanoids in Biology and Pathophysiology*, 2000; Vol. 905, pp 242-259.
14. Kachman, M. T.; Wang, H. X.; Schwartz, D. R.; Cho, K. R.; Lubman, D. M., A 2-D Liquid Separations/Mass Mapping Method for Interlysate Comparison of Ovarian Cancers. *Anal. Chem.* **2002,** 74, (8), 1779-1791.
15. Wang, H. X.; Kachman, M. T.; Schwartz, D. R.; Cho, K. R.; Lubman, D. M., A Protein Molecular Weight Map of ES2 Clear Cell Ovarian Carcinoma Cells Using a Two-Dimensional Liquid Separations/Mass Mapping Technique. *Electrophoresis* **2002,** 23, (18), 3168-3181.
16. Shin, B. K.; Wang, H.; Yim, A. M.; Le Naour, F.; Brichory, F.; Jang, J. H.; Zhao, R.; Puravs, E.; Tra, J.; Michael, C. W.; Misek, D. E.; Hanash, S. M., Global Profiling of the Cell Surface Proteome of Cancer Cells Uncovers an Abundance of Proteins with Chaperone Function. *J. Biol. Chem.* **2003,** 278, (9), 7607-7616.
17. Zhu, Y.; Wu, R.; Sangha, N.; Yoo, C.; Cho, K. R.; Shedden, K. A.; Katabuchi, H.; Lubman, D. M., Classifications of Ovarian Cancer Tissues by Proteomic Patterns. *Proteomics* **2006,** 6, (21), 5846-5856.
18. Page, J. S., Kelly, R.T., Tang, K., Smith, R.D., Ionization and Transmission Efficiency in an Electrospray Ionization-Mass Spectrometry Interface. *J. Am. Soc. Mass Spectrom.* **2007,** In Press.
19. Smith, R. D.; Loo, J. A.; Edmonds, C. G.; Barinaga, C. J.; Udseth, H. R., New Developments in Biochemical Mass-Spectrometry - Electrospray Ionization. *Anal. Chem.* **1990,** 62, (9), 882-899.
20. Kebarle, P.; Tang, L., From Ions in Solution to Ions in the Gas-Phase - the Mechanism of Electrospray Mass-Spectrometry. *Anal. Chem.* **1993,** 65, (22), A972-A986.
21. Tang, L.; Kebarle, P., Dependence of Ion Intensity in Electrospray Mass-Spectrometry on the Concentration of the Analytes in the Electrosprayed Solution. *Anal. Chem.* **1993,** 65, (24), 3654-3668.
22. Zook, D. R.; Bruins, A. P., On Cluster Ions, Ion Transmission, and Linear Dynamic Range Limitations in Electrospray (Ionspray) Mass Spectrometry. *Int. J. Mass Spectrom. Ion Phys* **1997,** 162, (1-3), 129-147.
23. Cech, N. B.; Enke, C. G., Practical Implications of Some Recent Studies in Electrospray Ionization Fundamentals. *Mass Spectrom. Rev.* **2001,** 20, (6), 362-387.
24. Zhou, L.; Yue, B. F.; Dearden, D. V.; Lee, E. D.; Rockwood, A. L.; Lee, M. L., Incorporation of a Venturi Device in Electrospray Ionization. *Anal. Chem.* **2003,** 75, (21), 5978-5983.
25. Hawkridge, A. M.; Zhou, L.; Lee, M. L.; Muddiman, D. C., Analytical Performance of a Venturi Device Integrated into an Electrospray Ionization Fourier Transform Ion Cyclotron Resonance Mass Spectrometer for Analysis of Nucleic Acids. *Anal. Chem.* **2004,** 76, (14), 4118-4122.
26. Yang, P. X.; Cooks, R. G.; Ouyang, Z.; Hawkridge, A. M.; Muddiman, D. C., Gentle Protein Ionization Assisted by High-Velocity Gas Flow. *Anal. Chem.* **2005,** 77, (19), 6174-6183.
27. Dixon, R. B.; Williams, D. K., Jr.; Sampson, J. S.; Edwards, J. R.; Hawkridge, A. M.; Muddiman, D. C., Integrating a Voltage-Assisted Air Amplifier

to Efficiently Sample Biological Specimens. In *The Asilomar Conference on Mass Spectrometry*, Pacific Grove, CA, 2007.
28. Dixon, R. B.; Muddiman, D. C.; Hawkridge, A. M.; Fedorov, A. G., Probing the Mechanism of an Air Amplifier using a LTQ-FT-ICR-MS and Fluorescence Spectroscopy. *J. Am. Soc. Mass Spectrom.* **2007,** 18, (11), 1909-1913.
29. Dixon, R. B.; Muddiman, D. C., Quantitative Comparison of a Flared and Standard Heated Metal Capillary Inlet with a Voltage-Assisted Air Amplifier on an Electrospray Ionization Linear Ion Trap Mass Spectrometer. *Rapid Commun. Mass Spectrom.* **2007,** 21, (19), 3207-3212.

572ba Optimization of a Bioreactor for Tissue Engineering Scaffolds

Mamatha Devarapalli and Sundararajan V. Madihally. School of Chemical Engineering, Oklahoma State University, 423 Engineering North, Stillwater, OK 74078

The goal of this study was to understand the flow distrbution in a reactor for growing large tissues (10 cm diameter and 2 mm thick) in vitro. For this purpose a parallel plate reactor would be ideal because of its construct which would give support to the scaffold and continuous flow ensures replenishment of the nutrients. However, rectangular reactors possess inherent problems related to dead spaces at the edges. Hence circular reactors which minimize the dead spaces within the reactor were used. First, fourteen differnt outlet and inlet configurations were assed for fluid distribution. From this, reactor with better fluid distribution was selected. Two aspects were considered for further analysis: (i) to ensure uniform distribution of the nutrients, residence time distribution (RTD) studies were carried out and (ii) to optimize the flow rate at which the nutrients are to be supplied for the cells to grow, oxygen and glucose consumptions were studied at varying flow rates using rate constants from literature. For nutrient consumption four different four differnt cell types were used. The simulations were done using the computational fluid dynamics packages CFX 11 (ANSYS Inc, Canonsburg, PA.) and/or Comsol 3.3 Multiphysics Package with Reaction Engineering Module (COMSOL, Inc., Burlington, MA). Experiments were performed using chitosan porous structures and reactors manufactured to the dimensions of the simulated condition. From the RTD studies it was observed that the RTD function E(t) peak shifted towards left showing that there is an increase in channeling with the porous structure than without the porous structure. The concentration profiles of oxygen and glucose were simulated for different cell types and it was observed that the optimum flow rate for all the cell types simulated were in between the range of 0.1-0.01ml/min and at around 0.001ml/min flow rate, almost all the nutrients were consumed. The pressure drop and shear stress distributions within the reactor were also studied. The reactor systems used here have distinct advantages for regenerating large tissue constructs. Scaffolds for skin, bladder, and cartilage can have very large surface area to volume ratios with very small thickness. Flow dynamics and mass transfer of the nutrients are two important concepts to be studied and understood well while designing a bioreactor for tissue engineering purposes. The computational and experimental techniques being developed show promise to further enhance the fundamental understanding of the processes for tissue regeneration.

572bb Stimulus Controlled Infectivity of Phage: A Novel Method for the Selection of Stimulus-Responsive Peptides

Mark Blenner and Scott Banta. Chemical Engineering, Columbia University, 500 W 120th St, New York, NY 10027

Here we present the development of a technique called Stimulus Controlled Infectivity of Phage (SCIP) and its application in the selection of ligand induced peptide conformational changes. Filamentous phage infect E. Coli expressing the F' pilus via interaction between the gene III coat protein and the TolA receptor on the bacterial surface. The gene III coat protein is comprised of three domains that have functional roles in the infection process: N1, which binds to the TolA receptor, N2, which binds to the F' pilus and CT, which is connected to the phage coat. Once the N2 domain binds, the pilus retracts and brings the N1 domain closer to the TolA receptor, allowing phage infection of E. Coli.. A flexible 19 amino acid linker (EGGGSEGGGSEGGGSEGGG) connects N1 and N2 and is believed to provide sufficient flexibility to allow a pilus bound phage to simultaneously bind the TolA receptor. Infectivity of phage with a flexible linker without glutamic acid residues is explored as well as phage with no linker, a cysteine constrained linker, and a structured tryptophan zipper linker. This system allows for the selection of peptide linkers exhibiting stimulus controlled conformational changes upon ligand binding from both a naïve and GPG constrained libraries using tandem positive and negative selections. During the positive selection, the phage population is allowed to infect E. Coli and propagate normally, removing any non-infective phage from the library. Phage are then recovered and incubated with ligand. Random peptide linkers that are able to bind the ligand and undergo a conformational change will lose their infective ability. The non-binders are removed by allowing them to infect E. Coli over a short time scale, and removing the E. Coli by centrifugation. The remaining phage are recovered, the ligand removed, and infectivity is once again restored. This technique can be used to develop allosteric ligand binders, protease resistant ligand binders, and stimulus-responsive elements for bionanotechnology applications.

572bc Simulation of Respiratory Flow In Human Lung with Simple Bronchial Model

Fukashi Kohori, Fuminori Hanaura, and Kiyotaka Sakai. Department of Chemical Engineering, Waseda University, 3-4-1 Okubo, Shinjuku-ku, Tokyo, Japan

Human lung consists of the bronchi in which the trachea continues to bifurcate and finally reaches 300-600 millions alveoli having a total surface area of approximately 100 m^2 at the end of the bronchi. Quantitative evaluation of respiratory air flow in the bronchus is capable of providing further information about pulmonary function and specifying respiratory disease. The objective of the present study is to reveal respiratory air flow velocity and pressure drop by a three-dimensional asymmetric human bronchial tree model. This model consists of three simplified elements and expresses a structure of five lobes of right upper lobe (RUL), right middle lobe (RML), right lower lobe (RLL), left upper lobe (LUL) and left lower lobe (LLL).

We analyzed the respiratory air flow by the three-dimensional asymmetric human bronchial tree model using the finite element method (ANSYS, PA). It was difficult to model all of the human bronchial air way so that we devised three simplified models such as asymmetric Y-bifurcation model, symmetric Y-bifurcation model and T-bifurcation model. Human bronchi were made up by these three models. The asymmetric Y-bifurcation model was used for the first bronchus that consisted of right lung at 30° and left lung at 45° connected to the mother bronchus based on anatomical data. We used the T-bifurcation model for the second and third bronchi in the right lung. The RUL was located perpendicularly to the second bronchus. The RML was located perpendicularly to the third bronchus and further the RLL was located axially to the third bronchus.

Maximum average air flow velocity was calculated to be 1140 mm/s in RUL, 1490 mm/s in RML, 2650 mm/s in RLL, 1330 mm/s in LUL, 1200 mm/s in LLL. When we assumed all pressure drops to be 100%, the pressure drop until the seventh bronchus was 80 % in RUL, 85 % in RML, 84 % in RLL, 79 % in LUL, 81 % in LLL respectively.

572bd Ewing's Sarcoma Spheroids as a 3-D Model of *In Vivo* Biology Useful for High-Throughput Preclinical Screening of Antineoplastic Drug Candidates

Emily Burdett[1], Antonios G. Mikos[1], and Joseph Ludwig[2]. (1) Bioengineering, Rice University, MS-142, P.O. Box 1892, Houston, TX 77251, (2) Sarcoma Medical Oncology, University of Texas M. D. Anderson Cancer Center, 1515 Holcombe Blvd, Houston, TX 77030

Pre-clinical screening for candidate drugs with potential antineoplastic activity typically relies upon high-throughput screens across cancer cell lines grown in monolayer culture, but this method often fails to accurately predict whether compounds eventually prove to be successful in later human clinical trials. In part this may be due to dedifferentiation and change in cellular phenotype that occurs upon removal from a native *in vivo* 3-D cellular microenvironment. Subsequent animal models do involve cell placement within a tissue microenvironment, but they are not a high-throughput platform, and do not allow for repeated biopsies for analysis during testing. A high-throughput screening method that can recreate this microenivoronment would offer greater predictive power, streamline the preclinical antineoplastic drug discovery process, and by extension, decrease the time required to move a drug from bench to bedside. In this study toward the development of such a screening model, we compare Ewing's Sarcoma cell lines grown in monolayers to 3-D cellular aggregates called spheroids in order to determine the effect that a 3-D architecture has upon cellular phenotype. The response of several different Ewing's Sarcoma cell lines to cancer therapeutic drugs was monitored through the use of cellular proliferation assays in order to evaluate the degree of therapeutic resistance conferred to cells by three dimensional growth. Drugs tested include several different chemotherapeutics that are currently used in standard treatment for Ewing's Sarcoma, including doxorubicin, vincristine, and etoposide. The inhibitory effects of insulin-like growth factor I receptor (IGF-1R) agonists were also tested to investigate the hypothesis that these multicellular Ewing's spheroids could serve as an intermediate model that is more representative of the *in vivo* signaling cascade. These tests revealed that growth within spheroids does confer resistance to many of these therapies; the drug concentration required to achieve 50 % growth inhibition within spheroids is about tenfold higher than that for monolayer growth. These results suggest cancer cells grown within a 3-D environment have a more clinically accurate cellular response to therapeutics, and show potential for development of a more predictive high-throughput preclinical screen for cancer therapies through the use of 3-D culturing techniques.

572be Two-Compartment Modeling and Computer Programs to Find the Optimal Intravenous Drug Dosage Regimens

Kwang S. Kim and Laurent Simon. Otto H. York Department of Chemical Engineering, New Jersey Institute of Technology, University Heights, Newark, NJ 07102

A user-friendly computer program was developed to determine optimal drug-dosage regimens using a two-compartment pharmacokinetic model. The results are applicable for intravenous bolus injections (i. v. bolus) followed by a continuous infusion (i.v. infusion). To decrease the time it takes to reach a desired therapeutic plasma concentration while satisfying constraints (i.e., therapeutic window), the algorithm estimates the sizes of a set number of bolus doses given at specific times and the infusion rate. Published pharmacokinetic parameters, i.e., intercompartment transfer rate coefficients and apparent Michaelis-Menten constants of theophylline, were used to illustrate the methodology. Drug concentration profiles in the central (plasma) and peripheral cell compartments were solved by orthogonal collocation method in the *Mathematica*® (Wolfram Research, Inc.) environment. This technique has the potential to assist in providing customized patient care by helping to identify individualized drug doses that are both efficacious and safe.

572bf A Multi-Scale Approach to Bionanosystem Simulation; Applications to Computer-Aided Vaccine Design, Nanocapsules for Drug Delivery and Medical Imaging

Parag Adhangale[1], Cheluvaraja Srinath[2], Yinglong Miao[2], Zeina Shreif[2], and Peter Ortoleva[2]. (1) Center for Cell and Virus Theory, Indiana University, 800 E Kirkwood Avenue, Bloomington, IN 47405, (2) Ccvt, Indiana University, 800 E Kirkwood Avenue, Bloomington, IN 47405

We present a novel multi-scale algorithm that make the fully atomistic simulation of bionanosystems over physiologically relevant timescales and conditions, feasible . Our method, termed "NanoX", is based on the automatic generation of order parameters that describe features of nanoparticle migration, rotation and structural transition. The dynamics equations describe the order parameter evolution over milliseconds or longer with all-atom resolution. NanoX provides insights into the cross-talk between the highly fluctuating atomic motions and the coherent nanoparticle dynamics that underlie many biophysical behaviors.

Viral threats to global health warrant the development of efficient methods to design antiviral drugs and vaccines, and to predict characteristics of mutants of known viruses. Viral structural transitions (STs) constitute a fundamental life-cycle event. They are key to viral infection of a host cell and the maturation of a self-assembling virus.The atomic-scale fluctuations provide the stochastic forces and entropy that drive viral structural transitions, while overall viral structure affects the statistics of atomic fluctuations. The million-atom scale of the viruses and the time and length scales involved in their interaction with host cells, present a grand computational challenge. We present the application of NanoX to the study of cowpea chlorotic mottle virus (CCMV) which is an icosahedral member of the bromovirus group of the Bromoviridae family.

The design of nanocapsules for targeted delivery of therapeutics and imaging systems, presents many, often seemingly self-contradictory, constraints. NanoX predicts the interactions of functionalized nanocapsules with cell surface receptors and with drug, siRNA, gene or other payload. NanoX allows for the simulation of phenomena, such as structural transitions and disassembly of the nanocapsule accompanying timed payload release or due to premature degradation. A novel "salt shaker" effect that underlies fluctuation-enhancement of payload delivery is presented. We also discuss the potential applications of our methodology to the design of nanoparticles for medical imaging.

Central to such a simulation is the interplay between the order parameters and the atomic scale variables which is captured via stochastic equations derived rigorously from the fundamental laws of molecular physics (I.e. Fokker-Planck or Smoluchowski and the corresponding Langevin equations). The thermal average forces and diffusion coefficients that appear in the order parameter dynamics equations, evolve along with the system and are automatically generated within the dynamics loop. NanoX is built on the CHARMM inter-atomic force-field which has been well developed and optimized thus avoiding the need for recalibration of the model for different systems, and is available as free open-source software.

Avenues for future development are also discussed.

572bg Effects of Perfusion Flow on Hmsc Construct Development In Chitosan-Based Biomimetic 3-D Scaffolds

Katelyn Sellgren and Teng Ma. Chemical and Biomedical Engineering, Florida State University, 2525 Pottsdamer Street, Tallahassee, FL 32310

Biomimetic scaffolds mimic the in vivo extracellular matrix environments of the target tissues and have been used to augment the therapeutic potential of human mesenchymal stem cells (hMSC) for a broad range of diseases. Previous studies from our lab have used hydroxyapatite, chitosan, and gelatin (HCG) to fabricate a three dimensional scaffold that mimics the biological composition of bone tissue and have demonstrated the potency of the HCG composite scaffold on hMSC growth, retention, and osteogenic differentiation.[1] However, the successful regeneration of functional 3-D constructs, for substantial bone defects, from HCG scaffolds and hMSCs requires uniform spatial cell distribution and growth throughout the entire 3-D scaffold in a controlled environment. To meet the demand, we incorporated the 3-D HCG scaffolds into the perfusion bioreactor system developed in our lab and investigated the effects of cell seeding, flow rate, and flow modes on hMSC development.[2,3] The in-house perfusion bioreactor system has multiple, individually controlled flow chambers that share the same media source and inoculum facilitating the comparison of various operation conditions. The perfusion system also has the ability for dynamic cell seeding by depth filtration that enhances seeding efficiency and initial cell distribution. During the cell growth phase, the perfusion flow in each chamber can be controlled as either parallel or perpendicular to the construct, meeting the varying nutrient demands over the course of construct development. In the current study we fabricated an HCG scaffold with an average pore size of 93 ± 27 ìm to fit within our current bioreactor system. Dynamically seeded human mesenchymal stem cells showed a notable 67 % seeding efficiency, significantly higher than the static seeding efficiency. A higher cell number increase compared to static culture was observed after 10 days of transverse perfusion growth at a flow rate of 0.1 mL/min. SEM has confirmed the presence of cells throughout the 1 mm thickness of the scaffold. Cell metabolic activities were monitored by measuring glucose consumption and lactate production. A steady increase in metabolic activity beyond day 15 was found to accompany the cell number increase. Currently, we are investigating the effects of flow rate under parallel and transverse conditions on CFU-F, alkaline phosphatase (ALP), and calcium levels of cells retained from perfusion samples for up to 30 days of perfusion culture. Our overall objective is to develop a functional bone construct from hMSC and the chitosan-based biomimetic scaffold capable of large bone defect repair.

1. F. Zhao, W. Grayson, T. Ma*, B. Bunnell, and W. Lu. Effects of Hydroxyapatite on 3-D Human Mesenchymal Stem Cell Tissue Development. Biomaterials, 2006, 27:1859-1867.

2. F. Zhao and T. Ma* Perfusion bioreactor system for human mesenchymal stem cell tissue engineering: Dynamic cell seeding and construct development, F Biotechnology and Bioengineering, Vol. 91, 482-493, 2005.

3. F. Zhao, R. Chella, and T. Ma* Effects of Shear Stress on 3-D Human Mesenchymal Stem Cell Construct Development in a Perfusion Bioreactor System: Experiments and Hydrodynamic Modeling. . Biotechnology and Bioengineering, 96:584-595, 2007

572bh A Facs-Based Approach to Characterizing Culture Heterogeneity Based on Accumulation of the Anti-Cancer Agent Paclitaxel In Plant Cell Suspensions

Vishal Gaurav, Rohan A. Patil, and Susan C. Roberts. Chemical Engineering, University of Massachusetts Amherst, 686 North Pleasant Street, 159 Goessmann Lab, Amherst, MA 01003

Plant cells culture provides an alternative production source for plant-derived pharmaceuticals in which environmental conditions can be more easily controlled, manipulated, and optimized to yield high quantities of these valuable natural products. Our laboratory focuses on development and optimization of bio-processes for production of the anti-cancer agent Paclitaxel (Taxol®) in *Taxus* cell suspension cultures with an emphasis on understanding cellular metabolism at both the molecular and cellular level. Most studies concerning metabolite production via cell culture technology rely on culture-average parameters, which are often insufficient to describe culture heterogeneity. We have developed novel methods to isolate single cells from aggregated *Taxus* cultures, thereby allowing analysis of single cell phenotypes in a culture population. In this talk, we will present our recent work aimed at characterizing *Taxus* cell suspension sub-populations with respect to Paclitaxel accumulation. By analyzing cell populations with varying levels of Paclitaxel accumulation, we can understand the inherent molecular and metabolic differences amongst cells in culture and begin to understand the phenomena that underlay culture heterogeneity and production variability.

We have established a novel live-cell based indirect immunofluorescence assay for Paclitaxel based on PE fluorescence. Paclitaxel is primarily stored in the plant cell wall, and because our studies necessitate the sorting and selection of live cells both for mRNA expression analysis and re-culturing, it is critical that cells remain intact and not permeabilized or damaged throughout the staining procedure. This immunoassay was sufficient to stain respective amounts of cell wall-associated Paclitaxel in different sub-populations. Concomitantly, cells were also stained using fluorescein diacetate (FDA) to monitor viability. Multiparameter flow cytometry was performed and different Taxus cell populations were examined based on both PE and FITC fluorescence. To further validate the efficacy of this method, time-based viability tests were performed and demonstrated that cell viability remained high both during and after the staining procedure. We detect a broad range of Paclitaxel accumulation amongst *Taxus* cells in culture, providing an excellent target for fluorescence-activated cell sorting (FACS).

FACS can be effectively used to sort cells based on a variety of parameters including size, growth rate, protein production and metabolite production. Plant cells are challenging to sort due to their aggregated nature in suspension and relatively large size when compared to microbial and mammalian cells. We developed a new technique for high-throughput plant cell sorting using the BD FACSVantage equipped with a MacroSort option to accommodate our larger *Taxus* cells. A larger 200 μ m nozzle with optimized optics and fluidics conditions was used in analysis and sorting of *Taxus* cells based on Paclitaxel accumulation. The instrument was calibrated with 25 micron beads and *Taxus* cells were first sorted on the basis of size. Following a successful (~80-90% purity) sort by size, cells were stained for Paclitaxel using the procedure mentioned above and sorted based on fluorescence. Cells were gated into two populations based on low and high Paclitaxel accumulation and sorted accordingly. We were able to attain a high level of purity (~90%) for both sorted populations, demonstrating that live *Taxus* cells can be effectively sorted according to Paclitaxel content. The new technologies developed here not only allow recovery of low and high accumulating cells for metabolic analysis, but also allow for re-culture and initiation of potentially superior cell lines with both enhanced Paclitaxel accumulation and stability.

572bi Multiparametric Profiles of Human Immune Responses Produced by Single-Cell Analysis

J. Christopher Love[1], Qing Han[1], Qing Song[1], Elizabeth M. Bradshaw[2], Sally C. Kent[2], and David A. Hafler[2]. (1) Chemical Engineering, Massachusetts Institute of Technology, 77 Massachusetts Ave., Bldg. 66-456, Cambridge, MA 02139, (2) Neurology, Brigham and Women's Hospital, 77 Avenue Louis Pasteur, NRB, Room 641, Boston, MA 02115

Generating highly-resolved quantitative profiles of human immune responses remains a significant challenge because clinical samples are often limited in size, and cells of particular interest (e.g., autoreactive T cells) can be rare. This talk will describe an approach for analyzing large populations of single cells for both phenotypic, surface-expressed markers and secreted factors such as cytokines and antigen-specific antibodies. A soft lithographic technique called microengraving is employed to allow concurrent detection of cytokines and antibodies from peripheral blood mononuclear cells. Application of the approach to examine B cells from a recent-onset Type I diabetic subject showed the frequency of proinsulin-reactive B cells to be ~0.5% of the circulating B cells tested. Extensions for detection of multiple analytes simultaneously will be presented.

572bj Quantifying the Contractile Behavior of Individual Fibroblasts within Three-Dimensional Collagen Scaffolds: A Cellular Solids Approach

Brendan A. Harley, Chemical and Biomolecular Engineering, University of Illinois at Urbana-Champaign, Roger Adams Lab, MC-712, Box C-3, 600 S. Mathews Ave., Urbana, IL 61801 and Lorna J. Gibson, Materials Science and Engineering, Massachusetts Institute of Technology, 77 Massachusetts Avenue, Cambridge, MA 02139.

Cell-mediated contraction plays a critical role in many physiological and pathological processes, notably organized contraction during wound healing. Implantation of an appropriately formulated (*i.e.*, mean pore size, chemical composition, degradation rate) three-dimensional scaffold into an *in vivo* wound site effectively blocks the majority of organized wound contraction and results in induced regeneration rather than scar formation [1]. Improved understanding of cell contraction within three-dimensional constructs therefore represents an important area of study in tissue engineering.

Studies of cell contraction within three-dimensional constructs typically calculate an average contractile force from the gross deformation of a macroscopic substrate by a large cell population [2]. In this study, cellular solids theory has been applied to conventional column buckling relationships in order to quantify the magnitude of individual cell contraction events within a three-dimensional, collagen-glycosaminoglycan (CG) scaffold. This new technique can be used for studying cell mechanics with a wide variety of porous scaffolds that resemble low-density, open-cell foams. It extends previous methods for analyzing cell buckling of two-dimensional substrates to three-dimensional constructs. The mean contractile force (F_c) generated by individual dermal fibroblasts within CG scaffolds was calculated to range between 11 and 41 nN ($F_c = 26 \pm 13$ nN, Mean ± StDev), with an upper bound of cell contractility estimated at 450 nN.

Future iterations of this work will attempt to provide a more comprehensive treatment of cell contraction within 3D scaffolds and ECM analogs by integrating three-dimensional cell tracking using fluorescent technologies in order to incorporate the effect of loading eccentricity to this system and to better visualize cytoskeletal reorganization during contraction. Comparing the individual and populational cell contractile capacity within the CG scaffold may allow aspects of cell cooperativity to be probed and may suggest techniques for modifying the scaffold microstructure or mechanics to prevent organized scaffold (and therefore wound) contraction during *in vivo* wound healing and regeneration studies. The individual cell contraction technique adds a powerful new technique to the methods available to study cell contractility, particularly due to the ability to study individual cell contractility within a bioactive 3D scaffold with direct *in vivo* applications for induced tissue regeneration where understanding cell contraction is critical [1].

References:

1. Yannas IV. Tissue and Organ Regeneration in Adults. New York: Springer; 2001.

2. Freyman TM, Yannas IV, Yokoo R, Gibson LJ. Fibroblast contraction of a collagen-GAG matrix. Biomaterials 2001;22(21):2883-2891.

Keywords: fibroblast; contraction; force generation; biomaterial; cellular solid

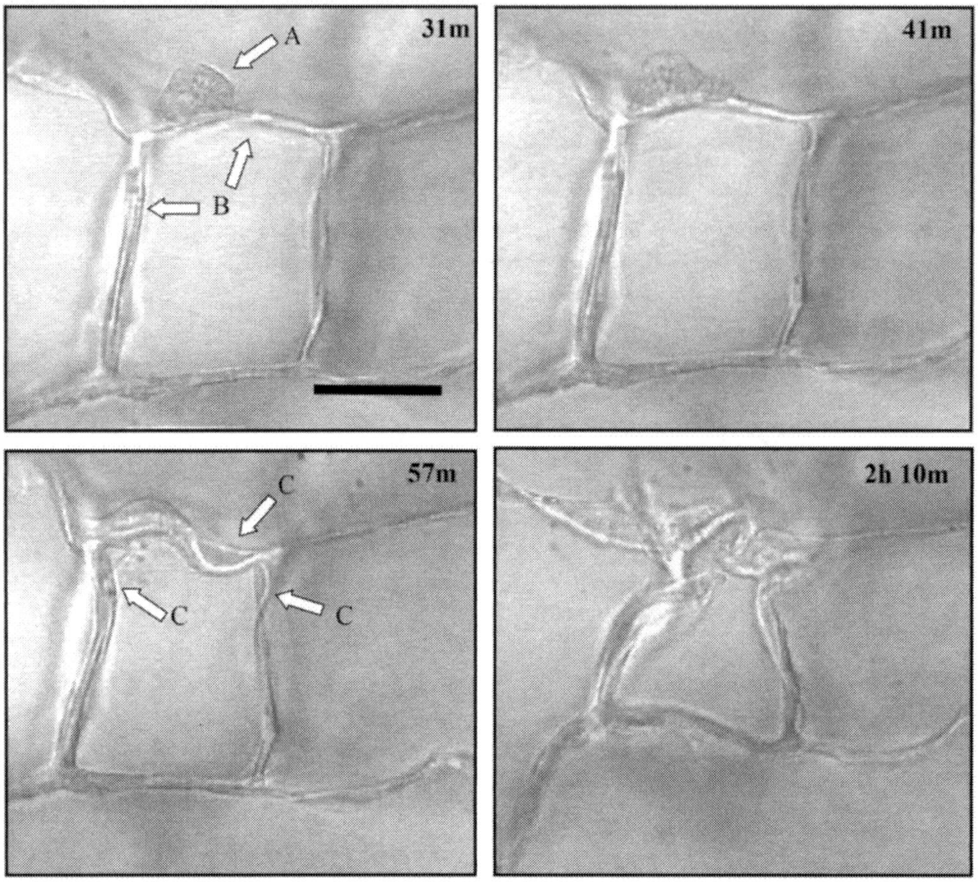

Figure 1. Time lapse images of an individual dermal fibroblast within the CG scaffold. The sequence of images shows a dermal fibroblast (arrow A) elongating and deforming the scaffold surrounding struts (arrows B). Several struts are deformed over time (arrows C). The number in the top right corner of each image indicates the time, in hours and minutes, after cell seeding. Scale bar: 50 μm.

ADDIN EN.REFLIST

572bk Modulating Endothelial Cell-Cell and Cell-Matrix Adhesion through Substrate Mechanics and Chemistry

Cynthia Reinhart-King and Joseph Califano. Department of Biomedical Engineering, Cornell University, Ithaca, NY 14853

While the effects of extracellular matrix chemistry have been well-studied, the role of substrate mechanics has only recently become of an area of intense interest. It is now well-known that substrate mechanics can alter many fundamental cell functions including cell adhesion, migration, contractility and gene expression. Using well-defined deformable substrates, we have recently found that cells can communicate mechanically through compliant substrates by exerting traction stresses that alter the strain field beneath adjacent cells. This exchange of mechanical information draws cells together and increases the stability of cell-cell adhesion between communicating cells. On softer substrates, where cells are less spread, cells tend to adhere to each other whereas on stiffer substrates, cells prefer to adhere and migrate on the substrate. We have recently extended these studies to explore the synergistic effects of substrate compliance and chemistry on the balance between cell-cell and cell-matrix adhesion. Our results indicate that decreasing the strength of cell-matrix adhesion by altering substrate mechanics or chemistry increases the stability of cell-cell adhesion. These changes in cell-cell adhesion result in increased endothelial cell aggregation and the formation of cellular networks. Our data underscore the importance of substrate mechanics in mediating the balance between cell-cell and cell-matrix adhesion and help to define material parameters that can enhance tissue formation on well-controlled biomaterial surfaces.

572bl Antisense Oligonucleotides (As ODNs) Based Leuckemia Therapy Using Therapeutic Antibody Conjugated Nano-Lipopolyplexes (NLPs)

Bo Yu, Department of Chemical and Biomolecular Engineering, The Ohio State University, 140 West 19th Avenue, Columbus, OH 43210, Yan Jin, NSF Center for Affordable Nanoengineering of Polymeric Biomedical Devices, The Ohio State University, 140 West 19th Avenue, Columbus, OH 43210, L. James Lee, Chemical and Biomolecular Engineering, The Ohio State University, 125 Koffolt Labs, 140 W 19th Ave, Columbus, OH 43210, and Robert J. Lee, College of Pharmacy, The Ohio State University, 542 Parks Hall, 500 West 12th Avenue, Columbus, OH 43210.

CD20 antibody is a FDA-approved therapeutic antibody that has been used in clinical treatment for leukemia. In this study, we conjugated antiCD20 on nano-lipopolyplexes (NLP) carrying Bcl-2 targeted antisense oligonucleotide (AS-ODN) for enhancing delivery to leukemia cells. In comparison to free ODN, the formulated ODN in antiCD20-NLP showed preferential uptake by B leukemia cells. The uptake of AS ODN correlated well with the CD20 expression levels on the cells. antiCD20-NLP mediated ODN delivery can enhance the intracellular Bcl-2 down-regulation in both B cell lines, such as Raji, and leukemia patient cells. The uptake and intracellular trafficking of ODN loaded antiCD20-NLP were characterized by confocal microscopy and Quantum Dot-FRET technology. We also found that the antiCD20 conjugated to NLPs can induce apoptosis on primary leukemia cells by the CD20 cross-linking mechanism. Our findings provide a new approach to improving the clinical efficacy of AS-ODNs therapy in treatment of B cell malignancies. The same approach can also be applied to therapeutic antibodies mediated small interfering RNA (siRNA) delivery for leukemia and other B cell related malignancies.

572bm Saccharide Modified Polyamines for the Prevention of Beta Amyloid Toxicity

James E. Henry, Cain Department of Chemical Engineering, Louisiana State University, 110 Chemical Engineering, South Stadium Rd, Baton Rouge, LA 70803

Alzheimer's disease (AD) is the leading cause of neurodegeneration in the United States, affecting approximately 4.5 million Americans in 2003, with an annual cost of care for these individuals estimated at over $100 billion. The cause of the neurotoxicity has been linked to the presence of Aβ fibrils. Aβ is a 39 to 43 amino acid long peptide that aggregates both in vivo and in vitro to form a variety of structures reported to be toxic and/or disease associated. These structures include a non-fibrillar species of approximately 17 to 42 kDa referred to as amyloid derived diffusible ligands (ADDLs), protofibrils species and fibril species. All the disease associated oligomers of Aβ have a high beta sheet content, in contrast to non-disease associated species. The currently most aggressively pursued Aβ oligomer target for both diagnostics and therapeutics is the ADDL.

In vivo, Aβ is known to accumulate in the cerebral cortex during disease, with deposition into mature amyloid plaques first in the entorhinal cortex and the hippocampus then later in the frontal cortex. During disease, it is apparent that Aβ accumulates in other parts of the body that are easier to sample for in vitro diagnostic measurements. Aβ circulates in both the blood and the cerebral spinal fluid. As disease progresses, there are significant changes in oligomeric Aβ in the cerebral spinal fluid.

While many investigators are developing structure specific antibodies for disease associated Aβ, there may be biomimetic materials that can be developed with antibody-like affinity but smaller molecular size that may be more appropriate for the demands of a neurotherapeutic. A variety of evidence indicates that Aβ may bind to cells via an interaction with surface sialic acids, and that the affinity of this interaction increases when the gangliosides or sialic acid molecules on the cell surface are clustered. Based on these data, it is hypothesized that biomimetic materials could be synthesized which would reproduce the clustered sialic acid structure of the cell surface, and therefore recreate the Aβ binding seen to occur on neuronal cell membranes.

It is proposed that targeting the fibrils will prevent toxicity by sequestering these fibrils. This technique has been demonstrated by previous work, but little has been done to improve the biomimicry of the sequestering agents. It is this area that will be explored. This technique relies on the attacking the theoretical "bottleneck" in the Alzheimer's process, the interaction of Aβ with neurons. This interaction is considered the bottleneck because there are several theorized environmental conditions that lead to the formation of Aβ fibrils. Additionally, the exact form of Aβ (3-mer, 5-mer, 12-mer, protofibril, fibril, etc) that interacts with the neuron is not concretely agreed upon. However, it is agreed that preventing neuronal interaction prevents toxicity.

A variety of evidence indicates that Aβ binds to cells via an interaction with surface glycolipids or glycoproteins with increased binding occurring in the presence of gangliosides or other clustered sialated molecules. Based on these data, it is hypothesized that membrane mimics could be synthesized which would reproduce the clustered sialic acid structure of the cell surface, and therefore compete with the cell surface for Aβ binding. While previous work has shown limited success, this work will greatly improve the knowledge toward intelligent molecule design and the efficacy of the sialated molecules. These results could have implications for the design of new agents that bind pathogenic Aβ peptides for the treatment of neurodegenerative disease. Based on our results, we will show the following results:

1. Production sialic acid modified polyamines.

2. Intrinsic toxicity of the new structures.

2. Efficacy of the molecules to prevent Aβ toxicity.

572bn The Prime Model for Protein Aggregation: Extension to All Twenty Amino Acids

Mookyung Cheon and Carol K. Hall. Chemical and Biomolecular Engineering, North Carolina State University, 911 Partners Way, Raleigh, NC 27695

We extend PRIME, an intermediate-resolution protein model previously applied to polyalanine and polyglutamine, to the description of the geometry and energetics of peptides containing all twenty amino acid residues. The PRIME model has been used previously in conjunction with discontinuous molecular dynamics, to simulate the self assembly of large systems of polyalanine peptides into fibrillar structures. The goal here is to develop a similar model but now applicable to all twenty amino acids, which would allow us to simulate the spontaneous fibrillization of heterogeneous proteins, including the Alzheimer's peptide A-beta. The parameters for extended PRIME include hydrogen-bonding , side-chain/side chain and side-chain/ main chain interaction range and strength, excluded volume parameters and side-chain bond lengths. These are obtained by applying a learning algorithm and linear programming technique that compares the known native states for a selection of proteins from the PDB with decoy structures generated by Levitt. The interactions are fine-tuned by seeking the optimal energy gap between native and decoy structures. Simulations on systems containing peptides known to form fibrils are being performed using the new parameter set.

Millions suffer from often debilitating bone injuries and diseases each year. Often, defects remain which must be treated to restore original function to the bone. Development of tissue engineering as a bone regeneration option requires identification of a suitable osteoconductive matrix. Hydrophobic scaffolds such as poly(propylene fumarates) and polyanhydrides have shown promise as osteoconductive matrices, and recently, hydrophilic polyethylene oxide (PEO) hydrogels to which cell adhesion ligands have been conjugated were demonstrated to provide an osteoconductive environment. Based on these results, scaffolds in which the level of hydrophobicity can be tuned may be particularly desirable for optimizing the formulation of bone tissue engineering matrices.

In the present study, we characterize the effects of novel inorganic-organic hydrogel scaffolds generated by the photo-cure of star polydimethylsiloxane ($PDMS_{star}$) (hydrophobic inorganic polymer) and linear hydrophilic PEO on rat osteoblast extracellular matrix (ECM) production, alkaline phosphatase expression, and calcium deposition. Initial studies have shown these hybrid scaffolds to display a range of hydrophobicities as well as microstructural, biochemical, and biomechanical properties that can be precisely tuned over a broad range.

By varying the ratio of $PDMS_{star}$ to PEO, the modulus of the hydrogels remained constant. However, gradual alterations in hydrophobicity and microscale scaffold morphology were introduced. These alterations in morphology impacted cellular ECM production and calcium deposition. Specifically, cells demonstrated a moderate increase in collagen type I, osteocalcein, and alkaline phosphatase as hydrophobic $PDMS_{star}$ levels increased from 0 to 1%. However, production of these same markers fell dramatically as hydrophobicity increased beyond this point. Similarly, the levels of collagen and GAG production varied significantly with hydrogel $PDMS_{star}$ concentration. Thus, by modifying the composition of novel inorganic-organic $PDMS_{star}$:PEO hydrogels, we are able to modulate the ECM production and differentiation of encapsulated rat osteoblasts. Future investigation of the impact of systematic alterations in the microstructural, biochemical, and biomechanical properties of these hybrid scaffolds on cell behavior should therefore yield profound insight into the dependence of cell behavior on material properties.

579b Cell Tracking and Cell Ratio Modulation for Cardiac Tissue Engineering

Rohin K. Iyer, Institute of Biomaterials and Biomedical Engineering (IBBME), University of Toronto, 164 College Street, Room 407, Toronto, ON M5S 3G9, Canada, Jane Chui, Division of Engineering Science, University of Toronto, 40 St. George Street, Toronto, ON M5S 2E4, Canada, and Milica Radisic, IBBME/Chemical Engineering and Applied Chemistry, University of Toronto, 164 College Street, Room 407, Toronto, ON M5S 3G9, Canada.

Introduction and Background

Recent work in the field of cardiac tissue engineering suggests that myocyte-nonmyocyte interactions may be crucial to engineering viable and functional heart tissues [1]. We previously showed that cultivation of cardiomyocytes (CM) on pre-formed networks (Pre-culture) of fibroblasts/endothelial cells (FB/EC), enhanced the structural and functional properties of cardiac "organoids" engineered at the microscale compared to simultaneously seeding the three cell types (Simultaneous Tri-culture), which resulted in non-beating organoids with a clustered morphology and lacking connexin-43 expression[2]. We hypothesized that these observations suggested a potential soluble factor-mediated mechanism involving vascular endothelial growth factor (VEGF) signaling. The goals of this study were to better visualize how these clusters formed with time and to further modulate the cell ratios used previously to select for better tri-culture conditions. To this end, we conducted cell tracking studies to get a visual picture of how cells organized both spatially and temporally into organoids when tri-cultured versus pre-cultured. We also assessed the effect of systematically varying the CM fraction (to improve the functional properties of the organoids) as well as the EC fraction (in order to improve capillary cord formation).

Materials and Methods

Microchannels were patterned by photopolymerization of liquid poly(ethylene glycol) diacrylate pre-polymer (PEG, 700 Da) around a polypropylene master, resulting in discs with microchannel features 100-200μm diameter, 4mm in length, as previously described [2]. The channels were coated with 5μL of Matrigel® and allowed to gel at 37°C prior to the addition of cells. Primary cardiac FB, an EC cell line (D4T), and primary neonatal rat CMs were used for tri-culture and pre-culture. For cell tracking studies, CellTracker™ Red, CellTracker™ Green and 4'-6-diamidino-2-phenylindole (DAPI) were used to label EC, FB, and CM, respectively. FB/EC were incubated in the dyes for 30 minutes at a concentration of 5μM while CM were incubated in DAPI at a concentration of 20μg/mL for 1 hour. For Pre-cultures, the FB/EC alone were mixed and seeded followed by CM two days later, while for Simultaneous Tri-cultures, CM/FB/EC were simultaneously mixed and seeded. The discs were flipped over and imaged under fluorescence microscopy at days 1 and 4 after seeding the CM. CM ratios were also varied as 40%, 60%, and 80% of the total cell number, which was fixed at $2x10^5$ cells/disc. In the case of the CM 100% group, all $2x10^5$ cells seeded were cardiomyocytes enriched by pre-plating the native heart isolate twice in T75 flasks to remove the more adherent non-myocytes, as previously described [2]. The percentages of each cell type seeded are given in Table 1 below.

Table 1: Percentages of cells seeded in Cell Tracking/CM ratio experiments

Group	%CM	%EC	%FB
Tri-/Pre-culture 40%	40%	47%	13%
Tri-/Pre-culture 60%	60%	31%	9%
Tri-/Pre-culture 80%	80%	16%	4%
CM 100%	100%	0%	0%

In the second part of the study, the fraction of EC was varied to promote capillary cord formation. From our cell tracking studies, we observed that the 60% myocyte fraction had the best structural and functional properties (elongated morphology and largest contraction amplitude). Thus, the myocyte fraction was fixed at 60% and the EC fraction was varied to span a range of densities. Two densities at the lower end of the spectrum (16×10^4 EC/disc and 31×10^4 EC/disc) were chosen because they were in agreement with seeding densities previously reported to support capillary cord formation. The latter value of 62×10^4 ECs/disc corresponded to the EC seeding density used in the Pre-culture 60% group (Table 1), and therefore served as a control for comparison to cell tracking studies. As the fraction of ECs increased, the fraction of fibroblasts seeded 24 hours later was correspondingly decreased to ensure the total cell density was maintained at $2\text{x}10^5$ cells/disc. Finally, two days after EC seeding, CM were seeded at a fraction of 60%. The final percentages of cells seeded in each group are given in Table 2 below.

Table 2 : Percentages of cells seeded in EC ratio / cord experiments

Group	%CM	%EC	%FB
Pre-culture 60% 16k EC	60%	8%	32%
Pre-culture 60% 31k EC	60%	15%	25%
Pre-culture 60% 40k EC	60%	20%	20%

Pre-culture 60% 62k EC	60%	31%	9%

Results and Discussion

Figure 1 shows the results of cell tracking studies performed on Pre-cultures grown at three different myocyte fractions (40%, 60%, and 80% of the total cell number). Cardiac FB are labelled green, D4T EC are labeled red, and neonatal CM appear as blue nuclei. In general cells appear more elongated in both the 60% and 80% groups than in the 40% group. As well, the cells appear more elongated on day 1 than on day 4. As expected, a higher proportion of FB (green) is seen in the 40% group compared to more EC (red) and CM (blue nuclei) in both the 60% and 80% groups. In the 60% cells appear very elongated and display evidence of cords (white arrows) consisting of both EC and FB, suggesting that this condition might be more conducive to formation of vascular precursors. Spontaneous contractions in the 60% and 80% were of a larger amplitude than in the 40% group.

Figure 2 shows the same data obtained from cell tracking studies on Simultaneous Tri-cultures. A comparison with Pre-cultures revealed marked differences in morphology. Cells appeared rounded and organized into spherical clusters (white arrows). Clusters were not apparent on day 1 but became more apparent on day 4, suggesting the cells organized into clusters over time. The clusters also appeared to increase in size with an increase in myocyte fraction. For the 60% CM fraction, fibroblasts (green) organized on the periphery of clusters while for the 80% CM fraction, both CMs (blue) and also fibroblasts (green) organized on periphery of clusters. In all three groups, EC (red) were primarily seen to organize within the centers of clusters, and this was more evident in the 60% and 80% groups. This localization of EC into clusters was also consistent with immunofluorescence data shown in previous studies[2]. We noted no spontaneous contractions in the Simultaneous Tri-culture 40% condition, consistent with our previous work [2], but were able to see sparse areas spontaneously contracting in the higher myocyte fractions (60%, 80%), suggesting that even in the case of Simultaneous Tri-culture it may be possible to restore contractile function by increasing the myocyte fraction.

To engineer vascularized organoids, we seeded D4T endothelial cells on Matrigel coated PEG microchannels to allow for capillary sprouts to form, a phenomenon reported to occur within 24 hours under these conditions. The ratio of CM was fixed at 60%, while the ECs were seeded at varying densities from 16×10^4 to 62×10^4 cells/disc. As shown in Figure 3 (top panel) 24 hours after seeding the endothelial cells, thin EC cords had formed in the 16k group (white arrows), but in the other groups, most of the cells had already proliferated into EC monolayers with only some evidence of cord formation at edges (single asterisks).

We then seeded cardiac FB 24 hours after cords had formed in order to stabilize them and allowed them to adhere for additional 24 hours. Finally, we seeded Enriched cardiomyocytes at the 60% fraction (total cell number = 2×10^5 cells/disc) and cultivated the cells for 7 days until organoids had formed. Figure 3 (bottom panel) shows phase contrast images of the resulting engineered organoids on Day 7 as a function of EC density. The 16k group formed three-dimensional organoids (white arrows) characterized by elongated cells. However in the higher EC fraction groups, we noted organoids that were extremely flat in morphology. As the EC density increased from 31k to 62k per disc, we also noted fewer spontaneous contractions and more rounded cells. We observed that the 16k group had a low excitation threshold (comparable to enriched CM alone) and higher overall viability as determined by live/dead staining.

References

1. Naito H, Melnychenko I, Didie M, et al. Optimizing engineered heart tissue for therapeutic applications as surrogate heart muscle. Circulation. 2006 Jul 4;114(1 Suppl):I72-8.

2. Iyer RK, Chiu, L. L. Y., Radisic, M. Microfabricated poly(ethylene glycol) templates enable rapid screening of tri-culture conditions for cardiac tissue engineering. Journal of Biomaterials Research A. (2008, in press.)

Acknowledgements

The author wishes to thank Dr. P. W. Zandstra

, Dr. M. V. Sefton,, and Dr. W. L. Stanford for cell lines, laboratory space, and use of their equipment. Funding was provided by NSERC (DG), CFI (LOF), ORDCF (ARTEC), OGSST, OGS, NSERC USRA, and NIH grant R01HL076485.

Figure 1: Cell Tracking on Pre-cultured organoids at three myocyte fractions (40%, 60%, 80%). EC appear red, FB appear green, and CM nuclei appear blue.

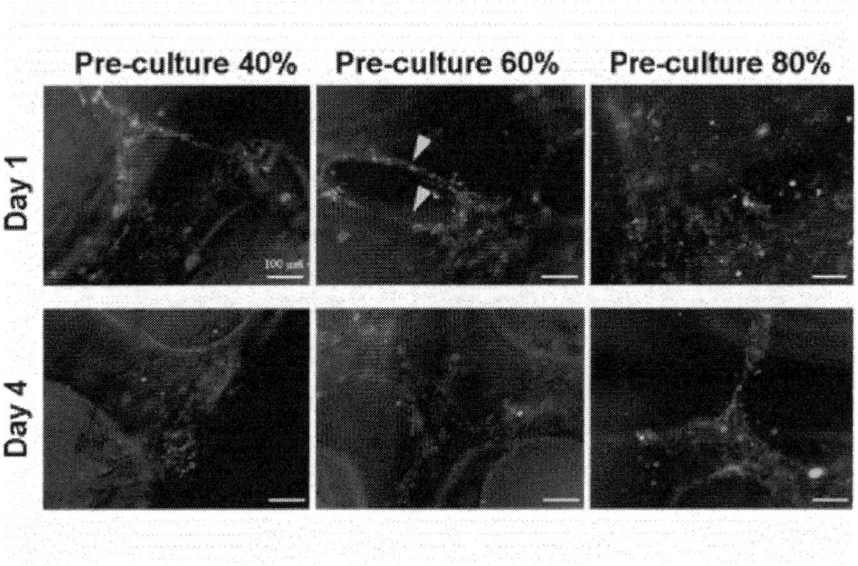

Figure 2: Cell Tracking on Simultaneously Tri-cultured organoids at three myocyte fractions (40%, 60%, 80%). EC appear red, FB appear green, and CM nuclei appear blue.

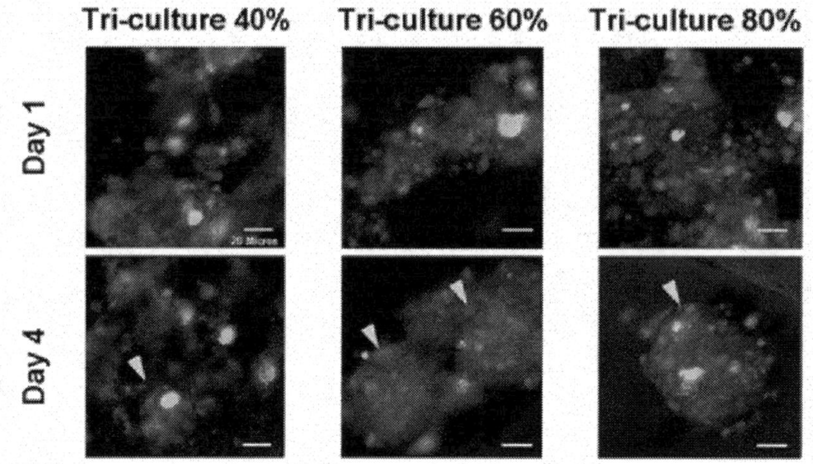

Figure 3: Phase contrast images of organoids cultivated at a fixed myocyte fractions (60%) at varying EC fractions (16k, 31k, 40k, and 62k EC/disc). The organoids were imaged at day 1 after EC seeding to assess cord formation and at day 7 after sequential cultivation of EC/FB/CM.

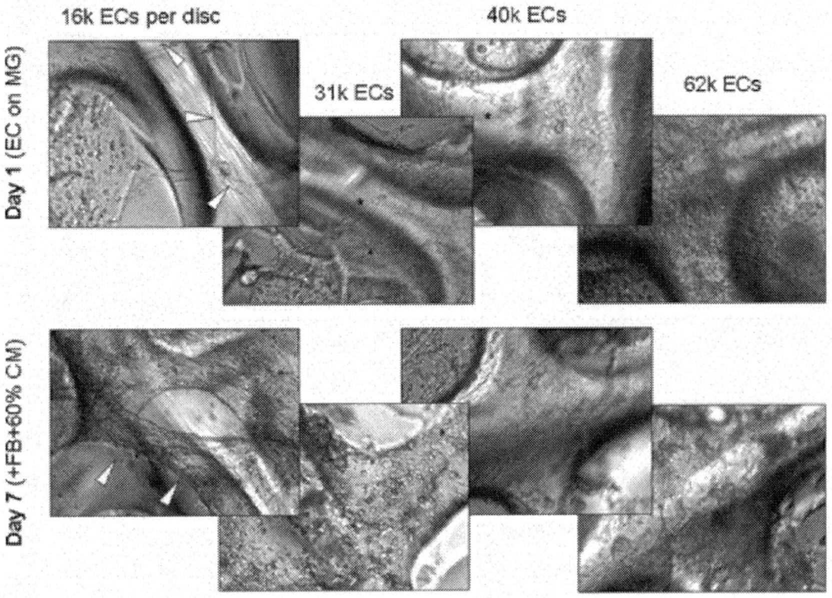

Web Page: chem-eng.utoronto.ca/~milica

579c Mineralized Extracellular Matrix Constructs Support the Osteogenic Differentiation of Mesenchymal Stem Cells In Vitro

F. Kurtis Kasper[1], Richard A. Thibault[1], and Antonios G. Mikos[2]. (1) Department of Bioengineering, Rice University, 6100 Main Street, MS-142, Keck Hall, Room 116, Houston, TX 77005-1892, (2) Departments of Bioengineering and Chemical and Biomolecular Engineering, Rice University, 6100 Main Street, MS-142, Keck Hall, Room 116, Houston, TX 77005-1892

A significant clinical need exists for synthetic scaffold materials capable of supporting and promoting bone regeneration *in vivo*. Previous studies from our laboratory have established that mesenchymal stem cells (MSCs) can be cultured on porous scaffolds under engineered conditions *in vitro* to generate a mineralized bioactive extracellular matrix coating of the scaffolds, which induces the osteogenic differentiation of subsequently cultured MSCs *in vitro*. The overall objectives of this study were to explore the application of biodegradable poly(ε-caprolactone) (PCL) scaffolds in generating mineralized bioactive extracellular matrix constructs and to assess the effect of the osteogenic culture medium supplement dexamethasone on the osteogenic differentiation of MSCs cultured on these constructs *in vitro*.

MSCs were harvested from the femora and tibiae of rats, expanded in osteogenic medium supplemented with dexamethasone for 6 days, seeded onto electrospun PCL fiber mesh scaffolds, and then cultured statically for 12 days in osteogenic medium supplemented with dexamethasone to produce mineralized bone-like extracellular matrix constructs. These constructs were then decellularized, seeded with freshly harvested and expanded MSCs, and cultured statically for 4, 8 and 16 days in osteogenic medium either with or without dexamethasone supplementation. MSCs were cultured in parallel on plain PCL scaffolds. Additionally, acellular PCL/ECM constructs were cultured in osteogenic medium supplemented with dexamethasone as a control. Following culture, the cellularity, alkaline phosphatase activity, and calcium content of scaffolds from each group were assessed.

This study demonstrates that MSCs cultured on PCL scaffolds in the presence of dexamethasone produce mineralized extracellular matrix constructs that direct the osteogenic differentiation of and mineral deposition by MSCs. Additionally, the PCL/ECM constructs support the osteogenic differentiation of MSCs (previously expanded in medium supplemented with dexamethasone), even in the absence of dexamethasone from the secondary culture medium. Finally, acellular PCL/ECM constructs alone induce mineral deposition from the culture medium. These novel bioactive mineralized constructs present tremendous potential for application as cell transplantation vehicles or as acellular scaffolds for bone tissue regeneration.

579d Modulating Endothelial Cell Response to Crosslinked Artificial Extracellular Matrix Proteins

Julie C. Liu, School of Chemical Engineering, Purdue University, 480 Stadium Mall Drive, West Lafayette, IN 47907 and David A. Tirrell, Division of Chemistry and Chemical Engineering and Joseph J. Jacobs Institute for Molecular Engineering for Medicine, California Institute of Technology, 1200 E. California Blvd., Pasadena, CA 91125.

A family of artificial extracellular matrix (aECM) proteins was designed for soft-tissue engineering applications. These proteins consist of i) elastin-based structural repeats to confer mechanical properties, ii) fibronectin cell-binding domains to promote cell adhesion, and iii) lysine residues that serve as crosslinking sites. Previous work established that cell-binding domain choice and context (i.e. amino acid choice more than 14 residues away from the cell-binding domain sequence) can be used to alter cell adhesion and spreading response to aECM proteins. This study demonstrates that ligand density of crosslinked aECM films provides another tool for controlling cell response.

We previously showed that cells adhere specifically to both the RGD and CS5 cell-binding sequences in adsorbed aECM proteins. This work focuses on aECM-RGD, a variant that contains the RGD sequence, and aECM-RDG, a negative control protein that contains a scrambled cell-binding domain. Covalent attachment of poly(ethylene glycol) (PEG) to crosslinked aECM substrates decreased nonspecific cell adhesion to aECM-RDG-PEG but did not preclude sequence-specific adhesion of endothelial cells to aECM-RGD-PEG. To alter cellular responses, ligand density was varied by mixing aECM-RGD and aECM-RDG prior to crosslinking. Increasing the RGD density resulted in more robust adhesion and spreading responses but did not affect cell motility rates. Control of cell-binding domain density in aECM proteins can thus be used to modulate cell adhesion and spreading, and will serve as an important design tool as these materials are developed further for use in surgery, tissue engineering and regenerative medicine.

579e a New Technique for Primary Hepatocyte Proliferation in Vitro

Cheul H. Cho[1], François Berthiaume[2], Arno W. Tilles[2], and Martin Yarmush[3]. (1) Biomedical Engineering, New Jersey Institute of Technology, 323 Martin Luther King Jr Blvd, Newark, NJ 07102, (2) Center for Engineering in Medicine, Massachusetts General Hospital, Harvard Medical School, 51 Blossom Street, Boston, MA 02114, (3) Research, Center for Engineering in Medicine, 51 Blossom Street, Boston, MA 02114

The current application for many potential cell-based treatments for liver failure is limited by the low availability of mature functional hepatocytes. Although adult hepatocytes have a remarkable ability to proliferate in vivo, attempts to proliferate adult hepatocytes in vitro have been less successful. In this study, we investigated the effect of coculture cell type on the proliferative response and the functional activities of hepatocytes. We show, for the first time, a robust proliferative response of primary adult rat hepatocytes when cocultured with mouse 3T3-J2 fibroblasts. Hepatocytes cultured at low density on growth-arrested 3T3-J2 fibroblast feeder layers underwent significantly higher proliferation rates than when cultured on feeder layers made of four other cell types. Increasing colony size correlated with an increase in hepatocellular functions. The proliferating hepatocytes retained their morphologic, phenotypic, and functional characteristics. To better understand the role of heterotypic cell-cell interactions on hepatocyte proliferation in cocultures, a cell patterning technique was used. Polydimethylsiloxane (PDMS) stencils with different circular hole sizes (300 or 1,000 μ m in diameter) were made by the procedure modified from our previous report [1]. Using a cell patterning technique, we found that 3T3-J2 fibroblasts stimulate DNA synthesis in hepatocytes by short-range heterotypic cell-cell interactions. When hepatocytes that proliferated in cocultures were harvested and further subcultured either on 3T3-J2 fibroblast feeders or in the collagen sandwich configuration, their behavior was similar to that of freshly isolated hepatocytes. We conclude that adult rat hepatocytes can proliferate in vitro in a coculture cell type-dependent manner, and can be serially propagated by coculturing with 3T3-J2 fibroblasts while maintaining their differentiated characteristics. Our results also suggest that one of the major reasons for the functional differences in hepatocyte cocultures may be due to the different proliferative responses of hepatocytes as a function of coculture cell type. This study provides new insights in the roles of coculture cell types and cell-cell interactions in the modulation of hepatic proliferation and function.

Reference:

[1] Jae-Sung Park, Cheul H. Cho, Natesh Parashurama, Yawen Li, François Berthiaume, Arno Tilles, Mehmet Toner, Martin Yarmush, "Microfabrication-based Modulation of Embryonic Stem Cell Differentiation", Lab on a Chip, 2007 (7): 1018-1028

579f Effect of Uniaxial Cyclic Stretch on Morphology and Gene Expression by Ligament Progenitor Cells Seeded on Electrospun Fibrous Meshes

Chris A. Bashur[1], Scott A. Guelcher[2], and Aaron S. Goldstein[1]. (1) Department of Chemical Engineering, 0211, Virginia Tech, Blacksburg, VA 24061-0211, (2) Chemical Engineering, Vanderbilt University, Olin Hall, Nashville, TN 37212

A tissue-engineered ligament is a promising alternative to autograft for anterior cruciate ligament (ACL) reconstruction, but strategies must be developed to form oriented tissues with high tensile strength. To accomplish this, we propose to seed ligament progenitor cells onto oriented fiber scaffolds and to use cyclic mechanical stretch on the cells to stimulate deposition of a ligament-like extracellular matrix (ECM). Here, electrospinning of a poly(esterurethane urea) (PEUUR) onto a rotating target will be used to create an oriented fiber mesh that will guide cell alignment and suppress cell reorientation in response to stretch. Further, our PEUUR is elastomeric, biocompatible, and degradable, making it a suitable substrate for both in vitro cyclic mechanical stretch and implantation in vivo.

For this study electrospun meshes will be formed with mean fiber diameters of 0.8 μm and orientation marked by an angular standard deviation of 30°, as our previous studies have shown that these characteristics ensure that ligament progenitor cells assume a spindle-shaped morphology and an orientation parallel to fiber alignment. Next, ligament progenitor cells – derived from bone marrow explants – will be cultured on our electrospun substrates and exposed intermittently to uniaxial cyclic stretch in a bioreactor that we have developed (0.5 Hz, 8% strain, 60 min daily). These experimental conditions have been shown to induce expression of the ECM proteins collagen types I and III, decorin, and tenascin-C. After performing cyclic stretch, cells will be analyzed for the expression of ECM proteins and for cell morphology (e.g., projected area, aspect ratio, and orientation relative to the axis of stretch). Control groups will include cells grown on fiber meshes but not exposed to stretch, and cells grown on smooth polydimethylsiloxane elastomer films. This study will serve as an initial condition for a series of systematic studies to probe the effect of mechanical stretch regimens on the formation of ligament-like tissues, and will provide materials for establishing a subcutaneous implantation model for characterizing integration and vascularization.

579g IGF-1 Signaling of Chondrocytes Embedded In Alginate-Hyaluronic Hydrogels

Diana M. Yoon, Department of Chemical and Biomolecular Engineering, University of Maryland, 2113 Chemical & Nuclear Engineering Bldg., College Park, MD 20742-2111, A. Hari Reddi, Department of Orthopaedic Surgery, University of California Davis School of Medicine - Center for Tissue Regeneration and Repair, 4635 Second Ave, Research Bldg I, Room 2000, Sacramento, CA 95817, and John P. Fisher, Fischell Department of Bioengineering, University of Maryland, 3238 Jeong H. Kim Building, College Park, MD 20742-2111.

Articular cartilage has a limited capacity for repair or regeneration. While there are a multitude of factors involved in cartilage function and repair, we propose that it is critical to understand the basic endogenous signaling mechanisms among chondrocytes in an engineered construct. An anabolic protein of particular interest is insulin-like growth factor-1 (IGF-1) because it aids in cartilage matrix production, specifically type II collagen expression. Therefore, we have begun to investigate how construct properties affect the insulin-like growth factor-1 (IGF-1) signaling pathway.

Chondrocytes embedded in alginate express IGF-1. Previous studies have demonstrated that in high concentrations of alginate, an up-regulation of IGF-1 expression is observed when decreasing paracrine signaling distances (increasing cell density) occurs. This suggests that high cell density and high matrix content enhances IGF-1 competition. Previous studies have also shown that by delivering exogenous IGF-1 to embedded chondrocytes in alginate leads to a decrease in endogenous IGF-1 expression and an increase in IGF-1 receptor (IGF-1R) expression. This result indicates that chondrocytes may increase the expression of IGF-1R to optimize IGF-1 signaling.

Our current work further extends our understanding of how the IGF-1 signaling pathway is altered by incorporating hyaluronic acid (HA) into alginate hydrogels. Hyaluronic acid is a major extracellular matrix that aids in proteoglycan production by chondrocytes through the receptor CD44. However, little is known about HA affects on the IGF-1 signaling pathway. Therefore, we investigated the effect of hyaluronic acid concentration (0, 0.5 mg/mL and 5.0 mg/mL) in alginate hydrogels upon IGF-1 and IGF-1R expression as well as its correlation with CD44 expression in both an in vitro model and in vivo model (subcutaneously implanted in SCID mice). Additionally, we induced the samples with exogenous IGF-1 (100 ng/mL) to further investigate how this delivery affects expression levels of IGF-1, IGF-1R and CD44 of chondrocytes embedded in the alginate-HA hydrogels. The results of this work indicate how extracellular matrix molecules impact the expression of IGF-1 signaling molecules by chondrocytes, demonstrating that the IGF-1 signaling pathway is affected by modifying the properties of an engineered construct.

587a In Vivo Monitoring of Viral Proteolytic Activity Using Luminescent Quantum Dot FRET-Based Probes

Divya Sivaraman[1], Marylynn V. Yates[2], Ashok Mulchandani[3], and Wilfred Chen[1]. (1) Chemical and Environmental Engineering, University of California Riverside, Riverside, CA 92521, (2) Department of Environmental Sciences, University of California Riverside, Riverside, CA 92521, (3) Department of Chemical and Environmental Engineering, University of California, Riverside, CA 92521

Proteases are enzymes that hydrolyze the peptide bonds of proteins. Many infectious microorganisms like bacteria and viruses use proteases as the key virulence factor for infection. Therefore proteases serve as an important target for characterization of disease development and can be potentially used to study various therapeutic agents. Here we are focusing on the use of a principle called fluorescence resonance energy transfer (FRET) to monitor the proteolytic activity of viruses in vivo. In particular, we seek to demonstrate the use of quantum dot bioconjugates to serve as sensitive probes in detecting viral protease activity in mammalian cells. Most commercially available quantum dots are not suitable for biological applications because of the presence of hydrophobic capping ligands on their surfaces. However, this can be overcome by exchanging the native TOPO capping with another surface ligand DHLA (dihydrolipoic acid), which renders them water soluble. In this work, the luminescent probes are based on FRET between a quantum dot that serves as an energy donor and rhodamine (an organic dye) that behaves as an energy acceptor. For the initial testing, we developed a substrate peptide sequence that contains the cleavage site which is recognized by the polio viral protease PV2A^{pro}. The rhodamine molecules are immobilized onto the surface of the quantum dots through the substrate peptide structure. The substrate peptide sequence acts as a spacer between the quantum dot and rhodamine thereby providing an optimal spacing for FRET to occur. Since our goal is to detect the presence of viral protease in living cells, a method that enables efficient intracellular delivery based on the use of the TAT peptide, which is derived from HIV-1 protein and has been shown to serve as excellent target carriers, was used. In this manner, the probes were introduced into BGMK (Buffalo green monkey kidney) cells and infected with different viral dosages. The linkage between the quantum dot donor and the rhodamine acceptor is cleaved during viral infection, due to the production of 2A^{pro} causing changes in fluorescence intensity. Utility of these new probes for monitoring viral activity and to screen for protease inhibitors will be discussed.

587b Facile Size Isolation and Peptide Conjugation to Lipid-Coated Microbubbles: Application to Molecular Imaging of Renal Tumor Models

Mark A. Borden, Chemical Engineering, Columbia University, 500 West 120th Street, New York, NY 10027, Jessica Kandel, Institute for Cancer Genetics, Columbia University, Babies & Children's Hospital North, Room 214, New York, NY 10032, and Darrell Yamashiro, Surgery, Columbia University, 161 Fort Washington Ave, Room 214, New York, NY 10032.

Ultrasound is rapidly emerging as a practical modality for molecular imaging this is capable of real-time, deep-tissue viewing and simultaneous intervention. Two important design criteria for probe development in ultrasound molecular imaging are ligand conjugation and contrast agent size. The former dictates to what degree the microbubble will selectively adhere to target tissue, while the latter determines the echo signature. More control of both will provide better quantification of biomarker expression. Lipid-coated microbubbles are well recognized for their biocompatibility and echogenicity. They can be fabricated by simple emulsification procedures that result in high yields. For example, over one billion microbubbles per milliliter of solution can be made within just a few seconds. Fortuitously, the resulting suspension is polydisperse in size with separate peaks centered at 1-2 micron and 4-5 micron diameter. Using chemical engineering principles, such as population balance modeling of migration in a gravitation field, these peaks can be rapidly and reproducibly isolated and characterized. Computational modeling and experimental validation of the technique will be presented. Following size isolation, peptide ligands can be efficiently conjugated using maleimide-thiol chemistry. This conjugation procedure will be presented as well.

Size-isolated, targeted microbubbles were used in a preliminary evaluation of ultrasound molecular imaging for angiogenesis in mouse model of a solid tumor in the kidney. Imaging with a high-frequency, small-animal scanner showed selective accumulation of RGD-peptide bearing microbubbles in the tumor vessels compared to no-ligand microbubbles. Tumor vessel architecture and perfusion also were evaluated. These early results show promise for ultrasound molecular imaging as a research tool for studying angiogenesis and the response to therapy in genetically driven renal tumor models.

587c Engineered Knottin Polypeptides. A New Class of In Vivo Molecular Imaging Agents

Richard H. Kimura[1], Aron M. Levin[2], Zhen Cheng[1], Sanjiv Sam Gambhir[1], and **Jennifer R. Cochran**[2]. (1) Radiology and Molecular Imaging Program, Stanford University Schools of Engineering and Medicine, 318 Campus Dr. West, MC5439, James H. Clark Center, W250, Stanford, CA 94305-5439, (2) Bioengineering, Stanford University Schools of Engineering and Medicine, 318 Campus Dr. West, MC5439, James H. Clark Center, W250, Stanford, CA 94305-5439

There is a critical need for molecular imaging probes that specifically target receptors that are overexpressed on tumors, and allow noninvasive characterization of tumors for patient-specific cancer treatment and disease management. We engineered cystine knot (knottin) polypeptides as a new class of in vivo molecular imaging agents of integrin receptor-expressing tumors. The knottin family includes protease inhibitors, toxins, and antimicrobials. These polypeptides are small (30-40 residues), very stable in vitro and in vivo, and their disulfide-constrained loops tolerate much sequence diversity. These qualities are desirable for drug design; however, knottins do not naturally bind to receptors expressed on tumors, rendering them useless in cancer applications.

Integrins are a family of extracellular matrix adhesion receptors that non-covalently associate into alpha/beta heterodimers with distinct ligand binding specificities. Several integrins, including alphavbeta3, alphavbeta5, and alpha5beta1, have generated clinical interest due to their expression on the surface of cancer cells and tumor neovasculature, as well as their proposed role in mediating angiogenesis, tumor growth, and metastasis. Despite the prevalence of integrin-binding peptides and peptidomimetics in the literature, suboptimal tumor targeting efficacy and pharmacokinetics have limited their clinical translation as molecular imaging agents.

We used the Ecballium elaterium trypsin inhibitor II (EETI-II), a knottin from the squash family of protease inhibitors, as a molecular scaffold for engineering integrin binding polypeptides. We used directed evolution to redirect the function of EETI-II from a protease inhibitor to a high affinity integrin-binding polypeptide. Yeast surface display was used to engineer EETI-II to bind to alphavbeta3 and alphavbeta5 integrin receptors with high (nanomolar) affinity. In the process, we also discovered the first known polypeptide that binds with high affinity to alphavbeta3, alphavbeta5, and alpha5beta1 integrins. Since all three of these integrins are co-expressed on tumors and contribute to angiogenesis, this knottin polypeptide has potential as a broad spectrum cancer therapeutic or imaging agent. Next, we showed that the engineered knottin polypeptides strongly inhibited tumor cell adhesion to the extracellular matrix protein vitronectin, and in some cases fibronectin, depending on their integrin binding specificity.

Finally, we used optical imaging and positron emission tomography to demonstrate that integrin binding affinity plays a critical role in tumor uptake in living organisms. High affinity knottin polypeptides exhibited a significant increase in tumor uptake in mouse human tumor xenograft models compared to weaker binding polypeptides, including a cyclic pentapeptide that is currently under clinical development for imaging of integrin-expressing tumors. Thus, engineered knottin polypeptides show great potential as clinical diagnostics for a variety of cancers.

Funded by: the NIH NCI Howard Temin Award 5K01 CA104706 and the Mallinckrodt Faculty Scholar Award (both to JRC), NCI ICMIC P50 CA114747 and NIH 5R25 CA118681 (both to SSG), a Stanford Molecular Imaging Scholars postdoctoral fellowship (to RHK), and a Stanford Dean's fellowship (to AML).

587d "Aquaporin-1: Its Presence, Distribution and Quantification In Arterial Endothelium and How These Change with Chronic Hypertension – with Relevance to Early Atherogenesis"

Jimmy Deon Toussaint[1], Tieuvi Nguyen[2], Stewart Russell[2], Kung-Ming Jan[3], and David Rumschitzki[1]. (1) Chemical Engineering, City College and the Graduate School and University Center (GSUC) of City Uinversity of New York, 140th Street &Convent Ave, New York, NY 10031, (2) Biomedical Engineering, City College and the Graduate School and University Center (GSUC) of City Uinversity of New York, 140th Street &Convent Ave, New York, NY 10031, (3) College of Physician and Surgeons, Columbia University, 3675 Riverdale Ave. Suite 5, Bronx, NY 10463

Atherosclerosis is responsible for over 50% of all deaths in the US and in all Western countries. The earliest events leading to atherosclerosis occur when low-density lipoproteins (LDLs) are transported, mainly by pressure driven convection, from the blood and deposited into the inner layers of high-pressure, large and intermediate sized arteries. Subsequent events narrow and harden these vessels' walls, leading possibly to strokes and heart attacks.

Permeability studies have shown that, at least over short time scales, macromolecular tracers appear to cross the endothelium of these atherosclerosis-prone vessels non-uniformly via rare focal leaks that are associated with transient leaky junctions of endothelial cells, some of which are either dying or dividing.

Pressure-driven transmural water transport advects these macromolecules across the endothelium into the subendothelial intimal space and initially spreads them in a direction parallel to the endothelium away from the leakage sites. The overall trans-wall water flux, not just the part through the few widened junctions, can dilute the local subendothelial macromolecular concentration, thereby causing the LDL to spread away from the focal leaky regions, eventually through the wall. Since this water flux affects the local LDL concentration in the intima, it very likely affects the kinetics of LDL attachment to the intima's extracellular matrix, which is believed to be one of the earliest and most critical events leading to lesion formations. The nature of this convective water flux is therefore a central factor in the prelesion events of atherosclerosis.

It has been accepted that water crosses the endothelium paracellularly, that is through tight and leaky junctions, the former, due to their numbers, being far more important than the latter to the overall trans-wall water flow. Since 1990, however, a family of ubiquitous transmembrane proteins called aquaporins (AQP) has been identified. AQPs are very specific and facilitate very efficient transcellular water transport. It is therefore natural to ask whether AQPs are present in arterial endothelial cells and, if so, if their presence implies a transcellular contribution in transmural water transport. Particularly, if water transport is not simply a passive process, that goes up and down with a changing transmural pressure, but is rather a protein-gated process, then a vessel might be able to actively regulate this transport by differential protein expression. Such transcellular transport, if it indeed is present, would play an important role in LDL transport once it has entered the subendothelial region through the leaky junctions and in its binding kinetics in the subendothelial portion of the vessel wall.

In our previous study, we used immunohistochemistry to identify and show the existence of Aquaporin-1 (AQP1) on both the luminal and abluminal membranes of rat aortic endothelial cells.

A known risk factor for atherosclerosis is hypertension and we postulated that one external condition that may influence a vessel's AQP expression is its transmural pressure. We previously confirmed this hypothesis in both genetically modified normotensive Wistar Kyoto rats and their hypertensive cousins Spontaneously Hypertensive rats (SHR) showing that endothelial cells from the chronically hypertensive rats appear to express far more AQP than their normotensive analogues.

In this study, we repeat our study in non- genetically modified normotensive (Sprague-Dawley rats without any operation, Sprague-Dawleys that have undergone a sham operation) and chronically hypertensive (Sprague-Dawleys that have undergone the Goldblatts procedure to induce hypertension by partially crimping one renal artery). We again show similar qualitative results to our previous genetically modified models study.

We also show evidence that blocking aquaporins, either chemically or by a knockdown experiment, reduces endothelial hydraulic conductivity, both in cultured monolayers and in whole vessels ex vivo.

This evidence begins to support the hypothesis that aortic endothelial cells may be able to actively regulate their aquaporin expression in response to chronic hypertensive conditions so as to control their transmural transport processes. Improved understanding of these processes may lead to novel therapies, not just in the case of atherosclerosis, but also for a wide variety of human diseases.

587e Quartz Crystal Microbalance Analysis of Growth Kinetics for Aggregation Intermediates of the Amyloid-β Protein

Joseph A. Kotarek, Kathryn C. Johnson, and Melissa A. Moss. Chemical Engineering, University of South Carolina, 2C02 Swearingen Engineering Center, Columbia, SC 29208

Alzheimer's disease (AD) is a neurodegenerative disorder that effects approximately 4.5 million Americans over the age of 65—a statistic that is set to triple by 2050. The widely accepted amyloid hypothesis proposes that the disease is caused by an accumulation of aggregating amyloid-beta protein, which eventually forms insoluble fibrils that deposit in the brain. This theory is supported by several genetic mutations linked to AD which increase the aggregation potential of the protein. Furthermore, experimental studies show that aggregated amyloid-beta is toxic in neuronal cultures, while its monomeric form causes no detrimental effects (Tickler et al. 2005).

While ample evidence supports this amyloid cascade hypothesis, the transformation of benign monomeric amyloid-beta into these toxic aggregates is not completely understood. The kinetics governing the nucleation driven reaction that produces these aggregates are also complex and their quantification has many associated difficulties. Predominant hurdles include both the isolation of specific mechanistic steps as well as the accurate quantification of aggregation in real time. The quartz crystal microbalance (QCM) is well suited as a biosensor for this study, as it provides a heterogeneous system that can measure mass changes in the nanogram range. QCM utilizes the piezoelectric effect in quartz crystals to detect changes in bound elastic mass as a variation in the frequency of oscillation. The methodology developed here excludes nucleation from the analysis and is based on the principle that only amyloid-beta aggregates, and not amyloid-beta monomers, are capable of supporting further aggregate growth. An avidin monolayer is deposited onto the gold surface of a crystal electrode via well established amino-coupling chemistry (Jung et al. 2006), upon which biotinylated amyloid-beta monomer is bound. This surface selectively recognizes aggregated forms of amyloid-beta and immobilizes these unlabeled aggregates, whereas monomer-monomer interactions are virtually non-existent in the system. Subsequent exposure to unlabeled monomer will initiate aggregate growth via aggregate-monomer interactions.

We show that this biosensor can selectively detect aggregates created in vitro and quantify their subsequent growth with sensitivity approaching the nanomolar range. Further quantification of soluble aggregate growth rates revealed a first order dependence on both monomer concentration adn immobilized aggregate density. The kinetic mechanisms isolated within this system compare well with previously developed models. An aggregate dissociation constant of 220 nm was determined, which is in agreement with data from studies of mature insoluble aggregates and indicates that the mechanisms of growth for soluble and insoluble aggregates are similar.

These parameters will allow for future optimization of the biosensor, as well as provide a better understanding of the aggregation process. Future work will explore the ability of the QCM sensor to selectively detect amyloid-beta aggregates in plasma and cerebral spinal fluid isolated from AD patients. Successful implementation of this technique could provide an effective means for the early diagnosis of AD.

Jung, H., Kim, J., Park, J., Lee, S., Lee, H., Kuboi, R., and Kawai, T. (2006). "Atomic force microscopy observation of highly arrayed phospholipid bilayer vesicle on a gold surface." Journal of Bioscience and Bioengineering, 102(1), 28-33.

Tickler, A. K., Wade, J. D., and Separovic, F. (2005). "The Role of Amyloid-beta Peptides in Alzheimer's Disease." Protein & Peptide Letters, 12(6), 513-519.

Engineering of Embryonic Stem Cell Neurogenic Differentiation for Neurodegenerative Diseases

Daniel E. Kehoe[1], **Ewa K. Stachowiak**[2], **Michal K. Stachowiak**[2] and **Emmanuel (Manolis) S. Tzanakakis**[1], (1)Chemical and Biological Engineering, State University of New York at Buffalo, Buffalo, NY USA, (2) Pathology and Anatomical Sciences, State University of New York at Buffalo, Buffalo, NY USA

ABSTRACT

Embryonic stem cells (ESCs) have the potential to become a limitless source of tissue for cell therapies and drug testing. Specifically, efficient generation of neuronal cells from human ESCs (hESCs) will have a major impact in transplantation therapies to replace aged or diseased neural tissue. This study focuses on better understanding the specific signals directing the differentiation of ESCs to neural progeny. We further examine the differentiation of hESCs in a bioreactor as a way of producing neuronal cells in quantities suitable for clinical therapies. The integrative nuclear fibroblast growth factor (FGF) signaling (INFS) has been show to mediate the activation of genes thereby coaxing the differentiation of umbilical cord blood cells and human neuronal precursors to neuronal cells. Here, we investigated if a similar mechanism can be utilized for the neurogenic commitment of mouse ESCs (mESCs). We found that treatment of mESCs with agents such as retinoic acid (RA) and cyclic adenosine monophosphate (cAMP) activates INFS, in part by nuclear translocation of an FGF receptor. Immunopositivity for glial fibrillary acidic protein (GFAP) revealed an increased propensity of mESCs for astrocyte differentiation upon stimulation with cAMP. In contrast, RA prompted the differentiation towards neurons as shown by enhanced expression of tyrosine hydroxylase and β3-tubulin. To that end, the localization of the RA receptor in differentiating stem cells in various stages of neuronal differentiation was also studied. These findings may lead to the development of methods for directed differentiation of ESCs to neuronal cells with high efficiency. For such methods to be utilized in clinical therapies, ESC-derived neurons must be generated in high quantities. Hence, we also examine if neuronal differentiation of hESCs can be performed in stirred suspension bioreactors. Cells formed neurosphere-like aggregates in the bioreactor and in a serum-free medium. A transition was noted to neuroepithelial cells and then radial glial cells. Finally, neuron-like cells emerged upon plating of the neurospheres. Differentiated cells displayed increased gene/protein levels of Pax6, Nestin, β3-tubulin, MAP2 and GABA (Fig. 1) compared to undifferentiated hESCs, and similar levels of ectoderm genes when compared to cells differentiated exclusively in static culture. This scalable system has the ability to produce mass quantities of neurospheres. In conjunction with an INFS-based method for directing the differentiation of these cells further to neuronal cells, this stem cell culture modality may facilitate the production of cellular material suitable for therapies against Parkinson and Alzheimer diseases.

INTRODUCTION

There are several neuronal diseases that afflict millions of Americans, including Alzheimer's, Huntington's, Parkinson's, Bell's Palsy and Neuromuscular disorders. Of these Parkinson's disease, which is a progressive disorder of the central nervous system affecting more than 1.5 million people in the United States, has a precedence of been treating using fetal stem cells.[1] Due to the inadequate amount and the moral issues of the use of fetal tissue, embryonic stem cells could perhaps be used to produce larger quantities of these cells for the treatment of Parkinson's disease one day. Since their creation, mouse and human embryonic stem cells have demonstrated the ability to proliferate in an undifferentiated state for long periods of time[2,3]. Due to this growth potential, these stem cells provide an ideal source for cell based therapies which require large quantities of cells, for a large number of patients. Producing a scalable process for the development of therapeutically effective tissue from embryonic stem cells could be very valuable in the future.

In addition to the ability to proliferate indefinitely, embryonic stem cells also possess the capability to differentiate to all three germ layers, including the ectoderm[4]. Functional neurons have been differentiated from mESCs for over a decade with the addition of RA and creating suspension aggregates of the mESCs termed embryoid bodies (EBs), and then let grow on surfaces[5]. The exact mechanism by which RA stimulates the differentiation is unknown, but the INFS pathway functions in association with cell differentiation and gene regulation[6], and could have a role. Cyclic AMP has shown the ability to elevate the gene expression levels of an important neuronal gene tyrosine hydroxylase[7,8] which may also coax the differentiation to mature neurons. Recently neurospheres have been produced by suspending hESCs aggregates on Petri dishes in a neuronal differentiation media and then placed on tissue culture surfaces for further maturation[9]. Bioreactors have been employed in the expansion of undifferentiated mESCs in a scalable process[10]. Human ESCs were recently differentiated hematopoietic progenitors via cultivation in stirred tank bioreactors[11].

This study delves into the mechanisms of neuronal differentiation using the mESC as a model, and then looks into creating neurons from hESCs in a more scalable process. The purpose is to better understand what is involved in the differentiation of ESCs to mature neurons, and create a system that can better produce large numbers of mature neurons to meet a large demand that might be needed for cell therapies for patients of Parkinson's disease, or other neuronal disorders.

RESULTS

The first section of this study focuses on the mechanisms, and factors that are important during ESC differentiation. For this work mESCs were coaxed to form neurons, and then investigated, and probed in a number of ways. The location of receptors for Fibroblast Growth Factor (FGF) and RA were studied over the differentiation time course. Different small molecules were added during the last stage of differentiation to see if it impacted cell differentiation. For the second segment of this paper, the differentiation of hESCs to neurons in a scalable bioreactor was assessed. The formation of neurospheres in the stirred suspension was assessed over time. More mature neurospheres were stained for neuronal markers and viewed under confocal microscopy. Finally gene expression was probed via real-time PCR for cells from neurospheres and then compared to undifferentiated hESCs.

Method of differentiation of ESCs to neurons

To produce neurons from mESCs a protocol was adapted from a paper discussed above by Fraichard[12]. Mouse ESCs were first grown undifferentiated on gel in the presence of LIF (Stage I), then they were trypsinized and placed in Petri dishes without LIF, and with 1 μM RA to form suspending EBs for 2 days, then the RA was removed for 2 additional days (Stage II). The EBs were then placed on a tissue culture surface, and let mature for 4 days (Stage III), and finally these cells are trypsinized and placed on slides for viewing, and additional treatment with stimulating molecules (Stage IV). Bright field images of each of these 4 stages are shown below in Figure 1, with the white bar indicating 200 μm.

Figure 1. Stages involved for the differentiation of mESCs to neurons (bars: 200 μm)

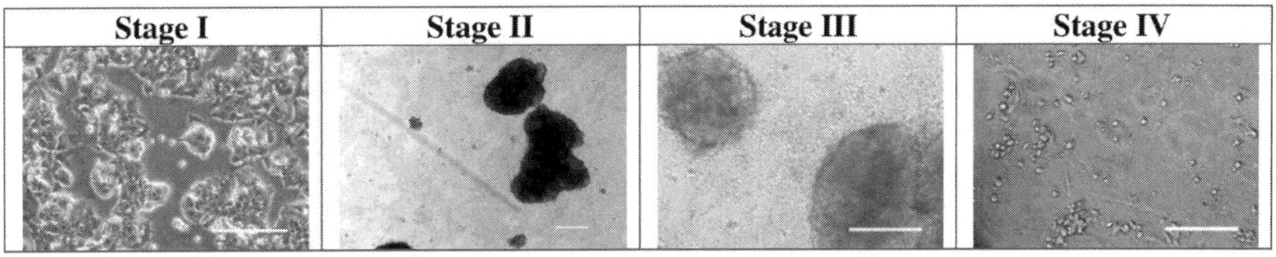

Investigation of the location of receptors of interest during differentiation

To delve into the mechanisms of the differentiations a closer examination of the location of the RA receptor (RAR), and another receptor of note, the FGF receptor 1 (FGFR1). Prior to the differentiation (Stage I), both receptors seem to be localized within the cytoplasm, but after differentiation and treatment with RA (Stage IV) the receptors were located in the nuclei of the cells, as shown in Figure 2 shown below.

Figure 2. Translocation of the FGF and RA receptors upon cell differentiation (bars: 50 μm)

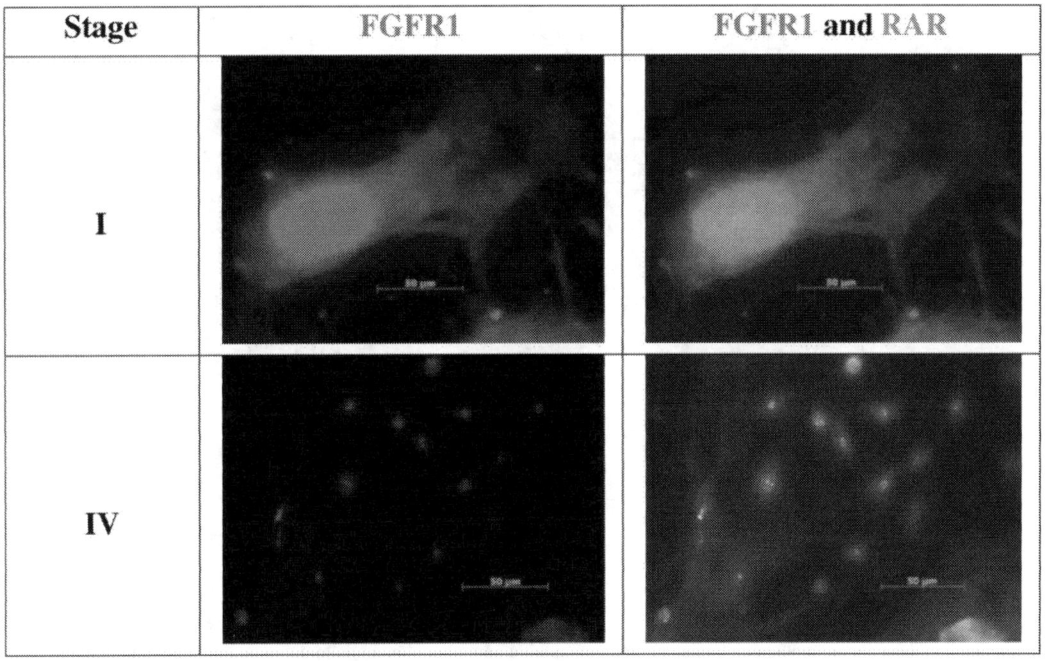

Coaxed differentiation of ESCs to neurons or astrocytes

The addition of small molecules cAMP and RA during the last stage of differentiation, for a duration of three days showed difference in the protein expression levels of several neuronal markers. Cyclic AMP treated stage IV differentiated mESCs showed higher levels of astrocytic marker GFAP, when compared to RA treatment (shown in the upper four images of Figure 3). On the other hand, RA stimulated differentiation caused an increase in the neuronal protein level of β-3-Tubilin, as shown below in the second half of Figure 3.

Figure 3. Stimulated differentiation of mESCs to astrocytes or neurons (bars = 50 um)

Treatment	Black and White	Merge with DAPI
cAMP	GFAP 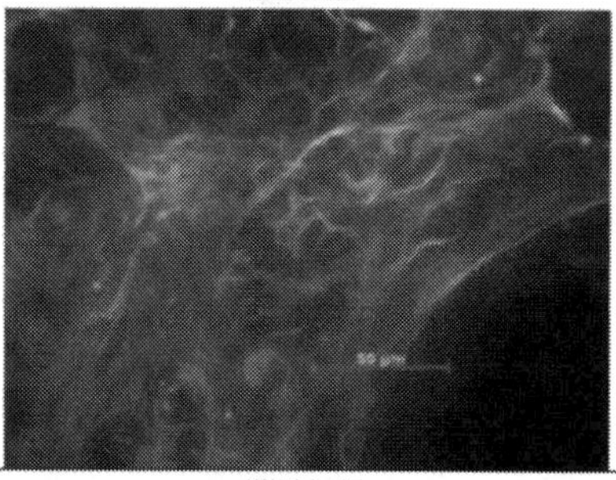	GFAP and DAPI 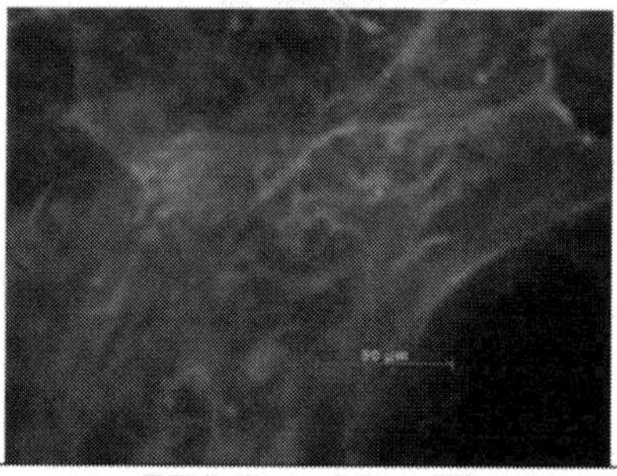
RA	GFAP 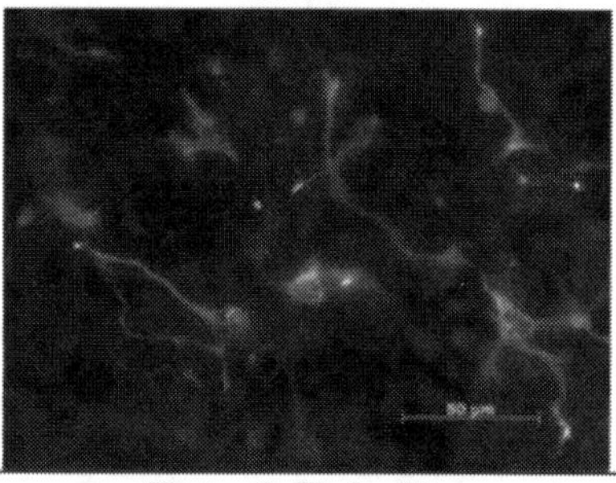	GFAP and DAPI 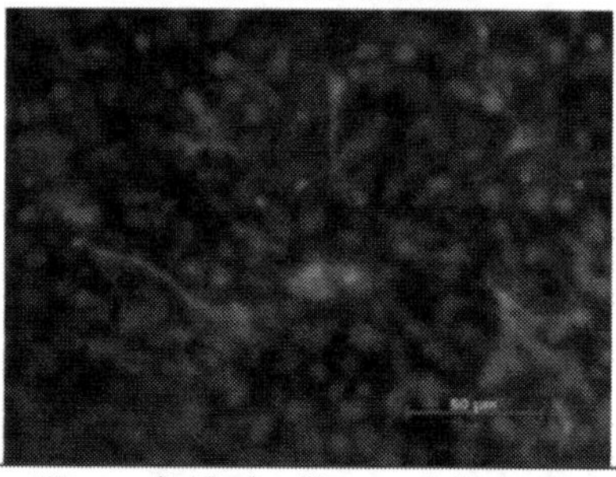
cAMP	Beta-3-Tubulin 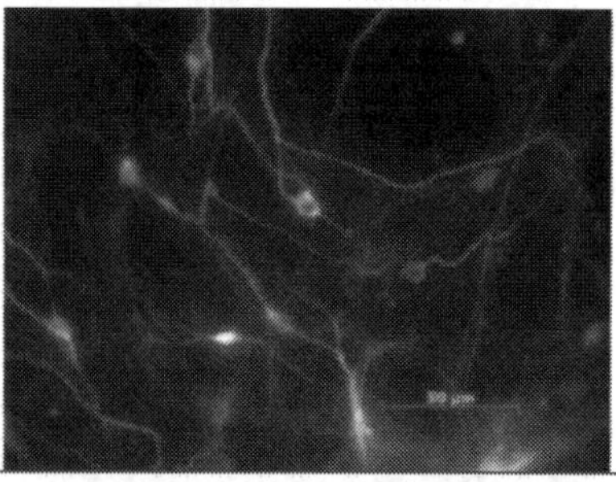	Beta-3-Tubulin and DAPI 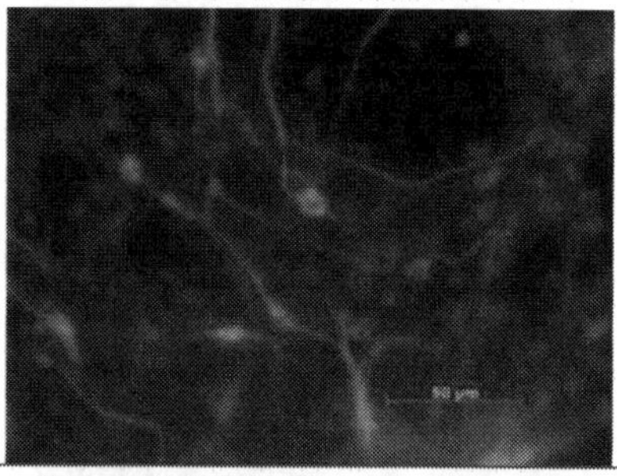
RA	Beta-3-Tubulin 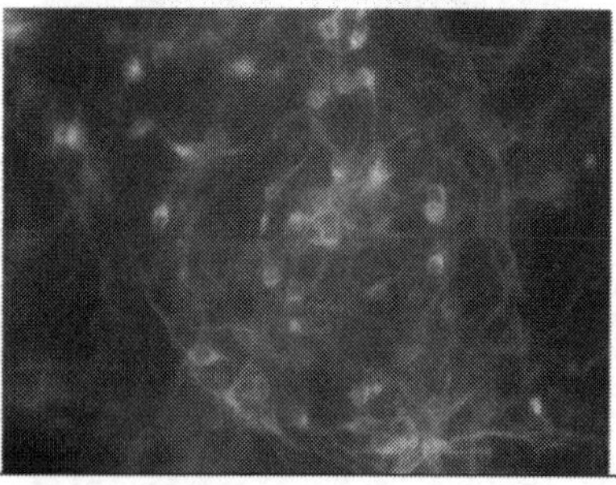	Beta-3-Tubulin and DAPI 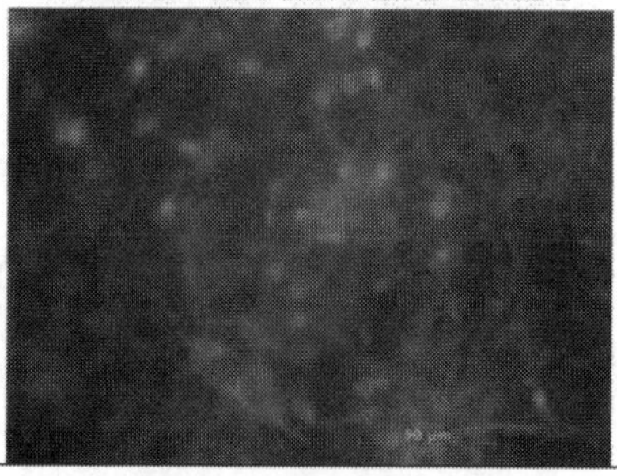

Method of differentiation of ESCs to neurospheres in the bioreactor

To produce neurospheres from hESCs in a bioreactor, a protocol was adapted from a study that produced neurospheres from hESCs in suspension[13]. Below in Figure 4, bright field images show how hESC aggregates from the bioreactor mature in the stirred tank. These cells were taken from matrigel coated tissue culture surfaces on day 0, and dissociated into small clumps of cells using collagenase, and then seeded into the bioreactor using neuronal proliferation media for the first four days, followed by neuronal differentiation media.

Figure 4. Bright field images of the maturation of neuropheres in a bioreactor

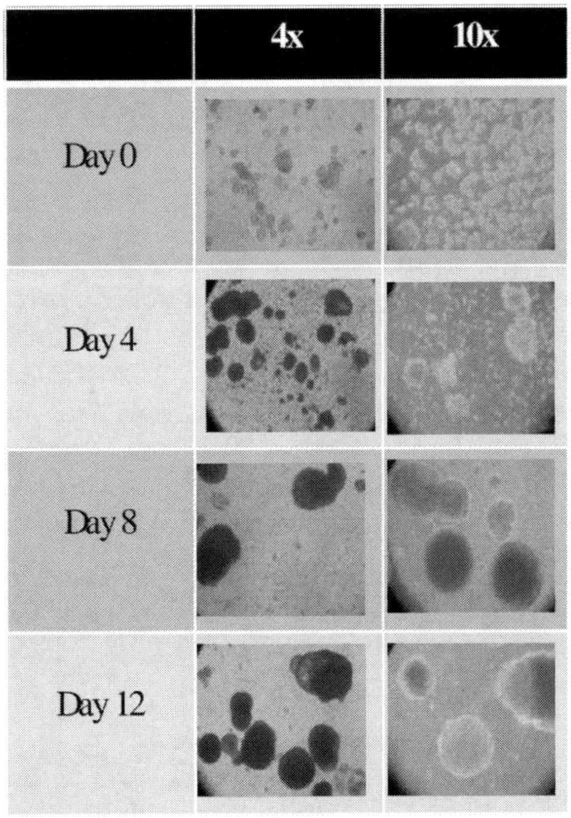

Analysis of neuronal proteins from neurospheres produced in the bioreactor

After 16 days in the bioreactor the neurospheres were removed and assessed for neuronal markers. Whole aggregates were stained with neuronal markers GABA, and MAP2, and as shown in Figure 5, when imaged on the confocal microscope, the neurosphere aggregates were stained positively for those markers.

Figure 5. Confocal images of neuropheres taken from bioreactor

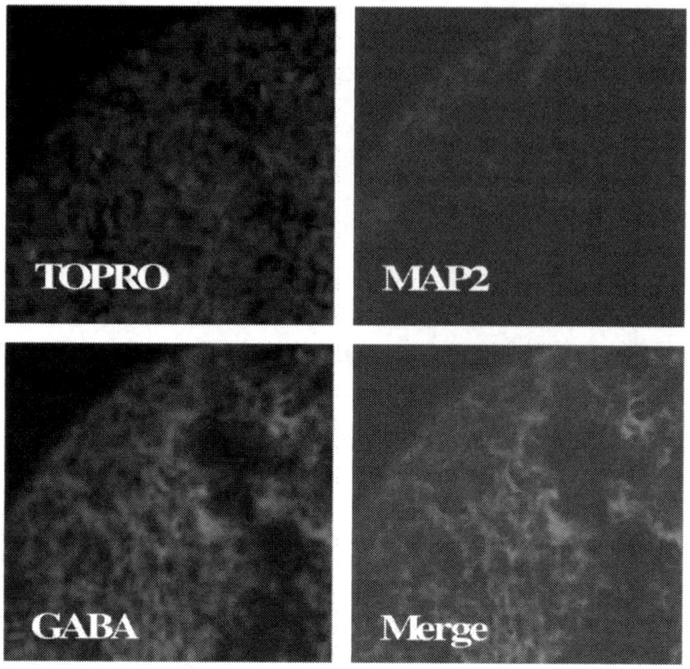

Analysis of neuronal RNA levels from neurospheres produced in the bioreactor

After 16 days in the bioreactor RNA was isolated from and then cDNA was produced and neuronal primers for PAX6 and β-3-Tubulin were tested along with stem cell marker NANOG. Levels of the stem cell marker decreased as the cells differentiated, while neuronal markers PAX6 and β-3-Tubilin increased on the log scale graph shown below in Figure 6.

Figure 6. Real-time PCR results of cDNA from Neurosphere produced in bioreactor.

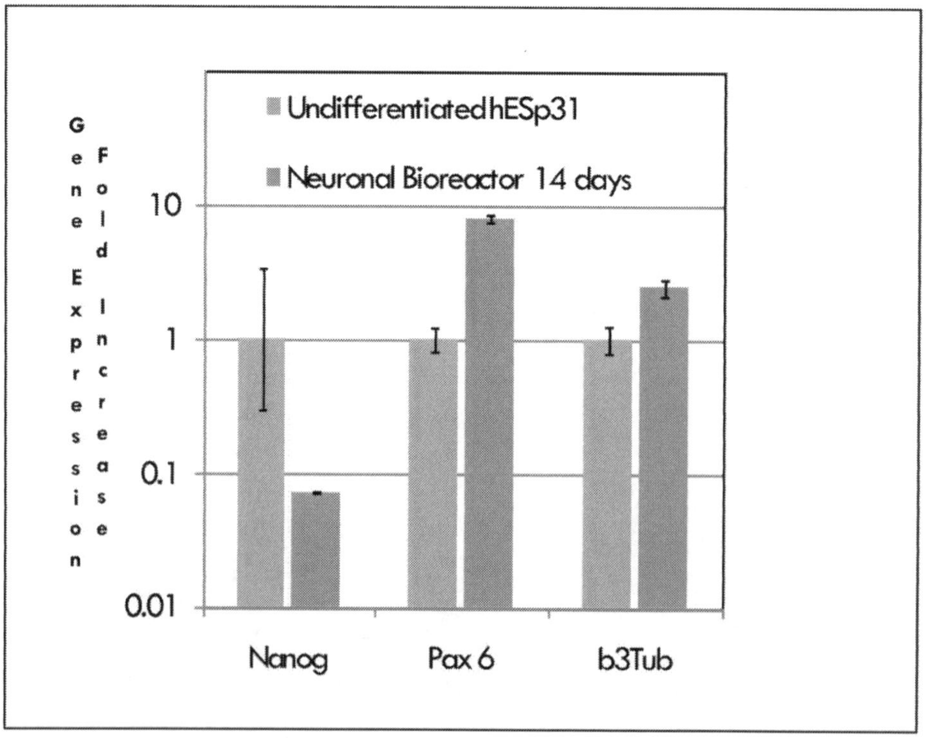

REFERENCES

[1] Olanow, C. W., J. H. Kordower, et al. (1996). "Fetal nigral transplantation as a therapy for Parkinson's disease." Trends Neurosci **19**(3): 102-9.

[2] Thomson, J. A., J. Itskovitz-Eldor, et al. (1998). "Embryonic stem cell lines derived from human blastocysts." Science **282**(5391): 1145-7.

[3] Evans, M. J. and M. H. Kaufman (1981). "Establishment in culture of pluripotential cells from mouse embryos." Nature **292**(5819): 154-6.

[4] Thomson, J. A., J. Itskovitz-Eldor, et al. (1998). "Embryonic stem cell lines derived from human blastocysts." Science **282**(5391): 1145-7.

[5] Fraichard, A., O. Chassande, et al. (1995). "In vitro differentiation of embryonic stem cells into glial cells and functional neurons." J Cell Sci **108 (Pt 10)**: 3181-8.

[6] Peng, H., J. Myers, et al. (2002). "Integrative nuclear FGFR1 signaling (INFS) pathway mediates activation of the tyrosine hydroxylase gene by angiotensin II, depolarization and protein kinase C." J Neurochem **81**(3): 506-24.

[7] Lewis, E. J., C. A. Harrington, et al. (1987). "Transcriptional regulation of the tyrosine hydroxylase gene by glucocorticoid and cyclic AMP." Proc Natl Acad Sci U S A **84**(11): 3550-4.

[8] Gueorguiev, V. D., R. J. Zeman, et al. (1999). "Differing temporal roles of Ca2+ and cAMP in nicotine-elicited elevation of tyrosine hydroxylase mRNA." Am J Physiol **276**(1 Pt 1): C54-65.

[9] Nat, R., M. Nilbratt, et al. (2007). "Neurogenic neuroepithelial and radial glial cells generated from six human embryonic stem cell lines in serum-free suspension and adherent cultures." Glia **55**(4): 385-99.

[10] Cormier, J. T., N. I. Nieden, et al. (2006). "Expansion of Undifferentiated Murine Embryonic Stem Cells as Aggregates in Suspension Culture Bioreactors." Tissue Eng.

[11] Cameron, C. M., W. S. Hu, et al. (2006). "Improved development of human embryonic stem cell-derived embryoid bodies by stirred vessel cultivation." Biotechnol Bioeng **94**(5): 938-48.

[12] Fraichard, A., O. Chassande, et al. (1995). "In vitro differentiation of embryonic stem cells into glial cells and functional neurons." J Cell Sci **108 (Pt 10)**: 3181-8.

[13] Nat, R., M. Nilbratt, et al. (2007). "Neurogenic neuroepithelial and radial glial cells generated from six human embryonic stem cell lines in serum-free suspension and adherent cultures." Glia **55**(4): 385-99.

587g Optimization of An Enrichment Process for Circulating Tumor Cells from the Blood of Head and Neck Cancer Patients through Depletion of Normal Cells

Liying Yang[1], James, Campbell Lang[2], Priya Balasubramanian[3], David Schuller[2], Amit Agrawal[4], Maciej Zborowski[5], and **Jeffrey J. Chalmers**[3]. (1) Department of Chemical and Biomolecular Engin, The Ohio State University, 140 W 19th Ave, Columbus, OH 43220, (2) College of Medicine and Public Heath, The Ohio State University, 300 West 10th Avenue, Columbus, OH 43210, (3) Chemical and Biomolecular Engineering, Ohio State University, 140 W. 19th Avenue, Columbus, OH 43210, (4) College of Medicine and Public Health, The Ohio State University, (5) Department of Biomedical Engineering, Cleveland Clinic Foundation, 9500 Euclid Ave, Cleveland, OH

The optimization of a purely negative depletion, enrichment process for circulating tumor cells, CTC's, in the peripheral blood of Head and Neck cancer patients is presented. The enrichment process uses a red cell lysis step followed by immunomagnetic labeling, and subsequent depletion, of CD45 positive cells. A number of relevant variables are quantified, or attempted to be quantified, which control the performance of the enrichment process. Six different immunomagnetic labeling combinations were evaluated as well as the significant difference in performance with respect to the blood source: buffy coats purchased from the Red Cross, fresh, peripheral blood from normal donors, and fresh peripheral blood from human cancer patients. After optimization, the process is able to reduce the number of normal blood cells in a cancer patient's blood from 4.05 x 109 to 8.04 x 103 cells/ml and still recover, on average, 2.32 CTC per ml of blood. For all of the cancer patient blood samples tested in which CTC were detected (20 out of 26 patients) the average recovery of CTCs was 21.7 per ml of blood, with a range of 282 to 0.53 CTC per ml of blood. Unlike a majority of other published studies, this study focused on quantifying as many factors as possible to facilitate both the optimization of the process as well as provide information for future performance comparisons. The authors are not aware any other reported study which has achieved the performance reported here (a 5.76 log10) in a purely negative enrichment mode of operation. Such a mode of operation of an enrichment process provides significant flexibility in that it has no bias with respect to what attributes define a CTC; thereby allowing the researcher or clinician to use any maker they choose to define whether the final, enrich product contains CTC's or other cell type relevant to the specific question (i.e. does the CTC have predominately epithelia or mesenchymal characteristics?).

ASAP: Accelerated Stability Assessment Program

Improved Protocol and Data Analysis for Accelerated Shelf-Life Estimation of Solid Dosage Forms

Kenneth C. Waterman

Research Fellow, Pfizer Inc

A more rapid and accurate accelerated aging protocol is used to determine the shelf-life of drugs in solid dosage forms based on different drug development phases, from excipient compatibility through late development. Reliable estimates for temperature and relative humidity effects on chemical stability are provided using the combination of an isoconversion paradigm and a humidity-corrected Arrhenius equation. Imprecision is incorporated into a Monte-Carlo simulation to propagate the variation inherent in the experiment. In early development phases, greater imprecision in predictions is tolerated to allow faster screening with reduced sampling. As programs progress, early development data are used to design appropriate test conditions for late stage stability programs. Parameters generated from the protocol also allow package selection without explicit screening. The Accelerated Stability Assessment Program, ASAP, is being used to define the critical degradation products for developing analytical methods, to select the appropriate packaging configuration, and to understand the stability impact of changes in the active pharmaceutical ingredient (API) or drug product manufacturing processes and ingredients.

In solution, chemical kinetics is characterized by a transformation from starting material to product following a reaction order. In the solid state, the situation is complicated by the multiplicity of molecular states. For example, molecules of an API in a formulation can exist in crystalline bulk, at the surface of the crystal, at defect sites as well as in amorphous states (either high energy or dissolved in the surrounding material). Each of these states often shows distinctly different reaction rates. Typically, the bulk crystalline API will react most slowly due to the lattice energy that must be overcome to cause reaction. The result of this heterogeneous kinetic situation is that the rate of API conversion to a degradation product will be a superposition of separate rates. This ultimately means that defining a rate constant for conversion of API to degradant in a classical sense can be futile.

The situation becomes still more complicated when temperature changes are evoked. The Arrhenius equation describes the effect of temperature on reaction rates in solution and gas phase:

$$\ln k = \ln A - E_a/(RT) \tag{1}$$

where k is the rate constant, A is an intercept related to the collision frequency, E_a is the activation energy for the chemical conversion, R is the gas constant and T is absolute temperature. As stated above, in the solid state, it may be difficult to impossible to determine the rate constant(s) describing the overall chemical conversion. Moreover, at different temperatures, the relative contribution of API in each of its different physical forms can be different. The result has generally been that many solid-state stability studies on drug products have been considered non-Arrhenius.

The situation can be operationally remedied with the introduction of an "isoconversion paradigm." In the isoconversion paradigm, a rate constant is defined as the reciprocal of the time needed to reach the specification limits of a degradant. This results in a rate constant that has potential contributions from different physical forms of the API; however, this proportion is kept constant across any accelerated storage conditions. By employing this isoconversion paradigm (at a constant relative humidity), the vast majority of API degradation processes appear to be Arrhenius. Exceptions occur when there is a phase transition such as a melt, glass transition or deliquescence. With these exceptions in mind, the isoconversion paradigm allows the general use of relatively high temperatures (to 80°C) to accurately determine ambient shelf-life by Arrhenius extrapolation.

To handle the range of relative humidities a drug product could be exposed to, we have found that chemical degradation universally follows a modified Arrhenius equation:

$$\ln k = \ln A - E_a/RT + B(ERH) \qquad (2)$$

where B is a fitting constant indicative of the moisture sensitivity of a particular system and ERH is the equilibrium relative humidity. The equation shows that relative humidity effects on API degradation are exponential. The moisture sensitivity of reactions does not indicate hydrolytic processes are involved. In fact, hydrolytic processes do not have significantly different B terms than other chemical mechanisms. The range of observed values for B is between 0 and 0.09. This exponential dependence can result in enormous differences in shelf-life depending on the relative humidity for storage or on the presence of desiccants. That this equation appears to be universal suggests that the dominant role of ERH on reaction rates is to affect mobility: ERH modifies the collision frequency but not the activation energy.

With the modified Arrhenius equation and the isoconversion paradigm, it is possible to define a design of experiment (DOE) protocol for stability. Using a five-point protocol varying temperature and ERH simultaneously, data can be generated that provide statistically meaningful fitting of the parameters in the equation. It should be noted that the times at each stability condition have to be adjusted to be consistent with the isoconversion paradigm. This means that average parameters are used in early studies, while the times are adjusted as data are available. It has generally be found that a two-week stability protocol provides reasonable data for accurately and precisely (propagating errors through a Monte-Carlo simulation procedure) projecting multiyear drug stability. A screening protocol for drug product is shown in Table 1.

Temperature (°C)	**Equilibrium Relative Humidity (%)**	**Number of Days at Storage Condition**
50	75	14
60	40	14
70	5	14
70	75	1
80	40	2

Table 1 General screening protocol for drug stability in drug products. Samples are stored at each condition exposed to the relative humidity of the environment (open bottle). Typically, an additional sample is stored at 5 °C to use as the initial control sample. For the best statistics in the experiment, all samples (including the control) should be analyzed as a single batch (i.e., samples removed from the stability chambers early should be held at 5 °C until all samples are ready to be analyzed).

Once the moisture sensitivity of drug stability is explicitly known for a system, stability in a package configuration becomes a matter of calculating the ERH inside a package as a function of time. This calculation can be accomplished using the initial relative humidity inside the package, the external relative humidity, the moisture adsorption isotherm of all internal components (including for example, any desiccants) and the moisture vapor transmission rate (MVTR) of the packaging. Using this model, we have successfully predicted the effect of such factors as the number of tablets in a bottle on the shelf-life of drug products. With the exponential dependence of stability with respect to relative humidity, one can sometimes observe what appears to be a threshold behavior: more rapid reaction as the relative humidity in a package increases. With this model validated, it is now possible to eliminate package screening since the results are accurately predicted from the science.

601b An Observed Correlation Between Flow and Electrical Properties of Pharmaceutical Blends

Kalyana Pingali, Department of Chemical and Biochemical Engineering, Rutgers University, 98 Brett Road, Piscataway, NJ 08854, Fernando J. Muzzio, Chemical and Biochemical Engineering, Rutgers University, 98 Brett Road, Piscataway, NJ 08854, and Troy Shinbrot, Biomedical Engineering, Rutgers University, 98 Brett Road, Piscataway, NJ 08854.

We study the relation between flow and conductivity of multiple formulations of pharmaceutical powder blends. Ten formulations were tested, consisting of two excipient sets, two active preparations, and a variety of food-grade additives including Magnesium Stearate (MgSt), and ionic and conductive materials such as ascorbic acid, talc, sodium carbonate, colloidal silica and TiO_2. Impedance, flow index and dilation were independently measured for all of the blends, and a strong correlation was found between every pair of these three properties. The relation between flow and dilation has been observed before; we find for the first time that there is a nearly linear relationship between flow index or dilation and impedance. This indicates that cohesive powder behavior depends on powder electrical properties, raising the question of whether additives such as MgSt affect friction and conductivity per se or whether these effects may be consequent to changes in electrical properties.

601c On the Role of Electrostatic Charge on the Drug Delivery through a Dry Powder Inhaler

Kewu Zhu, Wai Kiong Ng, and Reginald B. H. Tan. Crystallisation and Particle Sciences, Institute of Chemical and Engineering Sciences, 1, Pesek Road, Jurong Island, Singapore 627833, Singapore, Singapore

Electrostatic charge has been known to have great influence on the performance of dry powder inhalation (DPI) drug delivery. However our understanding on the electrostatic charge is still limited. The purpose of this work was to evaluate the role of electrostatic charge on the efficiency of drug emission and drug dispersion during inhalation drug delivery through a DPI. Salbutamol sulphate and inhalac 230 were chosen as model compounds, and the dry powder inhaler used was Rotahaler. A device for simultaneous measurements of electrostatic charge and aerosol aerodynamic diameter was developed in the current work. The accuracy of the measurement device was evaluated against an independent measurement using a Faraday Cage method. Aerosol emission efficiency, drug dispersion efficiency, and fine particle fraction (FPF) were characterized using a twin-stage-impinger (TSI) together with electrostatic charge characterization. The effect of electrostatic charge on the aerosol emission and drug dispersion was analyzed. It was found that electrostatic charge affected not only drug dispersion but also drug emission. The current study established a protocol for evaluating the role of electrostatic charge in the dry powder inhalation drug delivery, and could be useful for formulation and inhaler development.

601d In-Process Characterization of Coated Pharmaceutical Powder Blends Using NIR Spectroscopy

Lauren Beach[1], Jorge Ropero[2], Manel Alcalà[2], Rodolfo Romañach[2], and Rajesh Dave[3]. (1) Chemical Engineering, New Jersey Institute of Technology, 138 Warren Street, YCEES 208, Newark, NJ 07102, (2) Department of Chemistry, University of Puerto Rico Mayaguez, PO Box 9019, Mayagüez, PR 00681, (3) New Jersey Center for Engineered Particulates, New Jersey Institute of Technology, 138 Warren Street, YCEES 208, Newark, NJ 07102

Dry coating is an innovative technique in which nano particles are mechanically coated onto the surface of larger (from a micron to few microns) host particles to impart useful properties to the final product, which are engineered particles. Dry coated engineered particles can have improved flow and handling properties, and hence are very useful for a number of industrial applications. In this work, we present preparation, characterization and applications of such engineered particles to pharmaceutical applications.

A current need has arisen to characterize these dry coated powders, specifically active pharmaceutical ingredient powders. These powders were coated with nanoparticles using three different methods: the MAIC, the v-blender and the sieve. These powders, both before and after modification, need characterization, especially with respect to flow, to measure the improvement made by the dry coating. Powder flow is a critical parameter in both manufacturing and formulation.

The NIR is a viable method to do so with little trouble. NIR spectroscopy has been used to determine the flow of powders. By collecting spectra of voiding powders and determining the time per spectra, an accurate in line flow monitoring measurement can be taken and a qualitative flow rate measurement can be determined. Additionally, the NIR can give useful information, such as a qualitative flow rate, and can also differentiate the differences and consistency of the flow. This can help to compare the original uncoated powder and the nanocoated powder. The method can also compare the different methods used during the dry coating process. Other commercial characterization methods are not as useful, as they are offline tools and are often time consuming and do not give a direct flow measurement. Using this NIR method, spectra were obtained for blends of powder using non-coated API and then again using API coated with nanoparticles. Using the spectral data, the numerical flow rate of the voiding blends can be calculated.

601e Orientation Index of Tablets as a Means of Quantifying Intra-Tablet Coating Variation

Arjun V. Kalbag and Carl R. Wassgren. School of Mechanical Engineering, Purdue University, 585 Purdue Mall, West Lafayette, IN 47907-2088

Intra-tablet coating variability is the variation of coating mass over the surface of an individual tablet. Exposure times of different surfaces of a tablet (e.g., the cap and bands of a biconvex tablet) are directly related to the orientation of the tablet within the spray zone. We define the orientation index for a biconvex tablet as the ratio of the measured band-to-cap exposure areas to the ideal band-to-cap area ratio, with the ideal being the ratio of the band-to-cap surface areas for the given tablet. Therefore, a value of unity for the orientation index implies zero intra-tablet coating variability. Orientation index values larger than one indicate that more band is exposed to the spray than what is considered ideal, and values less than one indicate that more cap is exposed.

The different rates of coating mass accumulation on the caps and bands of biconvex tablets are examined using video-based techniques. Caps of directly compressed MCC tablets of different cap concavity are painted white. Three three pan speeds for two different biconvex shaped tablets and one flat faced cylindrical tablet are examined. The value of the orientation index measured from the video experiments are correlated to spray coating experiments where tablet weight gain is monitored by periodically sampling tablets and cutting the tablet along different planes to measure the coating film thickness. This goals of the work are to determine how tablet shape and pan speed influence intra-tablet coating variability.

601f Application of Multivariate Latent Variable Modeling to Pilot-Scale Spray Drying Monitoring and Fault Detection: Monitoring with Fundamental Knowledge

Salvador Garcia-Munoz, Process Modeling & Engineering Technology, Pfizer Global Research & Development, 445 Eastern Point Rd., Groton, CT 06340 and Dana Settell, Bend Research Inc., 64550 Research Road, Bend, OR 97701.

This work describes the use of multivariate latent variable modeling to enhance fundamental understanding of the common cause variability present in the operation of a pharmaceutical spray dryer; and provide a real time process monitoring and fault detection tool for this process. A latent variable model (LVM) was built and tested using commercially available software in a pilot-scale facility at Bend Research Pharmaceutical Process Development Inc. (BRPPD) in Bend, Oregon.

As part of this study, the pilot-scale spray dryer was modified to closely replicate control structure of the commercial-scale spray dryer since the system showed to also capture and monitor the performance of the controllers installed in the unit.

An LVM process monitoring system based on this model successfully identified seven process faults (faults were artificially introduced into the system) in real time during pilot-scale spray-drying trials at BRPPD. The performance of the system exceeded the expectations since it was not only able to detect those planned faults, but also unplanned events during the test.

A key advantage of the latent variable approach is that the variability that drives the process is easily understood in a fundamental way by interpreting the model parameters in the light of fundamental engineering knowledge (transport phenomena, thermodynamics, etc). The understanding of the common cause variability is greatly enhanced and enables the better understanding of the differences across scales for this unit. In monitoring the process, the faults are not only detected in a "statistical" way, but are also understood in a fundamental way by using the model to track down the driving forces that were involved in detecting such fault (e.g. a change in dryer thermodynamics, an abnormal behavior of the gas momentum across the unit).

The LVM process monitoring system can be used to detect spray drying process upsets in real time, potentially avoiding production of off-specification product and reducing process downtime. The system alerted the operator in much less time than would be possible using conventional univariate monitoring techniques, enabling the operators to address the upset before much product was lost. This system enables the operators to keep the process under tight operation, assuring the continuous quality of the end product.

Granule Strength Determination and Compactibility of Granules of Varying Density

Kevin Macias and Teresa Carvajal

Purpose: As in-line process analytics are becoming more widely accepted, it is important to illustrate that a measurement of interest will provide the expected outcome if that control point has reached a desired level. This methodology is especially true for novel applications. The purpose of this work was to investigate the role of densification during high shear granulation and the resulting impact on the compaction properties of the prepared granules. Additionally, the structural dependence of granule strength on the compaction phase was investigated to establish the validity of controlling granule density during high-shear wet granulation to ensure reproducible compaction properties.

Methods: Wet granulations were performed in a Diosna six-liter high-shear granulator equipped with a bottom driven impeller and side mounted chopper. Granulations consisted of 500 grams of Avicel PH 102 and utilized distilled water as the binder. To vary the density of prepared granules, granulations were carried out with impeller speeds of 175, 250 and 500 RPM at water levels corresponding to 50, 60, 70, 80, 90, 100 L/S%. At the 90 L/S% level, wet massing times were varied corresponding to 0, 500 and 7500 impeller revolutions. The chopper speed remained constant at 1000 RPM for all granulations. After drying the granulations, sieve fractions were prepared in order to minimize the effects of size distribution during further analysis. Envelope density of the sieve fractions (16, 20, 30, 40 mesh) was measured using the GeoPyc 1360 (Micromeritics, Norcross, GA). The 12.7mm ID sample chamber was chosen to perform all tests. The corresponding tight fitting piston was utilized with a compression load of 28N and a conversion factor of 0.1284 cm^3/mm. To determine the average granule strength a uni-axial confined compression method was employed as it simulates a compaction process that would be encountered during a compaction phase of tablet manufacture. Granule fractions were charged into the tooling with pre-lubricated die walls by volume. Uni-axial compression occurred using ½" flat-faced tooling (Natoli Engineering) on an MTS Sintech 30D load frame equipped with a 20,000lb load cell. Average compressive granule strength was derived using Kawakita analysis[1]. The compression stage was continued beyond the region necessary for Kawakita analysis to prepare compacts. After the compression stage, ejected compacts were allowed to elastically recover for a minimum of three days. Dimensions and weights were recorded prior to hardness testing using diametric compression. The tensile strength (σ) was derived from the compact hardness tests using Fell and Newton analysis[2] (Equation 1).

$$\sigma = \frac{2H}{\pi D t} \qquad \text{Eq. 1}$$

Where H is the force necessary to cause tensile failure (compact hardness), D is the diameter of the compact and t is the thickness of the compact.

Results: To model the relationship between granule density and granule strength, existing models were tested. The simplest model examined was the Ryshkewitch-Duckworth equation[3] (Equation 2).

$$\sigma = \sigma_0 \exp(-BP) \qquad \text{Eq. 2}$$

Where σ_0 is the strength of a non-porous sample, P is the sample porosity and B is the slope of the ln σ vs P line. This model was originally derived empirically to describe the relationship between the porosity of sintered ceramics and the compressive strength of samples. In the field of granulation, Rumpf proposed his famous model for relating the strength of agglomerates to the presence of pores in the sample[4] (Equation 3)

$$\sigma = \frac{1-\varepsilon}{\varepsilon}\frac{F}{d^2} \qquad \text{Eq. 3}$$

Where F is the mean tensile strength per unit cross-sectional area, d is the diameter of the primary particles and ε is the porosity of the agglomerate. By investigating Rumpf's model of granule strength, it can be found that the strength of a granule is affected by the strength of the bonds within the granule and the amount of void space in the granule. The strength of solid systems has been found to be reduced by flaws and all solids contain imperfections. It is the propagation of these flaws that allow materials to fail. The strength reduction of a material has been quantified by Griffith in his well known work[5] (Equation 4).

$$\sigma = \sqrt{\frac{2E\gamma}{\pi a}} \qquad \text{Eq. 4}$$

Where E is the elastic modulus, γ is the surface energy of the crack per unit area and a is the size of the flaw. Building on the model of Rumpf, Kendall[6] integrated the work of Griffith[5] to derive Equation 5.

$$\sigma = 3.7(1-\varepsilon)^4 \frac{F}{d\sqrt{da}} \qquad \text{Eq. 5}$$

Where a is the flaw size as stated in Griffith's Model (Equation 4), F, d and ε are the mean tensile strength per unit cross-sectional area, diameter of primary particles and the agglomerate porosity as stated in Rumpf's Model. Granule strength-density relationships were found to be well modeled using the Ryshkewitch-Duckworth model for all prepared granules[3] (Figure 1).

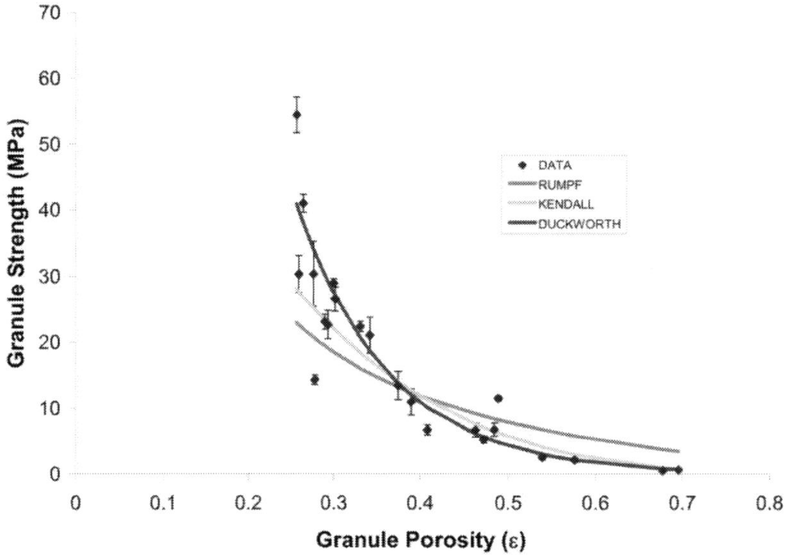

Figure 1 Application of models used to quantify granule strength dependence on granule density on 40 mesh granule data

A power law relationship was found between granule strength and compact strength for all prepared granules (Figure 2).

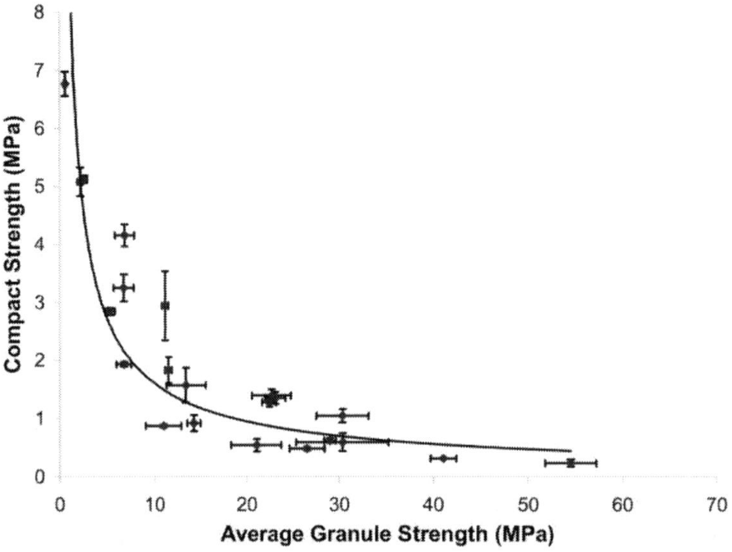

Figure 2 Power Law Model of Compact Strength versus Granule Strength (40 Mesh)

Upon truncation of granules which yielded poor compacts, Kendall's model was most applicable to describe the granule strength-density relationship and a narrowly linear relationship existed between granule strength and compact strength.

Conclusions: For the current system, the method of preparation did not influence the strength and compaction dependence on granule density. Illustrating that the desired compaction properties can be obtained solely by controlling granule density is quite significant. The method of granule density attainment did not influence the density-strength-compaction relationship. This illustrates the feasibility of controlling granule density during HSWG to arrive an end-point in terms of compaction characteristics. This systematic approach to process understanding is necessary to advance the transition from traditional, time-based granulation methods to the ideal platform of ensuring product quality during processing.

References

1. Kawakita K, Ludde K-H 1971. Some considerations on powder compression equations. Powder Technology 4(2):61.
2. Fell JT, Newton JM 1970. Determination of tablet strength by the diametral-compression test. Journal of Pharmaceutical Sciences 59:688-691.
3. Ryshkewitch E 1953. Compression Strength of Porous Sintered Alumina and Zirconia. 9th Communication to Ceramography. Journal of the American Ceramic Society 36(2):65-68.
4. Rumpf H. 1962. The Strength of Granules and Agglomerates. In W.A. Knepper. ed.: From Agglomeration, Wiley-Interscience. p 379- 418.
5. Griffith AA 1921. The Phenomena of Rupture and Flow in Solids. Philosophical Transactions of the Royal Society of London Series A 221:163-198.
6. Kendall K. 1987. Tribology in Particulate Technology. ed., Bristol and Philadelphia: IOP Publishing Ltd.

607a Overview of Models Used in Design Space Determination: A Regulatory Perspective

Sharmista Chatterjee, ONDQA (Office of New Drug Quality Assessment), CDER/FDA, 10903 New Hampshire Avenue, Silver Spring, MD 20993

A key element of the Quality by Design (QbD) paradigm of drug development is a delineation of a multi-dimensional design space for material and process parameters. Mathematical models can serve as a powerful tool in this approach. A design space could potentially include a combination of mechanistic and regression models. Mechanistic models mimic the underlying physical chemical phenomenon while the inherently empirical regression models are derived from statistically designed experiments. This talk would present an overview of role of mathematical models in the current paradigm of science based development and in addition illustrate FDA's perspective about this approach.

607b Reaction Kinetics for An API Process with Multiple Pathways and Hidden Intermediates

David Willcox[1], Deepak Ahuja[2], Tim Towne[1], Wenke Li[1], and Todd McDermott[1]. (1) Abbott Laboratories, 1401 Sheridan Road, R452-R13, North Chicago, IL 60091, (2) R452-R13, Abbott Laboratories, 1401 Sheridan Road, R452-R13, North Chicago, IL 60091

Variable results from a reaction often points to interactions between multiple steps in the reaction mechanism. This is complicated by reaction intermediates that are difficult to detect by HPLC or spectroscopy. An example of a step in an API process is discussed that illustrates these points. It was found that small changes in a reagent resulted in large changes in selectivity due to the presence of multiple reaction pathways controlled by an equilibrium reaction.

607c An Application of Reaction Engineering and Modeling In Quality by Design

Daniel Hallow, Justin Burt, Boguslaw Mudryk, Lucius Rossano, and Srinivas Tummala. Process Research and Development, Bristol-Myers Squibb, One Squibb Drive, New Brunswick, NJ 08903

Key aspects of a Quality by Design (QbD) approach include defining the design space and conducting a risk assessment of the process parameters. In this case study, we investigated a lactamization reaction using reaction modeling to perform a risk assessment and map the design space.

In the lactamization reaction, an organic base reacts with intermediate A to form intermediate B. In addition, the organic base and intermediate B can further react to form impurity C that is carried into the API product. A reaction model was built using data captured by FTIR and HPLC and then used to conduct numerous reaction simulations ("virtual experiments") at various conditions. The results of the virtual experiments were used to generate response surfaces of the potentially critical reaction parameters (i.e., temperature, net amount of base, and reaction time). This allowed the proposed operating space to be visualized and foster process risk assessment. In addition, these results demonstrated that the reaction requires a delicate balance of the critical parameters to allow reaction completion and avoid high impurity formation. The reaction temperature and organic base equivalents were then optimized to maximize reaction conversion while ensuring that impurity C remained below specifications. The model also showed that lowering the batch temperature during reaction endpoint analysis significantly reduced the rate of impurity C formation.

In conclusion, reaction modeling and simulation software was shown to be a very powerful tool to not only deepen our knowledge of reaction kinetics but also to provide a means to simulate numerous experiments, test alternative processing procedures, and develop a design space based upon a scientific and risk-based approach. This work culminated in the establishment of an alternative process procedure for the upcoming production campaign that mitigates risk, reduces the number of critical parameters, and ensures final product quality.

607d Ontology-Based Modeling Environment for Pharmaceutical Product Development: Milling Case Study

Pavan Kumar Akkisetty[1], Pradeep Suresh[2], Gintaras V. Reklaitis[2], and Venkat Venkatasubramanian[2]. (1) Chemical Engg, Purdue University, 480 Stadium Mall Drive, West Lafayette, IN 47907-2100, (2) Chemical Engineering, Purdue University, 480 Stadium Mall Drive, West Lafayette, IN 47907-2100

Milling process is one of the complex unit operations in the pharmaceutical domain. Its complexity arises due to the various breakage mechanisms that happen in a variety of mills. Traditionally population balance modeling (PBM) approach has been used to model the milling process. But certain important decisions such as selection of the mill type based on mill and material properties, scale up factors etc. have been heuristic in nature and are not obtained directly from PBM. Knowledge based tools and data driven models can be used to overcome these drawbacks. But the data driven (DD) model development depends on the knowledge of the expert who is building the model. The know-how of the product and DD model development are largely hidden in the expert's mind and in the development environment's syntax. Thus, this knowledge is not available for sharing across different platforms and for computer interpretation.

Our present work focuses on extracting the product and model development knowledge from the expert and making it computer interpretable. This knowledge includes the domain knowledge (e.g. milling) as well as data-driven modeling (e.g. statistical, neural nets) knowledge. This results in a transparent and standardized model development environment along the lines of the framework to manage mechanistic modeling knowledge proposed by Suresh et al [1]. Our ontology centric framework includes milling ontology and guidelines for mill selection, statistical ontology, guidelines for the statistical methods and a Java engine to integrate these with the Matlab solver. Product development for any pharmaceutical unit operation can be developed using our generic framework. We demonstrate our framework by developing a hybrid neural network model for the milling process.

References

1. Informatics Based Approach for Mathematical Knowledge Modeling in Process Operations, Pradeep Suresh, Chunhua Zhao, Girish Joglekar, Venkat Venkatasubramanian, 60f, AICHE Annual Meeting – 2007

MAPPING THE DROPLET DRYING PHENOMENA WITHIN A FLUIDIZED BED COATER AS A PREDICTION TOOL FOR COATING QUALITIES

Fesia L. Laksmana, University of Groningen, Groningen, The Netherlands
Paul J.A. Hartman Kok, Schering-Plough, Oss, The Netherlands
Herman Vromans, Schering-Plough, Oss, The Netherlands
Henderik W. J. Frijlink, University of Groningen, Groningen, The Netherlands
Kees Van der Voort Maarschalk, Schering-Plough, Oss, The Netherlands
Abstract ID: 126005

Introduction

For a high coating quality, agglomeration and excessive evaporation of the coating droplets during fluidized bed coating process should be avoided (1-3). This means that the coating process can often only be performed within a narrow operating window. Consequently, the selection of the process conditions and the process control are delicate steps to assure a good coating quality.

In spite of being an established technology, the process development of a fluidized bed coating process has often been performed in an empirical way. This approach obviously results in a large experimental program and the interpretation of the results stays far from first principles. In this manner, a rapid and continuous process development is difficult to be achieved. Process modeling is one of the methods that can aid the process understanding, as it enables a comprehensive parametric study of the involving sub-processes and further can be used as a process prediction tool once it is validated.

In this study, the physical phenomena occurring during a single droplet drying within a fluidized bed coater have been modeled, comprising of the changes in the droplet mass, water fraction, droplet size and droplet shape in time, before and after collision with the core particles. Using the models, the influence of various process conditions, i.e. inlet air temperatures, atomization pressures, and spraying rates could be studied comprehensively. The coatings obtained from the corresponding processes were characterized using a previously developed quantitative image analysis method (4, 5). In this way, the relationship between the extent of the droplet drying and the obtained coating quality could be assessed, which makes the model also usable as a development tool.

Heat and Mass Transfer Models

Inside a fluidized bed coater, the journey of droplets starts from being atomized from a spraying nozzle, colliding with core particles, spreading and further being dried and leaving deposit on the core surface as coating layer (3) (Figure 1). The drying conditions (temperature and vapor pressure) of the droplets were calculated using a mass and heat balance, where the total volume of the equipment was taken as the control volume and steady state mass and energy transfers were assumed. The mass and heat transfer between a single droplet and the hot-humid air can be expressed as Eq. 1. The heat transfer coefficient h was calculated using the established correlations (6). Different considerations were given during the modeling, according to the different stages that a droplet undergoes before it is completely dried.

During droplet's traveling period, the drying process of a droplet could be modeled quite straight-forwardly, assuming a spherical droplet wherein only changes in the water retained in the droplet and droplet diameter occur in time. Once the droplet hits the core particle, the droplet spontaneously spreads on the core surface, which extent depends on the impact (inertia spreading) and the surface energy (wet spreading) (7). Consequently, the changes in the droplet forms from spherical to spherical cap were taken into account in the drying model. Due to evaporation, the droplet contact line is receding (8). This occurs by keeping the contact angle more or less constant. Once the solute concentration increases to a critical value, pinning occurs. During pinning, the droplet diameter stays constant, while the droplet height decreases significantly (9). The transition between these two evaporation regimes is to be determined in this study, which influences the droplet deposit type i.e. dot-like or ring-like and therefore the coating quality. Eqs. 2-4 depict the drying models used to simulate the changes in the droplet mass in time at different periods, during i.e. droplet traveling, receding contact line period, and pinning period, respectively. In these models, shrinking control volume was assumed and linear correlation between the water fraction remained and the solution physical properties. These models were derived by substituting a set of correlations between droplet properties and its mass into Eq. 1.

$$\frac{dm}{dt} = -\frac{dQ/dt}{\lambda} = -\frac{hA(T_{a,in} - T_{wb})}{\lambda} \tag{1}$$

$$\frac{dm}{dt} = -\frac{(T_{a,in} - T_{wb})}{\lambda}\pi\, k(m)\left\{2\frac{6}{\rho\pi}^{1/3} m^{1/3} + 0.65\frac{\rho_a u_a}{\mu_a}^{0.5}\frac{\mu_a Cp_a}{k_a}^{1/3}\frac{6}{\rho\pi}^{0.5} m^{0.5}\right\} \tag{2}$$

$$\frac{dm}{dt} = -\frac{(T_{a,in} - T_{wb})}{4\lambda}\pi\, k(m)\left(1+tan^2\left(\frac{\theta}{2}\right)\right)\cdot$$

$$\left\{4\left(\frac{6}{\rho\pi\left(3tan\left(\frac{\theta}{2}\right)+tan^3\left(\frac{\theta}{2}\right)\right)}\right)^{1/3} m^{1/3} + 1.84\left(\frac{\rho_a u_a}{\mu_a}\right)^{0.5}\left(\frac{\mu_a Cp_a}{k_a}\right)^{1/3}\left(\frac{6}{\rho\pi\left(3tan\left(\frac{\theta}{2}\right)+tan^3\left(\frac{\theta}{2}\right)\right)}\right)^{0.5} m^{0.5}\right\} \tag{3}$$

$$\frac{dm}{dt} = -\frac{(T_{a,in} - T_{wb})}{4\lambda}\pi\, k(m)\left\{2d_{col} + 0.65\frac{\rho_a u_a}{\mu_a}^{0.5}\frac{\mu_a Cp_a}{k_a}^{1/3} d_{col}^{1.5}\right\}\cdot$$

$$\left(1+\left(\frac{6}{\rho\pi\left(3tan\left(\frac{\theta}{2}\right)+tan^3\left(\frac{\theta}{2}\right)\right)}\right)^{-2/3} m^{-2/3} h(m)\right) \tag{4}$$

The critical conditions used for the simulation at which the transition between the receding contact line and pinning regimes occurs were determined using the thermal gravimetric analysis (TGA) data. For each evaporation regime, a model was used, where the transition between the two regimes was optimized based on the goodness of the fit of the data to both of the models.

Material and Methods

Materials

Microcrystalline cellulose/MCC (Avicel PH102, FMC BioPolymer) was used as pellet excipient. Hydroxy-propyl methylcellulose/HPMC (Methocel E5 LV USP/ EP premium grade, Dow) was used as coating model material. Carmoisine (E122, Pomona BV) was used as pigment in the coating.

Experimental Methods

1. Pellets Production

The pellets were made in a high shear granulator (GRAL 10, Machines Colette, Belgium), with impeller and chopper rates of 600 and 3000 rpm, respectively. The pellets were made by mixing 500g of MCC with 500g of water, which was poured slowly (in 1 minute time) from the top of the GRAL. The granulation was performed for 20 minutes. The wet granules were further dried at 40°C for 8 hours in the vacuum drier (Elbanton, The Netherlands).

2. Particle Coating

The pellets with size fraction of 800-1000 micron were coated in the fluidized bed coater (Mycrolab, Hüttlin, Germany). 150 g of pellets was used for each batch. The coating solution can be fed either from the bottom or the top of the column. Furthermore, the spray rate, drying temperature, and atomization pressure were varied. The coating process was performed until 20% weight ratio of coating to core was sprayed. The sprayed HPMC solution was HPMC (5% w/w water) and carmoisine (0.1% w/w water) dissolved in cold water.

3. CSLM and Image Analysis

The carmoisine added to the coating fluoresces enabling the visualization of the coating layer under the CSLM and differentiate the coating layer from the core. A Leica inverted microscope DM IRE2 TCS SP2 (Leica microsystems, Germany) was used to acquire images of coated particles. Immersion oil type B (Cargille Lab, USA) was added to the coated particles prior to scanning. A green laser (HeNe 543nm/1.2mW) and a red laser (HeNe 633nm/10mW) were set at 100% and 37.66% intensity. The emission of the sample was collected at 652-716 nm. An HC Plan Fluotar 10x/0.30 objective was used with a zoom factor of 1 or 10 resulting in scanned area of 1500μm x 1500μm or 150μm x 150μm, used to acquire images for the characterization of coating thickness and pore structure, respectively. Image analysis was performed on the acquired images, developed previously, allowing the measurement of the coating thickness distribution around a pellet, coating porosity and the pore size distribution (4, 5).

4. Thermal Gravimetric Analysis (TGA)

The evaporation of a droplet of HPMC E5 solution (5% concentration) was followed using a TGA equipment (Mettler Toledo TGA/SDTA851^{e}, The Netherlands) at various the drying temperatures ranging from 40 to 80°C. For each temperature, the measurement was performed in three-fold. The droplet was dosed using Gilson micropipettes giving a 5 μl volume droplet and it was carefully dropped into a TGA aluminum 40 μl crucible. Continuous dry nitrogen gas flow at 40ml/min was purged above the droplet solution inside the TGA. Performing the droplet evaporation in TGA also simulates a single droplet drying inside a fluidized bed coater to some extent. The obtained polymer films were visually observed to confirm the droplet pinning theory used in this study. Furthermore, the data was used to derive the critical conditions at which pinning occurs as a function of drying temperature.

Results and Discussion

Figure 2 (top) shows the effect of drying temperature on the obtained coating porosity. Here, only the results on coating porosity are presented, as comparable results were obtained on the coating thickness and pore size. It is shown that the coating porosity decreases with the decrease of inlet air (drying) temperature used. Using current models, the effect of process conditions, in this case the drying temperature, on the evolution of the droplet properties during drying were able to be simulated. Important characteristics, such as droplet diameter, height, contact angle, water fraction retained, droplet spreading flux and pinning time could be determined, which are relevant to the film formation and therefore coating quality.

Figure 3 and Figure 4 show the typical evolution of the droplet diameter and contact angle during drying inside a fluidized bed coater. It can be seen that initially the droplet size decreases during the droplet traveling and then increases several-fold due to the droplet-particle impact and afterwards decreases again due to further drying. The higher the drying temperature used during coating process, the smaller is the diameter of droplet when it hits the core particle. Furthermore, the contact angle of the droplet upon collision with particle is higher at higher drying temperature, which is actually due to the impact of the droplet size on the contact angle. It was also found that at higher drying temperature, the pinning occurs earlier. As the wetting diameter is higher, the average polymer concentration is lower, leading to a ring-like polymer deposit, and therefore a poor film. On the other hand, if the pinning is delayed, the droplet will more likely form a dot-like film (9). The findings from our TGA experiments confirm the plausibility of this theory. Furthermore, a lower contact angle also results in a better spreading of the droplet on the core surface. Considering the effect of drying temperature on the characteristics of the droplet drying and the coating structures, it can be implied that the changes in the coating porosity is actually due to the variation in the pinning time and the droplet contact angle.

Increasing the atomization pressure reduces the initial droplet size, of which the pinning time is slightly lower and the contact angle is increased. Additionally, lowering the droplet size increases significantly the droplet spreading flux of the droplet on the particle, which was calculated as the ratio between a droplet area on the core surface upon collision multiplied by the number of droplets sprayed and divided by the total area of the core particles. In this case, the effect of the increase in the droplet spreading flux is much more significant on the porosity, considering also that the reduction in the pinning time due to the decrease in the droplet size is very little. As a result, the coating porosity becomes lower for coating process using high atomization pressure.

Using a high spraying rate during the coating process increases the bed relative humidity and the droplet size at the same time. This causes a more complex effect of the spraying rate on the characteristics of the droplet drying and therefore the coating quality. While delaying the pinning time, the increase in the spraying rate also reduces the droplet spreading flux, which effects on the coating quality are counteracting with each other. Consequently, altering the spraying rate does not vary greatly the coating porosity.

Figure 5 summarizes the effect of the droplet spreading flux and the droplet pinning time on the coating porosity. It is shown that the coating porosity is the highest when the droplet spreading flux and the pinning time are low. At low droplet spreading flux (due to high spraying rate/low atomization pressure), a very high pinning time is required to reduce the coating porosity. This can be achieved by using the lowest drying temperature possible. In this region, the droplet type-deposition significantly influences the obtained coating structure. At high droplet spreading flux, the influence of the pinning time becomes less significant on the coating porosity. The resulted coating structures in this regime are very much governed by the dispersion of the droplets through the bed of the core particles. Using the presented simulations, it is possible to find the suitable set of process conditions to assure a low coating porosity, i.e. a good coating quality.

Conclusions

From current droplet drying modeling results, the importance of the droplet spreading flux and the droplet pinning time on coating quality has been identified. A delayed pinning time leads to a dot-like polymer deposition resulting in a better coating film. A high droplet spreading flux allows a high distribution of droplets on core surfaces, giving also a coating with low porosity. In combination, it is shown that at low droplet spreading flux, the pinning phenomena of a droplet during drying becomes critical for the resulted coating structures, which is less significant at high droplet spreading flux. From the given modeling exercises, it has been demonstrated that it is possible using the presented models to better predict the coating qualities, by determining the droplet characteristics during drying at any given process conditions.

Acknowledgements

The authors would like to thank Mario Campana for his contribution on the derivation of the correlations for the hydrodynamic properties of the fluidized bed coater used for our study. We would also like to thank Friesland Food Research Centre Deventer, The Netherlands, especially Drs. Marcel Paques for letting us use their CSLM facility and Ing. Anno Koning for performing the CLSM measurements.

List of Symbols

A	Droplet area [=] m^2
Cp	Specific heat capacity [=] J/kgK
d_d	Droplet diameter [=] m
h_d	Droplet height [=] m
h	Heat transfer coefficient [=] W/m^2K
k	Thermal conductivity (of coating solution) [=] W/mK
m	Droplet mass [=] kg
Q	Heat [=] J
t	Time [=] s
T	Temperature [=] K
U	Velocity [=] m/s
λ	Specific heat of evaporation [=] J/kg
μ	Viscosity [=] Pa.s
θ	Contact angle ($2\tan^{-1}(2h_d/d_d)$) [=]
ρ	Density [=] kg/m^3

Subscript

a	Air
d	Droplet
in	Inlet air
max	Maximum
min	Minimum
o	Initial
p	Core particle
pin	Pinning
s	On core surface
t	In time
wb	Wet bulb

References

1. P. D. Hede, P. Bach, and A. D. Jensen. Small-scale top spray fluidized bed coating: Granule impact strength, agglomeration tendency and coating layer morphology, Powder Technology. 176:156-167 (2007).

2. N. Takei, K. Unosawa, and S. Matsumoto. Effect of the spray-drying process on the properties of coated films in fludized bed granular coaters, Advanced Powder Technology. 13:333-342 (2002).

3. K. Saleh, R. Cherif, and M. Hemati. An experimental study of fluidized-bed coating: influence of operating conditions on growth rate and mechanism, Advanced Powder Technology. 10:255-277 (1999).

4. Y.-S. Wu, L. J. Van Vliet, H. W. Frijlink, and K. Voort Maarschalk. Pore size distribution in tablets measured with a morphological sieve, International Journal of Pharmaceutics. 342:176-183 (2007).

5. Laksmana, F. L., Koning, A., Hartman Kok, P. J. A., Vromans, H., Frijlink, H. W., and Voort Maarschalk, K. Imaging and quantifying the effects of different process parameters on the quality of particle coating in fluidized bed coater in Proceedings of the 6th World Meeting on Pharmaceutics, Biopharmaceutics and Pharmaceutical Technology (2008).

6. W. E. Ranz and W. R. Marshall. Evaporation from drops, Chemical Engineering and Processing. 48:141-146 (1952).

7. M. Ikegawa and H. Azuma. Droplet behaviors on substrates in thin-film formation using ink-jet printing, JSME Int J Series B. 47:490-496 (2004).

8. S. Saritha and P. Neogi. Wetting kinetics of a thin film evaporating in air, Physics of Fluids. 19:1-6 (2007).

9. M. Kaneda, H. Ishizuka, Y. Sakai, J. Fukai, and S. T. A. Yasutake. Film formation from polymer solution using inkjet printing method, AIChE Journal. 53:1100-1108 (2007).

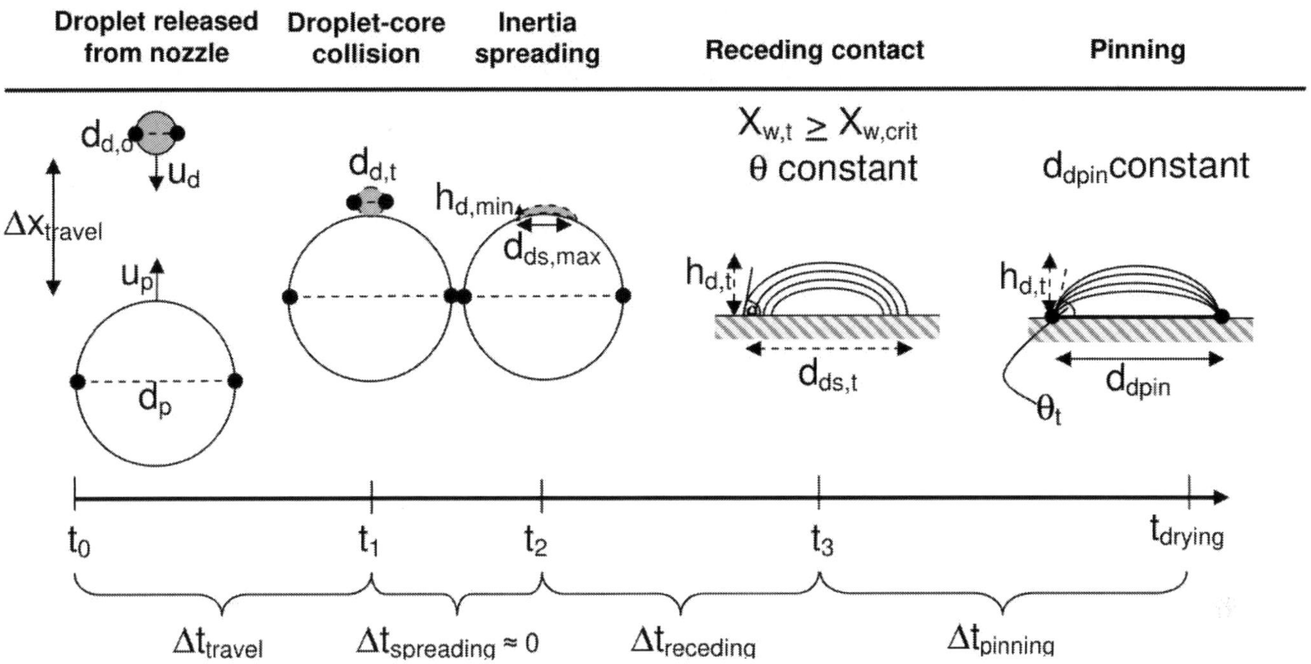

Figure 1. Schematic description of a droplet's journey inside a fluidized bed coater

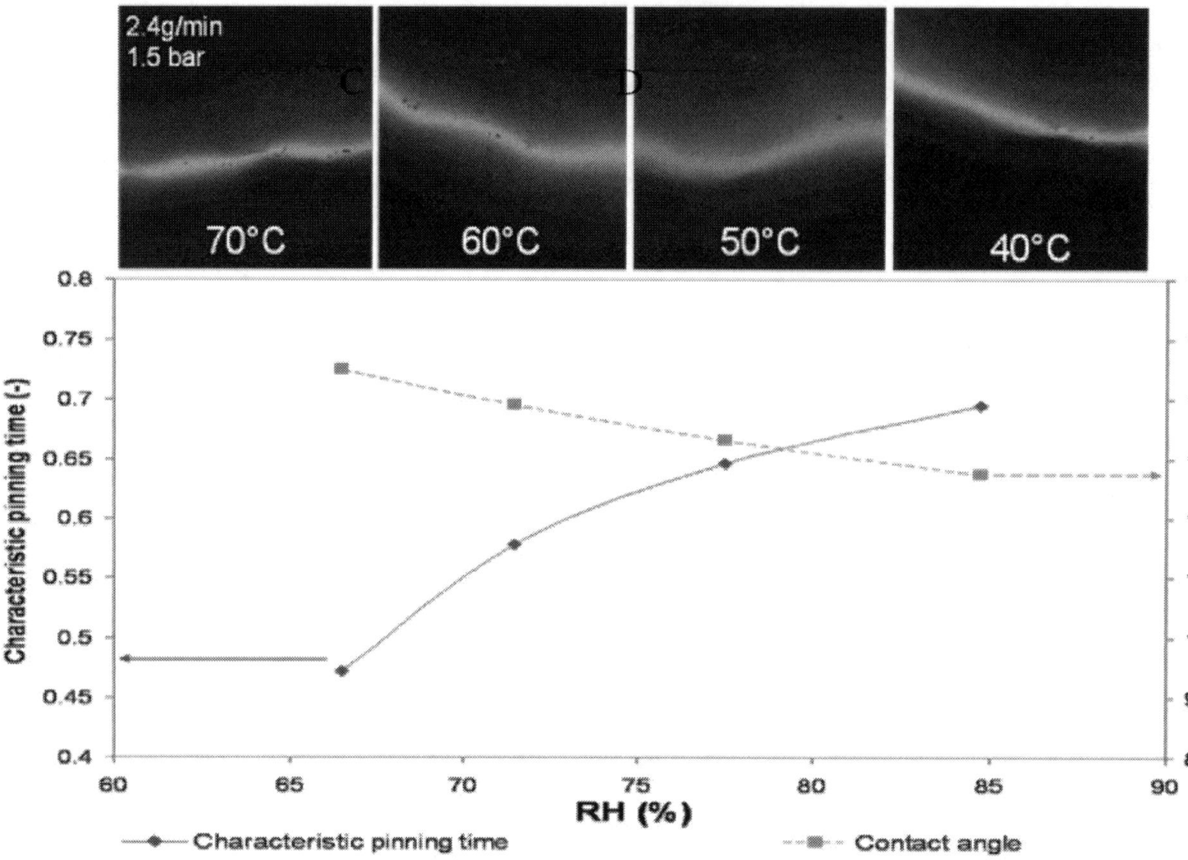

Figure 2. Influence of drying temperature on the characteristic pinning time and contact angle of the droplet upon collision and its impact on the coating structures.

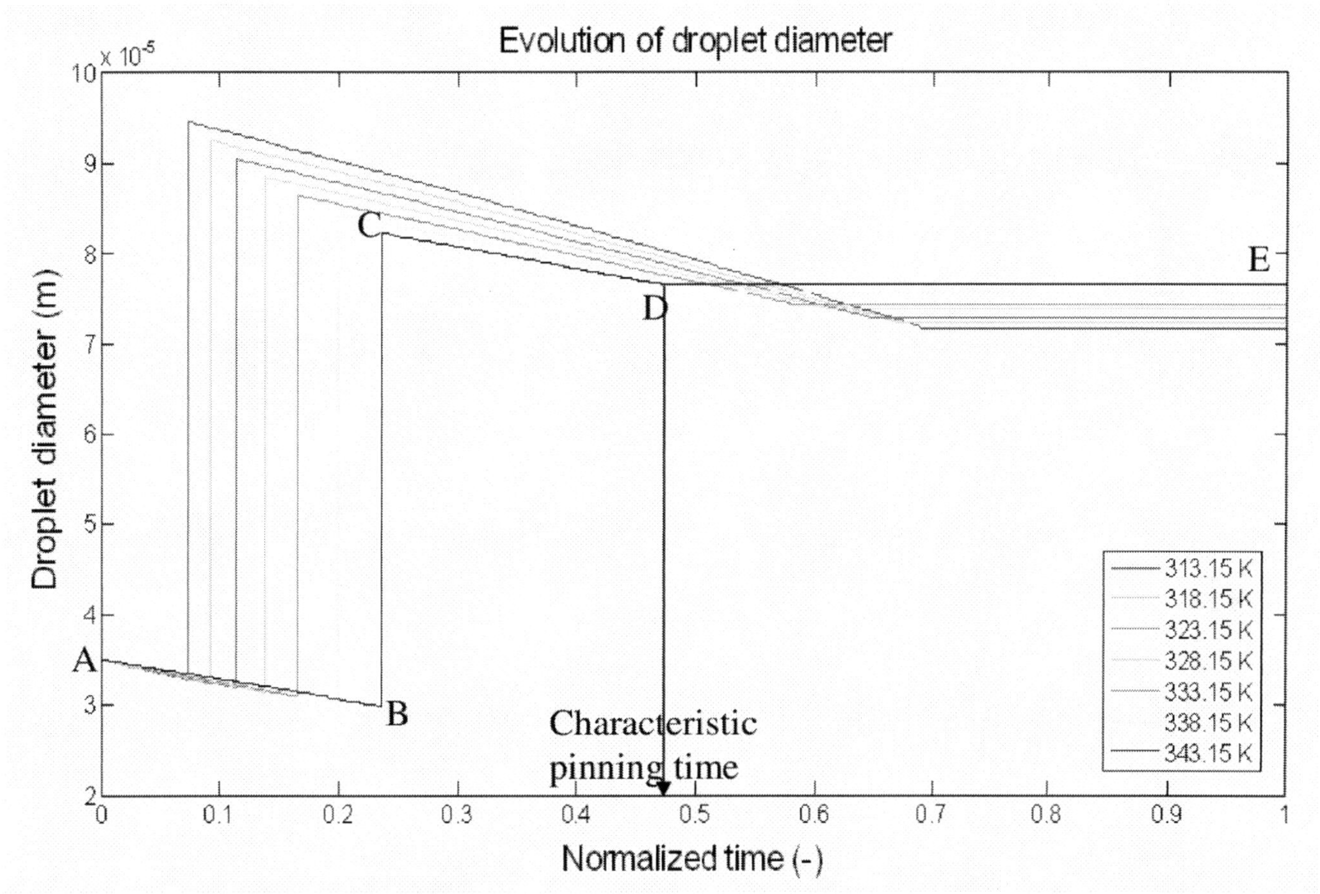

Figure 3. Evolution of droplet diameter during drying in a fluidized bed coater. Point A: Droplet is released from the nozzle; B: Droplet hits a core particle; C: Droplet maximally spreads on a core surface; C-D: Receding contact line evaporation regime, D-E: Pinning evaporation regime.

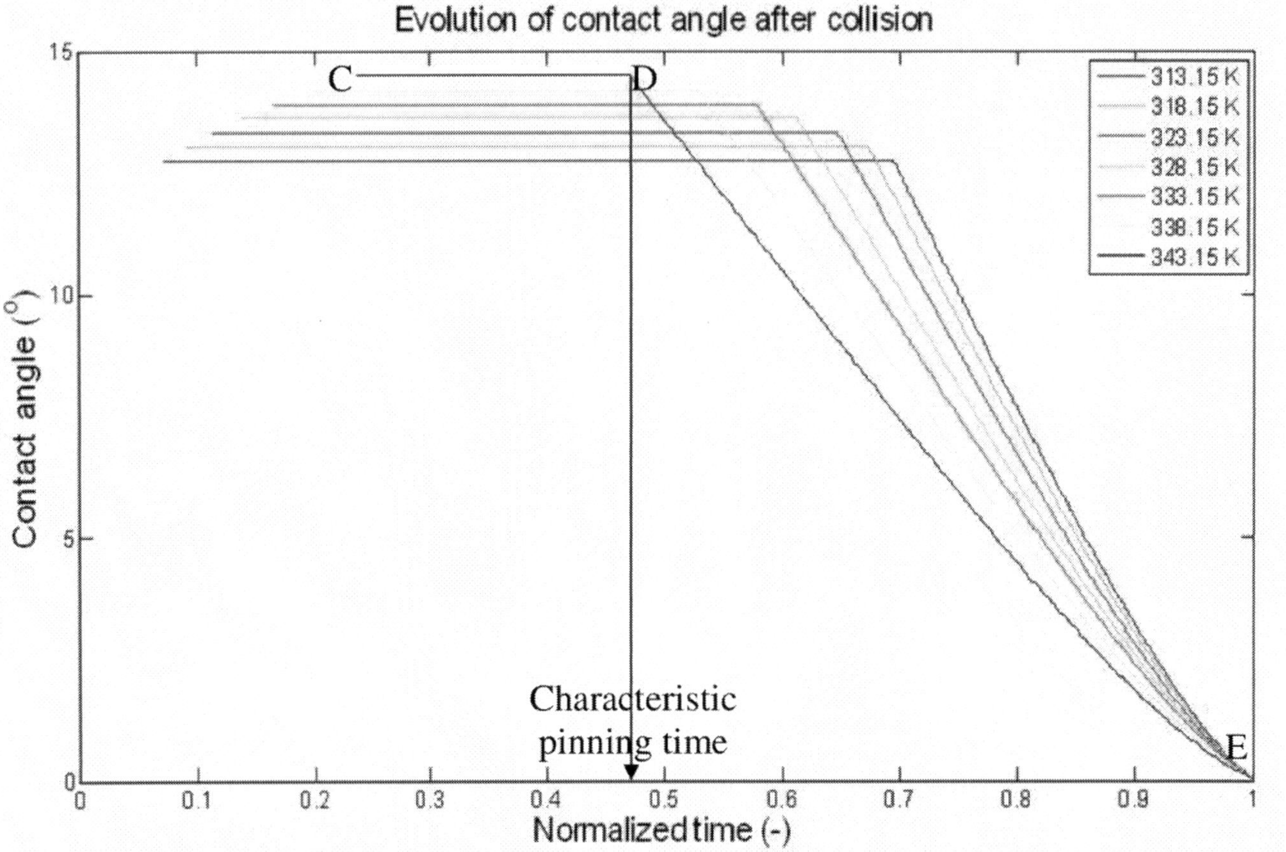

Figure 4. Evolution of droplet shape factor (ratio between droplet height and diameter) during drying in a fluidized bed coater. Point C: Droplet maximally spreads on a core surface; C-D: Receding contact line evaporation regime, D-E: Pinning evaporation regime.

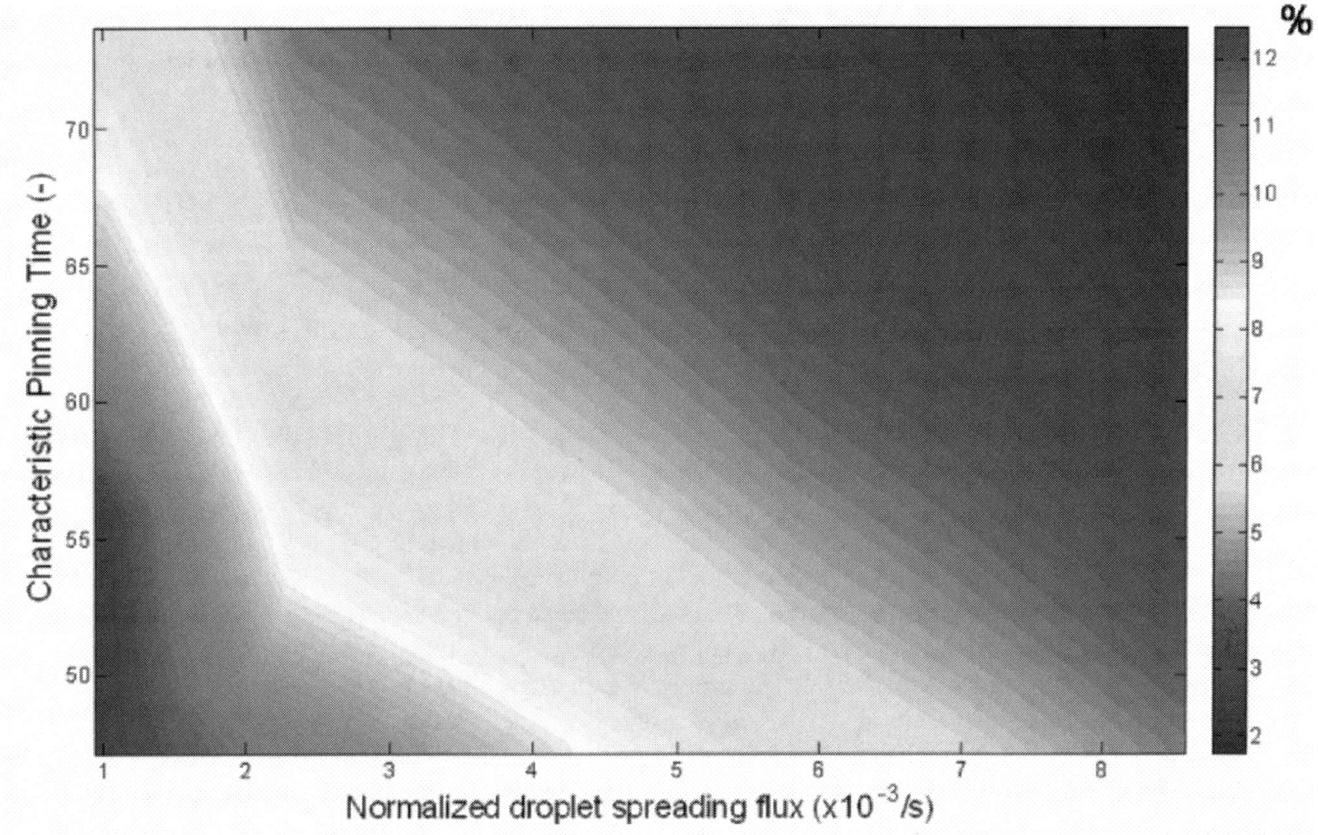

Figure 5. Influence of the characteristic pinning time and normalized droplet spreading flux on the coating porosity.

607f Use of Mathematical Models to Scale up Lyophilization Process

Venkat R. Koganti[1], Sumit Luthra[2], Evgenyi Shalaev[2], and Larry Gatlin[2]. (1) Parenteral Center of Emphasis, Pfizer Global Research and Development, Bldg. 156/Rm. 120U, Groton, CT 06340, (2) Parenteral Center of Emphasis, Pfizer, Groton, CT 06340

Lyophilization is the process of removing water from a solvent system at low temperatures to enhance the shelf life of a water labile product. This the most used process in pharmaceutical industry in the area of parenteral drug development. Even though the process has existed for years, scale up of the process has been using empirical approach in a majority of cases. Here we make an attempt to summarize in a concise manner the method that we follow to successfully scale the lyophilization process from lab scale (~200 vials) to commercial scale (~10000 vials) using mathematical models

Systematic approach using mathematical modeling to scale up lyophilization process will be presented. We will compare experimental results with modeling outputs using real time projects as well as model compounds e.g. sucrose. We compare a mathematically intense computational fluid dynamic (CFD) model (commercially available) which solves coupled heat and mass transfer equation to a simple one dimensional steady state model (based on literature). The predictions from both these models are compared to experimental results at different scales. Based on these results we conclude that we successfully applied mathematical models to scale up the lyophilization process. Most often we do obtain similar results from a simple one dimensional model and CFD model with respect to primary drying time and temperature predictions. CFD model does provide slightly more accurate predictions. It should be noted that using CFD model we can obtain additional estimates on water content present in the lyophile and can also perform secondary drying calculations.

Inputs to the models

• Heat transfer coefficient of the vial (can be estimated by a combination of experiments and theoretical calculations)

• Mass transfer resistance (we currently use the Manometric Temperature Measurement data to calculate resistance offered by dry layer as a function of dry layer thickness)

• Material properties of the product that is freeze dried and also vial dimensions

607g Magnetically Enhanced Centrifugation for Continuous Biopharmaceutical Processing

Fei Chen, Kenneth A. Smith, and T. Alan Hatton. Department of Chemical Engineering, Massachusetts Institute of Technology, 77 Massachusetts Avenue, Cambridge, MA 02139

The separation of biopharmaceutical products from production media continues to be a challenging task. High gradient magnetic Separation (HGMS) is a promising downstream processing technique as it integrates many batch operations into a single unit operation. However, so far, most HGMS processes are discontinuous and there is no commercially available one that can satisfy the specific needs of biopharmaceutical processing. We propose to exploit the interactions of functionalized magnetic particles with magnetic field gradients, forced convective flows and large centrifugal forces to facilitate the recovery of biopharmaceuticals from the media in which they are manufactured. We are developing various numerical tools, including, a particle trajectory model to track the particle movement and its capture onto magnetizable wires, a dynamic particle buildup growth model to study the dynamic process of particle accumulation and its effects on surrounding flow and magnetic fields, and a discrete element method (DEM) package to investigate the behavior of the particle sludge under a centrifugal field. Experiments were carried out to verify the validity of our models. We performed dimensional analysis and defined various dimensionless numbers in order to make the methods and results to be scalable. A small scale single-wire flow channel was made to test the dynamic buildup growth model. And a real magneto-centrifugal contactor was used to study the bulk movement of particle buildup under centrifugal force and compared with DEM simulation results. The models developed here will be used for the design, evaluation, and optimization of continuous magneto-centrifugal contactors that are suitable for biopharmaceutical downstream processing.

618a Tertiary Structure and Properties of a Glycoside Hydrolase Family 44 Endoglucanase from Clostridium Acetobutylicum

Christopher D. Warner[1], Julie A. Hoy[2], Taran C. Shilling[1], Micheal J. Linnen[1], Nathaniel D. Ginder[2], Clark F. Ford[3], Richard B. Honzatko[2], and Peter J. Reilly[1]. (1) Chemical and Biological Engineering, Iowa State University, Ames, IA 50014, (2) Department of Biochemistry, Biophysics, and Molecular Biology, Iowa State University, Ames, IA 50014, (3) Department of Food Science and Human Nutrition, Iowa State University, Ames, IA 50014

CAC0915, an endoglucanase whose genetic coding exists in *Clostridium acetobutylicum* but which is not normally expressed by it, was expressed by *Escherichia coli* after is gene was synthesized. It belongs to glycoside hydrolase family 44 and has been structurally and kinetically characterized. The crystal structure of CAC0915 was solved to 2.2 Å resolution, revealing a TIM barrel structure with a Greek key fold and b-sandwich domain. The catalytic acid, Glu180, and base, Glu352, are located 5.5 Å apart on the fourth and seventh beta strands of the TIM barrel. The catalytic acid is 4.5 Å from the glycosidic oxygen suggesting a rearrangement occurs upon ligand binding. This rearrangement maybe facilitated by a calcium atom, located behind the catalytic acid, which forms a kink in the peptide backbone. A catalytic triad composed of Glu180-His277-Ser351 raises the pK_a of the catalytic acid.

Substrates bind in a cleft composed of nine subsites, –4 through +5. Subsites –4 through –1 contain three aromatic residues (Trp58, Tyr65, and Trp385) which form stacking interactions with the substrate. Four hydrogen bonds are also present in the minus subsites, two of which help distort the pyranose ring in subsite –1. Subsites +1 to +5 contain one hydrogen bond, in subsite +2, and two aromatic residues (Trp320 and Trp324) in subsites +4 and +5. The topography of the binding cleft suggests the substrate will have higher affinity for subsites –1 to –4 than for subsites +1 to +5. This is consistent with the results of reaction product analysis on cellopentaose and cellohexaose substrates using thin layer chromatography. Cellopentaose produces cellotetraose and glucose and cellohexaose produces cellotriose and cellobiose, respectively, as primary products. However, both substrates also produce secondary reaction products, showing the substrate can bind in subsites –3 to +2 or –3 to +3.

CAC0915 hydrolyzes carboxylmethyl cellulose, birchwood xylan, larchwood xylan, cellopentaose, and cellohexaose. It is not active on mannan, laminarin, Avicel, pullulan, or lichenan. Michaelis–Menten kinetic parameters were determined on carboxymethyl cellulose (k_{cat} 18.9 U/mg, K_M 0.263 g/L), birchwood xylan (k_{cat} 31.6 U/mg, K_M 0.412 g/L) and larchwood xylan (k_{cat} 29.5 U/mg, K_M 0.278 g/L). CAC0915 is optimally active at pH 5.0 and has an activation energy for activity of 31.3 ± 2.3 kJ/mol-K and an activation energy for inactivation of 227 ± 20.5 kJ/mol-K.

SHEAR-INDUCED STRUCTURAL CHANGES IN BLOOD PROTEINS

Efrosyni Themistou[1], Indrajeet Singh[1], Sathy V. Balu-Iyer[2], Paschalis Alexandridis[1], and Sriram Neelamegham[1]

[1]Department of Chemical and Biological Engineering
[2]Department of Pharmaceutical Sciences
University at Buffalo - The State University of New York (SUNY)
Buffalo, NY 14260-4200, USA

Introduction

The physiological and pathological functions of blood proteins are imparted by changes in their structure. Von Willebrand factor (VWF) is a large, multimeric, multidomain glycoprotein in human blood [1]. VWF exists both as a dimer with molecular weight (MW) of ~500 kDa and as a multimer (polymer) with MW in the range of $10\text{-}20\times10^6$ Da [2]. It is commonly believed that conformation changes in VWF are important regulators of both the protein's physiological and pathological role [3-6]. While there is some biochemical evidence of VWF structural changes with shear [4,7-8], specific probes to rapidly detect conformational changes are yet unidentified. Such structural changes play a key role in regulating arterial thrombosis, by aiding platelet deposition at sites of vascular injury [3-4,6,9].

We are examining the biophysical and biological features regulating the size and structure of VWF in solution, both under static and fluid shear conditions, by using small angle neutron scattering (SANS) [10] and fluorescence methods. Here, we discuss the use of 4,4'-dianilino-1,1'-binaphthyl-5,5'-disulfonic acid dipotassium salt (bis-ANS), a fluorescence probe that partitions into protein hydrophobic domains [11-13], to investigate conformational changes in VWF solution structure by probing non-polar/hydrophobic domains of this protein that are exposed upon shear application.

In this study, we demonstrated that shear-dependent VWF changes in solution structure are accompanied by exposure of hydrophobic domains within the protein. More specifically, the fluorescence intensity due to probe-VWF interaction increased, suggesting that new hydrophobic domains were exposed upon fluid shear application at shear rates (G) greater than 2300-6000/s and at shear times (t_{sh}) greater than 1min. Such fluid shear-induced conformational changes are specific to VWF on the basis of control experiments on bovine serum albumin (BSA).

The importance of this work is two-folded: (1) Develop a new technique, bis-ANS fluorescence probing, to assess and quantify shear-induced structural changes of blood proteins. This new, easy to use, technique is able to quantify the effect of G and t_{sh} on protein conformation, by using small sample volumes. (2) To examine shear-mediated features in VWF that may regulate the cascade of events that contribute to arterial thrombosis and other blood coagulation processes. The findings emerging from this work may be applicable to other blood proteins.

Materials and Methods

Materials

Multimeric human VWF was purified from blood plasma cryoprecipitate obtained from Community Blood Bank (Erie, PA) [10,14]. Bis-ANS was from Invitrogen - Molecular Probes (Eugene, OR). BSA (96%) was purchased from Sigma-Aldrich (St. Louis, MO). The VWF concentration was measured using the Coomassie/Bradford protein assay kit (Pierce Biotechnology, Rockford, IL). Quantitative amino acid analysis was performed on non-sheared and sheared VWF samples at the Protein Chemistry Laboratory (Texas A&M University) to calibrate the Bradford assay.

Fluorescence Experiments

The fluorescent probe bis-ANS was dissolved in Na-K phosphate buffer and the probe solution was stored at 4°C for 3 days prior use. The optimum probe concentration for each experiment was found by measuring the fluorescence signal of the probe incubated with ~60 µg/mL protein. Thus, for studies with VWF and BSA (control protein), protein was incubated with 9.7 and 1.2 µg/mL of bis-ANS, respectively. Protein solutions were sheared using a cone-and-plate viscometer (Haake VT550, 0.5° cone angle, 5cm diameter) by varying (i) G=0-9600 /s and (ii) t_{sh}=1-5 min. 140µl of protein sample were incubated with 15 µl of bis-ANS stock solution prior to each measurement. The absolute fluorescence was measured using a QuantaMaster Fluorometer (PTI, Birmingham, NJ) at ambient temperature to detect probe-protein binding. The signal of the maximum peak of the emission spectrum was used to evaluate the extent of bis-ANS binding to VWF or BSA. The excitation wavelength used was 375 nm and an emission scan was collected between 420 and 620 nm. Experiments were performed using VWF (~60µg/ml) and BSA (~50µg/ml), both dissolved in Hepes buffer.

For sheared protein samples, the following parameters were measured or calculated: (i) $F^{(G)}$ (counts per second, cps): absolute fluorescence (raw data). (ii) $F^{(G)}_{max}$ (cps): peak emission fluorescence at fixed wavelength. (iii) $F_c^{(G)}$ (cps.mL/µg): concentration-normalized emission spectra after accounting for fluorescence due to free bis-ANS in Hepes buffer ($F^{(G)}_{Hepes+bis\text{-}ANS}$), intrinsic protein fluorescence in buffer ($F^{(G)}_{protein+Hepes}$), and fluorescence of Hepes buffer alone ($F^{(G)}_{Hepes}$): $F_c^{(G)} = (F^{(G)}_{protein+Hepes+bis\text{-}ANS} - F^{(G)}_{Hepes+bis\text{-}ANS} - F^{(G)}_{protein+Hepes} + F^{(G)}_{Hepes})$/concentration. (iv) $F_c^{(G)}{}_{max}$ (cps.mL/µg): Peak concentration-normalized emission fluorescence at fixed wavelength. All data are presented as mean ± standard deviation. The Students t-test was performed in some cases and a p value <0.05 was considered significant.

Results and Discussion

Shear-induced exposure of VWF hydrophobic domains

We used bis-ANS binding assay to detect shear-induced conformational changes in VWF by investigating the exposure of hydrophobic domains on its surface that is often correlated with alterations in protein function [11-12,15]. The effect of shear on the fluorescence intensity of the probe-protein solution was examined by comparing the fluorescence signal when either VWF or the control protein BSA was sheared at G=9600/s with that of the non-sheared protein.

Figures 1A and 1B are representative spectra of experiments performed with VWF and BSA, respectively. $F^{(G)}$ was measured for protein sheared at G=9600/s for 5 min (black patterned line) and this was compared to non-sheared sample (G=0/s, grey patterned line). In all cases, bis-ANS was added exactly 1 min after shear stoppage and data were recorded 1 min after that. The fluorescence intensity was low for Hepes buffer containing bis-ANS alone, Hepes buffer alone and protein alone samples. In Figure 1A, a shift towards lower wavelengths is observed upon addition of protein since $F^{(G)}_{max}$ appears at 490 nm in the presence of VWF and at 508 nm for dye alone in Hepes buffer. This suggests that the probe binds protein hydrophobic domains. $F^{(G)}_{max}$ increased by ~335% upon applying a G of 9600/s to VWF. This enhancement in fluorescence signal and probe binding is indicative of exposure of new hydrophobic domains in VWF subjected to shear. These new sites were likely inaccessible in the native protein. For the control protein BSA (Figure 1B), there was no significant change in $F^{(G)}_{max}$ intensity after applying high G (9600/s). Thus, BSA does not undergo conformation change and the signal change in Figure 1A is protein specific.

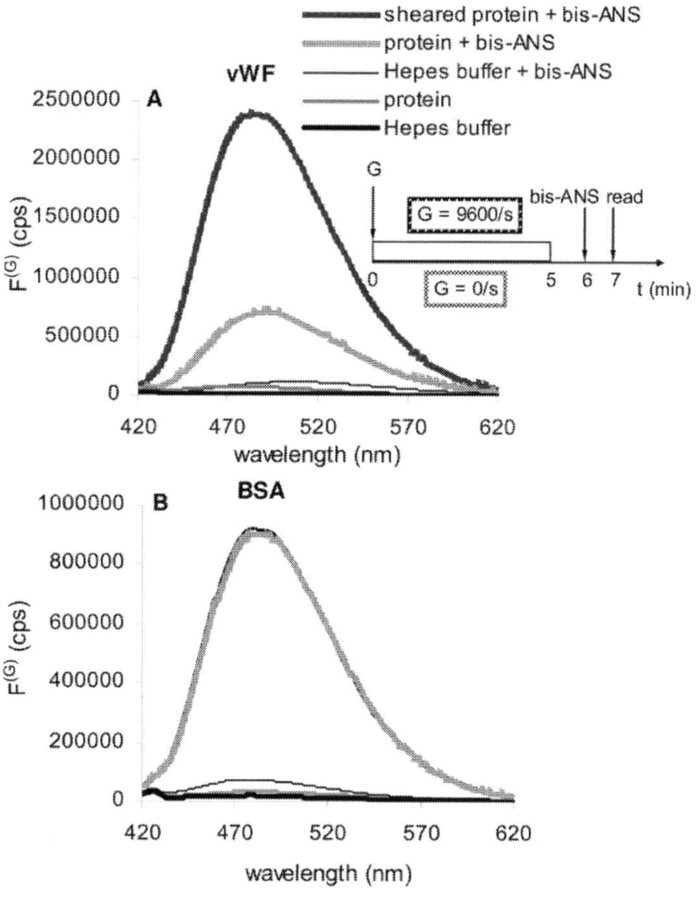

Figure 1. Shear-induced protein conformational changes detected by fluorescence of bis-ANS for (A) VWF and (B) BSA. The schematic presents the experimental protocol indicating the shear rate (G), the shear time, time of the probe addition and time of measurement.

Time and shear – dependent enhancement in bis-ANS binding

The physiological functions of VWF, including its binding interaction with blood platelets during atherothrombosis, are regulated by the magnitude of applied fluid shear. By using Annexin-V as a marker of platelet activation in a previous study of our group [14] significant platelet activation was mediated via GpIb-VWF interactions upon high shear (G=9600/s) even at the very low t_{sh} of 10 s. Higher G and t_{sh} were found to increase the extent of platelet activation [16]. By comparing the data from those studies, we conclude that VWF solution structure may undergo shear-induced conformational changes that occur at small time scales and are amplified at higher t_{sh}, relevant with the platelet activation.

Here, we apply the bis-ANS binding assay to find the G and t_{sh} at which conformational changes in VWF solution structure begin and also to quantify those changes.

Shear time effect

We examined the effect of t_{sh} on probe-binding and VWF conformation change. To this end, studies were performed at 9600/s with t_{sh} between 1 and 5 min (Figure 2A and 2B). $F^{(G)}$ was measured and normalized with respect to the protein concentration. The signal increases due to bis-ANS binding to VWF, but not to the control protein BSA, in a time-dependent manner. This change is observed in Figure 2A where there is an increase in the fluorescence intensity in the emission spectra with increasing the t_{sh}. Thus, the exposure of hydrophobic domains in VWF is enhanced with time. This increase in the fluorescence signal is also shown in Figure 2B that presents $F_c^{(G)}{}_{max}$ at different t_{sh}. No conformational change was observed for BSA since in Figure 2B the $F_c^{(G)}{}_{max}$ value is similar for t_{sh} of 1 and 5 min.

Shear rate effect

G values over the range of physiological and pathological conditions were applied to VWF to examine its solution structural changes. Upon step-increasing G, the bis-ANS fluorescence signal is increasing with shear, indicating a progressive increase in the extent of protein conformation change. Figures 2C and 2D describe the effect of varying G (0-9600 /s) over a 5 min time interval demonstrating that the level of bis-ANS binding to VWF increases with increasing G. Probe binding was not altered upon shearing the control protein BSA. Emission spectra of VWF samples are shown in Figure 2C. These same data are presented in Figure 2D that presents $F_c^{(G)}{}_{max}$ at different G. Thus, these studies measure the extent of hydrophobic domain exposure as a function of G. We observed that a minimum G of 6000/s is required for augmented probe binding, giving a 3-fold increase in the fluorescence signal. A higher G of 9600/s causes further enhancement of fluorescence intensity; a 4-fold increase of the probe signal is observed. These findings suggest that VWF specific conformational changes in solution that accompany hydrophobic domain exposure increase with shear above 6000/s in a shear-dependent manner.

Those results are in agreement with our SANS studies where the amount of neutron scattering also increases with G due to extended structural changes at the highest G that were used in these experiments. Both studies are indicative of the fact that VWF conformation changes are dependent on the magnitude of the applied shear. However, in our studies with bis-ANS as well as with SANS low levels of structural changes were observed at G below 2300/s, whereas at higher G high levels of VWF hydrophobic domain exposure were observed, giving an enhanced bis-ANS fluorescence signal.

It is important to compare our results with the literature, which shows that above a threshold G between 2000/s and 5000/s, fluorescently-tagged VWF structure changes from a compact to an elongated one [7]. From those results, it is suggested that at low, physiological G (<3000/s) small length scale changes occur due to rearrangement of domain level features, while at higher G (>2300/s) large length scale changes expose VWF hydrophobic domains, showing that the former ones may be a prerequisite for the second ones. Also, a study shows that the different domains of VWF show different stability since it was found that the VWF A2 domain may be most susceptible to unfolding than other VWF A domains [17].

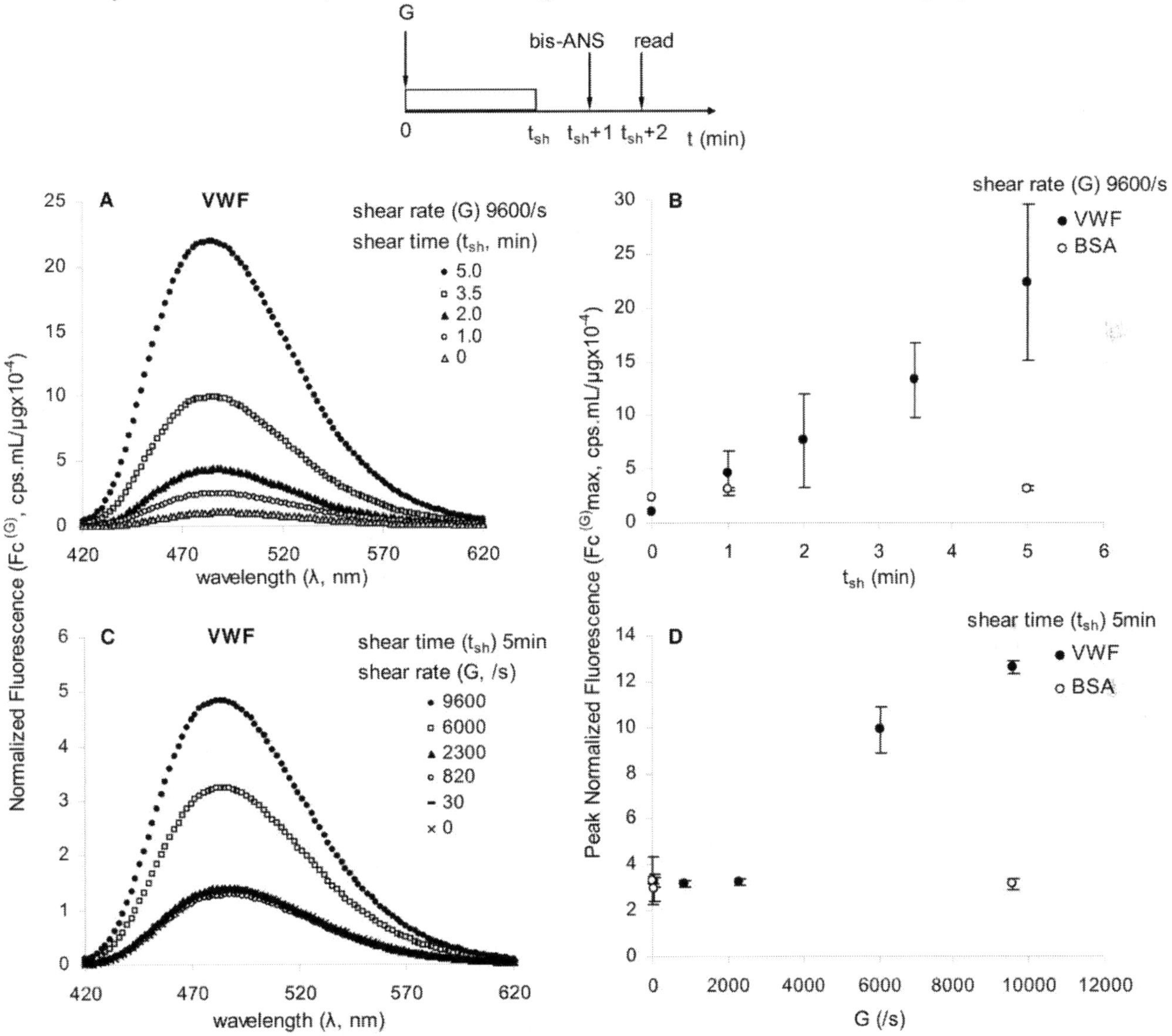

Figure 2. Shear and time dependence. A. Fluid shear was applied to VWF at 9600/s for various shear times (t_{sh}). Bis-ANS was added 1 min after shear stoppage and fluorescence spectra were measured 1 min thereafter as illustrated in the schematic. Normalized fluorescence due to bis-ANS VWF interaction $F_c^{(G)}$ was calculated as described in Experimental Part. **B.** Peak $F_c^{(G)}$ value at a fixed wavelength ($F_c^{(G)}{}_{max}$) is presented for 3 experiments. **C.** Studies similar to panel A were performed at a fixed t_{sh} (5 min) and the magnitude of fluid shear (G) was varied. **D.** Peak $F_c^{(G)}$ value at a fixed wavelength ($F_c^{(G)}{}_{max}$) is presented for 3 experiments.

Conclusions

A new, easy to use, technique is introduced here to investigate shear-induced conformational changes in large length scale features of VWF and measure/quantify their strength at different G and t_{sh}. Shear-induced exposure of VWF hydrophobic domains was observed at high G up to 9600/s and t_{sh} up to 5 min, indicated by the increase of the binding of the fluorescence probe bis-ANS to the sheared VWF compared to the non-sheared protein. The current work suggests that structural changes at large length scales in VWF are more obvious from physiological to high G and at t_{sh} higher than 1 min, whereas they are more dominant at higher t_{sh} and G. Such conformational changes may regulate alterations in VWF function upon shear [4,6,8], and enhance VWF binding to hydrophobic substrates and collagen [7,8], or VWF self-association [14]. The implications of these structural changes in VWF to arterial thrombosis are currently under investigation in our laboratory.

Acknowledgement

This work was supported by NIH grant HL77258.

References

1. Sadler, J. E. "Biochemistry and genetics of von Willebrand factor" *Annu. Rev. Biochem.* **1998**, *67*, 395-424.
2. Fowler, W. E.; Fretto, L. J.; Hamilton, K. K.; Erickson, H. P.; McKee, P. A. "Substructure of human von Willebrand factor" *J. Clin. Invest.* **1986**, *76*, 1491-1500.
3. Levy, G. G.; Nichols, W. C.; Lian, E. C.; Foroud, T.; McClintick, J. N.; McGee, B. M.; Yang, A. Y.; Siemieniak, D. R.; Stark, K. R.; Gruppo, R.; Sarode, R.; Shurin, S. B.; Chandrasekaran, V.; Stabler, S. P.; Sabio, H.; Bouhassira, E. E.; Upshaw, J. D.; Jr.; Ginsburg, D.; Tsai, H. M. "Mutations in a member of the *ADAMTS* gene family cause thrombotic thrombocytopenic purpura" *Nature* **2001**, *413*, 488-494.
4. Siedlecki, C. A.; Lestini, B. J.; Kottke-Marchant, K. K.; Eppell, S. J.; Wilson, D. L.; Marchant, R. E. "Shear-dependent changes in the three-dimensional structure of human von Willebrand factor" *Blood* **1996**, *88*, 2939-2950.
5. Tsai, H. M. "Physiologic cleavage of von Willebrand factor by a plasma protease is dependent on its conformation and requires calcium ion" *Blood* **1996**, *87*, 4235-4244.
6. Tsai, H. M.; Sussman I. I.; Nagel, R. L. "Shear stress enhances the proteolysis of von Willebrand factor in normal plasma" *Blood* **1994**, *83*, 2171-2179.
7. Schneider, S. W.; Nuschele, S.; Wixforth, A.; Gorzelanny, C.; Alexander-Katz, A.; Netz, R. R.; Schneider, M. F. "Shear-induced unfolding triggers adhesion of von Willebrand factor fibers" *Proc. Nat. Acad. Sci. USA* **2007**, *104*, 7899-7903.
8. Novak, L.; Deckmyn, H.; Damjanovich, S.; Harsfalvi, J. "Shear-dependent morphology of von Willebrand factor bound to immobilized collagen" *Blood* **2002**, *99*, 2070-2076.
9. Singh, I.; Shankaran, H.; Beauharnois, M. E.; Xiao, Z.; Alexandridis, P.; Neelamegham, S. "Solution structure of human von Willebrand factor studied using small angle neutron scattering" *J. Biol. Chem.* **2006**, *281*, 38266-38275.

10. Kasirer-Friede, A.; Cozzi, M. R.; Mazzucato, M.; De Marco, L.; Ruggeri, Z. M.; Shattil, S. J. "Signaling through GP Ib-IX-V activates aIIbβ3 independently of other receptors" *Blood* **2004**, *103*, 3403-3411.
11. Kosinski-Collins, M. S.; King, J. "In vitro unfolding, refolding, and polymerization of human gammaD crystallin, a protein involved in cataract formation" *Protein Sci.* **2003**, *12*, 480-490.
12. James, E. L.; Bottomley, S. P. "The mechanism of alpha 1-antitrypsin polymerization probed by fluorescence spectroscopy" *Arch. Biochem. Biophys.* **1998,** *356*, 296-300.
13. Balasubramanian, S. V.; Bruenn, J.; Straubinger, R. M. "Liposomes as Formulation Excipients for Protein Pharmaceuticals: A Model Protein Study" *Pharm. Res.* **2000**, *17,* 344-350.
14. Shankaran, H.; Alexandridis, P.; Neelamegham, S. "Aspects of hydrodynamic shear regulating shear-induced platelet activation and self-association of von Willebrand factor in suspension" Blood **2003**, *101*, 2637-2645.
15. Pierce, A.; deWaal, E.; Van Remmen, H.; Richardson, A. Chaudhuri, A. "A novel approach for screening the proteome for changes in protein conformation" *Biochemistry* **2006**, *45*, 3077-3085.
16. Zhang, J. N.; Bergeron, A. L.; Yu, Q.; Sun, C.; McIntire, L. V.; Lopez, J. A.; Dong, J. F. "Platelet aggregation and activation under complex patterns of shear stress" *Thromb. Haemost.* **2002**, *88*, 817-821.
17. Auton, M.; Cruz, M. A.; Moake J. "Conformational stability and domain unfolding of the Von Willebrand factor A domains" *J. Mol. Biol.* **2007**, *366*, 986-1000.

618c Expression and Characterization of a Computationally Designed Laccase-Like Enzyme

Doris J. Glykys[1], Geza Szilvay[1], Pablo Tortosa[2], Maria Suarez[2], Alfonso Jaramillo[3], and Scott Banta[1]. (1) Chemical Engineering, Columbia University, 500 W 120th St, New York, NY 10027, (2) Ecole Polytechnique, Route de Saclay, Palaiseau Cedex, 91128, France, (3) Laboratoire de Biochimie, Ecole Polytechnique, Route de Saclay, Palaiseau Cedex, 91128, France

Protein engineering of oxido-reductase enzymes, such as laccases, has attracted considerable interest due to their present and potential use in industrial applications. In an effort to improve the stability of laccases for various applications, computational methods were used to design a laccase-like enzyme with optimized folding characteristics and improved stability.

The laccase glycoprotein from the fungus *Trametes versicolor* (TvL) was used as a basis for designing a new laccase-like protein. The backbone structure of TvL was left intact while all the amino acids except for the ones around the trinuclear copper center were optimized for stability by computational methods. This newly designed laccase-like ~60 kDa amino acid chain was optimized for *E. coli* expression and was inserted in a pET vector with a C-terminal poly-His tag. Since this enzyme is expressed in bacteria, it will be non-glycosylated.

This laccase-like enzyme has been successfully expressed in the presence of copper following IPTG induction in *E. coli*. The majority of the expressed protein was found in the inclusion body fraction. In order to refold the protein, different strategies of denaturation and folding have been utilized. Under the right conditions, a small amount of the protein can be found in the soluble fraction and it has been purified using metal affinity, ion exchange and gel filtration chromatography methods.

The characterization of the final protein product includes molecular weight spectra (MALDI-TOF), conformational analysis (CD), thermal stability (CD) and copper content. Lastly, the kinetic activity of the enzyme under several conditions and with various substrates will be determined and compared to the original TvL laccase as well as to previously studied laccases from other sources.

618d The Role of the Two Active Sites of the Bacterial Disulfide Isomerase on the Catalysis of Oxidative Protein Folding

Silvia Arredondo and George Georgiou. Chemical Engineering, University of Texas at Austin, 1 University Station C0400, Austin, TX 78712

Protein disulfide isomerases are ubiquitous proteins that catalyze the proper formation of disulfide bonds during oxidative protein folding. The bacterial protein disulfide isomerase DsbC is a homodimeric V-shaped enzyme that consists of a dimerization domain, two α-helical linkers and two opposing thioredoxin fold catalytic domains. The functional significance of the two opposing catalytic sites is not well understood yet. To address this problem we fused the two polypeptides that compose the DsbC homodimer into a single protein. The resulting protein, mDsbC, was shown to exhibit high catalytic activity. Mutants containing only one active site or catalytic domain were constructed and their catalytic properties, including their ability to serve as a thiol oxidant and as a catalyst of disulfide bond rearrangement, were studied in detail and will be discussed. Apart from helping understand the catalytic cycle of disulfide isomerization, this study also provides insights onto the evolution of the eukaryotic enzymes (such as PDI) relative to their bacterial counterparts.

618e Solution Studies of Protein Folding Using a Site-Specific Probe of Cysteine Accessibility

J. David Pajerowski[1], Karthikan Rajagopal[2], and Dennis E. Discher[2]. (1) Biengineering, University of Pennsylvania, Philadelphia, PA 19104, (2) Chemical and Biomolecular Engineering, University of Pennsylvania, Philadelphia, PA 19104

Cysteine is a moderately hydrophobic amino acid that is under represented in most proteins, rendering it a useful probe of protein structure. In a reformulation of classical methods, we have developed a method to assess the accessibility of specific exposed and buried cysteines that provides a sensitive measure of the local structure of protein domains. The cysteine probes of choice are selectively reactive fluorophors, as they permit the extent of labeling to be interrogated using multiple measurement modalities including fluorescence microscopy, fluorescence of 1D and 2D protein gels, and spectroscopic analysis via HPLC. We have applied this method to study a muscular dystrophy relevant recombinant Ig domain from the Lamin A/C tail, using a 1-D gel system. By measuring the temperature dependant labeling of an endogenous cysteine, we calculate a change in the fractional accessibility that compares well with circular dichroism studies. A shortcoming of the gel based methodology is the inability to differentiate between the labeling of multiple cysteines in a given polypeptide. We address this by borrowing the proteomic approach of analyzing tryptic peptides via RP-HPLC. In this approach, proteins are labeled in their native state and then fragmented by trypsin. Labeled peptides of varying hydrophobicity are separated, and the extent of labeling quantified. The number of cysteines in a polypeptide that can be analyzed in this way is limited only by the ability of the HPLC to resolve the labeled tryptic fragments, providing a power site specific method for assessing local protein structure.

618f Application of a Continuously-Operated Enzyme Membrane Reactor In the Determination of the Operational Stability of An Enzyme Undergoing Reversible Denaturation

Thomas A. Rogers and Andreas Bommarius. Chemical and Biomolecular Engineering, Georgia Institute of Technology, 315 Ferst Drive, IBB, Atlanta, GA 30332

While many enzyme-catalyzed reactions offer clear advantages, such as high enantioselectivity and relatively mild process conditions, most enzymes suffer from denaturation when subjected to the elevated temperatures which are often used in commercially-viable processes. It is useful to model the combined effects of temperature on the activation (increased intrinsic kinetic rate) and deactivation (thermally induced degradation) of a biocatalyst and optimize the total turnover number in order to assess whether a biocatalyst will be cost-effective under a given set of process conditions. The correlation of short-term biochemical measurements to the accurate prediction of the productivity of an enzyme over its useful life has only recently been mentioned in the literature and has not yet been proven. Here, wild-type TEM-1 beta-lactamase was subjected to a temperature-time gradient in an enzyme membrane reactor (EMR), and the total turnover number was predicted by fitting conversion data to the Lumry-Eyring model. This has been attempted previously in the literature, but for an enzyme which deactivates irreversibly at high temperature. [1] The EMR results were correlated to kinetic data and thermal deactivation data collected in isothermal experiments. At temperatures well below the melting temperature of the enzyme, for which the population of unfolded protein in the reactor is very small compared to the population of native protein, it was demonstrated that the total turnover number is simply a function of the catalytic constant (kcat) and the observed deactivation rate (kD,obs), two quantities which are readily measured by experiment.

[1] P. R. Gibbs, et al. Biotechnology Progress (2005), 21, 762-774.

618g Refolding of Beta-Galactosidase In a Microfluidic Device

Sowmya Kondapalli, Sibley School of Mechanical and Aerospace Engineering, Cornell University, 245 Upson Hall, Ithaca, NY 14853 and Brian J. Kirby, Sibley School of Mechanical & Aerospace Engineering, Cornell University, Ithaca, NY 14853.

We have developed a microfluidic device that has the unique capability of automating the combinatorial process of finding solution conditions for protein refolding. The microfluidic approach offers the potential advantages of automation, cost-effectiveness, compatibility with optical detection, and a million-fold reduction in sample volumes as opposed to conventional techniques of hand-pipetting or using robotic systems. The device is a multi-layered Polydimethylsiloxane (PDMS) on glass device with automated controls for reagent aliquoting and mixing. Refolding experiments have been performed on the protein β-galactosidase, which is an essential enzyme in the human body that catalyzes the hydrolysis of β-galactosides like lactose into monosaccharides. The refolding yield has been quantified on-chip using fluorescein di-β-D-galactopyranoside (FDG), a caged-fluorescent molecule. The refolding protocol and the fluorometric assay are optimized for best on-chip signal intensity. The main impact of this work is on greatly reducing the cost of realizing protein pharmaceuticals.

620a Model-Guided Design of a Refactored Metabolic Pathway

Howard Salis[1], Zhirong Li[2], Krishna Niyogi[2], and Christopher Voigt[1]. (1) Department of Pharmaceutical Chemistry, UC San Francisco, 1700 4th St., 408 Byers Hall, San Francisco, CA 94158, (2) Department of Plant and Microbial Biology, UC Berkeley, 441 Koshland Hall, Berkeley, CA 94720-3102

Synthetic biology is an emerging field that applies bottom-up engineering principles to redesigning natural biological systems for biotech applications. We present an example of these principles by completely redesigning and optimizing a seven enzyme metabolic pathway extracted from Rhodobacter sphaeroides.

Using lower cost large-scale DNA synthesis, we may now rewrite (or refactor) the DNA sequences that encode useful metabolic pathways, extracted from any sequenced host organism, and optimize them for a genetically tractable host, such as *Escherichia coli*. In the process, we simplify the DNA sequence to more reliably engineer and optimize the product flux. Key steps include eliminating overlapping protein coding sequences, removing extraneous transcriptional or translational regulation, improving mRNA stability, and adding unique restriction sites. We also add synthetic gene regulation to control transcription and rationally select the translation initiation rate by designing synthetic 5' UTRs for each protein coding sequence. The synthetic 5' UTRs are generated by a newly developed model of translation initiation coupled with Monte Carlo sampling (see session 15C12).

We demonstrate this systematic procedure on the *crt*EBICFDA genes, extracted from Rhodobacter sphaeroides. The crt metabolic pathway produces visibly colored intermediates and products, making it a good test example. The final product of the pathway is the yellow carotenoid hydroxy-spheroidene. Using HPLC purification and absorbance detection, we quantify intermediate and product formation and evaluate the effectiveness of our design procedure. The final design results in a significant improvement in accumulation of the carotenoid hydroxy-spheroidene.

620b An Externally-Tunable Bacterial Band-Pass Filter for Enzymatic Activity

Takayuki Sohka[1], Richard A. Heins[1], Ryan M. Phelan[2], Jennifer Greisler[1], Craig A. Townsend[2], and **Marc Ostermeier**[1]. (1) Chemical and Biomolecular Engineering, Johns Hopkins University, 3400 North Charles Street, Baltimore, MD 21218, (2) Chemistry, Johns Hopkins University, 3400 North Charles Street, Baltimore, MD 21218

Synthetic biology aims at the assembly of complex biological systems using a network of robust, tunable, biological building blocks. The current paradigm for tuning synthetic biological systems is through reengineering system components. However, such an approach is tantamount to building a new device for each desired network behavior. Biological systems designed with the inherent ability to be tuned based on external stimuli will be more versatile. We engineered E. coli cells to behave as a band-pass filter for enzyme activity and small molecules. The band's location and band-width can be positioned within a four order of magnitude range simply by the addition of compounds to the growth media. Inclusion of an enzyme-substrate pair that functions as an attenuator in the network enabled this tunability. The system enabled bacteria growth to be patterned in response to chemical gradients in non-intuitive ways – linear patterns could be induced from radial concentration gradients. Mixtures of cells with different levels of enzyme activity could be induced to grow in a differentiated pattern corresponding to a rank order of enzyme activity. The system was used to isolate an engineered allosteric enzyme from a library of inactive and non-allosteric enzymes. The application of this strategy to other synthetic biological systems will increase their utility for biotechnological applications and their usefulness as a tool for gaining insight into biological network behavior and nature's underlying design principles.

620c Transcriptional Effects of Crp* Expression In *Escherichia Coli*

Reza Khankal[1], Debashis Ghosh[2], and Patrick C. Cirino[1]. (1) Chemical Engineering, Penn State University, 222 Fenske Lab, University Park, PA 16802, (2) Department of Statistics and Huck Institute of Life Sciences, Associate Professor, 514A WartikBuilding, University Park, PA 16802

Escherichia coli exhibits diauxic growth in sugar mixtures due to CRP-mediated catabolite repression and inducer exclusion related to phosphotransferase system enzyme activity. Replacement of the native *crp* gene with a catabolite repression mutant (referred to as *crp**) alleviates diauxic growth effects in *E. coli* and enables co-utilization of glucose and other sugars. While previous studies have examined the effects of expressing CRP* mutants on the expression of specific catabolic genes, little is known about the global transcriptional effects of CRP* expression. In this study, we compare the transcriptome of *E. coli* W3110 (expressing wild-type CRP) to that of mutant strain PC05 (expressing CRP*) in the presence and absence of glucose. While the simplest model of CRP*-mediated gene expression assumes insensitivity to glucose (or cAMP), our results show that gene expression in the context of CRP* is very different from that of wild-type in the absence of glucose, and is influenced by the presence of glucose. Approximately 400 genes were found to respond differently to glucose in PC05 compared to W3110. While most genes upregulated by glucose in W3110 are also upregulated by glucose in PC05, the glucose effect is significantly suppressed in PC05. We will present a detailed transcription analysis and relate these results to several important phenotypic differences between W3110 and PC05.

620d Regulation of Pentose Transport and Metabolism

Tasha Desai, Chemical and Biomolecular Engineering, University of Illinois, 600 South Matthews Ave., 210 Roger Adams Laboratory, Box C-3, Urbana, IL 61801 and Christopher V. Rao, Department of Chemical Engineering, University of Illinois, 600 South Matthews Ave., 211 Roger Adams Laboratory, Box C-3, Urbana, IL 61801.

Escherichia coli possess a large number of dedicated transporters for a wide range of carbohydrate substrates. These transporters are not constitutively expressed but rather induced in the presence of their cognate sugar. However, rarely does an organism such as E. coli see only one sugar in its natural environment as is common in traditional laboratory experiments. Rather, the organism is often confronted with multiple sugars. We hypothesize, therefore, that E. coli has evolved an additional layer of regulation that enables it to reprogram its metabolism in response to multiple sugars.

In this talk we discuss our efforts directed towards unraveling this regulation. Through the use of transcriptional reporters, we have observed a weak hierarchy in pentose utilization. For example, arabinose inhibits the expression of the xylose metabolic genes. Furthermore, deletion of the arabinose regulator, AraC, leads to increased xylose metabolic gene expression. Moreover, we also observe regulatory crosstalk by ribose and hexose sugars such as rhamnose and fucose. We have also systematically deleted all known pentose and hexose transporters both individually and in combination. Our results show significant cross-talk between the various transporters, both in terms of their specificity and also the regulation of their expression. Collectively, our results demonstrate that the regulation of individual pentose metabolism and transport pathways is interconnected to form a systems-level control system. We believe that these results will aid in the metabolic engineering of strains using pentose feedstocks, such as in biofuels production and natural product synthesis.

620e Knockout of Transcriptional Regulators for Decoupling Solvent Formation from Sporulation In *Clostridium Acetobutylicum*

Bryan P. Tracy, Chemical and Biological Engineering, Northwestern University, 2145 Sheridan Road, Evanston, IL 60208 and Eleftherios Terry Papoutsakis, Dept. of Chemical Engineering & Delaware Biotechnology Institute, University of Delaware, 15 Innovation Way, Newark, DE 19711.

Understanding and manipulating bacteria differentiation are intriguing goals motivated by pure scientific curiosity, medical microbiology, concerns of biological warfare, and industrial biotechnology. Of particular interest is endospore formation, which is well understood in the facultative anaerobe *Bacillus subtilis* (*B. subtilis*), but far less understood in related endospore formers such as clostridia. *Clostridium* is a genus of Gram-positive, obligate anaerobic, endospore-forming bacteria that includes many revered pathogenic species (*C. botulinum, difficile, tetani* and *perfringens*) and numerous commercially desired cellulolytic and/or solvent forming species (*C. acetobutylicum, thermocellum, cellulolyticum, beijerinckii*, etc.). Interestingly solvent formation and some toxin production are suggested to be genetically coupled with endospore formation. Thus we are interested in characterizing the genetic connections and controlling these coupled events.

To accomplish these tasks, we applied a novel gene disruption approach to individually knock out the function of *sigE* and *sigG*, which are suspected to be involved in either or both endospore and solvent formation in C. acetobutylicum. In both cases we successfully and absolutely decouple solvent formation from the completion of endospore differentiation. Both strains exhibited near wild-type (ATCC824) solvent production characteristics. Results also reinforced the suspected similarities between the transcriptional regulation events responsible for *B. subtilis* endospore formation by arresting endospore formation at stages II and III of asymmetric cell division for *sigE* and *sigG* knockouts respectively. This is the first report to purposefully decouple solvent formation from endospore differentiation. Results are significant for all clostridia and both strains serve as ideal commercial platforms for future optimization in converting cellulosic feedstocks to transportation fuels.

Web Page: www.papoutsakis.org/page.php?id=63

620f Improving Complex Phenotypes by Mutagenesis of the Bacterial RNA-Polymerase Holoenzyme: New Approaches for Transcriptional Engineering

Daniel Klein-Marcuschamer and Gregory Stephanopoulos. Chemical Engineering, Massachusetts Institute of Technology, 77 Massachusetts Ave. 56-422, Cambridge, MA 02139

Combinatorial or random methods for strain engineering have been extensively used for improvement of multigenic phenotypes and other traits for which the underlying mechanism is not fully understood. The preferred method has been mutagenesis and selection, but other methods have emerged. Our lab has successfully used mutant transcription factors, which direct the RNA polymerase (RNAP) during transcription, to engineer complex phenotypes in yeast and bacteria. These phenotypes encompass tolerance to harsh process conditions and overproduction of metabolites of interest in *Escherichia coli*, *Lactobacillus plantarum*, and *Saccharomyces cerevisiae*. Here, we show that it is also possible to impart new phenotypes by mutating the RNAP holoenzyme itself, in particular, the alpha subunit of the bacterial polymerase. Using a previously-developed statistical metric, we show that a library of mutants constructed by error-prone PCR can introduce phenotypic diversity, and we use this fact for guiding our experimental efforts. Finally, we present the use of this tool for improving tolerance of *E. coli* to butanol and other solvents, and for increasing the titers of two commercially-relevant products, tyrosine and hyaluronic acid.

620g Generation and Characterization of Novel Pathways to Degrade Xenobiotics

Stacey D. Finley[1], Linda J. Broadbelt[1], and Vassily Hatzimanikatis[2]. (1) Chemical and Biological Engineering, Northwestern University, 2145 Sheridan Road, E136, Evanston, IL 60208, (2) Laboratory of Computational Systems Biotechnology, EPFL, CH-1015, Lausanne, Switzerland

Anthropogenic compounds, called xenobiotics, are released into the environment in many different forms, including industrial pollutants and pesticides. Microorganisms are thought to be able to degrade these compounds; however, the reactions through which biodegradation occurs are largely unknown. We propose the use of the Biochemical Network Integrated Computational Explorer (BNICE), a computational framework for the discovery of biochemical reactions [1], in order to predict the biodegradation routes of xenobiotic compounds. The objective of this work was to demonstrate the applicability of this method and characterize the thermodynamic and cellular feasibility of the novel pathways generated by BNICE. A model of *E. coli* metabolism [2] was employed to determine how the novel pathways influence the existing metabolism of a reference organism. Using thermodynamic metabolic flux analysis (TMFA), thermodynamic constraints were combined with mass balance constraints and growth requirements of the cell to generate thermodynamically feasible flux profiles [3]. TMFA served as a tool to screen the thousands of novel pathways obtained from BNICE. As a result, we have obtained a set of pathways that are feasible under physiological conditions. We have compared the flux profiles for growth on glucose and growth on biphenyl in order to elucidate how growth using the xenobiotic as the sole carbon source impacts the central metabolic pathways. Additionally, we have explored how this analysis can be applied to *P. putida*, an organism more relevant to bioremediation.

References

1. Hatzimanikatis, V., *et al.*, Exploring the diversity of complex metabolic networks. Bioinformatics, 2005. 21(8): p. 1603-1609.

2. Reed, J.L., *et al.*, An expanded genome-scale model of *Escherichia coli* K-12 (iJR904 GSM/GPR). Genome Biology, 2003. 4: p. 54.1-54.12.

3. Henry, C.S., L.J. Broadbelt, and V. Hatzimanikatis, Thermodynamics-based metabolic flux analysis. Biophysical Journal, 2007. 92: p. 1792-1805.

Aspects of Screw Design for Wet Granulation Extrusion

J. Sun, M. R. Thompson
Department of Chemical Engineering, McMaster University, Hamilton, Ontario, Canada

ABSTRACT

This paper discusses our preliminary trials examining the particle size and particle shape of granules produced by different screw elements in a co-rotating twin-screw extruder. In general, conveying elements did not provide the necessary consolidation to allow large particles to grow in significant quantity, whereas the different mixing elements led to the large particles desired for tabletting. Granules from mixing elements varied from near-spherical to chains to grains to plate-like (disk) shapes depending on the type selected.

INTRODUCTION

In the development of twin-screw extrusion for granulation, current research has primarily focused on understanding the influence of process parameters and formulation variables on particle size and final tablet properties. Kleinebudde [1] studied the influence of water content and screw speed on the performance of the extruder. Remon's research group at Ghent University [2-4] studied the granulation of -lactose monohydrate in a twin screw extruder using an aqueous polyvinylpyrrolidone (PVP) solution as a binding agent. The authors found that granulation of lactose could be done with considerably less water in an extrusion process compared to a high shear mixer [2]. A group at Pfizer [5] studied wet granulation in a co-rotating twin-screw extruder and provided the first insight into the importance of screw design. They tested conveying elements and two mixing elements, kneading blocks (dispersive mixer) and discharging elements (distributive mixers). Kneading blocks tended to block the flow yet the discharging elements (otherwise known as comb elements) provided granulate discharge with evenly sized particles.

With twin screw extrusion as a new potential technology for particulate processing, it is likely that new screw element designs will have to be developed in order to satisfy the unique issues of processing granular matter; however, it is first necessary to establish the performance of available elements. To that ends, the aim of this paper is to discuss the granulation features of commercially available screw elements.

EXPERIMENTAL

MATERIALS - Lactose monohydrate was used as excipient (EMD Chemicals Inc., CA). Polyvinyl-pyrrolidone 100 GM (Alfa Aesar, M.W. 40000) was selected as the binder. No model active ingredient was included in the results discussed in this paper.

PROCEDURE - The experiments were performed in a 24 L/D Leistritz ZSE 27-HP co-rotating twin screw extruder using a simple screw profile, with an example shown in Figure 1. Different screw designs were examined which differed in the elements located 180 mm from the screw tips. The element types tested were: 1) conveying element with a 30mm or 40mm pitch, 2) kneading block with elements that were 30°, 60° or 90° offset from one another, and 3) discharging (comb) element, forwarding or reversing flow capacity. Figure 2 shows examples of each element type. The temperature of all zones was kept at 30°C by water cooling. The throughput rate was held constant at 2 kg/h while the screw speed was set to either 30 RPM or 80 RPM; the degree of channel fill was fixed as either 30% or 70% by these conditions.

The prepared binder solution consisted of 33 wt% PVP in distilled water. The binder solution was tumble

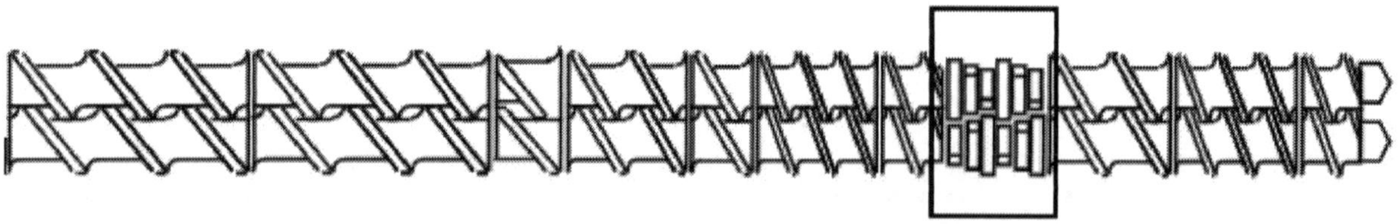

Fig 1. Example screw design with the highlighted section indicating the variable zone.

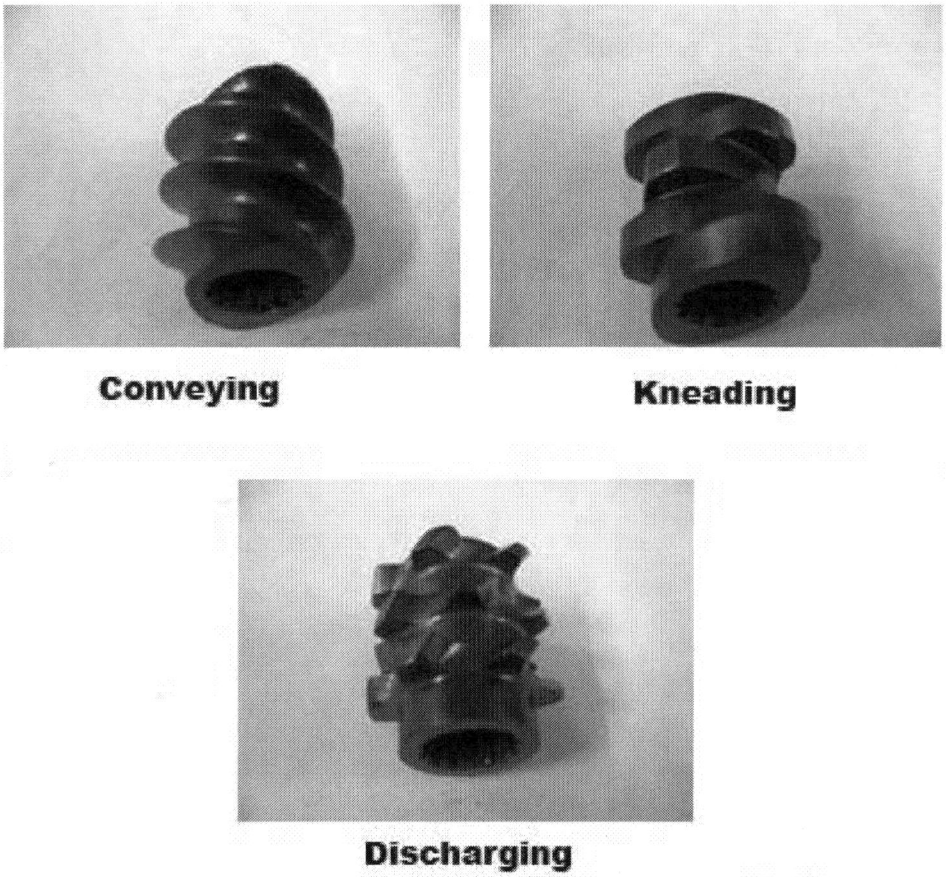

Fig 2. Examples of different screw elements

blended at 5-12% (w/w) with the excipient at 5-12% (w/w) for the trials. The pre-blend mixture was added to the extruder from a Brabender T20 twin-screw gravimetric feeder.

CHARACTERIZATION - After processing the extruded samples were air dried overnight at room temperature. For brevity the analysis discussed is limited to particle size distribution (PSD) and granule shape. The particle size distribution was determined by a mechanical sieve with screens of differing mesh size. The granule shape was inspected by reflective light optical microscope.

RESULTS AND DISCUSSION

For Sections 3.1-3.2 the binder solution was mixed at 7.5 wt% with the lactose, based on optimal operating conditions determined by Keleb [2, 3].

INFLUENCE OF SCREW DESIGN ON PSD

The effect that the different shear field of each element had on granulation, i.e. balancing particle growth versus rupture, was primarily assessed by particle size determination. The data disclosed pertains to the 30% degree of channel fill state and samples were taken at the element in question. Figure 3(a) compares the PSD for material entering the extruder from the feeder with samples from a 30mm or 40mm pitch conveying element. We see minor particle growth for the lactose/binder mixture in the 500-800μm region, preferentially for the element with the 40mm pitch. The greater free volume of the larger pitch will have comparatively lower shear on the particles and give less opportunity for rupture. Conversely, the kneading elements produced much larger particles as seen in Figure 3(b) which shows the PSD for the kneading block with elements at 30°, 60°, and 90° offset. Surprisingly there was little difference between the three blocks despite their different capacity for drag conveying versus pressure driven flow. Unlike the near-spherical granules found using the conveying elements, these particles formed plate-like shaped granules in the kneading elements. The lactose powder was compressed into thin disks within the nip region between two intermeshing elements. Finally, Figure 3(c) shows the PSD for the discharging element (two pairs of forwarding elements one after another or a single reversing pair). Both configurations resulted in particle growth, though the single pair exhibited more evidence of growth in balance with particle erosion. For the forwarding set (double pair), granules were nearly spherical whereas for the reversing pair, the exiting granules looked like reptating chains. With the reverse discharging element, too much backwards pressure occurred and the powder compacted too densely for the rupture mechanism

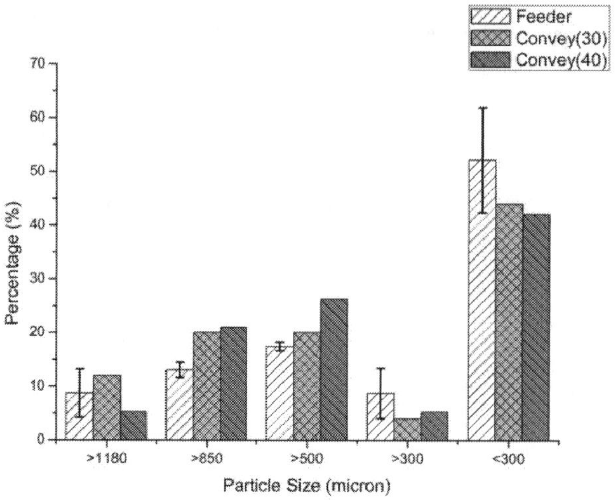

(a)

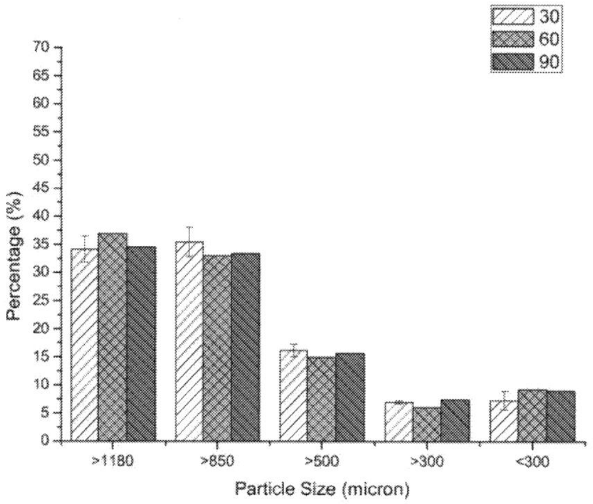

(b)

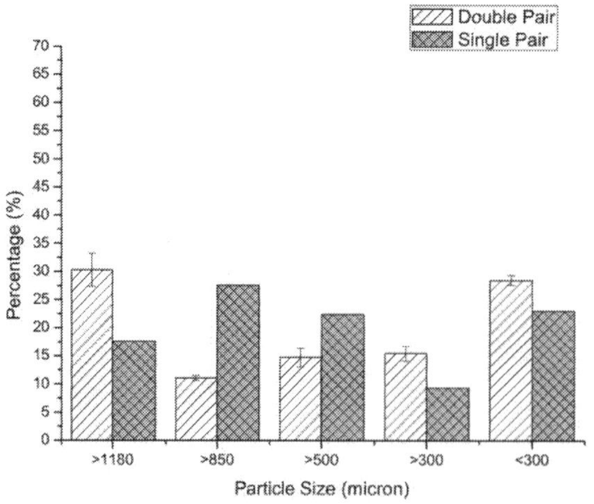

(c)

Fig 3. Effect of screw elements on PSD

to control particle size.

All of the subsequent discussion, the particle size distribution corresponded to the die exit of the extruder.

INFLUENCE OF SCREW SPEED ON PSD

For any twin screw extruder, both mass flow rate and screw speed are important controllable parameters since they affect the degree of channel fill which in turn affects the residence time and mixedness of a sample. In our case, the flow rate was fixed and screw speed was used to control the degree of fill. The PSD of granules exiting the machine at 30 RPM and 80 RPM for the 30mm conveying element or 60° offset kneading block are shown in Figure 4. For the conveying element, slower screw speed means more fill and more material exposed to the shear plane (i.e. the barrel); however, we see that higher shear (i.e. 80 RPM) was more important to induce particle rupture in this case. For the kneading block, the degree of fill bears little importance by itself since the element will almost always completely fill, and so we see higher shear by increased screw speed led to only a minor increase in particle rupture.

INFLUENCE OF BINDER CONCENTRATION ON PSD

The aqueous binder concentration is a critical factor on the morphology of granule. Granulation is only feasible within a certain water concentration range. Using the work of Keleb [2] to provide an operating window, four concentrations were examined in these experiments: 5%, 7.5%, 10%, and 12% (w/w). With 5% (w/w) concentration, granulation failed for all screw designs because of high frictional stresses between the lactose and the extruder surfaces. On the other hand, with 12% (w/w) concentration, the high proportion of water in the extrudate only resulted in a paste due to the efficient dispersion of the binder within the powder during extrusion. Figure 5 shows PSD for 7.5 and 10% (w/w) binder concentrations at 80 RPM. Across all screw elements examined, the higher binder concentration led to larger granules, in particular there was a notable loss of fine particles below 500 µm.

BINDER INJECTION VS PREBLENDING

As an alternative to the preblending method, the lactose was added by the feeder while the aqueous PVP solution was metered into the feed opening as liquid droplet by a peristaltic pump (Masterflex C/L). By droplet addition, granulation is controlled by immersion nucleation rather than distribution nucleation mechanism which results in poor binder spreading and large granules (>500 µm) in the both screw designs due to limited shear. Future work will compare the use of a dual manifold spray nozzle to the preblending method and droplet method.

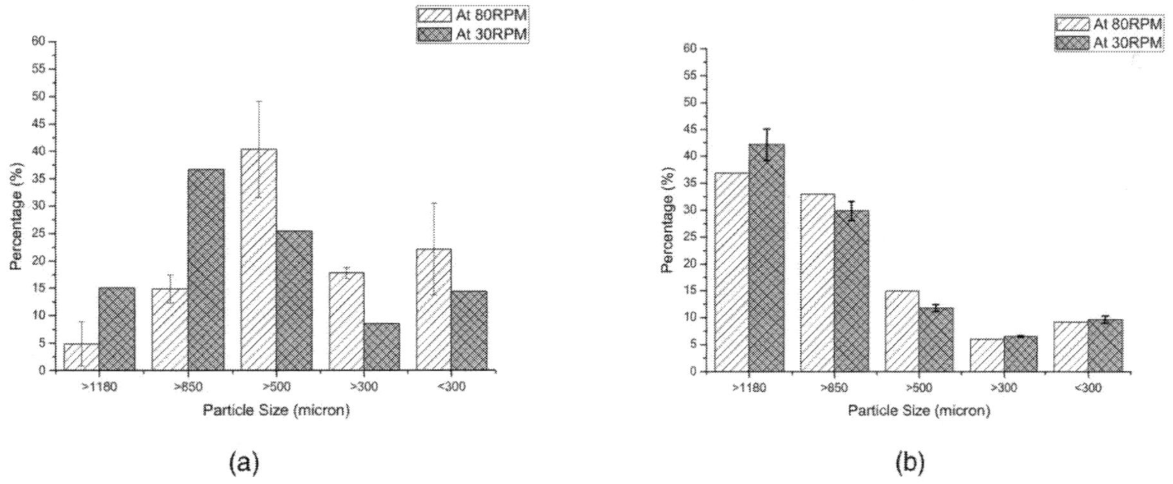

Fig 4. Effect of degree of channel fill on PSD, a) 30mm pitch conveying element and b) 60° offset kneading block.

CONCLUSION

The screw element selected determines the particle shape and particle size distribution of exiting granules in granulation extrusion. Granules varied from near-spherical to chains to grains to plate-like (disk) shapes depending on the screw element selected. Conveying elements did not provide the necessary consolidation to allow large particles to grow in significant quantity, whereas the different mixing elements led to the large particles desired for tabletting. Further experimentation will help to understand the rupture mechanism between the different mixing elements.

ACKNOWLEDGMENTS

The authors wish to extend their thanks to the Natural Sciences and Engineering Research Council (NSERC) of Canada for their funding of this work, and to the technical assistance of American Leistritz Extruder Corporation.

REFERENCES

1. P. Kleinebudde, H. Lindner, Experiments with an Instrumented Twin-Screw Extruder Using a Single-Step Granulation Extrusion Process. International Journal of Pharmaceutics, 1993. 94(1-3): p. 49-58.
2. E. I. Keleb et al., Continuous twin screw extrusion for the wet granulation of lactose. International Journal of Pharmaceutics, 2002. 239: p. 69-80.
3. E. I. Keleb et al., Twin screw granulation as a simple and efficient tool for continuous wet granulation. International Journal of Pharmaceutics, 2004. 273: p. 183-194.
4. A. Vermeire et al., Tablets Prepared by Single-Step Granulation/Tabletting: Interparticulate Binding Mechanism and Stability. Pharmaceutical Development and Technology,, 2005. 10: p. 397-403.
5. U. Shah, Use of a modified twin-screw extruder to develop a high-strength tablet dosage form. Pharmaceutical Technology, 2005. 29(6): p. 52-66.

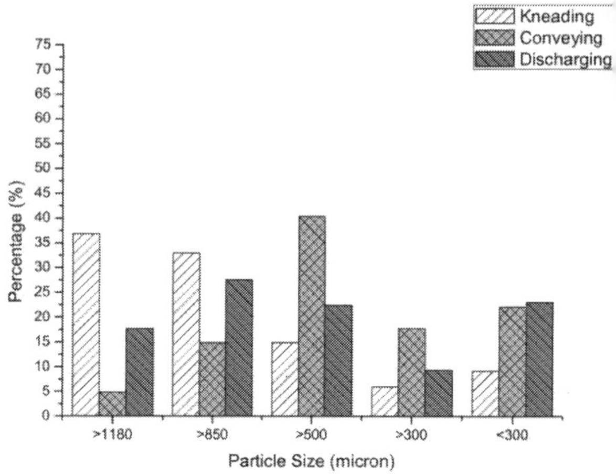

(a)

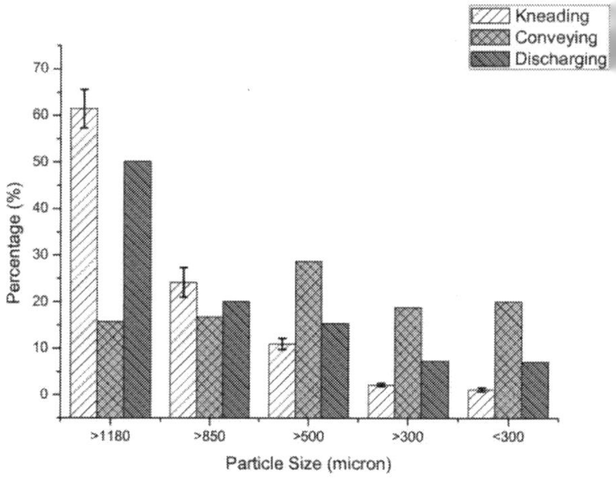

(b)

Fig. 5. Effect of aqueous binder concentration on PSD, a) 7.5 wt% binder and b) 10 wt% binder.

638b Continuous Production of Active Pharmaceutical Ingredients Using Recycle

Derek W. Griffin, Duncan A. Mellichamp, and Michael F. Doherty. UCSB, Department of Chemical Engineering, Santa Barbara, CA 93106

In the pharmaceutical and solids processing industry, it is important to consistently produce a product with specific properties such as size, shape, and morphology. Currently, the industry is dominated by batch processing where it can often be difficult to achieve the same product quality consistently from batch to batch. Thus, there has been a recent move to continuous processing because of the improved capacity usage, flexibility, and controllability offered by a continuous process compared to a batch one. A goal of solids processing is to make a product with a desired mean size and a certain size distribution. For certain active pharmaceutical ingredients (API's), such as paracetemol, the ideal product size is 20-50 microns to improve characteristics such as solubility and formulation. Since the size distribution from a crystallizer is generally larger than this target size there is the need for further processing, such as wet milling, to reduce the mean particle size. This study is aimed at the optimal operation of a continuous crystallization flowsheet in order to reduce the mean size of the crystal product and lower the operating costs of the entire plant. The proposed flowsheet is shown in Figure 1 which includes a cooling crystallizer followed by a solids separation step, such as a pusher centrifuge, that can be used for product classification. The solid product is taken off at a desired classification/cut size and part of the product stream is sent back the crystallizer as classified product removal. The remaining part of the solution is separated by filtration where the product is taken off and the mother liquor is recycled back to a dissolver to be mixed with the solute and make-up solvent. A fines destruction system (not shown) is also considered as part of the flowsheet which generally leads to higher levels of supersaturation, faster growth rates, and larger median crystal sizes at the expense of higher capital and operating costs. It is shown that it is possible to reduce the mean product size through the optimal operation of the mother liquor recycle and product classification system. The effects of various process operating parameters- such as crystallizer residence time, product classification cut size, and recycle flow rates- are investigated by tracking the crystal size distribution throughout the entire process.

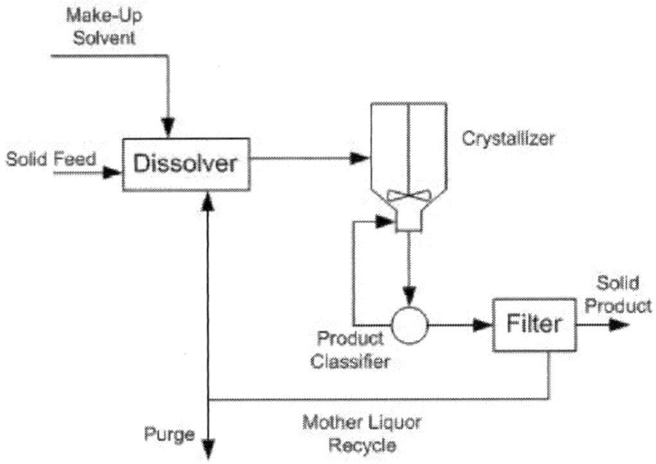

638c Development of Continuous Processes for the Production of Pharmaceutical Substances Using Heterogeneous Catalysis

Heidrun Gruber-Woelfler, Rafael J. P. Eder, Birgit Wilding, Iris Pflueger, Peter Feenstra, Gerburg Schider, Alexander Muhr, Christoph Kutschera, and Johannes G. Khinast. Institute for Process Engineering, Graz University of Technology, Inffeldgasse 21/A, Graz, A-8010, Austria

The pharmaceutical industry is currently attempting a transition towards continuous manufacturing in several areas, driven by the "PAT" and the "Critical-path" initiatives of the FDA (Federal Drug Administration). Currently, many catalysts used in pharmaceutical processing are homogeneous organometallic complexes, colloidal metals and biocatalysts. Tremendous achievements have been made in the implementation of these soluble complexes in industrial applications. However, use of homogeneous catalysts makes the implementation of continuous-flow processes more complex. Furthermore, homogeneous catalysts are usually in the same phase as the starting materials and products. Thus, one drawback is the difficulty to completely remove them from the reaction mixture, as mandated by regulatory bodies (FDA, EMEA = European Medicine Agency), in order to avoid metal carry-over. The immobilization of catalytic active compounds on solid supports is therefore a powerful method for overcoming the drawbacks of homogeneous catalysts, while maintaining their advantages.

In this work we present the development of heterogeneous catalysts and the investigation of their efficiency, long term stability and reusability. In a further step, these catalytic systems were implemented in continuous flow setups, which can be used for the production of pharmaceutical substances, such as

- preparation of amines by hydrosilylation of imines using heterogeneous Group 4 metallocenes

- application of heterogeneous Palladium-catalysts for the syntheses of arylpiperazines

- heterogeneous biocatalysis using enantioselective immobilized enzymes and

- particle formation using microfluidic devices

Preliminary results show that the developed setups lead to improved practicability and flexibility of the processes. Thus, these novel reaction systems constitute promising alternatives to existing batch applications.

638d Processing Pathway to Enhance Nanoparticle Stability

Varun Kumar and Robert K. Prud'homme. Chemical Engineering, Princeton University, A-217; Engineering Quadrangle, Olden Street, Princeton, NJ 08544

Formulation of hydrophobic drug compounds into nanoparticle form through the rapid precipitation technique has proven an attractive processing scheme for laboratory studies (Horn & Rieger, 2001; Hu et al., 2004; Johnson & Prud'homme, 2003) as well as for large scale operations (US Patent, BASF Aktiengesellschaft, 1999). In all of these approaches organic solvents must be employed in the processing to solubilize the active compound and then must be removed in the final dosage form for toxicological as well as nanoparticle stability reasons (Liu et al., 2007). The presence of organic solvent during the formulation often induces Ostwald ripening and secondary crystallization, which lead to the coarsening of the particle size distribution. Normal processing schemes include dialysis, which can require hours or days (Kim & Lee, 2001; Shuai et al., 2004) for solvent removal are too slow, and lyophilization is problematic for relatively high concentrations of organic solvent.

In this study we demonstrate the use of flash evaporation to rapidly remove tetrahydrofuran (THF) from a mixed solvent stream used to produce block copolymer stabilized nanoparticles by Flash NanoPrecipitation (Johnson & Prud'homme, 2003). The effectiveness of the solvent removal by flash evaporation is, in part, due to the strong non-ideality in the THF:water vapor-liquid equilibrium. The application has been demonstrated for the nanoparticle-formulation of model hydrophobic solute, β-carotene stabilized by PEO-b-PS (3k-b-1k) block copolymer. The dependence of stability of nanoparticles on the fraction of solvent present and degree of supersaturation have been analyzed.

Vacuum flash evaporation is especially well suited as the solvent removal process coupled with nanoparticle formation by Flash NanoPrecipitation. The process can easily scale from the laboratory-scale preparation of research samples, to large scale production.

638e Application of Data Reconciliation, Gross Error Detection and Real Time Optimization for Continuous Pharmaceutical Manufacturing

Michelle N. Chaffee, Gintaras V. Reklaitis, and Venkat Venkatasubramanian. Chemical Engineering, Purdue University, 480 Stadium Mall Drive, West Lafayette, IN 47907

Today pharmaceutical manufacturing still consists mainly of batch unit operations, which offer versatility and a means of tracking. However, recent focus has been on shifting the pharmaceutical manufacturing into a more continuous process in an effort to reduce the manufacturing costs and cycle times. Two of the automation tasks that must be addressed for effective continuous manufacturing are process monitoring and control. Monitoring and control are crucial to ensure consistent product quality and process economy. In order to operate at or near optimal conditions, an integrated control strategy which considers the entire production line is required.

The integrated control strategy will be encompassed in a real time process management (RTPM) system, which consists of two subsystems – real time optimization (RTO) and anomalous events management (AEM). These two subsystems will work together to identify gross errors, detect and diagnose faults, estimate and update model parameters, and suggest appropriate action to take to maintain optimal conditions all in real time. For the optimization, steady-state and, as appropriate, dynamic real time optimization (Tosukhowong, Lee et al. 2004) will be used in order to be sure that the process operates at optimal or near optimal conditions throughout the process life cycle.

In this work, an initial reduced set of functionalities is being explored for the monitoring and control of continuous pharmaceutical manufacturing. These functionalities include the use of data reconciliation, gross error detection and steady-state real time optimization. There have been numerous applications of these concepts in the past, in particular: the use of RTO in the petrochemical industry, which have proven to be successful (Yoon and Mijares 1996); an application of RTO to the pulp mill benchmark problem (Mercangöz and III 2008); and the use of RTO in an off-gas distribution system for iron and titanium (Woodward, Srinivasan et al. 2007). However, the application to continuous pharmaceutical manufacturing has not been developed to its fullest, especially for drug finishing steps such as a tableting line. This issue is addressed in this work by applying an integrated control strategy, which includes process models for optimization and control, to a continuous tablet manufacturing line – consisting of continuous feeders, blender, roll compactor, milling, and tablet press. We will present results of this approach for industrially inspired examples of anomalous events such as changes in the raw material properties. These feed changes may require changing the set points of the key variables associated with several of the downstream unit operations in order to insure that the tablet strength and other product properties remain within acceptable limits. The optimal values of these set points are thus determined be the RTO block.

References:

Mercangöz, M. and F. J. D. III (2008). "Real-time optimization of the pulp mill benchmark problem." Computers and Chemical Engineering 32: 789-804.

Tosukhowong, T., J. M. Lee, et al. (2004). "An introduction to a dynamic plant-wide optimization strategy for an integrated plant." Computers and Chemical Engineering 29: 199-208.

Woodward, L., B. Srinivasan, et al. (2007). "Real-time optimization of an off-gas distribution system of an iron and titanium plant." Computers and Chemical Engineering 31: 384-389.

Yoon, S. and G. Mijares (1996). "Real-time optimization boosts capacity of Korean olefins plant." Oil and Gas Journal 94(25): 36-41.

638f Analysis of Operating Conditions of a Continuous Powder Mixer Using Pept

Patricia M. Portillo[1], Andy Ingram[2], Jonathan P.K. Seville[3], M.G. Ierapetritou[4], and Fernando J. Muzzio[1]. (1) Chemical and Biochemical Engineering, Rutgers University, 98 Brett Road, Piscataway, NJ 08854, (2) Department of Chemical Engineering, University of Birmingham, Edgbaston, Birmingahm, United Kingdom, (3) School of Engineering, University of Warwick, Dean's Office, Coventry, CV4 7AL, United Kingdom, (4) Chemical & Biochemical Engineering Department, Rutgers University, the State University of New Jersey, 98 Brett Road, Piscataway, NJ 08854

In this work we examine the trajectory of particles within a continuous powder mixer using PEPT (Positron Emission Particle Tracking). Apart from experimental work examining mixing performance, no previous work exists investigating the particle trajectory from continuous powder mixing processes. The benefit of this approach is that the particle movement along the vessel can be measured non-invasively. PEPT has been reviewed in several existing papers and in this work we briefly summarize the method. A more detailed explanation of the technique can be found in Parker et al. [1] and in Jones et al. [2]. PEPT tracks a single radioactive tracer along a main flow of Lactose as a function of time. During radioactive decay, a high-energy positron is emitted, an electron nearby annihilates the positron, and two almost collinear gamma rays are generated. The PEPT camera consists of two symmetric plates that detect the gamma rays between them. If both gamma rays are absorbed by a detector element on each plate, then two lines can be drawn between the detector elements (one line for each gamma ray). After filtering scattered gamma rays, the intersection of the multiple pairs of lines gives the location of the tracer.

The continuous mixer examined in this work is manufactured by Buck Systems in Birmingham, England. The mixer is 0.31 meters long and the radius is 0.035 meters. The continuous mixer was inclined at an upward angle of 17„a. The mixer's rotation rate spans from 17 RPM to 200 RPM. The three agitator speeds examined in this work were 16, 75, and 170 rpm.

The effect of impeller rotation rate, powder flowrate, and powder cohesion on the particle trajectory, dispersive axial transport coefficient, and residence time are examined. It was found that the impeller rotation rate and powder hold-up affected the particle trajectory and residence time. Residence time is used to determine the strain and is an indirect indicator of the axial transport velocity the particle experiences within the vessel at varying processing conditions and bulk material properties. Several particle trajectories were recorded for each set of operating conditions (speed, powder, and flowrate). The average residence time was calculated using all trajectories. The findings were consistent with our previous experimental work (Portillo et al. [4]) using Danckwerts' RTD approach [3]. Experimental work already existed measuring residence time within a continuous mixer. However, the local residence time along the vessel axis had not been explored. We found that particles move along the axial length at fairly the same pace everywhere, even for the case that the vessel was inclined with a positive slope.

Cohesion is an important parameter affecting powder flow. The two materials examined were both common pharmaceutical ingredients that varied in cohesion: edible lactose (Dairy Crest) and free flowing lactose (Mallinckrodt). The effect of increasing cohesion (measured using the flow index, dilation, and the Hausner ratio) was longer average residence times. Interestingly, differences between the residence time profiles for the two materials, is greater at lower speeds and negligible at high speeds. Using the concepts of dispersion coefficient and Einstein's law, axial displacement was investigated. Experimentally the axial dispersion coefficient was related to residence time, consistent with theoretical derivations of the axial-dispersion model. Cohesion did not affect the axial dispersion coefficient significantly, but had considerable effects on the residence time and on the total particle path length.

Einstein [5] defined path length as the total distance covered by particles from the initial position (the entrance) to the final position. The path length is important for two reasons: (1) it provides a basis for measurement and prediction of the axial transport rate at the given fill level or powder hold-up; and (2) particle displacement is possibly impacted by electric charge which the particles may gain as they come in contact with other particles or the vessel. The time step interval is an important factor to consider, the work presented utilized an approximate time step of 0.017s, however we found that adjusting the time step altered the particle path length estimation. In particular a lower total length is obtained for longer time steps. Statistically it was found using an ANOVA that the total path length was shown to be affected by both cohesion and impeller rotation rate.

In summary, the effects of impeller rotation rate on the residence time, axial dispersion, and path length are complex. Therefore, in observations where the influence was questionable statistical analysis was applied in order to verify significance. We found that greater speeds resulted in lower residence times, higher axial dispersions, and lower total path length. The effect of powder hold-up was a bit more complicated since increasing the hold-up predominantly increased the residence time. Residence time showed an exponential trend with axial dispersion and a linear function with path length. In terms of the estimated particle total path length, longer path lengths occur at greater powder hold-ups. The effect of cohesion on the particles path length was highly significant. However, slightly higher axial dispersions were observed for less cohesive powders but this result was statistically not significant.

References:

[1] Parker D.J., Dijkstra A.E., Martin I.T.W., Seville J.P.K., 1997, Positron emission particle tracking studies of spherical particle motion in rotating drums, Chemical Engineering Science, 52, 13, 2011-2022.

[2] Jones J.R., Parker D.J., Bridgwater J., 2007, Axial mixing in a ploughshare mixer, Powder Technology, 178, 73-86.

[3] Danckwerts P.V., 1953, Continuous flow systems: Distribution of residence times, Chemical Engineering Science, 2, 1, 1-13.

[4] Portillo P.M., Ierapetritou M.G., and Muzzio F.J., 2008, Characterization of continuous Convective powder mixing processes, Powder Technology, 182, 368-378.

[5] Einstein, H.A., 1937. Bedload transport as a probability problem. Ph.D. Dissertation (English translation: In: H.W. Shen (Ed.), 1972. Sedimentation, Water Resources Publications, Fort Collins, CO, Appendix C. 105 pp.).

640a Polymeric Carrier-Mediated Intracellular Delivery of Phosphatidylinositol Trisphosphate to Overcome Insulin Resistance

Ilana Kachko[1], Tamar Traitel[1], Raz Jelinek[2], Assaf Rudich[3], and Joseph Kost[1]. (1) Chemical Engineering, Ben-Gurion University of the Negev, Ben Gurion University of the Negev, Beer-Sheva, Israel, (2) Chemistry, Ben-Gurion University of the Negev, Ben Gurion University of the Negev, Beer-Sheva, Israel, (3) Clinical Biochemistry, Ben-Gurion University of the Negev, Ben Gurion University of the Negev, Beer-Sheva, Israel

Background and aims: Production of phosphatidylinsoitol-3,4,5-trisphosphate (PIP3), induced by insulin is a major signaling step in the induction of insulin's metabolic actions in muscle and fat, and is impaired in various states of insulin resistance. Intracellular delivery of PIP3 into cells might therefore constitute an approach for bypassing such a signaling defect. However, this approach is challenged by the need to overcome permeability barriers through the plasma membrane, and to introduce the agent in well-defined spatial-temporal coordinates corresponding to endogenous PIP3 generation. Exogenous PIP3 may be introduced into cells using polycationic carriers which should mask its negative charges. However so far, these approaches only partially mimicked insulin's effects in normal, insulin sensitive cells, and displayed poor efficiency and low reproducibility. The overall aim of this work was to assess the intracellular transport of exogenous PIP3 to overcome cellular insulin resistance. **Materials and Methods:** Membrane-complex interactions and cellular localization were verified using biomimetic liposomal (lipid/polydiacetylene) membrane model and in L6 muscle cells in culture with fluorescent-PIP3 and live-cell microscopy. The ability of PIP3-carrier to induce insulin signaling and metabolic effects was evaluated in L6 myoblasts and in 3T3-L1 adipocytes electroporated with GFP-GLUT4myc plasmids. **Results:** A polymeric carrier, polyethyleneimine (PEI), was utilized at different molecular weights and branched/linear backbone to generate efficient electrostaticaly bound PIP3-carrier complexes. Fluorescent liposomes surface perturbation and ESR (Electron Spin Resonance) analyses revealed that PIP3-PEI formed complexes and penetrated the lipid bilayer. Branched PEI (25kDa) carrier was more efficient in membrane internalization than linear PEI of the same molecular weight. Maximal liposome-complex interaction occurred at 40 min. Live-cells kinetic studies revealed that the branched PEI-PIP3 complex enabled the retention of PIP3 at the cell periphery as compared to PIP3 alone. PIP3 delivered with PEI was found biologically available for binding to its intracellular ligand (GRP-PH) and for generating signaling effects, resulting in increased PKB phosphorylation, higher than that observed with neomycin as a carrier. Moreover, PIP3-carrier exposure increased Glut4 translocation and externalization in 3T3-L1 adipocytes. **Conclusions:** We demonstrated a polymeric system for exogenous PIP3 delivery that can induce insulin signaling and metabolic effect in muscle and fat cells based on biomimetic liposomes and cellular systems. Inducing insulin actions by intracellular PIP3 delivery in insulin resistant cells will shed light on the potential therapeutic use of this technology.

640b Novel Multifunctional Nanocarriers for Imaging and Therapy Enable Live Monitoring of Chemotherapy in a Rat Breast Tumor

Efstathios Karathanasis[1], Leslie Chan[1], Sri Balusu[1], Ananth Annapragada[2], Ioannis Sechopoulos[3], and Ravi Bellamkonda[1]. (1) W.H. Coulter Department of Biomedical Engineering, Georgia Institute of Technology, 313 Ferst Dr NW, Atlanta, GA 30332, (2) Health Information Sciences, University of Texas Health Science Center at Houston, 7000 Fannin St Suite 600, Houston, TX 77030, (3) Radiology, Emory University, 1701 Uppergate Drive NE Suite 5018, Atlanta, GA 30322

Introduction: Nanoscale control in fabricating nanocarriers allows for multifunctional carriers that can potentially enable personalized cancer therapy that facilitates diagnosing, treating and monitoring the progress of treatment for each individual patient. Our design criterions of such a multifunctional nanocarrier for breast cancer therapy included (1) high payload of iodinated contrast agent to allow visualization using x-ray imaging, (2) capability to probe tumor vascular permeability status by detection and quantification of the intratumoral extravascular accumulation of the agent using mammography, and (3) additional payload of a chemotherapeutic for treatment.

Methods: To address this need for monitoring therapy, we developed a long-circulating 100nm-scale liposomal nanocarrier co-encapsulating an X-ray contrast agent (iodixanol) for imaging and a chemotherapeutic (doxorubicin) for treatment. We tested the nanocarrier in a rat breast tumor model developed by inoculation of mammary adenocarcinoma cells (13762 MAT BIII) into the right flank of the animals. Following systemic administration, the intratumoral, extravascular deposition of the agent was imaged with a clinical digital mammography unit while the tumor response to the treatment was evaluated by measuring the tumor growth rate and survival rate of the animals.

Results: The multifunctional nanocarrier had diameters of 105 nm and contained 140 mg/mL iodine and 3 mg/mL doxorubicin (all encapsulated). In vitro cytotoxicity studies showed that the multifunctional nanocarrier co-encapsulating doxorubicin and iodine was equally potent as the conventional liposomal doxorubicin. Following administration of the agent at a dose of 350 and 7 mg/kg body weight of iodine and doxorubicin respectively, the mammography-based imaging allowed detection and quantification of the agent's uptake by the tumor. It was observed that specific tumors that exhibited high uptake of the nanocarrier as visualized by imaging were mostly benefited from the treatment showing low tumor growth and long survival.

Conclusions: In this work, a multifunctional nanocarrier co-loaded with an imaging agent and a chemotherapeutic was developed that satisfied the design criterions. We demonstrated that the nanocarrier was capable of predicting and monitoring the therapeutic outcome of the breast cancer treatment using mammography. *Such determination of the amount of chemotherapeutic reaching and remaining in the tumor during each cycle of treatment can potentially ensure an optimized personalized therapy regimen.*

640c Pharmacokinetic Model for the Distribution of Radiometal-Chelator-Dendrimer-PNA-IGF1 Analog Nanoparticles In Tumor-Bearing Mice

Armin W. Opitz[1], Nariman V. Amirkhanov[2], Kaijun Zhang[3], Mathew L. Thakur[3], Eric Wickstrom[4], and Norman J. Wagner[5]. (1) Chemical Engineering, University of Delaware, 150 Academy Street, Newark, DE 19716, (2) Laboratory of Nucleic Acids Chemistry, Institute of Chemical Biology & Fundamental Medicine, Novosibirsk, Russia, (3) Radiology, Thomas Jefferson University, Philadelphia, PA 19107, (4) Biochemistry & Molecular Biology, Thomas Jefferson University, Philadelphia, PA 19107, (5) Department of Chemical Engineering, University of Delaware, 150 Academy St., Newark, DE 19716

We hypothesized that overexpression of a mutated *KRAS* oncogene in a human pancreas cancer xenograft could be detected by external scintigraphic imaging of a *KRAS* peptide nucleic acid (PNA) hybridization probe, radiolabeled with a polydiaminopropanoyl (PDAP) dendrimer of ^{111}In ions, endocytosed into cancer cells that overexpress insulin-like growth factor 1 receptor (IGF1R) via an IGF1 peptide retro-inverso fragment, D(Cys-Ser-Lys-Cys). We assembled this molecular imaging nanoparticle (MIN) on a polymer support by solid-phase coupling starting from the C-terminus of the IGF1 peptide, purified the MIN by reverse-phase HPLC, and analyzed the MIN by mass spectrometry. The in vivo distribution of the *KRAS* MIN labeled with ^{111}In was established scintigraphically at 4 and 24 hours using human pancreas cancer xenografts in immunocompromised mice. The MIN distributed normally to the kidneys, livers, tumors, and other tissues. A two-compartment model of the tissue distribution was used to calculate an elimination rate constant of 1.64 ± 0.09/hr. The forward rate calculated for blood to tissue distribution was 0.26 ± 0.14/hr. The backward rate from tissue to blood was calculated as 0.089 ± 0.032/hr. Further it was determined that the MIN tends to reside in the vascular space of the mouse. An oligonucleotide pharmacokinetic model adapted from the literature fitted the distribution of the MIN in several tissues and in the blood by empirically adjusting the injected dose able to enter tissues and bind to its mRNA target at only 10-20% of the MIN concentration in blood. We speculate that the majority of the MIN is filtered out of the blood by the kidneys, and that a small fraction binds to the circulating insulin-like growth factor binding proteins (IGFBPs), which then aid in distributing the MIN into the different organs and tumor. This work was supported in part by NCI contract N01 CO27175 to E.W.

640d Co-Delivery of p53 and Paclitaxel with Cationic Oligopeptide Amphiphilic for Cancer Therapy

Nikken Wiradharma, Drug Gene and Protein Delivery; Chemical and Biomolecular Engineering, Institute of Bioengineering and Nanotechnology; National University of Singapore, 31 Biopolis Way, #04-01, The Nanos, Singapore 138669, 21 Lower Kent Ridge Road, Singapore 119077, Singapore, Singapore, Yen Wah Tong, Department of Chemical and Biomolecular Engineering, National University of Singapore, 4 Engineering Drive 4, Singapore, 117576, Singapore, and Yi Yan Yang, Drugs, Proteins and Genes Delivery, Institute of Bioengineering and Nanotechnology, 31 Biopolis Way #04-01, Singapore, Singapore.

Non-viral gene delivery has been proposed to carry therapeutic genes into mammalian cells with loss of gene functions. In this study, we focus on delivering p53-encoding plasmid DNA into human breast cancer PC3 cell lines, together with paclitaxel to achieve synergistic effects. This cell line has been reported to have loss of p53 function mutation, and as a control study, we provided the same study with human liver carcinoma HepG2 cell lines, which was known to have no such mutation. Cationic amphiphilic oligopeptide, containing Alanine, Phenylalanine, Histidine, and Lysine residues was designed and synthesized in this study to form self-assembled core-shell nanoparticles. By tuning the number of alanine and phenylalanine residues (in the hydrophobic core), we could adjust the CMC to be between 0.05 to 1.1 mg/mL, with the hydrodynamic radius ranging between 100-400 nm. The nanoparticles were able to encapsulate paclitaxel and at the same time bind p53 DNA on the surface of the nanoparticles, without significantly affecting the integrity of the particles. Paclitaxel encapsulation efficiency and loading levels were tested to be around 10.3% and 2.1%, respectively. The particle size of the peptide/DNA complexes were analysed by dynamic light scattering to be around 100-200 nm, while the zeta potential were characterized to be around +20mV. Cell cycle analysis of PC3 cell lines after the treatment with this co-delivery system has shown that synergistic effect could be achieved in arresting cell growth, while that of HepG2 cell lines did not show any synergy. This system may provide a smart strategy to induce cellular senescence in cancer cells with known p53 mutations by harvesting on the synergistic effects of p53 with anti-cancer drugs.

640e Comparison of In Vivo and In Vitro Delivery of Oligonucleotide Containing Lipopolyplexes for Cancer Therapy

Xulang Zhang[1], Chee Guan Koh[2], Bo Yu[2], Longzhu Piao[3], Shujun Liu[4], Guido I. Marcucci[4], Robert J. Lee[5], and J. James Lee[6]. (1) NSF Center for Affordable Nanoengineering of Polymeric Biomedical Devices, The Ohio State University, 140 West 19th Avenue, Columbus, OH 43210, (2) Department of Chemical and Biomolecular Engineering, The Ohio State University, 140 West 19th Avenue, Columbus, OH 43210, (3) Division of Pharmaceutics,, College of Pharmacy, The Ohio State University, 542 PARKS HALL, 500 W 12TH AVE, Columbus, OH 43210, (4) NCI Comprehensive Cancer Center (CCC),, The Ohio State University,, 2001 POLARIS PKWY, Columbus, OH 43210, (5) College of Pharmacy, The Ohio State University, 542 Parks Hall, 500 West 12th Avenue, Columbus, OH 43210, (6) The Ohio State University, Room 125A, Koffolt Labs., 140W. 19th Ave., Columbus, OH 43210

Abstract

Transferrin (Tf) conjugated lipopolyplexes (LPs) carrying oligonucleotides (ONs) such as antisense ODN and siRNA were synthesized and evaluated in Tf receptor positive cells and in a murine cancer xenograft model. In vitro, Tf-LP was more effective in down regulation of Bcl-2 in cells than non-targeted LP and free ON. The LPs were then evaluated in athymic mice carrying subcutaneous xenograft tumors by i.v. injection. Both treatment groups showed Bcl-2 down regulation by immunohistochemical staining and Western blotting. However, the Tf-LP was less effective than free ON, in contrast to the findings in vitro, probably due to the limited diffusion rate of LPs following extravasation. To investigate this possibility, an in vitro 3D cell cluster model was created by microcapsule technique and treated with both LPs and free ONs. The results indicated limited penetration of LPs into the cell cluster. The therapeutic efficacies of Tf-LP-ODN and free ODN were determined in the xenograft cancer model. Tf-LP-ODN showed slower plasma clearance, higher AUC, and greater accumulation in the tumor compared to free ODN. Tf-LP-ODN was found to be more effective in tumor growth inhibition and prolongation of survival than free ODN. This was associated with increased spleen weight and IL-12 production in the plasma. These results suggested a role of CpG effect to the immune system in the therapeutic response with the Tf-LPs.

640f Development of a Dry Powder Aerosol for the Dispersion and Eradication of Respiratory Biofilms

Emily Thomas, Department of Chemical and Biochemical Engineering, The University of Iowa, Iowa City, IA 52242, Buffy Stohs, Division of Pharmaceutics, The University of Iowa, Iowa City, IA 52242, and Jennifer Fiegel, Department of Chemical and Biochemical Engineering / Division of Pharmaceutics, The University of Iowa, Iowa City, IA 52242.

Biofilms caused by the bacteria Pseudomonas aeruginosa are a leading cause of persistent lung infections in patients who have chronic lung disease or are otherwise immunosuppressed. Mucoid strains of P. aeruginosa secrete polysaccharides to form an extracellular matrix which binds the bacteria together and protects them from antimicrobial agents and the patient's immune system. Because of this afforded protection by the biofilm matrix, current treatments are only minimally effective and result in alleviation of some symptoms but not a cure of the infection. Therefore, we are developing a new approach for the treatment of respiratory biofilms by combining a dispersion compound to break up bacterial colonies with antibiotics into a dry aerosol. These systems are aimed at enabling the use of traditional antibiotics for effective pathogen killing of biofilm-forming pathogens.

Inhalable dry powders have been formulated by spray drying aqueous solutions containing Ciprofloxacin HCl (antibiotic), glutamic acid (dispersion compound) and L-leucine (excipient). The effects of formulation and spray-drying process parameters on the aerodynamic diameter and chemical activity of the powder were examined using central composite design. Powder geometric diameter was determined by laser diffraction and particle density was estimated by tap density measurements. Powders were formulated with theoretical aerodynamic diameters, as estimated by the equation $d_{aero}=d_{geom}\sqrt{\rho}$, that ranged from 2-8 μm, appropriate for efficient deposition within the respiratory tract. The in vitro deposition profiles of the powders were evaluated using a Next Generation Impactor operated at 60 L/min. High deposition efficiencies (>40%) were observed for powders with optimized physical properties. Scanning electron microscopy showed that the particles have a collapsed spherical structure with a porous internal morphology. Powders were found to contain about 98% of the expected antibiotic load, indicating that the drug was not lost or harmed during the spray-drying process. Finally, the efficacy of the dry powder drug delivery system was tested on an in vitro respiratory biofilm model developed by culturing a mucoid producing strain of P. aeruginosa on Calu-3 human bronchial epithelial cells. Live/dead assay coupled with confocal microscopy showed that the drug delivery system was more effective at eliminating biofilms in vitro compared to traditional antibiotic treatment. This co-delivery system offers a new treatment strategy for bacterial biofilms which may improve the elimination of persistent infections in the lungs.

640g T Cell Proliferation Control with Small Molecule-Responsive RNA-Based Regulatory Systems

Yvonne Y. Chen[1], Christine E. Brown[2], Michael C. Jensen[2], and Christina D. Smolke[1]. (1) Chemical Engineering, California Institute of Technology, 1200 E. California Blvd., MC 210-41, Pasadena, CA 91125, (2) Division of Cancer Immunotherapeutics & Tumor Immunology, Beckman Research Institute, City of Hope National Medical Center, 1500 East Duarte Road, Duarte, CA 91010

Immunotherapy strategies utilizing engineered T cells hold great promise as treatments for diseases that are otherwise intractable to the natural immune system. However, the safety and efficacy of cellular immunotherapeutics critically depend on our ability to control the timing and extent of the proliferation of engineered cells used as therapeutic agents. Here we report on the development of a novel RNA-based strategy for small molecule-controlled transgene regulation and its application to T cell genetic engineering platforms for cancer immunotherapy.

Cytotoxic T Lymphocytes (CTLs) express highly specific receptor molecules that recognize disease markers and serve as the body's natural guard against infections. However, many diseases, notably cancer cells, have developed elaborate tactics to avoid detection by CTLs and evade immune responses. To address this challenge, T cells have been engineered to express synthetic receptors with specificity toward unnatural substrates, such as tumor-restricted antigens, thus redirecting the immune system against novel targets. While it has been demonstrated that engineered T cells can effectively target diseased cells, sustaining and regulating the growth of ex vivo expanded T cells in the tumor microenvironment remain major challenges. Our orthotopic model systems have demonstrated that without paracrine T cell growth factors, such as IL-2, CTL survival is of limited duration and insufficient for tumor control. Conversely, CTL genetic modification to constitutively express gamma-c cytokine transgenes such as IL-2 and IL-15 poses the risk of uncontrolled expansion. Therefore, a stringent regulatory system that is capable of fine-tuned control over growth factor production and the resultant T cell proliferation pattern is critical to the success of cellular immunotherapeutics.

We have developed a novel method to regulate T cell proliferation and survival by controlling cytokine transgene expression through small molecule-responsive RNA regulatory systems. The growth of IL-2 dependent CTLL-2 mouse T cells is controlled by the transgenic expression of cytokines such as IL-2 and IL-15, whose production levels are modulated by RNA-based regulatory elements implemented in the 3' untranslated region (UTR) of the cytokine transgenes. RNA regulatory elements are composed of three modular domains: (i) sensors consisting of RNA aptamers with high specificity for small-molecule ligands such as theophylline and tetracycline, (ii) actuators consisting of functional, noncoding RNA species such as microRNAs and self-cleaving ribozymes, and (iii) information transmitters consisting of rationally designed RNA sequences that transmit ligand-binding events from the sensor domain to changes in the activity of the actuator domain. By coupling small molecule-responsive RNA regulatory elements with target transcripts such as the egfp-t2a-il2 fusion gene, we construct systems in which transgene expression is stringently controlled and turned ON in the presence of the regulating small-molecule ligand. In a transient transfection system, we demonstrate ligand-dependent growth regulation through monitoring both fluorescence and cell culture viability. Stable integration of the regulatory systems into clonal T cell populations confirms their long-term ligand-responsive regulatory activities. Furthermore, we demonstrate the ability to fine-tune the response properties of these regulatory systems through sequence modifications targeted to the transmitter domains based on thermodynamic design strategies as well as combinatorial implementation of different classes of RNA-based regulatory elements. Progress is underway to examine small molecule control of the proliferation of engineered T cells harboring these regulatory systems in animal models, which will yield further insight into the in vivo functionality and possible improvements for RNA-based regulatory designs.

Coupled with technologies to generate new RNA sensors responsive to clinically relevant small molecule effectors, the platform reported here enables the design and construction of RNA-based regulatory systems tailored for applications that demand stringent control of cellular behavior and flexibility in the tuning of regulatory activities. This research highlights the exciting potential of new RNA-based regulatory systems in generating tightly-regulated, dynamic protein expression systems adaptable to pharmaceutical drugs with a high therapeutic index that are suitable for advanced clinical applications such as cellular therapeutics employing genetically modified T cells.

651a Application of Quality by Design Principles In the Model Predictions for Scale-up of Tablet Film Coating

Andrew Prpich[1], Mary T. Am Ende[1], Thomas Katzschner[2], Veronica Lubczyk[3], Holger Weyhers[4], and Georg Bernhard[3]. (1) Oral Products Center of Emphasis, Pfizer, Eastern Point Road, MS 8156-001, Groton, CT 06340, (2) Site Validation Committee, Pfizer Global Manufacturing, Heinrich Mack Str 35, Illertissen, 89257, Germany, (3) RFT Program Office, Pfizer Global Manufacturing, Heinrich Mack Str 35, Illertissen, 89257, Germany, (4) Site Leadership, Pfizer Global Manufacturing, Heinrich Mack Str 35, Illertissen, 89257, Germany

Two fundamental tablet film coating models were employed through a quality by design (QbD) approach to determine the operating parameters for scale-up of Varenicline IR (Chantix®/Champix®) tablet production in Illertissen, Germany. Due to high product demand, the tablet film coating production capacity was expanded from the existing Glatt GC1250 and IMA Perfima 200 coaters (~115 kg and 160 kg pan load, respectively) to a Glatt GC1500 coater (~350 kg pan load). The models were used in this study to establish an acceptable range of process parameters in the new coater to match the proven acceptable range (PAR) of operating conditions in the existing coaters. The established process parameters were then used to prioritize the experimental design to minimize the number of required trial runs and focus on optimization rather than validation.

A thermodynamic film coating model was used to determine the range of inlet air temperatures required to match the exhaust air conditions across the coaters at the target spray rates (600 – 1000 g/min) and pan air flow rate (6000 m3/hr). A film coating atomization model was used to determine the atomizing air pressure set point to achieve a mean coating solution droplet diameter of approximately 26 microns across the three coaters. Based on the model predictions, the following equipment set points were determined; (1) upper and lower inlet air temperatures of 75°C and 65°C, respectively, to maintain environmental similarity during film coating, and (2) atomization air pressure of 3.0 bar at the 600 g/min spray rate and 3.5 bar at the 1000 g/min spray rate to attain the desired droplet diameter. The recommendations were provided to the commercial site to guide the design of their scale-up trial runs. The results from the actual trials were compared against the corresponding model predictions to assess the validity of the guidance provided based on the two models. The results from this study illustrate how the thermodynamic and atomization film coating models can be used to improve the efficiency of the tablet film coating scale-up process.

651b Use of Molecular Modeling In the Formulation of Suspensions

Jamie T. Konkel, Medication Delivery Global R&D, Baxter Healthcare, Route 120 and Wilson Rd, WG3-3S, Round Lake, IL 60073 and Allan S. Myerson, Chemical Engineering, Illinois Institute of Technology, Chicago, IL 60616.

Presently, the formulation of stable crystalline nanosuspensions is largely an empirical process. Each new drug compound is screened with various surfactant systems to find the most stable formulation. This process can be time-consuming, and can require a prohibitive amount of drug in early development phases. This work was undertaken to develop a molecular modeling approach to surfactant screening, which could identify stabilizing surfactant systems for a particular drug, based on its crystal structure.

We have screened several model pharmaceutical drug crystals with surfactants, including the nonpolar surfactant, Polysorbate 80. The nanosuspensions were prepared by high-pressure piston gap homogenization of the drug in an aqueous surfactant mixture. Aliquots of each suspension were stress-tested to rapidly assess the physical stability of the formulation. The various combinations of model drugs and surfactants provided suspensions with a wide range of stabilities to model.

The molecular modeling was performed in Materials Studio (Accelrys). For each model drug, the adsorption of both the surfactant and a molecule of the drug were modeled on a face of the drug crystal. We calculated the binding energy of each drug with these additives, and the ratio of the surfactant binding energy to that of the drug correlated well with the rank order stabilization of the crystalline suspensions. A ratio of approximately 3 correlated to moderate stability; a ratio larger than 3, correlated to excellent stabilization. For the drugs and surfactants with poor stabilization, the ratio was near 1.

The use of molecular modeling for preliminary formulation of suspensions offers a way to visualize the interactions of surfactants on the surfaces of crystal particles. It can be less time-consuming than empirical formulation screening, and does not use valuable and limited drug material. The technique could help to prioritize the formulations tested in empirical laboratory experiments, so that only the most promising surfactant systems would need to be screened to find a stable formulation.

651c Use of Pharmacokinetic/pharmacodynamic (PK/PD) Modeling to Aid Formulation Selection

Larissa A. Wenning[1], Eseng Lai[2], Tami Crumley[2], Tian-Quan Cai[3], Fang Liu[4], M. G. Waters[3], Shaun Fitzpatrick[5], and Julie Stone[1]. (1) Clinical PK/PD, Merck & Co, Inc, WP75B-100, 770 Sumneytown Pike, West Point, PA 19486, (2) Clinical Pharmacology, Merck & Co, Inc, (3) Cardiovascular Diseases, Merck & Co, Inc, (4) Biostatistics, Merck & Co, Inc, (5) Pharm R&D, Merck & Co, Inc

Background: Niacin (nicotinic acid) has beneficial effects on lipids and reduces cardiovascular events in patients with dyslipidemia. PK/PD modeling was applied to project the probability that 3 candidate extended release niacin formulations would have comparable efficacy on lipids compared to a reference formulation. For the purpose of this projection, the beneficial effects of niacin were assumed to directly correlate with reduction in plasma free fatty acid (FFA) concentration.

Methods: A single dose crossover study was conducted in 32 healthy subjects to compare the pharmacokinetic (PK) profiles of 3 test extended release niacin formulations to a reference formulation. An Emax model was used to relate niacin plasma concentrations to percent inhibition of lipolysis as follows: Effect = Emax*C/(C+EC50); where Effect represents percent reduction in plasma FFA levels, C is niacin plasma concentration, Emax was fixed to 100% based on in vitro data, and EC50 was fixed to 12.4 ng/mL, again based on in vitro data. The PK/PD model was applied to project whether the PK differences between the 3 test formulations vs. the reference formulation, if there were any, would have a clinically meaningful impact on lipid efficacy (changes in LDL and HDL cholesterol) of ER niacin, which was postulated to derive from reduction of plasma FFA level.

Results: All three test formulations had different niacin plasma PK profiles compared to each other and the reference formulation. Despite the differences in PK profiles, all three formulations were projected to have similar effects on plasma FFA levels compared to the reference formulation based on the above PK/PD model. Two of these formulations were carried forward into a subsequent clinical study which showed similar effects on LDL and HDL cholesterol for both formulations compared to the reference.

Conclusions: PK/PD modeling is a useful tool that can aid interpretation of PK differences between formulations and selection of optimal formulations.

COMPUTER AIDED MODELLING IN PHARMACOKINETICS AS A TOPICAL ISSUE

Michael Wulkow* and Regina Telgmann, CiT GmbH,
Oldenburger Str. 200, Rastede, 26180, Germany, m.wulkow[at]cit-wulkow.de

Modeling and simulation in pharmacokinetics has turned into the focus of pharmaceutical companies and academic research, driven by the emerging consensus that *in silico* predictions, combined with *in vitro* data, have the potential to significantly increase insight into pharmacokinetic processes. Pharmacokinetics is the study of the time course of drug and metabolite levels in different fluids, tissues and excreta of the body. This includes the investigation and understanding of the processes of adsorption, distribution, metabolism and excretion (ADME) at the different stages of the drug discovery process:

- Early drug discovery: prioritize drug candidates and indicate critical pharmacokinetic properties
- Preclinical drug design: analyze in vivo data from animal studies, predict the PK in other species and in man and support dose finding for first-time-in-man studies
- Clinical phases: use modeling to understand and interpret clinical data and extrapolate to different scenarios like dosing schemes.

The pharmacokinetic profile of a drug strongly influences its delivery to biological targets, thereby affecting its efficacy and potential side effects. To support in silico methodology adequately, software tools are needed which do not only facilitate the solving of equations but also take over the ballast of repeated implementation of known and structurally equal working steps without preventing the input of open modeling ideas. In fact, such software (that does not mean purely equation based standard solvers) has to allow the modeling of ideas that go beyond the current knowledge – otherwise it would not be called innovative. The reasons for this claim will be explained in the lecture.

At first, there is no single established model underlying PK, and it will remain out of reach in the foreseeable future. The medical benefits of a drug depend not only on its biological effect at the target protein but also on its life-cycle within the organism – from its absorption into the blood, distribution to tissue and its eventual breakdown or excretion by the liver and kidneys. Pharmacokinetics (PK) is the study of the drug–organism interaction and comprises a number of physiological phenomena, which are modeled differently (i.e. using different model complexity and thus parameter dependency) at different stages. For example, at early stages (when not much data are available for candidates under consideration) the protein binding might be described by a fuP, B:P dependent basic model. Later it might be replaced by detailed description based on plasma and interstitial albumin concentrations or any other new insight from actual research. This structure holds for all processes in a PK model. Typically, following processes are modeled:

- Convection of drug molecules by the blood flow
- Binding to macromolecules in plasma and interstitial space
- Distribution into tissue
- Diffusion or active transport across the cellular membrane
- Metabolism or interaction with metabolic networks or signaling pathways etc.

A chosen model will be a compromise between detailed mechanistic description and required quality of the input parameters. At early stages of drug discovery, frequently measured *in vitro* parameters are used to parameterize early PBPK models. Either the parameter can directly be used in the model, or relevant model parameters are estimated through mechanistic equations from the measured *in vitro* parameters. Typically, the knowledge and the quality of parameters increases along the drug discovery and development process so that adaptation of the model to the current knowledge and parameter quality is possible.

Next, a physiologically based pharmacokinetic (PBPK) whole *body model* is a special type of compartmental model, in which the compartments represent anatomical volumes, such as organs or tissues. The compartments are connected in an anatomically meaningful way, to simulate drug exchange via the blood flow. The conceptual representation of a 14 organ PBPK model is shown in Fig. 1. Each compartment is further subdivided into the four (or more) sub compartments (phases): erythrocytes, plasma, interstitial and cellular space (also shown in Fig. 1).

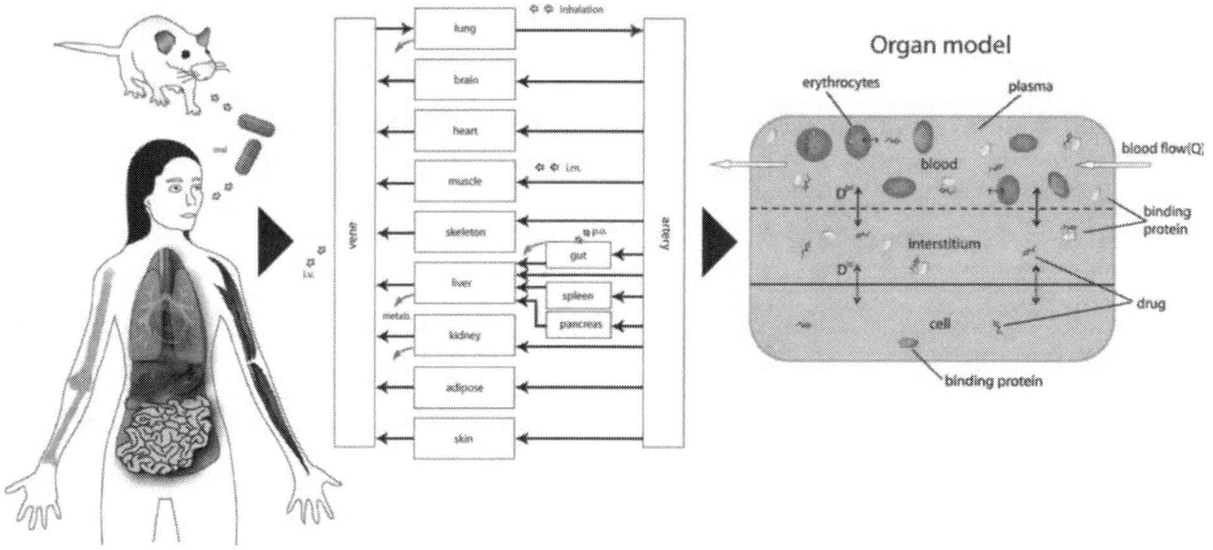

Figure 1: Organ structure of a physiologically based pharmacokinetic model

Such a given topology combined with a selection of models for the basic processes (one for each physiological process) builds a description of the PBPK model.

In mathematical terms, a PBPK model constitutes a large set of differential/algebraic equations describing the underlying processes. The current status of software development in pharmacokinetics is dominated by either a purely equation based approach — contradicting user-friendliness—or implementing a static model — contradicting flexibility. Instead, the requirements on user-friendliness and flexibility can be fulfilled by the use of sophisticated modular software concepts and structures and this has been done in MEDICI-PK.

During the development of the architecture of MEDICI-PK all parts of a PBPK model have been identified and examined in view of their connection. This has led to the orthogonal structure presented in Fig. 2. All structures principally defining the model equations but independent of concrete parameters are summarized as a so-called 'full body template'. The parameters to which the models refer can be classified as (i) compound dependent, (ii) species/individual dependent, (iii) dependent on the compound and species, (iv) general parameters. For a selection of compounds a given full body template can be (automatically) transformed into a simulation scenario now making possible simulation studies

- with different parameter sets
- with variations of basic processes
- with different dosing strategies
- for different individuals
- for different compounds
- with additional systems biology balances.

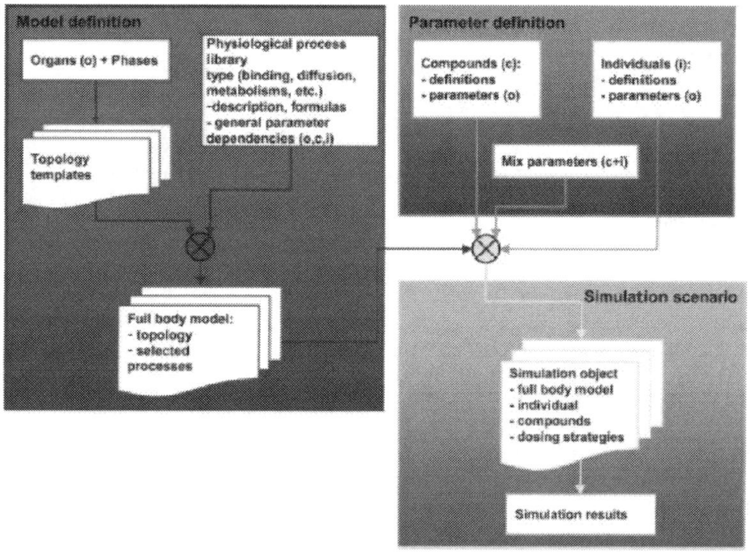

Figure 2: Schematic illustration of the orthogonal approach to software supported pharmacokinetic modeling [2]

When starting the simulation, the resulting differential equation system is automatically generated, including the assignment of all compound-specific parameter values and species-specific physiological parameter values to the respective processes.

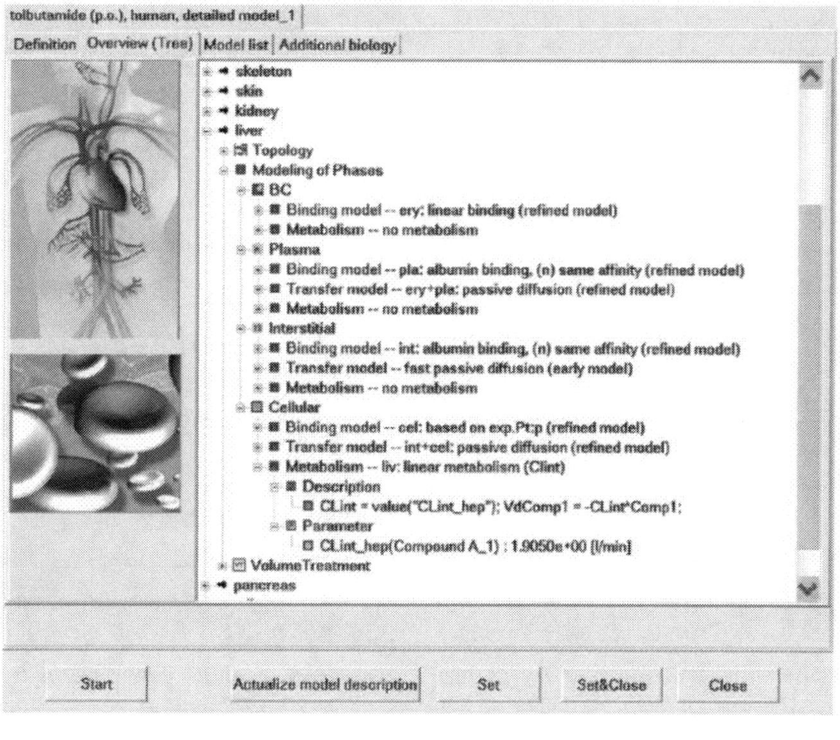

Figure 3: Tree view (among other overviews) of a MEDICI-PK simulation scenario

The development of MEDICI-PK was accompanied by several practical exercises with advanced students, in-house projects of companies and additional research activities. Actually it is used in a graduate research program. By that, from the beginning of the development, concept and architecture of the approach and its implementation were benchmarked. The feature matrix in Fig. 4 shows how the various design principles are realized in MEDICI-PK.

Important feature of the virtual laboratory MEDICI-PK

Property	Description	Realization in MEDICI-PK
User-friendliness	Assistance of the specific needs in physiologically based pharmacokinetic (PBPK) modeling.	Support of model building, hypothesis testing, user-defined output variables, data import from SBML, data export to Excel.
Application specificity	Software in terms of a PK-specific language.	Application-motivated design of underlying software structure; compounds, species, organs, dosing schemes, simulation objects as typical building blocks.
Modularity	Characterization of the overall PK model in terms of simple, independent modules that can be used in multiple contexts.	Extendable list of modules (physiological processes such as binding, transfer, metabolism, metabolic pathways, etc.) in a model basis. Definition of each full-body PK model in terms of the underlying physiological processes.
Orthogonal design and flexibility	Separation of model constituents that are independent of each other.	Independent definition of compound-specific data, species-specific physiological parameters and mathematical models to archive largest flexibility.
Openness	No limitation for adding new types of parameters, compounds, organs, species, physiological models, etc.	Easily extendable lists of parameters, compounds, organs, species, physiological models, dosing schemes, etc.
Transparency	No hidden calculations or parameters.	All parameters and models are directly assessable and editable.
Computational efficiency	Use of fast and accurate numerical techniques.	Integration of state-of-the-art numerical algorithms.

Figure 4: Design principles of Medici-PK [3]

Several different models and simulation studies have already been performed in MEDICI-PK in this way (e.g. a model for Cyclosporine A [5], a detailed model on the PK of tolbutamide and a generic PBPK model for early drug development [6]). Immediately, extensions of the existing models could be studied, for example incorporation of metabolites, comparison of different binding and tissue distribution models, drug–drug interaction studies, and comparison between different species (human, rat, and so on).

Based on our experience in other fields of chemistry and biology we see a high potential for in silico PK and PD approaches in the drug discovery and development process - if properly realized. Since even automatic parameter identification tools have been added to Medici-PK, new modeling ideas can directly be compared to experimental results.

Acknowledgement: We would like to thank Max von Kleist and Wilhelm Huisinga (Hamilton Institute, Dublin) for the fruitful cooperation during the development of MEDICI-PK and its foundations and for sharing ideas, models and graphics.

References:

[1] M. von Kleist, W. Huisinga, Hierarchical Approach to Physiologically based Pharmacokinetics: Refining Models with Increasing Knowledge, Submitted, 2007

[2] R. Telgmann, M. von Kleist, W. Huisinga, Software Supported Modelling in Pharmacokinetics, in 'Computational Life Sciences II', Lecture Notes in Computer Science 4216 (2006), M.R. Berthold, R. Glen, and I. Fischer (Eds.), Springer, pp. 216–225.

[3] W. Huisinga, R. Telgmann, M. Wulkow, The virtual lab approach to pharmacokinetics: design principles and concepts, Drug Discovery Today 11(2006) pp. 800-80

[4] R. Telgmann, Computer Aided Modeling, Dissertation FU Berlin, 2007, www.diss.fu-berlin.de/diss

[5[Kawai, R. et al. (1998), Physiologically based pharmacokinetics of cyclosporine A: extension to tissue distribution kinetics in rats and scale-up to human. J. Pharmacol. Exp. Ther. 287, 457–468

[6] Poulin, P. and Theil, F-P. (2002) Prediction of pharmacokinetics prior to in vivo studies. II. Generic physiologically based pharmacokinetic models of drug disposition. J. Pharm. Sci. 91, 1358–1370

[7] Pharmacometrics & Computational Disease Modeling, https://pharmacometrics.mi.fu-berlin.de/research-training-2.html

651e Modeling of Tablet Dissolution Attendant to Tablet Microstructure Using a Multi-Process F-D Method

Dan Braido, Rutgers University, 98 Brett rd., Piscataway, NJ 08854 and Alberto Cuitino, Mechanical and Aerospace Engineering, Rutgers University, 98 Brett rd., Piscataway, NJ 08854.

Purpose

The ultimate goal of this research is to develop a package of simulation tools which will be used in the design of delivery systems for pharmaceuticals. Current widespread models of tablet dissolution are not designed to investigate the effects of changes to the tablet micro-structure. The simulations herein focus on systems of compacted particles, paying particular attention to such factors as particle size, active particle distribution and porosity. Using this model, an algorithm has been designed which determines effective tablet parameters which will achieve a desired output.

Methods

An initial representation of the tablet is built or input into the workspace. This is essentially a 3-D model on a fixed Cartesian grid, implemented as a series of matrices storing cell based effective constants and values representing particle size, concentration, shape as well as information regarding the excipients. The dissolution of the tablet is considered for the case of said tablet completely submerged in a fluid. The tablet fluid interaction is calculated using a series of equations, each one modeling a particular process. The surface of the tablet is modeled using a level set method to track the moving boundary. This allows for a smooth representation of the interface existing between nodes, and is integral to the proper application of boundary conditions on the other equations. The penetration of solvent into the tablet is governed by a modified version of Fick's second law, the diffusion equation. This equation models how the surrounding fluid enters the tablet, thus controlling the actual amount of active particulate which is currently dissolving. Particle dissolution is considered to occur only in areas of the tablet where solvent is present. As the particles dissolve, the drug concentration of the solute increases. The diffusion of active drug in solution is then modeled using the same Fick's law equation as the solvent penetration. The drug is considered released as the solute diffuses through the tablet fluid interface. Most calculations are performed using a finite differencing technique. This model serves as the basis for the tablet design algorithm. The design algorithm iteratively simulates the dissolution of tablets within a given set of parameter ranges in an attempt to define a set of parameters which will produce a desired release profile. Both the release profile and tablet design constraints are input by the user, but all subsequent tablet design is performed by the algorithm.

Results

The model is able to produce curves representing the release of active drug into the fluid as well as track the concentration of solvent and volume of active particles inside the tablet. The tablet design algorithm is tested against physical dissolution profiles with reasonable agreement in tablet composition.

Conclusions

The current model provides an excellent basis for future extensions. The current framework is able to properly characterize a number of processes and is easily expandable to more. Even if a perfect representation of the true physical processes is unattainable, valuable insight into the importance of micro-scale factors on the overall dissolution of a tablet can still be determined.

651f Finite Element Analysis of Tablet Compaction Using a Modified Material Plasticity Model

Tuhin Sinha[1], Jennifer S. Curtis[2], Bruno C. Hancock[3], and Carl R. Wassgren[1]. (1) School of Mechanical Engineering, Purdue University, 585 Purdue Mall, Purdue University, West Lafayette, IN 47907-2088, (2) University of Florida, Chemical Engineering Department, P.O. Box 116005, Gainesville, FL 32611-6005, (3) Pfizer Inc., Groton, CT 06340

In this paper, finite element methods are used to investigate the compaction process of pharmaceutical tablets. The pharmaceutical powder is treated as a continuum with its elastic-plastic constitutive behavior described by the Drucker-Prager Cap (DPC) model. The model captures the shearing and densification of the powder during compaction and predicts the localized stress and relative density distributions. The DPC model was implemented for a cylindrical tablet geometry using the commercial software, Abaqus, for micro-crystalline cellulose, with the material parameters taken from the literature. The elastic-plastic material parameters of the constitutive model are strongly dependent on the relative density of the powder and evolve during the tablet compaction process. In the current study, a modified material model is used in which the DPC parameters are dependent on the current relative density of the powder and, hence, provide more accurate macroscopic property distributions within the tablet. The results show that significant differences in the final tablet stress and relative density distributions occur when constant material properties and relative density dependent properties are used.

651g Modeling of the Compaction of Confined Powder Beds Using a Modified Quasi-Continuum Approach

Athanas A. Koynov, Rutgers University, 98 Brett Rd., Piscataway, NJ 08854 and Alberto Cuitino, Mechanical Engineering, Rutgers University, 98 Brett Rd., Piscataway, NJ 08854.

The compaction of granular materials is a process encountered in a wide variety of systems – both naturally occurring and man-made. It is utilized for "green body" manufacturing by many industries from ceramics and metallurgy to the chemical and pharmaceutical ones. In fact, the currently most prevalent device for drug delivery – the solid tablet is produced through the process of powder compaction.

A clear understanding of the dependence of the mechanical properties of the solid product on both the characteristics of the starting materials and the system dynamics during the process itself is crucial for the effective design, control and scale-up of powder compaction. For this purpose, a quasicontinuum approach has been implemented to simulate the compaction of confined heterogeneous granular systems. The method focuses on the post-rearrangement stage of powder compaction where inter-particle voids have been filled-in and any further volume reduction can occur only via particle deformation. The developed tool is capable of simulating powder beds consisting of particles of different sizes, as well as material properties.

The quasicontinuum model is implemented using a finite element framework, in which the displacement of most constrained particles is computed based on the displacements of the nodes of an adaptive mesh. In this formulation inter-particle forces are resolved locally, while their relative displacement is represented using a constrained field. Equilibrium is enforced weakly using the principle of virtual displacement. The interactions between particles in different cells are handled in a way allowing for local operation and rendering symmetric tangent operators. History-dependent inter-particle bonding terms are used to simulate the mechanical strength of the finished compact. The initial configuration of a post-rearrangement powder bed is supplied by a static DEM model.

This method is particularly well suited for the dynamic computation of macroscopic variables, such as pressure and density, as well as the resolution of micro-level quantities, such as coordination number and loading paths. Since particle positions and inter-particle forces are computed explicitly, the model makes no assumptions regarding the homogeneity and isotropicity of the powder bed. It is, therefore, capable of providing local density distributions in the finished product. These can be used for the prediction of the product behavior during future loading, including the onset of fractures and instabilities caused by zones of higher and lower levels of compaction.

658a *E. Coli* Autolytic Vectors for High-Throughput Screening of Enzymes

Zhen Cai[1], Lihua Xu[2], Wanghui Xu[1], Rui Xue[3], and Zhanglin Lin[1]. (1) Department of Chemical Engineering, Tsinghua University, 1 Tsinghua Garden Road, Beijing, 100084, China, (2) DSM Innovation Center-White Biotechlogy, DSM China, 7 Dong San Huan Zhong Lu, Beijing, 100020, China, (3) Department of Chemistry, Tsinghua University, 1 Tsinghua Garden Road, Beijing, 100084, China

For high-throughput screening of enzymes expressed intracellularly in *E. coli*, a facile, consistent, reagentless and *in situ* cell lysis method was developed by combining the autolytic gene cassette *SRRz* from bacteriophage lamda with a series of UV- and heat-inducible promoters to construct *E. coli* autolytic vectors. Previously constructed UV- and heat-inducible *E. coli* autolytic vectors led to lysis efficienc of 60% and 90%, but restricted cell growth to be 30°C or below. Aiming to broaden the application of this heat-inducible autolytic vector series, three additional autolytic vectors with heat-inducible promoters *c*I857/p_R(T41C), *htpG*, and *clpB*, respectively were further tested. *E. coli* BL21 cells harboring these vectors grew well at 35°C and lysed efficiently (94%-98%) within 1-2 h after heat induction for 30 min at 42°C or 45°C. Promoter *c*I857/p_R(T41C) performed best with the highest lysis efficiency and lowest basal level lysis, while *htpG* showed a slightly more background lysis, and *clpB* preferred a higher heat induction temperature of 45°C instead of 42°C. Application of the autolytic vector using *c*I857/p_R(T41C) as promoter either in 96-well plates, or on nitrocellulose membranes, or on agar plates led to facile, efficient and consistent release of intracellular recombinant enzymes (*e.g.*, lysis efficiency of 91.8% ± 1.1% in 96-well plates). Further application in directed evolution was illustrated by improving the thermostability of amadoriase using this vector. This reagentless and *in situ* cell lysis method also has the potentials for lysis of miniaturized samples in clinical diagnosis and bioanalytical detection, and for lysis of cells in the microarray format.

658b Pawp (Protein A Well-plate Purification) – a High Throughput Method for Purification of Monoclonal Antibodies

Jennifer Hopp, Ross Pritchett, Maria Darlucio, Junfen Ma, and Judy H. Chou. Process Research & Development, Genentech, Inc., One Antibody Way, Oceanside, CA 92056

High throughput techniques for purification of monoclonal antibodies from harvested cell culture fluid can dramatically minimize feedstock requirements and accelerate process development decision-making during clone selection and cell culture process optimization. In this presentation, development of a scale-down high throughput antibody purification technique will be described. This method uses Protein A resin in a 96 well-plate format yielding sufficient quantity and concentration of antibody materials to perform multiple analyses of product quality per well. Protein A resins were selected and buffer conditions were optimized to obtain aggregate and charge variance comparability with antibodies purified by preparative Protein A chromatography. This comparability was demonstrated with three model monoclonal antibodies by size-exclusion chromatography, ion-exchange chromatography, imaged capillary isoelectric focusing, capillary zone electrophoresis, and glycan identification by capillary electrophoresis. The developed method can be performed in manual format and robotic format. It also maintains flexibility of resin types and amounts and offers cost-savings over commercially available products.

658c Influence of Enhancer Characteristics and Drug Hydrophobicity

K. Yerramsetty, V. K. Rachakonda, S. V. Madihally, R. L. Robinson Jr., and K. a. M. Gasem. School of Chemical Engineering, Oklahoma State University, 423 Engineering North, Stillwater, OK 74078

Altering the characteristics of the skin by using chemicals called penetration enhancers is one of the promising ways to breach the skin's barrier to drugs. In this work, several potential enhancers, menthone, 1-dodecyl-2-pyrrolidinone, 2-methyl-3-oxazolidinone, 1-methyl-2-pyrrolidone, nonanol, decanol, oleic acid and lauric acid, containing different functional groups were tested for their enhancement effect on three different classes of drugs viz. strongly hydrophilic insulin (Log Kow=-1.76), weakly hydrophobic melatonin (Log Kow=1.20) and strongly hydrophobic lidocaine (Log Kow=2.49). Initially, the experimental procedure was validated using permeability measurements of melatonin in the presence of four known CPEs with porcine skin, Franz Cell, and HPLC and using a similar procedure the permeabilities of the remaining drugs in the presence of CPEs is measured.

Further, three different characteristics of each CPE are measured: (i) The concentration of the CPE at the site of action, (ii) the permeability of the CPE alone through the skin and (iii) the toxic effect of the CPE on (a) Human Foreskin Fibroblasts (HFFs) using MTT assay at concentrations of three different orders of magnitude, and on (b) porcine abdominal skin by histology using H/E staining at the end of a 48-hour exposure period.

Our results suggest that (a) for a molecule to be a CPE, it has to be sufficiently hydrophobic and contain at least one hydrogen donating or accepting functional group; and (b) hydrophilic drugs respond better to CPE treatment than hydrophobic drugs. Further, the toxicity studies reveal that even though the cellular level toxicity studies are less laborious, they cannot be relied upon as a final conclusive indicator of the CPE's toxicity when compared to the more laborious but more accurate histological analysis.

658d Human In a Test Tube: Toward Drug Discovery Process In 3D Cell Scaffolds

Nicholas A. Kotov[1], Meghan J. Cuddihy[1], Jung Woo Lee[1], Richard C. Doty[2], and Joong Hwan Bahng[2]. (1) Chemical Engineering Department, Materials Science Department, Biomedical Engineering Department, University of Michigan, 2300 Hayward, Ann Arbor, MI 48109, (2) Nico Technologies Corp., Ann Arbor, MI 48109

Efficacy of in-vitro testing can be significantly improved provided that better ex-vivo models for different organs and tissues are developed. A large body of research indicates that cultured cells organized in three-dimensions (3D)behave a lot more closely to the original tissues and retain more natural functions than the cells in 2D cultures. However the currently available 3D scaffolds have either poor optical properties or impair cellular migration. Both of these factors are detrimental for scaffold utilization for rapid drug screening currently

used in industry. A new type of scaffold was developed based on inverted colloidal crystal (ICC) topology, which can resolve these issues and result in adequate

ex vivo models with 3D cellular organization resembling that of original organs. Additionally, the ICC scaffolds can be standardized exceptionally well,

which is critical for reproducibility of the drug screening assays. The ex vivo replicas of liver and bone marrow made in well plate format adaptable for drug

screening will be demonstrated.

658e A Network Based Approach for the Identification of Stress Responses In Lead Compound Transcriptome Profiles

Daniel R. Hyduke[1], Heng-Hong Li[1], Hyeon Ung Park[2], Sean P. Collins[2], Simeng Suy[3], Jiri Aubrecht[4], and Albert J. Fornace Jr.[1]. (1) Biochemistry and Molecular & Cellular Biology, Lombardi Comprehensive Cancer Center, Georgetown University Medical Center, 3970 Reservoir Rd. NW, Research Building E504, Washington, DC 20057, (2) Radiation Medicine, Lombardi Comprehensive Cancer Center, Georgetown University Medical Center, 3970 Reservoir Rd. NW, Research Building E504, Washington, DC 20057, (3) Drug Discovery Program, Lombardi Comprehensive Cancer Center, Georgetown University Medical Center, 3970 Reservoir Rd. NW, Research Building E504, Washington, DC 20057, (4) Global Research and Development, Pfizer, Groton, CT 06340

In addition to acting on a specific target, lead drug compounds possess the potential to initiate unexpected stress responses. Identification of these responses late in product development results in a significant loss on investment, and undetected effects may adversely affect the patient population. Additionally, the modes of action for promising drug candidates are often unknown, despite positive phenotypic effects. To facilitate the discovery of stress responses associated with lead compounds, we have developed a toxicogenomics database of the early transcriptomic response of a human lymphoblastoid cell line to a broad range of stress agents. Transcriptome measurements were taken for stress agent doses that elicited a robust response. The agents in this database include: a variety of anticancer compounds, endoplasmic reticulum stressors, antimetabolites, oxidizers, and other stressors. Because stress agents can induce a variety of overlapping responses, a straightforward comparison of a lead compound's transcriptome profile to the database will provide limited, if any, information about the compound's modes of action. To overcome this problem, we employed a biclustering approach to identify stable gene sets that are associated with the overlapping stress responses. These stress gene sets were assembled into a stress response network, and then the active stress components in a lead compound's transcriptome response were identified with network component analysis. We have validated our approach using independently generated transcriptome responses for positive and negative test cases. Additionally, we have used our approach to infer that a novel anticancer compound induces apoptosis by increasing oxidative stress--this inference was validated with independent biochemical assays.

658f High Throughput Determination of Gene Expression Levels Utilizing Mass Spectrometry and Solid-Phase Captureable Dideoxynucleotides

Daniel S. Duffield[1], Ashish Misra[1], Li Cai[2], and Sobin Kim[2]. (1) Chemical and Biochemical Engineering, Rutgers University, 98 Brett Rd, Piscataway, NJ 08854, (2) Biomedical Engineering, Rutgers University, 98 Brett Rd, Piscataway, NJ 08854

Here we present a rapid, precise and cost-effective method for determining gene expression levels. Current tools for expression analysis include DNA microarrays, Serial Analysis of Gene Expression (SAGE), and RT-PCR based methods. These offer efficient analysis of gene expression with a level of accuracy and throughput. However, there are notable limitations in these methods, such as long and laborious procedures (microarray and SAGE), and low throughput (RT-PCR). Thus new approaches that will allow more accurate and cost-effective analysis with multiplexing capability are highly desirable. To this end, we have developed a new method utilizing MALDI-TOF MS (matrix assisted laser desorption/ionization time-of-flight mass spectrometry) and biotinylated dideoxynucleotide triphosphate, the solid phase capturable terminators.

MALDI-TOF MS has been extensively studied for DNA analysis, i.e., DNA sequencing, SNP genotyping, and allele frequency determination, since it is accurate, suitable for quantification and easy for automation. Previously SPC-SBE, a multiplex genotyping method using solid phase capturable (SPC) dideoxynucleotides (ddNTPs) and single base extension (SBE) reaction, has been developed for MALDI-TOF based genetic variation detection. The method involves generating DNA fragments extended by a single base corresponding to the SNP of interest with SPC-ddNTPs such as biotinylated ddNTPs (biotin-ddNTPs). The DNA extension fragments containing biotin moiety at their 3' end are readily isolated from reaction mixture by utilizing a streptavidin-coated surface. The isolated DNA molecules are then analyzed with MALDI-TOF MS, allowing accurate DNA analysis for MALDI-TOF MS. The presented method further explores the use of biotin-ddNTPs and MALDI-TOF MS for gene expression analysis. Briefly, cDNAs are reversely transcribed from mRNAs isolated from breast tissue (normal and cancerous) and the cDNAs of interest and a known amount of internal standard (internal standard 1, IS-1) are PCR-amplified. A library of primers, one for each cDNA amplicon, are extended using biotin-ddNTPs with another internal standard (IS-2). Extension products are isolated by SPC on streptavidin-surface and analyzed with MALDI-TOF. Ratios between mass spectral peak areas are used to determine the relative quantity of each transcript. The system can enhance the accuracy in quantitative analysis of RNA samples, thus the level of gene expression. Additionally, our approach will increase the number of gene transcripts that can be analyzed simultaneously.

We have demonstrated the method in the analysis of gene expression in the mouse retinal tissue. Expression levels of five genes including, Rho, Lhx2, Hprt1, Sfrp2, and Nrl were determined utilitizing the method. The results strongly indicate that the method can provide fast processing time, great accuracy, and can be expanded to include high multiplexing capability.

As the proposed system utilizes biotin-ddNTPs and affinity capture of biotinylated DNA fragments, rapid capture, efficient washing and simple recovery of extension fragments is indispensable to further enhance the throughput provided by SPC in our gene expression analysis approach. Thus, we are constructing a novel microbead tip device for DNA isolation, which utilizes microbeads coated with biotin-binding proteins and a pipette tip. Together with the microbead tip system, our method will make a highly effective tool for the analysis of gene expression level changes.

658g Genomic Profiling: Amplification of Oligonucleotide Barcodes

Yusuf E. Murgha[1], Ayse Ozel[2], Onnop Srivannavit[2], Jean-Marie Rouillard[3], and Erdogan Gulari[4]. (1) Biomedical engineering, University of Michigan, 1107 Carl A. Gerstacker Building, 2200 Bonisteel, Blvd., Ann Arbor, MI 48109, (2) Chemical engineering, University of Michigan, 2300 Hayward St., 3074 H.H. Dow, Ann Arbor, MI 48109, (3) University of Michigan, 2300 Hayward St., 3074 H.H. Dow, Ann Arbor, MI 48109, (4) Chemical Engineering, The University of Michigan Ann Arbor, 2300 Hayward St., 3110 H.H. Dow Building, Ann Arbor, MI 48109

We describe a new method to synthesize and amplify cDNA for gene expression profiling on oligonucleotide microarrays. Currently, two methods used to amplify targets are polymerase chain reaction (PCR) or T7 polymerase based in vitro transcription. Conventional PCR due to its non-linear nature is prone to bias in amplification of multiple templates. Although, In Vitro transcription using T7-oligo(dT) or random primers amplify transcripts linearly, it has been reported that they fail to capture complete 5' information of transcripts and detect presence of rare transcripts. Both methods are further prone to cross hybridization effects on microarrays. To address these issues, we have developed a method using gene specific primers to capture information along any region of the gene.

Instead of cDNA amplification, a reporter sequence unique to a particular mRNA transcript is amplified and detected by complementary probe on microarray. The reporter tags (oligonucleotide barcodes) are incorporated with the gene specific primers used for reverse transcription. They are designed to have similar hybridization properties and minimal cross reactivity Instead of cDNA

Prior to detection, the tags are amplified by T7 based in vitro transcription and/or an emulsion based PCR method. These methods result in unbiased amplification of tags. Thus, all genes are represented in correct proportion for gene expression study.

670a A Bind-and-Lock VEGF Immobilization Strategy to Study VEGFR-2 Signaling

Sean M. Anderson and Tatiana Segura. Chemical and Biomolecular Engineering, University of California, Los Angeles, 420 Westwood Blvd., 5531 Boelter Hall, Los Angeles, CA 90095

Growth factors are a class of signaling proteins that direct cell fate through interaction with cell surface receptors. Although a myriad of possible cell fates stem from a growth factor binding to its receptor, the signaling cascades that result in one fate over another are still being elucidated. One possible mechanism by which nature modulates growth factor signaling is through the method of presentation of the growth factor - soluble or immobilized (matrix bound). For example, studies have shown that soluble VEGF, which can be internalized, enables endothelial cells to bypass confluent monolayer growth inhibition and continue to proliferate resulting in large, leaky, tumor-like blood vessels. In contrast, immobilized VEGF with a slower internalization rate causes endothelial cells to adhere to confluent monolayer growth inhibition, and results in the formation of small, structured capillaries. Here we present the methodology to study signaling of soluble versus immobilized VEGF through VEGFR-2. We have designed a strategy to covalently immobilize VEGF using its heparin-binding domain to orient the molecule (bind) and a secondary functional group to mediate covalent binding (lock). This bind-and-lock approach aims to allow VEGF to assume a bioactive orientation before covalent immobilization. Amine ($-NH_2$) and alcohol (-OH) terminated self-assembled monolayers of thiols on gold (SAMs) were used for immobilization. Heparin was oxidized to generate aldehyde groups to be used for Schiff base chemistry. Infrared (IR) spectroscopy confirmed aldehyde groups with the generation of a characteristic peak at 1730 cm^{-1}. Surface plasmon resonance (SPR) confirmed heparin and VEGF binding with surface densities of 44 $pmol/cm^2$ and 3 $fmol/cm^2$, respectively. Negligible amounts of heparin and VEGF were observed with surfaces containing no amines and negligible amounts of VEGF were observed with surfaces containing no heparin. VEGF immobilization was further confirmed with direct ELISA readings. To prepare the surfaces for cellular experiments, the surfaces were incubated with fibronectin. Cell studies showed that cells are able to bind to the heparin only surfaces, but showed enhanced binding to fibronectin-coated surfaces ($p < 0.05$). Further, cell spreading was most pronounced on surfaces that contained both heparin and fibronectin with a two-fold increase in cell area when compared to heparin alone. Last, endothelial cells plated on the VEGF-modified surfaces showed enhanced VEGFR-2 phosphorylation when compared to non-covalently bound VEGF and soluble VEGF ($p < 0.05$). This data suggest that immobilized VEGF signaling is different from soluble VEGF signaling. We believe that growth factor signaling is, at least, in part modulated by the physical nature of the growth factor - immobilized or soluble. The use of engineered surfaces such as the one described here are ideal for studying signaling and the resulting cell fates of immobilized growth factors, which could offer new insight into how nature regulates tissue formation and how to design scaffolds to guide tissue formation.

Web Page: tsegura.bol.ucla.edu/Site/home.html

670b EGFR Antagonists Based on the Fibronectin Domain

Benjamin J. Hackel and K. Dane Wittrup. Chemical Engineering, Massachusetts Institute of Technology, 50 Ames Street, E19-563, Cambridge, MA 02139

The epidermal growth factor receptor (EGFR) is a validated target in oncology. Yet the two FDA-approved antibodies against EGFR exhibit only modest therapeutic efficacy. We hypothesize that these deficiencies result from incomplete receptor inhibition that arises from autocrine ligand stimulation, untargeted EGFR variant III, insufficient receptor downregulation, and endocytic antibody consumption. We aim to engineer binders to multiple epitopes on the EGFR ectodomain for use in the study of these mechanisms and the development of improved therapies. The fibronectin type III domain is used as the scaffold because its single-domain, cysteine-free architecture facilitates the creation of bispecific constructs and enables biophysical FRET studies, and its small size (10 kDa) yields a pharmacokinetic profile that is well-suited for in vivo imaging.

Yeast surface display libraries were created with designed diversity in three solvent exposed loops of the tenth type III domain of human fibronectin. These libraries were screened for binders to the EGFR ectodomain using magnetic bead and flow cytometry sorting. Affinity maturation of lead clones was conducted using error-prone PCR and loop shuffling. Numerous clones were identified with affinities in the picomolar to nanomolar range. Antibody competition assays indicate that, collectively, these binders target multiple epitopes. The effect of internalization and downregulation of EGFR using both monovalent, bivalent, and bispecific constructs will be presented. The impact of affinity, binding epitope, and valency will be elucidated through the use of multiple binder combinations and formats. Implications for therapeutic development will be discussed.

670c EGFR Regulates Cell-Cell Adhesion and E-Cadherin Translocation through PKC-Delta

Raghvendra Singh, SUNY at Buffalo, 303 Furnas Hall, Dept. of Chemical Engg, University at Buffalo (North Campus), Amherst, NY 14260 and Stelios T. Andreadis, Chemical and Biological Engineering, SUNY-Buffalo, 908 Furnas Hall, Buffalo, NY 14260-4200.

EGF is known to affect adherens junctions and disrupt cell-cell adhesion in a variety of carcinomas but the underlying mechanisms are not completely understood. Using human tumor epithelial cells overexpressing EGFR we demonstrated that EGF-induced cell scattering was mediated by protein kinase C-delta (PKC-d). PKC-d knockdown by siRNA significantly inhibited EGF-induced internalization of E-cadherin into the cytoplasm and blocked cell scattering. EGF phosphorylated PKC-d at Y311 and ectopic expression of the mutant Y311F prevented PKC-d•binding to E-cadherin and EGF-induced cell-scattering. Moreover, blocking src or p38, but not p42/44 or JNK decreased EGF-induced PKC-d phosphorylation at Y311, consequently blocking its binding to E-cadherin and cell-scattering. Finally, EGF reduced expression of the tight junction protein, occludin, and this effect was also mediated by PKC-d. In summary, PKC-d mediated the effects of EGF on adherens and tight junctions thereby playing an important role in cell-cell adhesion with possible wider implications in tumor metastasis or epithelial-to-mesenchymal transition.

670d Activation of Endothelium In Alzheimer's Disease Brain Involves Soluble Aggregates of the Amyloid-Beta Protein

Francisco J. Gonzalez-Velasquez, Joseph A. Kotarek, and **Melissa A. Moss**. Department of Chemical Engineering, University of South Carolina, 2C02 Swearingen Engineering Center, Columbia, SC 29208

Alzheimer's disease (AD), the leading cause of dementia in the elderly, is characterized by the presence of amyloid plaques in the brain parenchyma and cerebrovasculature. These deposits are comprised primarily of the fibrillar form of the amyloid-beta protein. Monomeric amyloid-beta self-associates to form the fibrillar amyloid-beta that deposits as amyloid plaques. Vascular accumulation of fibrillar amyloid-beta occurs primarily in the basement membrane and initiates a cascade of events culminating in a weakening of the vessel wall that may lead to stroke as a result of cerebral hemorrhage. Amyloid-beta has been shown to elicit a number of responses in endothelial cells that contribute to these events; however, the mechanism for this stimulation remains unknown.

We have employed human brain microvasculature endothelial cell (HBMVEC) monolayers to assess the physiological effect of various amyloid-beta preparations. Amyloid-beta(1-40) monomer, fibril, and soluble aggregates were prepared and isolated using a combination of size exclusion chromatography (SEC) and centrifugation, and aggregate size was characterized using dynamic light scattering (DLS). Experimental data demonstrates that isolated soluble amyloid-beta(1-40) aggregation intermediates selectively activate endothothelial monolayers for both adhesion and subsequent transmigration of monocyte cells, as well as increased permeability, in the absence of endothelial cell death. In contrast, unaggregated amyloid-beta(1-40) monomer and mature amyloid-beta(1-40) fibril fail to induce any change in endothelial adhesion, transmigration, or permeability. Correlations between average amyloid-beta aggregate size and observed increases in both adhesion and monolayer permeability illustrate that smaller soluble aggregates are more potent activators of endothelium. Immunocytochemistry further reveals that NF-kappaB signaling cascades are involved in amyloid-beta stimulation of endothelium. These results support previous studies demonstrating heightened neuronal activity of soluble amyloid-beta aggregates, and further show that soluble aggregates also selectively exhibit activity in a vascular cell model.

As the contribution of vascular amyloid deposition in the progression of AD continues to gain support, therapeutic strategies should begin to encompass this vital facet of the disease. Resolution of the interactions between vascular amyloid-beta deposits and the endothelium will distinguish potential therapeutic targets for the reduction of vascular damage associated with AD.

670e Crosstalk Between Cell-Cell Contact and Soluble Factors In Regulating Epithelial Cell Proliferation

Anand R. Asthagiri, Chemical Engineering, California Institute of Technology, 1200 E. California Blvd MC 210-41, Pasadena, CA 91125 and Jin-Hong Kim, Bioengineering, California Institute of Technology, 1200 E. California Blvd MC 210-41, Pasadena, CA 91125.

Multicellular patterns and structures emerge as cells execute instructions received from cues in their microenvironment. Deciphering how cells integrate these cues to achieve an organized, functional structure is a fundamental question in biology with important biomedical implications in areas such as tissue engineering and regenerative medicine. An essential element in multicellular assembly involves direct cell-cell interactions. Cell-cell contacts are not only physical links between neighboring cells, but also sources of biochemical signals that instruct cellular behavior. Our lab is investigating how direct cell-cell interactions crosstalk with other environmental cues to affect individual cell fate choices, and thereby, generate multicellular patterns. In this talk, I will describe recent results that reveal a coupling between epidermal growth factor (EGF) and cell-cell contact in regulating the proliferation of human epithelial cells. This coupling between a long-range signal (EGF) and short-range cell-cell interactions gives rise to transient spatial heterogeneity in cell cycle activity. Single-cell measurements show that specific signaling pathways are influenced by cell-cell contact only when EGF concentration is below a critical concentration. These results reveal that quantitative competition between EGF and contact signaling pathways gives rise to spatiotemporal dynamics of epithelial cell proliferation.

670f Pkr Is Involved In Mediating Insulin Signaling and Inducing Insulin Resistance

Xuerui Yang, Department of Chemical Engineering and Materials Science, Michigan State University, 2527 Engineering Building, East Lansing, MI 48824 and Christina Chan, Chemical Engineering and Materials Science, Michigan State University, 2527 Engineering Building, East Lansing, MI 48824.

The major functions of insulin, e.g., regulation of the blood glucose level, amino acid uptake, lipid metabolism, are modulated by a signaling process called insulin signaling, which is initiated by the binding of insulin with its cellular membrane receptor protein, insulin receptor (IR) [1]. Upon binding with insulin, IR induces phosphorylation of its intracellular substrate, i.e., insulin receptor substrate (IRS), which plays a central role in triggering downstream pathways and thus signaling the effects of insulin on the cellular activities. The double-stranded RNA-dependent protein kinase (PKR) is best known for triggering cell defense responses by phosphorylating eIF-2α and thus activating cellular apoptosis during virus infection [2]. As a Ser/Thr protein kinase, PKR is also known for its role in mediating signaling pathways [3, 4] by interacting with downstream proteins such as NF-κB, MAPKs, and PP2A [5-8]. In the present study, we identified that in liver cells PKR suppressed the insulin signaling activity by inducing the inhibitory phosphorylation of IRS at its serine residues, which in its turn suppresses the tyrosine phosphorylation of IRS and blocks insulin signaling. It is the *first* time that PKR is shown to induce insulin resistance at the level of IRS. In addition, we also showed that insulin transiently inhibits the phosphorylation of PKR, which is potentially mediated by IRS and PP1. In summary, our results suggest, for the first time, that PKR is involved in mediating insulin signaling pathways and inducing insulin resistance, and we propose a feedback mechanism on the involvement of PKR in regulating the insulin signaling network.

1. Saltiel, A.R. and C.R. Kahn, Insulin signalling and the regulation of glucose and lipid metabolism. Nature, 2001. 414(6865): p. 799-806.

2. Barber, G.N., The dsRNA-dependent protein kinase, PKR and cell death. Cell Death Differ, 2005. 12 (6): p. 563-70.

3. Williams, B.R., PKR; a sentinel kinase for cellular stress. Oncogene, 1999. 18(45): p. 6112-20.

4. Williams, B.R., Signal integration via PKR. Sci STKE, 2001. 2001(89): p. RE2.

5. Kumar, A., et al., Double-stranded RNA-dependent protein kinase activates transcription factor NF-kappa B by phosphorylating I kappa B. Proc Natl Acad Sci U S A, 1994. 91(14): p. 6288-92.

6. Gil, J., J. Alcami, and M. Esteban, Activation of NF-kappa B by the dsRNA-dependent protein kinase, PKR involves the I kappa B kinase complex. Oncogene, 2000. 19(11): p. 1369-78.

7. Zhou, H.R., A.S. Lau, and J.J. Pestka, Role of double-stranded RNA-activated protein kinase R (PKR) in deoxynivalenol-induced ribotoxic stress response. Toxicol Sci, 2003. 74(2): p. 335-44.

8. Xu, Z. and B.R. Williams, The B56alpha regulatory subunit of protein phosphatase 2A is a target for regulation by double-stranded RNA-dependent protein kinase PKR. Mol Cell Biol, 2000. 20(14): p. 5285-99.

670g Dynamics and Regulation of Il-12 Receptor Signaling

Stacey D. Finley, Chemical and Biological Engineering, Northwestern University, 2145 Sheridan Road, E136, Evanston, IL 60208, Deepti Gupta, Biomedical Engineering, Northwestern University, Evanston, IL 60208, Linda J. Broadbelt, Northwestern University, Department of Chemical Engineering, 2145 Sheridan Road E136, Evanston, IL 60208-3120, Kathleen M. Brundage, Microbiology, Immunology & Cell Biology, West Virginia University, Morgantown, WV 26506, and David J. Klinke, Chemical Engineering, West Virginia University, Morgantown, WV 26506.

Engineering individualized immunotherapy is dependent on understanding the mechanistic basis for variations in disease susceptibility in outbred human populations. Inbred mouse strains provide model systems for quantifying functional differences among genetic variants of a species. The first objective of this study was to quantify dynamic differences in the Interleukin-12 receptor signaling (IL-12) pathway in two inbred mouse strains: C57Bl/6 and Balb/C. IL-12 is a key cytokine known to be involved in shaping cell-mediated immunity to infectious pathogens [1]. IL-12 initiates a cellular response via the IL-12 receptor, which is composed of two subunits: IL-12Rβ1 and IL-12Rβ2. The IL-12 signaling pathway is a member of the JAK/STAT family of signaling networks. Signaling via the JAK/STAT pathway culminates in the activation of STATs, a family of transcription factors that, upon activation, translocate to the nucleus and initiate protein synthesis. Suppressors of cytokine signaling (SOCS) proteins act as negative regulators of the JAK/STAT pathway. However, it has been shown that cellular context plays an important role in influencing the strength of signaling via particular reaction pathways [2]. The second objective of this study was to quantify how well the generalized JAK/STAT signaling mechanism represents the dynamics of STAT4 activation via the IL-12 pathway in naïve T helper cells, a cell critical for coordinating the adaptive immune response.

To quantify differences in IL-12 signaling, naïve $CD4^+$ $CD62L^+$ T-cells were isolated from murine Balb/C and C57Bl/6 splenocytes. The cells were activated for 44 hours using plate-bound anti-CD3 and anti-CD28 to provide nonspecific activation of the T cell receptor and co-stimulation, respectively. Activated naïve $CD4^+$ T cells were subdivided into three treatment groups: those receiving no stimulation, cells treated with anti-IL-4 only, and cells stimulated with anti-IL-4 and IL-12. Untreated cells and cells treated with anti-IL-4 only served as controls. Frequent samples of the cell populations were taken and stained with markers for IL-12Rβ1, IL-12Rβ2, SOCS3, and pSTAT4. Dynamic changes in protein expression were measured using flow cytometry and analyzed using R/Bioconductor. A mathematical model of the JAK/STAT pathway was created to quantify the differences in IL-12 signaling between mouse strains. Separate models incorporating various regulation mechanisms were created to test the strength of proposed negative regulators of the IL-12 signaling pathway.

IL-12R expression is dynamically regulated in naïve $CD4^+$ T-cells following activation by anti-CD3 and anti-CD28. Stimulation by IL-12 increased active STAT4 during the first 30 minutes and was followed by a steady decrease. Active STAT4 expression decreased to a greater extent in Balb/C cells compared to C57BI/6. The decrease in active STAT4 was correlated to a decrease in receptor number. Preliminary modeling results suggested that regulation of this pathway occurs via modulation of receptor expression rather than changes in activity of receptor. In summary, we have found dynamic differences between the two inbred mouse strains. These dynamic differences in IL-12 signaling appear to be regulated through different mechanisms than other members of the JAK/STAT signaling family. Reaction pathway analysis was helpful for discriminating between varying hypotheses regarding regulation of this particular pathway.

This work was supported in part by NIH RR016440 and NIH RR020866.

References

1. O'Garra, A., Cytokines induce the development of functionally heterogeneous T helper cell subsets. Immunity, 1998. 8: p. 275-283.

2. Klinke, D.J., *et al.*, Modulating temporal control of NF-kappaB activation: implications for therapeutic and assay selection. Biophysical Journal, 2008. In press.

673a Genomic and Proteomic Analysis of Lycopene-Overproducing *Escherichia Coli* Strains

Brian E. Mickus[1], Jeffrey C. Silva[2], Johannes P.C. Vissers[3], Gregory Stephanopoulos[1], and Charles L. Cooney[1]. (1) Chemical Engineering, Massachusetts Institute of Technology, 77 Massachusetts Avenue 56-454, Cambridge, MA 02139, (2) Cell Signaling Technology, Inc., 3 Trask Lane, Danvers, MA 01923, (3) Market Development Proteomics, Waters, Inc., Simonsway, Atlas Park, M22 5PP, Manchester, United Kingdom

Systems biology represents a powerful method to describe and manipulate phenotypes of interest by incorporating biological information from various levels of cellular organization. Such an approach is illustrated from a library of both rationally-directed and combinatorial gene knockout strains of *E. coli* containing the pAC-LYC plasmid which have been shown to produce elevated levels of the representative isoprenoid lycopene. Global genomic and proteomic expression changes associated with increased lycopene production of mutant *E. coli* constructs were discovered using whole-genome *E. coli* DNA microarrays and a novel LC-MS technique, respectively.

Following validation of the DNA microarray method by ANOVA analysis, transcriptional profiling was applied to five high lycopene-producing knockout mutants. One observation was greater than 4 times more differential expression for strains with a deletion in *hnr*, an important regulator of σ_s factor, as compared to mutants with metabolic targets. Additionally, the conservation of expression trends across a majority of distantly-related mutants for specific genes was seen. Within a specific mutant family of Δ*gdhA*, Δ*gdhA* Δ*aceE*, and Δ*gdhA* Δ*aceE* Δ_{P}*yjid* strains, *hisH* of the histidine biosynthetic pathway was down-regulated more than 10-fold while a gene encoding for a critical component of ATP synthase, *atpE*, was up-regulated more than 3-fold.

To supplement the genomic data, proteomic expression changes between the strains were analyzed using a novel LC-MS technique for simultaneous identification and label-free quantitation of proteins. The combined database search results from the LC-MS analysis of eight samples provided greater than 500 protein identifications. While a majority of proteins showed little expression change in the mutants relative to the pre-engineered strain, some key proteins were up- or down-regulated by nearly 10-fold. In particular, WrbA, a NADPH quinone oxidoreductase which reduces the quinone pool, was observed up-regulated over 5-fold while MdoG, a periplasmic protein which may affect the membrane storage capacity for lycopene, was down-regulated over 2-fold.

A simultaneous examination of both genomic and proteomic data revealed that the TCA cycle appears generally down-regulated while the PEP-glyoxylate cycle appears generally up-regulated across the various mutants. This is interesting due to the effect of the PEP-glyoxylate cycle upon cellular NADPH/NADH balance since lycopene biosynthesis requires large amounts of NADPH. Accordingly, a repressor of the glyoxylate pathway, IclR, and a membrane-bound pyridine nucleotide transhydrogenase, PntB, were targeted as potentially important to lycopene production.

Current work is progressing to overexpress and delete the *hisH, wrbA, iclR,* and *pntB* genes and determine resulting effects upon lycopene production.

673b Exploring Hypersecretion of Recombinant Proteins In *E. Coli* by Genome Sequencing

Prateek Gupta[1], Jeffrey C. Swanberg[2], and Kelvin H. Lee[2]. (1) Chemical and Biomolecular Engineering, Cornell University, 120 Olin Hall, Ithaca, NY 14853-5201, (2) Department of Chemical Engineering, University of Delaware, 150 Academy Street, Colburn Laboratory, Newark, DE 19716

Next-generation sequencing technologies are revolutionizing the pharmaceutical development. These technologies can not only be used in the drug discovery process but also for strain improvement in the manufacturing process. Using short read sequencing technology, we sequenced a parent strain and a hypersecreter strain (B41) of W3110 *E. coli*. B41 strain was created by chemically mutagenising the parent strain and secretes four-fold more active hemolysin protein (HlyA) relative to the parent strain via Type-I pathway. An average of 26.9x coverage of the two genomes resulted in a single G to T polymorphism in B41 strain at position 1,074,787 of the genome. The missense mutation results in translation termination of a transcriptional regulator RutR, coded by *ycdc* gene. The mRNA expression profiling data for the two strains also suggest significant expression change of the genes regulated by RutR, which is consistent with the observed mutation. We also tested the secretion phenotype in *ycdc* deletion mutant and observed a similar increase in active HlyA secretion relative to the parent strain.

673c Inter-Kingdom Signaling. A Signal-Mediated Framework of Host-Pathogen Interactions

Tarun Bansal, Thomas K. Wood, and Arul Jayaraman. Artie McFerrin Department of Chemical Engineering, Texas A&M University, 3122 TAMU, College Station, TX 77843

The human gastro-intestinal (GI) tract is colonized by approximately 10^{14} bacterial cells that belong to about 400 different species and co-exist with host cells. However, introduction of pathogens such as enterohemorrhagic *E. coli* (EHEC) disturbs this homeostasis and rapidly leads to colonization and infection. A characteristic feature of the GI tract is the abundance of signaling molecules, both of prokaryotic and eukaryotic origin, present in the lumen. Both pathogenic and non pathogenic bacteria secrete various quorum sensing and metabolic signals such as AI-2, AI-3, and indole. Similarly, host hormones like norepinephrine, dopamine and serotonin are also synthesized in the GI tract by the enteric nervous system. The close proximity of bacteria and the host cells, as well as the abundance of the signals they secrete, has led to a new signal-centric paradigm wherein the different GI tract signals are considered to be important mediators of homeostasis and infections through intra-kingdom (i.e., recognition of bacterial signals by other bacteria) and/or inter-kingdom (i.e., recognition of host signals by bacteria and vice-versa) signaling and communication [1, 2].

Work in our laboratory has focused on investigating intra- and inter-kingdom signaling in host cells, commensal bacteria, and the pathogen EHEC. Based on the structural similarity between indole and the human hormone melatonin, we hypothesized that indole acts as an inter-kingdom signal in intestinal epithelial cells. Gene expression studies with HCT-8 cells exposed to the bacterial signal indole showed that indole significantly up-regulated the expression of genes involved in apoptosis, immune response, interferons, and interleukins; thereby, suggesting that this commensal bacteria-secreted signal is used by intestinal epithelial cells in host cell inflammation and maintaining homeostasis. This observation was further validated through fluorescence microscopy analysis of the pro-inflammatory transcription factor NF-κB where the presence of indole decreased the induction TNF-α induced inflammatory response. Together, these results clearly demonstrate inter-kingdom recognition of bacterial signals by intestinal epithelial cells.

We also investigated inter-kingdom signaling in pathogenic bacteria by investigating the effect of intestinal epithelial cell secreted factors on EHEC pathogenicity. Gene expression studies with EHEC exposed to spent eukaryotic culture medium significantly increased the expression of 29 virulence genes belonging to LEE pathogenicity island and also of 22 genes involved in cell division and adhesion. Several AI-2 uptake genes and iron related genes were also upregulated. These observations are also corroborated by in vitro attachment assays where spent medium caused a 6-fold increase in EHEC attachment to HCT-8 cells. Together our results provide the basis of an inter-kingdom signaling-based framework where recognition of different signals by different elements underlies the balance between homeostasis and infection in the GI tract.

References

1. Bansal, T., D. Englert, J. Lee, M. Hegde, T. K. Wood, and A. Jayaraman. 2007. Differential Effects of Epinephrine, Norepinephrine, and Indole on *Escherichia coli* O157:H7 Chemotaxis, Colonization, and Gene Expression. Infect Immun 75:4597-4607.

2. Bansal, T., P. Jesudhasan, S. Pillai, T. K. Wood, and A. Jayaraman. 2008. Temporal Regulation of Enterohemorrhagic *Escherichia coli* Virulence Mediated by Autoinducer-2. Appl Microbiol Biotechnol 78:811-819.

Glycerol Kinase Overexpression Induces Metabolic Flux Alterations, Lipid Accumulation, and Increased Glucocorticoid Receptor Transcription Factor Activity In Rat Hepatoma Cells. A Systems Biology Study

Ganesh Sriram, Chemical and Biomolecular Engineering, University of Maryland, 1208D Chemical and Nuclear Engr. Bldg. 090, College Park, MD 20742, **Lilly Parr**, Human Genetics, University of California, Los Angeles, 695 Charles E. Young Dr. S. #5335, Los Angeles, CA 90095, **James C. Liao**, Chemical & Biomolecular Engineering, University of California, Los Angeles, 5531 Boelter Hall, 420 Westwood Plaza, Los Angeles, CA 90095 and **Katrina M. Dipple**, Human Genetics & Pediatrics - David Geffen School of Medicine, University of California, Los Angeles, 695 Charles E. Young Dr. S. #5335, Los Angeles, CA 90095

Glycerol kinase (GK) is an important lipogenic enzyme in liver, and exhibits several diverse ("moonlighting") cellular functions. GK overexpression has implications on insulin sensitivity, and GK deficiency is a single gene, yet complex genetic disorder whose complexities cannot be explained simply by lack of GK's biochemical activity. In the light of GK's multifaceted role, it is motivating to examine how perturbations in GK levels affect mammalian metabolic and regulatory networks. Therefore, we investigated metabolic flux, transcriptome, and transcription factor activity alterations caused by GK overexpression in a H4IIE rat hepatoma cell line.

13C labeling-based metabolic flux analysis revealed that GK-overexpressing cells had significantly higher flux through the pentose phosphate pathway (2.04 ± 0.10-fold) as compared to the wild type ($p < 0.01$). Comparable increases were also observed in the mRNA level and enzyme activity of glucose-6-phosphate dehydrogenase, the rate-limiting enzyme of this pathway, thus substantiating the flux analysis. cDNA microarray analysis of the GK-overexpressing and wild type cells revealed that GK overexpression alters the expression of several metabolic genes consistent with the observed metabolic flux changes. Lipid metabolism genes were differentially regulated in GK-overexpressing cells; this finding was corroborated by Oil Red O staining, which revealed larger lipid reserves in the GK-overexpressing cells.

We combined the microarray data with network component analysis to deduce that GK overexpression affected nine transcription factor activities, including that of the glucocorticoid-glucocorticoid receptor complex (GR). This result is particularly interesting because GR transcriptional activity is associated with GK's moonlighting role as ATP-stimulated translocation promoter. We further performed a glucocorticoid dose response experiment, which verified that the transcriptional activity of glucocorticoid receptor was indeed higher in the GK-overexpressing cells.

Thus, by using systems-level analyses, this work elucidates GK's central role in mammalian metabolism and transcription, and provides insights toward better understanding GK deficiency, a complex, yet single-gene inherited disease.

673e Metabolic Modeling of Ischemic Damage Treatment In Extracorporeal Perfusion

Korkut Uygun[1], Maria-Louisa Izamis[2], Nripen S. Sharma[3], Jack Milwid[4], Francois Berthiaume[4], and Martin Yarmush[5]. (1) Center for Engineering in Medicine, Harvard Med School, 51 Blossom Street, Boston, MA 02114, (2) Memp/hst, Harvard Med School/ MIT, 51 Blossom Street, Boston, MA 02114, (3) Center for Engineering in Medicine, Shriner Burns Hospital, Massachussets General Hospital, Harvard Medical School, 51 Blossom Street, Boston, MA 02114, (4) Center for Engineering in Medicine, Massachusetts General Hospital, Harvard Medical School, 51 Blossom Street, Boston, MA 02114, (5) Research, Center for Engineering in Medicine, 51 Blossom Street, Boston, MA 02114

Liver transplantation is currently the only established treatment for end-stage liver disease, but is limited by a severe shortage of viable donor organs; roughly 3,000 patients per year perish due to lack of transplantable donor organs. Livers obtained from donors after cardiac death (DCD) are currently an untapped donor source for liver transplantation and could increase the supply of donor livers by an estimated 6,000 per year. However, preservation of DCD livers by conventional methods, namely simple cold storage (SCS) in University of Wisconsin (UW) solution, is associated with a higher risk of primary non-function and delayed graft failure. We recently demonstrated that it is possible to resuscitate such livers using extracorporeal perfusion approaches. As compared to the current gold standard in organ preservation, cold storage, these methods provide the liver with oxygen and nutrition during storage; hence evidently enable a certain degree of repair by the organ. However, the state of the perfused liver is far from well understood, or characterized, hence a rational basis for further engineering the perfusion system is not available. Further, considering the expectable variability in these damaged livers, it is a critical necessity to evaluate their condition prior to transplantation, and optimize the repair during perfusion in order to maximize the probability of graft and patient survival.

Herein we describe the development of a metabolic flux model of ischemic reconditioning, with the ultimate goal engineering the perfusion systems in order to render marginal donor organs transplantable. Rat livers are subjected are harvested and subjected to 0, 60 or 90 minutes of warm ischemia in warm saline solution, as the model of donor after cardiac death. The livers are perfused for 6 hours in an extracorporeal perfusion system with erythrocyte supplemented WilliamsE medium. Perfusate samples obtained hourly from the inlet and outlet of the liver are analyzed for oxygen, amino acids and other metabolites to construct a metabolic flux model (46 metabolites and 72 fluxes). The differences between healthy livers (6/6 post-trasplant survival), ischemic livers reconditioned successfully (60 min, transplantation 11//11) and unsuccessful transplants (0/5) are analyzed, which reveals the dynamic progression of the metabolic processes during reconditioning, as well as pathways that cannot be adequately recovered with extensive ischemic damage. In addition, the results (fluxes and concentrations) are used to develop a viability score to predict the organ transplantability as a function of several metabolites during the perfusion, which can be used to quantitatively predict the probability of survival for a reconditioned liver. The use of the developed models and analysis for designing a second generation, portable perfusion system with a model-based metabolic feedback control is discussed.

673f Mathematical Modeling and Analysis of the Role of the BLR1 Protein and MAPK Activation In the Growth-Arrest and Differentiation Program of a Model Adult Stem-Cell

Ryan A. Tasseff[1], Satyaprakash Nayak[1], Damien Kudela[2], Jalal K. Siddiqui[1], Andrew Yen[3], and Jeffrey D. Varner[1]. (1) School of Chemical and Biomolecular Engineering, Cornell University, 120 Olin Hall, Ithaca, NY 14853, (2) School of Chemistry, Cornell University, Ithaca, NY 14853, (3) Department of Biomedical Sciences, Cornell University, Ithaca, NY 14853

Understanding the molecular basis of stem cell differentiation programs is one of the grand unmet challenges facing modern cell-biology. If these programs could be rationally manipulated, advanced stem-cell therapies could be developed for the treatment of a spectrum of human diseases, including cancer. For example, differentiation induction cancer chemotherapy has been used successfully to treat a sub-type of leukemia, Acute Promyelocytic Leukemia (APL) which occurs primarily in young adults. Differentiation induction chemotherapy uses Retinoic Acid (RA), a form of vitamin A, to cause growth arrest and differentiation of leukemic cells, thereby obviating the high toxicity typical of conventional chemotherapy. However, leukemias other than APL as well as other cancers are not-responsive to RA treatment. In this study, we integrated molecular network modeling and analysis techniques with molecular cell-biology to determine the critical regulatory molecules that interfere with differentiation induction therapies for cancers other than APL. The experimental system being analyzed was the HL-60 human myeloid leukemia cell-line. HL-60, an archetype experimental model for the study of differentiation programs since the late 1970s, is RA sensitive however it has a genetic background similar to RA-insensitive cells. Thus, HL-60 offers an opportunity to understand how the RA-response program functions for cancers other than APL.

A mathematical model of the molecular programs governing RA induced cell-cycle arrest and differentiation was formulated from literature and compared with protein- and Transcription Factor (TF)-array measurements taken from HL-60 cells exposed to RA. RA exposure induced the expression of two receptors, CD38 and BLR1, also known as CXCR5, early in the terminal differentiation process. Both CD38 and BLR1 contribute to MAPK signaling and propel RA-induced myeloid differentiation. A draft HL-60 mathematical model was constructed and tested against BLR1 wild-type (wt), knock-out and knock-in HL-60 cell-lines in the presence and absence of Retinoic Acid (RA). The initial HL-60 model architecture described the dynamics of 88 proteins which translated, when reversible interactions were considered, into 563 proteins and protein complexes interconnected by 951 interactions. A modular strategy was employed for the construction of the model. Separately identified MAPK, G1-cell-cycle, Calcium/Gq-protein and translation initiation modules were integrated into a master two-compartment (cytosol and nuclear) HL-60 network model. The master network described the regulated and constitutive expression of 17 proteins (Oct1, CREB2, ETS, BRN, IRF,SRPK, pRb, RhoGDI, P47phox, CD45, EIF2AK, SIIIp15, AAF, Stat5b, XRE and SAP) following a two-operator site promoter template. Activated transcription factors were assumed to reversibly bind operator site 1 (O1) which then allowed the recruitment of RNAp to operator site 2 (O2). Transcription of BLR1 was modeling by explicitly describing the reversible assembly of the CREB2:Oct1:NFATc3 transcription complex, the dissociation of CREB2 to activate the complex at O1 and then recruitment of RNAp at O2. The total concentration of the remaining 61 proteins was fixed by the initial condition of the simulation. Missing connectivity between pathway modules was estimated from a manual review of >100 publications and from the STRING and NETWORKIN databases. Also, when not experimentally derived, NETWORKIN was used to estimate kinase-target relationships and likely phosphorylation sites. The 563 Ordinary Differential Equations (ODEs) of the master model were generated using the UNIVERSAL code generation tool along with code to perform monte-carlo sensitivity analysis on the model parameters. Mass action kinetics were used to describe all protein-protein and protein-DNA interactions. The current HL-60 model has 951 parameters distributed among three types, association (kon), dissociation (koff) and catalytic (kcat). UNIVERSAL assigned heuristic values to unknown parameters where the mean binding affinity was micromolar and the average kcat was uniformly distributed around 500 s^-1.

In the absence of parameter tuning, simulations of MAPK activation following RA treatment in BLR1 wt, knock-out and knock-in HL-60 cells were compared with the recent study of Wang and Yen (J. Biol. Chem., 283:4375 - 4386, 2008). The objective of these initial simulation studies was to test whether or not the proper model structure (biological connectivity) was in place to capture qualitative trends in the Wang and Yen data. Thus, the heuristic model parameters generated by UNIVERSAL were used in these initial studies (however the initial conditions were adjusted). A single parameter set was used in all simulations. Consistent with the Wang and Yen data, the initial HL-60 network model was able to show up-regulation of BLR1 following RA exposure along with generation of ERKp. The also model correctly captured the direction of shifts in expression of BLR1 following RA-exposure in Raf+/Raf- HL-60 lines and the influence of a BLR1-/BLR1+ background upon MAPK activation (ERKp and MEKp) following RA-exposure. However, while the qualitative trends were consistent with data on a relative time-scale, the absolute time-scales were not in agreement. Moreover, some trends were not captured, e.g., the decrease of BLR1 mRNA at long exposure times. Interestingly, the assembly of the modular model components led to several testable linkages between BLR1 expression and MAPK. The route that was active in the initial simulations was Gq-protein activation of PKC via PLCg and DAG. BLR1 expression was predicted to drive IP3 and DAG formation through coupling with Gq-protein and PLCg activation. In the model DAG was assumed to be linked with the activation of Protein Kinase C (PKC) by helping recruit it to the plasma membrane. PKC was then assumed, consistent with literature, to be able to phosphorylate both Ras and Raf leading to MAPK activation. In the absence of parameter tuning, the BLR1-PKC connection could explain the activation of ERK following RA exposure but not the dependence of BLR1 expression upon MAPK activity. A second connection between MAPK and BLR1 was discovered in the initial analysis of the model. ERKp was found to activate the NFATc3 protein, a component in the transcription factor complex driving BLR1 expression. This connection could explain the dependence of BLR1 expression on Raf. Initial sensitivity analysis results confirm the critical role of Raf phosphorylation.

When taken together, the HL-60 modeling and sensitivity analysis results suggested experimentally testable hypothesis on the interconnections between BLR1 and MAPK activation. Understanding these relationships could be a critical first step to deciphering why RA-treatments fail in cancers other than APL. More generally, these studies demonstrate that dynamic molecular models of protein-protein and protein-DNA networks can be a useful tool to study complex programs in cell-biology.

673g Effect of Continuous Ultrasound on the Morphology and Cytoskeletal Architecture of Chondrocytes Maintained In 3-D Matrices

Anurandha Subramanian[1], **Sandra Noriega**[1], Gulnara Hassanova[1], Tarlan Mammedov[1], Joseph Turner[2], and Goutam Goshal[2]. (1) Chemical and Biomolecular Engineering, University of Nebraska - Lincoln, 207 L Othmer Hall, Lincoln, NE 68588, (2) Engineering Mechanics, University of Nebraska - Lincoln, W317.4 Nebraska Hall, Lincoln, NE 68588

Chondrocytes, the cellular component of the articular cartilage, have long been recognized as strain-sensitive cells, and have the ability to sense mechanical stimulation through surface receptors and intracellular signaling pathways. This stress-induced biological response of chondrocytes has been exploited to facilitate chondrocyte culture in in vitro systems and examples include the application of hydrostatic pressure, dynamic compression, hydrodynamic shear (i.e. rotating bioreactors) and application of low-intensity pulsed ultrasound (US). While the ability of US to influence chondrogenesis has been documented, the precise mechanisms of US induced stimulation continues to be investigated. As a consequence, there remains a critical need to evaluate the impact of US on chondrocytes in 3-D culture, which is a necessary microenvironment for maintaining the chondrocyte phenotype. In the present study an intermittent continuous ultrasound wave was employed to stimulate chondrocytes seeded in 3-D scaffolds. Non-US stimulated scaffolds served as the control. Both control and US stimulate groups were assayed for the total DNA content, morphology, and cartilage specific gene expression by RT-PCR. Our results show that chondrocytes when stimulated with continuous US for predetermined time intervals, possessed 1.2 to 1.4-times higher cellular viability than control and had higher levels of type-II collagen and aggrecan mRNA expression when compared to controls. Cell cytoskeleton structure was observed to be impacted by US and was reflected in actin, tubulin and vimentin structural distribution.

677a Dynamic Flux Balance Models for Simulation and Optimization of Saccharomyces Cerevisiae Fed-Batch Cultures

Jared L. Hjersted and Michael A. Henson. Department of Chemical Engineering, University of Massachusetts Amherst, 686 North Pleasant Street, Amherst, MA 01003

We have previously developed dynamic flux balance models for prediction of cellular growth and metabolic product formation rates in *Saccharomyces cerevisiae* batch and fed-batch cell cultures. These models couple steady-state stoichiometric balances on intracellular metabolites with dynamic extracellular balances on biomass, substrates, and metabolic byproducts through time-varying substrate uptake rates. In this contribution, we present the results of *Saccharomyces cerevisiae* fermentation experiments designed to evaluate the dynamic flux balance model predictions. We utilized a compartmentalized, genome-scale metabolic network to describe intracellular metabolism and simple Michaelis-Menten kinetics for the glucose and oxygen uptake rates. Kinetic parameters for the substrate uptake and the intracellular stoichiometric coefficients representing growth and non-growth associated energy requirements were estimated by nonlinear least squares optimization. A series of batch and fed-batch experiments demonstrated that the dynamic flux balance model was able to produce accurate substrate, biomass, and extracellular product concentration profile predictions over a wide range of fermentation conditions. The model was incorporated with a bi-level dynamic optimization scheme to compute fed-batch operating policies for optimal ethanol production in batch and fed-batch cultures. Experimental implementation of the optimal policies showed good agreement with the *in silico* predictions.

677b Optimization of Batch and Continuous Fermenters for Increasing Bioethanol Productivity Using Hybrid Cybernetic Models

Hyun-Seob Song, John A. Morgan, Nancy W. Y. Ho, Arvind Varma, and Doraiswami Ramkrishna. School of Chemical Engineering, Purdue University, Forney Hall of Chemical Engineering, 480 Stadium Mall Drive, West Lafayette, IN 47907

A process optimization study has been conducted using *cybernetic* models to explore ways to increase the productivity of fermenters using recombinant *Saccharomyces* yeast for converting glucose and xylose to bioethanol. For this purpose, a *hybrid* cybernetic model has been formulated for the Ho-Purdue strain 1400 (pLNH33) by incorporating elementary mode analysis into the cybernetic modeling framework. The hybrid cybernetic modeling approach models the distribution of uptake fluxes among different elementary modes based on pseudo-steady state hypotheses for the intracellular metabolites. The model shows excellent capability for describing and predicting various metabolic behaviors of the recombinant yeast as validated by several different sets of fermentation data.

The developed model has been used to optimize batch and continuous fermentation systems for maximum bioethanol productivity under the constraint that the unfermented sugar content at the end of fermentation for a batch culture or at the outlet for a continuous culture is below a certain threshold (e.g., 1 g/L). The theoretical maximum limit of ethanol productivity achievable by optimal operation has been estimated as a result. A comparison of two different cultivation methods shows that a batch fermenter outperforms a chemostat with respect to ethanol productivity in this example, i.e., the former shows the ethanol productivity within the range of 1.0 ~ 1.5 g/L/h by the optimization, while the latter 0.2 ~ 0.3 g/L/h. This might be ascribed to the strain characteristic of a strong preference for glucose to xylose as a carbon source. The theoretical implications of the obtained optimal operating strategies are discussed with a link to the internal flux distributions which can be estimated by the hybrid cybernetic model as well.

677c Indole and Quorum Sensing Signal in Escherichia Coli: Study of the Tryptophan Pathway by Means of a Metabolic Flux Analysis

Vanessa Ramirez Tovar and Andres Fernando Gonzalez Barrios. Chemical Engineering, Centro de Diseño de Productos y Procesos, Universidad de los Andes, Carrera 1a este No. 19 A 40 7° piso, Bogotá, Colombia

Bacteria are highly interactive and have extraordinary metabolic machinery for intracellular and extracellular communication. Nowadays, studies have been able to find a relationship between quorum sensing and many functions on *Escherichia coli*; such as operon *ftsQAZ* regulation which controls the expression of three genes, the expression of secretion systems type III, DNA replication inhibition, the activation of degradation pathways and biofilms formation [1,5]. In this work we study the signal indole which is a is a inter-species biofilm signal mediated trough the action of SdiA (Lee et al. 2007).

Here, we want to determine whether indole plays a more important role in the metabolic pathway or as a quorum sensing signal. For this, metabolic flux analysis of the tryptophan metabolic pathway was performed fixing the extracellular indole flux and synthesis of indole flux, and using two objective functions: minimization of redox potential and maximization of Acetyl-CoA flux. It was established that both, indole flux and synthesis of indole have a positive effect over the Acetyl-CoA flux, moreover, we found the cell finds it more efficient to minimize the use of energy rather than produce more of it; from which it is concluded that the best objective function for this system is the minimization of the redox potential. At high cell density and starving conditions cells minimize redox potential in order to decrease the number of oxidizing reactions that occur thus conserving its energy or using its energy in the most efficient way possible. It was proved that indole plays a more important role as a quorum sensing signal than it does in the cell metabolism; which was shown by the sensibility analysis.

Finally the phase diagram was constructed by calculating the shadow price for the redox potential over the concentration of extracellular indole. The diagram showed tree different zones: a irregular zone in which the shadow prize shows no tendency over a change in extracellular indole or extracellular indole; a second zone where the shadow price decreases to a minimum value; and the third zone in which the shadow price exhibit a regular tendency creating a plane figure.

677d Defining Design Space for Biotech Products: Case Study of Pichia Pastoris Fermentation

Jean Harms, Process Development, Amgen Inc, One Amgen Center Drive, MS 18S-1-A, Thousand Oaks, CA 91320 and Anurag Rathore, Process Development, Amgen Inc., One Amgen Center Drive, Thousand Oaks, CA 91320.

The concept of "design space" has been proposed in the ICH Q8 guideline and is gaining momentum in its application in the biotech industry. It has been defined as "The multidimensional combination and interaction of input variables (e.g., material attributes) and process parameters that have been demonstrated to provide assurance of quality." This paper presents a stepwise approach for defining design space for a biologic product. A case study, involving P. pastoris fermentation, is presented to facilitate this. First, risk analysis via Failure Modes and Effects Analysis (FMEA) is performed to identify parameters for process characterization. Second, small-scale models are created and qualified prior to their use in these experimental studies. Third, studies are designed using Design of Experiments (DOE) in order for the data to be amenable for use in defining the design space. Fourth, the studies are executed and the results analyzed for decisions on the criticality of the parameters, as well as, on establishing design space. For the application under consideration, it is shown that the fermentation unit operation is very robust with a wide design space and no critical operating parameters. This approach can be extended to other biotech unit operations and processes that can be scaled down and characterized at small scale.

Web Page: Biotechnol. Prog., article 10.1021/bp070338y S8756-7938(07)00338-4, Web release date: April 16, 2008

677e Strain Engineering and Fermentation Process Development toward Supplying Semi-Synthetic Artemisinin. A Key Component of Antimalarial Combination Therapies

Rika Regentin[1], Christopher J. Paddon[2], Patrick Westfall[2], Hiroko Tsuruta[1], Douglas J. Pitera[1], Diana J. Eng[1], Tizita Horning[1], Jacob R. Lenihan[1], Bonnie Lieu[2], Dae-Kyun Ro[3], Michelle C. Chang[4], Eric Paradise[5], Jay D. Keasling[6], Jack D. Newman[7], and Neil S. Renninger[7]. (1) Fermentation, Amyris Biotechnologies, Hayward, CA 94542, (2) Molecular Biology, Amyris Biotechnologies, Hayward, CA 94542, (3) Biological Sciences, University of Calgary, Calgary, T2N1N4, Canada, (4) Chemistry, University of California, Berkeley, Berkeley, CA 94720, (5) The Boston Consulting Group, Bethesda, MD 20814, (6) Berkeley Center for Synthetic Biology, UC, Berkeley, 717 Potter street, Berkeley, CA 94720, (7) Amyris Biotechnologies, Hayward, CA 94542

Malaria threatens 350-500 million lives every year, with over 1 million deaths, primarily children under the age of five. Artemisinin based Combination Therapies (ACT's) are recommended by World Health Organization for the treatment of uncomplicated malaria. Although currently more costly than traditional drug treatments, ACT's are preferred since *P. falciparum* resistance has been observed relative to almost all currently used antimalarials except for artemisinin and its deriviatives. Currently the only source of artemisinin is extraction from the plant *Artemisia annua*. Consequently, the supply of artemisinin for ACT's can be variable, due to a long planting cycles and climate-dependent yields. By cloning the genes from the *A. annua* artemisinin biosynthetic pathway into *Escherichia coli* and *Saccharomyces cerevisiae*, we have developed a process to produce semi-synthetic artemisinin through fermentation and chemical conversion. The biosynthetic pathways were constructed for two artemisinin precursors, amorpha-4,11-diene and artemisinic acid. Strain and process improvements increased production of amorpha-4,11-diene in *E. coli* to 25 g/L in small scale bioreactors. Artemisinic acid processes at the lab scale were developed for both organisms with titers > 1 g/L.This alternative artemisinin production option, if commercial scale production can be achieved, may be a good way to supplement the plant-derived supply since fermentation processes have a shorter cycle time, can be scaled to meet demand, and result in consistent quality.

677f Recombinant Protein Production In Escherichia Coli

Jane Gunson and Michael W Laird. Late Stage Cell Culture, Genentech, Inc, 1 DNA Way, MS 32, South San Francisco, CA 94080

Escherichia coli has been used for decades to produce recombinant proteins for basic research purposes and clinical indications. This bacterial production workhorse is rapid, versatile, economical, scalable and extremely well characterized. The genetics, physiology, molecular biology and biochemical analyses with this organism are straightforward and the regulatory hurdles are generally low. A survey of the published literature shows that numerous marketed protein drugs have been produced by from this bacterium including but not limited to human insulin, interferon-&alpha, interferon &beta-1b, interferon &gamma-1b, human growth hormone, bovine growth hormone, insulin growth factor-1 (IGF-1), keratinocyte growth factor, anti-VegF (Fab), human granulocyte colony stimulating factor (G-CSF) and interleukin-2.

Over the course of the last three decades here at Genentech, the Escherichia coli production platform has been used for a variety of research and development projects that includes growth factors, cytokines, ligands, antibody fragments and monovalent antibodies. An overview of the production technology and its current applications at Genentech will be described.

679a Embryonic Cardiomyocyte Beating on Mechanical Mimics of Normal and Fibrotic Heart Tissue

Christine I. Carag[1], Adam J. Engler[2], Colin Johnson[3], Hsin-Yao Tang[4], David W. Speicher[4], and Dennis E. Discher[1]. (1) Chemical and Biomolecular Engineering, University of Pennsylvania, Towne Bldg. Rm 311A, 220 S. 33rd St., Philadelphia, PA 19104, (2) Department of Molecular Biology, Princeton University, Princeton, NJ 08544, (3) Chemical and Biomolecular Engineering, Univ. Pennsylvania, 112 Towne Bldg, Philadelphia, PA 19104, (4) The Wistar Institute, 3601 Spruce St, Room 152, Philadelphia, PA 19104

Fibrotic rigidification following a myocardial infarct is known to impair cardiac output, and it is also known that cardiomyocytes on rigid culture substrates show a progressive loss of rhythmic beating. Here we study isolated embryonic cardiomyocytes on a series of flexible substrates and show that matrices which mimic the elasticity of the developing myocardial microenvironment are optimal for transmitting contractile work to the matrix and for promoting actomyosin striation and 1 Hz beating. On hard matrices that mechanically mimic a post-infarct fibrotic scar, cells over-strain themselves, lack striated myofibrils and stop beating; on softer matrices, cells preserve contractile beating for days in culture but exhibit fewer striations and do very little work. Optimal matrix produces a strain match between cell and matrix and suggests dynamic differences in intracellular protein structures. A "Cysteine Shotgun" method of labeling the in situ proteome reveals differences in several cytoskeletal proteins, including vimentin, filamin, and myosin. Combined with recent results that show stem cell differentiation is also highly sensitive to matrix elasticity, the results here highlight the need for greater attention to fibrosis and mechanical microenvironments in cell therapy.

679b Use of Genetic Engineering to Study Cell Behavior In a Synthetic Environment

James O. Blanchette, Chemical Engineering, University of South Carolina, 301 Main St, Room 2C02, Columbia, SC 29208 and Kristi Anseth, Department of Chemical and Biological Engineering, University of Colorado, Howard Hughes Medical Institute, ECCH 128, Campus Box 424, Boulder, CO 80309-0424.

INTRODUCTION

The combination of living tissue and a synthetic material holds great promise in the areas of tissue engineering and delivery of therapeutics. One of the challenges associated with this emerging field is controlling the behavior of the cells when they are no longer receiving the nutrients and signals that are present under physiological conditions. This study seeks to identify what stresses are placed on tissue in a polymeric environment focusing on encapsulated pancreatic islets.

Encapsulation of pancreatic islets has long been a strategy for treatment of diabetes[1]. A passive encapsulation system acts as a barrier to allow passage of nutrients and waste products, as well as glucose and insulin, while excluding immune cells. A successful barrier system will provide the encapsulated tissue with the nutrients they need while shielding them from the immune system allowing the cells to secrete insulin in response to elevated glucose concentrations. Despite the concerted efforts of numerous research groups, a viable encapsulation system leading to prolonged insulin independence has remained elusive.

This failure is the result of the large number of factors leading to loss of function for the transplanted tissue. Modifications to basic encapsulation systems that seek to address many of these factors individually have shown short-term success but failed to produce a clinically successful system for treatment of human patients. The most obvious factor is immune rejection, but studies comparing survival of encapsulated auto- and allograft tissue showed similar survival time indicating the role of other stresses in loss of function[2]. Some of these stresses include: insufficient diffusion of nutrients and waste products through the capsule resulting in hypoxia and oxidative stress, lack of biocompatibility of the capsule material and the lack of sufficient cell-cell and cell-ECM (extracellular matrix) interactions within the capsule.

EXPERIMENTAL PLAN

The central goal of the work proposed here is to improve the long term function and viability of encapsulated islet cells. The cells of particular interest within the islets are the beta cells which are responsible for the release of insulin. The intracellular signaling pathways that translate extracellular stresses into cell death (with particular attention given to anoikis) will be targeted for genetic modification to block these processes and in turn extend the viability and function of the graft tissue. To observe the role of hypoxia on cell survival, a hypoxia marker was developed to identify cells which have initiated a response to low oxygen.

To test the importance of cell-ECM interactions, islets were infected to overexpress integrin-linked kinase (ILK) and/or Bcl-2 prior to encapsulation in a poly(ethylene glycol) (PEG) matrix. We hypothesized that ILK (an integrin beta1 binding protein capable of phosphorylating the cytoplasmic subunit of b1 integrins) overexpression would prevent beta cell anoikis caused by the lack of physiological cell-ECM interactions within the PEG capsule. Bcl-2 (an integral protein found in the endoplasmic reticulum, nuclear envelope and mitochondria) is able to block apoptosis stimulated by a range of different stresses. Bcl-2 overexpression should also block anoikis but is less specific than ILK in this regard. Previous work using Bcl-2 overexpression in transplanted islets has shown promising results[3].

The encapsulation process isolates the islets from the blood supply which forces oxygen to diffuse a longer distance to reach the cells that compose each islet. To monitor whether encapsulated cells are receiving sufficient oxygen, an adenoviral marker was generated where a red fluorescent protein will be produced when a low oxygen stress response is initiated. Infected islets were encapsulated and exposed to different levels of oxygen to test the efficacy of this marker system.

RESULTS AND DISCUSSION

Initial studies were performed on a cell line (MIN6) to mimic the beta cells present in an islet. Previous work established that dispersed MIN6 cells show low viability in PEG gels after 5 days in culture. MIN6 cells infected to overexpress ILK, Bcl-2 or both showed considerably improved viability under these conditions compared to control transfected cells. Control cells showed $42 \pm 9\%$ viability which was similar to the results found for unmodified MIN6 cells encapsulated for 5 days and significantly lower than the experimental groups. After 5 days in the PEG capsule, Bcl-2-infected cells were $76 \pm 3\%$ viable, ILK-infected MIN6 cells were $95 \pm 3\%$ viable and cells infected with both ILK and Bcl-2 were $88 \pm 5\%$ viable. Statistical evaluation showed significant difference between the control cells and all three groups receiving Bcl-2 or ILK (p-values <0.001). The results suggest that anoikis is a cause of cell death observed in encapsulated dispersed MIN6 cells, and that survival may be improved by activating the pathway triggered by interactions of beta1 integrins with the ECM proteins lacking in the PEG hydrogel. We are, in effect, "hotwiring" this pathway to make the cells think they are interacting with the absent ECM matrix.

Similar studies were conducted with encapsulated islets obtained from Balb/c mice. When using the islets, insulin release in response to glucose stimulation was measured instead of viability as function a better metric than viability. Studies have shown a higher release of insulin after 4 weeks of encapsulation for Bcl-2 and ILK-infected islets compared to islets infected with a control virus. Compared to initial insulin release (release on day 1 set as 100%), Bcl-2 islets released 40%, ILK islets released 43% and control islets released 0% after 4 weeks of encapsulation. While 40 and 43% are not ideal values, they are certainly an improvement over the control cells and support the data obtained in the earlier studies.

Testing of the hypoxia marker showed that islets infected with the marker virus exhibited a fluorescent signal after 24 hours in a 1% oxygen environment. Control cells in this hypoxic environment showed no signal as did infected cells placed in a 21% oxygen environment. The intensity of the signal can also be controlled by the amount of virus used for the infection. This system will be used to study the impact of islet size and density as well as implantation site on oxygen diffusion to encapsulated islets.

CONCLUSIONS

The working hypothesis for these studies was that islets encapsulated in PEG networks were exhibiting loss of function due to apoptosis triggered by the absence of proper cell-ECM interactions and perhaps insufficient oxygen supply. Studies using the MIN6 beta cell line have shown that overexpression of signal transduction proteins involved in pro-survival pathways following interactions with ECM proteins can prolong survival in PEG gels. ILK seems to be a promising target for extending function of cell encapsulation systems for treatment of diabetes as well as other disease states. The decrease in insulin release after 4 weeks even in the modified cells indicates that other stresses are present on these cells and has motivated the development of the hypoxia marker. It is hoped that a combinatorial approach that seeks to address the range of stresses placed on encapsulated tissue can lead to longer term function. Since the challenges associated with islet encapsulation are common to many applications of cell-seeded biomaterials, we hope that these results will also be useful in other therapeutic delivery and tissue engineering applications.

REFERENCES:

[1] Lim F and Sun AM. Science; 210, 908 (1980).

[2] De Vos P et al. Diabetologia; 40, 262 (1997).

[3] Contreras JL et al. Transplantation; 71, 1015 (2001).

ACKNOWLEDGEMENTS

The viruses used in these studies were generated with support from the Howard Hughes Medical Institute. The HRE sequences used in the creation of the HRE-DsRed virus were generously donated by Dr Rachel Cowen. We would also like to thank Dr Stephen Langer for his assistance with the construction of the recombinant adenoviruses and Francisco Ramirez-Victorino and Philip Pratt for their assistance with isolating the Balb/c islets for these studies.

679c Directing Behavior of Neural Progenitor Cells by Altering Spatial Display of Peptide Signals in Self-Assembling Matrix

Qianqian Li and Ying Chau. Chemical Engineering, Hong Kong University of Science and Technology, Clear Water Bay, Kowloon, Hong Kong, Hong Kong

By applying a novel self-assembly peptide design template, we synthesized a nanofibrous bio-functional scaffold with a laminin-derived signal sequence isoleucine-lysine-valine-alanine-valine (IKVAV) inserting between repeating blocks of arginine-alanine-aspartate-alanine (RADA). The new material supported the growth of neural progenitor cells (PC12). Furthermore, cellular proliferation and differentiation could be controlled by the interplay of signal presentation format and matrix dimensionality. Enhanced and prolonged proliferation of PC12 cells as aggregates was observed in 3-D culture while pronounced neurite outgrowth was found in 2-D culture. Interestingly, cells previously exposed to 3-D culture were more prone to neural differentiation and synapse formation. In addition to characterizing the degree of differentiation by morphology and immunostaining, the stages of cell differentiation were detected by RT-PCR to better understand the cell-material interactions.

679d Cell Migration Studies at the Interface of a Dual Chemical-Mechanical Gradient

Yuzhen Yang[1], Albert Kwansa[2], Nathan Hale[3], Eric Rosenberg[4], and **Padma Rajagopalan**[3]. (1) Chemical Engineering, Lehigh University, Bethlehem, PA 18015, (2) School of Biomedical Engineering and Sciences, Virginia Polytechnic Institute and State University, Blacksburg, VA 24061, (3) Chemical Engineering, Virginia Polytechnic Institute and State University, Blacksburg, VA 24061, (4) Bioengineering Program, Lehigh University, Bethlehem, PA 18015

Cell migration is crucial in numerous physiological and physiological processes, such as wound healing, response to inflammation and cancer metastasis. In recent years, accumulating evidence indicates that cell movement is regulated not only by chemical factors but also by mechanical signals. In living organisms, cell migration is directed by complex chemical and mechanical stimuli. Thus, understanding the cellular response to a combination of chemical and mechanical stimuli will be critical for in vivo applications. In this study, the main purpose is to understand whether a chemical or mechanical stimulus plays the major role in directing cell migration. A polyacrylamide hydrogel that exhibits chemical and mechanical gradients on either side of an interface was synthesized. The chemical gradient was created by varying the collagen (Type I) concentration and the mechanical gradient was introduced by changing the concentration of the cross-linker. One side of the interface exhibited a high value for Young's modulus and low collagen concentration while the other side of the interface was soft and exhibited a high collagen concentration. Balb/c 3T3 fibroblasts were cultured on the hydrogels and their cell motility patterns were studied. Cell migration was observed by time lapse microscopy over a 8 hour period. AFM measurements were used to determine values for Young's modulus at the interface. Our results indicate that cells close to the interface moved towards the high collagen/low Young's modulus regions of the substratum, indicating a greater role for chemotaxis.

679e Parsing Stem Cell Behaviors on Complex Biomaterials Via High Content Imaging and Modeling

Matthew D. Treiser[1], Eric Yang[2], I.P. Androulakis[2], Joachim Kohn[3], Christopher S. Chen[4], and Prabhas V. Moghe[5]. (1) Biomedical Engineering, Rutgers, The State University of New Jersey, Piscataway, NJ 08854, (2) Biomedical Engineering, Rutgers University, Piscataway, NJ 08854, (3) Chemistry and Chemical Biology, Rutgers University and The New Jersey Center for Biomaterials, 145 Bevier Road, Piscataway, NJ 08854, (4) Department of Bioengineering, University of Pennsylvania School of Engineering and Applied Science, 210 S. 33rd St., 510 Skirkanich Hall, Philadelphia, PA 19104, (5) Chemical and Biochemical Engineering; Biomedical Engineering, Rutgers University, 599 Taylor Road, Piscataway, NJ 08854

The role of synthetic biomaterials on long-term stem cell differentiation is poorly understood and thus stem cell regenerative materials are difficult to rationally design. Recently, it has been demonstrated that the cytoskeletal proteins, particularly actin, are strong mediators of human mesenchymal stem cell (hMSC) differentiation, with early cell shape having large consequences on longer-term functions such as differentiation. As the cytoskeleton mediates the outside-in stem cell signaling, we sought to "profile" the effects of biomaterials based on the analysis of material induced changes in intracellular, quantitative descriptors of the cytoskeleton of stem cells. The ultimate goal is to predict long-term stem cell fates and lineage commitment based on the signature trends of early intracellular cytoskeletal organization correlated to biomaterial chemistry. Human mesenchymal stem cells were sub-cultured in 50:50 mixed osteogenic and adipogenic induction media on various polymeric biomaterial substrates. The substrates were spin-coated films from a combinatorially varied composition of a family of tyrosine-derived polycarbonates, (varied components included poly(ethylene glycol), anionic carboxylate groups, and a stiffening agent for radio-opacity, iodine) and from a family of combinatorially derived polymethacrylates. After 24 hours of culture, the cells were stained fluorescently for their actin cytoskeleton and high magnification (63X) and high resolution tile scan images spanning approximately 5 mm x 4.5 mm were produced using confocal imaging on a Leica TCS SP2 system. Utilizing these images, over 50 morphometric descriptors of actin were quantified for each substrate. Variations in the descriptor values were compared across different substrates and the role of material composition on stem cell descriptors was identified. Through multi-dimensional scaling (MDS) based modeling, cell populations were identified that best correlated with stem cell differentiation toward osteogenic vs. adipogenic lineages. This study highlights the possibility of using a combination of high content imaging and materials informatics toward predicting longer-term stem cell differentiation outcomes on complex biomaterials.

679f 3D Porous Scaffold Structure Design for Optimum Fluid Shear and Nutrient Transport to Osteoblastic Cells Cultured In a Perfusion Bioreactor

Roman S. Voronov, School of Chemical, Biological, and Materials Engineering, University of Oklahoma, 100 E. Boyd, SEC Room 311, Norman, OK 73019, Samuel VanGordon, Bioengineering, University of Oklahoma, 100 E. Boyd St., SEC T-335, Norman, OK 73019, Vassilios I. Sikavitsas, Chemical, Biological, and Materials Engineering, University of Oklahoma, 100 E. Boyd St., SEC Room T335, Norman, OK 73019, and Dimitrios Papavassiliou, Chemical Engineering, University of Oklahoma, Sarkeys Energy Center, 100 East Boyd Street, Norman, OK 73019.

In order to improve tissue formation in scaffolds, various mechanical parameters have been explored such as fluid flow, hydrostatic pressure, and substrate deformation. It is thought that fluid flow is the dominant mechanical stimulus for physiological bone cell behavior. Porous seeded cell constructs cultured in perfusion bioreactors have demonstrated increased improvement in cell expansion and tissue growth. Two distinct flow mechanisms are thought to be responsible for the increased response generated by cellular scaffolds cultured in perfusion bioreactors. Firstly, shear forces have shown to improve osteoblastic differentiation, cell proliferation, matrix deposition, extracellular matrix calcification and mechanical properties of the cellular construct. The second mechanism to which the success of flow perfusion cultured scaffolds is attributed to is enhanced nutrient and waste transport.

The variety of perfusion system arrangements, the inherently random internal architecture of porous scaffolds and the differences in cell-biomaterial interactions due to the various surface chemistry and cell types render any analytical estimation of the local shear forces and transport properties implausible, while the current scaffold architecture design is mostly empirical and trial-and-error based. The situation is further complicated by the transient nature of the tissue growth process within the scaffold while it is being cultured. Cell adhesion strengths change with time as cell coverage grows from an initial monolayer coverage of the scaffold to cells adhering to layers of other cells and excreted extracellular matrix. Expansion of cells and tissue during the culturing period within the porous network of the scaffold creates a continuously changing geometry. Thus the biological environment inside the scaffold can be visualized as a continuously changing 3D matrix of tissue that dynamically responds to mechanical stresses. Therefore, as a primary goal of this work the local shear force distributions as well as the nutrient transport are explored throughout the culturing process for various scaffold geometries. Based on the obtained results, the 3D porous scaffold structure is analyzed for optimum fluid shear and nutrient transport for osteoblastic cells cultured in a perfusion bioreactor.

Micro-Computed Tomography with 10 μm resolution is used to obtain 3D structure of the scaffolds. Histological staining is used for validation of the μCT images. Flows of osteogenic media through the cell seeded cylindrical scaffolds are modeled via fluid dynamics simulations (Lattice Boltzmann Method – LBM). Experimentally obtained pressure drops across the scaffolds are used as an input for the fluid dynamics simulations. High performance computing in conjunction with a house hybrid MPI/Open MP parallelized scheme is employed in order to take advantage inherent LBM parallelizability. The code is validated against analytical solutions for flows pressure driven flows in channels, pipes, ducts and an infinite array of spheres. Macroscopic mass transfer is modeled using the Lagrangian Scalar Tracking method (LST) in conjunction with the LBM algorithm. In LST, the motion of scalar markers is used to synthesize the scalar profile. The trajectories of these markers are composed by a convection part (obtained using the velocity field from the LBM simulations) and a diffusion part (i.e., Brownian motion obtained from a mesoscopic Monte-Carlo approach). This method is resourceful in terms of computational efficiency, in that it can be used to simulate various thermal boundary conditions and Schmidt number fluids with a single flow field resulting from an LBM simulation. Scaffolds are characterized based on their geometric characteristics (such as tortuosity, surface area to volume ratio, and porosity, pore size, and pore connectivity/anisotropy) in order to parameterize the obtained results as a function of different scaffold structures. Special attention is devoted to porosity of the scaffold since it has been shown to play a detrimental role in the internal geometry of the scaffolds. The study as performed at different time points throughout a two week culturing point. The theoretical framework developed allows for improved tissue generation via: maximum solid-fluid contact area for uniform cell adhesion, efficient nutrient and waste transport inside the scaffolds for promotion of cell survival and favorable mechanical stimulation of cell proliferation.

679g Multifunctional Tyrosine-Derived Polycarbonates: Effect of Chemistry on Smooth Muscle Cell Adhesion and Motility

Patrick A. Johnson[1], Arnold Luk[2], Alexei Demtchouk[2], Hiral Patel[3], Simon Gordonov[3], Larisa Sheihet[4], Das Bolikal[4], Joachim Kohn[4], and Prabhas V. Moghe[5]. (1) Chemical & Petroleum Engineering, University of Wyoming, Dept. 3295, 1000 E. University Ave., Laramie, WY 82071, (2) Department of Biomedical Engineering, Rutgers University, Piscataway, NJ 08854, (3) Biomedical Engineering, Rutgers University, 599 Taylor Rd, Piscataway, NJ 08854, (4) Chemistry and Chemical Biology, Rutgers University and The New Jersey Center for Biomaterials, 145 Bevier Road, Piscataway, NJ 08854, (5) Chemical and Biochemical Engineering; Biomedical Engineering, Rutgers University, 599 Taylor Road, Piscataway, NJ 08854

The performance of blood-contacting implants and vascular tissue engineering scaffolds depends heavily on the regulation of vascular cell responses. The goal of this study was to examine underlying substrate-smooth muscle cell responses, using a library of multifunctional, tyrosine-derived polycarbonates. Three chemical components within the polymer structure were selectively varied through copolymerization: 1) the content of iodinated tyrosine to achieve X-ray visibility; 2) the content of poly(ethylene glycol) (PEG) to decrease protein adsorption and cell adhesivity; and 3) the content of desaminotyrosyl-tyrosine (DT) which regulates the rate of polymer degradation. We quantified differential serum protein adsorption behavior due to the chemical components DT, iodinated tyrosine, and PEG using quartz crystal microbalance with dissipation. The complex interplay of the polymer components was tested on the adhesion, proliferation, and motility behavior cultured human aortic smooth muscle cells. The incorporation of PEG into the polymer reduced cell attachment, which was reversed in the presence of iodinated tyrosine. Further, we found that as little as 10% DT content was sufficient to negate the PEG effect in polymers containing iodinated tyrosine while in non-iodinated polymers the PEG effect on cell attachment was reversed. Cross-functional analysis of motility and proliferation revealed divergent substrate chemistry related cell response regimes. Additionally, our data suggests unique biorelevant properties following the incorporation of iodinated subunits in a polymeric biomaterial as a potential platform for X-ray visible devices.

693a Antibody Microarrays for Probing the Binding Cooperativity of Leukemia-Associated Cell Surface Antigens

Chaofang Yue, Chemical and Biomolecular Engineering, The Ohio State University, 125A Koffolt Laboratories, 140 west 19th Ave., Columbus, OH 43210, Yuan Yuan, Nanoscale Science and Engineering Center for Affordable Nanoengineering of Polymeric Biomedical Devices, The Ohio State University, 140 WEST 19TH AVE, Columbus, OH 43210, Bo Yu, Department of Chemical and Biomolecular Engineering, The Ohio State University, 140 West 19th Avenue, Columbus, OH 43210, J. James Lee, The Ohio State University, Room 125A, Koffolt Labs., 140W. 19th Ave., Columbus, OH 43210, and Michael E. Paulaitis, Department of Chemical engineering, The Ohio State University, 125 Koffolt Labs, 140, West 19th Ave, Columbus, OH 43210.

In the past several years, many novel targeted, immunosuppressive therapies have been developed for the treatment of leukemia. However, their therapeutic effects can deviate substantially from patient to patient due to variations in the levels of leukemia-associated cell surface antigens that are expressed across a diverse patient population. Efficient strategies for characterizing patient populations with heterogeneous clinical resistance to immunosuppressive therapies have not been established. One major challenge of targeted immunotherapies for leukemia treatment is designing targeting molecules or a repertoire of targeting molecules that can be applied across a heterogeneous population expressing a wide range of leukemia-associated cell surface antigens.

Recent studies have demonstrated that antibody microarrays can be effective tools for profiling leukemia-associated cell surface antigens [1,2]. In general, antibody microarrays quantify antigens expressed on a cell surface by measuring the density of cells captured on spots printed with the relevant antibodies. The level of cell capture depends primarily on the antigen concentration on surface of cells, and the antigen-antibody binding affinity/avidity. Most current strategies use only one targeting antibody for cell capture. The existence of positive cooperativity in the context of multivalent interactions is well recognized in the field of immunology. Thus, targeting strategies that can take advantage of multiple antibodies can potentially achieve higher targeting efficiency compared to single antibodies. The use of multiple targeting antibodies, however, requires the development of methods for rapidly and quantitatively characterizing cooperativity in binding cell surface antigens.

We have developing antibody microarrays to quantitatively characterize targeting efficiencies of multiple antibodies in a systematic, high-throughput fashion, which enables screening for optimal combinations of antibodies for targeted immunosuppressive therapies. Individual spots on the microarray are printed with different compositions of antibodies, and cell capture on each spot is measured to give the binding efficiency corresponding to that antibody composition. Comparing these results to cell capture on spots printed with the associated single antibodies gives the extent of binding cooperativity. Using a B-cell chronic lymphocytic leukemia cell line, we have shown that multiple antibodies can achieve much higher binding efficiencies compared to single antibodies printed at the same concentration. To investigate the kinetics of cell binding, a parallel-plate flow chamber has been designed to probe the kinetics of cell adhesion and the subsequent shear-induced detachment of bound cells. Time-tracked flow experiments can thus be carried out to estimate the dynamic range of cell binding to spots printed with multiple antibodies, compared to spots printed with single antibodies to provide important insights into the cooperative targeting of leukemia-associated cell surface antigens.

1. Belov, L., et al., Screening microarrays of novel monoclonal antibodies for binding to T-, B- and myeloid leukaemia cells. Journal of Immunological Methods, 2005. 305(1): p. 10-19.

2. Belov, L., et al., Analysis of human leukaemias and lymphomas using extensive immunophenotypes from an antibody microarray. British Journal of Haematology, 2006. 135(2): p. 184-197.

693b Antibody Microarrays for Immunoisolation of T-Cells and Immunodetection of Secreted Cytokines

He Zhu[1], Gulnaz Stybayeva[1], Erlan Ramanculov[2], Satya Dandekar[3], Michael George[3], and **Alexander Revzin**[1]. (1) Department of Biomedical Engineering, University of California at Davis, 451 East Health Sciences Dr., Davis, CA 95616, (2) National Center for Biotechnology, Astana, Kazakhstan, (3) Medical Microbiology and Immunology, University of California at Davis, 451 East Health Sciences Dr., Davis, CA 95616

Cytokine production by leukocytes in response to viral or bacterial pathogens correlates with disease outcomes and therefore carries significant diagnostic value. The goal of the present study was to develop a miniature device for detection of interleukin (IL)-2 and interferon (IFN)-gamma produced by a small population of CD4 and CD8 T-lymphocytes. Microarrays of CD4-, CD8- and cytokine-specific Ab spots were printed onto poly(ethylene glycol) (PEG) hydrogel-coated glass slides and then were enclosed inside a microfluidic device, creating a miniature (3 ul) immunoreaction chamber. Introduction of red blood cell (RBC) depleted whole human blood into the device followed by washing at a pre-defined shear stress resulted in isolation of pure (>95%) CD4+ and CD8+ T-cells on their respective Ab spots. Importantly, immune cells became localized next to anti-IL-2 and -IFN-gamma Ab spots. Mitogenic activation of captured T-cells was followed by immunofluorescent staining (all steps were carried out inside the microdevice) and revealed concentration gradients of IL-2 and IFN-gamma emanating from an array of immune cells. Concentration of IL-2 and IFN-gamma near CD4+ T-cells was estimated to be 30 ng/ml and 20 ng/ml respectively while CD8+ T-cells produced 15 ng/ml of IFN-gamma and no IL-2. The production of cytokines by T-cells was confirmed using intracellular cytokine staining and flow cytometry. This novel immunology tool is envisioned to have applications for HIV diagnosis, vaccine development and monitoring the effectiveness of antiretroviral therapy.

693c High Throughput Platelet Activation Kinetics

Manash S. Chatterjee and Scott L. Diamond. Chemical and Biomolecular Engineering, University of Pennsylvania, 5140 Vagelos Research Laboratories, 3340 Smith Walk, Philadelphia, PA 19104

Several agonists and antagonists acting together regulate the activation state of the platelet in vivo, allowing it to remain in a quiescent state during haemostasis and yet respond efficiently following vascular injury to form a stable thrombus.Using cytosolic Calcium concentration as a metric of platelet activation, we have developed a high throughput kinetic assay which allows us to rapidly study dose responses of agonists, antagonists and drugs as well as combinations of discrete states of these regulators. This assay allows us to probe 96 experimental conditions in replicates of 4 from a mere 15ml human blood draw. Furthermore, we have been able to phenotype 192 mouse platelet conditions using 2ml of pooled mouse blood, thereby greatly increasing throughput per knockout mice.Platelets were stimulated with the Collagen analogue Convulxin, Thrombin receptor agonist peptide (TRAP) and ADP in increasing concentrations to determine the IC50 for each agonist. Concentrations of 0, 0.1, 1 and 10x IC50 were selected, and all the 64 combinatorial combinations of the 4 states of the 3 agonists were studied simultaneously. We found that on short timescales the response is dominated by the characteristics of the GPCR mediated ADP/TRAP signal, whereas on longer timescales it becomes reminiscent of a Convulxin signal. At high concentrations of TRAP, ADP had little distinguishable effect, whereas the effect of Convulxin was apparent at all concentrations studied. Pharmacological treatments using the P2Y1 inhibitor MRS2179, the P2Y12 inhibitor MRS2395 and the COX inhibitor aspirin allowed us to quantify the differential contributions of each of these receptors and autocatalytically produced TxA2 to the overall ADP signal. Using the Calcium chelator EDTA in the medium and controlled recalcification protocols, we were able to distinguish the flux due to store operated extracellular Calcium entry. Using the SERCA pump inhibitor Thapsigargin and the PMCA inhibitor Carboxyeosin, we found evidence that the amplitude and duration of the signal could in part be determined by how quickly Calcium is removed from the cytoplasm by these pumps after agonist induced release from the stores. A Convulxin stimulus was found to produce a slow sustained elevation in cytoplasmic Calcium which was brought down very slowly using only the PMCA pumps. In contrast, a TRAP stimulus produced a rapid Calcium burst which was quickly brought down to baseline using both the PMCA and SERCA pumps.Given the central role of platelets in mediating cardiovascular pathology, this assay shows promise to perform a patient specific scan of 'Platelet Activation Space' and a control scan of a healthy donor done on the same plate with the objective of diagnosing efficiently undefined platelet defects.

Web Page: bloodsystemsbiology.org/

693d Screening Tools for the High-Throughput Analysis of CYP450-Mediated Drug Metabolism and CYP450 Inhibition

Sumitra Sukumaran, Chemical & Biological Engineering, Rensselaer Polytechnic Institute, Troy, NY 12180, Moo-Yeal Lee, Solidus Biosciences, 1223 Peoples Avenue, Troy, NY 12180, Benjamin Potsaid, The Center for Automation Technologies and Systems, Rensselaer Polytechnic Institute, Troy, NY 12180, Douglas S. Clark, Chemical Engineering, UC - Berkeley, Dept of Chemical Engineering, Berkeley, CA 94720 - 1462, and Jonathan S. Dordick, Chemical and Biological Engineering, Rensselaer Polytechnic Institute, 110 Eighth Street, Troy, NY 12180.

A common *in vivo* drug-drug interaction is the inhibition of Cytochrome P450 (CYP450) enzymes responsible for drug metabolism, resulting in higher systemic levels of a drug and possible toxicity. Drug-drug interactions represent a safety hurdle and a major expense in drug discovery. Hence it is vital to incorporate *in vitro* CYP450 inhibition assays early in the drug development process. Previous inhibition assays have been performed in 96-well plates, 1536-well plates, and on microfluidic platforms; however, truly high-throughput assays using nanovolumes of reagents have yet to be explored. We have developed a solid-phase alginate microarray consisting of up to 3,000 isolated CYP450 reactions on a single 25 X 75 mm functionalized glass slide for the high-throughput analysis of CYP450 inhibition. Individual CYP450 isoforms (CYP3A4, CYP2C9, CYP2D6, CYP2E1, and CYP1A2) were encapsulated in 20 nl alginate spots (prepared from a solution of 0.5% alginate and containing a total enzyme content < 50 pg) and used in fluorescence-based inhibition assays. For slide imaging and data extraction, the CYP450 microarrays were interfaced to a novel, wide-field, whole-slide imaging optics system, consisting of a telecentric composite lens coupled with traditional microscopic filters. Individual CYP50 isoforms encapsulated in alginate exhibited comparable and reproducible reactivity and storage stability, and increased linear reaction time, when compared to conventional CYP450 preparations in solution. We show that a single 1,080 or 2,856 spot array on a glass slide can simultaneously provide IC_{50} values for nine potential inhibitor compounds against several CYP450 isoforms. IC_{50} values and enzyme kinetics comparable to those in solution were obtained for CYP3A4, CYP2C9, CYP2D6, CYP2E1, and CYP1A2 with a number of compounds, including known inhibitor compounds such as sulfaphenazole, clotrimazole, quinidine, and erythromycin. Thus, a robust, sensitive, and accurate CYP450 inhibition assay has been developed that is high-throughput (nearly 2,000-fold miniaturization when compared to 96-well plate assays). The assay is now being validated by expanding to a broad set of test compounds. These advances represent critical steps toward a fully automated, high-throughput human pharmacokinetics (PK) technology. Potential long-term applications include very early-stage PK analysis in the drug discovery process and ultimately leading to uses that may partially enable personalized medicine.

693e Use of Fibrin Hydrogels for Localized and Cell-Controlled Lentiviral Gene Transfer

Shruti Raut, Roshan Padmashali, Pedro Lei, and Stelios T. Andreadis. Chemical and Biological Engineering, University at Buffalo, State University of New York, 916 Furnas Hall, Amherst, NY 14260

Recombinant lentiviruses are efficient gene delivery vehicles with the ability to infect non-dividing cells including neurons, hematopoietic and embryonic stem cells. In addition to high efficiency several applications may require localized and cell-controlled gene transfer to a subset of cells in a local microenvironment. To this end, we developed a strategy of lentivirus delivery using fibrin hydrogels. Our preliminary studies indicate that fibrin-mediated gene transfer is more efficient than traditional lentivirus transduction and that transduction efficiency depends on the properties of fibrin gels and the interaction of target cells with the matrix. Specifically, lentiviral transduction is optimum at intermediate concentrations of fibrinogen but independent of thrombin. Interestingly, fibrinolytic inhibitors decreased lentivirus transduction in a dose dependent manner, suggesting that matrix degradation is necessary for virus uptake. Surprisingly, at low fibrinogen concentrations lentiviral particles can diffuse out of the fibrin gels decreasing the transduction efficiency. To avoid this problem and obtain highly localized gene delivery we developed a novel method to covalently conjugate lentiviral particles into fibrin and achieve gene delivery in a cell-controlled manner. Using this strategy we engineered microarrays with immobilized lentivirus to express multiple genes in a single transduction step. This high throughput gene delivery strategy will have significant impact on many biological studies including profiling of adult and embryonic stem cells.

693f Parallel Gene Delivery for Cell-Based Activity Assays

Abigail D. Bellis and Lonnie D. Shea. Chemical and Biological Engineering, Northwestern University, 2145 Sheridan Road, Tech E136, Evanston, IL 60208

Many molecular markers have been inadequate for identifying therapeutic targets or predicting patient outcomes, which has motivated the development of cell-based assays that provide the physiological context in which to investigate molecular function. Gene delivery, in particular, is being employed with these assays to manipulate a specific factor within the cell. In previous proof of principle work, we have shown the ability to quantify cellular activity on a small scale within the format of a transfected cell array. Current efforts are focused on developing the methods to scale up this system to facilitate broader functional analysis. We are investigating the use of parallel gene delivery for high throughput cell based assays using either a slide or well-plate format. Achieving adequate transfection while maintaining high cell viability poses a significant challenge to automating this approach. Transfection of the luciferase encoding plasmid was analyzed using bioluminescent imaging. Bioluminescent imaging provides the ability to sensitively quantify transgene expression while limiting post-transfectional processing. For slides, plasmids are spotted at room temperature at a relative humidity of 85% in an array format onto the slide surface. The small volume of DNA/lipid complexes used during spotting complicates transfection, as uncontrolled surface adsorption or dehydration can diminish lipoplex activity. The addition of sucrose can protect and stabilize DNA/lipid complexes. Transfection in the presence of sucrose was increased four-fold relative to its absence. Slide incubation time before cell seeding was also investigated. Longer incubation times resulted in more complexes immobilized onto the spot surface. The correlation between transfection efficiency and incubation time was found to be dependent on initial DNA concentration within the spot. Multi-well plates were also investigated as an alternative to slides. Transfection within the plates was more consistent than on the slides yet requires more advanced automation for cell seeding and data acquisition. Because of the slightly lower transfection variability the plate format offered an important tool to determine proper experimental conditions for implementation within the transfected cell array. The ability to transfect cells with high efficiency in an array format is a fundamental tool that can be used in numerous cell-based screening assays.

693g Combinatorial Cell-Based Screening of Biodegradable Polymer Films and Microspheres: Implications for Vaccine Delivery

Latrisha K. Petersen[1], Andrew Adler[2], Jennifer H. Wilson[3], Michael Wannemuehler[3], Rajan Krishna[4], Xue Li[4], and Balaji Narasimhan[1]. (1) Chemical and Biological Engineering, Iowa State University, 2114 Sweeney Hall, Ames, IA 50011, (2) Bme, Northwestern University, Evanston, 60208, (3) Veterinary Microbiology & Preventive Medicine, Iowa State University, 2180 Vet Med, Ames, IA 50011, (4) Materials Science and Engineering, Iowa State University, 2220 Hoover Hall, Ames, IA 50011

Surface-erodible polyanhydrides have been thoroughly studied as carriers for drugs, proteins, and vaccines. In general, the focus of these studies has been on structure-function relationships carried out in a "one sample at a time" method. This costly, time consuming, error prone process for studying biomaterial structure-function relationships can be more efficiently investigated and optimized with the use of combinatorial techniques. The polyanhydrides of interest in this work were based on sebacic acid (SA), 1,6-bis(p-carboxyphenoxyhexane) (CPH), and 1,8-bis(p-carboxyphenoxy)-3,6-dioxaoctane (CPTEG) copolymers. Within a short amount of time, linearly varying discrete compositional libraries including 25 film chemistries and 11 microsphere chemistries of polyanhydride random copolymers (based on CPH:SA and CPTEG:CPH) were designed, synthesized, and characterized. Both libraries were rapidly screened for cytotoxicity, cytokine production and cell surface marker expression (the latter studies were only performed with microsphere libraries due to limited volume with film libraries). The libraries were incubated with primary antigen presenting cells of the immune system, dendritic cells and macrophages, and rapid screening with the MTT assay showed that these libraries had no pronounced cytotoxic effect for any CPH:SA or CPTEG:CPH composition at a concentration higher than that expected for in vivo biomedical applications. By performing flow cytometry (to assay for cell surface markers) and the Luminex assay (for cytokine production), we were able to clearly identify trends correlating polyanhydride chemistry to cell surface marker expression and cytokine production. There was an observable increase in MHC II, CD86, CD40 and TNFα with an increase in the CPH content of the copolymers (i.e., more hydrophobic polymers) and an evident increase in IL-6 with an increase in the SA content of the copolymers (i.e., less hydrophobic polymers). These trends are important because they dictate the pathway of the immune response to be either cell-mediated or humoral, which is essential knowledge for the rational design of biomaterials as adjuvants for vaccine delivery. We proceeded further to validate the data with the use of informatics and statistical methodologies by performing principal component analysis and model fitting which correlated with previous conventionally determined results. These findings add to the large body of evidence supporting the use of polyanhydrides as drug carriers and vaccine adjuvants. The techniques discussed are amenable to numerous applications in the area of high-throughput polymer synthesis and screening and will enable for future rapid and optimal design of biomaterials.

698a De Novo Protein Design Based on Binding Affinity Predictions for the Discovery of Novel Inhibitors for Complement 3 and HIV-1

Meghan L. Bellows, Ho Ki Fung, and Christodoulos A. Floudas. Department of Chemical Engineering, Princeton University, Engineering Quadrangle, Princeton, NJ 08544

Protein design, also known as the inverse folding problem, seeks the amino acid sequence that will fold into a given 3-dimensional template. The protein design problem exhibits degeneracy due to the fact that many amino acid sequences fold into a given template. It is therefore important to examine all the possible sequences for a given template and rank them based upon specific properties that are being designed (activity, specificity, etc.). A new de novo design framework based on binding affinity calculations is introduced. The framework consists of two stages: a sequence selection stage and a binding affinity calculation stage. The sequence selection stage produces a rank-ordered list of amino acid sequences with the lowest energies by solving an integer programming sequence selection model [1]. The sequence selection model incorporates backbone flexibility into the design process by utilizing the set of structures obtained from NMR experiments. In doing so, the pairwise energy between two residues can take on a range of values depending upon the range of distances obtained from the structures. The second stage employs Monte Carlo simulations to first predict the structure (using Rosetta Abinitio [2]) of the sequences from stage one and then to perform docking simulations (using Rosetta Dock [3]) between the new sequence and the target protein. A rotamerically-based ensemble of structures for the new peptide, the target protein, and the peptide-protein complex is generated using Rosetta Design [4], and is used to calculate approximate molecular partition functions of the new peptide, the target protein, and the peptide-protein complex. Using these approximate partition functions, a binding affinity is calculated [5]. The more accurate the partition functions are, the more precise the binding affinity will be. This new framework was applied to a complex of Complement C3c with compstatin variant E1 and a complex of HIV-1 gp120 with CCR5. The studies of the design of compstatin variants involve calculating the binding affinity of known sequences of compstatin obtained from previous designs in the literature. The binding affinities follow the trends of the activity of the designed sequences compared to the native sequence. The studies of the design of an inhibitor for HIV-1 gp120 based upon CCR5 look at sequences generated from the sequence selection stage using both the original selection model that does not incorporate backbone flexibility and sequences generated using the model that does incorporate backbone flexibility. The sequences examined so far demonstrate that the model incorporating the backbone flexibility provides more sequences with higher binding affinities compared to the native sequence of CCR5. The novel framework predicts improved binding affinities for a number of the candidate sequences designed based on the structure of compstatin for C3c inhibition and the structure of CCR5 for gp120 inhibition.

[1] H. K. Fung, M. S. Taylor, and C. A. Floudas. Novel Formulations for the sequence selection problem in de novo protein design with flexible templates. Optim. Methods & Software, 22:51-71, 2007.

[2] C. A. Rohl, C. E. M. Strauss, K. M. S. Misura, and D. Baker. Protein structure prediction using

Rosetta. Methods in Enzymology, 383:66-93, 2004.

[3] J. J. Gray, S. Moughon, C. Wang, O. Schueler-Furman, B. Kuhlman, C. A. Rohl, and D. Baker, Protein-protein docking with simultaneous optimization of rigid-body displacement and side-chain conformations. Journal of Molecular Biology, 331:281-299, 2003.

[4] Y. Liu and B. Kuhlman. RosettaDesign server for protein design. Nucleic Acids Research, 34:W235-W238, 2006.

[5] R. H. Lilien, B. W. Stevens, A. C. Anderson, and B. R. Donald. A novel ensemble-based scoring and search algorithm for protein redesign and its application to modify the substrate specificity of the Gramicidin Synthetase A Phenylalanine Adenylation enzyme. Journal of Computational Biology, 12:740-761, 2005.

Quantitative Measurement Technique for Transcription Factor Profiles

Zuyi (Jacky) Huang, Fatih Senocak, Arul Jayaraman and Juergen Hahn
Department of Chemical Engineering
Texas A& M University
College Station, TX 77843-3122

Introduction

Control of gene expression by transcription factors is an integral component of cell signaling and gene expression regulation (Hoffmann et al., 2007). Different transcription factors exhibit different expression and activation dynamics, and together govern the expression of specific genes and cellular phenotypes (Grove and Walhout, 2008). An important requirement for the development of these signal transduction models is the ability to quantitatively describe the activation dynamics of transcriptions so that parameters can be estimated for model development. The activation of transcription factors under different conditions have been conventionally monitored using protein binding techniques such as electrophoretic mobility shift assay or chromatin immunoprecipitation (Elnitski et al., 2006). While these techniques provide snapshots of activation at a small set of single time points, they can yield only qualitative or semi-quantitative data at best. This approach also requires the use of multiple cell populations for each time point at which transcription factor activation is to be measured, and often, the true dynamics of transcription factors are not captured due to limited sampling points and frequencies. Hence, these methods are not ideal for investigating time-dependent activation of transcription factors in a quantitative manner.

More recently, fluorescence-based reporter systems have been developed for continuous and non-invasive monitoring of transcription factors and elucidation of regulatory molecule dynamics. Recent studies (Thompson et al., 2004; Wieder et al., 2005, King et al., 2007) have used green fluorescent protein (GFP) as a reporter molecule for continuously monitoring activation of a panel of transcription factors, underlying the inflammatory response in hepatocytes for 24 h. These systems involve expressing GFP under the control of a minimal promoter such that GFP expression and fluorescence is observed only when a transcription factor is activated (i.e., when the transcription factor binds to its specific DNA binding sequence and induces expression from a minimal promoter). The dynamics of GFP fluorescence is used as the indicator for dynamics of the transcription factor being profiled. The primary drawback with this approach is that it does not provide direct activation rates of the transcription factors being investigated. Even though transcription factor dynamics influence GFP dynamics, the relationship between the two is non-trivial as the induction of

GFP fluorescence itself involves multiple steps (i.e., transcription of GFP mRNA, GFP protein translation, post-translational processing, etc) (Subramanian and Srienc, 1996), and not all of these steps contribute equally to regulation of GFP expression. The observed fluorescence dynamics in GFP reporter cell systems is the result of two different dynamics: (i) the dynamics of transcription factor activation by a soluble stimulus-mediated signal transduction pathway and (ii) the dynamics of GFP expression, folding, and maturation. Therefore, it is necessary to uncouple the effects of these independent systems in order to quantitatively determine transcription factor activation profiles underlying cellular phenotpyes.

In this work, we develop a strategy for determining transcription factor concentrations from fluorescence microscopy data. This technique is based upon the following steps:

1) The image analysis method based on K-means clustering and Principal Component Analysis (PCA) (Huang and Hahn, 2007) is implemented to obtain a fluorescence intensity profile from the fluorescence microscopy images.
2) Based on the model initially presented by Subramanian and Srienc (1996), a model for transcription, translation, and activation of GFP is derived to correlate transcription factor concentration with fluorescence intensity.
3) A procedure for solving an inverse problem involving the model developed in 2) and the fluorescence data derived from the image analysis from step 1) is presented. This procedure computes the transcription factor profiles from fluorescence intensity data.

The technique has been implemented to derive quantitative concentration of NF-kB from fluorescence microscopy images of hepatocytes stimulated by TNF-α with four different concentrations. Part of the derived NF-kB data is then used to develop and refine a model of the TNF-α signaling pathway. The refined model of the TNF-α signaling pathway is tested on data not used for parameter estimation and it is found that it can predict the dynamics of TNF-α signaling pathway very well.

Preliminaries

Image analysis based on K-means clustering and Principal Component Analysis (PCA)

This method distinguishes the regions of the image with similar brightness by PCA (Hotelling, 1933) and then groups them into the same cluster by K-means clustering (Kaufman and Rousseeuw, 1990). The outline of this method is shown in the following steps:

1) For each image from the time-series of images, the areas in the image representing cells where fluorescence can be seen are determined by PCA and K-means clustering.
2) Once the cell region has been determined it is possible to compute the average fluorescence intensity by the following formula:

$$I = \left(\frac{\sum_{k=1}^{N_f} I_{f,k}}{N_f} - \frac{\sum_{k=1}^{N_b} I_{b,k}}{N_b}\right)_{stimulation} \quad (1)$$

$I_{f,k}$ refers the fluorescent intensity of the k_{th} pixel in a fluorescent cell region, $I_{b,k}$ refers the fluorescent intensity of the k_{th} pixel belonging to the background, N_f is the total number of pixels in the fluorescent cell region, N_b is the total number of pixels in the background. For a RGB image, the fluorescent intensity I is defined as the sum of the values of red and green and blue of each pixel. The reason for subtracting the intensity of the pixels representing the background is to reduce measurement noise due to brightness variations.

3) The intensity for each image is combined together to get the fluorescent intensity profile for the whole set of time-series images.

Model development

Two models are involved in this work: (a) a model describing the dynamics of the proteins involved in TNF-α signaling and (b) a model describing the dynamics of the proteins of a green fluorescent protein reporter system. The first model has the TNF-α concentration as the input to the system and the output of the system is the dynamic profile of NF-kB that results from TNF-α stimulation. The second model uses the NF-kB concentration as the input and predicts the fluorescence intensity profile that can be measured. Using these two models it is possible to determine the NF-kB concentration during an experiment by solving an inverse problem of the second model. The generated data set can then be further used to adjust parameters of the first model. Figure 1 illustrates the relationship between these two models.

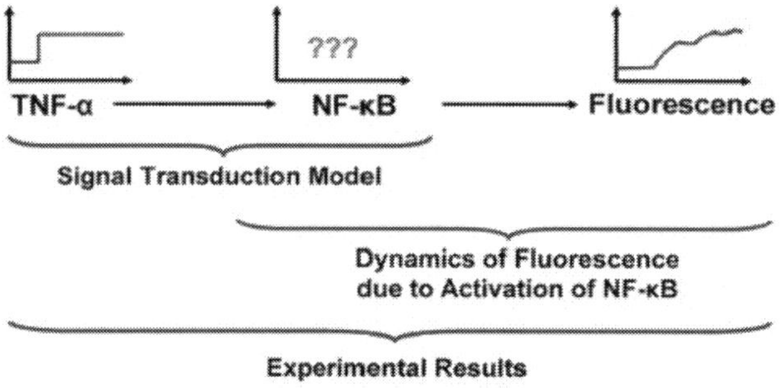

Figure 1: Relationship between input, output, and concentration of transcription factors with GFP-reporter systems.

The model describing TNF-α mediated signal transduction is shown in Figure 2 and the equations are given in 'Appendix 1'. This model is based upon the models described by Rangamani and Sirovich , 2007 and Lipniacki et al. , 2004 The model from Lipniacki et al. was used to describe signal transduction from IKKn to NF-kB whereas the model from Rangamani and Sirovich's work was used to describe signal transduction from TNF-α to IKKn. The reason for combining these two models is that the model from Lipniacki et al.'s work does not describe signal transduction from TNF-α to IKKn, while the paper by Rangamani and Sirovich states that the signal transduction from IKKn to NF-kB as described in their model should be updated as it represents a simplification of what is currently known about the signal transduction pathway. In order to combine these two models the assumption that c-IAP in the reaction "Caspase-3*+c-IAP->caspase-3*|c-IAP" from Rangamani and Sirovich's model can be replaced with $cgen_t$ from Lipniacki et al.'s model. The rationale behind this assumption is that c-IAP and $cgen_t$ are both involved in transcription of DNA.

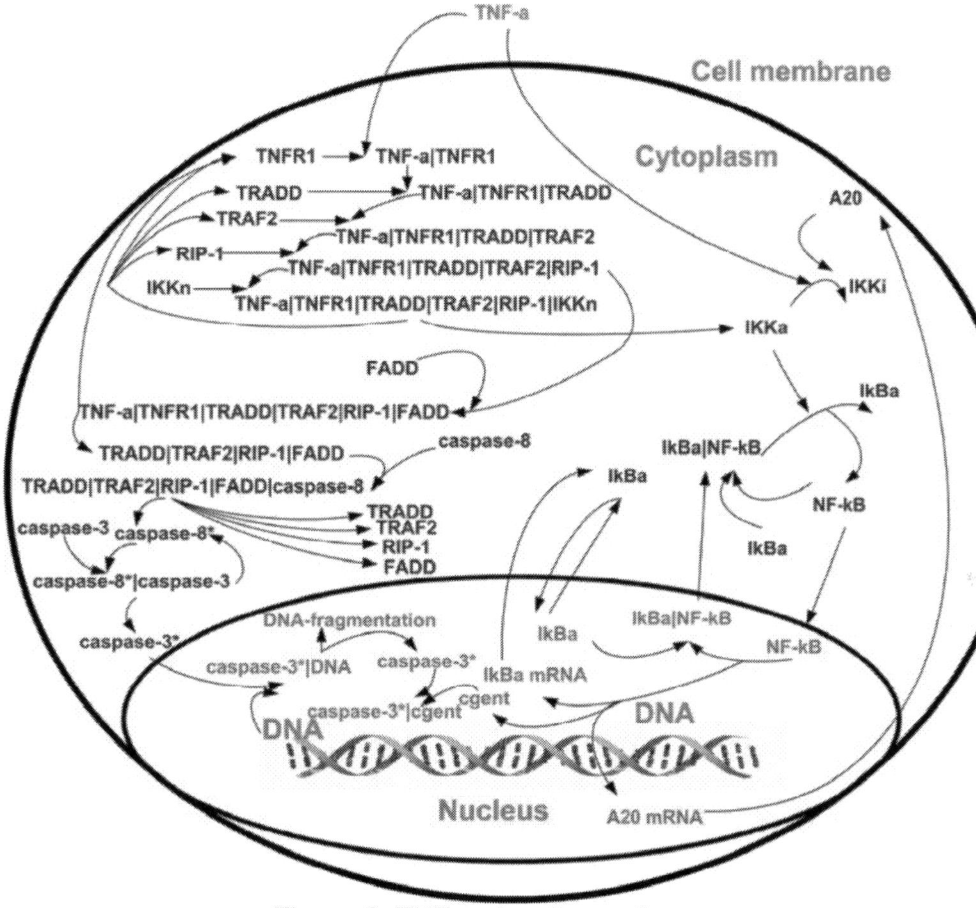

Figure 2: TNF-α signaling pathway

This integrated model, which consists of 37 differential equations and 60 parameters, can represent the dynamic behavior of the proteins involved in TNF-α-mediated NF-kB activation: TNF-α initiates the signal transduction by binding to its receptor TNFR1 and forming the complex TNF-α| TNFR1, which then recruits TRADD, TRAF2, RIP-1 to form the complex TNF-α| TNFR1|TRADD| TRAF2|RIP-1. This complex then activates two pathways: 1) it activates the apoptotic machinery by recruiting FADD; 2) it activates the NF-kB pathway by promoting the neutral form of IKK (IKKn) to the active form of IKK (IKKa). NF-kB is then released from the complex NF-kB|IkBα and translocates into the nucleus to initiate the transcription/translation process. Since the presence of NF-kB in the nucleus (i.e., activation of NF-kB) does not immediately lead to fluorescence seen in the images it is required to augment the developed model with equation describing transcription/translation as well as activation of GFP. The equations to be added are based upon the model described by Subramanian and Srienc , 1996 , where modifications are made to account for the constant reporter DNA levels in our experiments (i.e., due to stable integration of the reporter plasmid into the genomic DNA in our reporter cell line , Thompson et al. 2004) as well as to include the effect of transcription factor concentrations on the transcription rate. These changes result in the following model describing the measurement dynamics:

$$\frac{dm}{dt} = S_m p \frac{C_{NF-\kappa B}}{C + C_{NF-\kappa B}} - D_m m \tag{2}$$

$$\frac{dn}{dt} = S_n m - D_n n - S_f n$$

$$\frac{df}{dt} = S_f n - D_n f$$

where $C_{NF\text{-}kB}$ is the concentration of activated NF-kB in the nucleus, *m* is the mRNA concentration, *n* is the concentration of GFP, and *f* corresponds to the concentration of activated GFP. The values of the parameters shown in equation (2) are given in Table 1. The procedure for estimation of *C* is described below.

The experimental measurements consist of the fluorescence intensity, *I*, as seen on the images which is directly proportional to the concentration of activated green fluorescent protein:

$$f = \Delta I \tag{3}$$

Where Δ is the ratio between activated GFP and computed fluorescence intensity.

As *I* can be obtained from the fluorescence images that have been processed by the image analysis procedure described in the preliminary section, the dynamics of NF-kB can be computed by solving an inverse problem involving equations (2).

Table 1 - Parameters for the model shown in equation (2).

Parameter	Value	Parameter	Value
S_m	373 1/hr	S_f	0.347 1/hr
D_m	0.45 1/hr	C	108 nM
S_n	780 1/hr	p	5 nM
D_n	0.5 1/hr	$m(0)$, $n(0)$, $f(0)$	0 nM

A procedure to get the NF-kB profile from the fluorescent intensity data by solving an inverse problem

The activation of NF-kB in H35 reporter cells was investigated by stimulating with different TNF-α concentrations (6ng/ml, 10ng/ml, 13ng/ml, and 19 ng/ml) as described in the Methods section. The data was analyzed using the described image analysis procedure, resulting in the fluorescence intensity profiles shown (red line) in Figure 3. The error bars indicated +/- one standard deviation from the mean of the measurements taken for each time point.

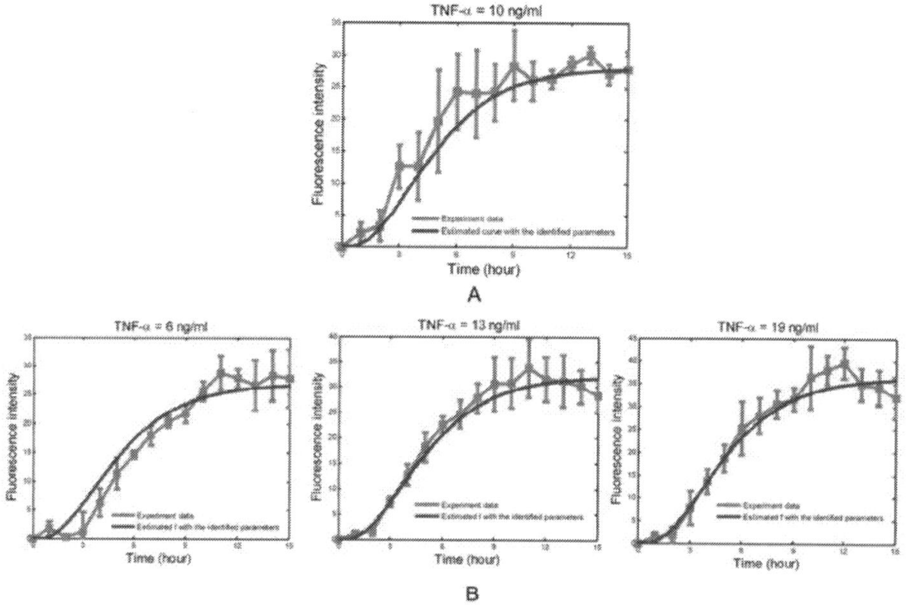

Figure 3: (A) Comparison of experimental data and the model predictions for f/Δ where the NF-kB concentration serves as the input to the model and is taken from Hoffman et al.'s paper (2002); (B) Experimental data and the fitted curve f/Δ for different TNF-α concentrations.

We developed a procedure that computes the NF-kB concentration profile from the experimental data by solving an inverse problem given by equations (2) and (3). In order to avoid a numerical solution of this inverse problem, we derived an analytical solution which computes $C_{NF\text{-}kB}$ from the fluorescence intensity profile *I*. This analytical solution treats equation (2) as a static nonlinearity

$$u = \frac{C_{\mathrm{NF-\kappa B}}}{C + C_{\mathrm{NF-\kappa B}}} \tag{4}$$

which is followed by a system of linear differential equations:

$$\begin{aligned} dm/dt &= S_m pu - D_m m \\ dn/dt &= S_n m - D_n n - S_f n \\ df/dt &= S_f n - D_n f \end{aligned} \tag{5}$$

Taking a Laplace transform of equation (5) results in $f(s)$ as a function of $u(s)$:

$$f(s) = \frac{S_f}{s + D_n} \cdot \frac{S_n}{s + D_n + S_f} \cdot \frac{S_m p}{s + D_m} u(s) \tag{6}$$

While it is possible to choose any function to describe $u(s)$, we opted for

$$u(s) = \frac{\omega_n^2}{s^2 + 2\varepsilon\omega_n s + \omega_n^2} \cdot \frac{T_\alpha}{s} \tag{7}$$

as $u(s)$ represents a concentration profile of $C_{\text{NF-kB}}$ that shows damped oscillatory behavior as has been reported in the literature (Hoffmann et al, 2002). Substituting equation (7) into equation (6) and performing an inverse Laplace transform results in:

$$f(t) = A_1 + A_2 e^{-D_n t} + A_3 e^{-(D_n + S_f)t} + A_4 e^{-D_m t} + A_7 e^{-\varepsilon\omega_n t} \sin(\omega_n \sqrt{1-\varepsilon^2}\, t + \varphi) \tag{8}$$

where A_1, A_2, A_3, A_4, A_7, and φ are constants with the values given in 'Appendix 2'. The values of the parameters ε, ω_n and T_α are estimated by fitting $f(t)$ to the experimental data for each experiment. The concentration of NF-kB is then given by:

$$C_{\text{NF-kB}} = \frac{CT_\alpha \sqrt{1-\varepsilon^2} - CT_\alpha e^{-\varepsilon\omega_n t} \sin(\omega_n \sqrt{1-\varepsilon^2}\, t + \phi)}{(1 - T_\alpha)\sqrt{1-\varepsilon^2} + T_\alpha e^{-\varepsilon\omega_n t} \sin(\omega_n \sqrt{1-\varepsilon^2}\, t + \phi)}, \text{ where} \tag{9}$$

$$\phi = \arctan \frac{\sqrt{1-\varepsilon^2}}{\varepsilon}$$

The values of C from equation (4) and Δ from equation (3) only need to be estimated once and can be assumed to be constant for all future experiments. We have chosen the concentration profile for NF-kB as reported in the paper by Hoffman et al., 2002, which corresponds to a stimulation with 10 ng/ml of TNF-α, as the input, and have estimated C and Δ from experimental data that we have collected for stimulation with 10 ng/ml of TNF-α. The value of C was determined to be 108 nM and Δ was found to be equal to 2.5562×10^4. It should be noted that some of the data derived from a stimulation with 10 ng/ml of TNF-α was used for determining these parameter values, while other data points will be used for testing model. Figure 3A shows the fit of equation (11) to the data generated by this experiment.

Figure 3B depicts the experimental data for stimulation with 6 ng/ml, 13 ng/ml, and 19 ng/ml of TNF-α as well as the results of the system identification using equation (8). The values for C and Δ are constant for these experiments, however, the values for ε, ω_n and T_α are estimated separately for each data set. The corresponding concentration profiles for NF-kB, as computed by equation (9) are shown in Figure 4. It can be seen that stimulation with higher concentrations of TNF-α results in larger long-term concentrations of NF-kB as well as in higher peak concentrations. One important aspect of this procedure is that the data obtained is quantitative (i.e., numerical values of the NF-kB profile at each time point are obtained) and not merely qualitative.

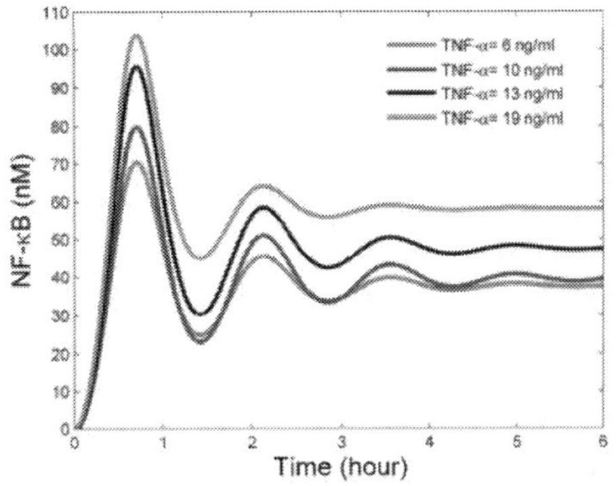

Figure 4: NF-kB profiles computed via solution of the inverse problem for TNF-α concentrations of 6 ng/ml, 10 ng/ml, 13 ng/ml, and 19 ng/ml.

These results for stimulation with 6 ng/ml, 13 ng/ml, and 19 ng/ml of TNF-α were used to estimate parameters of the signal transduction pathway model. Since the developed model contains many more parameters than can be estimated from three time series of data, it was required to use local sensitivity analysis to determine which parameters should be re-estimated. It was determined that the parameters c_3, k_{1p}, and k_r are good candidates for estimation. Nonlinear least square routines in MATLAB were then used to estimate these three parameters. The estimated values were found to be 0.0104, 0.0740 and 2.50, respectively. Since the data derived from the stimulation with 10 ng/ml of TNF-α was not used for estimating these parameters, this data set can be used for validating the accuracy of the updated model. Figure 5 shows the model prediction for 10 ng/ml of TNF-α together with the experimental results derived from the described image analysis procedure. It can be concluded that the updated model predicts experimental data very well.

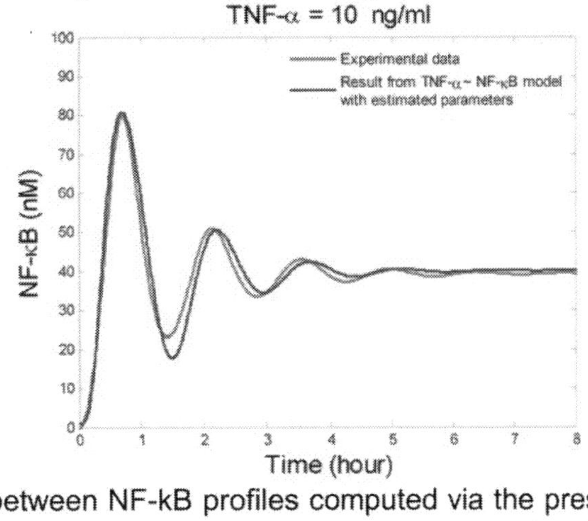

Figure 5: Comparison between NF-kB profiles computed via the presented technique for 10 ng/ml of TNF-α and simulation of the model where some parameters have been re-estimated.

Discussion and conclusion

In this study, we have demonstrated that transcription factor activation profiles can be quantitatively extracted from fluorescence reporter data. The proposed approach was effective in deriving transcription factor activation rates from GFP profiles generated from NF-kB reporter cells stimulated with 10 – 50 ng/mL of TNF-α, a concentration range that is commonly used in cell culture experiments (Damelin et al., 2007; King et al., 2007) and reported to result in strong activation of NF-kB (Wieder et al., 2005). However, predicting NF-kB activation at lower concentrations of TNF-α(< 10 ng/mL) was not as effective due to low levels of GFP signal. This is evident from Figure 3B which shows a better correlation between the model and experimental data at higher (13 and 19 ng/mL) than at lower (6 ng/mL) TNF-α concentrations. Therefore, while our method is effective for moderate-to-high levels of activation, further improvement (e.g., in the image analysis methods) is needed to increase the GFP signal/noise ratio for effectively predicting profiles of low abundance transcription factors.

Another discrepancy between the model and experimental data is predicting long-term NF-kB activation profiles. The data in Figure 3B shows that fluorescence decreases after ~ 11 h even though the stimulus (TNF-α) is continually present, with the decrease being more pronounced at the higher concentrations. However, this decrease is not reflected in Figure 3B which shows NF-kB levels being constant beyond 11 h as the assumed model structure from equation (7) cannot represent this decrease. It is possible to postulate a different profile for the transcription factor, resulting in differences in equation (7), e.g., one that can reflect such a decrease. However, it is not clear if the decrease in fluorescence observed after ~ 11 h of stimulation results from experimental artifacts (i.e., fluorescence photobleaching and cell death arising from cells being repeatedly exposed to UV light for imaging) or is a real biological phenomenon (i.e., consequence of change in gene expression arising due to constant stimulation with TNF-α). A better understanding of long-term activation is needed to evaluate this behavior.

In summary we have developed a methodology for quantitatively determining transcription factor profiles. This technique makes use of fluorescence microscopy images from a GFP reporter system for transcription factor activation and involves solving an inverse problem to determine the transcription factor profile from the fluorescence intensity dynamics. Data generated by this method can then be used to estimate parameters for signal transduction pathway models. This technique was applied to the activation of NF-kB by TNF-α, however, it can be used to determine transcription factor profiles for any system where limited qualitative knowledge about the transcription factor dynamics exists.

References

Damelin. LH., S. Coward, M. Kirwan, P. Collins, C. Selden, HJF. Hodgson (2007). "Fat-loaded HepG2 spheroids exhibit enhanced protection from Pro-oxidant and cytokine induced damage." J Cell Biochem, 101:723-734.

Elnitski, L., V. X. Jin, P. J. Farnham and S. J. Jones (2006). "Locating mammalian transcription factor binding sites: a survey of computational and experimental techniques." Genome Res 16, 1455-1464.

Grove, C. A. and A. J. M. Walhout (2008). "Transcription factor functionality and transcription regulatory networks." Molecular Biosystems, 4: 309-314.

Hotelling, H (1933). "Analysis of a Complex of Statistical Variables into Principal Components." Journal of Educational Psychology, 24, 417-441.

Hoffmann, A, A. Levchenko, M.L. Scott, and D. Baltimore (2002). "The IkB–NF-kB signaling module: temporal control and selective gene activation." Science, 298(8), 1241-1245.

Huang, Z and J. Hahn (2007). "Development and Comparison of Algorithms for Analysis of Fluorescent Images for Studying the Dynamics of Signal Transduction Pathways." AIChE Annual Conference at Salt Lake City, 2007.

Kaufman, L. and P. J. Rousseeuw (1990). "Finding Groups in Data: an Introduction to Cluster Analysis." John Wiley & Sons Press.

King, K.R., S. Wang, D. Irimia, A. Jayaraman, M. Toner, M.L. Yarmush (2007). "A high-throughput microfluidic real-time gene Expression living cell array." Lab Chip 7, 77-85.

Lipniacki, T., P. Paszek, A.R. Brasier, B. Luxon, and M. Kimmel (2004). "Mathematical model of NF-kB regulatory module." Journal of Theoretical Biology 228, 195–215.

Rangamani, P. and L. Sirovich (2007). "Survival and apoptotic pathways initiated by TNF-α: modeling and predictions." Biotechnology & Bioengineering 97(5), 1216-1229.

Subramanian, S. and F. Srienc (1996). "Quantitative analysis of transient gene expression in mammalian cells using the green fluorescent protein." J. Biotechnol. 49, 137-151.

Thompson, D. M., K. R. King, K. J. Wieder, M. Toner, M. L. Yarmush and A. Jayaraman (2004). "Dynamic gene expression profiling using a microfabricated living cell array." Anal Chem 76, 4098-4103.

Wieder, KJ., KR. King, DM. Thompson, C. Zia, ML. Yarmush, A. Jayaraman (2005). "Optimization of reporter cells for expression profiling in a microfluidic device." Biomed Microdevices, 7:213-222.

Appendix 1

This file describes the equations, the initial values of state variables, and parameters of the model describing TNF-α mediated signal transduction

$$\dot{x}_1 = -k_{1p}x_1\mathrm{u} + k_{2p}x_2 + k_{17p}x_{26} + k_{11p}x_{10}$$

$$\dot{x}_2 = k_{1p}x_1\mathrm{u} - k_{2p}x_2 - k_{3p}x_2x_3 + k_{4p}x_4$$

$$\dot{x}_3 = -k_{3p}x_2x_3 + k_{4p}x_4 + k_{11p}x_{10} + k_{20p}x_{29}$$

$$\dot{x}_4 = k_{3p}x_2x_3 - k_{4p}x_4 - k_{5p}x_4x_5 + k_{6p}x_6$$

$$\dot{x}_5 = -k_{5p}x_4x_5 + k_{6p}x_6 + k_{11p}x_{10} + k_{20p}x_{29}$$

$$\dot{x}_6 = k_{5p}x_4x_5 - k_{6p}x_6 - k_{7p}x_6x_7 + k_{8p}x_8$$

$$\dot{x}_7 = -k_{7p}x_6x_7 + k_{8p}x_8 + k_{11p}x_{10} + k_{20p}x_{29}$$

$$\dot{x}_8 = k_{7p}x_6x_7 - k_{8p}x_8 - k_{9p}x_8x_9 + k_{10p}x_{10} - k_{15p}x_8x_{25} + k_{16p}x_{26}$$

$$\dot{x}_9 = -k_{9p}x_8x_9 + k_{10p}x_{10} + k_{14p}x_{14}$$

$$\dot{x}_{10} = k_{9p}x_8x_9 - k_{10p}x_{10} - k_{11p}x_{10}$$

$$\dot{x}_{11} = k_{11p}x_{10} - k_3x_{11} - T_Rk_2x_{11}x_{17} - k_{deg}x_{11} - a_2x_{11}x_{19} + t_1x_{13} - a_3x_{11}x_{22} + t_2x_{14}$$

$$\dot{x}_{12} = k_3x_{11} + T_Rk_2x_{11}x_{17} - k_{deg}x_{12}$$

$$\dot{x}_{13} = a_2x_{11}x_{19} - t_1x_{13}$$

$$\dot{x}_{14} = a_3x_{11}x_{22} - t_2x_{14}$$

$$\dot{x}_{15} = c_{6a}x_{22} - a_1x_{15}x_{19} + t_2x_{14} - i_1x_{15}$$

$$\dot{x}_{16} = i_1k_vx_{15} - a_1x_{20}x_{16}$$

$$\dot{x}_{17} = c_4x_{18} - c_5x_{17}$$

$$\dot{x}_{18} = c_2 + c_1x_{16} - c_3x_{18}$$

$$\dot{x}_{19} = -a_2x_{11}x_{19} - a_1x_{19}x_{15} + c_{4a}x_{21} - c_{5a}x_{19} - i_{1a}x_{19} + e_{1a}x_{20}$$

$$\dot{x}_{20} = -a_1x_{16}x_{20} + i_{1a}k_vx_{19} - e_{1a}k_vx_{20}$$

$$\dot{x}_{21} = c_{2a} + c_{1a}x_{16} - c_{3a}x_{21}$$

$$\dot{x}_{22} = a_1x_{19}x_{15} - c_{6a}x_{22} - a_3x_{11}x_{22} + e_{2a}x_{23}$$

$$\dot{x}_{23} = a_1x_{20}x_{16} - e_{2a}k_vx_{23}$$

$$\dot{x}_{24} = c_{2c} + c_{1c}x_{16} - c_{3c}x_{24} - k_{28p}x_{24}x_{33}$$

$$\dot{x}_{25} = -k_{15p}x_8x_{25} + k_{16p}x_{26} + k_{20p}x_{29}$$

$$\dot{x}_{26} = k_{15p}x_8x_{25} - k_{16p}x_{26} - k_{17p}x_{26}$$

$$\dot{x}_{27} = k_{17p}x_{26} - k_{18p}x_{27}x_{28} + k_{19p}x_{29}$$

$$\dot{x}_{28} = -k_{18p}x_{27}x_{28} + k_{19p}x_{29}$$

$$\dot{x}_{29} = k_{18p}x_{27}x_{28} - k_{19p}x_{29} - k_{20p}x_{29}$$

$$\dot{x}_{30} = k_{20p}x_{29} - k_{21p}x_{30}x_{31} + k_{22p}x_{32} + k_{23p}x_{32}$$

$$\dot{x}_{31} = -k_{21p}x_{30}x_{31} + k_{22p}x_{32}$$

$$\dot{x}_{32} = k_{21p}x_{30}x_{31} - k_{22p}x_{32} - k_{23p}x_{32}$$

$$\dot{x}_{33} = k_{23p}x_{32} - k_{28p}x_{24}x_{33} - k_{24p}x_{33}x_{36} + k_{25p}x_{37} + k_{26p}x_{37}$$

$$\dot{x}_{34} = k_{26p}x_{37}$$

$$\dot{x}_{35} = k_{28p}x_{24}x_{33}$$

$$\dot{x}_{36} = -k_{24p}x_{33}x_{36} + k_{25p}x_{37}$$

$$\dot{x}_{37} = k_{24p}x_{33}x_{36} - k_{25p}x_{37} - k_{26p}x_{37}$$

$$y = x_{16} / k_r$$

State variables of the model and their initial values:

Name	Species	Initial values (μM)
x_1	TNFR1	0.1
x_2	TNF-α/TNFR	0
x_3	TRADD	0.15
x_4	TNF-α/TNFR1/TRADD	0
x_5	TRAF2	0.1
x_6	TNF-α/TNFR1/TRADD/TRAF2	0
x_7	RIP-1	0.1
x_8	TNF-α/TNFR1/TRADD/TRAF2/RIP-1	0
x_9	IKKn	0.2
x_{10}	TNF-α/TNFR1/TRADD/TRAF2/RIP-1/IKKn	0
x_{11}	IKKa	0
x_{12}	inactive IKK	0
x_{13}	cytoplasmic IKK|IkBa complex	0
x_{14}	cytoplasmic IKK|IkBa|NF-kB complex	0
x_{15}	free cxtoplasmic NF-kB	0.0003
x_{16}	free nuclear NF-kB	0.0023
x_{17}	cytoplasmic A20	0.0048
x_{18}	A20 transcription	0
x_{19}	free cytoplasmic IkBa	0.0025
x_{20}	free nuclear IkBan	0.0034
x_{21}	IkB transcription	0
x_{22}	cytoplasmic IkBa|NF-kB complex	0.0592
x_{23}	Nuclear IkBa|NF-kB complex	0.0001
x_{24}	Control gene mRNA level or c-IAP	0
x_{25}	FADD	0.1
x_{26}	TNF-α/TNFR1/TRADD/TRAF2/RIP-1/FADD	0
x_{27}	TRADD/TRAF2/RIP-1/FADD	0
x_{28}	Caspase-8	0.08
x_{29}	TRADD/TRAF2/RIP-1/FADD/caspase-8	0
x_{30}	Caspase-8*	0
x_{31}	Caspase-3	0.2
x_{32}	Caspase-8*/caspase-3	0
x_{33}	Caspase-3*	0
x_{34}	DNA-fragmentation	0
x_{35}	Caspase-3*/c-IAP	0
x_{36}	DNA intact	0.8
x_{37}	Caspase-3*/DNA	0

Note: u is the concentration of TNF-α, ng/ml. The molecule weight of TNF-α is 17 kDa. The unit ng/ml can be converted to μM by dividing by 17×10^3. y is the system output NF-kB after being scaled by k_r in units of μM.

Values of the parameters

Name	Value	Name	Value
k_v	5	k_{1p}	0.0740 (0.185)
AB^*	1	k_{15p}	0.185
c_1	$5\times10^{-7}\,AB$	k_{2p}	0.00125
c_2	0	k_{16p}	0.00125
c_3	0.0104 (0.0004)	k_{3p}	0.185
c_4	0.5	k_{17p}	0.37
c_5	0.0003	k_{4p}	0.00125
k_1	0.0025	k_{18p}	0.5
k_2	0.1	k_{5p}	0.185
k_3	0.0015	k_{19p}	0.2
k_{deg}	0.000125	k_{6p}	0.00125
a_2	0.2	k_{20p}	0.1
a_1	0.5	k_{7p}	0.185
a_3	1.	k_{21p}	0.1
t_1	0.1	k_{8p}	0.00125
t_2	0.1	k_{22p}	0.06
AA^*	1	k_{9p}	0.185
c_{1a}	$5\times10^{-7}\,AA$	k_{23p}	100
c_{2a}	0	k_{10p}	0.00125
c_{3a}	0.0004	k_{24p}	0.185
c_{4a}	0.5	k_{11p}	0.37
c_{5a}	0.0001	k_{25p}	0.00125
c_{6a}	0.00002	k_{12p}	0.014
i_1	0.0025	k_{26p}	0.37
e_{2a}	0.01	k_{13p}	0.00125
i_{1a}	0.001	k_{14p}	0.37
e_{1a}	0.0005	k_{28p}	0.5
c_{1c}	5×10^{-7}	p	1.75
c_{2c}	0	T_r^*	1
c_{3c}	0.0004	k_r	2.5

* Note: 1) $AA = 1$ refers to wt cell, while $AA = 0$ refers to IkBa deficient cell
2) $AB = 1$ refers to wt cell, while $AB = 0$ refers to A20 deficient cell
3) $T_r = 0$ when TNF-α is off, while $T_r = 1$ when TNF-α is on
4) Values in brackets refer to the model fit to the experimental data

Appendix 2

Equations for computing the values of the constants found in Equation (11)

$$A_1 = S_f S_n S_m p T_a / D_n /(D_n + S_f)/ D_m$$

$$A_2 = -S_n S_m p \omega_n^2 T_a /(D_m - D_n)/(D_n^2 - 2\varepsilon\omega_n D_n + \omega_n^2)/ D_n$$

$$A_3 = S_n S_m p \omega_n^2 T_a /(D_m - D_n - S_f)/((D_n + S_f)^2 - 2\varepsilon\omega_n (D_n + S_f) + \omega_n^2)/(D_n + S_f)$$

$$A_4 = S_f S_n S_m p \omega_n^2 T_a /(D_n - D_m)/(D_m - D_n - S_f)/(D_m^2 - 2\varepsilon\omega_n D_m + \omega_n^2)/ D_m$$

$$A_5 = -C_0 d_1 /(bd_1^2 + bd_0^2)$$

$$A_6 = C_0 (ad_1 + bd_0)/(bd_1^2 + bd_0^2)$$

$$A_7 = \sqrt{A_5^2 + (\frac{A_6 - A_5\varepsilon\omega_n}{\omega_n\sqrt{1-\varepsilon^2}})^2}$$

$$a = -\varepsilon\omega_n$$

$$b = \omega_n\sqrt{1-\varepsilon^2}$$

$$d_1 = -(a_3 + 4a)b^3 + (3a_3a^2 + 2a_2a + 4a^3 + a_1)b$$

$$d_0 = b^4 + a_3a^3 - (3a_3a + a_2 + 6a^2)b^2 + a_2a^2 + a^4 + a_1a$$

$$a_1 = (D_n^2 + D_n S_f) D_m$$

$$a_2 = D_n^2 + D_n S_f + 2D_n D_m + D_m S_f$$

$$a_3 = 2D_n + D_m + S_f$$

$$C_0 = S_f S_n S_m p \omega_n^2 T_\alpha$$

$$\varphi = \arctan\frac{A_5\omega_n\sqrt{1-\varepsilon^2}}{A_6 - A_5\varepsilon\omega_n}$$

698c Reconstructing Transcription Factor Networks Using the Living Cell Array

Eric Yang, Rutgers - The State University of New Jersey, Biomedical Engineering Department, Chemical & Biochemical Engineering Department, Room 108 BME Building - 617 Browser Road, Piscataway, NJ 08854, Ioannis Androulakis, Department of Biomedical Engineering, Rutgers - The State University of New Jersey, Biomedical Engineering Department, Chemical & Biochemical Engineering Department, Room 108 BME Building - 617 Browser Road, Piscataway, NJ 08854, and Martin Yarmush, Biomedical Engineering, Rutgers - The State University of New Jersey, Biomedical Engineering Department, Chemical & Biochemical Engineering Department, Room 108 BME Building - 617 Browser Road, Piscataway, NJ 08854.

Transcriptional interactions are one of the important components which govern gene expression. Their ability to alter gene expressions is dependent primarily upon the amount of activated transcription factor present in the system as well as the binding kinetics of said transcription factor. Therefore, the overall mechanism of transcriptional activation is that some ligand will activate a given transcription factor, this transcription factor will translocate into the nucleus, and then either up or down-regulate a given gene. The complexity of the system arises from the fact that transcription factors as well as their ligands can themselves be protein. Therefore, the activation of a given transcription factor may in turn alter the production of other transcription factors, which may eventually feedback and regulate the initial transcription factor via various means. Therefore, complexities arise from such a system because the activated levels of a given transcription factor may be regulated by the levels of many other transcription factors.

Thus, rather than thinking of transcriptional interactions as merely the interactions of an activated transcription factor with their various gene targets, it makes more sense to think of the interactions as a complex network in which transcription factors can interact with each other. However, to obtain such a network is important to quantify how the activation of one transcription factor can affect the level of activation of the other transcription factors.

The Living Cell Array (LCA)[1] is a novel microfluidics device which is able to determine the levels of an activated transcription factor in a high throughput manner. Not only is it able to obtain the levels of a set of transcription factors in parallel, but it is able to do this with a high temporal resolution as well under multiple stimulatory conditions. It accomplishes this by first utilizing a highly specific marker for the activity of a specific transcription factor. It accomplishes this task by first utilizing a plasmid which contains an unstable green fluorescence protein (GFP), a minimal promoter, and 4 repeats of a transcription factor consensus sequence obtained via TRANSFAC[2] which is highly specific for a given transcription factor.

When a transcription factor is activated by its specific external factor, it will then bind to the binding regions in the plasmid which then causes the transcription of the unstable GFP. This GFP due to its unstable nature is able to reflect the amount of transcription factor present at a relatively instantaneous point in time. Thus the fluorescence detected in a given cell functions as an accurate surrogate for a given transcription factor. The use of the microfluidics device then allows populations of identical cells to be stimulated in parallel with different stimulatory cocktails as well to be imaged in a high throughput manner.

The current experimental implementation of the LCA was used to explore first the response of hepatic cells to various inflammatory cytokines such as TNF-alpha, IL1, IL6, IFG-gamma, and the anti-inflammatory drug Dexamathasone. To quantify the effect of these factors upon inflammation, the transcription factor NFkB, STAT3, STAT1, HSE, GRE, ISRE were analyzed.

Because of the structure and quality of the data obtained via the LCA, it open up new opportunities for computational analysis, specifically through the determination of both network architectures as well as various clues as to the underlying mechanisms associated with transcription factor activation. We propose the use of a novel numerical technique to analyze the LCA data which we term Reverse Euler Deconvolution (RED)

Like many techniques, RED utilizes the general hypothesis that the system can be modeled via a set of ordinary differential equations in which the effect of multiple transcription factors can be treated in an additive sense. However, because of the nature of the data, in which there are more stimulatory profiles than there are transcription factors as well as the high temporal resolution of the data, we do not have to impose constraints to make the problem well-posed. In contrast to other techniques such as NCA[3], NIR [4], PLS[5], RED coupled with the LCA does not need to enforce any type of network architecture upon the data in the form of specific conditions a network must meet or in the number of connections that is required to fully define the system. Furthermore, RED treats the matrix A not as a matrix of constants as in the previous cases, but as time varying functions whose numerical response may provide insights as to the underlying process.

Utilizing a mixed integer linear programming formulation, we are not only able to solve for the time varying responses which are most likely with a given network architecture, but also the network architecture itself. Furthermore, we can show that this network architecture illustrates a significant conservation of interactions over multiple solutions as well as the fact that the interactions themselves show characteristic behavior of simple network motifs in response to a step input.

1. King KR, Wang S, Irimia D, Jayaraman A, Toner M, Yarmush ML: A high-throughput microfluidic real-time gene expression living cell array. Lab Chip 2007, 7:77-85.

2. Matys V, Fricke E, Geffers R, Gossling E, Haubrock M, Hehl R, Hornischer K, Karas D, Kel AE, Kel-Margoulis OV, et al: TRANSFAC: transcriptional regulation, from patterns to profiles. Nucleic Acids Res 2003, 31:374-378.

3. Liao JC, Boscolo R, Yang YL, Tran LM, Sabatti C, Roychowdhury VP: Network component

analysis: reconstruction of regulatory signals in biological systems. Proc Natl Acad Sci U S A 2003, 100:15522-15527.

4. Gardner TS, di Bernardo D, Lorenz D, Collins JJ: Inferring genetic networks and identifying compound mode of action via expression profiling. Science 2003, 301:102-105.

5. Boulesteix AL, Strimmer K: Predicting transcription factor activities from combined analysis of microarray and ChIP data: a partial least squares approach. Theor Biol Med Model 2005, 2:23.

698d Single-Cell Behavior and Population Heterogeneity: An Inverse Modeling Approach for Computing the Intrinsic Physiological State Functions

Konstantinos Spetsieris, Nikos V. Mantzaris, and Kyriacos Zygourakis. Chemical and Biomolecular Engineering, Rice University, Houston, TX 77005

Cell population balance models can account for the phenotypic heterogeneity that characterizes isogenic cell populations. To utilize the predictive power of these models, however, we must determine the single-cell reaction and division rates as well as the partition probability density function of the cell population. These functions (that are collectively called Intrinsic Physiological State or IPS functions) can be obtained through the Collins-Richmond (1962) inverse cell population balance modeling methodology, if we know the phenotypic distributions of (a) the overall cell population, (b) the dividing cell subpopulation and (c) the newborn cell subpopulation.

In earlier communications, we have described the development of a novel assay based on fluorescence microscopy and image processing that can accurately quantify the three aforementioned distributions and account for unequal partitioning at cell division. We have also developed a numerical procedure that can accurately recover the three IPS functions from simulated data and under the assumption that the three phenotypic distributions are known with arbitrarily high accuracy. In practice, however, our measurements will have uncertainty due to random errors. Moreover, the three distributions will be determined from a finite sample of the cell population. For example, our earlier studies with E. coli cells carrying the genetic toggle network with a GFP marker have shown that the fluorescence microscopy assay only needs about 3,000 cells to compute the overall phenotypic distribution of a cell population with accuracy comparable to that achievable with flow cytometry. A sample of this size yields about 300 dividing cells.

The objective of the present study is to develop a robust numerical procedure that can accurately recover the IPS functions of cell populations from finite samples and in the presence of the aforementioned sources of error. We have simulated finite sampling from a cell population by generating a finite number of random deviates following a range of pre-selected distributions. These random deviates represent the total cell property of interest for each individual cell measured. To simulate the effects of uncertainty in the experimental data, we added random errors to the data and evaluated different methods for noise filtering. We have also employed nonparametric methods to estimate the three densities. Additionally, we have compared the NDF to the CDF formulation for solving the integral equation that yields the partition probability density function.

The developed algorithm was tested by applying it to study a population of E. coli cells with an IPTG-inducible genetic toggle network equipped with a gfpmut3 gene that functions as a reporter of intracellular expression levels. Results from the inverse modeling will be analyzed to understand how IPTG affects the reaction and division rates as well as the partitioning mechanism at cell division, and to elucidate the interplay between cell population heterogeneity and single-cell behavior.

References

Collins JF, Richmond MH. 1962. Rate of growth of Bacillus cereus between divisions. Journal of General Microbiology, 28,15-33

STOCHASTIC SIMULATION AND SYSTEMS ANALYSIS OF INSULIN-STIMULATED GLUT4 TRANSLOCATION

Eric C. Kwei, Kevin R. Sanft, Jason E. Shoemaker, Linda R. Petzold, and Francis J. Doyle III
University of California, Santa Barbara, CA 93106

Introduction

The estimated number of people in the US with diagnosed diabetes has more than doubled in the last 15 years to 14.6 million, with associated annual medical costs of $174 billion; type 2 diabetes mellitus (T2DM) represents 90 to 95% of these cases [1].

T2DM is primarily linked to obesity and insulin resistance—a decreased sensitivity of glucose response to normal insulin levels—which have been linked to defects in the insulin signaling pathway. Analysis of detailed mathematical models of insulin signaling should yield a better understanding of the underlying mechanisms of insulin resistance and its subsequent progression to T2DM. This may be very relevant for maximizing treatment efficacy and minimizing side effects, which can ultimately improve the quality of life for those who suffer from T2DM.

We have brought a variety of engineering tools, including stochastic modeling, parameter sensitivity analysis, and robust performance analysis, to bear on an existing insulin signaling model, published by Sedaghat et al. [2]. This differential equation (DE) model, based primarily on mass action kinetics, largely reproduces the mechanisms in adipocytes (fat cells) that transduce an insulin input signal to movement of GLUT4, a glucose transporter responsible for glucose uptake, to the cell surface.

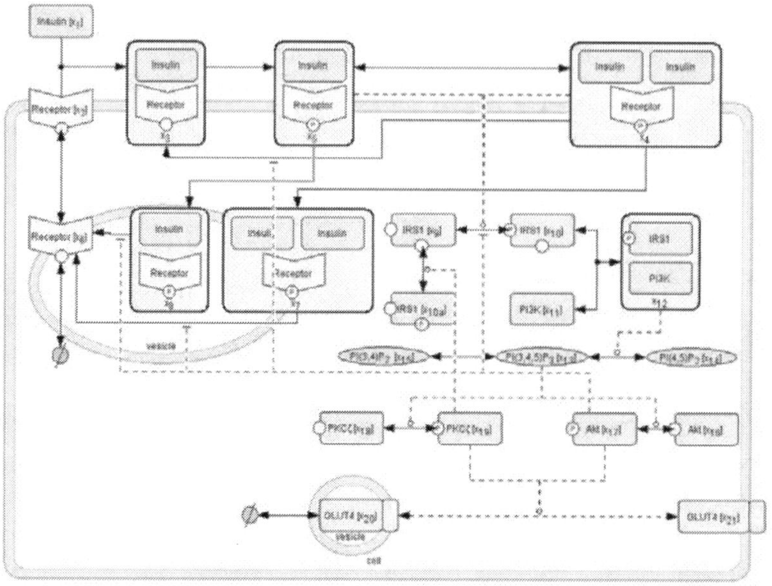

Figure 1. Sedaghat insulin-stimulated GLUT4 translocation model with feedback mechanisms

The Sedaghat model of insulin-stimulated GLUT4 translocation can be decomposed into 3 sub-models. The first sub-model describes insulin receptor dynamics—insulin binds to insulin receptor, causing subsequent receptor autophosphorylation; receptor recycling dynamics are also included. The second sub-model describes the phosphorylation cascade downstream from the insulin receptor: phosphorylated insulin receptor has tyrosine kinase activity, leading to the activation of a signaling cascadeconsisting of insulin receptor substrate 1 (IRS1), phosphatidylinositol 3-kinase (PI3K), phosphatidylinositol triphosphate (PIP_3), followed by protein kinases B (Akt) and C (PKC). The final sub-model describes the movement and fusion of specialized GLUT4 storage vesicles with the plasma membrane by Akt and PKC —the extent of GLUT4 translocation is quantified by the percentage of the total amount of GLUT4 that exists in the plasma membrane (surface GLUT4). Two versions of the model were proposed by Sedaghat et al., one incorporating both positive and negative feedback mechanisms with 21 states and one without these mechanisms with 20 states (Figure 1).

Upon scaling the deterministic model for a typical human adipocyte cell volume of 0.93 nL [3], we found that molecule counts revealed that several model species were present in small numbers (O(1) in some instances). When such small populations are involved, fluctuations may have significant effects on system behavior. To investigate this possibility, we have developed and simulated a stochastic version of the Sedaghat model.

While the Sedaghat model is the most complete available in the literature, it does not incorporate a number of known signaling components and has, to date, not been validated. Thus, we wanted to design an optimal experiment to better identify the model. To do this, we estimated the kinds of fluctuations one might observe in a reasonably-scaled experiment using our stochastic simulation of the Sedaghat model. Using these fluctuation estimations combined with parameter sensitivity analysis, we optimized experimental design for model identification.

We have also approached this model from a different but complementary perspective. It may not be terribly important to identify each model parameter precisely, as the signaling network structure may be robust to variations in that parameter. Conversely, the signaling network may be particularly fragile to perturbations in certain parameters, which may pinpoint potential causes for insulin resistance or even suggest potentially potent targets for drug therapy. Therefore, we performed a robustness analysis on the Sedaghat model to identify specific parameters that are crucial for robust performance of the GLUT4 response to insulin.

Stochastic Simulation of Sedaghat Model

The accuracy of a deterministic model fitted to data depends on measurement precision, which is limited by, among other factors, stochastic fluctuation magnitude. Because modern experimental methods can quantify the dynamic response of single target cells to insulin, we became interested in establishing whether a deterministic model can completely describe a system at this scale.

A discrete stochastic model can be derived from an ordinary differential equation (ODE) model by first representing the ODE's as individual chemical reactions. The rates of change in the state variables are then converted to *propensity functions* that describe the probability that a corresponding reaction will occur. When the ODE model is represented using only mass action kinetics and the chemical species are assumed to be well- mixed, the conversion to a stochastic model is straightforward and the resulting stochastic model has a time-dependent probability of being in any possible state that is described by the *chemical master equation* (CME). Due to its high dimensionality, the CME can rarely be solved exactly. However, using Gillespie's stochastic simulation algorithm (SSA) to run many realizations of the stochastic model, it is possible to get reasonable estimates of the solution of the CME [4].

A stochastic version of the Sedaghat model with feedback mechanisms was implemented using the StochKit software package. Custom code was added to modify the propensity functions when the deterministic model does not follow simple mass-action kinetics. Additional custom code was written to allow arbitrary insulin input functions, such as the pulse input used by Sedaghat et al. [5].

To estimate the effect of stochastic fluctuations, we used the SSA to generate 100 realizations of the stochastic model. A 15 minute pulse input of insulin was used, and state information was collected at every minute of simulation time up to 60 minutes.

Stochastic Simulation Results

For this model, appropriate metrics of the adipocyte response to insulin include the peak values of GLUT4 (glucose transporter) at the plasma membrane and the distribution of these values. There is qualitative agreement in the trajectory of the deterministic model and the mean trajectory of the stochastic models; however, the peak surface GLUT4 percentage is roughly 34% in the ODE model compared to 30% (1.51×10^6 molecules/cell) for the mean of the stochastic simulations (Figure 2). We attribute this difference to a non-mechanistic upper bound on the effect of Akt and PKC on GLUT4 translocation in the deterministic model (see [5] for a more complete explanation).

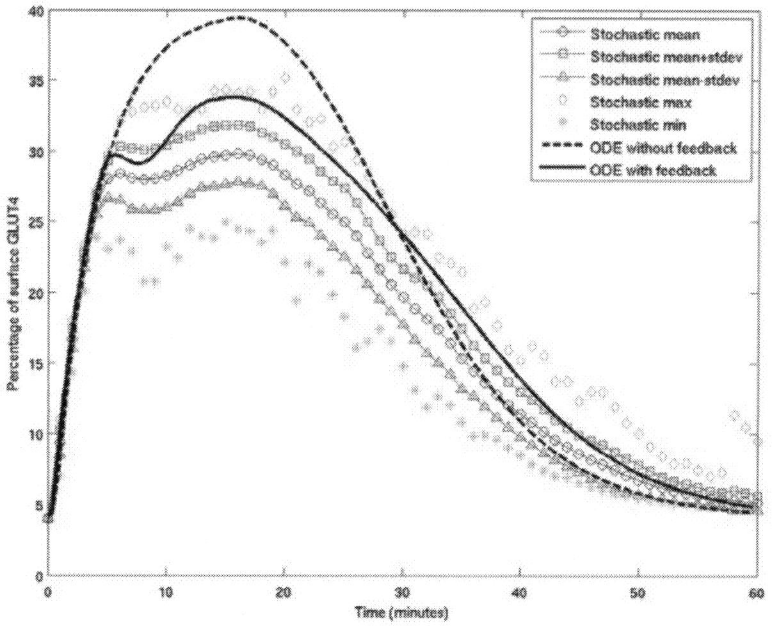

Figure 2. Means, standard deviations, and extreme values of percentage surface GLUT4 for 100 realizations of stochastic model versus Sedaghat model

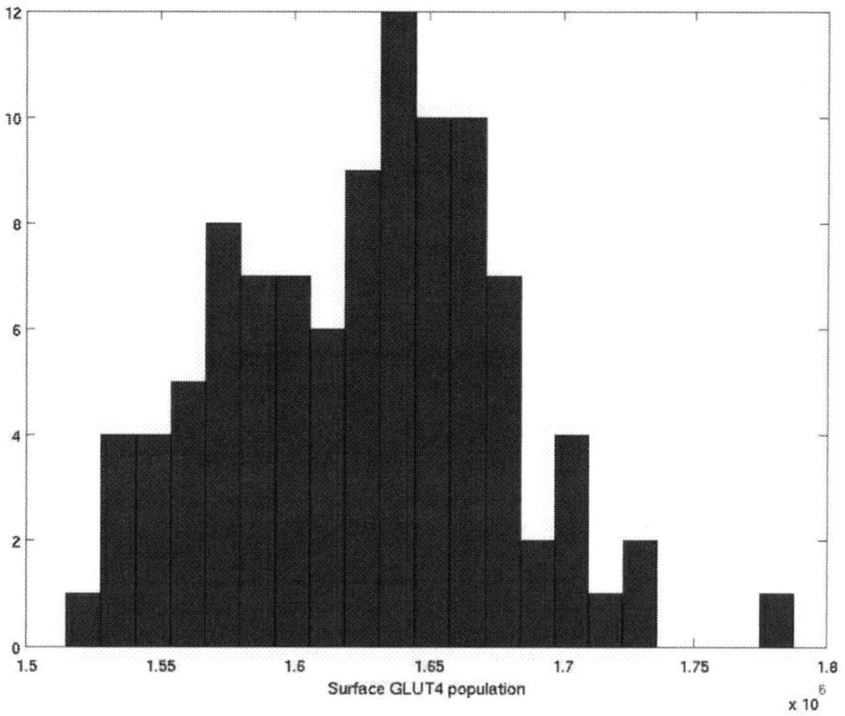

Figure 3. Histogram of the peak surface GLUT4 values for 100 realizations of the stochastic model.

The peak values of the individual stochastic realizations are summarized in Figure 3. It is of note that the peak of the mean values of surface GLUT4 percentage, 30%, does not equal the mean of the peak values, 32% (1.63×10^6 molecules/cell); this is due to fluctuations in time to reach peak value (mean: 15.48 minutes).

The stochastic fluctuations around peak surface GLUT4 are approximately Gaussian; the standard deviation is 2.1% in surface GLUT4 percentage—or 7% of the mean (Figure 2). It is difficult to assess how physiologically relevant these fluctuations are. In addition, non-mass action kinetics in the original deterministic model may distort the computed stochastic fluctuations unrealistically.

Experimental Design Using Sensitivity Analysis

Because experimental measurements can be sparse in biological systems, it is vital to develop an experimental design that can maximize the amount of information about model parameter values from limited and noisy measurements. We believe that there is utility in optimizing experimental protocols in this particular case because the Sedaghat model lacks experimental validation. By taking into account the model structure and estimating measurement noise (using stochastic fluctuation as a surrogate), we have used parameter sensitivity analysis to create experimental designs that are efficient for model identification, given constraints on measurement error and cost.

For an ODE model, a differential change in a parameter p_j from its nominal value may cause a change in state x_i, which is captured by a sensitivity coefficient, S_{ij} (where i ranges from 1 to the number of states N_x that will be measured and j ranges from 1 to the number of parameters N_p in the model that are of interest). For each measurement time point t_k (where k ranges from 1 to the number of timepoints N_t), a sensitivity coefficient matrix for each time point is generated (Equation 1).

$$S_{ij}(t_k) = \frac{\partial x_i(t_k)}{\partial p_j} \tag{1}$$

The sensitivity coefficient matrices are combined with estimations of measurement error (which are assumed to be Gaussian) for each state and timepoint to calculate the Fisher information matrix (FIM), where ${}_i(t_k)$ is the standard deviation for state x_i (Equation 2).

$$FIM = \sum_{k=1}^{N_t} S^T(t_k) V^{-1}(t_k) S(t_k)$$
$$V(t_k) = \begin{pmatrix} \sigma_1^2(t_k) & 0 & 0 \\ 0 & \ddots & 0 \\ 0 & 0 & \sigma_{N_x}^2(t_k) \end{pmatrix} \tag{2}$$

In the computations described below, we have selected ${}_i(t_k)$ to be the corresponding standard deviation from the stochastic simulations described above. The assumption that the stochastic trajectories have a normal distribution is not generally valid, particularly for states with O(1) copy count per cell volume.

A minimum measurement error of 1 molecule per cell volume is assumed, which allows $V(t_k)$ to be inverted in Equation 2. This minimum value is typically larger than the observed deviation from a normal distribution for states with O(1) copy count per cell volume; thus we can account somewhat for non-Gaussian behavior.

Finally, with the above assumptions, the FIM can be used to generate a lower bound the error that an unbiased estimation of parameter p_j, ${}_j$, would have if generated using measurements of a given choice of measured states, $\{x_i\}$, and measured timepoints, $\{t_k\}$ (Equation 3).

$$\sigma_j^2 \geq (FIM^{-1})_{jj} \tag{3}$$

For a more complete treatment, see [6].

To select between each choice of state and timepoint sets, we set a maximum allowable estimation error, the 95% confidence interval around the nominal parameter value, $1.96p_j$, for each "identified" parameter. For state and timepoint sets with identical numbers of identified parameters, the one with the smallest average normalized value of the 95% confidence interval for identified parameters (i.e., the most accurate estimation) is chosen.

Table 1. State measurement selection

# of states	Optimal states	# of identified p_j	Average $_j / p_j$
21	All	10	0.156
10	$x_2,x_5,x_6,x_8,x_9,x_{10},x_{13},x_{16},x_{20},x_{21}$	10	0.179
5	$x_2,x_5,x_9,x_{10},x_{20}$	8	0.236
3	x_2,x_9,x_{20}	4	0.288

Table 2. Timepoint measurement selection

# of timepoints	Optimal timepoints (min)	# of identified p_j	Average $_j / p_j$
60	All	10	0.156
15	1-5,10,19-24,46,56,57,60	9	0.179
10	1-5,10,24,46,56,57	8	0.150
5	1,3,4,46,56	8	0.200

Experimental Design Results

The sensitivity coefficient matrices were calculated for 31 parameters (after combining those that are *a priori* unidentifiable) in the Sedaghat model with feedback mechanisms over a 60 minute "experimental" duration using BioSens, a sensitivity analysis program [7]. As for the stochastic simulation, a 15-minute insulin pulse was used. Because the stochastic error estimations are in 1 minute intervals, the maximum number of timepoints that can be collected is 60. For any desired number of states and/or timepoints, a MATLAB routine was written that, first, maximizes the number of identified parameters and, second, minimizes the error in parameter estimation of identified parameters.

Selecting measured states that allow for maximum parameter estimation was the first experimental design undertaken, allowing all 60 timepoints to be measured. These results are summarized in Table 1 (Refer to Figure 1 for state numbers and signaling components). Overall, we observe that the model is not highly identifiable. Only 10 of 31 parameters can be identified with all state and timepoint measurements. More state measurements increases identified parameter count and accuracy. One would expect that directly measuring the states at the end of the signaling cascade would yield the highest amount of information, since later states would carry information about parameters that appear upstream. However, because different states have different measurement errors associated with them, this is not the case.

Selecting measured timepoints that allow for maximum parameter estimation was the next experimental design; these results are summarized in Table 2. Once again, taking more timepoints results in increased parameter identification and accuracy. The timepoints that contain the most parameter information seem to be clustered near the onset of insulin stimulation (1-5 minutes), with the remaining timepoints coming as the system relaxes back to a resting condition after the insulin stimulus is removed.

Table 3. Mixed state and timepoint selection

# of states	# of timepoints	# of identified p_i	Average $_i/p_i$
20	15	9	0.179
10	30	10	0.201
5	60	8	0.236

With the effects of state and timepoint measurement selection characterized separately, we turned our investigation to optimizing both state and timepoint measurement together. Thus, for a fixed total number of measurement datapoints (300), optimal state and timepoint measurements were determined and their effects on parameter identification are summarized in Table 3. For this example, an experimental design that tracks an intermediate number of states (10) and timepoints (30) would best identify model parameters. This method is easily adapted for modern high throughput methods of collecting biological data, including microarrays or phosphoproteomic methods, to select optimal measurement sets.

Insights from Robustness Analysis

Ultimately, it is hoped that the deterministic and stochastic models developed with the aid of the above techniques are useful in guiding modeling (in)validation and for (possibly combinatorial) drug targeting in insulin resistant cells. With this in mind, structured singular value (SSV or μ) analysis is performed to identify possible points of robust therapeutic intervention which minimize hazardous side effects. Fluctuations are not strictly limited to signaling pathways, as the intracellular environment is also subject to external noise; any therapeutic strategy applied must be robust to the uncertainty of the local environment as well as to the uncertainty in the therapy itself.

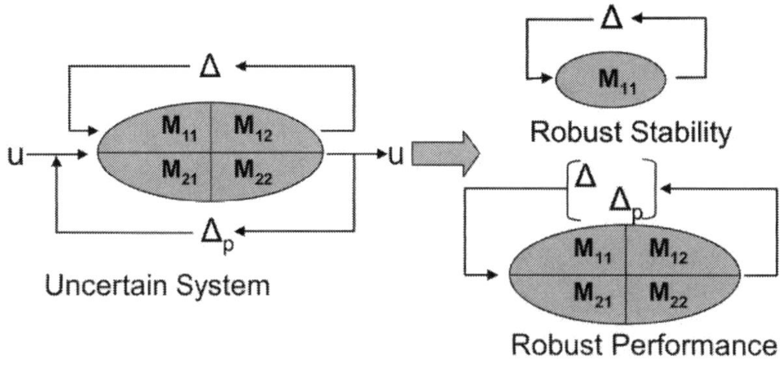

Figure 4. To test for robust stability of an uncertain system (left), one need only test the stability of the feedback of the uncertainty (top right). To extend the test to robust performance, the input/output channels are closed under feedback through a full uncertainty block, || p||•1. Reconfiguring the system (bottom left) and noticing the similarity, robust performance can be evaluated by evaluating the robust stability of the resulting block diagram.

We use SSV analysis is used to quantify the range of fluctuation a set of parameters can tolerate while maintaining robust insulin signaling performance. Briefly, SSV analysis identifies the smallest size perturbation, , which destabilizes the stable plant, **M**. The "size" of is measured in terms of its maximum singular value, $\bar{\sigma}$( ) (Figure 4). It is an extension of the Nyquist stability criterion, and μ is defined as

$$\mu(\mathbf{M}_{1,1}) = \min_{\Delta}\{\bar{\sigma}(\Delta) \mid \det(I - M_{1,1}\Delta) = 0 \text{ for structured } \Delta\}. \quad (4)$$

A full mathematical treatment of SSV analysis can found in [8].

The matrix is block-diagonal and is chosen to distribute parametric uncertainty to specific locations about the Jacobian, hence the uncertainty is structured. Furthermore, is generally weighted such that size of the system of uncertainty of interest is bounded, || ||•1; thus, when μ•1, the size of the perturbation to destabilize the system is greater than 1—outside of the uncertainty range of interest—and the system is robustly stable.

Extending robust stability to robust performance (RP) is achieved by applying a full uncertainty block, $_p$, to the input/output channels and closing the channels under negative feedback, as robust stability of this closed-loop structure ensures robust performance. The calculation of μ is an NP-hard problem; exact solutions can be found for small systems, but generally an upper and lower bound must be found for μ.

To evaluate combinatorial therapies in the insulin signaling pathway, the Sedaghat model without feedback mechanisms is linearized (the model with feedback mechanisms is not amenable to linearization). Robust performance is then defined as a tracking error about the nominal surface GLUT4 response. Several parameters from the nonlinear model do not appear in the linearized model, and many of the remaining parameters appear in related pairs which may be lumped into single parameters for the purposes of RP analysis, leaving 17 model parameters to be considered as possible therapeutic targets. For this study, 3 parameters (drug targets) are considered at a time, resulting in 680 possible parameter combinations.

First we sought to identify which sets of parameters, when perturbed, most effectively removed an individual cell from healthy performance. For each set considered, μ is calculated when each parameter is allowed to fluctuate within 5.0% of its nominal value (Figure 5). .

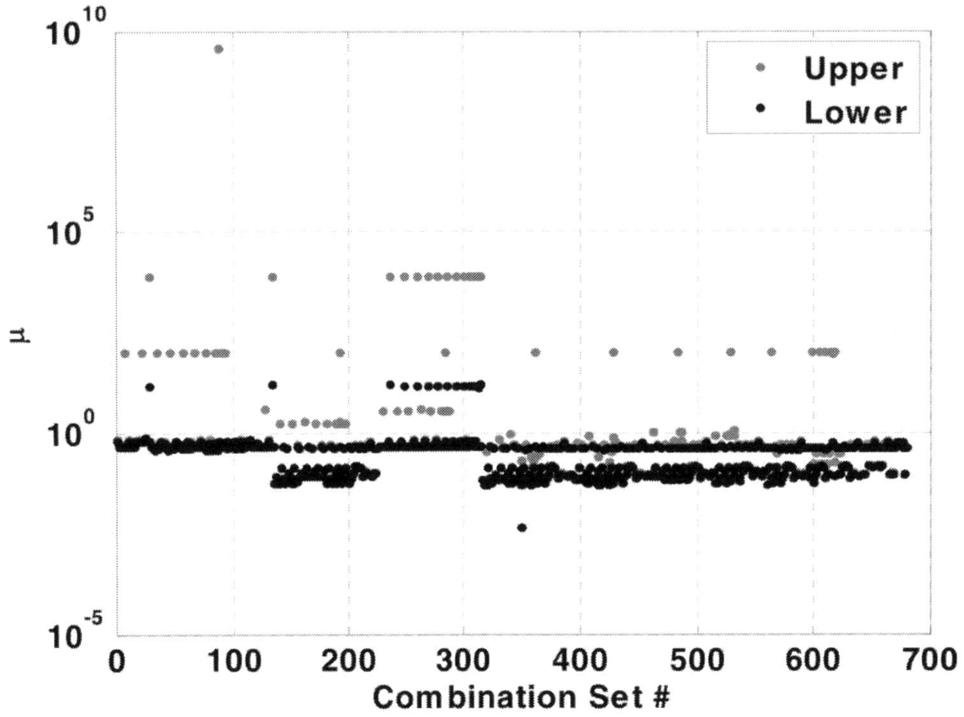

Figure 5. The upper and lower bound on μ for each of the possible 680 parameter combinations

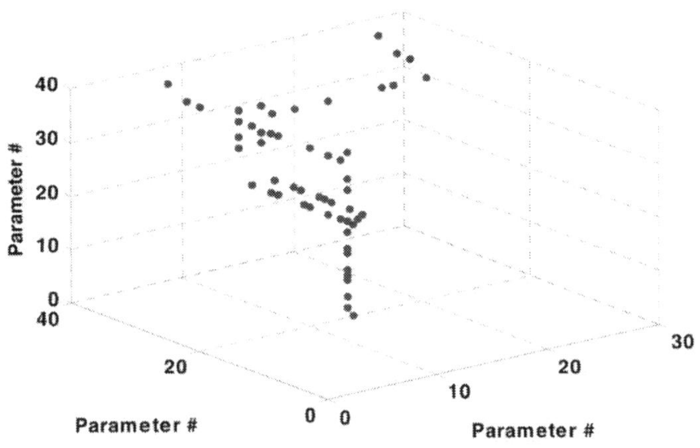

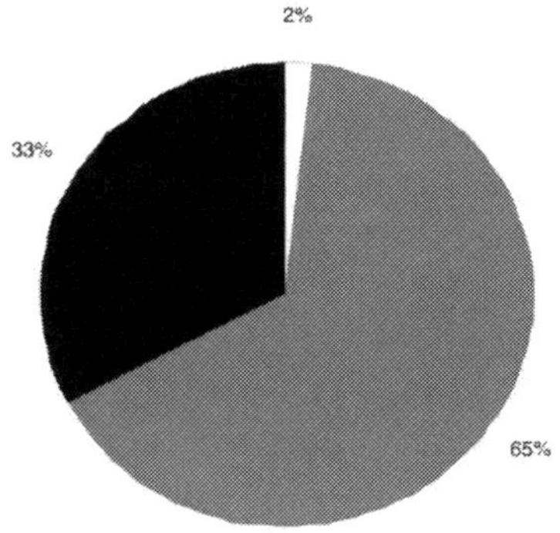

Figure 6. (Left) The set of all efficacious therapeutic combinations. (Right) The percentage of candidate therapeutic combinations that contain at least 1 (black), a combination of 2 (gray), or all three parameters PI3K, k_{-1}, and k_{9stim} (white).

Removing parameter sets for which the value of μ is poorly bounded (error difference> 0.1), 74.9% of the parameter sets remain to be analyzed for efficacy. Efficacy has been defined here as any remaining set with a μ value greater than 0.6. Applying this filter, only 8.1% of the original 680 possible combinations remain. Figure 6 shows how the remaining parameter combinations cluster about key parameters. Three parameters from the Sedaghat model appear in every candidate set, PI3K, k_{-1}, and k_{9stim}, and at least 2 of these three parameters appear in 65% of the possible therapeutic vectors. The parameter PI3K accounts for the amount of PI3K in the cell; k_{-1} is the dissociation rate of insulin from its receptors; and k_{9stim} represents the conversion of PIP_2 to PIP_3 by phosphorylated IRS1/activated PI3K complex. PI3K is involved in several cellular functions, has some known tumorigenic effects, and, as such, does not represent a suitable target vector. Removing strategies directly targeting PI3K, future work will focus on exploring the remaining 21 possible therapeutic combinations. These therapies will be applied to various models to test for efficacy *in silico.*

Conclusion

Stochastic modeling and simulation of the Sedaghat insulin-stimulated GLUT4 translocation network reveals fluctuations in concentrations that cannot be entirely captured in a deterministic model; these inherent fluctuations represent limits on measurement precision and are important to understanding the robustness and performance of the underlying biological system. We have used this understanding, with sensitivity analysis, to create optimized selections of states and timepoints for this network.

The SSV analysis we performed illustrates an effective preliminary *in silico* procedure for the rapid identification and screening of potential drug targets in this network. Given the large amount of cross-talk between the insulin signaling network and other critical, auxiliary networks, multi-drug therapies were identified that best regulate the insulin response while minimizing side effects, identifying a number of potential drug therapy sets.

References

1. Centers for Disease Control and Prevention, *Diabetes: Disabling Disease to Double by 2050.* 2008.
2. Sedaghat, A.R., A. Sherman, and M.J. Quon, *A mathematical model of metabolic insulin signaling pathways.* American Journal of Physiology-Endocrinology and Metabolism, 2002. **283**(5): p. E1084-E1101.
3. Leonhardt, W., H. Schneide, H. Haller, and M. Hanefeld, *Human Adipocyte Volumes - Maximum Size, and Correlation to Weight Index in Maturity Onset Diabetes.* Diabetologia, 1972. **8**(4): p. 287-291.
4. Gillespie, D.T., *General Method for Numerically Simulating Stochastic Time Evolution of Coupled Chemical-Reactions.* Journal of Computational Physics, 1976. **22**(4): p. 403-434.
5. Kwei, E.C., K.R. Sanft, L.R. Petzold, and F.J. Doyle III, *Systems Analysis of the Insulin Signaling Pathway*, in *Proc. 17th IFAC World Congress.* 2008: Seoul, Korea.
6. Zak, D.E., G.E. Gonye, J.S. Schwaber, and F.J. Doyle III, *Importance of input perturbations and stochastic gene expression in the reverse engineering of genetic regulatory networks: Insights from an identifiability analysis of an in silico network.* Genome Research, 2003. **13**(11): p. 2396-2405.
7. Taylor, S.R., R. Gunawan, K. Gadkar, and F.J. Doyle III, *BioSens Sensitivity Analysis Toolkit (UCSB's contribution to Bio-SPICE) v2* . 2005 [Available from: http://www.chemengr.ucsb.edu/~ceweb/faculty/doyle/biosens/BioSens.htm].
8. Skogestad, S. and I. Postlethwaite, *Multivariable Feedback Control: Analysis and Design.* 1996: Wiley.

698f Small World Network Models of Intercellular Coupling In the Mammalian Circadian Clock

Christina Vasalou[1], Mark Freeman[2], Erik D. Herzog[2], and **Michael A. Henson**[1]. (1) Department of Chemical Engineering, University of Massachusetts, 686 N. Pleasant Street, Amherst, MA 01003, (2) Department of Biology, Washington University, Box 1137 , 205 Monsanto Office, St. Louis, MT 63130-4899

The suprachiasmatic nucleus (SCN) of the hypothalamus is a multioscillator system that drives daily rhythms in mammalian behavior and physiology. Although intercellular communication within the SCN has been the focus of significant experimental effort, little is known about how SCN cells synchronize to each other to coordinate behavior. We previously developed a multicellular, molecular model of the mammalian circadian clock that incorporated recent data implicating the neurotransmitter vasoactive intestinal polypeptide (VIP) as the key synchronizing agent (To et al., 2007). This model assumed that cells were locally connected and that connection strengths were inversely proportional to the distance between cells due to the effects of VIP diffusion. In this contributon, we consider an alternative connectivity scheme based on small world networks that contain both short and long range synaptic connections for VIP mediated coupling. Individual SCN cells exhibit experimentally observed heterogeneities in uncoupled oscillator phenotype (sustained and damped oscillators) and VIP release characteristics (VIP producing and non-producing cells). We show that the small world network models are able to robustly synchronize despite heterogeneities in individual cells and network connectivity as long as the number of long range connections is sufficiently large. The model is used to investigate the hypothesis that age related circadian dysfunction is attributable to degeneration of the VIP coupling mechanism in the SCN rather than impairment of individual cell function. We find that partial removal of long range connections produces several experimentally observed behaviors including reduced VIP and CREB oscillation amplitudes in individual cells, a decrease in the mean period of the population, a higher percentage of non-oscillating cells, and reduced synchrony. The model thus predicts that loss of a few long range connections in the SCN can have large effects on circadian rhythmicity. Supported by NIH grant 78993.

To, T.-L., M. A. Henson, E. D. Herzog and F. J. Doyle III, "A Computational Model for Intercellular Synchronization in the Mammalian Circadian Clock," Biophysical Journal, 92, 3792-3803 (2007).

Microscale Analysis for quantifying Nitric Oxide induced Hypoxemia in Methemoglobinemia

*Srimoyee Bhattacharya, Saikat Chakraborty**

Indian Institute of Technology Kharagpur, Kharagpur 721302, INDIA

Abstract

Methemoglobinemia is a disorder of the blood caused by abnormally high levels of methemoglobin (MetHb) in the red blood cell (RBC) resulting from simultaneous uptake of oxygen (O_2) and nitric oxide (NO) by the human lungs. MetHb is produced in the RBC by irreversible oxidation of the oxygen carrying ferrous ion (Fe^{2+}) of the heme group of the hemoglobin (Hb) molecule to its non-oxygen binding ferric state (Fe^{3+}). The oxidation is induced by NO, which is directly inhaled by the patient or synthesized in excess by L-Arginine present in endothelial cells, or is produced from oxidizing nitrates ingested through drinking water or pharmaceutical agents.

In this work, we study the role of NO in the pathophysiology of methemoglobinemia and perform quantitative analysis of the relation between levels of NO inhaled by the patient and the severity of the disease. Reactions of NO occurring in the RBC with both Hb and oxyhemoglobin are considered in conjunction with the reaction between oxygen and Hb to form oxyhemoglobin and that for the reduction of MetHb to Hb.

Our dynamic simulations of NO and O_2 uptake in the RBC under continuous exposure to both gases reveal that at the end of the pulmonary transit time of 1 s, the oxygen saturation in the RBC equilibrate at 98% while breathing NO-free room air but decreases monotonically to 93% as the concentration of NO increases to 5 ppm and further down to 50% when the NO level in the blood increases to 80ppm. We show that an NO level of 10 ppm or higher while breathing in room air may be considered to be the critical NO concentration for Methemoglobinemia since it causes severe hypoxemia in patients by decreasing the oxygen saturation level to below its critical value of 91%.

We simulate the effects of oxygen therapy on MetHb and oxygen saturation levels in the blood and quantify the severity of hypoxemia by stratifying Methemoglobinemia patients who are oxygen responsive from those who fail to respond to pure oxygen. Our dynamic simulations reveal that ventilating patients with pure oxygen serves as an effective therapeutic strategy for NO levels of less than 10ppm in the RBC. Otherwise, Methemoglobinemia may be treated with methylene blue 1% solution (10mg/ml, 1-2 mg/kg) administered intravenously slowly over five minutes. This treatment is affected through the enzyme-inducing effect of methylene blue on levels of diaphorase II (NADPH methemoglobin reductase) which helps in the reduction of MetHb (Fe^{3+}) back to hemoglobin (Fe^{2+}) by providing an electron. Our kinetic modeling of this metabolic pathway reveals that methylene blue restores the iron in Hb to its oxygen-carrying state in less than a second of reaching the RBC.

Introduction

Methemoglobinemia is a disorder characterized by the presence of a higher than normal level of methemoglobin (MetHb) in the blood. It is a compound formed from hemoglobin (Hb) by oxidation of

* Corresponding author. Email: saikat@che.iitkgp.ernet.in Tel:+91-32222-83930 Fax: +91-32222-82250

iron atom from ferrous to ferric state. MetHb lacks the electron that is needed to form a bond with oxygen and hence, is incapable of oxygen transport. Generally, red blood cells (RBCs) are continuously exposed to various oxidant stresses and so MetHb is continually produced in humans. Typically, less than 1 percent of the total circulating hemoglobin in a healthy adult is present in the form of MetHb. However, injury or toxic agents convert a larger proportion of hemoglobin into MetHb. In healthy children, the ferric iron in MetHb is readily reduced to the ferrous state through the function of enzyme called as cytochrome b5 oxidase (also referred to as methemoglobin reductase), which is present in erythrocytes and other cells.

In this paper, we have concentrated in microscale modeling i.e., reactive uptake of NO in the RBC for the quantification of Methemoglobinemia. When a person is exposed continuously to NO, the reactions will attain equilibrium and the whole transport of NO through the lung will be diffusion controlled. Here spatial averaging is performed over the volume of the RBC. This model is used to obtain the increased levels of Methemoglobin inside RBC that is responsible for causing Methemoglobinemia and also estimate the levels of inhaled NO that could be fatal.

Reaction Chemistry

Nitric Oxide as previously discussed has rich reaction chemistry in Human Blood. The reactions that take place inside the RBC in presence of NO and O_2 are given as:

$$Hb + n.O_2 \underset{k_{-1}}{\overset{k_1}{\rightleftharpoons}} Hb(O_2) \tag{1}$$

$$NO + HbO_2 \underset{k_{-2}}{\overset{k_2}{\rightleftharpoons}} MetHb + NO_3^{-1} \tag{2}$$

$$NO + Hb \underset{k_{-3}}{\overset{k_3}{\rightleftharpoons}} HbNO \tag{3}$$

$$MetHb + b5_{Red} \underset{k_{-4}}{\overset{k_4}{\rightleftharpoons}} Hb + b5_{ox} \tag{4}$$

Reaction (1) shows that Oxygen-Hb equilibrium is assumed to follow Hill Kinetics with Hill constant, 'n' equal to 2.34. The equilibrium rate constant of Hill kinetics is calculated using

$$K_1 = \frac{H^n}{P_{50}^{\ n}}$$, where H= 7.4×10^5 Torr and P_{50}= 26 Torr. (1)

Absolute value of the reaction rate constant for NO induced oxidation of oxyhemoglobin to MetHb (2) was calculated by Eich and co-workers, (1996). Sheele et al., (1999) reported the reaction rate constant of the reaction between NO and deoxyhemoglobin as 2×10^7 $M^{-1}s^{-1}$. They measured the dissociation rates of Nitrosylhemoglobin (3). Also, Dissociation rate constant of MetHb is obtained to be 2.47×10^{-3} s^{-1} from Power et al., (2007). Reaction (4) shows the enzymatic MetHb reduction by reduced cytochrome *b5* back to normal hemoglobin. In normal RBCs, MetHb is maintained at a level of less than 1% of total hemoglobin through two metHb-reducing pathways. One of these systems is the redox cycle consisting of cytochrome b5 (cytb5) and cytochrome b5-metHb reductase (b5R), which uses NADH as an electron transfer to cytb5 ("cytb5-NADH system"). The other pathway uses flavin as an electron carrier for the reduction of MetHb coupled with NADPH oxidation, catalyzed by NADPH-dependent flavin reductase (FR) ("flavin-NADPH system"). Here we have considered first reduction pathway, because cytb5-NADH system is estimated to be responsible for more than 95% of metHb-reducing capacity under experimental conditions. Value of k_4 as $6.2 \times 10^3\ M^{-1}s^{-1}$; and k_{-4} as $0.583\ M^{-1}s^{-1}$ are given by Kuma , (1981) and Abe et al., (1979).

Mathematical Model

The diffusion-reaction equations in Lagrangian coordinates for a single RBC of any arbitrary geometry and volume Ω, and external surface area $\partial\Omega$ are given by,

$$\alpha_{O_2}\frac{\partial P_{O_2,rbc}}{\partial t} = D_{O_2}\alpha_{O_2}\nabla^2 P_{O_2,RBC} - R_1 \quad (6)$$

$$[Hb]_T\frac{\partial S_{HbO_2}}{\partial t} = D_{Hb}[Hb]_T\nabla^2 S_{HbO_2} + R_1 - R_2 \quad (7)$$

$$\alpha_{NO}\frac{\partial P_{NO,rbc}}{\partial t} = D_{NO}\alpha_{NO}\nabla^2 P_{NO,RBC} - (R_2 + R_3) \quad (8)$$

$$[Hb]_T\frac{\partial S_{MethHb}}{\partial t} = D_{Hb}[Hb]_T\nabla^2 S_{MethHb} + R_2 - R_4 \quad (9)$$

$$[Hb]_T\frac{\partial S_{HbNO}}{\partial t} = D_{Hb}[Hb]_T\nabla^2 S_{HbNO} + R_3 \quad (10)$$

where $P_{i,rbc}$ is the partial pressure of physically dissolved gas *'i'* (NO or O_2) in the RBC, S_j is the fractional hemoglobin *'j'* saturation, ∇^2 is the three dimensional Laplacian in the local coordinate in the RBC, D_i and D_{Hb} are the diffusion coefficients of gas *'i'* and hemoglobin inside the RBC, R_j is the net rate of conversion for corresponding reaction (j), $[Hb]_T$ is the total intra-erythrocytic hemoglobin (free+bound), and α_i is the solubility of gas *'i'* in the RBC. Since [b5R] is given to be 0.07 μM in Higasa et al. (1998), the scale of R_4 is very low as compared to R_2. Hence it can be ignored in Equation (9). Since the details of the NO concentration in the vasculature is not of interest to us, we assume that the total endothelial generated NO is consumed inside the RBC, i.e. it is not available for s-GMP activation. The uptake rate of oxygen depends on the alveolar partial pressure of oxygen, P_{AO2}, which is around 100 Torr.

Results and Discussion

We have taken initially deoxygenated blood i.e. arterial P_{O2} to be 40 torr corresponding to 75% S_{HbO2} for carrying out the simulation. This is because blood returning to the heart from the tissues has a low P_{O_2} (40 mmHg) and travels to the lungs via pulmonary arteries. The pulmonary arteries form pulmonary capillaries, which surround alveoli.

After oxygenation blood moves into the pulmonary veins which return to the left side of the heart to be pumped to the systemic tissues. Fig. 1(a) shows the dynamic simulation for S_{HbO2}. It can be noticed from the figure that when alveolar NO is nearly zero, oxygen saturation attains equilibrium at the value of 0.97 and the oxygen tension increases to 100 torr. Oxygen diffuses (moves through the membrane separating the air and the blood) from the high pressure in the alveoli (100 mmHg) to the area of lower pressure of the blood in the pulmonary capillaries (40 mmHg). It has been found that under this case oxygen saturation reaches a steady state within 0.8-1 sec of exposure of oxygen. Normally 1 sec is the transit time for oxygen consumption in red blood cell. Fig.1(b) shows the dynamic profile for S_{MetHb} in red blood cell. As discussed earlier 1% is known to be the level of MetHb in normal hemoglobin. But we can see that as the concentration of NO increases, MetHb saturation increases so much that it leads to the onset of methemoglobinemia. We have obtained the plot for 2 second but chopped it off in 1 second as this is the transit time of gas inside RBC. Moreover we have assumed initial condition that of deoxygenated blood inside RBC that is partial pressure of O_2 inside RBC to be of 40 torr.

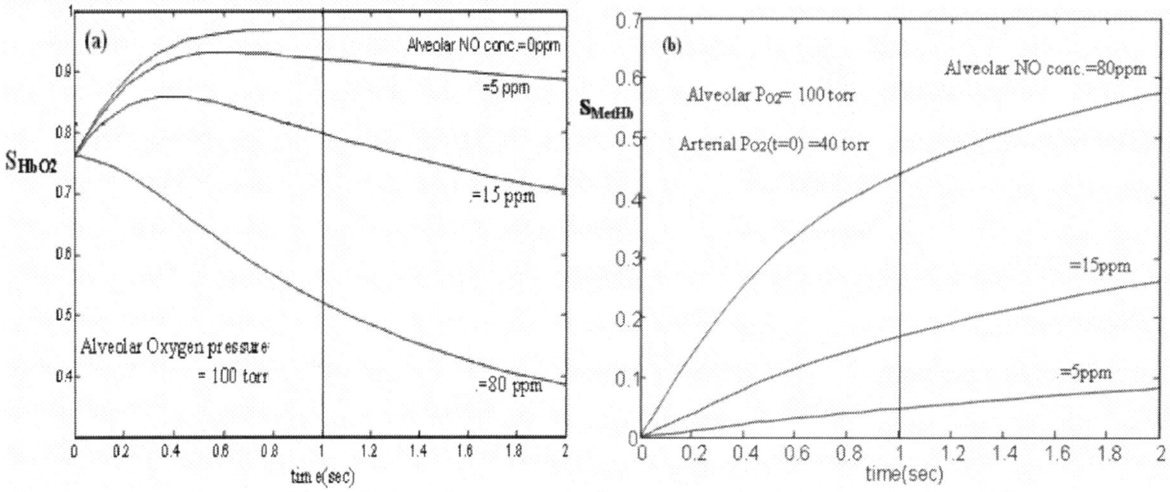

Fig.1. Dynamic Profile of (a) S_{HbO2} and (b) S_{MetHb} inside RBC for different values of alveolar NO concentration.

Therapeutic Strategy for Methemoglobinemia

Methemoglobinemia results from the oxidation of the heme moiety in the hemoglobin molecule from the ferrous to the ferric state. Auto-oxidation of Hb to MetHb occurs spontaneously at a slow rate in normal individuals, converting 0.5 to 3 percent of the available Hb to MetHb per day. The only physiologically important pathway for reducing MetHb back to Hb is the NADH-dependent reaction catalyzed by cytochrome b5 reductase (b5R). An alternative pathway is an enzyme utilizing NADPH generated by glucose-6-phosphate dehydrogenase (G6PD) in the hexose monophosphate shunt as a source of electrons. However, there is normally no electron carrier present in red blood cells to interact with NADPH methemoglobin reductase. . As a result, electron acceptors or redox dyes, such as methylene blue (MB), and flavin are required for this pathway to work. A congenital deficiency in the NADH-dependent cytochrome b5 methemoglobin reductase can be inherited in an autosomal recessive pattern. Patients who are homozygous for this enzymatic deficiency have congenital methemoglobinemia and may exhibit lifelong cyanosis. In order to treat this cyanosis, we decided to supply patient with oxygen therapy.

Fig. 2 shows the dynamic simulation of oxygen saturation with increasing oxygen supply for NO concentration of 5ppm. It shows that as oxygen supply increases above 250 torr, critical saturation increases and moves towards left till 0.2 sec and after that all follow the same trend with minute difference in values. From various plots for different NO concentration, we concluded that pretreatment with oxygen is useful for lower concentration of NO i.e. below 10ppm because supplying excess oxygen brings oxy-saturation to near about 91% which is the safe limit for normal human being. However when NO concentration is more than 10 ppm, excess oxygen supply does not play significant role. In such cases, we need to see other options available to us. Most significant and popular option for treatment of methemoglobinemia is Methylene Blue. Methylene blue is the first-line antidotal therapy. Methylene blue works as a cofactor for the enzyme NADPH-methemoglobin reductase. Methylene blue is oxidized into leukomethylene blue by accepting an electron from NADPH in the presence of NADPH-methemoglobin reductase. Leukomethylene blue then donates this electron to MetHb resulting in its conversion back to Hb. The initial dose is 1-2 mg/kg over 5 min. Its effects should be seen in approximately 20 min to 1 h. Fig.3 shows that MetHb saturation reaches to its minimum value within 0.1 second of methylene blue reaching RBC irrespective of alveolar NO concentration. The prophylactic preoperative methylene blue administration in a patient with congenital methemoglobinemia significantly decreased the MetHb level and increased the fractional oxygen saturation with a consequent increase of the safety margin against perioperative hypoxemia.

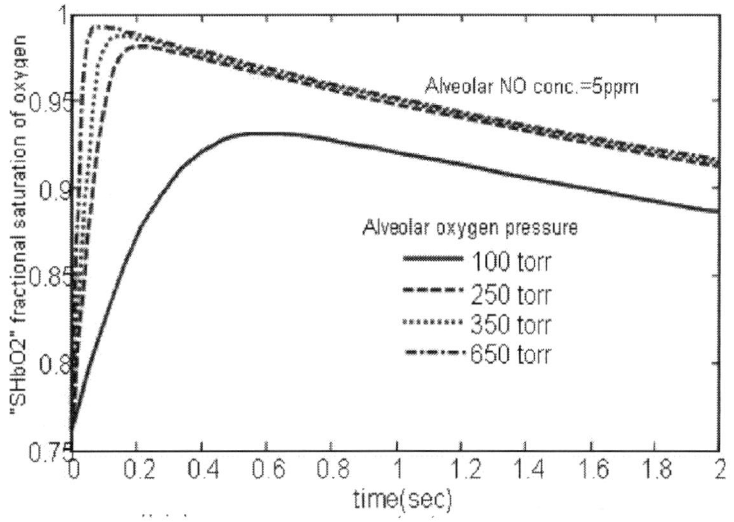

Fig. 2. Dynamic profile of oxygen saturation during oxygen therapy (alveolar NO conc. = 5 ppm).

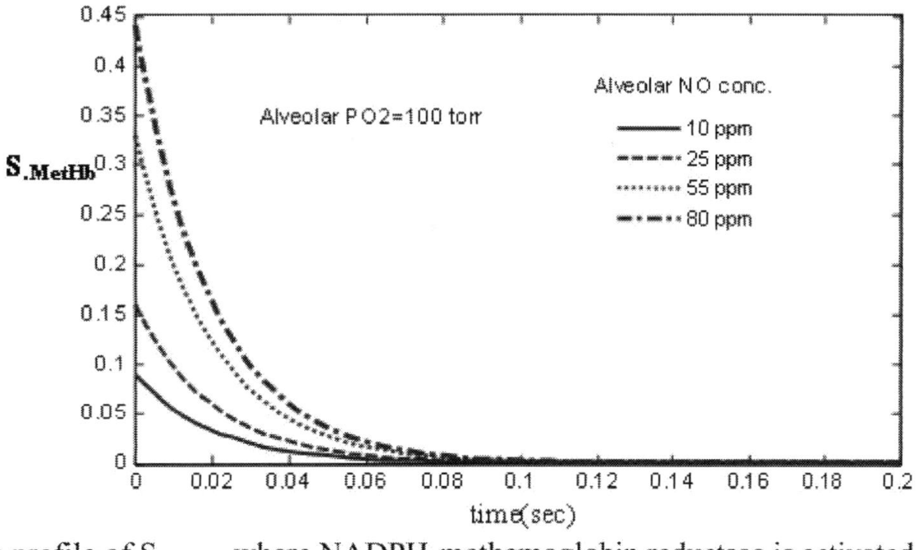

Fig. 3. Dynamic profile of S_{MetHb} where NADPH-methemoglobin reductase is activated by Methylene Blue for MetHb reduction to Reduced Hb for different NO concentration.

Conclusions

In this work, we have modeled the simultaneous uptake of Nitric Oxide and Oxygen by Red Blood Cell in order to quantify methemoglobinemia. The rate equations for reactions of NO with oxy- and deoxy-hemoglobin are developed and these equations are used to develop analytical expressions for O_2 and NO diffusing capacities of RBC for continuous exposure of NO. We have used LS based spatial averaging approach to solve these expressions. During continuous exposure to NO, we have assumed that the reactions achieve equilibrium and we have used it to obtain fractional saturation of various compounds. Dynamic simulation of our mathematical model shows that for higher NO concentration i.e. 10 ppm, oxy-saturation falls more rapidly going below the limit of 91% which is the safe limit for normal human being as discussed earlier. Also MetHb saturation can go upto 40% for 80 ppm of NO leading to fatal conditions of extreme methemoglobinemia. In order to cure this disease, we dealt with two kinds of treatment. First one is excess supply of oxygen which helps to cure cyanosis which occur upto 10 ppm of NO inside RBC. For NO concentration higher than this, first treatment does not work and so treatment with methylene blue is given more emphasis. Methylene blue works as a cofactor for the enzyme NADPH-methemoglobin reductase which helps in rapid conversion of MetHb back to Hb. It was observed from the plots that conversion process is activated within 0.1 second of methylene blue reaching inside RBC reducing MetHb to almost zero.

References

1. Abe K. and Sugita Y., 1979. Properties of cytochrome b5 and methemoglobin reduction in human erythrocytes, Eur J Biochem.101:423-8.
2. Azizia F., Kielbasaa J.E., Adeyigaa A.M., Mareea R.D., Fraziera M., Yakubuk M., Shieldsa H., Kingd S.B.and Kim-Shapiro A.E., 2005. Rates of nitric oxide dissociation from hemoglobin. Free Radical Biology & Medicine 39 145-151.
3. Boylston M. and Beer D., 2002. Methemoglobinemia: A Case Study. Crit Care Nurse; 22(4): 50-55.
4. Chakraborty S., Balakotaiah V., and Bidani A., 2004. Diffusing capacity reexamined: relative roles of diffusion and chemical reaction in red cell uptake of O2, CO, CO2 and NO. Journal of Applied Physiology, 97: 2284-2302.
5. Chakraborty S., Balakotaiah V., and Bidani A., 2007. Multiscale model for pulmonary oxygen uptake and its application to quantify hypoxemia in hepatopulmonary syndrome. Journal of Theoretical Biology 244 ,190–207.
6. Eich R.F., Li T., Lemon D.D., Doherty D.H., Curry S.R., Aitken J.F., Mathews A.J., Higasa K., Manabe J.I. and Yubisui T., 1998. Molecular basis of hereditary methemoglobinemia, types I and II: two novel mutations in the NADH-cytochrome b5 reductase gene, Br J Haematol. 103: 922-30.
7. Kreuzer K.H., 1970. Facilitated diffusion of oxygen and its possible significance: a review. Respir Physiol 9;1-30.
8. Kuma F.,1981. Properties of methemoglobin reductase and kinetic study of methemoglobin reduction, J Biol Chem.256:5518-23.
9. Lurie M.A., 1993. A Concise review: Methemoglobinemia. Am J Hematol; 42:7.
10. Peter R.2007. Health: The blue people. BrandweekNRX .
11. Power G.G., Bragg S.L., Oshiro B.T., Dejam A., Hunter C.J., and Blood A.B.,2007. A novel method of measuring reduction of nitrite-induced methemoglobin applied to fetal and adult blood of humans and sheep. Journal of Applied Physiology.
13. Simulation Study for Methemoglobin Reduction Pathways. http://www.ncbi.nlm.nih.gov/books/bv.fcgi?indexed=google&rid=eurekah.section.74876

710a Applications of NIR In Blend Monitoring and beyond: Case Studies from Development

Nancy E. Sever, Abbott Laboratories, 1401 North Sheridan Road, North Chicago, IL 60064

NIR has been established as a suitable tool for monitoring blend uniformity and determining blend endpoint. Although that has been the main emphasis for NIR, the NIR spectra contain additional information that can be exploited during development.

710b Optimization and Scale-up of High Shear Wet Granulation Processes Using Lasentec Fbrm and Pvm In-Situ Particle Characterization

Zane Arp[1], Benjamin Smith[2], and **Des O'Grady**[2]. (1) Pd-Most-Puc, GlaxoSmithKline, 709 Swedeland Rd, King of Prussia, PA 19406, (2) Particle Systems Characterization - Process Analytical Technology, Mettler Toledo, Redmond, WA 98052

High shear wet granulation is a common unit operation in solid dosage formulation, yet a mechanistic understanding of high shear wet granulation is not well known. One goal of high shear granulation is to yield a repeatable particle size and shape distribution at the batch endpoint. One measure of a successful granulation process is a repeatable endpoint particle distribution that achieves repeatable downstream drying, milling, flow properties, and content uniformity. However, inconsistencies often occur during granulation scale up due to changing raw materials or changing process dynamics such as segregation, agglomeration, or breakage. By designing a robust granulation process one can ensure consistent downstream processing from granulation through tablet dissolution. FBRM and PVM allow a user to directly link process control parameters to the particle distribution and the product quality. In this work bottom driven high shear wet granulation experiments at several volumes are compared with various formulation conditions and excipients. The application of in-situ particle size monitoring techniques, Lasentec FBRM and PVM is demonstrated to track a series of granulation batch process in the laboratory while varying the batch volume, raw materials, shear, and liquid addition. Target particle distributions are achieved by using FBRM® and PVM® to understand the critical process parameters and predict the effect during granulation scale up. FBRM® and PVM® are ultimately used to target a specific growth profile and endpoint particle size and shape distribution at the various scales and vessel configurations.

710c Characterizing In-Situ Particle Disintegration Mechanisms to Correlate Formulation Conditions to Tablet Dissolution Profiles

Des O'Grady[1], Eric Dycus[1], Eric J. Hukkanen[1], and Jeff Seely[2]. (1) Particle Systems Characterization, METTLER TOLEDO, Redmond, WA 98052, (2) Distek, North Brunswick, NJ 08902-4905

USP tablet dissolution testing is well established as the standard analytical test in the pharmaceutical industry. Yet dissolution testing often fails to detect the root cause of dissolution inconsistency. As the role of dissolution moves toward enabling quality by design and screening product development, one requires a more mechanistic understanding to why tablets release API with varying kinetics. The application of in-situ particle monitoring technology, Lasentec FBRM is demonstrated to track a series of tablet disintegration and dissolution profiles. By understanding the mechanism for particle disintegration and dissolution, one can correlate release inconsistencies to the upstream source (ie changes in raw materials, granulation, segregation, or tableting). One can use this information to quickly screen formulation development conditions to build quality into the formulation design and achieve a repeatable target release profile.

Web Page: www.mt.com/lasentec

710d High Throughput Product / Process Monitoring and Control Using Automated Td-NMR Technology

Supriyo Ghosh, minispec Division, Bruker Optics Inc., 2700 N Crescent Ridge Dr, The Woodlands, TX 77381

The manufacturing processes in chemical/pharmaceutical industries involve application of various unit operations on raw materials. These processing steps induce multiple physico-chemical changes leading to the desirable properties in the final product. During the process development as well as the industrial-scale implementation of the same, monitoring of the effects of each processing step need to be fast, accurate and precise. TD-NMR (time domain nuclear magnetic resonance) is a PAT (process analytical technology) which can be utilized to look into the physicochemical processes at a molecular level and precisely quantify them, using a small bench-top instrument. It obtains the signal from the atomic nucleus in a material. Mathematical analysis of the TD-NMR signal reveals the changes occurring in molecular level in a material. Measurements can be done in a matter of a few seconds to obtain quantitative characterization of materials undergoing physico-chemical changes. The entire measurement procedure can be done in a continuous fashion using either an on-line system or an automated robotic system. In this presentation, the basic principle of TD-NMR technology will be introduced. Various high throughput automated TD-NMR characterizations relevant to process development for pharmaceutical/biotech industries will be elaborated including but not limited to: tablet hardness/density, contactless weighing of medicine, moisture in powder/tablet, moisture in lyophilized materials, solvent/oil content, crosslink density, crystallization, copolymer ratio, hydration, impurity quantification, drying kinetics, fat-moisture-protein content, emulsion stability, droplet size distribution in emulsions, lean-fat analysis of live animals and tissue, porous media microstructure and transport. Probable new areas of pharmaceutical/biotech industry where this TD-NMR technology can be exploited in the future will also be discussed.

710e A Model-Based Framework for Systematic Product Quality Monitoring and Control

Ravendra Singh[1], Krist V. Gernaey[2], and Rafiqul Gani[1]. (1) CAPEC, Department of Chemical and Biochemical Engineering, Technical University of Denmark, Søltofts Plads, Building 229, 2800 Kgs. Lyngby, Denmark, (2) BioEng, Department of Chemical and Biochemical Engineering, Technical University of Denmark, DTU - Building 229, Søltofts Plads, 2800 Kgs. Lyngby, Denmark

Abstract

In the manufacturing industry, for example, the pharmaceutical industry, it is a goal to obtain consistently predefined end product qualities. A properly designed process monitoring and analysis system is needed to assure the predefined end product qualities. The design of process monitoring and analysis systems involve the identification of critical quality product parameters, selection of the appropriate monitoring tools and the determination of efficient control strategies to ensure that the selected critical quality parameters can be controlled and/or monitored. The framework for design of the monitoring/control system consists of three main parts: a knowledge base of methods and tools for monitoring/analysis, a model library (contains models for unit processes, sensors and controllers) to supplement the gaps in the knowledge base and an algorithm for design and analysis of the monitoring/control system. A set of design decisions need to be made in an integrated manner taking into account the interaction between product quality specifications, process operational constraints, cost of the monitoring and analysis system and the time needed for analysis.

The starting point of the design procedure for the monitoring/control system is the product quality specifications and process specifications (provided by designer of the system), followed by four analysis steps: process analysis, sensitivity analysis, interdependency analysis and performance analysis of monitoring/control tools. Process analysis provides an extended set of relevant process variables (identified through the use of a knowledge base) among which the critical quality parameters and the corresponding actuators have to be selected in subsequent analysis steps. It also provides a good understanding of the process-product rleationships. The critical quality parameters are identified through a sensitivity analysis (for this purpose the model library is used). The process variables, which violate the operational limits and have a major influence on the product quality are selected as the critical quality parameters. The pairing of critical quality parameters with the actuators is achieved through an interdependency analysis (model library is used). The process parameters, which are the most sensitive for the selected critical quality parameters are selected as the actuators. Appropriate tools for on-line measurement of the critical quality parameters are selected through performance analysis of the monitoring tools. In the process of selecting the monitoring tools, first the available monitoring tools together with specifications (e.g. accuracy, precision, operating range, time, cost etc.) for measurement of selected quality parameters are retrieved from a knowledge base and then their performances are compared to select (configure) the best monitoring/control system. In the final step, the most promising candidate systems are verified through simulation, ie., check for their monitoring/control performance. The objective of this presentation is to demonstrate the application of the systematic framework and its design methodology for product quality monitoring and control. The identification of the critical quality parameters, pairing with actuators and selection of monitoring tools will be illustrated through two case studies, an insulin production process and a tablet manufacturing process.

Towards a Robust, Greener API Process Via Cooling Crystallization

Dr. Evelina Kim (speaker)
Dr. James Marek

Abbott Laboratories, North Chicago, IL 60064

For the Nov., 2008 AIChE National Meeting in Philadelphia, PA

Abstract:

A late stage API candidate had been purified via antisolvent crystallization from IPA/MeOH. The antisolvent process was characterized by nucleation of undesirable crystal forms, large processing volumes, and small particle sizes (needles). A greener cooling crystallization from EtOH/water was developed to address these issues. FBRM and mid-IR PATs were used along with traditional HPLC methods to investigate supersaturation effects and gain insights into the nucleation and growth. The latter two phenomena were sensitive to mixing energy input. The cooling crystallization provides robust polymorph control, larger particles with smaller aspect ratio, lower processing volumes (higher throughput), and robust scalability as evidenced by runs at 200x and 2000x the lab scale (100 KG vs. 50 g.).

712a A Unified Approach to Mathematical Treatment of Multi-Particle Interactions during Size Reduction: Extending the Nonlinear Population Balance Model

Ecevit Bilgili, PhD, Pharmaceutical Commercialization Technology, Merck & Company, Inc., WP 78-110, P.O. Box 4, West Point, PA 19486

Breakage of particles is a complex phenomenon that is intended in size reduction processes for reducing particle size and enhancing specific surface area. As a unit operation, comminution or size reduction is widely used in a variety of industries including ceramics, composites, foods, minerals, paints and inks, pharmaceuticals, etc. The operation can be carried out under wet or dry conditions in batch or continuous modes. Among various approaches to the analysis of comminution processes, the use of population balance models (PBMs) provides a quantitative understanding at the process length scale. Not only can PBMs be used as a tool for simulation, design, and optimization, but also they can elucidate the breakage mechanism(s) such as massive fracture, cleavage, and/or attrition.

Although the traditional PBM has been described in a variety of mathematical forms (size-continuous form [1,2], size-discrete form [3], and matrix form [4]), it is always linear. The traditional model is based on the basic assumption that for particles with size x, the breakage probability p in discrete breakage events, as e.g. in roller milling, and specific breakage rate S during a non-discrete breakage process, as e.g. in ball or vibratory milling, are dependent only on the mass fraction of particles with size x at a given time and spatial location. In other words, the presence or absence of other particles with generic size z has no influence on the breakage probability or specific breakage rate of particles with size x. It is also assumed that the breakage (daughter) distribution function b is also independent of possible multi-particle interactions. These assumptions allowed researchers to find a simple yet elegant solution to the model equations. On the other hand, researchers have experimentally observed some interesting and non-intuitive breakage phenomena that cannot be explained by the traditional linear model (refer to the review in [5]). Gradual accumulation of fines in a batch mill is known to alter the specific breakage rate of the relatively coarse particles. Purposeful addition of fines to a particle assembly has a significant impact on the specific breakage rates. Particle bed breakage tests in the literature also show that multi-particle interactions among particles of different sizes lead to nonlinear breakage probability [6]. Bilgili [7] showed that the presence of multi-particle interactions during milling processes can lead to deviations from self-similar breakage behavior. All these phenomena can be labeled as nonlinear effects and arise from the multi-particle mechanical interactions during breakage events or non-discrete breakage processes [8].

Bilgili and Scarlett [5] and Bilgili et al. [9] have recently proposed a nonlinear population balance model for size reduction as a rate process, which decomposes the specific breakage rate into a size-dependent apparent specific rate function k and a population density dependent functional F. In other words, the birth and death terms (constitutive relations) in the population balance equation account for the multi-particle interactions phenomenologically. The population balance equations were derived for batch and continuous milling operations in view of the non-linear constitutive relations [5,7-10], and the numerical simulation results demonstrated the predictive capability of the novel nonlinear model in explaining the experimentally observed complex breakage behavior.

Although the recent efforts by Bilgili and co-workers [5,7-10] provide a sound mathematical framework for treating the multi-particle interactions or nonlinear effects during size reduction processes, the treatment of these interactions for single or multiple breakage events has been mainly semi-empirical. Baxter et al. [6,11] extended the breakage matrix approach of Broadbent and Calcott [4] and developed a sound experimental methodology to quantify the degree of multi-particle interactions. However useful the breakage matrix approach is for quantifying the extent of multi-particle interactions for a given feed size distribution, its predictive capability for any possible natural feed size distribution is questionable. This is because the fundamental parameters characterizing the breakage, i.e., the apparent breakage probability parameters, the daughter distribution parameters, and the multi-particle interaction parameters are lumped in the matrix elements in some unknown manner; they are not explicitly defined. Moreover, the breakage matrix approach, as used in [6,11], suggests an infinite number of matrix elements for infinitely different feed size distributions, which, to the least, creates a philosophical modeling dilemma. As described in [12], clearly the breakage matrix approach is no more than a matrix representation of the fundamental population balance equation usually described in some size-discrete form. This suggests that the breakage matrix can and should be derivable from the population balance equation with nonlinear constitutive relations for the birth and death terms.

In this paper, the population balance equation is formulated for breakage events starting with the matrix representation as proposed by Bilgili et al. [9]. After solving for the product mass fraction distribution as a function of feed mass fraction distribution and number of breakage events, we describe the nonlinear breakage matrix in terms of the apparent breakage probability matrix, daughter distribution matrix, and the multi-particle interaction matrix. For the case of single particle breakage, a simple experimental methodology is suggested for determining the elements of these matrices from a series of mono-dispersed breakage tests and natural feed breakage tests. In essence, the paper attempts to give the breakage matrix formalism [6,11] some predictive capability by deriving it from a fundamental population balance equation and to give it a more explicit mathematical structure. More importantly, a unified population modeling approach is presented for treating multi-particle interactions during both breakage events and rate processes.

1. Bass, L. Zur Theorie der Mahlvorgange. Zeits. Angew Maths. Phys. 1954, 5: 283–292.

2. Gaudin, A.M., Meloy, T.P. Model and comminution distribution equation for repeated fracture. Trans. AIME 1962, 223: 43–50.

3. Sedlatschek, K., Bass, L. Contribution to the theory of milling processes. Powder Metal. Bull. 1953, 6: 148–153.

4. Broadbent, R.A., Callcott, T.G. Coal breakage processes: II. A matrix representation of breakage. J.

Inst. Fuel 1956, 29: 528–539.

5. Bilgili, E., Scarlett, B. Population balance modeling of non-linear effects in milling processes. Powder Technol. 2005: 153, 59–71.

6. Baxter, J., Abu-Nahar, A., Tuzun, U. Understanding intra-mixture interactions in the breakage of dense particulate mixtures. Proc. 5th World Cong. Particle Technol., Paper No: 75e, Orlando, FL, Apr. 2006.

7. Bilgili, E. On the consequences of non-first-order breakage kinetics in comminution processes: absence of self-similar size spectra, Part. Part. Syst. Char. 2007, 24: 12–17.

8. Bilgili, E., Scarlett, B. Nonlinear effects in particulate processes. Nonlinear Anal. 2005, 63: e1131–e1141.

9. Bilgili, E. Yepes, J., Scarlett, B. Formulation of a non-linear framework for population balance modeling of batch grinding: beyond first-order kinetics. Chem. Eng. Sci. 2006, 61: 33–44.

10. Bilgili, E., Scarlett, B. Numerical simulation of open-circuit continuous mills using a non-linear population balance framework: incorporation of non-first-order effects. Chem. Eng. Technol. 2005, 28: 153–159.

11. Baxter, J., Abu-Nahar, A., Tuzun, U. The breakage matrix approach to inadvertent particulate degradation: dealing with intra-mixture interactions, Powder Technol. 2004, 143-144: 174-178.

12. Austin, L.G. A review: introduction to the mathematical description of grinding as a rate process. Powder Technol 1971, 5: 1–17.

712b On the Modeling of Milling In Pharmaceutical Industry

Nandkishor Nere[1], Ryan McCann[2], Kenneth Morris[2], and Doraiswami Ramkrishna[1]. (1) School of Chemical Engineering, Purdue University, 480 Stadium Mall Drive, West Lafayette, IN 47907-2100, (2) Department of Industrial and Physical Pharmacy, Purdue University, 480 Stadium Mall Drive, West Lafayette, IN 47907-2100

Milling of the active pharmaceutical ingredient is a significant unit operation practiced in the pharmaceutical and allied industry. Although it is simple in operation, the physics is complex and the milling behavior is difficult to predict from the first principles due to complicated interplay between various factors ranging from the breakage mechanisms and material properties in addition to operating conditions.

We present nonlinear breakage models to explain various observations reported in the literature based on the rational physics following the earlier work of Bilgili and Scarlett (2005). We show that the dynamics of the particle size distribution resulting due to certain nonlinear milling models also exhibits self-similarity that can very well be harnessed to extract models for both the breakage rates and daughter particle size distributions using the inverse problem approach developed by Sathyagal et al. (1994).

We show some interesting simulation results and validation of the adopted inverse problem approach for some model breakage rates and the daughter size distributions. It is also of interest to propose the methodology to relate the extracted breakage functions to the physics of the process and the operating parameters for two-stage hammer mill operated with screens in place that is a part of continuous tablet press.

References

1. Sathyagal, A. N.; Ramkrishna, D.; Narsimhan, G. Solutions of Inverse Problems in Population Balances II. Particle Break-Up. Comput. Chem. Eng. 19, (1995), 437.

2. Bilgili, E., Scarlett, B. Population balance modeling of non-linear effects in milling processes, Powder Technology 153, (2005), 59– 71.

712c Numerical Modeling of Two-Component Granulation

Themis Matsoukas and **Carl L. Marshall Jr.**. Chemical Engineering, The Pennsylvania State University, University Park, PA 16802

Granulation of pharmaceutical powders generally involves the mixing of an active pharmaceutical ingredient (API) and an excipient in the presence of a liquid binder. To apply population balance models, one must adapt the usual one-component algorithms to a *bivariate* distribution that takes into consideration both the size and composition of the granules. The numerical challenges of this problem are quite significant and the extension of one-component methods are not straightforward. These difficulties are magnified when one considers a realistic granulation environment in which the population balance equation (PBE) must be solved alongside with transport equations for momentum and heat in three-dimensional space. Though our work currently focuses on the two-component PBE alone, our goal is to perform realistic large-scale simulations of a fluidized bed granulator. Our short-term goal, and the focus of this work, is in developing numerical solutions of two-component PBE that are both sufficiently accurate and sufficiently fast to apply to real systems. In this study, we apply three different techniques with complementary advantages and limitations:

(i) Direct solution of the discrete, two-component PBE. This approach is rigorous and detailed but is computationally demanding and can only be used to track the very early stages of granulation

(ii) Constant-Number Monte Carlo (cNMC). This method allows the computation of the bivariate distribution over arbitrarily long times. Its chief disadvantage is that is not well-suited for systems that involve space and time gradients. This is due to the computational cost associated with increased dimensionality in space, and also due to a fundamental mismatch between event-driven Monte Carlo and time-driven process simulators.

(iii) Direct Quadrature Method of Moments (DQMOM): this methodology has proven to be very efficient in one-component systems and is currently the only viable option for interfacing the PBE with fluid dynamics. The method is applicable to two-component systems but such extensions are not straightforward, nor can they be implemented uniquely. Therefore, experimentation with DQMOM requires validation against known solutions.

(iv) Theory. Recent work from our group has resulted in the formulation of a theory of bicomponent aggregation, which, for special classes of kernels is capable of predicting the analytic form of the compositional distribution at long times. While exact predictions are not possible for all granulation kernels, theory is valuable in establishing known solutions and a basis for the interpretation of results.

To keep the formulation general, the aggregation kernel is expressed in the form $K_{12}=k_{12}\Psi_{12}$, where k_{12} depends on the size of the aggregating particles (granules) and Ψ_{12} depends on their composition. For

k_{12} we use the kinetic theory of a granular flow: $k_{12}=\beta\Psi_{12}d^{\gamma}_{43}(d_1+d_2)^2(1/d_1^3 + 1/d_2^3)^{0.5}$

where d_i is the diameter of particle i, which is assumed to be spherical. For Ψ_{12} we have chosen the expression $\Psi_{12}=exp[-a_{AB}(c_1+c_2-2c_1c_2)]$

where a_{AB} is a parameter and c_i is the mass. By choosing the value of a_{AB} we can simulate a wide range of behaviors. With $a_{AB}<0$, collisions between granules with dissimilar composition (solute-rich with excipient-rich) are favored, giving rise to enhanced mixing between components. With $a_{AB}>0$, the opposite is true: collisions between similar components are favored leading to poor mixing and possibly segregation of components. Finally, with $a_{AB}=0$ we recover the composition-independent kernel for which theory provides solutions.

We present results of simulations for various initial conditions in terms of amount of API and size ratio of the API and excipient particles. We find that Monte Carlo simulations with N=1000 to 10000 provides very good agreement with the rigorous (discrete) PBE and also with the predictions of the theory. Specifically, we find that the distribution of API in granules of mass v_i is a Gaussian function whose mean is equal to the overall mass fraction of API in the granulator and whose standard deviation is inversely proportional to the square root of v_i. Thus we conclude that the cNMC method produces accurate solutions over very long times. We present further results with both positive and negative values of the parameter a_{AB} and discuss the conditions under which the DQMOM method provides accurate solutions to the bicomponent problem.

712d Modelling and Simulation of the Crystallization of L-Glutamic Acid Polymorphs

Martin Hermanto[1], Nicholas Kee[2], Richard D. Braatz[3], and Min-Sen Chiu[1]. (1) Chemical and Biomolecular Engineering, National University of Singapore, 117576, Singapore, Singapore, (2) Department of Chemical and Biomolecular Engineering, National University of Singapore, 4 Engineering Drive 4, Singapore, 117576, Singapore, (3) Department of Chemical and Biomolecular Engineering, University of Illinois at Urbana-Champaign, 600 South Mathews Avenue, Box C-3, 293 Roger Adams Laboratory, Urbana, IL 61801-3602

Most pharmaceutical manufacturing processes include a series of crystallization processes where their product quality is often associated with the crystal final form (such as crystal habit, shape and size distribution). Recently, there is a rapid growth of interest in polymorphism (when a substance has multiple crystal forms) motivated by patent, operability, profitability, regulatory, and scientific considerations [1-3]. The different polymorphs can have orders-of-magnitude differences in properties such as solubility, chemical reactivity, and dissolution rate, which can have an adverse effect on downstream operability and performance of the crystal product [2]. As a result, controlling polymorphism to ensure consistent production of the desired polymorph is very critical in those industries, including in drug manufacturing where safety is of paramount importance.

This presentation considers the crystallization of L-Glutamic acid which has two polymorphic forms (the metastable alpha-form and the stable beta-form). A mathematical model based on population balance equations [4] is developed along with the fundamental equations describing the nucleation and growth kinetics for both polymorphs. Compared to past models for L-glutamic acid crystallization [5,6], this appears to be the first to estimate activation energies for the nucleation and growth processes, and the first to quantify the uncertainties in the kinetic parameters. Bayesian inference is used to determine the kinetic parameters from experimental data which includes on-line temperature measurement, in-situ concentration measurement by Attenuated Total Reflection Fourier Transform Infrared (ATR-FTIR) spectroscopy, and in-situ Chord Length Distribution (CLD) measurement by Focused Beam Reflectance Measurement (FBRM). While weighted least-squares methods are adequate for many problems, Bayesian inference is able to include prior knowledge in the statistical analysis which can produce models with higher predictive capability. Bayesian estimation produces a posterior distribution for the estimated parameters, which can be incorporated into robust control strategies for crystallization process [7].

Numerical simulation for polymorphic crystallizations enables the investigation of the effects of various operating conditions and can be used for optimal design and control [6,8,9]. Solving population balance equations is particularly challenging when the PDEs are hyperbolic with sharp gradients or discontinuities in the distribution [10]. Standard first-order methods require a very small grid size in order to reduce the numerical diffusion (i.e., smearing), whereas standard higher order methods introduce numerical dispersion (i.e., spurious oscillations), which usually results in a crystal size distribution with negative values. Efficient and sufficiently accurate computational methods for simulating the population balance equations are required to ensure the behaviour of the numerical solution is determined by the assumed physical principles and not by the chosen numerical method. This presentation considers a class of numerical algorithms known as weighted essentially non-oscillatory (WENO) methods [11-14] which were developed for especially accurate simulation of shock waves and provide much higher order accuracy than the previously considered methods for solving PBMs. These WENO methods are compared to the high resolution (HR) finite volume method [15] and a second-order finite difference (FD2) method, for the crystallization of L-glutamic acid polymorphs under conditions in which the distribution has sharp gradients.

References

[1] Fujiwara M, Nagy ZK, Chew JW, Braatz RD. First-principles and direct design approaches for the control of pharmaceutical crystallization. J Process Control 2005; 15:493-504.

[2] Blagden N, Davey R. Polymorphs take shape. Chem. Brit. 1999; 35:44-47.

[3] Brittain HG. The impact of polymorphism on drug development: A regulatory viewpoint. Amer. Pharm. Rev. 2000; 3:67-70.

[4] Randolph AD, Larson MA. Theory of particulate processes: analysis and techniques of continuous crystallization. San Diego: Academic Press. 1988.

[5] Ono T, Kramer HJM, Ter Horst JH, Jansens PJ. Process modeling of the polymorphic transformation of L-glutamic acid. Crystal Growth & Design 2004; 4(6):1161-1167.

[6] Scholl J, Bonalumi D, Vicum L, Mazzotti M. In situ monitoring and modeling of the solvent-mediated polymorphic transformation of L-Glutamic acid. Crystal Growth & Design 2006; 6(4):881-891.

[7] Nagy ZK, Braatz RD. Worst-case and distributional robustness analysis of finite-time control trajectories for nonlinear distributed parameter systems. IEEE Trans. Control Systems Technology 2003; 11(5):694-704.

[8] Hermanto MW, Chiu MS, Woo XY, Braatz RD. Robust optimal control of polymorphic transformation in batch crystallization. AIChE J. 2007; 53(10):2643-2650.

[9] Roelands CP, Jiang SF, Kitamura M, ter Horst JH, Kramer HJ, Jansens PJ. Antisolvent crystallization of the polymorphs of L-histidine as a function of supersaturation ratio and of solvent composition. Crystal Growth & Design 2006; 6(4):955-963.

[10] Sotowa KI, Naito K, Kano M, Hasebe S, Hashimoto I. Application of the method of characteristics

to crystallizer simulation and comparison with finite difference for controller performance evaluation. J. Process Control 2000; 10:203-208.

[11] Liu XD, Osher S, Chan T. Weighted essentially non-oscillatory schemes. J. Comput. Phys. 1994; 115:200-212.

[12] Jiang GS, Shu CW. Efficient implementation of weighted ENO schemes. J. Comput. Phys. 1996; 126:202-228.

[13] Henrick AK, Aslam TD, Powers JM. Mapped weighted essentially non-oscillatory schemes: Achieving optimal order near critical points. J. Comput. Phys. 2005; 207:542-567.

[14] Serna S, Marquina A. Power ENO methods: a fifth-order accurate weighted power ENO method. J. Comput. Phys. 2004; 194:632-658.

[15] Gunawan R, Fusman I, Braatz RD. High resolution algorithms for multidimensional population balance equations. AIChE J. 2004; 50(11):2738-2748.

Web Page: www.chee.nus.edu.sg/people/faculty_chiums.html

712e Breakage of Needle-Shaped Particle Assemblies : Estimation of Breakage Functions from Discrete Element Simulations

Pavol Rajniak[1], David Lamberto[2], Zdenek Grof[3], and **Frantisek Stepanek**[3]. (1) Merck & Co, Inc., WP78-110, PO Box 4, West Point, PA 19486, (2) Merck & Co., Inc., RY818-C108, P.O.Box 2000, Rahway, NJ 07065-0900, (3) Department of Chemical Engineering, Institute of Chemical Technology, Prague, Technicka 5, Prague 6, 166 28, Czech Republic

Crystals with a large aspect ratio (needle- or platelet-shaped) are a common occurrence in many pharmaceutical and fine chemicals processes. While relatively narrow particle size distribution (PSD) can be obtained in the crystallisation step by precise control of the crystal growth kinetics and hydrodynamic conditions, further fluid-solid separation such as filtration, filter washing, drying, and subsequent solids handling can often lead to uncontrolled changes in PSD due to breakage. In order to understand and eventually predict the effect of these unit operations on breakage patterns, population balance modelling is a useful tool – provided that correct values of input parameters are supplied. In particular, the so-called selection function (which specifies the probability that a particle will undergo a breakage event) and the breakage function (which specifies the size distribution of daughter particles once a breakage event does occur) need to be known. The breakage and selection functions can in principle be evaluated by retro-fitting experimental breakage data, but such procedure is not always reliable and requires considerable experimental effort.

In this contribution we present a new methodology for determining the breakage functions (breakage kernel and daughter distribution function) for a population balance model from Discrete Element (DEM) simulations of needle-shaped particle breakage. The main advantage of this methodology is that individual particles can be traced within a particle assembly during the simulation, and so the frequency of breakage events as well as the size distribution of daughter particles can be evaluated explicitly in a statistically robust way. The methodology consists of two parts: (i) the DEM simulation of needle-shaped particle breakage as described in [1], whereby the influence of parameters such as the aspect ratio, initial length distribution, and the intrinsic mechanical properties of particles and their distribution can be systematically investigated; (ii) post-processing of the DEM simulation outputs (i.e., broken and un-broken particles within a population as function of time and/or compression ratio) and the explicit evaluation of the breakage functions. Convoluted PSD data can be extracted both from DEM simulations and from subsequent population balance models of breakage, and used as an additional verification of the consistency of the two models and their comparison with experimental data.

References:

1. Grof Z., Kohout M., Štěpánek F., Multi-scale simulation of needle-shaped particle breakage under uni-axial compaction, Chem. Eng. Sci. 62, 1418-1429 (2007)

712f Multivariate Population Balance Equations for Functional Nanoparticle Formation Stabilized by Amphiphilic Block Copolymer Directed Assembly

Chungyin Cheng[1], R. Dennis Vigil, and R. O. Fox[2]. (1) Chemical and Biological Engineering, Iowa State University, Ames, IA 50011, (2) Department of Chemical & Biological Engineering, Iowa State University, 2114 Sweeney Hall, Ames, IA 50011-2230

Nanoprecipitation [1] is a process to produce functional nanoparticles protected by amphiphilic copolymer directed assembly. In this process, organic drug and copolymers are dissolved in solvent and rapidly mixed with nonsolvent in a customized micro-device to create high supersaturation. The growth of nanoparticles is kinetically arrested by block copolymer assembly on the surface of the organic drug, and therefore a tunable and narrow particle size distribution is achievable. In order to simulate the Nanoprecipitation process, the population balance equations (PBEs) are solved to obtain the particle size distribution (PSD) information in the particulate flow system. A bi-variate PBE model is presented to describe the particle stabilization process. Different particle compositions are represented by a bi-variate PSD with the first coordinate for the organic drug and the second for the copolymer. The detailed kinetics of micellization for the copolymer directed assembly are discussed. A two-region Brownian aggregation kernel is introduced to account for cases with larger aggregation number for micelles, where the hydrophilic block forms a corona region to prevent further aggregation. In this work, the bi-variate PBEs are solved using the quadrature method of moments (QMOM) [2] to obtain faster computations and the results (mean particle size) are compared with experimental data.

[1] B. K. Johnson and R. K. Prud'homme. Flash nanoprecipitation of organic actives and block copolymers using a confined impinging jets mixer. Australian Journal of Chemistry 56(10):1021–1024, 2003.

[2] Marchisio, D. L. & Fox, R. O. Solution of population balance equations using the direct quadrature method of moments Journal of Aerosol Science 36, 43-73, 2005

712g Component Mixing In Bicomponent Aggregation: Classification of Kernels

Themis Matsoukas[1], **Carl L. Marshall Jr.**[1], Kangtaek Lee[2], and Taehoon Kim[2]. (1) Chemical Engineering, The Pennsylvania State University, University Park, PA 16802, (2) Department of Chemical Engineering, Yonsei University, 134 Shinchon-dong, Seodaemun-gu, Seoul, 120-749, South Korea

Wet granulation is a size enlargement process in which a multicomponent mixture that contains an active compound, an excipient, and a binder, is lead to produce an agglomerated powder until desired properties are met with respect to mechanical strength of granules, flow characteristics and uniformity of composition. Over the past 15 years, there has been considerable effort towards the development of simulation tools for granulation that make use of increasingly sophisticated models to describe granule interactions and their coupling to the population balance equation (PBE). Virtually all of the previous studies consider granulation of a single component (univariate PBE). Yet more often that not, granulation is a multicomponent process. In pharmaceutical granulation, for example, an active pharmaceutical ingredient (API) is co-granulated with an inert excipient, in preparation for subsequent tablet formation. The purpose of granulation in this case is not only to increase the size of the granules, but also to improve mixing of components. Thus, in addition to changes in the size distribution, one needs to know the compositional distribution of components among granules of different sizes. Ideally, all granules should have the same composition that is equal to the overall ratio of the bulk amounts loaded in the unit. In reality, actual composition varies with granule size and time.

In this work, we present theoretical and experimental results of bicomponent granulation. We formulate a population balance approach to describe a bicomponent population of particles that contain an ingredient of interest ("solute") mixed with an inert compound ("excipient"). We classify granulation kernels into three categories: (i) kernels such that dissimilar components have higher rate of agglomeration compared to similar components; (ii) kernels such that dissimilar components have lower rate of agglomeration; and (iii) kernels that make no distinction between the components. Case (iii), which we refer to as "ideal" aggregation, leads to blending of components into a distribution that is Gaussian in the mass fraction of the solute. The mean composition of granules approaches quickly the desired composition and the variance of the distribution decreases inversely with granule size. Kernels of type (i) lead to quick blending of components. Kernels of the type (ii) inhibit blending and lead to a population of granules that is relatively segregated. Eventually, however, these kernels also produce Gaussian distributions but at a much slower rate compared to kernels (i) and (iii). The results of these simulations are compared with experimental granulation studies using mixtures of two excipients. We find that depending on the surface properties of the excipient, these systems exhibit behavior similar to that of kernels of type (ii) and (iii).

726a Recovery of Immunoglobulins from Bovine Serum through a Combination of Selective Precipitation and Crossflow Microfiltration

Georges Belfort and Adith Venkiteshwaran. Chemical and Biological Engineering, Rensselaer Polytechnic Institute (RPI), 110 8th St., Troy, NY 12180

A large proportion of protein therapeutics in the product pipelines of biotechnology firms are monoclonal antibodies (IgG). The high binding specificity of protein A affinity ligands for monoclonal antibodies is usually exploited in their downstream recovery. However, high costs associated with these ligands, along with requirements such as high pH for cleaning and problems involved with leaching of the ligand from the column, have prompted the development of alternative strategies for selective monoclonal antibody separation. Exploiting physicochemical properties with synthetic membrane processes to recover IgG from bovine serum has not proved very successful. As a result we have developed a novel method to selectively recover IgG from bovine serum. The key steps involve the selective precipitation of IgG, by addition of an optimum amount of ammonium sulfate, combined with crossflow membrane microfiltration to retain the precipitate and remove the undesirable soluble bovine serum albumin (BSA). High transmission of the soluble BSA was obtained with minimal fouling of the 0.1 micron hydrophilized poly(vinylidene flouride) (PVDF) membrane with retention of all the IgG. A subsequent diafiltration step resulted in the removal of about 97% of the BSA, and a yield and purity of 88% and 87%, respectively, for the IgG. The effects of wall-shear rate and membrane pore-size on the filtration of the above precipitate-containing feed were also investigated in total recycle mode to obtain a further understanding of the behavior of the system. The higher membrane pore-size resulted in higher concentration polarization and fouling due to the loss of precipitate particles within the membrane pores.

726b Ultra-Scale down Approaches for the Separation of Mammalian Cell Culture Broth In a Disc-Stack Centrifuge

Ferhana Zaman, Downstream Purification, Bioprocess Research & Development, Pfizer Inc., 700 Chesterfield Parkway West,, HH211-U, Chesterfield, MO 63017 and Sa V. Ho, Global Biologics, Pfizer, 700 Chesterfield Parkway West, Chesterfield, MO 63017.

Recovery of monoclonal antibodies produced extracellularly in mammalian cell culture systems typically requires centrifugation for the removal of cells and cell debris. Due to the complex nature of the centrifugation unit operation and the shear sensitiveness of the cell material, it is often difficult to obtain predictive laboratory scale data to closely mimic the performances of large scale centrifuges. The use of ultra scale-down studies to understand and better predict the clarification performance of pilot and industrial-scale continuous centrifuges for different monoclonal antibodies will be evaluated. With small material requirements, these approaches will allow rapid optimization of operating conditions.

726c Monoclonal Antibody Primary Recovery Clarification: Evaluating Product Quality Impact, Process Optimization and Scale-up

Bryan C. Dravis, Bioproduct R&D - Bioprocess Operations, Eli Lilly and Company, Lilly Corporate Center, DC 3515, Indianapolis, IN 46285 and Michael D. Hilgert, Bioproducts R&D - Bioprocess Operations, Eli Lilly and Company, Lilly Corporate Center, DC 3515, Indianapolis, IN 46285.

The primary recovery clarification of monoclonal antibodies (MAbs) from a mammalian cell culture is a straightforward solid-liquid separation of insoluble cellular components from a multifaceted cell culture broth. Consisting of centrifugation followed by a series of successive normal flow filtrations, the main objective of the clarification sequence is to properly clarify the starting complex cell culture for the subsequent series of chromatographic purifications and viral removal steps. Using a tailored design of experiments (DOE), investigations were conducted to assess the impact the primary recovery operations had on the overall downstream purification process and to facilitate the design of better scale down/up models for more accurate process definition and material supply.

A modified 2^3 factorial DOE was employed to explore a "representative" operating window for the primary recovery unit operations at pilot and manufacturing scales. The DOE incorporated Q/Σ theory boundaries for the centrifugation such that all results could be properly scaled and constrained across equipment sets. The DOE isolated the individual operating parameters for each unit operation that directly contributed to product quality, but the study also assessed how overall effect product stream characteristics impacted subsequent downstream unit operations. Product titers, quality, stability, clarity, throughput capacity, impurity removal, cell lysis and total protein analyses were all used as quantifiable measures in this process evaluation.

These primary recovery investigations were conducted across three separate manufacturing scales and evaluated the impact to several MAbs (IgG-1, IgG-4) and "MAb-like" therapeutic proteins. The study results have been used to properly scale, model and characterize primary recovery operations from lab to commercial scale, to identify safe process hold-locations, minimize impact to product quality and optimize process robustness. This work now serves to further enable "platform" approaches that facilitate better process understanding while also reducing process development time lines and permit manufacturing flexibility.

726d Downstream Scale-up Evaluation for High-Titer Mammalian-Based Processes: Platform Strategies for Improved Clarification and Purification Throughput

Joshua Walker, Process Biochemistry / Purification Scale up, Biogen Idec, 14 Cambridge Center, Cambridge, MA 02142

Previous work has demonstrated the control of mammalian cell broth buffer conditions during the production of high-titer therapeutic proteins to assist in clarification by TFF. However, there is limited understanding of the significant beneficial effects of harvest conditions on clarification performance by centrifugation and subsequent downstream chromatographic steps for processing high-titer products.

We will present a framework for describing the performance of disc-stack centrifugation in conjunction with depth filtration and capture column chromatography for cell and impurity removal using controlled harvest conditions. In combination, data from bench/pilot-scale to large-scale operations will be presented evaluating the impact of controlled parameters on downstream performance.

Pilot-scale centrifugation and depth filtration studies were then performed at a wide-range of centrifuge flowrates and RPMs using optimized harvest from various clinical products producing IgGs. The data shows a significant reduction in centrate clarity compared to clarifying with unadjusted harvest streams at all scales, with improved post-centrifuge filtration capacity and capture column eluate purity and clarity. Overall, this analysis allows for improved future design and optimization of large-scale centrifugal operations for processing mammalian cell culture broths.

726e Capturing of Cell Culture Derived Vaccinia Virus by Membrane Adsorbers

Michael W. Wolff[1], Corina Sievers[2], Sylvia Lehmann[1], Lars Opitz[1], Sara Post-Hansen[3], Rene Djurup[3], Rene Faber[4], and Udo Reichl[1]. (1) Bioprocess Engineering, Max Planck Institute for Dynamics of Complex Technical Systems, Sandtorstrasse 1, Magdeburg, 39106, Germany, (2) Bioprocess Engineering, Otto-von-Guericke Universität, Universitaetsplatz 2, Magdeburg, Germany, (3) Bavarian Nordic A/S, Hejreskovvej 10A, Kvistgård, Denmark, (4) Biotechnology Division, Sartorius Biotech GmbH, Weender Landstrasse 94-108, Goettingen, 37075, Germany

Smallpox is an acute, highly infectious viral disease unique to humans with a mortality rate around 25%. It is caused by the Variola virus that belongs to the family of Poxviruses. Smallpox was responsible for an estimated 300-500 million deaths in the 20th century. Following successful vaccination campaigns through the 19th and 20th centuries, the World Health Organization (WHO) certified the eradication of smallpox in 1980. After the eradication, the compulsory vaccination was abandoned – with the result that about half of the world's population is not vaccinated. This represents a potential threat in the case of a deliberate release of Variola virus as an act of bioterrorism. Consequently, several governments are ordering stock piles of smallpox vaccines to protect their populations from this remote, but extremely grave threat. MVA-BN® is a third generation smallpox vaccine based on the Modified Vaccinia Ankara (MVA) virus which demonstrates superior safety compared to traditional smallpox vaccines based on native Vaccinia virus (VV) strains. In addition, re-engineered VV represent as a robust vector a platform technology for vaccine delivery systems as e.g. in the case of HIV, Dengue fever, Japanese encephalitis and cancer.

Traditionally, VV- and MVA virus based vaccines have been grown in primary chicken embryo fibroblast cultures and purified either by sucrose cushion or sucrose gradient centrifugation, or by ultrafiltration. However, a potential shift from primary to continuous cell cultures would impose stricter requirements regarding the purity level of the vaccines, and a new generation of vaccine manufacturing processes is needed that include more sophisticated and innovative downstream techniques for purification.

Here, we report the development of an affinity chromatography of cell culture derived VV after an initial host cell homogenization and clearance centrifugation. The Vaccinia viral envelope protein A27L is known to bind to heparin.. Based on this, small scale chromatography experiments with heparinized polymer beads and cellufine sulfate beads in addition to heparinized cellulose membranes have been conducted. There we have found that membrane adsorbers are superior over bead based chromatography media in terms of efficiency and productivity.

Subsequent studies compared ion exchange membrane adsorbers with a heparinized membrane adsorber. The results indicate that the overall performance of the affinity chromatography in terms of virus capturing and contaminant removal is better than any of the tested ion exchange membrane adsorbers. Hence, membrane affinity chromatography represents a valuable choice to capture VV particles in addition to the general advantages of membrane chromatography.

726f Integrated IgG Fragmentation and Purification Using a Reactant Adsorptive Membrane Bioreactor Separator System

Deqiang Yu and Raja Ghosh. Department of Chemical Engineering, McMaster University, 1280 Main Street West, Hamilton, ON L8S 4L8, Canada

Immunoglobulin G fragments are widely used for different diagnostic and therapeutic applications. Fragmentation of immunoglobulin G is useful for characterizing antibodies. The enzymatic fragmentation of IgG is traditionally carried out by homogeneous liquid phase reaction and followed multi-step purification to remove the enzyme and byproducts. These processes tend to be complex with low overall productivity. The current work deals with use of a membrane chromatography system as a bioreactor and separator. In conventional bioreactors, the biocatalyst is immobilized within the reactor and the reactant is passed through it for carrying out the enzymatic reaction. In our bioreactor system the reactant i.e. IgG is reversibly immobilized and the enzyme is allowed to flow through the system. This approach allows the reaction to be carried out and multiple components i.e. reactant, impurities, enzyme, product and byproduct to be separated using a single device. In the present work hydrophobic interaction membrane chromatography (HIMC) was utilized to obtain F(ab')2 and Fab fragments directly from human serum using enzymes pepsin and papain. The reactant adsorptive HIMC system was able to selectively adsorb IgG from serum. When the bound IgG was pulsed with the enzyme under optimized condition, different fragments were obtained and these could be fractionated by use of appropriate gradient elution. The main advantage of our approach is the efficient integrated reaction and separation. Other anticipated advantages include low process time and scalability.

729a Petrobactin Biosynthesis: A Target for Antibiotics and a Platform for Producing Specialty Chemicals

Brian Pfleger[1], David Sherman[2], Tyler Nusca[2], Jaime Scaglione[2], and Jung-Yeop Lee[2]. (1) Chemical and Biological Engineering, University of Wisconsin Madison, 3629 Engineering Hall, 1415 Engineering Dr, Madison, WI 53706, (2) Life Sciences Institute, University of Michigan, Ann Arbor, MI 48109

In the ongoing fight against antibiotic resistance, iron acquisition, and more specifically siderophore biosynthesis, has become an attractive target for discovery of new antibiotics. Recent work with *Mycobacterium tuberculosis* has demonstrated the potential for small molecule inhibition of siderophore biosynthetic enzymes. Siderophore biosynthesis in *Bacillus anthracis* has been shown to be important for spore outgrowth and pathogenicity. Therefore, targeted inhibition of the biosynthesis of the essential siderophore, petrobactin, may hold promise as a potential therapeutic against anthrax. Our work has focused on characterizing the enzymes responsible for petrobactin biosynthesis and designing inhibitors for key biosynthetic steps. In the process, we have discovered that the substrate specificity of these enzymes is highly flexible and can be exploited for specialty chemical production.

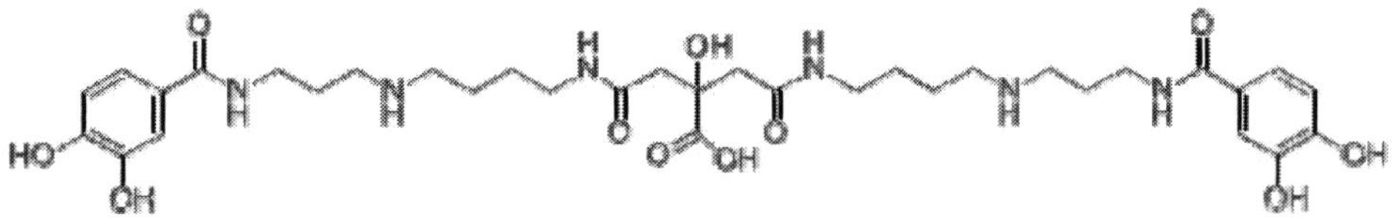

Petrobactin

We performed a series of studies with the objective of dissecting the catalytic function of each of the six Asb enzymes. Our initial studies focused on three enzymes, AsbC, AsbD, and AsbE, presumed to catalyze synthesis of the first intermediate in petrobactin biosynthesis (3, 4-dihydroxybenzoyl-spermidine). We have shown that AsbC is a 3, 4-dihydroxybenzoic acid-AMP ligase similar to EntE and VibE in enterobactin and vibriobactin biosynthesis. AsbD was shown to be an acyl/peptidyl/aryl carrier protein similar to thiolation domains found in polyketide, non-ribosomal peptide, and siderophore biosynthetic complexes. AsbE was shown to condense the AsbD-bound 3, 4-dihydroxybenzoic acid with spermidine, citryl-spermidine, and several polyamine analogs. Through substrate specificity assays, AsbC was shown to be capable of loading substrates with hydrogen bond donors at the meta and para positions of the benzyl ring onto holo-AsbD. This finding was supported by our initial inhibition studies where nonydrolyzable acylsulfate analogs of DHB-AMP containing hydroxyl groups at these positions were capable of inhibiting ATP-^{32}PPi exchange and those without did not.

Two condensation enzymes, AsbA and AsbB, have been shown to generate the amide bonds in the central portion of petrobactin. AsbA has been shown by others to generate citryl-spermidine through an ATP-dependent condensation mechanism. The order of the remaining condensations catalyzed by AsbB and AsbE is flexible, suggesting that these enzymes could be used to generate a variety of small molecules. The last enzyme in the pathway, AsbF, is responsible for generating protocatechuate from an intermediate in the shikimate pathway. Assays are currently being developed for use in high throughput screens of chemical libraries to indentify inhibitors of each enzyme. This presentation will demonstrate the novel petrobactin biosynthetic pathway, the flexibility of the participating enzymes, and progress towards identifying novel therapeutic leads against *Bacillus anthracis*.

729b A Polyketide Macrolactone Synthase from the Filamentous Fungus *Gibberella Zeae*

Hui Zhou, Chemical and Biomolecular Engineering, University of California, at Los Angeles, 420 Westwood Plaza, 5667 Boelter Hall, Los Angeles, CA 90034 and Yi Tang, Chemical and Biomolecular Engineering, University of California, Los Angeles, 420 Westwood Plaza, Los Angeles, CA 90095.

Abstract

Resorcylic acid lactones represent a unique class of fungal polyketides and display a wide range of biological activities, such as nanomolar inhibitors of Hsp90 and MAP kinase. The biosynthesis of these compounds is proposed to involve two fungal polyketide synthases (PKS) that function collaboratively to yield a 14-membered macrolactone with a resorcylate core. We report here the reconstitution of *Gibberella zeae* PKS13, which is the nonreducing PKS associated with zearalenone biosynthesis. Using a small molecule mimic of the natural hexaketide starter unit, we reconstituted the entire repertoire of PKS13 activities in vitro, including starter unit selection, iterative condensation, regioselective C2-C7 cyclization and macrolactone formation. PKS13 synthesized both natural 14-membered and novel 16-membered resorcylic acid lactones, indicating relaxed control in both iterative elongation and macrocyclization. Furthermore, PKS13 exhibited broad starter-unit specificities towards fatty acyl-CoAs ranging in sizes between C6 and C16. PKS13 was also proven to be able to interact with *Escherichia coli* fatty acid biosynthetic machinery and can be primed with fatty-acyl ACPp at low micromolar concentrations. A number of alkyl pyrones and alkyl resorcylic esters were produced by *E. coli* BL21(DE3) harboring PKS13 with a highest yield up to 5 mg/L. And we further demonstrated that BL21(DE3) harboring PKS13 can synthesize new polyketides at a yield of 2~3 mg/L, when supplemented with synthetic precursors, which showcases the utility of PKS13 in precursor-directed biosynthesis. PKS13 is therefore a highly versatile polyketide macrolactone synthase that is useful in the engineered biosynthesis of novel polyketides, including resorcylic acid lactones that are not found in nature.

Web Page: www.pnas.org/cgi/content/figsonly/105/17/6249

729c Studying the Structural Properties of Nesprin-1α By Using Engineered Protein Fragments

Zhixia Zhong[1], Siwei Chang[1], Katherine Wilson[2], and Kris Noel Dahl[1]. (1) Department of Chemical Engineering, Carnegie Mellon University, 5000 Forbes Avenue, Pittsburgh, PA 15213, (2) Johns Hopkins University, 725 N. Wolfe St., G10, Baltimore, MD 21205

Nesprins, encoded by three genes in humans, play key roles in mechanically linking the cytoskeleton to lamin intermediate filaments in the nucleus. One particular isoform, nesprin-1α (120 kDa), is an integral membrane protein of the inner nuclear membrane that binds directly to A-type lamins and emerin. Nesprin-1α plays a role in the pathogenesis of Emery-Dreifuss muscular dystrophy (EDMD), and heterozygous missense mutations in nesprin-1α lead to EDMD.Structurally, nesprin-1α has six spectrin like repeats (SLR) and a central rod domain, comprising 25% of its length. The SLR domains in nesprin-1α are weakly homologous to α- and β-spectrin (21-28%), but the structural properties of nesprin-1α, itself, were untested and the structural contributions of of the rod domain were unclear. Thermodynamic properties of other spectrin superfamily proteins involve cooperative coupling folding and unfolding of tandem three-helix bundles (SR domains). We hypothesized that the rod domain of nesprin-1α might confer unique molecular thermodynamics compared to other SR domain proteins. To test this model, we tested the thermodynamic stability of several nesprin-1α fragments including an engineered fragment of the full-length nesprin-1α without the rod domain. We measured the secondary structure and thermal unfolding properties of the constructed protein fragments of nesprin-1α fragments using circular dichroism. Despite the weak homology with spectrin, the predicted SLR domains in nesprin-1α have thermodynamic properties of typical SR domains based on their helicity, mean transition temperatures (Tm), ellipticity ratio $\Theta_{222}/\Theta_{208}$, multiple transition stages and cooperative coupling unfolding.

However, the protein fragments that included the rod domain were highly helical with a higher Tm than corresponding fragments lacking the rod domain. Thus, the rod domain overstabilizes the SLR domains by increasing the cooperative folding and by increasing overall helicity and Tm of the molecule. We also tested these protein fragments to look for *in vitro* binding to emerin. We found that the linker region is not necessary for protein binding, but our results suggest that the larger protein structure may be required for binding studies due to the cooperative folding of these SLR segments. Currently we are examining these engineered nesprin-1α fragments expressed in Hela cell to test the cell effects of the rod domain and SR domains. A combination of *in vitro* and cellular studies examining protein structure and interaction may lead to a better understanding of nuclear lamina organization as well as elucidate the mechanism of EDMD.

729d Characterization and Engineering of Peptide Binding to Class II Mhc by Yeast Co-Expression

Wei Jiang, Department of Chemical and Biomolecular Engineering, University of Pennsylvania, 220 South 33rd St., Philadelphia, PA 19104 and Eric T. Boder, Department of Chemical and Biomolecular Engineering, University of Tennessee, 1512 Middle Drive, 437 Dougherty Engineering Bldg., Knoxville, TN 37996-2200.

Class II major histocompatibility complex (MHC) -restricted T cell responses are related to a great number of diseases including autoimmunity, graft rejection, and atypical immune response. In depth characterization of peptide binding by class II MHC is critical to understanding issues in vaccine design, autoimmune disease, infectious disease progression, and transplantation rejection. The lack of an efficient, quantitative, robust, and widely applicable methodology for characterizing the peptide binding specificity and promiscuity of MHC alleles remains a bottleneck.

A yeast co-expression system, which can surface display peptide from influenza hemagglutinin (FLU) bound by soluble DR1 secreted by the same yeast cell, has been developed. In order to evaluate the performance of this system for studying the peptide binding properties of class II MHC molecules, initial studies focused on characterizing the peptide side chain preferences at the dominant anchor position (P1) of peptide/HLA-DR1 complexes. A set of yeast surface tethered FLU peptide analogues containing all natural amino acids (except for Cys) at P1 position have been tested for their ability to bind and anchor soluble HLA-DR1 to the cell surface. Flow cytometry analysis enabled discrimination of P1 amino acids supporting binding to the MHC and further allows evaluation of relative binding affinities. We find that , affinities follow the order Phe $\geq$ Tyr $\geq$ Trp $\approx$ Met $\approx$ Leu $\approx$ Ile > Val. All other amino acids tested yield no detectable binding to HLA-DR1, in agreement with binding preferences determined using phage display and/or other methods. We are extending this system to (1) study the peptide binding motif of HLA-DQ8, a major risk factor for susceptibility to type I diabetes, and (2) develop and validate an approach for retargeting defined MHC toward novel peptides for possible applications in vaccine technology. The yeast co-expression methodology enables both characterization and engineering of class II MHC peptide binding by manipulating both MHC molecules and peptides easily and efficiently in a format suitable for high throughput screening.

729e Prediction of siRNA Functionality Using Sequence and Thermodynamic Properties

Charles M. Roth[1], Pratik Sheth[2], Anik Jain[2], Vinit Shah[2], and Stanley M. Dunn[2]. (1) Rutgers University, 599 Taylor Road, Piscataway, NJ 08854, (2) Biomedical Engineering, Rutgers University, 599 Taylor Road, Piscataway, NJ 08854

Short interfering RNAs (siRNAs) are widely utilized in the laboratory for gene function studies and are being developed clinically to target viral genes, inflammatory targets and oncogenes. The selection of siRNA sequences is typically performed using software programs that rely upon heuristics and short motifs that have low predictive power. We have developed an approach to predicting siRNA functionality that combines mechanistic thermodynamic parameters with a novel representation of siRNA sequence. Specifically, the siRNA sequence is represented as a directed graph. We find that several graph metrics, including diameter and permanent, exhibit significantly different values in sequences found in a training set to be functional versus those found to be non-functional. We utilized these graph properties as features in a K nearest neighbors support vector machine classification algorithm, along with features representing the thermodynamic accessibility of the complementary region on the target mRNA and the kinetic accessibility of unwinding the antisense strand of the siRNA sequence, which is represented by the difference in thermodynamic stabilities of the 5'- and 3' termini. Applying this approach to a test set consisting of 103 functional and 93 non-functional sequences, we found that these features could be used to classify the functional sequences from the non-functional ones with a leave-one-out-cross-validation accuracy as high as 84.7%, a level significantly greater than that afforded by other classification approaches applied to the same data. Among the three types of features, the thermodynamics of target mRNA accessibility provided the greatest individual accuracy, but the three types of features were synergistic overall. Thus, the idea of combining mechanistic thermodynamic features with ones based on pure sequence representation is quite promising for the development of accurate siRNA prediction tools.

729f Engineering Vectors for Dual Gene Expression from Independent Promoters for High Throughput Studies

Jun Tian and Stelios T. Andreadis. Chemical and Biological Engineering, SUNY-Buffalo, 910 Furnas Hall, Buffalo, NY 14260-4200

Expression of multiple genes from the same target cell is required in several technological and therapieutic applications such as in vivo tracking of stem cells. In spite of such need, reaching independent, coordinate and high level dual-gene expression cannot be reliably accomplished by current gene transfer vehicles. To address this question, we designed and optimized a lentiviral vector carrying two internal promoters each driving expression of a single gene independent of the other. The two promoters were spaced by a synthetic polyA, as well as terminator and insulator sequences. A second insulator sequence was cloned downstream of the polyA of the second gene to alleviate the potential suppression of gene expression from the viral enhancer. All the internal transcription units were cloned in trans to avoid interference of viral mRNA production by the inserted polyAs. With this design, we found that more than 90% transduced cells by this vector co-expressed both genes and that the expression level of both genes was as high as that yielded from lentiviral vectros containing only a single transcriptional unit. Furthermore, the performance of the newly designed vector was independent of cell types and promoters used. Using this vector, we demonstrated tissue specific gene transfer from promoters specific for smooth muscle cells or epidermal keratinocytes. We also demonstrated that this vector can be used for dynamic, real-time monitoring of gene expression suggesting that it may be very useful in high throughput gene expression profiling studies.

729g Molecular Barcode-Labeled Aptamers for Biosensor Applications

Shengnan Xie, Department of Chemical Engineering and Materials Science, Michigan State University, 2527 Engineering Building, East Lansing, MI 48824 and S. Patrick Walton, Chemical Engineering and Materials Science, Michigan State University, East Lansing, MI 48824.

Techniques for parallel measurement of cellular molecules such as proteins and small molecules are important for complete understanding of cellular function and the development of diagnostic tools with greater accuracy and information content. Aptamers, single-stranded DNA or RNA molecules that bind to other molecules with high specificity and affinity, are gaining popularity as biosensing agents. Relative to antibodies, especially, aptamers are comparatively easy to modify and design for biosensing applications. In this presentation, we will describe our efforts at combining aptamers with molecular barcodes (MBs). MBs are a set of unique DNA labels that can be used to detect a single nucleic acid species from a mixed population. The MB serves as a unique identifier of the aptamer and the target molecule to which it binds. We are applying MB-labeled aptamers for the development of new parallel analytical techniques for biological molecules. We will describe our results for both small molecule and protein quantification.

IMPACT OF TWO COMMERCIAL ENZYMES APPLICATION ON SOME PHYSICOCHEMICAL PROPERTIES OF RED PITAYA JUICES

A.R. NUR 'ALIAA[1*], M.K. SITI MAZLINA[1] and F.S. TAIP[1]

[1]Department of Process and Food Engineering

Faculty of Engineering,Universiti Putra Malaysia,

43400 UPM Serdang, Selangor, Malaysia

*Corresponding Author

TEL: 00-603-89464315

FAX: 00-603-86567123

EMAIL: aliaa@eng.upm.edu.my

ABSTRACT

Application of enzyme in fruit juice clarification has been known widely in fruit processing industries. Clarification of red pitaya (*Hylocereus polyrhizus*) juice by enzymatic treatment has been done to obtain stable juice. The aim of this work was to observe the effectiveness of two commercial enzymes i.e. Pectinex CLEAR and Pectinex Ultra SP-L in clarifying red pitaya juices. The effects of both enzymes on some physicochemical properties of red pitaya juices were obtained. Enzymation leads to the juice with higher yield, higher acidity, and higher total soluble solids. It also increased the juice clarity and colour appearance. However, enzymation reduced the juice pH and viscosity. This study demonstrated the effect of pectinases application on several physicochemical properties of red pitaya juices with an attempt to obtain some interesting and valuable data for the fruit processing industries. The data obtained are helpful for the optimization of red pitaya juices production as a step towards promoting the potential of red pitaya juices as functional beverages.

Keywords: fruit juice clarification, red pitaya juice, *Hylocereus polyrhizus*, enzymatic treatment, physicochemical

1.0 INTRODUCTION

More than 500 edible fruits grow in the tropical and subtropical regions, but less than 15 are commercially processed. Only 15% of tropical fruits are used for juice production (Ramadan and Moersel, 2007). International markets for tropical fruits such as pineapple, mango, guava, banana and kiwi already exist. The most popular products made from tropical fruits are fruit juices, nectars, and drinks. The consumption of fruit juices from temperate zones (e.g. apple) is growing more slowly than that of citrus and tropical fruit juices altogether, which today account for about 70% of the fruit juice present in the market (Ramadan and Moersel, 2007). This trend has caused an upswing in the fruit juice industries of the fruit-growing countries, which endeavour to promote and improve production, to be competitive for both domestic demand and export markets (Askar, 1998).

Enzyme is an essential tool in juice processes, both in terms of quality improvement and cost saving (Ramadan and Moersel, 2007). Fruit and vegetable juice production is nowadays unthinkable without the use of enzymes (Baumann, 1981). The degradation of plant cell walls by exogenous enzymatic treatment results in easier release of the components contained in cells (Janser, 1997).

The cloudiness in the juices is mainly caused by the presence of polysaccharides such as pectin and starch. The pectin can be associated with plant polymers and the cell debris, which has a fiber-like molecular structure and makes the clarification process harder. Enzyme-catalyzed breakdown of the plant cell-wall matrix and middle lamella may first increase the immediate turbidity in the juice, which is generally assumed to be mainly due to the presence of pectin and other fractions of fruit cell-wall material (Grassin and Fauquembergue 1996). Therefore, enzymatic treatment by using pectinase is an effective way to reduce the pectin in the fruit juices because pectinase has the ability to hydrolyze pectin and cause pectin–protein complexes to flocculate (Rai *et al.* 2004; Lee *et al.* 2006; Liew Abdullah *et al.* 2006; Sin *et al.* 2006), which could be easily removed by filtration. A complete enzymatic breakdown of pectin is the key for producing clear and stable fruit juices. Though hard to believe, juice clarification is the oldest and still largest market for commercial pectinases (Baumann, 1981).

In fruit products, enzymes have been exploited for the following purposes: to break down all polymeric carbohydrates such as pectin, hemicelluloses and starches, thus increasing the yield of juice by enabling better pressing of the pulp; to improve the yield of substances contained in fruit, e.g., acids, coloring or aroma substances; to clarify juices; and to liquefy the entire fruit pulp for maximum yield (Gerhartz 1990). The resulting juice will have a much lower amount of pectin, lower viscosity and also lower turbidity, which is easier for filtration process (Isabella *et al.* 1995; Alvarez *et al.* 1998; Ceci and Lozano 1998; Rai *et al.* 2004; Lee *et al.* 2006; Sin *et al.* 2006).

The pitaya or dragon fruit is also known as pitahaya, strawberry pear, nanettikafruit or thanh long. It is a cactus species of the genus *Hylocereus* and *Stenocereus*. Generally, pitaya comes in three types: *Hylocereus undatus* (white flesh with pink skin), *Hylocereus polyrhizus* (red flesh with pink skin) and *Seleicereus megalanthus* (white flesh with yellow skin) (Wikipedia, 2007). Pitaya originated in Mexico and Central America and subsequently, the plant has been widely cultivated in countries such as Vietnam, Taiwan and Malaysia. Pitaya is mildly sweet and is low in calories, has an attractive flesh and is juicy with a subtle fruity flavor (melon-like flavor). Pitaya is round or oval, the skin is fuchsia pink or yellow depending on the species, leathery and slightly leafy, and the fruit usually weighs about 200–700 g. It is rich in fiber, vitamin C and minerals. Pitaya may be converted into juice, jam and wine. Pitaya juices are turbid, very viscous and contain colloidal suspension. Therefore, the juices need to be clarified in order to be commercialized. The use of pitaya as food colorant has also been reported (Stintzing *et al.* 2002; Mobhammer *et al.* 2005; Herbach *et al.* 2006; Esquivel *et al.* 2007).

Since no data have been available for enzymatic clarification of pitaya juices, thus in this work, some physicochemical parameters of red pitaya juices are reported for the first time. The data obtained are helpful for the optimization of red pitaya juices production as a step towards promoting the potential of pitaya juices as functional beverages.

2.0 Materials and Methods

2.1 Materials

Ripe red-fleshed pitayas were obtained from a local farm in Sepang (Selangor, Malaysia). Two commercial pectinase enzymes preparations were applied: Pectinex Ultra SP-L (from *Aspergillus niger*) and Pectinex CLEAR (from *Aspergillus aculeatus* and *Aspergillus niger*) both obtained from Novo Nordisk Ferment Ltd., Dittigen, Switzerland were used for enzymatic treatment of red pitaya juices and stored at 4°C. Pectinex Ultra SP-L and Pectinex CLEAR are commercial enzymes used to produce clear and stable fruit juices. The commercial enzyme contains mainly polygalacturonase, pectinesterase and pectin transeliminase. The activity of Pectinex Ultra SP-L and Pectinex CLEAR enzymes are 26 000 PG per ml (polygalacturonase activity per ml). It appears that such enzymatic preparations rupture the juice retaining particles of the pulp, such effect being apparently due in part to action of polygalacturonase in splitting the peptic chain and the action of pectinesterase in hydrolyzing the methyl ester groups of the pectin molecule (Wieland, 1972). The optimum enzyme reaction conditions are at pH 3.5 – 6.0 and temperature range below 50°C (Kashyap et al., 2001). These are ideal conditions for fruit processing.

2.2 Production of Clarified Red Pitaya Juices at Laboratory Scale

Pitayas were peeled and extracted using a juice extractor (Philips Juicer Model: HR 1858155; Royal Philips Electronics, Eindhoven, Netherlands) to remove the seeds and coarse cloud particles. Two commercial enzyme preparations were added separately to the pitaya pulp. The enzyme concentration, time and temperature of enzymatic treatment for Pectinex CLEAR enzyme have been obtained from trials done by Siti Mazlina *et al.* (2007). For Pectinex Ultra SP-L, the processing condition data for enzymatic clarification of red pitaya juice was also obtained from trials (data not published). For each sample, Pectinex Ultra SP-L and Pectinex CLEAR were added separately to the pulp and incubated (Wise Cube Shaking Incubator Model: WIS–S10; DAIHAN Scientific, Co., Ltd., Seoul, Korea) with continuous shaking (250 rpm). Table 2.1 shows the optimum conditions for enzymatic clarification of red pitaya pulp using both enzymes respectively.

Table 2.1 Optimum processing conditions for enzymatic clarification of red pitaya pulp

Types of enzymes	Enzyme concentration (%)	Incubation temperature (°C)	Incubation time (min)
Pectinex Ultra SP-L	0.10	40	45
Pectinex CLEAR	0.09	46	82

The treated pulp was pasteurized at 90°C for 5 min in a hot waterbath (Laboratory Waterbath Model: LMWB-6PC; PLT Scientific Sdn. Bhd., Selangor, Malaysia) and then cooled at 2°C using circulating waterbath (Model: 71; PolyScience, Illinois, USA). These steps were performed to the samples to inactivate the enzyme action before filtration. The treated red pitaya juices were centrifuged (Refrigerated Centrifuge 5800; Kubota Corporation, Fujioka, Japan) at 3000*g* for 10 minutes, and the supernatant was collected. The juice was filtered through a filter paper (Whatman No.1, Whatman International Ltd., Kent, England) using oil-less vacuum pump (Model: Rocker 300; Rocker Scientific Co., Ltd., Taipei County, Taiwan). The separation of the pulp from the juice by filtering the liquid pulp on a rotary vacuum filter avoids the costly pressing step and the juice obtained is clear (Baumann, 1971). Figure 2.1 illustrates the step for the production of clarified red pitaya juices at laboratory scales. A control sample was produced without enzymatic treatment step but was pasteurized at 90°C for 5 min followed by cooling at 2°C.

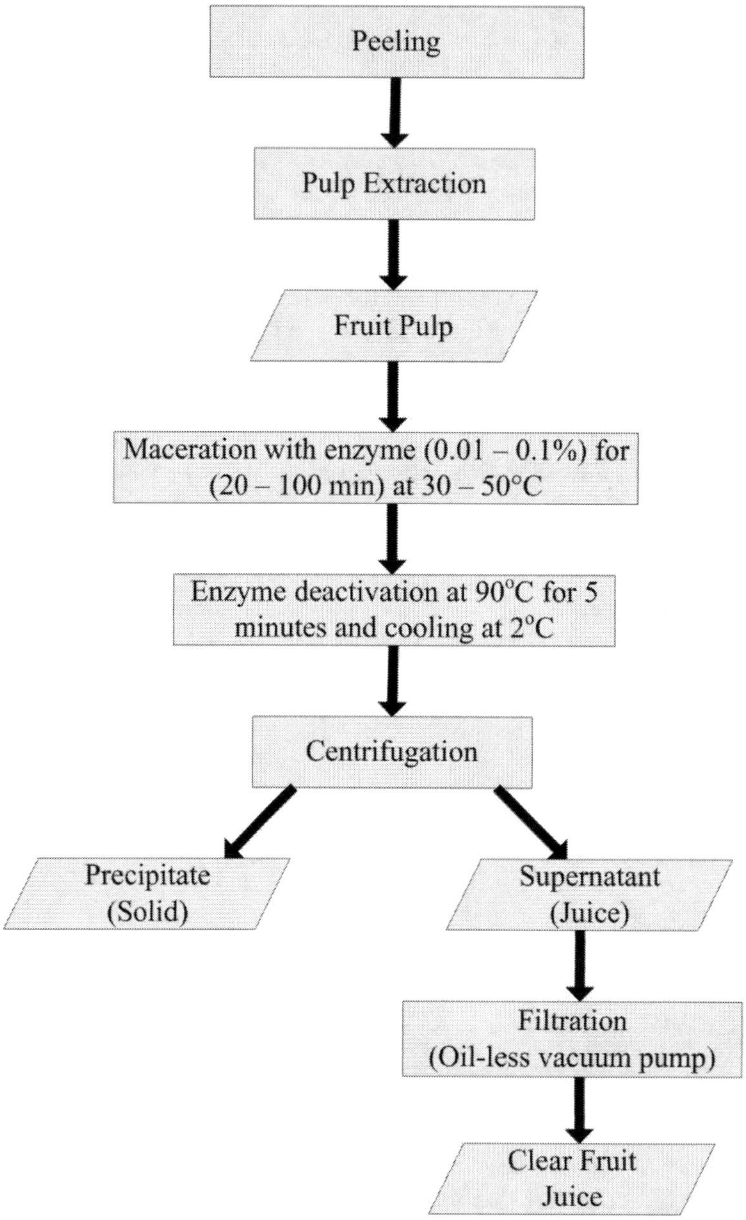

Figure 2.1: Steps for extraction and subsequent clarification by enzymatic treatment of red pitaya juice

2.3 Determination of some physicochemical properties of red pitaya juices

Total soluble solids (TSS) were determined using Atago Hand Refractometer (Model: Master-α; ATAGO Co., Ltd., Tokyo, Japan) and pH values were measured using Microprocessor-based Bench pH (Model: pH 211; Hanna Instruments (M) Sdn. Bhd., Selangor,

Malaysia). Juice yield, viscosity, clarity and colour measurement were obtained according to the methods reported by Nur 'Aliaa *et. al.* (2008).

3.0 Results and Discussion

Treatment of fruit pulp with appropriate enzyme preparations is a general practice in juice processing (Ramadan and Moersel, 2007). Enzymes are used in the juice industry to aid in the separation of juice from the fruit cells and to clarify the juice by the removal of pectin and naturally occurring starches that contribute to undesired viscosity, poor filtration, and a cloudy appearance (Enzyme Technical Association, 2002). Since juices extracted from ripe fruit contain substantial amount of pectins and starches, thus they lead to a cloudy appearance to the juices, making the juices less appealing by consumers due to the appearance and mouth-feel of the juices. The cloudiness that pectin causes is difficult to remove unless by enzymatic depectinization.

In this study, ripe red pitayas were selected from a local farm in Malaysia with considerably higher quality which is reflected by the fruit larger size, well-developed colour and very juicy pulp. Ripe fruits lead to juices freshness and quality which is crucial to the quality of the final product. Two pectinase enzymes application which were Pectinex Ultra SP-L and Pectinex CLEAR were examined to demonstrate the effects of pectinases on the quality of fruit juices.

3.1 Effect of enzyme application on some physicochemical properties of red pitaya juices

Table 3.1 presents some physicochemical properties of red pitaya juices. The yield of juice produced without enzymatic treatment is 78.53%. A dramatic increase can be observed in the recovery of juice produced by enzymatic treatments which were 83.2% for treatment using Pectinex Ultra SP-L and 92.38% using Pectinex CLEAR. Pectinases break down the fruit pulp resulting in increased yields of juices. It was observed that as the heating and enzymation continues, the pulp becomes markedly more fluid and the freeing of the pulp fragments is readily apparent (Wieland, 1972). This elevation might be of importance for industrial and commercial applications under large-scale juice processing (Ramadan and Moersel, 2007).

Table 3.1 Some physicochemical properties of red pitaya juices

Treatment	Yield (%v/v)	Viscosity (cps)	Absorbance value (abs)	Colour (L*)	Total acidity (meq/kg)	pH	TSS[a] (%Brix)
Fresh fruit pulp	-	7.23± 0.23	1.957± 0.07	2.93± 0.006	0.58± 0.03	4.8± 0.01	12.0± 0.00
Control juice (pasteurized)	78.53± 1.93	3.9± 0.00	0.179± 0.009	12.46± 0.02	0.70± 0.006	4.55± 0.01	12.0± 0.00
Juice treated with Pectinex Ultra SP-L	83.2± 1.04	3.0± 0.00	0.152± 0.03	22.4± 0.006	0.77± 0.01	4.41± 0.01	12.0± 0.00
Juice treated with Pectinex CLEAR	92.38± 2.26	3.0± 0.00	0.062± 0.000	24.76± 0.02	0.75± 0.06	4.45± 0.06	12.1± 0.06

Values given are the mean of three replicates ± standard deviation
[a]Total soluble solids

Application of enzymes reduces juices viscosity and improves the juices clarity. In the case of red pitaya juices, such enzymatic treatment is effective to reduce the relative viscosity of the pulp to less than half of its initial value. The absorbance values markedly decreased for juices treated with enzymes. This characteristic is more preferred by consumers when compared to the cloudy appearance showed by freshly extracted fruit pulps. L* values also markedly increased for enzyme-treated juices when compared to untreated samples. The total titratable acidity also increased in the juices treated with enzymes. This increase might be due to the formation of galacturonic acid by enzymatic breakdown of pectin (Acar, 1999). The total titratable acidity in red pitaya is in the range of 0.58-0.77%. Total acidity for other fruit juices such as goldenberry is 1.0% (Ramadan and Moersel, 2007), pear 0.3%, orange 0.8%, apple 0.9%, strawberry 0.9%, pineapple 1.1%, plum 2.2% and apricot 2.4% (Belitz, 1999).

Since the total acid content in the juice was slightly high, thus the pH of the juices was low, with values within 4.41-4.8. The decrease of pH values may be due to the release of carboxyl groups and galacturonic acid from pectin (Ramadan and Moersel, 2007). For red pitaya juices, there was no significant increase in TSS levels even after enzymatic treatment and the values remained at 12.0%Brix.

4.0 Conclusions

The differences between untreated juices when compared to enzyme-treated juices show that the mechanism of pectin degradation depends upon the types of enzymes used. Variety of enzymes available in fruit and vegetable processing contribute to some degree to the conversions, degradations and breakdown reactions desired (Wieland, 1972). From the study done, it can be seen that the effects of Pectinex Ultra SP-L and Pectinex CLEAR were different on red pitaya juices because pectinases act in different ways on the pectins. Pectins are situated in the primary cell walls and in the middle lamella. The mechanism of pectin degradation depends upon the pectic composition of the substrate. This study demonstrated the effect of pectinases application on several physicochemical properties of red pitaya juices. Moreover, this study attempted to obtain some data from the processing of fruit juices especially pitaya juices and to provide valuable and interesting information that would be useful for fruit processing industries. Utilization of pectinase as a processing aid in the production of pitaya juices proved to be a crucial step in order to enhance the efficiency of the whole system.

REFERENCES

ACAR, J., ALPER, N., and ESTÜRK, O. 1999. The production of cloudy apple nectar using total liquefaction enzymes. Fruit Process, *8*, 314-317.

ALVAREZ, S., ALVAREZ, R., RIERA, F.A., and COCA, J. 1998. Influence of depectinization on apple juice ultrafiltration. Colloids and Surfaces A: Physichochemical and Engineering Aspects, *138*, 377-382.

ASKAR, A. 1998. Enzymes in fruit juice processing. Fruit Process, *7*, 273-276.

BAUMANN, J.W. 1981. Application of enzymes in fruit juice technology. In *Enzymes and Food Processing,* (G.G. Birch, N. Blakebrough, & K.J. Parker, eds.) pp.129-147, London: Applied Science Publication.

BELITZ, H.D., and GROSCH, W. 1999. *Food Chemistry*, Springer, Berlin.

CECI, L., and LOZANO, J. 1998. Determination of enzymatic activities of commercial pectinases for the clarification of apple juice. Food Chemistry, *61*, 237-241.

ENZYME TECHNICAL ASSOCIATION, 2002, *Enzyme use in fruit juice processing*, Access on August 2008 through website: http://www.enzymetechnicalassoc.org/juice.html

ESQUIVEL, P., STINTZING, F.C., and CARLE, R. 2007. Pigment pattern and expression of colour in fruits from different *Hylocereus* sp. genotypes. Innovative Food Science & Emerging Technologies, *8*(3), 451-457.

GERHARTZ, W. 1990. *Enzymes in Industry*, VCH Publications, New York.

GRASSIN, C., and FAUQUEMBERGUE, P. 1996. Fruit juices. In *Industrial Enzymology,* (T.Godfrey & S. West, eds.) pp.225-264 (2nd ed.), London, UK:MacMillan Press.

HERBACH, K.M., ROHE, M., STINTZING, F.C., and CARLE, R. 2006. Structural and chromatic stability of purple pitaya (*Hylocereus polyrhizus* [Weber] Britton & Rose) betacyanins as affected by the juice matrix and selected additives. Food Research International, *39*, 667-677.

ISABELLA, M.B., GERALDO, A.M., and RAIMUNDO, W.F. 1995. Physical-chemical changes during extraction and clarification of guava juice. Food Chemistry, *54*(4), 388-386.

JANSER, E. 1997. Enzymes applications for tropical fruits and citrus. Fruit Process, *10*, 388-393.

KASHYAP, D.R., VOHRA, P.K., CHOPRA, S., and TEWARI, R. 2001. Applications of pectinases in the commercial sector: a review. Bioresource Technology, *77*, 215-227.

LEE, W.C, YUSOF, S., HAMID, N.S.A., and BAHARIN, B.S. 2006. Optimizing conditions for enzymatic clarification of banana juice using response surface methodology (RSM). Journal of Food Engineering, *73*, 55-63.

LIEW ABDULLAH, A.G., SULAIMAN, N.M., AROUA, M.K., and MEGAT MOHD NOOR, M.J. 2006. Response surface optimization of conditions for clarification of carambola fruit juice using a commercial enzyme. Journal of Food Engineering, *81*, 65-71.

MOBHAMMER, M.R., STINTZING, F.C., and CARLE, R. 2005. Colour studies on fruit juice blends from *Opuntia* and *Hylocereus* cacti and betalain-containing model solutions derived thereform. Food Research International, *38*, 975-981.

NUR 'ALIAA, A.R., SITI MAZLINA, M.K., TAIP, F.S., and LIEW ABDULLAH, A.G. 2008. Response surface optimization for clarification of white pitaya juice using response surface methodology, Journal of Food Process Engineering (*In Press*)

RAI, P., MAJUMDAR, G.C., DASGUPTA, S., and DE, S. 2004. Optimizing pectinase usage in pretreatment of mosambi juice for clarification by response surface methodology. Journal of Food Engineering, *64*, 397-403.

RAMADAN, M.F. and Moersel, J.T. 2007. Impact of enzymatic treatment on chemical composition, physicochemical properties and radical scavenging activity of goldenberry (*Physalis peruviana* L.) juice. Journal of the Science of Food and Agriculture, *87*, 452-460.

SIN, H.N., YUSOF, S., HAMID, N.S.A., and RAHMAN, R.A. 2006. Optimization of enzymatic clarification of sapodilla juice using response surface methodology. Journal of Food Engineering, *73*, 313-319.

SITI MAZLINA MUSTAPA KAMAL, NUR 'ALIAA ABD RAHMAN, and ABDUL GHANI LIEW ABDULLAH 2007. Clarification of red dragon fruit juice. Journal of Engineering Science and Technology (*In Press*)

STINTZING, F.C., SCHIEBER, A., and CARLE, R. 2002. Betacyanins in fruits from red-purple pitaya, *Hylocereus polyrhizus* (Weber) Britton Rose. Food Chemistry, *77*, 101-106.

WIELAND, H. 1972. *Enzymes in food processing and products*, pp.1-37, Noyes Data Corporation, New Jersey, USA.

WIKIPEDIA, 2007, Pitaya, Access on May 2007 through website: en.wikipedia.org/wiki/Dragon_fruit

730b The Changeability of Citrinin Production In Mutagenic Treatment of Monascus Purpureus Tistr 3541

Benjamaporn Wonganu, Department of Biotechnology, King Mongkut's University of Technology North Bangkok, 1518 Piboonsongkram Road, Bangsue, Bangkok, Thailand and Sasithorn Kongruang, Biotechnology, King Mongkut's University of Technology North Bangkok, 1518 Piboonsongkram Road, Bangsue, Bangkok, Thailand.

Citrinin is a nephrotoxic metabolite produced by nature filamentous fungi of the genera Monascus, which has been encountered as a natural contaminant in grains, food, feedstuffs, as well as biological fluids. However, it is recognized as an antibacterial metabolite. Citrinin was investigated in type culture of induced mutation Monascus. purpureus TISTR 3541 by HPLC to test citrinin producing ability along with pigment production. M. purpureus TISTR 3541 was mutated by ultrasonic wave inducing at 28°C, 45 kHz for 2 min to generate mutants for 5 generations. The results indicated that content of the citrinin was significantly increased as the induced mutation increasing from 12.72 % in G2 to 74.70 % in G5 compared with those from the wild type strain. Comparison result on the mutant generation effect in pigment production, statistical analysis showed that mutant TISTR 3541 G5 was able to produce significantly higher pigment than wild type (p-values < 0.0001). Red pigment stability was also evaluated. Results found that red color of TISTR 3541 G4 mutant was the most stable generation at pH when compared to G5, G1, G2 and G3 with the stability increase more than G0 (60.00, 47.30, 33.50, 24.43, and 2.31 time, respectively). Temperature stability of red pigment revealed that TISTR 3541 G1 mutant was more stable at 4°C than G3, G2, G5 and G4 with 124.00, 62.00, 50.71, 41.50 and 33.42 times, respectively when compared to G0. In the resistance to solvent, the result exhibited that all mutants in ethyl ether increased the stability more than G0.

730c Piezoelectric-Excited Millimeter Sized Cantilever (PEMC) Sensors for Highly Sensitive and Rapid Detection of Group a Streptococcus and Listeria Monocytogenes In Practical Matrixes

Harsh Sharma and Raj Mutharasan. Chemical and Biological Engineering, Drexel University, 3141 Chestnut Street, Philadelphia, PA 19104

The purpose of this study is to establish the feasibility of direct detection of infectious agents in practical matrixes (human serum for Group A Streptococcus [GAS] and chicken-wash for Listeria monocytogenes [LM]) and to determine the limit of detection for these antigen-matrix systems. We describe a biosensor-based approach for detection of the two infectious agents in practical matrixes in 15-20 min. The electrically-excited and electrically-sensed PEMC sensors show resonant frequency decrease as antigen binds to it. After immobilizing antibody to specific target pathogen, the sensor was exposed to test sample containing known concentrations of GAS or LM. Concentrations as low as 10 pathogens per ml show a decrease in resonant frequency of ~500 Hz while noise in measurement is ~ ±30 Hz. Because the sensors were shown in our earlier work to exhibit mass-change sensitivity of a few femtograms, it is feasible to detect at concentrations as low as one pathogen per ml. Current experiments at low concentrations will be described.

730d Elucidation of Biosensing Processes through Simulation

M. H. Akanda[1], Jin Wang[2], B. Chin[3], and Z.-Y. Cheng[3]. (1) Chemical Engineering, Auburn University, Auburn, AL 36849, (2) Department of Chemical Engineering, Auburn University, Auburn, AL 36849, (3) Materials Engineering, Auburn University, Auburn, AL 36849

In the US, more than ninety percent of foodborne illnesses are attributable to bacterial contamination of consumed food. Therefore, microbial detection is one of the most important steps to ensure enhanced safety in the food production and distribution network. Biological weapons have long been recognized by the United States as considerable threats to the country that could kill thousands of people and create an unparalleled medical, political and social crisis. Therefore, the need to improve our detection capabilities is also highlighted to prevent different forms of intentional bioterrorism. Biosensors are considered as having great potential for future pathogen detection due to their high sensitivity and near real time detection capability. The recently developed magnetoelastic biosensors have offered several advantages in terms of sensitivity, working environment, cost effectiveness etc over other existing biosensors. Thus, magnetoelastic biosensors show great promise to detect foodborne pathogens and hold huge potential to improve the safety of the food network. In order to detect the pathogens early, it is desirable to improve the biosensor's sensitivity, which can be achieved through proper sensor design. In this work, we perform simulations to investigate the effects of different factors such as geometry, flow condition etc on the sensor performance (such as detecting limit, and response time), as experiments alone are not sufficient due to practical limitations such as cost, time and technology.

In this work, we develop first principle models that incorporate fluid mechanics, mass transfer and surface reaction mechanism to describe stagnant as well as convective flow biosensing system. We consider both simplified geometries of biosensor such as hemisphere, disk, hemicylinder as well as more realistic cases such as cylindrical, rectangular bar. A set of partial differential equations with proper boundary conditions are derived based on first principle model (such as equation of continuity $DC_A/Dt=D_{AB}\nabla^2 C_A+r_A$ for mixtures) for different geometries and flowing conditions. Diffusion of pathogens from bulk fluid to biosensor surface may be diffusion controlled or adsorption/reaction controlled. Both conditions are considered for a set of specific biosensor geometries. Suitable coordinate systems are used to handle different sensor geometries. Numerical solutions are generated using COMSOL Multiphysics® to examine the effect of sensor shape and size on its performance. Simulation results are visualized with different plots to illustrate the dynamic behaviors of the biosensing system. In addition, analytical solutions for the simplified cases are derived and compared with numerical solutions. Model parameters such as diffusion coefficient, surface reaction/binding rate constant are estimated to fit available experimental results with appropriate regression techniques. Finally, model predictions are validated using experimental data. The simulation prediction enables one to analyze the performance of biosensors with different shapes and flowing conditions. Thus, the results obtained in this work will be used to guide sensor design in the future.

Reference:

Sheehan P. E., and Whitman L. J., "Detection limits for nanoscale biosensors", Nano Letters, 5(4), 803-807, 2005

Ziegler C., "Cantilever-based biosensors", Anal Bioanal Chem, 379, 946-959, 2004

Li, S. Q., Orona, L., Li, Z. M., and Cheng, Z.-Y., "Biosensor based on magnetostrictive microcantilever", Applied Physics Letters, 88 (7), 073507, 2006

Guntupalli R., Hu J., Lakshmanan R.S., Huang T.S., Barbaree J. M., and Chin B.A., "A magnetoelastic resonance biosensor immobilized with polyclonal antibody for the detection of Salmonella typhimurium", Biosensor and Bioelectronics 22, 1474-1479, 2007

Wan J., "Development and study of phage-coated magnetoelastic biosensor for the detection of Bacillus Anthracis sterne spores", PhD dissertation, Auburn University, Auburn, AL, 2008

730e Monte Carlo Simulation of Disinfection of Bacteria for Food Safety and Security: Non-Linear Approach

A. Argoti[1], L. T. Fan[1], and S. T. Chou[2]. (1) Department of Chemical Engineering, Kansas State University, 1005 Durland Hall, Manhattan, KS 66506, (2) Department of Finance and Banking, Kun Shan University, No. 949, Da-Wan Rd., Yung-Kang City, Taiwan

The analysis, modeling, and simulation of disinfection of bacteria, specifically, pathogenic bacteria, in foodstuff should belong to the repertoire of tools for quantitatively assessing strategies for food safety and security. Nevertheless, it is exceedingly difficult, if not impossible, to accurately quantify a population of bacteria throughout disinfection by resorting to any conventional deterministic approach: The bacteria are discrete and mesoscopic in size, thereby exhibiting random, or stochastic, fluctuations in their number concentration, especially at the tail-end of disinfection where their number is minute. Thus, it is indeed appropriate that the disinfection of bacteria in foodstuff be analyzed, modeled, and simulated according to a stochastic paradigm. Such stochastic simulation can be executed by means of the Monte Carlo method, which can deal with models based on linear as well as non-linear rate laws with nearly equal ease. For illustration, the mean, variance, and standard deviation of the number concentration of bacteria during disinfection of foodstuff have been estimated by Monte Carlo simulation. The numerical results are compared with both the analytical solutions and the available experimental data.

730f Characterization and Classification of Gamma-Irradiated Adipose Tissues Using Raman Spectroscopy and Chemometrics

Ramazan Kizil, Chemical Engineering, Istanbul Technical University, School of Chemical & Metallurgical Engineering, B-410 Chemical Engineering Department, Istanbul, 34469, Turkey and Joseph Irudayaraj, Agricultural & Biological Engineering, Purdue University, West Lafayette, IN 47907-2093.

Elucidation of irradiation damages to the structure of lipids is important to both cellular radiobiology and food irradiation fields. Lipid peroxidation is a major damaging process in membranes and liposomal dispersions and is responsible for the impairment of enzyme activities, chemically-induced toxicity and damages to DNA. Treatment of fats with ionizing radiation is believed to accelerate lipid oxidation via the action of various unstable free radicals.

Raman spectroscopy, a nondestructive analytical technique, was employed to characterize irradiation-induced changes in the chemical composition of adipose tissues at the molecular level. Adipose tissue is a connective tissue that functions as the major storage site for fat in the form of triglycerides.

We have investigated 1 kGy to 10 kGy gamma irradiation damages on the molecular structure of lamb, beef and porcine adipose tissues using Fourier transform Raman spectroscopy. Gamma irradiation-induced changes on the saturation level of fatty acids and the extent of lipid peroxidation were monitored by screening the C=C and C=O stretch related Raman scattering modes. The most irradiation sensitive vibrational modes was found to be the C-H stretch modes detected at 1302 cm-1 and 2800-3000 cm-1 region of Raman spectrum, probably due to the abstraction of hydrogen atoms from the aliphatic tail groups by the attack of free radicals. The irradiation perturbed Raman data was treated with a chemometric model, canonical variate analysis, to differentiate adipose tissues based on the extent of irradiation.

This presentation covers the radiochemistry of lipids and characterization of radiochemical changes upon irradiation at the molecular level. The potential of Raman spectroscopy along with a proper multivariate qualitative analysis method to discriminate adipose tissues with respect to the applied dose will be demonstrated. Since the proposed methodology investigates the radiation damages at the molecular level and can differentiate different dose irradiated tissues it has a potential for the rapid assessment of the safety and labeling of irradiated lipid containing food.

730g A Simple Model for Equilibrium Moisture Content In Food Products

Victor R. Vasquez, University of Nevada, Reno, Dept. of Chemical Engineering, Mail Stop 170, Reno, NV 89557-0136 and Charles J. Coronella, Chemical Engineering Dept., University of Nevada, Reno, Mailstop 170, Reno, NV 89557.

Equilibrium moisture content (EMC) plays a fundamental role on the properties of bio-derived products such as food, waste biomass, seeds, and many others. For foods, the moisture content affects flavor, texture, and storing conditions. In other applications, moisture affects drying conditions and energy requirements for food processing. Adequate modeling of EMC in food products as

a function of the water activity is very important for designing and understanding food engineering processes

as well for process monitoring and control. Most models used to correlate EMC with water activity are based on traditional adsorption models which take into account the adsorption energy of water with the substrate only and

do not consider other non-adsorption interactions of water within the substrate such as mixing and swelling. Many other models are empirical, which are commonly used for interpolation. Although a detailed description of this type of substrate is very complex, we present a simple model based on standard molecular-thermodynamic models and

standard statistical mechanics formulations to correlate the EMC with water activity that attempts to consider both adsorption and non-adsorption interactions of the water with the food product. Although the model is basic, it captures the behavior of water in this type of systems fairly well and it can be used to fit experimental data with parameters that provide better physical insight on the nature of substrate or food product. We illustrate the capabilities of the model with two case studies for food products.

737a A First-Principles Model for Eukaryotic DNA Synthesis Dynamics and the Implications for Cell Cycle Analysis

Marc R. Birtwistle, Peter G. Millili, and Babatunde A. Ogunnaike. Chemical Engineering, University of Delaware, 150 Academy St., Newark, DE 19716

Cell cycle analysis, an experimental tool used to quantify the fraction of cells in different phases of the cell cycle, is accomplished by measuring cellular DNA content: cells with one copy of the genome are in the G0 or G1 phases, cells with two copies of the genome are in the G2 or M phases, and cells with between one and two copies of the genome are in the S-phase. While the principles of cell cycle analysis are straightforward, in practice, measurement variability makes it difficult to obtain accurate estimates of the cell cycle phase fractions from data. There is a general consensus that such data should be analyzed by fitting them to a mixture of three probability distributions one each for (i) the combined G0/G1 phases, (ii) the S-phase, and (iii) the combined G2/M phases (Watson, 1992). While it is widely accepted that the DNA content of the G0/G1 and G2/M populations should be normally distributed, it remains unclear what the S-phase distribution should be (Watson, 1992). As the S-phase corresponds to the period of time when a cell is replicating its DNA, the probability of observing an S-phase cell with a particular DNA content is intimately related to the dynamics of DNA synthesis.

In the current work, we develop a first-principles, ordinary differential equation model for the dynamics of eukaryotic DNA synthesis, based on mechanisms of DNA replication that have been well-established over the past 50 years. The model is subsequently used to propose S-phase DNA content distributions. Using previously known replication fork rates and inter-replication origin distance distributions, our model is capable of describing the DNA synthesis dynamics in the fission yeast *S. Pombe*, the budding yeast *S. Cerevisiae*, the large frog eggs *Xenopus* oocytes, and the model mammalian cell line Chinese Hamster Ovary (CHO) cells. It has long been debated how *Xenopus* oocytes can replicate their large genome (6000 Mbp) in the observed short DNA synthesis phase (~20 min.) given the seemingly random placement of replication origins—a paradox that has been termed the "random replication problem". Our modeling results suggest that this random replication problem is solved in *Xenopus* oocytes by having (1) a short mean distance between potential replication origins, and (2) an extremely large number of replication units (foci) that all fire at least ~7.5 minutes before the scheduled end of the S-phase.

Our dynamic model shows that the S-phase DNA content distributions for a population of asynchronously cycling cells have a concave-up shape, with peaks at one and two copies of the genome. This concave-up distribution shape prescribed by our model is notable in that it is the exact opposite of all but one of the previously proposed S-phase distributions, all of which have a concave-down shape. The only exception is the empirically-based S-phase distribution model of Dean and Jett (1974), developed solely on the basis of goodness-of-fit to data, which indeed has a concave-down shape. Our results therefore provide a theoretical justification for this earlier purely empirical result and strongly suggest that fitting concave-down S-phase distributions to data from asynchronously cycling cell populations is inappropriate.

References

Watson JV. Flow Cytometry Data Analysis: Basic Concepts and Statistics. Cambridge University Press, 1992.

Dean PN and Jett JH. Mathematical analysis of DNA distributions derived from flow microfluorometry. J Cell Biol 60, 523-527 (1974).

737b A Kinetic Model Reveals the Effects of the Different Steps of the Translation Elongation Cycle on -1 Programmed Ribosomal Frameshifting

Pei-Yu Liao, School of Chemical and Biomolecular Engineering, Cornell University, Ithaca, NY 14853, Jonathan D. Dinman, Department of Cell Biology and Molecular Genetics, University of Maryland, College Park, College Park, MD 20742, and Kelvin H. Lee, Department of Chemical Engineering and Delaware Biotechnology Institute, University of Delaware, 15 Innovation Way, Newark, DE 19711.

Programmed ribosomal frameshifting (PRF) is the process by which ribosomes produce two different polypeptides from the same mRNA. In -1 PRF, stimulatory signals cause the ribosome to slip one nucleotide toward 5' end. In this study, we kinetically model -1 PRF through either incomplete translocation, or by slippage of P-site and A-site tRNAs either during or immediately after aa-tRNA accommodation. These models yield two possible -1 frameshift products: those incorporating zero frame A-site tRNA in the recoding site and products incorporating -1 frame A-site tRNA in the recoding site. Using known kinetic rate constants, we can determine the individual contributions of rates of translocation, GTP hydrolysis by elongation factors, and peptidyltransfer to -1 PRF. The modeling results are consistent with previous experimental observations. In addition, the model predicts that changes in the rate constants for each of these different steps can affect the ratio between two frameshift products. A dual fluorescence reporter system is currently being employed in *Escherichia coli* to empirically test the model.

737c A Hybrid off-Lattice Cell-Based Model of Tumor Growth

Junhwan Jeon and Peter T. Cummings. Department of Chemical Engineering, Vanderbilt University, Nashville, TN 37235-1604

We have developed a hybrid off-lattice cell-based version of Anderson et al. model [1]. We consider the hybridization on two levels: discrete space for molecules such as matrix-degradative enzymes (MDEs), nutrient or oxygen, and extracellular matrix (ECM) material, and continuous space for an individual cell. An important advantage of the off-lattice implementation is that there is no assumption made about whether the cells movements are in the diffusive regime (mean squared displacement ~ t) or the ballistic regime (mean squared displacement ~ t^2). According to the experimental results [2], the diffusive regime (in which the cell motion can be described by a random motility coefficient) takes a considerable time to develop (over 100 minutes for the mammalian cancer cells studied). We are going to present simulation results in terms of tumor shape and size which is strongly influenced by initial micro-environment.

References

1. Anderson, A.R.A.: A hybrid mathematical model of solid tumour invasion: the importance of cell adhesion. Math. Med. Biol. 22, 163-186 (2005).

2) Potdar, A. A., J. Lu, J. Jeon, A. M. Weaver, and P. T. Cummings: Bimodal Analysis of Mammary Epithelial Cell Migration in Two Dimensions, Annals of Biomedical Engineering, submitted (2007)

737d Diffusion-Enhanced Cellular Automata Viral Model

Jacob McGill and Mariajose Castellanos. Chemical and Biochemical Engineering, UMBC, 1000 Hilltop Circle, Baltimore, MD 21250

We present our work on the development of a dynamic population model of Influenzae A infection in epithelial tissue. In this model we took into account diffusional effects of the virus in vivo and integrated them with a cellular automata simulation to understand the effect of localized interactions on the dynamics of an Influenza infection on a host. We tested the impact of incorporating a virus diffusion model into the cellular automata simulation and what parameters were particularly important to the simulation. Finally, we explored how the diffusion of virus can influence complications in an infection by presenting a hypothetical case study of an immuno-deficient host. This case study shows a dramatic increase in the fraction of dead cells, as it almost doubles at its maximum to 81%. In this test case, the subject does recover, although it takes much longer than the 12 days needed to clear the normal virus in a normal host to fully recover. Additionally, by the 12th day a slight increase in infected cells can be observed. This re-infection is due to the virus lingering in the mucus layer above the epithelial cells. If we ignored the diffusional aspects of the infection, and only utilized cell to cell infection mechanisms, we would have been unable to observe this sort of dynamic.

737e Aging Studies. A Stochastic Approach

Suresh Kumar Poovathingal, Chemical and Biomolecular Engineering, National University of Singapore, National University of Singapore, Department of Chemical & Biomolecular Engineering, 4 Engineering Drive 4, Blk E5 #02-09, Singapore, 117576, Singapore, **Rudiyanto Gunawan**, Department of Chemical and Biomolecular Engineering, National University of Singapore, Singapore, Singapore, Jan Gruber, Department of Biochemistry, National University of Singapore, National University of Singapore, Department of Biochemistry, Neurobiology and Ageing Programme, Centre for Life Sciences (CeLS), 28 Medical Drive, #04-21 Lab 2, Singapore, 117456, Singapore, and Barry Halliwell, Department of Biochemistry, National University of Singapore, Yong Loo Lin School of Medicine, National University of Singapore, Yong Loo Lin School of Medicine, Department of Biochemistry, 8 Medical Drive, Blk MD7, #02-03, Singapore, 117597, Singapore.

1. Introduction:

Mitochondria are the powerhouse of eukaryotes. However, as a byproduct of cellular respiration they also produce mutagenic reactive oxygen species (ROS). Apart from the cell's nucleus, mitochondria are the only organelle which has its own genetic information. Mitochondrial DNA (mtDNA) is very compact with no introns or exons. Human mtDNA is made up of 16,565 base pairs, which encodes 13 polypeptides, 22 tRNAs and two rRNA's that are essential for the functioning of mitochondria [3]. The mitochondrial ROS can damage mtDNAs and this can lead to mtDNA mutations. Accelerated accumulation of mtDNA mutations has been postulated to occur during aging as a result of their proximity to the ROS source. The loss of mitochondrial functions due to these mutations has been associated with many metabolic and degenerative diseases, whose clinical symptoms progress with ageing [3]. The connection between mtDNA mutations and ageing is supported by experimental evidence showing that the amount of somatic mtDNA mutations increases with age [1] and mice with defective mtDNA polymerase show signs of accelerated ageing [2]. In this work, the consequence of inherent biological stochasticity on the maintenance of genomic stability of mitochondrial DNA (mtDNA) is investigated through the development of an *in silico* model

Mathematical models have been proposed to explain the accumulation of somatic mutations in mtDNAs. Kowald and Kirkwood have done pioneering work in the field of ageing modeling [3]. Their work was based on the hypothesis of ROS "vicious cycle". Vicious cycle theory postulates that ROS cause mtDNA mutations and in turn these mutations cause higher production of ROS. Most of the existing models however are deterministic (ODEs). Few exceptions include Langevin-type stochastic models due to Samuels and Chinnery [4], which were based on the assumption of relaxed replication of mtDNA. Here, mitochondrial fusion and fission were assumed to occur frequently enough to justify a single well-mixed mtDNA pool in the cell. This assumption is also used in the present model. One weakness of the existing models in the literature is the requirement of a large number of unknown parameters that need to be set or estimated. Quantitative estimation of most of the parameters needed for such elaborate modeling is challenging. To overcome this limitation, we have proposed a minimal chemical master equation model of mtDNA somatic mutation which can capture features of experimental data on mice [2]. Here, all the model parameters were obtained from published literature values.

2. **Model Description:**

The model captures two main processes that contribute to the maintenance of mtDNA genomic stability, namely mtDNA degradation and replication

mtDNA → Φ; Propensity $a=k_d*(W+M)$

mtDNA → 2mtDNA; Propensity $a=v_{max}*(1-(W^n/(W^n+K^n))$

where k_d, v_{max}, K and n are the model parameters. The a terms are known as the propensity functions, for which a•dt gives the probability that a given process (reaction) takes place in the time range of [t, t +dt). The model further tracks the number of wild type (W) and mutant (M) mtDNA molecules individually. In addition, there exists a probability (k_m) of wild type mtDNA replication that will produce one mutant molecule M. As the mtDNA replication is known to be regulated by nucleus, perhaps in response to the energetic needs of the cell, the propensity function for mtDNA replication depends on the number of wild-type mtDNAs according to a Hill-type function.

In order to deal with the stochastic nature inherent in cellular processes, the Stochastic Simulation Algorithm (SSA) was used to simulate this model [5]. A modified version of this algorithm has been used for the present work, which is not detailed here for brevity. The parameters were obtained from reported values based on experimental data for mice. The simulations were developed to represent the evolution of mtDNA mutations in the cardiomyocytes of mice, over its natural life span of 36 months. Two stages of the cell developments were considered in the modeling. The first stage involves the developmental stage in which there is a burst of mtDNA replication along with cell divisions. Here, mutations are primarily due to polymerase fidelity error during replication. In the second post-mitotic stage, mtDNA mutations primarily arise due to mtDNA turnover and oxidative damage.

3. **Results and Discussion:**

To validate the model, experimental results of age dependent accumulation of point mutations in wild type mice were used. In this experiment, point mutations on the *Taq*I restriction site (TCGA) in the gene encoding region of the 12S rRNA (bp 634-637) subunit of mice mtDNA were tracked using a highly sensitive RMC assay for a period of 36 months (maximum life span of a wild-type mice) [2]. Correspondingly, the simulation tracked the mtDNA population of both types (W and M) for the complete population of heart cells (~0.4 - 6×10^7 cells). The mutation burden predicted by the model and this experiment are in good agreement. An observation of the data from both simulations and experiments suggested that the mutational burden accumulated exponentially with time. However, the model predicts that the average of the mutation frequency increases in a linear fashion. The apparent exponential increase in the mutational burden was determined to be an artifact of small number of samples taken from a long-tailed distribution. To further validate the model, simulations were done for the mutator mice with a proof-reading deficient activity of DNA polymerase using data from [2, 6]. The model again demonstrated an excellent agreement with the experimental results for both the heterozygous and homozygous mutator mice (data not shown). The mutator model showed that the developmental stage makes a significant contribution to the accumulation of the mtDNA mutation, as seen in experiments [7]. The same observation was also made in the wild-type mice model, but to a much lesser degree. These results suggested that the mtDNA mutations were determined to some extent during its developmental phase.

4. **Conclusions:**

How cells maintains their mtDNA population in higher organisms is fundamentally important to our understanding of the relationship between mitochondrial genome, ageing, and diseases. In this work, a minimal model, based only on mtDNA turnover and mutations caused by damage and polymerase error, was developed. This model has shown to be in good agreement with experiments. Despite the apparent exponential increase of mtDNA mutation load with time, the model predicted that the average mutation frequency to grow linearly with age. This model provides a starting point for the development of more complex and realistic representations of mtDNA somatic mutation, including the role of mitochondrial fusion and fission

5. **References:**

1. Khaidakov, M., Heflich, M. H., Manjanatha, M. G., Myers, M. B., and Aidoo, A. 2003. Accumulation of point mutations in mitochondrial DNA of aging mice. Mutat. Res., 526(1-2):1-7.

2. Vermulst, M., Bielas, .J .H., Kujoth, G. C., Ladiges, W. C., Rabinovitch, P. S., Prolla, T. A., Loeb, A. 2007. Mitochondrial point mutations do not limit the natural lifespan of mice. Nat. Genet., 39(4):540-543.

3. Kowald, A. and wood, T. B. 1994. Towards a network theory of ageing: a model combining the free radical theory and the protein error theory. J. theor. Biol., 168(1):75-94.

4. Elson J.L., Samuels, D.C., Turnbull D.M., and Chinnery P.F. 2001. Random intracellular drift explains the clonal expansion of mitochondrial DNA mutations with age. Am. J. Hum. Genet., 68:802-806.

5. Gillespie D. T. 1977. Exact Stochastic simulation of coupled chemical reactions. J. Phys. Chem., 81 (25):2340-2361.

6. Trifunovic et al., 2004. Premature ageing in mice expressing defective mitochondrial DNA

polymerase. Nature. 429: pp. 417-423.

7. Trifunovic et al., 2005. Somatic mtDNA mutations cause aging phenotypes without affecting reactive oxygen species production. Proc. Natl. Acad. Sci. 102(50): pp. 17993-17998.

737f Refining a Kinetic Model of Mint Essential Oil Biosynthesis – Environmental and Genotypic Variation of Oil Quality and Quantity

Rigoberto Rios-Estepa[1], Iris Lange[2], Glenn W. Turner[2], James M. Lee[1], and B. M. Lange[2]. (1) School of Chemical Engineering and Bioengineering, Washington State University, P.O. BOX 642710, Pullman, WA 99164-2710, (2) Institute of Biological Chemistry, Washington State University, P.O. BOX 646340, Pullman, WA 99163

We have previously generated and tested a kinetic mathematical model for essential oil biosynthesis in peppermint [Rios-Estepa et al., 2008]. The inputs for our peppermint model were developmental profiles of gene expression, enzyme concentration, enzyme activity, metabolite levels, and the numbers of the specialized cells producing essential oils as variables, and kinetic properties of enzymes as static parameters. Our model predicted that, when plants were grown at low light intensities, the branch-point enzyme (+)-pulegone reductase was affected by (+)-menthofuran, a stress-induced side product, which acted as a weak competitive inhibitor of (+)-pulegone, the primary substrate for this enzyme. Further experiments with recombinant (+)-pulegone reductase demonstrated that this model prediction was correct, thus indicating the power of iterative approaches integrating mathematical modeling and experimentation.

Here we present a second generation model to describe changes in essential oil profiles under various environmental conditions and in different transgenic lines. Model adjustments include experimentally determined variations in the number of glandular trichomes (specialized essential oil-producing structures on leaf surfaces), the amount of essential oil stored per trichome, the distribution of glandular trichomes at different stages of leaf development, and modulation of gene expression patterns in transgenic lines with modified essential oil composition. The implications of these findings for transgenic approaches aimed at improving essential oil yield and composition are discussed. [Rios-Estepa et al., 2008. PNAS Feb 26;105(8):2818-23]

744a Linear Models for Prediction of Ibuprofen Crystal Morphology Based on Hydrogen Bonding Propensities

Charles Acquah, Chemical Engineering, University of Connecticut, Storrs, CT 06269, Arunprakash T. Karunanithi, U.S. Environmental Protection Agency, 26 W. Martin Luther King Drive, Cincinnati, OH 45268, Matthew Cagnetta, Department of Chemical, Materials and Biomolecular Engineering, University of Connecticut, 191 Auditorium Rd, U-3222, Storrs, CT 06269, Luke E. K. Achenie, Department of Chemical Engineering, Virginia Polytechnic Institute and State University, Randolph Hall 133, Blacksburg, VA 24061, and Steven Suib, Department of Chemistry, Chemical Engineering and Institute of Material Science, University of Connecticut, Storrs, CT 06269.

Solvents have a significant impact on the final crystal form of organic solids during solution crystallization. The use of polarity scales such as Hildebrand solubility parameter and dielectric constant for solvent selection often proves too generalized and do not provide enough insights into the solvent-solute intermolecular interactions directly affecting crystal growth and morphology. This paper addresses the challenging task of selecting an appropriate single component solvent property index that most accurately and sufficiently characterizes crystal morphology. Cooling crystallization experiments were carried out in a wide range of solvents using ibuprofen as a model pharmaceutical compound. Subsequently optical microscope images were used for quantitative characterization of morphology. Finally, linear models that correlate ibuprofen crystal morphology with pure solvent properties were developed. Our results show that, in general, there is a negative linear correlation between crystal aspect ratio (morphology) and a given solvent index. Some correlations revealed significant deviations which can be explained. Where appropriate, infrared spectroscopic measurements were used to analyze the deviations. The "acceptance number" has been identified as an index that significantly captures the ibuprofen-solvent hydrogen bonding intermolecular interactions. Predictions, using model based on acceptance number, were found to compare very well with experimentally determined aspect ratio data from the open literature. Finally, based on insights gained from this work, a flowchart to serve as a useful solvent selection guideline for crystallization of ibuprofen is proposed.

744b Mechanism of the Formation of Enantiotropic Polymorphs in Solution Crystallization

Wenju Wu, Pharmceutical Sciences, JJPRD, 1000 Route 202, OMP Building B241-D, Raritan, NJ 08869

From preclinical development through manufacture of the final drug product, the identification, selection and understanding of drug solid form will have an enormous practical and commercial impact. Crystallization is the critical step in solid form control. Physical and chemical properties denoted by cluster structure, size and shape affect the rate of crystallisation and therefore influence the polymorph selection process. The nucleation kinetics (i.e., induction time), of drug molecules in methanol, ethanol, 2-propanol and water was studied using Raman spectroscopy with multiple probes at 25°C at varying levels of supersaturation ranging from 1.3 to 2.3. Induction periods varied markedly with respect to relative supersaturation. While at the same relative supersaturation, the induction time in methanol was shortest and increased in order from ethanol, 2-propanol and water. The interfacial tensions, between drug crystal and methanol, ethanol, 2-propanol and water were estimated from their induction times based on nucleation theory and were found to be about 3.6, 4.1, 4.5 and 5.7 mJ/m^2 respectively. These values were of the same order of magnitude as those obtained from solubility data. The polymorphism of drug substance was studied in these solvents by maintaining a constant supersaturation (»1.6) and varying temperature from 15 to 45°C. Under these conditions the polymorph remained unchanged. The equation that displays the influence of interfacial tension, supersaturaton and temperature on crystallization kinetics was derived, and was found to be consistent with experimental observations. The mechanism of enantropic polymorphism in the solution crystallization is illustrated. Prediction of polymorphs in pure or mixtures of solvents based on interfacial tension values derived from solubility data is in agreement with the experimental findings.

744c Polymorph Selection In Solution Crystallization: Is Kinetics the Primary Determinant?

Sendhil K. Poornachary, Guangjun Han, Pui Shan Chow, and Reginald B. H Tan. Crystallization and Particle Science, Institute of Chemical & Engineering Sciences, 1 Pesek Road, Jurong Island, Singapore, 627833, Singapore

In determining the structural outcome of a solution crystallization process, there is a delicate interplay between the kinetics and thermodynamics controlling crystal nucleation. As a result, many a times, the stable crystalline form of an organic compound as predicted using lattice energy minimization approach may not be obtained experimentally. This was best illustrated by the case study of glycine, wherein it was shown that the thermodynamically stable γ-form grows 500 times slower than the metastable α-form at the same supersaturation, and hence, is difficult to crystallize from pure aqueous solution (Chew et al., *CrystEngComm*, 2007, 9, 128). Nevertheless, γ-glycine could be obtained from aqueous solutions at high or low pH conditions, as well as in the presence of certain salts and 'tailor-made' additives. Given this, we are confronted with the fundamental question - is kinetics or thermodynamics the primary determinant of polymorph selection?

In this work, the influence of tailor-made additives on the relative nucleation and growth behavior of the individual glycine polymorphs were studied. The observed changes in metastable zone widths and crystal morphology indicated growth inhibition of the metastable α-glycine by the additives (Poornachary et al., *Cryst. Growth Des.*, 2008, 8(1), 179). On the basis of the 'self-poisoning' (Towler et al., *J. Am. Chem. Soc.*, 2004, 126, 13347) and 'stereoselective nucleation inhibition' (Weissbuch et al., *Cryst. Growth Des.*, 2003, 3(2), 125) mechanisms, we rationalize that kinetics controls the appearance of γ-glycine. Furthermore, our data showed that NaCl and NH_4NO_3 can remarkably enhance γ-glycine growth, providing additional evidence that the γ-polymorph is kinetically favored in the presence of these salts (Han et al., *17th International Symposium on Industrial Crystallization, Maastricht, The Netherlands*, 2008). As an outcome of this investigation, we could conclude that the appearance of both the glycine polymorphs under these operating conditions is kinetically controlled.

744d Mechanistic Model for the Effect of Molecular Imposters on the Shape of Molecular Organic Crystals

Jacob P. Sizemore and M. F. Doherty. University of California Santa Barbara, Department of Chemical Engineering, Santa Barbara, CA 93106

The large majority of active pharmaceutical ingredients (APIs) are separated by crystallization from solution and later formulated into tablets as crystals. One important physical property of the crystal is its shape, which can impact the crystal's dissolution rate in the gut, its tendency to agglomerate, and its ability to be compacted into tablets. APIs are formed through a series of reactions, so there are always by-products in solution during crystallization, and these by-products can be quite similar in structure to the API itself. The structurally-related by-products can be occluded in the API crystal during crystallization and exert conspicuous effects on crystal shape. For example, Hendriksen et al. (1998) report the effect of several structurally related molecules on the morphology of paracetamol.

Models of crystal growth shape have moved away from those depending only on crystal geometry or thermodynamics to incorporate mechanistic ideas such as the incorporation of solute at kink sites. In particular, there exists a class of kinetic models based on step flow and the spiral growth of crystal surfaces which are founded on the work of Burton, Cabrera, and Frank (1951) and Chernov (1961); solvent effects in these models have been accounted for by Winn and Doherty (1998).

We furnish an extension of these crystal growth models to capture the effect of structurally-related by-products (here termed "imposters") based on a mechanistic description of the way a molecular imposter impacts a growing crystal surface. We argue that the imposter has no net effect on these growth spirals except at the point where the first turn of a spiral occurs. The model is applied to alpha-glycine growing from water in the presence of L-alanine, and our shape predictions are compared to experimental results (Weissbuch et al., 1983).

Hendriksen, Barry.A., David J.W. Grant, Paul Meenan, and Daniel A. Green. Crystallisation of paracetamol (acetaminophen) in the presence of structurally related substances. Journal of Crystal Growth, 183: 629—640, 1998.

Burton, W.K., N. Cabrera, and F.C. Frank. The Growth of Crystals and the Equilibrium Structure of their Surfaces. Phil. Trans. R. Soc., 243: 299—358, 1951.

Chernov, A.A. The Spiral Growth of Crystals. Soviet Physics Uspekhi, 4: 117—145, 1961.

Weissbuch, I., L. Addadi, Z. Berkovitch-Yellin, E. Gati, S. Weinstein, M. Lahav, and L. Leiserowitz. Centrosymmetric crystals for the direct assignment of the absolute configuration of chiral molecules: Application to the alpha-amino acids by their effect on glycine crystals. J. Am. Chem. Soc., 105: 6615—6621, 1983.

744e Precise Tailoring of the Crystal Size Distribution by Optimal Seeding Time Profiles

Nicholas Kee, Department of Chemical and Biomolecular Engineering, National University of Singapore, 4 Engineering Drive 4, Singapore, 117576, Singapore, Xing Yi Woo, Genome Institute of Singapore, 60 Biopolis Street, #02-01, Genome, Singapore, 138672, Singapore, Reginald B. H Tan, Crystallisation and Particle Science, Institute of Chemical & Engineering Sciences, 1 Pesek Road, Jurong Island, Singapore, 627833, Singapore, and Richard D. Braatz, Department of Chemical and Biomolecular Engineering, University of Illinois at Urbana-Champaign, 600 South Mathews Avenue, Box C-3, 293 Roger Adams Laboratory, Urbana, IL 61801-3602.

For active pharmaceutical ingredients (APIs), a major challenge has been in the development of a systematic engineering approach to design crystallization processes to produce a desired crystal size distribution to satisfy the API's bioavailability requirements. In this work, we explore the possibility to precisely achieve target crystal size distributions through the addition of seed crystals at various weights into an aging vessel for crystal growth for various times. In other words, this strategy optimizes a seeding time profile based on the target distribution, using seeds of narrow distribution, for the operation of a batch crystallizer. To simplify the formulation, a constant growth rate was used; experimentally, this can be easily achieved by constant supersaturation control [1, 2], which is optimal or near optimal for most crystallizers [3]. The optimal weights as a function of the drop time was derived using a nonlinear least-squares optimization solver, and appears as a "flipped" version of the target crystal size distribution. This approach has the capability of tailoring any crystal size distribution of any shape, for example, the flat-top and multi-modal distributions as demonstrated in this work, as long as the target size distributions do not have characteristics narrower than the narrowest size distribution of the seed crystals. Succinic acid was used as the model compound in this study; the nucleation, growth, and dissolution kinetics have been reported previously [4, 5].

References

1. Fujiwara, M.; Chow, P. S.; Ma, D. L.; Braatz, R. D. Crystal Growth & Design 2002, 2, 363-370.

2. Zhou, G. X.; Fujiwara, M.; Woo, X. Y.; Rusli, E.; Tung, H. H.; Starbuck, C.; Davidson, O.; Ge, Z.; Braatz, R. D. Crystal Growth & Design 2006, 6, 892-898.

3. Jones, A. G.; Mullin, J. W. Chemical Engineering Science 1974, 29 (1), 105.

4. Qiu, Y.; Rasmuson, A. C. AIChE Journal 1990, 36 (5), 665-676.

5. Qiu, Y.; Rasmuson, A. C. AIChE Journal 1991, 37 (9), 1293-1304.

744f Development and Scale-up of a Seeded Acidic Co-Addition Crystallization

Adam S. Butterbaugh, Kevin D. Seibert, and Paul C. Collins. Eli Lilly & Co., 1400 W. Raymond St., B110, Indianapolis, IN 46221

An isolation of a previously unisolated nucleoside HCl salt intermediate was developed using PAT in which the intermediate compound was recrystallized via an aqueous acidic and DMSO saturated solution seeded co-addition crystallization. Typical isolation via exclusively anti-solvent driven crystallization resulted in needle like morphology translating to poor isolation characteristics. Crystallization by seeded co-addition resulted in particles with markedly improved morphology and isolation characteristics. The process was successfully scaled-up in an accelerated manner from lab (0.5L) to pilot plant (200L) to production (8000L) in 4 months.

Systematic Design of Supersaturation Controlled Crystallization Processes

Z. K. Nagy*, E. Aamir, C. D. Rielly

Chemical Engineering Department, Loughborough University, Loughborough, LE11 3TU, United Kingdom
**e-mail: z.k.nagy@lboro.ac.uk*

The paper presents a novel methodology for the systematic design of the setpoint operating curves for supersaturation controlled crystallization processes, which produce a desired target crystal size distribution (CSD). The population balance model is solved using the method of characteristics under the assumptions of constant supersaturation and growth dominated process, yielding a simplified analytical expression for the size distribution. A design parameter for supersaturation controlled processes is introduced as a function of the supersaturation and time. Based on the design parameter and the simplified analytical model, the supersaturation setpoint and batch time are determined using an optimization approach to obtain a target distribution with a desired shape. A methodology is also described, which can be used to obtain the temperature profiles in the time domain, corresponding to a desired target CSD, providing a systematic direct design approach for practical applications and scale-up. Additionally, a method for designing the seed distribution is proposed, which can be used in conjunction with the supersaturation setpoint design, for shaping the product CSD. The proposed methods are exemplified for the model system of potash alum in water, for which the size dependent growth kinetic parameters have been identified based on industrial experimental data.

1. Introduction

Crystallization from solution is an industrially important unit operation due to its ability to provide a high purity separation. Batch cooling crystallization provides the advantages of being simple, flexible, and generally requires less process development and investment than many other separation/purification techniques. Many problems in downstream processes can be attributed to poor particle characteristics established in the crystallization step [1,2]. The shape of a crystal size distribution (CSD) produced from crystallisation, affects the efficiency of downstream operations such as filtration, drying and washing. Most of the product qualities are also directly related to the crystal size distribution [2] and the main difficulty in batch crystallization is to accomplish uniform and reproducible CSD [3]. One way to enhance the control of CSD is to use supersaturation control (SSC), which drives the process within the metastable zone to avoid nucleation [4]. Although this approach has proved to produce high quality crystals, the setpoint operating profiles for the supersaturation controller is usually chosen arbitrarily or by trial-and-error experimentation [4].

The paper presents a novel approach for the systematic design of the setpoint trajectory for the supersaturation controller so that a target CSD with a desired shape is obtained. In the case of seeded batch cooling crystallization processes controlled at constant supersaturation, the main governing phenomenon is growth. For these systems an analytical solution of the population balance equation can be obtained, which gives the entire CSD at any moment of the batch. A design parameter, as a function of the batch time and supersaturation, is introduced for supersaturation controlled crystallisation processes. The optimal design parameter is obtained by solving a constrained nonlinear optimization problem with the objective to achieve a desired shape of the CSD at the end of the batch, while maintaining a required minimum yield. The supersaturation setpoint and batch time to achieve the desired CSD can be obtained from the optimal design parameter taking into account the boundaries of the metastable zone, with additional uncertainty margins for robust operating profiles. A methodology is also illustrated, to derive the temperature versus time profiles from the optimal design parameter, which can be readily implemented in the case of industrial crystallizers based on classical temperature control systems. The methodology is extended for the design of the optimal seed distribution required to achieve the desired shape of CSD, which due to the characteristics of the growth kinetics may not be possible to achieve by optimizing the supersaturation only. The methodology is derived for processes with generic size dependent growth kinetics, and it is exemplified for the crystallisation of potash alum in water, using a mathematical model identified based on industrial experimental data [5].

2. Direct Design Approach

The traditional way of controlling the cooling crystallisation processes is to follow a predetermined temperature profile in time [4]. The direct design approach is based on the idea of operating the system within the metastable zone (Figure 1), which is bounded by the nucleation and the solubility curves [4]. The nucleation and solubility curves can be predetermined in automated experiments [6]. Operation close to the metastable limit (high supersaturation) results in excessive nucleation, lower purity and longer filtration times. Operation close to solubility curve (low supersaturation) leads to slow growth and long batch times. The setpoint supersaturation S_{sp} curve is a compromise between fast crystal growth and low nucleation rate. Therefore, in this technique, a supersaturation setpoint curve is chosen experimentally and is followed in the phase diagram using a supersaturation controller based on concentration measurement. In seeded crystallisation, the supersaturation is usually maintained at the desired constant value throughout the entire batch by application of properly designed control algorithms [4,7]. Since direct supersaturation measurement sensors are at prototype stage [3,4] and are often not available for industrial scale use, the supersaturation profiles can be redefined in terms of the temperature profiles in time, which are designed to maintain the supersaturation at a certain setpoint. Since these methodologies are becoming increasingly accepted in the pharmaceutical industries, it is important to have a systematic methodology for designing the supersaturation and temperature trajectories to produce products with the desired CSD.

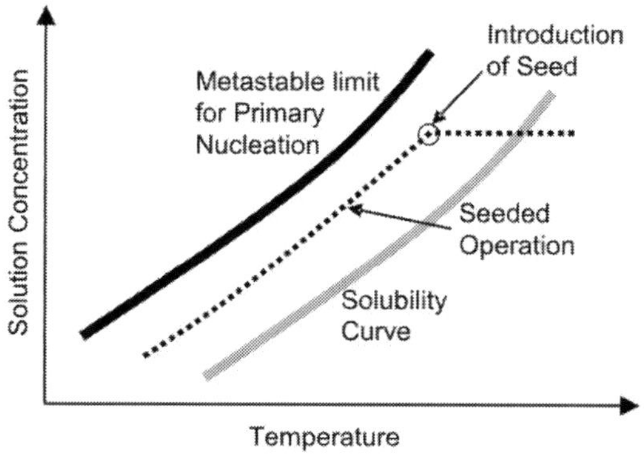

Fig 1: Operations of seeded batch cooling crystallizer.

3. Population Balance Modelling of Batch Crystallization Processes

Considering a single growth direction with one characteristic length L, and a well-mixed crystallizer with supersaturation control with growth as the only dominating phenomenon, the population balance equation (PBE) [1,3] has the form:

$$\frac{\partial f_n(L,t)}{\partial t} + \frac{\partial (G(S,L;\theta_g) f_n(L,t))}{\partial L} = 0, \quad (1)$$

where $f_n(L,t)$ is the crystal size distribution expressed as the number density function (number of crystal per unit mass of slurry), t is time, $G(S,L;\theta_g)$ is the rate of crystal growth, $S=(C-C_{sat})$ is the absolute supersaturation, C is the solute concentration, $C_{sat}=C_{sat}(T)$ is the saturation concentration, T the temperature, and θ_g is a vector of growth kinetic parameters. The generic PBE (1) can be reduced to a system of ODEs by applying the method of characteristics (MOCH). The aim of the MOCH is to solve the PBE by finding characteristic curves in the $L-t$ plane that reduce the partial differential equation to a system of ODEs [1]. If seed is added to suppress nucleation in the case of supersaturation controlled crystallisation, the process will be dominated by growth. For the generic case of size dependent growth, for which the kinetics are given by,

$$G=k_g S^g(1+\gamma L)^p, \tag{2}$$

where $\theta_g=[k_g,g,\gamma,p]$ is the growth parameter vector. Applying the MOCH equation (1) is reduced to the following system of two ODEs:

$$\frac{dL}{dt}=k_g S^g(1+\gamma L)^p, \tag{3}$$

$$\frac{df_n(L,t)}{dt}=-k_g S^g\gamma p(1+\gamma L)^{p-1}f_n(L,t). \tag{4}$$

In the case of well-controlled constant supersaturation, which follows the desired set-point value, S_{sp} the system (3)-(4) can be solved analytically [11] with the solution given by,

$$L=(((1+\gamma L_0)^{1-p}+k_g S^g t\gamma(1-p))^{\frac{1}{1-p}}-1)/\gamma, \tag{5}$$

$$f_n(L)=f_{n,0}(L_0)\left(1+\frac{k_g S^g t\gamma(1-p)}{(1+\gamma L_0)^{1-p}}\right)^{\frac{p}{p-1}}. \tag{6}$$

Discretizing the initial (seed) distribution $f_{n,0}(L_0)=f_{seed}(L_0)$ for different values of L_0, equations (5)-(6) can be used to compute the dynamic evolution of the CSD for a generic growth dominated process (the analytical solution is valid for $p\neq 1$ and $\gamma\neq 0$, however the methodology and equations (3)-(4) apply to the particular case of size independent growth $\gamma=0$, and/or $p=1$, for which simple analytical solutions also exist).

4. Systematic Design of Supersaturation Controlled Crystallisation

The CSD given by the system (5)-(6) is determined by the product between S^g and t, hence a design parameter (ϕ) can be defined as,

$$\phi=S^g t. \tag{7}$$

The optimal supersaturation control (SSC) design parameter $(\phi^\star)$ can be determined by minimizing the difference between the discretized target distribution and the predicted CSD obtained from the analytical estimator (5)-(6):

$$\min_{\phi}\{\sum_{i=1}^{N_d}(f_{v,i}-\tilde{f}_{v,i})^2\}, \tag{8}$$

$$\text{subject to:}\quad 0\le\phi\le\phi_{\max}, \tag{9}$$

$$C(t_{batch})\le C_{f,\max}, \tag{10}$$

where N_d is the number of discretization, $\tilde{f}_{v,i}$ is the discretized target CSD (volume particle density function), $f_{v,i} = f_{n,i}L_i^3 / \sum_{i=1}^{N_d}(f_{n,i}L_i^3\Delta L_i)$ is the discretized, estimated volumetric particle density, $C(t_{batch})$ is the solute concentration at the end of the batch and $C_{f,\max}$ is the maximum acceptable concentration at the end of the batch to achieve a required yield. The Matlab (MathWorks Inc.) function *fmincon* was used to solve the constrained nonlinear optimization problem (8)-(10). Once the design parameter ϕ is optimised for a desired target distribution, by using equation (7) it is possible to determine the supersaturation setpoint (S_{sp}) for a given batch time (t_{batch}), or to calculate the required batch time to achieve the desired distribution by controlling the process at a given supersaturation value. A batch processing unit may be a multi-purpose unit, which is used for several processing phases and may support multi-product manufacturing. In these case, the batch scheduling may become of key importance due to raw material and time constraints. The control design parameter ϕ, gives the flexibility to adjust the supersaturation for a fixed batch time, which can be calculated as,

$$S_{sp} = \left[\phi^\star / t_{batch}\right]^{1/g}, \qquad (11)$$

with $S_{sp} \leq S_{\max}$, where $S_{\max}$ is the boundary in which the supersaturation can be operated with confidence without producing nuclei, and is given by,

$$S_{\max} = S_{MSZW} - \Delta S_{MSZW}, \qquad (12)$$

where S_{MSZW} is the metastable zone width and ΔS_{MSZW} is safety back-off from the MSZW to provide robust performance (see Figure 2). The corresponding minimum batch time is given by

$$t_{batch,\min} = \phi^\star / S_{\max}^g. \qquad (13)$$

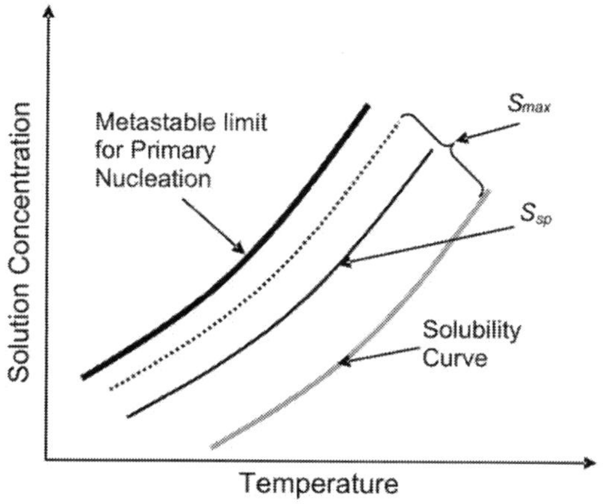

Fig 2: The supersaturation boundaries, in which the S_{sp} can be controlled with confidence without crossing the metastable zone to produce nuclei or solubility curve at which the dissolution starts.

In the cases of no scheduling limitations the batch time is the preferred design parameter to be adjusted, especially during the batch, since changes in the setpoint of the supersaturation controller may lead to control problems, such as undesired oscillations or overshoot.

In industrial practice, it is often difficult to control supersaturation due to the unavailability of installed sensors for measuring supersaturation. However temperature control systems are readily available for all crystallization systems. The temperature trajectory in the time domain can be designed for a desired supersaturation setpoint S_{sp} from the solubility curve, concentration and moments of the CSD during the batch. The solubility curve is given as a function of the temperature, $C_{sat}(T)$. For the system of potash alum in water used in this study, a second order polynomial is fitted to the experimental solubility data,

$$C_{sat}(T) = a_0 + a_1 T + a_2 T^2, \qquad (14)$$

where $a_0 = 3.63, a_1 = 2.43.10^{-2}, a_2 = 3.58.10^{-3}$, the temperature is in $°C$ and C_{sat} is in weight percentage expressed in the anhydrous potash alum [5]. The change in concentration with time is given by,

$$C(t) = C_0(0) - \rho_c k_v[\mu_3(t) - \mu_3(0)], \qquad (15)$$

where C_0 is the initial concentration in $wt\ \%$, ρ_c is the density of crystals in $kg\,\mu m^{-3}$, k_v is the volumetric shape factor, $\mu_3(0)$ is the initial third moment of the seed CSD and $\mu_3(t)$ is the third moment of the CSD at time t, with units of $\mu m^3\,(kg\,slurry)^{-1}$. The moments of crystal size distribution can be obtained from,

$$\mu_k(t) = \int_0^{\infty} f_n L^k dL \approx \sum_{i=1}^{N_d} (f_{n,i} L_i^k \Delta L_i), \text{ where } k = 0,1,2,\ldots\infty. \tag{16}$$

The temperature profile in the time domain used as the setpoint for the temperature controller, $T_{sp} = T(t)$, can be obtained by solving the nonlinear equation,

$$S_{sp}(t) - C(t) + a_2 T^2 + a_1 T + a_0 = 0. \tag{17}$$

The Matlab (Mathworks Inc.) function *fsolve* was used to solve equation (17).

4. Simulation Results and Discussion

Experimental investigations of the batch cooling crystallisation of potash alum $(KAl(SO_4)_2)$ in water were carried out. The experimental data were obtained using an industrial pilot crystallization system located at BASF (Ludwigshafen, Germany). The size dependent growth parameters ($\theta_g = [8.570\,\mu m\, s^{-1},\ 1,\ 0.005\,\mu m^{-1},\ 1.577]$) for the potash alum system in water were obtained using model-based estimation, assuming a well-mixed system and growth and secondary nucleation mechanisms. The model was solved using a combined approach based on the quadrature method of moments (QMOM) and method of characteristics (MOCH) [5]. A target CSD was obtained by experimental trial and error. The control design parameter (ϕ) was optimised to obtain the setpoint for the supersaturation controller and the batch time, in order to achieve the target experimental CSD. Results are shown in Figure 3. Since the target distribution was obtained experimentally, it represents a feasible setpoint for the system. Hence, the predicted CSD is in good agreement with the experimental CSD. Figures 3 (A and B) show the results using different pairs of supersaturation values and batch times, corresponding to a constant value of the design parameter ϕ.

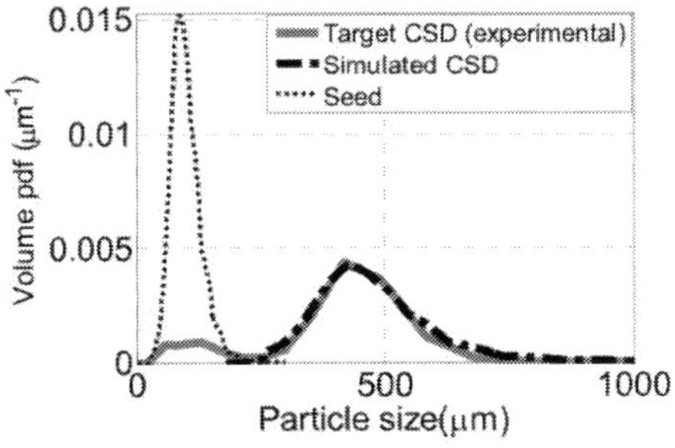

(A)

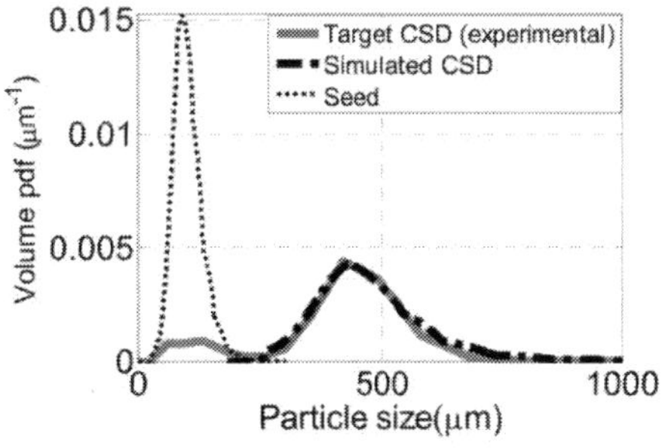

(B)

Fig 3: Results with the optimized design parameter $\phi = 0.206\ min$, designed to achieve the target experimental CSD. (A) fixed batch time of $t_{batch} = 80\ min$ and calculated $S_{sp} = 0.00257$; (B) fixed supersaturation setpoint $S_{sp} = 0.00215$ and calculated batch time $t_{batch} = 96\ min$.

The methodology was also evaluated for the design of crystallization systems with arbitrary target CSDs. A lognormal CSD was selected with mean $L_m = 420\ \mu m$ and standard deviation $\sigma = 0.22$. Figure 4(A) shows that the system with the optimized supersaturation design parameter of $\phi = 0.203\ min$ is able to achieve a product CSD with good agreement with the target CSD. The actual shape of the CSD, which can be achieved by designing the supersaturation level and/or batch time is limited, and is determined by the seed distribution and the growth kinetics of a particular system. Figure 4(B) illustrates the results of the optimal design when a narrow target distribution is used. The resulted design ($\phi = 0.198\ min$) provides a product with similar mean size, however the width of the distribution which can be achieved is limited by the system.

The designed supersaturation setpoints can be transformed into temperature profiles using the model inversion approach described in Section 3. Figure 5(A) shows the temperature profiles corresponding to the design parameter, $\phi = 0.206\ min$, obtained for different supersaturation setpoints and batch times in the case of the experimental target CSD. The corresponding concentration profiles can be observed in Figure 5(B). The yield produced is around 61 % in all cases, being higher than the minimum 50% required in the optimisation. The proposed simplified model-based direct design approach provides a systematic methodology to operate the crystallisation process at a constant supersaturation by controlling a temperature trajectory throughout the batch.

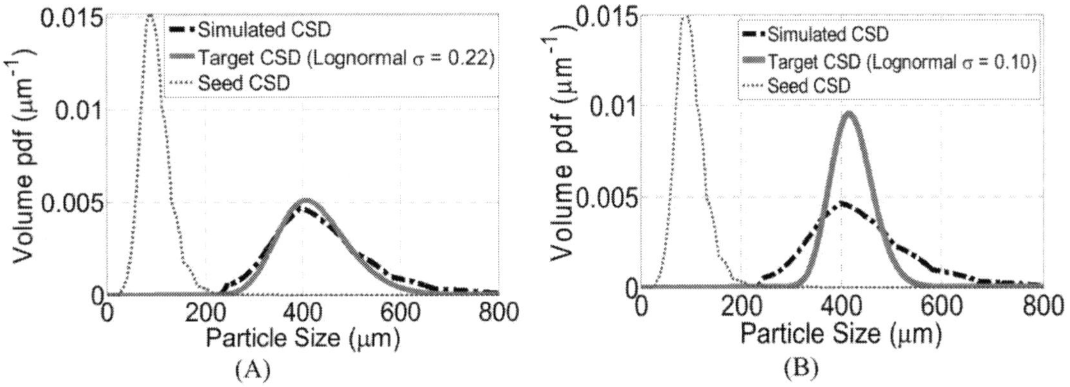

Fig 4: (A) Results with the optimized design parameter $\phi = 0.203\ min$ to achieve the target lognormal CSD with $L = 420\ \mu m$ and $\sigma = 0.22$; (B) results with the optimized design parameter $\phi = 0.198\ min$ for the target CSD represented as a narrow lognormal CSD with $L = 420\ \mu m$ and $\sigma = 0.10$.

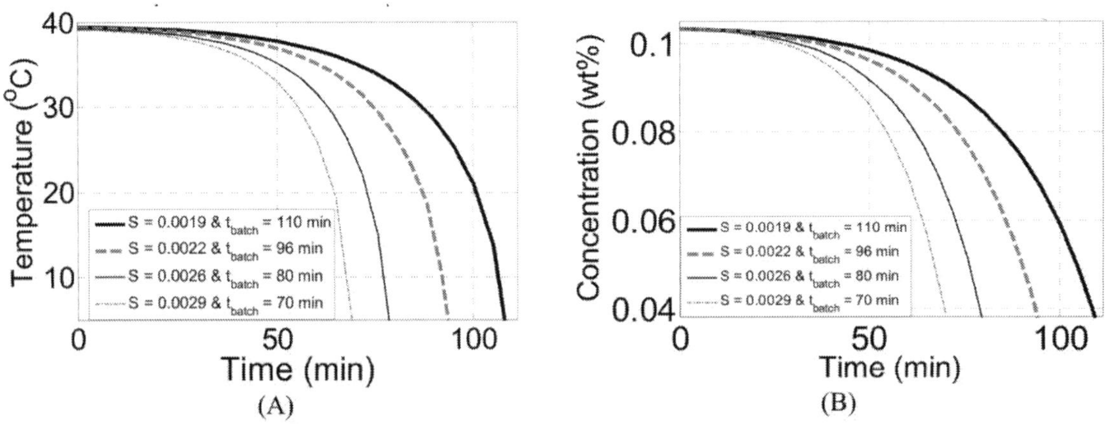

Fig 5: (A) Temperature profiles and (B) concentration profiles, obtained at different setpoint supersaturations, S_{sp} and batch times, t_{batch}, corresponding to the same design parameter $\phi = 0.203\ min$, optimized to achieve the experimental target CSD.

5. Shaping the CSD through Seed Recipe Design

In addition to the supersaturation and batch time (determined by the design parameter ϕ) the seed recipes can also be optimised [8] to obtain the desired target CSD, which may not be achieved by optimizing the ϕ only. A particular seed CSD can be obtained experimentally by mixing different amounts of seeds with different size distributions. The seed recipe was represented by a sum of Gaussians and the optimisation problem with the objective of shaping the distribution at the end of the batch was formulated as follows,

$$\min_{\beta}\left\{\sum_{i=1}^{N_d}(f_{v,i}-\tilde{f}_{v,i})^2\right\}, \tag{18}$$

Subject to:

$$0 \le \beta \le \beta_{\max}, \tag{19}$$

$$f_{n,seed}=\sum_{i=1}^{N_g} w_i N_i(L_{m,i},\sigma_i), \tag{20}$$

where $\beta=[w_1,L_{m,1},\sigma_1,\ldots,w_{N_g},L_{m,N_g},\sigma_{N_g}]$ is the seed design vector with N_g the number of Gaussians, w_i the weights, $L_{m,i}$ the mean sizes (μm) and σ_i (μm) the standard deviations of the respective Gaussians distributions, N_i, $i=1,2,\ldots,N_g$.

The seed can be designed for any target CSD, e.g., lognormal or bimodal. Figure 6(A) shows the results of the seed design, for the narrow lognormal target distribution shown in Figure 4(B) using the optimal supersaturation control design parameter, $\phi = 0.198\ min$. The seed required to produce the target CSD is narrower than the experimental seed used in the design of the supersaturation controller. Figures 6(B) and (C) show the results of the seed design when arbitrary bimodal target distributions were used in the case of fixed $\phi = 0.206\ min$. The predicted product CSDs are in good agreement with the target bimodal CSDs. Figure 7 shows the main steps of the proposed comprehensive and systematic methodology for shaping the product CSD for supersaturation controlled crystallisation processes, which combines seed recipe and operating policy designs.

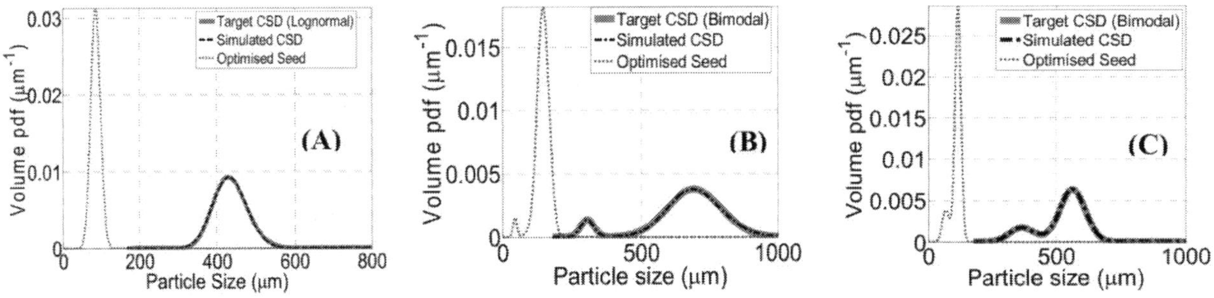

Fig 6: (A) Optimised seed ($\beta = [0.999, 84.9, 12.7, 0.001, 69.4, 28.3]$) for the target CSD shown in Figure 4(B); (B) optimised seed ($\beta = [0.47, 42.1, 7.6, 0.06, 62.4, 41.4, 0.47, 138.3, 21.7]$) for bimodal target CSD with separated peaks; (C) optimised seed ($\beta = [0.36, 57.8, 13.5, 0.17, 62.1, 26.1, 0.47, 113.3, 12.0]$) for bimodal target CSD with overlapping peaks.

6. Shaping the CSD via optimal dynamic seed addition

Achieving the shape of the final CSD by designing the initial size distribution as a mixture of seed with different size distributions can be difficult to apply at industrial level. Similar results can be achieved if a monomodal seed is introduced in the crystallizer during the duration of the crystallisation process. In this case the only requirement is that the seed CSD must be narrower than the target distribution. The narrower the seed distribution is the closer the final CSD will be to the target CSD. In this case the batch time is discretised in N intervals and the amount of seed dropped into the system at the beginning of each time interval is determined by solving the following optimisation problem:

$$\min_{w1,\ldots,w_N} \left\{ \sum_{i=1}^{N_d} (f_{v,i} - \tilde{f}_{v,i})^2 \right\} \tag{21}$$

Figure 7 shows the result of the optimised dynamic seed addition profile in the case of two target distributions. In both cases the seed distribution was Gaussian with a mean size of 60 μm and standard deviation of 5 μm. It can be seen that in the case of the lognormal target distribution seed addition was required during the entire duration of the batch to achieve the shape the form of the distribution in the small size range. In the case of the trapezoidal target distribution all crystals are larger than 200 μm. The results of the optimisation shows that this can be achieved if seed is dropt into the system during the first half of the batch. The error in the larger size range of the target distribution is due to the fact that the size dependent growth generates a widening of the seed distribution. Although the seed distribution is narrow, during the growth time required to achieve the larger size ranges desired for the target CSD, the distribution becomes wider.

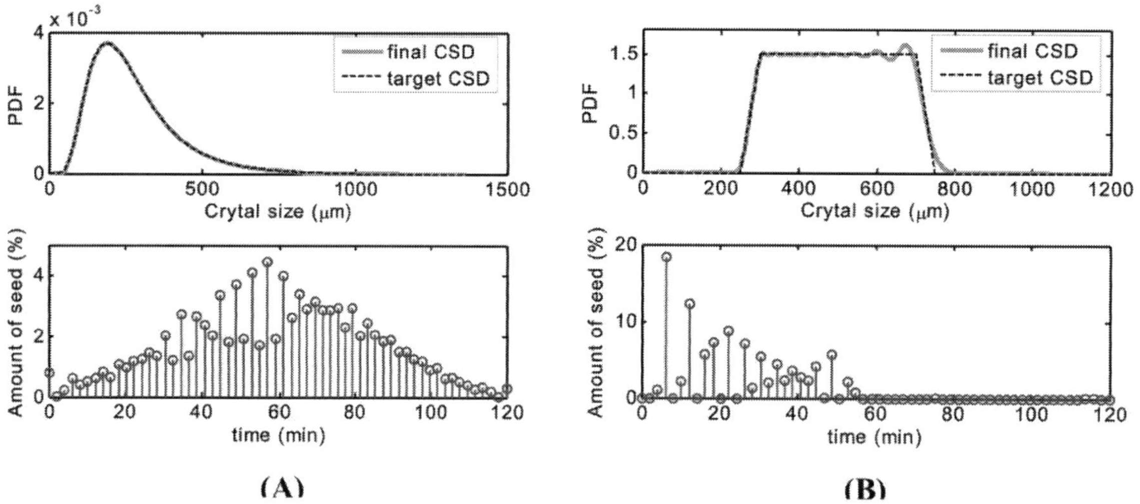

Fig. 7: Results of the dynamic seed addition for (A) lognormal and (B) trapezoidal target distribitions.

7. Conclusions

The paper describes a novel methodology for the systematic design of supersaturation controlled crystallisation prosesses. The approach is based on the idea that in the case of supersaturation controlled seeded crystallisation systems, the supersaturation is constant throughout the batch, and the process is dominated by growth only. A design parameter for supersaturation controlled batch crystallisation processes is introduced, and a simplified model based optimisation is used to derive the setpoint operating curve and batch time required to achieve the desired shape of the product CSD. The designed operating curve can be implemented in the phase diagram using supersaturation control, or in the time domain using classical temperature control. An approach to design an optimal seed recipe is also presented, which can be used to achieve a desired shape of the product CSD, in conjunction with the supersaturation control design. The methods are exemplified in the case of the batch cooling crystallisation of potash alum system in water, for which a model has been developed and identified based on industrial experimental data.

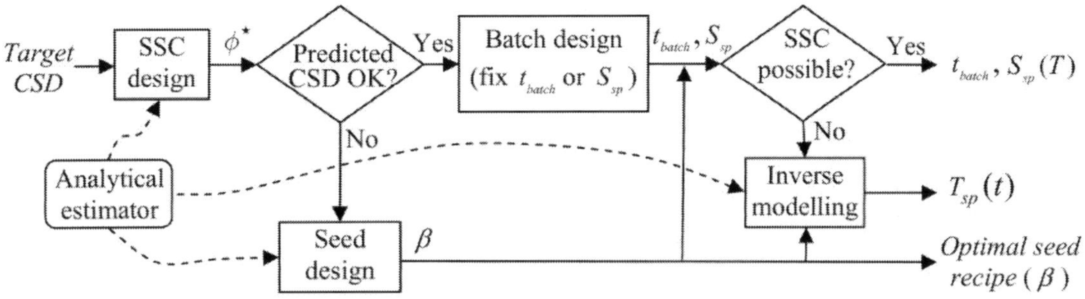

Fig 7: Flowchart of the systematic design of supersaturation controlled crystallisation processes, to achieve a desired target CSD.

8. References

[1] Hounslow, M. J., Reynolds G. K. Product engineering for crystal size distribution. AICHE Journal 52(2006), 2507-2517.

[2] Wibow, C., Ng, K. M. Design of integrated crystallisation systems. AICHE Journal 47(11), 2474-2492.

[3] Braatz, R. D. Advanced control of crystallisation processes. Annual Reviews in Control 26(2002), 87-99.

[4] Fujiwara, M., Nagy, Z. K., Chew, J. W., Braatz, R. D. First principles and direct design approaches for the control of pharmaceutical crystallisation. Journal of Process Control 15(2005), 493-504.

[5] Aamir, E., Nagy, Z. K., Rielly, C. D., Kleiner, T., Judat, B. Efficient crystal size distribution estimation approach for growth dominated crystallisation processes. Proceedings of the International Symposium on Industrial Crystallisation, Maastricht, The Netherlands, 2008.

[6] Glennon, B. Characterising the metastable zone and solubility curve using Lasentec FBRM and PVM. Chem. Eng. Res. Des. 80(2002),799-805.

[7] Zhang, G. P., Rohani, S. On-line optimal control of a seeded batch cooling crystalliser. Chemical Engineering Science 58(2003), 1887-1896.

[8] Chung, S. H., Ma, D. L. Braatz, R. D. Optimal seeding in batch crystallization. Canadian Journal of Chemical Engineering 77(3) (1999), 590-596.

Acknowledgements

Financial support provided by the Engineering and Physical Sciences Research Council (EPSRC), U.K., (grant EP/E022294/1) is gratefully acknowledged. BASF, Ludwigshafen, Germany, is also acknowledged for partial financial support and for providing the experimental data for model identification.

749a Disruption of Adipocyte-Endothelial Interactions Under Hyperglycemic Stress

Ning Lai and Kyongbum Lee. Chemical and Biological Engineering, Tufts University, 4 Colby Street, Science and Technology Center, Medford, MA 02155

Adipose tissue growth is essential to normal development and whole body metabolic homeostasis, but is also linked to many diseases. In particular, obesity is characterized by an excessive expansion of body fat (white adipose tissue, WAT) mass. The bulk of WAT is a loose association of white adipocytes held in a collagen matrix. The tissue is also highly vascularized. The vascular fraction contains locally resident precursor cells (preadipocytes) and endothelial cells that form the microcirculation. To date, the interactions between these cellular components in adipose tissue development remain poorly understood. As a first step, we investigated the reciprocal effects of adipocyte differentiation and endothelial cell growth using a collagen gel-based co-culture model. The co-culture consisted of human umbilical vein endothelial cells (HUVECs) and adipocytes derived from 3T3-L1 preadipocytes. We found that the co-culture synergistically enhanced the proliferation of the endothelial cells and the differentiation of adipocytes. Compared to a control culture seeded with only HUVECs, endothelial cell proliferation increased up to 6-fold. Compared to a control culture seeded with only 3T3-L1 cells, triglyceride (TG) accumulation, an important measure of differentiated adipocyte function, was increased up to 3.4-fold. The synergistic effects depended on both the total density and composition of the seeding mixture, with the largest increases observed at the highest total seeding density (2•106 cells/ml collagen) and preadipocyte:HUVEC ratio (90:10) tested in this study. To determine the metabolic basis for the enhanced TG accumulation, we compared the metabolic flux profiles of adipocytes with and without HUVEC co-culture. Results of metabolic flux analysis (MFA) showed that the co-culture up-regulated the activities of several carbohydrate pathways, including glycolysis and the pentose phosphate pathway (PPP). Interestingly, the up-regulation in adipocyte metabolic activity critically depended on endothelial cell viability. When the co-culture was subjected to a hyperglycemic stress (4.5 g/L medium glucose), HUVEC proliferation was abolished, and the TG content was decreased by 60 % compared to a control culture seeded with only 3T3-L1 cells. Taken together, our results demonstrate that the adipocyte-endothelial cell co-culture model captures a number of in vivo WAT features that are absent in conventional in vitro models, including endothelial cell-stress mediated attenuation in lipogenic activity. Prospectively, the co-culture system could serve as a useful platform to develop models of metabolic diseases that directly impact on WAT function, for example type 2 diabetes.

749b Identification of Novel Regulators of Innate Immune Responses Using High-Throughput Screening and Their Therapeutic Implications

Hwijin Kim and Brian Seed. Center for Computational and Integrative Biology & Department of Genetics, Massachusetts General Hospital, Harvard Medical School, 185 Cambridge Street, Boston, MA 02114

Virus infection rapidly induces type-I interferon (IFN-α/β) and proinflammatory cytokines, which further activates a subset of genes that confer an antiviral state on the host cell, thereby constituting the first line of defense against invading viruses. Despite major advances over the last decade in our understanding of the innate immune signaling pathways, the human gene components involved in the pathways have not yet been fully elucidated. Here, we have begun to systematically identify additional pathway components by high-throughput cell-based screening of human cDNA libraries that represent about 50% of the human coding transcriptome. Using luciferase-based transcriptional reporter assays that measure the activation of IFN-β and/or NF-κB promoters, combined with data mining, we have been able to identify a number of novel positive and negative regulators in the pathways. Hit validities were confirmed by various biochemical, genetic and biophysical methods, including shRNA knockdown and/or knockout mouse-based methods, RT-PCR, immunoblotting and co-immunoprecipitation. We expect our findings will offer novel targets for therapeutic manipulation of the innate immune signaling pathways, leading to the development of more effective agents to treat virus- and inflammation-related diseases.

749c Tumor-Induced T Cell Tolerance In a Prostate Cancer Model

Eileen M. Higham, Biological Engineering and Koch Institute for Integrative Cancer Research, Massachusetts Institute of Technology, 77 Massachusetts Ave., E17-132, Cambridge, MA 02139, K. Dane Wittrup, Chemical Engineering, Biological Engineering and Koch Institute for Integrative Cancer Research, Massachusetts Institute of Technology, 77 Massachusetts Ave., E19-551, Cambridge, MA 02139, and Jianzhu Chen, Biology and Koch Institute for Integrative Cancer Research, Massachusetts Institute of Technology, 77 Massachusetts Ave., E17-131, Cambridge, MA 02139.

Prostate cancer is the most prevalent malignancy among American men, leading to significant mortality each year. This is in part due to a lack of effective treatments for advanced prostate cancer, highlighting the critical need for novel therapies to treat this disease. The prostate is considered an ideal organ for cancer immunotherapy, because of its non-essential nature, and because of the presence of several prostate-specific antigens that could serve as targets for an immune response. However, such therapy is limited by tolerization of CD8+ T cells in the tumor local environment, essentially abrogating an anti-tumor response. The objective of this work is to understand the interactions between CD8+ T cells and tumor cells that lead to tolerance, and to design therapies to reinvigorate the immune response. A spontaneous prostate cancer model, the TRansgenic Adenocarcinoma of the Mouse Prostate (TRAMP) model (1), has been used to develop an in vivo system to explore tolerance. Tumor-induced tolerance has been characterized using TRAMP mice engineered to express the antigen SIY in their prostate tissue, TRAMP-SIY mice, coupled with T cells that recognize Kb-SIY, 2C T cells. We have demonstrated that 2C T cells are tolerized in the tumor local environment of TRAMP-SIY mice through both their inability to produce IFN-γ and their impaired cytotoxic killing functions. In vivo screens have also been used to define critical mechanisms of tolerance in this model, most notably tolerance mediated by cross-presentation. We have shown that systemic α-CD40 administration can initially overcome tolerizing mechanisms in TRAMP-SIY mice and can restore function to 2C T cells. Based on these findings, novel localized α-CD40 therapies are being rationally designed to break tolerance, while minimizing the autoimmune complications observed with systemic treatments.

1. Greenberg, N.,; DeMayo, F.; Finegold, M.; Medina, D.; Tilley, W.; Aspinall, J.; Cunha, G.; Donjacour, A.; Matusik, R.; Rosen, J.; Proc. Natl. Acad. Sci., 92, 3439-3443, 1995.

749d Using Innovative Microtechnology to Quantify the Effects of Precise Oxygen Delivery to In-Vitro Hypoxic Tumor Tissue

Bhushan J. Toley[1], Jaehyun Park[2], Byoungjin Kim[1], Michel M. Maharbiz[2], and Neil S. Forbes[1]. (1) Chemical Engineering, University of Massachusetts, Amherst, 686 North Pleasant Street, 159 Goessmna Lab, Amherst, MA 01003, (2) Electrical Engineering and Computer Science, University of California, Berkeley, 570 Cory Hall, Berkeley, CA 94720

Unlike normal tissues, tumors have hypoxic regions. Mathematical models suggest that cell survival in tumors is strongly dependent on and limited by the availability of oxygen, though no experimental evidence exists that eliminates the effect of other nutrient gradients and the acidic conditions that exist within the tumor. Our aim was to measure the rate and location of cell death and track cell migration in tumor tissue growing in a spatially and temporally defined oxygen microenvironment, to quantify the effect of modified oxygen gradients in tumors. We have developed an innovative microtechnology that allows spatial as well as temporal control over local oxygen delivery to in-vitro tissues. The gradient generating device consists of a single or an array of noble metal microelectrodes patterned over a glass substrate within a polydimethylsiloxane (PDMS) microchannel, which is filled with an electrolyte (deionized water). Electrolysis generates dissolved oxygen and can be used to generate a user defined oxygen gradient over the cell culture area. A single circular electrode of 10µ m diameter was used to generate oxygen. Cylindroids, which were formed by compressing multicellular spheroids in between two parallel surfaces, were positioned over the cell culture area so that their centers coincided with the electrode. A direct correlation between the electrolysis current and the oxygen generation rate was developed. Generation of oxygen at the center of cylindroids created oxygenated regions in otherwise hypoxic regions of the cylindroids. Fluorescent viability stains were used to determine viability profiles and viable regions were observed at the center of the cylindroids when oxygen was supplied for 60 hours. The corresponding controls had necrotic centers. Fluorescent nuclear stains were used to track cell migration and velocity profiles of the migrating cells were obtained. The nuclear morphology was monitored to determine the rate and location of apoptosing cells. A mathematical model for cylindroid growth was developed incorporating chemotaxis of individual cells towards a higher oxygen concentration. Experimental data was used to tune the parameters of the model to predict the relative contribution of chemotaxis (active) and growth (passive) to cell migration in tumor tissue. These experiments demonstrate the ability to study cell response in a spatially and temporally controlled oxygen microenvironment and could be used to predict microenvironment-specific tumor response.

749e Characterization of Resistance to Chemotherapy In Glioma Cell Lines

Salaheldin Hamed, Biomedical Engineering, Rutgers University, The State University of New Jersey, 599 Taylor Road, Piscataway, NJ 08554 and Charles M. Roth, Rutgers University, 599 Taylor Road, Piscataway, NJ 08854.

Gliomas are not responsive to chemotherapy because of tumor cell resistance, by which changes at the molecular level affect important cellular processes, such as cell cycle kinetics and apoptosis. A number of markers of resistance to chemotherapy in glioma models have been discovered; however, the mechanistic relationship between the expression of these markers and failure of chemotherapy remains to be elucidated. Morever, explaining resistance based on changes in a single genetic or molecular entity is inadequate, given the complexity of the system. To understand resistance to chemotherapy, identification of biological systems, and their interactions, is of great promise.

We have focused much of our attention to date on characterizing the response of glioma cell lines to chemotherapeutic drugs, carmustine and etoposide, at the level of cell cycle and apoptosis. In particular, we have developed a mathematical model of cell cycle kinetics that is able to reproduce the dynamics and dose response of cells to the two chemotherapeutic agents based on two parameters relating to cell cycle arrest and entry into apoptosis. We have shown that the model can be used to extract mechanistic information regarding the relative influence of these two processes upon tumor cells simply from pharamacological dose response curves, from which mechanism is not obtained using traditional analyses. The model suggests that carmustine elicits its effect by inducing apoptosis, while etoposide induces cell cycle arrest as a primary mode of action. Our current work seeks to apply this methodology to track the changes in tumor cells as they acquire resistance to chemotherapy.

We have generated glioma cell lines resistant to etoposide and carmustine by incremental stepwise exposure to sublethal doses of each drug. We have compared the effect of drug exposure to chemotherapy on important cellular processes, such as cell cycle distributions and apoptosis, between resistant and parent cell lines. We found that resistant cells have altered cell cycle distributions compared to parent cell lines at drug free conditions; moreover, changes in cell cycle distribution upon drug exposure are more drastic in parent cell lines compared to resistant cell lines. To deepen our understanding of resistant phenotype, we aim to use molecular profiling using microarray technology, which allows for global monitoring of changes in gene expression profiles. Microarrays exploring genes implicated in critical processes, such as cell cycle arrest and apoptosis, will be investigated in order to elucidate pathways implicated in resistance. We plan to construct regulatory networks from gene expression data in order to characterize the process of resistance development from a systems viewpoint.

749f The Role of the Flux of Platelet Agonists on Clot Formation and Stability

Keith B. Neeves and Scott L. Diamond. Chemical and Biomolecular Engineering, University of Pennsylvania, 3340 Smith Walk, 1150 Vagelos Laboratories, Philadelphia, PA 19104

The flux of platelet agonists into flowing blood is a critical event in thrombosis and hemostasis. However, few *in vitro* method exist for examining and controlling the role of platelet agonists on clot formation and stability under hemodynamic conditions. In this work, we describe a membrane-based method for introducing a solute into flowing blood at a defined flux. The device consisted of a track-etched polycarbonate membrane reversibly sealed between two microfluidic channels; one channel contained blood flowing at a physiologically relevant shear rate, and the other channel contained the agonist(s). As a proof-of-concept, ADP was introduced into whole blood flowing at 250 sec^{-1} at three fluxes (1.5, 2.4, and 4.4 x 10^{-18} mol μm^{-2} sec^{-1}). Platelet aggregation was monitored by fluorescence microscopy during the experiment and the morphology of aggregates was determined by *post hoc* confocal and electron microscopy. At the lowest flux (1.5 x 10^{-18} mol μm^{-2} sec^{-1}), we observed little to no aggregation. At the higher fluxes, we observed monolayer (2.4 x 10^{-18} mol μm^{-2} sec^{-1}) and multilayer (4.4 x 10^{-18} mol μm^{-2} sec^{-1}) aggregates of platelets and found that the platelet density within an aggregate increased with increasing ADP flux. This is the first demonstration of the threshold ADP flux required to trigger intense platelet deposition under flow. Preliminary results on the deposition of fibrin under flow show a similar dependence on thrombin flux. We expect this device to be a useful tool in unraveling the role of platelet agonists on clot formation and stability.

749g Pathological Biomineralization of Calcium Oxalate Kidney Stones

Jeffrey D. Rimer[1], Jeffrey A. Wesson[2], and Michael D. Ward[1]. (1) Molecular Design Institute, Department of Chemistry, New York University, 100 Washington Square East, New York, NY 10003, (2) Nephrology Division, Department of Veterans Affairs Medical Center and the Medical College of Wisconsin, 5000 West National Avenue, Milwaukee, WI 53295

The design of new technologies and pharmaceuticals for therapeutic drug delivery requires a fundamental understanding of the pathogenesis of physiological diseases; however, control and prevention of many common diseases, such as kidney stones, which affect 15 % of the U.S. population, are currently hindered by a knowledge gap pertaining to critical steps responsible for stone formation *in vivo*. Kidney stones derive predominantly from supersaturated calcium oxalate (CaOx), which crystallizes in renal tubules to produce two polymorphs, calcium oxalate monohydrate (COM) and dihydrate (COD). Adverse effects of the disease have been correlated to COM, which forms layered polycrystalline aggregates through a series of events involving nucleation, growth, aggregation, and crystal and/or aggregate attachment to epithelial cells. Our group [1] and others [2] have investigated interfacial aspects of these processes using atomic force microscopy (AFM), probing molecular-level events on the surfaces of kidney stone crystals (e.g. adhesion forces and growth kinetics) in an effort to identify the underlying mechanism(s) of their formation.

Lipids and proteins on cellular membranes can mediate crystal-crystal contact, while a multitude of urinary proteins, polysaccharides, lipids, and cellular debris adsorb to crystal surfaces, likely acting as adhesive glue among aggregates. Previous studies suggested that proteins with substantial anionic moieties serve as COM adhesives [1e]; however, a variety of urinary proteins have immerged as potential inhibitors of crystal nucleation, growth, and/or aggregation – most notably transferrin, human serum albumin, Tamm-Horsfall protein (THp), and osteopontin (OPN). Research efforts in our group have focused on (i) *in vitro* investigations of crystal growth to examine face-specific inhibitory effects in the presence of protein and macromolecular additives [1a,c], and (ii) adhesion force measurements using AFM probes modified with biologically relevant functional groups to study surface-adsorbate interactions involved in crystal adhesion and aggregation [3a]. We will present recent investigations that examine interactions between the most bioactive surface, COM (100), and functional groups that mimic L-arginine and L-glutamic acid residues. Experiments were performed with additives believed to influence crystal growth and aggregation, focusing on THp and OPN. A systematic study of THp was performed to identify the affect of sialylated and glycoslated side chains on the force of adhesion, while dynamic protein unfolding measurements were conducted on OPN – a protein that exhibits uncharacteristically strong adhesion to COM surfaces [3b]. Lastly, we have expanded the use of single amino-acid functionalized tips to directly measure protein-crystal and protein-protein interactions on COM surfaces using established protocols in the literature for binding proteins to AFM tips [4]. These collective results offer new insights on specific interactions that potentially regulate kidney stone formation, serving as a basis for future development of targeted drug delivery for COM stones.

1. (a) Guo et al., Langmuir 2002, 18, 4284, (b) Sheng et al., J. Am. Chem. Soc. 2003, 125, 2854, (c) Jung et al., Langmuir 2004, 20, 8587, (d) Sheng et al., Proc. Nat. Acad. Sci. 2005, 102, 267, (e) Ward and Wesson, Elements 2007, 3, 415.

2. (a) Qiu et al., Proc. Nat. Acad. Sci. 2004, 101, 1811, (b) Qiu et al., J. Am. Chem. Soc. 2005, 127, 9096, (c) Wang et al., Langmuir 2006, 22, 7279.

3. (a) Sheng et al., J. Am. Soc. Nephrol. 2005, 16, 1904, (b) Wesson et al., J. Am. Soc. Nephrol. 2003, 14, 139.

4. (a) Ebner et al., Current Nanoscience 2007, 3, 49, (b) Lee et al., Proc. Nat. Acad. Sci. 2006, 103, 12999, (c) Lee et al., Micron 2007, 38, 446, (d) Strunz et al., Proc. Nat. Acad. Sci. 1999, 96, 11277.

751a High Cell Density Fermentation of Deinococcus Radiodurans Under Optimized Conditions

Yi He, Biochemistry & Biophysics Ctr, National Heart, Lung, and Blood Institute, 50 South Dr,, Bethesda, MD 20892

Deinococcus radiodurans (DR) is the #1 candidate for bioremediating nuclear waste site, due to its unique ability to resist high doses of radiation. Its radiation resistance mechanism has been under intense investigation. However, studies of its growth media and growth conditions are scarce, and its large scale production has not been reported. In this study, a systematic investigation was carried out to produce DR at the 20-L fermentor scale. The phosphate buffer typically used was found to be inhibitory to DR growth, and caused cell grouping. HEPES and MOPS buffer helped DR growth. Several antifoams were investigated to support large scale production with submerged aeration, and KFO 673 defoamer was chosen based on its ability to prevent foaming without affecting DR growth. The traditional TGY media only support DR growth to 10 OD600. Using a Design of Experiments approach, glucose, Mg, Mn were found to be critical to support high density growth of DR. With addition of 10 g/L glucose, 0.5 g/L MgSO4•7H2O, 5 μM MnCl2 into TGY media, OD600 of 40 was achieved. The optimal pH and temperature for DR growth were found to be 7.0 and 37 oC, respectively. Fermentation was carried out in a 20-L fermentor using the newly developed media under the optimal conditions.

751b Production of L-Tyrosine by Recombinant E.Coli

Raymond R. Zolandz, Ranjan Patnaik, and David F. Kraynie. Central Research, Biochemical Sciences and Engineering, DuPont Co., 2309 Sunset Lake Road, Newark, DE 19702

A process has been developed for production of L-tyrosine. The process, demonstrated at 10 and 200 liter scale, uses overproducing Escherichia coli strains derived from an L-phenylalanine overproducing strain. Significant increase in L-tyrosine production was achieved via deletion of the chromosomal region encoding for the pheA gene, chorismate mutase/prephenate dehydratase, its leader peptide (pheL) and its associated promoter. A further increase in titer was achieved by overexpressing tyrA, encoding chorismate mutase/prephenate dehydrogenase, from a strong non-native trc promoter. Fermentation optimization studies were done, with particular emphasis on conditions to mitigate foaming issues associated with running a three-phase fermentation (media, air and tyrosine crystals). L-tyrosine titers of 55 g/l in 48 hours were demonstrated at the 200 L scale, which is the highest titer reported at that time..

Two primary separations schemes for isolation and purification of L-tyrosine from the fermentation broth were evaluated. The first uses physical separation of tyrosine crystals via selective centrifugation, while the second involves dissolving the tyrosine at high pH, filtering to remove biomass, then recovering the tyrosine by acid precipitation. L-tyrosine product purity of 98% was achieved with yields ranging from 90-95%.

751c Production of Natural Flavor Chemicals: Isoamyl Acetate Biosynthesis from Renewable Sugars Using Engineered Escherichia Coli

Praveen V. Vadlani, Grain Science and Industry, Kansas State University, 1980 Kimball Avenue, 202 BIVAP Building, Manhattan, KS 66506, Ka-Yiu San, Department of Bioengineering, Rice University, MS-142, PO Box 1892, Houston, TX 77005, and George N. Bennett, Biochemistry and Cell Biology, Rice University, 6100 Main Street MS#142, Houston, TX 77005.

Production of high-valued specialty chemicals using renewable resources is beneficial and is now cost-effective due to advances in metabolic engineering and fermentation techniques. Isoamyl acetate, produced via fermentation, is a natural flavor chemical with applications in food and other allied industries. A novel and cost-effective fermentation process was developed to produce isoamyl acetate using grain sorghum derived-glucose as feedstock. Two alcohol acetyltransferases from Saccharomyces cerevisiae (ATF1 and ATF2) were used in the esterification of isoamyl alcohol with acetyl coenzyme A. The respective genes were cloned and expressed in an appropriate ack-pta- mutant strain of Escherichia coli. Aerobic shake flasks experiments examined isoamyl acetate production over time, temperatures, and initial optical densities. The strain carrying the pBAD-ATF1 plasmid exhibited a high molar ester yield from glucose (1.13) after 48 hours of aerobic growth at 25¢ªC. Operational parameters have been optimized and low-cost media components, such as fusel oil, sorghum glucose and corn steep liquor was identified for high yield of isoamyl acetate. Using engineered Ecoli to produce natural ester is the first of its kind and an encouraging yield of 0.18 g isoamyl acetate/g of glucose was obtained with the low-cost fermentation medium.

751d Production of Organic Acid Esters by Extractive Fermentation and Enzymatic Esterification

Mingrui Yu, Hanjing Huang, and Shang-Tian Yang. Department of Chemical and Biomolecular Engineering, The Ohio State University, 140 West 19th Ave., Columbus, OH 43210

Organic acid esters such as ethyl lactate belong to a safe, biodegradable and relatively new family of solvents. They can be used as intermediates in chemical reaction and solvents in ink formulations and cleaning applications. An integrated fermentation, extraction and esterification process was developed to produce ethyl lactate ester from ethanol and lactic acid produced during fermentation. Lactic acid was first extracted to the organic solvent containing Alamine 336 and 2-octanol, and then reacted with ethanol to produce ethyl lactate in the extractive solvent using immobilized lipase from Candida antarctica or Yarrowia lipolytica. Ethyl lactate was continuously produced in a plug-flow reactor for 7 days without a decrease in the outlet concentration, suggesting the immobilized lipase was stable. The effects of water content, lactic acid concentration, different alcohols and molar ratio of alcohol to lactic acid on the conversion and productivity of esterification were investigated and will be reported in this paper. The integrated fermentation, extraction and esterification process can economically produce various organic acid esters from biomass.

751e Effectiveness of Shear Protective Additives in Enzyme Production by Pelleted Phanerochaete Chrysosporium

Zhimin Li[1], Yan Liu[2], Zhanyou Chi[1], and Shulin Chen[1]. (1) Biological Systems Engineering, Washington State University, 213 LJ Smith Hall, Pullman, WA 99164, (2) Department of Biosystems & Agricultural Engineering, Michigan State University, 203 Farrall Hall, East Lansing, MI 48824

Ligninolytic enzymes including lignin peroxidase, manganese peroxidase, and laccase play an important role in biodegradation and bioleaching. Phanerochaete chrysosporium is one of the well studied basidiomycetes to produce these three enzymes. The production of these enzymes is often conducted with solid state fermentation system as the strain's performance is compromised by shear stress caused by agitation and gas sparging. Limitations with solid state fermentation are relatively low efficiency and productivity, the difficulty for scaling up, and production inconsistency. In addition, P. chrysosporium needs oxygen to produce peroxidase. It is also very difficult to supply adequate oxygen to the interior of solid medium, leading to another serious limitation to enzyme production which has been significantly improved using pelletized P. chrysosporium culture in our previous studies. In this study, several shear protective additives including Pluronic F-68, BSA, Tween 80 and PEG8000 were compared for MnP production by pelleted P. chrysosporium. The results have shown that these additives increased the enzyme production significantly except BSA. After the optimization of concentration, Pluronic F-68 improved the productivity and enzyme level by 180% in 125mL flask. Moreover, the production in 5L stirred tank was achieved successfully at the agitation speed 300rpm. Additionally, some protective mechanisms were investigated in the study.

751f Study of a Two- Stage Culture of Docosahexaenoic Acid Producing Marine Algae Schizochytrium Limacinum SR21 with Shifting Dissolved Oxygen Level

Zhanyou Chi, Yan Liu, Craig Frear, and **Shulin Chen**. Department of Biological Systems Engineering, Washington State University, Pullman, WA 99163

Docosahexaenoic acid (DHA, C22:6), an important omega-3 fatty acid, is known as an essential nutrient for the development of brain and eye for the fetal and infant. Production of DHA from marine microalgae is taking more and more important role for providing the increasing demand of DHA, since fish oil will be not sustainable in the future, and there is potential pollutant from the environment. To decrease the production cost, cull potato and crude glycerol from biodiesel industry has been used to culture S. limacinum SR21 to produce DHA. However, difficulty has been encountered when pursuing a high density biomass production. Reviewing the life cycle of S. limacinum SR21 found that this alga exists as two statuses: zoospores and vegetative cells. A phenomena was observed that the zoospores, swimming rapidly in the culture broth, like to gathering at the surface of air-liquid surface. Based on this, a hypothesis was come up that the limitation of biomass production could be caused by the low oxygen concentration in the culture and produced too less cells in this condition. This was proved in this work, since the culture of DO controlled at 50% produced 181 million cells/ml , but the culture with 10% DO produced a cell density only 98.4 million cells/ml. Even lower density of 22.5 million cells/ml was produced in the culture with a fixed agitation rate of 150 rpm. It was also found that the size of new generated vegetative cells and lipid accumulated vegetative cells in the later phase of the culture is significantly different, due to the lipid accumulation inside the cell. The experiments also showed that high oxygen as 50% led to pH decreasing and be harmful to the lipid accumulation, and low oxygen concentration was obligatory for lipid accumulating. Thus, the culture of Schizochytrium limacnum SR21 can be divided into two stages: (1) cell number increasing stage: cell reproduction and cell number increasing with little increase in the size and weight of each cell and (2) cell size increasing stage: cells stopped reproduction but cell size enlarged due to lipids accumulation. A shifting oxygen control strategy was used to produce more cells with dissolved oxygen controlled at 50%, and then providing a limiting oxygen concentration for fatty acid accumulation. With this protocol, the production of algae biomass and DHA was improved to 37.9 g/L and 6.56 g/L, respectively. This work revealed an important feature of oleaginous microorganisms that high oxygen concentration is demanded for the reproduction while low or limiting oxygen concentration is obligatory for the lipid accumulation. Shifting the oxygen level is necessary to obtain high production of biomass and lipids, since the optimal condition for lipid accumulation conflicts with cell proliferation if one stage culture was used. This two-stage culture process could be used not only in the omega-3 fatty acids production process, but also more widely, in all single cell oil (SCO) producing processes, including biodiesel production from algae, which is the largest potential of SCO application.

ANALYSIS OF DIFFERENT AFFINITY MEMBRANES FOR THE PRIMARY CAPTURE STEP IN ANTIBODY MANUFACTURINGS

Cristiana Boi, Simone Dimartino and Giulio C. Sarti
Alma Mater Studiorum Università di Bologna, Bologna, Italy

Introduction

Protein A affinity chromatography is the most widespread method for the primary capture step in antibody manufacturing [1,2] . Due to the high cost of this natural ligand, produced with recombinant *Staphyloccucus aureus*, the risk of leaching and the low stability in sodium hydroxide, that is commonly used as sanitizing agent, the interest towards new synthetic ligands is of increasing importance [3].

In this work, new affinity membranes prepared by immobilizing two different protein A mimetic ligands are considered for the capturing step of a monoclonal antibody production process. The membranes have been extensively tested with pure polyclonal IgG solutions and with a cell culture supernatant containing IgG_1. The effects of flow rate and IgG concentration in the feed on the separation performance parameters like binding capacity, selectivity and process yield have been studied in detail.

The behavior and efficiency of the different affinity membranes will be presented and will be also compared with data available for a commercially available Protein A affinity membrane.

Materials and methods

Affinity membranes

Sartobind epoxy membranes (Sartorius Stedim Biotech GmbH, Göettingen, Germany) are stabilized reinforced cellulose membranes with active epoxy groups for ligand coupling; these membranes have a nominal pore size of 0.45 μm, an average thickness of 275 μm and a void fraction of 0.68.

Two different ligands have been immobilized on the membrane supports: A2P and the new B14, both ligands, designed to bind the Fc region of the antibody, are alternatives to protein A produced by Prometic Biosciences Ltd (Cambridge, UK). A2P was designed for the capture of polyclonal antibodies, whilst B14 for the capture of monoclonal antibodies. These synthetic ligands are stable in sodium hydroxide up to 1 M concentration, as a consequence they are easy to clean and sanitize.

Affinity membranes have been prepared by the Analytical Chemistry group at the University of Vienna by immobilizing A2P and B14 ligands on the Sartobind epoxy membranes generating an optimized spacer arm, thus obtaining A2P-TA-Sartoepoxy and B14-TA-Sartoepoxy, respectively

Sartobind Protein A® (Sartorius Stedim Biotech GmbH, Göettingen, Germany) was used as a benchmark affinity membrane.

Feed solutions

A human polyclonal antibody, namely Gammanorm (Octapharma, Stockholm, Sweden), was used as a pure IgG feed solution. A cell culture supernatant (Excell Gene, Switzerland) containing human IgG_1 at a concentration of about 0.12 mg/ml was used as a source of monoclonal antibody.

Phosphate Buffered Saline, PBS pH 7.4, was used as equilibration and washing buffer; 0.1 M glycine pH 2.8, 0.05 M citric acid pH 2.5 and 0.05 M acetic acid were used as elution buffers. Regeneration was accomplished using a solution of NaOH 0.6 M.

Experimental methods

Batch experiments have been performed by soaking the membranes in a IgG solution kept in gentle agitation using an orbital shaker until equilibrium was reached. The protein concentration was measured at the beginning and at the end of each experiment and the amount of protein adsorbed was calculated with a material balance.

Dynamic experiments have been performed using an FPLC AKTA Purifier 100 controlled by Unicorn software (GE Healthcare, Milan, Italy). In both experiments the membranes were cut into circles of 2.5 cm in diameter; for the dynamic runs a layered stack of membranes was housed in a flow cell.

The measure of IgG concentration of pure solutions was determined by UV readings at 280 nm (Shimadzu UV-1601). For complex solutions IgG concentration was determined by HPLC using a protein A affinity cartridge (PA ID, Applied Biosystems, Monza, MI, Italy) mounted on a liquid chromatography system (Alliance 2695 equipped with a dual wavelength UV detector 2487, Waters Milano, Italy). The purity of the eluted fractions was analyzed with size exclusion HPLC using a SEC column (Proteema 300, PSS, Mainz, Germany).

Results

Batch experiments

Equilibrium adsorption experimental data, obtained in a batch system, were fitted using the Langmuir model and the relevant thermodynamic parameters are reported in table 1. The new affinity membranes are endowed with a higher binding capacity than the benchmark Sartobind Protein A membrane, as it can be observed in Fig. 1, in which the experimental data for the three membranes are reported together with the relevant Langmuir isotherms.

Table 1. Thermodynamic equilibrium parameters of the different membranes considered

Affinity Membrane	q_{max}		K_d
	($\mu g/cm^2$)	(mg/ml)	(mg/ml)
Sartobind Protein A	56.2	2.04	0.278
A2P-TA-Sartoepoxy	96.2	3.24	0.0444
B14-TA-Sartoepoxy	71.8	2.61	0.0540

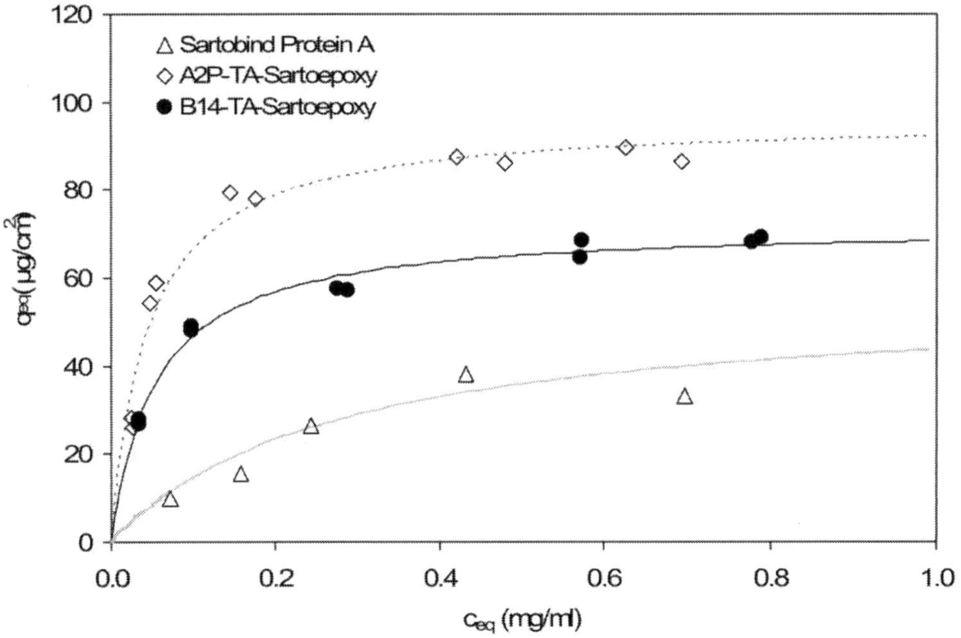

Figure 1. IgG equilibrium adsorption isotherms of three different affinity membranes

The recovery of the IgG adsorbed was very difficult from A2P-TA-Sartoepoxy membranes which are endowed with the highest binding capacity. Among the different elution buffers tested, acetic acid gave better performance, but only 59% of the IgG adsorbed was recovered. In Fig. 2 the amount of IgG eluted, using glycine and acetic acid, is reported versus the amount of IgG adsorbed.

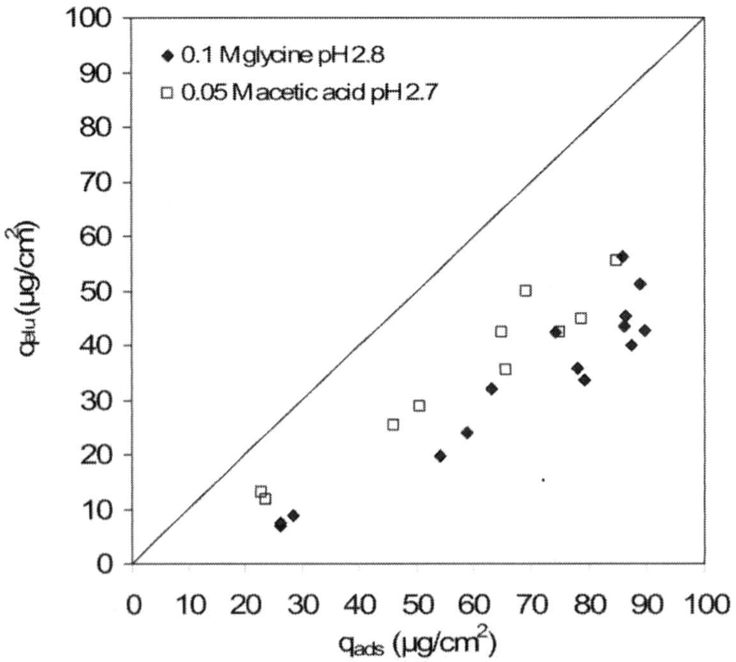

Fig. 2. Elution performance of A2P-TA-Sartoepoxy affinity membranes.

The strong binding between IgG and A2P ligand, has been already observed in previous work using different membranes [4] and is an important factor to consider in view of a possible industrial application of these membranes.

This behavior was not observed with membranes prepared with the new B14 ligand, for which good recoveries were obtained with all elution buffers, even if glycine, in this case, gave the best performance, with an average recovery of 92 %, as it can be observed in Fig. 3.

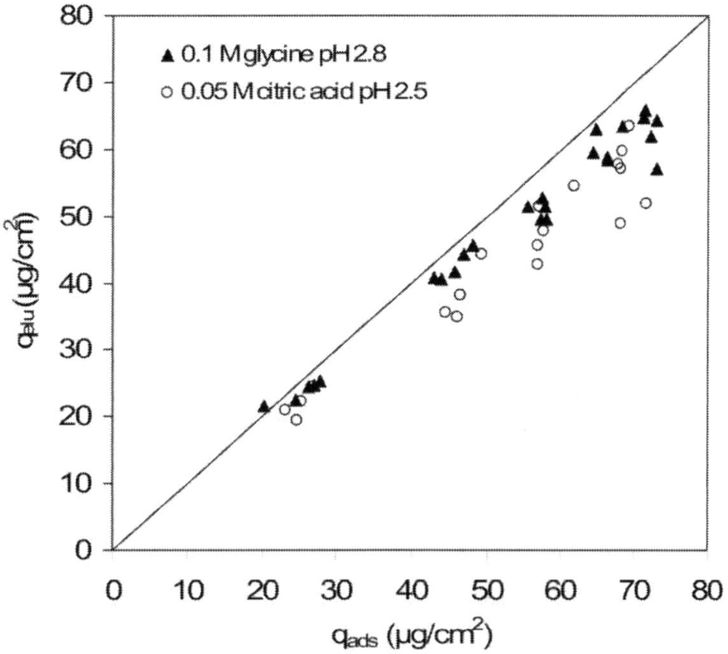

Fig. 3. Elution performance of B14-TA-Sartoepoxy affinity membranes.

Dynamic experiments

The affinity membranes were tested in dynamic experiments using pure IgG solutions and a cell culture supernatant which contains monoclonal IgG_1. With pure IgG solutions several runs were performed to study the effect of feed concentration and flow rate on membrane performance.

A comparison between A2P-TA-Sartoepoxy and B14-TA-Sartoepoxy membranes confirmed the results obtained in batch experiments, with A2P-TA-Sartoepoxy membranes endowed with higher dynamic binding capacity and lower recoveries, as it can be seen from the data reported in table 2.

Table 2. Results obtained in experiments with pure IgG at a feed concentration of 0.8 mg/ml

Affinity Membrane	$DBC_{100\%\ BT}$		Recovery
	(µg/cm^2)	(mg/ml)	(%)
A2P-TA-Sartoepoxy	70.8	2.57	57
B14-TA-Sartoepoxy	60.1	2.18	93

In experiments with the cell culture supernatant fractions were collected every milliliter and analyzed with both Protein A and SEC HPLC; in addition the most relevant fractions were analyzed also with SDS-PAGE gel electrophoresis under denaturing conditions. A typical run at a flow rate of 5 ml/min for a stack of five B14-TA-Sartoepoxy membranes is shown in Fig. 4.

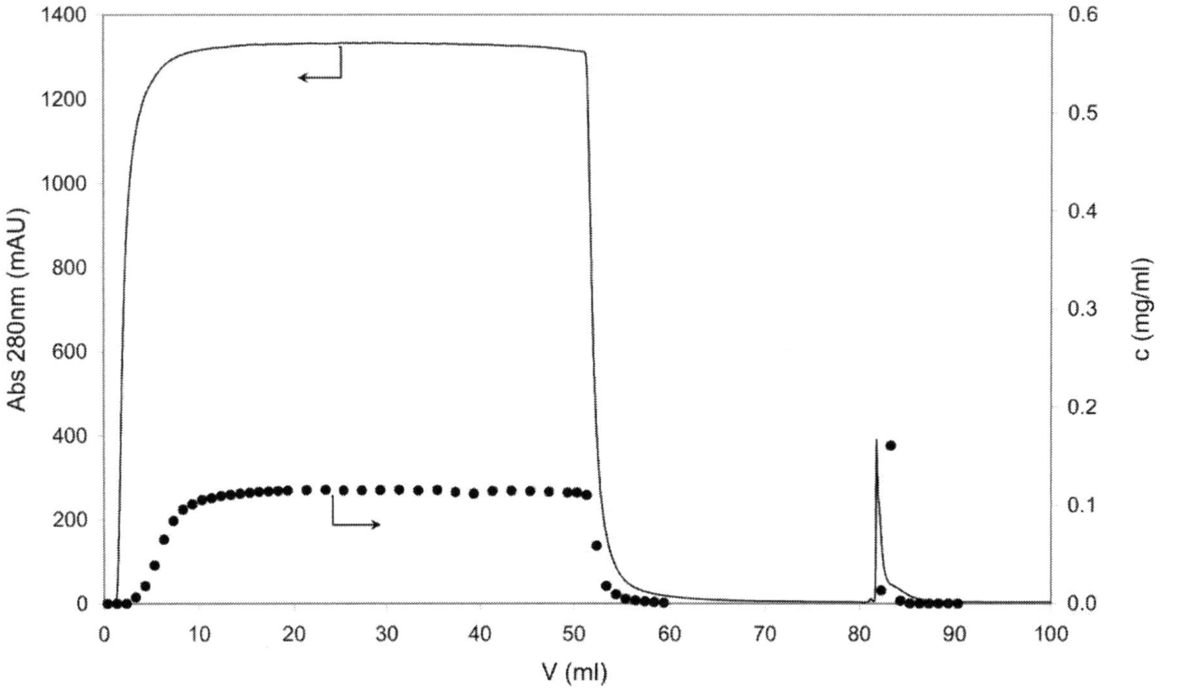

Fig. 4. Absorbance and IgG concentration profiles for experiments performed using the cell culture supernatant at a flow rate of 5 ml/min.

A direct comparison between the two membranes can be done by observing the SDS-PAGE electrophoresis that was run using fractions collected at the same volume for both membranes (Fig. 5). From the eluted fractions labeled E2, it can be observed that A2P-TA-Sartoepoxy membranes are not selective for IgG where IgG heavy chains can be barely detected, while for B14-TA-Sartoepoxy membranes both IgG heavy and light chains can be easily identified.

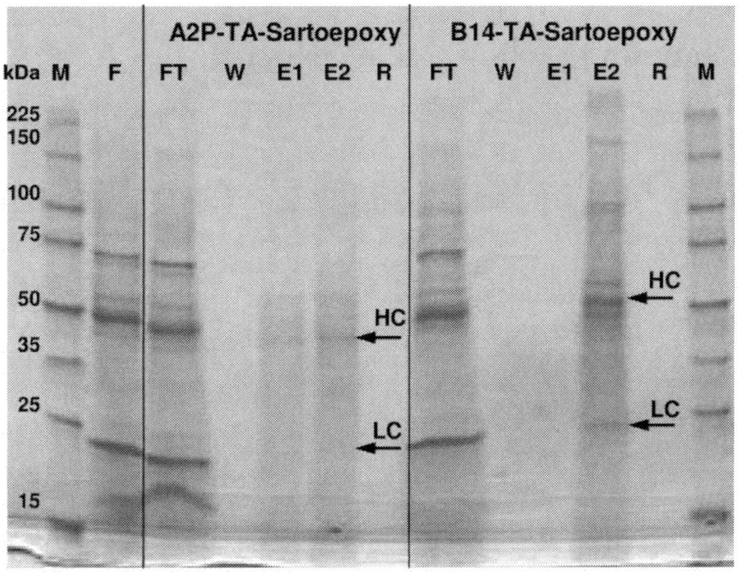

Fig. 5. SDS-PAGE electrophoresis under reducing conditions of the fractions collected during dynamic experiments performed with the cell culture supernatant. M: molecular weight markers; F: feed; FT: flow through; W: washing; E1 and E2: elution fractions; R: regeneration. HC labels IgG heavy chains and LC labels IgG light chains.

Conclusions

New affinity membranes have been experimentally characterized to determine their potential application in an IgG capture step of an antibody manufacturing process. The results obtained indicate that these new membranes are endowed with a good binding capacity for IgG, higher than the capacity of the commercially available Sartobind Protein A affinity membranes.

Although A2P-TA-Sartoepoxy membranes possess the highest binding capacity, B14-TA-Sartoepoxy membranes are superior in terms of IgG recovery and selectivity, thus they are worth further investigation to better evaluate their possible industrial application.

Acknowledgement

This work has been performed as part of the "Advanced Interactive Materials by Design" (AIMs) project, supported by the Sixth Research Framework Programme of the European Union (NMP3-CT-2004-500160).

References

1. Low, D.; O'Leary, R.; Pujar, N.S. (2007), "Future of Antibody purification". *J. Chromatogr. B 848*, 48-63.
2. Kelley, B. (2007), "Very large scale monoclonal antibody purification: The case for conventional unit operation", *Biotechnol. Prog., 23*, 995-1008.
3. Li R., Dowd V., Steward D.J., Burton S.J., Lowe C.R. (1998), "Design, synthesis and application of a protein A mimetic", *Nature biotech.*, 16, 190-195.
4. Boi C., Sarti G.C. (2007), "Development and characterisation of affinity membranes for immunoglobulin purification", *Separation Sci. Technol.*, 42, 2987 – 3001.

756b Polymeric Immunoglobulin Receptor-D1 Based Affinity Ligand for Igm Purification

Satyen Gautam and Kai Chee Loh. Department of Chemical and Biomolecular Engineering, National University of Singapore, 4 Engineering Drive 4, Singapore, 117584, Singapore

Recent application of immunoglobulin-M (IgM) for treatment of life-threatening diseases such as cancer and in stem cell research is based on the recognition between native or genetically engineered antibody and cell-surface receptors. Although naturally produced by the immune system, the need for IgM with unique specificities and designed for single application, has encouraged the search for novel IgM purification strategies. As purity requirements for the final antibody preparation are very stringent, particularly for in vivo applications, there has been the need to search for more efficient purification strategies. Among the available purification techniques, affinity chromatography is a promising technique as it allows for large purity gains in a single-step, reduction in non-specific interactions, increase in operational yields and facilitates the elimination of undesirable contaminants.

Our research aimed at investigating the suitability of peptidic ligands based on Domain 1 of the polymeric immunoglobulin receptor (pIgR-D1) for affinity purification of IgM molecules. pIgR is a glycosylated type I transmembrane protein, consisting of a 620 residue extracellular region, a 23 residue transmembrane region, and a 103 residue cytoplasmic tail. The extracellular region contains five domains (D1–D5) that share sequence similarity with Ig variable regions. Studies suggest that 1) binding between IgM and SC depends preferentially on a strong interaction with domain 1 of SC and 2) domain 1 of SC consists of three CDR (complimentary determining region) like loops and the region containing the CDR2-like loop (SSEGY) is most important in the pentameric IgM-binding process. These findings prompted us to investigate peptidic ligands incorporating the CDR-2 like loop, as affinity ligands for IgM purification.

In this regard, three peptides: pep12 (ITLISSEGYVSS), pep13 (ITLISSEGYVSSK) and pep14 (CITLISSEGYVSSK), incorporating the CDR-2 region of domain 1 were synthesized. Qualitative characterization of the binding of synthesized peptides to IgM, IgG1 and bovine serum albumin (BSA) was performed by Surface Plasmon resonance (SPR) biosensor assays on a CM5 chip. SPR results showed that pep14 bound specifically to IgM with no binding observed for IgG1 and BSA. On the other hand, pep12 and pep13 showed no binding to IgM, IgG1 and BSA. Circular dichroism (CD) studies were performed to ascertain the role of structural conformations for the differential binding behavior exhibited by the three peptides. CD results suggested a random coil structure for pep12 and pep13 while pep14 exhibited a mix of beta sheet and random coil conformation. The above study gives an insight to the crucial role of structural conformation of pep14 for IgM binding and establishes pep14 as an affinity ligand for IgM purification.

756c A Rational Approach for the Design and Synthesis of Selective Displacers for the Chromatographic Purification of Proteins

Srinavya Vutukuru[1], Sandesh D. Kate[1], Scott A. McCallum[2], Christopher J. Morrison[1], Steven M. Cramer[1], and Ravi S. Kane[1]. (1) Chemical and Biological Engineering, Rensselaer Polytechnic Institute, 110 8th Street, Troy, NY 12180, (2) Department of Biology, Rensselaer Polytechnic Institute, 110 8th Street, Troy, NY 12180

The development of effective strategies for designing selective displacers is critical for the successful implementation of displacement chromatography – a powerful technique for the purification of biomolecules. We have developed an affinity-based approach for designing selective displacers, based on the identification of a protein-binding moiety and its attachment to a resin-binding moiety via a linker. In particular, we synthesized displacers based on biotin, which selectively retained avidin as compared to aprotinin on SP Sepharose High Performance resin. In addition, we have extended this approach to develop an affinity-peptide-based displacer that discriminates between lysozyme and cytochrome c. Our approach may be broadly applicable for the design of selective displacers for many proteins of biological or pharmaceutical interest. Combinatorial methods such as phage display are well-established and enable the rapid identification of affinity ligands; the attachment of these ligands to resin-binding moieties allows the immediate design of selective displacers.

756d A Non-Chromatographic Elp-Intein Based Protein Purification Technique: Application to Mammalian Expression System

Wan-Yi Wu, Department of Chemical Engineering, Princeton University, Rm A217 Engineering Quadrangle, Princeton, NJ 08544 and David Wood, Department of Chemical Enginnering, Princeton University, Rm A417 Engineering Quadrangle, Princeton, NJ 08544.

Inteins are self-splicing protein elements that can be engineered for different purposes, one of which is the development of self-cleaving purification tags. In previous work, we combined a pH-sensitive self-cleaving intein with a thermally-responsive elastin-like polypeptide (ELP), to create a very simple method for non-chromatographically purifying target proteins expressed in E. coli.

For complex human therapeutics protein productions, however, mammalian expression systems are typically required for more complicated post-translational processing ability. In this work, we demonstrate that ELP-tagged target proteins can be successfully secreted by Chinese Hamster Ovary (CHO) cells. Premature intein cleavage was observed as expected since the pH and temperature required for CHO expression is permissive for the cleaving reaction. Thus finding an alternative to solve the premature cleaving problem is a key goal of our current work. The strategies will be discussed, as well as the potential for the ELP-intein-based technique to provide a rapid, flexible, and inexpensive approach to purify CHO-expressed recombinant proteins.

756e Programmable Isoelectric Chromatofocusing and Its Application to Targeted Proteomics and Preparative Purification of Proteins

Derek Y.C. Choy, Chemical and Biological Engineering, Michael Smith Laboratories, University of British Columbia, 2185 East Mall, Vancouver, BC V6T 1Z4, Canada and Charles Haynes, Michael Smith Laboratories, University of British Columbia, 2185 East Mall, Vancouver, BC V6T 1Z4, Canada.

Chromatofocusing (CF), which primarily separates analytes on the basis of differences in isoelectric point (pI), theoretically achieves peak capacities greater than ion exchange or size exclusion chromatography, making it attractive both as a first separation dimension in multidimensional liquid chromatography platforms for resolving proteomes and as an efficient mode for high-resolution protein purification at the preparative scale. However, reaching this theoretical limit with complex protein mixtures requires improvements in chromatofocusing theory and related methodologies. Toward this goal, we have developed a model and new method for designing CF loading and elution buffers such that the buffering components can be added in a programmed manner to sculpt elution pH gradients over the range of pH 10 to 3.5, and thereby maximize peak capacity on a strong anion exchange column. The method uses small volatile buffering species as opposed to larger polyampholyte molecules, allowing for the facile removal of salts that may interfere with downstream analytical steps such as mass spectrometry. The use of a strong anion exchange column permits application to both bottom-up and top-down proteomics applications. When this chromatofocusing technique is applied with a linear pH gradient to the non-denaturing separation of the blood plasma proteome, superior fractionation performance is observed relative to previously reported chromatofocusing strategies and to traditional anion exchange chromatography. High resolution is also achieved in the non-denaturing fractionation of other proteomes, including cell extracts from E. coli, mouse C2C12, human HEPG2, MMAN and platelet cells; as well as culture media and vesicle extracts for various strains of Cryptococcus, demonstrating the power of optimized CF for protein purification, targeted proteomics or general protein inventory studies. Finally, our programmable CF technology can be applied to individual proteins, providing a means to selectively capture and purify therapeutic products from cell culture on the basis of small differences in pI and gradients in net change with pH near the target protein's pI. Linear velocities as high as 20m/h can be applied to these separations, as demonstrated by our studies on the purification of recombinant human transferrin from cultures of the methylotrophic yeast Pichia pastoris.

Post-Viral Inactivation Filtration for Impurity Removal in Biopharmaceutical Processes

Jeffrey E. Galvin, David Maheu & Sushil Abraham
Amgen Inc., 4000 Nelson Rd., Longmont, CO 80503

Abstract

Viral inactivation is a critical bioprocessing step that employs chemical treatments such as pH titration or addition of surfactants to inactivate enveloped viral contaminants. A filtration step often follows the viral inactivation to reduce the turbidity that results from the chemical treatment. This paper describes the development and characterization of a post-viral inactivation filtration step for a biopharmaceutical process. In addition to reducing turbidity, the filtration step also removes impurities such as host cell proteins and nucleic acids. Although the purification process already shows sufficient capability to remove these impurities, the post-viral inactivation filtration significantly reduces the burden on downstream steps, thereby increasing overall process robustness and consistency.

Introduction

Several types of filters have been characterized for removal of turbidity resulting from neutralization of the viral inactivated pool in biopharmaceutical purification. In early development of a monoclonal antibody (MAb) purification process, it was observed that a single-stage membrane filter quickly became clogged with impurity particles, leading to the introduction of a two-stage filtration train consisting of a prefilter followed by a membrane filter. Several pre-filter types were characterized, and depth filter media was found to be the most effective pre-filter for this stream. In particular, composite filters containing both a pre-filter depth media (i.e. glass fiber or silica) followed by a membrane filter were found to provide the most robust performance in terms of throughput.

Filter characterization studies focused on identifying the filter area required to complete the filtration in a reasonable processing time while avoiding replacement of the filters mid-process due to high pressure drop. These studies were conducted using traditional filter sizing methods such as constant pressure (Vmax) and constant flow (Pmax). Additional studies looked at the ability of the two-stage filtration train to remove turbidity as well as reduce process impurities such as DNA, host cell (Chinese hamster ovary) proteins (CHOP), and high molecular weight species (HMWS).

One aspect of the step that has not been fully characterized is the mechanism for removal of impurities by the filtration train. Two potential mechanisms are removal of impurity particles by size exclusion and adsorption of impurities on the surface of the filter media. The purpose of this paper is to investigate impurity removal mechanisms by comparing a characterization case study to historical data.

The case study focused on DNA and CHOP because of the importance of the VI filtration step in removing these impurities. Although about 98% of DNA in the harvest filtrate is removed by Column 1, an additional 4-5 logs of DNA clearance is normally achieved across the VI filtration step (Figure 1). CHOP is also reduced by 1-2 logs across the step (Figure 2). Subsequent chromatography steps, however, have been shown in a worst-case study to be able to remove an additional 1-2 logs of both CHOP and DNA, adding robustness to the overall purification process.

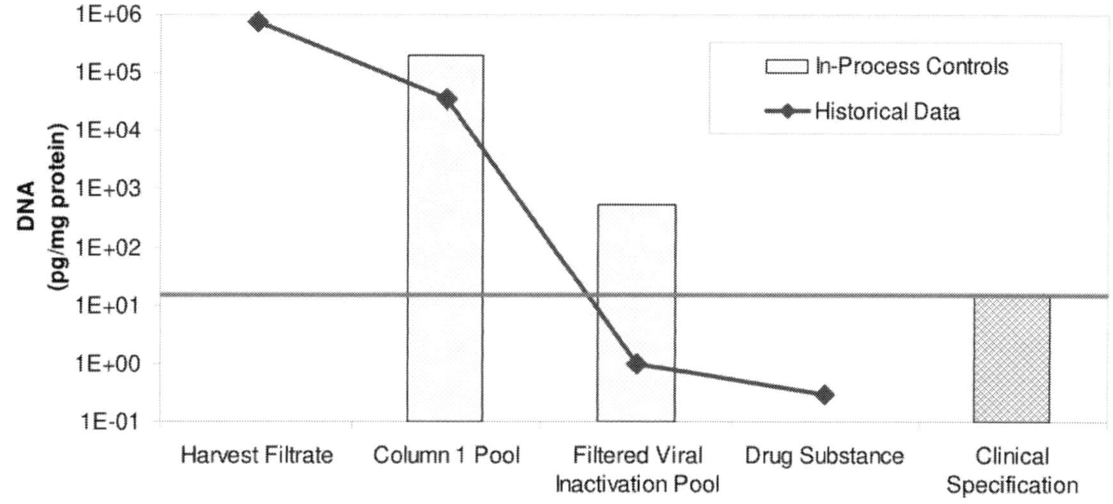

Figure 1. Process Clearance of DNA [pg/mg protein]

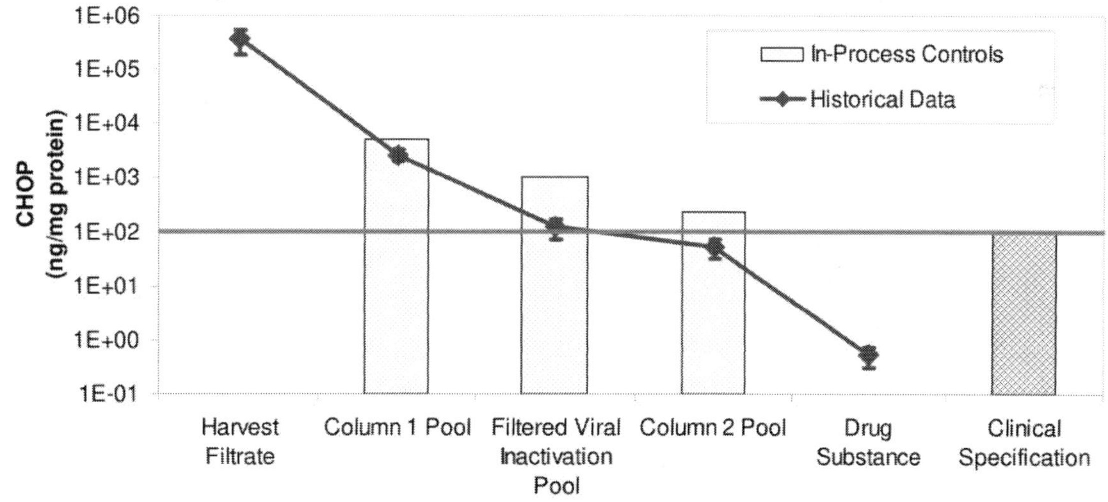

Figure 2. Process Clearance of CHOP [ng/mg protein]

Materials and Methods

Column 1 pool generated in this experiment was virally inactivated according to the process summarized in Figure 3. A buffer addition was performed after acid titration to simulate the buffer chase after transfer from titration tank to hold tank. However, in a departure from the standard process, a buffer with higher ionic strength was used for the chase dilution. The use of a higher ionic strength buffer increased the ionic strength and conductivity of the pre-filtered pool above what is normally observed.

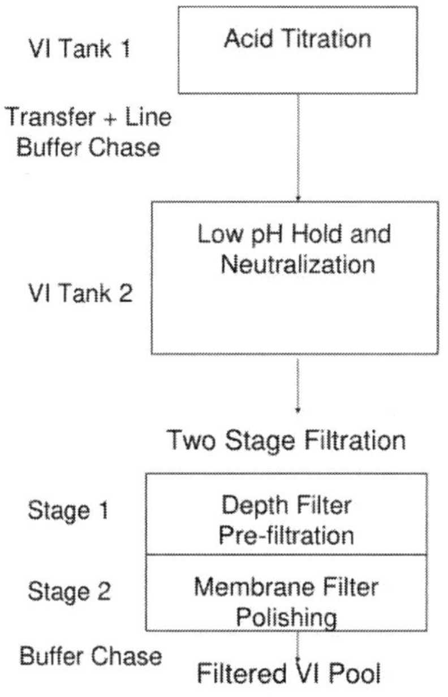

Figure 3. Viral Inactivation Process Flow Diagram

The resulting neutralized viral inactivated pool was filtered over three separate filter trains in parallel. Train 1 used a smaller version of the stage 1 depth filter, while Train 2 used a different stage 1 depth filter with similar materials of construction. Train 3 was a control. All three trains used a dual-layer membrane filter as the stage 2 filter. Flow through samples were taken every 2 liters and sent for CHOP and DNA analysis.

Results

Table 1 summarizes the post-VI filtration performance results. In all trains, CHOP and DNA (Figures 4 and 5) were significantly higher than representative historical data. Turbidity for the pre-filtered pool was lower than typically observed prior to filtration, while post-filtration turbidity was comparable (Table 1). Pool conductivity was elevated due to the use of the high ionic strength buffer in the chase step. Results from the different stage 1 filters used in Trains 1 and 2 were comparable.

Table 1 Summary of Post-VI Filtration Results vs. Historical Data

	Turbidity [NTU]		Conductivity [mS/cm]	CHOP [ng/mg]	DNA [pg/mg]
	Pre-filtration	Post-filtration (FVIP)	FVIP	FVIP	FVIP
Train 1	67.8	5.88	5.53	1529	472
Train 2		7.17	5.46	1610	304
Train 3		3.61	5.53	685	42
Historical Data mean ± SD	189 ± 33	1.99 ± 0.20	2.02 ± 0.27	290 ± 86	<1

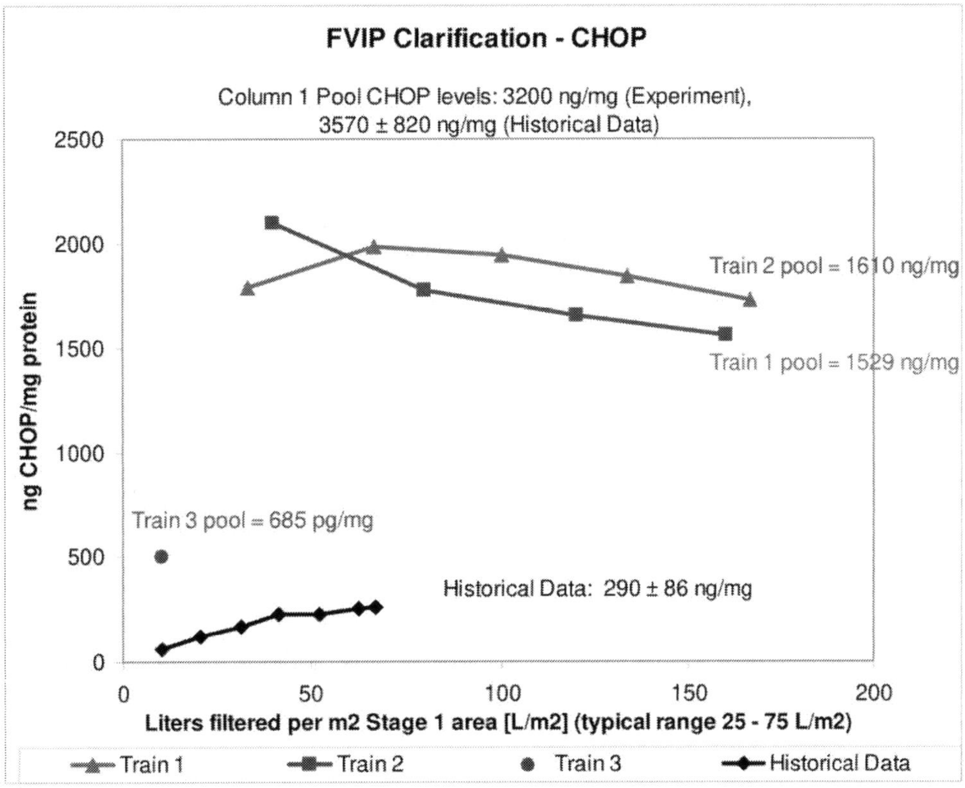

Figure 4. CHOP Flow Through for Experiment vs. Historical Data

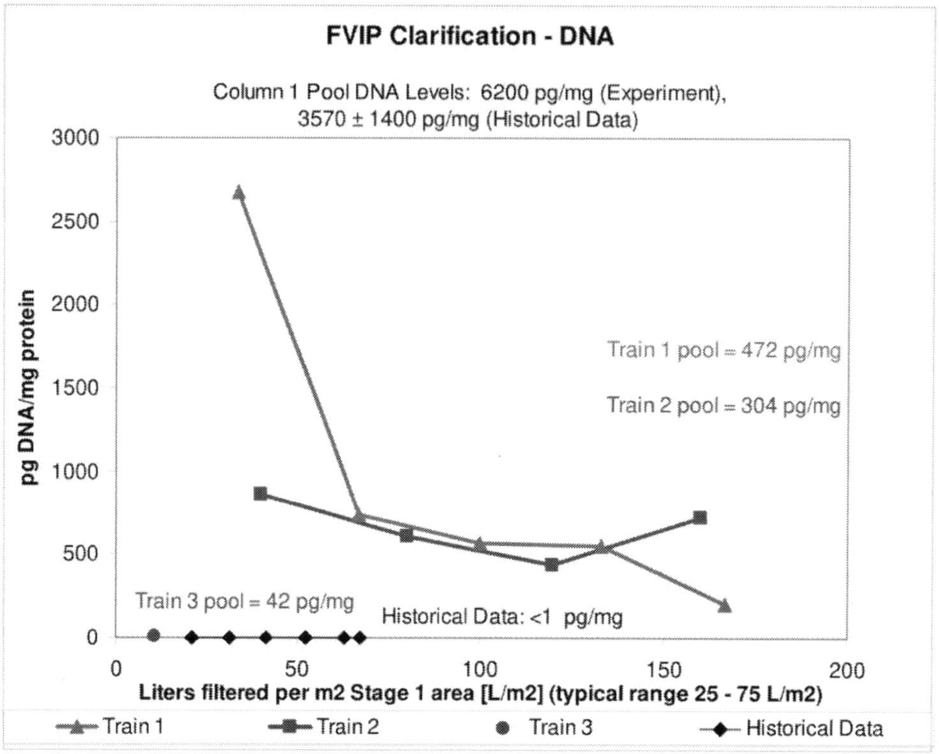

Figure 5. DNA Flow Through for Experiment vs. Historical Data

Discussion

As shown in Figures 4 and 5, removal of CHOP and DNA was much less effective in this experiment compared with historical data. Flow through sampling indicated that filter overloading was not the root cause. The pre-filtered material was less turbid (Table 1) than usual, and the pool conductivity was higher than usual. The low turbidity suggested a smaller particle size distribution, which may have reduced size exclusion removal efficiency. The high ionic strength, indicated by high conductivity, may have reduced the extent of adsorption of small impurity particles on the filter media. Ionic strength could have also impacted the extent of precipitation of impurity particles.

Although some impurity removal may be realized at this step, in general the viral inactivation clarification filter area requirement should not be based on achieving a target impurity removal. The adsorptive properties of the filters are not sufficiently controlled by the filter vendors to be considered reproducible. Therefore, the process should be developed to minimize the reliance on this step for impurity removal. For a robust biopharmaceutical process, the downstream chromatography steps should have sufficient capacity to remove elevated levels of CHOP and DNA, as is the case for this process (Tables 2 and 3).

Table 2. CHOP Levels for Downstream Steps vs. Historical Data and Worst Case Study

	CHOP [ng/mg protein]		
	FVIP	Column 2 Pool	Column 3 Pool
Experiment	1424	147	
Historical Data mean ± SD	290 ± 86	111 ± 26	
Worst Case Study	72	45	< 0.52

Table 3. DNA Levels for Downstream Steps vs. Historical Data and Worst Case Study

	DNA [pg/mg protein]		
	FVIP	Column 2 Pool	Column 3 Pool
Experiment	578	91	
Historical Data	<1.22	<0.85	
Worst Case Study	554	142	< 1

Conclusions

Impurity removal in the post-viral inactivation filtration can significantly reduce the burden on downstream steps, but performance in this step can be sensitive to process conditions. However, a robust purification process will demonstrate sufficient capability to remove impurities without relying on the filtration step, therefore aligning with best practices for MAb processes.

756g Sulphated Cellulose Membranes – An Economic Virus Capturing Method for Influenza Virus Purification

Lars Opitz, Nadine Solf, Udo Reichl, and Michael W. Wolff. Bioprocess Engineering, Max Planck Institute for Dynamics of Complex Technical Systems, Sandtorstrasse 1, Magdeburg, 39106, Germany

Influenza is one of the most worldwide spread diseases. Besides antiviral medical treatments, prophylactic vaccinations are crucial to control seasonal influenza epidemics. Hence, every year large amounts of vaccine doses have to be produced. Conventionally, embryonated chicken eggs are used for human influenza vaccine production. This production process has only a limited scalability. In addition, these vaccines contain egg derived proteins which may cause allergic reactions. Hence, cell culture based vaccine production processes have been developed, which require an adapted downstream processing strategy. One unit operation, which is often used in influenza virus purification, is column chromatography based on cellufine® sulphate. The main disadvantage of this method for this application is the limited flow rate due to the high back pressure, leading to a suboptimal productivity rate of the process.

The present study describes a capturing method for cell culture derived influenza viruses based on sulphated reinforced cellulose membranes. Purification studies have been done using different MDCK cell derived influenza virus strains: A/Wisconsin/67/2005 (H3N2), A/Puerto Rico/8/34 (H1N1) and B/Malaysia/2506/2004. Purification efficiency concerning viral yield as well as total protein and host cell dsDNA reduction using sulphated membranes was directly compared to commercially available cation exchange adsorbers and to column based cellufine® sulphate resin. The modified membranes achieved high product recoveries and contaminant reduction. Due to a fast binding kinetic and a low back pressure, membrane adsorbers enable to operate the capturing process at an increased flow rate. Hence, the productivity can be significantly enhanced making them to a valuable choice for industrial influenza vaccine production processes.

765a In Silico Modeling of Uptake and Metabolism of Arsenicals In Hepatocytes with Implications for Hepatotoxicity

S.K. Stamatelos, C.J. Brinkerhoff, A.F. Sasso, S.S. Isukapalli, and P.G. Georgopoulos. Environmental and Occupational Health Sciences Institute, University of Medicine and Dentistry of New Jersey (UMDNJ)/Robert Wood Johnson Medical School (RWJMS), 170 Frelinghuysen Road, Piscataway, NJ 08854

Humans are continuously exposed to arsenic, as it is very often present in drinking water (typically as an inorganic ion) as well as in various foods (mostly in organic forms) [1]. Arsenic is a known carcinogen and its ability to induce various types of cancer, such as skin, liver, bladder and lung cancer has been demonstrated in various epidemiological studies involving the inhalation and ingestion exposure routes. Arsenic causes oxidative stress by increasing production of Reactive Oxygen Species (ROS) [2]. Furthermore, low doses of arsenic in mice during gestation have been associated with gene expression changes related to steroid pathways as well as with liver cancer in the adult stage [3].

Inorganic arsenic in water is mainly present in the form of arsenate (iAsV), which is uptaken slowly by cells, as it is negatively charged at physiological pH. Arsenate in the blood is quickly converted to arsenite (iAsIII) [4].

The goal of the effort presented here is to examine in silico specific hepatic effects of arsenic exposures. The first step is the quantitative characterization of the time-course dynamics of different chemical forms of arsenic within human hepatocytes. This is pursued through the development of a "cellular level pharmacokinetic (PK) model" that simulates the kinetics of various "arsenic species," i.e. arsenite (iAsIII), monomethylated As (MMA) and dimethylated As (DMA), inside hepatocytes. The PK model considers iAsIII uptake by hepatocytes via aquaporin isozymes 9 (AQP9's) [5] and metabolism through a series of methylation and glutathione conjugation reactions [6]. MMAIII is effluxed via AQP9's [7] while DMA simply diffuses across the cell membrane. Glutathione adducts of arsenicals are effluxed via multidrug resistant proteins 1/2 (MRP's) [8]. The parameterization of the model utilizes primarily experimental data from human hepatocytes exposed to arsenite in vitro [9]. However, the kinetic structure employed in this model is also able to replicate experimental results involving mouse hepatocytes exposed to arsenic [10]. Bayesian Markov Chain Monte Carlo (MCMC) simulations and systematic sensitivity runs have been conducted in order to develop estimates of the distributions of model parameters and to characterize essential features of the behavior of the model under different conditions. These parameterizations have then been used in evaluating the PK model with respect to dose-extrapolations, and for incorporation into whole-body Physiologically Based Toxicokinetic (PBTK) models of arsenic.

On-going and future components of this research effort involve the coupling of this PK model with various endpoint-specific pharmacodynamic (PD) models for the cellular-level effects of arsenic species. One specific initial objective within this effort involves the development of hypotheses and corresponding computational models, such as indirect response PD models [11], that will link the levels of arsenic species within hepatocytes, and especially of the methylated arsenic forms, with ROS profiles and induced DNA damage. This PK/PD modeling approach is going to provide linkages among exposure to a toxicant, target tissue dosimetry and effects in the subcellular level.

References

1. Georgopoulos, P.G., et al., Biologically based modeling of multimedia, multipathway, multiroute population exposures to arsenic. J Expo Sci Environ Epidemiol, 2007.

2. Qian, Y., et al., Cdc42 regulates arsenic-induced NADPH oxidase activation and cell migration through actin filament reorganizatlion. Journal of Biological Chemistry, 2005. 280(5): p. 3875-3884.

3. Liu, J., et al., Global gene expression associated with hepatocarcinogenesis in adult male mice induced by in utero arsenic exposure. Environmental Health Perspectives, 2006. 114(3): p. 404-411.

4. Vahter, M. and E. Marafente, Reduction and binding of arsenate in marmoset monkeys. Archives of Toxicology, 1985. 57(2): p. 119-124.

5. Liu, Z.J., et al., Arsenite transport by mammalian aquaglyceroporins AQP7 and AQP9. Proceedings of the National Academy of Sciences of the United States of America, 2002. 99(9): p. 6053-6058.

6. Hayakawa, T., et al., A new metabolic pathway of arsenite: arsenic-glutathione complexes are substrates for human arsenic methyltransferase Cyt19. Archives of Toxicology, 2005. 79(4): p. 183-191.

7. Liu, Z.J., M. Styblo, and B.P. Rosen, Methylarsonous acid transport by aquaglyceroporins. Environmental Health Perspectives, 2006. 114(4): p. 527-531.

8. Kala, S.V., et al., Formation and urinary excretion of arsenic triglutathione and methylarsenic diglutathione. Chemical Research in Toxicology, 2004. 17(2): p. 243-249.

9. Styblo, M., et al., Metabolism of arsenic in primary cultures of human and rat hepatocytes. Chemical Research in Toxicology, 1999. 12(7): p. 560-565.

10. Kedderis, G.L., et al., Kinetics of arsenic methylation by freshly isolated B6C3F1 mouse hepatocytes. Chemico-Biological Interactions, 2006. 161(2): p. 139-145.

11. Sharma, A. and W.J. Jusko, Characteristics of indirect pharmacodynamic models and applications to clinical drug responses. British Journal of Clinical Pharmacology, 1998. 45(3): p. 229-239.

765b Cell Migration Paths of Epithelial Cells Resemble Levy Modulated Correlated Random Walk Pattern

Alka A. Potdar[1], Junhwan Jeon[1], Alissa M. Weaver[2], and Peter T. Cummings[1]. (1) Department of Chemical Engineering, Vanderbilt University, Nashville, TN 37235-1604, (2) Cancer Biology, Vanderbilt University Medical Center, 448 Preston Research Building, Nashville, TN 37232

Cellular tracks of MCF-10A pbabe, neuN and neuT random migration on 2-D plastic substrates, analyzed using *Bimodal analysis*, were found to reveal a Levy modulated correlated random work pattern. We find two types of Levy flights (corresponding to the directional and re-orientation phases) having a specific form of turn angle distribution representing correlation between move step lengths within flights. The Levy exponent, μ , for both the power-law distributed flights was found to be close to 2. This is reminiscent of foraging strategies adopted by various organisms from sea creatures, bumble-bees to micro-zooplankton in randomly distributed nutrient conditions where Levy walk has been found to be an optimum strategy to enhance encounter with food. The presence of non-uniform turn angle distribution of move-step lengths within a flight and the power-law distributed flights indicate that the epithelial cells might be exhibiting a Levy modulated correlated random walk (LMCRW). Moreover, the early super diffusive nature of mean squared displacement of the cells confirm the Levy modulation.

765c Predicting Translation Initiation Rate from RNA Sequence

Howard Salis[1], Ethan Mirsky[2], and Christopher Voigt[1]. (1) Department of Pharmaceutical Chemistry, UC San Francisco, 1700 4th St., 408 Byers Hall, San Francisco, CA 94158, (2) Biophysics graduate program, UC San Francisco, 1700 4th St., 408 Byers Hall, San Francisco, CA 94158

The reliable genetic engineering of bacterial systems would be greatly aided by a more quantitative and predictive understanding of gene expression. The production rate of a protein can be precisely tuned by varying the mRNA sequence of the 5' UTR (untranslated region), including the ribosome binding site (RBS), that precedes the protein's coding sequence. However, there is currently no quantitative model that can predict an RNA sequence that yields a desired production rate of protein.

We have developed and experimentally validated a statistical thermodynamic model of translation initiation in Escherichia coli. We combine the quantitative model with Monte Carlo sampling to create a design method that generates synthetic 5' UTR RNA sequences. The user inputs a protein coding sequence and a desired translation initiation rate and the design method will generate a RNA sequence accordingly. We measure the accuracy of the design method by generating numerous synthetic 5' UTRs that drive the expression of the fluorescent protein RFP in a simplified test system. We then compare the design method's predictions with flow cytometry data. The design method is capable of accurately generating RNA sequences that yield the desired translation rate, with rates varying across three orders of magnitude.

The presented design method enables the rational tuning of the production rates of one or more proteins in a synthetic or natural biological system and speeds up the optimization of metabolic pathways or genetic networks. With the emergence of lower cost large-scale DNA synthesis and the rapid assembly of genetic systems, the development bottleneck now shifts towards identifying the DNA sequence that ultimately yields a desired system behavior. These and other quantitative and predictive models aim to provide that much-needed missing link.

765d Design Principles for Riboswitch Function

Chase L. Beisel, Chemical Engineering, California Institute of Technology, Pasadena, CA 91125 and Christina Smolke, Chemistry and Chemical Engineering, California Institute of Technology, 236A Spalding, Mail Code 210-41, Pasadena, CA 91125.

Modular composition frameworks support the rapid assembly of biological components into devices that can be implemented toward the most pressing research challenges facing society. Implicit in these frameworks is the ability to modify individual component behavior to predictably tune and program device performance. However, design principles are needed to guide component modification and accurately predict their effect on device performance. We were interested in delineating design principles for the rapid tuning and programming of RNA-based devices. RNA is a versatile building block for device construction based on a variety of inherent properties: a manageable size of sequence space, predictable folding and sequence recognition through base pairing interactions, diverse regulatory mechanisms, catalytic activity, and specific and tight binding of various biomolecules. By utilizing these properties, researchers have successfully constructed RNA-based information processing devices called riboswitches that program cellular behavior through detection of user-defined signals in the intracellular environment. As the identity of the inputs and outputs are only loosely constrained, synthetic riboswitches have tremendous potential as in vivo biosensors and autonomous control systems. Composition frameworks for RNA device assembly from modular functional components are currently being developed, where recent studies have offered a few design principles. A critical factor that remains to be evaluated is the role of kinetics and thermodynamics in riboswitch performance. Inherent processes such as ligand binding or transcription rate have a profound impact as shown experimentally, although kinetic factors are not currently considered as part of device construction. To address this issue, we investigated riboswitch behavior in silico to develop a quantitative link between the dynamics of riboswitch behavior and overall performance toward gaining further insights into the design and programming of riboswitch function.

To draw general conclusions regarding riboswitch function, we considered three representative regulatory mechanisms: transcriptional termination, translational repression, and mRNA destabilization. Modeling results suggest that riboswitch performance is highly dependent on the competition between irreversible rates such as mRNA degradation or transcriptional termination and reversible rates such as conformational switching or ligand binding. When reversible rates dominate (thermodynamically-driven), riboswitch performance is independent of the selected regulatory mechanism. When the rates are balanced (kinetically-driven), riboswitch performance generally decreases, transcriptional folding becomes important, and performance depends on the selected regulatory mechanism. When irreversible rates dominate (non-functional), riboswitch performance is insensitive to the input signal. In addition, imposing an upper limit on the input signal reduces the maximal dynamic range and establishes a bounded parameter space for optimal riboswitch performance. Model predictions are supported by published experimental data and physical modification of a synthetic riboswitch.

From our modeling efforts, we arrived at multiple design principles to guide the construction and modification of synthetic riboswitches. These principles address selection and modification of device components as well as formulation of composition frameworks. Current work is focused on applying these principles to improving the performance of a synthetic riboswitch actively used in our group for biomedical and biotechnological applications.

765e Intravital Quantification of Tissue Fluid Balance Changes In Lymphedema

Joseph M. Rutkowski and Melody A. Swartz. Institute of Bioengineering, École Polytechnique Fédérale de Lausanne (EPFL), Batiment AAB; Station 15, Lausanne, 1015, Switzerland

Hydraulic conductivity, as the physical tissue property that governs interstitial flow and transport in all tissues, is a potentially controlling factor in processes of tumor and immune cell trafficking, drug and vaccine delivery, and cell and tissue morphogenesis. Quantification of tissue hydraulic conductivity is therefore useful when exploring these processes, but, to date, determining this parameter in vivo has been challenging. Primary lymphedema is a congenital pathology of dysfunctional lymphatic drainage characterized by swelling of the limbs, thickening of the dermis, and fluid and lipid accumulation in the underlying tissue. We hypothesized that lymphedema presents an excellent model to demonstrate how matrix remodeling dictates the hydraulic conductivity of the tissue. We utilized two mouse models of lymphedema, the Chy mouse and the K14-VEGFR-3-Ig mouse, which exhibit dermal lymphedema symptomatically similar to the human condition resulting from deficient vascular endothelial growth factor receptor (VEGFR)-3 signaling and a subsequent lack of dermal lymphatic capillaries. Using a simple, reproducible, intravital technique, we infused a fluorescently-labeled dextran intradermally in the mouse tail at low pressures and quantitatively determined the functional interstitial transport implications resulting from these mutations. We found that despite similarities in increased skin hydration and elevated interstitial fluid pressure, dermal tissue adaptations were significantly different in the two mouse models. Chy mouse skin possessed much higher levels of collagen and fat, effectively densifying the matrix, minimizing interstitial flow, and normalizing hydraulic conductivity. Conversely, while the fat and collagen content in K14-VEGFR-3-Ig skin was relatively normal, the tissue conductivity was greatly increased due to the increased hydration. Thus, we hypothesize that the Chy mouse has adapted to a lack of initial lymphatics by decreasing interstitial transport, while the K14-VEGFR3-Ig mouse has adapted to increase or facilitate interstitial flow over longer distances. These responses to reduced lymphatic drainage (via missing initial lymphatics) suggest that tissue remodeling in lymphedema is not simply a consequence, but an adaption to limit tissue swelling by attempting to normalize the interstitial fluid balance.

765f Validating a Dimensionless Number for Glucose Homeostasis in Humans

David Klinke, Department of Chemical Engineering, West Virginia University, Morgantown, WV 26506-6102

Background: Understanding type 2 diabetes is challenged by the diversity of patient phenotypes. Translating data across species and among individuals is a barrier for understanding the genetic loci that underpin this multifactorial disease. Dynamic scaling, based upon dimensional analysis, is a common technique in engineering used to translate data among different systems. The objective of this study was to gain qualitative insight into underlying biological phenomena responsible for glucose homeostasis using dimensional analysis.

Methods: A dimensionless number was derived using variables involved in the production of insulin and in the sensitivity of glucose metabolism to insulin. The resulting dynamic scaling relationship was validated against patient data obtained for over 2000 individuals that range in phenotype from normal to severe type 2 diabetes. Individuals were identified in the third National Health and Nutrition Evaluation Survey.

Results: Patient groups clustered in different regions based upon the severity of clinical symptoms. The cross-sectional comparison among patient groups shows that progression from normal to clinical onset of type 2 diabetes is consistent with a decrease in the sensitivity to insulin. However, a non-linear change in insulin production capacity was associated with an increase in severity in diabetic symptoms (i. e., normal to clinical diagnosis of type 2 diabetes).

Conclusions: This dimensionless number provides a method for discriminating between patient groups from first principles. By analogy with other dimensionless numbers, this number may be used to monitor basic physiological variables responsible for glucose homeostasis. In addition, similarity in dynamic trajectories could provide a criterion for selecting relevant animal models for diabetes.

767a Electrochemical Hydrogenation of Soybean Oil Using Transition Metal Carbide Based Electrocatalysts

Adam Lausche, Saemin Choi, Kanako Okada, and Levi Thompson. Chemical Engineering, University of Michigan, 3074 H.H. Dow, 2300 Hayward, Ann Arbor, MI 48105

More than 8,300,000 metric tons of partially hydrogenated vegetable oil are produced in the United States annually. Most of this oil is processed using a high temperature, high pressure thermochemical process with hydrogen gas and a solid nickel catalyst. Partially hydrogenated vegetable oils produced using these conventional methods typically contain significant amounts of trans-fatty acids (TFA). These TFAs have been linked to coronary artery disease. Recently an electrochemical approach has been reported for the partial hydrogenation of soybean oil. This process was carried out at low temperatures using a Pd black cathode electrocatalyst and resulted in a significant reduction in the formation of TFAs. Research described in this paper investigated the use non-noble metal catalysts for the electrochemical hydrogenation of soybean oil. Early transition metal carbides have demonstrated good activities and selectivities for this reaction. The experiments were carried out using a reactor that resembled a proton exchange membrane fuel cell. A Pt-black anode was used with Nafion 117 to produce the membrane electrode assemblies (~5 cm2), along with an early transition metal carbide or Pd-black based cathode. The activities and selectivities were evaluated at 60-80 °C, with a current density of 0.16 A/cm2 and oil flow rate of 100 mL/min. The oils were characterized using gas chromatography (according to American Oil Chemists' Society standards of fatty acid methyl esters) and iodine value measurements. The electrocatalysts were characterized using x-ray powder diffraction, BET surface area analysis, and x-ray photoelectron spectroscopy. The product chemistries and current efficiencies varied significantly with the type of carbide electrocatalyst. The NbC electrocatalyst was most active among the carbides and was nearly as active as the Pd black electrocatalyst. As shown in Figure 1, the NbC cathode exhibited better selectivities than the Pd-black cathode, producing lower concentrations of stearic acid (C18:0). Both materials produced a lower TFA content (10-15% at an iodine value of 110) when compared to a commercial reactor with Raney nickel catalyst (up to 40%). The reaction products were consistent with a Horiuti-Polanyi mechanism for the hydrogenation of triacylglycerols, which involves the hydrogenation of unsaturated fats (e.g. linolenic, linoleic, and oleic acids) and isomerization to produce geometric and positional isomers (e.g. elaidic acid). These and other results will be presented in this paper.

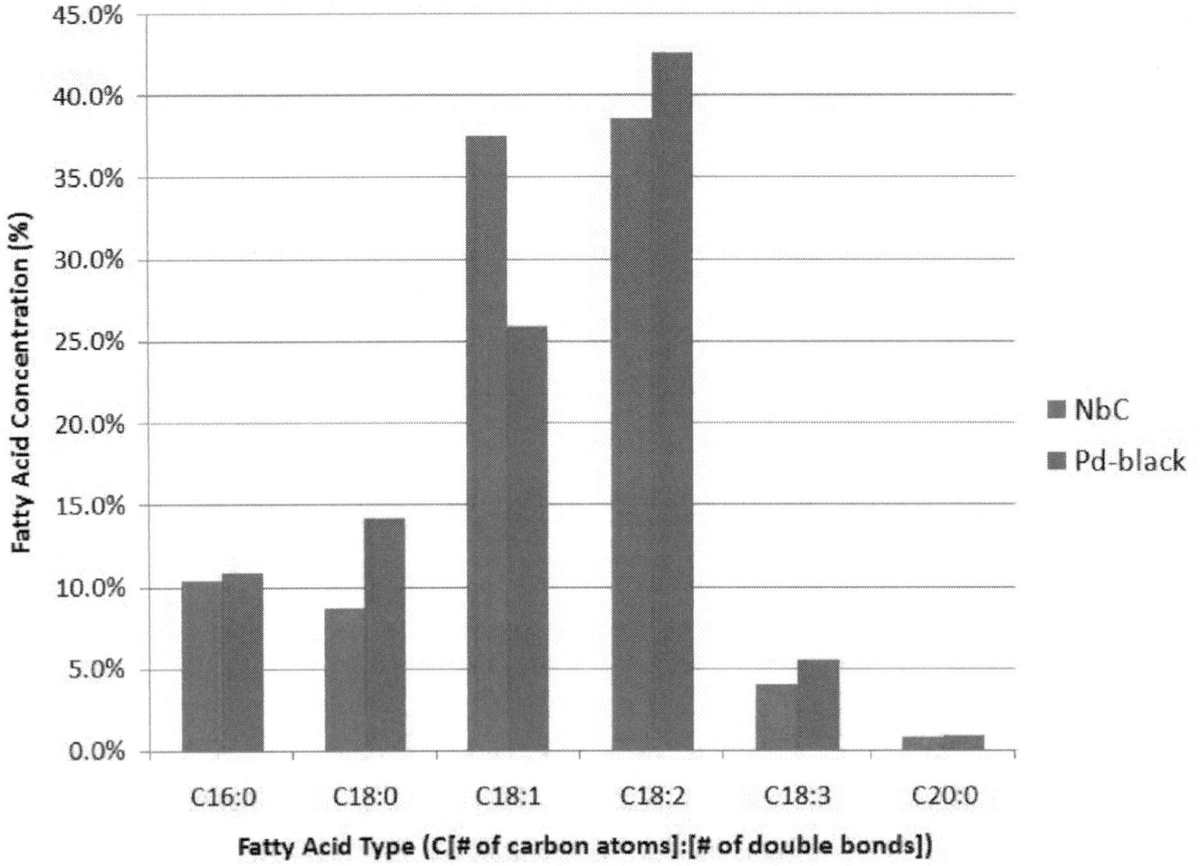

767b Nanoscalar Structures of Zein Microcapsules Modulate Release Kinetics of Encapsulated Lysozyme

Qixin Zhong, Department of Food Science and Technology, University of Tennessee, 2509 River Drive, 115B McLeod Building, Knoxville, TN 37996-4539 and Minfeng Jin, Food Science and Technology, The University of Tennessee, 2509 River Drive, Knoxville, TN 37996.

Numerous studies have been reported to encapsulate drugs or other bioactive compounds in biopolymeric micro- or nanocapsules to achieve a controlled release of the bioactives. Food biopolymers are less studied as carriers, albeit abundant and generally-recognized-as-safe. Further, many of the processes used to encapsulate bioactives are questionable for food applications that require very large production capacities and appropriate solvents. This work reports release characteristics of lysozyme, a naturally-occurring antimicrobial, from corn zein microcapsules produced using spray drying. We show that the additives of surfactant and essential oil profoundly changed the internal structures of capsules and that nanostructures of these capsules controlled the mass transfer of the encapsulated lysozyme. Without additives, micrometer-sized pores existed in microcapsules, which corresponded to quick equilibria of lysozyme release profiles at pH 2-8, with a smaller amount of lysozyme released at a higher pH. With the addition of Tween 40, the release characteristics was not fundamentally changed, although the matrix of microcapsules changed to packed nanoparticles. When thymol was used, the wall structure became continuous at the nanoscale observed and a sustained release was observed at pH 6 and 8 over 36 days. This work illustrated that molecular interactions and nanoscale structures both determined release profiles of lysozyme. The technology being developed may be used to produce food grade antimicrobial delivery systems to enhance microbial food safety.

Effect of concentration on rheological properties of media-milled cellulose

Ta-Yuan Yu and An-I Yeh*
Graduate Institute of Food Science and Technology
National Taiwan University
1 Roosevelt Road, Section 4, Taipei, Taiwan
e-mail: yehs@ntu.edu.tw
Phone: +886-2-33664121
Fax: +886-2-23620849
*: Corresponding author

Abstract

Cellulose is the most common organic compound on Earth. It is not digestible by human and is often referred to as 'dietary fiber' or 'roughage', acting as a hydrophilic bulking agent for feces. Nevertheless, the pliable but strong characteristics limit the applications of cellulose in food. Size reduction would be an attractive method to broaden the utilization of cellulose. Nano/submicron scale cellulose particles have been prepared by using media mill. The rheological behavior media-milled cellulose-suspension has been investigated using a dynamic rheometer.

Cotton cellulose, with a volume mean diameter 33.19 μm, was used as the raw material in this study. The diameter of cellulose was reduced to nano/submicron scale as determined by dynamic laser light measurement and illustrated by TEM. Both concentration and milling time affected the average particle size. In general, increasing milling time reduced the particle size. After being milled for 180 min, the highest concentration of 11% exhibited an average particle size (in volume) of 0.769 μm with the volume percentage of nano/submicron particles greater than 68%.

The viscosity of suspension increased with the concentration and exhibited shear-thinning behavior, which can be described by Williamson model. The milling resulted in viscoelastic solid properties when the content of cellulose was 5 to 11%.

Nevertheless, for low concentration of 1 and 3%, the suspension exhibited viscoelastic fluid behavior. The results showed that the size reduction to nano/submicron scale induced changes in rheological properties.

767d The Efficiency of Medicinal Plants Extract Inhibited Plant Disease Bacteria

Sumonthip Kongtun Janphuk, Department of Biotechnology, Faculty of Applied Science, King Mongkut's University of Technology North Bangkok, 1518 Piboonsongkram Road, Bangsue, Bangkok,10800, Thailand

The efficiency of medicinal plants extract for inhibition Xanthomonas campestris and Bacillus cereus, the causal agents of soft rot of vegetable and food poison. The extract of 10 medicinal plants were tested with Hibiscus sabdariffa Linn., Punicagranatum Linn., Ziinggiber zerumber(L) Smith., Citrus hystrix DC., Tammaridus indica Linn., Phyllanthus emblica Linn., Phyllanthus acidus Skeels., Momordica Charantia Linn., Garcinia manggostana Linn., Allium ascalonicum Linn. Ninety five percent of alcohol was used in the extraction process and diluted to serial dilution from 10-10-6μg/ml. Result was found that the minimum medicinal plant of Garcinia mangostana Linn. (mangosteen) extract inhibited the growth of B. cereus in dilution 10 , that have clear zone 25.50 millimeters. For other dilutions 10-1, 10-2 , 10-3 ,10-4 ,10-5 and 10-6 μg/ml, the clear zones decrease to 17.57, 9.88, 11.98, 4.75, 4.25 and 6.00 millimeters respectively. Punica granatum Linn. was the best antibacterial activity toward Xanthomonas campestris. Result indicated clear zone 26.00 of millimeters at dilution 10 and 23.63, 15.88, 8.65, 8.55, 9.00 and 8.68 at dilution 10-1 ,10-2 , 10-3 , 10-4, 10-5 and 10-6 μg/ml respectively.

767e Catalytic Membranes for Partial Hydrogenation of Vegetable Oil

Devinder Singh, Peter Pfromm, and Mary Rezac. Chemical Engineering, Kansas State University, 105 Durland Hall, Manhattan, KS 66506

Annual U.S. consumption of partially hydrogenated vegetable oil is nearly 14 billion pounds. The conventional hydrogenation process used produces trans fatty acids (TFA), the content of which can be as high as 45-50%. Dietary TFAs are associated with increased risk of ischemic heart disease, colon cancer, type-2 diabetes and allergic diseases in children. FDA regulations require listing of the TFA content on food labels. So the design of a commercial hydrogenation process/technology that produces minimum amounts of TFA becomes essential.

In the current technology catalyst particles are immersed in oil that contains small amounts of dissolved hydrogen which results in catalyst surface being hydrogen starved and thus promotes the isomerisation to TFAs. The isomerization to TFA can be reduced by having a high hydrogen concentration at the surface of the catalyst. Our membrane reactor approach relies on adding hydrogen directly near the catalytic sites of known hydrogenation catalysts located on the skin of a polymeric high-performance asymmetric gas separation membrane. A membrane capable of selectively transporting hydrogen while acting to prevent any loss of liquid phase is incorporated in the reactor housing. Oil is pumped on one surface of the membrane where it comes into contact with the catalytic metal surface supported on the polymeric membrane support. The metal catalyst has a high hydrogen coverage that has diffused through the membrane due to an imposed chemical potential driving force. High concentrations of hydrogen on the catalyst surface, and the resulting decrease in temperature, promote the hydrogenation reaction at the expense of the cis to trans isomerization. The approach studied is also compatible with existing hydrogenation technologies and requires no additional utilities.

Experiments were performed with membranes having different membrane properties and catalyst loadings using platinum, palladium, and Pd-Pb alloy as catalyst. At an iodine value of 100 the membrane reactor (70C, 3.4 atm H2) produced less than 2 wt% TFA while the conventional system produced nearly 10 wt% TFA (Pt catalyst). The presentation will discuss the concept and the influence of membrane properties, catalyst type and catalyst loadings on TFA formation and hydrogenation selectivity.

767f Induced Mutation of Monascus Purpureus by Ultrasonic Wave for Food Coloring

Sasithorn Kongruang, Department of Biotechnology, King Mongkut's University of Technology North Bangkok, 1518 Piboonsongkram Road, Bangsue, Bangkok, 10800, Thailand

Natural food coloring is biopigment produced by Monascus purpureus. It can be widely used in many industries such as foods, pharmaceutics and cosmetics. Induced mutation of Monascus purpureus TISTR 3002 and TISTR 3385 was experimented by using ultrasonic wave at 28 oC, 45 kHz for 2 min to develop an efficiency food coloring mutant for 4 generations. Red pigment stability to temperatures, pH range and solvents was also evaluated. Preliminary study of the growth rates on PDA was found that TISTR 3180, TISTR 3090, TISTR 3002, TISTR 3385 and TISTR 3541 were 5.241, 4.950, 4.863, 4.746 and 4.670 mm/day, respectively. Results showed that these fungi produced yellow, orange and red pigment at the maximum wavelength of 340, 440 and 480 nm in that order. Temperature stability of red pigment exhibited that a G2 mutant of TISTR 3002 was the most stable generation at 78 oC when compared to G3, G1 and G4 with the decrease of stability more than G0 : 3.27, 1.94, 1.93 and 1.44 times, respectively. Red color of TISTR 3002 G4 mutant was more stable at pH 7 than G3 and G1 with the values of 8.62, 3.10 and 1.50 times. The resistance to solvent, distilled water, degradation results revealed that TISTR 3385 G1, G2, G3 and G4 mutant when compared to G0 at the value equivalent to 1 time. In terms of the comparison on the generation effect to color production, statistical analysis showed that TISTR 3002 G2 mutant produced significantly higher color than wild type about 2 times ($p = 0.007$), while TISTR 3385 G4, G3, G1 and G2 mutants yielded 2.17, 1.71, 1.58 and 1.49 times with the p-values of 0.003, 0.016, 0.025 and 0.033, respectively.

767g Nickel-Boron Alloy Nanocatalysts Reduce the Formation of Trans Fatty Acids In Hydrogenated Soybean Oils

Qixin Zhong[1], Tonglin Li[1], Weinong Zhang[2], and Roy Zhenhu Lee[3]. (1) Department of Food Science and Technology, University of Tennessee, 2509 River Drive, 115B McLeod Building, Knoxville, TN 37996-4539, (2) Food Science and Technology, University of Tennessee, 2509 River Drive, 115B McLeod Building, Knoxville, TN 37996-4539, (3) Johnson Matthey, Inc., 435 Devon Park Drive, Wayne, PA 19087

The edible oils and fats industry is still looking for novel solutions that can reduce the unsaturation of lipids and at the same time maintain low concentrations of trans- and saturated fatty acids. In this work, we adapted a simple method to synthesize amorphous alloy nanocatalysts of nickel and boron (Ni-B), which was used in literature to hydrogenate unsaturated hydrocarbons and the hydrogenated products were reported to have a low concentration of trans-isomers. Soybean oil was hydrogenated with a group of Ni-B alloy catalysts synthesized at various conditions. We report that some Ni-B alloy catalysts showed activities similar to conventional nickel and palladium catalysts, while some did not have activities. The trans-fat contents in products hydrogenated by Ni-B were approximately half of those when the commercial nickel catalyst was used. The selectivities of Ni-B catalysts were also different from the commercial nickel catalyst. We further showed that, although the hydrogenation reaction rate was strongly affected by reaction conditions, the production of trans-fat was primarily affected by the catalyst and secondarily by the studied reaction conditions. Because of the simplicity to synthesize amorphous catalysts, this work may lead to a group of novel nanocatalysts that integrate different functions of various transition metals to synergistically reduce the formation of both trans- and saturated fatty acids in partially hydrogenated edible oils and fats.

769a In Process Wireless Coating Bed Temperature Monitoring: Application of a Novel Commercial off the Shelf Solution for PAT and Qbd Applications

Zane Arp and Christine McDade. Pd-Most-Puc, GlaxoSmithKline, 709 Swedeland Rd, King of Prussia, PA 19406

In order to meet GSKs DFM initiatives and meet requirements set forth by the FDA for process understanding, in depth understanding of critical quality attributes (CQA's) is imperative. One of the most important CQA's that must be controlled and understood in coating applications is the temperature of the product bed. This temperature is very important as overheating can negatively affect product and underheating can cause inadequate drying of the coat, coating quality problems, and substantially longer coating times. Recently a new wireless temperature monitoring technology has been identified by GlaxoSmithKline that will allow direct bed temperature measurement. The CorTemp® temperature probe is a wireless, tablet based temperature probe about the size of a large multi-vitamin. This tablet based temperature probe can be placed directly into the coating bed during the coating process allowing direct measurement of the coating pan bed temperature from the pharmaceutical tablets perspective. The nature of this technology makes it a substantial improvement over any other currently used method of temperature monitoring such as exhaust temperature and infrared guns. In this presentation work is reported showing proof of concept work on temperature monitoring in multiple coaters with the CorTemp® temperature probe. Results from these studies clearly show that the CorTemp® temperature probe is a valid method for real-time monitoring and improving process knowledge in coating applications that can be utilized in the future for process control of these unit operations.

769b Rapid Characterization of Virus Solutions for Filter Challenge Studies

Leonard F. Pease III[1], De-Hao Tsai[2], Kurt A. Brorson[3], Kenneth D. Cole[4], Michael Zachariah[2], and Michael J. Tarlov[5]. (1) Process Measurements Division, National Institute of Standards and Technology, 100 Bureau Drive, Gaithersburg, MD 20899, (2) University of Maryland and NIST, College Park, MD 20742, (3) Center for Drug Evaluation and Research (CDER), Food and Drug Administration (FDA), 10903 New Hampshire Ave., Silver Spring, MD 20903, (4) Biochemical Science Division, National Institute of Standards and Technology, 100 Bureau Dr, Gaithersburg, MD 20899, (5) National Institute of Standards and Technology, Gaithersburg, MD 20899-8362

Here we describe the use of electrospray differential mobility analysis (ES-DMA) as a process analytical technology (PAT) to characterize the size, concentration, and suitability for filter testing of virus containing solutions. Cell cultures typically used in the production of monoclonal antibodies (mAbs) and therapeutic recombinant proteins (TRPs) produce endogenous type C retrovirus particles and can potentially become infected by adventitious viruses. Biopharmaceutical manufacturing schemes typically include process steps to decrease the viral load including filters, which are membrane-based devices that remove large viruses (e.g. retroviruses) and/or small viruses (e.g. parvoviruses) from product by size exclusion. In 2002, the Parenteral Drug Association (PDA) organized the virus filter task force to develop a common nomenclature and a standardized test method for classifying and identifying viral-retentive filters. In support of this effort, we determined the sizes of several bacteriophages ranging from ~20 nm to ~80 nm including pp7, ΦX174, PR772, and MS2 using ES-DMA. Virus concentration can be assessed with ES-DMA by including an internal particle standard such as gold nanoparticles at a known concentration. This talk will describe the theory of operation of ES-DMA, its use to determine the size and concentration of viral preparations, and its potential application for characterizing vaccines and virus-like particles.

769c Anomaly Detection and Diagnosis In Continuous Pharmaceutical Manufacturing: An Ontological Approach

Intan Munirah Hamdan, Gintaras V. Reklaitis, and Venkat Venkatasubramanian. Chemical Engineering, Purdue University, 480 Stadium Mall Drive, Forney Hall of Chemical Engineering, West Lafayette, IN 47907

The development and manufacturing of pharmaceutical products are governed by strict safety regulations since defective products would have a devastating consequence in public healthcare. As a result, pharmaceutical products undergo rigorous product testing to meet strict product quality specifications. With the push towards continuous manufacturing to reduce production costs, the identification and immediate correction of deviations from desired operating conditions become more important. Timely anomaly detection and diagnosis can potentially avoid the progression of abnormal events and thus reduce productivity loss without compromising the quality and safety of the products. Additionally, an automated fault detection and diagnosis system reduces the reliance of plant operation on human operators, which according to industrial statistics results in 70% of industrial accidents [1].

While several fault detection and diagnosis frameworks have been proposed in the literature, an approach that combines intelligent trend monitoring using Qualitative Trend Analysis (QTA) with causal reasoning based on Signed Digraphs (SDG) has been shown to be quite successful in the past. in diagnosing incipient faults [2]. However, these frameworks generally require the user to be familiar with programming languages and are difficult to customize. To facilitate the application of SDG and QTA to detect and diagnose faults a framework is being developed within the Purdue Ontology for Pharmaceutical Engineering (POPE) environment which allows users to perform fault detection and diagnosis more easily.. The vision is to call upon the framework developed in [3], through POPE, to perform fault diagnosis without direct user involvement. Thus, the user need only provide mechanistic models (represented in ontologies) that describe the unit operation along with qualitative relationships within model variables, and a stream of data to classify faults, and subsequently detect and diagnose the faults.

The incorporation of the fault detection and diagnosis framework into POPE would allow direct access to a built-in library of control strategies; this would then enable rapid fault detection and diagnosis and immediate identification of the appropriate control methodology to correct the fault(s). For example, if a higher than desired ribbon density is detected; the possible causes are deviation in pressure, feed speed or motor malfunction. If through diagnosis, the most likely cause is determined to be a malfunction of the pressure transducer. The fault-tolerant control strategy would then provide new controller parameters to bring density back to desired value so as not to interrupt operations. We will present results of this approach for industrially inspired case studies.

References

1. Venkatasubramanian, V., R. Rengaswamy, K. Yin, and S. N. Kavuri (2003). A review of process fault detection and diagnosis Part I: Quantitative model-based methods. Computers and Chemical Engineering, 27,293-311

2. Maurya, M.R., R. Rengaswamy, V. Venkatasubramanian (2007). A signed directed graph and qualitative trend analysis – based Framework for incipient fault diagnosis. Chemical Engineering Research and Design, 85(A10) 1407-1422

769d Monitoring and Control of Cell Culture Bioreactors Using on-Line Scanning Dielectric Spectroscopy

Cary Opel and Ashraf Amanullah. Oceanside Process R&D, Genentech, Inc., 1 Antibody Way, Oceanside, CA 92056

Achieving reliable on-line cell density measurements is one of the most valuable, but difficult goals in process monitoring. Capacitance probes offer one solution to this problem by measuring the enclosed viable biovolume of a culture. One limitation to this method is that the relationship between biovolume and cell density changes during batch and fed-batch cultures, making it difficult to predict cell density from capacitance readings. Recent advances in scanning dielectric spectroscopy show promise alleviating this limitation by providing information on the size and size distribution of a culture. With this knowledge, one can theoretically calculate the number of cells in a given volume. Capacitance probes improve the online characterization of a cell culture process, contributing to the Process Analytical Technology Initiative

Batch and fed-batch cultures using CHO cells were performed and analyzed using a Fogale Biomass System. Offline samples were analyzed using a Nova Biomedical Flex, a Beckman-Coulter Z2 Cell Analyzer and a Beckman-Coulter ViCell. Packed cell volumes were also measured. Several methods of analyzing the set of capacitance data were used to derive useful culture parameters such as viable cell density, viable packed cell volume, integrated viable cell density, growth rates, average cell diameter and cell size distribution. First, a simple model was created using a capacitance-to-parameter ratio that was corrected each offline sample. Second, Partial Least Squares (PLS) regression was used to create a prediction model based on a finite training set of batch and fed-batch cultures. Finally, theoretical calculations based on the Beta-Dispersion curve were used to predict the parameters of interest. Growth rates from these methods were calculated then used to test the accuracy of viable cell density predictions.

Finally, some practical applications of capacitance probes were demonstrated through the integration of the signal with a DeltaV bioreactor control system. This includes an automated batch re-feed for the maintenance of inoculum cells, monitoring of base addition via conductivity (an additional measurement provided by the capacitance probe), exact target densities for inoculum trains, online indication of glucose and lactate exhaustion, and the implementation of a capacitance controlled continuous bioreactor.

AIChE 2008 Annual Meeting

Abstract

Chiajen Lai*, Mario Hubert, Gregory Lane, Thomas LaPorte, Paul Lobben, Lori Spangler

Lasentec FBRM IPC-PAT method development and validation for wet milling process

Lasentec FBRM in-situ measurement is used to monitor particle size in many different operations such as crystallization and wet milling. It can be installed in recirculation loops, or directly in vessels of different sizes ranging from 50 mL to thousands of gallons. Despite the popularity of the technique, it has never been used as an In-Process-Control (IPC) tool in the pharmaceutical industry. Its common use is to understand and troubleshoot processes or what is generally termed "information gathering". One reason that the FBRM technique has not been used as an IPC or "release" test is the higher probability of variable results. The variation normally encountered is due to difficulties in controlling operating conditions to the level that is possible for off-line tests. However, there is evident advantage to the use of in-line process analytical techniques (PAT) such as Lasentec FBRM to speed up and streamline API manufacturing and potentially allow superior process control which is not possible with traditional off-line measurements.

This presentation describes efforts towards developing and validating the Lasentec FBRM as an IPC method for a wet milling process of an active pharmaceutical ingredient (API). Results from a pilot plant campaign suggest up to 10% variation with different flow rates in the recirculation loop in which the Lasentec probe was placed. Additional laboratory scale studies reveal other contributing factors to variability when the Lasentec probe is installed in the crystallizer - rpm of the agitator, location of the probe, and scanning speed of the laser. Results from the robustness studies will be discussed in this presentation.

*presenting author

769f Process Analytical Technology Based Robust Scale-up of Pharmaceutical Crystallization Processes

Zoltan K. Nagy[1], **Levente L. Simon**[2], and Konrad Hungerbuehler[2]. (1) Chemical Engineering Department, Loughborough University, Loughborough, LE11 3TU, England, (2) Institute for Chemical and Bioengineering, Safety and Environmental Technology Group, ETH Zurich, G-137, HCI, ETH Hoenggerberg, Zurich, 8093, Switzerland

Crystallization is of prime importance in the pharmaceutical industry as more than 80% of the active pharmaceutical ingredients are produced in crystalline form. The crystallization step has a strong effect on the desired polymorphic form, crystal size distribution and shape of crystals. These critical properties then affect the downstream processing as well as the therapeutic response of the drug. Nucleation and crystal growth are the governing phenomena characterising most of the crystallization processes. To control a desired balance between nucleation and growth in industry normally an open loop control strategy is employed when the system follows a preset cooling profile [1]. The quality of product obtained as a result of this is often not satisfactory and is characterised by large batch-to-batch variability as the control does not account for any dynamic changes in the system due to disturbances or scale-up. A key objective in the design of crystallization operating recipes is to obtain consistent product quality and avoid excessive secondary nucleation. The industrial practice to achieve this objective is to use seeded operation. However, the production of seed with consistent properties, such as size distribution or polymorphic form is difficult. In the case of unseeded operation when the "seed" is generated in situ the initial size distribution at the onset of nucleation is uncertain, which will lead to variability in the product quality. Due to the advent of process analytical technology and increased availability of sophisticated equipment which can be used in situ, the control of crystallization processes has improved significantly in recent years [2, 3].

Supersaturation control (SSC) is more and more commonly used today for improved consistency in the final crystalline product quality for cooling and antisolvent crystallization systems [1,2,4]. The idea of supersaturation control is to control the process within the metastable zone so that secondary nucleation or the formation of undesired polymorphs is avoided. The design of the SSC requires the choice of a supersaturation setpoint or profile within the metastable zone. The performance of the SSC depends on the correct determination of the MSZW. However the boundaries of the MSZW can change significantly due to changing mixing conditions during scale-up. Therefore a supersaturation operating profile designed on smaller scale may not be effective on the industrial scale. The cloud point (nucleation) can be detected by modern process analytical equipment. However the nucleation boundary of the MSZW is the combined result of the actual nucleation (formation of critical cluster) and growth to the detectable size. This contribution evaluates the use of different PAT for the detection of the MSZW at two different scale reactors. Focused beam reflectance measurement (FBRM), attenuated total reflectance (ATR) Fourier Transform Infrared (FTIR) and UV spectroscopy, together with calorimetric measurements and image analysis will be corroborated for the detection of MSZW. The advantages and caveats of each method will be presented. For the determination of the robust operating zone in the phase diagram a series of experiments with all the aforementioned PAT tools are conducted under different operating conditions to estimate the uncertainties on the boundaries of the MSZW. The small scale experiments are performed in a 50 mL patented calorimetric reactor [5]. The reactor temperature is controlled using the principle of power compensation. In order to compensate changes of the heat transfer through the reactor wall, Peltier elements are incorporated between the jacket and the surrounding heat exchanger. Combining the principles of power compensation and heat balance calorimetry, the reactor allows the direct measurement of the heat release without any additional calibration.

According to our knowledge this is the first time when crystallization experiments are conducted with simultaneous monitoring of five different in situ signals from the system (FBRM, ATR-UV, ATR-FTIR, in situ image analysis, and calorimetric measurement).

Larger scale MSZW experiments are also performed on a 100L scale reactor and the effect of the scale up on the MSZW is evaluated. A systematic methodology is proposed based on PAT based MSZW experiments to determine a robust operating zone which can be used as the operating envelop for supersaturation controlled crystallization systems, to design the operating curves which are robust against scale-up and minimize variability in the final product quality.

References

[1] Nagy, Z. K. , J.W. Chew, M. Fujiwara, and R. D. Braatz, Comparative performance of concentration and temperature controlled crystallizations, Journal of Process Control, 18 (3-4), 399-407, 2008.

[2] M. Fujiwara, Z. K. Nagy, J. W. Chew, R. D. Braatz, First-principles and direct design approaches for the control of pharmaceutical crystallization, Journal of Process Control, 15, 493-504, 2005.

[3] Doki, N., Seki, H., Takano, K., Asatani, H., Yokota, M. and Kubota, N., Process Control of Seeded Batch Cooling Crystallization of the Metastable -Form Glycine Using an In-Situ ATR-FTIR Spectrometer and an In-Situ FBRM Particle Counter, Crystal Growth & Design, 4 (5), 949-953, 2004.

[4] Zhou, G. X., M. Fujiwara, X. Y. Woo, E. Rusli, H.-H. Tung, C. Starbuck, O. Davidson, Z. Ge, and R. D. Braatz, Direct design of pharmaceutical antisolvent crystallization through concentration control, Crystal Growth & Design, 6 (4), 892 -898, 2006.

[5] Zogg A., Fischer U., Hungerbühler K. A new small scale reaction calorimeter that combines the principles of power compensation and heat balance. Industrial and Engineering Chemistry Research (2003) 42: 767-776

Optimization of API Drying Via Hemi-solvate Conversion Kinetics

Dr. Kimberley Allen (speaker)
Dr. Nandkishor Nere
Dr. James Marek

Abbott Laboratories, North Chicago, IL 60064

For the Nov., 2008 AIChE National Meeting in Philadelphia, PA

Abstract:

Drying of an early stage API candidate required excessively long cycle times in the pilot plant filter dryer. Investigation of the drying curve revealed the rate-limiting step was de-solvation, as opposed to removal of unbound ethanol solvent. Intrinsic kinetics were then developed for the ethanol hemi-solvate as a function of temperature, and compared to the apparent kinetics derived from the pilot scale drying. Mass Spectrometry and Raman Spectroscopy PATs supplemented the kinetic study to elucidate a more fundamental understanding of the form conversion. The results were combined to optimize drying conditions on scale and model the time needed to reach an API dryness of 0.1wt%.

AUTHOR INDEX

AUTHOR INDEX

AUTHOR INDEX

AUTHOR INDEX

AUTHOR INDEX

AUTHOR INDEX

AUTHOR INDEX

AUTHOR INDEX

AUTHOR INDEX

AUTHOR INDEX

AUTHOR INDEX

AUTHOR INDEX

AUTHOR INDEX

AUTHOR INDEX

AUTHOR INDEX

AUTHOR INDEX

AUTHOR INDEX

AUTHOR INDEX

AUTHOR INDEX

AUTHOR INDEX

AUTHOR INDEX

AUTHOR INDEX

AUTHOR INDEX

AUTHOR INDEX

AUTHOR INDEX

AUTHOR INDEX